静态与动态成像
电子光学
——周立伟学术论文选

Static and Dynamic Electron Optics of Wide Beam Focusing
—Selected Academic Papers by Zhou Liwei

周立伟　著

北京理工大学出版社
BEIJING INSTITUTE OF TECHNOLOGY PRESS

版权专有 侵权必究

图书在版编目(CIP)数据

静态与动态成像电子光学：周立伟学术论文选 / 周立伟著. -- 北京：北京理工大学出版社，2022.9
　ISBN 978-7-5763-1692-6

Ⅰ. ①静… Ⅱ. ①周… Ⅲ. ①电子光学-文集 Ⅳ. ①O463

中国版本图书馆 CIP 数据核字（2022）第 163357 号

责任编辑：陈莉华　　　　**文案编辑**：陈莉华　辛丽莉
责任校对：刘亚男　　　　**责任印制**：李志强

出版发行 / 北京理工大学出版社有限责任公司
社　　址 / 北京市丰台区四合庄路 6 号
邮　　编 / 100070
电　　话 / (010) 68944439（学术售后服务热线）
网　　址 / http://www.bitpress.com.cn

版 印 次 / 2022 年 9 月第 1 版第 1 次印刷
印　　刷 / 三河市华骏印务包装有限公司
开　　本 / 787 mm × 1092 mm　1/16
印　　张 / 45.5
字　　数 / 1015 千字
定　　价 / 196.00 元

图书出现印装质量问题，请拨打售后服务热线，负责调换

夫科学上成大事者，不唯有追求真理的激情、求索质疑的胆识，亦必有坚忍不拔之大志。

科学研究是创造性的劳动，它探索未知，追求卓越，需要独立的精神、自由的思想、开放的头脑和舒展的心灵。

使静态与动态成像电子光学有一个较为完善的逻辑结构与理论体系是我研究工作的出发点和一生奋斗的目标。

周立伟

献给亲爱的父母亲
　　——智慧、力量和勇气的源泉

前　　言

我从事宽束电子光学的研究当自1962年在苏联列宁格勒（现俄罗斯圣彼得堡）乌里扬诺夫（列宁）电工学院电物理系读研究生开始。在此以前，我在北京工业学院辅导和讲授电子光学课程，但对这一门分支学科仅是一知半解，脑海中存在着许多疑问和科学问题，这些问题推动我去思索。

1962年11月，我被派遣到苏联列宁格勒乌里扬诺夫电工学院读研究生，将变像管与像增强器的电子光学理论与计算作为自己的研究方向，这是出国前校系领导交给我从事夜视领域研究的光荣任务。但在当时，成像电子光学的理论与技术涉及军事应用，在苏联也是保密的，列宁格勒电工学院给我分配的指导教师又是研究超高频电子光学的，中苏两国之间的关系正处于低潮，人地两疏，困难重重。尽管自己心中对研究没有底，但小卒过河，义无反顾，只有努力往前冲，奋力一搏了。这样，我在列宁格勒谢德林图书馆和苏联科学院图书馆里，在寂寞的学案上整整地读了几年书，虽摸索着前进，却打下了较为扎实的基础。终于在静电聚焦同心球系统的电子光学的理论研究上有所突破，顺利地通过了苏联物理数学副博士学位论文答辩。

当我回国时，即1966年5月，正值"文化大革命"开始。最初的2年，浑浑噩噩，被卷入群众运动。后来我才意识到自己思想认识的糊涂，便转向思考成像电子光学中的科学问题，积极投入夜视会战，研究电子光学系统的计算与设计等。像我国绝大多数知识分子一样，光和热真正发挥出来是在党的十一届三中全会之后。1978年以来，有好几年的时间，自己感到年青时代思考过的种种科学问题和想法在脑海中翻滚，抑制不住对于科学研究的冲动和创造的渴望。自那时起，我和我的团队的同志与研究生们一起，完成了多项课题的研究，得到了国内外学术界的肯定和承认，其中一些研究成果在实际应用中也取得了成功，获得了显著的社会与经济效益。

回顾自己的一生，我的科学研究生涯也像西天取经，真是需要信徒一样的虔诚，能经受无穷无尽的磨难百折不回。但是，能在科学上另辟蹊径，有所发现，有所创造，何尝不是一种快乐和幸福呢？不过，此中的苦楚与欢乐，只有亲身经历才能体会。就我个人来说，我是幸福的，能在科学长途的跋涉中，历尽磨难坚持下来；我也是幸运的，我始终受到前辈老师、各级领导、国内外学术界同行和亲人们的关怀、支持和鼓励。

我是学工出身的，数理基础薄弱，但我愿作科学长途百折不回的跋涉者。"路漫漫其修远兮，吾将上下而求索"。我的信心是坚定的，目标是明确的，我立志不但要给"成像电子光学"学科理论有一个较为圆满的物理解释，创造较为完整的理论体系，建立属于自己的科学学派；还要使所创造的理论与技术在生产实践与国防建设中得到应用。60多年来，我一直朝着这个方向在努力。

因为时常有一些国内外友人来函索要文章，研究生们也希望系统地了解我们在成像电子光学领域的研究与贡献。为此，我整理了所发表的有关成像电子光学的文章，觉得研究的思路与脉络还是比较清楚的，读者可以从中看出我们的宽电子束聚焦与成像的研究体系是如何建立和发展起来的。于是，在1993年出版《宽电子束聚焦与成像》学术论文选的基础上，编成了这本《静态与动态成像电子光学——周立伟学术论文集》，共选入63篇文章，分成11个专题。如果这本学术论文集能对国内外同行，特别是我国的科学工作者、大学教师以及年轻学人们有所裨益，这将是我极大的荣幸和愉快。

在学术论文集即将出版之际，我要强调的是，这本论文集不仅是我60多年来科学活动的部分记录，也是我领导的科研团队集体研究的成果。它包含了与我长期合作的西安近代化学研究所方二伦高级工程师呕心沥血的贡献，我的研究生倪国强、金伟其、张智诠教授等悉心研究的成就，以及艾克聪、潘顺臣研究员等历届研究生卓有成效的工作。其中也包括我国电子光学先行者、已故的北京大学西门纪业教授的杰出贡献。他们功不可没，在这里谨向他们表示衷心的感谢。

在本论文集出版之际，我深深怀念已故的清华大学孟昭英院士、中国科学院王大珩院士、北京大学吴全德院士，也感谢他们对我的指导、关怀和鼓励；我深深怀念在苏联求学期间列宁格勒乌里扬诺夫电工学院电子医疗器件教研室主任О. Б. 茹里叶教授，也感谢他对我的鼓励和关爱。我衷心感谢北京理工大学光学工程系李振沂主任以及系内老师们对我的支持、关爱和帮助。我十分想念已故的妻子吕素芹，是她的爱一直陪伴着我、鼓舞着我，使我能有所成就。我也十分感谢大学时代北京工业学院8531班的同学们，那份兄弟情谊让我深深怀念。同时，我也要感谢中科院力学所"钱学森科学和教育思想研究会"的老师们，感谢他们给予我的友情鼓励。北京理工大学与光电学院领导以及光电子成像技术教研室的同事和学生们对本论文集的出版给予了热情支持和帮助，我谨向他们表示由衷的感激之情。

借此机会，我还要特别感谢中国工程院、国防科学技术工业委员会、国家教育委员会、工业和信息化部、中国兵器工业总公司、中国兵器装备总公司，以及国家自然科学基金会所给予的物质与精神上的鼓励和支持。

最后，我要深深地感谢为出版这本学术论文集付出辛勤劳动的北京理工大学出版社的责任编辑辛丽莉、陈莉华和其他编辑同志们。

周立伟
于北京

导　　言

众所周知，控制电子在电场与磁场作用下在真空中的运动藉以聚焦、成像、偏转以形成电子束是带电粒子光学（电子光学）的基本问题。只有掌握电子在电场和磁场中运动的规律才能设计、制造直到创造新的电子光学器件与仪器。

自 1926 年 B. Busch 研究旋转对称磁场中电子运动以来，经过近百年的发展，电子光学作为技术物理的一门分支学科，它所研究的对象及其内涵已与当年大有差异。例如，出现了与电子显微镜相联系的电子显微学，与摄像管、显像管相联系的电视电子光学，与变像管、像增强器相联系的成像电子光学，与微波器件相联系的强流电子光学，考虑电子波动性的波动电子光学，与高能加速器、回旋管相联系的相对论电子光学和束流传输理论，与行波管相联系的超高频电子光学，与质谱仪、能谱仪相联系的离子束光学等。尽管就其出发点与基本原理来说有许多是共同的，但不少物理概念、研究途径、设计方法与处理手段却是很不相同的。

本论文集所涉及的内容主要是研究真空光电子成像器件中的宽电子束聚焦与成像，即成像电子光学，亦称宽束电子光学。

宽电子束成像系统的电子光学的主要特点是电子光学系统利用本身的光阴极所发射的光电子流，在电子接收器（屏靶）上形成图像或一定形状的电子信号。由于阴极面本身就是物面，并且直接处于场中，电子以不同的初角度和初速度自阴极面各点处射出，逸出电子的初速度很小，但其斜率可趋于无穷。若束电流很小，通常空间电荷效应可不予考虑。这一类系统在电子光学术语中通常称为阴极透镜。

作为聚焦成像的电子光学系统，一般要求有大的视场，小的几何失真，高的鉴别率和良好的调制传递特性。也就是说，越来越要求系统能形成足够清晰（高空间分辨、高时间分辨和高对比传递）和畸变极小的图像。

由此可见，成像电子光学的基本问题乃是研究在各种聚焦方式下大物面宽电子束成像的问题。即如何使光阴极上的电子像尽可能不走样地转移到接收器（屏靶）上，并满足其他要求（如时空分辨的要求，高速摄影中分幅的要求等）。电子光学工作者研究的任务不但要探讨成像系统的基本理论问题，如近轴光学、像差理论、曲轴光学、动态光学、电子光学传递函数与像质评定，而且要对计算场与轨迹的方法进行探讨，设计各种适应不同需求的成像器件，进而探讨电子光学系统最优化设计与电子光学逆设计等问题。

自 20 世纪 70 年代以来，本着发展我国光电子成像技术的需要，本人领导了一个科研团队研究宽电子束聚焦成像理论与成像电子光学系统设计，得到了我校与兵器工业集团有限公司以及国防科学技术工业委员会、国家教育委员会、国家科学技术委员会、工业和信息化部、兵器科学研究院、国家自然科学基金委员会的支持与资助。在这一领域中，我们开展了多方面的探索与研究，发表学术论文，推广研究成果，并与国内外研究所合作交流等。兹选出本人领衔的63篇文章，记载着中国学人对电子光学宝库的贡献，现分为如下11个专题以飨读者：

（1）静电聚焦同心球系统的成像电子光学；

（2）电磁复合聚焦同心球系统的成像电子光学；

（3）倾斜型电磁聚焦系统的成像电子光学；

（4）电磁聚焦移像系统的成像电子光学；

（5）曲轴宽电子束聚焦理论；

（6）阴极透镜的几何横向像差与近轴横向像差理论；

（7）成像系统的电子光学传递函数；

（8）成像电子光学系统的计算与设计；

（9）成像电子光学系统的逆设计；

（10）静态与动态成像电子光学；

（11）动态成像电子光学的时间像差理论。

现就这些文章的研究背景、科学问题的提出、所取得的进展以及对科学的贡献做一些简要的说明。

第一组文章分为4类专题，共19篇文章，研究成像电子光学系统的理想模型，如静电聚焦同心球系统、电磁复合聚焦同心球系统、倾斜型电磁聚焦系统和电磁聚焦移像系统。

这一组文章，主要研究一些理想的成像电子光学系统，具体表现在其电位和磁场分布以及电子轨迹都能写成解析形式，可定量地研究系统的电子光学性质和像差。本专题主要探讨和分析这类理想系统聚焦成像的基本问题以及所包含的具体矛盾，解剖它作为阴极透镜所具有的矛盾的特殊性，寻求它们对于研究阴极透镜具有普遍意义和规律性的线索。

通过这四类专题的研究找到了静电聚焦同心球系统与电磁聚焦同心球系统自阴极逸出的电子轨迹的成像位置的精确解以及近轴解，并把它表达成级数展开的形式，解决了宽束电子光学理想成像等概念和定义电子光学近轴像差与几何像差等问题。从静电聚焦与电磁复合聚焦同心球系统电子光学理想模型所具有的矛盾的特殊性出发，进而研究宽电子束成像矛盾的普遍性。对理想电子光学系统的几何横向像差与近轴横向像差理论的探索，对进一步研究轴对称阴极透镜具有实际意义。

更为重要的是，通过对同心球系统电子光学理论的研究，初学者能建立起宽束电子光学成像与像差的一些基本概念，如解析解、精确解、近轴解、实际成像、理想成像、近轴像差与几何像差等，熟悉成像电子光学系统的横向像差的基本概念，它乃是近轴横向像差与几何横向像差的合成，而不仅仅只有几何横向像差。尤其深刻理解了，无论成像系统是静电聚焦或是电磁聚焦，决定系统鉴别率的因素主要是二级近轴横向色差，以Recknagel-Artimovich公式表示，它仅与逸出电子的初能、阴极面上的场强以及系统的线性放大率有关，而与系统的具体电极结构及轴上电位分布无关。顺便指出，第一组文章的论述为研究成像电子光学提供了坚实的理论基础，是深入学习成像电子光学最好的入门内容。这一研究是作者在20世纪60—70年代对静电与电磁聚焦同心球系统的电子光学理论的科学阐释，被国际上誉为独树一帜，创立了自己的科学学派。

第二组文章分为3类专题，共19篇文章，涉及曲轴宽电子束聚焦理论、阴极透镜的几何横向像差与近轴横向像差理论以及成像系统的电子光学传递函数。

关于第一类专题。在探讨大物面、宽电子束聚焦与成像的科学问题时，我们的研究表明，用张量分析方法研究主轨迹为曲光轴下电子在曲线坐标系中的运动，是一条十分有效的途径。我们从曲线坐标系下的普遍轨迹方程出发探讨大物面曲轴宽电子束聚焦的普遍理论，推导主轨迹方程和曲近轴轨迹方程，研究曲近轴系统的正交条件。这一类专题的探索对于确定成像系统的场曲与像散，以及研究成像系统的电子光学传递函数具有实际意义。

关于第二类专题。长期以来，阴极透镜的三级几何横向像差理论，是电子光学学者们研究和关心的对象和主题，但对它的认识不够全面而且偏颇。我们的研究证明了，对于阴极透镜，除了几何横向像差外，还必须考虑由于逸出初能不同引起的色像差，即近轴横向像差。阴极透镜的横向像差应是近轴横向像差与几何横向像差的合成。成像电子光学系统的鉴别率应由二级近轴横向色差即Recknagel-Artimovich表示式所确定。此外，我们还提出了由近轴方程渐近解来研究成像电子光学系统近轴横向像差的途径，推导了近轴横向像差普遍形式的表示式。

关于第三类专题。对于成像系统的像质评定，自20世纪70年代始，我们主要研究电子光学传递函数（简称ETF），它能更全面地评价电子光学系统的成像品质。由于二级近轴横向色差即Recknagel-Artimovich表示式对于评定系统的像质很重要，于是我们建立确定了近轴横向色差半径的三维曲面图，用形象的手段，严密地探讨了成像系统的点传递函数（PSF）、调制传递函数（MTF）和均方根半径（RMS）的确定。我们的研究证明，以指数函数形式表示的调制扩散函数解析式 $T(f) = \exp[-(\pi\rho f)^2]$ 足以精确地表达成像系统的MTF，这里 ρ 是成像系统在某一个像面上的均方根半径，它反映了逸出电子束在像面上落点密集的程度可用来表征系统的调制传递特性。

在曲轴宽电子束聚焦理论、阴极透镜的近轴像差理论以及成像系统的电子光学传递函数的研究上，都有我们独特的贡献。

第三组文章分为两类专题，共9篇文章，涉及成像电子光学系统的计算、正设计、优化设计与逆设计。

正设计与优化设计乃是在已给系统场的所有结构与电磁参数下求系统的电子光学成像特性、像差与电子光学传递函数。前者采用逐次试算的方法，人为地调整结构参量与电参量来逼近所期望的设计结果。后者由程序自动寻求一组参数值，使之在满足一定的（约束）条件下，某些电子光学参量如像面位置、放大率达到所需值时，其成像质量等指标达到最佳值。我们的研究详细叙述了求解电场和磁场的计算方法，确定在已给场下的系统的电子光学聚焦成像性质。采用以多重网格法计算场与以约束变尺度法作为优化算法进行静电像管优化设计的工作。由于成像电子光学系统的正设计与优化设计涉及所研制像管的水平和质量，工作量十分巨大，团队的同志们以巨大的精力和热情投入这一工作，尤其是作为程序设计的负责人方二伦高级工程师功不可没，他长年累月对软件包不断地进行改进，为成像电子光学系统的正设计与优化设计做出了不可磨灭的贡献。

成像电子光学系统的逆设计是电子光学系统设计的一个极为艰难的课题。其实质是按所给定的成像特性，由满足此特性的电子轨迹出发，反算电极系统的形状和所施加的电压，线圈绕组的形状位置以及通过线圈的电流等。逆设计的问题极为困难与复杂，国际电子光学学术界对该问题也没有找到真正解决的途径，本课题组仅进行了一些探索与尝试。

第四组文章分为两类专题，共16篇文章，涉及动态电子光学与时间像差理论，这是将静态宽束电子光学的研究扩展到动态成像电子光学领域。

第一类专题基于电子运动方程，对两电极静电同心球系统求解由光阴极逸出的运动电子的空间和时间轨迹，推导成像位置和电子飞行时间的精确和近似表达式，并由空间和时间轨迹的解析解出发讨论系统的空间与时间特性。对空间像差和时间像差给出统一的定义，结果证实：时间色差和空间色差正是初始条件参量$(\varepsilon_r, \varepsilon_z)$与$(\varepsilon_r, \varepsilon_{z1})$的两条近轴电子轨迹与实际电子轨迹在成像位置处的时间和空间上的差异。

第二类专题主要考察俄罗斯科学家提出的"τ-变分法"与我们提出的"直接积分法"研究成像电子光学系统时间像差的理论。"τ-变分法"由变分原理出发，"直接积分法"由牛顿运动方程出发，分别导出了动态电子光学成像系统时间像差系数的微分表示式和积分表示式。我们利用静电聚焦同心球系统理想模型的解析解检验了"直接积分法"与"τ-变分法"给出的成像系统时间像差系数的表达式，证明了两种途径的正确性与等价性，殊途同归。但"直接积分法"给出的二级几何时间像差系数以积分形式表示，计算更为简便，更适合于成像系统的实际计算与设计。由这一研究，我们建立了较为完整的静态与动态成像电子光学理论体系。我们的理论严格证明了，无论是静态还是动态成像电子光学，空间像差是近轴横向像差和几何横向像差的合成，而时间像差是近轴时间像差和几何时间像差的合成；无论是系统的空间分辨率或是时间分辨率，都与阴极面的场强直接相关。

本学术论文集在编辑时对所收入的文章一般不做变动，但文章的作者、标题与摘要均用中英两种文字刊出。对原文中某些印刷错误或个别不适之处做了一些修正和删改。

目 录

一、静电聚焦同心球系统的成像电子光学

Imaging Electron Optics of Electrostatic Focusing Concentric Spherical Systems ……（ 1 ）
    1.1 A 章：电子轨迹方程 …………………………………………（ 3 ）
    1.2 B 章：近轴横向色差与几何横向球差 ……………………（ 15 ）
    1.3 C 章：多电极同心球系统的电子光学 ……………………（ 24 ）
    1.4 D 章：最小弥散圆与最佳像面位置的确定 ………………（ 35 ）

二、电磁复合聚焦同心球系统的成像电子光学

Imaging Electron Optics of Combined Electromagnetic Focusing Concentric Spherical Systems ……（ 47 ）
    2.1 Part A：Paraxial Optics ………………………………………（ 49 ）
    2.2 Part B：Paraxial Aberrations …………………………………（ 63 ）
    2.3 Part C：Approximate Solutions of Paraxial Equation …………（ 75 ）
    2.4 Part D：Asymptotic Solutions of Paraxial Equation ……………（ 83 ）
    2.5 电磁聚焦同心球系统的电子光学 ……………………………（ 95 ）
    2.6 Electron Optics of Concentric Spherical Electromagnetic Focusing Systems ……（ 110 ）
    2.7 电磁聚焦同心球系统的精确解 ……………………………（ 122 ）

三、倾斜型电磁聚焦系统的成像电子光学

Imaging Electron Optics of Oblique Electromagnetic Focusing Systems ……（ 127 ）
    3.1 Electron Optics of Oblique Electromagnetic Focusing Systems …（ 129 ）
    3.2 倾斜型像管的电子光学像差 …………………………………（ 142 ）
    3.3 倾斜型像管电子光学系统的计算与分析 ……………………（ 151 ）

四、电磁聚焦移像系统的成像电子光学

Imaging Electron Optics of Electromagnetic Focusing Systems for Image Transference ……（ 163 ）
    4.1 电磁聚焦移像系统理论的研究 ………………………………（ 165 ）

4.2 图像无旋转的电磁聚焦移像系统的研究 …………………………………………（ 176 ）
4.3 Electrostatic and Magnetic Imaging Without Image Rotation …………………（ 185 ）
4.4 图像无旋转的曲轴电磁聚焦成像 …………………………………………………（ 196 ）
4.5 一种新型的放大率 $M \neq 1$ 电磁聚焦移像系统 ……………………………………（ 203 ）

五、曲轴宽电子束聚焦理论

Theory of Wide Electron Beam Focusing Having Curvilinear Axes …………（ 215 ）

5.1 Tensor Analysis of Electron Motion in Curvilinear Coordinate Systems
（Part Ⅰ） ……………………………………………………………………（ 217 ）
5.2 Tensor Analysis of Electron Motion in Curvilinear Coordinate Systems
（Part Ⅱ） ……………………………………………………………………（ 237 ）
5.3 曲轴宽电子束聚焦的普遍理论 ……………………………………………………（ 247 ）
5.4 A Generalized Theory of Wide Electron Beam Focusing ………………………（ 262 ）
5.5 曲轴宽电子束聚焦的理论研究 ……………………………………………………（ 275 ）
5.6 宽电子束聚焦的变分理论 …………………………………………………………（ 288 ）
5.7 相对论修正下的宽电子束聚焦的普遍理论 ………………………………………（ 296 ）
5.8 曲近轴方程组确定静电聚焦成像系统的场曲和像散的研究 ……………………（ 304 ）

六、阴极透镜的几何横向像差与近轴横向像差理论

Theory of Geometrical Lateral Aberrations and Paraxial Lateral Aberrations for Cathode Lenses ………………………………………………………………（ 315 ）

6.1 关于电磁复合聚焦阴极透镜的几何横向像差理论 ………………………………（ 317 ）
6.2 Variation Theory of Geometrical Aberrations in Cathode Lenses ……………（ 329 ）
6.3 静电成像电子光学近轴横向像差理论 ……………………………………………（ 340 ）
6.4 On Verification for Paraxial Lateral Aberrations of Imaging Electrostatic Electron Optics Based on Asymptotic Solutions ………………………………（ 349 ）
6.5 Theory of Paraxial Lateral Aberrations in Imaging Electrostatic Electron Optics Based on Asymptotic Solutions ………………………………………（ 365 ）
6.6 Test and Verification of Theory for Paraxial Lateral Aberrations by a Bi-electrode Electrostatic Concentric Spherical System Model ………………（ 376 ）

七、成像系统的电子光学传递函数

Electron Optical Transfer Function of Imaging Systems ………………………（ 391 ）

7.1 成像系统的电子光学调制传递函数与均方根半径的研究 ………………………（ 393 ）
7.2 On Determination of Poly-energetic Point Spread Function and Modulation Transfer Function in Electron-optical Imaging Systems Using Three Dimensional Coordinates ………………………………………………………………………（ 406 ）
7.3 电子光学成像系统全色点扩散函数的研究 ………………………………………（ 419 ）

| 7.4 | On Modulation Transfer Function of Cathode Lenses in Image Tubes ……… (429)
| 7.5 | 一种用曲轴轨迹确定轴外点扩散分布的方法……………………………………… (444)

八、成像电子光学系统的计算与设计

Computation and Design of Imaging
Electron Optical Systems ……………………………………………………… (449)

| 8.1 | 变像管及像增强器电子光学系统的计算机辅助设计……………………… (451)
| 8.2 | Optimization Design of Image Tubes with Electrostatic Focusing ……… (461)
| 8.3 | Some Problems of Mathematical Simulation in Optimization Design of Electrostatic Image Tubes ……………………………………………………… (469)
| 8.4 | A Study on Electron Optical Systems for Conical Immersion Lenses ……… (484)
| 8.5 | A Multigrid Method for the Computation of Rotational Symmetrical Electrostatic Fields …………………………………………………………… (496)
| 8.6 | 一种计算轴对称磁场的边界元 – 有限元混合法的研究……………………… (503)

九、成像电子光学系统的逆设计

Inverse Design of Imaging Electron Optical Systems ……………… (511)

| 9.1 | 静电聚焦成像系统电子光学逆设计的研究……………………………………… (513)
| 9.2 | An Inverse Design of Magnetic Focusing Coil for Electrostatic and Magnetic Imaging ……………………………………………………………… (527)
| 9.3 | An Inverse Design of Electrostatic Focusing Field for Electrostatic and Magnetic Imaging ……………………………………………………………… (537)

十、动态成像电子光学

Dynamic Imaging Electron Optics ……………………………………………… (549)

| 10.1 | On Paraxial Chromatic Aberrations Limited Temporal Resolution and Spatial Resolution in Image Tubes …………………………………………… (551)
| 10.2 | 论两电极静电同心球系统的近轴电子光学及其空间 – 时间像差 ……… (556)
| 10.3 | Paraxial Imaging Electron Optics and Its Spatial-temporal Aberrations for a Bi-electrode Concentric Spherical System with Electrostatic Focusing …… (564)
| 10.4 | Static and Dynamic Imaging Electron Optics and Spatial-temporal Aberrations in a Bi-electrode Spherical Concentric System with Electrostatic Focusing …… (575)
| 10.5 | 两电极静电同心球系统的成像电子光学及其空间 – 时间像差 ………… (590)
| 10.6 | 静电聚焦同心球系统的静态和动态近轴光学 …………………………… (602)
| 10.7 | On Electron-optical Spatial-temporal Aberrations in a Bi-electrode Spherical Concentric System with Electrostatic Focusing …………………………… (608)
| 10.8 | Paraxial Imaging Electron Optics and Its Spatial-temporal Aberrations for a Bi-electrode Concentric Spherical System with Electrostatic Focusing …… (622)

十一、动态成像电子光学的时间像差理论

Temporal Aberration Theory of Dynamic Imaging Electron Optics ………… （633）

- 11.1 关于动态电子光学时间像差理论的研究 ………………………… （635）
- 11.2 关于 τ - 变分法研究动态电子光学成像系统的时间像差理论 ………… （643）
- 11.3 Theory of Temporal Aberrations for Cathode Lenses ……………… （653）
- 11.4 静电聚焦同心球系统验证电子光学成像系统的时间像差理论 ……… （669）
- 11.5 静电同心球电子光学系统时间像差系数的计算 …………………… （678）
- 11.6 On the Theory of Temporal Aberrations for Dynamic Electron Optics ……… （682）
- 11.7 直接积分法研究电子光学成像系统的时间像差理论 ……………… （694）
- 11.8 静电同心球系统验证直接积分法的时间像差系数的研究 ………… （703）

一、静电聚焦同心球系统的成像电子光学

Imaging Electron Optics of Electrostatic Focusing Concentric Spherical Systems

1.1 A 章：电子轨迹方程
Electron Trajectory Equations

摘要：在成像电子光学中，静电聚焦同心球系统具有一系列宝贵的性质，其电位分布与电子轨迹能以解析形式表示，可以定量地研究系统的电子光学成像特性与横向像差。尽管前人已有不少研究，但都限于零级近似成像的认识，且其中存在着不少谬误。本系列文章将全面研究静电聚焦两电极与多电极同心球系统中电子的运动轨迹、电子光学成像特性与横向像差，探讨电子束在成像段形成的图像弥散，给出一些新的结论和认识，纠正了文献中存在的一些谬误，建立自己的理论体系。本系列文章的第一篇主要探讨电子在两电极同心球系统中电子在静电场作用下的运动轨迹，导出了两电极静电聚焦同心球系统中自阴极面逸出的电子轨迹在极坐标系下的表示式 $\rho = f(\varphi)$ 与圆柱坐标系下新的轨迹方程的表示式 $r = r(z)$，给出了自光阴极逸出的电子在成像段的行进轨迹的交轴位置及其斜率的近似和精确表示式。本文为全面研究静电聚焦同心球系统的电子光学性质及其像差奠定了基础。

Abstract: In imaging electron optics, the electrostatic focusing concentric spherical system has a series of valuable properties, its potential distribution and electron trajectory can be represented by analytical form, and the electron optical imaging characteristics and its lateral aberrations can be quantatively studied. Although there are many literatures that have been studied, they are limited to the understanding of the zeroth order approximation of imaging, and a lot of fallacies exist in it. This series of articles fully study the electron motion trajectory, electron optical imaging characteristics and lateral aberrations for the electrostatic focusing concentric spherical system composed of two electrodes or multiple electrodes, investigate the image dispersion and abberations, correct some fallacies in literatures, put forward some new idea and recognition, and establish my own theoretical system. The first article of this series mainly explores the moving electron trajectory under the electrostatic field in the concentric spherical electrostatic focusing system composed of two electrodes, deduces new representations for the electron trajectory emitted from the photocathode in the polar coordinate system $\rho = f(\varphi)$ and in the cylindrical coordinate system $r = r(z)$, explores the exact and approximate formulae of axial intersection position and its angle of inclination at the image section for the electron trajectory emitted from the photocathode. This paper lays a foundation in studying the electron optical properties and lateral aberrations for the electrostatic focusing concentric spherical system.

周立伟. 北京理工大学. 光学学报（Acta Optica Sinica），V. 42，No. 8，2022，0811001 – 1 – 8.

本文初载：周立伟. 北京理工大学. 工程光学（Engineering Optics）（内部），1，1978，71 – 87.

引 言

由两个具有共同曲率中心的球面电极所组成的球形电容器系统简称两电极同心球静电聚焦系统。研究同心球系统作为静电聚焦像管的成像系统从电子光学观点来看是颇有意义的。这是因为，系统的电位分布和电子轨迹都可写成解析形式，可以定量地研究系统的电子光学性质和像差。无论是严格精确的或是近似实现的同心球系统，都能得到良好的像质。这类系统本身具有一系列宝贵的性质：对于凹面阴极－凸面栅状阳极系统，如果接收图像的荧光屏制成球面形状，其曲率半径等于成像位置到系统中心的距离，则除了色差与球差外，场曲与屏面曲率一致，其他类型的像差——彗差、畸变和像散不复存在。因此，不少像管设计都采用接近于同心球系统的结构。20世纪60年代开始发展的第一代级联像增强器的电子光学系统就其本质来说属于两电极静电聚焦同心球系统的一种变型。

按照辩证唯物主义观点，在考察事物时，我们必须从本质上认识到世界上不存在纯粹的普遍性，要使普遍性广泛得到承认，必须由特殊性来加以证实。因此，由静电阴极透镜的理想模型——同心球系统出发，分析和探讨这类系统聚焦成像所包含的具体矛盾，解剖它作为阴极透镜所具有的矛盾的特殊性，从中找出一些对于阴极透镜具有普遍意义和规律性的线索，这不但对于研究同心球型像管的电子光学提供理论基础。而且，由于理想模型成像的矛盾的特殊性中包含静电阴极透镜宽电子束成像的矛盾的普遍性，故对于进一步研究轴对称阴极透镜具有实际意义。

本系列文章将着重研究两电极静电聚焦同心球电子光学系统中电子行进的轨迹方程，探讨电子光学成像性质及其像差，研讨影响电子光学成像质量的最小弥散圆，确定最佳像面的位置，并将结果推广到任意多个电极静电聚焦同心球电子光学系统中。

关于两电极静电聚焦同心球系统的电子光学，虽然 E. Ruska（1933）[1]、P. H. Schagen（1952）[2]、О. И. 谢曼（1958）[3]等人先后曾进行了研究，导出了轨迹方程，讨论了系统的成像，但他们的工作主要是研究电子在同心球内部的行进轨迹，对于电子光学成像的考察仅停留在零级近似上[1,2]。本文将推导两电极静电同心球电子光学系统自阴极物面以初速度 v_0、初角度 α_0 射出的电子轨迹新的表示式，以及逸出电子的交轴位置及其倾角的精确与近轴表示式，为全面分析两电极和多电极同心系统的电子光学成像性质和像差奠定了基础[4,5]。

为了揭示规律性，本文借助几何光学通常采用的符号规则导出同心球系统的轨迹方程普遍形式解，它无论对于凹面阴极－凸面阳极，或是凸面阴极－凹面阳极构成的系统都是适用的，由此可极方便地确定轨迹的交轴位置及其斜率，研讨系统的电子光学成像性质。

1 符号规则

图1所示的是两种不同的静电同心球系统：凹面阴极－凸面阳极系统和凸面阴极－凹面阳极系统。设球面阴极 C 和栅状球面阳极 A 的曲率半径为 R_c 和 R_a，系统的共同曲率中心为 O。由于系统的球对称性，故可用极坐标 (ρ, φ) 来描述电子轨迹。令极坐标的原点位于系统的曲率中心 O，并设阴极 C 的电位 $\phi_c = 0$，栅状阳极 A 对于阴极 C 的电位为 ϕ_{ac}，电子以初速度 v_0、初角度 α_0 自阴极上的原点处射出。

一、静电聚焦同心球系统的成像电子光学

为了推导适合上述两种不同的同心球系统的普遍公式,兹规定符号规则如下。

(1) 线段。设射线行进方向自左至右,故规定线段由左向右为正,由下向上为正,反之为负。为了定出某一线段参量的符号,还需规定线段的计算起点。图1中各线段与角度参量的计算起点规定如下:

R_c、R_a、ρ、$R_{i\alpha}$、R_i^* 均以曲率中心 O 算起到相应球面的顶点,其中星号(*)对应近轴轨迹。纵向像差 Δz 自理想像面位置 I^* 点算起,横向像差 Δr 由 $\varphi=0$ 轴线算起。

(2) 角度。一律以锐角来度量,规定逆时针转为正,顺时针转为负。与线段参量要规定计算起点一样,角度也要规定起始轴。图1中各角度参量的起始轴为:

α_0、φ、γ、α_i 均以 $\varphi=0$ 轴线为起始轴,其中,γ 为 D 点的角度坐标,α_i 为电子轨迹在交点 I_α 处与 $\varphi=0$ 的轴线的交角;ζ、τ 均以过该点与中心 O 的连线为起始轴。ζ 为极坐标 (ρ, φ) 轨迹的切线与过该点与曲率中心 O 之间的夹角,τ 为电子轨迹在 D 点的切线与矢径 OD 之间的夹角。

图1上所标注的全部是绝对值。因此,图中在负的角度或线段前加一负号,使它变为正值。

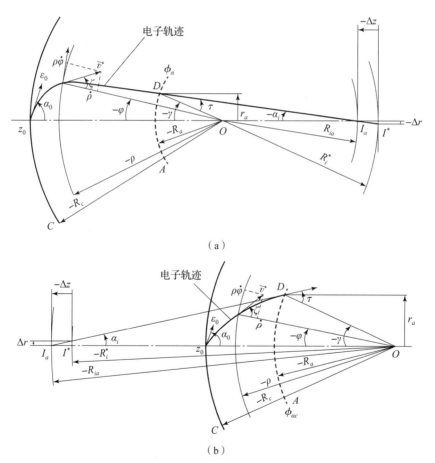

图1 两电极静电聚焦同心球系统的电子轨迹

(a) $n>2$ $\left(n=\dfrac{R_c}{R_a}\right)$;(b) $2>n>1$

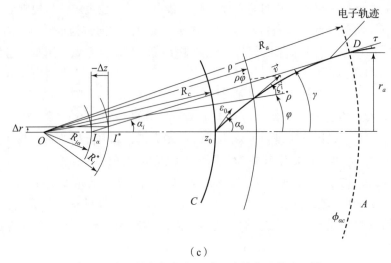

(c)

图1 两电极静电聚焦同心球系统的电子轨迹（续）

(c) $n < 1$

2 极坐标系下的实际轨迹($\rho = f(\varphi)$)

同心球系统两电极之间的场等效于集中的球心的点电荷所产生的场，故不难求得矢径坐标 ρ 处相对于阴极 C 的电位 $\phi_{\rho c}$ 的表达式

$$\phi_{\rho c} = E_c R_c \left(\frac{R_c}{\rho} - 1 \right) \tag{1}$$

式中，E_c 为阴极面上的电场强度。当 $\rho = R_a$ 时，有

$$E_c = \frac{\phi_{ac}}{R_c(n-1)} \tag{2}$$

式中，$n = \frac{R_c}{R_a}$，对于凹面阴极-凸面阳极系统，$n > 1$，R_c 为负值；对于凸面阴极-凹面阳极系统，$n < 1$，R_c 为正值。由此可见，无论 $n > 1$ 或 $n < 1$，E_c 永为负值，指向球面阴极。

在同心球系统中，电子在初速度 v_0、初角度 α_0 的逸出方向与系统轴线所构成的平面内的运动遵循能量守恒定律和有心力场角动量守恒定律：

$$\frac{1}{2} m_0 v^2 = \frac{1}{2} m_0 v_0^2 + e\phi_{\rho c} \tag{3}$$

$$\rho^2 \dot\varphi = R_c^2 \dot\varphi_0 = \text{const} \tag{4}$$

式中，m_0、e 为电子的质量与电荷，e 取绝对值；v 为电子在极坐标 (ρ, φ) 处的速度；$\dot\varphi = \frac{d\varphi}{dt}$，$\varphi_0$ 为电子转动的初角度。

方程（4）中的常数决定于初始条件

$$R_c \dot\varphi_0 = v_0 \sin\alpha_0 \tag{5}$$

于是有

$$\rho^2 \dot\varphi = R_c v_0 \sin\alpha_0 \tag{6}$$

在极坐标 (ρ, φ) 中，电子速度 v 可表示为

$$v^2 = \dot{\rho}^2 + (\rho\dot{\varphi})^2 \tag{7}$$

令 $\mu = \dfrac{R_c}{\rho}$，则

$$d\mu = -\dfrac{R_c}{\rho^2}d\rho \tag{8}$$

利用关系式 $\dfrac{d\varphi}{d\rho} = \dfrac{\dot{\varphi}}{\dot{\rho}}$，由式（3）、式（5）、式（7）并考虑式（1）、式（2），便得到

$$d\varphi = \dfrac{-\sin\alpha_0 d\mu}{[1 + p(\mu-1) - \mu^2\sin^2\alpha_0]^{1/2}} \tag{9}$$

式中，参量 p 虽然仅是一个数学符号，但它与 $(n-1)$ 的乘积具有明显的物理意义，表示系统的加速电位与初电位的比值，即 $p(n-1) = \dfrac{\phi_{ac}}{\varepsilon_0}$。其中，$\varepsilon_0 = \dfrac{1}{2}\dfrac{m_0}{e}v_0^2$，$\varepsilon_0$ 为静止的光电子为获得其发射能量所要求的加速电位，称为初电位。通常，$p(n-1)$ 的值是很大的。在一般的像管中，$p(n-1) = 10^4 \sim 10^5$。

由积分式（9），考虑初始条件 $\rho = R_c$ 时，$\mu = 1$，$\varphi = 0$，便有

$$\rho = f(\varphi) = \dfrac{R_c}{\left(1 - \dfrac{p}{2\sin^2\alpha_0}\right)\cos\varphi - \dfrac{1}{\tan\alpha_0}\sin\varphi + \dfrac{p}{2\sin^2\alpha_0}} \tag{10}$$

式（10）是 E. Ruska 首先推导得到的[1]。

3 极坐标系下实际轨迹的另一种表示（$\varphi = \psi(\rho)$）

由图 1，可求得极坐标 (ρ, φ) 处轨迹的切线与过该点与中心 O 之连线间的夹角 ζ。因为 $\tan\zeta$ 等于 (ρ, φ) 处垂直于矢径方向的速度 $\rho\dot{\varphi}$ 与沿着矢径方向的速度 $\dot{\rho}$ 的比值：

$$\tan\zeta = \dfrac{\rho\dot{\varphi}}{\dot{\rho}} = \rho\dfrac{d\varphi}{d\rho} = -\mu\dfrac{d\varphi}{d\mu} \tag{11}$$

因此，由式（9）可得到

$$\tan\zeta = \dfrac{\mu\sin\alpha_0}{[1 + p(\mu-1) - \mu^2\sin^2\alpha_0]^{1/2}} \tag{12}$$

而由式（10）得

$$\tan\zeta = \dfrac{\mu}{\left(1 - \dfrac{p}{2\sin^2\alpha_0}\right)\sin\varphi + \dfrac{1}{\tan\alpha_0}\cos\varphi} \tag{13}$$

联立式（12）、式（13）和式（10），则可解出 $\sin\varphi$、$\cos\varphi$ 以及 $\tan\zeta$ 的表达式[4]：

$$\sin\varphi = -\dfrac{(\mu-1)(c + d_0)}{bd_0 + b_0 c}\sin\alpha_0 \tag{14}$$

$$\cos\varphi = \dfrac{bc + b_0 d_0}{bd_0 + b_0 c} \tag{15}$$

$$\tan\zeta = \dfrac{\mu}{b}\sin\alpha_0 \tag{16}$$

式中，

$$b_0 = \cos \alpha_0,$$

$$d_0 = 1 - 2(n-1)\frac{\varepsilon_0}{\phi_{ac}}\sin^2 \alpha_0,$$

$$b = \left[1 + \frac{\phi_{ac}(\mu-1)}{\varepsilon_0(n-1)} - \mu^2 \sin^2 \alpha_0\right]^{1/2}, \tag{17}$$

$$c = 1 - 2\mu(n-1)\frac{\varepsilon_0}{\phi_{ac}}\sin^2 \alpha_0$$

式（14）~式（16）表示电子在两电极同心球系统中的行进轨迹 $\varphi = \psi(\rho)$，是本文作者在俄罗斯留学期间导出的。它的优点是对于多电极同心球系统的电子追迹具有类推性。

4 极坐标下实际轨迹以半角的正切表示 $\left(\tan\left(\dfrac{\varphi}{2}\right) = f(\rho)\right)$

如果我们将在下列的三角公式

$$\tan \frac{\varphi}{2} = \frac{\sin \varphi}{1 + \cos \varphi} \tag{18}$$

中代入式（14）、式（15）的表示式，则不难得到轨迹以半角的正切的表示式[4]

$$\tan \frac{\varphi}{2} = -\frac{(\mu-1)(c+d_0)}{b(c+d_0) + b_0(c+d_0)} \sin \alpha_0 \tag{19}$$

顺便指出，1951 年，对两电极同心球系统的电子轨迹，P. H. Schagen 由

$$\tan \frac{\varphi}{2} = \frac{1 - \cos \varphi}{\sin \varphi} \tag{20}$$

出发，首先导得以正切的半角表示的轨迹表示式[2]

$$\tan \frac{\varphi}{2} = \frac{-\cos \alpha_0 + [1 + p(\mu-1) - \mu^2 \sin^2 \alpha_0]^{1/2}}{-\dfrac{p}{\sin \alpha_0} + (\mu+1)\sin \alpha_0} \tag{21}$$

可以看出，本文导出的两电极同心球系统的轨迹表示式（14）~式（16），与 E. Ruska 的公式（10）和 P. H. Schagen 的公式（21）以及与作者本人曾给出的表示式（19）相比较，对于确定系统的电子光学成像性质与像差更为简捷方便。而且，无论对于 $n > 1$ 或 $n < 1$ 的系统，它们都是普遍适用的。此外，对于多电极静电同心球系统，式（14）~式（16）具有类推性。必须强调的是，在使用以上公式时，线段和角度都必须遵循上述的符号规则。

5 圆柱坐标系下的实际轨迹 ($r = r(z)$)

对于轴对称电子光学系统，用极坐标 $\rho = f(\varphi)$ 或 $\varphi = \psi(\rho)$ 等形式表示的电子轨迹在使用上颇为不便。因此，我们研究以圆柱坐标 $r = r(z)$ 的形式描述自两电极同心球系统的阴极面原点以初速度 v_0、初角度 α_0 逸出的电子轨迹。图 2 为圆柱坐标系下静电聚焦两电极同心球系统的实际电子轨迹。

令圆柱坐标系的原点取在阴极面上逸出电子的起始点上，z 轴与极坐标 (ρ, φ) 的起始轴线相重合，坐标 (r, z) 的符号规则同前。显然，两个坐标系之间存在着下列关系：

$$r = \rho \sin \varphi, \quad z = \rho \cos \varphi - R_c \tag{22}$$

由此解出 $\sin \varphi$ 和 $\cos \varphi$，代入式（10），便有

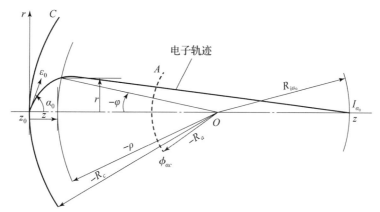

图 2 圆柱坐标系下静电聚焦两电极同心球系统的实际电子轨迹 $r=r(z)$

$$\rho = z + R_c + \frac{-2z\sin^2\alpha_0 + 2r\sin\alpha_0\cos\alpha_0}{p} \tag{23}$$

将式（23）代入关系式 $r^2 = \rho^2 - (z+R_c)^2$ 中，并利用 $p(n-1) = \phi_{ac}/\varepsilon_0$，则可得到两电极静电同心球系统中圆柱坐标系下由阴极面原点逸出的实际电子轨迹的解析表示式[6]

$$r(z) = \frac{2(n-1)}{1 - 4(n-1)^2\frac{\varepsilon_r\varepsilon_z}{\phi_{ac}^2}}\left(\frac{\varepsilon_r}{\phi_{ac}}\right)^{1/2} \times$$

$$\left\{(z+R_c)\left(\frac{\varepsilon_z}{\phi_{ac}}\right)^{1/2} - \frac{2z(n-1)\varepsilon_z^{1/2}\varepsilon_r}{\phi_{ac}^{3/2}} - (z+R_c)\left[\frac{-z}{(n-1)(z+R_c)} + \frac{(z+R_c)^2\varepsilon_z + z^2\varepsilon_r}{(z+R_c)^2\phi_{ac}}\right]^{1/2}\right\}$$

$$\tag{24}$$

式（24）是两电极静电同心球系统中实际轨迹在圆柱坐标系下的精确解，没有进行任何简化，这是作者自俄罗斯留学回国后导得的，公开发表在本人1993年出版的专著《宽束电子光学》[6]一书中，如该书的120页的式（5-64）所示。

为了与国外同样的研究做一个对比，作者在这里列出了俄罗斯科学家在1983年对同一问题的研究所推导的圆柱坐标系下两电极同心球系统中自阴极面逸出的电子轨迹的表示式[7,8]

$$\vec{r} = \frac{2R_k^3 n_1 e^{i\beta_0}(\cos\omega + i\sin\omega\cos\gamma)}{\phi_c^2 R_k^2 \cos^2\omega(1+\tan^2\omega\cos^2\gamma) - \Theta^2 R_k}\frac{1}{} \left\{ (\phi_c - 2\varepsilon_t n_1)\left(r_0\varepsilon_t\cos\omega - R_k\varepsilon_n^{1/2}\varepsilon_t^{1/2}w - \frac{\varepsilon_t r_0 w\cos\omega}{\cos\gamma}\right) + \right.$$

$$\frac{\phi_c}{2n_1}\left\{4\varepsilon_t n_1^2(R_k\varepsilon_n^{1/2}w + \varepsilon_t^{1/2}r_0\cos\omega)^2 + R_k^2[(\phi_c - 2\varepsilon_t n_1)w + 2\varepsilon_t n_1\cos\gamma]^2 - R_k^2\phi_c^2 w^2\right\}^{1/2} -$$

$$\left. 2\varepsilon_n^{1/2}\varepsilon_t^{3/2}R_k n_1\cos\gamma - \frac{\varepsilon_t r_0 w\phi_c}{\cos\gamma}\cos\omega \right\} + R_k w\tan\gamma e^{i\beta_0} \tag{25}$$

式中，β_0 为自阴极面逸出电子的方位角，r_0 为电子逸出的初高度，且

$$\Theta = 2R_k\varepsilon_n^{1/2}\varepsilon_t^{1/2}n_1\cos\gamma + r_0\cos\omega(2\varepsilon_t n_1 - \phi_c), \quad n_1 = \frac{R_k}{R_c} - 1, \quad \tan\gamma = \frac{r_0}{R_k - z}, \quad w = 1 - \frac{z}{R_k}$$

应该指出，俄罗斯科学家提出的公式（25）中符号的意义与本文有一些不同。在他们的表示式中，R_k、R_c 分别表示球面阴极和球面阳极的曲率半径，ϕ_c 表示阳极对于阴极的电位，ε_t、ε_n 分别表示逸出电子的切向初电位和法向初电位，其他均是电子自球面阴极逸

出时所对应的角度坐标。我们可以清晰地看出，与本文的圆柱坐标系下两电极静电同心球系统中电子轨迹表示式（24）相比较，以矢量形式表示的式（25）是如此繁复，所有的参量都混杂在一起，读者根本看不出哪些参量和因素在起主要作用，哪些是根本不重要的。因而难于总结其中存在的规律，更谈不上探讨理想成像和横向像差了。

作者认为，这原本是一个很简单的几何问题，也就是将球坐标系下表示的轨迹，换成圆柱坐标系下表示的轨迹而已。问题在于，俄罗斯的研究者们不是由同心球的球面的顶点出发，求逸出电子在子午面内行进轨迹，而是想寻找一个由球面阴极上的任一点向任意方向逸出的电子在圆柱坐标系下的解，从而把一个十分简单的问题复杂化了。本来，对于静电同心球电子光学系统，任何过系统中心的线都可以视为系统的轴线。这样，研究人员完全可以简单地选择圆柱坐标系的系统中心轴（它也是球对称系统的对称轴）作为研究的出发点，来寻找逸出电子在子午面内的行进轨迹。这样的处理，问题很容易解决，无非是将球坐标系换成圆柱坐标系而已，所得到的表示式是很简单而清晰的，物理意义是清晰的，可以给予精确的表述。

在这里，作者绝对无意嘲笑我们的俄罗斯同行。因为，作为一个科学家，在思考科学问题时常会想到，是直扑主题，还是追求更普遍形式的解。"把一个简单的科学问题复杂化"也常是我进行科学研究时的一个误区，以为是自己在追求这一科学问题的更为广泛的普适解。

式（24）是本文作者首先导得的，它较现有的电子光学文献[9]给出的表示式更为简洁。下面，作者将进一步考察式（24），抛弃一些更微小的量，进一步考察电子轨迹的性状，从中归纳出具有普遍意义的结论。

6　圆柱坐标系下的近轴轨迹（$r^* = r^*(z)$）[6]

如果我们在式（24）中引入近轴条件，略去分母中和大括弧内较1小得多的 ε_r/ϕ_{ac} 及更高的阶次项，便得到以圆柱坐标表示的自原点以初始条件参量 $(\varepsilon_0, \alpha_0)$ 射出的近轴轨迹，以 $r^*(z)$ 表示为

$$r^*(z) = 2(n-1)\left(\frac{\varepsilon_r}{\phi_{ac}}\right)^{1/2}(z+R_c)\left\{\left(\frac{\varepsilon_z}{\phi_{ac}}\right)^{1/2} - \left[\frac{-z}{(n-1)(z+R_c)} + \frac{\varepsilon_z}{\phi_{ac}}\right]^{1/2}\right\} \quad (26)$$

若在上式中引入两电极同心球系统的轴上电位分布表示式

$$\phi(z) = \phi_{ac}\frac{-z}{(n-1)(z+R_c)} \quad (27)$$

则可把式（24）表示成如下极为简明的形式

$$r^*(z) = 2z\sqrt{\frac{\varepsilon_r}{\phi(z)}}\left\{\sqrt{1+\frac{\varepsilon_z}{\phi(z)}} - \sqrt{\frac{\varepsilon_z}{\phi(z)}}\right\} \quad (28)$$

式（28）乃是实际轨迹表达式（24）的抽象与简化，$r^*(z)$ 之所以被称为近轴轨迹，它的轨迹即式（26）正是圆柱坐标系下静电阴极透镜的近轴轨迹方程

$$r''(z) + \frac{\phi'(z)}{2[\phi(z)+\varepsilon_z]}r'(z) + \frac{\phi''(z)}{4[\phi(z)+\varepsilon_z]}r(z) = 0 \quad (29)$$

在同心球系统的轴上电位分布式（27）的解，故我们称它为"近轴解"。这样，我们首次

证实了，若在一个具体的电子光学系统的实际轨迹表达式（24）中，舍弃高于电子径向初电位与加速电位的比值 ε_r/ϕ_{ac} 及其相关的高阶幂次项，即所谓满足近轴条件时，此轨迹即式（28）便可称为近轴轨迹，而它确实是近轴轨迹方程的解析解。这可以说是静电聚焦成像电子光学中极为难得的一例，是本文作者首先发现并证实的。

由式（28），不难求得行进中的近轴轨迹的斜率以及近轴电子的径向和轴向速度的表达式：

$$r'^* = \frac{2}{R_c+z}\left(\frac{\varepsilon_r}{\phi(z)}\right)^{1/2}\left\{z\left[\left(1+\frac{\varepsilon_z}{\phi(z)}\right)^{1/2}-\left(\frac{\varepsilon_z}{\phi(z)}\right)^{1/2}\right]+\frac{R_c}{2}\left(1+\frac{\varepsilon_z}{\phi(z)}\right)^{-1/2}\right\} \quad (30)$$

$$\dot{r}^* = \frac{1}{R_c+z}\left(\frac{2e}{m_0}\varepsilon_r\right)^{1/2}\left\{R_c+2z\left(1+\frac{\varepsilon_z}{\phi(z)}\right)^{1/2}\left[\left(1+\frac{\varepsilon_z}{\phi(z)}\right)^{1/2}-\left(\frac{\varepsilon_z}{\phi(z)}\right)^{1/2}\right]\right\} \quad (31)$$

$$\dot{z}^* = \left\{\frac{2e}{m_0}[\phi(z)+\varepsilon_z]\right\}^{1/2} \quad (32)$$

由此可见，当电子自轴上原点离开阴极后，其轴向速度 $\dot{z}^*$ 迅速增大，斜率 r'^* 急剧下降；在离阴极不远的某一位置处，轨迹便能处处满足条件 $r'^2 \ll 1$，近轴轨迹实际变为傍轴轨迹。在 $z = -\frac{1}{2}R_c$ 处，$(\dot{r}^*)^2$ 和 $(r'^*)^2$ 都趋于零。近轴电子的径向速度 $\dot{r}^*$ 在径向初速 v_{0r} 正值的情况下满足如下的条件式：

$$\frac{v_{0r}}{M} < \dot{r}^* < v_{0r} \quad (33)$$

式中，M 为线放大率。由此可见，自光阴极逸出的电子，其径向初速的变化是极其微小的。

这里，我们再一次强调，所谓"近轴解"，是指由近轴轨迹方程或近轴运动方程解得的解。计算表明，"近轴解"是足够精确代表所谓的"精确解"或"实际解"。因此，自轴上点逸出的近轴电子轨迹与实际电子轨迹几乎完全重合在一起，作图上是难以区分的。

表 1 给出了 $R_c = -25$ mm，$R_a = -5$ mm 的两电极同心球系统在 $\phi_{ac} = 10\ 000$ V 下以 $\varepsilon_0 = 1$ V，$\alpha_0 = 90°$ 逸出的实际电子轨迹与近轴电子轨迹的计算结果。由表 1 可见，实际轨迹与近轴轨迹的解是何等接近，基本重合在一起，以致误差难于用作图表示。

表 1　两电极静电聚焦同心球系统中的实际电子轨迹与近轴电子轨迹

同心球型像管参数	离阴极面之距离 z/mm	轴上电位分布 ϕ/V	实际轨迹		近轴轨迹	
			$r(z)$ /mm	$r'(z)$	$r^*(z)$ /mm	$r^{*'}(z)$
$R_c = -25$ mm	4	476.170 5	0.366 620 0	0.037 133 7	0.366 605 6	0.037 147 0
$R_a = -5$ mm	8	1 176.470 5	0.466 520 0	0.015 470 6	0.466 475 2	0.015 434 8
$\phi_{ac} = 10\ 000$ V	12	2 307.692 3	0.499 692 1	0.001 620 5	0.499 598 4	0.001 601 3
$\varepsilon_0 = 1$ V	16	4 444.444 4	0.480 120 6	-0.011 682 7	0.480 000 0	-0.0116 666
$\alpha_0 = 90°$	20	1 0000.000 0	0.400 319 8	-0.030 088 6	0.400 000 0	-0.0300 000

7 实际轨迹交轴位置及其斜率的确定[6]

以上关于轨迹方程的推导是为研究电子光学成像及像差服务的。现在我们确定由轴上点逸出的电子通过阳极后轨迹的性状。由式（14）~式（17），不难求得自阴极上原点以初电位 ε_0、初角度 α_0 射出到达阳极 A 上 D 点的电子轨迹，由此可以确定电子轨迹在 D 点的切线与矢径 OD 间之夹角 τ 和 D 点的角度坐标 γ。

假定阳极是对于电子透明的带电的理想网——等电位的栅状电极，在栅状阳极后面的空间为无场空间。当电子经过带电的等电位理想网时，假定电子速度的大小和方向都不变，轨迹的解及其斜率都是连续的。因此，通过栅状阳极后的电子轨迹将是一条切于过 D 点轨迹的直线。下面我们将证明，对于凹面阴极 - 凸面阳极系统，当 $n > 2$ 时，轨迹是会聚的；它交于 $\varphi = 0$ 的轴线于 I_α 点，形成阴极面的实像（图 1-1（a））。对于 $1 < n \leqslant 2$ 的凹面阴极 - 凸面阳极系统或 $n < 1$ 的凸面阴极 - 凹面阳极系统，轨迹是发散的；则切于 D 点的轨迹向后延长线亦交于 I_α 点，形成阴极面的虚像（图 1（b）和图 1（c））。

按照图 1 所给出的三角形 $\triangle ODI_\alpha$，不难求得交点 I_α 位置的表达式

$$R_{i\alpha} = R_a \frac{\sin \tau}{\sin \alpha_i} = R_a \frac{\tan \tau}{\sin \gamma + \tan \tau \cos \gamma} \tag{34}$$

式中，α_i 为电子轨迹在交点 I_α 处与 $\varphi = 0$ 轴线的交角；$R_{i\alpha}$ 为曲率中心 O 至交点 I_α 的距离。

显然，若令 $\rho = R_a$，则 $\mu = n$，$\varphi = \gamma$，$\zeta = \tau$，$b = b_1$，$c = c_0$；于是式（14）~式（16）可写成如下形式

$$\sin \gamma = -\frac{(n-1)(c_0 + d_0)}{b_1 d_0 + b_0 c_0} \sin \alpha_0 \tag{35}$$

$$\cos \gamma = \frac{b_1 c_0 + b_0 d_0}{b_1 d_0 + b_0 c_0} \tag{36}$$

$$\tan \tau = \frac{n}{b_1} \sin \alpha_0 \tag{37}$$

式中，

$$\begin{aligned} b_0 &= \cos \alpha_0, \\ c_0 &= 1 - 2n(n-1)\frac{\varepsilon_0}{\phi_{ac}} \sin^2 \alpha_0, \\ d_0 &= 1 - 2(n-1)\frac{\varepsilon_0}{\phi_{ac}} \sin^2 \alpha_0, \\ b_1 &= \left[1 + \frac{\phi_{ac}}{\varepsilon_0} - n^2 \sin^2 \alpha_0 \right]^{1/2} \end{aligned} \tag{38}$$

将式（35）~式（37）三个表达式代入式（34）中，便得到两电极同心球系统中以初角度 α_0、初电位 ε_0 自阴极物点射出的电子轨迹交轴位置的精确表达式

$$R_{i\alpha} = -R_a \frac{n(b_1 d_0 + b_0 c_0)}{b_1(n-1)(c_0 + d_0) - n(b_1 c_0 + b_0 d_0)} \tag{39}$$

同样，由

$$\tan\alpha_i = \tan(\gamma+\tau) = \frac{\tan\gamma+\tan\tau}{1-\tan\gamma\tan\tau} \tag{40}$$

不难求得轨迹在 I_α 处的斜率的精确表达式

$$\tan\alpha_i = -\frac{b_1(n-1)(c_0+d_0)-n(b_1c_0+b_0d_0)}{b_1(b_1c_0+b_0d_0)+n(n-1)(c_0+d_0)\sin^2\alpha_0}\sin\alpha_0 \tag{41}$$

由式（39）和式（41）可见，如果给出系统的结构参量 R_a、R_c 和电参量 ϕ_{ac}，则不难求得电子自阴极面以初始条件参量 (ε_0,α_0) 射出时电子轨迹的交轴位置 $R_{i\alpha}$ 及其斜率 $\tan\alpha_i$ 的精确值。

现简化 $R_{i\alpha}$ 和 $\tan\alpha_i$ 的表达式。为此，用二项式定理按 $\left(\frac{\varepsilon_0}{\phi_{ac}}\right)^{1/2}$ 乘幂展开式（39）和式（41）中各项，略去较 1 小得多的 $\left(\frac{\varepsilon_0}{\phi_{ac}}\right)^{3/2}$ 等以上高阶乘幂项，便得到轨迹交轴位置 $R_{i\alpha}$ 及其斜率 $\tan\alpha_i$ 的表示式分别为

$$R_{i\alpha} = -R_a\frac{n}{n-2}\left[1+\frac{2(n-1)}{(n-2)}\left(\frac{\varepsilon_0}{\phi_{ac}}\right)^{1/2}\cos\alpha_0+\frac{2n(n-1)}{(n-2)^2}\frac{\varepsilon_0}{\phi_{ac}}\cos^2\alpha_0-\frac{2(n-1)^2}{(n-2)}\frac{\varepsilon_0}{\phi_{ac}}\sin^2\alpha_0\right] \tag{42}$$

$$\tan\alpha_i = -(n-2)\left(\frac{\varepsilon_0}{\phi_{ac}}\right)^{1/2}\sin\alpha_0\left[1-\frac{2(n-1)}{(n-2)}\left(\frac{\varepsilon_0}{\phi_{ac}}\right)^{1/2}\cos\alpha_0+\frac{3n-2}{2(n-2)}\left(\frac{\varepsilon_0}{\phi_{ac}}\right)\cos^2\alpha_0+\right.$$
$$\left.\frac{(n-1)(n^2-n+2)}{2(n-2)}\left(\frac{\varepsilon_0}{\phi_{ac}}\right)\sin^2\alpha_0\right] \tag{43}$$

由此出发，可以研究静电聚焦同心球系统的电子光学性质及其像差。

结束语

本文研究了两电极静电聚焦同心球系统的逸出电子在极坐标系与圆柱坐标系下的运动轨迹，导出了新的轨迹表示式，它能方便地确定自阴极面逸出的电子轨迹的实际交轴位置及其倾角。本文导出了该系统在圆柱坐标系下由阴极面原点逸出的实际电子轨迹的解析表示式（24）；纠正了一些文献中相关表示式的谬误，证明了若实际轨迹满足近轴条件，其电子轨迹的解可以由近轴轨迹方程求得。本文给出了两电极静电聚焦同心球系统自阴极面逸出的电子轨迹落点及其斜率的精确解表示式（39）与式（41），为研究电子光学系统的理想成像与横向像差奠定基础。

参 考 文 献

[1] RUSKA E. Zur fokussierbarkeit von kathoden-strahlbundeln grosser ausgangsquerchnitte [J]. Z. Angew. Physik. 1933,83(9):684-687 (in German).

[2] SCHAGEN P H, BRUINING H, FRANCKEN J C. A simple electrostatic electron-optical system with only one voltage [J]. Philips Research Report, 1952, 7(2):119-132.

[3] 谢曼 О И. 电子光学理论基础[M]. 朱宜, 译. 北京:高等教育出版社,1958.

[4] ЧЖОУ Л В (ZHOU LIWEI). Исследование электростатических фокусирующих

систем электронно-оптических преобразователей изобржении с шаровой и осевой симметрией [D]. Ленинград:Диссертация ЛЭТИ. ,1966.

[5] ЧЖОУ Л В. Об аберрации в электронной оптической системе с шаровой и осевой симметрией[C]. Тезисы докладов XX1 Научно-Технической Конференции, Ленинград, 1966:137－138.

[6] 周立伟. 宽束电子光学[M]. 北京理工大学出版社, 1993.

[7] ИГНАТЬЕВ А Н, КУЛИКОВ Ю В. Математическая модель катодной линзы типа сферический конденсатор—Новые методы расчета электронно-оптических систем [M]. Москва：Издателъство（Наука）, Сибирское отделение, 1983:131－133.

[8] ИЛЬИН В П, КАТЕШОВ В А, КУЛИКОВ Ю В, МОНАСТЕРСКИЙ М А. Численные методы оптимизации эмиссионных электронно-оптических систем [M]. Москва：Издателъство（Наука）, Сибирское отделение,1987.

[9] КЕЛЬМАН В М, САПАРГАЛИЕВ А А, ЯКУШЕВ Е М. Теория катодных линз, II—Электростатическая линза с вращательной симметрией поля[J]. Ж. Т. Ф. ,1973, 43 (1):52－56.

1.2 B章：近轴横向色差与几何横向球差
Paraxial Lateral Chromatic Aberration and Geometrical Lateral Spherical Aberration

摘要：在前文的基础上，本文进一步研究两电极静电聚焦同心球系统的轴上点电子光学横向色差。研究表明，对于成像电子光学系统，由光阴极发射的电子所形成的图像弥散，其轴上点的图像弥散系由近轴横向色差与几何横向球差两部分组成，证明了在静电聚焦同心球系统中，阴极透镜的二级近轴横向色差即 Recknagel-Арцимович 公式普遍成立。文中研究了宽电子束与细电子束之间轴上物点横向色差之差异，最后讨论了两电极同心球系统向近贴聚焦系统过渡的特例。

Abstract: On the basis of the previous paper, this paper goes further to study the electron optical lateral aberration of axial point for the bi-electrode electrostatic focusing concentric spherical system. It shows that the image diffusion of axial point formed by the electrons emitted from the photocathode in an electron optical system is composed of two parts, the paraxial lateral chromatic aberration and the geometric lateral spherical aberration. Confirming that in the electrostatic focusing concentric spherical system, the second order paraxial lateral chromatic aberration of cathode lens, that is, the Recknagel-Арцимович formula, is generally valid. This paper also studies the differences of lateral aberrations of axial object point between the wide electron beam optics and the narrow electron beam optics. Finally, a special case of transit of proximity focusing system from the bi-electrode electrostatic focusing concentric spherical system is also investigated.

引 言

在上一篇文章中，我们导出了两电极静电聚焦同心球系统中自阴极面逸出的电子轨迹在极坐标系与圆柱坐标系下新的表示式：$\rho = f(\varphi)$ 与 $r = r(z)$。研究发现，若自阴极面逸出的电子轨迹的表示式中舍弃电子径向初电位与加速电位的比值 ε_r/ϕ_{ac} 及其相关的高阶幂次项，则自光阴极逸出的电子束，如果其逸出电子的轴向初能量都相同，就能实现理想聚焦。本文在此基础上研讨两电极静电聚焦同心球系统中电子光学的理想成像、轴上物点的横向色差，以及它所组成的近轴横向色差与几何横向球差。

周立伟（Zhou Liwei）北京理工大学. 光学学报（Acta Optica Sinica），V. 42, No. 8, 2022, 0811002 – 1 – 7.

1 两电极同心球系统的成像电子光学性质

1.1 近轴条件下的理想成像

定义

$$\varepsilon_z = \varepsilon_0 \cos^2\alpha_0, \quad \varepsilon_r = \varepsilon_0 \sin^2\alpha_0 \tag{1}$$

为电子的轴向和径向初速对应的初电位。若在前面"电子轨迹方程"给出的两电极同心球系统的实际轨迹表示式（42）与式（43）中，考虑成像电子光学系统的近轴条件，略去含有较1小得多的 ε_r/ϕ_{ac} 的各项，便得到自阴极面上原点以初电位 ε_0、初角度 α_0 逸出的近轴轨迹到达成像位置 I^* 处的距离为 R_i^* 和斜率为 $\tan\alpha_i^*$ 的表达式

$$R_i^* = -R_a \frac{n}{n-2}\left[1 + \frac{2(n-1)}{(n-2)}\left(\frac{\varepsilon_z}{\phi_{ac}}\right)^{1/2} + \frac{2n(n-1)}{(n-2)^2}\left(\frac{\varepsilon_z}{\phi_{ac}}\right)\right] \tag{2}$$

$$\tan\alpha_i^* = -(n-2)\left(\frac{\varepsilon_r}{\phi_{ac}}\right)^{1/2}\left[1 - \frac{2(n-1)}{(n-2)}\left(\frac{\varepsilon_z}{\phi_{ac}}\right)^{1/2} + \frac{3n-2}{2(n-2)}\frac{\varepsilon_z}{\phi_{ac}}\right] \tag{3}$$

上标星号（*）表示满足近轴条件的情况。

式（2）表明，由阴极面轴上点以相同的轴向初速对应的初电位 ε_z 射出的电子束（其初角度 α_0 和初电位 ε_0 都不相同）经历同心球系统场后将穿过透明的栅状阳极后会聚于同一点（当 $n>2$ 时，实像情况），或者好像是从阴极后面某一点发出一般（当 $n<2$ 时，虚像情况）。因此，在近轴条件下，只有 ε_z 相同的电子才能理想聚焦；当 ε_z 改变时，理想成像的位置也随着变动。

1.2 系统放大率

由于系统的球对称性，阴极球面上的弧段成像同样亦是曲率半径为 R_i^* 的弧段，此时横向放大率（简称放大率）M 由式（2）可表示为

$$M = \frac{R_i^*}{R_c} = -\frac{1}{n-2}\left[1 + \frac{2(n-1)}{(n-2)}\left(\frac{\varepsilon_{z_1}}{\phi_{ac}}\right)^{1/2} + \frac{2n(n-1)}{(n-2)^2}\frac{\varepsilon_{z_1}}{\phi_{ac}}\right] \tag{4}$$

很明显，在 $n>2$ 时，形成倒立的实像；在 $n<2$ 时，形成正立的虚像。

可以看出，在两电极情况下，系统的放大率以及像面位置主要取决于系统的结构参量，电参量与初始条件参量的影响则是次要的。因此，两电极的成像电子光学系统亦称为定焦系统或自聚焦系统。

当 $\varepsilon_z=0$，即 $\alpha_0=90°$ 时，切于阴极面的近轴电子射线将会聚于极限像面位置 I_t^* 处。于是，对应于极限像面的 R_{it}^* 和 M_t 由下式确定：

$$R_{it}^* = -R_a \frac{n}{n-2} \tag{5}$$

$$M_t = -\frac{1}{n-2} \tag{6}$$

式（5）、式（6）就是通常文献[1~4]中见到的零级近似式。值得注意的是，R_{it}^* 和 M_t 与阳极对于阴极的加速电位 ϕ_{ac} 无关。自然，令 $\varepsilon_z=\varepsilon_0$，$\alpha_0=0°$，则可确定对应于高斯像面 I_g^* 处的 R_{ig}^* 和 M_g 值。

1.3 物像空间不变式

物像空间不变式也就是几何光学中一般所说的 Lagrange-Helmholtz（拉格朗日 – 亥姆霍兹）不变式，它代表光学系统在理想条件下的成像特性，也包括电子光学系统在近轴条件下理想成像的普遍特性。对于两电极静电聚焦同心球系统，在近轴成像下，如果定义角放大率 Γ 为

$$\Gamma = \frac{\tan \alpha_i^*}{\tan \alpha_0} \tag{7}$$

则由式（3）、式（4）和式（7），可得角放大率 Γ 与横向放大率 M 的乘积为

$$M\Gamma = \left(\frac{\varepsilon_z}{\phi_{ac}}\right)^{1/2} \left[1 - \frac{\varepsilon_z}{2\phi_{ac}} + \frac{3}{8}\left(\frac{\varepsilon_z}{\phi_{ac}}\right)^2\right] = \sqrt{\frac{\varepsilon_z}{\phi_{ac} + \varepsilon_z}} \tag{8}$$

式中，与 M、Γ 相关的式（3）、式（4）内包括 $\left(\frac{\varepsilon_z}{\phi_{ac}}\right)^2$ 项的考虑。式（8）就是几何光学中极为重要的拉格朗日 – 亥姆霍兹不变式，说明电子光学系统与几何光学系统的理想成像是类似的。因此，在阴极透镜中，当已知物方各量 $\sqrt{\varepsilon_z}$、$\tan \alpha_0$、r_0，则对任意一个像方空间来说，其像方各量 $\sqrt{\phi_i + \varepsilon_z}$、$\tan \alpha_i^*$、$r_i^*$ 三者的乘积为常数，由式（8）可得

$$r_i^* \tan \alpha_i^* \sqrt{\phi_i + \varepsilon_z} = r_0 \tan \alpha_0 \sqrt{\varepsilon_z} = 常数 \tag{9}$$

2 两电极同心球系统的近轴横向色差与几何横向球差

在阴极透镜中，色差与球差通常是互相联系、密切相关的。人们经常用名词色球差概括由于逸出电子不同的初角度和不同的初速度综合作用所引起的像差。但是，这个名词较为笼统，人们并不清楚究竟是哪一类像差（色差还是球差）对于图像的模糊所起的作用。在本文中，作者建议用近轴色差表示在近轴条件下由于逸出电子轴向初能的差异所导致的像差，几何球差表示因逸出电子几何量的差异所导致的像差。

对于成像电子光学系统，定义从阴极面原点逸出的轴向初电位为 ε_{z_1} 的电子按近轴条件聚焦处 I^* 为理想像面位置，而从同一点发出的初电位 ε_0、初角度 α_0 的实际电子轨迹交于 I_{α_0} 处，则产生像的位移，即纵向像差，以 Δz 表示，如图 1 所示。

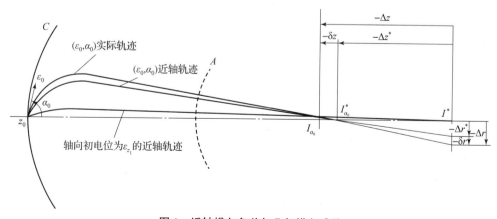

图 1 近轴横向色差与几何横向球差

显然

$$\Delta z = R_{i\alpha} - R_i^* \tag{10}$$

式中，$R_{i\alpha}$ 即上一文章导出的式 (39)；R_i^* 以式 (2) 表示，不过其中的 ε_z 以 ε_{z_1} 代之。将它们代入式 (10) 中，经过整理，便得到如下的纵向像差 Δz 表示式：

$$\Delta z = \Delta z^* + \delta z = \Delta z_1^* + \Delta z_2^* + \delta z_2 \tag{11}$$

式中，

$$\Delta z_1^* = -\frac{2M^2\sqrt{\phi_{ac}}}{E_c}(\sqrt{\varepsilon_z} - \sqrt{\varepsilon_{z_1}}) \tag{12}$$

$$\Delta z_2^* = -\frac{2M^2}{E_c}\left[\frac{n}{n-2}(\varepsilon_z - \varepsilon_{z_1}) + \frac{4(n-1)}{(n-2)}\sqrt{\varepsilon_{z_1}}(\sqrt{\varepsilon_z} - \sqrt{\varepsilon_{z_1}})\right] \tag{13}$$

$$\delta z_2 = \frac{2M^2(n-1)}{E_c}\varepsilon_r \tag{14}$$

同样，在轴向初能为 ε_{z_1} 所确定的近轴成像位置处的轴上点横向像差 Δr 可表示为

$$\Delta r = -\Delta z \tan \alpha_i \tag{15}$$

式中，$\tan \alpha_i$ 表示对应于初始条件参量 $(\varepsilon_0, \alpha_0)$ 的实际电子轨线在成像点 I^* 处与 $\varphi = 0$ 轴线所造成的斜率，它由上一文章导出的式 (43) 确定。

将式 (11) 和式 (43) 代入式 (15) 中，则轴上点的横向像差 Δr 可表示如下：

$$\Delta r = \Delta r^* + \delta r = \Delta r_2^* + \Delta r_3^* + \delta r_3 \tag{16}$$

式中，

$$\Delta r_2^* = \frac{2M}{E_c}\sqrt{\varepsilon_r}(\sqrt{\varepsilon_z} - \sqrt{\varepsilon_{z_1}}) \tag{17}$$

$$\Delta r_3^* = -\frac{2M}{E_c\sqrt{\phi_{ac}}}\sqrt{\varepsilon_r}(\varepsilon_z - \varepsilon_{z_1}) \tag{18}$$

$$\delta r_3 = -\frac{2M}{E_c\sqrt{\phi_{ac}}}(n-1)(\varepsilon_r)^{3/2} \tag{19}$$

上述诸式中各参量意义如下：

E_c——阴极面的电场强度，$E_c = \dfrac{\phi_{ac}}{R_c(n-1)}$；

M——系统的横向放大率，以本文的式 (4) 表示；

Δz^*、Δr^*——近轴纵向色差与近轴横向色差，即初始条件参量 $(\varepsilon_0, \alpha_0)$ 的近轴轨迹在对应于某一轴向初能 ε_{z_1} 所确定的近轴成像位置的纵向与横向偏离；

Δz_j^* $(j = 1, 2)$——一级与二级近轴纵向色差；

Δr_j^* $(j = 2, 3)$——二级与三级近轴横向色差；

δz，δr——几何纵向球差与几何横向球差，即实际轨迹在对应的近轴成像位置上的纵向与横向偏离。其中，δz_2 为二级几何纵向球差；δr_3 为三级几何横向球差。

这里，Δz^*、Δr^* 与 δz、δr 均自成像点算起，自左至右为正，自下至上为正。

在本文中，轴上点的纵向像差和横向像差用阶次分类以区别于细束电子光学通常所用的三级和高级像差。这是因为在阴极透镜中，从阴极面发出的电子除了初能量的差别外，

还具有不同的逸出角,且轨迹的斜率可能趋于无穷,加之理想像面位置与所取的轴向初电位 ε_{z_1} 有关,用细束电子光学通常的表示法划分纵向像差和横向像差已不可能。在这里,纵向像差按照 $(\varepsilon_0/\phi_{ac})^{j/2}$ 的阶次分类,横向像差则按照 $(\varepsilon_0/\phi_{ac})^{(j+1)/2}$ 的阶次分类。

考察式(16)及其式(19),可得出以下结论:

轴上点的横向像差 Δr 可视为近轴横向色差 Δr^*(即由轴向初能 ε_{z_1} 与 ε_z 不同的两条近轴轨迹所形成的像差)与几何横向球差 δr(即由同一初始条件参量的实际轨迹与近轴轨迹所形成的像差)的合成,如图1所示。其中,式(17)表示的二级近轴横向色差 Δr_2^*,以两位研究电子光学先驱者名字 Recknagel 和 Арцимович 命名,亦称 R – A 公式,它对于确定系统的成像质量即鉴别率起着主要的、决定的作用。它较式(18)的三级近轴横向色差 Δr_3^* 和式(19)的三级几何横向球差 δr_3 要大一个数量级。几何横向球差 δr_3 正是通常电子光学中的所谓三级几何球差,它与 $\left(\dfrac{\varepsilon_r}{\phi_{ac}}\right)^{3/2}$ 成比例。通过三级近轴横向色差 Δr_3^*,我们便把通常的二级近轴横向色差 Δr_2^* 与三级几何横向球差 δr_3 联系了起来。

应该指出,本文所导得的二级近轴横向色差 Δr_2^* 被谢曼[1]称为 Recknagel-Арцимович 公式[2,3],简称 R – A 公式,是成像电子光学中最重要的一个表示式。它所形成的图像弥散,我们将在后面"最小离散圆与最佳像面位置的确定"中详细讨论。

在这里,我想指出,虽然本文中的个别结论曾在一些学者关于同心球系统电子光学的研究中得出过,例如,Крупп[4]证实了高斯像面上的横向像差为 $\Delta r_g = \dfrac{2M\varepsilon_0}{E_c}$,Schagen[5] 给出了极限像面上的横向像差为 $\Delta r_t = \dfrac{M\varepsilon_0}{E_c}$,Арцимович[3] 给出了最佳像面上的横向像差为 $\Delta r_m = 0.6\dfrac{M\varepsilon_0}{E_c}$。谢曼[1]则是最早注意到在阴极透镜中,其图像弥散值在极限像面、最佳像面与高斯像面存在着 1 : 0.6 : 2 关系。本文作者的贡献是在具体的静电聚焦同心球系统中证实了这一规律。

顺便指出,一些文献对于轴上点所构成的横向像差乃是近轴横向色差与几何横向球差的合成,并不严格区分,其认识也是含混的。而且,在 Кельман(1973)[6]等电子光学学者关于静电阴极透镜像差的著述中,所关心的仍是三级几何横向球差 δr_3 等,虽然它是处处存在的。实际上,它较二级近轴横向色差 Δr_2^* 要小一个数量级,对于成像质量的影响是较为次要的,甚至可以不予考虑。

3 细电子束下的横向球差[7]

下面用几何电子光学通常采用的方法,表示细电子束下高斯像面上的横向球差。当阴极面物点以很小的倾角 α_0 射出电子束时,角度的余弦可用下式表达:

$$\cos\alpha_0 = 1 - \frac{1}{2}\sin^2\alpha_0 - \frac{1}{8}\sin^4\alpha_0 \tag{20}$$

把它代入式(14)中,即

$$\Delta r = \frac{2M}{E_c}\left[\sqrt{\varepsilon_r}(\sqrt{\varepsilon_z} - \sqrt{\varepsilon_{z_1}}) - \frac{1}{\sqrt{\phi_{ac}}}\sqrt{\varepsilon_r}(\varepsilon_z - \varepsilon_{z_1}) - \frac{1}{\sqrt{\phi_{ac}}}(n-1)(\varepsilon_r)^{3/2}\right] \tag{21}$$

令 $\varepsilon_{z_1} = \varepsilon_0 \cos\delta_0$，$\varepsilon_z = \varepsilon_0 \cos\alpha_0$，展开上式，便可得到

$$\Delta r = \frac{2M}{E_c}\varepsilon_0 \sin\alpha_0 \left[(1-\cos\delta_0) - \frac{1}{2}\sin^2\alpha_0 - \frac{1}{8}\sin^4\alpha_0 - \frac{\sqrt{\varepsilon_0}}{\sqrt{\phi_{ac}}}(1-\cos^2\delta_0) - \frac{\sqrt{\varepsilon_0}}{\sqrt{\phi_{ac}}}(n-2)\sin^2\alpha_0\right] \tag{22}$$

在细电子束情况下，若像面置于高斯像面处，即令 $\cos\delta_0 = 1$，于是，其横向球差（包括三级与五级）可表示为

$$\Delta r_g^{3+5} = -\frac{M\varepsilon_0}{E_c}\sin^3\alpha_0 \left(1 + \frac{1}{4}\sin^2\alpha_0\right) \tag{23}$$

通常，只需考虑三级横向球差项，即

$$\Delta r_g^3 = -\frac{M\varepsilon_0}{E_c}\sin^3\alpha_0 \tag{24}$$

这与通常的细束电子光学的结论是一样的。

若令 $\sin\alpha_0 = 0.14$，$\alpha_0 = 8°$，则三级横向球差在全部球差中占 99% 以上。由此可见，细束电子光学中研究高斯像面上的像差只须考虑三级球差项。

若 $\delta_0 \neq 0$ 时，则出现一级横向球差。当 $\delta_0 \leqslant \alpha_0$ 时，一级横向球差与三级、五级球差有抵消的作用；随着成像面离高斯面愈远，一级横向球差在全部球差中表现愈显著。在极限像面时，一级横向球差达到最大值。令极限像面上的横向球差为 Δr_t^s，则有

$$\Delta r_g^s \simeq \frac{\sin^2\alpha_0}{2}\Delta r_t^s \tag{25}$$

这表明，细束下高斯像面上的球差远小于极限像面上的球差。必须指出，尽管细电子束下的理想成像面通常指的是高斯像面，但细束最小弥散圆对应的最佳像面并不位于高斯像面处，而是位于离高斯像面不远处，它与细束下被限制的逸出角 α_0 所对应的位置相靠近。其值可估计如下：

上面已经求出，高斯像面上的散射圆半径为

$$\Delta r_g = -\frac{M\varepsilon_0}{E_c}\sin^3\alpha_0 \left(1 + \frac{1}{4}\sin^2\alpha_0\right) \tag{26}$$

对式（22）求极值条件，并利用求宽束下最小弥散圆半径式相似的方法[7]，可求得细束下的最小弥散圆半径为

$$\Delta r_{\min}^s = -\frac{1}{4}\frac{M\varepsilon_0}{E_c}\sin^3\alpha_0 \left(1 + \frac{5}{16}\sin^2\alpha_0\right) \tag{27}$$

由式（26）和式（27）可估计细束下 Δr_g^s 和 $\Delta r_{\min}^s$ 的比值：

$$\Delta r_{\min}^s = \frac{1}{4}\frac{1}{1-\frac{1}{16}\sin^2\alpha_0}\Delta r_g^s \tag{28}$$

因此，很明显，用 Крупп[4] 所提出的如下表达式

$$\Delta r_{\min}^s = \frac{1}{1+\frac{1}{\sin\alpha_0}}\Delta r_g^s \tag{29}$$

来估计细束下最小弥散圆半径是不能成立的。

4 纵向均匀电场的电子光学——同心球系统向近贴聚焦系统的过渡[7]

对于两电极同心球系统，若令 $n=1$，即 R_c、R_a 都趋于无穷大。于是球面阴极、栅状球面阳极都变成平面电极，同心球系统就变成近贴聚焦系统，如图 2 所示。

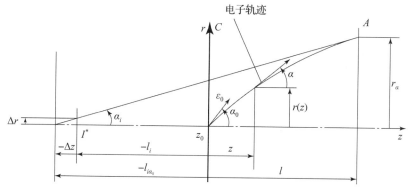

图 2 近贴聚焦系统（$n=1$）的电子轨迹与近轴横向色差

按照图，平面阴极至平面阳极的距离 l 可表示为

$$l = (R_a - R_c) \Big|_{\substack{R_a \to \infty \\ R_c \to \infty}} \tag{30}$$

l 的符号自阴极面算起，自左至右为正。于是

$$n = \frac{R_c}{R_a} = \left(1 - \frac{l}{R_a}\right)\Big|_{R_a \to \infty} \to 1 \qquad \mu = \frac{R_c}{\rho} = \frac{1}{1+\dfrac{z}{R_c}}\Big|_{R_c \to \infty} \to 1$$

$$p(\mu - 1) = \frac{\phi_{ac}}{\varepsilon_0} \frac{(\mu-1)}{(n-1)} = \frac{\phi_{ac}}{\varepsilon_0} \frac{\dfrac{-z}{z+R_c}}{\dfrac{-l}{R_a}}\Bigg|_{\substack{R_c \to \infty \\ R_a \to \infty}} = \frac{\phi_{ac}}{\varepsilon_0} \frac{z}{l} \tag{31}$$

将上述关系式代入"电子轨迹方程"中的式（12），则纵向均匀电场中电子轨迹的斜率为

$$r'(z) = \frac{\sqrt{\varepsilon_r}}{\sqrt{\dfrac{\phi_{ac}}{l}z + \varepsilon_z}} \tag{32}$$

其电子轨迹表达式为

$$r(z) = \frac{2\sqrt{\varepsilon_r}}{\dfrac{\phi_{ac}}{l}}\left(\sqrt{\dfrac{\phi_{ac}}{l}z + \varepsilon_z} - \sqrt{\varepsilon_z}\right) \tag{33}$$

式（33）表示实际的电子轨线，若令 $\phi(z) = \dfrac{\phi_{ac}}{l}z$，则式（33）给出的近轴轨迹与"电子轨迹方程"中的式（28）在形式上是完全一样的，而且，纵向均匀电场中实际轨迹与近轴轨迹是完全重合的。

现讨论系统的电子光学性质。首先研究虚像面的位置。设 l_i 为阴极至虚像面的距离，它自阴极面算起自左至右为正。故可表示为

$$l_i = (R_{i\alpha} - R_c)\Big|_{\substack{R_c \to \infty \\ R_a \to \infty}} \tag{34}$$

将"电子轨迹方程"中的式（42）的 $R_{i\alpha}$ 代入式（34）中，将其中各项经过类似式（31）的交换，如其展开式中的第一项可表示为

$$\left(-R_a \frac{n}{n-2} - R_c\right)\Big|_{\substack{R_c \to \infty \\ R_a \to \infty}} = -l \tag{35}$$

最后可得

$$l_i = -l\left[1 - 2\left(\frac{\varepsilon_0}{\phi_{ac}}\right)^{1/2}\cos\alpha_0 + 2\frac{\varepsilon_0}{\phi_{ac}}\cos\alpha_0 - \left(\frac{\varepsilon_0}{\phi_{ac}}\right)^{3/2}\cos^3\alpha_0\right] = -l\left[\sqrt{1 + \frac{\varepsilon_z}{\phi_{ac}}} - \sqrt{\frac{\varepsilon_z}{\phi_{ac}}}\right]^2 \tag{36}$$

由式（36）出发，可讨论近贴聚焦系统的电子光学性质：

（1）由阴极面上原点以相同的 ε_z 射出的电子射线经过纵向均匀场后好像是自阴极面后某一点发出一般，式（36）表明，只有 ε_z 相同的电子才能理想聚焦。

（2）由式（4），可以得到 $M = +1$，表明是正立的虚像。同样可以证明 Lagrange-Helmholtz 不变式式（8）成立。

（3）极限（虚）像平面位置由式（36）可得

$$l_{it} = -l \tag{37}$$

这表明，它较高斯（虚）像面离阴极的距离远。

（4）平面阳极上的最大散射圆半径及其斜率可表示为

$$r_a\Big|_{\alpha_0 = \frac{\pi}{2}} = 2l\sqrt{\frac{\varepsilon_0}{\phi_{ac}}}, \quad \tan\alpha_i\Big|_{\alpha_0 = \frac{\pi}{2}} = \sqrt{\frac{\varepsilon_0}{\phi_{ac}}} \tag{38}$$

由此可见，切于阴极面发出的电子轨线在经过纵向均匀场后与阴极法线成很小的倾斜角。阴极前面的电场虽然把从阴极某点射出的电子迅速会聚成细束，但在阳极平面上不能形成点状的像，而是半径为 r_a 的散射圆。因此，它实际上是"投射成像"。

由于近贴聚焦系统，$M = +1$，同心球系统的像差公式（11）和公式（16）可完全推广到近贴聚焦的情况，如一级与二级纵向近轴色差可表示如下：

$$\Delta z_1^* = \frac{-2\sqrt{\phi_{ac}}}{E_c}(\sqrt{\varepsilon_z} - \sqrt{\varepsilon_{z_1}}), \quad \Delta z_2^* = \frac{2}{E_c}(\varepsilon_z - \varepsilon_{z_1}) \tag{39}$$

对于二级与三级横向近轴色差，则有

$$\Delta r_2^* = \frac{2\sqrt{\varepsilon_r}}{E_c}(\sqrt{\varepsilon_z} - \sqrt{\varepsilon_{z_1}}), \quad \Delta r_3^* = -\frac{2\sqrt{\varepsilon_r}}{E_c\sqrt{\phi_{ac}}}(\varepsilon_z - \varepsilon_{z_1}) \tag{40}$$

此处 $E_c = \frac{-\phi_{ac}}{l}$。由于实际轨迹与近轴轨迹合二为一，故此系统只剩下近轴纵向色差 Δz^* 与近轴横向色差 Δr^*，几何纵向球差 δz 和几何横向球差 δr 均等于零。因此，不论是两电极同心球系统或是近贴聚焦系统，二级近轴横向色差公式是普遍成立的。

结束语

本文指出，静电聚焦同心球系统的轴上点横向像差由近轴横向色差与几何横向球差组

成。证实了在静电聚焦同心球系统中,阴极透镜的二级近轴横向色差公式即 R - A 公式(17)普遍成立,探讨了系统的近轴横向色差与几何横向球差,以及宽电子束与细电子束之间的相互关系。

文中对 $n<1$ 和 $n>1$ 时球面阴极的两电极同心球系统和 $n=1$ 时平面阴极的近贴聚焦系统的成像与像差都进行了考察,所得的结论对于研究定焦型像管和近贴聚焦型像管具有实际意义,对于研究阴极透镜的近轴光学及其横向像差具有指导意义。

参 考 文 献

[1] 谢曼 О И. 电子光学理论基础[M]. 朱宜,译. 北京:高等教育出版社,1958.

[2] RECKNAGEL A. Theorie des elektrisohen elektronen miktroskops fur selbstrakler[J]. Z. Angew. Physik,1941,117: 689.

[3] АРЦИМОВИЧ Л А. Электростатические свойства эмисионных систем[J]. Изд, АН СССР Сер. Физ. ,1944,8(6):313.

[4] КРҮПП Д М. Кружок сферической аберрации в катодной линзе с шаровой симметрией[J]. Труды ГОИ, 1963, 30(159):22 - 29.

[5] SCHAGEN P,H BRUINING H, FRANCKEN J C. A simple electrostatic electron-optical system with only one voltage [J]. Philips Research Report, 1952, 7(2):119 - 132.

[6] КЕЛЬМАН В М, САПАРГАЛИЕВ А А, ЯКУШЕВ Е М. Теория катодных линз II—Электростатическая линза с вращательной симметрией поля[J]. Ж. Т. Ф. , 1973, 43(1):52 - 56.

[7] 周立伟. 宽束电子光学[M]. 北京:北京理工大学出版社,1993.

1.3 C章：多电极同心球系统的电子光学
On Electron Optics of Concentric Spherical System Composed of Multiple Electrodes

摘要：鉴于两电极静电聚焦同心球系统的定焦性质，当该系统的几何尺寸相对关系给定后，像面位置与放大率也就随之确定，改变系统的电参量对于像面位置和放大率的变动是极其微小的。本文研究在光阴极与栅状阳极组成的两电极同心球系统之间插入任意多个栅极后，该系统的电子光学成像特性及其横向像差的变化规律。本文再次证实了在多电极静电聚焦同心球系统中，成像电子光学系统的二级近轴横向色差即 Recknagel-Арцимович 公式依然成立。文中着重讨论了三电极静电聚焦同心球系统的电子光学成像特性。

Abstract：In view of the fixed-focus nature of imaging for the bi-electrode electrostatic focusing concentric spherical system, when the geometry of the system is relatively fixed, the image plane position and the magnification of system are immediately determined, and if the electrical parameters of the system are changed, the image plane position and its magnification will be changed extremely tiny. This paper studies the variation of the electron optical imaging characteristics and its lateral aberrations when inserting a large number of gate electrodes into the bi-electrode electrostatic focusing concentric spherical system between the photocathode and the anode. This paper once again confirms that for the multi-electrode electrostatic focusing concentric spherical system, the second order paraxial lateral chromatic aberration of the cathode lens, i. e., the Recknagel- Арцимович formula, is still valid. The electron optical imaging characteristics of the tri-electrode electrostatic focusing concentric spherical system is emphatically discussed in the text.

引 言

通过对两电极静电聚焦同心球系统的电子光学性质的进一步考察，不难发现，其成像面的位置取决于系统的结构参量。如果电极系统制造得不精确，或者在装配时不能保持给定的几何尺寸及相对关系，则阴极物面上的图像便不能很好地聚焦在给定的成像平面上，而且也无法改善聚焦质量。也就是说，当两电极同心球系统的几何尺寸相对关系给定后，像面位置与放大率也就随之确定。改变系统的电参量对于像面位置和放大率的变动是极其微小的。若我们在两电极同心球系统中插入一个中间栅极，此栅极不仅能够控制电子流（如在高速摄影中作为电子快门），而且还起着调焦电极的作用。当改变中间栅极的电位时，可以得到最佳的聚焦。本文将两电极静电聚焦同心球成像电子光学的研究推广到更普遍的情况——多电极静电聚焦同心球系统成像电子光学。

1 参量意义与符号规则

如图 1 所示，若在阴极 C 和栅状阳极 A 所组成的两电极静电聚焦同心球系统中，插入 j 个具有共同球心的曲率半径为 R_{s_j}、电位为 $\phi_{s_j,c}$ 的栅状球面电极 $S_j(j=1,2,\cdots,m)$，形成多电极静电聚焦同心球系统。

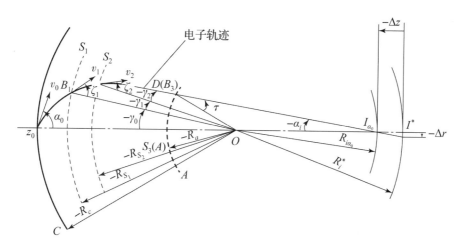

图 1 多电极同心球系统 ($m=2$)

兹规定参量意义与符号规则如下。

几何参量如下。

线段：R_c——阴极面半径；

R_a——阳极半径；

R_{s_j}——第 j 个中间栅极 S_j 的半径 $(j=1,2,\cdots,m)$；

R_i^*——与 ε_{z_1} 相对应的理想像面位置 I^* 处的曲率半径。

以上各线段值均自曲率中心 O 算起，自左至右为正，反之为负。

Δz——纵向像差，自像面位置 I^* 算起，自左为右为正，反之为负；

Δr——横向像差，像面位于 I^* 处，由 $\varphi=0$ 的轴线算起，自下而上为正，反之为负。

星号（*）代表满足近轴条件下的情况。

角度：一律以锐角计算。

γ_0——电子轨迹在第一中间栅极 S_1 上的角度坐标；

γ_j——电子轨迹在两个相邻电极 S_j 和 S_{j+1} 之间的角度坐标；

ζ_j——中间栅极 S_j 上轨迹落点处的切线与过该点矢径之间的夹角；

τ——逸出电子在栅状阳极 A 上落点处的切线与过该点矢径之间的夹角；

α_i——初始条件为 $(\varepsilon_0, \alpha_0)$ 的电子轨迹在成像位置 $I_{\alpha_0}^*$ 处与 $\varphi=0$ 轴线之夹角。

这里，α_i、γ_0 以轴线为轴，γ_j、ζ_j 和 τ 以过点 B_j 和 D 的矢径为轴，逆时针为正，反之为负。

电参量：

ϕ_{ac}——栅状阳极 A 相对于阴极 C 的电位；

ϕ_{s_jc}——第 j 个中间栅极 S_j 相对于阴极 C 的电位；

$\phi_{s_{j+1}s_j}$——第 $j+1$ 个中间栅极 S_{j+1} 相对于第 j 个中间栅极 S_j 的电位；

ε_0——以初速度 v_0 自阴极 C 射出的电子所对应的初电位；

ε_j——由第 j 个中间栅极 S_j 以速度 v_j 射出的电子所对应的电位，$\varepsilon_j = \varepsilon_0 + \phi_{s_jc}$。

辅助参量：

$$n = \frac{R_c}{R_a}, y_j = \frac{R_c}{R_{s_j}}, x_{j,j+1} = \frac{R_{s_j}}{R_{s_{j+1}}}, r_{j+1,j}^2 = \frac{\phi_{s_{j+1}c}}{\phi_{s_jc}},$$

$$r_{m+1,1}^2 = \frac{\phi_{s_{m+1}c}}{\phi_{s_1c}} = \frac{\phi_{ac}}{\phi_{s_1c}}, r_{j,1}^2 = \frac{\phi_{s_jc}}{\phi_{s_1c}}, r_{1,1}^2 = 1 \quad (j = 1, 2, \cdots, m)$$

2 多电极静电聚焦同心球系统的轨迹交轴位置及其斜率的精确表达式[1,2,3]

多电极同心球系统的电子追迹计算将按以下的次序进行：

首先考察阴极 C 和第一中间栅极 S_1 的系统（简称 $C-S_1$ 系统）。设自阴极面原点 z_0 发出初角度为 α_0、初速度为 v_0 的电子，则在此系统内的电子轨迹如"电子轨迹方程"文中所述，可以很容易被确定。于是可求得逸出电子在栅极 S_1 的落点 B_1 的位置以及相应的角度 ζ_1 和 γ_0。

其次研究 S_1-S_2 系统。很明显，如果假定栅网电极是透明的，当电子通过时，其速度和斜率都不发生变化。我们可以把第一中间栅极 S_1 看作阴极 C，而从 B_1 点发出角度为 ζ_1、速度为 v_1 的电子，很容易用类似 $C-S_1$ 系统的处理方法确定 B_2 点的位置以及相应的角度 ζ_2 和 γ_1。

以此类推，不难求得 S_m-A 系统中电子到达阳极 A 上的 D 点的位置以及角度 τ 和 γ_m。若假定栅状阳极后面为无场空间，便可以求得自阴极面 z_0 点发出的初角度为 α_0、初速度为 v_0 的电子，经过 m 个中间电极和阳极后与 $\varphi = 0$ 轴线的交点 I_α 的位置及其斜率。

对多电极静电聚焦同心球系统，自光阴极逸出的电子经过系统后，其轨迹的最终交轴点 I_α 处的曲率半径 $R_{i\alpha}$ 和斜率 $\tan\alpha_i$ 可以表述如下：

$$R_{i\alpha} = R_a \frac{\sin\tau}{\sin\alpha_i} = R_a \tan\tau \frac{\cos\tau}{\sin\left[(\gamma_0 + \sum_{j=1}^{m}\gamma_j) + \tau\right]}$$

$$= R_a \tan\tau \{\sin\gamma_0 \cos(\sum_{j=1}^{m}\gamma_j) + \cos\gamma_0 \sin(\sum_{j=1}^{m}\gamma_j) +$$

$$\tan\tau[\cos\gamma_0 \cos(\sum_{j=1}^{m}\gamma_j) - \sin\gamma_0 \sin(\sum_{j=1}^{m}\gamma_j)]\}^{-1} \quad (1)$$

$$\tan\alpha_i = \tan\left[\left(\sum_{j=0}^{m}\gamma_j\right)+\tau\right] = \frac{\tan\gamma_0 + \tan\left(\sum_{j=1}^{m}\gamma_j\right) + \tan\tau\left[1-\tan\gamma_0\tan\left(\sum_{j=1}^{m}\gamma_j\right)\right]}{1-\tan\gamma_0\tan\left(\sum_{j=1}^{m}\gamma_j\right)-\tan\tau\left[\tan\gamma_0+\tan\left(\sum_{j=1}^{m}\gamma_j\right)\right]} \quad (2)$$

式（1）和式（2）中各量具体计算如下：对于 $C-S_1$ 系统，按照前面"电子轨迹方程"中推导的式（35）～式（37），则下列关系式成立。

令 $n \to y_1$，$\tau \to \zeta$，$\gamma_1 \to \gamma_0$，$\phi_{ac} \to \phi_{s_1c}$，便有

$$\sin\gamma_0 = -\frac{(y_1-1)(c_0+d_0)}{b_1 d_0 + b_0 c_0}\sin\alpha_0 \quad (3)$$

$$\cos\gamma_0 = \frac{b_1 c_0 + b_0 d_0}{b_1 d_0 + b_0 c_0} \quad (4)$$

$$\tan\zeta_1 = \frac{y_1}{b_1}\sin\alpha_0 \quad (5)$$

式中，

$$b_0 = \cos\alpha_0,\ b_1 = \left(1+\frac{\phi_{s_1c}}{\varepsilon_0}-y_1^2\sin^2\alpha_0\right)^{1/2},\ c_0 = 1-2y_1(y_1-1)\frac{\varepsilon_0}{\phi_{s_1c}}\sin^2\alpha_0$$

$$d_0 = 1-2(y_1-1)\frac{\varepsilon_0}{\phi_{s_1c}}\sin^2\alpha_0 \quad (6)$$

类似地，对于由第 j 和第 $j+1$ 中间电极构成的两电极系统，则以下列参量进行变换：

$$\alpha_0 \to \zeta_j,\ y_1 \to x_{j,j+1},\ \varepsilon_0 \to \varepsilon_j,\ \gamma_0 \to \gamma_j,\ \zeta_1 \to \zeta_{j+1},\ \phi_{s_jc} \to \phi_{s_{j+1}s_j}$$

下面逐个确定 γ_j 和 $\zeta_{j+1}(j=1,2,\cdots,m)$ 的值。为此，先确定 S_1-S_2 系统的 $\tan\zeta_2$、$\sin\gamma_1$ 和 $\cos\gamma_1$ 各值。对于 $\tan\zeta_2$，按式（5），有

$$\tan\zeta_2 = \frac{x_{1,2}\sin\zeta_1}{\left(1+\frac{\phi_{s_2s_1}}{\varepsilon_1}-x_{1,2}^2\sin^2\zeta_1\right)^{1/2}} \quad (7)$$

式中，

$$\sin\zeta_1 = \frac{y_1\sin\alpha_0}{\left(1+\frac{\phi_{s_1c}}{\varepsilon_0}\right)^{1/2}},\ x_{1,2} = \frac{y_2}{y_1},\ \frac{\phi_{s_2s_1}}{\varepsilon_1} = \frac{\phi_{s_2c}-\phi_{s_1c}}{\varepsilon_0+\phi_{s_1c}} = \frac{\frac{\phi_{s_2c}}{\varepsilon_0}-\frac{\phi_{s_1c}}{\varepsilon_0}}{1+\frac{\phi_{s_1c}}{\varepsilon_0}}$$

将以上各式代入式（7）中，则 $\tan\zeta_2$ 具有下列简明形式：

$$\tan\zeta_2 = \frac{y_2\sin\alpha_0}{\left(1+\frac{\phi_{s_2c}}{\varepsilon_0}-y_2^2\sin^2\alpha_0\right)^{1/2}} = \frac{y_2}{b_2}\sin\alpha_0 \quad (8)$$

同样，求 $\sin\gamma_1$、$\cos\gamma_1$ 值时，式（3）、式（4）中各量需进行下列替换：

$$y_1 \to x_{1,2} = \frac{y_2}{y_1},\quad \sin\alpha_0 \to \sin\zeta_1 = \frac{y_1\sin\alpha_0}{\left(1+\frac{\phi_{s_1c}}{\varepsilon_0}\right)^{1/2}},$$

$$b_0 = \cos\alpha_0 \to \cos\zeta_1 = \frac{b_1}{\left(1+\frac{\phi_{s_1c}}{\varepsilon_0}\right)^{1/2}},\ b_1 \to \left(1+\frac{\phi_{s_2s_1}}{\varepsilon_1}-x_{1,2}^2\sin^2\zeta_1\right)^{1/2} = \frac{b_2}{\left(1+\frac{\phi_{s_1c}}{\varepsilon_0}\right)^{1/2}}$$

$$c_0 \to 1 - 2x_{1,2}(x_{1,2}-1)\frac{\varepsilon_1}{\phi_{s_2s_1}}\sin^2\zeta_1 = 1 - 2y_2(y_2-y_1)\frac{1}{r_{2,1}^2 - r_{1,1}^2}\frac{\varepsilon_0}{\phi_{s_1c}}\sin^2\alpha_0 = c_1 \quad (9)$$

$$d_0 \to 1 - 2(x_{1,2}-1)\frac{\varepsilon_1}{\phi_{s_2s_1}}\sin^2\zeta_1 = 1 - 2y_1(y_2-y_1)\frac{1}{r_{2,1}^2 - r_{1,1}^2}\frac{\varepsilon_0}{\phi_{s_1c}}\sin^2\alpha_0 = d_1$$

于是，$\sin\gamma_1$、$\cos\gamma_1$ 有类似式（3）、式（4）的形式：

$$\sin\gamma_1 = -\frac{(y_2-y_1)(c_1+d_1)}{b_2d_1+b_1c_1}\sin\alpha_0 \quad (10)$$

$$\cos\gamma_1 = \frac{b_2c_1+b_1d_1}{b_2d_1+b_1c_1} \quad (11)$$

式中，

$$b_2 = \left(1+\frac{\phi_{s_2c}}{\varepsilon_0}-y_2^2\sin^2\alpha_0\right)^{1/2}, \quad c_1 = 1 - \frac{2y_2(y_2-y_1)}{r_{2,1}^2-r_{1,1}^2}\frac{\varepsilon_0}{\phi_{s_1c}}\sin^2\alpha_0,$$

$$d_1 = 1 - \frac{2y_1(y_2-y_1)}{r_{2,1}^2-r_{1,1}^2}\frac{\varepsilon_0}{\phi_{s_1c}}\sin^2\alpha_0$$

由此可见，对于 $\tan\zeta_{j+1}$、$\cos\gamma_j$ 和 $\sin\gamma_j$，显然可以类推成如下的表达式：

$$\tan\zeta_{j+1} = \frac{y_{j+1}}{b_{j+1}}\sin\alpha_0 \quad (12)$$

$$\sin\gamma_j = -\frac{(y_{j+1}-y_j)(c_j+d_j)}{b_{j+1}d_j+b_jc_j}\sin\alpha_0 \quad (13)$$

$$\cos\gamma_j = \frac{b_{j+1}c_j+b_jd_j}{b_{j+1}d_j+b_jc_j} \quad (14)$$

式中，

$$b_j = \left(1+\frac{\phi_{s_jc}}{\varepsilon_0}-y_j^2\sin^2\alpha_0\right)^{1/2}, \quad c_j = 1 - \frac{2y_{j+1}(y_{j+1}-y_j)}{r_{j+1,1}^2-r_{j,1}^2}\frac{\varepsilon_0}{\phi_{s_1c}}\sin^2\alpha_0,$$

$$d_j = 1 - \frac{2y_j(y_{j+1}-y_j)}{r_{j+1,1}^2-r_{j,1}^2}\frac{\varepsilon_0}{\phi_{s_1c}}\sin^2\alpha_0$$

当 $j=m$ 时，则由式（12），得

$$\tan\tau = \tan\zeta_{m+1} = \frac{n}{b_{m+1}}\sin\alpha_0 \quad (15)$$

关于式（2）中的 $\tan\gamma_0$、$\tan\gamma_j$ 值，则可由式（3）、式（4）以及式（13）、式（14）求得

$$\tan\gamma_0 = -\frac{(y_1-1)(c_0+d_0)}{b_1c_0+b_0d_0}\sin\alpha_0 \quad (16)$$

$$\tan\gamma_j = -\frac{(y_{j+1}-y_j)(c_j+d_j)}{b_{j+1}c_j+b_jd_j}\sin\alpha_0 \quad (17)$$

将以上导得的表达式代入式（1）和式（2）中，便可得到多电极同心球系统中以初始条件参量（ε_0，α_0）逸出的电子轨迹交轴位置 $R_{i\alpha}$ 及其斜率 $\tan\alpha_i$ 的精确表达式。由式（1）和式（2）可见，$R_{i\alpha}$ 和 $\tan\alpha_i$ 取决于几何参量 R_c、$R_j(j=1,2,\cdots,m)$、R_a，电参量 $\phi_{s_jc}(j=1,2,\cdots,m)$ 和 ϕ_{ac}，以及初始条件参量（ε_0，α_0）。如果给定以上参量，便能确定 $R_{i\alpha}$ 与 $\tan\alpha_i$ 的精确值。

3　多电极静电聚焦同心球系统的轨迹交轴位置及其斜率的近似表示式[1,3,4]

由式（1）和式（2）的 $R_{i\alpha}$ 和 $\tan\alpha_i$ 的精确表示式，我们很难看出各参量之间的相互关系。为此，我们将这两表达式展开，略去较 1 小得多的 $\left(\dfrac{\varepsilon_r}{\phi_{s_jc}}\right)^{3/2}$ 等高阶乘幂项，经过繁复的运算和整理，便可得到多电极静电聚焦同心球系统下电子轨迹的交轴位置 $R_{i\alpha}$ 及其斜率 $\tan\alpha_i$ 的近似表达式：

$$R_{i\alpha} = -R_a n A \left\{ 1 + 2r_{m+1,1}(y_1-1)A\sqrt{\dfrac{\varepsilon_z}{\phi_{s_1c}}} + \left[AC + 4A^2 r_{m+1,1}^2(y_1-1)^2\right]\dfrac{\varepsilon_z}{\phi_{s_1c}} + A\left[C - 2Br_{m+1,1}(y_1-1)\right]\dfrac{\varepsilon_r}{\phi_{s_1c}} \right\} \tag{18}$$

$$\tan\alpha_i = -\dfrac{1}{Ar_{m+1,1}}\sqrt{\dfrac{\varepsilon_r}{\phi_{s_1c}}}\left[1 - 2r_{m+1,1}(y_1-1)A\sqrt{\dfrac{\varepsilon_z}{\phi_{s_1c}}} + DA\dfrac{\varepsilon_z}{\phi_{s_1c}} + (D+E)A\dfrac{\varepsilon_r}{\phi_{s_1c}}\right] \tag{19}$$

上述两式足以阐明同心球系统的电子光学成像性质，并推导出电子光学横向像差。
公式中的符号 A，B，C，D，E 可表示如下。

3.1　两电极静电聚焦同心球系统（$m=0$）

令 $y_1 = n$，$r_{1,1} = 1$，$\phi_{s_1c} = \phi_{ac}$，则有

$$A = \dfrac{1}{n-2}, B = n-2, C = 2(n-1), D = \dfrac{3n-2}{2}, E = \dfrac{n^2(n-2)}{2} \tag{20}$$

将式（20）各表达式代入式（18）和式（19），便得到"电子轨迹方程"文章中式（42）、式（43）的结果。其成像特性与横向像差已在"近轴横向色差与几何横向球差"文章中叙述过。

3.2　三电极静电聚焦同心球系统（$m=1$）

若在两电极同心球系统中，插入一中间栅极，其半径为 R_s，电位为 ϕ_{s_1c}，便形成三电极同心球系统。令

$$y_1 = y = \dfrac{R_c}{R_s}, y_2 = n = \dfrac{R_c}{R_a}, r_{2,1} = r = \left(\dfrac{\phi_{ac}}{\phi_{s_1c}}\right)^{1/2}$$

则有

$$A = \dfrac{r+1}{r(n-2) + 2r^2(y-1) - n} \tag{21}$$

$$B = \dfrac{2(y-1)n - r(y-2)^2}{2r} + \dfrac{2(n-y)^2[n - 2r(y-1)] + (n-y)(n^2 + ry^2)}{2r(r+1)^2(y-1)} - \dfrac{(n-y)^3}{(y-1)(r+1)^3} - \dfrac{n^2[(n-1) + r(y-1)]}{2r^2(y-1)} + \dfrac{2(n-y)[n - r(y-1)]}{r(r+1)} \tag{22}$$

$$C = -r(y-1) + \dfrac{n-y}{r+1} - \dfrac{(n-1) + r(y-1)}{r(r+1)} \tag{23}$$

$$D = r(y-1) + \frac{n-y}{r+1} + \frac{n}{2r^2} \qquad (24)$$

$$E = r(y-1)(3y^2 - 4y) - 4n(y-1)^2 - \frac{n^3}{2r^2} + \frac{2n^2[r(y-1)+(n-1)]}{r(r+1)} + \frac{2r(n-y)^3}{(r+1)^3} +$$

$$\frac{(n-y)(n^2+ry^2) + 4(n-y)^2[2r(y-1)-n]}{(r+1)^2} \qquad (25)$$

式(21)的 A 表示式曾被 Linden 和 Snell 导出过[4]。但他们两位研究的仅是三电极同心球系统的零级近似成像,故他们的文章中,上述 B、C、D、E 的表示式都没有给出。

3.3 多电极同心球系统($m \geqslant 2$)

由式(18)、式(19)可见,在近轴条件下,与 $\frac{\varepsilon_r}{\phi_{s_1c}}$ 相关的较为复杂的 B 和 E 可以略去,于是,系数 A、C、D 可以表述如下:

$$A = \frac{1}{r_{m+1,1}\left[2(y_1-1) + \sum_{i=1}^{m}\frac{2(y_{i+1}-y_i)}{r_{i+1,1}+r_{i,1}} - \frac{n}{r_{m+1,1}}\right]} \qquad (26)$$

$$C = -r_{m+1,1}(y_1-1) + r_{m+1,1}\sum_{i=1}^{m}\frac{y_{i+1}-y_i}{r_{i+1,1}r_{i,1}(r_{i+1,1}+r_{i,1})} - \frac{nA+1}{2Ar_{m+1,1}^2} \qquad (27)$$

$$D = r_{m+1,1}(y_1-1) + \frac{n}{2r_{m+1,1}^2} - r_{m+1,1}\sum_{i=1}^{m}\frac{y_{i+1}-y_i}{r_{i+1,1}r_{i,1}(r_{i+1,1}+r_{i,1})} \qquad (28)$$

关于系数 B、E 的表示式可参阅文献[1,3],这里不再列出了。由此,可以研究具有任意多个中间栅极的多电极静电聚焦同心球系统的理想成像与像差。

4 多电极静电聚焦同心球系统的电子光学成像与横向像差[1,3]

4.1 理想成像

如果我们考虑阴极透镜的近轴条件,则可略去式(18)中大括弧内和式(19)中中括号内含有 $\varepsilon_r/\phi_{s_1c}$ 的项,且令 $\varepsilon_z = \varepsilon_{z_1}$,则可得到自阴极面上物点 z_0 以某一轴向初电位逸出 ε_{z_1} 的近轴电子轨迹到达理想成像位置离曲率中心 O 之距离为 R_i^*,即

$$R_i^* = -R_a nA\left\{1 + 2r_{m+1,1}(y_1-1)A\sqrt{\frac{\varepsilon_{z_1}}{\phi_{s_1c}}} + [AC + 4A^2 r_{m+1,1}^2(y_1-1)^2]\frac{\varepsilon_{z_1}}{\phi_{s_1c}}\right\} \qquad (29)$$

而轨迹在 I_0^* 处的斜率 $\tan\alpha_i^*$ 为

$$\tan\alpha_i^* = -\frac{1}{Ar_{m+1,1}}\sqrt{\frac{\varepsilon_r}{\phi_{s_1c}}}\left\{1 - 2r_{m+1,1}(y_1-1)A\sqrt{\frac{\varepsilon_{z_1}}{\phi_{s_1c}}} + DA\frac{\varepsilon_{z_1}}{\phi_{s_1c}}\right\} \qquad (30)$$

式中,上标"*"号表示近轴条件下的情况。

式(29)表明,由阴极面上原点 z_0 以相同的轴向初电位 ε_{z_1} 射出的电子束,在经过同心球系统场穿过透明的栅状阳极后会聚于同一点($A>0$,实像情况)或者好像是从阴极后面某一点发出一般($A<0$,虚像情况)。很明显,只有 ε_{z_1} 相同的近轴电子束才能理想聚

焦；当 ε_{z_1} 改变时，理想成像的位置亦随着变动。式（29）中上标"*"号表示相应于近轴条件的情况。

由于系统的球对称性，与 ε_{z_1} 相应的电子束自阴极球面上的弧段射出，其成像同样为曲率半径为 R_i^* 的弧段。此时，线放大率 M 可表示为

$$M = -A\left\{1 + 2r_{m+1,1}(y_1 - 1)A\sqrt{\frac{\varepsilon_{z_1}}{\phi_{s_1 c}}} + [AC + 4A^2 r_{m+1,1}^2 (y_1 - 1)^2]\frac{\varepsilon_{z_1}}{\phi_{s_1 c}}\right\} \qquad (31)$$

显然，系统形成的是倒立的实像或正立的虚像，取决于 A 值。由式（21）的表示式可以看出，A 仅是结构参量和电参量的函数。故对多电极静电聚焦同心球系统，线放大率 M 不仅取决于系统的几何参量，而且与电参量有关，而初始条件的参量的影响则是高阶小的。

若令 $\varepsilon_{z_1}=0$，即逸出电子切于阴极面射出，此时可定出极限像面的位置 R_{it}^* 以及相应的放大率 M_t：

$$R_{it}^* = -R_a nA, \quad M_t = -A \qquad (32)$$

式（32）乃是多电极同心球静电聚焦系统的零级近似式，由此可以看出系数 A 的明显的物理意义。

由线放大率 M 和角放大率 Γ 的乘积，得

$$M\Gamma = \frac{\sqrt{\varepsilon_{z_1}}}{r_{m+1,1}\sqrt{\phi_{s_1 c}}}\left[1 + A(C+D)\frac{\varepsilon_{z_1}}{\phi_{s_1 c}}\right] \qquad (33)$$

因

$$A(C+D) = -\frac{1}{2r_{m+1,1}^2} \qquad (34)$$

故可证明拉格朗日－亥姆霍兹关系式

$$M\Gamma = \frac{\sqrt{\varepsilon_{z_1}}}{\sqrt{\phi_{ac}}}\left(1 + \frac{\varepsilon_{z_1}}{2\phi_{ac}}\right) = \frac{\sqrt{\varepsilon_{z_1}}}{\sqrt{\phi_{ac} + \varepsilon_{z_1}}} \qquad (35)$$

依然成立。因此，不论对于两电极同心球系统，或是在阴极和栅状阳极之间插入多个中间电极所组成的多电极同心球系统，尽管电位分布发生了变化，拉格朗日－亥姆霍兹关系式始终成立。

4.2 三电极静电聚焦同心球系统的成像特性[4]

我们在前面内容中已给出三电极同心球系统电子光学特性的表示式，由式（21）

$$A = \frac{r + 1}{r(n - 2) + 2r^2(y - 1) - n}$$

不难看出有如下的变化规律。

（1）欲使像面位置位移、放大率之模增大，可以有如下途径：（a）提高 $\phi_{s_1 c}$ 或降低 ϕ_{ac} 使 r^2 下降；（b）中间电极 S 靠近阴极 C 使 y 下降；（c）阳极 A 靠近阴极 C 使 n 下降。

（2）像面位置与放大率值取决于电极系统的相对位置和电压比，故系统放大 K 倍或缩小到原来的 $\frac{1}{K}$，像面位置亦相应放大 K 倍或缩小到原来的 $\frac{1}{K}$，而放大率不变；当电压比不变时，像面位置与放大率基本不变。

(3) 由系统成实像的条件 $A > 0$，可以导得以下条件式：

$$n > y > 1 + \frac{n - r(n-2)}{2r^2} \quad (36)$$

在两电极情况下，当 $n = 2$ 时，系统是发散的，但当插入中间栅极时，使 y 满足条件

$$2 > y > 1 + \frac{1}{r^2}, \quad r^2 > 1 \quad (37)$$

便能形成实像，且 r^2 越大，中间栅极位置越能靠近阴极。这说明，在两电极系统中加一中间电极且加上一个较阳极为低的电位，有可能使系统更为紧凑。

当然，亦可令 $r^2 < 1$，即 $\phi_{ac} < \phi_{sc}$。此时若 $n = 2$，则无论如何配置中间电极，条件式 (36) 不复成立，只有当 $n > 2$ 时才有可能。现假定 $n = 3$，则有

$$3 > y > 1 + \frac{3-r}{2r^2} \quad (38)$$

由此可得 $r^2 > \frac{9}{16}$ 时才能形成实像。例如：令 $r^2 = \frac{16}{25}$，则 $3 > y > 2.71$。

因此，一般来说，中间电极比阳极电位高，将使系统纵向尺寸增大，以满足给定的像面位置和放大率的要求。从物理意义上来说，并非一定要 $\phi_{ac} > \phi_{sc}$，电子才能会聚；而当 $\phi_{ac} > \phi_{sc}$ 时，只要电子在 $C-S$ 系统内获得了足够的会聚，就可以抵消 $S-A$ 系统内的发散。这样，电子仍然会聚，形成实像。

(4) 如果在三电极系统中，使电压比与结构参数满足

$$r^2 = \frac{n-1}{y-1}$$

则

$$A = \frac{1}{n = 2} \quad (39)$$

这表明，当在两电极同心球系统的等位面上安置一中间栅极，其 ϕ_{sc} 值即取等位面的电位值，且电极的形状与等位面形状一致，则系统的特性保持不变。通常，三电极系统的设计往往是在两电极系统的等位面处安置一个形状与等位面形状相同的中间栅极，调节此中间栅极的电位值即可适当改变成像面的位置和放大率。

4.3 多电极同心球系统的纵向和横向像差

对于多电极同心球系统，"多电极同心球系统的电子光学"中关于两电极同心球系统的纵向像差与横向像差的定义是完全适用的。其纵向像差 Δz 以 "近轴横向色差与几何横向球差"中的式 (11) 表示，即

$$\Delta z = \Delta z^* + \delta z = \Delta z_1^* + \Delta z_2^* + \delta z_2$$

式中，

$$\Delta z_1^* = -\frac{2M^2 \sqrt{\phi_{ac}}}{E_c}(\sqrt{\varepsilon_z} - \sqrt{\varepsilon_{z_1}})$$

$$\Delta z_2^* = -\frac{2M^2}{E_c}\left[\frac{C + 4Ar_{m+1,1}(y_1-1)^2}{2(y_1-1)}(\varepsilon_z - \varepsilon_{z_1}) - 4Ar_{m+1,1}(y_1-1)\sqrt{\varepsilon_{z_1}}(\sqrt{\varepsilon_z} - \sqrt{\varepsilon_{z_1}})\right]$$

(40)

$$\delta z_2 = \frac{2M^2}{E_c}\left[\frac{-C + 2B(y_1 - 1)r_{m+1,1}}{2(y_1 - 1)}\varepsilon_r\right] \tag{41}$$

横向像差 Δr 以"近轴横向色差与几何横向球差"中的式（16）表示，即

$$\Delta r = \Delta r^* + \delta r = \Delta r_2^* + \Delta r_3^* + \delta r_3$$

式中，

$$\Delta r_2^* = \frac{2M}{E_c}\sqrt{\varepsilon_r}(\sqrt{\varepsilon_z} - \sqrt{\varepsilon_{z_1}}),$$

$$\Delta r_3^* = -\frac{2M}{E_c}\frac{\sqrt{\varepsilon_r}}{\sqrt{\phi_{ac}}}\left[\frac{C}{2(y_1 - 1)}(\varepsilon_z - \varepsilon_{z_1})\right] \tag{42}$$

$$\delta r_3 = -\frac{2M}{E_c}\frac{\sqrt{\varepsilon_r}}{\sqrt{\phi_{ac}}}\left[\frac{2Br_{m+1,1}(y_1 - 1) - C}{2(y_1 - 1)}\varepsilon_r\right] \tag{43}$$

这里各参量的意义同"近轴横向色差与几何横向球差"中的。其中，E_c 为阴极面处的电场强度，它可表示为

$$E_c = \frac{\phi_{s_1c}}{R_c(y_1 - 1)} \tag{44}$$

在本系列文章中，纵向和横向像差用阶次分类以区别于细束电子光学通常所用的三级和高级像差。这是因为，在阴极透镜中，从阴极面发出的电子除电子初能量的差别外，还具有不同的逸出角，且轨迹的斜率可能趋于无穷，加之理想像面位置与所取的轴向初电位 ε_z 有关，用细束电子光学通常的表示法划分纵向和横向像差已不可能，在这里，纵向像差按照 $(\varepsilon_0/\phi_{ac})^{j/2}$ 的阶次分类，横向像差则按照 $(\varepsilon_0/\phi_{ac})^{(j+1)/2}$ 的阶次分类。

对多电极与两电极同心球静电聚焦系统的横向像差表示式进行数值分析，可得到如下结论。

（1）对同心球型电子光学系统，轴上点的横向像差 Δr 可视为近轴横向色差 Δr^*（即由轴向初能 ε_{z_1} 与 ε_z 两条不同的近轴轨迹所形成的像差）与几何横向球差 δr（即由同样的初始条件参量的实际轨迹与近轴轨迹所形成的像差）的合成，如图1所示。其中，"近轴横向色差与几何横向球差"中的式（17）表示的二级近轴横向色差 Δr_2^* 即 R-A 公式成立。它对于系统的成像质量起着主要的、决定的作用。它较式（42）的三级近轴横向色差 Δr_3^* 与式（43）的三级几何横向球差 δr_3 要大一个数量级。故更高级次的像差可完全忽略不计。这个结论对于任何轴对称的静电电子光学成像系统也是成立的。

（2）我们从具体的系统（两电极与多电极同心球系统）中证实了阴极透镜的二级近轴横向色差公式即 R-A 公式普遍成立。"近轴横向色差与几何横向球差"中的式（17）表明，阴极透镜的近轴横向色差 Δr_2^* 在二级近似下仅与电子逸出的初电位、初角度、阴极面上的场强以及系统的线放大率有关，而与电极结构和轴上电位分布无关。

（3）由横向像差公式可以看出，三级几何横向球差 δr_3 实际上正是通常电子光学中所谓三级几何球差的主要项，它与 $\left(\sqrt{\dfrac{\varepsilon_r}{\phi_{ac}}}\right)^3$ 成比例。这样，通过三级近轴横向色差 Δr_3^* 我们便把通常的二级近轴横向色差 Δr_2^* 与三级几何横向球差 δr_3 联系起来了。

结束语

本文将两电极静电聚焦同心球成像电子光学的研究推广到更普遍的情况——多电极静电聚焦同心球系统成像电子光学。研究表明，无论对于两电极或是多电极静电聚焦同心球系统，其轴上点横向像差系由近轴横向色差与几何横向球差所组成，证实了对于静电聚焦电子光学系统，其二级近轴横向色差公式即 R-A 公式普遍成立。本文对于研究成像电子光学中实际的三电极调焦系统与四电极变倍系统的设计具有理论指导意义。

参 考 文 献

［1］ЧЖОУ Л В. Исследование электростатических фокусирующих систем электронно-оптических преобразователей изобржении с шаровой и осевой симметрией［D］. Ленинград：Диссертация ЛЭТИ. ,1966.

［2］ЧЖОУ Л В. Об аберрации в электронной оптической системе с шаровой и осевой симметрией［C］. Тезисы докладов XX1 Научно-Технической Конференции, Ленинград, 1966:137-138.

［3］周立伟. 宽束电子光学［M］. 北京：北京理工大学出版社,1993.

［4］LINDEN S R, SNELL P A. Shutter image converter tubes［J］. Proceedings of the IRE. , 1957,45:513-517.

1.4 D章：最小弥散圆与最佳像面位置的确定
On Determination of Minimum Diffusing Circle and Optimal Image Position

摘要：本文由所导出的阴极透镜中普遍成立的二级近轴横向色差即著名的 Recknagel-Арцимович 表示式出发，研究电子束形成的最小弥散圆以及最佳成像面的位置的确定，考察由物面原点发射的电子在栅状阳极上形成的散射圆以及整个阴极面发射所形成的交叉颈，描绘由物面原点逸出的电子射束在成像段所形成的电子射线的包络。本文有助于读者理解电子光学像管中电子行进的轨迹以及成像段的电子轨迹的发散与会聚。

Abstract: Starting from the second order paraxial lateral chromatic aberration for the cathode lens, that is, from the well-known Recknagel-Арцимович formula in the imaging electron optical system, this paper will investigate the minimum dispensing circle formed by the electron beam, study the determination of the minimum circle of image diffusion in which the optimal image plane is located, investigate the contour of crossover formed by the electron beam emitted from the whole cathode surface, as well as the diffusion circle formed by the electron beam emitted from the whole photocathode surface, determine the optimal imaging plane, depict the envelope of the electron rays formed in the imaging section. The present paper will help the readers to understand the moving electron trajectory in the image tubes, as well as its beam convergence and divergence in the imaging section.

引 言

对成像电子光学系统，电子束自光阴极逸出，在器件中行进，其会聚与发散的状况，以及它在成像面上图像清晰的程度是一个普遍关心的问题。尤其是对于从事变像管与像增强器的设计者，需要清晰了解电子束在器件中是如何行进的，如何使成像面位于电子束斑最密集处，以获得最佳的像质。本文将研究成像电子光学系统由物面原点发射的电子在栅状阳极上形成的散射圆以及整个物面发射的电子所形成的交叉颈，并由所导出的阴极透镜中普遍成立的二级近轴横向色差即 Recknagel-Арцимович 表示式出发，描绘由物面原点逸出的电子射束在成像段所形成的电子射线的包络，研究电子束形成的最小弥散圆，确定最佳成像面的位置。本文的研究有助于理解电子光学系统中成像段的电子的行进轨迹及其形成的像差。

周立伟（Zhou Liwei）. 北京理工大学. 光学学报（Acta Optica Sinica），V. 42, No. 8, 2022, 0811004-1-9.

1 交叉颈与栅状阳极上的散射圆[1]

在静电聚焦同心球系统中，可以由下面两个参量看出行进中的电子束的会聚状况：

一是由阴极面物点以初始条件参量（ε_0，α_0）射出的电子轨迹在栅状阳极上的散射圆半径 r_a。这是考察由物点逸出的电子束的会聚状况。二是由阴极物面射出的电子束在交叉点处形成的散射圆半径（通常称为交叉颈）r_c，它是考察整个物面逸出的电子束的会聚状况。

关于栅状阳极上的散射圆半径 r_a 的大小，这是设计者关心的一个问题。它关系到由阴极物点出发的电子束的会聚状况。由本系列文章的"电子轨迹方程"中的式（35）出发，可估计其值之大小：

$$r_a = R_a \sin\gamma = -R_a \frac{(n-1)(c_0+d_0)}{b_1 d_0 + b_0 c_0} \sin\alpha_0 \tag{1}$$

式中，b_1、b_0、c_0、d_0 以"电子轨迹方程"中的式（38）表示。将上式展开并引入近轴条件，于是有

$$r_a^* = 2(R_a - R_c)\sqrt{\frac{\varepsilon_r}{\phi_{ac}}}\left[\left(1+\frac{\varepsilon_z}{\phi_{ac}}\right)^{1/2} - \left(\frac{\varepsilon_z}{\phi_{ac}}\right)^{1/2}\right] \tag{2}$$

当 $\alpha_0 = 90°$ 时，则 r_a^* 取最大值，它可以表示为

$$r_a^*\bigg|_{\alpha_0=\frac{\pi}{2}} = 2(R_a - R_c)\sqrt{\frac{\varepsilon_0}{\phi_{ac}}} = 2l\sqrt{\frac{\varepsilon_0}{\phi_{ac}}} \tag{3}$$

式（3）实质上即是"近轴横向色差与几何横向球差"中"纵向均匀电场的电子光学"中的式（33）。这说明，阳极上的散射圆半径的大小仅与光阴极至阳极的距离有关，而与电极的形状基本无关。这个结论对于成像电子光学系统的设计者具有重要的参考价值。

关于交叉颈，这是考察整个物面逸出的电子束的会聚状况。令 r_{cr} 表示电子束会聚所形成的交叉颈半径，如图 1 中所示的 OH 线段；ψ_0 表示半个阴极面上有效尺寸（物面高度 r_0）与中心 O 所对应的角度，ψ_0 以系统轴线 Oz_0 为起始轴，逆时针为正。按照图中所示的三角形 $\triangle OHI_{\alpha_0}$，则有

$$\frac{r_{cr}}{\sin(-\alpha_i)} = \frac{R_{i\alpha_0}}{\sin[90°+(\psi_0+\alpha_i)]} \tag{4}$$

故可求得交叉颈半径 r_{cr} 的表示式为

$$r_{cr} = -R_{i\alpha_0}\frac{\tan\alpha_i}{\cos\psi_0 - \sin\psi_0 \tan\alpha_i} \tag{5}$$

式中，轨迹交轴位置 $R_{i\alpha_0}$ 及其斜率 $\tan\alpha_i$ 皆以本系列文章"近轴横向色差与几何横向球差"中式（38）、式（39）表示。

若在式（5）中代入 $R_{i\alpha_0}$、$\tan\alpha_i$ 的表示式（见"电子轨迹方程"中的式（39）、式（40）），取零级近似，则可得到用于电子枪估计交叉颈半径 r_{cr} 的如下表示式：

$$r_{cr} \approx -R_c \frac{1}{\cos\psi_0}\left(\frac{\varepsilon_r}{\phi_{ac}}\right)^{1/2} \tag{6}$$

由此可以估计交叉颈与半个阴极面有效尺寸的比值 η_0：

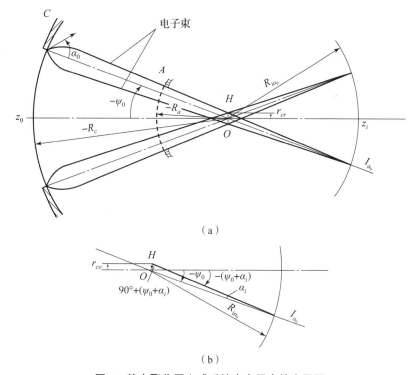

图 1　静电聚焦同心球系统中电子束的交叉颈

（a）交叉颈的形成；（b）交叉颈部分放大图

$$\eta_0 = \frac{r_{cr}}{-R_c \sin \psi_0} = \frac{2}{\sin 2\psi_0} \sqrt{\frac{\varepsilon_r}{\phi_{ac}}} \tag{7}$$

当 $\alpha_0 = 90°$，$\psi_0 = -45°$，$\dfrac{\phi_{ac}}{\varepsilon_0} = 1.6 \times 10^4$，则有 $\eta_0 = 1.58\%$。

由式（7）可见，在零级近似下，η_0 与系统的结构形状无关，仅与电子的逸出初能与加速电位的比值有关。上述结论对于研究电子枪的交叉颈具有意义。

现对 Д. М. Крупп[2] 建议加限制光阑的问题进行讨论。由式（3），令极间距离等于 20 mm，$\dfrac{\varepsilon_0}{\phi_{ac}} = 10^{-4}$，则逸出电子束在阳极孔阑的半径最大将不超过 0.2 mm。如果把如此小孔径的限制光阑放在阳极位置上，则不但能截获中心电子束，更能截获轴外逸出的电子束，妨碍了大尺寸阴极面图像的形成。由此可见，在像管中，自阴极面以任意角度射出的电子束在强场作用下迅速聚成细束，Д. М. Крупп 试图用光阑限制成像电子束是没有任何实际意义的。

但是，在某些像管中，可以在电子束行进途径的交叉点位置上安置附加光阑，使其孔径等于或大于交叉颈。但它的作用并不是切除从整个阴极面射出的电子束，而是切除杂散电子以及尽可能地减少屏上发出光线对于光阴极的反馈作用。

2　最佳像面位置、最小弥散圆的确定[1,4]

在本系列文章"近轴横向色差与几何横向球差"中，我们证明了，对同心球型电子光学系统，轴上点像差主要是由二级近轴横向色差 Δr_2^* 所确定。由此我们从具体的系统中证明了影响系统成像质量的 Recknagel-Арцимович 公式普遍成立。它表明，影响电子光学系统图像清晰度的二级近轴横向色差，与电极的具体结构、静电电位的具体分布无关，而与逸出电子的初电位和初角度、阴极面上的场强以及系统的放大率有关。

下面，我们由二级近轴横向色差的 Recknagel-Арцимович 表示式出发[2,3]，确定系统的最佳像面位置及最小弥撒圆。首先，令 $\varepsilon_z = \varepsilon_0 \cos^2\alpha_0$，$\varepsilon_r = \varepsilon_0 \sin^2\alpha_0$，将文章"近轴横向色差与几何横向球差"中的式（17）表示为

$$\Delta r_2^* = \frac{2M\varepsilon_0}{E_c}\sin\alpha_0\left(\cos\alpha_0 - \sqrt{\frac{\varepsilon_{z_1}}{\varepsilon_0}}\right) \tag{8}$$

在上式中代入置换式

$$\Delta\rho = \frac{E_c}{M\varepsilon_0}\Delta r_2^* \tag{9}$$

则有

$$\Delta\rho = 2\sin\alpha_0\left(\cos\alpha_0 - \sqrt{\frac{\varepsilon_{z_1}}{\varepsilon_0}}\right) \tag{10}$$

式中，$\Delta\rho$ 为无量纲的近轴横向色差。

由式（10）可见，当 α_0 改变时，则与 $\frac{\sqrt{\varepsilon_{z_1}}}{\sqrt{\varepsilon_0}}$ 相对应的每一个像面上，$\Delta\rho$ 就有一个确定的弥散圆半径的极大值。如果给定一个 $\frac{\sqrt{\varepsilon_{z_1}}}{\sqrt{\varepsilon_0}}$ 值，即给定一个像平面位置，就可以确定在什么样的角度 α_0 下，$\Delta\rho$ 取极大值。为此，我们需要求 $\Delta\rho\left(\frac{\sqrt{\varepsilon_{z_1}}}{\sqrt{\varepsilon_0}}, \alpha_0\right)$ 的极值条件。因为 $\Delta\rho$ 只是作为 α_0 的函数，在某个 α_0 下，取 $\Delta\rho$ 的极大值 $\Delta\rho_{1v}$。现先求 $\Delta\rho$ 对 α_0 的偏微商。

$$\frac{\partial(\Delta\rho)}{\partial(\alpha_0)} = 2\left(\cos 2\alpha_0 - \sqrt{\frac{\varepsilon_{z_1}}{\varepsilon_0}}\cos\alpha_0\right) \tag{11}$$

由此可确定极值的条件

$$\sqrt{\frac{\varepsilon_{z_1}}{\varepsilon_0}} = \frac{\cos 2\alpha_v}{\cos\alpha_v} \tag{12}$$

即

$$2\cos^2\alpha_v - \sqrt{\frac{\varepsilon_{z_1}}{\varepsilon_0}}\cos\alpha_v - 1 = 0 \tag{13}$$

对式（13）取极值，我们便可确定弥散圆半径的大小。此时，角度 α_v 对应于像面位置的关系式为

$$\cos\alpha_v = \frac{\sqrt{\frac{\varepsilon_{z_1}}{\varepsilon_0}} \pm \sqrt{\frac{\varepsilon_{z_1}}{\varepsilon_0} + 8}}{4} \tag{14}$$

由物理条件 $\cos\alpha_v \geq 0$，故上式的 ± 号应取正值。

如果我们以 α_v 来表示 $\Delta\rho_{1v}$，则由式（10）和式（12），得

$$\Delta\rho_{1v} = -2\sin\alpha_v\left(\frac{\cos 2\alpha_v}{\cos\alpha_v} - \cos\alpha_v\right) = 2\frac{\sin^3\alpha_v}{\cos\alpha_v} \tag{15}$$

由式（12）和式（15）可见，$0 \leq \frac{\sqrt{\varepsilon_{z_1}}}{\sqrt{\varepsilon_0}} \leq 1$，$\alpha_v$ 被限制于 $-\frac{\pi}{4} \leq \alpha_v \leq \frac{\pi}{4}$ 的范围内，于是，$\Delta\rho_{1v}$ 在 $[-1, +1]$ 之间变化。

由式（10），对于极限像面，即令 $\sqrt{\varepsilon_{z_1}} = 0$，则有

$$\Delta\rho_t = \sin 2\alpha_v \tag{16}$$

$\Delta\rho_t$ 为极限像面上的弥散圆半径。

当 $\alpha_v = \pm(\pi/4)$ 时，$|\Delta\rho_t|$ 取极大值，而且 $|\Delta\rho_t|_{90°-\alpha_v} = |\Delta\rho_t|_{\alpha_v}$，$|\tan\alpha_i|_{90°-\alpha_v} > |\tan\alpha_i|_{\alpha_v}$。这说明，如果将 α_v 限制在 $-\frac{\pi}{4} \leq \alpha_v \leq \frac{\pi}{4}$ 之内，则在极限像面上，逸出角 α_v 的电子射线在该面上的交点与逸出角 $(90°-\alpha_v)$ 的电子射线相同，但斜率的绝对值要小。在过极限像面后，在轨迹与轴相交之前，α_v 的轨迹则在 $(90°-\alpha_v)$ 的轨迹之上。由图2可见，密集电子轨迹的包络线（焦散线，实质上是回转的焦散面）由 $[-\pi/4, \pi/4]$ 的电子轨迹所决定。而式（15）正是焦散线的表示式，$\Delta\rho_{1v}$ 表示 α_0 在 $[-\pi/4, \pi/4]$ 的电子束与轴相交前在某一像面上的散射圆半径。

因此，在 $[-\pi/4, \pi/4]$ 的电子射束与轴相交之前，弥散圆半径由焦散线决定；而在轨迹与轴相交之后，自极限像面开始，散射圆半径的大小由 $\alpha_0 = \pm 90°$ 的轨迹所决定。由式（10），当 $\alpha_0 = \pm 90°$ 时，$\Delta\rho$ 亦取极大值，我们将它以 $\Delta\rho_{2v}$ 表示：

$$\Delta\rho_{2v} = \pm 2\sqrt{\frac{\varepsilon_{z_1}}{\varepsilon_0}} \tag{17}$$

于是，当 $0 \leq \sqrt{\frac{\varepsilon_{z_1}}{\varepsilon_0}} \leq 1$ 时，若 $\alpha_0 = \pm 90°$，则 $\Delta\rho_{2v}$ 在 $[-2, +2]$ 范围内变化。根据上面的讨论，我们可以看出，当相应的像面位置 $0 \leq \sqrt{\frac{\varepsilon_{z_1}}{\varepsilon_0}} \leq 1$ 变化时，$\Delta\rho_{1v}$ 和 $\Delta\rho_{2v}$ 按相反的趋势单调地变化。

显然，对应于 $\sqrt{\frac{\varepsilon_{z_1}}{\varepsilon_0}}$ 的像面上散射圆半径将由 $\Delta\rho_{1v}$、$\Delta\rho_{2v}$ 中较大者决定。上面已经指出，$\Delta\rho_{1v}$ 与 $\Delta\rho_{2v}$ 是按相反的趋势单调地改变。因此不难发现，弥散圆半径取最小值的像平面位置应该由下述条件决定：

$$\Delta\rho_{1v} = \Delta\rho_{2v} \tag{18}$$

于是，由式（15）和式（17），则可表示为

$$-2\sqrt{\frac{\varepsilon_{z_1}}{\varepsilon_0}} = 2\frac{\sin^3\alpha_v}{\cos\alpha_v} \tag{19}$$

将极值条件式（12）代入上式中，则得出 α_v 应满足如下条件：

$$\frac{\sin^3\alpha_v}{\cos 2\alpha_v} = -1 \tag{20}$$

式（20）亦可写成如下形式：
$$\sin^3\alpha_v - 2\sin^2\alpha_v + 1 = (\sin\alpha_v - 1)(\sin^2\alpha_v - \sin\alpha_v - 1) = 0 \quad (21)$$
不难得到 $\sin\alpha_v$ 的三个解：
$$\sin\alpha_{v_1} = 1, \quad \sin\alpha_{v_{2,3}} = \frac{1 \pm \sqrt{5}}{2} \quad (22)$$
对于我们所讨论的情况：$-\dfrac{\pi}{4} \leqslant \alpha_v \leqslant 0$，故只剩下一个合理的解：
$$\sin\alpha_v = \frac{1-\sqrt{5}}{2} = -0.618 \quad (23)$$
由此可以定出 α_v：
$$\alpha_v = -38°10' \quad (24)$$
由此我们得到这样的结论：最小弥散圆的位置位于 $\alpha_0 = +\left(\dfrac{\pi}{2}\right)$ 的电子轨线与 $\alpha_v = -38°10'$ 的电子轨线相交处（也可以说在 $\alpha_0 = -\dfrac{\pi}{2}$ 与 $\alpha_v = +38°10'$ 两条电子轨线的相交处），如图 2 所示。

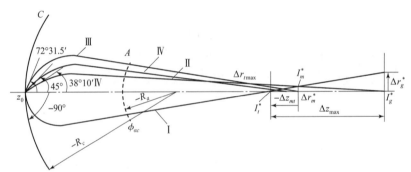

图 2　最佳像面位置与最小弥散圆的确定

在图 2 中，假定逸出电子射线均是 $\varepsilon_0 = \varepsilon_m$ 的单色束情况，射线 I 表示由阴极原点 z_0 以 $\alpha_0 = -90°$ 的初角度射出的近轴电子轨线，它在高斯像面上形成最大弥散圆 Δr_g^*；射线 II 表示由阴极上同一原点 z_0 以 $\alpha_v = +38°10'$ 的初角度射出的近轴电子轨线，最佳像平面即垂直于轴并过射线 I 和射线 II 交点的平面，从而形成最小弥散圆 Δr_m^*；射线 III 表示以 $72°31.5'$ 逸出的电子轨线，它与光轴交于最佳像面位置 I_m^* 处；射线 IV 表示以 $+45°$ 逸出的电子轨线，它在极限像面 I_t^* 处形成弥散圆 Δr_t^*。

我们把决定最小弥散圆的位置称为最佳像面位置，它由下式
$$\sqrt{\frac{\varepsilon_{z_1}}{\varepsilon_0}} = \frac{\cos 2\alpha_v}{\cos\alpha_v} = -\frac{\sin^3\alpha_v}{\cos\alpha_v} = 0.30028 \quad (25)$$
来决定，若令 $\varepsilon_{z_1} = \varepsilon_0 \cos^2\delta_0$，则与最佳像面位置相对应的逸出角 δ_0 为
$$\delta_0 = \pm 72°31.5' \quad (26)$$
故由式（17）可得
$$\Delta\rho_{\min} = \mp 2\sqrt{\frac{\varepsilon_{z_1}}{\varepsilon_0}} = \mp 0.60056 \quad (27)$$

因此，由式（9），可得最小弥散圆的半径为

$$\Delta r_{\min} = \pm 0.6 \frac{M\varepsilon_0}{E_c} \tag{28}$$

折合到阴极上的最小弥散圆直径为

$$2\left|\frac{\Delta r_{\min}}{M}\right| = 1.2\left|\frac{\varepsilon_0}{E_c}\right| \tag{29}$$

通常，ε_0 取光电发射时电子的最大初电位 $\varepsilon_{\max} = \varepsilon_m$。式（29）表明，无论是具体的同心球型电子光学系统或是阴极透镜，折合到阴极上的最小弥散圆的直径仅与电子的最大初电位 ε_m 和阴极面上的场强 E_c 的比值有关。这个结果对于成像器件的分辨本领的研究具有重要的意义。

由文章"近轴横向色差与几何横向球差"中式（12），可确定位于 $\sqrt{\frac{\varepsilon_{z_1}}{\varepsilon_0}} = 0.3$ 的最佳像面与位于 $\sqrt{\frac{\varepsilon_{z_1}}{\varepsilon_0}} = 0$ 的极限像面之间的距离：

$$\Delta z_{mt} = 0.6 \frac{M^2 \sqrt{\phi_{ac}\varepsilon_0}}{E_c} \tag{30}$$

图 2 表示两电极同心球系统中各个像面对应的位置，I_g^*、I_m^* 和 I_t^* 分别对应于高斯像面、最佳像面和极限像面的位置。显然，极限像面 I_t^* 至高斯像面 I_g^* 之距离可用一级纵向近轴色差表示

$$\Delta z_{tg} = \Delta z_{\max} = -\frac{2M^2 \sqrt{\phi_{ac}\varepsilon_0}}{E_c}\frac{1}{2} = -\frac{M^2 \sqrt{\phi_{ac}\varepsilon_0}}{E_c} \tag{31}$$

式中，$\Delta z_{\max}$ 为极限像面至高斯像面的距离。极限像面、最佳像面与高斯像面上的散射圆半径存在着如下的关系：

$$\Delta r_t^* : \Delta r_m^* : \Delta r_g^* = 1 : 0.6 : 2 \tag{32}$$

式（32）正是谢曼 О.И. 当初提出 Recknagel-Арцимович 表示式的依据。

3 成像段区域电子轨迹形成的弥散圆与焦散面[1,4,5]

下面我们讨论成像段区域电子轨迹弥散所形成的圆锥面与焦散面。由上所述，从阴极面原点 z_0 射出的电子束并不聚焦于一点，但总可以找到电子轨线十分密集的那部分包络曲面，这就是所谓的焦散面。显然，单色电子束的焦散面由 $\Delta\rho = \Delta\rho_{1v}$ 绕光轴旋转而得。从阴极面上 z_0 点射出的平面电子束在极限像面与高斯像面之间的光轴上密集而形成焦散线。图上的粗线表示焦散线和焦散曲面的形状。此回转的焦散面在子午面上的截线可以由式（15）的如下表示式确定：即

$$\Delta\rho_{1v} = 2\frac{\sin^3\alpha_v}{\cos\alpha_v}$$

式中，$\cos\alpha_v$ 的表示式已由式（14）给出，若该式中的 $\sqrt{\frac{\varepsilon_{z_1}}{\varepsilon_0}}$ 以 $\cos\delta_0$ 表示，它可表示为

$$\cos\alpha_v = \frac{\sqrt{\cos^2\delta_0 + 8} + \cos\delta_0}{4} \tag{33}$$

由此可以解得

$$\sin\alpha_v = \frac{\sqrt{8 - 2\cos^2\delta_0 - 2\cos\delta_0 \sqrt{\cos^2\delta_0 + 8}}}{4} \tag{34}$$

$$\sin^3\alpha_v = \frac{4 - \cos^2\delta_0 - \cos\delta_0 \sqrt{\cos^2\delta_0 + 8}}{32}\sqrt{8 - 2\cos^2\delta_0 - 2\cos\delta_0 \sqrt{\cos^2\delta_0 + 8}} \tag{35}$$

将以上三式代入式（15），由式（9）便得到折合到阴极面上的弥散半径值：

$$\Delta r_{1v} = \pm \frac{M\varepsilon_0}{8E_c}\left(\sqrt{\cos^2\delta_0 + 8} - 3\cos\delta_0\right)\sqrt{8 - 2\cos^2\delta_0 - 2\cos\delta_0 \sqrt{\cos^2\delta_0 + 8}} \tag{36}$$

式（36）描绘了成像电子光学系统由光阴极物面原点逸出的电子束在成像段区间由极限像面 $\cos\delta_0 = 0$ 至高斯像面 $\cos\delta_0 = 1$ 形成的焦散面的形状，该式表示逸出角为 α_0 在 $[-\pi/4, \pi/4]$ 的电子束与轴相交之前在某一像面上的散射圆半径，如图3所示，其数据如表1中的 $\Delta\rho_{1v}$ 所示。

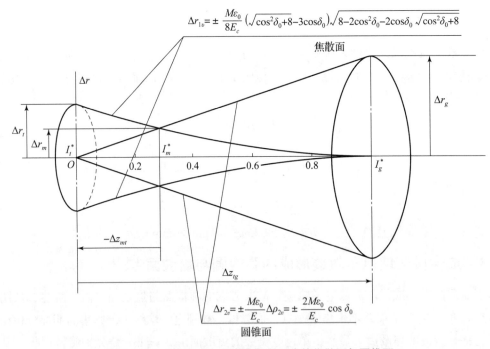

图3　轴上点的电子束在成像段所形成的焦散面与圆锥面

其次，当逸出的电子轨迹与轴相交之后，自极限像面开始，散射圆半径由逸出角 α_0 为 $[-\pi/2, \pi/2]$ 的轨迹所决定。由式（17），$\Delta\rho$ 亦取极大值，我们将它以 $\Delta\rho_{2v}$ 表示。于是，由式（9）可得

$$\Delta r_{2v} = \mp \frac{2M\varepsilon_0}{E_c}\cos\delta_0 \tag{37}$$

式（37）描绘了电子束由极限像面 $\cos\delta_0 = 0$ 至高斯像面 $\cos\delta_0 = 1$ 所形成的圆锥面的形状，

如图 3 所示。其数据如表 1 的 $\Delta\rho_{2v}$ 所示。

表 1　轴上物点逸出的电子束在极限像面至高斯像面间形成的横向像差（束斑）

$\cos\delta_0$ / $\Delta\rho_v$	0	0.1	0.2	0.3	0.4	0.5	0.6	0.7	0.8	0.9	1
$\Delta\rho_{1v}$	1	0.861 17	0.727 92	0.600 9	0.480 96	0.369 01	0.266 30	0.174 44	0.095 76	0.034 14	0
$\Delta\rho_{2v}$	0	0.2	0.4	0.6	0.8	1	1.2	1.4	1.6	1.8	2

应该指出，以上的结论是基于单色束且不考虑初角度分布求得的。当考虑逸出电子的初角度和初能量分布时，则最小弥散圆的大小及其对应的最佳像面位置将会有相应的变化。

结束语

本文研究了成像电子光学系统由阴极物面原点发射的电子在栅状阳极上形成的散射圆以及整个物面发射的电子所形成的交叉颈，并由所导出的成像电子光学系统中普遍成立的二级近轴横向色差即 Recknagel-Арцимович 表示式，描绘了由物面原点逸出的电子射束在成像段所形成的电子射线的包络，确定了电子束形成的最小弥散圆以及最佳成像面的位置。本文有助于理解电子光学系统中成像段的电子的行进轨迹以及电子光学横向像差的形成。

本系列文章的总结

（1）本系列文章最初思考的问题。

在研究静电聚焦成像系统的电子光学之初，主要思考两个问题：一是，傍轴电子光学理论和方法不能用来解决成像电子光学问题；近轴电子光学理论只能解决理想成像，而且只适合解决邻近对称轴区域的电子光学问题。二是，成像电子光学系统的横向像差是不是当时学术界普遍认为仅有三级几何横向像差。

（2）成像电子光学研究的切入点。

科学研究从何处切入是解决问题的关键。当时的思考是，如果能找到一个可以求得解析解的成像电子光学系统的理想模型，从它入手，把它的成像规律、实际轨迹和近轴轨迹等都研究透了，便能对理想成像、横向像差等有一个正确理解和把握。由此出发，可指导一般静电成像电子光学系统理论与设计。最终，选择两电极同心球静电聚焦系统中电子运动的轨迹作为研究的切入点。

（3）由前人研究中找存在的问题。

静电聚焦球对称系统的电子光学，尽管德、英、俄等国的一些科学家也曾研究过，但他们的重点是研究电子在同心球系统内部运动的轨迹，得到的仅是电子初速为零的轨迹的近似解。这样的处理对于研究系统的成像特性是远远不够的，需要研究的是如何求得电子从阴极面轴上点逸出后通过同心球系统后与轴相交的精确着陆点。

（4）找电子轨迹精确解与近轴解是本研究的关键。

本研究找到了两电极与多电极同心球静电聚焦系统自阴极逸出的电子轨迹的成像位置

的精确解以及近轴解，并把它表达成级数展开的形式，由此解决了理想成像等电子光学性质概念和定义电子光学像差的问题。由静电聚焦同心球系统电子光学理想模型所具有的矛盾的特殊性出发，进而研究宽电子束成像矛盾的普遍性，对成像电子光学的深入研究具有指导意义。

（5）研究的新发现与新认识。

本课题的研究得到了如下的新进展、新结论和新认识：

一是，导出了自阴极面逸出的电子在静电聚焦同心球系统中的新的轨迹表示式。它如同光线光学由一个折射面过渡到下一个折射面的追迹公式，不但能用于研究两电极同心球系统的电子光学，而且也适用于研究任意多个电极同心球系统的电子追迹。

二是，找到了电子束自阴极面逸出经过同心球系统后最终会聚的精确落点，即电子轨迹的精确解，这为研究理想成像，讨论电子光学性质和定义横向像差打下坚实基础。与此同时，给出了两电极静电同心球系统中圆柱坐标系下由阴极面原点逸出的实际电子轨迹的解析表示式，纠正了一些文献中存在的谬误。

三是，由精确解导出了静电聚焦同心球系统中电子轨迹的近轴解，它正是电子运动方程或轨迹方程的解析解，证明了所提出的理论之无误。

四是，给出了成像系统的电子光学横向像差新的定义，提出了成像电子光学系统的横向像差乃是近轴横向像差与几何横向像差的合成，而不仅仅是当时电子光学学术界普遍认为只有几何横向像差。

五是，证实了不论是两电极或是多电极同心球系统，还是一般的静电聚焦成像电子光学系统，决定系统鉴别率的是二级近轴横向色差，它仅与逸出电子的初电位、初角度、阴极面上的场强以及系统的线放大率有关，而与系统的具体电极结构及轴上电位分布无关。本研究证实了谢曼提出的确定成像电子光学系统鉴别率的 Recknagel-Арцимович 表示式普遍成立。

六是，考察了成像系统中电子束形成的最小弥散圆以及最佳成像面的位置的确定，形象地展示了成像段所形成的电子射线的包络。

……

（6）评价。

1993 年，作者出版了一部学术专著《宽束电子光学》，详细总结了作者与所领导的团队在成像电子光学领域探索的研究成果，其中包括本系列文章"静电聚焦同心球系统成像电子光学"和 2019 年"电磁聚焦同心球系统成像电子光学"系列文章的一些内容。这部专著得到了国内外 12 位光电子领域科学家的高度评价与赞誉。2000 年，诺贝尔奖金获得者普洛霍罗夫院士高度赞扬了作者在成像电子光学研究中的科学成果，认为这是关乎创建科学"学派"的工作。他在作者当选为俄罗斯联邦工程科学院外籍院士的贺电中称"您是您的科学学派的创立者"。

致　谢

作者对清华大学张海涛教授在本系列文章中有关计算与绘图给予的帮助表示衷心的感谢，对女儿周霞给予计算机制图上的帮助也表示感谢。

参 考 文 献

[1] 周立伟.宽束电子光学[M].北京:北京理工大学出版社,1993.

[2] КРУПП Д М. Кружок сферической аберрации в катодной линзе с шаровой симметрией[J]. Труды ГОИ, 1963,30(159):22-29.

[3] RECKNAGEL A. Theorie des elektrisohen elektronen miktroskops fur selbstrakler[J]. Z. Angew. Physik, 1941,117:689.

[4] АРЦИМОВИЧ Л А. Электростатические свойства эмисионных систем[J]. А Н СССР Сер. физ., 1944,8(6):313.

[5] 谢曼 О И.电子光学理论基础[M].北京:高等教育出版社,1958.

[6] ЧЖОУ Л В. Исследование электростатических фокусирующих систем электронно-оптических преобразователей изобржении с шаровой и осевой симметрией[D]. Ленинград: Диссертация ЛЭТИ.,1966.

二、电磁复合聚焦同心球系统的成像电子光学

Imaging Electron Optics of Combined Electromagnetic Focusing Concentric Spherical Systems

2.1 Part A: Paraxial Optics
近轴光学

Abstract: The ray solutions of paraxial equation for imaging electron optics have been explored by an ideal model of combined electromagnetic concentric spherical system in the present paper. The exact analytical expressions of rotational angle as well as two special solutions of paraxial equation in this system have been firstly derived, and the paraxial imaging properties have also been investigated. The results have been generalized to the bi-electrode electrostatic concentric spherical system, the homogeneous and parallel combined electromagnetic system, and the electrostatic proximity system.

摘要：本文通过复合电磁同心球系统的理想模型，对成像电子光学近轴方程的轨迹求解进行探讨，首次推导了该系统的近轴电子轨迹的转角及近轴方程的两个特解的解析表达式，探讨了近轴成像性质，并将结果推广到两电极静电同心球系统、均匀平行复合电磁系统和静电近贴系统中。

Introduction

A spherical condenser system composed of two concentric spherical electrodes with a common curvature center is called a bi-electrode concentric spherical electrostatic focusing system. If we add to this system a concentric radial magnetic field parallel to the electrostatic field, we can get a bi-electrode combined electromagnetic concentric spherical focusing system, which is simply called as a combined electromagnetic concentric spherical system. In other words, the system has concentric radial electric and magnetic fields, which are always mutually parallel, the direction of the electric field intensity always points to the cathode surface, and the direction of the magnetic induction intensity is identical to or opposite to the direction of electric field intensity[1]. Similarly to a bi-electrode electrostatic concentric spherical system, study on a combined electromagnetic concentric spherical system as an imaging system has always been an object of interest for the researchers in electron optical study with wide beam focusing. From the viewpoint of electron optics, their magnetic and electric field distributions can be expressed as analytical forms, and the system can achieve both magnifying or de-magnifying requirements that the magnification is greater than 1 or less than 1. If the analytical form representation of the electron trajectory can be obtained, the electron optical properties and aberrations can be quantitatively studied. No matter whether it is strictly accurate or approximately realized (its special case is a bi-electrode electrostatic concentric spherical system), a very good image quality can be achieved.

If the imaging screen is a spherical one, due to the spherical symmetry, the system has no aberrations other than the paraxial chromatic aberration and the geometrical spherical aberration. Especially when the combined electromagnetic focusing image tube is numerically calculated, the analytical solution derived from the model can be used as a solution of initial ray tracing or to verify the error of the initial trajectory calculation.

In general, for a combined electromagnetic imaging system, it is a very tricky problem for solving the analytical solutions of the electron trajectory from the electron optical paraxial equation at the known electrostatic potential and magnetic induction distributions. Early in 1970s, we had solved the analytical solution of the electron trajectory under the initial axial electron potential $\varepsilon_z \neq 0$ (where ε_z is the initial axial potential corresponding to the initial axial energy of emitted electrons) in a bi-electrode electrostatic concentric spherical system. Later, we obtained the analytical solution of the practical trajectory and the exact solution, approximate solution and asymptotic solution of the paraxial equation in this system, and systematically studied the lateral aberration and the temporal aberration[2-5]. For the combined electromagnetic imaging system, because the electric field and the magnetic field are converged, the electric field accelerates the trajectory, and the magnetic field rotates the trajectory. No matter the ray equation or the motion equation, the solution is very complicated. People have been trying to find the analytical solutions of their paraxial equations in order to more accurately understand the status of electron motion and analyze the constitution of paraxial aberrations and geometric aberrations. Because of this, it has always been a problem to be solved in the academia of electron optics.

The combined electromagnetic concentric spherical system has been studied as an imaging system[6-7], but its electron trajectory has been solved by a computer numerical method. In Ref. [8], the imaging properties and its aberrations have been explored, but they were studied only for a simple homogeneous and parallel combined electromagnetic system. In 1978, we firstly reported his study on the combined electromagnetic concentric spherical system by analytical method, deduced the analytical expression of a special solution of $v(z,\varepsilon_z)$ for the paraxial electron ray by using analogy method, derived the focusing condition of system, investigated the paraxial spherical-chromatic aberration, and generalized the results to several special cases[1]. Later, Russian scientists N. A. Smirnov, M. A. Monastyrski, and Y. V. Kulikov also studied the imaging properties and aberrations of the combined electromagnetic concentric spherical system, in which expressions of the zeroth order approximation of two special solutions of paraxial equation at $\varepsilon_z = 0$ were given, and on this basis they had defined the paraxial lateral aberrations[9].

Fortunately, for the combined electromagnetic concentric spherical system, although its potential distribution and magnetic field distribution are intertwined with each other, through a clever treatment, we may obtain an exact analytical representation of two special solutions of the paraxial equation at $\varepsilon_z \neq 0$.

It should be pointed out that obtaining the exact analytical expressions of two special solutions of paraxial equation in a combined electromagnetic concentric spherical system is an extremely rare case in electron optics with wide beam focusing. In the present paper, we shall describe in detail how to get the exact expressions of two special solutions of the paraxial equation in the combined electromagnetic concentric spherical system, and generalize the results obtained to some special cases—the bi-electrode electrostatic concentric spherical system, the homogeneous and parallel combined electromagnetic system, and the electrostatic proximity system. In the following sections, we will discuss approximate solutions and asymptotic solutions of two special solutions and the paraxial lateral aberrations.

1 General description of paraxial imaging for the combined electromagnetic concentric spherical systems

In the combined electromagnetic concentric spherical systems (Fig. 1), the paraxial ray $u^*(z)$ in the rotational coordinate system is the solution of following equation[10]:

$$u^{*''}(z) + \frac{\phi'(z)}{2[\phi(z)+\varepsilon_z]}u^{*'}(z) + \frac{\phi''(z)}{4[\phi(z)+\varepsilon_z]}\left[\phi''(z) + \frac{e}{2m_0}B^2(z)\right]u^*(z) = 0 \quad (1)$$

where $u^*(z)$ is the position vector of electrons in the rotational coordinate system (x, y); $\phi(z)$ and $B(z)$ are the axial potential distributions and axial magnetic induction intensity distribution respectively; ε_z is the initial axial potential corresponding to the initial axial energy of emitted electrons; e/m_0 is the ratio of electron charge to mass; $' = d/dz$ expresses the differential with respect to z; the star mark "$*$" represents that it is under the paraxial condition.

Let $r^*(z)$ be the position vector of electrons in the fixed coordinate system (x, y), which has the following relationship with $u^*(z)$:

$$r^*(z) = u^*(z)\exp[j\chi(z,\varepsilon_z)] \quad (2)$$

where $j = \sqrt{-1}$; $\chi(z,\varepsilon_z)$ is a rotational angle of electron trajectory, which is an angle of the rotational coordinate system $u(x,y)$ rotated with respect to the fixed coordinate system $r(x,y)$:

$$\chi(z,\varepsilon_z) = \int_{z_0}^{z} \sqrt{\frac{e}{8m_0}} \frac{B(z)}{\sqrt{\phi(z)+\varepsilon_z}} dz + \chi_0 \quad (3)$$

where $\chi_0 = \chi(z_0 = 0)$ is the initial angle of the rotational coordinate system. In general, we suppose that $\chi_0 = 0$.

Eq. (3) shows that $\chi(z,\varepsilon_z)$, the rotational angle of electron trajectory emitted from cathode surface is independent of the height of the object r_0 and the initial radial velocity $\dot{r}_0$ of the electrons and it is only related to ϕ_{ac}, $B(z)$, and the axial velocity $\dot{z}$ of the electrons. As long as the axial initial velocity $\dot{z}$ of electrons is the same, all electrons, whether on axis or off axis, turn at the same angle.

Eq. (1) is a second order linear homogeneous differential equation, in which two special solutions $v = v(z)$ and $w = w(z)$ should satisfy the following initial conditions:

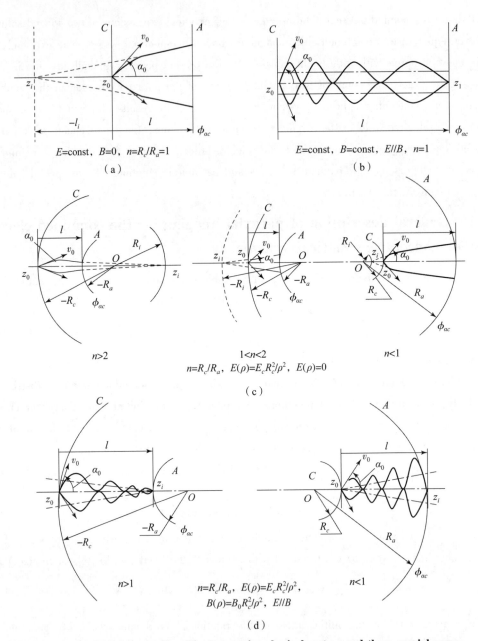

Fig. 1 Combined electromagnetic concentric spherical system and three special cases

$$v(z=0)=0, \quad v'(z=0)=\frac{1}{\sqrt{\varepsilon_z}} \tag{4}$$

$$w(z=0)=1, \quad w'(z=0)=1 \tag{5}$$

And these two special solutions should satisfy the following Wronski determinant:

$$\sqrt{\phi(z)+\varepsilon_z}(v'w-vw')=1 \tag{6}$$

The general solution of paraxial Eq. (1) can be written as

$$u^*(z)=r_0 w(z)+\left[\sqrt{\frac{m_0}{2e}}\dot{r}_0-(k\times r_0)\sqrt{\frac{m_0}{8e}}B_0\right]v(z) \tag{7}$$

From Eq. (2) we may have

$$r^*(z) = r_0 w(z) \exp[j\chi(z)] + \left[\sqrt{\frac{m_0}{2e}}\dot{r}_0 - (k \times r_0)\sqrt{\frac{m_0}{8e}}B_0\right]v(z)\exp[j\chi(z)] \quad (8)$$

where k is a unit vector along the positive direction of axis z, r_0 and $\dot{r}_0$ are the initial position vector and the radial initial velocity vector of electron emitted from the cathode surface, respectively.

In the ideal image plane at $z = z_i$, which corresponds to ε_{z_1}, let $v(z, \varepsilon_{z_1}) = 0$, $\chi(z, \varepsilon_{z_1}) = i\pi$, where i is the number of loops. Then, we have

$$r^*(z_i, \varepsilon_{z_1}) = r_0 w(z_i, \varepsilon_{z_1}) = r_0 M \quad (9)$$

where M is the linear magnification of the system.

2 Solution of paraxial ray for the combined electromagnetic concentric spherical system

2.1 Analytical expressions of fields

For the combined electromagnetic concentric spherical system, the direction of the electric field intensity E coincides with the normal of the cathode surface, and the magnetic induction intensity B is parallel to E everywhere. According to the spherical symmetry condition, in the direction of vector radius ρ, we have

$$\begin{aligned}E_\rho &= E_c \frac{R_c^2}{\rho^2} \\ B_\rho &= B_0 \frac{R_c^2}{\rho^2}\end{aligned} \quad (10)$$

where E_c is the electric field intensity at the cathode surface, and B_0 is the magnetic induction intensity at the cathode surface.

Starting from Eq. (10), we can derive the expressions of the axial potential distribution $\phi(z)$ and the axial distribution of magnetic induction intensity $B(z)$ as

$$\begin{aligned}\phi(z) &= (-E_c)\frac{R_c z}{R_c + z} \\ B(z) &= B_0 \frac{R_c^2}{(R_c + z)^2}\end{aligned} \quad (11)$$

They can be expressed as

$$\phi(z) = \frac{z}{nl - (n-1)z}\phi_{ac} \quad (12)$$

$$B(z) = \frac{n^2 l^2}{[nl - (n-1)z]^2}B_0 \quad (13)$$

where $n = R_c/R_a$, R_c and R_a are respectively the radius of curvature of the spherical cathode and the spherical screen (anode) of the concentric spherical system, its value is calculated from the center O of the curvature, the direction of electron travelling is positive, and vice versa; l is the distance between the cathode and the anode, $l = R_a - R_c$; ϕ_{ac} is the electrostatic potential of the anode relative to the cathode; E_c is the electric field intensity at the cathode surface, can be expressed by

$$E_c = \frac{\phi_{ac}}{R_c(n-1)} = \frac{\phi_{ac}}{-nl} \quad (14)$$

The direction of E_c always points to the cathode, whether $n > 1$ or $n < 1$. Therefore, E_c takes a negative value forever. For the combined electromagnetic concentric spherical system, the image on the spherical cathode is transferred to the spherical anode (screen). As a result, $n = 1/M$.

2.2 Exact analytical solution for the rotation angle $\chi(z, \varepsilon_z)$ of electron ray

Now we shall seek the expressions of two special solutions of paraxial Eq. (1) under the axial potential distribution [Eq. (12)] and the axial magnetic induction distribution [Eq. (13)]. At first, we shall solve the rotation angle $\chi(z, \varepsilon_z)$ of electron trajectory in the rotating coordinate system. Differentiating Eq. (12), we obtain

$$\phi'(z) = (-E_c) \frac{(nl)^2}{[nl - (n-1)z]^2} \quad (15)$$

By transforming Eq. (13), $B(z)$ can be expressed in the following form

$$B(z) = \phi'(z) \frac{B_0}{-E_c} \quad (16)$$

Eq. (16) shows the relationship of transformation between axial potential distribution and axial magnetic induction distribution in the case of concentric spherical system. It is precisely because of the transformation of the critical step of Eq. (16) that the integration of Eq. (3) and solving the electro-optical paraxial Eq. (1) becomes possible. Substituting Eq. (16) into Eq. (3), we may obtain the angle $\chi(z, \varepsilon_z)$ that the electron ray rotates with the rotating coordinate system:

$$\chi(z, \varepsilon_z) = \sqrt{\frac{e}{2m_0}} \frac{B_0}{-E_c} \int_{z_0}^{z} \frac{1}{2} \frac{\phi'(z)}{\sqrt{\phi(z) + \varepsilon_z}} dz \quad (17)$$

Integrating it, we have

$$\chi(z, \varepsilon_z) = \sqrt{\frac{e}{2m_0}} \frac{B_0}{-E_c} [\sqrt{\phi(z) + \varepsilon_z} - \sqrt{\varepsilon_z}] \quad (18)$$

in which we suppose that $\chi_0 = \chi(z_0 = 0) = 0$, $\phi(z_0 = 0) = 0$.

Eq. (18) may be written as a more concise form. Let $k^2 = \left[\frac{e}{2m_0} \frac{B_0^2}{(-E_c)} \right]$, and its dimension is $1/l$. Substituting it to Eq. (18), we may obtain the exact analytical expression of rotational angle $\chi(z, \varepsilon_z)$ as

$$\chi(z, \varepsilon_z) = \frac{k}{\sqrt{-E_c}} [\sqrt{\phi(z) + \varepsilon_z} - \sqrt{\varepsilon_z}] \quad (19)$$

2.3 Exact analytical expression for the special solution $\chi(z,\varepsilon_z)$

Now we seek the exact analytical expression of the first special solution $v(z,\varepsilon_z)$ of the paraxial Eq. (1). As we know, the special solution $v(z,\varepsilon_z)$ is a helix rotated with the rotating coordinate system. Since the angle $v(z,\varepsilon_z)$ of electron ray rotated along with the rotating coordinate system is known by Eq. (19), we may suppose that

$$v(z,\varepsilon_z) = v_1(z,\varepsilon_z)\sin\chi(z,\varepsilon_z) \tag{20}$$

where $v_1(z,\varepsilon_z)$ is a function to be solved.

Substituting Eq. (20) to the scalar form of Eq. (1):

$$v''(z) + \frac{\phi'(z)}{2[\phi(z)+\varepsilon_z]}v'(z) + \frac{\phi''(z)}{4[\phi(z)+\varepsilon_z]}\left[\phi''(z) + \frac{e}{2m_0}B^2(z)\right]v(z) = 0 \tag{21}$$

we may obtain items which are related to $\sin\chi(z,\varepsilon_z)$ and $\cos\chi(z,\varepsilon_z)$:

$$\left\{2v'_1(z,\varepsilon_z)\chi'(z,\varepsilon_z) + \left[\chi''(z,\varepsilon_z) + \frac{\phi'(z)}{2[\phi(z)+\varepsilon_z]}\chi'(z,\varepsilon_z)\right]v_1(z,\varepsilon_z)\right\}\cos\chi(z,\varepsilon_z) = 0 \tag{22}$$

$$\left\{v''_1(z,\varepsilon_z) + \frac{\phi'(z)v'(z,\varepsilon_z)}{2[\phi(z)+\varepsilon_z]} + \left[\frac{\phi''(z)+\frac{k^2}{-E_c}\phi'^2(z)}{4(\phi(z)+\varepsilon_z)} - \chi'^2(z,\varepsilon_z)\right]v_1(z,\varepsilon_z)\right\}\sin\chi(z,\varepsilon_z) = 0 \tag{23}$$

Obviously, if Eq. (22) and Eq. (23) are tenable, the coefficients in the brackets should be equal to zero, respectively.

Now, we substitute Eq. (19) of rotational angle $\chi(z,\varepsilon_z)$ and its derivatives

$$\chi'(z,\varepsilon_z) = \frac{k}{\sqrt{-E_c}}\frac{\phi'(z)}{2\sqrt{\phi(z)+\varepsilon_z}} \tag{24}$$

$$\chi''(z,\varepsilon_z) = \frac{k}{\sqrt{-E_c}}\frac{\phi''(z)}{2\sqrt{\phi(z)+\varepsilon_z}} - \frac{k}{\sqrt{-E_c}}\frac{\phi'^2(z)}{4[\phi(z)+\varepsilon_z]^{3/2}} \tag{25}$$

into the brackets of Eq. (22), through a series of simplified treatment, and we may have

$$2v'_1(z,\varepsilon_z)\phi'(z) + v_1(z,\varepsilon_z)\phi''(z) = 0 \tag{26}$$

Seeing that the axial potential distribution $\phi(z)$ of the concentric spherical system expressed by Eq. (12) satisfies the following relationship:

$$2\left[1 - \frac{\phi'(z)}{\phi(z)}\right] + \frac{\phi''(z)}{\phi'(z)}z = 1 \tag{27}$$

we can transform Eq. (26) into the following form:

$$\frac{v'_1(z,\varepsilon_z)}{v_1(z,\varepsilon_z)}z + \frac{\phi'(z)}{\phi(z)}z = 1 \tag{28}$$

From Eq. (28) we may get the solution

$$v_1(z,\varepsilon_z) = \frac{2z\sqrt{-E_c}}{k\phi(z)} \tag{29}$$

Substituting the expression of $v_1(z,\varepsilon_z)$ of Eq. (29) into the coefficients in the brackets of

Eq. (23), we can also prove that Eq. (23) is tenable.

Therefore, from Eq. (20), we may obtain the exact analytical expression of the first special solution $v(z,\varepsilon_z)$:

$$\frac{2z\sqrt{-E_c}}{k\phi(z)}\sin\left\{\frac{k}{\sqrt{-E_c}}\left[\sqrt{\phi(z)+\varepsilon_z}-\sqrt{\varepsilon_z}\right]\right\} \tag{30}$$

It is not difficult to prove that Eq. (30) satisfies the initial condition [Eq. (4)] of the first special solution $v(z,\varepsilon_z)$.

2.4 Exact analytical expression for the special solution $w(z,\varepsilon_z)$

For the second special solution $w(z,\varepsilon_z)$, similar to above-mentioned method, its analytical solution can be modeled on the following supposition

$$w_2(z,\varepsilon_z) = w_1(z,\varepsilon_z)\cos\chi(z,\varepsilon_z) \tag{31}$$

where $w_1(z,\varepsilon_z)$ is a function to be solved.

Substituting Eq. (31) to the scalar form of Eq. (1), we may obtain coefficient items which are related to $\sin\chi(z,\varepsilon_z)$ and $\cos\chi(z,\varepsilon_z)$, which should be equal to zero, respectively:

$$-2w_1'(z,\varepsilon_z)\chi'(z,\varepsilon_z) - w_1(z,\varepsilon_z)\times\left[\chi''(z,\varepsilon_z)+\frac{1}{2}\frac{\phi'(z)}{\phi(z)+\varepsilon_z}\chi'(z,\varepsilon_z)\right] = 0 \tag{32}$$

$$w_1''(z,\varepsilon_z)+\frac{1}{2}\frac{\phi'(z)}{\phi(z)+\varepsilon_z}w_1'(z,\varepsilon_z)+\left\{-\chi'^2(z,\varepsilon_z)+\frac{1}{4[\phi(z)+\varepsilon_z]}\times\left[\phi''(z)+\frac{e}{2m_0}B^2(z)\right]\right\}w_1(z,\varepsilon_z) = 0 \tag{33}$$

Through the simplified treatment to Eq. (32), we may obtain an expression which is similar to Eq. (26):

$$2w_1'(z,\varepsilon_z)\phi'(z) + w_1(z,\varepsilon_z)\phi''(z) = 0 \tag{34}$$

Seeing that the axial potential distribution $\phi(z)$ of the concentric spherical system expressed by Eq. (12) satisfies Eq. (27), we can transform Eq. (34) into the following form:

$$\frac{w_1'(z,\varepsilon_z)}{w_1(z,\varepsilon_z)}z + \frac{\phi'(z)}{\phi(z)}z = 1 \tag{35}$$

If we suppose that

$$w_1(z,\varepsilon_z) = 1 + \frac{z}{R_c}$$
$$w_1'(z,\varepsilon_z) = \frac{1}{R_c} \tag{36}$$

it is not difficult to prove that Eq. (35) is tenable. Therefore, the solution of $w_2(z,\varepsilon_z)$ can be obtained as

$$\left(1+\frac{z}{R_c}\right)\cos\left\{\frac{k}{\sqrt{-E_c}}\left[\sqrt{\phi(z)+\varepsilon_z}-\sqrt{\varepsilon_z}\right]\right\} \tag{37}$$

but it satisfies the following initial condition:

$$w_2(z=0) = 1, \quad w_2'(z=0) = \frac{1}{R_c} \tag{38}$$

Under this assumption, we can also prove that the Wronski determinant [Eq. (6)] is tenable, i.e.,

$$\sqrt{\phi(z) + \varepsilon_z}(v'w_2 - vw'_2) = 1 \tag{39}$$

Although $w_2(z,\varepsilon_z)$ in Eq. (37) is a special solution of paraxial Eq. (1), it does not satisfy the initial condition [Eq. (5)] as we required. In order to make the special solution $w(z,\varepsilon_z)$ satisfy both the assumptions of the initial condition [Eq. (5)] and the Wronski determinant [Eq. (6)], we may express the second special solution $w(z,\varepsilon_z)$ of the paraxial Eq. (1) as follows:

$$w(z,\varepsilon_z) = w_2(z,\varepsilon_z) - \frac{\sqrt{\varepsilon_z}}{R_c} v(z,\varepsilon_z) \tag{40}$$

We have proved that the special solutions $w_2(z,\varepsilon_z)$ and $v(z,\varepsilon_z)$ satisfy the paraxial Eq. (1), so that their combination into a new special solution $w(z,\varepsilon_z)$ will inevitably satisfy the paraxial Eq. (1). Examining Eq. (40), we found that the initial condition [Eq. (5)] is satisfied.

Similarly, it is not difficult to prove that the two special solutions $v(z,\varepsilon_z)$ and $w(z,\varepsilon_z)$, expressed by Eq. (30) and Eq. (40), satisfy the initial conditions [Eq. (4) and Eq. (5)], and the Wronski determinant [Eq. (6)]

$$\sqrt{\phi(z) + \varepsilon_z}(v'w - vw') = \sqrt{\phi(z) + \varepsilon_z}\left[v'\left(w_2 - \frac{\sqrt{\varepsilon_z}}{R_c}v\right) - v\left(w'_2 - \frac{\sqrt{\varepsilon_z}}{R_c}v'\right)\right] = 1 \tag{41}$$

is also tenable. Thus, from Eq. (40), the second special solution $w(z,\varepsilon_z)$ can be expressed by

$$w(z,\varepsilon_z) = \left(1 + \frac{z}{R_c}\right)\cos\left\{\frac{k}{\sqrt{-E_c}}[\sqrt{\phi(z) + \varepsilon_z} - \sqrt{\varepsilon_z}]\right\} - \frac{\sqrt{\varepsilon_z}}{R_c}\frac{2z\sqrt{-E_c}}{k\phi(z)}\sin\left\{\frac{k}{\sqrt{-E_c}}[\sqrt{\phi(z) + \varepsilon_z} - \sqrt{\varepsilon_z}]\right\} \tag{42}$$

Therefore, we first strictly derived the exact analytical Eq. (30) and Eq. (42) of the two special solutions $v(z,\varepsilon_z)$ and $w(z,\varepsilon_z)$ for the combined electromagnetic concentric spherical systems under the assumption $\varepsilon_z \neq 0$.

When $\varepsilon_z = 0$, the two special solutions $v(z,\varepsilon_z)$ and $w(z,\varepsilon_z)$ expressed by Eq. (30) and Eq. (42), as well as the rotational angle $\chi(z)$ expressed by Eq. (19) can be simplified as

$$v(z) = \frac{2z\sqrt{-E_c}}{k\phi(z)}\sin\left[\frac{k}{\sqrt{-E_c}}\sqrt{\phi(z)}\right],$$

$$w(z) = \left(1 + \frac{z}{R_c}\right)\cos\left[\frac{k}{\sqrt{-E_c}}\sqrt{\phi(z)}\right] \tag{43}$$

$$\chi(z) = \frac{k}{\sqrt{-E_c}}\sqrt{\phi(z)} \tag{44}$$

Eq. (43) is the same as Eq. (1.99) and Eq. (1.100) given in Ref. [11]. It is actually a zeroth order approximate solution of the paraxial trajectory, which is a special case of the results of this paper.

3 Paraxial imaging properties for the combined electromagnetic concentric spherical system

From Eq. (30), it can be seen that for a combined electromagnetic concentric spherical system, the electron trajectory emitted from the cathode surface is a conic helix with unequal pitches. From Eq. (19), it is not difficult to determine the various image position $z_m(m=1,2,3,\cdots)$ which corresponds to $\varepsilon_z = \varepsilon_{z_1}$. The imaging condition is obviously as $\sin\chi(z_m, \varepsilon_{z_1}) = 0$, i.e.,

$$\frac{k}{\sqrt{-E_c}}\left[\sqrt{\frac{z_m}{nl-(n-1)z_m}\phi_{ac} + \varepsilon_{z_1}} - \sqrt{\varepsilon_{z_1}}\right] = m\pi, \quad m = 1,2,3,\cdots,i \tag{45}$$

where m is the number of loops.

From Eq. (45), it can be seen that regardless of the initial radial potential ε_r of the emitted electrons, they can be ideally focused as long as their initial axial potential ε_z of electron along the principal trajectory is equal to ε_{z_1}.

Obviously, when $\chi = [(2m-1)/2]\pi$, the conical helix of electron trajectory takes extreme value $r_{j_m}^*$ (maximum or minimum) at z_j, that is

$$r_{j_m}^* = r^*(z_j) = 2\sqrt{\frac{2m_0}{e}}\frac{\sqrt{\varepsilon_r}}{B_0}\left(1 - \frac{n-1}{n}\frac{z_j}{l}\right), \quad j = \frac{2m-1}{2} \tag{46}$$

where the value of z_j can be obtained, provided that m in Eq. (45) is substituted by $(2m-1)/2$. It follows from Eq. (46) that $r_{j_m}^*$ depends on l, n, z_j, and the initial radial potential ε_r of the electron, but is inversely proportional to B_0, and it decreases (when $n > 1$) or increases (when $n < 1$) the value of z_j, thus forming a conical helix, as shown in Fig. 1 (d).

Let the electron be emitted from the center of spherical photocathode, 2θ denotes the angle between the central axis and a connection line which joins neighboring amplitudes of the conical helix. We may have

$$\tan 2\theta = \frac{r_{j+1}^* - r_j^*}{z_{j+1} - z_j} = -2\sqrt{\frac{2m_0}{e}}\frac{\sqrt{\varepsilon_r}}{B_0}\frac{n-1}{nl} = 2\sqrt{\frac{2m_0}{e}}\frac{\sqrt{\varepsilon_r}}{B_0}\frac{1}{R_c} = \text{const} \tag{47}$$

It follows that when $n > 1$ or R_c takes negative value, $\tan 2\theta$ will be negative. It means that the radius of gyration is gradually reduced, but when $n < 1$ or R_c takes a positive value, $\tan 2\theta$ will be positive, indicating that the radius of gyration is gradually increased.

The distance between two adjacent amplitudes of the conical helix is

$$\Delta z_j = z_{j+1} - z_j = \frac{4m\dfrac{m_0}{e}\dfrac{-E_c}{B_0^2}\pi^2}{\left\{1+(n-1)\dfrac{4m^2+1}{4}\left[\dfrac{1}{\phi_{ac}}\dfrac{2m_0}{e}\dfrac{(-E_c)^2}{B_0^2}\pi^2\right]\right\}^2} \tag{48}$$

It means that when $n > 1$ or R_c takes negative value, Δz_j is decreased, but when $n < 1$ or R_c takes the negative value, Δz_j is increased.

From Eq. (45), the ideal imaging position z_i corresponding to ε_{z_1} at i loops can be obtained. The distance l between the photocathode and the screen can be expressed as

$$z_i = l = \sqrt{\frac{2m_0}{e}}\frac{i\pi}{nB_0}(\sqrt{\phi_{ac} + \varepsilon_{z_1}} - \sqrt{\varepsilon_{z_1}}) \tag{49}$$

In fact, this type of imaging system tends to use a very strong accelerating electric field, so the electron velocity reaches the image plane far faster than the axial initial velocity. If we omit the term of ε_{z_1} in Eq. (49), the distance l between photocathode and screen can be approximated by the limiting image position of z_i of i loops focusing.

$$z_i = l \approx i\pi \sqrt{\frac{2m_0}{e}} \frac{\sqrt{\phi_{ac}}}{nB_0} \tag{50}$$

If n, l, i are given, the value of $\sqrt{\phi_{ac}}/B_0$, will be completely determined.

At the image plane position $z_i = l$, $\sin\chi(z_i, \varepsilon_{z_1}) = 0$, $\cos\chi(z_i, \varepsilon_{z_1}) = (-1)^i$. Therefore, from Eq. (30), we may obtain

$$v'(z_i = l) = \frac{n}{\sqrt{\phi_{ac} + \varepsilon_{z_1}}}(-1)^i \tag{51}$$

From the linear magnification $M = 1/n$ and the angular magnification

$$\Gamma = \frac{v'(z_i)}{v'(z_0)}(-1)^i \tag{52}$$

we may prove that the Lagrange-Helmhotz relationship

$$M\Gamma = \frac{\sqrt{\varepsilon_{z_1}}}{\sqrt{\phi_{ac} + \varepsilon_{z_1}}} \tag{53}$$

is still valid in the combined electromagnetic concentric spherical system.

In Ref. [1] we have given an example of computations for a combined electromagnetic focusing image tube (a demagnifying tube, $M = 1/5$; a magnifying tube, $M = 5$).

4 Some special cases

4.1 Homogeneous and parallel combined electromagnetic system

For the combined electromagnetic concentric spherical system, let both R_c and R_a tend to be infinite, that is $n = 1$. The spherical cathode and spherical screen will become flat cathode and flat screen, and this system will be converted into a normal homogeneous and parallel combined electromagnetic system with unity magnification which is also called a long magnetic cathode lens.

In this case, $\phi(z) = \phi_{ac}z/l = -E_c z$, $E_c = \phi_{ac}/l = $ const, $B(z) = B_0 = $ const, $1/R_c = 0$.

Therefore, in a combined homogeneous and parallel electromagnetic system, from Eq. (19), Eq. (30) and Eq. (42), it is not difficult to get the exact analytical expression of the rotational angle $\chi(z, \varepsilon_z)$ and the exact analytical expressions of two special solutions $v(z, \varepsilon_z)$ and $w(z, \varepsilon_z)$:

$$\chi(z, \varepsilon_z) = \frac{k}{\sqrt{-E_c}}(\sqrt{-E_c z + \varepsilon_z} - \sqrt{\varepsilon_z}) \tag{54}$$

$$v(z, \varepsilon_z) = \frac{2}{k\sqrt{-E_c}} \sin\left[\frac{k}{\sqrt{-E_c}}(\sqrt{-E_c z + \varepsilon_z} - \sqrt{\varepsilon_z})\right] \tag{55}$$

$$w(z, \varepsilon_z) = \cos\left[\frac{k}{\sqrt{-E_c}}(\sqrt{-E_c z + \varepsilon_z} - \sqrt{\varepsilon_z})\right] \tag{56}$$

It is well known that in this system, the electrons move with a constant acceleration in the axial direction, and make a circular motion with constant radial velocity $\dot{r}_0$. The transverse orbit is a circle returning to the axis in a period $T = 2\pi m_0/(eB_0)$, its gyration radius is $r_j = 2(m_0/e) \times (\dot{r}_0/B_0)$, and therefore, its electron trajectory is depicted as a cylindrical helical line whose thread pitch gradually increases.

The image positions of various stages correspond to

$$z_m = \dot{z}_0(mT) + \frac{1}{2}\frac{-E_c}{m_0}(mT)^2, \quad m = 1,2,3\cdots \tag{57}$$

The ideal image position (screen position) can be expressed by

$$z_i = l = \frac{i\pi}{k\sqrt{-E_c}}(\sqrt{\phi_{ac} + \varepsilon_{z_1}} - \sqrt{\varepsilon_{z_1}}) \tag{58}$$

4.2 Bi-electrode electrostatic concentric spherical system

Regarding to special solutions of a bi-electrode electrostatic concentric spherical system, it has been studied already in our monograph and articles[2-5,10]. Here we study it as a special case of combined electromagnetic concentric spherical system, and the same results can also be obtained.

If in the combined electromagnetic concentric spherical system, the magnetic field is not present, $B(z) = 0$, i.e., $k = 0$, this system becomes a purely electrostatic concentric spherical system. It will have three cases: $n < 1$, $1 < n < 2$, and $n > 2$, as shown in Fig. 1 (c).

Its axial distribution is $\phi(z) = (-E_c)R_c z/(R_c + z)$. Therefore, from Eq. (19) we know that $\chi(z) = 0$, $\sin\chi(z) = 0$ and $\cos\chi(z) = \pm 1$. Its electron trajectory will be a plane curve.

Because

$$\lim_{\chi(z)=0}\frac{\sin\chi(z)}{\chi(z)} = 1 \tag{59}$$

it is not difficult to get the analytical solutions of two special solutions from Eq. (30) and Eq. (42):

$$v(z,\varepsilon_z) = \frac{2z}{\phi(z)}[\sqrt{\phi_{ac} + \varepsilon_z} - \sqrt{\varepsilon_z}] \tag{60}$$

$$w(z,\varepsilon_z) = 1 + \frac{z}{R_c} - \frac{\sqrt{\varepsilon_z}}{R_c}\frac{2z}{\phi(z)}[\sqrt{\phi_{ac} + \varepsilon_z} - \sqrt{\varepsilon_z}] \tag{61}$$

The derivatives are

$$v'(z,\varepsilon_z) = \frac{2}{\phi(z)}\frac{z}{R_c + z}[\sqrt{\phi(z) + \varepsilon_z} - \sqrt{\varepsilon_z}] + \frac{R_c}{(R_c + z)\sqrt{\phi(z) + \varepsilon_z}} \tag{62}$$

$$w'(z,\varepsilon_z) = \frac{1}{R_c} - \frac{\sqrt{\varepsilon_z}}{R_c}\left\{\frac{2}{\phi(z)}\frac{z}{R_c + z}[\sqrt{\phi(z) + \varepsilon_z} - \sqrt{\varepsilon_z}] + \frac{R_c}{(R_c + z)\sqrt{\phi(z) + \varepsilon_z}}\right\} \tag{63}$$

It is not difficult to prove that these two special solutions Eq. (60) and Eq. (61) satisfy the initial conditions Eq. (4) and Eq. (5), and the Wronski determinant [Eq. (6)] is also tenable.

It will be seen from Eq. (62) that only when $R_c < 0$ and $z > -R_c/2$, i.e., $n > 2$, $v'(z,\varepsilon_z) <$

0, or the electron trajectory emitted from the axial point is deflected and inclined to the axis, this system will have a possibility to form a real image.

4.3 Electrostatic proximity focusing system

For a bi-electrode electrostatic concentric spherical system, if both R_c and R_a tend to infinity $1/R_c = 0$, $1/R_a = 0$, i.e., $n = 1$ the spherical cathode and spherical screen turn into the plane cathode and plane screen, and this system becomes an electrostatic proximity focusing system. In this case, the axial potential distribution is $\phi(z) = (-E_c)z$, $\phi'(z) = -E_c = \phi_{ac}/l = $ const.

Since in this system, electron moving in the radial direction has not been affected by the uniform field, the velocity of electron in the radial direction remains unchanged. However, the electron moving in the axial direction is affected by the uniform field, so the velocity of electron along the axial direction can be expressed by

$$\dot{z} = \sqrt{\frac{2e}{m_0}(-E_c z + \varepsilon_z)} \qquad (64)$$

The velocity in axial direction is increased along with square root of z, which is the distance from the plane cathode, and the electron trajectory is a parabola. From Eq. (60) and Eq. (61), we may obtain the expressions of two special solutions as

$$v(z, \varepsilon_z) = \frac{2}{-E_c}(\sqrt{\phi(z) + \varepsilon_z} - \sqrt{\varepsilon_z}) \qquad (65)$$

$$w(z, \varepsilon_z) = 1 \qquad (66)$$

The further description concerning the electrostatic proximity focusing and its electron optics may consult the related references[10].

Summary

For the combined electromagnetic concentric spherical system, seeking for solutions of the electron trajectory emitted from the cathode surface is always a problem of common interest. Through the mutual transformation between the electrostatic potential distribution and the magnetic induction distribution of this system, we have subtly solved the difficult problem of seeking for analytical solutions of two special solutions by solving paraxial equation of electron optics, which is a second order homogeneous linear differential equation. The exact expressions of rotational angle and two special solutions of electron ray have been firstly derived, the paraxial imaging properties have also been investigated, and the results have been generalized to the bi-electrode electrostatic concentric spherical system, the homogeneous and parallel combined electromagnetic system, and the electrostatic proximity system. Results of the present part have practical meaning for the design of combined electromagnetic image tube, and lay a foundation for further study of the paraxial aberration theory for the combined electromagnetic imaging system.

References

[1] CHOU L W (ZHOU L W). Electron optics of concentric spherical electromagnetic

focusing systems [J]. Advances in Electronics and Electron Physics, 1979, 52: 119 – 132.

[2] ZHOU L W, GONG H, ZHANG Z Q, et al. On electron-optical spatial and temporal aberrations in a bi-electrode spherical concentric system with electrostatic focusing [J]. Proceedings of SPIE, 2009, 7384:738435.

[3] ZHOU L W, GONG H, ZHANG Z Q, et al. Static and dynamic imaging electron-optics and spatial-temporal aberrations in a bi-electrode spherical concentric system with electrostatic focusing[J]. Optik, 2011,122(4):287 – 294.

[4] ZHOU L W, GONG H, ZHANG Z Q, et al. Paraxial imaging electron optics and its spatial-temporal aberrations for a bi-electrode concentric spherical system with electrostatic focusing [J]. Optik, 2011,122(4):295 – 299.

[5] ZHOU L W, GONG H. Theory of paraxial lateral aberrations of electrostatic imaging electrostatic electron optics based on asymptotic solutions and its verification[J]. Optik, 2011, 122(4): 300 – 306.

[6] BEURLE R L, WREATHALL W M. Aberration in magnetic focus systems [J]. Advances in Electronics and Electron Physics, 1962, 16: 333 – 340.

[7] ZACHROV B. A demagnifying image tube for nuclear physics applications [J]. Advances in Electronics and Electron Physics, 1962, 16: 99 – 104.

[8] PICAT J P. Optical property of electromagnetically focused electronic cameras [J]. Astronomy and Astrophysics, 1971,11:257 – 267.

[9] SMIRNOV N A, MONASTYRSKI M A, KULIKOV Y V. Some problems on electron optics for cathode lenses of combined fields with spherical symmetry [J]. Journal of Technical Physics, 1979, 48(2): 2590 – 2595.

[10] ZHOU L W. Electron optics with wide beam focusing [M]. Beijing: Beijing Institute of Technology Press, 1993.

[11] ELIN V P, KATESHOWV V A, KULIKOV Y V, et al. Numerical methods of optimization for emissive electron optical systems [M]. Siberia: Press Science,1987.

2.2 Part B: Paraxial Aberrations
近轴像差

Abstract: Starting from the two special solutions of the paraxial equation of the combined electromagnetic concentric spherical system, the paraxial longitudinal and lateral aberrations of the combined electromagnetic concentric spherical system have been explored by expanding the representation of the image rotation angle at the ideal imaging position. The main contribution of this paper is to obtain the analytical expression of the two special solutions of paraxial equation at ideal image plane, and it has proved that the paraxial spherical-chromatic aberration of the second order that determines the limited spatial resolution of imaging can still be characterized in terms of the Recknagel-Artimovich formula. Expressions of the paraxial lateral aberration are composed of paraxial spherical-chromatic aberration, paraxial magnification chromatic aberration, and paraxial anisotropic chromatic aberration have deduced.

摘要：本文由复合电磁同心球系统的近轴方程两个特解出发，通过展开理想成像位置处图像转角表达式的途径，探讨了复合电磁同心球系统的近轴纵向像差和近轴横向像差，主要贡献是求得了近轴方程的两个特解在理想像面位置处的解析表达式，由此证明了决定极限空间分率的二级近轴色球差能以 Recknagel-Artimovich 公式描述，导出了近轴横向像差的表达式，该像差由近轴色球差、近轴放大率色差和近轴各向异性色差等组成。

Introduction

For a bi-electrode electrostatic concentric spherical system, it has been demonstrated that the paraxial lateral spherical-chromatic aberration of the second order that determines the limited spatial resolution of the system can be expressed by the Recknagel-Artimovich formula[1-3] which can be simplified as R-A formula. Similarly, we have already proved that the R-A formula is still established in the general electrostatic electron optical imaging system[3]. This shows that it has universal significance in evaluating image quality for the electrostatic imaging systems. However, in the combined electromagnetic electron optical imaging system, what is the influence of paraxial spherical-chromatic aberration and other special types of paraxial lateral aberration on the image quality, and how does the introduction of a magnetic field affect the paraxial lateral aberration of the system? How big it is, whether the R-A formula is established, etc., are all worth considering and need to give a clear answer.

Zhou Liwei (周立伟). 北京理工大学. Acta Optica Sinica (光学学报), V. 39, No. 4, 2019, 0411002-1-8.

In the previous paper[1], we have given a general expression of paraxial trajectory $r^*(z,\varepsilon_z)$ in the combined electromagnetic imaging system as

$$r^*(z,\varepsilon_z) = r_0 w(z,\varepsilon_z)\exp[j\chi(z,\varepsilon_z)] + \left[\sqrt{\frac{m_0}{2e}}\dot{r}_0 - (k\times r_0)\sqrt{\frac{e}{8m_0}}B_0\right]v(z,\varepsilon_z)\exp[j\chi(z,\varepsilon_z)] \tag{1}$$

where $j = \sqrt{-1}$, r_0 and $\dot{r}_0$ are the initial position vector and the initial radial velocity vector of electron emitted from the cathode surface, respectively, k is a unit vector along the positive direction of axis z, ε_z is the initial axial potential corresponding to the initial axial energy of emitted electrons; e/m_0 is the ratio of electron charge to mass; B_0 is the magnetic induction intensity at the cathode surface, and the star mark "*" represents that it is under the paraxial condition.

In Eq. (1), $v(z,\varepsilon_z)$ and $w(z,\varepsilon_z)$ are two special solutions of electron optical paraxial equation, which should satisfy the following initial conditions

$$v(z=0) = 0, \quad v'(z=0) = \frac{1}{\sqrt{\varepsilon_z}} \tag{2}$$

$$w(z=0) = 1, \quad w'(z=0) = 0 \tag{3}$$

and these two special solutions should satisfy the following Wronski determinant:

$$\sqrt{\phi(z) + \varepsilon_z}(v'w - vw') = 1 \tag{4}$$

Let $\chi(z,\varepsilon_z)$ indicate the rotating angle of electron trajectory emitted from the cathode surface, that is, the angle at which the rotating coordinate system (x, y) rotates relatively to the fixed coordinate system (x, y):

$$\chi(z,\varepsilon_z) = \int_{z_0}^{z}\sqrt{\frac{e}{8m_0}}\frac{B(z)}{\sqrt{\phi(z)+\varepsilon_z}}dz + \chi_0 \tag{5}$$

In general, we suppose that $\chi_0 = \chi(z_0 = 0) = 0$.

In the ideal image plane at $z = z_i$, which corresponds to ε_{z_1}, let $v(z_i,\varepsilon_{z_1}) = 0$, $\chi(z_i,\varepsilon_{z_1}) = i\pi$, where i is the number of loops for the rotation of emitted electrons. Then, from Eq. (1), we have

$$r^*(z,\varepsilon_{z_1}) = r_0 w(z,\varepsilon_{z_1}) = r_0 M \tag{6}$$

where M is the linear magnification of system in the ideal image plane at $z = z_i$.

The paraxial lateral aberration $\Delta r^*(z,\varepsilon_{z_1})$ in the ideal image plane at $z = z_i$ is defined as[4]

$$\Delta r^*(z_i,\varepsilon_z) = r^*(z_i,\varepsilon_z) - r^*(z_i,\varepsilon_{z_1}) \tag{7}$$

Therefore, from Eq. (1) it can be formulated as

$$\Delta r^*(z_i,\varepsilon_z) = \sqrt{\frac{m_0}{2e}}\dot{r}_0 v(z_i,\varepsilon_z)\exp[j\chi(z_i,\varepsilon_z)] + r_0\{w(z_i,\varepsilon_z)\exp[j\chi(z_i,\varepsilon_z)] - M\} -$$

$$(k\times r_0)\sqrt{\frac{e}{8m_0}}B_0 v(z_i,\varepsilon_z)\exp[j\chi(z_i,\varepsilon_z)]$$

$$= \Delta r_v^*(z_i,\varepsilon_z) + \Delta r_w^*(z_i,\varepsilon_z) + \Delta r_u^*(z_i,\varepsilon_z) \tag{8}$$

where $\Delta r_v^*(z_i,\varepsilon_z)$; $\Delta r_w^*(z_i,\varepsilon_z)$ and $\Delta r_u^*(z_i,\varepsilon_z)$ are called as paraxial spherical-chromatic aberration, paraxial magnification chromatic aberration, and paraxial anisotropic chromatic aberration, respectively.

Eq. (8) is the starting point for studying the paraxial lateral aberrations of a combined electromagnetic concentric spherical system. In the previous paper[1], we have derived the exact analytical expressions of two special solutions $v(z,\varepsilon_z)$, $w(z,\varepsilon_z)$ and the rotating angle $\chi(z,\varepsilon_z)$. Therefore, this paper attempts to obtain the analytical solution of the paraxial lateral aberrations for the combined electromagnetic concentric spherical system so that we can accurately analyze the effects of various types of paraxial lateral aberrations on the imaging quality of the system.

Similar to the study of the bi-electrode electrostatic concentric spherical system, the formation regularity of paraxial imaging and paraxial lateral aberrations derived from the combined electromagnetic concentric spherical system will have universal guiding significance for studying general combined electromagnetic imaging systems. This is just the purpose of the present research.

1 Basic formulae

For the combined electromagnetic concentric spherical system, suppose that the electric field intensity **E** is perpendicular to the spherical cathode, and the magnetic induction intensity **B** is parallel to the electric field intensity **E** everywhere. The axial potential distribution $\phi(z)$ and the axial distribution of magnetic induction intensity $B(z)$ can be expressed as[5]

$$\phi(z) = \frac{z}{nl - (n-1)z}\phi_{ac} \tag{9}$$

$$B(z) = \frac{n^2 l^2}{[nl - (n-1)z]^2}B_c \tag{10}$$

where $n = R_c/R_a$, and R_c and R_a are the curvature radii of spherical cathode and spherical screen (anode) of the combined electromagnetic concentric spherical system respectively, its value is calculated from the center of curvature, and the direction of the electrons is positive, and vice versa; l is the distance between cathode and anode $l = R_a - R_c$; ϕ_{ac} is the electrostatic potential of the anode relative to the cathode; B_0 is the magnetic induction intensity at the cathode surface.

In the previous paper[1], we have derived exact expressions of two special solutions of $v(z,\varepsilon_z)$ and $w(z,\varepsilon_z)$ from the electron optical paraxial equation of this system as follows:

$$v(z,\varepsilon_z) = \frac{2z\sqrt{-E_c}}{k\phi(z)}\sin\left\{\frac{k}{\sqrt{-E_c}}[\sqrt{\phi(z)+\varepsilon_z} - \sqrt{\varepsilon_z}]\right\} \tag{11}$$

$$w(z,\varepsilon_z) = \left(1 + \frac{z}{R_c}\right)\cos\left\{\frac{k}{\sqrt{-E_c}}[\sqrt{\phi(z)+\varepsilon_z} - \sqrt{\varepsilon_z}]\right\} - \frac{\sqrt{\varepsilon_z}}{R_c}\frac{2z\sqrt{-E_c}}{k\phi(z)}\sin\left\{\frac{k}{\sqrt{-E_c}}[\sqrt{\phi(z)+\varepsilon_z} - \sqrt{\varepsilon_z}]\right\} \tag{12}$$

where $k^2 = \frac{e}{2m_0}\frac{B_0^2}{-E_c}$, and its dimension is $\frac{1}{l}$. E_c the electric field intensity at the cathode surface can be expressed by

$$E_c = \frac{\phi_{ac}}{R_c(n-1)} = \frac{\phi_{ac}}{-nl} \qquad (13)$$

The direction of E_c points to the cathode surface, whether $n > 1$ or $n < 1$, and it takes negative value forever. These two special solutions [Eq. (11) and Eq. (12)] satisfy initial conditions [Eq. (2) and Eq. (3)]. Meanwhile, its Wronski determinant [Eq. (4)] is also satisfied.

Due to the influence of the magnetic field, $\chi(z, \varepsilon_z)$, the rotating angle of electron emitted from the cathode surface during the course of travelling can be expressed as

$$\chi(z, \varepsilon_z) = \frac{k}{\sqrt{-E_c}} [\sqrt{\phi(z) + \varepsilon_z} - \sqrt{\varepsilon_z}] \qquad (14)$$

From Eq. (8), it can be seen that solving the paraxial lateral aberrations $\Delta r^*(z_i, \varepsilon_z)$ can be attributed to finding the value of two special solutions $v(z, \varepsilon_z)$, $w(z, \varepsilon_z)$ and the rotating angle $\chi(z, \varepsilon_z)$ of electron at the imaging position $z = z_i$. From the analytical expressions of each type of paraxial lateral aberrations $\Delta r_v^*(z_i, \varepsilon_z)$, $\Delta r_w^*(z_i, \varepsilon_z)$ and $\Delta r_u^*(z_i, \varepsilon_z)$, we may quantitatively analyze the effect of each special type of paraxial lateral aberrations on image quality.

2 Paraxial longitudinal chromatic aberration

Before discussing the paraxial lateral aberration $\Delta r^*(z_i, \varepsilon_z)$, we first study the analytical expression of the paraxial longitudinal chromatic aberration Δz^*. In electron optics, study of paraxial longitudinal chromatic aberration is of significance, not only useful for studying the depth of field, but also for understanding the axial dispersion of electrons emitted from initial axial object point of cathode surface with different axial initial energies. Let an electron emitted from the initial object point of cathode surface, whose initial axial potential is $\varepsilon_z = \varepsilon_{z_1}$, under the common action of electric and magnetic fields, be accelerated along the axis and rotate round the axis, go through i loops moving as a conical helix, and finally intersect at the position of ideal image plane $z_i = l$ which is determined by $v(z_i, \varepsilon_{z_1}) = 0$. From Eq. (14), letting $\chi(z_i, \varepsilon_{z_1}) = i\pi$, we may obtain the expression of the ideal image position:

$$z_i = l = i\pi \frac{\sqrt{\phi_{ac}}}{kn\sqrt{-E_c}} \left(\sqrt{1 + \frac{\varepsilon_{z_1}}{\phi_{ac}}} - \sqrt{\frac{\varepsilon_{z_1}}{\phi_{ac}}} \right) \qquad (15)$$

Thus, the other paraxial trajectory emitted from the same initial object point with initial axial potential ε_z (but $\varepsilon_z \neq \varepsilon_{z_1}$), intersects at the point z_i^*, which is adjacent to the image position $z_i = l$, and it can be determined by[4-5]

$$\frac{k}{\sqrt{-E_c}} \left[\sqrt{\frac{z_i^*}{nl - (n-1)z_i^*} \phi_{ac} + \varepsilon_z} - \sqrt{\varepsilon_z} \right] = i\pi \qquad (16)$$

From Eq. (16) and Eq. (15), we can obtain the expression of z_i^*:

$$z_i^* = i\pi \frac{\sqrt{\phi_{ac}}}{k\sqrt{-E_c}} \frac{\sqrt{1+\frac{\varepsilon_{z_1}}{\phi_{ac}}} - \sqrt{\frac{\varepsilon_{z_1}}{\phi_{ac}}} + 2\sqrt{\frac{\varepsilon_z}{\phi_{ac}}}}{1+(n-1)\left(\sqrt{1+\frac{\varepsilon_{z_1}}{\phi_{ac}}} - \sqrt{\frac{\varepsilon_{z_1}}{\phi_{ac}}}\right)\left(\sqrt{1+\frac{\varepsilon_{z_1}}{\phi_{ac}}} - \sqrt{\frac{\varepsilon_{z_1}}{\phi_{ac}}} + 2\sqrt{\frac{\varepsilon_z}{\phi_{ac}}}\right)} \quad (17)$$

The paraxial longitudinal chromatic aberration Δz^* can be defined as

$$\Delta z^* = z_i^* - z = z_i^* - l \quad (18)$$

Expanding Eq. (15) and Eq. (17) and omitting higher order small quantities which are above the second order of $\sqrt{\varepsilon_z/\phi_{ac}}$, we may express $z_i = l$ and z_i^* as

$$z_i = l = i\pi \frac{\sqrt{\phi_{ac}}}{nk\sqrt{-E_c}} \left(1 + \sqrt{\frac{\varepsilon_{z_1}}{\phi_{ac}}} + \frac{\varepsilon_{z_1}}{2\phi_{ac}}\right) \quad (19)$$

$$z_i^* = i\pi \frac{\sqrt{\phi_{ac}}}{nk\sqrt{-E_c}} \Bigg[1 - \frac{2(n-1)}{n}\left(\sqrt{\frac{\varepsilon_z}{\phi_{ac}}} - \sqrt{\frac{\varepsilon_{z_1}}{\phi_{ac}}}\right) - \frac{4(n-1)}{n^2}\left(\sqrt{\frac{\varepsilon_z}{\phi_{ac}}} - \sqrt{\frac{\varepsilon_{z_1}}{\phi_{ac}}}\right)^2 - \sqrt{\frac{\varepsilon_{z_1}}{\phi_{ac}}} + 2\sqrt{\frac{\varepsilon_z}{\phi_{ac}}} + \frac{\varepsilon_{z_1}}{2\phi_{ac}} \Bigg] \quad (20)$$

From Eq. (13), we can transform Eq. (15) into

$$\frac{i\pi}{k} = \frac{k}{\sqrt{-E_c}}\left[\sqrt{\phi_{ac}+\varepsilon_{z_1}} - \sqrt{\varepsilon_{z_1}}\right] \quad (21)$$

Eq. (21) shows the relationship between magnetic induction intensity B_0, electric field intensity E_c, and electrostatic potential ϕ_{ac} when the ideal imaging is formed. Substituting Eq. (21) into Eq. (20) and expanding it, omitting the small quantities higher than the second order of $\sqrt{\varepsilon_z/\phi_{ac}}$ and $\sqrt{\varepsilon_{z_1}/\phi_{ac}}$, and taking into account the linear magnification $M = 1/n$, from Eq. (18), we may obtain the expression of the paraxial longitudinal chromatic aberration Δz^* as follows:

$$\Delta z^* = \Delta z_1^* + \Delta z_2^* \quad (22)$$

where

$$\Delta z_1^* = \frac{2M^2}{E_c}\sqrt{\phi_{ac}}(\sqrt{\varepsilon_{z_1}} - \sqrt{\varepsilon_z}) \quad (23)$$

$$\Delta z_2^* = \frac{-4M^2(1-M)}{E_c}\sqrt{\varepsilon_z}(\sqrt{\varepsilon_{z_1}} - \sqrt{\varepsilon_z}) \quad (24)$$

And Δz_1^* and Δz_2^* are the so-called paraxial longitudinal chromatic aberration of the first order and the second order, respectively. Although they are derived from the analytical solution of the combined electromagnetic concentric spherical system, the magnetic induction intensity B_0 term in these expressions does not appear. In addition, the expression of the paraxial longitudinal chromatic aberration of the first order is exactly the same as that of the bi-electrode electrostatic concentric spherical system[2,4]. This result deepens people's understanding of the imaging characteristics of electron optical system with wide electron beam focusing, and it is of great importance for the designers when choosing which type of electron optical imaging system is required.

It is not to be wondered at if we analyze it from a physical point of view. As we know, in a combined electromagnetic concentric spherical system, the formation of a conical helix is mainly a combination of an electric field that accelerates the electrons in the axial direction and a magnetic field that rotates electrons around an axis of symmetry. Electrons traveling in the axial direction due to the difference in the initial axial energy are accelerated by the electric field, and the divergence is produced resulting in the formation of paraxial longitudinal chromatic aberration. The magnetic field only acts to rotate the electron orbit around a certain axis, so it is natural that there is little contribution to the paraxial longitudinal chromatic aberration.

3 Paraxial lateral aberration

3.1 Paraxial lateral chromatic aberration

On the basis of paraxial longitudinal chromatic aberration Δz^*, we shall first discuss the paraxial lateral spherical-chromatic aberration $\Delta r_v^*(z_i, \varepsilon_z)$. From Eq. (8), it has been defined as

$$\Delta r_v^*(z_i, \varepsilon_z) = \sqrt{\frac{m_0}{2e}} \dot{r}_0 v(z_i, \varepsilon_z) \exp[j\chi(z_i, \varepsilon_z)] \tag{25}$$

From Eq. (11) we know that when $\varepsilon_z = \varepsilon_{z_1}$, at the ideal image plane $z_i = l$, we have

$$v(z_i, \varepsilon_{z_1}) = \frac{2}{nk\sqrt{-E_c}} \sin\chi(z_i, \varepsilon_{z_1}) = \frac{2}{nk\sqrt{-E_c}} \sin\left\{\frac{k}{\sqrt{-E_c}}\sqrt{\phi_{ac}}\left(\sqrt{1 + \frac{\varepsilon_{z_1}}{\phi_{ac}}} - \sqrt{\frac{\varepsilon_{z_1}}{\phi_{ac}}}\right)\right\} \tag{26}$$

If $\varepsilon_z \neq \varepsilon_{z_1}$, the special solution $v(z, \varepsilon_z)$ at the ideal image plane $z_i = l$ will be not equal to zero, i. e.,

$$v(z_i, \varepsilon_z) = \frac{2}{nk\sqrt{-E_c}} \sin\chi(z_i, \varepsilon_{z_1}) = \frac{2}{nk\sqrt{-E_c}} \sin\left\{\frac{k\sqrt{\phi_{ac}}}{\sqrt{-E_c}}\left(\sqrt{\frac{z_i}{nl-(n-1)z_i}} + \frac{\varepsilon_z}{\phi_{ac}} - \sqrt{\frac{\varepsilon_z}{\phi_{ac}}}\right)\right\} \neq 0 \tag{27}$$

Thus, all of these electrons that have same $\varepsilon_z (\varepsilon_z \neq \varepsilon_{z_1})$ will be intersected at an adjacent position z_i^*, which is close to the image plane $z_i = l$, that is

$$v(z_i^*, \varepsilon_z) = \frac{2}{nk\sqrt{-E_c}} \sin\chi(z_i^*, \varepsilon_{z_1})$$

$$= \frac{2}{nk\sqrt{-E_c}} \sin\left\{\frac{k\sqrt{\phi_{ac}}}{\sqrt{-E_c}}\left(\sqrt{\frac{z_i^*}{nl-(n-1)z_i^*}} + \frac{\varepsilon_z}{\phi_{ac}} - \sqrt{\frac{\varepsilon_z}{\phi_{ac}}}\right)\right\} \neq 0 \tag{28}$$

From Eq. (27), we know that the key to finding the special solution $v(z_i, \varepsilon_z)$ lies in seeking the value of the rotational angle of image: $\sin\chi(z_i, \varepsilon_z)$. From Eq. (18), we have

$$z_i = l = z_i^* - \Delta z^* \tag{29}$$

Substituting it into the expression of $\sin\chi(z_i, \varepsilon_z)$ in Eq. (27), we have

$$\sin\chi(z_i,\varepsilon_z) = \sin\left\{\frac{k}{\sqrt{-E_c}}\sqrt{\phi_{ac}}\left(\sqrt{\frac{z_i^* - \Delta z^*}{nl-(n-1)(z_i^*-\Delta z^*)} + \frac{\varepsilon_z}{\phi_{ac}}} - \sqrt{\frac{\varepsilon_z}{\phi_{ac}}}\right)\right\} \quad (30)$$

Expanding Eq. (30), using Eq. (28) and the values of l, z_i^*, and Δz^* given in Eq. (19), Eq. (20), and Eq. (22), and omitting all of the small quantities higher than the second order of $\sqrt{\varepsilon_z} = /\sqrt{\phi_{ac}}$ and $\sqrt{\varepsilon_{z_1}} = /\sqrt{\phi_{ac}}$, through a lengthy and tedious calculation, we may obtain

$$\sin\chi(z_i,\varepsilon_z) \approx \chi(z_i,\varepsilon_z) = \frac{k}{\sqrt{-E_c}}\sqrt{\phi_{ac}}\left[-\left(\sqrt{\frac{\varepsilon_z}{\phi_{ac}}} - \sqrt{\frac{\varepsilon_{z_1}}{\phi_{ac}}}\right) + \frac{1}{2}\left(\frac{\varepsilon_z}{\phi_{ac}} - \frac{\varepsilon_{z_1}}{\phi_{ac}}\right)\right] \quad (31)$$

Substituting it into Eq. (27), we may obtain the expression of special solution $v(z_i,\varepsilon_z)$ at the ideal image plane

$$v(z_i,\varepsilon_z) = \frac{2M}{E_c}\left[(\sqrt{\varepsilon_z} - \sqrt{\varepsilon_{z_1}}) - \frac{1}{2\sqrt{\phi_{ac}}}(\varepsilon_z - \varepsilon_{z_1})\right] \quad (32)$$

Eq. (32) is a very important expression in imaging electron optics. It conveys an unusual message that the special solution $v(z_i,\varepsilon_z)$ at the ideal image plane is only related to M, the linear magnification, E_c the electric field strength at cathode surface, and ϕ_{ac}, the electrostatic potential at image plane, but it has nothing to do with the magnetic induction intensity B_0. Eq. (32) shows that its value is a small quantity first order approximation. Although the special solution $v(z,\varepsilon_z)$ will take different values at various locations z, but at the image position $z = z_i$, it takes a definite value, as shown in Eq. (32). In particular, it should be pointed out that Eq. (32) is generally established in imaging electron optics. The well-known R-A formula derives from this, and it is of great significance for the study of imaging electron optics. This is an important conclusion we first made explicitly in the study of Imaging electron optics.

Therefore, the paraxial lateral spherical chromatic aberration $\Delta r_v^*(z_i,\varepsilon_z)$ of a combined electromagnetic concentric spherical system expressed by Eq. (25) can be written as

$$\Delta r_v^*(z_i,\varepsilon_z) = \frac{2M}{E_c}\sqrt{\frac{m_0}{2e}}\dot{r}_0\left[(\sqrt{\varepsilon_z} - \sqrt{\varepsilon_{z_1}}) - \frac{1}{2\sqrt{\phi_{ac}}}(\varepsilon_z - \varepsilon_{z_1})\right]\exp[j\chi(z_i,\varepsilon_z)] \quad (33)$$

Now we consider the effect of image rotation on the paraxial lateral spherical-chromatic aberration. Expanding the exponential function $\exp[j\chi(z_i,\varepsilon_z)]$, we have

$$\exp[j\chi(z_i,\varepsilon_z)] = 1 - \frac{jk}{\sqrt{-E_c}}(\sqrt{\varepsilon_z} - \sqrt{\varepsilon_{z_1}}) \quad (34)$$

and substituting it into Eq. (33), we may have

$$\Delta r_v^*(z_i,\varepsilon_z) = \frac{2M}{E_c}\sqrt{\frac{m_0}{2e}}\dot{r}_0\left[(\sqrt{\varepsilon_z} - \sqrt{\varepsilon_{z_1}}) - \frac{1}{2\sqrt{\phi_{ac}}}(\varepsilon_z - \varepsilon_{z_1}) - \frac{jk}{\sqrt{-E_c}}(\sqrt{\varepsilon_z} - \sqrt{\varepsilon_{z_1}})^2\right] \quad (35)$$

The associated term related to $j = \sqrt{-1}$ appears in Eq. (35), indicating that this term is in a direction perpendicular to the vector $\dot{r}_0$. Thus, when the electrons emitted from the object point of cathode surface with the initial velocity vector $\dot{r}_0$, but $\varepsilon_z \neq \varepsilon_{z_1}$, reach the image plane $z = z_i = l$, the dispersion of the image is composed of two parts. The role of the electrostatic field is the main one, which determines the second order term of the spherical-chromatic aberration. The less important part is given by the magnetic field, which causes a very small image rotation $\chi(z_i = l, \varepsilon_z)$ at the image plane. Its effect is only limited to the third order term of spherical-chromatic aberration.

The module of paraxial spherical-chromatic aberration of the second order can be expressed as

$$\Delta r_{v2}^* = \frac{2M}{E_c}\sqrt{\varepsilon_r}(\sqrt{\varepsilon_z} - \sqrt{\varepsilon_{z_1}}) \tag{36}$$

which is the so-called R-A formula in imaging electron optics. Eq. (36) has been given by Ref. [5] in 1979.

It should be noted that the R-A formula [Eq. (36)] is universally established in imaging electron optical systems. This shows that it actually originates from Eq. (32). I would like to point out that as early as in the 1940s, the German scholar Recknagel[6] and the Soviet scholar Artimovich[7] separately gave expressions of diffusion circle at the limited image plane and at the optimal image plane in studying electrostatic imaging electron optics. Therefore, Eq. (36) was named R-A formula. Actually, regardless of domestic and foreign literature the strict identification of R-A formula has not been given. It is our contribution that we have strictly deduced the R-A formula [Eq. (36)] from the study of imaging electron optics with electrostatic concentric spherical system and combined electromagnetic concentric spherical system.

Eq. (36) shows that for a combined electromagnetic concentric spherical system, the electrostatic field plays a major and decisive role in the formation of the paraxial spherical-chromatic aberration. The term of magnetic field appearing in Eq. (35) occurs only in paraxial spherical chromatic aberration of the third order, and it does not play a decisive role in image quality.

From the above mentioned description, we have proved that for these two different types of concentric spherical systems in imaging electron optics, electrostatic or electromagnetic, the expressions of the paraxial longitudinal chromatic aberration of the first order Δz_1^* [Eq. (23)] and of the paraxial lateral chromatic aberration of the second order Δr_{v2}^* [Eq. (36)] are exactly the same. They are only related to the field strength at the cathode E_c and the linear magnification of system M, but not related to the magnetic field and the specific structure of the system. Obviously, this conclusion is not accidental, and it reveals the law with universal significance in wide beam electron optical imaging.

3.2 Paraxial magnification chromatic aberration

It should be pointed out that for a bi-electrode electrostatic concentric spherical system, if we talk about the geometric image formed by the cathode object, it is located somewhere behind the anode, at a screen with a radius of curvature R_i. Similarly, for a combined electromagnetic concentric spherical system, the geometric image is located at an anode-screen with a radius of curvature R_a. Therefore, only paraxial spherical chromatic aberration is present, and any other form of aberrations no longer exists.

However, in electron optics with wide beam focusing, when we talk about the paraxial imaging, both the object plane that emits photoelectrons and the image plane that accepts the photoelectrons are abstracted as flat planes, and the paraxial lateral aberration is on this flat image plane formed and measured.

Thus, when we study the paraxial imaging and its aberrations of the off-axis object point, besides paraxial spherical chromatic aberration, other special kinds of paraxial aberrations will appear at the flat image plane, for example, one of these is the paraxial magnification chromatic aberration $\Delta r_v^*(z_i, \varepsilon_z)$. From Eq. (8), it has been defined as

$$\Delta r_w^*(z_i, \varepsilon_z) = r_0 \{ w(z_i, \varepsilon_z) \exp[j\chi(z_i, \varepsilon_z)] - M \} \tag{37}$$

From Eq. (12), when $z_i = l$, $w(z_i, \varepsilon_z)$ can be expressed as

$$w(z_i, \varepsilon_z) = (1 + l/R_c)\cos\chi(z_i, \varepsilon_z) - \frac{2l}{R_c k \phi_{ac}}\sqrt{-E_c}\sqrt{\varepsilon_z}\sin\chi(z_i, \varepsilon_z) \tag{38}$$

where $\sin\chi(z_i, \varepsilon_z)$ can be expressed by Eq (31). Then, $\cos\chi(z_i, \varepsilon_z)$ takes the following form:

$$\cos\chi(z_i, \varepsilon_z) = 1 - \frac{1}{2}\frac{k^2}{-E_c}\phi_{ac}\left(\sqrt{\frac{\varepsilon_z}{\phi_{ac}}} - \sqrt{\frac{\varepsilon_{z_1}}{\phi_{ac}}}\right)^2 \tag{39}$$

Substituting Eq. (31) and Eq. (39) into Eq. (38), we may get an approximate expression of $w(z_i, \varepsilon_z)$:

$$w(z_i, \varepsilon_z) = M\left[1 - \frac{2(M-1)}{M}\sqrt{\frac{\varepsilon_z}{\phi_{ac}}} \times \left(\sqrt{\frac{\varepsilon_z}{\phi_{ac}}} - \sqrt{\frac{\varepsilon_{z_1}}{\phi_{ac}}}\right) - \frac{1}{2}\frac{k^2}{-E_c}(\sqrt{\varepsilon_z} - \sqrt{\varepsilon_{z_1}})^2\right] \tag{40}$$

Similar to the first special solution $v(z, \varepsilon_z)$ at the image plane z_i, i.e., $v(z_i, \varepsilon_z)$, the second special solution $w(z, \varepsilon_z)$ at the image plane z_i, $w(z_i, \varepsilon_z)$, which is related to the magnification M, is also a very important parameter in imaging electron optics. It differs from $v(z_i, \varepsilon_z)$ that $w(z_i, \varepsilon_z)$ is not only related to the electrostatic field, but also related to the magnetic field. However, regardless of the electrostatic field or magnetic field, their effect on $w(z_i, \varepsilon_z)$ is only limited to small quantities of the second order.

From Eq. (37), the paraxial magnification chromatic aberration $\Delta r_w^*(z_i, \varepsilon_z)$ can be expressed as

$$\Delta r_w^*(z_i, \varepsilon_z) = r_0 \left(\left\{ -2(M-1)\sqrt{\frac{\varepsilon_z}{\phi_{ac}}}\left(\sqrt{\frac{\varepsilon_z}{\phi_{ac}}} - \sqrt{\frac{\varepsilon_{z_1}}{\phi_{ac}}}\right) + M\left[1 - \frac{1}{2}\frac{k^2}{-E_c}(\sqrt{\varepsilon_z} - \sqrt{\varepsilon_{z_1}})^2\right] \right\} \exp[j\chi(z_i, \varepsilon_z)] - M \right) \tag{41}$$

Substituting the expression of rotational angle of the exponential function $\exp[j\chi(z_i, \varepsilon_z)]$ of Eq. (34) into the above mentioned Eq. (41), we may obtain

$$\Delta r_w^*(z_i, \varepsilon_z) = \Delta r_{w2}^*(z_i, \varepsilon_z) + \Delta r_{w3}^*(z_i, \varepsilon_z) = r_0 M\left[-\frac{jk}{\sqrt{-E_c}}(\sqrt{\varepsilon_z} - \sqrt{\varepsilon_{z_1}})\right] +$$

$$r_0 M \left[-\frac{2(M-1)}{M\phi_{ac}} \sqrt{\varepsilon_z} (\sqrt{\varepsilon_z} - \sqrt{\varepsilon_{z_1}}) - \frac{jk}{2\sqrt{-E_c}\sqrt{\phi_{ac}}} (\varepsilon_z - \varepsilon_{z_1}) + \frac{jk}{-E_c} (\sqrt{\varepsilon_z} - \sqrt{\varepsilon_{z_1}})^2 \right] \quad (42)$$

In Eq. (42) we have found that some quantities related to $j = \sqrt{-1}$ appear, indicating that these terms are located along the direction perpendicular to the vector r_0.

The module of paraxial magnification chromatic aberration of the second order can be expressed as

$$\Delta r_{w2}^*(z_i, \varepsilon_z) = |r_0| M \left[\frac{-k}{\sqrt{-E_c}} (\sqrt{\varepsilon_z} - \sqrt{\varepsilon_{z_1}}) \right] \quad (43)$$

It follows from Eq. (42) that the paraxial magnification chromatic aberration $\Delta r_w^*(z_i, \varepsilon_z)$ is caused by the combination of the magnetic field and electrostatic field, but mainly it is related to the magnetic field, and the influence of the electrostatic field is secondary. Eq. (43) shows that the paraxial magnification chromatic aberration of the second order $\Delta r_{w2}^*(z_i, \varepsilon_z)$ is entirely caused by the magnetic field, but its value is in the direction perpendicular to the vector r_0. It dominates the main part in the paraxial magnification chromatic aberration $\Delta r_w^*(z_i, \varepsilon_z)$, Eq. (43) shows that $\Delta r_{w2}^*(z_i, \varepsilon_z)$, the paraxial magnification chromatic aberration of the second order, is a quantity that cannot be neglected when we design a general combined electromagnetic imaging system.

3.3 Paraxial anisotropic chromatic aberration

The other special kind paraxial chromatic aberration of off-axis object point is the paraxial anisotropic chromatic aberration $\Delta r_u^*(z_i, \varepsilon_z)$. From Eq. (8), it is defined as

$$\Delta r_u^*(z_i, \varepsilon_z) = -(k \times r_0) \times \sqrt{\frac{e}{8m_0}} B_0 v(z_i, \varepsilon_z) \exp[j\chi(z_i, \varepsilon_z)] \quad (44)$$

Substituting the expressions of $v(z_i, \varepsilon_z)$ and $\exp[j\chi(z_i, \varepsilon_z)]$ represented by Eq. (32) and Eq. (34) into Eq. (44), we may obtain the expression of paraxial anisotropic chromatic aberration:

$$\Delta r_u^*(z_i, \varepsilon_z) = -(k \times r_0) \frac{kM}{\sqrt{-E_c}} \left[(\sqrt{\varepsilon_z} - \sqrt{\varepsilon_{z_1}}) - \frac{1}{2\sqrt{\phi_{ac}}} (\varepsilon_z - \varepsilon_{z_1}) - \frac{jk}{-E_c} (\sqrt{\varepsilon_z} - \sqrt{\varepsilon_{z_1}})^2 \right] \quad (45)$$

The module of paraxial anisotropic chromatic aberration of the second order $\Delta r_{u2}^*(z_i, \varepsilon_z)$ can be written as

$$\Delta r_{u2}^*(z_i, \varepsilon_z) = |r_0| M \frac{k}{\sqrt{-E_c}} (\sqrt{\varepsilon_z} - \sqrt{\varepsilon_{z_1}}) \quad (46)$$

From Eq. (45), the paraxial anisotropic chromatic aberration $\Delta r_u^*(z_i, \varepsilon_z)$ is completely caused by the image rotation induced by the magnetic field. The paraxial anisotropic chromatic aberration of the second order $\Delta r_{u2}^*(z_i, \varepsilon_z)$ shown in Eq. (46) has the same order of magnitude as the paraxial magnification chromatic aberration $\Delta r_{w2}^{*}(z_i, \varepsilon_z)$ shown in Eq. (43), but the former occurs in the negative direction of initial object vector $k \times r_0$, and the latter occurs in the direction perpendicular to initial object vector r_0.

Summary

For the study of the electron optical paraxial aberration of the combined electromagnetic concentric spherical system, this paper derives analytical expressions of paraxial longitudinal aberration Δz^* and paraxial lateral aberration $\Delta r^*(z_i, \varepsilon_z)$ by expanding the expression of image rotation angle at the ideal image plane position. The latter includes the paraxial spherical chromatic aberration $\Delta r_v^*(z_i, \varepsilon_z)$, the paraxial magnification chromatic aberration $\Delta r_w^*(z_i, \varepsilon_z)$, and the paraxial anisotropic chromatic aberration $\Delta r_u^*(z_i, \varepsilon_z)$. This work makes it possible to quantitatively analyze the effects of the electron optical aberrations on the image quality system.

From this presentation, we can make the following conclusions.

(1) We have presented for the first time that the analytical expression of $v(z_i, \varepsilon_z)$, which is the special solution $v(z, \varepsilon_z)$ of paraxial equation at ideal image plane, is only related to M—the linear magnification, E_c—the electric field intensity, and ϕ_{ac}—the electrostatic potential of the anode relative to the cathode, but it is not related to B_0—the magnetic induction intensity at the cathode surface. On this basis we have proven that the paraxial spherical chromatic aberration of the second order which determines the ultimate spatial resolution of the system can still be characterized by the R-A formula, thus providing a theoretical basis for applying this formula to design the electrostatic image tube and the combined electromagnetic image tube.

(2) The paraxial spherical-chromatic aberration $\Delta r_v^*(z_i, \varepsilon_z)$ is the only aberration for the imaging of axial object point. Its formation is due to the dispersion of the initial axial energy of the electrons under the action of the accelerating electric field. Its aberration pattern is a diffuse circle. The paraxial longitudinal aberration Δz^* is defined as an axial divergence of different initial energies of electrons. This research shows that these two aberrations are completely caused by the electrostatic field. The effect of the magnetic field is only to make the circle image have a slight rotation. It has no substantial contribution to the paraxial spherical chromatic aberration and the paraxial longitudinal chromatic aberration.

(3) Under the action of magnetic field, the paraxial magnification chromatic aberration $\Delta r_w^*(z_i, \varepsilon_z)$ makes the image of an off-axis object point have a slight deformation along the direction of radius vector, and the paraxial anisotropic chromatic aberration $\Delta r_u^*(z_i, \varepsilon_z)$ makes the image of an off-axis object point have a slight deformation along the negative direction perpendicular to initial object vector. The module of $\Delta r_{w2}^*(z_i, \varepsilon_z)$ and $\Delta r_{u2}^*(z_i, \varepsilon_z)$ has a simple form, which is only related to electric field intensity E_c, magnetic induction intensity B_0, and linear magnification M. The combination of Eq. (43), Eq. (46), and Eq. (36) can be used to evaluate the image quality of off-axis object points in general combined electromagnetic imaging systems. This conclusion is of great significance for the design of combined electromagnetic image tubes.

In summary, compared to the bi-electrode electrostatic concentric spherical system, in spite of introduction of a magnetic field, for the imaging of an axial object point, the central aberration, i.e., the spherical chromatic aberration, in the combined electromagnetic concentric spherical system has not given substantial change, and it can still be described by the R-A formula. However, in the combined electromagnetic concentric spherical system, for the imaging of an off-axis object point, the paraxial anisotropic chromatic aberration has arisen, which does not appear in the bi-electrode electrostatic concentric spherical system. Furthermore, due to the influence of the magnetic field, the order of magnitude of the paraxial magnification chromatic aberration is also greater than that of the bi-electrode electrostatic concentric spherical system. This paper gives an extremely concise form of these two kinds of paraxial lateral aberrations—the paraxial anisotropic chromatic aberration and the paraxial magnification chromatic aberration, which can easily evaluate the image quality of axis object point imaging. All of these conclusion and understanding have been firstly given, and they will be of importance for the researchers and designers to design combined electromagnetic image tubes.

References

[1] ZHOU L W. Imaging electron optics of a combined electromagnetic concentric spherical system. Part A paraxial optics[J]. Acta Optica Sinica, 2019, 39(4):0411001.

[2] ZHOU L W, GONG H, ZHANG Z Q, et al. Static and dynamic imaging electron-optics and spatial-temporal aberrations in a bi-electrode spherical concentric system with electrostatic focusing[J]. Optik, 2011, 122(4): 287-294.

[3] ZHOU L W, GONG H, ZHANG Z Q, et al. Paraxial imaging electron optics and its spatial-temporal aberrations for a bi-electrode concentric spherical system with electrostatic focusing[J]. Optik, 2011, 122(4):295-299.

[4] ZHOU L W. Electron optics with wide beam focusing[M]. Beijing: Beijing Institute of Technology Press, 1993.

[5] ZHOU L W. Electron optics of concentric spherical electromagnetic focusing systems [J]. Advances in Electronics and Electron Physics, 1979, 52: 119-132.

[6] RECKNAGEL A. Theorie des elektrischen elektronenmikroskops fur selbststrahler [J]. Zeitschrift Fur Physik, 1941, 117(11/12): 689-708.

[7] ARTIMOVICH A. Electrostatic properties of emission system [J]. Information of Academy of Science USSR, Series of Physics, 1941, 8(6): 313-328.

2.3 Part C: Approximate Solutions of Paraxial Equation
近轴方程近似解

Abstract: The approximate expressions of special solutions of the paraxial equation and its paraxial lateral ions have been explored by an ideal model of a combined electromagnetic concentric spherical system in this paper. The approximate expressions of two special solutions of the paraxial equation in a combined electromagnetic concentric spherical system have been derived, and on this basis, some special types of paraxial lateral aberrations such as paraxial spherical-chromatic aberration, paraxial magnification chromatic aberration, and paraxial anisotropic chromatic aberration have been deduced. The results show that the paraxial lateral aberration deduced from the approximation of two special solutions is exactly the same as the exact solution, which proves the feasibility of the approximation to solve the paraxial lateral aberration.

摘要：本文通过复合电磁同心球系统的理想模型，探讨了近轴方程特解的近似表示及其近轴横向像差的求解，导出了复合电磁同心球系统近轴方程两个特解的近似表达式，在此基础上导出了一些特殊类型的近轴横向像差的表达式，如近轴色球差、近轴放大率色差和近轴各向异性色差。结果表明，由两个特解的近似解推导得到的近轴横向像差与使用精确解的结果完全一致，由此证明近似解求解近轴横向像差的方法是可行的。

Introduction

In two previous articles[1-2], we have deduced the exact analytical expressions of the two special solutions of paraxial equation, the rotational angle of electron ray, and the paraxial lateral aberrations at an arbitrary image plane where $\varepsilon_{z_1} \neq 0$ in a combined electromagnetic concentric spherical system, under the assumption that the initial axial energy of electron emitted from the cathode surface is not equal to zero. On this basis, we shall deduce the approximate expressions of two special solutions, as well as the paraxial lateral aberration through this approach. The result shows that in electron optics with wide beam focusing, expansion approximate expressions of two special solutions only needs accuracy to the second order of small quantities, and its accuracy is good enough to investigate the paraxial lateral aberration that affects the imaging quality of the system.

1 Approximate expressions of two special solutions of paraxial equation

In the previous papers on paraxial optics of combined electromagnetic concentric spherical systems[1-3], we have derived analytical expressions of two special solutions $v(z,\varepsilon_z)$, $w(z,\varepsilon_z)$ and the rotational angle of electron ray $\chi(z,\varepsilon_z)$ from the paraxial equation as follows:

$$v(z,\varepsilon_z) = \frac{2z\sqrt{-E_c}}{k\phi(z)}\sin\chi(z,\varepsilon_z) \tag{1}$$

$$w(z,\varepsilon_z) = \left(1 + \frac{z}{R_c}\right)\cos\chi(z,\varepsilon_z) - \sqrt{\varepsilon_z}\frac{2z\sqrt{-E_c}}{R_c k\phi(z)}\sin\chi(z,\varepsilon_z) \tag{2}$$

$$\chi(z,\varepsilon_z) = \frac{k}{\sqrt{-E_c}}[\sqrt{\phi(z)+\varepsilon_z} - \sqrt{\varepsilon_z}] \tag{3}$$

where ε_z is the initial axial potential corresponding to the initial axial energy of emitted electrons; $k^2 = \frac{e}{2m_0}\frac{B_0^2}{-E_c}$, and its dimension is $1/l$; e/m_0 is the ratio of electron charge to mass; B_0 is the magnetic induction intensity at the cathode surface; $\phi(z)$ is the axial potential distribution, and $B(z)$ is the axial distribution of magnetic induction intensity. For the combined electromagnetic concentric spherical system, they can be expressed as

$$\phi(z) = \frac{z}{nl-(n-1)z}\phi_{ac} \tag{4}$$

$$B(z) = \frac{n^2 l^2}{[nl-(n-1)z]^2}B_0 \tag{5}$$

where $n = R_c/R_a$, R_c and R_a are the radii of curvature of the spherical cathode and the spherical screen (anode), respectively. Its value is calculated from the curvature center O, and the direction of electron travelling is positive, and vice versa. l is the distance between the cathode and the anode $l = R_a - R_c$, ϕ_{ac} is the electrostatic potential at the anode relative to the cathode. E_c is the electric field intensity at the cathode surface, can be expressed

$$E_c = \frac{\phi_{ac}}{R_c(n-1)} = \frac{\phi_{ac}}{-nl} \tag{6}$$

Two special solutions $v = v(z,\varepsilon_z)$ and $w = w(z,\varepsilon_z)$ of paraxial equation should satisfy the following initial conditions:

$$v(z=0) = 0, \quad v'(z=0) = \frac{1}{\sqrt{\varepsilon_z}} \tag{7}$$

$$w(z=0) = 1, \quad w'(z=0) = 0 \tag{8}$$

and these two special solutions should satisfy the following Wronski determinant:

$$\sqrt{\phi(z)+\varepsilon_z}[v'(z,\varepsilon_z)w(z,\varepsilon_z) - v(z,\varepsilon_z)w'(z,\varepsilon_z)] = 1 \tag{9}$$

As can be seen from Eq. (1) and Eq. (2), the terms of the main part occupied by these two special solutions are not dearly shown. Now according to the power of $\sqrt{\varepsilon_z/\phi(z)}$ such as

$(\sqrt{\varepsilon_z/\phi(z)})^0$, $(\sqrt{\varepsilon_z/\phi(z)})^1$ and $(\sqrt{\varepsilon_z/\phi(z)})^2$, we expand Eq. (1) and Eq. (2), omitting terms higher than $(\sqrt{\varepsilon_z/\phi(z)})^2$, and arranging the expressions one by one in order of power, and thus we can see which one plays a major role.

First, we expand the analytical expressions which contain the trigonometric function of $\sin\chi(z,\varepsilon_z)$ and $\cos\chi(z,\varepsilon_z)$ in Eq. (1) and Eq. (2) to obtain the approximate expressions arranged by the power of $\sqrt{\varepsilon_z/\phi_{ac}}$:

$$\sin\chi(z,\varepsilon_z) = \sin\left[\frac{k\sqrt{\phi(z)}}{\sqrt{-E_c}}\right] - \sqrt{\varepsilon_z}\frac{k}{\sqrt{-E_c}}\cos\left[\frac{k\sqrt{\phi(z)}}{\sqrt{-E_c}}\right] - $$
$$\varepsilon_z\left\{\frac{1}{2}\frac{k^2}{-E_c}\sin\left[\frac{k\sqrt{\phi(z)}}{\sqrt{-E_c}}\right] - \frac{k}{2\sqrt{\phi(z)}\sqrt{-E_c}}\cos\left[\frac{k\sqrt{\phi(z)}}{\sqrt{-E_c}}\right]\right\} \quad (10)$$

$$\cos\chi(z,\varepsilon_z) = \cos\left[\frac{k\sqrt{\phi(z)}}{\sqrt{-E_c}}\right] + \sqrt{\varepsilon_z}\frac{k}{\sqrt{-E_c}}\sin\left[\frac{k\sqrt{\phi(z)}}{\sqrt{-E_c}}\right] - $$
$$\varepsilon_z\left\{\frac{k}{2\sqrt{-E_c}\sqrt{\phi(z)}}\sin\left[\frac{k\sqrt{\phi(z)}}{\sqrt{-E_c}}\right] + \frac{k^2}{2(-E_c)}\cos\left[\frac{k\sqrt{\phi(z)}}{\sqrt{-E_c}}\right]\right\} \quad (11)$$

Substituting them into Eq. (1) — Eq. (3) we may obtain the expressions of two special solutions $v(z,\varepsilon_z)$, $w(z,\varepsilon_z)$ and the rotation angle $\chi(z,\varepsilon_z)$ which are arranged by the power of $\sqrt{\varepsilon_z/\phi(z)}$ as follows:

$$v(z,\varepsilon_z) = \frac{2z}{\phi(z)}\frac{\sqrt{-E_c}}{k}\sin\left[\frac{k\sqrt{\phi(z)}}{\sqrt{-E_c}}\right] - \sqrt{\varepsilon_z}\frac{2z}{\phi(z)}\cos\left[\frac{k\sqrt{\phi(z)}}{\sqrt{-E_c}}\right] + $$
$$\varepsilon_z\left\{\frac{z}{[\phi(z)]^{3/2}}\cos\left[\frac{k\sqrt{\phi(z)}}{\sqrt{-E_c}}\right] - \frac{z}{\phi(z)}\frac{k}{\sqrt{-E_c}}\sin\left[\frac{k\sqrt{\phi(z)}}{\sqrt{-E_c}}\right]\right\} \quad (12)$$

$$w(z,\varepsilon_z) = (-E_c)\frac{z}{\phi(z)}\cos\left[\frac{k\sqrt{\phi(z)}}{\sqrt{-E_c}}\right] + \sqrt{\varepsilon_z}\left\{\sqrt{-E_c}k\left(1 - \frac{2}{k^2 R_c}\right)\frac{z}{\phi(z)}\sin\left[\frac{k\sqrt{\phi(z)}}{\sqrt{-E_c}}\right]\right\} - $$
$$\varepsilon_z\left\{k\sqrt{-E_c}\frac{z}{2[\phi(z)]^{3/2}}\sin\left[\frac{k\sqrt{\phi(z)}}{\sqrt{-E_c}}\right] + \left(\frac{k^2}{2} - \frac{2}{R_c}\right)\frac{z}{\phi(z)}\cos\left[\frac{k\sqrt{\phi(z)}}{\sqrt{-E_c}}\right]\right\} \quad (13)$$

$$\chi(z,\varepsilon_z) = \frac{k}{\sqrt{-E_c}}\sqrt{\phi(z)}\left[1 - \sqrt{\frac{\varepsilon_z}{\phi(z)}} + \frac{\varepsilon_z}{2\phi(z)}\right] \quad (14)$$

It should be pointed out that the Russian scientists had worked on solving the electron trajectories both in a bi-electrode electrostatic concentric spherical system and in a combined electromagnetic concentric spherical system[4-6]. The expressions of two special solutions $v(z,\varepsilon_z)$ and $w(z,\varepsilon_z)$ given in Ref. [5] are only the first term of Eq. (12) and Eq. (13) of this paper, that is, the zeroth order approximations of $v(z,\varepsilon_z)$ and $w(z,\varepsilon_z)$. Their work had given the analytical expressions of two special solutions in which the initial energy of electron emitted from the cathode surface is equal to zero ($\varepsilon_z = 0$). Similarly, the work on paraxial lateral aberrations by Russian scientists is also investigated and discussed only at the limited image plane ($\varepsilon_{z_1} = 0$). Of course, such treatment really simplifies the problem, but only zeroth order approximation of special solutions can be obtained. So it is difficult to investigate the formation and variation of paraxial lateral aberrations at different image planes.

2 Approximate expressions of special solutions under some special conditions

The approximate expressions of special solutions of the combined electromagnetic concentric spherical system given above are general solutions which can be generalized to some special cases.

2.1 Bi-electrode electrostatic concentric spherical system

For a bi-electrode electrostatic concentric spherical system, let $B_0 = 0$, i.e., $k = 0$. Therefore, we have $\chi(z,\varepsilon_z) = 0$. From Eq. (12) and Eq. (13), it is not difficult to get the approximate expressions of two special solutions $v(z,\varepsilon_z)$ and $w(z,\varepsilon_z)$, which are arranged by the power of $\sqrt{\varepsilon_z/\phi_{ac}}$:

$$v(z,\varepsilon_z) = \frac{2z}{\sqrt{\phi(z)}} - \sqrt{\varepsilon_z}\frac{2z}{\phi(z)} + \varepsilon_z\frac{z}{[\phi(z)]^{3/2}} \tag{15}$$

$$w(z,\varepsilon_z) = 1 + \frac{z}{R_c} - \sqrt{\varepsilon_z}\frac{2z}{R_c\sqrt{\phi(z)}} + \varepsilon_z\frac{2z}{R_c\phi(z)} \tag{16}$$

in which the following equation is satisfied.

$$(-E_c)\frac{z}{\phi(z)} = 1 + \frac{z}{R_c} \tag{17}$$

We have given Eq. (15) and Eq. (16) for studying the bi-electrode electrostatic concentric spherical system[7-8]. Its paraxial imaging and paraxial aberrations can be referred to Refs. [7-10].

2.2 Homogeneous and parallel combined electro-magnetic imaging system

For a homogeneous and parallel combined electromagnetic imaging system, letting $\phi(z) = (-E_c)z$, $B(z) = B_0$, and $1/R_c = 0$, and from Eq. (12) and Eq. (13), we may get the approximate expressions of two special solutions $v(z,\varepsilon_z)$, $w(z,\varepsilon_z)$ and the rotational angle $\chi(z,\varepsilon_z)$ as follows:

$$v(z,\varepsilon_z) = \frac{2}{k\sqrt{-E_c}}\sin(k\sqrt{z}) - \sqrt{\varepsilon_z}\frac{2}{-E_c}\cos(k\sqrt{z}) + \varepsilon_z\left[\frac{\cos(k\sqrt{z})}{(-E_c)^{3/2}\sqrt{z}} - \frac{k\sin(k\sqrt{z})}{(-E_c)^{3/2}}\right] \tag{18}$$

$$w(z,\varepsilon_z) = \cos(k\sqrt{z}) + \sqrt{\varepsilon_z}\frac{k}{\sqrt{-E_c}}\sin(k\sqrt{z}) - \varepsilon_z\left[\frac{k^2\cos(k\sqrt{z})}{2(-E_c)} + \frac{k\sin(k\sqrt{z})}{2(-E_c)\sqrt{z}}\right] \tag{19}$$

$$\chi(z,\varepsilon_z) = k\left[\sqrt{z} - \sqrt{\frac{\varepsilon_z}{-E_c}} + \frac{\varepsilon_z}{2(-E_c)\sqrt{z}}\right] \tag{20}$$

Eq. (18) and Eq. (19) have been derived by Monastyrski[4] when the asymptotic solutions of paraxial equation were studied.

2.3 Electrostatic proximity system

For an electrostatic proximity system, from Eq. (15) and Eq. (16), letting $\phi(z) = (-E_c)z$, we may get the approximate expressions of two special solutions $v(z,\varepsilon_z)$ and $w(z,\varepsilon_z)$:

$$v(z,\varepsilon_z) = \frac{2\sqrt{z}}{\sqrt{-E_c}} - \sqrt{\varepsilon_z}\frac{2}{-E_c} + \varepsilon_z \frac{1}{(-E_c)^{3/2}\sqrt{z}} \tag{21}$$

$$w(z,\varepsilon_z) = 1 \tag{22}$$

If we carefully examine the exact and approximate expressions of the special solutions $v(z,\varepsilon_z)$ and $w(z,\varepsilon_z)$, as well as their derivatives in the above four cases, it is not difficult to see that the exact expression fully satisfies the initial condition of the special solution (whether it is the initial position or the initial slope). Although the approximate expressions are simplified from the exact expressions, they are arranged according to the power of $\sqrt{\varepsilon_z/\phi_{ac}}$, the expression expanding to ε_z/ϕ_{ac} is enough, and they are indeed accurate enough to describe the trajectories of the electrons emitted from the cathode surface, but they may not necessarily satisfy the initial slope or initial position of the special solutions. By the way, I would like to point out that in the previous studies of electron optics with wide beam focusing, when studying the approximate solution of the electron trajectory, people took great efforts to make various assumptions to simulate the initial electron trajectory emitted from the cathode surface and the initial conditions and to satisfy the special solutions that are considered as the primary considerations, but the results are insignificant. The reason actually lies in not understanding the essential difference between the exact and approximate expressions.

3 Paraxial lateral aberrations of combined electromagnetic concentric spherical systems derived from approximate solutions

Now, we discuss the paraxial lateral aberrations $\Delta r^*(z_i,\varepsilon_z)$ of combined electromagnetic concentric spherical systems by using approximate expressions of special solutions of paraxial equation. The paraxial lateral aberrations $\Delta r^*(z_i,\varepsilon_z)$ have been defined as[2]

$$\Delta r^*(z_i,\varepsilon_z) = \sqrt{\frac{m_0}{2e}}\dot{r}_0 v(z_i,\varepsilon_z)\exp[j\chi(z_i,\varepsilon_z)] + r_0\{w(z_i,\varepsilon_z)\exp[j\chi(z_i,\varepsilon_z)] - M\} -$$

$$(k \times r_0)\sqrt{\frac{e}{8m_0}}B_0 v(z_i,\varepsilon_z)\exp[j\chi(z_i,\varepsilon_z)]$$

$$= \Delta r_v^*(z_i,\varepsilon_z) + \Delta r_w^*(z_i,\varepsilon_z) + \Delta r_u^*(z_i,\varepsilon_z) \tag{23}$$

where $j = \sqrt{-1}$; $z_i = l$ is the image position determined by the initial electron energy of ε_{z_1}, $\Delta r_v^*(z_i,\varepsilon_z)$, $\Delta r_w^*(z_i,\varepsilon_z)$ and $\Delta r_u^*(z_i,\varepsilon_z)$ are the so-called paraxial spherical-chromatic aberration, paraxial magnification chromatic aberration, and paraxial anisotropic chromatic aberration respectively.

Firstly, we shall find the approximate solution of $v(z,\varepsilon_z)$. Letting $z_i = l$, $\phi(z_i) = \phi_{ac}$, and using the equation $l/\phi_{ac} = M/(-E_c)$ at $v(z_i, \varepsilon_{z_1}) = 0$, from Eq. (12), we have

$$v(z_i, \varepsilon_z) = -(\sqrt{\varepsilon_z} - \sqrt{\varepsilon_{z_1}})\frac{2M}{-E_c}\cos\left(\frac{k}{\sqrt{-E_c}}\sqrt{\phi_{ac}}\right) +$$
$$(\varepsilon_z - \varepsilon_{z_1})\frac{2M}{-E_c}\left[\frac{1}{2}\frac{1}{\sqrt{\phi_{ac}}}\cos\left(\frac{k}{\sqrt{-E_c}}\sqrt{\phi_{ac}}\right) - \frac{k}{2}\frac{1}{\sqrt{-E_c}}\sin\left(\frac{k}{\sqrt{-E_c}}\sqrt{\phi_{ac}}\right)\right] \quad (24)$$

Since the paraxial lateral aberrations are measured the ideal image plane which is located at the image position $z_i = l$, determined by the electron initial energy, we may have

$$\frac{k}{\sqrt{-E_c}}(\sqrt{\phi_{ac} + \varepsilon_z} - \sqrt{\varepsilon_{z_1}}) = i\pi, \quad i = 1, 2, \cdots \quad (25)$$

Its solution is

$$\frac{k}{\sqrt{-E_c}}\sqrt{\phi_{ac}} = i\pi\left(1 + \sqrt{\frac{\varepsilon_{z_1}}{\phi_{ac}}} + \frac{\varepsilon_{z_1}}{2\phi_{ac}}\right) \quad (26)$$

Therefore, we have

$$\sin\left(\frac{k}{\sqrt{-E_c}}\sqrt{\phi_{ac}}\right) \approx i\pi\sqrt{\frac{\varepsilon_{z_1}}{\phi_{ac}}}$$
$$\cos\left(\frac{k}{\sqrt{-E_c}}\sqrt{\phi_{ac}}\right) \approx 1 - \frac{1}{2}(i\pi)^2\frac{\varepsilon_{z_1}}{\phi_{ac}} \quad (27)$$

Substituting Eq. (27) into Eq. (24), we may obtain

$$v(z_i, \varepsilon_z) = \frac{2M}{E_c}\left[(\sqrt{\varepsilon_z} - \sqrt{\varepsilon_{z_1}}) - \frac{1}{2}\frac{1}{\sqrt{\phi_{ac}}}(\varepsilon_z - \varepsilon_{z_1})\right] \quad (28)$$

Eq. (28) is a very important conclusion in imaging electron optics. Although we do not know the specific characteristics of the electron trajectory carried with $(\varepsilon_z, \varepsilon_r)$ emitted from the origin of the cathode surface, we know that the electron trajectories will intersect on the identified image plane defined by ε_{z_1} and form a diffusion circle of some size.

Secondly, to find the solution of $w(z, \varepsilon_z)$ from Eq. (1), we may have

$$\chi(z, \varepsilon_z) = -\frac{k}{\sqrt{-E_c}}\left[(\sqrt{\varepsilon_z} - \sqrt{\varepsilon_{z_1}}) - \frac{1}{2}\frac{1}{\sqrt{\phi_{ac}}}(\varepsilon_z - \varepsilon_{z_1})\right] \quad (29)$$

$$\cos\chi(z, \varepsilon_z) = 1 - \frac{k^2}{2(-E_c)}(\sqrt{\varepsilon_z} - \sqrt{\varepsilon_{z_1}})^2 \quad (30)$$

Substituting them into Eq. (2), we may obtain

$$w(z_i, \varepsilon_z) = M\left[1 + \frac{2(M-1)}{M\phi_{ac}}\sqrt{\varepsilon_z}(\sqrt{\varepsilon_z} - \sqrt{\varepsilon_{z_1}}) - \frac{k^2}{2(-E_c)}(\sqrt{\varepsilon_z} - \sqrt{\varepsilon_{z_1}})^2\right] \quad (31)$$

Considering the effect of image rotation angle on the paraxial lateral aberrations, and expanding the exponential function $\exp[j\chi(z_i, \varepsilon_z)]$, we have

$$\exp[j\chi(z_i, \varepsilon_z)] = 1 - \frac{jk}{\sqrt{-E_c}}(\sqrt{\varepsilon_z} - \sqrt{\varepsilon_{z_1}}) \quad (32)$$

Substituting Eq. (28), Eq. (31), and Eq. (32) into Eq. (23), we may obtain the paraxial lateral aberration of combined electromagnetic concentric spherical system derived by approximate expressions of special solutions.

The paraxial spherical-chromatic aberration is

$$\Delta r_v^*(z_i,\varepsilon_z) = \frac{2M}{E_c}\sqrt{\frac{m_0}{2e}}\dot{r}_0\left[(\sqrt{\varepsilon_z}-\sqrt{\varepsilon_{z_1}}) - \frac{1}{2\sqrt{\phi_{ac}}}(\varepsilon_z-\varepsilon_{z_1}) - \frac{jk}{\sqrt{-E_c}}(\sqrt{\varepsilon_z}-\sqrt{\varepsilon_{z_1}})^2\right] \tag{33}$$

The paraxial magnification chromatic aberration is

$$\Delta r_w^*(z_i,\varepsilon_z) = r_0 M\left[-\frac{jk}{\sqrt{-E_c}}(\sqrt{\varepsilon_z}-\sqrt{\varepsilon_{z_1}}) - \frac{k^2}{-E_c}(\sqrt{\varepsilon_z}-\sqrt{\varepsilon_{z_1}})^2 - \frac{2(M-1)}{\phi_{ac}}\sqrt{\varepsilon_z}(\sqrt{\varepsilon_z}-\sqrt{\varepsilon_{z_1}}) + \frac{jk}{2\sqrt{-E_c}\sqrt{\phi_{ac}}}(\varepsilon_z-\varepsilon_{z_1})\right] \tag{34}$$

The paraxial anisotropic chromatic aberration is

$$\Delta r_u^*(z_i,\varepsilon_z) = -(k \times r_0)\frac{kM}{\sqrt{-E_c}}\left[(\sqrt{\varepsilon_z}-\sqrt{\varepsilon_{z_1}}) - \frac{1}{2\sqrt{\phi_{ac}}}(\varepsilon_z-\varepsilon_{z_1}) - \frac{jk}{-E_c}(\sqrt{\varepsilon_z}-\sqrt{\varepsilon_{z_1}})^2\right] \tag{35}$$

It can be seen that for the paraxial lateral aberration, which is obtained by approximate expressions of special solutions, or by expanding the expression of the image rotation angle at the ideal image position, the result is completely consistent[2].

Summary

Based on the exact analytical expressions of special solutions of paraxial equation, the approximate expressions of two special solutions of the paraxial equation have been deduced, for a combined electromagnetic concentric spherical system, a bi-electrode electrostatic concentric spherical system, a homogeneous and parallel combined electromagnetic imaging system, and an electrostatic proximity system. The paraxial lateral aberrations have been derived, including paraxial spherical-chromatic aberration $\Delta r_v^*(z_i,\varepsilon_z)$, paraxial magnification chromatic aberration $\Delta r_w^*(z_i,\varepsilon_z)$, and paraxial anisotropic chromatic aberration $\Delta r_u^*(z_i,\varepsilon_z)$. Research has shown that in electron optics with wide beam focusing the paraxial lateral aberrations can be solved through the approximate expressions of two special solutions, which needs only accuracy within the second order of small quantities, and its accuracy is absolutely satisfied.

References

[1] ZHOU L W. Imaging electron optics of a combined electromagnetic concentric spherical system. Part A paraxial optics[J]. Acta Optica Sinica, 2019, 39(4):0411001.

[2] ZHOU L W. Imaging electron optics of a combined electromagnetic concentric spherical system. Part B: paraxial aberration[J]. Acta Optica Sinica, 2019, 39(4): 0411002.

[3] ZHOU L W. Electron optics of concentric spherical electromagnetic focusing systems[J]. Advances in Electronics and Electron Physics, 1979, 52: 119-132.

[4] MONASTYRSKI M A. On asymptotic solutions of paraxial equation of electron optics [J]. Journal of Technical Physics, 1978, 48(6): 1117 – 1122.

[5] SMIRNOV N A, MONASTYRSKI M A, KULIKOV Y V. Some problems on electron optics for cathode lenses of combined fields with spherical symmetry [J]. Journal of Technical Physics, 1979, 48(2): 2590 – 2595.

[6] ELIN V P, KATESHOWV V A, KULIKOV Y V, et al. Numerical methods of optimization for emissive electron optical systems [M]. Siberia: Press Science, 1987.

[7] ZHOU L W, GONG H, ZHANG Z Q, et al. Static and dynamic imaging electron-optics and spatial-temporal aberrations in a bi-electrode spherical concentric system with electrostatic focusing[J]. Optik, 2011, 122(4): 287 – 294.

[8] ZHOU L W, GONG H, ZHANG Z Q, et al. Paraxial imaging electron optics and its spatial-temporal aberrations for a bi-electrode concentric spherical system with electrostatic focusing[J]. Optik, 2011, 122(4): 295 – 299.

[9] ZHOU L W. Electron optics with wide beam focusing[M]. Beijing: Beijing Institute of Technology Press, 1993.

[10] ZHOU L W, GONG H. Theory of paraxial lateral aberrations of electrostatic imaging electrostatic electron optics based on asymptotic solutions and its verification[J]. Optik, 2011, 122(4):300 – 306.

2.4 Part D: Asymptotic Solutions of Paraxial Equation
近轴方程渐近解

Abstract: The asymptotic solutions of the paraxial equation for a combined electromagnetic concentric spherical system have been explored in this paper. Expressions of each kind of coefficients of two special solutions given by the asymptotic method for solving the paraxial equation have been deduced. By verifying the exact solution of two special solutions of the combined electromagnetic concentric spherical system, this paper proves that the approach based on asymptotic solutions given by Monastyrski[6] to solve the two special solutions of paraxial equation in electron optics is basically correct and feasible, and only a few aspects need to be improved.

摘要：本文首次探讨了复合电磁同心球系统近轴方程的渐近解，推导了复合电磁同心球系统中近轴方程两个特解的渐近解中各类系数的表达式，通过复合电磁同心球系统两个特解精确解的验证，证明了 Monastyrski[6] 提出的用渐近解求解成像电子光学近轴方程两个特解的方法正确且可行，仅个别之处需要改进。

Introduction

In the process of studying electron optics with wide beam focusing, solving the two special solutions of paraxial equation in an electrostatic focusing imaging system and a combined electromagnetic imaging system is always an important problem which should be paid more attention to. From this, we can study electron trajectories emitted from the photocathode and its imaging properties and aberrations in an electron optical imaging system. In a series of papers[1-4], we have investigated problems concerning the paraxial imaging, the spatial and temporal aberrations in an ideal model of the bi-electrode electrostatic concentric spherical system. In Ref. [5], we have proved that the asymptotic solution may be an effective way to solve the special solutions of the axial equation In the electrostatic imaging system. This paper has induced interest and attention in the academic communities of electron optics. The outstanding advantage the asymptotic solution for solving the paraxial equation is that the paraxial trajectory can be decomposed into expressions that are arranged in terms of powers of various orders, which is very favourable for the pursuit of paraxial lateral aberrations at all levels. However, whether or not the approach of asymptotic solutions can effectively solve two special solutions of paraxial equation of electron

---

Zhou Liwei (周立伟). 北京理工大学. Acta Optica Sinica (光学学报), V.39, No.4, 2019, 0411004 - 1 - 7.

optics in the combined electromagnetic imaging system has not been sufficiently proved so far.

As is known, in a combined electromagnetic concentric spherical system, the electric field and the magnetic field overlap, in which the electric field causes electron acceleration along the axial direction, and the magnetic field rotates the electrons around the axis. Therefore, the electron trajectory emitted from the photocathode is not a planar trajectory, but a rotational space trajectory, which shows the difficulty of its solution. Up to now, only in Refs. [5-6] two special solutions of the paraxial equation in imaging electron optics using the asymptotic solution method have been discussed. Ref. [5] discusses a bi-electrode concentric spherical system with electrostatic focusing, Ref. [6] discusses a combined homogeneous and parallel electromagnetic system and both obtain the asymptotic solutions of their paraxial equation. However, it should be pointed out that in these two articles, the latter model is too simple: its electric field intensity and magnetic induction intensity are uniform and parallel everywhere; and the former model only studies a pure electrostatic concentric spherical system without any magnetic focusing. So it is not sufficient to prove that the asymptotic solution method is universal and effective for solving the electron optical imaging systems.

In Refs. [7-9], based on the electron-optical paraxial equation of the imaging system, we have deduced the exact and approximate analytical expressions of two special solutions of paraxial equation in a combined electromagnetic concentric spherical system and discussed the paraxial lateral aberrations. In this paper, the ideal model of a combined electromagnetic concentric spherical system is studied to solve two special solutions of the paraxial equation of electron optics by an asymptotic solution method. It is proved that the asymptotic solution method is really an effective approach that can be tried, but the existing methods and procedures for finding the special solutions of the paraxial equation by asymptotic solution method have to be further improved.

1 General description of solving paraxial equations by asymptotic solution method

In Ref. [7], we have given a general expression for the paraxial trajectory $r(z)$ in a combine electromagnetic imaging system as follows:

$$r(z) = r_0 w(z)\exp[j\chi(z)] + \left[\sqrt{\frac{m_0}{2e}}\dot{r}_0 - (k \times r_0)\sqrt{\frac{e}{8m_0}}B_0\right]v(z)\exp[j\chi(z)] \quad (1)$$

Where r_0 an $\dot{r}_0$ represent initial position vector and initial radial velocity vector of electrons emitted from the cathode surface, respectively; k is a unit vector along the positive direction of axis z; e/m_0 is the ratio of electron charge to mass; B_0 is the magnetic induction intensity at the cathode surface; $\chi(z)$ is an angle of the rotational coordinate system $u(x,y)$ with respect to the fixed coordinate system $r(x,y)$:

$$\chi(z) = \int_{z_0}^{z}\sqrt{\frac{e}{8m_0}}\frac{B(z)}{\sqrt{\phi(z)+\varepsilon_z}}dz + \chi_0 \quad (2)$$

In general, we suppose that $\chi_0 = \chi(z=0) = 0$. Here, $\phi(z)$ is the initial axial potential distribution. $B(z)$ is the axial distribution of magnetic induction intensity, and ε_z is the initial axial potential corresponding to the initial axial energy of emitted electrons.

Suppose that $v(z, \varepsilon_z)$ and $w(z, \varepsilon_z)$ are two special solutions of paraxial equation in imaging electron optics, then we have

$$u''(z) + \frac{1}{2}\frac{\phi'(z)}{\phi(z)+\varepsilon_z}u'(z) + \frac{1}{4}\left[\frac{\phi''(z)}{\phi(z)+\varepsilon_z} + \frac{e}{2m_0}B^2(z)\right]u(z) = 0 \quad (3)$$

in the rotating coordinate system, which satisfies the following initial conditions:

$$v(z=0) = 0, \quad v'(z=0) = \frac{1}{\sqrt{\varepsilon_z}} \quad (4)$$

$$w(z=0) = 1, \quad w'(z=0) = 0 \quad (5)$$

And these two special solutions should satisfy the following Wronski determinant:

$$\sqrt{\phi(z)+\varepsilon_z}(v'w - vw') = 1 \quad (6)$$

An approach to solve the special solutions $v(z, \varepsilon_z)$ and $w(z, \varepsilon_z)$ of paraxial equation by using asymptotic solutions has been given in Ref. [6], which can be described as

$$v(z,\varepsilon_z) = \sqrt{\phi(z)}\xi_0(z) + \sqrt{\varepsilon_z}\eta_0(z) + \varepsilon_z\left[\frac{\xi_0(z)}{2\sqrt{\phi(z)}} + \frac{1}{2}\xi_1(z)\sqrt{\phi(z)}\right] + \cdots \quad (7)$$

$$w(z,\varepsilon_z) = \omega_0(z) + \sqrt{\varepsilon_z}\sqrt{\phi(z)}\zeta_0(z) + \varepsilon_z\omega_1(z) + \cdots \quad (8)$$

where coefficients ξ_0, η_0, ω_0, ξ_1 and ω_1 are the solutions of following differential equations:

$$L_0\begin{pmatrix}\xi_K\\\zeta_K\end{pmatrix} = -\begin{pmatrix}\xi''_{k-1}\\\zeta''_{k-1}\end{pmatrix}, \quad M_0\begin{pmatrix}\eta_K\\\omega_K\end{pmatrix} = -\begin{pmatrix}\eta''_{k-1}\\\omega''_{k-1}\end{pmatrix} \quad (9)$$

where $\xi_{-1} = \zeta_{-1} = \eta_{-1} = \omega_{-1} = 0$. L_0 and M_0 are linear differential operators, which can be expressed as

$$L_0 = \phi(z)\frac{d^2}{dz^2} + \frac{3}{2}\phi'(z)\frac{d}{dz} + \left[\frac{3}{4}\phi''(z) + \frac{e}{8m_0}B^2(z)\right] \quad (10)$$

$$M_0 = \phi(z)\frac{d^2}{dz^2} + \frac{1}{2}\phi'(z)\frac{d}{dz} + \left[\frac{1}{4}\phi''(z) + \frac{e}{8m_0}B^2(z)\right] \quad (11)$$

The initial condition of Eq. (9) can be determined according to the following procedure:

$$\frac{\xi_0(0)\phi'(0)}{2} = 1 \quad \rightarrow \quad \xi_0(0) + \eta_0(0) = 0,$$
$$\downarrow \quad\quad\quad\quad (12)$$
$$\frac{\xi_1(0)\phi'(0)}{2} = +\xi'_0(0) + \eta'_0(0) = 0 \rightarrow \cdots$$

$$\omega_0(0) = 1 \quad \xi_0(0) + \omega_1(0) = 0,$$
$$\downarrow \quad\quad \uparrow \quad\quad \downarrow \quad\quad\quad (13)$$
$$\frac{\zeta_0(0) + \phi'(0)}{2} + \omega'_0(0) = 0 \rightarrow \cdots$$

where the arrow indicates the sequence of calculation.

It must be pointed out that from the author's practice the expression of the first special solution $v(z,\varepsilon_z)$ of paraxial equation given in Ref. [6], which is solved in the form of asymptotic solutions, seems to be incorrect. The coefficient of term $\xi_1(z)\sqrt{\phi(z)}$, associated with the variable ε_z, is not 1, but 1/2. Eq. (7) is the result after correction.

2 Analytical expressions of coefficients for asymptotic solutions in a combined electromagnetic concentric spherical system

2.1 Relationship between potential distribution and magnetic induction distribution

In a combined electromagnetic concentric spherical system, any straight line passing through the center of curvature O can be considered as an axis. For simplicity, we study the situation of the central axis of the system. Therefore, the axial potential distribution $\phi(z)$ and the axial magnetic induction distribution $B(z)$ can be written as[7]

$$\phi(z) = \frac{z}{nl-(n-1)z}\phi_{ac} \tag{14}$$

$$B(z) = \frac{n^2 l^2}{[nl-(n-1)z]^2}B_0 \tag{15}$$

where $n = R_c/R_a$, R_c and R_a are the curvature radii of spherical cathode and spherical screen (anode), respectively. Its value is calculated from the center of curvature O, with the direction of electron travelling positive, and vice versa. l is the distance between the cathode and the anode, $l = R_a - R_c$. B_0 is the magnetic induction intensity at the cathode. ϕ_{ac} is the electrostatic potential of the anode relative to the cathode. E_c is the electric field intensity at the cathode surface, which can be expressed as

$$E_c = \frac{\phi_{ac}}{R_c(n-1)} = \frac{\phi_{ac}}{-nl} \tag{16}$$

It should be pointed out that the coefficients of the asymptotic solution expression are solved by using the second order homogeneous linear differential Eq. (9), but the linear differential operators [Eq. (10) and Eq. (11)] of the equation have arisen from not only the parameter of electrostatic field, but also the parameter of magnetic field, so seeking the analytical expression for the asymptotic solution coefficient is by no means an easy task. Fortunately, there is an intrinsic link between the potential distribution $\phi(z)$ and the magnetic induction intensity distribution $B(z)$ represented by Eq. (14) and Eq. (15), making it possible to solve analytical expressions for asymptotic solution coefficients.

From Eq. (14), differentiating $\phi(z)$, we obtain

$$\phi'(z) = (-E_c)\frac{(nl)^2}{[nl-(n-1)z]^2} \tag{17}$$

Therefore, the expression of $B(z)$ represented by Eq. (15) can be rewritten as

$$B(z) = \frac{\phi'(z)}{-E_c}B_0 \tag{18}$$

Thus, the linear differential operators of Eq. (10) and Eq. (11), in which the magnetic field parameters are converted by Eq. (18) can be rewritten by the following form:

$$\frac{3}{4}\phi''(z) + \frac{e}{8m_0}B^2(z) = \frac{3}{4}\phi''(z) + \frac{k^2}{4(-E_c)}\phi'^2(z) \tag{19}$$

$$\frac{1}{4}\left[\phi''(z) + \frac{e}{2m_0}B^2(z)\right] = \frac{1}{4}\left[\phi''(z) + \phi'^2(z)\frac{k^2}{-E_c}\right] \tag{20}$$

where $k^2 = \frac{e}{2m_0}\frac{B_0^2}{-E_c}$. In this way, by Eq. (18) we represent the quantity of the magnetic field $B(z)$ in the linear differential operator as the parameter of the electrostatic field $\phi(z)$, thus making it possible to solve Eq. (9). Eq. (18) has been given in Ref. [7].

2.2 Solving the coefficients of asymptotic solutions in the first special solution

Now we shall seek asymptotic solutions of the first special solution $v(z)$ given by Eq. (7). The coefficients are $\xi_0(z)$, $\eta_0(z)$, and $\xi_1(z)$.

Firstly, we explore the coefficient $\eta_0(z)$. From Eq. (9) and Eq. (20), the corresponding second order linear homogeneous differential equation can be pressed as

$$M_0\{\eta_0(z)\} = \phi(z)\frac{d^2}{dz^2}\eta_0(z) + \frac{1}{2}\frac{d}{dz}\eta_0(z) + \left[\frac{1}{4}\phi''(z) + \phi'^2(z)\frac{k^2}{-E_c}\right]\eta_0(z) = -\eta''_{-1} = 0 \tag{21}$$

If we assume that the coefficient $\eta_0(z)$ can be written as

$$\eta_0(z) = -\frac{2z}{\phi(z)}\cos\left[\frac{k^2}{\sqrt{-E_c}}\sqrt{\phi(z)}\right] \tag{22}$$

and put it into Eq. (21), through a series differential operations and putting it in order, we will have

$$\left[\phi'(z) + \frac{z\phi''(z)}{2} - \frac{z\phi'^2(z)}{\phi(z)}\right] \times \left\{\frac{2k}{\sqrt{-E_c}\sqrt{\phi(z)}}\sin\left[\frac{k\sqrt{\phi(z)}}{\sqrt{-E_c}}\right] + \frac{3}{\phi(z)}\cos\left[\frac{k\sqrt{\phi(z)}}{\sqrt{-E_c}}\right]\right\} = 0 \tag{23}$$

It is not difficult to prove that the value in the first bracket in Eq. (23) is equal to zero, that is

$$\phi'(z) + \frac{z\phi''(z)}{2} - \frac{z\phi'^2(z)}{\phi(z)} = 0 \tag{24}$$

Thus, Eq. (23) is tenable, so the differential Eq. (21) is satisfied. In addition, when $z = 0$, $\eta_0(z=0)$ satisfies the requirement of the initial condition proposed by Eq. (12), that is, $\eta_0(0) = -2/\phi'_0$. It can be seen that Eq. (22) is just the solution of the coefficient $\eta_0(z)$.

Secondly, we shall find the coefficient $\xi_0(z)$. From Eq. (9) and Eq. (19), we obtain

$$L_0[\xi_0(z)] = \phi(z)\frac{d^2\xi_0(z)}{dz^2} + \frac{3}{2}\phi'(z)\frac{d\xi_0(z)}{dz} + \left[\frac{3}{4}\phi''(z) + \frac{k^2}{4(-E_c)}\phi'^2(z)\right]\xi_0(z) = -\xi''_{-1} = 0 \tag{25}$$

If we assume that the coefficient $\xi_0(z)$ can be written as

$$\xi_0(z) = \frac{2z}{[\phi(z)]^{3/2}} \frac{\sqrt{-E_c}}{k} \sin\left[\frac{k}{\sqrt{-E_c}} \sqrt{\phi(z)}\right] \tag{26}$$

and put it into Eq. (25), we will have the following expression similar to Eq. (23):

$$\left[\phi'(z) + \frac{z\phi''(z)}{2} - \frac{z\phi'^2(z)}{\phi(z)}\right] \times \left\{\frac{3}{k[\phi(z)]^{3/2}} \frac{\sqrt{-E_c}}{\sqrt{-E_c}} \sin\left[\frac{k\sqrt{\phi(z)}}{\sqrt{-E_c}}\right] + \frac{2}{\phi(z)} \cos\left[\frac{k\sqrt{\phi(z)}}{\sqrt{-E_c}}\right]\right\} = 0 \tag{27}$$

It can be seen from Eq. (24) that Eq. (27) is also tenable. When $z = 0$, satisfies the requirement of the initial condition proposed by Eq. (12), that is, $\xi_0(0) = 2/\phi'_0$. Thus, it can be seen that Eq. (26) is just one of the solutions of the coefficient $\xi_0(z)$.

It should be noted that in Eq. (7) of the asymptotic solution of the first special solution $v(z)$, for the coefficient $\xi_0(z)$, the differential Eq. (25) also has another solution.

If we suppose that

$$\xi_0(z) = \frac{2z}{\phi(z)} \cos\left[\frac{k}{\sqrt{-E_c}} \sqrt{\phi(z)}\right] \tag{28}$$

and put it into Eq. (25), we will have the following expression similar to Eq. (23):

$$\left[\phi'(z) + \frac{z\phi''(z)}{2} - \frac{z\phi'^2(z)}{\phi(z)}\right] \times \left\{-\frac{1}{\phi(z)} \cos\left[\frac{k\sqrt{\phi(z)}}{\sqrt{-E_c}}\right] - \frac{1}{\phi(z)} \frac{2k\sqrt{\phi(z)}}{\sqrt{-E_c}} \sin\left[\frac{k\sqrt{\phi(z)}}{\sqrt{-E_c}}\right]\right\} = 0 \tag{29}$$

From Eq. (24), we can see that Eq. (29) is also satisfied. Meanwhile, Eq. (28) also satisfies the requirement of the initial condition proposed by Eq. (12), that is, $\xi_0(0) = 2/\phi'_0$ at $z = 0$. Therefore, Eq. (28) is the second solution of coefficient $\xi_0(z)$. But we choose Eq. (26) as a solution of the coefficient $\xi_0(z)$ needed by Eq. (25). The reason is that it has passed the theoretical testing of a combined electromagnetic concentric spherical system.

Thirdly, we shall find the coefficient $\xi_1(z)$. First, find the value of $\xi_1(z)$ at $z = 0$ on the basis of procedure in Eq. (12) and Eq. (13), for searching initial conditions of asymptotic solutions according to Ref. [6].

At $z = 0$, from Eq. (26) or Eq. (28) and from Eq. (22), we may obtain

$$\begin{aligned} \xi_0(0) = \xi_0(z=0) = \frac{2}{\phi'_0} \\ \eta_0(0) = \eta_0(z=0) = -\frac{2}{\phi'_0} \end{aligned} \tag{30}$$

Obviously, Eq. (30) satisfies the requirements of initial condition of Eq. (12) when asymptotic solutions are solved as follows

$$\frac{\xi_0(0)\phi'(0)}{2} = 1 \rightarrow \xi_0(0) + \eta_0(0) = 0 \tag{31}$$

Similarly, differentiating $\xi_0(z)$ of Eq. (26) and $\eta_0(z)$ of Eq. (22) at $z = 0$, we may obtain

$$\eta'_0(z=0) = \frac{k^2}{\phi'_0}, \quad \xi'_0(z=0) = 0 \tag{32}$$

Thus, from the initial condition in Eq. (11)

$$\frac{\xi_0(0)\phi'(0)}{2} + \xi'_0(0) + \eta'_0(0) = 0 \tag{33}$$

we may obtain the relationship in which the coefficient $\xi_1(z)$ should be satisfied at $z = 0$:

$$\xi_1(0) = -\frac{2k^2}{\phi'^2_0} \tag{34}$$

From Eq. (9) and Eq. (19), we now solve the differential equation

$$L_0[\xi_1(z)] = \phi(z)\frac{d^2\xi_1(z)}{dz^2} + \frac{3}{2}\phi'(z)\frac{d\xi_1(z)}{dz} + \left[\frac{3}{4}\phi''(z) + \frac{k^2}{4(-E_c)}\phi'^2(z)\right]\xi_1(z) = -\xi''_0 \tag{35}$$

The equation to be solved $L_0[\xi_1(z)]$ is exactly the same as $L_0[\xi_0(z)]$, where the latter is a second order linear homogeneous differential equation, while the former is a second order non-homogeneous differential equation. Due to the two solutions of $L_0[\xi_0(z)]$ we have obtained above solutions of the linear homogeneous differential equation $L_0[\xi_0(z)]$ must have a similar form, and there is only a difference in the constant coefficient.

Thus, for the coefficient $\xi_1(z)$, we can assume that

$$\xi_1(z) = -\frac{2z}{[\phi(z)]^{3/2}}\frac{k}{\sqrt{-E_c}}\sin\left[\frac{k}{\sqrt{-E_c}}\sqrt{\phi(z)}\right] \tag{36}$$

It satisfies the initial condition in Eq. (34), that is

$$\xi_0(0) = \xi_1(z)\big|_{z=0} = -\frac{2z}{[\phi(z)]^{3/2}}\frac{k}{\sqrt{-E_c}}\frac{k}{\sqrt{-E_c}}\sqrt{\phi(z)}\bigg|_{z=0} = -\frac{2k^2}{\phi'^2_0} \tag{37}$$

So Eq. (36) is one of special solutions of Eq. (35). The second special solution of Eq. (35) can be assumed as

$$\xi_1(z) = -\frac{2z}{\phi(z)}\frac{k^2}{-E_c}\cos\left[\frac{k}{\sqrt{-E_c}}\sqrt{\phi(z)}\right] \tag{38}$$

It also satisfies the initial condition in Eq. (34) that is

$$\xi_1(0) = \xi_1(z)\big|_{z=0} = -\frac{2z}{\phi(z)}\frac{k^2}{-E_c}\bigg|_{z=0} = -\frac{2k^2}{\phi'^2_0} \tag{39}$$

It can be seen that the two special solutions of the obtained coefficients $\xi_1(z)$ satisfy the requirements second order linear homogeneous differential equations $L_0[\xi_1(z)]$ and the initial conditions specified in Eq. (12) and Eq. (13). However, from the results of the verification, only the asymptotic solution of Eq. (36) is reasonable and what we need. But the procedure for solving two special solutions of paraxial equation by asymptotic solution proposed in Ref. [6] does not address the choice of solutions for second order linear homogeneous differential equations. In addition, according to the differential Eq. (35), we can of course obtain from Eq. (26) or Eq. (28) the solution $\xi_0(z)$ to determine the right end term ξ''_0 of the non-homogeneous differential Eq. (35) and to obtain the general solution of the equation. However, according to the author's opinion, it is not necessary to resort to the general solution of the second order linear non-homogeneous differential equations to find the paths of asymptotic solution coefficients. From this point of view, the approach of asymptotic solutions suggested by Ref. [6] for solving the paraxial equation seems to need some additions and corrections.

2.3 Solving the coefficients of asymptotic solutions in the second special solution

Now we shall seek asymptotic solutions of the second special solution $w(z)$ given by Eq. (8). Their coefficients are $w_0(z)$, $\zeta_0(z)$, and $\omega_0(z)$. First, we explore the coefficient $\omega_0(z)$. From Eq. (9) and Eq. (20), its corresponding differential equation can be expressed as

$$M_0[\omega_0(z)] = \phi(z)\frac{d^2\omega_0(z)}{dz^2} + \frac{1}{2}\phi'(z)\frac{d\omega_0(z)}{dz} + \frac{1}{4}\left[\phi''(z) + \frac{k^2}{-E_c}\phi'^2(z)\right]\omega_0(z) = -\omega''_{-1} = 0 \tag{40}$$

From the special solution $w(z=0) = 1$, we can obtain the initial condition of $\omega_0(z)$: $\omega_0(z=0) = 1$.

Let

$$\omega_0(z) = (-E_c)\frac{z}{\phi(z)}\cos\left[\frac{k}{\sqrt{-E_c}}\sqrt{\phi(z)}\right] \tag{41}$$

Substituting it into Eq. (40), we have

$$\left[\phi'(z) + \frac{z\phi''(z)}{2} - \frac{z\phi'^2(z)}{\phi(z)}\right] \times \left\{\frac{3E_c}{2\phi(z)}\cos\left[\frac{k}{\sqrt{-E_c}}\sqrt{\phi(z)}\right] - \frac{k}{\sqrt{\phi(z)}}\sqrt{-E_c}\sin\left[\frac{k}{\sqrt{-E_c}}\sqrt{\phi(z)}\right]\right\} = 0 \tag{42}$$

From Eq. (24), we can see that Eq. (41) is a special solution of $\omega_0(z)$.

In view of

$$(-E_c)\frac{z}{\phi(z)} = 1 + \frac{z}{R_c} \tag{43}$$

Eq. (41) can be expressed as

$$\omega_0(z) = \left(1 + \frac{z}{R_c}\right)\cos\left[\frac{k}{\sqrt{-E_c}}\sqrt{\phi(z)}\right] \tag{44}$$

Thus, it can be seen that the initial condition, $\omega_0(z=0) = 1$ is satisfied.

Next, find the coefficients $\zeta_0(z)$. Like the method described above, from Eq. (9) and Eq. (19), we have

$$\phi(z)\frac{d^2\zeta_0}{dz^2} + \frac{3}{2}\phi'(z)\frac{d\zeta_0}{dz} + \left[\frac{3}{4}\phi''(z) + \frac{e}{8m_0}B^2(z)\right]\zeta_0 = -\zeta_{-1} = 0 \tag{45}$$

Then, $\zeta_0(z=0)$ must meet the following initial conditions:

$$\frac{\zeta_0(0)\phi'(0)}{2} + \omega'_0(0) = 0 \tag{46}$$

We first evaluate $\omega'_0(z=0) = \omega'_0(0)$. Differentiating Eq. (44), we may obtain

$$\omega'_0(z=0) = \frac{1}{R_c} - \frac{1}{2}k^2 \tag{47}$$

Then, from Eq. (46), we can get the value of coefficient $\zeta_0(z)$ at $z=0$:

$$\zeta_0(0) = -\frac{2\omega'_0(0)}{\phi'(0)} = \frac{2}{E_c}\left(\frac{1}{R_c} - \frac{k^2}{2}\right) \tag{48}$$

If we assume that the coefficient $\zeta_0(z)$ satisfying Eq. (45) has the form

$$\zeta_0(z) = \left[\sqrt{-E_c}k\left(1-\frac{2}{k^2 R_c}\right)\right] \times \frac{z}{[\phi(z)]^{3/2}}\sin\left[\frac{k}{\sqrt{-E_c}}\sqrt{\phi(z)}\right] \tag{49}$$

it not only satisfies Eq. (45) but also satisfies the initial condition of Eq. (48), i.e.,

$$\zeta_0(0) = \zeta_0(z=0) = \left[k_2\left(1-\frac{2}{k^2 R_c}\right)\right] \times \frac{z}{\phi(z)} \frac{\sin\left[\frac{k}{\sqrt{-E_c}}\sqrt{\phi(z)}\right]}{\frac{k}{\sqrt{-E_c}}\sqrt{\phi(z)}}\bigg|_{z=0} = \frac{2}{E_c}\left(\frac{1}{R_c} - \frac{k^2}{2}\right) \tag{50}$$

Thus, Eq. (49) is the solution of the coefficient $\zeta_0(z)$.

Now we find the coefficient $\omega_1(z)$ associated with the variable ε_z in Eq. (8), which must satisfy Eq. (9) and Eq. (19):

$$M_0[\omega_1(z)] = \phi(z)\frac{d^2\omega_1(z)}{dz^2} + \frac{1}{2}\phi'(z)\frac{d\omega_1(z)}{dz} + \frac{1}{4}\left[\phi''(z) + \frac{k^2}{-E_c}\phi'^2(z)\right]\omega_1(z) = -\omega''_0 \tag{51}$$

It should be pointed out that Ref. [6] does not indicate whether $\omega_1(z)$ is a single solution or a combination solution of the differential Eq. (50). However, regardless of the solution of the equation, its initial condition should satisfy the requirement in Eq. (13). From Eq. (48), we have

$$\omega_1(0) = -\zeta_0(0) = -\frac{2}{E_c}\left(\frac{1}{R_c} - \frac{k^2}{2}\right) \tag{52}$$

Now we assume that the coefficient $\omega_1(z)$ has the following form, i.e.,

$$\omega_1(z) = -k\sqrt{-E_c}\frac{z}{2[\phi(z)]^{3/2}}\sin\left[\frac{k\sqrt{\phi(z)}}{\sqrt{-E_c}}\right] - \left(\frac{k^2}{2} - \frac{2}{R_c}\right)\frac{z}{\phi(z)}\cos\left[\frac{k\sqrt{\phi(z)}}{\sqrt{-E_c}}\right] \tag{53}$$

We have proved above that the sine and cosine solutions in Eq. (53) both are special solutions of the second order homogeneous linear Eq. (51), so the sum is also necessarily the solution of Eq. (51) if they can both requirements of the initial conditions in Eq. (52) at $z=0$.

We divide the coefficient $\omega_1(z)$ of Eq. (52) into two terms and find their values at $z=0$, i.e.,

$$-k\sqrt{-E_c}\frac{z}{2[\phi(z)]^{3/2}}\sin\left[\frac{k\sqrt{\phi(z)}}{\sqrt{-E_c}}\right]\bigg|_{z=0} = -k^2\frac{z}{2\phi(z)}\frac{\sin\left[\frac{k\sqrt{\phi(z)}}{\sqrt{-E_c}}\right]}{\frac{k\sqrt{\phi(z)}}{\sqrt{-E_c}}}\bigg|_{z=0} \tag{54}$$

$$= k^2\frac{1}{2E_c} - \left(\frac{k^2}{2} - \frac{2}{R_c}\right)\frac{z}{\phi(z)}\cos\left[\frac{k\sqrt{\phi(z)}}{\sqrt{-E_c}}\right]\bigg|_{z=0} = \frac{1}{E_c}\left(\frac{1}{R_c} - \frac{k^2}{2}\right)$$

Therefore

$$\omega_1(z=0) = \omega_1(0) = \frac{k^2}{2E_c} + \frac{1}{E_c}\left(\frac{k^2}{2} - \frac{2}{R_c}\right) = -\frac{2}{E_c}\left(\frac{1}{R_c} - \frac{k^2}{2}\right) \tag{55}$$

Thus, we have proved that the assumed solution [Eq. (53)] of the coefficient $\omega_1(z)$ not only satisfies the requirement of Eq. (51), but also satisfies the requirement of the initial condition of Eq. (52). So Eq. (53) is just the solution of the coefficient $\omega_1(z)$.

We now substitute the coefficients $\eta_0(z)$, $\xi_0(z)$ and $\xi_1(z)$ of asymptotic solutions expressed by Eq. (22), Eq. (26), and Eq. (36) into Eq. (7) of the first special solution $v(z)$, and substitute the coefficients $\omega_0(z)$, $\xi_0(z)$, and $\omega_1(z)$ of asymptotic solutions expressed by Eq. (41), Eq. (49), and Eq. (53) into Eq. (8) of the second special solution $w(z)$. One can obtain the expressions of asymptotic solutions of $v(z)$ and $w(z)$:

$$v(z,\varepsilon_z) = \frac{2z}{\phi(z)}\frac{\sqrt{-E_c}}{k}\sin\left[\frac{k\sqrt{\phi(z)}}{\sqrt{-E_c}}\right] - \sqrt{\varepsilon_z}\frac{2z}{\phi(z)}\cos\left[\frac{k\sqrt{\phi(z)}}{\sqrt{-E_c}}\right] +$$
$$\varepsilon_z\left\{\frac{z}{[\phi(z)]^{3/2}}\cos\left[\frac{k\sqrt{\phi(z)}}{\sqrt{-E_c}}\right] - \frac{z}{\phi(z)}\frac{k}{\sqrt{-E_c}}\sin\left[\frac{k\sqrt{\phi(z)}}{\sqrt{-E_c}}\right]\right\} \tag{56}$$

$$w(z,\varepsilon_z) = (-E_c)\frac{z}{\phi(z)}\cos\left[\frac{k\sqrt{\phi(z)}}{\sqrt{-E_c}}\right] + \sqrt{\varepsilon_z}\sqrt{-E_c}\,k\left(1 - \frac{2}{k^2 R_c}\right)\frac{z}{\phi(z)}\sin\left[\frac{k\sqrt{\phi(z)}}{\sqrt{-E_c}}\right] -$$
$$\varepsilon_z\left\{k\sqrt{-E_c}\frac{z}{2[\phi(z)]^{3/2}}\sin\left[\frac{k\sqrt{\phi(z)}}{\sqrt{-E_c}}\right] + \left(\frac{k^2}{2} - \frac{2}{R_c}\right)\frac{z}{\phi(z)}\cos\left[\frac{k\sqrt{\phi(z)}}{\sqrt{-E_c}}\right]\right\} \tag{57}$$

Eq. (56) and Eq. (57) are exactly the same as the expansion of the exact solution to the paraxial equation of the combined electromagnetic concentric spherical system given in Ref. [9]. This illustrates the feasibility of the asymptotic solution approach for solving the electron-optical paraxial equation.

Summary

In 1978, the Russian scholar M. A. Monastyrski proposed the asymptotic solution to solve the special solutions of the paraxial equation. From then on, it has been involved to solve the solution of the second order linear homogeneous differential equation, and the difficulty is great. At that time, as an example, he proposed using the asymptotic solution to solve a very simple electron optical problem in which a homogeneous and parallel combined electromagnetic focusing system is suggested, i.e., assuming the electric field intensity $E(z) = E(0) = E_c = $ const, the axial potential distribution $\phi(z) = -E_c z$, and the axial magnetic induction intensity $B(z) = B(0) = B_0 = $ const. Obviously, it naturally and greatly reduces the difficulty of solving differential equations. For more than 30 years, there had been no better example of the asymptotic solutions to solve the problem for two special solutions of paraxial equation in imaging electron optics. In 2011, we first solved the electron optical problem of a bi-electrode electrostatic concentric spherical system by using the asymptotic solution method[5], in which the axial potential distribution $\phi(z)$ is shown in Eq. (14). Of course, the solution was a bit more difficult. In this paper, we have discussed the asymptotic solution method to solve the electron optical problem of a combined electro-magnetic concentric spherical system. That is to say, an axial magnetic induction $B(z)$ expressed by Eq. (15) is superimposed on the axial potential distribution $\phi(z)$ expressed by Eq. (14). Of course, solving such a problem is more difficult. Our purpose is to prove that the asymptotic solution can indeed be used to solve the special solutions of paraxial equation in imaging electron optics.

It should be pointed out that we are more fortunate. But, if we did not study the electron optics of electrostatic and electromagnetic combined concentric spherical systems, in other words, if we have not found the special solutions of paraxial equation for these two systems in advance, our mathematical skills cannot solve such a problem of asymptotic solutions by solving the second order linear homogeneous differential equations. That is to say, we have already known in advance what are the correct analytical solutions for the special solutions of $v(z)$ and $w(z)$, then reviewed the theory and approach of asymptotic solution method. There is nothing shameful to make this clear.

Our conclusion is that through the tests of our electrostatic and electromagnetic concentric spherical electron optical systems, the approach and procedure for solving two special solutions of paraxial equation by asymptotic solution method proposed by M. A. Monastyrski are proved correct and feasible.

If we want to pick up some shortcomings of M. A. Monastyrski's work, we have pointed out above that the expression of $v(z, \varepsilon_z)$ given in Ref. [6] seems to be a written error, in which the coefficient of the term $\xi_1(z) \sqrt{\phi(z)}$ associated with the variable ε_z is not 1, but rather 1/2 [see Eq. (7)]. What we do not understand is that his article has proposed that for some coefficients, the second order non-homogeneous differential equations need to be solved, that is, a general solution of the differential equation must be considered. Practice shows that this is completely unnecessary. Two examples that we have solved show that the asymptotic solution to be solved for the two special solutions of the paraxial equation requires only a single special solution or a combination of two special solutions of the second order linear homogeneous differential equation, since it only needs accuracy to small quantities of the second order, and thus it does not require to solve the second order non-homogeneous differential equations. It means that we can suppose that the right hand term of Eq. (35) is equal to zero, i. e., $-\xi''_0 = 0$.

References

[1] ZHOU L W, GONG H, ZHANG Z Q, et al. Paraxial imaging electron optics and its spatial-temporal aberrations for a bi-electrode concentric spherical system with electrostatic focusing[J]. Optik, 2011,122(4):295 - 29.

[2] ZHOU L W, GONG H, ZHANG Z Q, et al. Static and dynamic imaging electron optics and spatial-temporal aberrations in a bi-electrode spherical concentric system with electrostatic focusing[J]. Optik, 2011, 122(4): 287 - 294.

[3] ZHOU L W. Electron optics with wide beam focusing[M]. Beijing: Beijing Institute of

Technology Press, 1993.

［4］ZHOU L W. Electron optics of concentric spherical electromagnetic focusing systems [J]. Advances in Electronics and Electron Physics, 1979, 52:119 – 132.

［5］ZHOU L W, GONG H. Theory of paraxial lateral aberrations of electrostatic imaging electrostatic electron optics based on asymptotic solutions and its verification[J]. Optik, 2011, 122(4): 300 – 306.

［6］MONASTYRSKI M A. On asymptotic solutions of paraxial equation of electron optics [J]. Journal of Technical Physics, 1978, 48(6): 1117 – 1122.

［7］ZHOU L W. Imaging electron optics of a combined electromagnetic concentric spherical system. Part A: paraxial optics[J]. Acta Optica Sinica, 2019, 39(4): 0411001.

［8］ZHOU L W. Imaging electron optics of a combined electromagnetic concentric spherical system. Part B: paraxial aberration[J]. Acta Optica Sinica, 2019, 39(4): 0411002.

［9］ZHOU L W. Imaging electron optics of a combined electromagnetic concentric spherical system. Part C: approximate solutions of paraxial equation[J]. Acta Optica Sinica, 2019, 39(4): 0411003.

2.5 电磁聚焦同心球系统的电子光学
Electron Optics of Electromagnetic Focusing Concentric Spherical Systems

摘要：本文由电磁聚焦曲近轴方程出发，导出了电磁聚焦同心球系统的近轴轨迹的解析表达式，讨论了系统的聚焦性质与电子光学横向色差和球差。本文研究了由电磁聚焦同心球系统向具有均匀平行的电场和磁场的电磁聚焦系统、静电聚焦同心球系统和近贴聚焦系统过渡的三种特殊情况，最后考察了加上栅网电极的三电极同心球静电聚焦系统和均匀平行的电场和磁场的三电极电磁聚焦系统的电子光学。

Abstract: Based on the electromagnetic focusing curvilinear paraxial equation, the analytical solutions for solving the paraxial trajectories of concentric spherical electromagnetic focusing system have been deduced, the electron optical properties and the electron optical lateral chromatic and spherical aberrations have been discussed. In this paper, the concentric spherical electromagnetic focusing system together with three other systems, namely, the electromagnetic focusing system using parallel homogeneous magnetic and electric fields, the concentric electrostatic focusing system and proximity focusing system are studied. Finally, the electron optics of the concentric spherical electrostatic focusing system with spherical mesh and the electromagnetic focusing system with planar mesh using parallel homogeneous magnetic and electric fields are also discussed.

引言

由两个共同曲率中心的球面电极加上相应的电位所组成的系统称为两电极同心球静电聚焦系统；若在此系统加上同心的辐射磁场，使它与电场的方向处处平行，则构成两电极同心球电磁聚焦系统，简称电磁聚焦同心球系统。也就是说，系统具有球对称的径向电场和磁场。

和众所周知的静电聚焦同心球系统一样，研究电磁聚焦同心球系统作为像管的成像系统从电子光学观点来看是颇有意义的。这是因为系统可实现放大率大于1或小于1的要求，电场和磁场分布以及电子近轴轨迹都可写成解析形式，可以定量地研究系统的聚焦性质和像差。无论是严格精确或近似实现的电磁聚焦同心球系统，都能得到好的像质，对于理想的电磁聚焦同心球系统模型，除了色球差外，其他类型的像差不复存在。

电磁聚焦同心球系统作为成像系统曾在1962年光电成像器件会议上叙述过（B. Zacharov[1]，R. L. Beurle 和 W. M. Wreathall[2]）。但是，电子轨迹是用数字计算机求解的，

周立伟. 北京理工大学. 兵工学报（Acta Armentarii）（试刊），No.1, 1979, 66–81.

没有得到轨迹的解析表达式，对于系统的聚焦性质与像差并没有充分讨论过。

本文进一步发展作者在静电聚焦同心球系统电子光学的研究工作。我们将电磁聚焦同心球系统连同其他三类典型系统，即具有平行均匀的磁场和电场的电磁聚焦系统、静电聚焦同心球系统和近贴聚焦系统（图1）用符号规则统一起来综合地进行考察。由普遍的情况——电磁聚焦同心球系统出发，探讨求解轨迹的方法，导出近轴电子轨迹的解析表达式、聚焦条件和色像差，验算 B. Zacharov 给出的缩小管型以及与此相对应的放大管型。最后讨论加上栅网电极的三电极同心球静电聚焦系统和平行均匀磁场和电场的三电极电磁聚焦系统的情况。

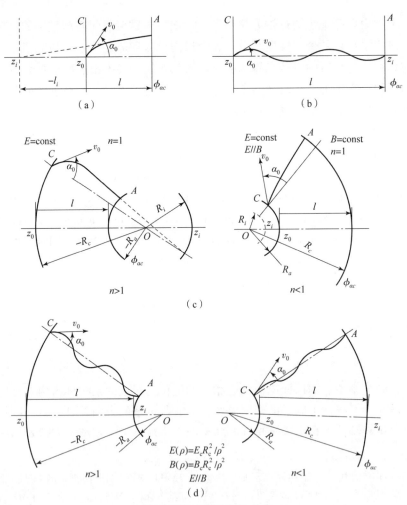

图1　四种典型的聚焦系统

（a）近贴系统；（b）平行均匀电场和磁场的电磁聚焦系统；（c）静电聚焦同心球系统；（d）电磁聚焦同心球系统

1　电磁聚焦同心球系统

1.1　近轴电子轨迹的解与聚焦条件

对于电磁聚焦同心球系统，主轨迹为一条垂直于阴极面的直线。现相对于主轨迹确定其相邻轨迹的位置。如图2所示，过主轨迹的某点 N 作主轨迹的垂直平面，此平面与主轨迹的相邻轨迹的交点 N^* 的位置将由点 N 到点 N^* 的向量 p 表达。令 s 为由初始点 P_0 算起沿着主轨迹的直线距离，一撇表示对 s 的微分。现引入正交坐标系，其坐标轴指向主轨迹点 N 处的切线、主法线和副法线的方向 t、n、b。于是由阴极面逸出的初角度为 α_0，初速度为 v_0 的电子轨迹可以用其参考轴为曲率和挠率都等于零的主轨迹的曲近轴方程进行求解：

$$(p_2' \sqrt{\phi_*})' = -\frac{1}{4} \frac{\phi''}{\sqrt{\phi_*}} p_2 - \sqrt{\frac{e}{2m_0}} \left(\frac{B'}{2} p_3 + B p_3' \right) \quad (1)$$

$$(p_3' \sqrt{\phi_*})' = -\frac{1}{4} \frac{\phi''}{\sqrt{\phi_*}} p_3 + \sqrt{\frac{e}{2m_0}} \left(\frac{B'}{2} p_2 + B p_2' \right) \quad (2)$$

式中，p_2、p_3、ϕ、B 是独立变量 s 的函数，"'" 表示对 s 的微分，B 和 ϕ 为沿主轨迹的磁感应和电位分布，m_0、e 为电子的质量和电荷，$\phi_* = \phi + \varepsilon_0 \cos^2 \alpha_0$，$\varepsilon_0 = m_0 v_0^2 / (2e)$ 为静止的光电子获得其发射能量所要求的加速电位，称为初电位，p_2、p_3 乃是 p 在主法线 n 和副法线 b 上的投影。

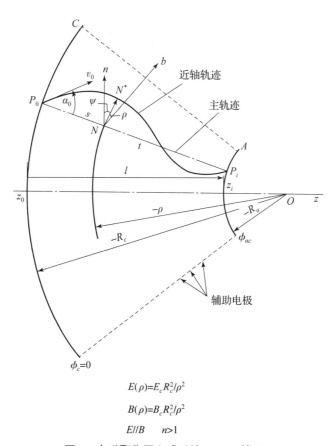

$E(\rho) = E_c R_c^2 / \rho^2$
$B(\rho) = B_c R_c^2 / \rho^2$
$E // B \quad n > 1$

图 2　电磁聚焦同心球系统（$n > 1$ 情况）

方程式（1）和式（2）与 Г. А. Гринберг 和 P. A. Sturrock 的曲轴方程相似。但是，在我们的推导中，对于成像系统，其 $\phi + \varepsilon_0$ 应以 $\phi + \varepsilon_0 \cos^2 \alpha_0$ 替代。因此，方程式（1）和

式（2）对于参考轴为中心轴的轴对称阴极透镜也是有效的。

若我们以 p 表示向量 $\boldsymbol{p}$ 的模，$\boldsymbol{p}$ 和 $\boldsymbol{n}$ 之间的夹角，即子午面的旋转角以 ψ 表示，并令 $p_2 = p\cos\psi$，$p_3 = p\sin\psi$，则方程式（1）和式（2）可以写为

$$\psi' = \sqrt{\frac{e}{2m_0}} \frac{1}{\sqrt{\phi_*}} \left(\frac{B}{2} + \frac{C}{p^2} \right) \tag{3}$$

$$\sqrt{\phi_*} (\sqrt{\phi_*} p')' = -p\left(\frac{1}{4}\phi'' + \frac{e}{8m_0} B^2\right) + \frac{e}{2m_0} \frac{C^2}{p^3} \tag{4}$$

式中，C 为积分常数。

由于系统的球对称性，故阴极面的任一法线都可看成主轨迹。设电子逸出点取在原点上，以 z 表示离阴极原点的距离，并以 $r(z)$ 和 $B(z)$ 代替 $p(s)$、$\phi(s)$ 和 $B(s)$，这样得到的结果并不失去普遍性。

将同心球系统的场表达式

$$\phi(z) = \frac{z}{nl - (n-1)z} \phi_{ac} \tag{5}$$

$$B(z) = \frac{n^2 l^2}{[nl - (n-1)z]^2} B_c \tag{6}$$

代入方程式（3）和式（4）中，便可得到近轴电子轨迹在初始条件 v_0、α_0、$z_0 = 0$，$r_0 = 0$，$\psi_0 = 0$ 下的解为

$$\phi(z) = \sqrt{\frac{e}{2m_0}} \frac{B_c}{-E_c} \times \left[\sqrt{\frac{z}{nl-(n-1)z} \phi_{ac} + \varepsilon_z} - \sqrt{\varepsilon_z} \right] \tag{7}$$

$$r(z) = 2\sqrt{\frac{2m_0}{e}} \sqrt{\frac{\varepsilon_r}{B_c}} \times \left(1 - \frac{n-1}{n} \frac{z}{l}\right) \sin\psi(z) \tag{8}$$

式中，$n = R_c/R_a$，其中 R_c、R_a 乃是球面阴极 C 和球面靶 A 的曲率半径，它们自曲率中心 O 算起，自左至右为正。B_c 和 E_c($-\phi_{ac}/(nl)$) 乃是光阴极处的磁感应强度和电场强度，l 是阴极到靶的距离，$\varepsilon_z = \varepsilon_0 \cos^2\alpha_0$，$\varepsilon_r = \varepsilon_0 \sin^2\alpha_0$ 称为轴向和径向初电位。

当 $\sin\psi = 0$，$\psi = m\pi$ ($m = 1, 2, \cdots, i$，i 为聚焦圈数)，由式（7）可得到各级成像位置 z_m：

$$z_m = \frac{2\sqrt{\frac{2m_0}{e}} \frac{m\pi}{B_c} \sqrt{\varepsilon_z} - \frac{2m_0}{e} m^2 \pi^2 \frac{E_c}{B_c^2}}{1 + \frac{2(n-1)}{\phi_{ac}} \left(\frac{m_0}{e} m^2\pi^2 \frac{E_c^2}{B_c^2} - \sqrt{\frac{2m_0}{e}} m\pi \frac{E_c}{B_c} \sqrt{\varepsilon_z} \right)} \tag{9}$$

由 $\psi = (2m-1)\pi/2$，方程式（8）的 $r(z)$ 在 $z = z_j$ 处取极大值，此处 z_j 值可由式（9）定出，只需将 m 以 $(2m-1)/2$ 代之。由此可见，电子轨迹是一条圆锥螺线，其锥度可表示为：

$$\tan\theta = -2\sqrt{\frac{2m_0}{e}} \frac{\sqrt{\varepsilon_z}}{B_c l} \frac{n-1}{n} \tag{10}$$

式中，θ 为圆锥螺线相邻波腹的连线对于主轨迹的倾角。

很明显，圆锥螺线的旋转半径与 B_c、l、n 以及逸出电子的径向初速度有关，并与 B_c

或 E_c 成反比，且当 $n<1$ 时逐渐增大，而当 $n>1$ 时，则逐渐减小。

令 $m=i$，则可求得与逸出电子轴向初电位相 ε_{z_1} 对应的成像距离即极间距离的关系式：

$$z_i = l = \frac{i\pi}{nB_c}\sqrt{\frac{2m_0}{e}}(\sqrt{\phi_{ac}+\varepsilon_{z_1}}+\sqrt{\varepsilon_{z_1}}) \tag{11}$$

实际使用时，往往采用很强的加速电场，电子到达像面的速度较轴向初速度大得多，即 $\phi_{ac}\gg\varepsilon_{z_1}$，则极间距离 l 可以用下式近似地表示：

$$l \approx i\pi\sqrt{\frac{2m_0}{e}}\frac{\sqrt{\phi_{ac}}}{nB_c} \tag{12}$$

式（12）即系统的聚焦条件，如果 l、n、i 给定，则 $\sqrt{\phi_{ac}}/B_c$ 便可以完全确定了。

同样，在 $z_i=l$ 处，由式（8）和式（11）不难证明成像系统中拉格朗日-亥姆霍兹不变式

$$M_y M_u = \frac{\sqrt{\varepsilon_{z_1}}}{\sqrt{\phi_{ac}+\varepsilon_{z_1}}} \tag{13}$$

普遍成立，此处 $M_y=M=1/n$ 为系统的线放大率，而 $M_u=r'(z_i)/r'(z_0)$ 为系统的角放大率。

1.2 成像电子光学系统的纵向和横向色差

下面我们用色球差的名词概括成像系统由于不同的初角度和不同的初速度综合作用引起的像差。

定义从阴极面原点发出与 ε_{z_1} 相对应的近轴电子聚焦的位置 $z_i=l$ 为理想像面位置，而从同一点发出的初电位为 ε_0、初角度为 α_0 的另一条近轴电子轨迹交于 $z_{i\alpha_0}$ 处，则产生像的位移，即纵向近轴色差 Δz，如图 3 所示。

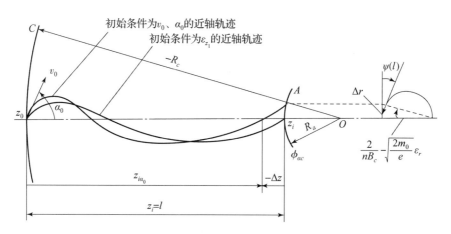

图 3 纵向近轴色差与横向近轴色差

显然：

$$\Delta z = z_{i\alpha_0} - l \tag{14}$$

这里 l 用式（11）表示，而 $z_{i\alpha_0}$ 利用式（9）并考虑到式（11），便有

$$z_{i\alpha_0} = \sqrt{\frac{2m_0}{e}} \frac{i\pi}{nB_c} \sqrt{\phi_{ac}} \times \frac{\sqrt{1+\frac{\varepsilon_{z_1}}{\phi_{ac}}} - \sqrt{\frac{\varepsilon_{z_1}}{\phi_{ac}}} + 2\sqrt{\frac{\varepsilon_z}{\phi_{ac}}}}{1 - \frac{2(n-1)}{n}\left(\sqrt{1+\frac{\varepsilon_{z_1}}{\phi_{ac}}} - \sqrt{\frac{\varepsilon_{z_1}}{\phi_{ac}}}\right)\left(\sqrt{\frac{\varepsilon_{z_1}}{\phi_{ac}}} - \sqrt{\frac{\varepsilon_z}{\phi_{ac}}}\right)} \quad (15)$$

由式（14），经过整理，略去$(\varepsilon_0/\phi_{ac})^{3/2}$的高阶项，便得到：

$$\Delta z = \Delta z_1 + \Delta z_2 \quad (16a)$$

这里，

$$\Delta z_1 = -\frac{2M^2}{E_c}\sqrt{\phi_{ac}}(\sqrt{\varepsilon_z} - \sqrt{\varepsilon_{z_1}}) \quad (16b)$$

$$\Delta z_2 = -\frac{2M^2}{E_c}\left[\frac{n}{n-2}(\varepsilon_z - \varepsilon_{z_1}) + \frac{4(n-1)}{(n-2)}\sqrt{\varepsilon_{z_1}}(\sqrt{\varepsilon_z} - \sqrt{\varepsilon_{z_1}})\right] \quad (16c)$$

同样，以初始条件参量(v_0, α_0)的近轴电子轨迹在$z_i = l$处的像面上转了$\psi(l)$角，它可以表示为

$$\psi(l) = i\pi + \frac{1}{2}\sqrt{\frac{e}{2m_0}}\frac{B_c}{E_c}\frac{n}{l}\sqrt{\phi_{ac}} \times \frac{\Delta z}{\left[n - n(-1)\frac{z_{i\alpha_0}}{l}\right]^2 \left[\frac{z_{i\alpha_0}}{nl - n(-1)z_{i\alpha_0}} + \frac{\varepsilon_z}{\phi_{ac}}\right]^{1/2}} -$$

$$\frac{1}{8}\sqrt{\frac{e}{2m_0}}\frac{B_c}{E_c}\frac{n^2}{l^2}\sqrt{\phi_{ac}} \times \frac{\Delta z^2}{\left[n - n(-1)\frac{z_{i\alpha_0}}{l}\right]^3 \left[\frac{z_{i\alpha_0}}{nl - n(-1)z_{i\alpha_0}} + \frac{\varepsilon_z}{\phi_{ac}}\right]^{1/2}} +$$

$$\frac{1}{2}\sqrt{\frac{e}{2m_0}}\frac{B_c}{E_c}\frac{n(n-1)}{l^2}\sqrt{\phi_{ac}} \times \frac{\Delta z^2}{\left[n - n(-1)\frac{z_{i\alpha_0}}{l}\right]^4 \left[\frac{z_{i\alpha_0}}{nl - n(-1)z_{i\alpha_0}} + \frac{\varepsilon_z}{\phi_{ac}}\right]^{3/2}} + \cdots$$

$$(17)$$

由图3，电子轨迹离此像面上光轴的距离即横向色差Δr可表示为

$$\Delta r = 2\sqrt{\frac{2m_0}{e}} \frac{\sqrt{\varepsilon_r}}{nB_c} \sin\psi(l) \quad (18)$$

将式（17）代入式（18），略去$(\varepsilon_0/\phi_{ac})^2$的高阶项，便得到横向近轴色差$\Delta r$的表达式：

$$\Delta r = \Delta r_2 + \Delta r_3 \quad (19a)$$

式中，

$$\Delta r_2 = \frac{2M}{E_c}\sqrt{\varepsilon_r}(\sqrt{\varepsilon_z} - \sqrt{\varepsilon_{z_1}}) \quad (19b)$$

$$\Delta r_3 = -\frac{M}{E_c}\frac{\sqrt{\varepsilon_r}}{\sqrt{\phi_{ac}}}(\varepsilon_z - \varepsilon_{z_1}) \quad (19c)$$

在式（16a）和式（19a）中，Δz_1、Δz_2为一级和二级近轴纵向色差，Δr_2、Δr_3为二级和三级近轴横向色差，它们是按照$(\varepsilon_0/\phi_{ac})^{1/2}$的阶次进行划分的。应该指出，如果我们能找到实际轨迹表达式，则可求出系统的几何纵向和横向色球差δz和δr，即以同一初始条件参量(v_0, α_0)的近轴轨迹与实际轨迹之间的纵向偏离以及此二轨迹在像面上的横向偏离。数值计算表明，一级近轴色差Δz_1和二级近轴横向色差Δr_2在全部色差中起着主要的

作用；Δz_2 和 Δr_3 与 Δz_1 和 Δr_2 相比较是高阶小量。

应该指出的是，在展开式（17）时，必须使转角 $\psi(l)$ 满足下列条件：

$$i\pi - \frac{\pi}{4} < \psi(l) < i\pi + \frac{\pi}{4} \tag{20}$$

于是，由式（18）可导得条件式：

$$\frac{B_c}{-E_c} < \frac{1}{2}\sqrt{\frac{m_0}{2e}}\frac{\pi}{\sqrt{\varepsilon_0}} \tag{21}$$

由式（16b）和式（19b）不难证明，由阴极面原点以初电位逸出的各方向的近轴单色电子束在 $\varepsilon_{z_1} = 0$ 的像面至 $\varepsilon_{z_1} = \varepsilon_0$ 的像面的距离为

$$\Delta z_{\min} = -\frac{-2M^2\sqrt{\phi_{ac}\varepsilon_0}}{E_c} \tag{22}$$

的区间上密集形成焦散面与焦散线。其最小弥散圆所在的最佳像面位置与 $\varepsilon_{z_1} = 0.3\varepsilon_0$ 相对应。最小弥散圆半径为

$$\Delta r_{\min} = 0.6\frac{M}{-E_c}\varepsilon_0 = 0.6\frac{l}{\phi_{ac}}\varepsilon_0 \tag{23}$$

由此可见，无论是放大型像管或是缩小型像管，其弥散圆的大小只与所施加的电位 ϕ_{ac}、初电位 ε_0 和极间距离 l 有关，而与放大率 M 无关。从物理意义上来说，缩小管的电子轨迹螺旋线的回转半径是逐步减小的，但由于 B_c 较小，故起始回转半径大；而对于放大管，尽管电子轨迹螺旋线的回转半径是逐步增大的，但由于 B_c 较大，起始回转半径小；二者最后在靶面上得到的弥散圆的大小是一样的。

1.3 缩小管和放大管的计算举例

利用上述的表达式计算 B. Zacharov 建议的缩小管以及放大管，可以得到很明晰的结果，如表 1 所示。放大管与缩小管在结构上完全相似，只不过光阴极与靶互易位置［见图 1（d）］。

2 均匀平行电场和场的电磁聚焦系统

若 R_c 和 R_a 同时都趋于无穷大，即球面阴极和球面靶都变为平面阴极和平面靶，$n = 1$，$B = $ 常数，$E = $ 常数，则以上导得的公式完全能适用于均匀平行电场和磁场的电磁聚焦系统，如图 1（b）所示。

众所周知，在此系统中，电子在轴向做等加速运动，横向轨道乃是周期为 $T = 2\pi m_0/(eB)$ 且回到轴上的一个圆圈，其回转半径为 $m_0 \dot{r}_0/(eB)$。因此，电子轨迹描绘的是一条螺距逐渐增大的圆柱螺线。

很明显，对于均匀平行电场和磁场的电磁聚焦系统，由于实际轨迹与近轴轨迹的解是一致的，故几何色差 δz 和 δr 等于零。

3 静电聚焦同心球系统

在不存在磁场的情况下，即 $B(z) = 0$，于是电磁聚焦同心球系统就变成纯粹的静电聚

焦同心球系统，如图 1（c）所示。因 $\psi(z) = 0$，故电子轨迹便变为平面曲线。因为

$$\lim_{\psi(z) \to 0} \frac{\sin \psi(z)}{\psi(z)} = 1 \tag{24}$$

故由式（7）和式（8）可得初始条件 (v_0, α_0) 下近轴轨迹的解为

$$r(z) = 2[nl - (n-1)z] \frac{\sqrt{\varepsilon_r}}{\phi_{ac}} \times \left[\sqrt{\frac{z}{nl-(n-1)z}\phi_{ac} + \varepsilon_z} - \sqrt{\varepsilon_z}\right] \tag{25}$$

由此不难求得电子近轴轨迹在阳极 D 处的斜率：

$$r'(z=l) = -\frac{n-2}{\sqrt{\phi_{ac}}} \sqrt{\varepsilon_r} \left[1 - \frac{2(n-1)}{n-2} \times \frac{\sqrt{\varepsilon_z}}{\phi_{ac}} + \frac{3n-2}{2(n-2)} \frac{\varepsilon_z}{\phi_{ac}}\right] \tag{26}$$

表 1 电磁聚焦缩小管和放大管的计算

数据与计算结果		缩小管（$M=1/5$）	放大管（$M=5$）
初始数据	R_c	-0.25 m	0.05 m
	R_a	-0.05 m	0.25 m
	n	5	1/5
	l	0.20 m	0.20 m
	ϕ_{ac}	20 000 V	20 000 V
	ε_0	1 V	1 V
计算数值	E_c	-2×10^4 V/m	-5×10^5 V/m
	$B_{\min}$	$B_{\min} = B_c = 0.010\ 486$ Wb/m²	$B_{\min} = B_a = 0.010\ 486$ Wb/m²
	$B_{\max}$	$B_{\max} = B_a = 0.262\ 15$ Wb/m²	$B_{\max} = B_c = 0.262\ 15$ Wb/m²
	M	1/5	5
	$\tan \theta$	-2.572×10^{-3}	2.058×10^{-5}
	i	7	7
场分布	$\phi(z)$	$\phi(z) = 20\ 000 \dfrac{z}{1-4z}$	$\phi(z) = 500\ 000 \dfrac{z}{1+20z}$
	$B(z)$	$B(z) = \dfrac{0.010\ 486}{(1-4z)^2}$	$B(z) = \dfrac{0.262\ 15}{(1+20z)^2}$
电子近轴轨迹	$\psi(z)$	$\psi(z) = 0.155\ 5 \times$ $\left(\sqrt{\dfrac{2 \times 10^4 z}{1-4z} + \varepsilon_0 \cos^2 \alpha_0} - \sqrt{\varepsilon_0 \cos^2 \alpha_0}\right)$	$\psi(z) = 0.155\ 5 \times$ $\left(\sqrt{\dfrac{5 \times 10^5 z}{1+20z} + \varepsilon_0 \cos^2 \alpha_0} - \sqrt{\varepsilon_0 \cos^2 \alpha_0}\right)$
	$r(z)$	$r(z) = 6.43 \times 10^{-4} \times$ $\sqrt{\varepsilon_0} \sin \alpha_0 \times (1-4z) \sin \psi(z)$	$r(z) = 2.572 \times 10^{-5} \times$ $\sqrt{\varepsilon_0} \sin \alpha_0 \times (1+20z) \sin \psi(z)$

续表

数据与计算结果		缩小管（$M=1/5$）	放大管（$M=5$）
各级成像位置	z_1	0.018 869 7	0.000 830 0
	z_2	0.061 541 2	0.003 493 8
	z_3	0.105 888 7	0.008 613 5
	z_4	0.141 599 3	0.017 682 0
	z_5	0.167 790 9	0.034 489 2
	z_6	0.186 534 4	0.071 305 1
	z_7	0.200 000 0	0.200 000 0
波腹位置 (z_j, r_j) $j=\dfrac{2(m-1)}{2}$ 单位(m)	$z_{0.5}, r_{0.5}$	0.005 000 6, $6.303\ 18 \times 10^{-4}$	0.000 204 9, $2.582\ 54 \times 10^{-5}$
	$z_{1.5}, r_{1.5}$	0.038 796 4, $5.432\ 16 \times 10^{-4}$	0.001 906 9, $2.670\ 09 \times 10^{-5}$
	$z_{2.5}, r_{2.5}$	0.084 465 2, $4.257\ 56 \times 10^{-4}$	0.005 682 4, $2.864\ 30 \times 10^{-5}$
	$z_{3.5}, r_{3.5}$	0.125 006 5, $3.214\ 83 \times 10^{-4}$	0.012 561 5, $3.215\ 07 \times 10^{-5}$
	$z_{4.5}, r_{4.5}$	0.155 775 3, $2.423\ 46 \times 10^{-4}$	0.024 691 5, $3.842\ 47 \times 10^{-5}$
	$z_{5.5}, r_{5.5}$	0.177 946 4, $1.853\ 23 \times 10^{-4}$	0.048 800 2, $5.082\ 28 \times 10^{-5}$
	$z_{6.5}, r_{6.5}$	0.193 811 8, $1.445\ 16 \times 10^{-4}$	0.111 221 2, $8.293\ 22 \times 10^{-5}$
电子光学色球差	Δz_1	$\Delta z_1 = -5.656\ 85 \times 10^{-4} \times (\sqrt{\varepsilon_{z_1}} - \sqrt{\varepsilon_0}\cos\alpha_0)$	$\Delta z_1 = -0.014\ 142\ 1 \times (\sqrt{\varepsilon_{z_1}} - \sqrt{\varepsilon_0}\cos\alpha_0)$
	Δr_1	$\Delta r_1 = 1 \times 10^{-5}\sqrt{\varepsilon_0}\sin\alpha_0 \times (\sqrt{\varepsilon_{z_1}} - \sqrt{\varepsilon_0}\cos\alpha_0)$	$\Delta r_1 = 1 \times 10^{-5}\sqrt{\varepsilon_0}\sin\alpha_0 \times (\sqrt{\varepsilon_{z_1}} - \sqrt{\varepsilon_0}\cos\alpha_0)$
	$\Delta r_{\min}$	$\Delta r_{\min} = 6 \times 10^{-6}$ m	$\Delta r_{\min} = 6 \times 10^{-6}$ m
	$\Delta r_{\max}$	$\Delta r_{\max} = 5.656\ 85 \times 10^{-4}$ m	$\Delta r_{\max} = 0.014\ 142$ m

如果我们假定阳极是对于电子透明的带电的理想网——等电位的栅状阳极，在栅状阳极后面的空间为无场空间。当电子经过带电的理想网时，其斜率不变。于是由式（25）和式（26），对于 ε_{z_1} 的电子，可得到成像位置 R_i 的表达式：

$$R_i = -R_c \frac{1}{n-2}\left[1 + \frac{2(n-1)}{n-2}\frac{\sqrt{\varepsilon_{z_1}}}{\phi_{ac}} + \frac{2n(n-1)}{(n-2)^2}\frac{\varepsilon_{z_1}}{\phi_{ac}}\right] \tag{27}$$

式中，R_i 自曲率中心 O 算起，自左至右为正。由此可见，只有当 $n>2$ 时才能获得实像。

当 $\varepsilon_{z_1}=0$ 时，则可得 P. Schagen[3] 等人推导的成像位置近似式：

$$R_i \approx \frac{-R_c}{n-2} \tag{28}$$

这实际乃是极限像面位置的表达式。

下面讨论系统的电子光学色球差。如图 4 所示，对于初始条件为 (v_0, α_0) 的电子，由 $\tan\alpha_i = r'(l)$ 以及正弦定理 $\sin\tau = R_{i\alpha_0}\sin\alpha_i/R_a$，可求得角度 τ 和 γ，即

$$\tan\tau = n\sqrt{\frac{\varepsilon_r}{\phi_{ac}}}\left(1 - \frac{1}{2}\frac{\varepsilon_z}{\phi_{ac}}\right) \tag{29}$$

$$\sin\gamma = -2(n-1)\sqrt{\frac{\varepsilon_r}{\phi_{ac}}} \times \left(1 - \sqrt{\frac{\varepsilon_z}{\phi_{ac}}} + \frac{1}{2}\frac{\varepsilon_z}{\phi_{ac}}\right) \tag{30}$$

此处 τ 为轨迹在阳极处的逸出角，α_i 为轨迹在 $z_{i\alpha_0}$ 处与轴的交角，γ 为 OD 的连线与轴的交角。角度一律以锐角表示，γ、α_i 以中心轴为轴，τ 以 OD 为轴，逆时针为正。

由式（27）、式（29）和式（30），不难求得纵向近轴色差 Δz 的表达式为

$$\Delta z = \Delta z_1 + \Delta z_2 \tag{31a}$$

式中

$$\Delta z_1 = -\frac{2M^2}{E_c}\sqrt{\phi_{ac}}(\sqrt{\varepsilon_z} - \sqrt{\varepsilon_{z_1}}) \tag{31b}$$

$$\Delta z_2 = -\frac{2M^2}{E_c}\left[\frac{n}{n-2}(\varepsilon_z - \varepsilon_{z_1}) - \frac{4(n-1)}{n-2}\sqrt{\varepsilon_{z_1}}(\sqrt{\varepsilon_z} - \sqrt{\varepsilon_{z_1}})\right] \tag{31c}$$

和横向近轴色差 Δr 的表达式

$$\Delta r = \Delta r_2 + \Delta r_3 \tag{32a}$$

式中，

$$\Delta r_2 = -\frac{2M}{E_c}\sqrt{\varepsilon_r}(\sqrt{\varepsilon_z} - \sqrt{\varepsilon_{z_1}}) \tag{32b}$$

$$\Delta r_3 = -\frac{2M}{E_c}\frac{\sqrt{\varepsilon_r}}{\sqrt{\phi_{ac}}}(\varepsilon_z - \varepsilon_{z_1}) \tag{32c}$$

这里 M 为系统的线放大率，以 $M = R_i/R_c$ 表示。

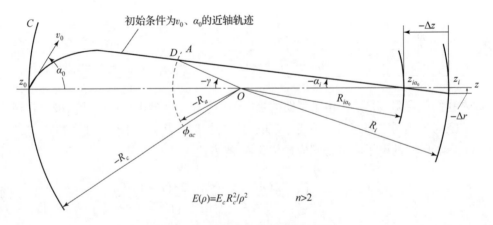

图 4 两电极同心球静电聚焦系统

如果我们利用作者导出的静电聚焦同心球系统自阴极面逸出初电位为 ε_0、初角度为 α_0 的实际轨迹表达式

$$r(z) = \frac{2(n-1)\sqrt{\varepsilon_r}}{\sqrt{\phi_{ac}}\left[1-4(n-1)^2\dfrac{\varepsilon_r\varepsilon_z}{\phi_{ac}^2}\right]} \times \left\{(z+R_c)\sqrt{\dfrac{\varepsilon_z}{\phi_{ac}}} - \dfrac{2z(n-1)\sqrt{\varepsilon_z}\varepsilon_r}{\phi_{ac}^{3/2}} - \right.$$

$$\left.(z+R_c)\left[\dfrac{-z}{(n-1)(z+R_c)} + \dfrac{(z+R_c)^2\varepsilon_z + z^2\varepsilon_r}{(z+R_c)^2\phi_{ac}}\right]^{1/2}\right\} \quad (33)$$

则应在式（31a）和式（32a）的像差公式中附加以下两项，即纵向几何球差 δz 和横向几何球差 δr：

$$\delta z = \frac{2M^2}{E_c}(n-1)\varepsilon_r \quad (34)$$

$$\delta r = \frac{2M}{E_c\sqrt{\phi_{ac}}}(n-1)\varepsilon_r^{3/2} \quad (35)$$

表 2 中给出了 $R_c = -25$ mm，$R_a = -5$ mm 和 $R_c = 5$ mm，$R_a = 25$ mm 的两种静电聚焦同心球系统在 $\phi_{ac} = 10\,000$ V 以 $\alpha_0 = 90°$ 逸出的电子近轴轨迹与实际轨迹的计算结果。由表 2 可见，由磁场 $B = 0$ 的近轴方程式（4）所导出的近轴轨迹表达式（25）与实际轨迹表达式（33）是非常接近的[4]。

表 2　静电聚焦同心球系统的近轴轨迹与实际轨迹[4]

管子参数	离阴极面的距离 z/mm	电位分布 φ(z)/V	近轴轨迹 r(z)/mm	近轴轨迹 r′(z)	实际轨迹 r(z)/mm	实际轨迹 r′(z)
$R_c = 25$ mm	4	476.170 5	0.366 605 6	0.037 097 0	0.366 620 0	0.037 133 7
$R_a = -5$ mm	8	1 176.470 5	0.466 475 2	0.015 434 8	0.466 520 0	0.013 170 6
$\phi_{ac} = 10\,000$ V	12	2 307.692 3	0.499 598 4	0.001 601 3	0.499 692 1	0.001 620 5
$\varepsilon_0 = 1$ V	16	4 444.444 4	0.480 000 0	-0.011 666 6	0.480 120 6	-0.011 682 7
$\alpha_0 = 90°$	20	10 000.000 0	0.400 000 0	-0.030 000 0	0.400 319 8	-0.030 088 6
$R_c = 5$ mm	4	5 555.555 5	0.107 331 2	0.019 379 2	0.107 333 5	0.019 377 1
$R_a = 25$ mm	8	7 692.307 6	0.182 427 2	0.018 418 1	0.182 432 6	0.018 410 7
$\phi_{ac} = 10\,000$ V	12	8 823.529 4	0.255 499 2	0.018 160 4	0.255 506 7	0.018 152 9
$\varepsilon_0 = 1$ V	16	9 523.809 5	0.327 900 8	0.018 054 0	0.327 912 4	0.018 043 7
$\alpha_0 = 90°$	20	10 000.000 0	0.400 000 0	0.018 000 0	0.400 012 8	0.017 981 0

4　近贴聚焦系统

对于静电聚焦同心球系统，我们假定 R_c、R_a 都趋于无穷大，即 $n = 1$，于是式（25）就变成近贴聚焦系统（如图 1（a）所示）的解。很明显，电子轨迹是一条抛物线，自阴极面发出的纵向初电位为 ε_{z_1} 的电子将在阴极后面形成虚像。

令 l_i 为阴极至虚像面的距离，它自阴极面算起，自右至左为负，故可表示为

$$l_i = (R_i - R_c)|_{R_c \to \infty, R_a \to \infty} \tag{36}$$

将式（27）代入式（36）中，并利用 $R_a - R_c = l_i$，$n = R_c/R_a = (1 - l/R_a)|_{R_a \to \infty}$，则有

$$l_i = -l\left(\sqrt{1 + \frac{\varepsilon_{z_1}}{\phi_{ac}}} - \sqrt{\frac{\varepsilon_{z_1}}{\phi_{ac}}}\right)^2 \tag{37}$$

静电聚焦同心球系统的色差公式完全适用于近贴聚焦系统，只须令 $n = 1$。在此条件下，式（34）的 δz 和式（35）的 δr 都等于零，说明近贴聚焦系统的实际轨迹的解与近轴轨迹的解是一致的。

5 带有栅网的三电极聚焦系统

5.1 带有栅网的三电极同心球静电聚焦系统

如果我们在两电极同心球静电聚焦系统的光阴极与阳极之间插入一个曲率半径为 R_s 的球形栅网，于是就变成三电极同心球静电聚焦系统，如图 5 所示。

系统的计算可按以下的次序进行。

首先考察阴极 C – 栅网 S 系统，自阴极原点 z_0 发出的初角度为 α_0、初速度为 v_0 的电子将落在栅网 S 的 B 点上，其角度 ζ 和 γ_0 可以由上面导得的式（29）和式（30）来表示，即：

$$\tan\zeta = y\sqrt{\frac{\varepsilon_r}{\phi_{sc}}}\left(1 - \frac{1}{2}\frac{\varepsilon_z}{\phi_{sc}}\right) \tag{38}$$

$$\sin\gamma_0 = -2(y-1)\sqrt{\frac{\varepsilon_r}{\phi_{sc}}} \times \left(1 - \sqrt{\frac{\varepsilon_z}{\phi_{sc}}} + \frac{1}{2}\frac{\varepsilon_z}{\phi_{sc}}\right) \tag{39}$$

式中，$y = R_c/R_s$，ϕ_{sc} 是栅网 S 对于阴极 C 的电位。

其次，我们研究由栅网 S 和阳极 A 组成的系统。我们仍假定，轨迹通过栅网和阳极时，其斜率不变。很明显，如果把栅网看作阴极，则自网射出发射角为 ζ，速度为 v_1 的电子将落到阳极 A 的 D 点上。按照上述，D 点的位置即与其相对应的角度 τ 和 γ_1 很容易确定，只需对式（38）和式（39）进行以下变换：

$$\alpha_0 \to \zeta, \quad y \to \frac{n}{y}, \quad \varepsilon_0 \to \varepsilon_0 + \phi_{sc}, \quad \gamma_0 \to \gamma_1, \quad \zeta \to \tau, \quad \phi_{sc} \to \phi_{ac} - \phi_{sc}$$

通过上述变换，我们可以求得 $\tan\tau$ 和 $\sin\gamma_1$ 的表达式：

$$\tan\tau = \frac{n}{r}\left(\frac{\varepsilon_r}{\phi_{sc}}\right)^{1/2}\left(1 - \frac{1}{2r^2}\frac{\varepsilon_z}{\phi_{sc}}\right) \tag{40}$$

$$\sin\gamma_1 = \frac{-2(n-y)}{r+1}\sqrt{\frac{\varepsilon_r}{\phi_{sc}}} \times \left(1 - \frac{1}{2r}\frac{\varepsilon_z}{\phi_{sc}}\right) \tag{41}$$

式中，$r^2 = \phi_{ac}/\phi_{sc}$。

同样，假定阳极后面为等位空间，由图 5 轨迹交轴位置 $R_{i\alpha_0}$ 可以按正弦定理求得

$$R_{i\alpha_0} = R_a \frac{\sin\tau}{\sin\alpha_i} = R_a \frac{\tan\tau}{\sin(\gamma_0 + \gamma_1) + \cos(\gamma_0 + \gamma_1)\tan\tau} \tag{42}$$

将式（39）、式（40）和式（41）代入上式中，便得到

$$R_{i\alpha_0} = R_c M_0 \left\{ 1 - 2r(y-1)M_0 \sqrt{\frac{\varepsilon_z}{\phi_{sc}}} + [M_0 C_0 + 4M_0^2 r^2 (y-1)^2] \frac{\varepsilon_z}{\phi_{sc}} \right\} \quad (43)$$

式中，M_0 为零级近似线放大率，它可表示为

$$M_0 = -\frac{r+1}{r(n-2) + 2r^2(y-1) - n} \quad (44)$$

C_0 为与结构参量的电参量有关的参数：

$$C_0 = r(y-1) - \frac{n-y}{r+1} + \frac{(n-1) + r(y-1)}{r(r+1)} \quad (45)$$

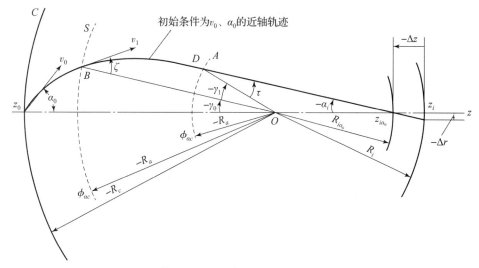

图 5 带有栅网的三电极同心球静电聚焦系统

在式（43）中，若令 $\varepsilon_z = 0$，则我们就可得到 S. R. Linden 和 P. A. Snell[5]所导得的零级近似表达式，由式（43）可以讨论系统的聚焦特性和像差，例如，由式（44）可以导得系统成实像的条件：

$$n > y > 1 + \frac{n - r(n-2)}{2r^2} \quad (46)$$

此外，我们也可仿照 P. Schagen 以及 S. R. Linden 和 P. A. Snell 来讨论阳极带有小孔的静电聚焦同心球系统的计算，这里就不细述了。

5.2 带有栅网的平行均匀电场和磁场的三电极电磁聚焦系统

对于带有平面栅网的平行均匀电场和磁场的三电极电磁聚焦系统，如图 6 所示，如前所述，若要在靶位置处得到良好的聚焦，则转角必须为 π 的整数倍，即 $\psi = i\pi$。

由式（7），在栅网处的转角为

$$\psi(l_{cs}) = \sqrt{\frac{e}{2m_0}} \frac{B}{\sqrt{\phi_{sc}}} l_{cs} \left(\sqrt{1 + \frac{\varepsilon_{z_1}}{\phi_{sc}}} - \sqrt{\frac{\varepsilon_{z_1}}{\phi_{sc}}} \right) \quad (47)$$

对于在靶处的转角，则式（47）中各值以下列诸值代替：

$$l_{cs} \to l_{sa}, \quad \phi_{sc} \to \phi_{as}, \quad \varepsilon_{z_1} \to \phi_{sc} + \varepsilon_{z_1}$$

于是

$$\psi(l_{sa}) = \sqrt{\frac{e}{2m_0}} \frac{B}{\sqrt{\phi_{as}}} l_{sa} \left(\sqrt{1 + \frac{\phi_{sc} + \varepsilon_{z_1}}{\phi_{as}}} - \sqrt{\frac{\phi_{sc} + \varepsilon_{z_1}}{\phi_{as}}} \right) \quad (48)$$

式中，ϕ_{sc} 是栅网对于阴极的电位，ϕ_{as} 是靶对于栅网的电位，l_{cs} 是阴极到栅网的距离，l_{sa} 是栅网到靶的距离。

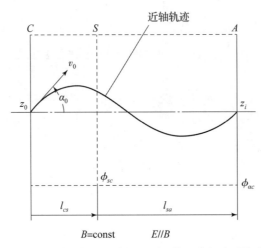

图 6 带有栅网的均匀平行电场和磁场的三电极电磁聚焦系统

由 $\psi(l_{cs}) + \psi(l_{sa}) = i\pi$，便可得到系统的聚焦条件，磁感应强度 B 可表示为

$$B = i\pi \sqrt{\frac{2m_0}{e}} \left[\frac{l_{cs}}{\sqrt{\phi_{sc}}} \left(\sqrt{1 + \frac{\varepsilon_{z_1}}{\phi_{sc}}} - \sqrt{\frac{\varepsilon_{z_1}}{\phi_{sc}}} \right) + \frac{l_{sa}}{\sqrt{\phi_{as}}} \left(\sqrt{1 + \frac{\phi_{sc} + \varepsilon_{z_1}}{\phi_{as}}} - \sqrt{\frac{\phi_{sc} + \varepsilon_{z_1}}{\phi_{as}}} \right) \right]^{-1} \quad (49)$$

若令 $\varepsilon_{z_1} = 0$，并把 B 的单位 Wb/m^2 变为高斯，其他各量仍用 SI 制单位表示，于是我们得到

$$B(\text{高斯}) = \frac{0.106i}{\frac{l_{cs}}{\sqrt{\phi_{sc}}} + \frac{l_{sa}}{\sqrt{\phi_{as}}} \left(\sqrt{1 + \frac{\phi_{sc}}{\phi_{as}}} - \sqrt{\frac{\phi_{sc}}{\phi_{as}}} \right)} \quad (50)$$

式（49）曾为 S. R. Linden 和 P. A. Snell[5] 证明过。

附带指出，无论对于静电聚焦系统或是电磁聚焦系统，当在两电极同心球系统中插入栅网电极时，仍可证明，一级纵向和二级横向近轴色差具有式（16b）和式（19b）的形式。

结束语

本文由普遍的情况——电磁聚焦同心球系统出发，导出了近轴电子轨迹的解析解，确定了各级成像点的位置并给出了系统的结构参数与电磁参数之间的关系式，研究了以阶次分类的近轴色差与几何球差，计算了 B. Zacharov 的缩小管以及与此结构上相似的放大管，再一次证明了成像系统的二级近轴色差公式普遍成立。文中无论对 $n > 1$、$n < 1$ 的球面电极系统或是 $n = 1$ 的平面电极系统，对有无磁场 B 的情况以及对带有栅网的三电极聚焦系统都进行了考虑，所得的结论对于研究静电聚焦型像管和电磁聚焦型像管具有实际意义，对于研究阴极镜的电子光学及其像差具有指导意义。

参 考 文 献

[1] ZACHAROV B. A demagnifying image tube for nuclear physics applications [J]. Advances in Electronics and Electron Physics, 1962, 16: 99 – 104.

[2] BEURLE R L, WREATHALL W M. Aberration in magnetic focus systems [J]. Advances in Electronics and Electron Physics, 1962, 16: 333.

[3] SCHAGEN P, BRUINING H, FRANCKEN J C. A simple electrostatic electron-optical system with only one voltage [J]. Philips Research Report, 1952, 7(2): 119 – 130.

[4] ЧЖОУ Л В. Об аберрации в электронной оптической системе с шаровой и осевой симметрией [C]. Тезисы докладов XX1 Научно-Технической Конференции, Ленинград, 1966: 137 – 138.

[5] LINDEN S R, SNELL P A. Shutter image converter tubes [J]. Proceedings of the IRE., 1957, 45: 513 – 517.

2.6 Electron Optics of Concentric Spherical Electromagnetic Focusing Systems
同心球电磁聚焦系统的电子光学

Abstract: The concentric spherical electromagnetic focusing system as an imaging system was described at the 2nd Symposium on Photo-electronic Image Devices, held in Imperial College, London, in 1978, it has been paid great attention to, but up to now the analytic solution for the electron trajectories in this system has not been given. Based on the electron optical paraxial equation, the analytical solutions for solving the "paraxial" trajectories of a bi-electrode concentric spherical electromagnetic focusing system have been given. In this paper, the concentric spherical electromagnetic focusing system together with three other systems, namely, the electromagnetic focusing system using parallel homogeneous magnetic and electric fields, the concentric electrostatic focusing system and proximity focusing system are studied. The analytic solution for "paraxial" trajectories, the focusing conditions and the sphero-chromatic aberrations are given, and the tri-electrode concentric spherical electrostatic focusing system with spherical mesh and the tri-electrode electromagnetic focusing system with planar mesh using parallel homogeneous magnetic and electric fields are also discussed.

摘要：同心球电磁聚焦系统作为成像系统在1978年伦敦帝国理工学院举行的第二届光电子成像器件会议上报告过，已引起了广泛的注意，但直到现在并没有给出其系统的电子轨迹的解析解。本文由电子光学近轴方程出发，导出了同心球电磁聚焦系统的近轴轨迹的表达式，讨论了系统的聚焦性质与电子光学色球差，探讨了由同心球电磁聚焦系统向具有均匀平行的电场和磁场的电磁聚焦系统、同心球静电聚焦系统和近贴聚焦系统过渡的三种特殊情况，最后研究了加上栅网电极的三电极同心球静电聚焦系统和均匀平行的电场和磁场的三电极电磁聚焦系统的电子光学。

Introduction

A system composed of two concentric spherical electrodes and their corresponding potentials is called a bi-electrode concentric spherical electrostatic focusing system. If we add to this system a concentric radial magnetic field parallel to the electric field, we get a concentric spherical electromagnetic focusing system, in other words, the system has spherically symmetric radial electric and magnetic fields.

Chou Liwei (Zhou Liwei). Beijing Institute of Technology. Advances in Electronics and Electron Physics (电子学与电子物理学的进展). 1979, 52, 119−132.

From the viewpoint of electron optics, it is interesting to study the concentric spherical electromagnetic focusing system as an imaging system, as well as the more familiar concentric spherical electrostatic focusing system. This is because the system can meet both magnifying and demagnifying requirements. The magnetic field and electric field distributions and the "paraxial" electron trajectories can be written as analytic forms and the focusing characteristics and the aberrations can be investigated quantitatively. With a concentric spherical electromagnetic focusing system, whether realized accurately or approximately, a good image quality can be obtained. Only spherical and chromatic aberrations are present.

The concentric spherical electromagnetic focusing system as an imaging system was described at the 2nd Symposium on Photo-electronic Image Devices[1,2], but the electron trajectories were solved by digital computers. An analytic solution for the electron trajectories in this system has not yet been given, and the focusing characteristics and the aberrations have not been fully discussed.

In this paper, the concentric spherical electromagnetic focusing system is studied together with three other systems, namely, the electromagnetic focusing system using parallel homogeneous magnetic and electric fields, the concentric spherical electrostatic focusing system and the proximity focusing system (Fig. 1). The method for solving the electron trajectories in these systems is explored; the analytic solution for "paraxial" trajectories, the focusing conditions and the sphero-chromatic aberrations are given; the demagnifying tube suggested by Zacharov[1] and the corresponding magnifying tube are verified; and the case of the tri-electrode concentric spherical electrostatic focusing system and the tri-electrode electromagnetic focusing system using parallel, homogeneous electric and magnetic fields are discussed.

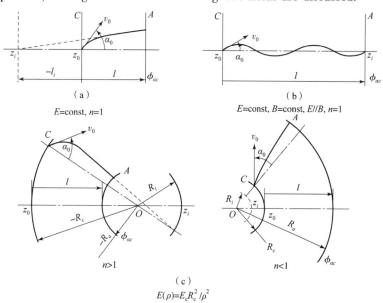

Fig. 1 Four focusing systems

(a) The proximity focusing system;

(b) The electromagnetic focusing system using parallel, homogeneous electric and magnetic fields;

(c) The concentric spherical electrostatic focusing system

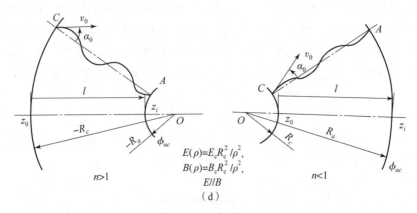

Fig. 1 Four focusing systems（续）
(d) The concentric spherical electromagnetic focusing system

1 Solution for "paraxial" trajectories and its focusing condition in the concentric spherical electromagnetic focusing system

For the concentric spherical electromagnetic focusing system using spherically symmetric magnetic and electric fields (Fig. 2), the principal trajectories are straight lines perpendicular to the photocathode surface. Now we determine the position of neighboring trajectories relative to the principal trajectory $P_0 P_i$. A plane normal to the principal trajectory at a point N intersects a neighboring trajectory at a point N^*. The position of N^* can be expressed by the vector $\boldsymbol{p}$ from N to N^*. Let s be the linear distance along the principal trajectory from the initial point P_0 on the photocathode and a prime denote differentiation with respect to s. We introduce an orthogonal coordinate system with coordinate axes assigned to the tangential, normal and bi-normal directions $\boldsymbol{t}, \boldsymbol{n}, \boldsymbol{b}$ at point N. Then, the trajectory of an electron from the photocathode with angle of emission α_0 and velocity of emission v_0 can be solved from the curvilinear "paraxial" equations, in which the reference axis is the principal trajectory whose curvature and torsion are equal to zero:

$$(p_2' \sqrt{\phi_*})' = -\frac{1}{4} \frac{\phi''}{\sqrt{\phi_*}} p_2 - \sqrt{\frac{e}{2m_0}} \left(\frac{B'}{2} p_3 + B p_3' \right) \tag{1}$$

$$(p_3' \sqrt{\phi_*})' = -\frac{1}{4} \frac{\phi''}{\sqrt{\phi_*}} p_3 - \sqrt{\frac{e}{2m_0}} \left(\frac{B'}{2} p_2 + B p_2' \right) \tag{2}$$

where p_2, p_3, and ϕ are functions of the independent variable s; B and ϕ are the magnetic induction and the electrostatic potential along the principal trajectory respectively; e and m_0 are the charge and mass of the electron; $\phi_* = \phi + \varepsilon_0 \cos^2 \alpha_0$, ε_0 is the so called "initial potential" of the emitted electron, $\varepsilon_0 = m_0 v_0^2 / 2e$; p_2 and p_3 are the projections of the neighbouring trajectory vector $\boldsymbol{p}$ on the normal $\boldsymbol{n}$ and binormal $\boldsymbol{b}$ of the principal trajectory.

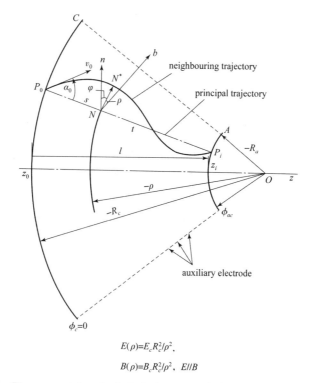

$$E(\rho) = E_c R_c^2/\rho^2,$$
$$B(\rho) = B_c R_c^2/\rho^2, \quad E /\!/ B$$

Fig. 2 The concentric spherical electromagnetic focusing system ($n > 1$)

Eq. (1) and Eq. (2) are similar to those derived by G. A. Greenberg[3] and P. A. Sturrock[4], but in our derivation $\phi + \varepsilon_0$ is replaced by $\phi + \varepsilon_0 \cos^2 \alpha_0$, and therefore Eq. (1) and Eq. (2) are also valid for cathode lenses with axially symmetric fields whose reference axis is the central axis of the system.

If we express the modulus of vector $\boldsymbol{p}$ by p, the angle of rotation between $\boldsymbol{p}$ and $\boldsymbol{n}$ by ψ, and let $p_2 = p\cos\psi$, $p_3 = p\sin\psi$, then Eq. (1) and Eq. (2) can be written as

$$\psi' = \sqrt{\frac{e}{2m_0}} \frac{1}{\sqrt{\phi_*}} \left(\frac{B}{2} + \frac{C}{p^2} \right) \tag{3}$$

$$\sqrt{\phi_*} (\sqrt{\phi_*} p')' = -p\left(\frac{1}{4}\phi'' + \frac{e}{8m_0} B^2 \right) + \frac{e}{2m_0} \frac{C^2}{p^3} \tag{4}$$

where C is the constant of integration. Because of the spherical symmetry of the system, any of the normal to the photocathode surface may be taken as a principal trajectory; we take z_0 as the initial point of emitting electrons and let z be the distance from the initial point z_0. In this case, the replacement of $p(s)$, $\phi(s)$ and $B(s)$ by $r(z)$, $\phi(z)$ and $B(z)$ respectively will not affect the generality of the results.

Substituting the field equations of the said system:

$$\phi(z) = \frac{z}{nl - (n-1)z} \phi_{ac} \tag{5}$$

$$B(z) = \frac{n^2 l^2}{[nl - (n-1)z]^2} B_c \tag{6}$$

into Eq. (3) and Eq. (4), we can obtain the solution for the "paraxial" electron trajectory with initial conditions ν_0, α_0, $z_0 = 0$, $r_0 = 0$, $\psi_0 = 0$:

$$\psi(z) = \sqrt{\frac{e}{2m_0}} \frac{B_c}{-E_c} \left[\sqrt{\frac{z}{nl-(n-1)z}\phi_{ac} + \varepsilon_z} - \sqrt{\varepsilon_z} \right] \tag{7}$$

$$r(z) = 2\sqrt{\frac{2m_0}{e}} \frac{\sqrt{\varepsilon_r}}{B_c} \left(1 - \frac{n-1}{n}\frac{z}{l}\right) \sin\psi(z) \tag{8}$$

where $n = R_c/R_a$, R_c and R_a are the radii of the spherical photocathode C and the spherical anode (target) A, R_c and R_a are taken as positive from left to right, starting from the centre of curvature 0; B_c and $E_c (= -\phi_{ac}/nl)$ are the magnetic induction and the electric field intensity at the photocathode respectively; l is the distance from the photocathode to target; $\varepsilon_z = \varepsilon_0 \cos^2\alpha_0$ and $\varepsilon_r = \varepsilon_0 \sin^2\alpha_0$ are the so called initial axial and radial potentials.

From $\sin\psi = 0$, namely, $\psi = m\pi$ ($m = 1, 2 \cdots i$, i = number of loops), for the various stages of the imaging positions z_m, from Eq. (7) we have

$$z_m = \frac{2\sqrt{\frac{2m_0}{e}}\frac{m\pi}{B_c}\sqrt{\varepsilon_z} - \frac{2m_0}{e}m^2\pi^2\frac{E_c}{B_c^2}}{1 + \frac{2(n-1)}{\phi_{ac}}\left(\frac{m_0}{e}m^2\pi^2\frac{E_c^2}{B_c^2} - \sqrt{\frac{2m_0}{e}}m\pi\frac{E_c}{B_c}\sqrt{\varepsilon_z}\right)} \tag{9}$$

From $\psi = [(2m-1)\pi]/2$, $r(z)$ in Eq. (8) tends to a minimum r_i at z_i, for which we may substitute $(m-1)/2$ for m in Eq. (9). It follows that the electron trajectory is a conical helix, and the radius of rotation of the conical helix, inversely proportional to B_c or E_c gradually decreases when $n > 1$, and increases when $n < 1$.

Let $m = i$, for the imaging distance z_i or the distance l from the photocathode to the target corresponding to the initial axial potential ε_{zl} of electrons emitted from the photocathode, we obtain

$$z_i = l = \frac{i\pi}{nB_c}\sqrt{\frac{2m_0}{e}}(\sqrt{\phi_{ac} + \varepsilon_{zl}} + \sqrt{\varepsilon_{zl}}) \tag{10}$$

In actual practice, owing to the fact that a very strong accelerating electric field is usually applied, the velocity of electrons arriving at the target is much greater than the initial electron velocity, i.e., $\phi_{ac} \gg \varepsilon_{zl}$. Then, l can be expressed approximately by

$$l \approx i\pi\sqrt{\frac{2m_0}{e}}\frac{\sqrt{\phi_{ac}}}{nB_c} \tag{11}$$

This is the focusing condition for the system. From Eq. (11), the value of $\sqrt{\phi_{ac}}/B_c$ can be fully determined when l, n, i are given.

2 "Paraxial" lateral chromatic aberrations

In what follows, the aberrations of the imaging system caused by different emission angles and different emission velocities are called "spherical and chromatic aberrations".

We define the focusing position $z_i = l$ expressed by Eq. (10) as an ideal image position corresponding to the initial axial potential ε_{zl}. Another "paraxial" electron with angle of emission α_0 and velocity of emission v_0 intersects the axis at point $z_{i\alpha_0}$. The axial deviation from the ideal image position is the so called longitudinal "paraxial" chromatic aberration Δz given by

$$\Delta z = z_{i\alpha_0} - l \tag{12}$$

From Eq. (9) and considering Eq. (10), $z_{i\alpha_0}$ can be expressed by

$$z_{i\alpha_0} = \sqrt{\frac{2m_0}{e}} \frac{i\pi}{nB_c} \sqrt{\phi_{ac}} \frac{\sqrt{1+(\varepsilon_{z_1}/\phi_{ac})} - \sqrt{\varepsilon_{z_1}/\phi_{ac}} + 2\sqrt{\varepsilon_z/\phi_{ac}}}{1 - [2(n-1)/n][\sqrt{1+(\varepsilon_{z_1}/\phi_{ac})} - \sqrt{\varepsilon_{z_1}/\phi_{ac}}](\sqrt{\varepsilon_{z_1}/\phi_{ac}} - \sqrt{\varepsilon_z/\phi_{ac}})} \tag{13}$$

Substituting Eq. (10) and Eq. (13) in Eq. (12) and neglecting the higher order terms of $(\varepsilon_0/\phi_{ac})^{3/2}$, we have

$$\Delta z = \Delta z_1 + \Delta z_2 \tag{14}$$

where

$$\Delta z_1 = -\frac{2M^2}{E_c} \sqrt{\phi_{ac}} (\sqrt{\varepsilon_z} - \sqrt{\varepsilon_{z_1}}) \tag{14a}$$

and

$$\Delta z_2 = -\frac{2M^2}{E_c} \left\{ -\frac{2(n-1)}{n}(\sqrt{\varepsilon_z} - \sqrt{\varepsilon_{z_1}})^2 - \sqrt{\varepsilon_{z_1}}(\sqrt{\varepsilon_z} - \sqrt{\varepsilon_{z_1}}) \right\} \tag{14b}$$

where $M = 1/n$ is the linear magnification.

In can be seen from Fig. 3 that if "paraxial" electron with initial conditions (v_0, α_0) in the ideal image plane at $z_i = l$ is rotated through an angle $\psi(l)$, the transverse deviation of the electron trajectory from the ideal image position in the image plane, i.e., the so called lateral "paraxial" chromatic aberration Δr can be written as

$$\Delta r = 2\sqrt{\frac{2m_0}{e}} \frac{\sqrt{\varepsilon_r}}{nB_c} \sin\psi(l) \tag{15}$$

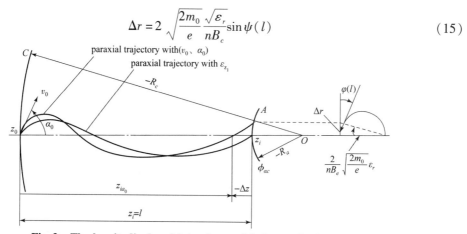

Fig. 3 The longitudinal and lateral paraxial chromatic aberrations

Neglecting the higher order terms of $(\varepsilon_0/\phi_{ac})^2$, from Eq. (15) we have

$$\Delta r = \Delta r_2 + \Delta r_3 \tag{16}$$

where

$$\Delta r_2 = \frac{2M}{E_c} \sqrt{\varepsilon_r} (\sqrt{\varepsilon_z} - \sqrt{\varepsilon_{z_1}}) \tag{16a}$$

$$\Delta r_3 = -\frac{M}{E_c\sqrt{\phi_{ac}}}\sqrt{\varepsilon_r}(\varepsilon_z - \varepsilon_{z_1}) \tag{16b}$$

In Eq. (14) and Eq. (16), Δz_1, Δz_2 are the first order and the second order "paraxial" longitudinal chromatic aberrations, Δr_2, Δr_3 are the second order and the third order "paraxial" lateral chromatic aberrations, respectively, which are divided according to the order of $(\varepsilon_0/\phi_{ac})^{1/2}$. Numerical analysis shows that Δz_1 and Δr_2 account for most of the total chromatic aberrations, while Δz_2 and Δr_3 are higher order terms compared with Δz_1 and Δr_2.

It should be noted that when we expand Eq. (15), we must ensure that the angles of over or under rotation $\psi(l)$ are less than $\pi/4$. From Eq. (15), we obtain the conditional expression

$$\frac{B_c}{-E_c} < \frac{1}{2}\sqrt{\frac{m_0}{2e}}\frac{\pi}{\sqrt{\varepsilon_0}} \tag{17}$$

It can be proved form Eq. (14a) and Eq. (16a), that the electrons emitted from photocathode with monochromatic ε_0 form a dense defocusing surface in the region between the image planes with $\varepsilon_{z_1} = 0$ and $\varepsilon_{z_1} = \varepsilon_0$ separated by

$$\Delta z_{max} = \frac{-2M^2\sqrt{\phi_{ac}\varepsilon_0}}{E_c} \tag{18}$$

In the plane of the best focus of electrons, which corresponds to $\varepsilon_{z_1} = 0.09\varepsilon_0$, the minimum radius Δr_{min} of the circle of confusion in the electromagnetic focusing system can be expressed by

$$\Delta r_{min} = 0.6\frac{M}{-E_c}\varepsilon_0 = 0.6\frac{l}{\phi_{ac}}\varepsilon_0 \tag{19}$$

It may be seen that whether for the magnifying type or the demagnifying type of electromagnetic focusing system, the size of the circle of confusion only depends on l, ε_0 and ϕ_{ac}. This can be explained as follows. In the demagnifying system, as B_c is small, the initial radius of rotation is rather big, while the radius of rotation of the helix gradually decreases. In the magnifying system, as B_c is large, the initial radius of rotation is rather small, while the radius of rotation of the helix gradually increases. Finally, the size of the circle of confusion at the target surface in the two cases is the same.

The demagnifying tube suggested by B. Zacharov[1] was verified and the corresponding magnifying tube was calculated from the said equations. The magnifying tube resembles the demagnifying tube in structure, but the photocathode and the target exchange positions [Fig. 1(d)].

It is clear that we can follow K. F. Hartley[5] in calculating the modulation transfer function of the system, but it is suggested that one more term Eq. (16b) may be added. In addition, we also can follow J. P. Picat[6] and R. F. Beurle and W. M. Wreathall[2] in calculating the resolution of the system.

3 The electromagnetic focusing system using parallel homogeneous magnetic and electric fields

If both R_a and R_c tend to infinity, i.e., $n = 1$, $E = $ const and $B = $ const, all the formulae mentioned above can be applied to the case of the electromagnetic focusing system using parallel homogeneous magnetic and electric fields [Fig. 1 (b)]. It is well known that in this system, the electrons move with a constant acceleration in the axial direction, the transverse orbit is a circle returning to the axis in a period $T = 2\pi m_0/eB$, and the radius of rotation is $(m_0/e)(\dot{r}_0/B)$. Therefore, the electron trajectory describes a cylindrical helix whose pitch gradually increases. It is obvious that the solution for the real trajectories in this system is in accordance with the solution for the "paraxial" trajectories and the "geometric" chromatic aberrations δz and δr are equal to zero.

4 The concentric spherical electrostatic focusing system

In the absence of a magnetic field, i.e., $B(z) = 0$, the concentric spherical electromagnetic focusing system becomes a purely electrostatic focusing system [Fig. 1 (c)]. As $\psi(z) = 0$, the electron trajectories are plane curves. Because

$$\lim_{\psi(z)=0} \frac{\sin \psi(z)}{\psi(z)} = 1$$

from Eq. (7) and Eq. (8) we obtain the solution for the "paraxial" trajectories under the initial condition (v_0, α_0) as

$$r(z) = 2[nl - (n-1)z] \frac{\sqrt{\varepsilon_r}}{\phi_{ac}} \left(\sqrt{\frac{z}{nl-(n-1)z} \phi_{ac} + \varepsilon_z} - \sqrt{\varepsilon_z} \right) \tag{20}$$

Suppose the anode is an ideal "electron-transparent" mesh, and the space behind the anode is free from fields, when electrons pass through the mesh, their slope is unchanged. Then, from Eq. (20) the distance from the center of curvature to the image position for electrons with ε_{z_1} can be obtained:

$$R_i = -R_c \frac{1}{n-2} \left[1 + \frac{2(n-1)}{n-2} \sqrt{\frac{\varepsilon_{z_1}}{\phi_{ac}}} + \frac{2n(n-1)}{(n-2)^2} \frac{\varepsilon_{z_1}}{\phi_{ac}} \right] \tag{21}$$

where R_i is taken as positive from left to right, starting from center O. It may be seen from Eq. (21) that real image can be obtained only when $n > 2$. If $\varepsilon_{z_1} = 0$, we can get the approximate expression for the image position

$$R_i \cong \frac{-R_c}{n-2} \tag{22}$$

derived by P. Schagen, et al[7].

For the longitudinal "paraxial" chromatic aberration Δz, we obtain

$$\Delta z = \Delta z_1 + \Delta z_2 \tag{23}$$

where

$$\Delta z_1 = -\frac{2M^2}{E_c} \sqrt{\phi_{ac}} (\sqrt{\varepsilon_z} - \sqrt{\varepsilon_{z_1}}) \tag{23a}$$

$$\Delta z_2 = -\frac{2M^2}{E_c} \left(\frac{n}{n-2} (\varepsilon_z - \varepsilon_{z_1}) - \frac{4(n-1)}{n-2} \sqrt{\varepsilon_{z_1}} (\sqrt{\varepsilon_z} - \sqrt{\varepsilon_{z_1}}) \right) \tag{23b}$$

and for the lateral "paraxial" chromatic aberration Δr, we get
$$\Delta r = \Delta r_2 + \Delta r_3 \tag{24}$$
where
$$\Delta r_2 = \frac{2M}{E_c}\sqrt{\varepsilon_r}(\sqrt{\varepsilon_z} - \sqrt{\varepsilon_{z_1}}) \tag{24a}$$

$$\Delta r_3 = -\frac{2M}{E_c\sqrt{\phi_{ac}}}\sqrt{\varepsilon_r}(\varepsilon_z - \varepsilon_{z_1}) \tag{24b}$$

where M is the linear magnification, $M = R_i/R_c$.

If we use the solution for the real trajectories $r(z)$ under the initial condition (v_0, α_0), we have

$$r(z) = \frac{2(n-1)\sqrt{\varepsilon_r}}{\sqrt{\phi_{ac}}\left(1 - 4(n-1)^2\frac{\varepsilon_r\varepsilon_z}{\phi_{ac}^2}\right)}\left\{(z+R_c)\sqrt{\frac{\varepsilon_z}{\phi_{ac}}} - \frac{2z(n-1)\sqrt{\varepsilon_z\varepsilon_r}}{\phi_{ac}^{3/2}} - \right.$$
$$\left.(z+R_c)\left[\frac{-z}{(n-1)(z+R_c)} + \frac{(z+R_c)^2\varepsilon_z + z^2\varepsilon_r}{(z+R_c)^2\phi_{ac}}\right]^{1/2}\right\} \tag{25}$$

The following "geometric" spherical aberrations δz and δr, which are the axial deviation and the transverse deviation in the image plane of the real trajectory from the "paraxial" trajectory with the same initial conditions, respectively, should be added to Eq. (23) and Eq. (24) respectively[8].

$$\delta z = \frac{2M^2}{E_c}(n-1)\varepsilon_r \tag{26}$$

$$\delta r = \frac{2M}{E_c\sqrt{\phi_{ac}}}(n-1)\varepsilon_r^{3/2} \tag{27}$$

It is clear that δz, Δz_2 and δr, Δr_2 are of the same order of magnitude respectively.

In Table 1, the calculated results for the "paraxial" trajectory and the real trajectory in two electrostatic focusing systems ($R_c = -25$ mm, $R_a = -5$ mm, $\phi_{ac} = 10,000$ V and $R_c = 5$ mm, $R_a = 25$ mm, $\phi_{ac} = 10,000$ V) with the initial conditions ($\alpha_0 = 90°$, $\varepsilon_0 = 1$ V) are given. Numerical calculation shows that the solution for the "paraxial" trajectories [Eq. (20)] and solution for the real trajectories [Eq. (25)] are very close to each other.

Table 1 Solutions for the "paraxial" and real trajectories in two concentric spherical electrostatic focusing systems

Tube parameters	Distance from the photocathode z/mm	Potential along axis $\phi(z)$/V	"Paraxial" trajectory		Real trajectory	
			$r(z)$/mm	$r'(z)$	$r(z)$/mm	$r'(z)$
$R_c = -25$ mm	4	476.170,5	0.366,605,6	0.037,097,0	0.366,620,0	0.037,133,7
$R_a = -5$ mm	8	1,176.470,5	0.466,475,2	0.015,434,8	0.466,520,0	0.015,470,6
$\phi_{ac} = 10,000$ V	12	2,307.692,3	0.499,598,4	0.001,601,3	0.499,692,1	0.001,620,5
$\varepsilon_0 = 1$ V	16	4,444.444,4	0.48	-0.011,666,6	0.480,120,6	-0.011,682,7

续表

Tube parameters	Distance From the photocathode z/mm	Potential along axis $\phi(z)$/V	"Paraxial" trajectory		Real trajectory	
			$r(z)$/mm	$r'(z)$	$r(z)$/mm	$r'(z)$
$\alpha_0 = 90°$	20	10,000.000,0	0.4	−0.03	0.400,319,8	−0.030,088,6
$R_c = 5$ mm	4	5,555.555,5	0.107,331,2	0.019,379,2	0.107,333,2	0.019,377,1
$R_a = 25$ mm	8	7,692.307,6	0.182,427,2	0.018,418,1	0.182,432,6	0.018,410,7
$\phi_{ac} = 10,000$ V	12	8,823.529,4	0.255,499,2	0.018,160,4	0.255,506,7	0.018,152,9
$\varepsilon_0 = 1$ V	16	9,523.809,5	0.327,900,8	0.018,054,0	0.327,912,4	0.018,043,7
$\alpha_0 = 90°$	20	10,000.000,0	0.4	0.018,000,0	0.400,012,8	0.017,981,0

5 The proximity focusing system

For the concentric spherical electrostatic focusing system, we assume that both R_a and R_c tend to infinity, i.e., $n = 1$. Eq. (20) then reduces to the solution for the proximity focusing system [Fig. 1 (a)]. It is obvious that the electron trajectory is a parabola and the electrons emitted from the photocathode with ε_{z_1} form a virtual image behind the cathode. Let l_i be the distance from the photocathode to the virtual image point, we have

$$l_i = (R_i - R_c)_{R_c \to \infty, R_a \to \infty} \tag{28}$$

where l_i is taken as negative from right to left, starting from the photocathode.

Substituting Eq. (21) into Eq. (28), and using $R_a - R_c = l$, $n = R_c/R_a = (1 - l/R_a)_{R_a \to \infty}$, we obtain

$$l_i = -l \left(\sqrt{1 + \frac{\varepsilon_{z_1}}{\phi_{ac}}} - \sqrt{\frac{\varepsilon_{z_1}}{\phi_{ac}}} \right)^2 \tag{29}$$

The chromatic aberration formulae, Eq. (23) and Eq. (24) of the concentric spherical electrostatic focusing system are entirely suitable to the proximity focusing system with $n = 1$. Under this condition, δz in Eq. (26) and δr in Eq. (27) will both equal zero. This means that the solution for real trajectories in this system is in accordance with the solution for "paraxial" trajectories.

6 The tri-electrode focusing systems with grid

6.1 The tri-electrode concentric spherical electrostatic focusing system with spherical grid

If we insert a spherical grid (mesh) with radius of curvature R, in front of the photocathode of the bi-electrode concentric spherical electrostatic focusing system, it becomes a tri-electrode concentric spherical electrostatic focusing system. Based on the above discussions, the distance R_i from the center of curvature to the image positions for electrons with ε_{z_1} can be written as

$$R_i = R_c M_0 \left\{ 1 - 2r(y-1)M_0 \sqrt{\frac{\varepsilon_{z_1}}{\phi_{sc}}} + [M_0 C_0 + 4M_0^2 r^2 (y-1)^2]\frac{\varepsilon_{z_1}}{\phi_{sa}} \right\} \quad (30)$$

where $y = R_c/R_s$; $r^2 = \phi_{ac}/\phi_{sc}$; ϕ_{sc} is the potential of the grid relative to the photocathode; M_0 is the linear magnification of the zeroth order, derived by S. R. Linden and P. A. Snell[9] as

$$M_0 = \frac{r+1}{r(n-2) + 2r^2(y-1) - n} \quad (31)$$

and C_0 is parameter dependent on the structure and applied potentials given by

$$C_0 = r(y-1) - \frac{n-y}{r+1} + \frac{(n-1) + r(y-1)}{r(r+1)} \quad (32)$$

From Eq. (30) we can study the focusing characteristics and the aberrations. For example, from Eq. (31) we can derive the following condition, in which a real image will be formed:

$$n > y > 1 + \frac{n - r(n-2)}{2r^2} \quad (33)$$

In addition, we can prove that the first order "paraxial" spherochromatic aberrations Δ_{z_1} [Eq. 14 (a)] and Δ_{r_1} [Eq. 16 (a)] are also valid in the tri-electrode electrostatic focusing system.

Finally, it should be pointed out that we can follow P. Schagen, et al[7] discuss the case of electrostatic focusing systems in which the anode has an aperture.

6.2 The tri-electrode electromagnetic focusing system using parallel, homogeneous magnetic and electric fields with a plane grid

For the tri-electrode electromagnetic focusing system using parallel, homogeneous magnetic and electric fields with a plane grid (mesh), as mentioned above, good focus may be obtained at the target if the angle of rotation ψ is an integral multiple of π, i.e., $\psi(l_{cs}) + \psi(l_{sa}) = i\pi$. Thus, we have the focusing condition for this system. The magnetic induction B can be written as:

$$B = i\pi \sqrt{\frac{2m_0}{e}} \left[\frac{l_{cs}}{\sqrt{\phi_{sc}}} \left(\sqrt{1 + \frac{\varepsilon_{z_1}}{\phi_{sc}}} - \sqrt{\frac{\varepsilon_{z_1}}{\phi_{sc}}} \right) + \frac{l_{sa}}{\sqrt{\phi_{as}}} \left(\sqrt{1 + \frac{\phi_{sc} + \varepsilon_{z_1}}{\phi_{as}}} - \sqrt{\frac{\phi_{sc} + \varepsilon_{z_1}}{\phi_{as}}} \right) \right]^{-1} \quad (34)$$

where ϕ_{sc} is the potential of the mesh relative to the photocathode, ϕ_{as} is the potential of the anode (i.e., target) relative to the mesh, l_{cs} is the distance from the photocathode to the mesh and l_{sa} is the distance from the mesh to the target. Let $\varepsilon_{z_1} = 0$, transform the unit of B from Wb/m^{-2} to Gauss, and leave the other values in SI units, then we get

$$B = \frac{0.106i}{\frac{l_{cs}}{\sqrt{\phi_{sc}}} + \frac{l_{sa}}{\sqrt{\phi_{as}}} \left(\sqrt{1 + \frac{\phi_{sc}}{\phi_{as}}} - \sqrt{\frac{\phi_{sc}}{\phi_{as}}} \right)} \quad (35)$$

as was derived by S. R. Linden and P. A. Snell[9].

It can be proved that the first order and the second order "paraxial" chromatic aberrations, Eq. (14a) and Eq. (16a), are also valid for this system.

Conclusions

The analytic solution of the "paraxial" trajectories and the expressions of the sphero-chromatic aberrations for four typical electrostatic and electromagnetic focusing systems are derived. The form of the second order "paraxial" lateral chromatic aberrations for electrostatic and electromagnetic focusing imaging systems are proved to be the same.

The results of this paper can be used to design electromagnetic focusing image tubes and to calculate the initial electron trajectories in image tubes.

Acknowledgements

The author wishes to thank the director of the North Industries Corporation for permission to publish this paper. Thanks are also given to the Optical Engineering Department of Beijing Institute of Technology for the support and cooperation.

References

[1] ZACHAROV B. A demagnifying image tube for nuclear physics applications [J]. Advances in Electronics and Electron Physics, 1962, 16: 99 – 104.

[2] BEURLE R. L, WREATHALL W M. Aberration in magnetic focus systems [J]. Advances in Electronics and Electron Physics, 1962,16:333 – 340.

[3] ГРИНБЕРГ Г А. Основы обшей теории фокусируюшего действия электростатических и магнитных полеи II. пространственные электростатические поля [J]. Изд, А Н СССР, 1942,37(9):295 – 303.

[4] STURROCK P A. The imaging properties of electron beams in arbitrary static electromagnetic fields. [J]. Philosophical Transactions, 1952, A245:155 – 187.

[5] HARTLEY K F. On the electron optics of magnetically focused image tubes [J]. J. Phys. D, 1974,7:1612 – 1620.

[6] PICAT J P. Optical Property of electromagnetically focused electronic cameras [J]. Astron. & Astrophys,1971,11:257 – 267.

[7] SCHAGEN P, BRUINING H, FRANCKEN J C. A simple electrostatic electron-optical system with only one voltage. [J]. Philips Research Report,1952,7(2):119 – 130.

[8] ЧЖОУ Л В. Об аберрации в электронной оптической системе с шаровой и осевой симметрией [C]. Тезисы докладов XX1 Научно-Технической Конференции, Ленинград,1966:137 – 138.

[9] LINDEN S R, SNELL P A. Shutter image converter tubes [J]. Proceedings of the IRE, 1957,45:513 – 517.

2.7 电磁聚焦同心球系统的精确解
On Exact Solutions of a Concentric Spherical Electromagnetic Focusing System

摘要：电磁聚焦同心球系统的电子轨迹近轴解曾被作者详细研究过，但迄今为止尚未求得此系统的电子轨迹的精确解。本文由球坐标系下电子光学广义拉格朗日函数出发，导出了电磁聚焦同心球系统的电子运动方程，求得了电子轨迹的精确解。这对于进一步研究系统的电子光学像差具有实际意义。

Abstract: The "paraxial" solution of electron trajectories for the concentric spherical electromagnetic focusing systems was derived by the author, but up to now the exact solutions of electron trajectories have not been deduced. Based on the general Lagrange function of electron optics in the spherical coordinate system, we have derived the electron motion's equations and then the exact solutions of trajectory for the concentric spherical electromagnetic focusing system. It has the practical meaning for further studies of the electron optical aberrations of imaging systems.

引言

电磁聚焦同心球系统近年来逐渐引起人们广泛注意。周立伟[1]于1978年首先用解析方法研究了这一普遍形式的系统，求得了该系统的电子轨迹的近轴解，推导了系统的聚焦条件与电子光学色球差，由此考察了其他几种特殊情况。1979年年底，СМИРНОВ N A 等人[2]对此系统进行研究，求得了近轴电子轨迹解，并由此推导了电磁复合聚焦阴极透镜的像差表达式及色差系数。

正如两电极静电聚焦同心球系统的精确解对于确定静电聚焦系统的电子光学性质与像差具有实际意义一样[3]，以电磁聚焦同心球系统作为物理模型，求其精确解也是很令人鼓舞的，因为借助于所求得的精确解可以检验电磁复合聚焦系统成像参量与像差表达式以及评价所选择的数值方法的精确性。

但是，应该指出，以往的文献都并没有给出电磁聚焦同心球系统的电子轨迹精确解的表示式。本文的任务就是试图求得轨迹的精确解，以便确定系统的几何横向像差与近轴横向像差，并利用此理论结果来考察目前所得到的电磁复合聚焦阴极透镜的像差两种典型表达式的精确性。

周立伟，倪国强. 北京理工大学. 电子管技术 (Technology of Electron Tubes), No. 2, 1985, 50–53.

1 同心球系统的轨迹方程

在以同心球的球心为原点的球坐标系 (ρ, θ, ψ) 中，电场强度 E 和磁感应强度 B 的分量 $(E_\rho, E_\theta, E_\psi)$ 和 $(B_\rho, B_\theta, B_\psi)$ 分别为 $\left(E_c \dfrac{R_c^2}{\rho^2}, 0, 0\right)$ 和 $\left(B_0 \dfrac{R_c^2}{\rho^2}, 0, 0\right)$，其中 E_c、B_0 分别为阴极面处的电场强度与磁感应强度，R_c 为阴极面的曲率半径（见图1）。由此可导得电位 $\phi_{\rho c} = \phi(\rho)$ 的表达式

$$\phi_{\rho c} = E_c R_c \left(\frac{R_c}{\rho} - 1\right) \tag{1}$$

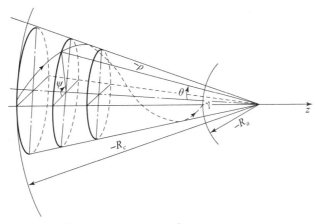

图1 (ρ, θ, ψ) 球坐标系及电子轨迹

及磁矢位矢量 A 的分量表达式

$$A_\psi = \frac{B_0 R_c^2}{\rho} \frac{1-\cos\theta}{\sin\theta}, \quad A_\theta = 0, \quad A_\rho = 0 \tag{2}$$

按电子光学的广义拉格朗日函数及相应的拉格朗日-欧拉方程出发，可以求得电子运动方程如下：

$$m_0(\ddot{\rho} - \rho\dot{\theta}^2 - \rho\sin^2\theta\,\dot{\psi}^2) = -\frac{eE_c R_c^2}{\rho^2} \tag{3a}$$

$$m_0(2\rho\dot{\rho}\dot{\theta} + \rho^2\ddot{\theta} - \rho^2\sin\theta\cos\theta\,\dot{\psi}^2) = -eB_0 R_c^2 \sin\theta\,\dot{\psi} \tag{3b}$$

$$m_0 \frac{\mathrm{d}}{\mathrm{d}t}(\rho^2 \sin^2\theta\,\dot{\psi}) = eB_0 R_c^2 \sin\theta\,\dot{\theta} \tag{3c}$$

由此出发，可以证明

$$-m_0(\rho^2 \sin\theta\cos\theta\cos\psi\,\dot{\psi} + \rho^2\dot{\theta}\sin\psi) + eB_0 R_c^2 \sin\theta\cos\psi = C_1 \tag{4a}$$

$$m_0(-\rho^2 \sin\theta\cos\theta\sin\psi\,\dot{\psi} + \rho^2\dot{\theta}\cos\psi) + eB_0 R_c^2 \sin\theta\sin\psi = C_2 \tag{4b}$$

$$m_0 \rho^2 \sin^2\theta\,\dot{\psi} + eB_0 R_c^2 \cos\theta = C_3 \tag{4c}$$

其中 C_1、C_2、C_3 是由场参量及电子逸出初始条件所确定的常数，它们乃是下列常矢量

$$N = m_0(\rho \times \dot{\rho}) + eB_0 R_c^2 \frac{\rho}{\rho} \tag{5}$$

在 (x, y, z) 直角坐标系中的投影，即

$$N_x = m_0[y\dot{z} - (z+R_c)\dot{y}] + eB_0R_c^2\frac{x}{\rho} = C_1 \tag{6a}$$

$$N_y = m_0[(z+R_c)\dot{x} - x\dot{z}] + eB_0R_c^2\frac{y}{\rho} = C_2 \tag{6b}$$

$$N_z = m_0(x\dot{y} - \dot{x}y) + eB_0R_c^2\frac{z+R_c}{\rho} = C_3 \tag{6c}$$

由此可见，常矢量 N 乃是动量矩矢量 $m_0(\rho \times \dot{\rho})$ 与矢量 $eB_0R_c^2\dfrac{\vec{\rho}}{\rho}$ 的几何合成。当逸出电子的初始条件给定后，N 在大小、方向上均不变。故如果已知轨迹在某点的位置和电子的速度，则不难确定矢量 N，而 N 与电子位置矢量 $\vec{\rho}$ 之间的夹角 γ 恒定不变，并由下式确定：

$$\cos\gamma = \frac{eB_0R_c^2}{N} \tag{7}$$

这说明电子将沿圆锥体表面运动，此圆锥体之轴与矢量 N 重合，而其顶点位于系统的球心。

应该指出，N 虽是一个常矢，但其指向除了与所考察的电子逸出点有关外，还取决于电子逸出的初速度 v_0、初角度 α_0 以及初方位角 β_0。当电子从直角坐标系的原点（$x_0 = 0$，$y_0 = 0$，$z_0 = 0$）以 (v_0, α_0, β_0) 逸出时，它与曲率半径为 R_c 的球面的交点的坐标为 $\bar{x}_0 = -R_c\sin\gamma\sin\beta_0$，$\bar{y}_0 = R_c\sin\gamma\cos\beta_0$，（$\gamma$ 取正值）。

当我们将球面坐标系的轴 $\theta = 0$ 与恒定矢量 N 的方向重合，在这样坐标系安排下，运动将按锥体表面 $\theta = \gamma$ 进行，此时可以证明电子的回转角速度由下式确定：

$$\dot{\psi} = \frac{N}{m_0\rho^2} \tag{8}$$

由上式以及 $\theta = $ 常数可见，电子矢径在单位时间内扫过的面积 $\dfrac{1}{2}\rho^2\sin^2\theta\dot{\psi} = $ 常数。这正是角动量守恒：

$$\frac{m_0}{2}(\dot{\rho}^2 + \rho^2\dot{\theta}^2 + \rho^2\sin^2\theta\,\dot{\psi}^2) - eE_cR_c\left(\frac{R_c}{\rho} - 1\right) = \frac{1}{2}m_0v_0^2 \tag{9}$$

由 $\theta = \gamma$，$\dot{\theta} = 0$ 及 $\mathrm{d}t = \dfrac{m_0\rho^2}{N}\mathrm{d}\psi$，对上式作变换，并积分，得

$$\rho = \frac{2\varepsilon_r^*R_c}{1 + p\sin[(\psi - \psi_c)\sin\gamma]} \tag{10}$$

这就是电磁聚焦同心球系统电子轨迹精确解的表达式。其中

$$\psi_c = \frac{1}{\sin\gamma}\arcsin\frac{1 - 2\varepsilon_r^*}{p} \tag{11a}$$

$$p = \sqrt{1 - 4\varepsilon_r^*(1 - \varepsilon_0^*)} \tag{11b}$$

$$\varepsilon_r^* = \frac{\varepsilon_r}{E_cR_c}, \quad \varepsilon_0^* = \frac{\varepsilon_0}{E_cR_c} \tag{11c}$$

$$\psi|_{t=0} = 0 \tag{11d}$$

这里 ε_0、ε_r 分别是逸出电子的初电位及径向初电位。

十分容易验证,对静电场,$B_0 = 0$,则 $\sin\gamma = 1$,上述解就是静电场两电极同心球系统电子轨迹的精确解[3]。

2 电子轨迹的精确解

对上述球坐标系进行坐标变换,以上述锥体上通过电子逸出的母线为 z 轴,并以该点为 z 轴的起始点,采用柱坐标系 (r, ψ^*, z),以电子逸出速度所在的子午面为电子矢径 r 的方位角 ψ^* 的起始面(见图2),这时电子在该坐标系中的逸出点坐标为 $(r_0 = 0, z_0 = 0)$,逸出初始条件为 $(v_0, \alpha_0, \beta_0 = 0)$、$(\psi_0^* = \beta_0 = 0)$,则可得到在该圆柱坐标系中电子轨迹以球坐标系中的经度角 ψ 为参变量的精确解表达式:

$$z = -R_c \left\{ 1 - \frac{2\varepsilon_r^* [1 - (1-\cos\psi)\sin^2\gamma]}{1 + p\sin[(\psi-\psi_c)\sin\gamma]} \right\} \tag{12a}$$

$$r = -(R_c + z)\sin\gamma \frac{\sqrt{\sin^2\psi + (1-\cos\psi)^2 \cos^2\gamma}}{1 - (1-\cos\psi)\sin^2\gamma} \tag{12b}$$

$$\tan(\psi^* + m\pi) = \tan\frac{\psi}{2}\cos\gamma \tag{12c}$$

式中,$m = \left[\dfrac{\psi^*}{\pi}\right]$,即 m 取 ψ^*/π 的整数部分。

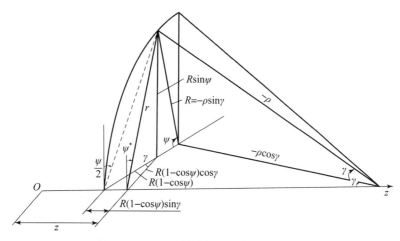

图2 (r, ψ^*, z) 柱坐标系及电子运动位置

由此出发,就可以进一步深入讨论实际轨迹精确解与电子光学成像性质、成像参量及电子光学像差,并进而将其与轨迹的近轴解作比较,研究和检验阴极透镜的电子光学像差理论。

参 考 文 献

[1] 周立伟. 电磁聚焦同心球系统的电子光学[J]. 兵工学报,1979(1):66 – 81.

[2] СМИРНОВ N A, КУЛИКОВ Ю В, МОНАСТЕРСКИЙ M A. Некоторые вопросы электронной оптики катодных линз с комбинированными полями, обладающими сферической симметией[J]. Ж. Т. Ф.,1979,49(12):2590 – 2595.

[3] 周立伟. 两电极同心球系统的电子光学[J]. 工程光学,1978(1):71 – 87.

三、倾斜型电磁聚焦系统的成像电子光学

Imaging Electron Optics of Oblique Electromagnetic Focusing Systems

3.1 Electron Optics of Oblique Electromagnetic Focusing Systems

倾斜型电磁聚焦系统的电子光学

Abstract: Equations for the electron motion referred to movable coordinate system in an oblique electromagnetic focusing system are given, the aberration expressions of the system and the root mean square radius characterizing the resolution of system are derived. Based upon the results obtained, a general purpose computer program for designing oblique electromagnetic focusing systems is suggested. Constructional dimensions and electromagnetic parameters of the system, the electron optical transfer function and the root mean square radius are calculated. The result shows that the new method of calculating electron optical transfer function and root mean square radius for oblique image tubes will save about 3/4 – 5/6 of the total time of computation, retaining the same degree of accuracy as compared with the traditional methods.

摘要：本文给出了倾斜型电磁聚焦系统在随动坐标系下的电子运动方程、导出了系统的像差表达式，探讨了表征该系统的鉴别率特性的均方根半径，编制了设计倾斜型电磁聚焦系统的通用程序，计算了电子光学外形尺寸，电子光学传递函数与均方根半径，检验并比较了所采用的横向像差表达式的精度。结果表明，利用像差方程计算倾斜型像管的传递函数较通常由运动方程计算落点的传统法省3/4～5/6的时间并具有足够的精度。

Introduction

The oblique electromagnetic focusing system is the form among other varieties of focusing systems that has seen rapid developments in recent years. It can make full use of opaque photocathodes, the ultraviolet photocathode and Ⅲ – Ⅴ compound photocathodes in particular. The electron optical system of the oblique image tubes is in fact made up of an uniform cross-field, that is, the direction of uniform magnetic field B deviates a certain angle θ with respect to the direction of the uniform electrostatic field E which is parallel to the optical axis. It differs from the normal image tube in that the opaque photocathode is located at the back of the tube, while the screen or target is at the front of it. They are thus mutually exchanged in their position, with an angle θ of deviation between them. When the incident image is projected on the opaque photocathode through an optical system, the emitted photoelectrons, under the electric field, are accelerated and rotated along the parabola deviating from the direction of magnetic field. They are

---

Zhou Liwei[a], Fang Erlun[b]. a) Beijing Institute of Technology, b) Xi'an Research Institute of Modern Chemistry. Journal of Beijing Institute of Technology（北京理工大学学报）, English Edition V. 10 S1, 1990, 19 – 32.

then focused on the screen or target on the output side. However, the screen (target) and input window are not symmetrically located on both sides of the tube axis, but deviates with respect to each other through an angle θ, as shown in Fig. 1.

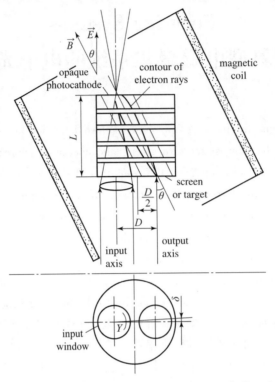

Fig. 1 Image tube with oblique electromagnetic focusing

For this system, J. P. Picat[1] first put forward the new problem and proved the possibility of forming a real image. C. B. Johnson et al[2-4], deduced the electron motion equation, explicated the imaging of electron trajectory at zero initial velocity of emergent electrons, and discussed constructional design of the system. J. L. Lowrance[5,6] investigated the point spread function (PSF) and the modulation transfer function (MTF) of the system directly from the equations of motion of the electrons, and made use of CCD as a receiving target of the system. But none of the above researchers analyzed the aberrations of this system.

Starting from the electron motion equation under uniform crossed electrostatic and magnetic fields, we here attempt to introduce a movable coordinate system and deduce another expression of electron trajectory and aberrations for such a system. Based on the results obtained, a general purpose computer program for designing oblique electromagnetic focusing systems is suggested, whereby the constructional dimensions and electromagnetic parameters of the system, the electron optical transfer function (ETF) and the root mean square (RMS) radius are calculated.

1 Electron trajectories in movable coordinate system

Taking B and E to be constant and suppose they intersect at an angle of θ. Using Cartesian coordinate system (x, y, z), we take the direction of the magnetic field as the z axis and the plane containing vectors $\boldsymbol{B}$ and $\boldsymbol{E}$ as the xz plane. Then, introducing a movable coordinate system $(\bar{x}, \bar{y}, \bar{z})$, we have

$$\begin{aligned}\bar{x} &= x, \\ \bar{y} &= y - \frac{V_{ac}}{BL}\sin\theta \cdot t, \\ \bar{z} &= z - \dot{z}_0 t - \frac{e}{m_0}\frac{V_{ac}}{2L}\cos\theta \cdot t^2\end{aligned} \tag{1}$$

where e/m_0 is the charge to mass ratio of the electron, L and V_{ac} are the distance from photocathode to the target and their applied potential respectively.

The origin of the movable coordinates ($\bar{x}=0$, $\bar{y}=0$, $\bar{z}=0$) moves along the parabola in the yz plane. From Eq. (1), we can get

$$y = \frac{m_0}{e}\frac{\tan\theta}{B}\left[\left(\frac{2e}{m_0}\frac{V_{ac}}{L}\cos\theta \cdot z + \dot{z}_0^2\right)^{1/2} - \dot{z}_0\right] \tag{2}$$

Equations for electron motion in a movable coordinate system will take the form

$$\begin{aligned}\bar{x} &= \rho\cos(\omega t + \varphi) - \frac{m_0}{eB}\left(\dot{y}_0 - \frac{V_{ac}\sin\theta}{BL}\right), \\ \bar{y} &= \rho\sin(\omega t + \varphi) + \frac{m_0}{eB}\dot{x}_0, \\ \bar{z} &= 0\end{aligned} \tag{3}$$

ρ and φ in Eq. (3) are determined from the initial conditions

$$\rho = \frac{m_0}{eB}\left[\dot{x}_0^2 + \left(\dot{y}_0 - \frac{V_{ac}\sin\theta}{BL}\right)^2\right]^{1/2}, \quad \tan\varphi = \frac{-\dot{x}_0}{\dot{y}_0 - \frac{V_{ac}\sin\theta}{BL}} \tag{4}$$

It is obvious that Eq. (3) describes a circle with radius ρ located in the $\bar{x}\bar{y}$ plane, and the emerging electrons move along a cylindrical helix having its axis moving along a parabola in the yz plane.

Now, expressions for the electron trajectories of the system can be derived in term of $r(\bar{z}=0)$ and $\psi(\bar{z}=0)$. Let r be the distance of the electron trajectory in the $\bar{x}\bar{y}$ plane from $\bar{x}=0$, $\bar{y}=0$, ψ be the angle between the line segment r and the tangential of the described circle at the initial point of the movable coordinate system, we then have

$$r = 2\rho\sin\psi \tag{5}$$

Using Eq. (3) and Eq. (4), the electron ray equations in terms of r and ψ becomes

$$\begin{aligned}r(\bar{z}=0) &= 2\frac{m_0}{eB}\left[\dot{x}_0^2 + \left(\dot{y}_0 - \frac{V_{ac}\sin\theta}{BL}\right)^2\right]^{1/2}\sin\psi, \\ \psi(\bar{z}=0) &= \frac{1}{2}\frac{BL}{V_{ac}\cos\theta}\left[\left(\frac{2e}{m_0}\frac{V_{ac}\cos\theta}{L}z + \dot{z}_0^2\right)^{1/2} - \dot{z}_0\right]\end{aligned} \tag{6}$$

The electron trajectory moving in a movable coordinate system can be expressed by

$$\bar{x} = r\cos\left(\psi + \varphi - \frac{\pi}{2}\right),$$
$$\bar{y} = r\sin\left(\psi + \varphi - \frac{\pi}{2}\right)$$
(7)

$\dot{x}_0$, $\dot{y}_0$, $\dot{z}_0$ have the following relations with the initial velocity v_0 and initial angle ($\alpha_0 = 0 - 90°$), ($\beta_0 = 0 - 360°$) of the emerging electrons from the cathode surface:

$$\dot{x}_0 = v_0\cos\alpha_0\sin\theta + v_0\sin\alpha_0\cos\beta_0\cos\theta,$$
$$\dot{y}_0 = v_0\sin\alpha_0\sin\beta_0,$$
$$\dot{z}_0 = v_0\cos\alpha_0\cos\theta - v_0\sin\alpha_0\cos\beta_0\sin\theta$$
(8)

where θ is the oblique angle between the axes x and X, which belongs to another Cartesian coordinate system X, Y, Z, and α_0, β_0 are the angles of emission and the azimuth angle of the electron respectively (Fig. 2).

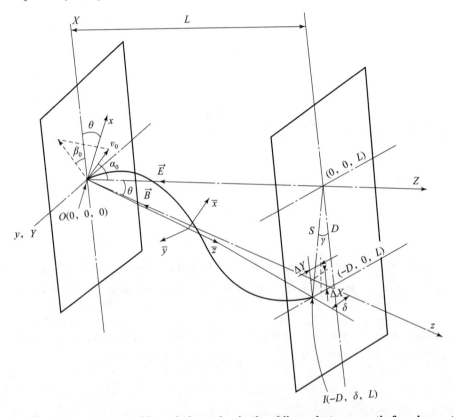

Fig. 2 Electron trajectory and lateral aberration in the oblique electromagnetic focusing system

2 Determination of constructional dimensions and electromagnetic parameters

In the following discussions, we assume that the ideal image position corresponds to the situation when electrons have their initial velocity $\dot{z}_1$ along z axis. Then, the expressions for constructional dimensions and electromagnetic parameters in this system can be obtained as

$$L = N\pi \left(\frac{2m_0}{e}\right)^{1/2} \frac{\sqrt{V_{ac}}}{B} \left[\left(1 + \frac{m_0}{2e}\frac{\dot{z}_1^2}{V_{ac}}\right)^{1/2} + \left(\frac{m_0}{2e}\right)^{1/2} \frac{\dot{z}_1}{\sqrt{V_{ac}}}\right] \cos\theta \tag{9}$$

$$\frac{V_{ac}}{B^2} = L^2 \left[\left(1 + \frac{m_0}{2e}\frac{\dot{z}_1^2}{V_{ac}}\right)^{1/2} - \left(\frac{m_0}{2e}\right)^{1/2} \frac{\dot{z}_1}{\sqrt{V_{ac}}}\right]^2 \left(2\frac{m_0}{e}\pi^2 N^2 \cos^2\theta\right)^{-1} \tag{10}$$

$$\frac{\delta}{L} = \frac{\tan\theta}{\pi N \cos\theta}\left[\left(1 + \frac{m_0}{2e}\frac{\dot{z}_1^2}{V_{ac}}\right)^{1/2} - \left(\frac{m_0}{2e}\right)^{1/2}\frac{\dot{z}_1}{\sqrt{V_{ac}}}\right]^2 \tag{11}$$

$$\frac{S}{L} = \tan\theta \left\{1 + \frac{1}{\pi(N\cos\theta)^2}\left[\left(1 + \frac{m_0}{2e}\frac{\dot{z}_1^2}{V_{ac}}\right)^{1/2} - \left(\frac{m_0}{2e}\right)^{1/2}\frac{\dot{z}_1}{\sqrt{V_{ac}}}\right]^4\right\}^{1/2} \tag{12}$$

$$\frac{D}{L} = \tan\theta \tag{13}$$

$$\gamma = \arctan\left\{1 + \frac{1}{\pi N \cos\theta}\left[\left(1 + \frac{m_0}{2e}\frac{\dot{z}_1^2}{V_{ac}}\right)^{1/2} - \left(\frac{m_0}{2e}\right)^{1/2}\frac{\dot{z}_1}{\sqrt{V_{ac}}}\right]^2\right\} \tag{14}$$

$$\theta = \arccos\left[\frac{1}{2\pi^2 N^2 (1 + S^2/L^2)}\left\{\pi^2 N^2 - \left[\left(1 + \frac{m_0}{2e}\frac{\dot{z}_1^2}{V_{ac}}\right)^{1/2} - \left(\frac{m_0}{2e}\right)^{1/2}\frac{\dot{z}_1}{\sqrt{V_{ac}}}\right]^4 + \left\{\left[\left(1 + \frac{m_0}{2e}\frac{\dot{z}_1^2}{V_{ac}}\right)^{1/2} - \left(\frac{m_0}{2e}\right)^{1/2}\frac{\dot{z}_1}{\sqrt{V_{ac}}}\right]^3 + 2\pi^2 N^2\left[\left(1 + \frac{m_0}{2e}\frac{\dot{z}_1^2}{V_{ac}}\right)^{1/2} - \left(\frac{m_0}{2e}\right)^{1/2}\frac{\dot{z}_1}{\sqrt{V_{ac}}}\right]^4 + 4\pi^2 N^2 \left(\frac{S}{L}\right)^2\left[\left(1 + \frac{m_0}{2e}\frac{\dot{z}_1^2}{V_{ac}}\right)^{1/2} - \left(\frac{m_0}{2e}\right)^{1/2}\frac{\dot{z}_1}{\sqrt{V_{ac}}}\right]^4 + \pi^4 N^4\right\}^{1/2}\right\}\right]^{1/2} \tag{15}$$

Designations of letters used for geometrical and electromagnetic parameters will be clear from Fig. 2. In the above equations, if $\dot{z}_1 = 0$, then we may have the expressions given by C. B. Johnson[4]. In general, $\dot{z}_1 = 0$ may be taken into account in determining the initial geometrical and electromagnetic parameters of the system.

3 The lateral aberration

The lateral aberration of the emerging electrons having initial velocity $(\dot{x}_0, \dot{y}_0, \dot{z}_0)$ with respect to the ideal image point $(X = -D, Y = \delta, Z = L)$ corresponding to $\dot{z}_1$ can be derived with higher accuracy

$$\Delta X = \frac{m_0}{e}\frac{L}{V_{ac}\cos^2\theta}(\dot{z}_1 - \dot{z}_0)\dot{x}_0 + \frac{m_0}{2e}\frac{L\tan\theta}{V_{ac}\cos^2\theta}(\dot{z}_1 - \dot{z}_0)^2 -$$

$$\left(\frac{m_0}{e}\right)^{3/2}\frac{N\pi L}{\sqrt{2}V_{ac}^{3/2}}\left\{(\dot{z}_1 - \dot{z}_0)^2\dot{y}_0 + \frac{(\dot{z}_1 - \dot{z}_0)}{\pi N}\times(\dot{x}_0^2\tan\theta + \dot{x}_0\dot{z}_0\frac{\cos 2\theta}{\cos\theta}) + \frac{1}{2\pi N\cos^2\theta}\dot{x}_0(\dot{z}_1 - \dot{z}_0)^2\right\} +$$

$$\left(\frac{m_0}{e}\right)^{3/2}\frac{L}{\sqrt{2}V_{ac}^{3/2}}\left\{(\dot{z}_1 - \dot{z}_0)\dot{z}_0^2\tan\theta + \frac{1}{2\cos^2\theta}(\dot{z}_1 - \dot{z}_0)^2\dot{z}_0\tan\theta\right\}$$

(16a)

$$\Delta Y = \frac{m_0}{e}\frac{L}{V_{ac}\cos\theta}(\dot{z}_1-\dot{z}_0)\dot{y}_0 + \left(\frac{m_0}{e}\right)^{3/2}\frac{\pi NL}{\sqrt{2}V_{ac}^{3/2}\cos\theta}\bigl[(\dot{z}_1-\dot{z}_0)^2\dot{x}_0 - $$
$$\frac{(\dot{z}_1-\dot{z}_0)\dot{y}_0}{\pi N}(\dot{x}_0\tan\theta + \dot{z}_0) - \frac{1}{2\pi N\cos^2\theta}(\dot{z}_1-\dot{z}_0)^2\dot{y}_0\bigr] \quad (16b)$$

By analyzing the aberration formulae, it goes without saying that the first terms in Eq. (16a) and Eq. (16b) related to ε_0/V_{ac} is similar to the chromatic aberration of the uniform parallel electrostatic and magnetic focusing system. However, the second term in (16a) which is independent of $\dot{x}_0$, $\dot{y}_0$ appearing in X direction corresponds to an image displacement. It results from the asymmetry of the system. The final second order terms in Eq. (16a) and Eq. (16b) related to $(\varepsilon_0/V_{ac})^{3/2}$ are those similar to higher order aberrations.

Neglecting the second order terms in Eq. (16a) and Eq. (16b), the abridged expression of lateral aberration with sufficient accuracy for oblique electromagnetic focusing systems can be written as

$$\Delta X = \frac{m_0}{e}\frac{L}{V_{ac}\cos^2\theta}(\dot{z}_1-\dot{z}_0)\dot{x}_0 + \frac{m_0}{2e}\frac{L\tan\theta}{V_{ac}\cos^2\theta}(\dot{z}-\dot{z}_0)^2 \quad (17a)$$

$$\Delta Y = \frac{m_0}{e}\frac{L}{V_{ac}\cos\theta}(\dot{z}_1-\dot{z}_0)\dot{y}_0 \quad (17b)$$

It will be observed for a number of computations that the lateral aberration pattern of this kind of focusing system is rather complicated. On the one hand, it is related to the position of image plane corresponding to $\dot{z}_1$; on the other hand, $\dot{x}_0$, $\dot{y}_0$ relate not only to the angles, α_0, β_0 of the emerging electrons, but also to the angle θ between $\boldsymbol{E}$ and $\boldsymbol{B}$ of the system. So even taking into account the first terms in Eq. (16a) and Eq. (16b), the pattern will not be a perfect circle.

Let $\theta = 0$, we can obtain from Eq. (16a) and Eq. (16b) expressions for the lateral aberration expressions for the focusing system using uniform parallel magnetic and electric fields. Their derivation can be seen in Ref. [7].

4 The RMS (root mean square) radius

If we calculate the RMS radius $\overline{\Delta r}$ of the oblique electrostatic and magnetic focusing system with the angle of emission $\alpha_0(0-90°)$, azimuth angle $\beta_0(0-360°)$ and initial energy $\varepsilon_0(0-\varepsilon_{0\max})$, as follows:

$$\overline{\Delta r} = \sqrt{\overline{\Delta X^2} + \overline{\Delta Y^2}} \quad (18)$$

then the results obtained can be used as a measure of the resolution of the system.

For a Lambert angular distribution of electrons and a beta distributions of electron emission velocities, the two components of RMS radius can be expressed as follows:

$$\overline{\Delta X^2} = \int_0^1\int_0^{2\pi}\int_0^{\pi/2}(\Delta X)^2\frac{\sin 2\alpha_0}{2\pi}N(\xi)\,\mathrm{d}\alpha\mathrm{d}\beta\mathrm{d}\xi$$

$$\overline{\Delta Y^2} = \int_0^1 \int_0^{2\pi} \int_0^{\pi/2} (\Delta Y)^2 \frac{\sin 2\alpha_0}{2\pi} N(\xi) \mathrm{d}\alpha \mathrm{d}\beta \mathrm{d}\xi \qquad (19)$$

where
$$N(\xi) = \beta_{m,n}(\xi) = \frac{(m+n+1)!}{m!\ n!} \xi^m (1-\xi)^n \qquad (20)$$

$\xi = \dfrac{\varepsilon_0}{\varepsilon_{0m}}$, $0 \leqslant \xi \leqslant 1$, $\varepsilon_{0m} = \varepsilon_{0\max}$ is the maximum value of initial energy; m, n can take any value from 1, 2, $\cdots$, 8, depending on what is needed in practice.

Substituting Eq. (17) for ΔX, ΔY and Eq. (20) for $N(\xi)$ in Eq. (19), and integrating it, we have expressions of $\overline{\Delta X^2}$ and $\overline{\Delta Y^2}$. The RMS radius $\overline{\Delta r}$ of Eq. (18) then can be written as

$$\overline{\Delta r} = \frac{\varepsilon_{0m}}{(V_{ac}/L)\cos\theta} \left\{ \frac{(m+n+1)!}{m!\ n!} \left[\left(\frac{\varepsilon_{z_1}}{\varepsilon_{0m}}\right)^2 \frac{\tan^2\theta}{\cos^2\theta} \alpha_1 + \right.\right.$$
$$\frac{\varepsilon_{z_1}}{\varepsilon_{0m}} \left[\frac{2}{\cos^2\theta} \left(1 + \frac{3}{4}\sin^2\theta\tan^2\theta\right) \right] \alpha_2 - \left(\frac{\varepsilon_{z_1}}{\varepsilon_{0m}}\right)^{1/2} \frac{16}{15} \left(\frac{\cos\theta+1}{\cos\theta}\right) \alpha_{5/2} + \qquad (21)$$
$$\left. \left. \left(\frac{2}{3} + \frac{1}{3}\tan^2\theta\sin^2\theta + \frac{1}{8}\tan^4\theta\sin^2\theta\right)\alpha_3 \right] \right\}^{1/2}$$

where
$$\alpha_j = \sum_{i=0}^{n} (-1)^i \frac{1}{i!(n-i)!} \frac{1}{i+m+j} \quad (j=1,2,5/2,3) \qquad (22)$$

For parallel uniform electrostatic and magnetic fields, and assuming only Lambertian angular distribution of electrons, the RMS can be written as[8]

$$\overline{\Delta r} = \sqrt{2}\frac{L}{V_{ac}}\varepsilon_{0m} \left\{ \frac{\varepsilon_{z_1}}{\varepsilon_{0m}} - \frac{16}{15}\left(\frac{\varepsilon_{z_1}}{\varepsilon_{0m}}\right)^{1/2} + \frac{1}{3} \right\}^{1/2} \qquad (23)$$

5 Calculations of the constructional dimensions, electron trajectories and lateral aberrations

As an example of a study on typical oblique image tubes, let us take the initial data given in Ref. [4]. With the aid of the electron motions Eq. of (6) and Eq. (7) in a movable coordinate system, the electron trajectory in the oblique electromagnetic focusing system can be expressed clearly by a coplanar curve (a coplanar parabola in general) in the yz plane together with a circular motion, representing rotation around this coplanar curve. Fig. 3 shows the geometrical pattern for one of the electron trajectories ($\alpha_0 = 61.5°$, $\beta_0 = 0°$, $\varepsilon_0 = 4$ V), which the constructional dimensions are given below in the next page.

When the typical data given, we can investigate the accurate position for the point of fall of the electrons, the initial parameters of which are (ε_0, α_0, β_0), at a distance L from the photocathode in an image plane which is perpendicular to the axis z.

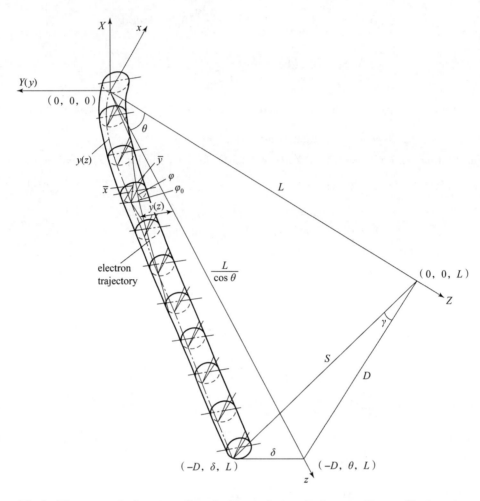

Fig. 3 The geometrical pattern of an electron trajectory in the movable coordinate system

System parameters:	Geometrical dimensions:	Parameters for the circle:
$BV = 0.157 \times 10^{-1}$ T	$\delta = 0.164,916 \times 10^{-1}$ m	$x_c = 0.262,471 \times 10^{-1}$ m
$V_{ac} = 20,000$ V	$L = 0.838,583 \times 10^{-1}$ m	$y_c = 0.429,549 \times 10^{-1}$ m
$\theta = 28.5°$	$D = 0.455,313 \times 10^{-1}$ m	$\rho = 0.265,963 \times 10^{-1}$ m
$N = 1$	$S = 0.484,259 \times 10^{-1}$ m	$\varphi = 9.294,332,3°$
$\varepsilon_0 = 4$ V	$\gamma = 19.191,052,0°$	

Using Eq. (17a) and Eq. (17b), we calculate ΔX, ΔY in two image planes, the positions of which correspond to $\dot{z}_1 = 0$, and $\dot{z}_1 = 0.25 v_0$ under the following initial conditions: $\varepsilon_0 = 4$ V, $\Delta \alpha = 15°$, $\Delta \beta = 15°$ (see Fig. 4). Thus, from Fig. 4, we can have an idea about the change of geometrical pattern. It is not difficult to calculate the lateral aberrations with higher accuracy Eq.

(16a), Eq. (16b), Eq. (17a) and Eq. (17b), and to compare them with the accurate deviation. It is estimated from computations that the relative error will be less than 0.5% [using Eq. (16a) and Eq. (16b)] and 1% [using Eq. (17a) and Eq. (17b)] respectively. Aberrations formed by bundles of electron beam with different angles of emergence in an image plane, we are also allowed to investigate the geometrical pattern of aberrations in different image planes.

It is evident from Fig. 4 that the aberration patterns have a remarkable asymmetry with respect to the axis Y, but only a slight asymmetry with respect to the axis X. When $\beta_0 = 0°$ or $\beta_0 = 180°$, the electrons emitted from the photocathode with angles $30°$, $45°$, $60°$ and $90°$ do not fall on the axis X, but deviate slightly from the axis X. It is worth noticing that the aberration pattern and the spot diagram are crowded densely together between the image planes, which correspond to $\dot{z}_1 = 0$ and $\dot{z}_1 = 0.25 v_0$, while they tend to diverge when they lie off the two positions to the left or to the right. From this, it can be seen that the dense distribution of the aberration pattern and its asymmetry with respect to the axes X and Y will have noticeable influence on the MTF and PTF of the system.

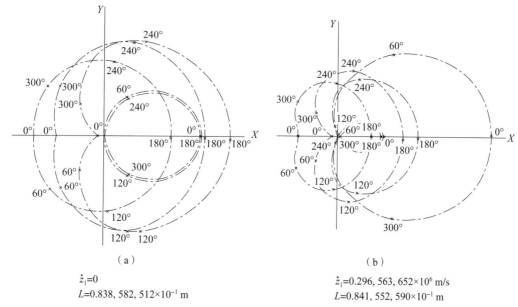

(a) $\dot{z}_1 = 0$
$L = 0.838, 582, 512 \times 10^{-1}$ m

(b) $\dot{z}_1 = 0.296, 563, 652 \times 10^6$ m/s
$L = 0.841, 552, 590 \times 10^{-1}$ m

Fig. 4 The geometrical patterns of lateral aberrations at two image positions

($B = 0.157 \times 10^{-1}$ T, $V_{ac} = 20,000$ V, $N = 1$, $\theta = 28.5°$, $\varepsilon_0 = 4$ V)

—··—··— $\alpha_0 = 30°$;   — · — · — $\alpha_0 = 45°$;

— · — · — $\alpha_0 = 60°$;   ——— $\alpha_0 = 90°$

6 Computations for ETF and RMS radius

As is known, the ETF consists of the MTF and the phase transfer function (PTF). In fact, the ETF computation includes computation of the spot diagram (i.e., PSF), the line spread function (LSF), the MTF and the C as will be described along the following steps:

(1) Choice of initial energy and initial angle distributions;

(2) Definition for the position of an ideal image point;
(3) Construction of a rectangular frame for the spot diagram;
(4) Determination of electron distribution's function;
(5) Computation for the spot diagram;
(6) Demonstration of LSF in directions x and y;
(7) Computation of the MTF and PTF (a Fourier transformation for the LSF).

Fig. 5(a) and Fig. 5(b) show the patterns of LSF in two directions taken from the spot diagram. The MTF and PTF are shown in Fig. 5(c) and Fig. 5(d). The image plane corresponds to $\dot{z}_1 = 0.2v_{0m} (L = 0.840,957,7 \times 10^{-1}$ m) and the Lambert distribution and beta distribution $\beta_{1.8}$ are used.

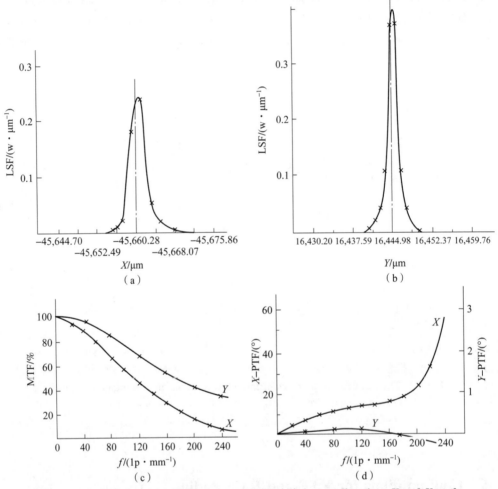

Fig. 5 The LSF (a, b), MTF (c) and PTF (d) in two directions X and Y at the image position, corresponding to $\dot{z}_1 = 0.2v_{0m} = 0.237,25 \times 10^{-1}$ m/s, $L = 0.840,957,7 \times 10^{-1}$ m, $(-D, \delta) = (-0.456,602,8 \times 10^{-1}$ m, $0.164,449,8 \times 10^{-1}$ m)

It may be seen from this that the PTF in the Y direction is close to zero owing to the symmetry of LSF with respect to the axis X, but the LSF in X direction looks much larger.

Fig. 6 gives the graphs of MTF for the same system at several different image positions. It is obvious from the $\beta_{1.8}$ distribution that the optimum image plane, i.e., the image plane corresponding to the optimum MTF, is located $\dot{z}_1 = 0.2v_{0m}$.

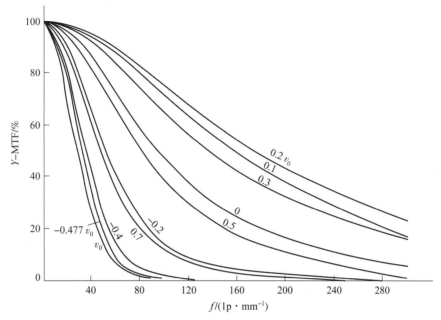

Fig. 6 **The MTF curves at different image positions ($\beta_{1.8}$ distribution, at Y direction)**

Along with the computation of spot diagram and ETF at a certain image position, the RMS radius can also be calculated with the following equation

$$\overline{\Delta r} = \left[\sum_{i=1}^{n} (\Delta r_i)^2 \frac{W_i}{W_{sum}} \right]^{1/2} \tag{24}$$

where Δr_i is the distance from the point of fall of electrons to the ideal image point in the image plane, when electrons are emitted from the initial point with different initial angles and initial energies; n is the number of points of fall; W_i is the corresponding weight of emitted electrons and W_{sum} is the sum of W_i. On the basis of our definition, W_{sum} has been normalized so that it is taken as a unity.

Fig. 7 shows the graphs of RMS radius as a function of the image position L for different oblique angles: 0°, 15°, 30°, 40°. It is obvious that the minimum RMS radius increases while the angle of obliquity θ becomes larger, and at the same time, the image plane with the best focus is moving towards the image plane, which corresponds to $\dot{z}_1 = 0$.

Furthermore, it is discovered that the size of RMS radius corresponds to the characteristics of MTF at each image plane. For example, the minimum RMS radius corresponds to the best MTF at the image position $\dot{z}_1 = 0.2v_{0m}$. Such a correspondence ensures that the parameter RMS radius can be used as an evaluating index for the MTF of the system. So it is convenient for designers to evaluate the image quality of a system by using the RMS radius. This conclusion has thus a practical meaning for calculation and evaluating the electron optical characteristics of image tubes.

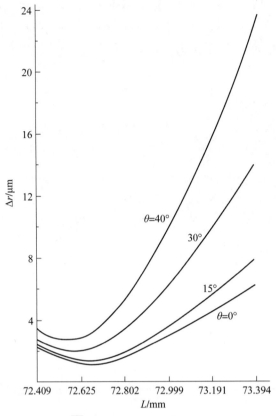

Fig. 7 The RMS radius $\overline{\Delta r}$ versus image positions at different oblique angles

($V_{ac} = 21,500$ V, $N = 1$, $\varepsilon_{0m} = 4$ V)

1) $\theta = 0°$, $B = 0.216, 5 \times 10^{-1}$ T; 2) $\theta = 15°$, $B = 0.209, 1 \times 10^{-1}$ T;
3) $\theta = 30°$, $B = 0.187, 5 \times 10^{-1}$ T; 4) $\theta = 40°$, $B = 0.165, 8 \times 10^{-1}$ T

Conclusions

(1) The description of electron trajectories with a movable coordinate system appears to be neat in presentation and elucidating in conception, and in these respects it has its own merit compared with other forms of expression thus far used.

(2) The expressions of aberration derived for designing the oblique image tubes and calculating MTF give enough accuracy in computation, saving however about 3/4 – 5/6 of the time spent in computation.

(3) It has been found that the RMS radius appears to correspond well with the resolution or MTF, so it can be used as an evaluating index for the image quality of electron optical systems.

(4) A general purpose computer program is proposed for designing the oblique image tubes. It can also be used for designing image tubes using uniform parallel magnetic and electric fields and for the MTF analysis of the concentric spherical systems with electromagnetic focusing.

Acknowledgement

We wish to acknowledge the helpful discussions and earnest assistance of Dr. J. L. Lowrance, Princeton University, USA and Prof. P. R. Yan, Beijing Institute of Technology, China.

References

[1] PICAT J P, COMBES M, FELENBOK P, et al. The electronic camera used in a reflection mode[J]. Advances in Electronics and Electron Physics, 1972, 33:557-562.

[2] JOHNSON C B, HALLAM K L. Correction to the oblique electron lens [J]. IEEE Transactions on Electron Devices, 1974, 21(1):131.

[3] JOHNSON C B, WILLIAMS J T. The Oblique electron lens [C]. The 7th Symposium on Photo-electronic Imaging Devices (PEID), London, Imperial College, 1978: 407.

[4] JOHNSON C B, HALLAM K L. The oblique image converter [J]. Advances in Electronics and Electron Physics, 1976, 40:69-82.

[5] LOWRANCE J L, ZUCCHINO P, RENDA G, et al. ICCD development at Princeton [J]. Advances in Electronics and Electron Physics, 1979, 52:441.

[6] LOWRANCE J L. Oblique magnetic focus point spread profiles and MTFs [J]. Advances in Electronics and Electron Physics, 1985, 64B:591.

[7] CHOU L W. Electron optics of concentric spherical electromagnetic focusing systems [J]. Advances in Electronics and Electron Physics, 1979, 52:119-132.

[8] BEURLE R L, WREATHALL W M. Aberration in magnetic focus systems [J]. Advances in Electronics and Electron Physics, 1962, 16:333.

3.2 倾斜型像管的电子光学像差
Electron Optical Aberrations of Oblique Image Tube

摘要：本文给出了倾斜型电磁聚焦系统在随动坐标系下的电子运动方程，导出了系统的像差表达式，探讨了表征该系统的鉴别率特性的均方根像差，由此研究系统的点列图与电子光学传递函数。

Abstract: In this paper, equations for the electron motion referred to movable coordinate system in an oblique electromagnetic focusing system are given, the aberration expressions of the system and the root mean square radius characterizing the resolution of system are derived, the root mean square (RMS) radius and electron optical transfer function are investigated.

引言

倾斜型像管是为了有效地使用不透明光阴极（特别是Ⅲ~Ⅴ族光阴极），在近年来迅速发展的一种新管型。实质上，这种像管的电子光学系统乃是一种均匀交叉场，即均匀磁场 B 相对于与光轴平行的均匀电场 E 错开或倾斜一个角度。与一般的像增强器不同的是，不透明光阴极位于像管的后面而屏靶在前面，两者恰好颠倒了一下，并且还错开一个位置。当入射光子通过光学系统聚焦成像在不透明极上，逸出的光电子受电场加速，围着与磁场方向偏离的抛物线旋转，从而聚焦到输出端的屏靶上。不过此时屏靶输出窗与入射窗的位置并不位于与管轴相对称的两侧，而是错开一个角度，如图1所示。这种倾斜型系统，首先由 Picat[1] 等人提出，并给出了电子运动的表达式，此后 Johnson 等人[2][3][4] 讨论了逸出电子初速度为零时电子轨迹的成像，探讨了系统的外形尺寸等。后来，Lowrance 等人直接由运动方程出发，研究了此系统中电子在屏靶上的点列图，并利用 CCD 器件作为系统的接收屏靶。但以上这些工作对轨迹的性状、系统的电子光学像差并未进行深入的分析。本文由交叉的均匀复合场下的运动方程出发，引入随动坐标系，导出描写系统中电子运动轨迹的另一种表达式。由所导出的像差表达式探讨"点列图"与电子光学传递函数。

1 随动坐标系下的电子运动方程

设 B 为常数，E 为常数，B 与 E 以某一角度 θ 相交。取直角坐标系 (x, y, z)，选沿着磁场 B 的方向为 z 轴，而经过矢量 E、B 的平面为 xz 平面。于是自物点 $(0, 0, 0)$、初速度为 $(\dot{x}_0, \dot{y}_0, \dot{z}_0)$ 逸出的电子运动方程可表示为

$$x = \rho\cos(\omega t + \varphi) - \frac{1}{\omega}\left(\dot{y}_0 - \frac{a_x}{\omega}\right) \tag{1a}$$

周立伟[a]，方二伦[b]．a) 北京工业学院，b) 西安现代应用化学研究所．工程光学（Engineering Optics），No.1, 1980, 1–13.

三、倾斜型电磁聚焦系统的成像电子光学

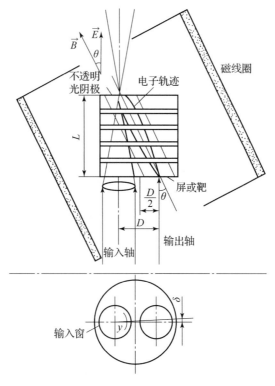

图1 倾斜型电磁聚焦像管

$$y = \rho \sin(\omega t + \varphi) + \frac{1}{\omega}(\dot{x}_0 + a_x t) \tag{1b}$$

$$z = \dot{z}_0 t + \frac{1}{2} a_z t^2 \tag{1c}$$

式中，$\omega = \frac{e}{m_0} B$ 为电子在磁场中的圆频率，$\frac{e}{m_0}$ 为电子的荷质比；$a_x = \frac{e}{m_0} \frac{\phi_{ac}}{L} \sin\theta$、$a_z = \frac{e}{m_0} \cdot \frac{\phi_{ac}}{L} \cos\theta$ 分别是电子在 x 方向和 z 方向所受的加速度；L 和 ϕ_{ac} 是光阴极至屏靶的距离和屏靶相对于光阴极的电位。

式（1）中的 ρ、φ 值可由初始条件定出：

$$\rho = \frac{m_0}{eB} \left[\dot{x}_0^2 + \left(\dot{y}_0 - \frac{\phi_{ac} \sin\theta}{BL} \right)^2 \right]^{1/2} \tag{2}$$

$$\tan\varphi = \frac{-\dot{x}_0}{\dot{y}_0 - \frac{\phi_{ac} \sin\theta}{BL}} \tag{3}$$

展开式（1），并利用 $t = 0$ 时由初始条件所得的关系式：

$$\rho \sin\varphi = -\frac{\dot{x}_0}{\omega}, \quad \rho\omega \cos\varphi = \dot{y}_0 - \frac{a_x}{\omega}$$

可导得

$$x = \frac{1}{\omega}\left[\dot{x}_0 \sin\omega t + (\cos\omega t - 1)\left(\dot{y}_0 - \frac{a_x}{\omega}\right)\right] \qquad (4a)$$

$$y = \frac{1}{\omega}\left[\dot{x}_0(1-\cos\omega t) + \left(\dot{y}_0 - \frac{a_x}{\omega}\right)\sin\omega t + \frac{1}{\omega}a_x t\right] \qquad (4b)$$

$$z = \dot{z}_0 t + \frac{1}{2}a_z t^2 \qquad (4c)$$

上式曾为 Picat 等人导出过[1]。

当 $t = NT = \frac{2N\pi}{\omega}$ 时，则电子将聚于某处，对应的位置为

$$x_1 = 0 \qquad (5a)$$

$$y_1 = \delta = \frac{2m_0}{e}(\pi N \sin\theta/L)\frac{\phi_{ac}}{B^2} \qquad (5b)$$

$$z_1 = \frac{L}{\cos\theta} = N\pi\sqrt{\frac{2m_0}{e}}\frac{\sqrt{\phi_{ac}}}{B}\left(\sqrt{1+\frac{m_0}{2e}\frac{\dot{z}_0^2}{\phi_{ac}}}+\sqrt{\frac{m_0}{2e\phi_{ac}}}\dot{z}_0\right) \qquad (5c)$$

由此可见，不管逸出电子的初速度分量 $\dot{x}_0$、$\dot{y}_0$ 如何，只要 $\dot{z}_0$ 相同，系统便能理想聚焦。

由式（1）或式（4c），可导得

$$t = \frac{m_0}{e}\frac{L}{\phi_{ac}\cos\theta}\left(\sqrt{\frac{2e}{m_0}\frac{\phi_{ac}\cos\theta}{L}z + \dot{z}_0^2} - \dot{z}_0\right)$$

必须指出，用式（1）或式（4）还是难于清晰地表达系统中电子的运动轨迹，现我们引入随动坐标系（$\bar{x}$，$\bar{y}$，$\bar{z}$）：

$$\bar{x} = x \qquad (6a)$$

$$\bar{y} = y - \frac{\phi_{ac}}{BL}\sin\theta \cdot t \qquad (6b)$$

$$\bar{z} = \dot{z}_0 t - \frac{e}{m_0}\frac{\phi_{ac}}{2L}\cos\theta \cdot t^2 \qquad (6c)$$

由式（6）可见，随动坐标的原点（$\bar{x}=0$，$\bar{y}=0$，$\bar{z}=0$）在 yz 的平面上做抛物线运动。其方程为

$$y = \frac{m_0}{e}\frac{\tan\theta}{B}\left(\sqrt{\frac{2e}{m_0}\frac{\phi_{ac}}{L}\cos\theta \cdot z + \dot{z}_0^2} - \dot{z}_0\right) \qquad (7)$$

于是，式（1）变为

$$\bar{x} = \rho\cos(\omega t + \varphi) + \frac{m_0}{eB}\left(\dot{y}_0 - \frac{\phi_{ac}\sin\theta}{BL}\right) \qquad (8a)$$

$$\bar{y} = \rho\sin(\omega t + \varphi) + \frac{m_0}{eB_0}\dot{x}_0 \qquad (8b)$$

$$\bar{z} = 0 \qquad (8c)$$

由式（8）可以得出

$$\left[\bar{x} + \frac{m_0}{eB}\left(\dot{y}_0 - \frac{\phi_{ac}}{BL}\sin\theta\right)\right]^2 + \left(\bar{y} - \frac{m_0}{eB}\dot{x}_0\right)^2 = \rho^2 \qquad (9)$$

显然，式（9）是描述位于 $\bar{x}\bar{y}$ 平面内半径为 ρ 的圆。圆心位于

$$\bar{x} = -\frac{m_0}{eB}\left(\dot{y}_0 - \frac{\phi_{ac}}{BL}\cos\theta\right), \qquad \bar{y} = \frac{m_0}{eB}\dot{x}_0$$

这表明，逸出电子以基线位于 yz 平面的一条抛物线做圆柱螺旋运动。

现推导以 $r(\bar{z}=0)$、$\psi(\bar{z}=0)$ 表示系统的电子轨迹，令 $r(\bar{z}=0)$ 为电子在 $\bar{x}\bar{y}$ 平面上的落点离 $\bar{x}=0$、$\bar{y}=0$ 的距离，$\psi(\bar{z}=0)$ 为直线 $r(\bar{z}=0)$ 与过随动坐标系原点处圆的切线间的夹角。在 $\bar{x}=0$、$\bar{y}=0$ 处该切线的斜率为 $\left(\dot{y}_0 - \frac{\phi_{ac}}{BL}\sin\theta\right)/\dot{x}_0$，于是有

$$r(\bar{z}=0) = 2\rho\sin\psi \tag{10}$$

由式（9），可得

$$\bar{x}^2 + \bar{y}^2 + 2\frac{m_0}{eB}\left(\dot{y}_0 - \frac{\phi_{ac}}{BL}\sin\theta\right)\bar{x} - 2\frac{m_0}{eB}\dot{x}_0\bar{y} = 0$$

将式（10）的 $r^2 = \bar{x}^2 + \bar{y}^2 = 4\rho^2\sin^2\psi$ 以及式（8）的 $\bar{x}$、$\bar{y}$ 值代入上式，便得到

$$\psi(\bar{z}=0) = \frac{eB}{2m_0}t$$

于是，以随动坐标系（$\bar{x}$，$\bar{y}$，$\bar{z}$）表示的电子轨迹方程为

$$r(\bar{z}=0) = 2\frac{m_0}{eB}\left[\dot{x}_0^2 + \left(\dot{y}_0 - \frac{\phi_{ac}\sin\theta}{BL}\right)^2\right]^{1/2}\sin\psi(\bar{z}=0) \tag{11a}$$

$$\psi(\bar{z}=0) = \frac{BL}{2\phi_{ac}\cos\theta}\left(\sqrt{\frac{2e}{m_0}\frac{\phi_{ac}\cos\theta}{L} + \dot{z}_0^2} - \dot{z}_0\right) \tag{11b}$$

于是，电子在随动坐标系中的运动轨迹可表示为

$$\bar{x} = r\cos\left(\psi + \psi_0 - \frac{\pi}{2}\right) \tag{12a}$$

$$\bar{y} = r\sin\left(\psi - \psi_0 - \frac{\pi}{2}\right) \tag{12b}$$

由式（11b），令 $\psi(\bar{z}=0) = N\pi$，同样可求得理想聚焦位置位于与式（5）对应的各值处。

现讨论 $\dot{x}_0$、$\dot{y}_0$、$\dot{z}_0$ 与自阴极面逸出电子的初速度 v_0、初角度（α_0，β_0）的关系。作（X，Y，Z）坐标系，这里 Z 轴与 E 平行，Y 轴与 y 轴重合，X 轴与 x 轴夹角为 θ，逸出角 α_0 为逸出电子与 Z 轴的夹角（$\alpha_0 = 0\sim90°$），方位角 β_0 为逸出电子在 XY 平面上的投影与 X 轴的夹角（$\beta_0 = 0\sim360°$）。于是有

$$\dot{x}_0 = v_0\cos\alpha_0\sin\theta + v_0\sin\alpha_0\cos\theta\cos\beta_0 \tag{13a}$$

$$\dot{y}_0 = v_0\sin\alpha_0\sin\beta_0 \tag{13b}$$

$$\dot{z}_0 = v_0\cos\alpha_0\cos\theta - v_0\sin\alpha_0\sin\theta\cos\beta_0 \tag{13c}$$

2　外形尺寸与电磁参数的确定

在以下讨论中，设理想成像位置与逸出电子的 $\bar{z}_1$ 相对应，于是可得系统的外形尺寸与电磁参数关系式：

$$L = N\pi \sqrt{\frac{2m_0}{e}} \frac{\sqrt{\phi_{ac}}}{B} \left(\sqrt{1 + \frac{m_0 \dot{z}_1^2}{2e\phi_{ac}}} - \sqrt{\frac{m_0}{2e}} \frac{\dot{z}_1}{\sqrt{\phi_{ac}}} \right) \cos\theta \tag{14}$$

$$\frac{\phi_{ac}}{B^2} = L^2 \left(\sqrt{1 + \frac{m_0 \dot{z}_1^2}{2e\phi_{ac}}} - \sqrt{\frac{m_0}{2e}} \frac{\dot{z}_1}{\sqrt{\phi_{ac}}} \right)^2 \left(2 \frac{m_0}{e} \pi^2 N^2 \cos^2\theta \right)^{-1} \tag{15}$$

$$\frac{\delta}{L} = \frac{\tan\theta}{\pi N \cos\theta} \left(\sqrt{1 + \frac{m_0 \dot{z}_1^2}{2e\phi_{ac}}} - \sqrt{\frac{m_0}{2e}} \frac{\dot{z}_1}{\sqrt{\phi_{ac}}} \right)^2 \tag{16}$$

$$\frac{S}{L} = \tan\theta \left[1 + \frac{1}{(\pi N \cos\theta)^2} \left(\sqrt{1 + \frac{m_0 \dot{z}_1^2}{2e\phi_{ac}}} - \sqrt{\frac{m_0}{2e}} \frac{\dot{z}_1}{\sqrt{\phi_{ac}}} \right)^4 \right]^{1/2} \tag{17}$$

$$\frac{D}{L} = \tan\theta \tag{18}$$

$$\phi = \arctan\left[\frac{1}{\pi N \cos\theta} \left(\sqrt{1 + \frac{m_0 \dot{z}_1^2}{2e\phi_{ac}}} - \sqrt{\frac{m_0}{2e}} \frac{\dot{z}_1}{\sqrt{\phi_{ac}}} \right)^2 \right] \tag{19}$$

$$\theta = \arccos\left\{ \frac{1}{2\pi^2 N^2 (1 + S^2/L^2)} \left[\pi^2 N^2 - \left(\sqrt{1 + \frac{m_0 \dot{z}_1^2}{2e\phi_{ac}}} - \sqrt{\frac{m_0}{2e}} \frac{\dot{z}_1}{\sqrt{\phi_{ac}}} \right)^4 + \left(\sqrt{1 + \frac{m_0 \dot{z}_1^2}{2e\phi_{ac}}} - \sqrt{\frac{m_0}{2e}} \frac{\dot{z}_1}{\sqrt{\phi_{ac}}} \right)^8 + \right. \right.$$
$$\left. \left. 2\pi^2 N^2 \left(\sqrt{1 + \frac{m_0 \dot{z}_1^2}{2e\phi_{ac}}} - \sqrt{\frac{m_0}{2e}} \frac{\dot{z}_1}{\sqrt{\phi_{ac}}} \right)^4 + 4\pi^2 N^2 \left(\sqrt{1 + \frac{m_0 \dot{z}_1^2}{2e\phi_{ac}}} - \sqrt{\frac{m_0}{2e}} \frac{\dot{z}}{\sqrt{\phi_{ac}}} \right)^4 + \pi^4 N^4 \right]^{1/2} \right\}^{1/2} \tag{20}$$

以上公式中几何参量的符号意义如图 2 所示。在上述公式中,若令 $\dot{z}_1 = 0$,便得到 Johnson[4] 所导出的关系式。一般,在计算外形尺寸时,可考虑 $\dot{z}_1 = 0$。

3 横向像差

由于横向像差应在位于与 $\boldsymbol{E}$ 相垂直的屏靶平面上衡量,故引入 (X, Y, Z) 坐标系,如图 2 所示。于是以初速度 $(\dot{x}_0, \dot{y}_0, \dot{z}_0)$ 逸出的电子相对于 $\dot{z}_1$ 所确定的理想像点 $(X = -D, Y = \delta, Z = -L)$ 的横向像差可表示为

$$\Delta x = \left[-\frac{1}{\omega} \left(\dot{y}_0 - \frac{a_x}{\omega} \right)(1 - \cos\omega t_L) \right] \cos\theta - \left(\dot{z}_0 t_L + \frac{1}{2} a_z t_L^2 \right) \sin\theta + D, \tag{21a}$$

$$\Delta y = \left(\frac{\dot{y}_0}{\omega} - \frac{a_x}{\omega^2} \right) \sin\omega t_L - \frac{\dot{x}_0}{\omega} \cos\omega t_L + \frac{1}{\omega}(a_x t_L + \dot{x}_0) - \delta \tag{21b}$$

这里 t_L 即电子到达位于 $Z = L$ 像面上的 t 值。它可由下式求得:

$$L = \left[-\frac{1}{\omega} \left(\dot{y}_0 - \frac{a_x}{\omega} \right)(1 - \cos\omega t_L) + \frac{\dot{x}_0}{\omega} \sin\omega t_L \right] \sin\theta + \left(\dot{z}_0 t_L + \frac{1}{2} a_z t_L^2 \right) \cos\theta \tag{22}$$

由式 (22) 解得 t_L 值,代入式 (21) 中,便得到系统的横向像差值。

式 (21) 表示的横向像差含有时间参量 t_L,且与系统的几何参量和电参量的关系不明显。因此,进一步将式 (21) 化简。

三、倾斜型电磁聚焦系统的成像电子光学

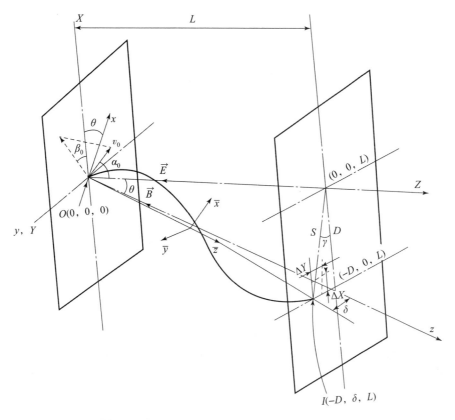

图 2 倾斜型像管的坐标系、电子轨迹与横向像差

实际，以 $\dot{z}_0$ 逸出的电子将相对于 $\dot{z}_1$ 所对应的理想像点多转或少转了 ψ_L 角，于是可令

$$\omega t_L = 2N\pi + \psi_L \tag{23}$$

代入式（22）后，略去高阶小项，便可求得

$$\psi_L = \frac{LB}{\phi_{ac}}(\dot{z}_1 - \dot{z}_0)\frac{1}{\cos\theta}\left(1 - \frac{\dot{x}_0\tan\theta + \dot{z}_0}{2\pi N\cos\theta}\frac{LB}{\phi_{ac}} - \frac{\dot{z}_1 - \dot{z}_0}{4\pi N\cos^3\theta}\frac{LB}{\phi_{ac}}\right) \tag{24}$$

将式（23）及其所求得的 ψ_L 值代入式（21）中，略去高于 $\frac{L^2B}{(\phi_{ac})^2}$ 之项，便得到

$$\Delta x = \frac{m_0}{e}\frac{L}{\phi_{ac}}(\dot{z}_1 - \dot{z}_0)\dot{x}_0\frac{1}{\cos^2\theta} + \frac{m_0}{e}\frac{L}{\phi_{ac}}(\dot{z}_1 - \dot{z}_0)^2\frac{\tan\theta}{2\cos^2\theta} - $$

$$\frac{m_0}{e}\frac{L^2B}{(\phi_{ac})^2}\left[\frac{(\dot{z}_1 - \dot{z}_0)^2\dot{y}_0}{2\cos\theta} + \frac{(\dot{z}_1 - \dot{z}_0)}{2\pi N\cos\theta}\left(\dot{x}_0^2\tan\theta + \dot{x}_0\dot{z}_0\frac{\cos 2\theta}{\cos^2\theta}\right) + \frac{\dot{x}_0(\dot{z}_1 - \dot{z}_0)^2}{4\pi N\cos^3\theta}\right] + $$

$$\frac{m_0}{e}\frac{L^2B}{(\phi_{ac})^2}\left[\frac{(\dot{z}_1 - \dot{z}_0)^2\dot{z}_0^2\tan\theta}{2\pi N\cos\theta} + \frac{(\dot{z}_1 - \dot{z}_0)^2\dot{z}_0\tan\theta}{4\pi N\cos^3\theta}\right] \tag{25a}$$

$$\Delta y = \frac{m_0}{e}\frac{L}{\phi_{ac}}(\dot{z}_1 - \dot{z}_0)\dot{y}_0\frac{1}{\cos\theta} + \frac{m_0}{e}\frac{L^2B}{(\phi_{ac})^2}\left[\frac{(\dot{z}_1 - \dot{z}_0)^2\dot{x}_0}{2\cos^2\theta} - \frac{(\dot{z}_1 - \dot{z}_0)\dot{y}_0}{2\pi N\cos^2\theta}(\dot{x}_0\tan\theta + \dot{z}_0) - \frac{(\dot{z}_1 - \dot{z}_0)^2\dot{y}_0}{4\pi N\cos^4\theta}\right] \tag{25b}$$

若在上式中，$\frac{LB}{\sqrt{\phi_{ac}}}$ 以 $N\pi\sqrt{\frac{2m_0}{e}}\cos\theta$ 代替，则可得

$$\Delta x = \frac{m_0}{e}\frac{L}{\phi_{ac}\cos^2\theta}(\dot{z}_1-\dot{z}_0)\dot{x}_0 + \frac{m_0}{2e}\frac{L\tan\theta}{\phi_{ac}\cos^2\theta}(\dot{z}_1-\dot{z}_0)^2 -$$
$$\left(\frac{m_0}{e}\right)^{3/2}\frac{N\pi L}{\sqrt{2}(\phi_{ac})^{3/2}}\left[(\dot{z}_1-\dot{z}_0)^2\dot{y}_0 + \frac{(\dot{z}_1-\dot{z}_0)}{\pi N}\left(\dot{x}_0^2\tan\theta + \dot{x}_0\dot{z}_0\frac{\cos 2\theta}{\cos\theta}\right) + \frac{\dot{x}_0(\dot{z}_1-\dot{z}_0)^2}{2\pi N\cos^2\theta}\right] +$$
$$\left(\frac{m_0}{e}\right)^{3/2}\frac{L}{\sqrt{2}(\phi_{ac})^{3/2}}\left[(\dot{z}_1-\dot{z}_0)\dot{z}_0^2\tan\theta + \frac{1}{2\cos^2\theta}(\dot{z}_1-\dot{z}_0)^2\dot{z}_0\tan\theta\right]$$
(26a)

$$\Delta y = \frac{m_0}{e}\frac{L}{\phi_{ac}\cos\theta}(\dot{z}_1-\dot{z}_0)\dot{y}_0 + \left(\frac{m_0}{e}\right)^{3/2}\frac{\pi N L}{\sqrt{2}(\phi_{ac})^{3/2}\cos\theta} \times$$
$$\left[(\dot{z}_1-\dot{z}_0)^2\dot{x}_0 - \frac{(\dot{z}_1-\dot{z}_0)\dot{y}_0}{\pi N}(\dot{x}_0\tan\theta + \dot{z}_0) - \frac{(\dot{z}_1-\dot{z}_0)^2\dot{y}_0}{2\pi N\cos^2\theta}\right]$$
(26b)

分析像差公式可见，第一项类似于均匀平行复合电磁聚焦系统的色差项，不过在 X 方向还出现了与 $\dot{x}_0$、$\dot{y}_0$ 无关的类似于畸变的像差项，它仍是由于系统的不对称性所引起的。式（25）或式（26）中最后的二阶小项乃是与上述像差相类似的高级像差项。

当在式（25）或式（26）中略去二阶小项时，便得到倾斜型像管具有足够精确度的横向像差表达式：

$$\Delta x = \frac{m_0}{e}\frac{L}{\phi_{ac}\cos^2\theta}(\dot{z}_1-\dot{z}_0)\dot{x}_0 + \frac{m_0}{e}\frac{L\tan\theta}{2\phi_{ac}\cos^2\theta}(\dot{z}_1-\dot{z}_0)^2 \quad (27a)$$

$$\Delta y = \frac{m_0}{e}\frac{L}{\phi_{ac}\cos\theta}(\dot{z}_1-\dot{z}_0)\dot{y}_0 \quad (27b)$$

必须指出，这种类型像管的横向像差的图形是复杂的。一方面，它与 $\dot{z}_1$ 所对应的像面位置有关；另一方面，$\dot{x}_0$、$\dot{y}_0$ 并不单纯与逸出电子的初角度（α_0，β_0）有关，而且还与系统的 **E** 和 **B** 间的夹角 θ 有关，故即使就式（27a）、式（27b）中第一项而言，图形也不是纯粹的圆。

当 $\theta=0$ 时，倾斜型像管就变成轴对称的均匀复合场电磁聚焦系统。我们由式（26a）、式（26b）可得到系统的横向像差表达式。只须令 $\theta=0$，由于系统的轴对称性，故可令横向像差 Δr 为

$$\Delta r = \sqrt{(\Delta x)^2+(\Delta y)^2} = \Delta r_2 + \Delta r_3 \quad (28)$$

当把 $\dot{x}_0$、$\dot{y}_0$ 的合成 $\dot{r}_0$ 以及 $\dot{z}_0$、$\dot{z}_1$ 化为初电位分量 ε_r、ε_z、ε_{z_1} 后，将式（26a）、式（26b）代入式（28）中，经过简化，便得到

$$\Delta r_2 = -\frac{2}{E}\sqrt{\varepsilon_r}(\sqrt{\varepsilon_{z_1}}-\sqrt{\varepsilon_z}) \quad (29a)$$

$$\Delta r_3 = \frac{2}{E}\frac{\sqrt{\varepsilon_r}}{\sqrt{\phi_{ac}}}(\varepsilon_{z_1}-\varepsilon_z) \quad (29b)$$

式中，$E=-\frac{\phi_{ac}}{L}$。式（29a）、式（29b）曾在文献［6］中导出过。

利用式（14）~式（20），可对倾斜型电磁聚焦系统进行初步设计。由式（7）和式（12），可得系统的电子轨迹，并可由式（25）、式（26）或式（27）得到各像面上的像差图形。最后，由这些公式出发，通过电子数字计算机，不难求得像面上的点列图与系统的调制传递函数（MTF）和相位传递函数（PTF），有关计算结果将在下文叙述。

4　均方根横向像差

如果我们计算逸出角 α_0 自 0 到 90°，方位角 β_0 自 0 到 360° 和能量 ε_0 自 0 到 ε_{max} 变化下系统的均方根横向像差 $\overline{\Delta r}$ 值：

$$\overline{\Delta r} = \sqrt{(\overline{\Delta x})^2 + (\overline{\Delta y})^2} \tag{30}$$

则式（30）可用来作为评价系数鉴别率特性的一个尺度。

首先我们仅考虑光电子的初角度呈朗伯分布，于是均方根横向像差的两个分量可表示为：

$$\overline{\Delta x^2} = \int_0^{2\pi}\int_0^{\frac{\pi}{2}} (\Delta x)^2 \frac{\sin 2\alpha_0}{2\pi} d\alpha_0 d\beta \tag{31a}$$

$$\overline{\Delta y^2} = \int_0^{2\pi}\int_0^{\frac{\pi}{2}} (\Delta y)^2 \frac{\sin 2\alpha_0}{2\pi} d\alpha_0 d\beta \tag{31b}$$

代入式（27a）、式（27b），并积分得到

$$\overline{\Delta x^2} = \left(\frac{m_0}{e}\frac{L}{\phi_{ac}\cos^2\theta}\right)^2 \left(\frac{v_0^2}{2}\right)^2 \left[\left(\frac{\dot{z}_1}{v_0}\right)^4 \tan^2\theta + \left(\frac{\dot{z}_1}{v_0}\right)^2 \left(1+\sin^2\theta+\frac{3}{2}\sin^2\theta\tan^2\theta\right) - \frac{16}{15}\left(\frac{\dot{z}_1}{v_0}\right)\cos\theta + \frac{1}{3}\left(1-\frac{1}{2}\sin^2\theta+\frac{1}{2}\sin^4\theta+\frac{3}{8}\tan^2\theta\sin^4\theta\right)\right] \tag{32a}$$

$$\overline{\Delta y^2} = \left(\frac{m_0}{e}\frac{L}{\phi_{ac}\cos^2\theta}\right)^2 \left(\frac{v_0^2}{2}\right)^2 \left[\left(\frac{\dot{z}_1}{v_0}\right)^2 - \frac{16}{15}\left(\frac{\dot{z}_1}{v_0}\right)\cos\theta + \frac{1}{3}\cos^2\theta + \frac{1}{6}\sin^2\theta\right] \tag{32b}$$

对式（32a）和式（32b）求极值，则可得 $\overline{\Delta x^2}|_{min}$ 约位于

$$\frac{\dot{z}_1}{v_0} \doteq 0.533\cos\theta - 1.216\cos\theta\sin^2\theta$$

处，而 $\overline{\Delta y^2}|_{min}$ 对应

$$\frac{\dot{z}_1}{v_0} \doteq 0.533\cos\theta \tag{33}$$

如果令 $\theta = 0$，则仅考虑朗伯分布的均方根横向像差值可由式（32）和式（31）求得，即

$$\overline{\Delta r^2} = 2\left(\frac{m_0}{e}\frac{L}{\phi_{ac}}\right)^2 \left(\frac{v_0^2}{2}\right)^2 \left[\left(\frac{\dot{z}_1}{v_0}\right)^2 - \frac{15}{16}\frac{\dot{z}_1}{v_0} + \frac{1}{3}\right] \tag{34}$$

最小值对应于 $\frac{\dot{z}_1}{v_0} \doteq 0.533$ 处，其值为

$$\overline{\Delta r^2}|_{min} = \frac{22}{225}\left(\frac{\frac{m_0}{2e}v_0^2}{\frac{\phi_{ac}}{L}}\right)^2 \tag{35}$$

式（34）、式（35）曾为 Beurle 与 Wreathall[6] 导出过。

关于像差公式的精度分析、点列图以及各种初能分布下的传递函数将在下文叙述。

参 考 文 献

[1] PICAT J P, COMBES M, FELENBOK P, et al. The electronic camera used in a reflection mode[J]. Advances in Electronics and Electron Physics, 1972, 33:557-562.

[2] JOHNSON C B, HALLAM K L. Correction to the oblique electron lens[J]. IEEE Transactions on Electron Devices, 1974, 21(1):131.

[3] JOHNSON C B, WILLIAMS J T. The oblique electron lens [C]. The 7th Symposium on Photo-electronic Imaging Devices (PEID), London, Imperial College, 1978: 407.

[4] JOHNSON C B, HALLAM K L. The oblique image converter[J]. Advances in Electronics and Electron Physics, 1976, 40:69-82.

[5] CHOU L W. Electron optics of concentric electromagnetic focusing systems[J]. Advances in Electronics and Electron Physics,1979,52:119-132.

[6] BEURLE R L, WREATHALL W M. Aberration in magnetic focus systems [J]. Advances in Electronics and Electron Physics, 1962, 16(1110):333-340.

3.3 倾斜型像管电子光学系统的计算与分析
Electron Optical Computation and Analysis of Oblique Image Tube

摘要：本文由前文所导出的随动坐标系下的电子轨迹与像差方程出发编制了设计倾斜型像管的通用程序，计算了电子光学外形尺寸、电子光学传递函数与均方半径，检验并比较了所采用的横向像差表示式的精度。结果表明，利用像差表示式计算倾斜型像管的传递函数较目前由运动方程计算落点的传统方法节省 3/4～5/6 的时间，并具有足够的精度。文中对一系列像面位置上的电子落点分布及其传递函数进行了考察，对各种不同初能分布下的传递函数进行了分析与比较，对初角度与初能分布的划分进行了考察。

Abstract: Start from the electron trajectory referred to movable coordinate system and aberration equation given by previous paper, in this article, we have compiled a universal program for designing the oblique image tube. We have also examined and compared the electron optical contour dimension, electron optical transfer function and root mean square (RMS) radius, deduced the precision of expressions for the lateral aberrations. Result shows that the computation for the transfer function by using the lateral aberration expressions can save time by 3/4 - 5/6, comparing to the spot diagram by using the traditional method. In this article, the spot diagram and transfer function at a series of image plane have been studied, the transfer function under different initial distributions have been compared and analyzed, and how to divide the initial emission energy and initial emission angle have also been investigated.

引言

关于倾斜型电磁聚焦系统的电子光学传递函数的计算，J. L. Lowrance 在他的文章与报告中[2,3]首先做了叙述，其出发点是基于众所周知的均匀交叉场下的电子运动方程。一般来说，这样的传递函数计算需要花费大量的上机时间。因此，如何节省计算时间而不降低计算精度是一个亟待解决的问题。

本文将从另外一个角度，即利用作者在文献［1］中所推导的横向像差表达式对传递函数与均方半径进行研究，并对该公式的计算精度和计算速度进行考察。在此基础上编制设计倾斜型像管的通用程序。程序通过计算外形尺寸以及不同像面位置下的电子光学调制传递函数与相位传递函数以及均方半径，从而给出管子在最佳聚焦下的成像特性。计算结果表明，程序对于倾斜型电磁聚焦系统的计算具有足够的精度并节省大量时间。

周立伟[a]，方二伦[b]，艾克聪[a]. a) 北京工业学院，b) 西安现代化学研究所. 工程光学（Engineering Optics）（内部），No. 2，1980，1-16.

1 外形尺寸与随动坐标系下的轨迹计算

在以下计算中，取国际单位制，上文[1]中长度参量如 L、S、D、δ 以米（m）表示，电位 ϕ_{ac} 与初电位 ε_0 以伏（V）表示，磁感应强度 B 以特斯拉（T）表示，速度 v_0 及其分量（$\dot{x}_0$，$\dot{y}_0$，$\dot{z}_0$）以米/秒（m/s）表示，ϕ 和 θ 以角度（°）表示。

现取 C. B. Johnson[4] 的初始数据，作为研究典型倾斜型像管的一个实例：
$$B = 0.157 \times 10^{-1} \text{ T}, \phi_{ac} = 0.2 \times 10^6 \text{ V}, \theta = 28.5°, N = 1, \varepsilon_{0m} = 4 \text{ V}$$
若我们对逸出电子的初角度取以下三组数据：

(a) $\alpha_0 = 61.5°$，$\beta_0 = 0°$；(b) $\alpha_0 = 69°$，$\beta_0 = 45°$；(c) $\alpha_0 = 90°$，$\beta_0 = 90°$

则由上文的式 (20) 可见，其沿轴的初速度设为 $\dot{z}_0 = 0$。于是，按上文式 (14) 与式 (19)，可求得其外形尺寸如下：

$$L = 0.838\ 582\ 512 \times 10^{-1} \text{ m}$$
$$\delta = 0.164\ 915\ 630 \times 10^{-1} \text{ m}$$
$$D = 0.455\ 313\ 154 \times 10^{-1} \text{ m}$$
$$S = 0.484\ 259\ 469 \times 10^{-1} \text{ m}$$
$$\phi = 19.910\ 520\ 99°$$

利用随动坐标系所表示的电子运动方程，可以通过在 yz 平面内的平面基线（一般呈抛物线形式）以及沿此基线旋转的圆方程及其参数，将轨迹明晰地表达出来。由于上述 (a)、(b)、(c) 三组数据使 $\dot{z}_0 = 0$，故由原点逸出的这三条轨迹虽都绕同一 $y(z)$ 基线旋转，并最终会聚于同一点 $(-D, \delta, L)$ 处，不过其回转半径却是不同的，图 1 乃是 $\alpha_0 = 61.5°$，$\beta_0 = 0°$，$\varepsilon_0 = 4$ V 的电子轨迹以抛物线的基轴而旋转的图形。

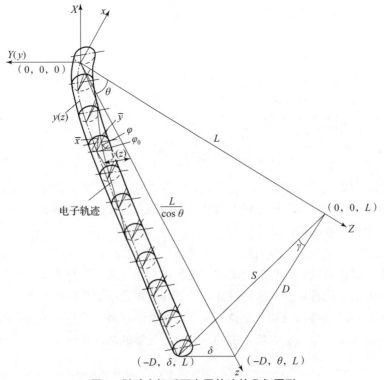

图 1　随动坐标系下电子轨迹的几何图形

2 像差方程的精度与像差的几何图形

2.1 精度验算

按照所给的典型数据，可求得以初始条件参量（ε_0，α_0，β_0）在离阴极距离为 L 处且垂直于 z 轴的像面上的精确落点位置以及它与理想像点（$-D$，δ）的差异，即上文公式（21）。同样也不难求得精确到二级近似的像差表达式（25）与式（27）。我们曾计算了初始条件为 $\varepsilon_0 = 4$ V，$\alpha_0 = 30°$、$45°$、$60°$、$90°$，$\beta_0 = 0$、$30°$、$60°$、$90°$、…、$330°$ 在 $\dot{z}_1 = 0$，即相应 $L = 0.838\,582\,5 \times 10^{-1}$ m 的像面上的各表达式所对应的横向像差计算值。计算表明，无论用上文的式（25）或是用式（27）代替精确的式（21），其差别已难于用作图方法表示，若令 $\Delta r = \sqrt{(\Delta x)^2 + (\Delta y)^2}$，并通过比较 Δr 值的大小来检验误差，其相对误差分别小于 1% 和 5%。由此可见，式（27）表示的像差公式，具有足够的精确度。

顺便指出，在计算表示像差的精确式时，须先用近似方法（如逐次接近法）求出下式

$$L = \left[-\frac{1}{\omega}\left(\dot{y}_0 - \frac{a_x}{\omega}\right)(1 - \cos\omega t_L) - \frac{\dot{x}_0}{\omega}\sin\omega t_L\right]\sin\theta + \left(\dot{z}_0 t_L + \frac{a_x}{2}t_L^2\right)\cos\theta \tag{1}$$

中的 t_L 值以及 ωt_L 值，然后代入上文的式（21）中，可求得 Δx、Δy 的精确值。

2.2 像差的几何图形

为研究整个电子束射线和像平面交点的分布情况，与几何光学技法相类似，我们采用围绕着阴极面的法线的同心圆锥，从整个电子束中取出位于这些圆锥面上的电子射线组，每组这样的电子射线，称为单元电子束。它在像平面上的交点形成一封闭曲线。根据这些曲线的形状和位置就能了解整个电子束中射线在像平面上的分布情况。我们用 $\varepsilon_0 = 4$ V，$\Delta\alpha = 15°$，$\Delta\beta = 15°$ 和上文的 Δx、Δy 公式计算像面位置位于

$$\dot{z}_1 = -v_0\sin\theta = -0.477v_0, \quad \dot{z}_1 = -\frac{1}{2}v_0\sin\theta = -0.238v_0, \quad \dot{z}_1 = 0, \quad \dot{z}_1 = 0.25v_0,$$

$$\dot{z}_1 = 0.5v_0, \quad \dot{z}_1 = 0.75v_0, \quad \dot{z}_1 = v_0$$

7 个像面上的像差值。这里 $\dot{z}_1 = -0.477v_0$ 与 $\dot{z}_1 = v_0$ 是逸出电子有可能到达的最近与最远的像面位置。这样，我们不但可以了解不同逸出角 α_0 的圆锥射出的单元电子束在像面上的图形变化，而且亦可考察不同像面上的像差图形变化的情况。

计算结果表明，用作图技法已难于表现这三套像差表达式之间的差异。因此，我们只示出了用上文式（25）描绘此 7 个像面上的像差图形，如图 2 所示。

由图 2 可以清晰地看出，像差图形对于 Y 轴显著不对称，而对于 X 轴略微有些不对称。由方位角 $\beta = 0°$，或 $\beta = 180°$，逸出角 α_0 为 $30°$、$45°$、$60°$、$90°$，逸出的电子落点并不落在 X 轴上，而是交在 X 轴略有偏离的某处，它们在 $\dot{z}_1 = 0$ 和 $\dot{z}_1 = 0.25v_0$ 所对应的像面上较为密集，而左右偏离此二位置的图形都显著扩散。由下面研究可见，像差图形的密集程度与它对 X、Y 轴的对称状况，将对系统的电子光学调制传递函数与相位传递函数产生显著的影响。

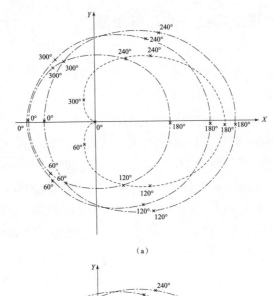

(a)

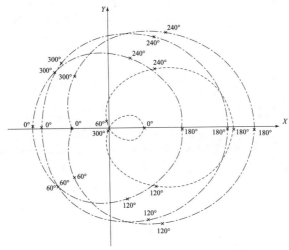

(b)

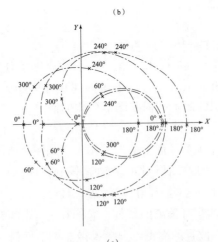

(c)

图 2 像面位于不同位置的像差图形

(a) $\begin{cases} \xi = -0.566\,031\,779 \times 10^6 \text{ m/s} \\ L = 0.832\,942\,013 \times 10^{-1} \text{ m} \end{cases}$; (b) $\begin{cases} \xi = -0.283\,015\,889 \times 10^6 \text{ m/s} \\ L = 0.835\,757\,889 \times 10^{-1} \text{ m} \end{cases}$; (c) $\begin{cases} \xi = 0 \\ L = 0.838\,582\,512 \times 10^{-1} \text{ m} \end{cases}$

三、倾斜型电磁聚焦系统的成像电子光学

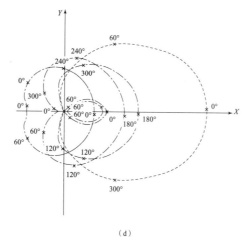

(d)

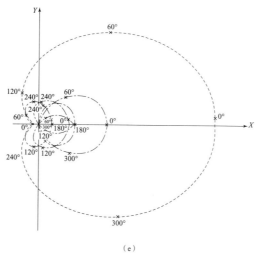

(e)

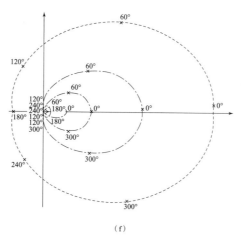

(f)

图 2　像面位于不同位置的像差图形（续）

(d) $\begin{cases} \xi = 0.296\ 563\ 652 \times 10^{6}\ \text{m/s} \\ L = 0.841\ 552\ 590 \times 10^{-1}\ \text{m} \end{cases}$; (e) $\begin{cases} \xi = 0.593\ 127\ 305 \times 10^{6}\ \text{m/s} \\ L = 0.844\ 533\ 15 \times 10^{-1}\ \text{m} \end{cases}$; (f) $\begin{cases} \xi = 0.889\ 000\ 957 \times 10^{6}\ \text{m/s} \\ L = 0.847\ 524\ 191 \times 10^{-1}\ \text{m} \end{cases}$;

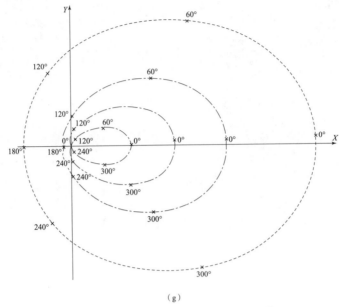

(g)

图 2 像面位于不同位置的像差图形（续）

(g) $\begin{cases} \xi = 0.118\,625\,400 \times 10^6 \text{ m/s} \\ L = 0.850\,525\,713 \times 10^{-1} \text{ m} \end{cases}$

($B = 0.157 \times 10^{-1}$ T，$\phi_{ac} = 20\,000$ V，$N = 1$，$\theta = 28.5°$，$\varepsilon_0 = 4$ V)

$\alpha_0 = 30°$，$\alpha_0 = 45°$，$\alpha_0 = 60°$，$\alpha_0 = 90°$

3 点列图、调制传递函数与均方半径的计算

点列图（即点扩散函数 PSF）、线扩散函数（LSF）、调制传递函数（MTF）与相位传递函数（PTF）的计算过程如下。

（1）选择初角度与初能分布。

一般，初角度按朗伯分布，初能分布可选择贝塔分布，也可选择其他类型的分布，如余弦分布和瑞利分布等，取决于所采用的光阴极以及入射辐射的波长。程序除包括上述三种形式的电子初能理论分布外，还可对实验分布进行处理。

（2）确定理想像点位置。

由 $\dot{z}_0$ 确定像面位置，于是可求得作为理想像点的坐标 $(-D, \delta)$ 值，在下面的计算中我们将以 $(-D, \delta)$ 为原点。

（3）为点列图构成矩形网络。

取 $\varepsilon_0 = \varepsilon_{0m}$，然后在 $\Delta\alpha = 15°$，$\Delta\beta = 15°$ 的间隔依次求 Δx、Δy，并确定 Δx、Δy 的正负最大值。然后将它们在 X 方向和 Y 方向各分为 16 或 32 等份，构成大量矩阵网格，一般 X 方向和 Y 方向的分格是不同的。通常格子的划分应对于 X 轴对称，而对于 Y 轴可以是不对称的。但在我们给出的点列图中，格子的划分对于 X 轴和 Y 轴都是对称的。

（4）$\Delta\xi$、$\Delta\alpha$、$\Delta\beta$ 的划分与电子分布函数。

按照所需要的计算精度要求，对初能量与初角度进行划分，定出 $\Delta\xi$、$\Delta\alpha$、$\Delta\beta$ 值。于是电子分布函数 W 即逸出电子的权数可由下式求出：

$$W(\bar{\xi}_i, \bar{\alpha}_j, \bar{\beta}_k) = \Delta N(\bar{\xi}_i) \frac{1}{2} (\cos 2\alpha_{j-1} - \cos 2\alpha_j) \frac{\beta_k - \beta_{k-1}}{2\pi} \quad (2)$$

式中，$\bar{\xi}_i = \frac{\xi_{i-1} + \xi_i}{2}$，$\bar{\alpha}_j = \frac{\alpha_{j-1} + \alpha_j}{2}$，$\bar{\beta}_k = \frac{\beta_{k-1} + \beta_k}{2}$，$\Delta N(\xi_i)$ 是对 $N(\xi_i)$ 直接积分而求出与各个单能区所对应的电子数。即

$$\Delta N(\bar{\xi}_i) = \int_{\xi_{i-1}}^{\xi_i} N(\xi) \mathrm{d}\xi \quad (3)$$

（5）点列图的计算。

按 $\bar{\varepsilon}_i = \bar{\xi}_i \varepsilon_m$、$\bar{\alpha}_j$、$\bar{\beta}_k$ 值，由上文式（25）或式（27）计算 Δx、Δy 值。在与 Δx、Δy 相对应的点列图格子上记下所携带的权数，点列图的显示以 16×16 的形式给出。按照我们的定义，点列图上所示出的电子总权数等于 1。

（6）调制传递函数（MTF）和相位传递函数（PTF）的计算。

$X(Y)$ 方向的线扩散函数（LSF）乃是点列图上 $Y(X)$ 轴上权电子迭加除以间隔。其单位现以电子权数/微米表示，按照我们的安排，两个方向的线扩散函数以 32 等份的形式给出。

（7）线扩散函数（LSF）的计算。

对上述的线扩散函数进行傅里叶变换，便得到点像沿 X 方向和 Y 方向的调制传递函数（MTF）与相位传递函数（PTF）。

对 C. B. Johnson 给出的典型倾斜型像管，我们曾经画出在某一像面处（例如对应于 $\dot{z}_1 = 0.2 v_{0m}$，$L = 0.840\,957\,7 \times 10^{-1}$ m）的点列图。图 3（a）、图 3（b）乃是由此点列图上在两个方向上取下的线扩散函数的图形，其调制传递函数与相位传递函数如图 3（c）、图 3（d）所示。

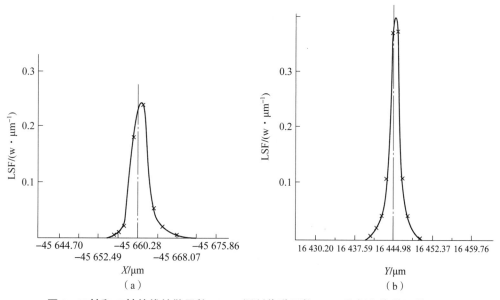

图 3　X 轴和 Y 轴的线扩散函数 LSF、调制传递函数 MTF 和相位传递函数 PTF

(a) X 轴的线扩散函数；(b) Y 轴的线扩散函数

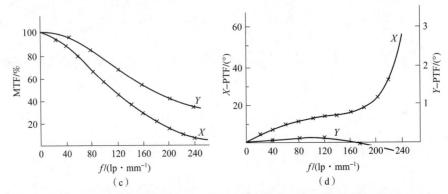

图 3 X 轴和 Y 轴的线扩散函数 LSF、调制传递函数 MTF 和相位传递函数 PTF（续）
（c）调制传递函数 MTF；（d）相位传递函数 PTF

由点列图与线扩散函数的图形可见，由于沿 Y 方向的线扩散函数基本上对 X 轴是对称的，其 PTF 接近于零，而且沿 X 方向的线扩散函数对 Y 轴极不对称，故其 PTF 较为显著。

图 4 给出了像面位于不同位置上的调制传递函数，由 $\beta_{1,8}$ 分布可见，最佳调制传递函数的像面位置约对应于 $\dot{z}_1 = 0.2 v_{0m}$ 处。

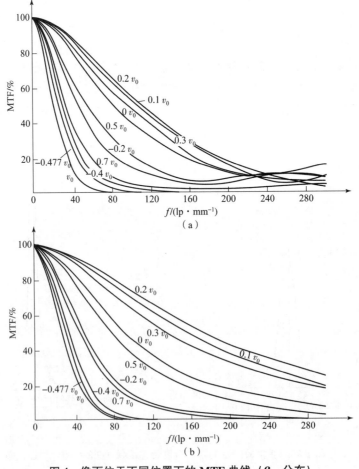

图 4 像面位于不同位置下的 MTF 曲线（$\beta_{1,8}$ 分布）
（a）x 方向；（b）y 方向

在计算某一像面上的点列图（点扩散函数）以及传递函数的同时，不难求出表征系统鉴别率特性的均方半径值：

$$\overline{\Delta r} = \left[\sum_{i=1}^{n} (\Delta r_i)^2 \frac{\omega_i}{\omega_{\text{total}}} \right]^{1/2} \tag{4}$$

式中，Δr_i 为初角度为 $(\bar{\alpha}_j, \bar{\beta}_k)$，初能量 $\bar{\xi}_0$ 自原点处逸出且携带权数为 ω_i 的电子在像面的落点离理想像点的距离，落点数 ω_i 为逸出电子对应的权，ω_{total} 为 ω_i 之和，在我们的计算标中，该值已归一化，故令 $\omega_{\text{total}} = 1$。

图 5 示出了均方半径随不同倾斜角 θ（磁感应强度 B 随 θ 改变以保证像面位置维持相同的距离）与所处的像面位置 L 的变化曲线。由图可见，当倾斜角度变大时，最小均方半径随 $\overline{\Delta r}_{\min}$ 增大，它所对应的像面位置向着极限像面方向移动。

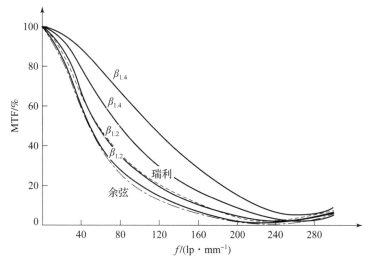

图 5　不同的能量分布（贝塔分布、余弦分布、瑞利分布）的 MTF 曲线

4　电子光学传递函数计算中若干问题的讨论

4.1　用像差公式求传递函数的计算精度与计算速度

考察上文以二级近似的横向像差公式（27）替代精确的落点表示式（21）求传递函数的计算精度与计算速度。计算表明，式（27）所得到的调制传递函数值与精确式所得的值仅在有效数字第三位或第四位上有差异，已难于用作图方法表示。二级近似与精确表示式所得到的传递函数的误差在 $f = 200$ lp/mm 以内分别为 1% 与 2% 以下。但计算时间却节省 3/4 或 5/6 之多，由此可见所导出的像差方程的优越性。

4.2　初能量与初角度的划分

在计算传递函数时，问题在于：究竟怎样划分 $\Delta\xi$、$\Delta\alpha$、$\Delta\beta$，即取多少个落点才能使计算传递函数的时间短、精度高、收效大。表 2 中示出了在不同的 $\Delta\xi$、$\Delta\alpha$、$\Delta\beta$ 划分下（由 360 点到 129 600 点）对应 $f = 20$ lp/mm、50 lp/mm、100 lp/mm、150 lp/mm、200 lp/mm

处的调制传递函数值，由表 1 可见：

（1）当所取的点数使调制传递值达到足够的精度时，更精细地划分 $\Delta\xi$、$\Delta\alpha$、$\Delta\beta$ 使落点数成倍增加将在计算精度上得益甚微，只是徒然耗费计算时间，如表中 6 480 点、129 600 点等项，点数成 20 倍增大，而计算结果改进不大。

（2）初能 $\Delta\xi$ 的划分较 $\Delta\alpha$ 和 $\Delta\beta$ 划分更为重要些，在相同的点数下，$\Delta\xi$ 取得小些，其收益较 $\Delta\alpha$ 和 $\Delta\beta$ 划分得小要更大些。如表 2 中，同样是 21 600 点，（$\Delta\xi=0.05$，$\Delta\alpha=3°$ 和 $\Delta\beta=10°$）的调制传递函数值较（$\Delta\xi=0.10$，$\Delta\alpha=3°$ 和 $\Delta\beta=5°$）要更接近于精确值，而（$\Delta\xi=0.05$，$\Delta\alpha=22.5°$ 和 $\Delta\beta=30°$）960 点较（$\Delta\xi=0.10$，$\Delta\alpha=10°$ 和 $\Delta\beta=30°$）1 080 点的计算精度要更高一些。

（3）一般来说，计算点数在 1 000 点以上便具有足够的精度，如（$\Delta\xi=0.10$，$\Delta\alpha=15°$ 和 $\Delta\beta=15°$）1 440 点直到 1 200 lp/mm 宽广的范围内都有很高的精度，故选取 1 440 点无论对于精度或计算速度的要求都是合适的。如果对 MTF 仅要求在低频范围有一定的精度，如允许误差在 3% ~ 5%，则（$\Delta\xi=0.10$，$\Delta\alpha=22.5°$，$\Delta\beta=45°$）320 点或（$\Delta\xi=0.10$，$\Delta\alpha=15°$，$\Delta\beta=45°$）480 点亦就足够了。

表 1　三种像差公式计算的传递函数与计算时间

$f/(\text{lp}\cdot\text{mm}^{-1})$	MTF（子午）			MTF（弧矢）		
	式（21a）	式（26）	式（27）	式（21a）	式（26）	式（27）
0	1.000 000 0	1.000 000 0	1.000 000 0	1.000 000 0	1.000 000 0	1.000 000
20	0.969 831 1	0.969 696 5	0.969 480 7	0.986 809 9	0.986 822 2	0.986 777 6
40	0.889 794 5	0.889 322 7	0.888 594 7	0.949 209 1	0.949 255 8	0.949 106 3
60	0.782 683 7	0.781 807 0	0.780 524 7	0.892 484 9	0.892 582 3	0.892 339 3
80	0.668 646 7	0.667 419 9	0.665 695 4	0.823 673 2	0.823 834 6	0.823 590 2
100	0.558 955 8	0.557 518 4	0.555 475 6	0.749 747 3	0.749 988 3	0.749 891 5
120	0.458 143 7	0.456 644 9	0.454 373 0	0.676 258 6	0.676 603 3	0.676 819 0
140	0.367 425 9	0.365 944 5	0.363 548 0	0.606 773 2	0.607 254 6	0.607 927 9
160	0.286 730 0	0.285 269 4	0.282 856 7	0.543 016 4	0.543 672 5	0.544 908 0
180	0.215 545 8	0.214 105 8	0.211 700 4	0.485 398 0	0.486 264 8	0.488 121 8
200	0.153 433 0	0.152 064 1	0.149 682 7	0.433 587 1	0.434 689 9	0.437 187 9
250	0.048 190 6	0.047 810 9	0.047 611 5	0.325 248 7	0.326 914 1	0.330 931 5
300	0.081 670 1	0.082 420 6	0.084 817 0	0.240 053 7	0.241 901 9	0.247 211 6
PSF、LSF、MTF、PTF 的计算			式（21）	式（25）		式（27）
三种像差公式与式（21）计算时间的比值			100%	22%		16%

表 2　不同 $\Delta\varepsilon$、$\Delta\alpha$、$\Delta\beta$ 划分下的 MTF 值

($B = 0.157 \times 10^{-1}$ T，$\theta = 28.5°$，$\phi_{ac} = 20\,000$ V，$N = 1$，$\varepsilon_{0m} = 4$ V)

$f/(\text{lp} \cdot \text{mm}^{-1})$ ($\Delta\xi$，$\Delta\alpha$，$\Delta\beta$) 点数	20	50	100	150	200
(0.025，2°，5°) 129 600 点	0.969 654 553	0.837 382 361	0.558 062 991	0.325 490 449	0.153 272 011
(0.05，2°，5°) 64 800 点	0.969 690 554	0.837 514 145	0.558 170 995	0.326 745 260	0.153 655 378
(0.025，3°，10°) 43 200 点	0.969 617 935	0.837 145 133	0.557 234 145	0.324 252 029	0.152 220 277
(0.05，3°，10°) 21 600 点	0.969 696 499	0.837 495 633	0.557 518 391	0.324 383 249	0.152 064 055
(0.10，3°，5°) 21 600 点	0.969 098 533	0.834 841 169	0.553 450 837	0.321 783 041	0.152 187 939
(0.05，5°，20°) 6 480 点	0.969 589 203	0.837 116 726	0.557 550 552	0.328 079 697	0.159 662 099
(0.025，10°，30°) 4 320 点	0.969 582 253	0.837 494 260	0.560 633 849	0.330 723 400	0.155 137 815
(0.10，15°，15°) 1 440 点	0.968 349 046	0.832 656 671	0.554 862 610	0.330 861 220	0.153 437 930
(0.10，10°，30°) 1 080 点	0.968 830 631	0.833 010 090	0.546 137 441	0.317 627 872	0.144 962 032
(0.05，22.5°，30°) 960 点	0.969 486 897	0.839 761 503	0.569 591 995	0.333 547 000	0.148 006 448
(0.10，30°，30°) 360 点	0.967 591 611	0.827 760 262	0.550 392 82	0.336 629 344	0.171 271 521

5　倾斜型像管通用程序的功能

按照上述分析，编制了设计电磁聚焦倾斜型像管的通用程序，它有以下功能：

(1) 按照给定的初始条件 B、ϕ_{ac}、θ、N 和 ε_{0m} 值，计算对应于所取 $\dot{z}_1$ 值下的外形尺寸 L、δ、D、S、ϕ。

(2) 按照给定的条件，计算随动坐标系下的单条或成组电子轨迹。

(3) 可选择任一套像差公式计算给定像面上的像差值，以 $\Delta\alpha = 15°$、$\Delta\beta = 15°$ 的形式给出，亦可求位于 $\dot{z}_1 = -v_{0m}\sin\theta$，$\dot{z}_1 = v_{0m}$ 之间成组像面位置上的像差值。

(4) 按照所选择的初能分布计算任一给定像面或成组像面上的点列图 PSF、LSF、MTF 和 PTF。

(5) 如令 $\theta = 0°$，则可用来设计和计算均匀平行复合电磁聚焦像管。

(6) 如令 $\theta = 0°$，在上文式（25）中以阴极透镜电场强度 $-E_c$ 代替 $\dfrac{\phi_{ac}}{L}$ 并乘以静电系统的线放大率 M，则可用来分析静电阴极透镜二级横向像差所构成的传递函数。

结论

(1) 以随动坐标系表示的电子轨迹较迄今为止的各种表达式概念明确，表达简洁。

(2) 利用像差方程计算倾斜型像管的传递函数具有足够的计算精度，且节省 3/4 ~ 5/6 的计算时间。

（3）编制了适用于设计倾斜型电磁聚焦像管以及均匀平行电磁聚焦像管的通用程序，并可用来对同心球电磁聚焦系统与静电阴极透镜的传递函数进行计算分析。

致谢

作者致谢北京工业学院严沛然教授、普林斯顿大学天文台 J. L. Lowrance 博士对于本文给予的热忱帮助与有益的讨论。

参 考 文 献

[1] 周立伟,方二伦. 倾斜型电磁聚焦系统的电子光学[J]. 工程光学,1980,2:1-16.

[2] LOWRANCE J L. A review of solid state image sensors [J]. Advances in Electronics and Electron Physics, 1979,52:441.

[3] LOWRANCE J L. Study of oblique magnetic focusing [C]. Interim report-Grant NSG-5277 Princeton University Observatory,1979.

[4] JOHNSON C B. The oblique image converter [J]. Advances in Electronics and Electron Physics,1976,40:79.

[5] CSORBA I P. Modulation transfer function of image tube lenses[J]. Applied Optics, 1977, 16(10): 2647-2650.

[6] ZACHAROV B. A Demagnifying image tube for nuclear physics applications [J]. Advances in Electronics and Electron Physics, 1962,16:76.

[7] CHOU L W. Electron optics of concentric spherical electromagnetic focusing systems [J]. Advances in Electronics and Electron Physics, 1979,52:119-132.

四、电磁聚焦移像系统的成像电子光学

Imaging Electron Optics of Electromagnetic Focusing Systems for Image Transference

4.1 电磁聚焦移像系统理论的研究
On the Theory of Combined Electrostatic and Magnetic Focusing Systems for Image Transference

摘要：本文探讨了 И. И. Цуккерман 关于电磁聚焦移像系统的理论，指出该理论尚有值得商榷与补正之处，并证明了若主轨迹方程与相邻轨迹方程以矢量形式表示，则其求解将较 И. И. Цуккерман 建议的方程简便。最后，文中结合实例对一种典型的电磁聚焦移像系统进行了考察。

Abstract：Theory of combined electrostatic and magnetic focusing system has been investigated by I. I. Tsukkerman, but his theory is suitable only for systems with narrow electron beam focusing. This paper vertifies the theory of combined electrostatic and magnetic system with wide electron beam focusing for image transference, gives a comment on the Tsukkerman's theory. The article shows that if the principal trajectory equation and its neighboring trajectory equations are expressed in vector form, it will be more convenient and simpler to find out the solutions of trajectories then those trajectory equations expressed in scalar form suggested by. I. I. Tsukkerman. Finally, using the theory obtained, a practical example is given for combined electrostatic and magnetic systems with wide electron beam focusing for image transference.

引言

早在 20 世纪 50 年代，苏联学者 И. И. Цуккерман 提出了一种研究电磁聚焦移像系统的电子光学特性的新方法[1,2]，它的实质是，由某一物点逸出的电子轨迹 $r(z)$，可用非线性方程解得的主轨迹 $r_主(z)$ 和某一位置 z 平面内连接主轨迹的相邻轨迹的矢量 $p(z)$ 表示（见图1）。由此出发，可探讨系统的聚焦成像性质。

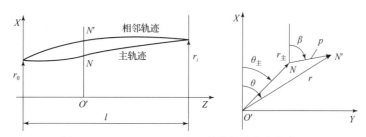

图 1　И. И. Цуккерман 定义的主轨迹与相邻轨迹

周立伟. 北京工业学院. 北京工业学院学报（Journal of Beijing Institute of Technology），No. 3, 1983, 12 – 24.

И. И. Цуккерман 的工作推进了移像系统的研究，引起了人们的重视与兴趣。但是，在他的著作中，所给出的主轨迹方程与相邻轨迹方程以及所定义的放大率表达式并不适用于分析复合电磁聚焦阴极透镜的成像系统。其次，对所给出的主轨迹方程与相邻轨迹方程的求解也缺乏深入的探讨。此外，文献 [1] 中相邻轨迹的幅角公式并没有考虑逸出电子的初始方位角 β_0，以致其公式仅适用于 $\beta_0 = 0$ 的情况。

本文由旋转坐标系下轴对称复合场的近轴方程出发，给出以矢量形式表示的主轨迹方程与相邻轨迹方程，于是主轨迹与相邻轨迹的模与方向可由此方程矢量形式的解来确定。由于主轨迹方程与相邻轨迹方程乃是具有相同系数的二阶线性齐次微分方程，故具有相同的特解，而通解可以通过特解 v、w 的组合并结合主轨迹与相邻轨迹的初始条件来确定。因此，大大简化了求解。文中进一步证明了以矢量形式表达的主轨迹方程与相邻轨迹方程可以转化为经过补正后的 И. И. Цуккерман 方程，并对系统放大率与图像转角的确定进行了讨论。最后，以一理想的电磁聚焦移像系统为例对 И. И. Цуккерман 的方法与本文的方法进行比较与考察。

1 И. И. Цуккерман 移像系统理论评述

按照 И. И. Цуккерман[1,2]，轴对称电磁聚焦系统的主轨迹方程可表示为

$$\sqrt{\phi}\frac{\mathrm{d}}{\mathrm{d}z}\left(\sqrt{\phi}\frac{\mathrm{d}r_{\mathrm{prin.}}}{\mathrm{d}z}\right) + \frac{1}{4}\frac{\mathrm{d}^2\phi}{\mathrm{d}z^2}r_{\mathrm{prin}} + \frac{eB^2(0)}{8m_0}\left[\frac{B^2(z)}{B^2(0)} - \frac{r_0^4}{r_{\mathrm{prin}}^4}\right]r_{\mathrm{prin}} = 0 \tag{1}$$

$$\sqrt{\phi}\frac{\mathrm{d}\theta_{\mathrm{prin.}}}{\mathrm{d}z} = \sqrt{\frac{e}{8m_0}}B(0)\left[\frac{B(z)}{B(0)} - \frac{r_0^2}{r_{\mathrm{prin.}}^2}\right] \tag{2}$$

主轨迹的相邻轨迹可以用以下的复数形式表示：

$$\boldsymbol{p}(z) = p(z)\mathrm{e}^{\mathrm{i}\beta(z)} \tag{3}$$

这里 $p(z)$ 是 $\boldsymbol{p}(z)$ 的模，$\beta(z)$ 是矢量 $\boldsymbol{p}$ 相对固定坐标系 X 轴转过的角度，它由下式确定[2]：

$$\beta(z) = \sqrt{\frac{e}{8m_0}}\int_0^z \frac{B(z)}{\sqrt{\phi(z)}}\mathrm{d}z \tag{4}$$

而 $p(z)$ 满足以下方程：

$$\sqrt{\phi}\frac{\mathrm{d}}{\mathrm{d}z}\left(\sqrt{\phi}\frac{\mathrm{d}p}{\mathrm{d}z}\right) + \frac{1}{4}\frac{\mathrm{d}^2\phi}{\mathrm{d}z^2}p + \frac{eB^2(z)}{8m_0}p = 0 \tag{5}$$

式中，e/m_0 为电子的荷质比，$\phi = \phi(z)$，$B = B(z)$ 分别为电位与磁感应强度的沿轴分布，$B(0) = B_0$ 为物平面处的磁感应强度，$r_{\mathrm{prin.}} = r_{\mathrm{prin.}}(z)$，$\theta_{\mathrm{prin.}} = \theta_{\mathrm{prin.}}(z)$，$p = p(z)$，$\beta = \beta(z)$ 的意义如图 1 所示。

И. И. Цуккерман 假定相邻轨迹方程式 (5) 的两个特解 u_α、u_β 满足以下初始条件：

$$\begin{aligned}u_\alpha(0) &= 0, \quad u_\alpha'(0) = 1, \\ u_\beta(0) &= 1, \quad u_\beta'(0) = 0\end{aligned} \tag{6}$$

并由文献 [3] 给出以下初始条件：

$$r(0) = r_0, \quad r'(0) = r_0', \\ \theta(0) = 0, \quad \theta'(0) = \theta_0' \tag{7}$$

则 $r(z)$ 与 $\theta(z)$ 的表达式为

$$r(z) = r_0 \sqrt{\left(u_\beta + \frac{r_0'}{r_0} u_\alpha\right)^2 + \left(\theta_0' - \sqrt{\frac{e}{8m_0\phi(0)}} B(0)\right)^2 u_\alpha^2} \tag{8}$$

$$\theta(z) = \sqrt{\frac{e}{8m_0}} \int_0^z \frac{B(z)}{\sqrt{\phi(z)}} \mathrm{d}z + \arctan \frac{\theta_0' - \sqrt{\frac{e}{8m_0\phi(0)}} B(0)}{\frac{u_\beta}{u_\alpha} + \frac{r_0'}{r_0}} \tag{9}$$

设 $z = z_i$ 时，$u_\alpha(z_i) = 0$。И. И. Цуккерман 给出放大率 M 与图像转角 $\theta(z_i)$ 的表达式为

$$M = u_\beta(z_i) = \frac{1}{u_\alpha'(z_i)} \sqrt{\frac{\phi(0)}{\phi(z_i)}} \tag{10}$$

$$\theta(z_i) = \sqrt{\frac{e}{8m_0}} \int_0^{z_i} \frac{B(z)}{\sqrt{\phi(z)}} \mathrm{d}z - n\pi \tag{11}$$

应该指出，И. И. Цуккерман 上述理论尚有若干值得商榷之处：

（1）由于给出的主轨迹方程与相邻轨迹方程所得到的 $r(z)$、$\theta(z)$ 的解的表达式（8）、式（9）以及所定义的放大率表达式（10）在 $\phi(0) = 0$ 时出现不确定性，故此理论并不适用于分析复合电磁聚焦阴极透镜的成像系统。

（2）$r(z)$、$\theta(z)$ 的解表达式（8）、式（9）中所定义的 u_α、u_β 在文献 [3] 中是旋转坐标系下复合电磁聚焦傍轴方程的两个特解，而 И. И. Цуккерман 著作[1]中，u_α、u_β 却被定义为固定坐标系下 p 方程的特解，尽管两者的特解在形式上是一样的，但其含义是截然不同的。此外，И. И. Цуккерман 没有注意到放大率应在固定坐标系内定义，其式（10）中所用到的 $u_\alpha'(z_i)$ 或 $u_\beta(z_i)$ 均是对旋转坐标系而言。众所周知，此二坐标系之间有一角度 χ。

（3）相邻轨迹的幅角公式（4）并没有考虑逸出电子的初始方位角 β_0，导致该式仅适用于 $\beta_0 = 0$ 的情况。

为此，我们在下面对 И. И. Цуккерман 移像系统的理论进行改进和补充。

2 矢量形式的主轨迹方程与相邻轨迹方程及其求解

对于电磁复合聚焦移像系统（阴极位于场中 $\phi(0) = 0$）的电子轨迹的研究，出发点仍然是西门纪业[4]导出的矢量形式的线性方程：

$$\boldsymbol{u}'' + \frac{\phi'(z)}{2[\phi(z) + \varepsilon_z]} \boldsymbol{u}' + \frac{1}{4[\phi(z) + \varepsilon_z]} \left[\phi''(z) + \frac{e}{2m_0} B^2(z)\right] \boldsymbol{u} = 0 \tag{12}$$

这里 ε_z 为逸出电子对应的轴向初电位，$\boldsymbol{u} = \boldsymbol{u}(z)$ 为旋转坐标系 (x, y) 中的轨迹矢量，它与固定坐标系中的轨迹矢量 $\boldsymbol{r}$ 有如下关系：

$$\boldsymbol{r} = \boldsymbol{u} \mathrm{e}^{\mathrm{i}\chi} \tag{13}$$

式中 $\chi = \chi(z)$ 为旋转坐标系绕 z 轴旋转相对于固定坐标系转过的角度，它是 $\boldsymbol{u}$ 所在平面位置 z 的函数[4]：

$$\chi = \sqrt{\frac{e}{8m_0}} \int_0^z \frac{B(z)}{\sqrt{\phi(z) + \varepsilon_z}} \mathrm{d}z + \chi_0 \tag{14}$$

式中，χ_0 为旋转坐标系的初始转角，通常 χ_0 可假定等于零。

由式（13）、式（14）可见，旋转坐标系的转角χ与逸出电子的初角度（α_0，β_0）无关，仅与逸出电子的轴向初电位ε_z有关；只要ε_z相同，都转过相同的转角。因此，当任意轨迹矢量r与主轨迹矢量$r_{prin.}$取相同的ε_z时，则主轨迹的相邻轨迹矢量可用下式表示：

$$p = r - r_{prin.} \tag{15}$$

按定义，$r = ue^{i\chi}$，$r_{prin.} = u_{prin.}e^{i\chi}$，若令

$$q = u - u_{prin.} \tag{16}$$

则p和q的关系同样可表示为

$$p = qe^{i\chi} \tag{17}$$

由方程式（12）与式（16），不难证明，旋转坐标系内以$u_{prin.}$表示的主轨迹方程或是以q表示的相邻轨迹方程都可用方程式（12）的矢量形式的线性方程来表示，只须以$u_{prin.}$或q代替u即可。由此方程结合$r_{prin.}$与p的初始条件，便可解出$u_{prin.}$和q值以及相应的$r_{prin.}$和p值。$u_{prin.}$、$r_{prin.}$与p、q等矢量如图2所示。

由方程式（12）可见，无论是$u_{prin.}$或是q方程都与u方程具有相同的系数，故其标量形式方程的特解都相同，设特解v、w的初始条件为

$$v(0) = 0, \quad v'(0)\sqrt{\varepsilon_z} = 1$$
$$w(0) = 1, \quad w'(0)\sqrt{\varepsilon_z} = 0 \tag{18}$$

逸出电子的初始条件如前文[5]所述，令

$$r_0 = r_0 e^{i\theta_0}, \quad r_0' = e^{i\beta_0}\tan\alpha_0 \tag{19}$$

式中，θ_0为逸出电子的初始位置r_0对X轴的幅角，α_0为初始电子的逸出角，β_0已如上述。

于是，对同一高度r_0发出的主轨迹，其初始条件可表示为

$$r_{0prin.} = r_0 e^{i\theta_0}, \quad r_{0prin.}' = 0 \tag{20}$$

而主轨迹的相邻轨迹p的初始条件可表示为

$$p_0 = r_0 - r_{0prin.} = 0,$$
$$p_0' = r_0' - r_{0prin.}' = e^{i\beta_0}\tan\alpha_0 \tag{21}$$

由此可导得主轨迹与相邻轨迹的通解为：

在旋转坐标系（x，y）内

$$u_{prin.} = r_0 e^{i(\theta_0 - \chi_0)}w - ir_0\sqrt{\frac{e}{8m_0}}B_0 e^{i(\theta_0 - \chi_0)}v,$$
$$q = \varepsilon_r^{1/2} e^{i(\beta_0 - \chi_0)}v \tag{22}$$

在固定坐标系（X、Y）内

$$r_{prin.} = r_0 e^{i(\theta_0 + \chi - \chi_0)}w - ir_0\sqrt{\frac{e}{8m_0}}B_0 e^{i(\theta_0 + \chi - \chi_0)}v,$$
$$p = \varepsilon_r^{1/2} e^{i(\beta_0 + \chi - \chi_0)}v \tag{23}$$

显然，由式（22），不难求得方程式（12）的通解u为

$$u = \left[\varepsilon_r^{1/2} e^{i(\beta_0 - \chi_0)} - i\sqrt{\frac{e}{8m_0}}B_0 r_0 e^{i(\theta_0 - \chi_0)}\right]v + r_0 e^{i(\theta_0 - \chi_0)}w \tag{24}$$

于是电子在固定坐标系（X、Y）内的位置矢量 r 不难表示为

$$r = \left[\varepsilon_r^{1/2} \mathrm{e}^{\mathrm{i}(\beta_0 + \chi - \chi_0)} - \mathrm{i}\sqrt{\frac{e}{8m_0}} B_0 r_0 \mathrm{e}^{\mathrm{i}(\theta_0 + \chi - \chi_0)}\right] v + r_0 \mathrm{e}^{\mathrm{i}(\theta_0 + \chi - \chi_0)} w \tag{25}$$

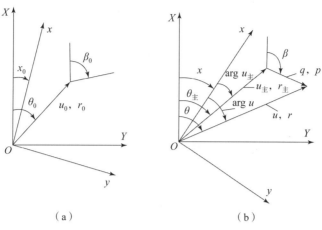

图2 固定坐标系与旋转坐标系下主轨迹与相邻轨迹

(a) 在 $Z=0$ 物面上；(b) 在 $Z=z$ 平面上

这样，我们便得到了主轨迹与近轴轨迹的矢量形式的解的表达式（23），而式（25）正表示此二轨迹的合成。

现确定系统的放大率与图像转角，若假定在像面 $z_i = l$ 处，与轴向初电位 ε_{z_1} 相对应的电子能理想聚焦，即

$$v(z_i, \varepsilon_{z_1}) = 0 \tag{26}$$

则由式（23）或式（25），可求得放大率 M 的表达式为

$$M = \frac{r(z_i, \varepsilon_{z_1})}{r_0} = \mathrm{e}^{\mathrm{i}[\chi_i(z_i, \varepsilon_{z_1}) - \chi_0(\varepsilon_{z_1}) - n\pi]} |w(z_i, \varepsilon_{z_1})| \tag{27}$$

由此可见，放大率的模等于 $|w(z_i, \varepsilon_{z_1})|$，但 r_i 与 r_0 之间还有一个夹角为 $\chi_i(z_i, \varepsilon_{z_1}) - \chi_0(\varepsilon_{z_1}) - n\pi$。由朗斯基行列式

$$\sqrt{\phi(z) + \varepsilon_{z_1}}(v'w - w'v) = 1 \tag{28}$$

式（27）还可以表示为

$$M = \mathrm{e}^{\mathrm{i}[\chi_i(z_i, \varepsilon_{z_1}) - \chi_0(\varepsilon_{z_1}) - n\pi]} \left|\frac{1}{v'(z_i, \varepsilon_{z_1})\sqrt{\phi(z_i) + \varepsilon_{z_1}}}\right| \tag{29}$$

式中，χ_i 乃是旋转坐标系在 $z = z_i$ 处相对于固定坐标系转过的角度，式（14）中当 $\varepsilon_z = \varepsilon_{z_1}$，$z = z_i$ 时，$\chi = \chi_i$，而图像转角 $\theta(z_i) - \theta_0$，即 r_i 矢量相对于 r_0 转过的角度，则可由式（25）表示为

$$\theta(z_i) - \theta_0 = \chi_i - \chi_0 - n\pi \quad (n = 1, 2, 3\cdots) \tag{30}$$

式中，n 为聚焦圈数。

若在 $z = z_i$ 处，旋转坐标系相对于初始位置转动了 π 的整数倍，即

$$\chi_i - \chi_0 = n\pi, \quad (n = 1, 2, 3\cdots) \tag{31}$$

于是

$$M = (-1)^n w(z_i, \varepsilon_{z_1}) = (-1)^n \frac{1}{v'(z_i, \varepsilon_{z_1}) \sqrt{\phi(z_i) + \varepsilon_{z_1}}} \quad (32)$$

$$\theta_i = \theta_0 \quad (33)$$

式（29）和式（32）便避免了文献［1］中所定义的放大率当 $\phi(0) = 0$ 时的不确定性问题，它无论对于 $\varepsilon_{z_1} = 0$ 或是 $\varepsilon_{z_1} \neq 0$ 都是适用的。

诚然，由于 $u_主$ 和 q 方程与 u 方程具有相同的特解，只是初始条件不同，通解才有差异。看来似乎并不需要经过 $u_主$ 和 q 的转换来求 u 值。但是从物理意义上来说，$u_主(r_主)$ 和 $q(p)$ 方程表示的轨迹比 $u(r)$ 方程给出的轨迹更为直观和清晰。这样，自同一物点逸出的电子轨迹可视为转动的主轨迹与围绕主轨迹做螺旋运动的相邻轨迹的合成。

3 对 И. И. Цуккерман 方程的改进与补充[4,5]

下面我们将旋转坐标系下矢量形式的主轨迹与相邻轨迹方程化为 И. И. Цуккерман 方程的形式，但使之适用于计算 $\phi(0) = 0$ 情况的成像系统。

由式（12）、式（13），不难求得固定坐标系下矢量形式的 r 方程为

$$[\phi(z) + \varepsilon_z] r'' + \frac{1}{2}\phi'(z) r' + \frac{1}{4}\phi''(z) r + i\sqrt{\frac{e}{2m_0}} \sqrt{\phi(z) + \varepsilon_z} \left[\frac{1}{2} B'(z) r + B(z) r'\right] = 0 \quad (34)$$

令 $r = re^{i\theta}$，将 r 及其导数代入上式，便有

$$[\phi(z) + \varepsilon_z] r'' + \frac{1}{2}[\phi(z) + \varepsilon_z] r' + \left\{\frac{1}{4}\phi''(z) + \sqrt{\phi(z) + \varepsilon_z} \sqrt{\frac{e}{2m_0}} B(z)\theta' - [\phi(z) + \varepsilon_z]\theta'^2\right\} r + $$
$$i\left\{\left[2\theta'(\phi(z) + \varepsilon_z) - \sqrt{\frac{e}{2m_0}} \sqrt{\phi(z) + \varepsilon_z} B(z)\right] r' + \left[(\phi(z) + \varepsilon_z)\theta'' + \frac{1}{2}\phi'(z)\theta' - \frac{1}{2}\sqrt{\frac{e}{2m_0}} \sqrt{\phi(z) + \varepsilon_z} B'(z)\right] r\right\} = 0 \quad (35)$$

若令

$$\theta' = \sqrt{\frac{e}{8m_0}} \frac{1}{\sqrt{\phi(z) + \varepsilon_z}} \left[B(z) + \frac{2m_0}{e} \frac{C}{r^2}\right] \quad (36)$$

式中，

$$C = \sqrt{\frac{2e}{m_0}} r_0^2 \left[\sqrt{\varepsilon_z} \theta_0' - \sqrt{\frac{e}{8m_0}} B(0)\right] \quad (37)$$

则方程式（35）中带有虚数 i 的大括弧项等于零，故式（35）变为

$$\sqrt{\phi(z) + \varepsilon_z} \frac{d}{dz}\left[\sqrt{\phi(z) + \varepsilon_z} \frac{dr}{dz}\right] + \left[\frac{1}{4}\phi''(z) + \frac{e}{8m_0} B^2(z)\right] r - \frac{m_0}{2e} \frac{C^2}{r^3} = 0 \quad (38)$$

在方程式（38）与式（36）中，若以 $r_{prin.}$ 代替 r，$\theta_{prin.}$ 代替 θ，且 $r_{prin.}(0) = r_0$，$\theta_{prin.}(0) = \theta_0$，$\theta'_{prin.}(0) = 0$，则由式（37），$C = -\frac{e}{2m_0} r_0^2 B(0)$，便得到主轨迹方程。同样，若以 p 代替 r，β 代替 θ，$p(0) = 0$，$\beta(0) = \beta_0$，于是 $C = 0$，便得到相邻轨迹方程。现将它们表达成以下形式：

主轨迹方程为

$$\sqrt{V(z)+\frac{\varepsilon_z}{\phi_m}}\frac{d}{dz}\left(\sqrt{V(z)+\frac{\varepsilon_z}{\phi_m}}\frac{dr_{\text{prin.}}}{dz}\right)+\frac{V''(z)}{4}r_{\text{prin.}}+\lambda r_{\text{prin.}}\left(G^2(z)-\frac{r_0^4}{r_{\text{prin.}}^4}\right)=0 \quad (39)$$

$$\sqrt{V(z)+\frac{\varepsilon_z}{\phi_m}}\frac{d\theta_{\text{prin.}}}{dz}=\sqrt{\lambda}\left(G^2(z)-\frac{r_0^2}{r_{\text{prin.}}^2}\right) \quad (40)$$

相邻轨迹方程为

$$\sqrt{V(z)+\frac{\varepsilon_z}{\phi_m}}\frac{d}{dz}\left(\sqrt{V(z)+\frac{\varepsilon_z}{\phi_m}}\frac{dp}{dz}\right)+\frac{V''(z)}{4}p+\lambda G^2(z)p+0 \quad (41)$$

$$\frac{d\beta(z)}{dz}=\sqrt{\lambda}\frac{G(z)}{\sqrt{V(z)+\frac{\varepsilon_z}{\phi_m}}} \quad (42)$$

式中，$V(z)=\phi(z)/\phi_m$，$G(z)=B(z)/B(0)$，ϕ_m 为系统轴上电位的最大值，λ 为特征值，它可表示为

$$\lambda=\frac{e}{8m_0}\frac{B_0^2}{\phi_m} \quad (43)$$

这便是对 И. И. Цуккерман 理论经过修正后适应于成像系统的固定坐标系下的主轨迹方程与相邻轨迹方程，它与旋转坐标系下矢量形式的 $u_{\text{prin.}}$ 和 q 方程是等效的。但是，这两种轨迹方程表达式对于求解来说是大有差异的。以矢量形式表示的 $u_\pm$ 和 q 方程乃是具有相同系数的二阶线性齐次方程，具有共同的特解。通解可以通过特解线性组合求得，这是比较方便的。而以式（39）、式（40）给出的主轨迹方程为非线性方程，式（41）、式（42）表示的相邻轨迹方程为线性齐次方程，它们的求解自然要复杂与困难得多。

幸运的是，我们可以通过上节中导出的主轨迹与相邻轨迹的矢量解，来求方程式（39）~式（42）的解。

由旋转坐标系内的主轨迹矢量 $u_\pm$ 与相邻轨迹矢量 q，确定轨迹在固定坐标系内的位置（模与幅角）。于是，对于主轨迹，则有

$$r_{\text{prin.}}=|u_{\text{prin.}}|=r_0\sqrt{w^2+\frac{e}{8m_0}B_0^2v^2} \quad (44)$$

而 $\theta_{\text{prin.}}=\chi+\arg u_{\text{prin.}}$（见图2），前者则可以用式（14）表示，而后者为

$$\arg u_{\text{prin.}}=\arctan\frac{w\sin(\theta_0-\chi_0)-\sqrt{\frac{e}{8m_0}}B_0v\sin(\theta_0-\chi_0)}{w\cos(\theta_0-\chi_0)+\sqrt{\frac{e}{8m_0}}B_0v\sin(\theta_0-\chi_0)}=\theta_0-\chi_0-\arctan\frac{\sqrt{\frac{e}{8m_0}}B_0v}{w}$$

于是

$$\theta_{\text{prin.}}=\theta_0+\sqrt{\lambda}\int_0^z\frac{G(z)}{\sqrt{V(z)+\frac{\varepsilon_z}{\phi_m}}}dz-\arctan\frac{\sqrt{\frac{e}{8m_0}}B_0v}{w} \quad (45)$$

对于相邻轨迹，则有

$$p = |\boldsymbol{q}| = \varepsilon_r^{1/2} |v| \tag{46}$$

$$\beta = \sqrt{\lambda} \int_0^z \frac{G(z)}{\sqrt{V(z) + \dfrac{\varepsilon_z}{\phi_m}}} \mathrm{d}z + \beta_0 + (m-1)\pi \tag{47}$$

当 $\chi - \chi_0 \leqslant m\pi$ 时，取 $m = 1, 2, \cdots, n$（n 为总聚焦圈数）。

由此可见，只须知道特解 $v(z)$ 和 $w(z)$ 值，即可方便地求得在固定坐标系下主轨迹与相邻轨迹的模及其幅角。

将式（44）、式（45）与式（46）、式（47）代入式（39）、式（40）与式（41）、式（42）中，不难证明它们正是标量形式的主轨迹方程与相邻轨迹方程的解。

同样，不难确定以初始条件式（19）逸出的电子轨迹在固定坐标系 (X, Y) 内的落点位置。

$$r = |\boldsymbol{u}| = \left\{ \left(\varepsilon_r + \frac{e}{8m_0} B_0^2 r_0^2 \right) v^2 + r_0^2 w^2 + \sqrt{\frac{e}{2m_0}} B_0 r_0 \varepsilon_r^{1/2} v^2 \sin(\theta_0 - \beta_0) + 2\varepsilon_r^{1/2} r_0 vw \cos(\theta_0 - \beta_0) \right\}^{1/2} \tag{48}$$

$$\theta = \theta_0 + \sqrt{\lambda} \int_0^z \frac{G(z)}{\sqrt{V(z) + \dfrac{\varepsilon_z}{\phi_m}}} \mathrm{d}z - \arctan \frac{r_0 \sqrt{\dfrac{e}{8m_0}} B_0 v - \varepsilon_r^{1/2} v \sin(\beta_0 - \theta_0)}{r_0 w + \varepsilon_r^{1/2} v \cos(\beta_0 - \theta_0)} \tag{49}$$

当 $v(z_i, \varepsilon_{z_1}) = 0$ 时，放大率可表示为

$$M = \frac{\boldsymbol{r}_i(z_i, \varepsilon_{z_1})}{\boldsymbol{r}_0} = \frac{r_i \mathrm{e}^{\mathrm{i}\theta_{\mathrm{prin.}}(z_i, \varepsilon_{z_1})}}{r_0 \mathrm{e}^{\mathrm{i}\theta_0}} = (-1)^n w(z_i, \varepsilon_{z_1}) \mathrm{e}^{\mathrm{i}[\chi_i(z_i, \varepsilon_{z_1}) - \chi_0 - n\pi]} \tag{50}$$

同样，图像转角表示式（45）仍可用式（30）表示。由此可见，这两种方法的一致性得到了彻底的证明。

4 电磁聚焦移像系统的二级横向像差

设理想聚焦的像点 z_i 与逸出电子的轴向初电位 ε_{z_1} 相对应，即

$$\boldsymbol{r}_i(z_i, \varepsilon_{z_1}) = r_0 M \mathrm{e}^{\mathrm{i}\theta_0} \tag{51}$$

当同一物点以初始条件 $(\varepsilon_0, \alpha_0, \beta_0)$ 逸出的近轴电子，其 $\varepsilon_z \neq \varepsilon_{z_1}$，则旋转坐标系的 x 轴在该位置处将多转 $\chi_i - \chi_0$ 的角度。于是近轴像差或称二级横向像差 $\Delta \boldsymbol{r}$ 可定义为

$$\Delta \boldsymbol{r} = \boldsymbol{r}_i(z_i, \varepsilon_{z_1}) - \boldsymbol{r}_i(z_i, \varepsilon_z) \tag{52}$$

按式（25），则上式可表示为

$$\Delta \boldsymbol{r} = \varepsilon_r^{1/2} v_i \mathrm{e}^{\mathrm{i}(\beta_0 + \chi_i - \chi_0)} + [w_i \mathrm{e}^{\mathrm{i}(\chi_i - \chi_0)} - M] r_0 \mathrm{e}^{\mathrm{i}\theta_0} - \mathrm{i} \sqrt{\frac{e}{8m_0}} B_0 r_0 v_i \mathrm{e}^{\mathrm{i}(\theta_0 + \chi_i - \chi_0)} \tag{53}$$

式中，$w_i = w(z_i, \varepsilon_z)$，$v_i = v(z_i, \varepsilon_z)$，$\chi_i = \chi(z_i, \varepsilon_z)$，$\chi_0 = \chi(z_0, \varepsilon_z)$。式中第一项乃是中心近轴色差（即二级近轴色差），它表示当 $r_0 = 0$ 时，以初始条件 $(\varepsilon_0, \alpha_0, \beta_0)$ 逸出的 $\varepsilon_z \neq \varepsilon_{z_1}$ 的近轴电子在 ε_{z_1} 所决定的像面上与中心像点的偏离。在一阶近似下，它可用众所周知的 R-A 公式表示，后两项乃是与 ε_z 相对应的轨迹在 ε_{z_1} 像面上由于放大率的改变以及磁场旋转所引起的位移项。

5 实例考察

下面我们通过一个实例来对求解主轨迹与相邻轨迹的两种方法进行考察。

我们可以证明，若移像系统的电位与磁感应强度沿轴分布为

$$\phi(z) = \frac{Mz}{l-(1-M)z}\phi_{ac} \tag{54}$$

$$B(z) = \frac{l^2}{[l-(1-M)z]^2}B(0) \tag{55}$$

则可得到放大率为 M 的无旋转的图像。这里 l 为阴极物面至成像面的距离，M 为所希求的放大率，ϕ_{ac} 为成像面相对于阴极物面的电位。

按照文献 [5]，我们不难给出特解 v、w 可表示为

$$v(z) = \sqrt{\frac{8m_0}{e}}\frac{1}{B_0}\left(1-\frac{1-M}{l}z\right)\sin(\chi-\chi_0) \tag{56}$$

$$w(z) = \left(1-\frac{1-M}{l}z\right)\cos(\chi-\chi_0) + \frac{1-M}{l}\sqrt{\frac{8m_0}{e}}\frac{\varepsilon_z^{1/2}}{B_0}\left(1-\frac{1-M}{l}z\right)\sin(\chi-\chi_0) \tag{57}$$

式中，

$$\chi = k\sqrt{\frac{lz}{l-(1-M)z}+\frac{\varepsilon_z}{\phi_0'}}, \quad \chi_0 = k\sqrt{\frac{\varepsilon_z}{\phi_0'}}$$

$$k = \sqrt{\frac{e}{2m_0}}\frac{B_0}{\phi_0'^{1/2}}, \quad \phi_0' = -E_c = \frac{M\phi_{ac}}{l}$$

由式 (22)，旋转坐标系下的主轨迹与相邻轨迹可表示为

$$\boldsymbol{u}_{\text{prin.}} = r_0\left(1-\frac{1-M}{l}z\right)e^{i(\theta_0-\chi)} + r_0\left(\frac{1-M}{l}\right)\sqrt{\frac{8m_0}{e}}\frac{\varepsilon_z^{1/2}}{B_0}\left(1-\frac{1-M}{l}z\right)\sin(\chi-\chi_0)e^{i(\theta_0-\chi_0)} \tag{58}$$

$$\boldsymbol{q} = \sqrt{\frac{8m_0}{e}}\frac{\varepsilon_r^{1/2}}{B_0}\left(1-\frac{1-M}{l}z\right)\sin(\chi-\chi_0)e^{i(\beta_0-\chi_0)} \tag{59}$$

在固定坐标系下，则

$$\boldsymbol{r}_{\text{prin.}} = r_0\left(1-\frac{1-M}{l}z\right)e^{i\theta_0} + r_0\left(\frac{1-M}{l}\right)\sqrt{\frac{8m_0}{e}}\frac{\varepsilon_z^{1/2}}{B_0}\left(1-\frac{1-M}{l}z\right)\sin(\chi-\chi_0)e^{i(\theta_0+\chi-\chi_0)} \tag{60}$$

$$\boldsymbol{p} = \sqrt{\frac{8m_0}{e}}\frac{\varepsilon_r^{1/2}}{B_0}\left(1-\frac{1-M}{l}z\right)\sin(\chi-\chi_0)e^{i(\theta_0+\chi-\chi_0)} \tag{61}$$

由式 (49) 与式 (50)，我们不难求得主轨迹的模及其幅角：

$$r_{\text{prin.}} = r_0\left(1-\frac{1-M}{l}z\right)\left[1-\left(\frac{1-M}{l}\right)\sqrt{\frac{8m_0}{e}}\frac{\varepsilon_z^{1/2}}{B_0}\sin 2(\chi-\chi_0) + \left(\frac{1-M}{l}\right)^2\frac{8m_0}{e}\frac{\varepsilon_z}{B_0^2}\sin^2(\chi-\chi_0)\right]^{1/2} \tag{62}$$

$$\theta_{\text{prin.}} = \theta_0 + \chi - \chi_0 - \arctan\frac{\sin(\chi-\chi_0)}{\cos(\chi-\chi_0)+\left(\frac{1-M}{l}\right)\sqrt{\frac{8m_0}{e}}\frac{1}{B_0}\varepsilon_z^{1/2}\sin(\chi-\chi_0)}\frac{n!}{r!(n-r)!} \tag{63}$$

相邻轨迹的模及其幅角可由式（46）、式（47）求得

$$p = \varepsilon_r^{1/2} \left| \sqrt{\frac{8m_0}{e}} \frac{1}{B_0} \left(1 - \frac{1-M}{l}z\right) \sin(\chi - \chi_0) \right| \tag{64}$$

$$\beta = \chi - \chi_0 + \beta_0 + (m-1)\pi \tag{65}$$

这里，当 $\chi - \chi_0 \leq m\pi$ 时，取 $m = 1, 2, \cdots, n$（n 为总聚焦圈数）。

由式（62）、式（63）可见，当 $\varepsilon_z = 0$ 时，$\theta_{\text{prin.}} = \theta_0$，主轨迹为一条直线。

由式（56）、式（57）可见，当 $\varepsilon_z = \varepsilon_{z_1}$, $z = z_i = l$ 时，$\chi_i - \chi_0 = n\pi$，则

$$v(z_i, \varepsilon_{z_1}) = 0, \quad w(z_i, \varepsilon_{z_1}) = (-1)^n M \tag{66}$$

这与式（32）的定义是一致的。

由式（14），我们不难求得在 $z_i = l$ 的像面上，ε_z 电子多转的 $\chi_i - \chi_0$ 角为

$$\chi_i - \chi_0 = n\pi + \frac{k}{\sqrt{\phi_0'}}(\sqrt{\varepsilon_{z_1}} - \sqrt{\varepsilon_z}) - \frac{1}{2} \frac{k^2}{n\pi \phi_0'}(\varepsilon_{z_1} - \varepsilon_z) \tag{67}$$

于是由式（53），便可求得近轴横向像差为

$$\Delta r = \left[\frac{2M}{\phi_0'} \varepsilon_r^{1/2}(\sqrt{\varepsilon_{z_1}} - \sqrt{\varepsilon_z}) - \frac{M}{\phi_0'\sqrt{\phi_{ac}}} \varepsilon_r^{1/2}(\varepsilon_{z_1} - \varepsilon_z)\right] \times$$
$$\exp\left\{i\left[n\pi + \beta_0 + \frac{k}{\sqrt{\phi_0'}}(\sqrt{\varepsilon_{z_1}} - \sqrt{\varepsilon_z}) - \frac{1}{2} \frac{k^2}{n\pi \phi_0'}(\varepsilon_{z_1} - \varepsilon_z)\right]\right\} -$$
$$r_0 \frac{M-1}{l}\left[\frac{2M}{\phi_0'} \varepsilon_z^{1/2}(\sqrt{\varepsilon_{z_1}} - \sqrt{\varepsilon_z}) - \frac{M}{\phi_0'\sqrt{\phi_{ac}}} \varepsilon_z^{1/2}(\varepsilon_{z_1} - \varepsilon_z)\right] \times$$
$$\exp\left\{i\left[n\pi + \theta_0 + \frac{k}{\sqrt{\phi_0'}}(\sqrt{\varepsilon_{z_1}} - \sqrt{\varepsilon_z}) - \frac{1}{2} \frac{k^2}{n\pi \phi_0'}(\varepsilon_{z_1} - \varepsilon_z)\right]\right\} \tag{68}$$

这与文献［6］在 $r_0 = 0$ 时轴上点色差公式是一致的。

结束语

（1）本文对 И. И. Цуккерман 理论进一步改进与补充，使之适用于研究阴极透镜的移像系统，并由旋转坐标系下矢量方程的解导出 И. И. Цуккерман 方程的主轨迹与相邻轨迹的解的表达式。

（2）用旋转坐标系下矢量形式的主轨迹方程与相邻轨迹方程研究移像系统实质上是与 И. И. Цуккерман 建议的标量形式的方程一致的，前者的求解较后者简便。

（3）本文研究了求解某一典型移像系统的实例，给出了其主轨迹与相邻轨迹的解以及近轴横向像差的表达式。

参 考 文 献

［1］ЦУККЕРМАН И И. Электронная Оптика В Телевидении［M］. Москва-Ленинград，ГЭИ，1958.

［2］ЦУККЕРМАН И И. О Магнитной фокусировке в передающих трубках с перенсом изобрaжния[J]. Ж. Т. Ф.,1953,23:1228-1238,

[3] COSSLETT V E. Introduction to electron optics[M]. Oxford：Oxford University Press,1946.

[4] 西门纪业. 复合浸没物镜的电子光学性质和像差理论[J]. 物理学报,1957,13(4)：339-356.

[5] 周立伟,方二伦. 一种新型的放大率 $M \neq 1$ 的电磁聚焦移像系统[J]//夜视技术论文集. 北京：北京理工大学出版社,1982,1：206-219.

[6] CHOU L W. Electron optics of concentric spherical electromagnetic focusing systems[J]. Advances in Electronics and Electron Physics,1979,52：119-132.

4.2 图像无旋转的电磁聚焦移像系统的研究
A Study of Electrostatic and Magnetic Focusing Image Transference System Without Image Rotation

提要：本文对 И. И. Цуккерман[1] 的电磁聚焦移像系统的理论进行补正和补充，放宽了实现图像无转角的条件，导出了能获得放大率 $M=1$ 或 $M\neq1$ 且无转角的图像移像系统中轴上相对电位 $V(z)$ 和磁感应强度分布 $G(z)$ 的关系式，以及特征值 λ 应满足的条件，提出了设计图像无转角且放大率可变的移像系统的新途径。文中给出了解析实例与数值计算实例。

Abstract: A method for designing electrostatic and magnetic focusing image transference systems is introduced with variable magnification but without image rotation and two conditions, the eigenvalue λ as well as the potential $V(z)$ and the magnetic induction $G(z)$ along the axis are derived. As soon as these two conditions are satisfied in the system, an image with magnification $M=1$ or $M\neq1$ but without image rotation can be obtained. Thus, the method for designing image transference system without image rotation appears to be an improvement and development for the existing one. Some practical examples are also given.

引言

И. И. Цуккерман 曾对电磁复合聚焦移像系统进行了一系列的研究，提出了研究该系统聚焦成像特性的理论和方法，给出了用电子光学方法实现图像像倍且无转角的移像系统的条件。其理论的实质，一是用主轨迹矢量 $r_{\pm}(z)$ 及其相邻轨迹矢量 $\rho(z)$ 来描述电子轨迹矢量 $r(z)$；二是按所给条件，将边值问题下标量形式的傍轴方程化为 Sturm-Liouville 方程，从而获得在图像无旋转下实现聚焦的电场和磁场的分布。应当指出，其轨迹方程中的电场和磁场均采用旋转对称系统的轴上分布。

И. И. Цуккерман 的工作推进了移像系统的研究，但其理论尚有值得商榷与补正之处；一是他在著作中[1~4]应用的是傍轴方程，而不是近轴方程，故其主轨迹方程与相邻轨迹方程以及所定义的放大率表达式并不适用于宽电子下电磁复合聚焦移像系统的分析，且对主轨迹非线性方程的求解探讨得不够[5]；二是他所提出的系统获得无旋转图像的条件过于苛刻，难以实现，因而限制了该理论的适用范围。文献 [5] 针对前一个问题，进一步考察了宽电子束下电磁复合聚焦移像系统的电子光学理论，由矢量形式的近轴轨迹方程出发并采用轴上准规范化电位 $\phi(z)+\varepsilon_z$，使主轨迹方程与相邻轨迹方程适用于阴极透镜成像系统；并导出了旋转坐标系中以特解 v、w 表示的主轨迹和相邻轨迹的通解及系统放大率的表达式。

周立伟ª，倪国强ª，方二伦ᵇ. a) 北京工业学院），b) 西安近代化学研究所. 电子学报, V. 12, No. 6, 1984, 33－34.

本文在此基础上，进一步探讨了系统实现图像无旋转的新途径，以及相应的设计方法。作者认为，只要将电磁复合聚焦移像系统的标量形式的近轴方程作为磁聚焦的特征值问题来处理，使所给出的相对电位与磁感应强度的沿轴分布满足某一条件式（从而放宽了И. И. Цуккерман 所提出的条件），则可确定磁聚焦的特征值，并构成图像无旋转且放大率 $M=1$ 或 $M \neq 1$ 的电磁聚焦移像系统。

1 移像系统的基本理论

对于轴对称电磁聚焦移像系统，主轨迹 $\boldsymbol{r}_\pm(z)$ 与相邻轨迹 $\boldsymbol{\rho}(z)$ 可用以下复数形式表示

$$\begin{aligned}\boldsymbol{r}_\pm(z) &= r_\pm(z) e^{i\theta_\pm(z)} \\ \boldsymbol{\rho}(z) &= p(z) e^{i\beta(z)}\end{aligned} \tag{1}$$

于是有

$$r(z) = \boldsymbol{r}_\pm(z) + \boldsymbol{\rho}(z) \tag{2}$$

式中，$r_\pm(z)$、$p(z)$ 分别是 $\boldsymbol{r}_\pm(z)$ 和 $\boldsymbol{\rho}(z)$ 的模；$\theta_\pm(z)$、$\beta(z)$ 是 $r_\pm(z)$ 和 $p(z)$ 相对于固定坐标系 x 轴所转过的角度，如图 1 所示。

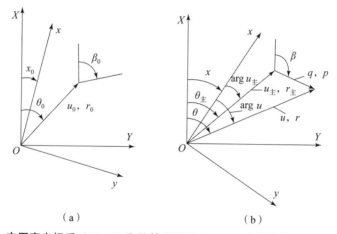

图 1 在固定坐标系 (X, Y) 和旋转坐标系 (x, y) 中的主轨迹与相邻轨迹

(a) 在 $Z=0$ 物面上；(b) 在 $Z=z$ 平面上

由矢量形式的近轴轨迹方程出发[6]，不难分别导得主轨迹方程和相邻轨迹方程为[5]

$$\sqrt{V(z)+\varepsilon_z^*}\frac{\mathrm{d}}{\mathrm{d}z}\left(\sqrt{V(z)+\varepsilon_z^*}\frac{\mathrm{d}r_\pm}{\mathrm{d}z}\right) + \left\{\frac{V''(z)}{4} + \lambda\left[G^2(z) - \frac{r_0^4}{r_\pm^4}\right]\right\}r_\pm = 0 \tag{3}$$

$$\sqrt{V(z)+\varepsilon_z^*}\frac{\mathrm{d}\theta_\pm}{\mathrm{d}z} = \sqrt{\lambda}\left[G(z) - \frac{r_0^2}{r_\pm^2}\right] \tag{4}$$

$$\sqrt{V(z)+\varepsilon_z^*}\frac{\mathrm{d}}{\mathrm{d}z}\left(\sqrt{V(z)+\varepsilon_z^*}\frac{\mathrm{d}p}{\mathrm{d}z}\right) + \left[\frac{V''(z)}{4} + \lambda G^2(z)\right]p = 0 \tag{5}$$

$$\frac{\mathrm{d}\beta}{\mathrm{d}z} = \sqrt{\lambda}\frac{G(z)}{\sqrt{V(z)+\varepsilon_z^*}} \tag{6}$$

式中，z 是以阴极物平面到像面的距离 l 为系统单位长度的相对轴向位置坐标，$0 \leq z \leq 1$；撇号表示对 z 求导；$V(z)=\phi(z)/\phi_i$；$G(z)=B(z)/B_0$，$\phi(z)$、$B(z)$ 分别是电位与磁感应

强度的沿轴分布; $\phi_i = \phi(z_i = 1)$, $B_0 = B(z_0 = 0)$, 且 $V(z_0 = 0) = 0$, $V(z_i = 1) = 1$, $G(z_0 = 0) = 1$; r_0 为逸出电子的离轴高度; $\varepsilon_z^* = \varepsilon_z/\phi_i$; 而 $\varepsilon_z = \varepsilon_0 \cos^2\alpha_0$, ε_0、ε_z 分别为电子初能与轴向初能所对应的初电位, α_0 则为电子的初始逸出角; λ 为无量纲的特征值, 它可表示为

$$\lambda = \frac{e}{8m_0} \frac{B_0^2 l^2}{\phi_i} \tag{7}$$

式中, e/m_0 为电子的荷质比。

在求解式 (3) ~式(6) 时, 我们引入了如图 1 所示的旋转坐标系 (x, y), 它与固定坐标系有一个夹角 $\chi(z)$。不难证明[5]

$$\chi(z) = \sqrt{\lambda} \int_0^z \frac{G(z)}{\sqrt{V(z) + \varepsilon_z^*}} dz + \chi_0 \tag{8}$$

式中, χ_0 是旋转坐标系的初始转角。在旋转坐标系下, 矢量形式的主轨迹方程和相邻轨迹方程具有相同的形式, 因而有相同的特解[6]。若矢量方程对应的标量方程的两个特解 v、w 的初始条件为

$$\begin{aligned} v(z_0) &= 0, \quad v'(z_0) \sqrt{\varepsilon_z^*} = 1; \\ w(z_0) &= 1, \quad w'(z_0) \sqrt{\varepsilon_z^*} = 0 \end{aligned} \tag{9}$$

则矢量轨迹方程的通解可由 v、w 的组合并结合主轨迹和相邻轨迹的初始条件来确定。

当电子以初始条件 $\boldsymbol{r}_0 = r_0 e^{i\theta_0}$ 和 $\boldsymbol{r}'_0 = l\tan\alpha_0 \cdot e^{i\beta_0}$ 逸出时, 式 (3) 和式 (4) 的解可表示为

$$r_{主}(z) = r_0 \sqrt{w^2 + \lambda v^2} \tag{10}$$

$$\theta_{主}(z) = \theta_0 + \sqrt{\lambda} \int_0^z \frac{G(z)}{\sqrt{V(z) + \varepsilon_z^*}} dz - \arctan\frac{\sqrt{\lambda} v}{w} \tag{11}$$

而相邻轨迹则可表示为

$$p(z) = \sqrt{\varepsilon_r^*} l |v| \tag{12}$$

$$\beta(z) = \sqrt{\lambda} \int_0^z \frac{G(z)}{\sqrt{V(z) + \varepsilon_z^*}} dz + \beta_0 + (m-1)\pi \tag{13}$$

式中, $\varepsilon_z^* = \varepsilon_z/\phi_i$, 而 $\varepsilon_r = \varepsilon_0 \sin^2\alpha_0$, 为与逸出电子径向初能对应的初电位; θ_0、β_0 分别为 $\theta_{主}(z)$、$\beta(z)$ 的初始值。在式 (13) 中, 当等号右边第一项等于或小于 $m\pi$ 时, 应取 $m = 1, 2, \cdots, n$ (n 为总聚焦圈数)。

若在像面 $z_i = l$ 处, 具有相对轴向初电位 $\varepsilon_{z_1}^*$ 的电子能够理想聚焦, 即

$$V(z_i, \varepsilon_{z_1}^*) = 0 \tag{14}$$

则在此像面上, 系统"放大率矢量"可表示为

$$M = \frac{r(z_i, \varepsilon_{z_1}^*)}{r_0} = w(z_i, \varepsilon_{z_1}^*) e^{i(\chi_i - \chi_0)} = \frac{1}{v'(z_i, \varepsilon_{z_1}^*) \sqrt{1 + \varepsilon_{z_1}^*}} e^{i(\chi_i - \chi_0)} \tag{15}$$

其模

$$M = |w(z_i, \varepsilon_{z_1}^*) e^{i(\chi_i - \chi_0)}| \tag{16}$$

而图像转角 $[\theta_{\pm}(z_i) - \theta_0]$，即 $\theta_{\pm}(z_i, \varepsilon_{z_1}^*)$ 相对于 r_0 所转过的角度可表示为

$$\theta_{\pm}(z_i) - \theta_0 = \chi_i - \chi_0 - n\pi \tag{17}$$

式中，$\chi_i = \chi(z_i)$，可由式（8）确定（令其中 $\varepsilon_z^* = \varepsilon_{z_1}^*$）。由此可见，若在像面 z_i 处，旋转坐标系相对于初始位置转动的角度为 π 的整数倍时，即当

$$\chi_i - \chi_0 = n\pi \tag{18}$$

时，有 $\theta_{\pm}(z_i) = \theta_0$。式（18）就是图像无旋转的条件。这时"放大率矢量"的方向将与固定坐标系的 X 轴重合，并可用标量值表示为

$$M = (-1)^n w(z_i, \varepsilon_{z_1}^*) = (-1)^n \frac{1}{v'(z_i, \varepsilon_{z_1}^*)\sqrt{1 + \varepsilon_{z_1}^*}} \tag{19}$$

式（15）和式（19）就避免了文献［1］所定义的放大率当规范化电位 $\phi(z_0) = 0$ 时的不确定性问题，对于 $\varepsilon_{z_1}^* = 0$ 或 $\varepsilon_{z_1}^* \neq 0$ 的情况，也都是适用的。此外，由式（19）可见，在式（18）的条件下，当 n 为奇数时，$w(z_i, \varepsilon_{z_1}^*)$ 与 $v'(z_i, \varepsilon_{z_1}^*)$ 为负值，当 n 为偶数时，$w(z_i, \varepsilon_{z_1}^*)$ 与 $v'(z_i, \varepsilon_{z_1}^*)$ 为正值，故 M 将恒取正值。

2 实现图像无旋转的电磁聚焦移像系统的途径

文献［1］、［3］、［4］指出，当系统满足

$$G(z_i) = r_0^2 / r_{\pm}^2(z) \tag{20}$$

的条件时，便能实现图像无转角的要求。应当指出，式（20）的条件过于苛刻。按此条件，$r_{\pm}(z)$ 便以式（3）在纯电场下的形式，由 $r_{\pm}(z_0) = r_0$、$r'_{\pm}(z_0) = 0$ 作为初值问题来确定，再由 $r_{\pm}(z_i)/r_0$ 来确定放大率 M；主轨迹乃是位于子午面内的平面曲线；磁场的作用是使电子束聚焦，并使主轨迹处处与磁力线一致。这在实际上是较难做到的。

通常，主轨迹在磁场的作用下，已不再是平面曲线。只要主轨迹电子逸出的速度方向和阴极面处的磁力线的夹角不为零，主轨迹就是一条空间螺旋曲线。因此，有必要对图像无旋转的电磁聚焦移像系统的理论重新加以考察。

根据式（14）和式（18），我们可把理想聚焦与图像无旋转的问题归结为求电磁聚焦移像系统的标量形式近轴轨迹方程

$$\frac{\mathrm{d}}{\mathrm{d}z}\left(\sqrt{V(z) + \varepsilon_{z_1}^*}\frac{\mathrm{d}v}{\mathrm{d}z}\right) + \frac{1}{\sqrt{V(z) + \varepsilon_{z_1}^*}}\left[\frac{V''(z)}{4} + \lambda G^2(z)\right]v = 0 \tag{21}$$

在边界条件

$$v(z_0) = 0, \quad v(z_i, \varepsilon_{z_1}^*) = 0 \tag{22}$$

下的解，并使其满足图像无旋转的条件式（18）。

对式（21）作变量代换，令

$$u = \sqrt{G}v, \quad \psi = \int_0^z \frac{G(z)}{\sqrt{V(z) + \varepsilon_{z_1}^*}}\mathrm{d}z$$

则式（21）和式（22）可分别写为

$$\frac{\mathrm{d}^2 u}{\mathrm{d}\psi^2} + \lambda u + \frac{u}{4G^2}\left\{\left(\frac{\mathrm{d}G}{\mathrm{d}\psi}\right)^2 - 2G\frac{\mathrm{d}^2 G}{\mathrm{d}\psi^2} + \frac{G}{V + \varepsilon_{z_1}^*}\left[G\frac{\mathrm{d}^2 V}{\mathrm{d}\psi^2} + \frac{\mathrm{d}G}{\mathrm{d}\psi}\frac{\mathrm{d}V}{\mathrm{d}\psi} - \frac{G}{2(V + \varepsilon_{z_1}^*)}\left(\frac{\mathrm{d}V}{\mathrm{d}\psi}\right)^2\right]\right\} = 0 \tag{23}$$

$$u(0) = 0, \quad u(\psi_i, \varepsilon_{z_1}^*) = 0 \tag{24}$$

式中，$\psi_i = \psi(z_i)$。若使式（23）中大括号内的关系式为零，则该式便简化为

$$\frac{d^2 u}{d\psi^2} + \lambda u = 0 \tag{25}$$

这便是众所周知的 Sturm-Liouville 方程。按照特征值及其相应的方程解——特征函数存在的基本定理，在满足某些实际要求的情况下，总可以选择特征值，使式（25）可解。于是，在解得的场分布下，可以选择 B_0、ϕ_i、l，使之在 z_i 处实现聚焦。

由式（25）和边界条件式（24），可解得

$$u = \sin(\sqrt{\lambda}\psi) \tag{26}$$

和

$$\sin(\sqrt{\lambda}\psi_i) = 0 \tag{27}$$

显然，式（27）与式（18）是完全一致的。由此可确定与 n 级像相关的特征值 λ_n。而正因为 λ 可取各个离散值，才有可能实现多次聚焦，并在物面与像面之间存在 $(n-1)$ 级中间像。这些中间像分别对应于 λ_1、λ_2、…、λ_{n-1}。

把式（23）中大括号内的关系式和式（26）中的 ψ 变换为 z 的函数，相应地可得

$$\frac{3(V+\varepsilon_{z_1}^*)}{G^2}\left(\frac{dG}{dz}\right)^2 - \frac{2(V+\varepsilon_{z_1}^*)}{G}\frac{d^2 G}{dz^2} - \frac{1}{G}\frac{dV}{dz}\frac{dG}{dz} + \frac{d^2 V}{dz^2} = 0 \tag{28}$$

$$\lambda_n = \frac{e}{8m_0}\frac{B_0^2 l^2}{\phi_i} = \frac{n^2\pi^2}{\left[\int_0^1 \frac{G(z)}{\sqrt{V(z)+\varepsilon_{z_1}^*}} dz\right]^2} \tag{29}$$

这说明，只要把式（21）作为磁聚焦的特征值问题来处理，就可得到上面的两个条件式，使移像系统能在像面上获得聚焦且无转角的图像。由此可见，我们不必像文献 [1] 那样要求主轨迹的 $\theta_\pm(z)$ 处处等于 θ_0，而仅要求在终端处 $\theta_\pm(z_i) = \theta_0$。至于特解 v，可由式（26）求得，特解 w 可根据常微分方程理论求得。若考虑式（9），它们可表示为：

$$v(z,\varepsilon_{z_1}^*) = \frac{1}{\sqrt{\lambda}}\frac{1}{\sqrt{G(z)}}\sin\left(\sqrt{\lambda}\int_0^z \frac{G(z)}{\sqrt{V(z)+\varepsilon_{z_1}^*}} dz\right) \tag{30}$$

$$w(z,\varepsilon_{z_1}^*) = -\sqrt{\varepsilon_{z_1}^*}v(z,\varepsilon_{z_1}^*)\frac{d}{dz}\left(\frac{1}{\sqrt{G(z)}}\right) - v(z,\varepsilon_{z_1}^*)\int \frac{1}{v^2(z,\varepsilon_{z_1}^*)}\frac{1}{\sqrt{V(z)+\varepsilon_{z_1}^*}} dz \tag{31}$$

最后可由式（30）和式（19）求得放大率 M 的表达式为

$$M = \frac{1}{\sqrt{G(z_i)}} \tag{32}$$

这表明：在满足条件式（28）和式（29）的情况下，M 仅取决于 $B(z_i)/B_0$ 的值，在形式上和 $\phi(z)$、$B(z)$ 的分布以及 ϕ_i 值无关。

文献 [4]、文献 [1] 中虽也有与式（32）相同的结果，但它是由理想化的条件式（20）求取的，而不是由所定义的放大率的一般表达式证得的。从物理意义上看，文献 [4]、文献 [1] 要求磁力线在阴极物平面处垂直于该平面，即 $G'(z_0) = 0$，以保证位于子午平面内的主轨迹电子处处沿磁力线运动；而在本文中，只要满足我们提出的条件，主轨迹电子就在以通过逸出点的磁力线为母线的某曲面上做螺旋式运动，最后仍在该磁力线上聚焦。因此，两者在放大率的表达式上虽然一致，但物理图像是明显不同的。

3 图像无旋转移像系统的设计

综上所述，图像无旋转移像系统的设计可归纳为如下：

(1) 假定 $V(z)$ 或 $G(z)$ 分布。

(2) 将条件式（28）作为二阶常微分方程的边值问题，求取相应的 $G(z)$ 或 $V(z)$ 分布。

(3) 确定聚焦圈数 n，由条件式（29）求取 λ 值；根据需要选择 ϕ_i、B_i 及 l 的值。

(4) 由式（32）求放大率 M。

(5) 由式（30）、式（31）求特解 v、w，或将式（21）作为二阶齐次常微分方程的初值问题，结合初始条件式（9）求取 v、w。

最后，由电子运动的初始条件，根据式（10）~式（13）及式（2），确定电子运动的轨迹。然后，根据 $V(z)$、$G(z)$ 分布及 ϕ_i、B_0 值，再进行电极和磁线圈的设计。

显然，对式（28）来说，只要给定的 $G(z)$ 或 $V(z)$ 是连续的，且其一阶、二阶导数连续有界，并在 $[0, 1]$ 上有 $G(z) \neq 0$，或在 $[0, 1]$ 上有 $V(z) + \varepsilon_{z_1}^* \neq 0$，则必定存在满足初始条件 $V(0) = 0$，$V'(0) = V'_0$ 的 $V(z)$ 的唯一解，或满足初始条件 $G(0) = 1$，$G'(0) = G'_0$ 的 $G(z)$ 的唯一解。

实际上，无论是由 $G(z)$ 求 $V(z)$，还是由 $V(z)$ 求 $G(z)$，我们都是作为边值问题来处理的。其边值条件分别是

$$V(z_0) = 0, \quad V(z_0) = 1, \tag{33}$$

$$G(z_0) = 1, \quad G(z_0) = 1/M^2 \tag{34}$$

这在实质上扩展了 И. И. Цуккерман 的方法：不但可在已知 $G(z)$ 时求相应的 $V(z)$；而且，特别是在由 $V(z)$ 求 $G(z)$ 时，作为边值问题，用所企求的各个放大率 M 来求取与 $V(z)$ 相对应的各个 $G(z)$，而无须作为初值问题，不必规定 $G'(0) = 0$，使得在假定情况下，只产生唯一解。我们将图像无旋转的条件放宽，这给场分布的选择提供了很大的灵活性，为真正实现图像无旋转而放大率可变的电磁复合聚焦移像系统提供了可能性。

4 图像无旋转的移像系统的解析实例

文献 [8] 中曾给出一种理想的 $M \neq 1$ 的移像系统场分布的表达式

$$G(z) = \frac{1}{[1 - (1-M)z]^2} \tag{35}$$

$$V(z) = \frac{Mz}{[1 - (1-M)z]} \tag{36}$$

不难证明，它们满足条件式（28），并由式（30）、式（31），特解 v、w 可表示为

$$v(z) = \frac{1}{\sqrt{\lambda}} [1 - (1-M)z] \sin(\chi - \chi_0),$$

$$w(z) = [1 - (1-M)z] \cos(\chi - \chi_0) + \sqrt{\frac{\varepsilon_z^*}{\lambda}} (1-M)[1 - (1-M)z] \sin(\chi - \chi_0)$$

式中，

$$\chi = \frac{2\sqrt{\lambda}}{M}\sqrt{V(z)+\varepsilon_z^*},\ \chi_0 = \frac{2\sqrt{\lambda}}{M}\sqrt{\varepsilon_z^*}$$

由式（10）~式（13）便可确定电子主轨迹以及相邻轨迹的表达式：

$$r_{\text{主}}(z) = r_0\left[1-(1-M)z\right]\left[1+(1-M)\sqrt{\frac{\varepsilon_z^*}{\lambda}}\sin 2(\chi-\chi_0)+(1-M)^2\left(\frac{\varepsilon_z^*}{\lambda}\right)\sin^2(\chi-\chi_0)\right]^{1/2} \tag{37}$$

$$\theta_{\text{主}}(z) = \theta_0 + \chi - \chi_0 - \arctan\frac{\sin(\chi-\chi_0)}{\cos(\chi-\chi_0)+(1-M)\sqrt{\frac{\varepsilon_z^*}{\lambda}}\sin(\chi-\chi_0)} \tag{38}$$

$$p(z) = \sqrt{\frac{\varepsilon_z^*}{\lambda}} \cdot l \cdot [1-(1-M)z]|\sin(\chi-\chi_0)| \tag{39}$$

$$\beta(z) = \chi - \chi_0 + \beta_0 + (m-1)\pi \tag{40}$$

当 $M=1$ 时，式（37）~式（40）便是均匀平行复合电磁聚焦系统的解[7]。

关于在上述场分布下系统的一阶横向像差的分析、场有微扰下获得不失真图像的途径，以及场内电子轨迹、像差与调制传递函数的计算，我们将另文叙述。

图 2 给出了 $M=1.4$、$n=1$ 时的 $G(z)$、$V(z)$、$v(z)$、$w(z)$、$r_{\text{主}}(z)$、$p(z)$ 的曲线。图 3 则是对应于不同放大率 M 的上述场分布曲线图。图 3 中过坐标原点的会聚曲线族是 $V(z)$ 分布；过 (0, 1) 点的曲线族是 $G(z)$ 分布。

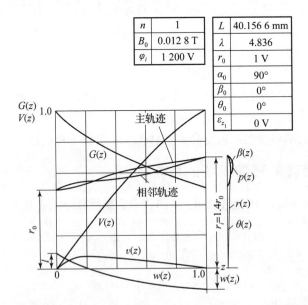

图 2 $M=1.4$ 时的 $G(z)$、$V(z)$、$v(z)$、$w(z)$ 等曲线

应当指出，对于式（35）中的 $G(z)$ 而言，式（28）的 $V(z)$ 只有唯一解，即式（36）。但对于式（36）中的 $V(z)$ 来说，式（28）将对应不同的放大率 M 有多个 $G(z)$ 的解，而式（35）的解只是其中之一。

5 计算实例

以如上研究为基础，我们编制了计算图像无旋转移像系统的轴上相对电磁场分布和其他有关数据的源程序。该程序可在已知函数或数值形式的 $G(z)$ 或 $V(z)$ 的情况下，求取相应的 $V(z)$ 或 $G(z)$。图 4 即是针对同一个 $V(z)$ 所算得的对应于各个不同放大率 M 的 $G(z)$ 分布。图中

$$V(z) = \frac{\gamma z}{1-(1-\gamma)z} \tag{41}$$

其中参变量 γ 取 0.6。由图 4 可以看出，在计算范围内，式（28）的解是稳定的。

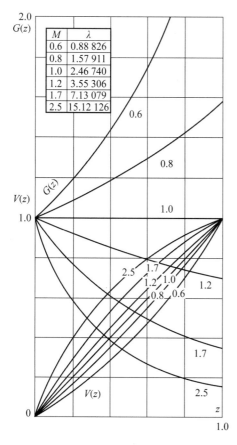

图 3　各种不同的 $V(z)$ 对应的 $G(z)$ 场分布曲线

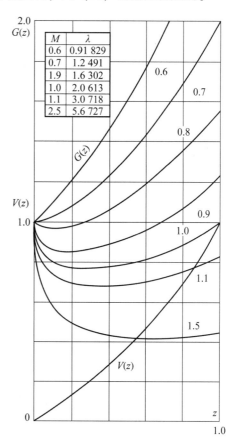

图 4　$V(z)$ 与其对应的不同 $G(z)$ 场分布曲线

结束语

本文补正了 И. И. Цуккерман 的理论；导出了实现图像无旋转的电磁聚焦移像系统的相对电位与磁感应强度沿轴分布的关系式，以及特征值应满足的条件；给出了移像系统轨迹方程的特解表示式；提出了实现图像无旋转且放大率可变的移像系统的新的设计方法。本文结果对电磁聚焦移像系统的研究、计算和设计将具有实际意义。

参 考 文 献

[1] ЦУККЕРМАН И И. Электронная Оптика В Телевидении[M]. Москва-Ленинград, ГЭИ,1958.

[2] ЦУККЕРМАН И И. О магнитной фокусировке в передающих трубках с пиренсом изображния[J]. Ж. Т. Ф. , 1952,23:1228 – 1238.

[3] ЦУККЕРМАН И И. Магнитные электронно-оптические системы с переменным увеличением без поворота изображения. [J]. Ж. Т. Ф. , 1955, 25:950 – 952.

[4] ЦУККЕРМАН И И. Электронно-оптический метод изменения масштаба электронного изображения[J]. Радиотехника, 1957,12(3): 4 – 9.

[5] 周立伟. 电磁聚焦移像系统理论的研究[J]. 北京理工大学学报,1983,3(9): 12 – 24.

[6] 西门纪业. 复合浸没物镜的电子光学性质和像差理论[J]. 物理学报,1957,13(4): 339 – 356.

[7] CHOU L W. Electron optics of concentric spherical electromagnetic focusing systems [J]. Advances in Electronics and Electron Physics, 1979,52:119 – 132.

4.3 Electrostatic and Magnetic Imaging Without Image Rotation
图像无旋转的电磁聚焦移像

Abstract: A method is discussed for designing electrostatic and magnetic imaging systems with $M=1$ or $M\neq 1$, without image rotation in this paper. Tsukkerman's condition for an imaging system free of rotation does not appear to be applicable to practical design. We have modified his theory to obtain two other conditions for the normalized axial electrostatic potential $V(z)$, the normalized axial magnetic induction $G(z)$, and a lens strength parameter λ which can be interpreted as an eigenvalue. Some analytical examples and numerical results are given.

摘要：本文对 Tsukkerman 的电磁聚焦移像系统的理论进行了修正和补充，放宽了实现图像无转角的条件，导出了能获得放大率 $M=1$ 或 $M\neq 1$ 且无转角的图像下移像系统的轴上相对电位和磁感应强度分布的关系式，以及特征值 λ 应满足的条件，提出了设计图像无转角且放大率可变的移像系统的新途径。文中给出了解析实例与数值计算实例。

Introduction

Tsukkerman's theory [2-6] can be outlined as follows.

(1) The vector of position $r(z) = r_*(z) + p(z)$ of an electron trajectory is written as the sum of that of a principal trajectory $r_*(z)$ *and an increment* $p(z)$ for which simpler equations are obtained.

(2) The axial potential $V(z)$ and the magnetic induction $G(z)$ are subject to conditions simplifying the "paraxial" ray equation. For a given object plane $z = z_0$ and given image plane $z = z_i$, the simplified ray equation can be treated as an eigenvalue problem.

Tsukkerman's theory does not appear to be applicable to the case where $\phi(z_0) = 0$, i.e., where the object plane is a cathode surface. Zhou Liwei[7] has modified Tsukkerman's theory to make it applicable to that case. In the present paper, we have applied this modified theory to establish two other conditions for $V(z)$, $G(z)$ and λ.

1 Fundamental theory of the system

For an axial-symmetric cathode lens, we express the complex coordinate $r_*(z)$ of the principal ray and $p(z)$ of the increment by their moduli and arguments:

Zhou Liwei（周立伟）[a], Ni Guoqiang（倪国强）[a], Fang Erlun（方二伦）[b], a) 北京工业学院. b) 西安现代化学研究所. Conferences on Electron Optical Systems for Microscopy Microanalysis and Microlithography. SEM Inc., AMF O'Hare, 1984: 37-43.

$$r_*(z) = r_*(z)e^{i\theta_*(z)}, \quad p(z) = p(z)e^{i\beta(z)} \tag{1}$$

Their sum

$$r(z) = r_*(z) + p(z) \tag{2}$$

describes a ray neighboring to the principal ray (Fig. 1).

List of symbols

$B(z)$	Axial magnetic induction
B_0	Magnetic induction at the cathode surface
E	Electron charge
$e\varepsilon_0$	Initial energy of emission electron, $e\varepsilon_0 = (m_0/2)v_0^2$
$e\varepsilon_r^*$	Relative transversal initial energy of emission electron, $e\varepsilon_r^* = e\varepsilon_0 \sin^2\alpha_0/\phi_i$
$e\varepsilon_z^*$	Relative axial initial energy of emission electron, $e\varepsilon_z^* = e\varepsilon_0 \cos^2\alpha_0/\phi_i$. $e\varepsilon_{z_1}^*$ can be focused ideally at the image plane
$G(z)$	Relative axial magnetic induction, $G(z) = B(z)/B_0$
L	Axial distance from cathode surface to image plane
$\boldsymbol{M}$	Magnification vector
M	Magnification
m_0	Electron mass
m	Integer number, $0 < m \leq n$
n	Total of loops
$\boldsymbol{p}(z)$	Vector of increment of electron ray
$p(z)$	Module of $\boldsymbol{p}(z)$
$\boldsymbol{q}(z)$	Vector of increment of electron ray in a twisted coordinate system (x, y)
$\boldsymbol{r}(z)$	Vector of electron ray
$r(z)$	Module of $\boldsymbol{r}(z)$
r_0	Off-axis height from axis z at the cathode surface
$\boldsymbol{r}_*(z)$	Vector of principal electron ray
$r_*(z)$	Module of $\boldsymbol{r}_*(z)$
$\boldsymbol{u}(z)$	Vector of electron ray in twisted coordinate system (x, y)
$\boldsymbol{u}_*(z)$	Vector of principal electron ray in a twisted coordinate system (x, y)
U	Variable in Eq. (24)
$V(z)$	Relative axial electric potential, $V(z) = \phi(z)/\phi_i$
$v(z), w(z)$	Two independent special solutions of the "paraxial" ray equation

v_0	Initial velocity of electrons
(X, Y, Z)	Fixed Cartesian coordinate system
(x, y, z)	Twisted Cartesian coordinate system
Z	Axial coordinate measured in the units of the distance l between cathode surface and image plane
z_0, z_i	Coordinates of cathode surface and image plane, respectively
α_0	Emission angle of electrons
$\beta(z)$	Angle of $\boldsymbol{p}(z)$ turned around to the axis X
β_0	Initial value of $\beta(z)$
γ	Parameter in Eq. (46)
$\theta(z)$	Angle of $\boldsymbol{r}(z)$ turned around to the axis X
$\theta_*(z)$	Angle of $\boldsymbol{r}_*(z)$ turned around to the axis X
θ_0	Initial value of $\theta_*(z)$
λ	Dimensionless lens strength parameter
$\phi(z)$	Axial electrostatic potential
ϕ_i	Electrostatic potential at the image plane
$\chi(z)$	Angle of rotation of the twisted coordinate system (x, y) with respect to the fixed coordinate system (X, Y)
χ_0	Initial value of $\chi(z)$
ψ	Variable in Eq. (25)

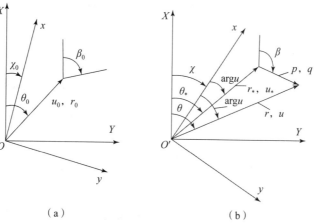

Fig. 1 The principal trajectory and its increment in fixed Cartesian coordinate system (x, y) and twisted Cartesian coordinate system (x, y) at $z = 0$ object plane and $z = z$ plane

Form the "paraxial" ray equation for $r(z)$, two equations for the principal ray

$$\sqrt{V(z)+\varepsilon_z^*}\frac{\mathrm{d}}{\mathrm{d}z}\left[\sqrt{V(z)+\varepsilon_z^*}\frac{\mathrm{d}r_*}{\mathrm{d}z}\right]+\left\{\frac{1}{4}V''(z)+\lambda\left[G^2(z)-\frac{r_0^4}{r_*^4}\right]\right\}r_*=0 \quad (3)$$

$$\sqrt{V(z)+\varepsilon_z^*}\frac{\mathrm{d}\theta_*}{\mathrm{d}z}=\sqrt{\lambda}\left[G(z)-\frac{r_0^2}{r_*^2}\right] \quad (4)$$

and two others for the increment

$$\sqrt{V(z)+\varepsilon_z^*}\frac{\mathrm{d}}{\mathrm{d}z}\left[\sqrt{V(z)+\varepsilon_z^*}\frac{\mathrm{d}p}{\mathrm{d}z}\right]+\left[\frac{1}{4}V''(z)+\lambda G^2(z)\right]p=0 \quad (5)$$

$$\frac{\mathrm{d}\beta}{\mathrm{d}z}=\sqrt{\lambda}\frac{G(z)}{\sqrt{V(z)+\varepsilon_z^*}} \quad (6)$$

are derived [7], where z is the axial coordinate measured in the units of the distance l between cathode and image plane, i.e., $z_0=0$, $z_i=l$. Primes denote differentiation with respect to z, $\phi(z)$ is the axial electrostatic potential,

$$\phi(0)=0, \quad \phi_i=\phi(l), \quad B_0=B(0), \quad V(z)=\phi(z)/\phi_i, \quad G(z)=B(z)/B_0, \quad r_0=r(0),$$

$e\varepsilon_0=\frac{1}{2}m_0 v_0^2$ is the initial energy of the electron, α_0 is the angle between its initial tangent and the $z-$axis, $\varepsilon_z^*=\varepsilon_0\cos^2\alpha_0/\phi_i$, $\varepsilon_r^*=\varepsilon_0\sin^2\alpha_0/\phi_i$ and

$$\lambda=\frac{e}{8m_0}\frac{B_0^2 l^2}{\phi_i} \quad (7)$$

is dimensionless lens strength parameter.

In order to solve the Eq. (3) to Eq. (6), we introduce twisted coordinate system by using the complex coordinate $\boldsymbol{u}=\boldsymbol{r}\exp(-\mathrm{i}\chi(z))$ where

$$\chi(z)=\sqrt{\lambda}\int_0^z\frac{G(z)}{\sqrt{V(z)+\varepsilon_z^*}}\mathrm{d}z+\chi_0 \quad (8)$$

In the twisted coordinate system the differential equations for the principal ray and the increment assume the same form [7]:

$$[V(z)+\varepsilon_z^*]u''+\frac{1}{2}V'(z)u'+\left[\frac{1}{4}V''(z)+\lambda G^2(z)\right]u=0 \quad (9)$$

Their solutions can be expressed using two linearly independent solutions $v(z)$ and $w(z)$ of this ray equation satisfying the initial conditions:

$$v(z_0=0)=0, \quad v'(z_0=0)\sqrt{\varepsilon_z^*}=1,$$
$$w(z_0=0)=1, \quad w'(z_0=0)\sqrt{\varepsilon_z^*}=0 \quad (10)$$

For electrons emitted from the cathode surface with the initial conditions:

$$r_0=r_0\mathrm{e}^{\mathrm{i}\theta_0}, \quad r'_0=l\tan\alpha_0\mathrm{e}^{\mathrm{i}\beta_0} \quad (11)$$

the solution of the principal trajectory equation can be written as

$$r_*(z)=r_0\sqrt{w^2+\lambda v^2} \quad (12)$$

$$\theta_*(z) = \theta_0 + \sqrt{\lambda} \int_0^z \frac{G(z)}{\sqrt{V(z) + \varepsilon_z^*}} dz - \arctan\frac{\sqrt{\lambda} v}{w} \quad (13)$$

whereas the solution for the increment becomes

$$p(z) = \sqrt{\varepsilon_z^*} l |v| \quad (14)$$

$$\beta(z) = \sqrt{\lambda} \int_0^z \frac{G(z)}{\sqrt{V(z) + \varepsilon_z^*}} dz + \beta_0 + (m-1)\pi \quad (15)$$

with $\theta_0 = \theta_*(z_0 = 0)$, $\beta_0 = \beta(z_0 = 0)$. The integer number m which depends on z is defined by the inequality:

$$(m-1)\pi < \sqrt{\lambda} \int_0^z \frac{G(z)}{\sqrt{V(z) + \varepsilon_z^*}} dz \leq m\pi$$

its value for $z_i = l$ is $m = n$. If

$$v(z_i, \varepsilon_{z_1}^*) = 0, \quad (16)$$

then all electrons having the same initial axial velocity component, the same value of $\varepsilon_z^* = \varepsilon_{z_1}^*$, are ideally focused at the image plane $z = l$. The complex magnification becomes

$$M = \frac{r_*(z_i, \varepsilon_{z_1}^*)}{r_0} = w(z_i, \varepsilon_{z_1}^*) e^{i(\chi_1 - \chi_0)} = \frac{1}{v'(z_i, \varepsilon_{z_1}^*)\sqrt{1 + \varepsilon_{z_1}^*}} e^{i(\chi_1 - \chi_0)} \quad (17)$$

and its module

$$M = |w(z_i, \varepsilon_{z_1}^*)|$$

For the angle of image rotation we obtain

$$\theta_*(z_i) - \theta_0 = \chi_i - \chi_0 - n\pi \quad (18)$$

where $\chi_i = \chi(l)$ follows from Eq. (8) for $z = l$. The system is free of image rotation if $\theta_*(z_i) = \theta_0$, i.e.,

$$\chi_i - \chi_0 = n\pi \quad (19)$$

The module of the magnification becomes

$$M = (-1)^n w(z_i, \varepsilon_{z_i}^*) = (-1)^n \frac{1}{v'(z_i, \varepsilon_{z_i}^*)\sqrt{1 + \varepsilon_{z_1}^*}} \quad (20)$$

Eq. (17) and Eq. (20) can also be applied to the case $V(0) = 0$ and $\varepsilon_{z_1}^* = 0$.

2 A new approach to realize an imaging system without image rotation

Tsukkerman has pointed out that a system without image rotation can be realized if the condition

$$G(z) = \frac{r_0^2}{r_*^2(z)} \quad (21)$$

is satisfied, then Eq. (3) assumes the form of the "paraxial" ray equation for a purely electrostatic field. It can be solved with the initial conditions $r_*(0) = r_0$, $r'_*(0) = 0$, and the magnification becomes $M = r_*(l)/r_0$. The principal trajectory is then meridional ray following a magnetic line of force.

Since we think that the condition Eq. (21) can hardly be realized, we suggest another solution of the problem. In order to solve the "paraxial" ray equqtion in the twisted coordinate system

$$\frac{d}{dz}\left(\sqrt{V(z)+\varepsilon_{z_1}^*}\frac{dv}{dz}\right)+\left(\frac{V''(z)}{4\sqrt{V(z)+\varepsilon_{z_1}^*}}+\lambda\frac{G^2(z)}{\sqrt{V(z)+\varepsilon_{z_1}^*}}\right)v=0 \quad (22)$$

for $v(z)$ with the boundary condition

$$v(z_0=0)=0, \quad v(z_i=l,\varepsilon_{z_1}^*)=0 \quad (23)$$

we introduce the new coordinates

$$u=\sqrt{G}v \quad (24)$$

and

$$\psi=\int_0^z \frac{G(z)}{\sqrt{V(z)+\varepsilon_{z_1}^*}}dz \quad (25)$$

The ray equation in terms of the new coordinates can be written as

$$\frac{d^2u}{d\psi^2}+\lambda u+\frac{u}{4G^2}\left\{\left(\frac{dG}{d\psi}\right)^2-2G\frac{d^2G}{d\psi^2}+\frac{G}{V+\varepsilon_{z_1}^*}\left[G\frac{d^2V}{d\psi^2}+\frac{dG}{d\psi}\frac{dV}{d\psi}-\frac{G}{2(V+\varepsilon_{z_1}^*)}\left(\frac{dV}{d\psi}\right)^2\right]\right\}=0 \quad (26)$$

and

$$u(0)=0, \quad u(\psi_i,\varepsilon_{z_1}^*)=0 \quad (27)$$

where

$$\psi_i=\psi(z_i=l)$$

If the contents of the curved brackets in Eq. (26) vanish, Eq. (26) is reduced to the Sturm-Liouville equation:

$$\frac{d^2u}{d\psi^2}+\lambda u=0 \quad (28)$$

The solution which satisfies the condition (27) is

$$u=\sin(\sqrt{\lambda}\psi) \quad (29)$$

where

$$\sin(\sqrt{\lambda}\psi_i)=0 \quad (30)$$

Eq. (30) is equivalent to the condition of Eq. (19). It is allowed to define the n-th eigenvalue λ_n for which $\sqrt{\lambda_n}\psi_i=n\pi$. For $\lambda=\lambda_n$, there are $(n-1)$ intermediate foci between the cathode surface $z_0=0$ and the image plane $z_i=l$.

If the contents of the curved brackets in Eq. (26) are expressed as a function of z and set equal to zero, we obtain

$$\frac{3(V+\varepsilon_{z_1}^*)}{G^2}\left(\frac{dG}{dz}\right)^2-\frac{2(V+\varepsilon_{z_1}^*)}{G}\frac{d^2G}{dz^2}-\frac{1}{G}\frac{dV}{dz}\frac{dG}{dz}+\frac{d^2V}{dz^2}=0 \quad (31)$$

The n-th eigenvalue must satisfy the condition:

$$\lambda_n = \frac{e}{8m_0}\frac{B_0^2 l^2}{\phi_i} = \frac{n^2\pi^2}{\left[\int_0^1 \frac{G(z)}{\sqrt{V(z)+\varepsilon_{z_1}^*}}\mathrm{d}z\right]^2} \quad (32)$$

If these two conditions are satisfied, the cathode surface $z_0 = 0$ will be imaged onto the image plane $z_i = l$ without image rotation. Our conditions are more general than Tsukkerman's condition of Eq. (21) since they do not imply that $\theta_*(z) = \theta_0$ for the principal ray but only that $\theta_*(z_i) = \theta_0$ in the image plane.

From Eq. (24) and Eq. (28) together with the initial conditions of Eq. (10), it follows that

$$v(z,\varepsilon_{z_1}^*) = \frac{1}{\sqrt{\lambda G(z)}}\sin\left[\sqrt{\lambda}\int_0^z \frac{G(z)}{\sqrt{V(z)+\varepsilon_{z_1}^*}}\mathrm{d}z\right] \quad (33)$$

$$w(z,\varepsilon_{z_1}^*) = -\sqrt{\varepsilon_{z_1}^*}\,v(z,\varepsilon_{z_1}^*)\left[\frac{\mathrm{d}}{\mathrm{d}z}\frac{1}{\sqrt{G(z)}}\right]_{z=0} + \frac{1}{\sqrt{G(z)}}\cos\left[\sqrt{\lambda}\int_0^z \frac{G(z)}{\sqrt{V(z)+\varepsilon_{z_1}^*}}\mathrm{d}z\right]$$

(34)

For $z = z_i$ we found the magnification

$$M = \frac{1}{\sqrt{G(z_i)}} = \sqrt{B_0/B(z_i)} \quad (35)$$

The same result has been obtained by Tsukkerman [3,4,6], but we think that our treatment is more general and more realistic.

3 Design of an imaging system without image rotation

The design procedure is as follows:

(1) $V(z)$ or $G(z)$ is given;

(2) Use condition of Eq. (31) to determine $G(z)$ given $V(z)$ or $V(z)$ from given $G(z)$. $V(z)$ and $G(z)$ must satisfy the conditions:

$$V(z_0 = 0) = 0, \quad V(z_i = 1) = 1,$$
$$G(z_0 = 0) = 1, \quad G(z_i = 1) = \frac{1}{M^2}$$

(3) Determine M;

(4) Choose suitable values of ϕ_i, B_0, l and n which must be in agreement with Eq. (32), and evaluate $\lambda = \lambda_n$;

(5) The solutions v and w of Eq. (22) are then given by Eq. (33) and Eq. (34);

(6) Determine $\boldsymbol{r}(z) = \boldsymbol{r}_*(z) + \boldsymbol{p}(z)$ using the solutions $v(z)$ and $w(z)$;

(7) Design electrodes and magnetic circuit in a way that they generate the fields $V(z)$ and $G(z)$ with the constants V_i and B_0.

We think that our method is more general and flexible than Tsukkerman's since his method is restricted to the case that $V(z)$ is given and that $G'(0) = 0$, and it yields only one solution $G(z)$ and one value for the magnification for the given $V(z)$. On the other hand, for a given $V(z)$, we can choose an arbitrary value for the magnification M, and then solve Eq. (31) for $G(z)$ with the boundary conditions $G(0) = 1$; $G(l) = 1/M^2$.

4 An analytic example for the design of an imaging system without image rotation

If we choose

$$G(z) = \frac{1}{[1-(1-M)z]^2} \tag{36}$$

and

$$V(z) = \frac{Mz}{1-(1-M)z} \tag{37}$$

then we have

$$G\frac{d^2 G}{dz^2} - \frac{3}{2}\left(\frac{dG}{dz}\right)^2 = 0,$$
$$G\frac{d^2 V}{dz^2} - \frac{dV}{dz}\frac{dG}{dz} = 0 \tag{38}$$

i. e., Eq. (31) is satisfied[1,7].

From Eq. (33) and Eq. (34) we obtain

$$v(z) = \frac{1}{\sqrt{\lambda}}[1-(1-M)z]\sin(\chi - \chi_0) \tag{39}$$

$$w(z) = [1-(1-M)z]\left[\cos(\chi - \chi_0) + \sqrt{\frac{\varepsilon_z^*}{\lambda}}(1-M)\sin(\chi - \chi_0)\right] \tag{40}$$

where

$$\chi = \frac{2\sqrt{\lambda}}{M}\sqrt{V(z)+\varepsilon_z^*}, \quad \chi_0 = \frac{2\sqrt{\lambda}}{M}\sqrt{\varepsilon_z^*} \tag{41}$$

with v and w from Eq. (39) and Eq. (40) we obtain from Eq. (12) to Eq. (15):

$$r_*(z) = r_0[1-(1-M)z]\left[1+(1-M)\sqrt{\frac{\varepsilon_z^*}{\lambda}}\sin_2(\chi-\chi_0)+(1-M)^2\frac{\varepsilon_z^*}{\lambda}\sin^2(\chi-\chi_0)\right]^{1/2} \tag{42}$$

$$\theta_*(z) = \theta_0 + \chi - \chi_0 - \arctan\frac{\sin(\chi-\chi_0)}{\cos(\chi-\chi_0)+(1-M)\sqrt{\frac{\varepsilon_z^*}{\lambda}}\sin(\chi-\chi_0)} \tag{43}$$

$$p(z) = \sqrt{\frac{\varepsilon_r^*}{\lambda}} l [1 - (1 - M)] |\sin(\chi - \chi_0)| \tag{44}$$

$$\beta(z) = \chi - \chi_0 + \beta_0 + (m-1)\pi \tag{45}$$

For $M = 1$, the example describes the well-known system using parallel homogeneous electric and magnetic fields. Fig. 2 shows graphs for the case $M = 1.4$, $n = 1$. In Fig. 3, $V(z)$ and $G(z)$ are plotted for a number of different magnifications in the range $0.6 \leq M \leq 2.5$.

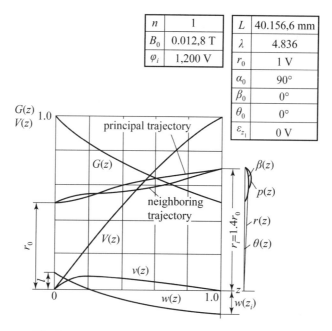

Fig. 2 An analytic example of imaging system with
$M = 1.4$, $n = 1$ (graphs of $G(z)$, $V(z)$, $v(z)$, $w(z)$, $r(z)$, $\theta(z)$, $p(z)$, $\beta(z)$)

5 A computational example for the design of an imaging system with variable magnification

For a given $V(z)$, an arbitrary value for the magnification M may be chosen, and then Eq. (31) may be solved numerically for $G(z)$ with the boundary conditions $G(0) = 1$, $G(l) = 1/M^2$. We have worked out a computer program to do this. As an example, Fig. 4 shows the given function

$$V(z) = \frac{\gamma z}{1 - (1 - \gamma)z} \tag{46}$$

with $\gamma = 0.6$, a number of solutions $G(z)$ for different values of the given magnification in the range $0.6 \leq M \leq 2.5$.

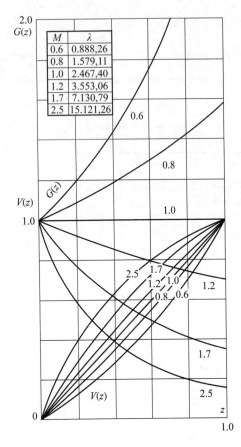
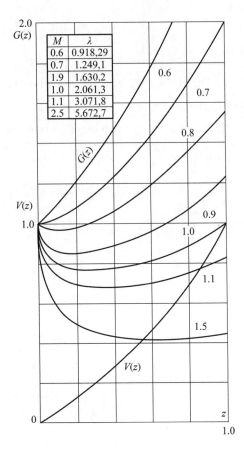

Fig. 3 Some analytic curves $V(z)$ and $G(z)$ for different magnification

Fig. 4 A computational example for a system with variable magnification

Summary

Tsukkerman's method of finding combined electrostatic and magnetic imaging systems without image rotation is generalized in order to make it applicable to cathode lenses where the electric potential in the cathode surface vanishes and to magnetic fields which are not necessarily perpendicular to the cathode surface. A few analytical and numerical examples are given.

Acknowledgement

We would like to thank Prof. F. Lenz from University of Tubingen, Germany, for a critical review of the manuscript.

References

[1] CHOU L W. Electron optics of concentric electromagnetic focusing systems [J]. Advances in Electronics and Electron Physics, 1979, 52: 119 – 132.

[2] TSUKKERMAN I I. On the "non-optical" theory of focusing in magnetic field of rotation [J]. Zh. Tekh. Fiz. ,1952,22:1843 - 1847(in Russian).

[3] TSUKKERMAN I I. On the magnetic focusing in camera tubes with image section[J]. Zh. Tekh. Fiz. ,1953,23:1228 - 1238(in Russian).

[4] TSUKKERMAN I I. Magnetic electron optical systems with variable magnification but without image rotation[J]. Zh. Tekh. Fiz,1955,25:950 - 952 (in Russian).

[5] TSUKKERMAN I I. Electron-optical method for scaling television picture[J]. Radiotekh. Electron,1957, 12:3 - 9 (in Russian).

[6] TSUKKERMAN I I. Electron optics in television[M]. Oxford: Pergamon Press, 1961.

[7] ZHOU L W. On the theory of combined electrostatic and magnetic focusing imaging system[J]. Journal of Beijing Institute of Technology, 1983,3:12 - 24(in Chinese).

4.4 图像无旋转的曲轴电磁聚焦成像
Electrostatic and Magnetic Imaging Having Curvilinear Axes Without Image Rotation

摘要：在曲轴宽电子束聚焦理论和电磁聚焦成像系统理论的基础上，本文进一步发展了图像无旋转的曲轴电磁聚焦移像理论，给出了曲轴系统中获得图像无旋转的充要条件，并讨论了可获得图像无旋转系统的曲轴电场和磁场分布的若干特例。

Abstract: In this paper, a theory on electrostatic and magnetic imaging having curvilinear axis without image rotation has been further developed on the basis of wide electron beam focusing and theory of electrostatic and magnetic imaging. The full and necessary conditions for producing an image without image rotation in the electrostatic and magnetic focusing system having curvilinear axis have been given. Some particular examples of electrostatic and magnetic fields with curvilinear axis for obtaining an imaging without image rotation have been discussed.

引 言

长期以来，苏联学者 И. И. Цуккерман 关于电磁聚焦成像系统的理论受到人们普遍的重视[1-3]。其理论的实质，一是用主轨迹 $r_{\pm}(z)$ 及其相邻轨迹 $p(z)$ 来描述电子轨迹 $r(z)$；二是按所给的条件，将边值问题下标量形式的傍轴方程转化为 Sturm-Liouville 方程，从而获得在图像无旋转条件下实现聚焦的电场和磁场分布。近年来周立伟等利用矢量形式的线性近轴轨迹方程，对 И. И. Цуккерман 的理论进行了修正和进一步的发展[4-11]，不仅使之适用于宽电子束聚焦成像系统，而且放宽了获得图像无旋转的条件，使 $M \neq 1$ 的电磁聚焦成像系统的实现成为可能，并用逆设计的方法设计出了实际的图像无旋转系统，从而建立了电磁聚焦移像系统的理论和有效的设计计算方法。

以上研究是基于直轴电子光学的傍轴或近轴轨迹方程进行的，电场和磁场均采用旋转对称的沿轴分布，其结果只适用于旋转对称系统的情况，严格来说仅对于近轴区域是有效的。本文在近年研究工作的基础上，利用所提出的曲轴宽电子束聚焦的普遍理论[2-12]，研究曲轴条件下图像无旋转的电磁聚焦移像理论。

1 曲轴电磁聚焦移像的基本理论

如图 1 所示，$\overparen{O_1O_2}$ 是一条曲光轴（它也是一条初始逸出的电位 $\varepsilon_s = \varepsilon_{s_1}$ 的电子的主轨迹），在曲光轴上我们建立随动的 Frenet 局部坐标系。设在曲光轴 $r_{\pm}(s)$ 外，从物平面 $s =$

金伟其，周立伟，倪国强．北京理工大学．光电子学技术（Optoelectronics Technology），No. 3, 1990, 18-24.

s_0 上作 $p_{\text{主}}(s) + q(s)$，于是有

$$r(s) = r_{\text{主}}(s) + p(s) = r_{\text{主}}(s) + p_{\text{主}}(s) + q(s) \tag{1}$$

$$p(s) = p_{\text{主}}(s) + q(s) \tag{2}$$

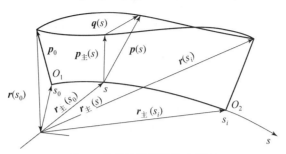

图 1 曲光轴及其轴外电子束

我们所要研究的问题就是以沿曲光轴的电场和磁场分布来描述轴外电子束，并寻求曲光轴上电场和磁场的某种分布，使电子束在像面 $s = s_i(\varepsilon_{s_1})$ 上形成放大率为 M 的像，即 $q(s_i) = 0$，且图像无旋转。

所谓图像无旋转，在直轴情况下，其含义是在固定坐标系中，相对于物点，像点没有转角；在曲轴情况下，则表示在随动的 Frenet 坐标系中，像点相对于物点没有转角。

如上所述，由于曲轴系统应具有点聚焦性质，按文献［11］、文献［12］有

$$F - R = 0, \quad N - Q = 0 \tag{3}$$

式中，F、R、N、Q 均为曲光轴上场的函数，在式（3）的条件下，其定义为[12]

$$n = \sqrt{\phi(s) + \varepsilon_{s_1}}, \quad \eta = \sqrt{e/(2m_0)},$$

$$F = R = n(\chi^2 - k^2) - \frac{1}{4n}\phi'' + \frac{1}{2}\eta\left(-2\chi B_{t00} + \frac{\varphi_{010} + 2n^2 k}{2n^2}B_{t00} - \frac{\varphi_{001}}{2n^2}B_{n00}\right),$$

$$N = Q = (n\chi)' - \frac{k\varphi_{001}}{2n} + \frac{1}{2}\eta\left(B_{b01} + B_{n10} + \frac{\varphi_{001}}{2n^2}B_{b00} + \frac{\varphi_{010}}{2n^2}B_{n00}\right), \tag{4}$$

$$K = n\chi - \frac{1}{2}\eta B_{t00},$$

$$K' = N$$

式中，k、χ 分别为曲光轴的曲率和挠率；e/m_0 为电子静止荷质比；φ_{0ij} 和 B_{tij}、B_{nij}、B_{bij}（$i, j = 0, 1, 2 \cdots$）分别为电场和磁场分量沿曲轴展开的系数；$\varphi_{000} = \phi(s)$ 为曲光轴上的电位分布；"′" 表示对弧长 s 的微商。

于是，系统的曲轴方程变为矢量形式的曲近轴轨迹的方程：

$$\frac{\mathrm{d}}{\mathrm{d}s}\left(n\frac{\mathrm{d}p}{\mathrm{d}s}\right) = Fp - \mathrm{i}\left(Np + 2K\frac{\mathrm{d}p}{\mathrm{d}s}\right) \tag{5}$$

考虑到 $p = pe^{\mathrm{i}\theta}$，则式（5）可变成曲轴坐标系下的分量形式：

$$n\frac{\mathrm{d}}{\mathrm{d}s}\left(n\frac{\mathrm{d}p}{\mathrm{d}s}\right) = nFp + \left(\frac{C_k^2}{p^4} - K^2\right)p, \quad n\frac{\mathrm{d}\theta}{\mathrm{d}s} = -K + C_k/p^2 \tag{6}$$

式中，

$$C_k = p_0^2\left(\sqrt{\varepsilon_{s_1}}\theta_0' + K_0\right) \tag{7}$$

为了消除式 (6) 中的非线性项，引入旋转坐标系：

$$p = u\mathrm{e}^{\mathrm{i}\gamma}; \quad p_{\text{主}} = u_{\text{主}}\mathrm{e}^{\mathrm{i}\gamma}; \quad q = v\mathrm{e}^{\mathrm{i}\gamma} \tag{8}$$

$$\gamma(s) = -\int_{s_0}^{s}(K/n)\mathrm{d}s + \gamma_0 \tag{9}$$

在旋转坐标系下轨迹矢量 u、$u_{\text{主}}$、v 均满足如下微分方程：

$$\frac{\mathrm{d}}{\mathrm{d}s}\left(n\frac{\mathrm{d}w}{\mathrm{d}s}\right) = Uw \tag{10}$$

式中，

$$U = F - K^2/n \tag{11}$$

根据文献 [6]，设式 (10) 表示的标量形式方程的特解 w_α、w_β 满足下列初始条件：

$$\begin{aligned} w_\alpha(s_0) &= 0, \quad \sqrt{\varepsilon_{s_1}}w'_\alpha(s_0) = 1, \\ w_\beta(s_0) &= 1, \quad \sqrt{\varepsilon_{s_1}}w'_\beta(s_0) = 0 \end{aligned} \tag{12}$$

逸出电子的初始条件为

$$p_0 = p_0\mathrm{e}^{\mathrm{i}\theta_0}, \quad p'_0 = \tan\alpha_0 \cdot \mathrm{e}^{\mathrm{i}\beta_0} \tag{13}$$

式中，θ_0 为逸出电子的初始位置 p_0 与 u_0 的夹角；α_0、β_0 分别为 p_0 处电子的初始逸出角和方位角。

由此，可导出曲光轴的主轨迹和相邻轨迹的通解。

在旋转坐标系内：

$$\begin{aligned} u_{\text{主}} &= p_0\mathrm{e}^{\mathrm{i}(\theta_0-\gamma_0)}w_\beta + \mathrm{i}p_0K_0\mathrm{e}^{\mathrm{i}(\theta_0-\gamma_0)}w_\alpha, \\ v &= \varepsilon_\perp^{1/2}\mathrm{e}^{\mathrm{i}(\beta_0-\gamma_0)}w_\alpha, \\ u &= u_{\text{主}} + v = [\varepsilon_\perp^{1/2}\mathrm{e}^{\mathrm{i}(\beta_0-\gamma_0)} + \mathrm{i}K_0p_0\mathrm{e}^{\mathrm{i}(\theta_0-\gamma_0)}]w_\alpha + p_0\mathrm{e}^{\mathrm{i}(\theta_0-\gamma_0)}w_\beta \end{aligned} \tag{14}$$

在 Frenet 局部坐标系内：

$$\begin{aligned} p_{\text{主}} &= p_0\mathrm{e}^{\mathrm{i}(\theta_0+\gamma-\gamma_0)}w_\beta + \mathrm{i}p_0K_0\mathrm{e}^{\mathrm{i}(\theta_0+\gamma-\gamma_0)}w_\alpha, \\ q &= \varepsilon_\perp^{1/2}\mathrm{e}^{\mathrm{i}(\beta_0+\gamma-\gamma_0)}w_\alpha, \\ p &= (\varepsilon_\perp^{1/2}\mathrm{e}^{\mathrm{i}\beta_0} + \mathrm{i}K_0p_0\mathrm{e}^{\mathrm{i}\beta_0})\mathrm{e}^{\mathrm{i}(\gamma-\gamma_0)}w_\alpha + p_0\mathrm{e}^{\mathrm{i}(\theta_0+\gamma-\gamma_0)}w_\beta \end{aligned} \tag{15}$$

利用式 (15) 就可确定式 (6) 的解。式中，$\varepsilon_\perp = \varepsilon_0\sin^2\alpha_0$，$\varepsilon_0$ 为电子的初始逸出电位。

假定在像平面 s_i 处，ε_{s_1} 的电子能理想聚焦，即满足

$$w_\alpha(s_i, \varepsilon_{s_1}) = 0 \tag{16}$$

则可求得放大率矢量 M 为

$$M = \frac{p(s_i, \varepsilon_{s_1})}{p_0} = \mathrm{e}^{\mathrm{i}(\gamma_i-\gamma_0-n\pi)}|w_\beta(s_i, \varepsilon_{s_1})| = \mathrm{e}^{\mathrm{i}(\gamma_i-\gamma_0-n\pi)}\left|\frac{1}{w'_\alpha(s_i,\varepsilon_{s_1})\sqrt{\phi(s_i)+\varepsilon_{s_1}}}\right| \tag{17}$$

式中，γ_i 是旋转坐标系在 s_i 处相对于 Frenet 坐标系旋转的角度，$n = 1, 2, 3\cdots$ 为聚焦圈数，而在 Frenet 坐标系中，像点 $p(s_i, \varepsilon_{s_1})$ 相对于物点 p_0 转过的角度为

$$\theta(s_i) - \theta_0 = \gamma_i - \gamma_0 - n\pi \tag{18}$$

由此可见，若使图像无旋转，$\theta_i - \theta_0 = 0$，必须使

$$\gamma_i - \gamma_0 = n\pi \tag{19}$$

即在 s_i 处旋转坐标系相对于初始位置旋转了 $n\pi$。

2 实现图像无旋转的条件

如上所述，实现图像无旋转的问题可归结为求解式（10）的标量

$$\frac{\mathrm{d}}{\mathrm{d}s}\left(n\frac{\mathrm{d}w_\alpha}{\mathrm{d}s}\right)=\left(F-\frac{K^2}{n}\right)w_\alpha \tag{20}$$

在边界条件 $w_\alpha(0)=0$、$w_\alpha(s_i)=0$ 和图像无旋转的条件 $\gamma_i-\gamma_0=n\pi$ 下的解。

与文献［6］同理，对式（20）作变量代换

$$\xi=\sqrt{|K|}w_\alpha,\quad \psi=-\int_{s_0}^{s}(K/n)\mathrm{d}s \tag{21}$$

当

$$\frac{3n^2}{K^2}\left(\frac{\mathrm{d}K}{\mathrm{d}s}\right)^2-\frac{2n^2}{K}\frac{\mathrm{d}^2K}{\mathrm{d}s^2}-\frac{1}{K}\frac{\mathrm{d}\phi}{\mathrm{d}s}\frac{\mathrm{d}K}{\mathrm{d}s}-4nF=0 \tag{22}$$

$$\psi_i=\psi(s_i)=-\int_{s_0}^{s_i}(K/n)\mathrm{d}s=\gamma_i-\gamma_0=n\pi \tag{23}$$

式（20）可变为 Sturm-Liouville 方程：

$$\frac{\mathrm{d}^2\xi}{\mathrm{d}\psi^2}+\xi=0,\quad \xi(0,\varepsilon_{s_1})=\xi(\psi_i,\varepsilon_{s_1})=0 \tag{24}$$

由此，式（20）的解可变为

$$w_\alpha(s,\varepsilon_{s_1})=-\sqrt{\frac{K_0}{K}}\cdot\frac{1}{K_0}\sin\psi(s) \tag{25}$$

因此式（22）、式（23）就是曲轴情况下图像无旋转的充要条件。利用式（4），还可将式（22）写成

$$2n^2N^2-2n^2KN'-KN\phi'-4n^2K^2F=0 \tag{26}$$

由此可见，曲轴情况下的图像无旋转充要条件与直轴时有类似的形式，但由于 K、F 中还包含曲光轴的曲率、挠率及其他场系数，而不是简单的沿轴电位 $\phi(s)$ 和磁场 $B(s)$ 的分布，因此情况要复杂得多。

利用参量变易法，还可求得另一个特解 w_β：

$$w_\beta(s,\varepsilon_{s_1})=-\frac{\mathrm{d}}{\mathrm{d}s}\left(\sqrt{\frac{K_0}{K}}\right)\bigg|_{s_0}\varepsilon_{s_1}^{1/2}w_\alpha+\sqrt{\frac{K_0}{K}}\cos\psi \tag{27}$$

当 $\psi_i=n\pi$ 时，由式（17）可得

$$M=\sqrt{K_0/K_i} \tag{28}$$

这表明，对于图像无旋转的曲轴移像系统，其放大率 M 取决于 K 在物面、像平面上的比值 $\sqrt{K_0/K_i}$，而与这两个平面之间的曲轴性质、电场和磁场的分布无直接关系。由于 K 中包含 n、χ 和 B_{t00}，在一般情况下，放大率与所加的电场和磁场都有关系，但当 $\chi=0$ 时，曲轴系统与直轴一样，仅与物面和像面上 B_{t00} 的比值有关。

3 图像无旋转的移像系统举例

对于 $k=\chi=0$ 的旋转对称直轴系统，本文结果与文献［6］是一致的，在文献［6］中已给出了一些实例，这里不再详述。

众所周知，$k(s)$、$\chi(s)$是曲线的两个不变量，当$k(s)$、$\chi(s)$确定后，曲线形状就唯一地确定。实际上，式（22）中给出了$k(s)$、$\chi(s)$、$\phi(s)$、$\phi_{010}(s)$、$\phi_{001}(s)$、$B_{t00}(s)$、$B_{n00}(s)$和$B_{b00}(s)$等8个参数之间的关系，考虑到曲光轴本身满足主轨迹方程[13]，因此，在满足图像无旋转的条件时，还有5个参数是独立的，故远比直轴情况复杂。下面讨论几种曲轴上的电位和磁场分布，它们可在像面$s_i = l(s_0 = 0)$上获得放大率为M的无旋转图像。

（1）首先，考虑无主轨迹色球差的平面圆弧曲光轴，它满足的条件[12]为

$$\varphi_{010} = -2kn^2, \quad \varphi_{001} = 0 \tag{29}$$

$$k = \text{常数}, \quad \chi = 0 \tag{30}$$

由式（4）可得

$$F = -nk^2 - \frac{\phi''}{4n}, \quad K = -\frac{1}{2}\eta B_{t00}(s) \tag{31}$$

① 在式（22）中，令

$$\frac{3}{2}\left(\frac{dK}{ds}\right)^2 - K\frac{d^2K}{ds^2} = 0, \quad \frac{1}{K}\frac{d\phi}{ds}\frac{dK}{ds} + 4nF = 0 \tag{32}$$

则可得到轴上电位和磁场的分布：

$$\phi_*(s) = \phi(s) + \varepsilon_{s_1} = \frac{M(\phi_i + \varepsilon_{s_1})\sin 2ks + \varepsilon_{s_1}\sin 2k(l-s)}{\sin 2kl\left(1 - \frac{1-M}{l}s\right)} \tag{33}$$

$$B_{t00}(s) = \frac{B_0}{\left(1 - \frac{1-M}{l}s\right)^2} \tag{34}$$

式中，$B_0 = B_{t00}(0)$，$\phi_i = \phi(s=l)$分别为物面上的轴向磁感应强度和像面电位；kl为曲轴旋转的角度。由式（34）可见$M = \sqrt{B_{t00}(0)/[B_{t00}(s_i)]}$。

② 令$B_{t00}(s) = B_0 e^{+2bs}$，代入式（22），可得轴上电位分布应满足方程：

$$\phi_*'' - 2b\phi_*' + (4b^2 + k^2)\phi_* = 0 \tag{35}$$

求解得（$w = \sqrt{k^2 + 3b^2}$）：

$$\phi_*(s) = \frac{e^{bs}}{\sin \omega l}[\varepsilon_{s_1}\sin \omega(l-s) + (\phi_i + \varepsilon_{s_1})e^{-bl}\sin \omega s] \tag{36}$$

求出$B_{t00}(s)$和$\phi_*(s)$之后，由式（9）就可确定$\gamma(s)$，进而求出待解确定整个轴外电子束的运动规律。

（2）考虑另一类特殊的空间曲线，它除满足无主轨迹色球差条件式（29）外，还应满足

$$n(\chi^2 - k^2) - \chi\eta B_{t00} = 0 \tag{37}$$

于是

$$F = -\frac{\phi''}{4n}, \quad K = n\chi - \frac{1}{2}\eta B_{t00}(s) \tag{38}$$

可以证明，若用K/K_0代替$G(s)$，则文献[6]给出的电位和磁感应强度分布可直接作为沿曲轴的电位和K/K_0的分布。由于只满足式（37），故k、χ中还有一个量是独立的，

可根据情况选取。

（3）一般来说，由已知的曲光轴及其曲轴场系数求解，能够实现具体的电磁场系统是极为困难的，但是在某些特殊条件下进行研究还是有吸引力的，这也是今后曲轴电子光学系统逆设计的一个较难的研究课题。这里，我们给出一个共轴圆筒电极所形成的静电场，它可实现曲光轴为 $k = -\chi =$ 常数的螺旋线，且实现 $M = 1$ 的图像无旋转的点聚焦。

如图2所示，设曲轴所在柱面 $\rho = a$ 上电位为 ϕ（常数），则在圆柱坐标系 (ρ, θ, z) 中系统的电位分布为

$$\varphi(\rho, z) = \phi + C_1 \ln(\rho/a) \quad (C_1 \text{ 为常数}) \tag{39}$$

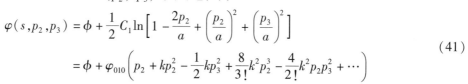

图 2 共轴圆筒电极中的螺旋线电子束

螺旋线曲光轴的曲率和挠率为

$$k = -\chi = 1/(2a) \tag{40}$$

不难证明，在曲轴 s 平面上 (p_2, p_3) 点的电位为

$$\begin{aligned}\varphi(s, p_2, p_3) &= \phi + \frac{1}{2}C_1 \ln\left[1 - \frac{2p_2}{a} + \left(\frac{p_2}{a}\right)^2 + \left(\frac{p_3}{a}\right)^2\right] \\ &= \phi + \varphi_{010}\left(p_2 + kp_2^2 - \frac{1}{2}kp_3^2 + \frac{8}{3!}k^2 p_2^3 - \frac{4}{2!}k^2 p_2 p_3^2 + \cdots\right)\end{aligned} \tag{41}$$

即

$$\varphi_{010} = -(C_1/a) = 2kn^2, \quad \varphi_{020} = 2k\varphi_{010}, \quad \varphi_{002} = -k\varphi_{010}, \quad \varphi_{030} = 8k^2\varphi_{010},$$
$$\varphi_{012} = -4k^2\varphi_{010}, \quad \varphi_{001} = \varphi_{011} = \varphi_{003} = \varphi_{021} = 0 \tag{42}$$

由式（4）得

$$F = R = 0, \quad N = Q = 0, \quad K = n\chi = \text{常数} \tag{43}$$

满足点聚焦条件式（3）和图像无旋转条件式（22）。旋转坐标系下的曲近轴轨迹方程为

$$\boldsymbol{u}'' = -\chi^2 \boldsymbol{u} \tag{44}$$

特解为

$$w_\alpha = \frac{1}{k}\frac{1}{\sqrt{\varepsilon_{s_1}}}\sin ks, \quad w_\beta = \cos ks \tag{45}$$

通解为

$$\boldsymbol{u} = \frac{1}{k}\boldsymbol{u}_0' \sin ks + \boldsymbol{u}_0 \cos ks \tag{46}$$

$$\gamma(s) = \gamma_0 - \int_0^s \frac{K}{n}\mathrm{d}s = \gamma_0 + ks \tag{47}$$

当 $ks_i = n\pi$ 时，实现 $M = 1$ 的点聚焦。各级像点 $s_i(n)$ 对应的 z 方向坐标为

$$z_i(n) = \sqrt{2}n\pi \quad (C_n = 1, 2, 3\cdots) \tag{48}$$

由式（15）可得 Frenet 坐标系中电子束的轨迹。

若已知系统在 $\rho = R_1$ 的电极筒上的电位为 U_1，则由式（39）和式（42）可得

$$a = R_1 \exp\left(\frac{U_1 - \phi}{\phi + \varepsilon_{s_1}}\right) \tag{49}$$

由于物、像点均为相同电位，因此这一系统并不适用于物面为阴极面的成像系统，但

作为一个前置预加速系统的曲轴细电子束成像系统仍是有意义的。

结束语

本文研究了曲轴情况下图像无旋转的电磁聚焦移像系统的理论，给出了以矢量形式线性轨迹方程特解表示的主轨迹和相邻轨迹的解，推导了实现图像无旋转的充要条件及特解，证明了在满足图像无旋转条件时，系统放大率等于$\sqrt{K_0/K_i}$，与物面、像面之间曲光轴的形状、电场和磁场沿曲轴分布无直接关系。这些结果对曲近轴区域系统的成像性质的描述是有效的。文中还讨论了满足图像无旋转条件下电场和磁场沿曲轴分布的若干实例。

参 考 文 献

［1］ЦУККЕРМАН И И. Электронная Оптика В Телевидении［M］. Москва-Ленинград，ГЭИ,1958.

［2］ЦУККЕРМАН И И. О Магнитной фокусировке в передающих трубках с перенсом изображния［J］. Ж. Т. ф.,1953,23:1228 – 1238.

［3］ЦУККЕРМАН И И. Электронно-оптический метод изменения масштаба электронного изображения［J］. Радиотехника,1957,12(3):4 – 9.

［4］ЦУККЕРМАН И И. Магнитные электронно-оптические системы с переменным увеличением без поворота изображния.［J］. Ж. Т. ф., 1955, 25:950 – 952.

［5］ЦУККЕРМАН И И. Электронно-оптический метод изменения масштаба электронного изображения［J］. Радиотехника, 1957,12(3):4 – 9.

［6］ZHOU L W, NI G Q, FANG E L. Electrostatic and magnetic imaging without image rotation［J］. Electron Optical Systems, SEM. Inc., 1984:37 – 43.

［7］周立伟,倪国强,方二伦. 图像无旋转的电磁聚焦移像系统的研究［J］. 电子学报, 1984,12(3):33 – 44.

［8］周立伟,方二伦. 一种新型的放大率$M \neq 1$的电磁聚焦移像系统［J］//夜视技术论文集. 1982,1:206 – 219.

［9］周立伟,仇伯仓,倪国强. 电磁聚焦移像系统中聚焦磁场的逆设计［J］. 电子学报, 1989,17(2):21 – 27.

［10］倪国强,周立伟,金伟其,等. 电磁聚焦移像系统中静电聚焦场的逆设计［J］. 电子科学学刊,1989,11(3):236 – 243.

［11］倪国强,周立伟,金伟其,等. 一种新的设计变倍电磁聚焦像增强器方法的研究［C］. 国际光电子科学与工程学术会议,北京,1989(ICOESE'89).

［12］ZHOU L W. A generalized theory of wide electron beam focusing［J］. Advances in Electronics and Electron Physics, 1985:575 – 589.

［13］周立伟,金伟其,倪国强. 曲轴宽电子束聚焦理论的研究［J］. 光电子学技术,1988, 8(4):8 – 22.

4.5 一种新型的放大率 $M \neq 1$ 电磁聚焦移像系统
A New Combined Electrostatic and Magnetic Focusing System for Image Transference with Magnification $M \neq 1$

摘要：本文提出一种电位与磁感应强度的沿轴分布，它能构成理想的放大率 $M \neq 1$ 的新型电磁聚焦移像系统。文中由电磁聚焦系统的矢量形式的近轴方程出发，分析这一类电磁聚焦移像系统的电子轨迹、聚焦条件与色球差；讨论当场有微扰时即所谓的准电磁聚焦 $M \neq 1$ 移像系统电子轨迹的变化以及获得不失真图像的途径。其 $M=1$ 的均匀平行复合电磁聚焦系统可视为本情况的一个特例。

Abstract: This article describes a new combined electrostatic and magnetic focusing system for image transference with magnification $M \neq 1$. This system is constructed on the basis of theoretical electrostatic potential and magnetic induction along axis, and it will realize the requirements without image rotation. Based on the "paraxial" ray equations in vector form for the combined electrostatic and magnetic fields, the electron trajectories, the focusing condition and the spherochromatic aberrations of such an electrostatic and magnetic focusing system are analyzed, the variance of electron trajectories and the means of obtaining a true image to the original will also be discussed in case the fields have to be slightly disturbed. The uniform parallel combined electrostatic and magnetic focusing system with $M = 1$ can be regarded as a special case of the abovementioned system.

引 言

均匀平行复合电磁聚焦系统作为移像系统广泛应用于光电成像器件与电视摄像器件中。特别是近年来，由于它在半导体集成电路制板技术的应用受到格外地重视，其研究是比较充分的。但是，对于放大率 $M \neq 1$ 的电磁聚焦移像系统的研究是很不够的。迄今为止，场和轨迹能以解析形式表示的，只有 $M=1$ 的均匀平行复合电磁聚焦系统。从理论上还没有找到一种合适的电位与磁感应沿轴分布，它能满足 $M>1$ 或 $M<1$ 且图像无旋转的要求。

本文的目的是试图建立一种电位与磁感应强度的沿轴分布，构成理想的放大率 $M \neq 1$ 新型移像系统。文中由复合场的矢量形式的近轴方程出发求轨迹的解析解，由此研究系统的近轴像差，并讨论当场有微小变化时的电子轨迹以及获得不失真图像的途径。

周立伟[a]，方二伦[b]．a) 北京工业学院，b) 西安现代化学研究所．夜视技术论文集（Proceedings on Technology of Night Vision），V. 1, 1982, 206 – 219.

周立伟，金伟其．一种放大率 $M \neq 1$ 的倒像式电磁聚焦成像 [J]．激光与光电子学进展，1995（A01）：1.

1 由复合场矢量形式的近轴方程求解 $M \neq 1$ 电磁聚焦移像系统

对于电磁复合聚焦系统的电子轨迹,可以由西门纪业[1]导出的矢量形式的线性方程

$$u'' + \frac{\phi'(z)}{2[\phi(z)+\varepsilon_z]}u' + \frac{1}{4[\phi(z+\varepsilon_z)]}\left[\phi''(z) + \frac{e}{2m_0}B^2(z)\right]u = 0 \quad (1)$$

出发进行研究。里 $u = x + iy$,x、y 为旋转坐标系的坐标,旋转坐标系绕 z 轴转动的角度为 χ,它与固定坐标系的 X、Y 有如下关系

$$r = X + iY = u e^{i\chi} \quad (2)$$

方程式(1)中的 u 是在旋转坐标系中的矢量,引入的目的是除去电磁场所引起近轴轨迹的旋转。旋转坐标系相对于固定坐标系的转角 $\chi(z)$ 是 $u(z)$ 所在平面的位置函数,它可表示为

$$\chi = \sqrt{\frac{e}{8m_0}} \int_0^z \frac{B(z)}{\sqrt{\phi(z)+\varepsilon_z}} dz + \chi_0 \quad (3)$$

式中,χ_0 为旋转坐标系的初始转角。

为了求解式(1),应当给出初始条件。按照定义

$$r_0 = r_0 e^{i\theta_0}, \quad r_0' = e^{i\beta_0} \tan \alpha_0 \quad (4)$$

式中,$r_0 = \sqrt{X_0^2 + Y_0^2}$,$\theta_0 = \arctan \frac{Y_0}{X_0}$,$\beta_0$ 为逸出电子的方位角,即电子初速度矢量 v_0 在 XOY 平面上的投影与 X 轴的夹角,如图1所示。

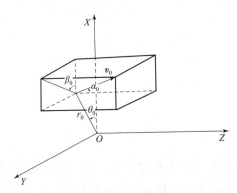

图1 逸出电子在固定坐标系中的初始位置和初始方向

对于电磁聚焦移像系统,令 l 为阴极物面至成像面的距离,M 为所希求的放大率,ϕ_{ac} 为成像面相对于阴极物面的电位,B_0 为阴极面处的磁感应强度,由文献[2],我们对电位与磁感应强度的沿轴分布做如下的假设:

$$\phi(z) = \frac{Mz}{l-(1-M)z}\phi_{ac} \quad (5)$$

$$B(z) = \frac{l^2}{[l-(1-M)z]^2}B_0 \quad (6)$$

如果按照我们所假定的电位与磁感应强度沿轴分布式(5)和式(6),则可求得线性方程式(1)的通解如下:

$$u = C_1 v + C_2 w \quad (7)$$

式中，v、w 为标量形式的线性方程式（1）的两个特解，它可以表示为

$$v = \sqrt{\frac{8m_0}{e}} \frac{1}{B_0} \left(1 - \frac{1-M}{l}z\right) \sin(\chi - \chi_0) \tag{8}$$

$$w = \left(1 - \frac{1-M}{l}z\right) \cos(\chi - \chi_0) \tag{9}$$

这里 χ、χ_0 可由式（3）求得

$$\chi = k\sqrt{\frac{lz}{l-(1-M)z} + \frac{\varepsilon_z}{\phi_0'}}, \quad \chi_0 = k\sqrt{\frac{\varepsilon_z}{\phi_0'}} \tag{10}$$

式中，$k = \sqrt{\frac{e}{2m_0}} \frac{B_0}{\phi_0'^{1/2}}$，$\phi_0' = -E_c = \frac{M\phi_{ac}}{l}$。

显然，特解 v、w 应满足如下初始条件：

$$v(0) = 0, \quad v'(0) = 1/\varepsilon_z^{1/2}, \tag{11}$$
$$w(0) = 1, \quad w'(0) = (M-1)/l$$

通解式（7）中的常数 C_1、C_2 可由式（4）与式（11）求得：

$$C_1 = \varepsilon_r^{1/2} e^{i(\beta_0 - \chi_0)} - i\sqrt{\frac{e}{8m_0}} B_0 r_0 e^{i(\theta_0 + \chi_0)} - r_0 \left(\frac{M-1}{l}\right) \varepsilon_z^{1/2} e^{i(\theta_0 - \chi_0)}, \tag{12}$$
$$C_2 = r_0 e^{i(\theta_0 - \chi_0)}$$

式中，$\varepsilon_r = \varepsilon_0 \sin^2 \alpha_0$。于是式（7）可表示为

$$\boldsymbol{u} = \sqrt{\frac{8m_0}{e}} \frac{1}{B_0} \varepsilon_r^{1/2} \left(1 - \frac{1-M}{l}z\right) e^{i(\beta_0 - \chi_0)} \sin(\chi - \chi_0) - $$
$$\sqrt{\frac{8m_0}{e}} \frac{1}{B_0} r_0 \varepsilon_z^{1/2} \left(\frac{M-1}{l}\right) \left(1 - \frac{1-M}{l}z\right) e^{i(\theta_0 - \chi_0)} \sin(\chi - \chi_0) + r_0 \left(1 - \frac{1-M}{l}z\right) e^{i(\theta_0 - \chi)} \tag{13}$$

在固定坐标系内，解将取如下形式：

$$\boldsymbol{r} = \sqrt{\frac{8m_0}{e}} \frac{1}{B_0} \varepsilon_r^{1/2} \left(1 - \frac{1-M}{l}z\right) e^{i(\beta_0 + \chi - \chi_0)} \sin(\chi - \chi_0) - $$
$$\sqrt{\frac{8m_0}{e}} \frac{1}{B_0} r_0 \varepsilon_z^{1/2} \left(\frac{M-1}{l}\right) \left(1 - \frac{1-M}{l}z\right) e^{i(\theta_0 + \chi - \chi_0)} \sin(\chi - \chi_0) + r_0 \left(1 - \frac{1-M}{l}z\right) e^{i\theta_0} \tag{14}$$

通常，轨迹 $\boldsymbol{r}$ 的模 r 及其幅角 θ 可直接由矢量 $\boldsymbol{u}$ 求得，由图 2 可见，

$$\theta = \chi + \arg \boldsymbol{u} = k\sqrt{\frac{lz}{l-(1-M)z} + \frac{\varepsilon_{z_1}}{\phi_0'}} + \arctan \frac{u_y}{u_\chi} \tag{15}$$

$$r = |\boldsymbol{u}| = \sqrt{u_\chi^2 + u_y^2} \tag{16}$$

式中，

$$u_\chi = \left[\varepsilon_r^{1/2} \cos(\beta_0 - \chi_0) + \sqrt{\frac{e}{8m_0}} B_0 r_0 \sin(\theta_0 - \chi_0) + r_0 \frac{M-1}{l} \varepsilon_z^{1/2} \cos(\theta_0 - \chi_0)\right] v + r_0 \cos(\theta_0 - \chi_0) w,$$

$$u_y = \left[\varepsilon_r^{1/2} \sin(\beta_0 - \chi_0) - \sqrt{\frac{e}{8m_0}} B_0 r_0 \cos(\theta_0 - \chi_0) + r_0 \frac{M-1}{l} \varepsilon_z^{1/2} \sin(\theta_0 - \chi_0)\right] v + r_0 \sin(\theta_0 - \chi_0) w$$

$$\tag{17}$$

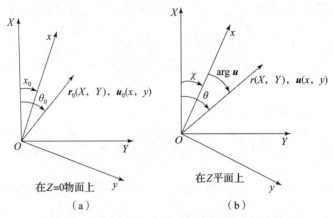

在 Z=0 物面上 　　　　　在 Z 平面上
　　（a）　　　　　　　　（b）

图 2　固定坐标系与旋转坐标系下的电子轨迹 $r(X,Y)$ 与 $u(x,y)$

由 $\sin(\chi - \chi_0) = 0$，即 $\chi - \chi_0 = j\pi$（$j = 1, 2, \cdots, m$，m 为聚焦圈数），便可由式（10）、式（14）确定各级成像点的位置：

$$z_j = \frac{2\sqrt{\dfrac{2m_0 j\pi}{e}} \dfrac{1}{B_0} + \dfrac{2m_0}{e} j^2 \pi^2 \dfrac{\phi_0'}{B_0^2}}{1 + \dfrac{2(1-M)}{M\phi_{ac}}\left(\dfrac{m_0}{e} j^2 \pi^2 \dfrac{\phi_0'^2}{B_0^2} + \sqrt{\dfrac{2m_0}{e}} j\pi \dfrac{\phi_0'}{B_0}\sqrt{\varepsilon_z}\right)}, \tag{18}$$

$$r_j = r_0\left(1 - \frac{1-M}{l}z_j\right)e^{i\theta_0} = \left(1 - \frac{1-M}{l}z_j\right)r_0$$

由此可见，此电子轨迹类似一圆锥螺线，其回转半径与 B_0 成反比，当 $M<1$ 时，半径逐渐减小，反之则增大。

令 $j = m$，则与逸出电子轴向初电位 ε_{z_1} 相对应的成像距离 z_i 即 l 便可表示为

$$z_i = l = \frac{m\pi}{B_0}M\sqrt{\frac{2m_0}{e}}\left(\sqrt{\phi_{ac} + \varepsilon_{z_1}} + \sqrt{\varepsilon_{z_1}}\right) \tag{19}$$

此即系统的聚焦条件，当 $z_i = l$ 时，$\sin(\theta_0 - \chi_0) = 0$，由式（14），便有

$$r_i = Mr_0 e^{i\theta_0} = Mr_0$$

这说明，不管逸出电子的初始方向（α_0, β_0）如何，只要 ε_{z_1} 相同，便能理想聚焦，且 r_i 与 r_0 同向，放大率为 M。

图 3 示出了 $M = 2$ 下电位与磁感应强度沿轴分布时主轨迹与相邻轨迹的一个实例。

2　近轴横向像差的确定

设理想聚焦的像点与逸出电子的轴向初电位 ε_{z_1} 相对应，即

$$r_i(z_i, \varepsilon_{z_1}) = r_0 M e^{i\theta_0} \tag{20}$$

当同一物点以初始条件（$\varepsilon_0, \alpha_0, \beta_0$）逸出的近轴电子，其 $\varepsilon_z \neq \varepsilon_{z_1}$，则旋转坐标系的 x 轴将不与逸出电子的初始坐标轴重合，而是多转了 $\chi - \chi_0$ 角。于是近轴像差可定义为

$$\Delta r = r_i(z_i, \varepsilon_z) - r_i(z_i, \varepsilon_{z_1}) \tag{21}$$

如图 4 所示。

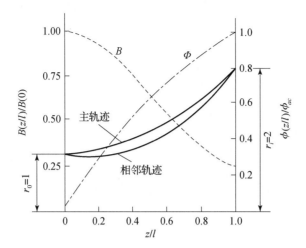

图 3 当 $M=2$ 时系统的电位与磁感应沿轴分布以及主轨迹与相邻轨迹

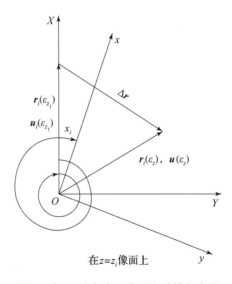

在 $z=z_i$ 像面上

图 4 在 ε_{z_1} 对应的 z_i 像面上的横向像差

由式（10），不难求得多转的 $\chi - \chi_0$ 角为

$$\chi - \chi_0 = m\pi + \frac{k}{\sqrt{\phi_0'}}(\sqrt{\varepsilon_{z_1}} - \sqrt{\varepsilon_z}) - \frac{1}{2}\frac{k^2}{m\pi\phi_0'}(\varepsilon_{z_1} - \varepsilon_z) \tag{22}$$

于是由式（14）和式（21），便可求得近轴横向像差

$$\Delta \boldsymbol{r} = \left[\frac{2M}{\phi_0'}\varepsilon_r^{1/2}(\sqrt{\varepsilon_{z_1}} - \sqrt{\varepsilon_z}) - \frac{M}{\phi_0'\sqrt{\phi_{ac}}}\varepsilon_r^{1/2}(\varepsilon_{z_1} - \varepsilon_z)\right] \times \exp\left\{\mathrm{i}\left[\beta_0 + \frac{k}{\sqrt{\phi_0'}}(\sqrt{\varepsilon_{z_1}} - \sqrt{\varepsilon_z}) - \right.\right.$$

$$\left.\left.\frac{1}{2}\frac{k^2}{m\pi\phi_0'}(\varepsilon_{z_1} - \varepsilon_z)\right]\right\} - r_0\frac{M-1}{l}\left[\frac{2M\varepsilon_z^{1/2}}{\phi_0'}(\sqrt{\varepsilon_{z_1}} - \sqrt{\varepsilon_z}) - \frac{M\varepsilon_z^{1/2}}{\phi_0'\sqrt{\phi_{ac}}}(\varepsilon_{z_1} - \varepsilon_z)\right] \times$$

$$\exp\left\{\mathrm{i}\left[\theta_0 + \frac{k}{\sqrt{\phi_0'}}(\sqrt{\varepsilon_{z_1}} - \sqrt{\varepsilon_z}) - \frac{1}{2}\frac{k^2}{m\pi\phi_0'}(\varepsilon_{z_1} - \varepsilon_z)\right]\right\} \tag{23}$$

上式第一项是中心近轴横向色差（二级色差，它表示当 $r_0 = 0$ 时以初始条件 $(\varepsilon_0, \alpha_0, \beta_0)$ 逸出的 $\varepsilon_z \neq \varepsilon_{z_1}$ 近轴电子轨迹在 ε_{z_1} 所决定的像面上的偏离）；第二项是与 ε_z 相对应的主轨迹在 ε_{z_1} 像面上由于放大率的改变与磁场旋转所引起的均匀偏移项。

3 场有微扰情况下的散焦特性

当上述聚焦场的分布有微小变化时，如 $\phi(z)$ 变为 $\phi(z) + \Delta\phi(z)$，$B(z)$ 变为 $B(z) + \Delta B(z)$，若在所考察的区间内，即 $0 \leq z \leq l$，$\Delta\phi/\phi(z) \ll 1$，$\Delta B/B(z) \ll 1$，则我们称之为准电磁聚焦移像系统。

将 $\Delta\phi$、ΔB 在物面处对 z 作泰勒展开，并限制到二级项，便有

$$\Delta B = \Delta B_0 + \Delta B_0' z + \frac{1}{2} \Delta B_0'' z^2,$$
$$\Delta\phi = \Delta\phi_0' z + \frac{1}{2} \Delta\phi_0'' z^2 \tag{24}$$

式中，ΔB_0 为物面处磁感应强度的变化，$\Delta B_0'$、$\Delta\phi_0'$、$\Delta B_0''$、$\Delta\phi_0''$ 乃是 ΔB、$\Delta\phi$ 在物面处对 z 的一阶与二阶导数。

为了研究微扰场引起轨迹的变化，我们将式（13）按 $\varepsilon_z^{1/2}$ 的幂次展开，略去含 ε_z 等高阶幂次项，并令

$$\xi = \frac{lz}{l - (1-M)z} \tag{25}$$

于是有

$$u = \sqrt{\frac{8m_0}{e}} \frac{1}{B_0} \varepsilon_r^{1/2} \left(1 - \frac{1-M}{l}z\right) e^{i\beta_0} \sin k\sqrt{\xi} + \frac{2\varepsilon_r^{1/2} \varepsilon_z^{1/2}}{\phi_0'} \left(1 - \frac{1-M}{l}z\right) e^{i(\beta_0 + k\sqrt{\xi})} (\delta - 1) +$$
$$r_0 \left(1 - \frac{1-M}{l}z\right) e^{i(\theta_0 - k\sqrt{\xi})} \left(1 - i\delta k\sqrt{\frac{\varepsilon_z}{\phi_0'}}\right) - \sqrt{\frac{8m_0}{e}} \frac{r_0}{B_0} \varepsilon_z^{1/2} \left(\frac{M-1}{l}\right) \left(1 - \frac{1-M}{l}z\right) e^{i\theta_0} \sin k\sqrt{\xi} \tag{26}$$

按照式（7），我们可得

$$v = \sqrt{\frac{8m_0}{e}} \frac{1}{B_0} \left(1 - \frac{1-M}{l}z\right) \sin k\sqrt{\xi} \tag{27}$$

$$w = \left(1 - \frac{1-M}{l}z\right) \cos k\sqrt{\xi} \tag{28}$$

$$C_1 = \varepsilon_r^{1/2} e^{i\beta_0} - i\sqrt{\frac{e}{8m_0}} B_0 r_0 e^{i\theta_0} - i\frac{k\varepsilon_r^{1/2} \varepsilon_z^{1/2}}{\phi_0'^{1/2}} e^{i\beta_0} - \frac{k^2 \varepsilon_z^{1/2}}{2} \delta r_0 e^{i\theta_0} - r_0 \left(\frac{M-1}{l}\right) \varepsilon_z^{1/2} e^{i\theta_0} \tag{29}$$

$$C_2 = r_0 e^{i\theta_0} + (\delta - 1) \frac{2\varepsilon_r^{1/2} \varepsilon_z^{1/2} e^{i\beta_0}}{\phi_0'} - i\frac{\varepsilon_z^{1/2} k r_0 e^{i\theta_0}}{\phi_0'^{1/2}} \delta \tag{30}$$

这里引入系数 δ 是为了考虑初始条件式（4）：当 $z = 0$ 时，$\delta = 1$；当 $z \neq 0$ 时，$\delta = 0$。

同样，轨迹 $r(z)$ 可表示如下：

$$r(z) = \sqrt{\frac{8m_0}{e}} \frac{1}{B_0} \varepsilon_r^{1/2} \left(1 - \frac{1-M}{l}z\right) e^{i(\beta_0 + k\sqrt{\xi})} \sin k\sqrt{\xi} + \frac{2\varepsilon_r^{1/2} \varepsilon_z^{1/2}}{\phi_0'} e^{i(\beta_0 + 2k\sqrt{\xi})} \left(1 - \frac{1-M}{l}z\right) (\delta - 1) +$$
$$\left(1 - \frac{1-M}{l}z\right) r_0 e^{i\theta_0} - \sqrt{\frac{8m_0}{e}} \frac{1}{B_0} \varepsilon_z^{1/2} r_0 \left(\frac{M-1}{l}\right) \left(1 - \frac{1-M}{l}z\right) e^{i(\theta_0 + k\sqrt{\xi})} \sin k\sqrt{\xi} \tag{31}$$

四、电磁聚焦移像系统的成像电子光学

由式（27）与式（28）可见，函数 v、w 乃是 $\varepsilon_z = 0$ 下方程式（1）的两特解。因此，式（26）与式（31）乃是以 $\varepsilon_z = 0$ 下的特解来表达 $\varepsilon_z \neq 0$ 情况下的轨迹。

在准电磁聚焦 $M \neq 1$ 移像系统的情况下，则方程式（1）的通解可表示为

$$\boldsymbol{u} = \boldsymbol{C}_1 (v + \Delta v) + \boldsymbol{C}_2 (w + \Delta w) \tag{32}$$

显然，Δv、Δw 乃是由于式（24）中 $\Delta \phi$、ΔB 的微扰项所引起的。当将 $\phi(z) + \Delta \phi$、$B(z) + \Delta B$ 代替 $\phi(z)$、$B(z)$ 代入方程式（1）后，将所有含有与 $\Delta \phi$、ΔB 相关的项移到方程的右端，从而构成非齐次线性方程，后面部分比之原以 $\phi(z)$、$B(z)$ 代入的方程的解可视为一微扰。于是该方程便可用变动任意常数法求解。若我们求 $z = l$ 处聚焦位置的值，以及仅考虑 M 接近于 1 的情况，则经过繁复的运算，可得

$$\Delta v = (-1)^{m+1} \left\{ \frac{\Delta \phi_0''}{\phi_0'^{3/2}} \left(\frac{l}{M}\right)^{3/2} \left[\frac{1}{6} + \frac{1}{m^2 \pi^2} + \frac{M-1}{M}\left(\frac{1}{5} - \frac{2}{m^2 \pi^2} + \frac{3}{m^4 \pi^4}\right)\right] + \right.$$

$$\frac{\Delta \phi_0'}{\phi_0'^{3/2}} \left(\frac{l}{M}\right)^{1/2} \left[1 + \frac{M-1}{M}\left(\frac{1}{3} - \frac{2}{m^2 \pi^2}\right)\right] - \frac{2\Delta B_0}{B_0 \phi_0'^{1/2}} \left(\frac{l}{M}\right)^{1/2} \left[1 + \frac{M-1}{M}\left(\frac{2}{3} - \frac{1}{m^2 \pi^2}\right)\right] -$$

$$\frac{2\Delta B_0'}{B_0 \phi_0'^{1/2}} \left(\frac{l}{M}\right)^{3/2} \left[\left(\frac{1}{3} - \frac{1}{2m^2 \pi^2}\right) + \frac{M-1}{M}\left(\frac{3}{5} - \frac{3}{m^2 \pi^2} + \frac{9}{2m^4 \pi^4}\right)\right] - \frac{2\Delta B_0'}{B_0 \phi_0'^{1/2}} \left(\frac{l}{M}\right)^{5/2} \times$$

$$\left. \left[\left(\frac{1}{10} - \frac{1}{2m^2 \pi^2} + \frac{3}{4m^4 \pi^4}\right) + \frac{M-1}{M}\left(\frac{2}{7} - \frac{3}{m^2 \pi^2} + \frac{15}{m^4 \pi^4} - \frac{45}{2}\frac{1}{m^6 \pi^6}\right)\right] \right\} \mathrm{sgn}(\boldsymbol{B}) \tag{33}$$

$$\Delta w = (-1)^{m+1} M \left\{ \frac{\Delta \phi_0''}{\phi_0'} \frac{l}{M}\left(\frac{3}{2} \frac{M-1}{M} \frac{1}{m^2 \pi^2}\right) + \frac{\Delta B_0}{B_0} \frac{M-1}{M} + \frac{\Delta B_0'}{2B_0} \frac{l}{M} \left[1 + 3\frac{M-1}{M}\left(1 - \frac{3}{m^2 \pi^2}\right)\right] - \right.$$

$$\left. \frac{1}{4} \frac{\Delta B_0''}{B_0} \left(\frac{l}{M}\right)^2 \left[\left(\frac{3}{m^2 \pi^2} - 1\right) - 4\frac{M-1}{M}\left(1 - \frac{15}{2}\frac{1}{m^2 \pi^2} + \frac{45}{2}\frac{1}{m^4 \pi^4}\right)\right] \right\} \tag{34}$$

这里，由于假定 M 接近于 1，故 $\frac{M-1}{M}$ 的高次项便被略去，式（34）中不出现 $\frac{\Delta \phi_0'}{\phi_0'}$ 项乃是由于它与 $\frac{M-1}{M}$ 的高次项相关系，表现为高阶小项。

由此二公式出发，可以研究存在微扰项时图像的转角、畸变与散焦特性。

由式（34）可见，与电场和磁场的微扰项相关的放大率的改变量可以确定。我们知道，当 $\Delta w = 0$ 时，系统的放大率将保持不变，当 $\Delta w > 0$ 且 m 为偶数时，微扰项将使系统的放大率增大，m 为奇数时则减少。$\Delta w < 0$ 时则反之。由该式可见，若选择适当的 B_0、B_0' 和 ϕ_0' 值，则可使系统的畸变减至最小。

其次，微扰场引起的图像转角可由下式确定：

$$\Delta \theta = \Delta \chi + \Delta \psi \tag{35}$$

式中，$\Delta \chi$ 乃是由于式（3）中引入微扰项时而出现旋转坐标系的补充旋转。将式（24）代入式（3）中，略去与上述一致的高阶小项，便得到

$$\Delta \chi = m\pi \left[\frac{\Delta B_0}{B_0}\left(1 + \frac{2}{3}\frac{M-1}{M}\right) + \frac{\Delta B_0'}{3B_0}\frac{l}{M}\left(1 + \frac{9}{5}\frac{M-1}{M}\right) + \frac{\Delta B_0''}{10B_0}\left(\frac{l}{M}\right)^2\left(1 + \frac{20}{7}\frac{M-1}{M}\right) - \right.$$

$$\left. \frac{1}{2}\frac{\Delta \phi_0'}{\phi_0'}\left(1 + \frac{1}{3}\frac{M-1}{M}\right) - \frac{1}{12}\frac{\Delta \phi_0''}{\phi_0'}\frac{l}{M}\left(1 + \frac{12}{5}\frac{M-1}{M}\right) \right] \mathrm{sgn}(\boldsymbol{B}) \tag{36}$$

由于转角的方向与 $\boldsymbol{B}$ 的方向有关，故令

$$\mathrm{sgn}(\boldsymbol{B}) = \begin{cases} 1, & \text{当 } \boldsymbol{B} \text{ 与沿轴单位矢量 } \boldsymbol{k} \text{ 同向时} \\ -1, & \text{当 } \boldsymbol{B} \text{ 与 } \boldsymbol{k} \text{ 反向时} \end{cases}$$

而 $\Delta\psi$ 被定义为

$$\Delta\psi = \arg(\boldsymbol{u} + \Delta\boldsymbol{u}) - \arg\boldsymbol{u} \tag{37}$$

这里 $\arg\boldsymbol{u}$ 或 $\arg(\boldsymbol{u}+\Delta\boldsymbol{u})$ 乃是旋转坐标系 χ 轴与矢量 $\boldsymbol{u}$ 或 $\boldsymbol{u}+\Delta\boldsymbol{u}$ 之间的夹角，于是 $\psi = \arg\boldsymbol{u} = \arctan\dfrac{u_y}{u_\chi}$，按多变量函数的泰勒公式，上式可展开为

$$\Delta\psi = \frac{\partial}{\partial v}\left(\arctan\frac{u_y}{u_\chi}\right)\Delta v + \frac{\partial}{\partial w}\left(\arctan\frac{u_y}{u_\chi}\right)\Delta w \tag{38}$$

利用式 (17)，并注意到 $v(z_i \ne l) = 0$，$w(z_i = l) = (-1)^m M$，并略去高阶小项，则 $\Delta\psi$ 可表示为

$$\Delta\psi = (-1)^{m+1}\frac{k\phi_0'^{1/2}}{2M}\Delta v \tag{39}$$

由式 (33)，便可求得 $\Delta\psi$ 为

$$\Delta\psi = m\pi\left\{\frac{\Delta\phi_0''}{\phi_0'}\left(\frac{l}{M}\right)\left[\left(\frac{1}{12} - \frac{1}{2m^2\pi^2}\right) + \frac{M-1}{M}\left(\frac{1}{10} - \frac{1}{m^2\pi^2} - \frac{3}{2m^4\pi^4}\right)\right] + \frac{1}{2}\frac{\Delta\phi_0'}{\phi_0'}\left[1 + \frac{M-1}{M} \times \right.\right.$$
$$\left(\frac{1}{3} - \frac{2}{m^2\pi^2}\right)\right] - \frac{\Delta B_0}{B_0}\left[1 + \frac{M-1}{M}\left(\frac{2}{3} - \frac{1}{m^2\pi^2}\right)\right] - \frac{\Delta B_0'}{B_0}\left(\frac{l}{M}\right)\left[\left(\frac{1}{3} - \frac{1}{2m^2\pi^2}\right) + \right.$$
$$\left.\frac{M-1}{M}\left(\frac{3}{5} - \frac{3}{m^2\pi^2} + \frac{3}{m^4\pi^4}\right)\right] - \frac{\Delta B_0''}{B_0}\left(\frac{l}{M}\right)^2\left[\left(\frac{1}{10} - \frac{1}{2m^2\pi^2} + \frac{3}{4m^4\pi^4}\right) + \right.$$
$$\left.\left.\frac{M-1}{M}\left(\frac{2}{7} - \frac{3}{m^2\pi^2} + \frac{15}{m^4\pi^4} - \frac{45}{2}\frac{1}{m^6\pi^6}\right)\right]\right\}\mathrm{sgn}(\boldsymbol{B}) \tag{40}$$

于是 $\Delta\theta$ 可由式 (36) 和式 (40) 求得

$$\Delta\theta = m\pi\left\{\frac{\Delta\phi_0''}{\phi_0'}\frac{l}{M}\left[-\frac{1}{2m^2\pi^2} - \frac{M-1}{M}\left(\frac{1}{10} + \frac{1}{m^2\pi^2} + \frac{3}{2m^4\pi^4}\right)\right] + \right.$$
$$\frac{\Delta\phi_0'}{\phi_0'}\left[-\frac{M-1}{M}\frac{1}{m^2\pi^2}\right] + \frac{\Delta B_0}{B_0}\left(\frac{M-1}{M}\frac{1}{m^2\pi^2}\right) + \frac{\Delta B_0'}{B_0}\left(\frac{l}{M}\right)\left[\frac{1}{2m^2\pi^2} + \frac{M-1}{M}\left(\frac{3}{m^2\pi^2} - \frac{9}{2m^4\pi^4}\right)\right] +$$
$$\left.\frac{\Delta B_0''}{B_0}\left(\frac{l}{M}\right)^2\left[\left(\frac{1}{2m^2\pi^2} - \frac{3}{4m^4\pi^4}\right) + \frac{M-1}{M}\left(\frac{3}{m^2\pi^2} - \frac{15}{m^4\pi^4} + \frac{45}{2}\frac{1}{m^6\pi^6}\right)\right]\right\}\mathrm{sgn}(\boldsymbol{B}) \tag{41}$$

由式 (41) 可见，当场偏离原聚焦系统时，在最终的图像上便引起图像的旋转。转角的大小与场的微扰值有关。但如果能适当地选择 $\Delta\phi_0'$、$\Delta\phi_0''$、ΔB_0、$\Delta B_0'$ 和 $\Delta B_0''$ 值，则可使图像转角减到最小程度直至消失。

由于微扰场的引入，故系统中心图像除了正常情况下式 (23) 第一项的一级色球差外，还会引起系统的散焦，即对应于原像点的偏离，该值可以近似地用下式表示：

$$\Delta\boldsymbol{r} = \varepsilon_r^{1/2}\Delta v\mathrm{e}^{\mathrm{i}(\beta_0+m\pi)} - \frac{\mathrm{i}k\varepsilon_r^{1/2}\varepsilon_z^{1/2}}{\phi_0'^{1/2}}\Delta v\mathrm{e}^{\mathrm{i}(\beta_0+m\pi)} - \frac{2\varepsilon_r^{1/2}\varepsilon_z^{1/2}}{\phi_0'}\Delta w\mathrm{e}^{\mathrm{i}(\beta_0+m\pi)} \tag{42}$$

由此可估计磁感应强度与电位分布有微扰时对散焦图像的影响，并且连同二、三级中心色差一起，以确定系统达到锐焦的条件，这里就不细述了。关于 $M \ne 1$ 电磁聚焦移像系统的像差将另文叙述。

4 均匀平行复合电磁聚焦系统作为移像系统

在以上公式中，若令 $M=1$，则可得到均匀平行复合电磁聚焦移像系统的表达式，现列举如下。

（1）特解与通解：

$$v = \sqrt{\frac{8m_0}{e}}\frac{1}{B_0}\sin(\chi-\chi_0), \qquad (43)$$

$$w = \cos(\chi-\chi_0)$$

$$\boldsymbol{C}_1 = \varepsilon_r^{1/2}\mathrm{e}^{\mathrm{i}(\beta_0-\chi_0)} - \mathrm{i}\sqrt{\frac{e}{8m_0}}B_0 r_0 \mathrm{e}^{\mathrm{i}(\theta_0-\chi_0)}, \qquad (44)$$

$$\boldsymbol{C}_2 = r_0 \mathrm{e}^{\mathrm{i}(\theta_0-\chi_0)}$$

$$\boldsymbol{u} = \sqrt{\frac{8m_0}{e}}\frac{1}{B_0}\varepsilon_r^{1/2}\mathrm{e}^{\mathrm{i}(\beta_0-\chi_0)}\sin(\chi-\chi_0) + r_0 \mathrm{e}^{\mathrm{i}(\theta_0-\chi)}, \qquad (45)$$

$$\boldsymbol{r} = \sqrt{\frac{8m_0}{e}}\frac{1}{B_0}\varepsilon_r^{1/2}\mathrm{e}^{\mathrm{i}(\beta_0+\chi-\chi_0)}\sin(\chi-\chi_0) + r_0 \mathrm{e}^{\mathrm{i}\theta_0}$$

（2）近轴横向像差：

$$\Delta \boldsymbol{r} = \left\{\frac{2}{\phi_0'}\varepsilon_r^{1/2}(\sqrt{\varepsilon_{z_1}} - \sqrt{\varepsilon_z}) - \frac{1}{\phi_0'\sqrt{\phi_{ac}}}\varepsilon_r^{1/2}(\varepsilon_{z_1} - \varepsilon_z)\right\} \times$$

$$\exp\left\{\mathrm{i}\left[\beta_0 + \frac{k}{\sqrt{\phi_0'}}(\sqrt{\varepsilon_{z_1}} - \sqrt{\varepsilon_z}) - \frac{1}{2}\frac{k^2}{m\pi\phi_0'}(\varepsilon_{z_1} - \varepsilon_z)\right]\right\} \qquad (46)$$

由上可见，此系统只剩下横向色球差，均匀偏移项消失。

（3）轨迹 $\boldsymbol{u}$、$\boldsymbol{r}$ 即式（26）与式（31）对 $\varepsilon_z^{1/2}$ 展开的表示式：

$$\boldsymbol{u} = \sqrt{\frac{8m_0}{e}}\frac{1}{B_0}\varepsilon_r^{1/2}\mathrm{e}^{\mathrm{i}\beta_0}\sin k\sqrt{z} + \frac{2\varepsilon_r^{1/2}\varepsilon_r^{1/2}}{\phi_0'}\mathrm{e}^{\mathrm{i}(\beta_0+k\sqrt{z})}(\delta-1) + r_0 \mathrm{e}^{\mathrm{i}(\theta_0-k\sqrt{z})}\left(1 - \mathrm{i}\delta k\sqrt{\frac{\varepsilon_z}{\phi_0'}}\right) \qquad (47)$$

$$\boldsymbol{r} = \sqrt{\frac{8m_0}{e}}\frac{1}{B_0}\varepsilon_r^{1/2}\mathrm{e}^{\mathrm{i}(\beta_0+k\sqrt{z})}\sin k\sqrt{z} + \frac{2\varepsilon_r^{1/2}\varepsilon_r^{1/2}}{\phi_0'}\mathrm{e}^{\mathrm{i}(\beta_0+2k\sqrt{z})}(\delta-1) + r_0 \mathrm{e}^{\mathrm{i}\theta_0} \qquad (48)$$

此时，v、w、$\boldsymbol{C}_1$ 和 $\boldsymbol{C}_2$ 可表达如下：

$$v = \sqrt{\frac{8m_0}{e}}\frac{1}{B_0}\sin k\sqrt{z}, \qquad (49)$$

$$w = \cos k\sqrt{z}$$

$$\boldsymbol{C}_1 = \varepsilon_r^{1/2}\mathrm{e}^{\mathrm{i}\beta_0} - \mathrm{i}\sqrt{\frac{e}{8m_0}}B_0 r_0 \mathrm{e}^{\mathrm{i}\theta_0} - \mathrm{i}\frac{k\varepsilon_r^{1/2}\varepsilon_z^{1/2}}{\phi_0'^{1/2}}\mathrm{e}^{\mathrm{i}\beta_0} - \frac{k^2\varepsilon_z^{1/2}}{2}\delta r_0 \mathrm{e}^{\mathrm{i}\theta_0}$$

$$\boldsymbol{C}_2 = r_0 \mathrm{e}^{\mathrm{i}\theta_0} + (\delta-1)\frac{2\varepsilon_r^{1/2}\varepsilon_z^{1/2}}{\phi_0'^{1/2}}\mathrm{e}^{\mathrm{i}\beta_0} - \mathrm{i}\frac{\varepsilon_z^{1/2}kr_0 \mathrm{e}^{\mathrm{i}\theta_0}}{\phi_0'^{1/2}}\delta \qquad (50)$$

（4）场有微扰下 Δv、Δw、$\Delta \chi$、$\Delta \psi$ 和 $\Delta \theta$ 的表达式：

$$\Delta v = (-1)^{m+1}\left[\frac{\Delta\phi_0''}{\phi_0'^{3/2}}l^{3/2}\left(\frac{1}{6} - \frac{1}{m^2\pi^2}\right) + \frac{\Delta\phi_0'}{\phi_0'^{3/2}}l^{1/2} - \frac{2\Delta B_0}{B_0\phi_0'^{1/2}}l^{1/2} - \frac{2\Delta B_0'}{B_0\phi_0'^{1/2}}l^{3/2}\left(\frac{1}{3} - \frac{1}{2m^2\pi^2}\right) - \right.$$

$$\frac{\Delta B_0''}{B_0 \phi_0'^{1/2}} l^{5/2} \left(\frac{1}{5} - \frac{1}{m^2 \pi^2} + \frac{3}{2m^4 \pi^4} \right) \bigg] \operatorname{sgn}(\boldsymbol{B}) \tag{51}$$

$$\Delta w = (-1)^{m+1} \left[\frac{\Delta B_0'}{2B_0} l - \frac{1}{4} \frac{\Delta B_0''}{B_0} l^2 \left(\frac{3}{m^2 \pi^2} - 1 \right) \right] \tag{52}$$

$$\Delta \chi = m\pi \left(\frac{\Delta B_0}{B_0} + \frac{\Delta B_0'}{3B_0} l + \frac{\Delta B_0''}{10 B_0} l^2 - \frac{1}{2} \frac{\Delta \phi_0'}{\phi_0'} - \frac{1}{12} \frac{\Delta \phi_0''}{\phi_0'} l \right) \operatorname{sgn}(\boldsymbol{B}) \tag{53}$$

$$\Delta \psi = m\pi \left[\frac{\Delta \phi_0''}{12 \phi_0'} l \left(\frac{1}{3} - \frac{6}{m^2 \pi^2} \right) + \frac{1}{2} \frac{\Delta \phi_0'}{\phi_0'} - \frac{\Delta B_0}{B_0} - \frac{\Delta B_0'}{3 B_0} l \left(1 - \frac{3}{2 m^2 \pi^2} \right) - \right.$$

$$\left. \frac{\Delta B_0''}{10 B_0^2} l^2 \left(1 - \frac{5}{m^2 \pi^2} + \frac{15}{2 m^4 \pi^4} \right) \right] \operatorname{sgn}(\boldsymbol{B}) \tag{54}$$

$$\Delta \theta = \left[-\frac{1}{2m\pi} \frac{\Delta \phi_0''}{\phi_0'} l + \frac{1}{2m\pi} \frac{\Delta B_0'}{B_0} l + \frac{\Delta B_0''}{2m\pi B_0} l^2 \left(1 - \frac{3}{2 m^2 \pi^2} \right) \right] \operatorname{sgn}(\boldsymbol{B}) \tag{55}$$

(5) 微扰场引起的中心色球差的附加项 Δr 仍以式（42）表示，只需令 $M=1$。

以上这些公式与 Ю. В. Куликов[3] 研究均匀平行复合电磁聚焦系统所导得的公式基本符合，但文献［3］与本文式（43）～式（55）所对应的各式中，具有不少印刷、遗漏甚至推导错误，例如与式（52）相对应的 Δw 表达式中就多了 $-\dfrac{\Delta \phi_0'}{\phi_0'} \dfrac{l}{8}$ 项，该项本是与高阶小联系的。此外，由于对横向色球差的理解与定义不同，式（46）与文献［3］对应的公式则大有差异，但本文的结果更符合实际。

我们由式（52）可以看出，磁场微扰的影响对于放大率的改变量远较电场微扰的影响来得大。

若将均匀平行复合电磁聚焦系统用于作为电子制板装置时，总是尽可能希望得到无畸变无旋转的图像，式（52）与式（55）指出了这种可能性。

设电场是均匀的，$\Delta \phi_0' = 0$，$\Delta \phi_0'' = 0$，磁感应强度分布相对于某一平面（例如 $z=0$ 平面）对称，故 $B(z)$ 的展开式在 $z=0$ 处取以下形式

$$B(z) = B(0) = \frac{1}{2} B_0'' z^2 \tag{56}$$

令式（52）中 $\Delta w = 0$，利用由式（56）求得的 $B'(z)$ 和 $B''(z)$，便可得到

$$z = \frac{l}{2} \left(\frac{3}{m^2 \pi^2} - 1 \right)$$

当 $m=1$ 时，则有

$$z = -0.348 l$$

这表明，如果物面（阴极）位于磁场对称平面之左距离为 $0.348l$ 处，则当图像旋转时，其放大率不变。同时，若希望 $\Delta \theta = 0$，则图像无旋转。若设 $\Delta \phi_0' = \Delta \phi_0'' = 0$，则可由式（55）求得

$$z = -\left(1 - \frac{3}{2 m^2 \pi^2}\right) l$$

这说明，当 $m=1$ 时，若把物面（阴极）置于磁场对称平面之左的 $0.847l$ 处，则可得到无旋转的图像。此二结论曾为 Ю. В. Куликов 指出[3]，并与美国专利[4]是一致的。

参 考 文 献

[1] 西门纪业. 复合浸没物镜的电子光学性质和像差理论[J]. 物理学报,1957,13(4):339-356.

[2] CHOU L W. Electron optics of concentric electromagnetic focusing systems [J]. Advances in Electronics and Electron Physics,1979,52:119-132.

[3] КУЛИКОВ Ю В. Об оценке качества изображения катодных электронно-оптической систем с учетом аберации третьего порядка. [J]. Радиотехника и Электроника 1975,20 (6):1249-1254.

[4] VINE J. The design of electrostatic zoom image intensifiers. [J] Advances in Electronics and Electron Physics, 1969,28A:537-543.

五、曲轴宽电子束聚焦理论

Theory of Wide Electron Beam Focusing Having Curvilinear Axes

5.1 Tensor Analysis of Electron Motion in Curvilinear Coordinate Systems (Part I)

曲线坐标系中电子运动的张量分析（第Ⅰ部分）

Abstract: Study the of charged particle's motion using a curvilinear coordinate system along a curved optical axis has long been one of the fundamental problems in electron optics. From then on, one could investigate and explore almost the whole field of problems in electron optics such as narrow and wide electron beam focusing, Gaussian imaging and paraxial imaging properties, electron optical aberrations, and electron optical transfer function. Based on the previous articles[4,5], Part Ⅰ of the present paper takes a further step to investigate the electron motion around a curved optical axis in the electrostatic and magnetic fields described by electrostatic potential and scalar or vector magnetic potential, using the contravariant and covariant forms of Newton's equation and Lorentz force, in which the Frenet local coordinate system along the principal trajectory is used. Furthermore, we will apply the variation method to study the electron motion and electron trajectories on the assumption of coordinate system mentioned above. Part Ⅰ of the present paper has given the different forms of equations for the electron motion and electron trajectories in the most general case, and fully verifies no differences between the variation method and the electrodynamic method in studying electron optics.

摘要：以主轨迹为曲光轴研究曲线坐标系中带电粒子的运动长期以来是电子光学的一个基本问题。由此出发，人们可以研究与考察电子光学领域中，如细电子束与宽电子束聚焦，高斯成像与近轴成像性质，电子光学像差和电子光学传递函数等几乎一切问题。在作者研究曲轴宽电子束聚焦理论的基础上[4,5]，本文这一部分进一步使用 Newton 方程和 Lorentz 力的逆变形式和协变形式以及广义变分函数来探讨曲线坐标系中以静电电位、磁标位或磁矢位表示的静电场和磁场中的电子轨迹。本文这一部分是在最普遍的情况下给出了各种形式的电子运动方程和电子轨迹方程，并充分证实了用变分方法和电动力学方法研究电子运动的一致性。

Introduction

Study of electron beam focusing along curved principal trajectory has long been one of the most fundamental problems in optics of charged particles, and great attention has been paid in it. As early as the beginning of the 1949s, Г. А. ГРИНБЕРГ[1] deduced the paraxial trajectory equations with an arbitrary curved optical axis, based on the fundamental equation of electron motion. At the 1950s, P. A. Sturrock[2] investigated electron optical systems with a curved optical axis, where the orthogonality condition was satisfied, applying the variation principle. P. P. Kas'yankov[3] derived the paraxial trajectory equation on the basis of general trajectory equations in a curvilinear coordinate system. Since the 1980s, Zhou LiWei[4,5] has explored the general forms of electron motion's equation and trajectory equation using tensor analysis, and approached a subject of wide electron beam focusing, in which the object (cathode) is immersed directly in the electrostatic and magnetic fields and the cathode surface is to be imaged with wide electron beam, so that electrons leave the cathode with small velocities and large inclinations. Investigation of papers show that the tensor analysis is a very efficient tool in studying electron optics. Using the tensor method, it is very convenient to derive the general forms of electron motion's equation and electron trajectories equation under the condition of an arbitrary curved optical axis, which is the starting point for studying the focusing properties of electron optical systems with a curved axis or a straight axis.

Part I of this paper takes a further step to investigate the electron motion around a curved optical axis in the electrostatic and magnetic fields described by the electrostatic potential and the scalar or vector magnetic potential, using the contravariant and covariant forms of Newton's equation and Lorentz force, in which the Frenet local coordinate system moving along the principal trajectory is used. Furthermore, we will apply the variation method to study the electron motion and electron trajectory on the assumption of coordinate system mentioned above.

Study of the electron motion in the twisted Frenet local orthogonal coordinate system (s, u, v) along the principal trajectory will be described in part II of this paper.

1 Frenet local coordinate system and metric tensors

Let $r_N = r_N(x^1)$ be a natural representation of a space curve, consisting of the curved axis of an electron beam (principal trajectory), emitted from point A at the object surface, where x^1 is the arc length along the curved axis from the initial point A. The normal plane through an arbitrary point N on the curved axis intersects the neighboring trajectory at point N^*. It follows that the position of N^* can be determined by a vector p from N to N^*. If the expression of $p = p(x^1)$ is known, then the curvilinear neighboring trajectory will be completely defined.

Introduce Frenet local coordinate system (x^1, x^2, x^3) (indices 1, 2, 3 are here used as superscripts, not exponents) and denote the unit vector at the point N by t, n, b for the tangent, principal normal and binormal directions, respectively. As the vector p is placed in the normal plane orthogonal to the tangent of the principal trajectory, therefore, its components at the principal normal and binormal directions are x^2 and x^3, respectively, as shown in Fig. 1.

The curvature and torsion of space curve (principal trajectory) r_N expressed by $k = k(x^1)$ and $\chi = \chi(x^1)$, respectively, can be given as

$$k = \frac{|r'_N \times r''_N|}{|r'_N|^3} \qquad (1-1)$$

$$\chi = \frac{|r'_N \times r''_N| \cdot r'''_N}{|r'_N \times r''_N|^2} \qquad (1-2)$$

Here and below the prime indicates derivatives with respect to the arc length x^1.

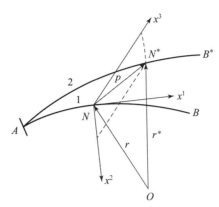

Fig. 1 The Frenet local coordinate system, principal trajectory and its neighboring trajectories

(1—principal trajectory; 2—neighboring trajectory)

From analytical geometry, the system of unit vectors in the Frenet triad at the unit vectors in the Frenet triad at the point N is defined by

$$t = r'_N, \text{(tangent)}$$
$$n = r''_N/k, \text{(normal)}$$
$$b = (r'_N \times r''_N)/k \text{(binormal)} \qquad (1-3)$$

The vector $\boldsymbol{r}^*$ at point N^* on the neighboring trajectory can be expressed by

$$r^* = r_N + p = r_N + x^2 n + x^3 b \qquad (1-4)$$

Now, let us calculate the element of arc length ds^* in the selected curvilinear coordinate system. Differentiating Eq. (4), using Frenet-Serret's formulae:

$$\frac{dt}{dx^1} = kn, \quad \frac{dn}{dx^1} = -kt + \chi b, \quad \frac{db}{dx^1} = -\chi n \qquad (1-5)$$

we obtain

$$dr^* = (1 - kx^2)dx^1 t + (dx^2 - \chi x^3 dx^1)n + (dx^3 + \chi x^2 dx^1)b \qquad (1-6)$$

Considering that $(ds^*)^2 = (dr^*)^2$, and expressing it in tensor form, we may have

$$(ds^*)^2 = g_{ij}dx^i dx^j \qquad (1-7)$$

where g_{ij} is the fundamental metric tensor, which can be written as follows:

$$(g_{ij}) = \begin{pmatrix} (1-kx^2)^2 + (\chi x^2)^2 + (\chi x^3)^2 & -\chi x^3 & \chi x^2 \\ -\chi x^3 & 1 & 0 \\ \chi x^2 & 0 & 1 \end{pmatrix} \qquad (1-8)$$

The conjugate metric tensor g^{ij} can be derived by g_{ij} given in Eq. (1-8)[6]:

$$g^{ij} = G^{ij}/g \qquad (1-9)$$

where g denotes the value of determinant of the matrix (g_{ij}):

$$g = \det|g_{ij}| = (1-kx^2)^2 \qquad (1-10)$$

and G^{ij} is the cofactor of g_{ij} in the determinant $|g_{ij}|$. This quantity can be written in the form

$$G^{ij} = \begin{vmatrix} g_{ps} & g_{pt} \\ g_{rs} & g_{rt} \end{vmatrix}$$

where i, p, r and j, s, t are both cyclic permutations of 1, 2, 3. Therefore, we have

$$(g^{ij}) = \begin{pmatrix} \dfrac{1}{g} & \dfrac{\chi x^3}{g} & \dfrac{-\chi x^2}{g} \\ \dfrac{\chi x^3}{g} & \dfrac{(1-kx^2)^2 + (\chi x^3)^2}{g} & \dfrac{-(\chi)^2 x^2 x^3}{g} \\ \dfrac{-\chi x^2}{g} & \dfrac{-(\chi)^2 x^2 x^3}{g} & \dfrac{(1-kx^2)^2 + (\chi x^2)^2}{g} \end{pmatrix} \quad (1-11)$$

2 Electron motion's equation of "contravariant" type in Frenet local coordinate system

In the non-relativistic case, the contravariant form of Newton's equation can be written as[7]:

$$m_0 \left\{ \ddot{x}^\lambda + \begin{Bmatrix} \lambda \\ j \; k \end{Bmatrix} \dot{x}^j \dot{x}^k \right\} = F^\lambda \quad (\lambda = 1,2,3) \quad (1-12)$$

where the dot indicates derivatives with respect to the time t; m_0 is the electron mass; $\begin{Bmatrix} \lambda \\ j \; k \end{Bmatrix}$ are the Christoffel symbols of the second kind, will have the form[6]:

$$\begin{Bmatrix} \lambda \\ j \; k \end{Bmatrix} = \frac{1}{2} g^{\lambda i} \left(\frac{\partial g_{ij}}{\partial x^k} + \frac{\partial g_{ik}}{\partial x^j} - \frac{\partial g_{jk}}{\partial x^i} \right) \quad (1-13)$$

Using the fundamental metric tensor g_{ij} and the conjugate metric tensor g^{ij}, we can express $\dfrac{\partial g_{ij}}{\partial x^R}$ and $\begin{Bmatrix} \lambda \\ j \; k \end{Bmatrix}$ in the form of matrix with 3 marks, as shown in Table 1.1 and Table 1.2, respectively.

In the electrostatic and magnetic field the contravariant form of Lorentz force F acting on an electron can be written as:

$$F^\lambda = e(\nabla \varphi)^\lambda - e(v \times B)^\lambda \quad (\lambda = 1,2,3) \quad (1-14)$$

where e is the magnitude of electron charge; φ is the electrostatic potential; B is the magnetic induction; v is the vector of velocity of the electron motion; $(\nabla \varphi)^\lambda$ is the contravariant component of gradient of a scalar potential φ, can be expressed by

$$(\nabla \varphi)^\lambda = g^{\lambda k} \frac{\partial \varphi}{\partial x^k} \quad (\lambda = 1,2,3) \quad (1-15)$$

Using the conjugate metric tensor g^{ij} given by Eq. (1-11), we obtain $(\nabla \varphi)^\lambda$, as shown in Table 1.3.

Table 1.1 $\dfrac{\partial g_{ij}}{\partial x^k}$ (i, j, $k = 1$, 2, 3)

(i, j, k denote the row, column and page, respectively)

$k = 1$

$\dfrac{\partial g_{11}}{\partial x^1}$	$\dfrac{\partial g_{12}}{\partial x^1}$	$\dfrac{\partial g_{13}}{\partial x^1}$
$2(1-kx^2)(-k'x^2) + 2\chi\chi'[(x^2)^2+(x^3)^2]$	$-\chi' x^3$	$\chi' x^2$
$\dfrac{\partial g_{21}}{\partial x^1}$	$\dfrac{\partial g_{22}}{\partial x^1}$	$\dfrac{\partial g_{23}}{\partial x^1}$
$-\chi' x^3$	0	0
$\dfrac{\partial g_{31}}{\partial x^1}$	$\dfrac{\partial g_{32}}{\partial x^1}$	$\dfrac{\partial g_{33}}{\partial x^1}$
$\chi' x^2$	0	0

$k = 2$

$\dfrac{\partial g_{11}}{\partial x^2}$	$\dfrac{\partial g_{12}}{\partial x^2}$	$\dfrac{\partial g_{13}}{\partial x^2}$
$2(1-kx^2)(-k) + 2(\chi^2)x^2$	0	χ
$\dfrac{\partial g_{21}}{\partial x^2}$	$\dfrac{\partial g_{22}}{\partial x^2}$	$\dfrac{\partial g_{23}}{\partial x^2}$
0	0	0
$\dfrac{\partial g_{31}}{\partial x^2}$	$\dfrac{\partial g_{32}}{\partial x^2}$	$\dfrac{\partial g_{33}}{\partial x^2}$
χ	0	0

$k = 3$

$\dfrac{\partial g_{11}}{\partial x^3}$	$\dfrac{\partial g_{12}}{\partial x^3}$	$\dfrac{\partial g_{13}}{\partial x^3}$
$2(\chi)^2 x^3$	$-\chi$	0
$\dfrac{\partial g_{21}}{\partial x^3}$	$\dfrac{\partial g_{22}}{\partial x^3}$	$\dfrac{\partial g_{23}}{\partial x^3}$
$-\chi$	0	0
$\dfrac{\partial g_{31}}{\partial x^3}$	$\dfrac{\partial g_{32}}{\partial x^3}$	$\dfrac{\partial g_{33}}{\partial x^3}$
0	0	0

Table 1.2 Christorffel symbols of the second kind $\begin{Bmatrix} \lambda \\ j \quad k \end{Bmatrix}$ (j, k, $\lambda = 1, 2, 3$)

(j, k, λ denote the row, column and page, respectively)

$\lambda = 1$

$\begin{Bmatrix} 1 \\ 1 \quad 1 \end{Bmatrix}$	$\begin{Bmatrix} 1 \\ 1 \quad 2 \end{Bmatrix}$	$\begin{Bmatrix} 1 \\ 1 \quad 3 \end{Bmatrix}$
$\dfrac{k\chi x^3 - k'x^2}{1 - kx^2}$	$\dfrac{-k}{1-kx^2}$	0
$\begin{Bmatrix} 1 \\ 2 \quad 1 \end{Bmatrix}$	$\begin{Bmatrix} 1 \\ 2 \quad 2 \end{Bmatrix}$	$\begin{Bmatrix} 1 \\ 2 \quad 3 \end{Bmatrix}$
$\dfrac{-k}{1-kx^2}$	0	0
$\begin{Bmatrix} 1 \\ 3 \quad 1 \end{Bmatrix}$	$\begin{Bmatrix} 1 \\ 3 \quad 2 \end{Bmatrix}$	$\begin{Bmatrix} 1 \\ 3 \quad 3 \end{Bmatrix}$
0	0	0

$\lambda = 2$

$\begin{Bmatrix} 2 \\ 1 \quad 1 \end{Bmatrix}$	$\begin{Bmatrix} 2 \\ 1 \quad 2 \end{Bmatrix}$	$\begin{Bmatrix} 2 \\ 1 \quad 3 \end{Bmatrix}$
$\dfrac{k - k^2 x^2 - (\chi)^2 x^2 - \chi' x^3 + (\chi x^3)^2 k - \chi k' x^2 x^3}{1 - kx^2}$	$-\dfrac{k\chi x^3}{1-kx^2}$	$-\chi$
$\begin{Bmatrix} 2 \\ 2 \quad 1 \end{Bmatrix}$	$\begin{Bmatrix} 2 \\ 2 \quad 2 \end{Bmatrix}$	$\begin{Bmatrix} 2 \\ 2 \quad 3 \end{Bmatrix}$
$-\dfrac{k\chi x^3}{1-kx^2}$	0	0
$\begin{Bmatrix} 2 \\ 3 \quad 1 \end{Bmatrix}$	$\begin{Bmatrix} 2 \\ 3 \quad 2 \end{Bmatrix}$	$\begin{Bmatrix} 2 \\ 3 \quad 3 \end{Bmatrix}$
$-\chi$	0	0

$\lambda = 3$

$\begin{Bmatrix} 3 \\ 1 \quad 1 \end{Bmatrix}$	$\begin{Bmatrix} 3 \\ 2 \quad 1 \end{Bmatrix}$	$\begin{Bmatrix} 3 \\ 1 \quad 3 \end{Bmatrix}$
$\dfrac{\chi' x^2 - (\chi)^2 x^3 + k'\chi(x^2)^2 - k(\chi)^2 x^2 x^3}{1 - kx^2}$	$-\dfrac{\chi}{1-kx^2}$	0
$\begin{Bmatrix} 3 \\ 2 \quad 1 \end{Bmatrix}$	$\begin{Bmatrix} 3 \\ 2 \quad 2 \end{Bmatrix}$	$\begin{Bmatrix} 3 \\ 2 \quad 3 \end{Bmatrix}$
$-\dfrac{\chi}{1-kx^2}$	0	0
$\begin{Bmatrix} 3 \\ 3 \quad 1 \end{Bmatrix}$	$\begin{Bmatrix} 3 \\ 3 \quad 2 \end{Bmatrix}$	$\begin{Bmatrix} 3 \\ 3 \quad 3 \end{Bmatrix}$
0	0	0

Table 1.3 $(\nabla\varphi)^\lambda$ $(\lambda=1,2,3)$

$(\nabla\varphi)^1$	$\dfrac{1}{(1-kx^2)^2}\left(\dfrac{\partial\varphi}{\partial x^1}+\chi x^3\dfrac{\partial\varphi}{\partial x^2}-\chi x^2\dfrac{\partial\varphi}{\partial x^3}\right)$
$(\nabla\varphi)^2$	$\dfrac{1}{(1-kx^2)^2}\left\{\chi x^3\dfrac{\partial\varphi}{\partial x^1}+[(1-kx^2)^2+(\chi x^3)^2]\dfrac{\partial\varphi}{\partial x^2}-(\chi)^2 x^2 x^3\dfrac{\partial\varphi}{\partial x^3}\right\}$
$(\nabla\varphi)^2$	$\dfrac{1}{(1-kx^2)^2}\left\{-\chi x^2\dfrac{\partial\varphi}{\partial x^1}-(\chi)^2 x^2 x^3\dfrac{\partial\varphi}{\partial x^2}+[(1-kx^2)^2+(\chi x^2)^2]\dfrac{\partial\varphi}{\partial x^3}\right\}$

From Eq. (1–8), Eq. (1–10) and Eq. (1–15), we may obtain $\dot{x}_j$ and $(v\times\nabla\Omega)^\lambda$, as shown in Table 1.4 and Table 1.5, respectively.

Table 1.4 $\dot{x}_j$ $(j=1,2,3)$

$\dot{x}_1$	$[(1-kx^2)^2+(\chi x^3)^2+(\chi x^2)^2]\dot{x}^1-\chi x^3\dot{x}^2+\chi x^2\dot{x}^3$
$\dot{x}_2$	$-\chi x^3\dot{x}^1+\dot{x}^2$
$\dot{x}_3$	$\chi x^2\dot{x}^1+\dot{x}^3$

Table 1.5 $(v\times\nabla\Omega)^\lambda$ $(\lambda=1,2,3)$

$(v\times\nabla\Omega)^1$	$\dfrac{1}{1-kx^2}\left\{(\dot{x}^2-\chi x^3\dot{x}^1)\dfrac{\partial\Omega}{\partial x^3}-(\dot{x}^3+\chi x^2\dot{x}^1)\dfrac{\partial\Omega}{\partial x^2}\right\}$
$(v\times\nabla\Omega)^2$	$\dfrac{1}{1-kx^2}\left((\dot{x}^3+\chi x^2\dot{x}^1)\dfrac{\partial\Omega}{\partial x^1}-\{[(1-kx^2)^2+(\chi x^3)^2+(\chi x^2)^2]\dot{x}^1-\chi x^3\dot{x}^2+\chi x^2\dot{x}^3\}\dfrac{\partial\Omega}{\partial x^3}\right)$
$(v\times\nabla\Omega)^3$	$\dfrac{1}{1-kx^2}\left(\{[(1-kx^2)^2+(\chi x^3)^2+(\chi x^2)^2]\dot{x}^1-\chi x^3\dot{x}^2+\chi x^2\dot{x}^3\}\dfrac{\partial\Omega}{\partial x^2}-(\dot{x}^2-\chi x^3\dot{x}^1)\dfrac{\partial\Omega}{\partial x^1}\right)$

If the magnetic field is expressed by scalar magnetic potential Ω, $B=\mu_0 H$, $H=-\nabla\Omega$, where H is the magnetic field intensity, μ_0 is the magnetic permeability in vacuum, then we have

$$(v\times B)^\lambda=-\mu_0(v\times\nabla\Omega)^\lambda=-\dfrac{\mu_0}{\sqrt{g}}\left(\dot{x}_j\dfrac{\partial\Omega}{\partial x^k}-\dot{x}_k\dfrac{\partial\Omega}{\partial x^j}\right)(\lambda=1,2,3) \quad (1-16)$$

where λ, j, k is the cyclic permutation of 1, 2, 3; $\dot{x}_j$ is the covariant components of v, which can be expressed by the contravariant components of v

$$\dot{x}_j=g_{ji}\dot{x}^i \quad (j=1,2,3) \quad (1-17)$$

Substituting the quantities of Table 2.2, Table 1.3 and Table 1.5 in Eq. (1–12) and Eq. (1–14) (for convenience, we use symbols (s, p_2, p_3) instead of (x^1, x^2, x^3)), we obtain the electron motion's equation in Frenet local coordinate system (s, p_2, p_3):

$$\dfrac{m_0}{e}\left(\ddot{s}+\dfrac{k\chi p_3-k'p_2}{1-kp_2}\dot{s}^2-\dfrac{2k}{1-kp_2}\dot{s}\dot{p}_2\right)=\dfrac{1}{(1-kp_2)^2}\left(\dfrac{\partial\varphi}{\partial s}+\chi p_3\dfrac{\partial\varphi}{\partial p_2}-\chi p_2\dfrac{\partial\varphi}{\partial p_3}\right)+I_1 \quad (1-18\text{a})$$

$$\dfrac{m_0}{e}\left[\ddot{p}_2+\left(k-k^2p_2-\chi^2 p_2-\chi'p_3+\dfrac{k\chi^2 p_3^2-\chi k'p_2 p_3}{1-kp_2}\right)\dot{s}^2-\dfrac{2k\chi p_3}{1-kp_2}\dot{s}\dot{p}_2-2\chi\dot{s}\dot{p}_3\right]=$$

$$\dfrac{1}{(1-kp_2)^2}\left\{\chi p_3\dfrac{\partial\varphi}{\partial s}+[(1-kp_2)^2+\chi^2 p_3^2]\dfrac{\partial\varphi}{\partial p_2}-p_2 p_3\chi^2\dfrac{\partial\varphi}{\partial p_3}\right\}+I_2 \quad (1-18\text{b})$$

$$\frac{m_0}{e}\left[\ddot{p}_3 + \left(\chi'p_2 - \chi^2 p_3 + \frac{\chi k'p_2^2 - \chi^2 kp_2p_3}{1-kp_2}\right)\dot{s}^2 + \frac{2\chi\dot{s}\dot{p}_2}{1-kp_2}\right] =$$
$$\frac{1}{(1-kp_2)^2}\left\{-\chi p_2 \frac{\partial\varphi}{\partial s} - \chi^2 p_2 p_3 \frac{\partial\varphi}{\partial p_2} + [(1-kp_2)^2 + \chi^2 p_2^2]\frac{\partial\varphi}{\partial p_3}\right\} + I_3 \quad (1-18c)$$

where

$$I_1 = \frac{\mu_0}{1-kp_2}\left[(\dot{p}_2 - \chi p_3 \dot{s})\frac{\partial\Omega}{\partial p_3} - (\dot{p}_3 + \chi p_2 \dot{s})\frac{\partial\Omega}{\partial p_2}\right] \quad (1-19a)$$

$$I_2 = \frac{\mu_0}{1-kp_2}\left((\dot{p}_3 + \chi p_2 \dot{s})\frac{\partial\Omega}{\partial s} - \{[(1-kp_2)^2 + (\chi p_3)^2 + (\chi p_2)^2]\dot{s} - \chi p_3\dot{p}_2 + \chi p_2\dot{p}_3\}\frac{\partial\Omega}{\partial p_3}\right) \quad (1-19b)$$

$$I_3 = \frac{\mu_0}{1-kp_2}\left(\{[(1-kp_2)^2 + (\chi p_2)^2 + (\chi p_3)^2]\dot{s} - \chi p_3\dot{p}_2 + \chi p_2\dot{p}_3\}\frac{\partial\Omega}{\partial p_2} - (\dot{p}_2 - \chi p_3\dot{s})\frac{\partial\Omega}{\partial s}\right) \quad (1-19c)$$

where the magnetic field is expressed by scalar magnetic potential Ω. Eqs. (1−18 a, b, c) are called the electron motion's equation of "contravariant" type.

If the magnetic field is expressed by vector magnetic potential A, $B = \nabla \times A$, then we have

$$(v \times B)^\lambda = (v \times (\nabla \times A))^\lambda = \frac{1}{\sqrt{g}}[\dot{x}_j(\nabla \times A)_k - \dot{x}_k(\nabla \times A)_j] \quad (\lambda = 1,2,3) \quad (1-20a)$$

where

$$(\nabla \times A)_l = g_{li}(\nabla \times A)^i = g_{li}\frac{1}{\sqrt{g}}\left(\frac{\partial A_k}{\partial x^j} - \frac{\partial A_j}{\partial x^k}\right) \quad (l=1,2,3) \quad (1-20b)$$

where (λ, j, k), (i, j, k) are both cyclic permutation of 1, 2, 3. The quantities of $(\nabla \times A)_j$ and $[v \times (\nabla \times A)^\lambda]$ are shown in Table 1.6 and Table 1.7, respectively.

Table 1.6 $(\nabla \times A)_j$ $(i=1, 2, 3)$

$(\nabla \times A)_1$	$\frac{(1-kx^2)^2 + (\chi x^2)^2 + (\chi x^3)^2}{1-kx^2}\left(\frac{\partial A_3}{\partial x^2} - \frac{\partial A_2}{\partial x^3}\right) - \frac{\chi x^3}{1-kx^2}\left(\frac{\partial A_1}{\partial x^3} - \frac{\partial A_3}{\partial x^1}\right) + \frac{\chi x^2}{1-kx^2}\left(\frac{\partial A_2}{\partial x^1} - \frac{\partial A_1}{\partial x^2}\right)$
$(\nabla \times A)_2$	$\frac{-\chi x^3}{1-kx^2}\left(\frac{\partial A_3}{\partial x^2} - \frac{\partial A_2}{\partial x^3}\right) + \frac{1}{1-kx^2}\left(\frac{\partial A_1}{\partial x^3} - \frac{\partial A_3}{\partial x^1}\right)$
$(\nabla \times A)_3$	$\frac{\chi x^2}{1-kx^2}\left(\frac{\partial A_3}{\partial x^2} - \frac{\partial A_2}{\partial x^3}\right) + \frac{1}{1-kx^2}\left(\frac{\partial A_2}{\partial x^1} - \frac{\partial A_1}{\partial x^2}\right)$

Table 1.7 $[v \times (\nabla \times A)]^\lambda$ $(\lambda =1, 2, 3)$

$[v \times (\nabla \times A)]^1$	$\frac{1}{(1-kx^2)^2}\left[(\chi x^2 \dot{x}^2 + \chi x^3 \dot{x}^3)\left(\frac{\partial A_3}{\partial x^2} - \frac{\partial A_2}{\partial x^3}\right) + (-\chi x^3 \dot{x}^1 + \dot{x}^2)\left(\frac{\partial A_2}{\partial x^1} - \frac{\partial A_1}{\partial x^2}\right) - (\chi x^2 \dot{x}^1 + \dot{x}^3)\left(\frac{\partial A_1}{\partial x^3} - \frac{\partial A_3}{\partial x^1}\right)\right]$

$[v \times (\nabla \times A)]^2$	$\dfrac{1}{(1-kx^2)^2} \{ \{[(1-kx^2)^2 + (\chi x^3)^2]\dot{x}^3 + (\chi)^2 x^2 x^3 \dot{x}^2 \} \left(\dfrac{\partial A_3}{\partial x^2} - \dfrac{\partial A_2}{\partial x^3} \right) - $ $[(\chi)^2 x^2 x^3 \dot{x}^1 + \chi x^3 \dot{x}^3] \left(\dfrac{\partial A_1}{\partial x^3} - \dfrac{\partial A_3}{\partial x^1} \right) + \{\chi x^3 \dot{x}^2 - [(1-kx^2)^2 + (\chi x^3)^2]\dot{x}^1 \} \left(\dfrac{\partial A_2}{\partial x^1} - \dfrac{\partial A_1}{\partial x^2} \right) \}$
$[v \times (\nabla \times A)]^3$	$\dfrac{1}{(1-kx^2)^2} \{ -\{[(1-kx^2)^2 + (\chi x^2)^2]\dot{x}^2 + (\chi)^2 x^2 x^3 \dot{x}^3 \} \left(\dfrac{\partial A_3}{\partial x^2} - \dfrac{\partial A_2}{\partial x^3} \right) + $ $[(\chi)^2 x^2 x^3 \dot{x}^1 - \chi x^2 \dot{x}^2] \left(\dfrac{\partial A_2}{\partial x^1} - \dfrac{\partial A_1}{\partial x^2} \right) + \{[(1-kx^2)^2 + (\chi x^2)^2]\dot{x}^1 + \chi x^2 \dot{x}^3 \} \left(\dfrac{\partial A_1}{\partial x^3} - \dfrac{\partial A_3}{\partial x^1} \right) \}$

Substituting the quantities of Table 1.2, Table 1.3, and Table 1.7 in Eq. (1−12) and Eq. (1−14), we also obtain the electron motion's equation of "contravariant" type, which can also be written in the same form as Eq.18 (a, b), provided that

$$I_1 = -\frac{1}{(1-kp_2)^2} \left\{ (\chi p_2 \dot{p}_2 + \chi p_3 \dot{p}_3) \left(\frac{\partial A_3}{\partial p_2} - \frac{\partial A_2}{\partial p_3} \right) + \right.$$
$$\left. (\dot{p}_2 - \chi p_3 \dot{s}) \left(\frac{\partial A_2}{\partial s} - \frac{\partial A_1}{\partial p_2} \right) - (\chi p_2 \dot{s} + \dot{p}_3) \left(\frac{\partial A_1}{\partial p_3} - \frac{\partial A_3}{\partial s} \right) \right\} \quad (1-21\text{a})$$

$$I_2 = -\frac{1}{(1-kp_2)^2} \Big(\{[(1-kp_2)^2 + (\chi p_3)^2]\dot{p}_3 + (\chi)^2 p_2 p_3 \dot{p}_2 \} \left(\frac{\partial A_3}{\partial p_2} - \frac{\partial A_2}{\partial p_3} \right) -$$
$$[(\chi)^2 p_2 p_3 \dot{s} + \chi p_3 \dot{p}_3] \left(\frac{\partial A_1}{\partial p_3} - \frac{\partial A_3}{\partial s} \right) + \{\chi p_3 \dot{p}_2 - [(1-kp_2)^2 + (\chi p_3)^2]\dot{s} \} \left(\frac{\partial A_2}{\partial s} - \frac{\partial A_1}{\partial p_2} \right) \Big)$$
$$(1-21\text{b})$$

$$I_3 = -\frac{1}{(1-kp_2)^2} \Big(-\{[(1-kp_2)^2 + (\chi p_2)^2]\dot{p}_2 + (\chi)^2 p_2 p_3 \dot{p}_3 \} \left(\frac{\partial A_3}{\partial p_2} - \frac{\partial A_2}{\partial p_3} \right) +$$
$$[(\chi)^2 p_2 p_3 \dot{s} - \chi p_2 \dot{p}_2] \left(\frac{\partial A_2}{\partial s} - \frac{\partial A_1}{\partial p_2} \right) + \{[(1-kp_2)^2 + (\chi p_2)^2]\dot{s} + \chi p_2 \dot{p}_3 \} \left(\frac{\partial A_1}{\partial p_3} - \frac{\partial A_3}{\partial s} \right) \Big)$$
$$(1-21\text{c})$$

where the magnetic field is expressed by vector magnetic potential A.

3 The electron motion's equation of "covariant" type in Frenet local coordinate system

In the non-relativistic case, the covariant form of Newton's equation will have the form[7]:

$$m_0 \left(\ddot{x}_\lambda - \begin{Bmatrix} i \\ \lambda \quad k \end{Bmatrix} \dot{x}_i \dot{x}^k \right) = F_\lambda \quad (\lambda = 1, 2, 3) \quad (1-22)$$

Using Eq. (1−17), we have

$$\ddot{x}_\lambda = \frac{\mathrm{d}}{\mathrm{d}t}(\dot{x}_\lambda) = \frac{\mathrm{d}}{\mathrm{d}t}(g_{\lambda k} \dot{x}^k) = g_{\lambda k} \ddot{x}^k + \dot{x}^k \dot{x}^m \frac{\partial g_{\lambda k}}{\partial x^m} \quad (1-23)$$

then Eq. (1−22) can be written as

$$m_0 \left(g_{\lambda k} \ddot{x}^k + \dot{x}^k \dot{x}^m \frac{\partial g_{\lambda k}}{\partial x^m} - \begin{Bmatrix} i \\ \lambda \quad k \end{Bmatrix} g_{ij} \dot{x}^j \dot{x}^k \right) = F_\lambda \qquad (1-24)$$

Using Eq. (1 – 8) and Table 1.1 and Table 1.2, we may obtain the quantities of $g_{\lambda k} \ddot{x}^k$, $\dot{x}^k \dot{x}^m \frac{\partial g_{\lambda k}}{\partial x^m}$ and $\begin{Bmatrix} i \\ \lambda \quad k \end{Bmatrix} g_{ij} \dot{x}^j \dot{x}^k$, as shown in Table 1.8 – Table 1.10, respectively.

Table 1.8 $g_{\lambda k} \ddot{x}^k$ ($\lambda = 1, 2, 3$)

$g_{1k}\ddot{x}^k$	$[(1-kx^2)^2 + (\chi x^2)^2 + (\chi x^3)^2] \ddot{x}^1 - \chi x^3 \ddot{x}^2 + \chi x^2 \ddot{x}^3$
$g_{2k}\ddot{x}^k$	$-\chi x^3 \ddot{x}^1 + \ddot{x}^2$
$g_{3k}\ddot{x}^k$	$\chi x^2 \ddot{x}^1 + \ddot{x}^3$

Table 1.9 $\dot{x}^k \dot{x}^m \frac{\partial g_{\lambda k}}{\partial x^m}$ ($k, m, \lambda = 1, 2, 3$; k, m, λ denote the row, column and page, respectively)

$\lambda = 1$

$(\dot{x}^1)^2 \frac{\partial g_{11}}{\partial x^1}$	$\dot{x}^1 \dot{x}^2 \frac{\partial g_{11}}{\partial x^2}$	$\dot{x}^1 \dot{x}^3 \frac{\partial g_{11}}{\partial x^3}$
$(\dot{x}^1)^2 [-2k'x^2 - 2kk'(x^2)^2 + 2\chi\chi'(x^2)^2 + 2\chi\chi'(x^3)^2]$	$\dot{x}^1 \dot{x}^2 [-2k(1-kx^2) + 2(\chi)^2 x^2]$	$\dot{x}^1 \dot{x}^3 [2(\chi)^2 x^3]$
$\dot{x}^2 \dot{x}^1 \frac{\partial g_{12}}{\partial x^1}$	$(\dot{x}^2)^2 \frac{\partial g_{12}}{\partial x^2}$	$\dot{x}^2 \dot{x}^3 \frac{\partial g_{12}}{\partial x^3}$
$\dot{x}^2 \dot{x}^1 [-\chi' \dot{x}^3]$	0	$\dot{x}^2 \dot{x}^3 (-\chi)$
$\dot{x}^3 \dot{x}^1 \frac{\partial g_{13}}{\partial x^1}$	$\dot{x}^3 \dot{x}^2 \frac{\partial g_{13}}{\partial x^2}$	$(\dot{x}^3)^2 \frac{\partial g_{13}}{\partial x^3}$
$\dot{x}^3 \dot{x}^1 (\chi' x^2)$	$\dot{x}^3 \dot{x}^2 (\chi)$	0

$\lambda = 2$

$(\dot{x}^1)^2 \frac{\partial g_{21}}{\partial x^1}$	$\dot{x}^1 \dot{x}^2 \frac{\partial g_{21}}{\partial x^2}$	$\dot{x}^1 \dot{x}^3 \frac{\partial g_{21}}{\partial x^3}$
$(\dot{x}^1)^2 (-\chi' x^3)$	0	$\dot{x}^1 \dot{x}^3 (-\chi)$
$\dot{x}^2 \dot{x}^1 \frac{\partial g_{22}}{\partial x^1}$	$(\dot{x}^2)^2 \frac{\partial g_{22}}{\partial x^2}$	$\dot{x}^2 \dot{x}^3 \frac{\partial g_{22}}{\partial x^3}$
0	0	0
$\dot{x}^3 \dot{x}^1 \frac{\partial g_{23}}{\partial x^1}$	$\dot{x}^3 \dot{x}^2 \frac{\partial g_{23}}{\partial x^2}$	$(\dot{x}^3)^2 \frac{\partial g_{23}}{\partial x^3}$
0	0	0

$\lambda = 3$

$(\dot{x}^1)^2 \frac{\partial g_{31}}{\partial x^1}$	$\dot{x}^1 \dot{x}^2 \frac{\partial g_{31}}{\partial x^2}$	$\dot{x}^1 \dot{x}^3 \frac{\partial g_{31}}{\partial x^3}$
$(\dot{x}^1)^2 (\chi' x^2)$	$\dot{x}^1 \dot{x}^2 (\chi)$	0
$\dot{x}^2 \dot{x}^1 \frac{\partial g_{32}}{\partial x^1}$	$(\dot{x}^2)^2 \frac{\partial g_{32}}{\partial x^2}$	$\dot{x}^2 \dot{x}^3 \frac{\partial g_{32}}{\partial x^3}$
0	0	0
$\dot{x}^3 \dot{x}^1 \frac{\partial g_{33}}{\partial x^1}$	$\dot{x}^3 \dot{x}^2 \frac{\partial g_{33}}{\partial x^2}$	$(\dot{x}^3)^2 \frac{\partial g_{33}}{\partial x^3}$
0	0	0

Table 1.10 $\begin{Bmatrix} i \\ \lambda \ \ k \end{Bmatrix} g_{ij} \dot{x}^j \dot{x}^k$ ($i = 1, 2, 3$)

(j, k, $\lambda = 1, 2, 3$; j, k, λ denote the row, column and page, respectively.)

$\lambda = 1$

$\begin{Bmatrix} i \\ 1 \ \ 1 \end{Bmatrix} g_{i1} \dot{x}^1 \dot{x}^1$	$\begin{Bmatrix} i \\ 1 \ \ 2 \end{Bmatrix} g_{i1} \dot{x}^1 \dot{x}^2$	$\begin{Bmatrix} i \\ 1 \ \ 3 \end{Bmatrix} g_{i1} \dot{x}^1 \dot{x}^3$
$-[k'x^2 - kk'(x^2)^2 - \chi\chi'(x^3)^2 - \chi\chi'(x^2)^2](\dot{x}^1)^2$	$-[k(1-kx^2) - (\chi)^2 x^2]\dot{x}^1\dot{x}^2$	$(\chi)^2 x^3 \dot{x}^1 \dot{x}^3$
$\begin{Bmatrix} i \\ 1 \ \ 1 \end{Bmatrix} g_{i2} \dot{x}^2 \dot{x}^1$	$\begin{Bmatrix} i \\ 1 \ \ 2 \end{Bmatrix} g_{i2} \dot{x}^2 \dot{x}^2$	$\begin{Bmatrix} i \\ 1 \ \ 3 \end{Bmatrix} g_{i2} \dot{x}^2 \dot{x}^3$
$-[-k(1-kx^2) + (\chi)^2 x^2 + \chi' x^3]\dot{x}^2 \dot{x}^1$	0	$-\chi \dot{x}^2 \dot{x}^3$
$\begin{Bmatrix} i \\ 1 \ \ 1 \end{Bmatrix} g_{i3} \dot{x}^3 \dot{x}^1$	$\begin{Bmatrix} i \\ 1 \ \ 2 \end{Bmatrix} g_{i3} \dot{x}^3 \dot{x}^2$	$\begin{Bmatrix} i \\ 1 \ \ 3 \end{Bmatrix} g_{i3} \dot{x}^3 \dot{x}^3$
$-[(\chi)^2 x^3 - \chi' x^2]\dot{x}^3 \dot{x}^1$	$\chi \dot{x}^3 \dot{x}^2$	0

$\lambda = 2$

$\begin{Bmatrix} i \\ 2 \ \ 1 \end{Bmatrix} g_{i1} \dot{x}^1 \dot{x}^1$	$\begin{Bmatrix} i \\ 2 \ \ 2 \end{Bmatrix} g_{i1} \dot{x}^1 \dot{x}^2$	$\begin{Bmatrix} i \\ 2 \ \ 3 \end{Bmatrix} g_{i1} \dot{x}^1 \dot{x}^3$
$-[k - k^2 x^2 - (\chi)^2 x^2](\dot{x}^1)^2$	0	0
$\begin{Bmatrix} i \\ 2 \ \ 1 \end{Bmatrix} g_{i2} \dot{x}^2 \dot{x}^1$	$\begin{Bmatrix} i \\ 2 \ \ 2 \end{Bmatrix} g_{i2} \dot{x}^2 \dot{x}^2$	$\begin{Bmatrix} i \\ 2 \ \ 3 \end{Bmatrix} g_{i2} \dot{x}^2 \dot{x}^3$
0	0	0
$\begin{Bmatrix} i \\ 2 \ \ 1 \end{Bmatrix} g_{i3} \dot{x}^3 \dot{x}^1$	$\begin{Bmatrix} i \\ 2 \ \ 2 \end{Bmatrix} g_{i3} \dot{x}^3 \dot{x}^2$	$\begin{Bmatrix} i \\ 2 \ \ 3 \end{Bmatrix} g_{i3} \dot{x}^3 \dot{x}^3$
$\chi x^3 \dot{x}^1$	0	0

$\lambda = 3$

$\begin{Bmatrix} i \\ 3 \ \ 1 \end{Bmatrix} g_{i1} \dot{x}^1 \dot{x}^1$	$\begin{Bmatrix} i \\ 3 \ \ 2 \end{Bmatrix} g_{i1} \dot{x}^1 \dot{x}^2$	$\begin{Bmatrix} i \\ 3 \ \ 3 \end{Bmatrix} g_{i1} \dot{x}^1 \dot{x}^3$
$(\chi^2) x^3 x^1 x^1$	0	0
$\begin{Bmatrix} i \\ 3 \ \ 1 \end{Bmatrix} g_{i2} \dot{x}^2 \dot{x}^1$	$\begin{Bmatrix} i \\ 3 \ \ 2 \end{Bmatrix} g_{i2} \dot{x}^2 \dot{x}^2$	$\begin{Bmatrix} i \\ 3 \ \ 3 \end{Bmatrix} g_{i2} \dot{x}^2 \dot{x}^3$
$-\chi \dot{x}^2 \dot{x}^1$	0	0
$\begin{Bmatrix} i \\ 3 \ \ 1 \end{Bmatrix} g_{i3} \dot{x}^3 \dot{x}^1$	$\begin{Bmatrix} i \\ 3 \ \ 2 \end{Bmatrix} g_{i3} \dot{x}^3 \dot{x}^2$	$\begin{Bmatrix} i \\ 3 \ \ 3 \end{Bmatrix} g_{i3} \dot{x}^3 \dot{x}^3$
0	0	0

In Table 1.10, $\left\{\begin{array}{c}i\\ \lambda\ k\end{array}\right\}g_{ij}\dot{x}^j\dot{x}^k$ is still described by the form of matrix with 3 marks, but each term expresses the sum from $i=1$ to $i=3$. Here we apply the Einstein's summation agreement, for example,

$$\left\{\begin{array}{c}i\\ 1\ 1\end{array}\right\}g_{i1}\dot{x}^1\dot{x}^1 = \sum_{i=1}^{3}\left\{\begin{array}{c}i\\ 1\ 1\end{array}\right\}g_{i1}\dot{x}^1\dot{x}^1 \quad (1-25)$$

In the electrostatic and magnetic fields, the covariant form of Lorentz force F acting on an electron can be written as

$$F_\lambda = e(\nabla\varphi)_\lambda - e(v\times B)_\lambda \quad (\lambda=1,2,3) \quad (1-26)$$

where $(\nabla\varphi)_\lambda$, the covariant components of gradient of a scalar potential φ, is just the partial derivative with respect to the contravariant coordinate x^λ, i.e.,

$$(\nabla\varphi)_1 = \frac{\partial\varphi}{\partial x^1},\ (\nabla\varphi)_2 = \frac{\partial\varphi}{\partial x^2},\ (\nabla\varphi)_3 = \frac{\partial\varphi}{\partial x^3} \quad (1-27)$$

If the magnetic field is expressed by scalar magnetic potential Ω, then

$$(v\times B)_\lambda = -\mu_0(v\times\nabla\Omega)_\lambda = -\mu_0\sqrt{g}[\dot{x}^j(\nabla\Omega)^k - \dot{x}^k(\nabla\Omega)^j] \quad (\lambda=1,2,3) \quad (1-28)$$

where λ, j, k is a cyclic permutation of 1, 2, 3, the expressions of $(\nabla\Omega)^k$ still is the same as $(\nabla\varphi)^\lambda$ in Table 3, only φ will be substituted by Ω. Therefore, $(v\times\nabla\Omega)_\lambda$ will be given in Table 1.11.

Substituting the quantities of Table 1.8 – Table 1.11 and Eq. (1-27) in Eq. (1-24) and Eq. (1-26), where symbols, x^1, x^2, x^3 are replaced by s, p_2, p_3 too, then we obtain the electron motion's equation of "covariant" type in Frenet local coordinate system (s, p_2, p_3):

$$\frac{m_0}{e}\{[(1-kp_2)^2 + (\chi p_2)^2 + (\chi p_3)^2]\ddot{s} - \chi p_3\ddot{p}_2 + \chi p_2\ddot{p}_3 + [-k'p_2 + kk'(p_2)^2 +$$

$$\chi\chi'(p_2)^2 + \chi\chi'(p_3)^2]\dot{s}^2 + 2(-k+k^2p_2+\chi^2p_2)\dot{s}\dot{p}_2 + 2\chi^2p_3\dot{s}\dot{p}_3\} = \frac{\partial\varphi}{\partial s} + J_1 \quad (1-29a)$$

$$\frac{m_0}{e}\{-\chi p_3\ddot{s} + \ddot{p}_2 + (k-k^2p_2-\chi^2p_2-\chi'p_3)\dot{s}^2 - 2\chi\dot{s}\dot{p}_3\} = \frac{\partial\varphi}{\partial p_2} + J_2 \quad (1-29b)$$

$$\frac{m_0}{e}\{\chi p_2\ddot{s} + \ddot{p}_3 + (\chi'p_2 - \chi^2p_3)\dot{s}^2 + 2\chi\dot{s}\dot{p}_2\} = \frac{\partial\varphi}{\partial p_3} + J_3 \quad (1-29c)$$

where

$$J_1 = -\frac{\mu_0}{1-kp_2}\Big\{(\chi p_2\dot{p}_2 + \chi_3\dot{p}_3)\frac{\partial\Omega}{\partial s} + \{[(1-kp_2)^2 + (\chi p_3)^2]\dot{p}_3 + \chi^2 p_2 p_3\dot{p}_2\}\frac{\partial\Omega}{\partial p_2} -$$

$$\{[(1-kp_2)^2 + (\chi p_2)^2]\dot{p}_3 + \chi^2 p_2 p_3\dot{p}_3\}\frac{\partial\Omega}{\partial p_3}\Big\}, \quad (1-30a)$$

$$J_2 = \frac{\mu_0}{1-kp_2}\Big\{(\dot{p}_3 + \chi p_2\dot{s})\frac{\partial\Omega}{\partial s} + [\chi p_3\dot{p}_3 + (\chi)^2 p_2 p_3\dot{s}]\frac{\partial\Omega}{\partial p_2} - \{[(1-kp_2)^2 + (\chi p_2)^2]\dot{s} + \chi p_2\dot{p}_3\}\frac{\partial\Omega}{\partial p_3}\Big\}$$

$$(1-30b)$$

$$J_3 = -\frac{\mu_0}{1-kp_2}\left\{(\dot{p}_2 - \chi p_3 \dot{s})\frac{\partial \Omega}{\partial s} - \{[(1-kp_2)^2 + (\chi p_3)^2]\dot{s} + \chi p_3 \dot{p}_2\}\frac{\partial \Omega}{\partial p_2} - \right. \quad (1-30c)$$
$$\left. [\chi p_2 \dot{p}_2 - (\chi)^2 p_2 p_3 \dot{s}]\frac{\partial \Omega}{\partial p_3}\right\}$$

where the magnetic field is expressed by scalar magnetic potential Ω.

If the magnetic field is expressed by vector magnetic potential $\boldsymbol{A}$, $\boldsymbol{B} = \nabla \times \boldsymbol{A}$, then we have

$$[v \times B]_\lambda = [v \times (\nabla \times A)]_\lambda = \sqrt{g}[\dot{x}^j (\nabla \times A)^k - \dot{x}^k (\nabla \times A)^j] \quad (\lambda = 1, 2, 3) \quad (1-31a)$$

where

$$(\nabla \times \boldsymbol{A})^i = \frac{1}{\sqrt{g}}\left(\frac{\partial A_k}{\partial x^j} - \frac{\partial A_j}{\partial x^k}\right) \quad (i = 1, 2, 3) \quad (1-31b)$$

where λ, j, k and i, j, k are both cyclic permutation of 1, 2, 3. From this, $[v \times (\nabla \times A)]_\lambda$ will be given in Table 1.12.

Table 1.11 $(v \times \nabla \Omega)_\lambda$ $(\lambda = 1, 2, 3)$

$(v \times \nabla \Omega)_1$	$-\frac{1}{1-kx^2}\{(\chi x^2 \dot{x}^2 + \chi x^3 \dot{x}^3)\frac{\partial \Omega}{\partial x^1} + \{[(1-kx^2)^2 + (\chi x^3)^2]\dot{x}^3 + (\chi)^2 x^2 x^3 \dot{x}^2\}\frac{\partial \Omega}{\partial x^2} - \{[(1-kx^2)^2 + (\chi \dot{x}^2)^2]\dot{x}^2 + (\chi)^2 x^2 x^3 \dot{x}^3\}\frac{\partial \Omega}{\partial x^3}\}$
$(v \times \nabla \Omega)_2$	$\frac{1}{1-kx^2}\{(\dot{x}^3 + \chi x^2 \dot{x}^1)\frac{\partial \Omega}{\partial x^1} + [\chi x^3 \dot{x}^3 + (\chi)^2 x^2 x^3 \dot{x}^1]\frac{\partial \Omega}{\partial x^2} - \{[(1-kx^2)^2 + (\chi x^2)^2]\dot{x}^1 + \chi x^2 \dot{x}^3\}\frac{\partial \Omega}{\partial x^3}\}$
$(v \times \nabla \Omega)_3$	$-\frac{1}{1-kx^2}\{(\dot{x}^2 - \chi x^3 \dot{x}^1)\frac{\partial \Omega}{\partial x^1} - \{[(1-kx^2)^2 + (\chi x^3)^2]\dot{x}^1 - \chi x^3 \dot{x}^2\}\frac{\partial \Omega}{\partial x^2} - [\chi x^2 \dot{x}^2 - (\chi)^2 x^2 x^3 \dot{x}^1]\frac{\partial \Omega}{\partial x^3}\}$

Table 1.12 $[v \times (\nabla \times A)]_\lambda$ $(\lambda = 1, 2, 3)$

$[v \times (\nabla \times A)]_1$	$\left(\frac{\partial A_2}{\partial x^1} - \frac{\partial A_1}{\partial x^2}\right)\dot{x}^2 - \left(\frac{\partial A_1}{\partial x^3} - \frac{\partial A_3}{\partial x^1}\right)\dot{x}^3$
$[v \times (\nabla \times A)]_2$	$\left(\frac{\partial A_3}{\partial x^2} - \frac{\partial A_2}{\partial x^3}\right)\dot{x}^3 - \left(\frac{\partial A_2}{\partial x^1} - \frac{\partial A_1}{\partial x^2}\right)\dot{x}^1$
$[v \times (\nabla \times A)]_3$	$\left(\frac{\partial A_1}{\partial x^3} - \frac{\partial A_3}{\partial x^1}\right)\dot{x}^1 - \left(\frac{\partial A_3}{\partial x^2} - \frac{\partial A_2}{\partial x^3}\right)\dot{x}^2$

Similarly to derivation of Eq. (1-29a, b, c), substituting Table 1.12 for Table 1.11, we obtain the electron motion's equation of "covariant" type, which can also be written in the same form as Eq. (1-29a, b, c), provided that

$$J_1 = -\dot{p}_2\left(\frac{\partial A_2}{\partial s} - \frac{\partial A_1}{\partial p_2}\right) + \dot{p}_3\left(\frac{\partial A_1}{\partial p_3} - \frac{\partial A_3}{\partial s}\right) \quad (1-32a)$$

$$J_2 = -\dot{p}_3\left(\frac{\partial A_3}{\partial p_2} - \frac{\partial A_2}{\partial p_3}\right) + \dot{s}\left(\frac{\partial A_2}{\partial s} - \frac{\partial A_1}{\partial p_2}\right) \tag{1-32b}$$

$$J_3 = -\dot{s}\left(\frac{\partial A_1}{\partial p_3} - \frac{\partial A_3}{\partial s}\right) + \dot{p}_2\left(\frac{\partial A_3}{\partial p_2} - \frac{\partial A_2}{\partial p_3}\right) \tag{1-32c}$$

where the magnetic field is expressed by vector magnetic potential $\boldsymbol{A}$.

It will be seen from Eqs. (1-18a, b, c), Eqs. (1-19a, b, c), Eq. (1-21a, b, c) and Eqs. (1-29a, b, c), Eqs. (1-30a, b, c), Eqs. (1-32a, b, c) that the electron motion's equation of "covariant" type is more complicated than the "contravariant" one, because it contains derivatives of the second order of two or more variables with respect to time t. Applying elimination of unknown to Eqs. (1-29a, b, c), so that only one variable's derivative of the second order with respect to time t is involved, then we may transform the electron motion's equation of "covariant" type into the "contravariant" one.

4 Deduction of the electron motion's equation using generalized Lagrange's function

As is well known, in electron optics, the generalized Lagrange's function L of charged particles usually can be expressed by

$$L = \frac{m_0}{2}(v \cdot v) + e\varphi - e(\boldsymbol{A} \cdot v) \tag{1-33}$$

In Frenet local coordinate system, the tensor form of Eq. (1-33) may be written as

$$L = \frac{1}{2}m_0 g_{jk}\dot{x}^j\dot{x}^k + e\varphi - e(A_k\dot{x}^k) \tag{1-34}$$

which makes the functional

$$\int_{t_0}^{t_1} L\,\mathrm{d}t \tag{1-35}$$

an extreme value.

The Euler's equation of this variation problem may be written as

$$\frac{\mathrm{d}}{\mathrm{d}t}\left(\frac{\partial L}{\partial \dot{x}^i}\right) - \frac{\partial L}{\partial x^i} = 0 \quad (i=1,2,3) \tag{1-36}$$

this is just the electron motion's equation to be derived.

From Eq. (1-34), we have

$$\frac{\partial L}{\partial x^i} = \frac{1}{2}m_0 \frac{\partial g_{jk}}{\partial x^i}\dot{x}^j\dot{x}^k + e\frac{\partial \varphi}{\partial x^i} - e\frac{\partial A_k}{\partial x^i}\dot{x}^k \tag{1-37}$$

$$\frac{\partial L}{\partial \dot{x}^i} = \frac{1}{2}m_0(g_{ik}\dot{x}^k + g_{ij}\dot{x}^j) - eA_i \tag{1-38}$$

$$\frac{\mathrm{d}}{\mathrm{d}t}\left(\frac{\partial L}{\partial \dot{x}^i}\right) = \frac{1}{2}m_0\left(g_{ik}\ddot{x}^k + \frac{\partial g_{ik}}{\partial x^j}\dot{x}^j\dot{x}^k + g_{ij}\ddot{x}^j + \frac{\partial g_{ij}}{\partial x^k}\dot{x}^k\dot{x}^j\right) - \frac{\mathrm{d}}{\mathrm{d}t}(eA_i) \tag{1-39}$$

Substituting them in Eq. (1-36), we may obtain

$$m_0\{g_{ij}\ddot{x}^j + [i,jk]\dot{x}^j\dot{x}^k\} = e\frac{\partial \varphi}{\partial x^i} - e\frac{\partial A_k}{\partial x^i}\dot{x}^k + e\frac{\partial A_i}{\partial x^j}\dot{x}^j \quad (i=1,2,3) \tag{1-40}$$

where $g_{ij}\ddot{x}^j$ has been given in Table 1.8, $[i, jk]$ is the Christoffel function of the first kind:

$$[i,jk] = \frac{1}{2}\left(\frac{\partial g_{ij}}{\partial x^k} + \frac{\partial g_{ik}}{\partial x^j} - \frac{\partial g_{jk}}{\partial x^i}\right) \quad (1-41)$$

which can also be expressed in the form of matrix with 3 marks, as shown in Table 1.13.

Table 1.13 $[i, j, k]$ ($i, j, k = 1, 2, 3; j, k, i$) **denote the row, column and page, respectively**

$i = 1$

[1, 1, 1]	[1, 1, 2]	[1, 1, 3]
$(1-kx^2)(-k'x^2) + \chi\chi'(x^2)^2 + \chi\chi'(x^3)^2$	$(1-kx^2)(-k) + (\chi')^2 x^2$	$(\chi')^2 x^3$
[1, 2, 1]	[1, 2, 2]	[1, 2, 3]
$(1-kx^2)(-k) + (\chi')^2 x^2$	0	0
[1, 3, 1]	[1, 3, 2]	[1, 3, 3]
$(\chi')^2 x^3$	0	0

$i = 2$

[2, 1, 1]	[2, 1, 2]	[2, 1, 3]
$k - k^2 x^2 - (\chi')^2 x^2 - \chi' x^3$	0	$-\chi'$
[2, 2, 1]	[2, 2, 2]	[2, 2, 3]
0	0	0
[2, 3, 1]	[2, 3, 2]	[2, 3, 3]
$-\chi'$	0	0

$i = 3$

[3, 1, 1]	[3, 1, 2]	[3, 1, 3]
$\chi' x^2 - (\chi')^2 x^3$	χ'	0
[3, 2, 1]	[3, 2, 2]	[3, 2, 3]
χ'	0	0
[3, 3, 1]	[3, 3, 2]	[3, 3, 3]
0	0	0

From Eq. (1-40), we can also derive the electron motion's equation of "covariant" type in Frenet local coordinate system (s, p_2, p_3), in which the magnetic field is expressed by vector magnetic potential $\boldsymbol{A}$, as it was derived in Eqs. (1-29a, b, c) and Eqs. (1-32a, b, c).

5 Electron trajectory equation in Frenet local coordinate system

To transform the general electron motion's equation with respect to independent variable t, the time, into the electron trajectory equation with respect to independent variable s, the arc length along principal trajectory, let $v^* = \mathrm{d}s^*/\mathrm{d}t$, the velocity of electron at the neighboring trajectory. From the standpoint of energy, we have

$$\frac{\mathrm{d}s^*}{\mathrm{d}t} = \sqrt{\frac{2e}{m_0}\varphi_*} \qquad (1-42)$$

where $\varphi_* = \varphi + \varepsilon_0$, $e\varepsilon_0$ is the initial energy of electrons emitted from the object surface, φ_* is usually called the normalized potential. The elements of arc length $\mathrm{d}s$ at the principal trajectory and $\mathrm{d}s^*$ at its neighboring trajectory are related such that

$$\left(\frac{\mathrm{d}s^*}{\mathrm{d}s}\right)^2 = (1-kp_2)^2 + \chi^2(p_2^2 + p_3^2) - 2\chi p_3 p_2' + 2\chi p_2 p_3' + p_2'^2 + p_3'^2 \qquad (1-43)$$

Transforming each of the parameters $\dot{s}$, $\dot{p}_2$, $\ddot{p}_2$, we have

$$\dot{s} = \frac{\mathrm{d}s}{\mathrm{d}s^*}\frac{\mathrm{d}s^*}{\mathrm{d}t} = \sqrt{\frac{2e}{m_0}\varphi_*}\frac{\mathrm{d}s}{\mathrm{d}s^*},$$

$$\dot{p}_2 = \sqrt{\frac{2e}{m_0}\varphi_*}\frac{\mathrm{d}s}{\mathrm{d}s^*}p_2',$$

$$\ddot{p}_2 = \frac{e}{m_0}\left\{2\varphi_*\left(\frac{\mathrm{d}s}{\mathrm{d}s^*}\right)^2 p_2'' + p_2'\frac{\mathrm{d}}{\mathrm{d}s}\left[\varphi_*\left(\frac{\mathrm{d}s}{\mathrm{d}s^*}\right)^2\right]\right\} \qquad (1-44)$$

Similar expressions can be obtained for $\dot{p}_3$ and $\ddot{p}_3$. Substituting Eq. (43) and Eq. (1-44) in Eqs. (1-18b, c), Eqs. (1-19b, c), Eqs. (1-21b, c) and Eqs. (1-29b, c), Eqs. (1-30b, c), Eqs. (1-32b, c), we can easily derive the electron trajectory equations of both "contravariant" type and "covariant" type in Frenet local coordinate system, in which the electrostatic and magnetic fields are expressed by electrostatic potential, scalar or vector magnetic potential.

The electron trajectory equation of "contravariant" type in Frenet local coordinate system can be obtained as follows:

$$\frac{2\varphi_*}{(1-kp_2)^2 + \chi^2(p_2^2+p_3^2) - 2\chi p_3 p_2' + 2\chi p_2 p_3' + p_2'^2 + p_3'^2}p_2'' +$$

$$p_2'\frac{\mathrm{d}}{\mathrm{d}s}\left[\frac{\varphi_*}{(1-kp_2)^2 + \chi^2(p_2^2+p_3^2) - 2\chi p_3 p_2' + 2\chi p_2 p_3' + p_2'^2 + p_3'^2}\right] +$$

$$\frac{2\varphi_*}{(1-kp_2)^2 + \chi^2(p_2^2+p_3^2) - 2\chi p_3 p_2' + 2\chi p_2 p_3' + p_2'^2 + p_3'^2} \times$$

$$\left(k - k^2 p_2 - \chi^2 p_2 - \chi' p_3 + \frac{k\chi^2 p_3^2 - \chi k' p_2 p_3}{1-kp_2} - \frac{2k\chi p_3}{1-kp_2}p_2' - 2\chi p_3'\right)$$

$$= \frac{1}{(1-kp_2)^2}\left\{\chi p_3 \frac{\partial \varphi}{\partial s} + \left[(1-kp_2)^2 + (\chi p_3)^2\right]\frac{\partial \varphi}{\partial p_2} - \chi^2 p_2 p_3 \frac{\partial \varphi}{\partial p_3}\right\} + M_1 \qquad (1-45\mathrm{a})$$

$$\frac{2\varphi_*}{(1-kp_2)^2 + \chi^2(p_2^2+p_3^2) - 2\chi p_3 p_2' + 2\chi p_2 p_3' + p_2'^2 + p_3'^2}p_3'' +$$

$$p_3'\frac{\mathrm{d}}{\mathrm{d}s}\left[\frac{\varphi_*}{(1-kp_2)^2 + \chi^2(p_2^2+p_3^2) - 2\chi p_3 p_2' + 2\chi p_2 p_3' + p_2'^2 + p_3'^2}\right] +$$

$$\frac{2\varphi_*}{(1-kp_2)^2 + \chi^2(p_2^2+p_3^2) - 2\chi p_3 p_2' + 2\chi p_2 p_3' + p_2'^2 + p_3'^2} \times$$

$$\left(\chi' p_2 - \chi^2 p_3 + \frac{\chi k' p_2^2 - \chi^2 k p_2 p_3}{1-kp_2} - \frac{2\chi p_2'}{1-kp_2} \right)$$

$$= \frac{1}{(1-kp_2)^2} \left\{ -\chi p_2 \frac{\partial \varphi}{\partial s} - \chi^2 p_2 p_3 \frac{\partial \varphi}{\partial p_2} + [(1-kp_2)^2 + (\chi p_2)^2] \frac{\partial \varphi}{\partial p_3} \right\} + M_2 \quad (1-45\text{b})$$

where

$$M_1 = \sqrt{\frac{2e}{m_0}} \mu_0 \frac{1}{1-kp_2} \left(\frac{\varphi_*}{(1-kp_2)^2 + \chi^2(p_2^2+p_3^2) - 2\chi p_3 p_2' + 2\chi p_2 p_3' + p_2'^2 + p_3'^2} \right)^{1/2} \times$$

$$\left[(p_3' + \chi p_2) \frac{\partial \Omega}{\partial s} - \{[(1-kp_2)^2 + \chi^2(p_2^2+p_3^2)] - \chi p_3 p_2' + \chi p_2 p_3'\} \frac{\partial \Omega}{\partial p_3} \right] \quad (1-46\text{a})$$

$$M_2 = \sqrt{\frac{2e}{m_0}} \mu_0 \frac{1}{1-kp_2} \left(\frac{\varphi_*}{(1-kp_2)^2 + \chi^2(p_2^2+p_3^2) - 2\chi p_3 p_2' + 2\chi_2 p_2 p_3' + p_2'^2 + p_3'^2} \right)^{1/2} \times$$

$$\left[\{[(1-kp_2)^2 + \chi^2(p_2^2+p_3^2)] - \chi p_3 p_2' + \chi p_2 p_3'\} \frac{\partial \Omega}{\partial p_2} - (p_2' - \chi p_3) \frac{\partial \Omega}{\partial s} \right] \quad (1-46\text{b})$$

where the magnetic field is expressed by scalar magnetic potential Ω.

The electron trajectory equation of "contravariant" type can also be written in the same form as Eq. (1-45a), Eq. (1-45b), provided that

$$M_1 = -\sqrt{\frac{2e}{m_0}} \frac{1}{(1-kp_2)^2} \left(\frac{\varphi_*}{(1-kp_2)^2 + \chi^2(p_2^2+p_3^2) - 2\chi p_3 p_2' + 2\chi p_2 p_3' + p_2'^2 + p_3'^2} \right)^{1/2} \times$$

$$\left\{ \{[(1-kp_2)^2 + (\chi p_3)^2] p_3' + \chi^2 p_2 p_3 p_2'\} \left(\frac{\partial A_3}{\partial p_2} - \frac{\partial A_2}{\partial p_3} \right) - \right.$$

$$\left. (\chi^2 p_2 p_3 + \chi p_3 p_3') \left(\frac{\partial A_1}{\partial p_3} - \frac{\partial A_3}{\partial s} \right) + \{\chi p_3 p_2' - [(1-kp_2)^2 + (\chi p_3)^2]\} \left(\frac{\partial A_2}{\partial s} - \frac{\partial A_1}{\partial p_2} \right) \right\}$$

$$(1-47\text{a})$$

$$M_2 = -\sqrt{\frac{2e}{m_0}} \frac{1}{(1-kp_2)^2} \left(\frac{\varphi_*}{(1-kp_2)^2 + \chi^2(p_2^2+p_3^2) - 2\chi p_3 p_2' + 2\chi_2 p_3' + p_2'^2 + p_3'^2} \right)^{1/2} \times$$

$$\left(-\{[(1-kp_2)^2 + (\chi p_2)^2] p_2' + \chi^2 p_2 p_3 p_3'\} \left(\frac{\partial A_3}{\partial p_2} - \frac{\partial A_2}{\partial p_3} \right) - \right.$$

$$\left. (\chi^2 p_2 p_3 - \chi p_2 p_2') \left(\frac{\partial A_2}{\partial s} - \frac{\partial A_1}{\partial p_2} \right) + \{[(1-kp_2)^2 + (\chi p_3)^2] + \chi p_2 p_3'\} \left(\frac{\partial A_1}{\partial p_3} - \frac{\partial A_3}{\partial s} \right) \right)$$

$$(1-47\text{b})$$

in which the magnetic field is expressed by vector magnetic potential A.

The electron trajectory equation of "covariant" type in Frenet local coordinate system can be obtained as follows:

$$\frac{2\varphi_*}{(1-kp_2)^2 + \chi^2(p_2^2+p_3^2) - 2\chi p_3 p_2' + 2\chi p_2 p_3' + p_2'^2 + p_3'^2} p_2'' +$$

$$(p_2' - \chi p_3) \frac{\mathrm{d}}{\mathrm{d}s} \left[\frac{\varphi_*}{(1-kp_2)^2 + \chi^2(p_2^2+p_3^2) - 2\chi p_3 p_2' + 2\chi p_2 p_3' + p_2'^2 + p_3'^2} \right] +$$

$$\frac{2\varphi_*}{(1-kp_2)^2+\chi^2(p_2^2+p_3^2)-2\chi p_3 p_2'+2\chi p_2 p_3'+p_2'^2+p_3'^2}$$
$$(k-k^2p_2-\chi^2p_2-\chi'p_3-2\chi p_3')=\frac{\partial\varphi}{\partial p_2}+N_1 \qquad (1-48\text{a})$$

$$\frac{2\varphi_*}{(1-kp_2)^2+\chi^2(p_2^2+p_3^2)-2\chi p_3 p_2'+2\chi p_2 p_3'+p_2'^2+p_3'^2}p_3''+$$
$$(p_3'-\chi p_2)\frac{\mathrm{d}}{\mathrm{d}s}\left[\frac{\varphi_*}{(1-kp_2)^2+\chi^2(p_2^2+p_3^2)-2\chi p_3 p_2'+2\chi p_2 p_3'+p_2'^2+p_3'^2}\right]+$$
$$\frac{2\varphi_*}{(1-kp_2)^2+\chi^2(p_2^2+p_3^2)-2\chi p_3 p_2'+2\chi p_2 p_3'+p_2'^2+p_3'^2}(\chi'p_2-\chi^2p_3-2\chi p_2')=\frac{\partial\varphi}{\partial p_3}+N_2$$
$$(1-48\text{b})$$

where

$$N_1=\sqrt{\frac{2e}{m_0}}\mu_0\frac{1}{1-kp_2}\left[\frac{\varphi_*}{(1-kp_2)^2+\chi^2(p_2^2+p_3^2)-2\chi p_3 p_2'+2\chi p_2 p_3'+p_2'^2+p_3'^2}\right]^{1/2}\times$$
$$\left\{(p_3'+\chi p_2)\frac{\partial\Omega}{\partial s}+(\chi p_3 p_3'+\chi^2 p_2 p_3)\frac{\partial\Omega}{\partial p_2}-[(1-kp_2)^2+\chi^2 p_2^2+\chi p_2 p_3']\frac{\partial\Omega}{\partial p_3}\right\} \quad (1-49\text{a})$$

$$N_2=-\sqrt{\frac{2e}{m_0}}\mu_0\frac{1}{1-kp_2}\left[\frac{\varphi_*}{(1-kp_2)^2+\chi^2(p_2^2+p_3^2)-2\chi p_3 p_2'+2\chi p_2 p_3'+p_2'^2+p_3'^2}\right]^{1/2}\times$$
$$\left\{(p_2'+\chi p_3)\frac{\partial\Omega}{\partial s}-[(1-kp_2)^2+\chi^2 p_3^2+\chi p_3 p_2']\frac{\partial\Omega}{\partial p_2}-(\chi p_2 p_2'-\chi^2 p_2 p_3)\frac{\partial\Omega}{\partial p_3}\right\} \quad (1-49\text{b})$$

in which the magnetic field is expressed by scalar magnetic potential Ω.

The electron trajectory equation of "covariant" type can also be written in the same form as Eq. (1-48a), Eq. (1-48b), provided that

$$N_1=-\sqrt{\frac{2e}{m_0}}\left[\frac{\varphi_*}{(1-kp_2)^2+\chi^2(p_2^2+p_3^2)-2\chi p_3 p_2'+2\chi p_2 p_3'+p_2'^2+p_3'^2}\right]^{1/2}\times$$
$$\left[\left(\frac{\partial A_1}{\partial p_2}-\frac{\partial A_2}{\partial s}\right)+\left(\frac{\partial A_3}{\partial p_2}-\frac{\partial A_2}{\partial p_3}\right)p_3'\right] \qquad (1-50\text{a})$$

$$N_2=-\sqrt{\frac{2e}{m_0}}\left[\frac{\varphi_*}{(1-kp_2)^2+\chi^2(p_2^2+p_3^2)-2\chi p_3 p_2'+2\chi p_2 p_3'+p_2'^2+p_3'^2}\right]^{1/2}\times$$
$$\left[\left(\frac{\partial A_1}{\partial p_3}-\frac{\partial A_3}{\partial s}\right)+\left(\frac{\partial A_2}{\partial p_3}-\frac{\partial A_3}{\partial p_2}\right)p_2'\right] \qquad (1-50\text{b})$$

in which the magnetic field is expressed by vector magnetic potential $\mathbf{A}$.

Obviously, it is not difficult to prove that the electron trajectory equations of both "contravariant" type and "covariant" type can be transformed each other.

6 Deduction of electron trajectory equation using generalized variation function

According to the principle of minimum action, in the generalized field of conservative forces, the action of particles moving along an actual trajectory from point P_0 to point P_1 takes an extreme value, i.e.,

$$\delta \int_{P_0}^{P_1} P \cdot \mathrm{d}r^* = 0 \tag{1-51}$$

where r^* is the radius vector of trajectory of particle's motion; P is the generalized momentum of particles.

As is known, the generalized momentum P of electrons moving in the electromagnetic field can be expressed by

$$P = m_0 \dot{r}^* - eA \tag{1-52}$$

If Eq. (1-52) is written in tensor form, then its covariant components can be expressed as in the following:

$$P_i = \frac{\partial L}{\partial \dot{x}^i} = m_0 \dot{x}_i - eA_i \quad (i = 1, 2, 3) \tag{1-53}$$

where L is the generalized Lagrange's function in form of tensor, as given in Eq. (1-34).

To make a variable substitution, we replace the element $\mathrm{d}\dot{r}^*$ in the integral of Eq. (1-53) by the element of arc length $\mathrm{d}s$ along the principal trajectory. Considering that $|\mathrm{d}\dot{r}^*| = \mathrm{d}s^*$ and applying Eq. (1-42) and Eq. (1-43), we have the variation function in Frenet local coordinate system (s, p_2, p_3):

$$F = \varphi_*^{1/2} [(1 - kp_2)^2 + \chi^2(p_2^2 + p_3^2) - 2\chi p_3 p_2' + 2\chi p_2 p_3' + p_2'^2 + p_3'^2]^{1/2} - \sqrt{\frac{e}{2m_0}}(A_1 + A_2 p_2' + A_3 p_3') \tag{1-54}$$

which makes the functional

$$\int_0^s F(s, p_2, p_3, p_2', p_3') \mathrm{d}s$$

an extreme value. The Euler's equations of this variation problem may be written as

$$\begin{aligned}\frac{\mathrm{d}}{\mathrm{d}s}\left(\frac{\partial F}{\partial p_2'}\right) - \frac{\partial F}{\partial p_2} &= 0, \\ \frac{\mathrm{d}}{\mathrm{d}s}\left(\frac{\partial F}{\partial p_3'}\right) - \frac{\partial F}{\partial p_3} &= 0\end{aligned} \tag{1-55}$$

From this, we can derive the electron trajectory equation with respect to independent variable s, the arc length along the principal trajectory, in Frenet local coordinate system, as given in Eq. (1-48a), Eq. (1-48b) and Eq. (1-50a), Eq. (1-50b).

Conclusions

In the present paper, on the basis of the contravariant and covariant forms of Newton's equation and Lorentz force, we have separately derived the electron motion's equations and electron trajectory equations of both "contravariant" types and "covariant" types in Frenet local coordinate system, in which the electrostatic and magnetic fields are expressed by electrostatic potential and scalar or vector magnetic potential. Although the two types of equations have different forms in appearance, it was proved to be no essential difference between them, as they can be transformed with each other. From the generalized Lagrange's function and generalized variation function in tensor form, we can also derive the electron motion's equation of "covariant" type and the electron trajectory equation of "covariant" type, the forms of which coincide with those equations deduced from the covariant form of Newton's equation and Lorentz force. It verifies no difference between the variation method and the electrodynamic method in a general case.

As for the electron motion's equation of "contravariant" type and of "covariant" type, it has been found that only one variable's derivative of the second order with respect to t is involved in the electron motion's equation of "contravariant" type, which is much simpler than equation of "covariant" type. But the electron trajectory equation of "covariant" type appears to simpler than equation of "contravariant" type.

References

[1] ГРИНБЕРГ Г А. Избранные Вопросы Математической Теории Электрических и Магмитных Явлений[M]. Изд, А Н СССР,1948:507-535.

[2] STURROCK P A. Static and dynamic electron optics[M]. Cambridge: Cambridge University Press, 1955.

[3] KAS'YANKOV P P. Theory of electromagnetic system with curvilinear axis[M]. Leningrad:Press of Leningrad University, 1956:50-57 (in Russian).

[4] ZHOU L W. Optics of wide electron beam focusing[J]. Electron Optical Systems, SEM, Inc. ,1984:45-62.

[5] ZHOU L W. A generalized theory of wide electron beam focusing[J]. Advances in Electronics and Electron Physics, 1985,64B:575-589.

[6] BORISENKO A I, TARAPOV I E. Vector and tensor analysis with applications[M]. New York: Dover Publications, Inc. , 1968.

[7] MCCONNELL A J. Application of tensor analysis[M]. New York: Dover Publications, Inc. ,1957.

5.2 Tensor Analysis of Electron Motion in Curvilinear Coordinate Systems (Part II)

曲线坐标系中电子运动的张量分析（第II部分）

Abstract: This part of present paper continues study of the electron motion in curvilinear coordinate system using tensor analysis. A twisted Frenet local coordinate system (s, u, v) moving along a principal trajectory is introduced. Therefore, no matter what tensor forms of Newton's equation and Lorentz force are used, both the electron motion's equation and the electron trajectory equation of contravariant and covariant forms derived from the assumption of coordinate system mentioned above will take the same form. As has been treated in the preceding part of present paper, we also investigate the electron motion in twisted local coordinate systems using generalized Lagrange's function and generalized variation function, and again verify that there is no difference between the variation method and electrodynamic method in studying electron optics.

摘要：本文的这一部分继续用张量分析方法研究在曲线坐标系中电子的运动，文中引入沿主轨迹运动的转动的 Frenet 局部正交坐标系 (s, u, v)。这样，不管 Newton 方程和 Lorentz 力采用什么张量形式，在正交坐标系的假设下所导得的逆变或协变型电子运动方程和电子轨迹方程将取相同的形式。如本文的上一部分处理的那样，我们还利用广义 Lagrange 函数和广义变分函数研究在转动的局部正交坐标系中电子的运动，并再一次证实了在电子光学用变分方法和电动力学方法研究电子运动的一致性。

Introduction

In part I of present paper, on the basis of contravariant and covariant forms of Newton's equation and Lorentz force, as well as on the basis of the generalized Lagrange's function and generalized variation function, we have derived the electron motion's equation and electron trajectory equation, in which the electrostatic and magnetic fields are expressed by electrostatic potential, scalar or vector magnetic potential.

To obtain a local orthogonal coordinate system moving along the principal trajectory, R. G. E. Hutter[1] introduced a twisted Frenet local coordinate system (s, u, v). This part of present paper will continue a further study of the electron motion using tensor analysis on the assumption of this coordinate system (s, u, v).

Zhou Liwei, Ni Guoqiang, Qiu Bocang. Beijing Institute of Technology. Optik, V. 78, No. 2, 1988, 101–107.

1 Twisted Frenet local orthogonal coordinate system and metric tensors

A twisted Frenet local orthogonal coordinate system (s, u, v) moving along the principal trajectory was introduced in Ref. [1], which rotates with respect to the Frenet local coordinate system (s, p_2, p_3) described in the first part of present paper. The two coordinate systems have relations with

$$p_2 = u\cos\gamma + v\sin\gamma$$
$$p_3 = -u\sin\gamma + v\cos\gamma \qquad (2-1)$$

where γ is the angle between the old p_2-axis and the new u-axis (see Fig. 1), which is defined by

$$\gamma(s) = \int_0^s \chi(s)\,\mathrm{d}s \qquad (2-2)$$

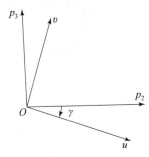

Fig. 1 The coordinate system (p_2, p_3) and the twisted coordinate system (u, v)

where $\chi = \chi(s)$ is the torsion of principal trajectory.

From Eq. (1-6) and Eq. (1-7), we have

$$(\mathrm{d}s^*)^2 = (\mathrm{d}\boldsymbol{r}^*)^2 = [(1-kp_2)^2 + \chi^2(p_2^2+p_3^2)](\mathrm{d}s)^2 - 2\chi p_3 \mathrm{d}s\mathrm{d}p_2 + 2\chi p_2 \mathrm{d}s\mathrm{d}p_3 + (\mathrm{d}p_2)^2 + (\mathrm{d}p_3)^2 \qquad (2-3)$$

Differentiating Eq. (2-1), and substituting it in Eq. (2-3), then we have

$$(\mathrm{d}s^*)^2 = (\mathrm{d}\boldsymbol{r}^*)^2 = [(1-ku\cos\gamma-kv\sin\gamma)^2\mathrm{d}s^2 + \mathrm{d}u^2 + \mathrm{d}v^2] \qquad (2-4)$$

The fundamental metric tensor g_{ij} can be written as follows:

$$(g_{ij}) = \begin{pmatrix} (1-ku\cos\gamma-kv\sin\gamma)^2 & 0 & 0 \\ 0 & 1 & 0 \\ 0 & 0 & 1 \end{pmatrix} \qquad (2-5)$$

The conjugate metric tensor g^{ij} will have the form:

$$(g^{ij}) = \begin{pmatrix} \dfrac{1}{(1-ku\cos\gamma-kv\sin\gamma)^2} & 0 & 0 \\ 0 & 1 & 0 \\ 0 & 0 & 1 \end{pmatrix} \qquad (2-6)$$

2 The electron motion's equation in the twisted Frenet local orthogonal coordinate system

Now, we derive the electron motion's equation from Eq. (1-12) and Eq. (1-14). In the case of twisted Frenet local coordinate systems, $\begin{Bmatrix} \lambda \\ j\ k \end{Bmatrix}$, the Christoffel symbols of the second kind, can be listed in Table 2.1. According to Eq. (1-15), $(\nabla\varphi)^{\lambda}$, the contravariant components of gradient of a scalar potential φ, will have representations, as shown in Table 2.2.

Table 2.1 $\begin{Bmatrix} \lambda \\ j\ k \end{Bmatrix}$ in twisted Frenet local coordinate system (j, k, $\lambda = 1, 2, 3$)

(j, k, λ denote the row, column and page, respectively)

$\lambda = 1$

$\begin{Bmatrix} 1 \\ 1\ 1 \end{Bmatrix}$	$\begin{Bmatrix} 1 \\ 1\ 2 \end{Bmatrix}$	$\begin{Bmatrix} 1 \\ 1\ 3 \end{Bmatrix}$
$\dfrac{1}{1-ku\cos\gamma-kv\sin\gamma} \times [-k'(u\cos\gamma+v\sin\gamma)+k\chi(u\sin\gamma-v\cos\gamma)]$	$\dfrac{-k\cos\gamma}{1-ku\cos\gamma-kv\sin\gamma}$	$\dfrac{-k\sin\gamma}{1-ku\cos\gamma-kv\sin\gamma}$
$\begin{Bmatrix} 1 \\ 2\ 1 \end{Bmatrix}$	$\begin{Bmatrix} 1 \\ 2\ 2 \end{Bmatrix}$	$\begin{Bmatrix} 1 \\ 2\ 3 \end{Bmatrix}$
$\dfrac{-k\cos\gamma}{1-ku\cos\gamma-kv\sin\gamma}$	0	0
$\begin{Bmatrix} 1 \\ 3\ 1 \end{Bmatrix}$	$\begin{Bmatrix} 1 \\ 3\ 2 \end{Bmatrix}$	$\begin{Bmatrix} 1 \\ 3\ 3 \end{Bmatrix}$
$\dfrac{-k\sin\gamma}{1-ku\cos\gamma-kv\sin\gamma}$	0	0

$\lambda = 2$

$\begin{Bmatrix} 2 \\ 1\ 1 \end{Bmatrix}$	$\begin{Bmatrix} 2 \\ 1\ 2 \end{Bmatrix}$	$\begin{Bmatrix} 2 \\ 1\ 3 \end{Bmatrix}$
$k\cos\gamma(1-ku\cos\gamma-kv\sin\gamma)$	0	0
$\begin{Bmatrix} 2 \\ 2\ 1 \end{Bmatrix}$	$\begin{Bmatrix} 2 \\ 2\ 2 \end{Bmatrix}$	$\begin{Bmatrix} 2 \\ 2\ 3 \end{Bmatrix}$
0	0	0
$\begin{Bmatrix} 2 \\ 3\ 1 \end{Bmatrix}$	$\begin{Bmatrix} 2 \\ 3\ 2 \end{Bmatrix}$	$\begin{Bmatrix} 2 \\ 3\ 3 \end{Bmatrix}$
0	0	0

$\lambda = 3$

$\begin{Bmatrix} 3 \\ 1\ 1 \end{Bmatrix}$	$\begin{Bmatrix} 3 \\ 2\ 1 \end{Bmatrix}$	$\begin{Bmatrix} 3 \\ 1\ 3 \end{Bmatrix}$
$k\sin\gamma(1-ku\cos\gamma-kv\sin\gamma)$	0	0
$\begin{Bmatrix} 3 \\ 2\ 1 \end{Bmatrix}$	$\begin{Bmatrix} 3 \\ 2\ 2 \end{Bmatrix}$	$\begin{Bmatrix} 3 \\ 2\ 3 \end{Bmatrix}$
0	0	0
$\begin{Bmatrix} 3 \\ 3\ 1 \end{Bmatrix}$	$\begin{Bmatrix} 3 \\ 3\ 2 \end{Bmatrix}$	$\begin{Bmatrix} 3 \\ 3\ 3 \end{Bmatrix}$
0	0	0

Table 2.2 $(\nabla\varphi)^{\lambda}$ in twisted Frenet local coordinate system (s, u, v) $(\lambda = 1, 2, 3)$

$(\nabla\varphi)^1$	$\dfrac{1}{(1-ku\cos\gamma-kv\sin\gamma)^2}\dfrac{\partial\varphi}{\partial s}$
$(\nabla\varphi)^2$	$\dfrac{\partial\varphi}{\partial u}$
$(\nabla\varphi)^3$	$\dfrac{\partial\varphi}{\partial v}$

If the magnetic field is expressed by scalar magnetic potential Ω, then $(v \times \nabla\Omega)^{\lambda}$ can be easily found by Eq. (16), as listed in Table 2.3.

Table 2.3 $(v \times \nabla\Omega)^{\lambda}$ in twisted Frenet local coordinate system (s, u, v) $(\lambda = 1, 2, 3)$

$(v \times \nabla\Omega)^1$	$\dfrac{1}{1-ku\cos\gamma-kv\sin\gamma}\left(\dot{u}\dfrac{\partial\Omega}{\partial v}-\dot{v}\dfrac{\partial\Omega}{\partial u}\right)$
$(v \times \nabla\Omega)^2$	$\dfrac{1}{1-ku\cos\gamma-kv\sin\gamma}\left[\dot{v}\dfrac{\partial\Omega}{\partial s}-\dot{s}(1-ku\cos\gamma-kv\sin\gamma)^2\dfrac{\partial\Omega}{\partial v}\right]$
$(v \times \nabla\Omega)^3$	$\dfrac{1}{1-ku\cos\gamma-kv\sin\gamma}\left[\dot{s}(1-ku\cos\gamma-kv\sin\gamma)^2\dfrac{\partial\Omega}{\partial u}-\dot{u}\dfrac{\partial\Omega}{\partial s}\right]$

Substituting the values of Table 2.1 – Table 2.3 in Eq. (1–12) and Eq. (1–14), we may derive the electron motion's equation in twisted Frenet local coordinate system (s, u, v):

$$\frac{m_0}{e}\left\{\ddot{s}+\frac{1}{1-ku\cos\gamma-kv\sin\gamma}[-k'(u\cos\gamma+v\sin\gamma)+k\chi(u\sin\gamma-v\cos\gamma)]\dot{s}^2-\frac{2k\cos\gamma}{1-ku\cos\gamma-kv\sin\gamma}\dot{s}\dot{u}-\frac{2k\sin\gamma}{1-ku\cos\gamma-kv\sin\gamma}\dot{s}\dot{v}\right\}=\frac{1}{(1-ku\cos\gamma-kv\sin\gamma)^2}\frac{\partial\varphi}{\partial s}+R_1$$
(2–7a)

$$\frac{m_0}{e}\{\ddot{u}+k\cos\gamma(1-ku\cos\gamma-kv\sin\gamma)\dot{s}^2\}=\frac{\partial\varphi}{\partial u}+R_2 \qquad (2-7b)$$

$$\frac{m_0}{e}\{\ddot{v}+k\sin\gamma(1-ku\cos\gamma-kv\sin\gamma)\dot{s}^2\}=\frac{\partial\varphi}{\partial v}+R_3 \qquad (2-7c)$$

where

$$R_1=\frac{\mu_0}{1-ku\cos\gamma-kv\sin\gamma}\left(\dot{u}\frac{\partial\Omega}{\partial v}-\dot{v}\frac{\partial\Omega}{\partial u}\right) \qquad (2-8a)$$

$$R_2=\frac{\mu_0}{1-ku\cos\gamma-kv\sin\gamma}\left[\dot{v}\frac{\partial\Omega}{\partial s}-\dot{s}(1-ku\cos\gamma-kv\sin\gamma)^2\frac{\partial\Omega}{\partial v}\right] \qquad (2-8b)$$

$$R_3=\frac{\mu_0}{1-ku\cos\gamma-kv\sin\gamma}\left[\dot{s}(1-ku\cos\gamma-kv\sin\gamma)^2\frac{\partial\Omega}{\partial u}-\dot{u}\frac{\partial\Omega}{\partial s}\right] \qquad (2-8c)$$

where the magnetic field is expressed by scalar magnetic potential Ω.

If the magnetic field is expressed by magnetic vector potential A, then $(v \times \nabla \times A)^{\lambda}$ can be easily found by Eq. (1–20a, b), as listed in Table 2.4:

Table 2.4 $(v \times \nabla \times A)^\lambda$ in twisted Frenet local coordinate system (s, u, v) $(\lambda = 1, 2, 3)$

$[v \times \nabla \times A)]^1$	$\dfrac{1}{(1-ku\cos\gamma-kv\sin\gamma)^2}\left[\dot{u}\left(\dfrac{\partial A_u}{\partial s}-\dfrac{\partial A_s}{\partial u}\right)-\dot{v}\left(\dfrac{\partial A_s}{\partial v}-\dfrac{\partial A_v}{\partial s}\right)\right]$
$[v \times \nabla \times A)]^2$	$\dot{v}\left(\dfrac{\partial A_v}{\partial u}-\dfrac{\partial A_u}{\partial v}\right)-\dot{s}\left(\dfrac{\partial A_u}{\partial s}-\dfrac{\partial A_s}{\partial u}\right)$
$[v \times \nabla \times A)]^3$	$\dot{s}\left(\dfrac{\partial A_s}{\partial v}-\dfrac{\partial A_v}{\partial s}\right)-\dot{u}\left(\dfrac{\partial A_v}{\partial u}-\dfrac{\partial A_u}{\partial v}\right)$

The substitution of $(v \times \nabla \times A)^\lambda$ in Table 2.4 and for $(v \times \nabla \Omega)^\lambda$ in Table 2.3 will give the electron motion's equation in the twisted Frenet local coordinate system (s, u, v), which can also be written in the same form, provided that

$$R_1 = \frac{1}{1-ku\cos\gamma-kv\sin\gamma}\left[\dot{u}\left(\frac{\partial A_u}{\partial s}-\frac{\partial A_s}{\partial u}\right)-\dot{v}\left(\frac{\partial A_s}{\partial v}-\frac{\partial A_v}{\partial s}\right)\right] \quad (2-9\text{a})$$

$$R_2 = -\left[\dot{v}\left(\frac{\partial A_v}{\partial u}-\frac{\partial A_u}{\partial v}\right)-\dot{s}\left(\frac{\partial A_u}{\partial s}-\frac{\partial A_s}{\partial u}\right)\right] \quad (2-9\text{b})$$

$$R_3 = -\left[\dot{s}\left(\frac{\partial A_s}{\partial v}-\frac{\partial A_v}{\partial s}\right)-\dot{u}\left(\frac{\partial A_v}{\partial u}-\frac{\partial A_u}{\partial v}\right)\right] \quad (2-9\text{c})$$

in which the magnetic field is expressed by vector magnetic potential A. It should be noted that here and below A_s, A_u, A_v are covariant components of A.

It is easy to prove that in twisted Frenet local orthogonal coordinate system, the electron motion's equation derived from the Newton's equation and Lorentz force, whether they are in contravariant form or in covariant form, will have the same form.

3 The electron trajectory equation in the twisted Frenet local orthogonal coordinate system

Similarly to part I of the present paper, we can also transform the general electron motion's equation with respect to variable t, the time, into the general electron trajectory equation with respect to variable s, the arc length along principal trajectory.

In an twisted Frenet local coordinate system, the element of arc length ds at the principal trajectory and ds^* at the its neighboring trajectory are related as following:

$$\left(\frac{ds^*}{ds}\right)^2 = (1-ku\cos\gamma-kv\sin\gamma)^2 + u'^2 + v'^2 \quad (2-10)$$

From Eq. (1-44), the parameters $\dot{s}$, $\dot{u}$, $\ddot{u}$ can be transformed to

$$\dot{s} = \sqrt{\frac{2e}{m_0}\varphi_*}\frac{ds}{ds^*},$$

$$\dot{u} = \sqrt{\frac{2e}{m_0}\varphi_*}\frac{ds}{ds^*}u',$$

$$\ddot{u} = \frac{e}{m_0}\left\{2\varphi_*\left(\frac{ds}{ds^*}\right)^2 u'' + u'\frac{d}{ds}\left[\varphi_*\left(\frac{ds}{ds^*}\right)^2\right]\right\} \quad (2-11)$$

Similar expressions can be obtained for $\dot{v}$ and $\ddot{v}$. Substituting Eq. (2-10) and Eq. (2-11) in Eqs. (2-7a, b) and Eqs. (2-8a, b), we can derive the electron trajectory equation in a twisted Frenet local coordinate system as follows:

$$\frac{2\varphi_*}{(1-ku\cos\gamma-kv\sin\gamma)^2+u'^2+v'^2}u'' + u'\frac{d}{ds}\left[\frac{\varphi_*}{(1-ku\cos\gamma-kv\sin\gamma)^2+u'^2+v'^2}\right] +$$

$$2k\cos\gamma(1-ku\cos\gamma-kv\sin\gamma)\frac{\varphi_*}{(1-ku\cos\gamma-kv\sin\gamma)^2+u'^2+v'^2} = \frac{\partial\varphi}{\partial u} + S_1 \quad (2-12a)$$

$$\frac{2\varphi_*}{(1-ku\cos\gamma-kv\sin\gamma)^2+u'^2+v'^2}v'' + v'\frac{d}{ds}\left[\frac{\varphi_*}{(1-ku\cos\gamma-kv\sin\gamma)^2+u'^2+v'^2}\right] +$$

$$2k\sin\gamma(1-ku\cos\gamma-kv\sin\gamma)\frac{\varphi_*}{(1-ku\cos\gamma-kv\sin\gamma)^2+u'^2+v'^2} = \frac{\partial\varphi}{\partial v} + S_2 \quad (2-12b)$$

where

$$S_1 = \sqrt{\frac{2e}{m_0}}\mu_0 \left[\frac{\varphi_*}{(1-ku\cos\gamma-kv\sin\gamma)^2+u'^2+v'^2}\right]^{1/2} \times$$

$$\left[\frac{v'}{-ku\cos\gamma-kv\sin\gamma}\frac{\partial\Omega}{\partial s} - (1-ku\cos\gamma-kv\sin\gamma)\frac{\partial\Omega}{\partial v}\right] \quad (2-13a)$$

$$S_2 = \sqrt{\frac{2e}{m_0}}\mu_0 \left[\frac{\varphi_*}{(1-ku\cos\gamma-kv\sin\gamma)^2+u'^2+v'^2}\right]^{1/2} \times$$

$$\left[(1-ku\cos\gamma-kv\sin\gamma)\frac{\partial\Omega}{\partial u} - \frac{u'}{1-ku\cos\gamma-kv\sin\gamma}\frac{\partial\Omega}{\partial s}\right] \quad (2-13b)$$

where the magnetic field is expressed by scalar magnetic potential Ω.

From Eqs. (2-7b, c) and Eqs. (2-9b, c), the electron trajectory equation can also be written in the same form as Eqs. (2-9a, c), provided that

$$S_1 = -\sqrt{\frac{2e}{m_0}}\left[\frac{\varphi_*}{(1-ku\cos\gamma-kv\sin\gamma)^2+u'^2+v'^2}\right]^{1/2}\left[v'\left(\frac{\partial A_v}{\partial u}-\frac{\partial A_u}{\partial v}\right)-\left(\frac{\partial A_u}{\partial s}-\frac{\partial A_s}{\partial u}\right)\right]$$
$$(2-14a)$$

$$S_2 = -\sqrt{\frac{2e}{m_0}}\left[\frac{\varphi_*}{(1-ku\cos\gamma-kv\sin\gamma)^2+u'^2+v'^2}\right]^{1/2}\left[\left(\frac{\partial A_s}{\partial v}-\frac{\partial A_v}{\partial s}\right)-u'\left(\frac{\partial A_v}{\partial u}-\frac{\partial A_u}{\partial v}\right)\right]$$
$$(2-14b)$$

where the magnetic field is expressed by vector magnetic potential $\mathbf{A}$.

4 Deduction of electron motion's equation and electron trajectory equation using generalized Lagrange's function and generalized variation function

In twisted Frenet local coordinate system (s, u, v), the generalized Lagrange's function of tensor form Eq. (1-34) can be transformed to

$$L = \frac{1}{2}m_0\left[(1-ku\cos\gamma-kv\sin\gamma)^2\dot{s}^2 + \dot{u}^2 + \dot{v}^2\right] + e\varphi - e(A_s\dot{s} + A_u\dot{u} + A_v\dot{v}) \quad (2-15)$$

In that case, $[i, j, k]$, the Christoffel function of the first kind, which is expressed by Eq.

(41), can also be written in the form of matrix with 3 marks, as shown in Table 2.5.

Table 2.5 $[i, j, k]$ **in twisted Frenet local coordinate system** (s, u, v) $(i, j, k = 1, 2, 3)$; $(j, k, i$ denote the row, column and page, respectively)

$i = 1$

[1, 1, 1]	[1, 1, 2]	[1, 1, 3]
$(1 - ku\cos\gamma - kv\sin\gamma) \times$ $[(-k'(u\cos\gamma + v\sin\gamma) +$ $k\chi(u\sin\gamma - v\cos\gamma)]$	$(1 - ku\cos\gamma - kv\sin\gamma) \times$ $(-k\cos\gamma)$	$(1 - ku\cos\gamma - kv\sin\gamma)$ $\times(-k\sin\gamma)$
[1, 2, 1]	[1, 2, 2]	[1, 2, 3]
$(1 - ku\cos\gamma - kv\sin\gamma) \times$ $(-k\cos\gamma)$	0	0
[1, 3, 1]	[1, 3, 2]	[1, 3, 3]
$(1 - ku\cos\gamma - kv\sin\gamma) \times$ $(-k\sin\gamma)$	0	0

$i = 2$

[2, 1, 1]	[2, 1, 2]	[2, 1, 3]
$(1 - ku\cos\gamma - kv\sin\gamma) \times$ $(-k\cos\gamma)$	0	0
[2, 2, 1]	[2, 2, 2]	[2, 2, 3]
0	0	0
[2, 3, 1]	[2, 3, 2]	[2, 3, 3]
0	0	0

$i = 3$

[3, 1, 1]	[3, 1, 2]	[3, 1, 3]
$(1 - ku\cos\gamma - kv\sin\gamma) \times$ $k\sin\gamma$	0	0
[3, 2, 1]	[3, 2, 2]	[3, 2, 3]
0	0	0
[3, 3, 1]	[3, 3, 2]	[3, 3, 3]
0	0	0

Substituting the components of g_{ij} [Eq. (2-5)] and $[i, j, k]$, the Christoffel function of the first kind (Table 2.5), in Eq. (1-40) of Euler's equation, which is satisfied by Eq (2-15) of the generalized Lagrange's function, we can get the electron motion's equation, whose form coincides with Eq. (2-7a, b, c) and Eq. (2-9a, b, c).

Similar to it from Eq. (1-54), we can also obtain the generalized variation function in a twisted Frenet local coordinate system:

$$F = \varphi_*^{1/2} [(1 - ku\cos\gamma - kv\sin\gamma)^2 + u'^2 + v'^2]^{1/2} - \sqrt{\frac{e}{2m_0}}(A_s + A_u u' + A_v v') \quad (2-16)$$

which make the functional

$$\int_0^F F(s,u,v,u',v')\,\mathrm{d}s$$

an extreme value. The Euler's equation of this variation problem may be written as

$$\frac{\mathrm{d}}{\mathrm{d}s}\left(\frac{\partial F}{\partial u'}\right) - \frac{\partial F}{\partial u} = 0,$$
$$\frac{\mathrm{d}}{\mathrm{d}s}\left(\frac{\partial F}{\partial v'}\right) - \frac{\partial F}{\partial v} = 0 \tag{2-17}$$

From this, we can derive the electron trajectory equation with respect to independent variable s, the arc length along the principal trajectory, in twisted Frenet local coordinate system, whose form coincides Eqs. (2–12a, b) and Eq. (2–14a, b).

It must be pointed out when A_s, A_u, A_v are represented by physical components of A, then covariant component A_s in Eq. (2–16) should be replaced by A_s $(1 - ku\cos\gamma - kv\sin\gamma)$, but the physical components A_u, A_v and the covariant components A_u, A_v are just the same.

5 Some special cases

5.1 When the principal trajectory is straight axis, then $k=0$, $\chi=0$, $u=p_2$, $v=p_3$. Now we shall discuss two special cases.

1) In rectangular Cartesian coordinate system (x, y, z)

In that case, let $s=z$, $u=p_2=x$, $v=p_3=y$ and suppose $A_s=A_z=A_1$, $A_u=A_x=A_2$, $A_v=A_y=A_3$; then from Eq. (1–48a), Eq. (1–48b) and Eq. (1–50a) – Eq. (1–50a) or Eq. (2–12a), Eq. (2–12b) and Eq. (2–14a), Eq. (2–14b), it is not difficult to derive the electron trajectory equation in rectangular Cartesian coordinate system (x, y, z):

$$\frac{\mathrm{d}}{\mathrm{d}z}\left[\frac{\varphi_*^{1/2}}{(1+x'^2+y'^2)^{1/2}}x'\right] - \frac{(1+x'^2+y'^2)^{1/2}}{2\varphi_*^{1/2}}\frac{\partial\varphi}{\partial x} + \sqrt{\frac{e}{2m_0}}\left\{y'\left(\frac{\partial A_y}{\partial x}-\frac{\partial A_x}{\partial y}\right)-\left(\frac{\partial A_x}{\partial z}-\frac{\partial A_z}{\partial x}\right)\right\}=0 \tag{2-18a}$$

$$\frac{\mathrm{d}}{\mathrm{d}z}\left[\frac{\varphi_*^{1/2}}{(1+x'^2+y'^2)^{1/2}}y'\right] - \frac{(1+x'^2+y'^2)^{1/2}}{2\varphi_*^{1/2}}\frac{\partial\varphi}{\partial y} + \sqrt{\frac{e}{2m_0}}\left\{x'\left(\frac{\partial A_x}{\partial y}-\frac{\partial A_y}{\partial x}\right)-\left(\frac{\partial A_y}{\partial z}-\frac{\partial A_z}{\partial y}\right)\right\}=0 \tag{2-18b}$$

where the prime indicates derivatives with respect to z.

The above equations can also be found from generalized variation function Eq. (1–54) or Eq. (2–16), which in the rectangular Cartesian coordinate system will take the following form:

$$F = \varphi_*^{1/2}(1+x'^2+y'^2)^{1/2} - \sqrt{\frac{e}{2m_0}}(A_x x' + A_y y' + A_z) \tag{2-19}$$

The Euler's equations of this variation problem will give the electron trajectory equation, as written in Eq. (2–18a, b).

2) In cylindrical coordinate system (r, θ, z)

In that case, let $s=z$, $u=p_2=r\cos\theta$, $v=p_3=r\sin\theta$. Suppose $A_z=A_1$, $A_r=A_2\cos\theta+A_3\sin\theta$,

$A_\theta = -A_2 \sin\theta + A_3 \cos\theta$; then Eq (1-54) or Eq (1-16) of the generalized variation function takes the following form:

$$F = \varphi_*^{1/2}(1 + r'^2 + r^2\theta'^2)^{1/2} - \sqrt{\frac{e}{2m_0}}(A_z + A_r r' + A_\theta r\theta') \qquad (2-20)$$

which makes the functional

$$\int_0^s F(z,r,\theta,r',\theta')\,\mathrm{d}z$$

an extreme value. The Euler's equation of this variation problem may be written as:

$$\frac{\mathrm{d}}{\mathrm{d}z}\left(\frac{\partial F}{\partial r'}\right) - \frac{\partial F}{\partial r} = 0 \qquad (2-21\mathrm{a})$$

$$\frac{\mathrm{d}}{\mathrm{d}z}\left(\frac{\partial F}{\partial \theta'}\right) - \frac{\partial F}{\partial \theta} = 0 \qquad (2-21\mathrm{b})$$

From this, we can derive the electron trajectory equations in cylindrical coordinate system:

$$\frac{\mathrm{d}}{\mathrm{d}z}\left[\frac{\varphi_*^{1/2} r'}{(1+r'^2+r^2\theta'^2)^{1/2}}\right] - \frac{(1+r'^2+r^2\theta'^2)^{1/2}}{2\varphi_*^{1/2}}\frac{\partial \varphi}{\partial r} -$$

$$\frac{r\theta'^2 \varphi_*^{1/2}}{(1+r'^2+r^2\theta'^2)^{1/2}} - \sqrt{\frac{e}{2m_0}}\left[\theta'\left(\frac{\partial(rA_\theta)}{\partial r} - \frac{\partial A_r}{\partial \theta}\right) + \left(\frac{\partial A_z}{\partial r} - \frac{\partial A_r}{\partial z}\right)\right] = 0 \qquad (2-22\mathrm{a})$$

$$\frac{\mathrm{d}}{\mathrm{d}z}\left[\frac{\varphi_*^{1/2} r^2 \theta'}{(1+r'^2+r^2\theta'^2)^{1/2}}\right] - \frac{(1+r'^2+r^2\theta'^2)^{1/2}}{2\varphi_*^{1/2}}\frac{\partial \varphi}{\partial \theta} +$$

$$\sqrt{\frac{e}{2m_0}}\left[r'\left(\frac{\partial A_r}{\partial \theta} - \frac{\partial(rA_\theta)}{\partial r}\right) + \left(\frac{\partial A_z}{\partial \theta} - \frac{\partial(rA_\theta)}{\partial z}\right)\right] = 0 \qquad (2-22\mathrm{b})$$

5.2 When the principal trajectory is a planar curve, then $\chi = 0$, $u = p_2$, $v = p_3$. In that case, both types of electron trajectory equation will take the same form. From Eqs. (1-48 a, b) and Eqs. (1-49a, b) or Eqs. (2-12a, b) and Eqs. (2-13a, b), the electron trajectory equation can be deduced:

$$\frac{2\varphi_*}{(1-kp_2)^2 + p_2'^2 + p_3'^2}p_2'' + p_2'\frac{\mathrm{d}}{\mathrm{d}s}\left(\frac{\varphi_*}{(1-kp_2)^2 + p_2'^2 + p_3'^2}\right) +$$

$$\frac{2\varphi_*}{(1-kp_2)^2 + p_2'^2 + p_3'^2}(k - k^2 p_2) = \frac{\partial \varphi}{\partial p_2} + Q_1 \qquad (2-23\mathrm{a})$$

$$\frac{2\varphi_*}{(1-kp_2)^2 + p_2'^2 + p_3'^2}p_3'' + p_3'\frac{\mathrm{d}}{\mathrm{d}s}\left(\frac{\varphi_*}{(1-kp_2)^2 + p_2'^2 + p_3'^2}\right) = \frac{\partial \varphi}{\partial p_3} + Q_2 \qquad (2-23\mathrm{b})$$

where

$$Q_1 = \sqrt{\frac{2e}{m_0}}\mu_0 \frac{1}{1-kp_2}\left[\frac{\varphi_*}{(1-kp_2)^2 + p_2'^2 + p_3'^2}\right]^{1/2}\left[p_3'\frac{\partial \Omega}{\partial s} - (1-kp_2)^2\frac{\partial \Omega}{\partial p_3}\right] \qquad (2-24\mathrm{a})$$

$$Q_2 = \sqrt{\frac{2e}{m_0}}\mu_0 \frac{1}{1-kp_2}\left[\frac{\varphi_*}{(1-kp_2)^2 + p_2'^2 + p_3'^2}\right]^{1/2}\left[(1-kp_2)^2\frac{\partial \Omega}{\partial p_2} - p_2'\frac{\partial \Omega}{\partial s}\right] \qquad (2-24\mathrm{b})$$

where the magnetic field is expressed by scalar magnetic potential Ω.

Using Eqs. (1-48a, b) and Eqs. (1-50a, b) or Eqs. (2-12a, b) and Eqs. (2-14 a, b), the electron trajectory equation can also be written in the same form, as Eqs. (2-23a, b), provided that

$$Q_1 = -\sqrt{\frac{2e}{m_0}} \left[\frac{\varphi_*}{(1-kp_2)^2 + p_2'^2 + p_3'^2} \right]^{1/2} \left[p_3' \left(\frac{\partial A_3}{\partial p_2} - \frac{\partial A_2}{\partial p_3} \right) - \left(\frac{\partial A_2}{\partial s} - \frac{\partial A_1}{\partial p_2} \right) \right] \quad (2-25a)$$

$$Q_1 = -\sqrt{\frac{2e}{m_0}} \left[\frac{\varphi_*}{(1-kp_2)^2 + p_2'^2 + p_3'^2} \right]^{1/2} \left[p_2' \left(\frac{\partial A_3}{\partial p_2} - \frac{\partial A_2}{\partial p_3} \right) - \left(\frac{\partial A_1}{\partial p_3} - \frac{\partial A_3}{\partial s} \right) \right] \quad (2-25b)$$

in which the magnetic field is expressed by vector magnetic potential **A**.

Conclusions

In the present paper, we have investigated the electron motion in twisted Frenet local orthogonal coordinate system moving along the principal trajectory. From the contravariant form or covariant form of Newton's equation and Lorentz force, as well as from the generalized Lagrange's function and generalized variation function, we have derived the electron motion's equations Eqs. (2-7a, b, c), Eqs. (2-8a, b, c) or Eq. (2-9a, b, c) and electron trajectory equations Eqs. (2-12a, b), Eqs. (2-13a, b) or Eq. (2-14a, b), in which the electrostatic and magnetic fields are expressed by electrostatic potential and scalar or vector magnetic potential. We once again verify that there is no difference between the variation method and electrodynamic method in studying electron optics.

At first appearance, both the electron motion's equation and the electron trajectory equation in twisted Frenet local coordinate system are much simpler than the preceding equations derived in part I of this paper. But these equations are related with the angle γ, which is the integral of torsion χ along the principal trajectory. It is better for us to use the equations derived in part I of the present paper, than to use the equations here, as has been mentioned in previous paper [2].

Two special cases have been discussed when the principal trajectory is a straight axis or a planar curve. The electron trajectory equation and electron motion's equation deduced above have practical meaning for studying electron optical systems with axial symmetry or planar symmetry.

The result and these conclusions given in part I and part II of present paper may be of theoretical and practical importance for studying narrow and wide electron beam focusing, Gaussian imaging properties, aberrations and electron optical transfer function.

References

[1] HUTTER R G E. Deflection defocusing effects in cathode ray tubes at large deflection Angles[J]. IEEE—ED,171970:1022-1031.

[2] ZHOU L W. Optics of wide electron beam focusing[J]. Electron Optical Systems, SEM, Inc.,1984:45-52.

5.3 曲轴宽电子束聚焦的普遍理论
A Generalized Theory of Wide Electron Beam Focusing Having Curvilinear Axes

摘要：本文用张量分析的方法推导了 Frenet 局部曲线坐标系下电子运动方程和轨迹方程。由此出发，考虑阴极物面 $\varphi(0)=0$ 且自阴极面逸出的电子速度很小而斜率可能很大的宽电子束的情况，推导了 Frenet 曲线坐标系下的主轨迹方程与曲近轴轨迹方程，研究了满足正交条件下曲近轴系统的特性。文中还着重讨论了满足正交条件下宽电子束聚焦的若干实例。曲线坐标系下细电子束聚焦的问题与轴对称系统的宽电子束聚焦的问题均可视为本文的特例。

Abstract: Equations of electron motion and electron trajectory in a curvilinear coordinate system have been derived using tensor analysis in this paper. Considering that the object (cathode) is placed directly in the electrostatic and magnetic fields and the cathode surface is to be imaged with wide electron beam, thus the electrons leave the cathode with small velocities and large inclinations, equations have been deduced for the principal trajectory and the curvilinear "paraxial" trajectories in the Frenet local coordinate system, and the orthogonality conditions for systems with wide electron beam focusing are studied. Some special cases, cases in which systems satisfy the orthogonality conditions of wide electron beam focusing, have emphatically been discussed. The problem involving narrow electron beam focusing around a curvilinear axis can be regarded as a special case of this generalized treatment. The identity and difference between narrow and wide electron beam focusing have also been discussed.

引 言

带电粒子束的聚焦是电子光学基本问题之一，长期以来受到人们普遍的重视。曲线坐标系下的细电子束聚焦的普遍理论是由 Г. А. Гринберг[1~3] 和 P. A. Sturrock[4~5] 在 20 世纪 40—50 年代分别建立的。前者由电子运动的基本方程出发导出了曲线坐标系下的曲傍轴轨迹方程，后者由变分理论出发研究了满足正交条件下曲傍轴系统的电子光学性质。近几十年来，在此基础上，开展了大量的研究工作。

严格说来，满足曲傍轴条件出发所导得细电子束聚焦的轨迹方程，并不适用于研究阴极物面 $\varphi(s=0)=0$ 且自阴极逸出的电子速度很小而斜率可能很大的宽电子束聚焦的问题（如阴极透镜）。因此，进一步探讨宽电子束聚焦的普遍理论是很有必要的。必须指出，为了将本文的结果与 Г. А. Гринберг 的相比较，本文采用 CGSM 单位制，并采用 Г. А.

周立伟（Zhou Liwei）. 北京工业学院. 工程光学（Engineering Optics 内部），No. 2, 1983, 37–54.

Гринберг 文章的符号。

在研究宽束电子光学时，与探讨轴对称阴极透镜问题相类似[6]，我们可假定相邻轨迹处处满足如下的曲近轴条件：

$$p^2(s) \approx 0, \quad p(s) \ll \rho(s), \quad p(s) \ll \tau(s) \tag{1}$$

$$\dot{p}^2(s) \ll 1 \tag{2}$$

假定

$$\frac{\frac{m_0}{2e}(\dot{p}_0^2 - \dot{p}^2)}{\varphi(s) + \varepsilon_s} \ll 1$$

而且 $\varphi(0) = 0$，但是并不一定处处要满足 $\dot{p}^2(s) \ll 1$ 的条件。这里 s 为曲线轴自物点算起的弧长，$p(s)$ 为相邻轨迹相对于曲线轴的高度。点号表示对时间 t 的导数，$\rho(s)$、$\tau(s)$ 分别为曲线轴的曲率半径与挠率半径。实际上，所谓满足曲近轴条件，是将 $p(s)$、$\dot{p}(s)$ 作为一阶小量，而把 p^2、$\dot{p}^2$、$p\dot{p}$ 等量视为高阶小。

本文由曲线坐标系下的劳伦茨方程出发，利用张量表示法，导出了普遍形式的轨迹方程。在此基础上，引入曲近轴条件，推导了主轨迹方程与曲近轴轨迹方程，研究了曲近轴系统的正交条件，并讨论了满足正交条件下宽电子束聚焦的若干实例。

曲线坐标系下细电子束聚焦的问题和轴对称系统的宽电子束聚焦的问题，均可视为本文的特例。

1 曲线坐标系下电子运动方程与轨迹方程

设空间任意曲线 $r = r(x^1)$ 为自物面上 A 点逸出的电子束的曲线轴（主轨迹）的矢径，其中 x^1 为由起始点算起沿着该曲线轴的弧长。在点 N 处作主轨迹曲线的垂直平面，此平面与相邻轨迹 AB^* 的交点 N^* 的位置将由点 N 到点 N^* 的矢量 p 来表达。如果给出 $p = p(s)$ 表达式，则相邻轨迹的曲线便可完全确定。

引入曲线坐标系 (x^1, x^2, x^3)（注意！这里 1、2、3 不是指数，而是上标），其单位向量 ξ_1、ξ_2、ξ_3 指向主轨迹点 N 处的切线、主法线和副法线方向。空间任意曲线 r 的曲率半径与挠率半径以 $\rho = \rho(x^1)$ 和 $\tau = \tau(x^1)$ 来表达。因为 p 是位于主曲线的垂直平面内，它在主法线和副法线上的分量分别为 x^2 和 x^3，如图 1 所示。

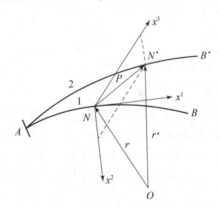

图 1 曲线坐标系、主轨迹与相邻轨迹

于是点 N^* 处相邻轨迹的矢径 r^* 可表示为

$$r^* = r + p = r + x^2 \xi_2 + x^3 \xi_3 \qquad (3)$$

现计算在所选择的曲线坐标系内的弧长元 ds^*。由微分式（3），应用微分几何中 Frenet–Serret 公式，并考虑到 $(ds^*)^2 = (dr^*)^2$，将它以张量形式表示，便有[7]

$$(ds^*)^2 = g_{ij} dx^i dx^j \qquad (4)^{[注2]}$$

式中，g_{ij} 是二阶二重协变张量的元素。二阶二重协变张量或称基本测度张量 (g_{ij}) 可表达如下：

$$(g_{ij}) = \begin{bmatrix} \left(1 - \dfrac{1}{\rho}x^2\right)^2 + \left(\dfrac{1}{\tau}x^2\right)^2 + \left(\dfrac{1}{\tau}x^3\right)^2 & -\dfrac{1}{\tau}x^3 & \dfrac{1}{\tau}x^2 \\ -\dfrac{1}{\tau}x^3 & 1 & 0 \\ \dfrac{1}{\tau}x^2 & 0 & 1 \end{bmatrix} \qquad (5)$$

非相对论下劳伦兹方程的逆变分量具有以下形式[8][9]：

$$m_0(\ddot{x}^l + \Gamma^l_{ij}\dot{x}^i\dot{x}^j) = e(A^l + B^l),\ (l = 1, 2, 3) \qquad (6)$$

式中 m_0、e 分别为电子的静止质量和电荷，Γ^l_{ij} 乃是二类克里斯托弗符号，它可表示为

$$\Gamma^l_{ij} = \dfrac{1}{2} g^{kl} \left(\dfrac{\partial g_{kj}}{\partial x^i} + \dfrac{\partial g_{ik}}{\partial x^j} - \dfrac{\partial g_{ij}}{\partial x^k} \right) \qquad (7)$$

这里 g^{ij} 是二阶二重逆变张量的元素，它可由下列公式求得

$$g^{ij} = \dfrac{G(i,j)}{g} \qquad (8)$$

式（8）中 $G(i,j)$ 是对于元素 g_{ij} 的代数余子式，g 表示由元素 g_{ij} 所组成的行列式。由式（5）可得

$$g = \det|g_{ij}| = \left(1 - \dfrac{1}{\rho}x^2\right)^2 \qquad (9)$$

利用式（5）和式（8），并考虑到 $\Gamma^l_{ij} = \Gamma^l_{ji}$，便可求得各个 Γ^l_{ij} 值。

式（6）中的 A^l 和 $B^l(l = 1, 2, 3)$ 可表示如下：

$$A^l = g^{li} \dfrac{\partial \varphi}{\partial x^i} \qquad (10)$$

$$B^1 = \dfrac{1}{\sqrt{g}} \left(\dot{x}_2 \dfrac{\partial \Omega}{\partial x^3} - \dot{x}_3 \dfrac{\partial \Omega}{\partial x^2} \right),$$

$$B^2 = \dfrac{1}{\sqrt{g}} \left(\dot{x}_3 \dfrac{\partial \Omega}{\partial x^1} - \dot{x}_1 \dfrac{\partial \Omega}{\partial x^3} \right), \qquad (11)$$

$$B^3 = \dfrac{1}{\sqrt{g}} \left(\dot{x}_1 \dfrac{\partial \Omega}{\partial x^2} - \dot{x}_2 \dfrac{\partial \Omega}{\partial x^1} \right)$$

$$\dot{x}_i = g_{ij} \dot{x}^j \qquad (12)$$

式中，φ 为电场的电位，磁场 H 可通过磁标位 Ω 确定，$H = -\text{grad}\,\Omega$。

下面，为了方便起见，我们用另一种坐标符号 (s, p_2, p_3) 代替 (x^1, x^2, x^3)。利用以上表达式，由式（6），便可求得电子在 (s, p_2, p_3) 坐标系内的运动方程：

$$\frac{m_0}{e}\left[\ddot{s} + \frac{\frac{1}{\rho\tau}p_3 + \frac{1}{\rho^2}\rho' p_2}{1-\frac{1}{\rho}p_2}\dot{s}^2 + \frac{2\frac{1}{\rho}}{1-\frac{1}{\rho}p_2}\dot{s}\dot{p}_2\right] = \frac{1}{\left(1-\frac{1}{\rho}p_2\right)^2}\left(\frac{\partial\varphi}{\partial s} + \frac{1}{\tau}p_3\frac{\partial\varphi}{\partial p_2} - \frac{1}{\tau}p_2\frac{\partial\varphi}{\partial p_3}\right) +$$

$$\frac{1}{1-\frac{1}{\rho}p_2}\left[\left(\dot{p}_2 - \frac{1}{\tau}p_3\dot{s}\right)\frac{\partial\Omega}{\partial p_3} - \left(\dot{p}_3 + \frac{1}{\tau}\dot{s}p_2\right)\frac{\partial\Omega}{\partial p_2}\right] \quad (13\text{a})$$

$$\frac{m_0}{e}\left[\ddot{p}_2 + \left(\frac{1}{\rho} - \frac{1}{\rho^2}p_2 - \frac{1}{\tau^2}p_2 + \frac{1}{\tau^2}\tau' p_3 + \frac{\frac{1}{\rho\tau^2}p_3^2 + \frac{1}{\rho^2\tau}\rho' p_2 p_3}{1-\frac{1}{\rho}p_2}\right)\dot{s}^2 - \frac{2\frac{1}{\rho\tau}p_3}{1-\frac{1}{\rho}p_2}\dot{s}\dot{p}_3 - 2\frac{1}{\tau}\dot{s}\dot{p}_3\right]$$

$$= \frac{1}{\left(1-\frac{1}{\rho}p_2\right)^2}\left\{\frac{1}{\tau}p_3\frac{\partial\varphi}{\partial s} + \left[\left(1-\frac{1}{\rho}p_2\right)^2 + \frac{1}{\tau^2}p_3^2\right]\frac{\partial\varphi}{\partial p_2} - \frac{1}{\tau^2}p_2 p_3\frac{\partial\varphi}{\partial p_3}\right\} +$$

$$\frac{1}{1-\frac{1}{\rho}p_2}\left\{\left(\dot{p}_3 + \frac{1}{\tau}p_2\dot{s}\right)\frac{\partial\Omega}{\partial s} - \left\{\left[\left(1-\frac{1}{\rho}p_2\right)^2 + \frac{1}{\tau^2}p_2^2 + \frac{1}{\tau^2}p_3^2\right] - \frac{1}{\tau}p_3\dot{p}_2 + \frac{1}{\tau}p_2\dot{p}_3\right\}\frac{\partial\Omega}{\partial p_3}\right\} \quad (13\text{b})$$

$$\frac{m_0}{e}\left[\ddot{p}_3 + \left(-\frac{1}{\tau^2}\tau' p_2 - \frac{1}{\tau^2}p_3 - \frac{\frac{1}{\rho\tau^2}\rho' p_2^2 + \frac{1}{\tau^2\rho}p_2 p_3}{1-\frac{1}{\rho}p_2}\right)\dot{s}^2 + \frac{2\frac{1}{\tau}\dot{s}\dot{p}_2}{1-\frac{1}{\rho}p_2}\right] =$$

$$\frac{1}{\left(1-\frac{1}{\rho}p_2\right)^2}\left\{\frac{1}{\tau}p_2\frac{\partial\varphi}{\partial s} - \frac{1}{\tau^2}p_2 p_3\frac{\partial\varphi}{\partial p_2} + \left[\left(1-\frac{1}{\rho}p_2\right)^2 + \frac{1}{\tau^2}p_2^2\right]\frac{\partial\varphi}{\partial p_3}\right\} +$$

$$\frac{1}{1-\frac{1}{\rho}p_2}\left\{\left\{\left[\left(1-\frac{1}{\rho}p_2\right)^2 + \frac{1}{\tau^2}p_2^2 + \frac{1}{\tau^2}p_3^2\right]\dot{s} - \frac{1}{\tau}p_3\dot{p}_2 + \frac{1}{\tau}p_2\dot{p}_3\right\}\frac{\partial\Omega}{\partial p_2} - \left(\dot{p}_2 - \frac{1}{\tau}\dot{s}p_3\right)\frac{\partial\Omega}{\partial s}\right\} \quad (13\text{c})$$

方程式（13）曾为 П. П. Касьянков[9]导出过，不过形式上与上述略有不同，注意现在式（13）中上标表示指数。

现将式（13）化为轨迹方程的形式。设 v 为相邻轨迹上电子的速度，$v = \frac{ds^*}{dt}$。按能量守恒定律，便可求得

$$\frac{ds^*}{dt} = \sqrt{\frac{2e}{m_0}\varphi_*} \quad (14)$$

式中，$\varphi_* = \varphi + \varepsilon_0$，$\varepsilon_0$ 是逸出电子初能对应的初电位；φ_* 通常称为规范化电位。按式（4），坐标曲线的弧长元 ds 与 ds^* 有以下关系：

$$\left(\frac{ds^*}{ds}\right)^2 = \left(1-\frac{1}{\rho}p_2\right)^2 + \frac{1}{\tau^2}(p_2^2 + p_3^2) - 2\frac{1}{\tau}p_3 p_2' + 2\frac{1}{\tau}p_2 p_3' + p_2'^2 + p_3'^2 \quad (15)$$

变换式（13）中 $\dot{s}$、$\dot{p}_2$、$\dot{p}_3$、$\ddot{p}_2$、$\ddot{p}_3$ 各量，于是

$$\dot{s} = \frac{\mathrm{d}s}{\mathrm{d}t} = \frac{\mathrm{d}s}{\mathrm{d}s^*} \frac{\mathrm{d}s^*}{\mathrm{d}t} = \sqrt{\frac{2e}{m_0}\varphi_*} \frac{\mathrm{d}s}{\mathrm{d}s^*},$$

$$\dot{p}_2 = \frac{\mathrm{d}p_2}{\mathrm{d}t} = \frac{\mathrm{d}p_2}{\mathrm{d}s} \frac{\mathrm{d}s}{\mathrm{d}s^*} \frac{\mathrm{d}s^*}{\mathrm{d}t} = \sqrt{\frac{2e}{m_0}\varphi_*} \frac{\mathrm{d}s}{\mathrm{d}s^*} p_2', \tag{16}$$

$$\ddot{p}_2 = \sqrt{\frac{2e}{m_0}\varphi_*} \frac{\mathrm{d}s}{\mathrm{d}s^*} \frac{\mathrm{d}}{\mathrm{d}s}\left(\sqrt{\frac{2e}{m_0}\varphi_*} \frac{\mathrm{d}s}{\mathrm{d}s^*} p_2'\right) = \frac{e}{m_0}\left\{2\varphi_*\left(\frac{\mathrm{d}s}{\mathrm{d}s^*}\right)^2 p_2'' + p_2' \frac{\mathrm{d}}{\mathrm{d}s}\left[\varphi_*\left(\frac{\mathrm{d}s}{\mathrm{d}s^*}\right)^2\right]\right\}$$

$\dot{p}_3$、$\ddot{p}_3$ 有类似表达式。将它们代入方程式（13b）、式（13c）中，便可导得曲线坐标系下的轨迹方程：

$$\frac{2\varphi_*}{\left(1-\frac{1}{\rho}p_2\right)^2 + \left(\frac{1}{\tau}p_2\right)^2 + \left(\frac{1}{\tau}p_3\right)^2 - 2\frac{1}{\tau}p_3 p_2' + 2\frac{1}{\tau}p_2 p_3' + p_2'^2 + p_3'^2} p_2'' +$$

$$p_2' \frac{\mathrm{d}}{\mathrm{d}s}\left[\frac{\varphi_*}{\left(1-\frac{1}{\rho}p_2\right)^2 + \left(\frac{1}{\tau}p_2\right)^2 + \left(\frac{1}{\tau}p_3\right)^2 - 2\frac{1}{\tau}p_3 p_2' + 2\frac{1}{\tau}p_2 p_3' + p_2'^2 + p_3'^2}\right] +$$

$$\frac{2\varphi_*}{\left(1-\frac{1}{\rho}p_2\right)^2 + \left(\frac{1}{\tau}p_2\right)^2 + \left(\frac{1}{\tau}p_3\right)^2 - 2\frac{1}{\tau}p_3 p_2' + 2\frac{1}{\tau}p_2 p_3' + p_2'^2 + p_3'^2} \times$$

$$\left(\frac{1}{\rho} - \frac{1}{\rho^2}p_2 - \frac{1}{\tau^2}p_2 + \frac{1}{\tau^2}\tau' p_3 + \frac{\frac{1}{\rho\tau^2}p_3^2 + \frac{1}{\rho^2\tau}\rho' p_2 p_3}{1-\frac{1}{\rho}p_2} - \frac{\frac{2}{\rho\tau}p_3 p_2'}{1-\frac{1}{\rho}p_2} - 2\frac{1}{\tau}p_3'\right)$$

$$= \frac{1}{\left(1-\frac{1}{\rho}p_2\right)^2}\left\{\frac{1}{\tau}p_3 \frac{\partial\varphi}{\partial s} + \left[\left(1-\frac{1}{\rho}p_2\right)^2 + \frac{1}{\tau^2}p_3^2\right]\frac{\partial\varphi}{\partial p_2} - \frac{1}{\tau^2}p_2 p_3 \frac{\partial\varphi}{\partial p_3}\right\} +$$

$$\frac{1}{1-\frac{1}{\rho}p_2}\sqrt{\frac{2e\varphi_*/m_0}{\left(1-\frac{1}{\rho}p_2\right)^2 + \left(\frac{1}{\tau}p_2\right)^2 + \left(\frac{1}{\tau}p_3\right)^2 - 2\frac{1}{\tau}p_3 p_2' + 2\frac{1}{\tau}p_2 p_3' + p_2'^2 + p_3'^2}} \times$$

$$\left\{\left(p_3' + \frac{1}{\tau}p_2\right)\frac{\partial\Omega}{\partial s} - \left\{\left[\left(1-\frac{1}{\rho}p_2\right)^2 + \frac{1}{\tau^2}p_2^2 + \frac{1}{\tau^2}p_3^2\right] - \frac{1}{\tau}p_3 p_2' + \frac{1}{\tau}p_2 p_3'\right\}\frac{\partial\Omega}{\partial p_3}\right\} \tag{17a}$$

$$\frac{2\varphi_*}{\left(1-\frac{1}{\rho}p_2\right)^2 + \left(\frac{1}{\tau}p_2\right)^2 + \left(\frac{1}{\tau}p_3\right)^2 - 2\frac{1}{\tau}p_3 p_2' + 2\frac{1}{\tau}p_2 p_3' + p_2'^2 + p_3'^2} p_3'' +$$

$$p_3' \frac{\mathrm{d}}{\mathrm{d}s}\left[\frac{\varphi_*}{\left(1-\frac{1}{\rho}p_2\right)^2 + \left(\frac{1}{\tau}p_2\right)^2 + \left(\frac{1}{\tau}p_3\right)^2 - 2\frac{1}{\tau}p_3 p_2' + 2\frac{1}{\tau}p_2 p_3' + p_2'^2 + p_3'^2}\right] +$$

$$\frac{2\varphi_*}{\left(1-\frac{1}{\rho}p_2\right)^2 + \left(\frac{1}{\tau}p_2\right)^2 + \left(\frac{1}{\tau}p_3\right)^2 - 2\frac{1}{\tau}p_3 p_2' + 2\frac{1}{\tau}p_2 p_3' + p_2'^2 + p_3'^2} \times$$

$$\left(-\frac{1}{\tau^2}\tau' p_2 - \frac{1}{\tau^2}p_3 - \frac{\frac{1}{\tau\rho^2}\rho' p_2^2 + \frac{1}{\tau^2\rho}p_2 p_3}{1-\frac{1}{\rho}p_2} + \frac{2\frac{1}{\tau}p_2'}{1-\frac{1}{\rho}p_2}\right)$$

$$= \frac{1}{\left(1 - \frac{1}{\rho}p_2\right)^2}\left\{\frac{1}{\tau}p_2\frac{\partial\varphi}{\partial s} - \frac{1}{\tau^2}p_2 p_3\frac{\partial\varphi}{\partial p_2} + \left[\left(1 - \frac{1}{\rho}p_2\right)^2 + \frac{1}{\tau^2}p_2^2\right]\frac{\partial\varphi}{\partial p_3}\right\} +$$

$$\frac{1}{1 - \frac{1}{\rho}p_2}\sqrt{\frac{2e\varphi_*/m_0}{\left(1 - \frac{1}{\rho}p_2\right)^2 + \left(\frac{1}{\tau}p_2\right)^2 + \left(\frac{1}{\tau}p_3\right)^2 - 2\frac{1}{\tau}p_3 p_2' + 2\frac{1}{\tau}p_2 p_3' + p_2'^2 + p_3'^2}} \times$$

$$\left\{\left\{\left[\left(1 - \frac{1}{\rho}p_2\right)^2 + \frac{1}{\tau^2}p_2^2 + \frac{1}{\tau^2}p_3^2\right] - \frac{1}{\tau}p_3 p_2' + \frac{1}{\tau}p_2 p_3'\right\}\frac{\partial\Omega}{\partial p_2} - \left(p_2' - \frac{1}{\tau}p_3\right)\frac{\partial\Omega}{\partial s}\right\} \quad (17b)$$

以式（17）所表示的轨迹方程是本文首次给出的，这种形式无论对于研究宽电子束或是细电子束的聚焦和像差都是十分方便的。

2 主轨迹方程与曲近轴轨迹方程

在普遍形式的轨迹方程中，我们引入曲近轴条件式（1）、式（2），将 p_2、p_3、$\dot{p}_2$、$\dot{p}_3$ 作为一阶小量，采用文献 [10]、文献 [11] 的方法，来考察式（17）中各系数项。

将 N^* 点处的电位 φ 和磁标位 Ω 在曲线轴的 N 点处作泰勒展开，便有

$$\varphi(s, p_2, p_3) = \phi(s) + p_2\varphi_2(s) + p_3\varphi_3(s) + \frac{1}{2}p_2^2\varphi_{22}(s) + p_2 p_3\varphi_{23}(s) + \frac{1}{2}p_3^2\varphi_{33}(s) + \cdots \quad (18)$$

$$\Omega(s, p_2, p_3) = \Omega(s) + p_2\Omega_2(s) + p_3\Omega_3(s) + \frac{1}{2}p_2^2\Omega_{22}(s) + p_2 p_3\Omega_{23}(s) + \frac{1}{2}p_3^2\Omega_{33}(s) + \cdots \quad (19)$$

这里引入的符号为

$$\varphi\Big|_{\substack{p_2=0\\p_3=0}} = \varphi(s,0,0) = \phi(s),\ \Omega\Big|_{\substack{p_2=0\\p_3=0}} = \Omega(s,0,0) = \Omega(s) \quad (20)$$

$$\frac{\partial\varphi}{\partial p_i}\Big|_{\substack{p_2=0\\p_3=0}} = \varphi_i,\quad \frac{\partial\Omega}{\partial p_i}\Big|_{\substack{p_2=0\\p_3=0}} = \Omega_i,\quad \frac{\partial^2\varphi}{\partial p_i \partial p_j}\Big|_{\substack{p_2=0\\p_3=0}} = \varphi_{ij},\quad \frac{\partial^2\Omega}{\partial p_i \partial p_j}\Big|_{\substack{p_2=0\\p_3=0}} = \Omega_{ij} \quad (21)$$

此外，$\varphi_* = \varphi(s, p_2, p_3) + \varepsilon_0$，$\varphi(0,0,0) = \phi(0) = 0$，$\varphi_1 = \frac{d\varphi}{ds}$。

现考察轨迹方程式（17）第一项的系数，它与主轨迹的切向速度 $\dot{s}$ 有关，不难看出，在电子逸出的起始点处，它可表示为

$$\left[\frac{\varphi_*}{\left(1 - \frac{1}{\rho}p_2\right)^2 + \left(\frac{1}{\tau}p_2\right)^2 + \left(\frac{1}{\tau}p_3\right)^2 - 2\frac{1}{\tau}p_3 p_2' + 2\frac{1}{\tau}p_2 p_3' + p_2'^2 + p_3'^2}\right]_{s=0, p_2=0, p_3=0}$$

$$= \left(\frac{\varepsilon_0}{1 + p_2'^2 + p_3'^2}\right)_{s=0, p_2=0, p_3=0} = \varepsilon_s \quad (22)$$

因此，第一项系数的 0 + 1 级近似，便可表示为

$$\left[\frac{\varphi_*}{\left(1 - \frac{1}{\rho}p_2\right)^2 + \left(\frac{1}{\tau}p_2\right)^2 + \left(\frac{1}{\tau}p_3\right)^2 - 2\frac{1}{\tau}p_3 p_2' + 2\frac{1}{\tau}p_2 p_3' + p_2'^2 + p_3'^2}\right]_{0+1} = \varepsilon_s + \phi +$$

$$p_2\varphi_2 + p_3\varphi_3 + (\varepsilon_s + \phi)\frac{2}{\rho}p_2 \quad (23)$$

考察该系数的 0 + 1 + 2 级近似，不难看出，式（23）成立的条件是

$$p^2 \approx 0,$$
$$\frac{\frac{m_0}{2e}(\dot{p}^2 - \dot{p}_0^2)}{\phi + \varepsilon_s} \ll 1 \tag{24}$$

这便是所谓曲近轴条件。因此，在研究宽电子束聚焦时，是将 p 与 $\sqrt{\frac{m_0}{2e}}\dot{p}\Big/\sqrt{\phi + \varepsilon_s}$ 作为一级小量，而其本身与相互的乘积视为二级小量，可以略去。

按此，对式（17）其他各项进行同样处理，经过相当繁复的运算，便可得到如下形式的方程：

$$\left(\frac{2}{\rho}(\phi + \varepsilon_s) - \varphi_2 + \sqrt{\frac{2e}{m_0}}\sqrt{\phi + \varepsilon_s}\,\Omega_3\right)_0 + \left\{2(\phi + \varepsilon_s)p_2'' + p_2'\frac{\mathrm{d}}{\mathrm{d}s}(\phi + \varepsilon_s) - \right.$$
$$\left[\frac{4}{\tau}(\phi + \varepsilon_s) + \sqrt{\frac{2e}{m_0}}\sqrt{\phi + \varepsilon_s}\,\Omega_1\right]p_3' + \left[\frac{2}{\rho}\varphi_2 + 2\left(\frac{1}{\rho^2} - \frac{1}{\tau^2}\right)(\phi + \varepsilon_s) - \varphi_{22} + \right.$$
$$\sqrt{\frac{2e}{m_0}}\sqrt{\phi + \varepsilon_s}\left(\Omega_{23} + \frac{\varphi_2}{2(\phi + \varepsilon_s)}\Omega_3 - \frac{1}{\tau}\Omega_1\right)\right]p_2 + \left[\frac{2}{\rho}\varphi_3 - \varphi_{23} + \frac{2}{\tau^2}\tau'(\phi + \varepsilon_s) - \frac{1}{\tau}\varphi_1 + \right.$$
$$\left.\left.\sqrt{\frac{2e}{m_0}}\sqrt{\phi + \varepsilon_s}\left(\Omega_{33} + \frac{\varphi_3}{2(\phi + \varepsilon_s)}\Omega_3\right)\right]p_3\right\}_1 = 0, \tag{25a}$$

$$-\left(\varphi_3 + \sqrt{\frac{2e}{m_0}}\sqrt{\phi + \varepsilon_s}\,\Omega_2\right)_0 + \left\{2(\phi + \varepsilon_s)p_3'' + p_3'\frac{\mathrm{d}}{\mathrm{d}s}(\phi + \varepsilon_s) + \right.$$
$$\left[\frac{4}{\tau}(\phi + \varepsilon_s) + \sqrt{\frac{2e}{m_0}}\sqrt{\phi + \varepsilon_s}\,\Omega_1\right]p_2' + \left[-\frac{2}{\tau^2}\tau'(\phi + \varepsilon_s) + \frac{1}{\tau}\varphi_1 - \varphi_{23} - \right.$$
$$\left.\sqrt{\frac{2e}{m_0}}\sqrt{\phi + \varepsilon_s}\left(\Omega_{22} + \frac{\varphi_2\Omega_2}{2(\phi + \varepsilon_s)}\right)\right]p_2 - \left[\varphi_{33} + \frac{2}{\tau^2}(\phi + \varepsilon_s) + \right.$$
$$\left.\left.\sqrt{\frac{2e}{m_0}}\sqrt{\phi + \varepsilon_s}\left(\Omega_{23} + \frac{\varphi_3\Omega_2}{2(\phi + \varepsilon_s)} + \frac{1}{\tau}\Omega_1\right)\right]p_3\right\}_1 = 0 \tag{25b}$$

我们在这里假定坐标线段 s 是一条由物面上某点射出的空间曲线轴。现在令这条曲线轴是一条电子轨迹。因为上述方程乃是自物面某点射出的任意轨迹的方程，为了要使坐标线段也是电子轨迹之一，其必要与充分条件是上述方程满足 $p_2 \equiv p_3 \equiv 0$。

显然，此时以下标为 0 表示的括弧项必须为零。由此可导得

$$\frac{2(\phi + \varepsilon_s)}{\rho} = \varphi_2 - \sqrt{\frac{2e}{m_0}}\sqrt{\phi + \varepsilon_s}\,\Omega_3 \tag{26a}$$

$$\varphi_3 = -\sqrt{\frac{2e}{m_0}}\sqrt{\phi + \varepsilon_s}\,\Omega_2 \tag{26b}$$

式（26）称为主轨迹方程（或称主曲线方程、主电子方程）。由此解得的坐标线段 s 表示的电子轨迹称为主轨迹（它可以是电子束的轴线，也可以是系统的轴线）。

当主轨迹方程满足式（26）时，则以下标为 1 表示的大括弧项等于零时所导得的相邻轨迹方程可表达成类似文献 [1]、文献 [12] 所给出的如下形式：

$$\frac{\mathrm{d}}{\mathrm{d}s}\left(n\frac{\mathrm{d}p_2}{\mathrm{d}s}\right) = Fp_2 + Np_3 + 2K\frac{\mathrm{d}p_3}{\mathrm{d}s} \tag{27a}$$

$$\frac{\mathrm{d}}{\mathrm{d}s}\left(n\frac{\mathrm{d}p_3}{\mathrm{d}s}\right) = Rp_3 - Qp_2 - 2K\frac{\mathrm{d}p_2}{\mathrm{d}s} \tag{27b}$$

式 n、K、F、R、N 和 Q 表示为主轨迹曲线 s 的如下函数：

$$n = \sqrt{\phi + \varepsilon_s}, \quad K = \frac{\sqrt{\phi + \varepsilon_s}}{+} \sqrt{\frac{e}{8m_0}} \Omega_1 \tag{28a}$$

$$F = \sqrt{\phi + \varepsilon_s}\left(\frac{1}{\tau^2} - \frac{1}{\rho^2}\right) + \frac{\varphi_{22}}{2\sqrt{\phi + \varepsilon_s}} - \frac{\varphi_2}{\rho\sqrt{\phi + \varepsilon_s}} + \sqrt{\frac{e}{2m_0}}\left[\frac{\Omega_1}{\tau} - \Omega_{23} - \frac{\Omega_3\varphi_2}{2(\phi + \varepsilon_s)}\right] \tag{28b}$$

$$R = \frac{\sqrt{\phi + \varepsilon_s}}{\tau^2} + \frac{\varphi_{33}}{2\sqrt{\phi + \varepsilon_s}} + \sqrt{\frac{e}{2m_0}}\left[\frac{\Omega_1}{\tau} + \Omega_{23} + \frac{\Omega_2\varphi_3}{2(\phi + \varepsilon_s)}\right] \tag{28c}$$

$$N = \frac{\varphi_1}{2\tau\sqrt{\phi + \varepsilon_s}} - \frac{\sqrt{\phi + \varepsilon_s}}{\tau^2}\frac{\mathrm{d}\tau}{\mathrm{d}s} - \frac{\varphi_3}{\rho\sqrt{\phi + \varepsilon_s}} + \frac{\varphi_{23}}{2\sqrt{\phi + \varepsilon_s}} - \sqrt{\frac{e}{2m_0}}\left[\Omega_{33} + \frac{\Omega_2\varphi_3}{2(\phi + \varepsilon_s)}\right] \tag{28d}$$

$$Q = \frac{\varphi_1}{2\tau\sqrt{\phi + \varepsilon_s}} - \frac{\sqrt{\phi + \varepsilon_s}}{\tau^2}\frac{\mathrm{d}\tau}{\mathrm{d}s} - \frac{\varphi_{23}}{2\sqrt{\phi + \varepsilon_s}} - \sqrt{\frac{e}{2m_0}}\left[\Omega_{22} + \frac{\Omega_2\varphi_3}{2(\phi + \varepsilon_s)}\right] \tag{28e}$$

不难证明，式（28）各分式中系数 N、Q 和 K 之间存在着下列关系式

$$\frac{N+Q}{2} = \frac{\mathrm{d}K}{\mathrm{d}s} \tag{29}$$

方程式（27）能描写自阴极物面 $\phi(0)=0$ 逸出的初始斜率可能很大的相邻轨迹，我们将之命名为曲近轴轨迹方程，以有别于 Г. А. Гринберг 等导得的细电子束下的曲（傍）轴轨迹方程。

再一次强调指出，求解方程式（27）时所取的系数 n、F、N、R、Q、K 等值均是主轨迹上的值。因此，通常是将式（26）和式（27）联立求解。此外，主轨迹方程式（26）和曲近轴轨迹方程式（27）的各系数，必须取相同的 ε_s 值。

还应该指出，在求解方程式（26）和式（27）时其电位和磁标位展开式的系数之间还应满足以下补充关系式：

$$\varphi_{11} + \varphi_{22} + \varphi_{33} = 0, \quad \Omega_{11} + \Omega_{22} + \Omega_{33} = 0 \tag{30a}$$

$$\varphi_1 = \phi' = \frac{\mathrm{d}\phi}{\mathrm{d}s}, \quad \varphi_{11} + \frac{1}{\rho}\varphi_2 = \frac{\mathrm{d}^2\phi}{\mathrm{d}s^2} \tag{30b}$$

$$\Omega_1 = \frac{\mathrm{d}\Omega}{\mathrm{d}s}, \quad \Omega_{11} + \frac{1}{\rho}\Omega_2 = \frac{\mathrm{d}^2\Omega}{\mathrm{d}s^2} \tag{30c}$$

$$\frac{\mathrm{d}}{\mathrm{d}s}\left(\frac{2(\phi + \varepsilon_s)}{\rho}\right) = \varphi_{12} - \frac{1}{\rho}\varphi_1 + \frac{1}{\tau}\varphi_3 - \sqrt{\frac{2e}{m_0}}\sqrt{\phi + \varepsilon_s}\left(\Omega_{31} - \frac{\Omega_2}{\tau}\right) - \sqrt{\frac{e}{2m_0}}\frac{1}{\sqrt{\phi + \varepsilon_s}}\varphi_1\Omega_3 \tag{30d}$$

$$\varphi_{31} - \frac{1}{\tau}\varphi_2 = -\sqrt{\frac{2e}{m_0}}\sqrt{\phi + \varepsilon_s}\left(\Omega_{12} - \frac{\Omega_1}{\rho} + \frac{\Omega_3}{\tau}\right) - \sqrt{\frac{e}{2m_0}}\frac{1}{\sqrt{\phi + \varepsilon_s}}\varphi_1\Omega_2 \tag{30e}$$

Г. А. Гринберг 曾经指出[1]。带电粒子束的聚焦问题可以由两种方式提出，通常由已给定的电场和磁场分布出发，研究确定轨迹的微分方程，求带电粒子的轨迹，从而确定其

聚焦性质。但是，也可以反其道而行之，先给出粒子束的形状，然后求得保证实现此束的形状的场分布。方程式（26）和式（27）将 Г. А. Гринберг 理论发展于宽电子束情况，同样也包括了这两种可能性。诚然，式（27）这两个二阶线性微分方程的待解函数 p_2、p_3 是耦合的。由此可见，研究任意电场和磁场下宽电子束的聚焦性质甚至连一级近似轨迹也是不容易的。

3 曲近轴系统的正交条件

曲傍轴系统的正交条件是由 P. A. Sturrock 首先给出的[4]。下面我们由方程式（27）出发讨论曲近轴系统的正交条件。我们首先使方程式（27）中不出现 p'_3 项，而方程式（27）中不出现 p'_2 项，使方程的形式简化。

引入坐标变换

$$p_2 + \mathrm{i} p_3 = (u + \mathrm{i} v)\mathrm{e}^{\mathrm{i}\chi} \tag{31}$$

且令

$$\chi' = \frac{\mathrm{d}\chi}{\mathrm{d}s} = -\frac{K}{n} \tag{32}$$

对式（27）进行变换，便可得到以旋转坐标系 (u, v) 表示的曲近轴轨迹方程：

$$\frac{\mathrm{d}}{\mathrm{d}s}\left(n\frac{\mathrm{d}u}{\mathrm{d}s}\right) = a_{11}u + a_{12}v \tag{33a}$$

$$\frac{\mathrm{d}}{\mathrm{d}s}\left(n\frac{\mathrm{d}v}{\mathrm{d}s}\right) = a_{21}u + a_{22}v \tag{33b}$$

式中，

$$a_{11} = F\cos^2\chi + R\sin^2\chi + (N-Q)\sin\chi\cos\chi - \frac{K^2}{n} \tag{34a}$$

$$a_{12} = N\cos^2\chi + Q\sin^2\chi + (F-R)\sin\chi\cos\chi - K' \tag{34b}$$

$$a_{21} = -N\sin^2\chi - Q\cos^2\chi - (F-R)\sin\chi\cos\chi + K' \tag{34c}$$

$$a_{22} = F\sin^2\chi + R\cos^2\chi - (N-Q)\sin\chi\cos\chi - \frac{K^2}{n} \tag{34d}$$

由式（33）可见，此两方程的变量仍未分离。但是，由式（34）和式（29）不难证明 $a_{12} = a_{21}$，说明这两个方程是自共轭方程。

但是，要使变量分离，还必须满足条件 $a_{12} = a_{21} = 0$，如果此条件成立，则 $a_{12} + a_{21} = 0$，由此可得

$$\tan 2\chi = \frac{N-Q}{F-R} \tag{35}$$

用式（35）变换式（34）中 a_{11} 和 a_{22} 表达式，并令 $U = a_{11}$，$V = a_{22}$，则可导得

$$\frac{\mathrm{d}}{\mathrm{d}s}\left(n\frac{\mathrm{d}u}{\mathrm{d}s}\right) = Uu \tag{36a}$$

$$\frac{\mathrm{d}}{\mathrm{d}s}\left(n\frac{\mathrm{d}v}{\mathrm{d}s}\right) = Vv \tag{36b}$$

式中

$$U = \frac{1}{2}(F+R) + \frac{1}{2}\sqrt{(F-R)^2 + (N-Q)^2} - \frac{K^2}{n} \qquad (37a)$$

$$V = \frac{1}{2}(F+R) - \frac{1}{2}\sqrt{(F-R)^2 + (N-Q)^2} - \frac{K^2}{n} \qquad (37b)$$

由式（35）和式（32），我们可进一步导得使变量分离时 n、K、F、R、N 和 Q 等系数所应满足的条件式：

$$n[(N-Q)(F'-R') + (F-R)(N'-Q')] = 2K[(F-R)^2 + (N-Q)^2] \qquad (38)$$

这便是曲近轴系统的正交条件，我们将满足正交条件式（38）的系统称为正交曲近轴系统，以有别于 P. A. Sturrock 所建立的正交（曲傍轴）系统。

正如 P. A. Sturrock[4] 所指出的，满足正交条件的曲轴系统，可以形成点聚焦或线聚焦的图像。

式（36）分别为 u 和 v 的二阶线性齐次微分方程。尽管方程在阴极物面处存在着奇点，但求解方法与轴对称阴极透镜的线性方程并无根本差别。因此，u、v 的解可以用两个线性独立的特解来表示。令 u_α、u_β 和 v_α、v_β 为方程式（36）的两组特解，它们满足如下初始条件：

$$u_\alpha(0) = v_\alpha(0) = 0,\ \sqrt{\varepsilon_s}u'_\alpha(0) = \sqrt{\varepsilon_s}v'_\alpha(0) = 1 \qquad (39a)$$

$$u_\beta(0) = v_\beta(0) = 0,\ \sqrt{\varepsilon_s}u'_\beta(0) = \sqrt{\varepsilon_s}v'_\beta(0) = 0 \qquad (39b)$$

为了使系统获得无像散点聚焦的像，显然必须使式（36）中 $U = V$，这只有在

$$F - R = N - Q = 0 \qquad (40)$$

时才有可能。于是在 $\varepsilon_s = \varepsilon_{s_i}$ 对应的像面 $s = s_i$ 处，将有

$$u_\alpha(\varepsilon_{s_1}, s_i) = u_\alpha(\varepsilon_{s_1}, s_i) = 0 \qquad (41a)$$

$$u'_\alpha(\varepsilon_{s_1}, s_i) = v'_\alpha(\varepsilon_{s_1}, s_i) \qquad (41b)$$

其放大率之模可表示为

$$M_u = M_v = M = \left| \frac{1}{u'_\alpha(\varepsilon_{s_1}, s_i)\sqrt{\phi(s_i) + \varepsilon_{s_1}}} \right| \qquad (42)$$

由式（36）和式（38）可见，满足条件式（40）的系统必须同时满足正交条件。而条件式（40）对应于下列两个关系式：

$$\frac{\sqrt{\phi + \varepsilon_s}}{\rho^2} + \frac{\varphi_{33} - \varphi_{22}}{2\sqrt{\phi + \varepsilon_s}} + \frac{\varphi_2}{\rho\sqrt{\phi + \varepsilon_s}} + \sqrt{\frac{2e}{m_0}}\left(\Omega_{23} + \frac{\Omega_3\varphi_2 + \Omega_2\varphi_3}{4\sqrt{\phi + \varepsilon_s}}\right) = 0 \qquad (43a)$$

$$\frac{\varphi_{23}}{\sqrt{\phi + \varepsilon_s}} - \frac{\varphi_3}{\rho\sqrt{\phi + \varepsilon_s}} + \sqrt{\frac{2e}{m_0}}\left(\Omega_{22} - \Omega_{33} - \frac{\Omega_3\varphi_3 - \Omega_2\varphi_2}{2\sqrt{\phi + \varepsilon_s}}\right) = 0 \qquad (43b)$$

条件式（40）或条件式（43）说明，以 $\varepsilon_s = \varepsilon_{s_1}$ 逸出的电子束将能理想聚焦（点聚焦），系统将具有类似轴对称电子光学系统的性质。

若曲近轴系统仅满足正交条件式（38），但并不满足条件式（40）时，$U \neq V$，此时，特解 u_α、u_β 和 v_α、v_β 是不同的。在此情况下，放大率各向异性。由式（39）和式（36），不难求得在 $\varepsilon_s = \varepsilon_{s_1}$ 下两个方向的放大率之模为：

$$M_u = \left| \frac{1}{u'_\alpha(\varepsilon_{s_1}, s_{1i})\sqrt{\phi(s_{1i}) + \varepsilon_{s_1}}} \right| \qquad (44a)$$

$$M_v = \left| \frac{1}{v'_\alpha(\varepsilon_{s_1}, s_{2i}) \sqrt{\phi(s_{2i}) + \varepsilon_{s_1}}} \right| \tag{44b}$$

即使在同一 $\varepsilon_s = \varepsilon_{s_1}$ 下，聚焦像的位置 s_{1i} 与 s_{2i} 并不重合。这说明，通过正交电子光学场，可能有两个理想聚焦像，它们与 $\varepsilon_s = \varepsilon_{s_1}$ 相对应。在 $s = s_{1i}$ 处，在 $\boldsymbol{\xi}_3$ 方向形成（弧矢）焦线，而在 $s = s_{2i}$ 处，却在 $\boldsymbol{\xi}_2$ 方向形成（子午）焦线。而此两者位置之差，可用来表征曲近轴系统的像散。关于静电阴极透镜中宽电子束形成的场曲与像散的确定，我们将另文叙述。

4 宽电子束聚焦举例

4.1 主轨迹为直轴且旋转对称下宽电子束聚焦[10][13]

当主轨迹为直轴时，$\frac{1}{\rho} = \frac{1}{\tau} = 0$。考虑场的旋转对称性，电位在通过轴时取极值，因此

$$\varphi_2 = \varphi_3 = 0, \quad \varphi_{23} = 0 \tag{45a}$$
$$\Omega_2 = \Omega_3 = 0, \quad \Omega_{23} = 0 \tag{45b}$$

同样，由于主法线方向与副法线方向没有区别，则由式（30）可得

$$\varphi_{22} = \varphi_{33} = -\frac{1}{2}\varphi_{11} = -\frac{1}{2}\frac{d^2\varphi}{ds^2} \tag{46a}$$

$$\Omega_{22} = \Omega_{33} = -\frac{1}{2}\Omega_{11} = -\frac{1}{2}\frac{d^2\Omega}{ds^2} \tag{46b}$$

由式（43）可见，此时电子束聚焦不但满足正交条件式（38）且满足条件式（40），可见系统具有理想聚焦（点聚焦）的性质。

对于轴对称情况下，令 $p = p_2 + \mathrm{i}p_3$，$s = z$，$\varepsilon_s = \varepsilon_z$，$\phi(s) = \phi(z)$，$\Omega_1 = \frac{d\Omega}{ds} = -H(z)$，便可得到

$$n = \sqrt{\phi(z) + \varepsilon_z}, \quad K = -\sqrt{\frac{e}{8m_0}}H(z),$$
$$F = R = -\frac{1}{4}\frac{\phi''}{\sqrt{\phi(z) + \varepsilon_z}}, \quad N - Q = 0 \tag{47}$$

于是有

$$U = V = -\frac{1}{4}\frac{\phi''}{\sqrt{\phi(z) + \varepsilon_z}} - \frac{e}{8m_0}\frac{H^2(z)}{\sqrt{\phi(z) + \varepsilon_z}} \tag{48}$$

令

$$\boldsymbol{u} = u + \mathrm{i}v = \boldsymbol{p}\mathrm{e}^{-\mathrm{i}\chi} \tag{49}$$

$$\chi' = \sqrt{\frac{e}{8m_0}}\frac{H(z)}{\sqrt{\phi(z) + \varepsilon_z}} \tag{50}$$

则方程式（36）便化为如下形式

$$[\phi(z)+\varepsilon_z]u'' + \frac{1}{2}\phi'(z)u' + \frac{1}{4}\left[\phi'' + \frac{e}{2m_0}H^2(z)\right]u = 0 \qquad (51)$$

这便是场具有旋转对称且主轨迹为直轴下宽电子束聚焦的线性方程[10]，电磁聚焦同心球系统自阴极任一点逸出的电子束聚焦亦可用本方程进行求解[13]。

4.2 静电阴极透镜的宽电子束聚焦

对于轴对称静电阴极透镜，自阴极面逸出的轴外主轨迹将是一条垂直于阴极面的平面曲线，此时

$$\Omega = 0, \quad \frac{1}{\tau} = 0, \quad \varphi_{23} = 0$$

于是式（28）变为

$$n = \sqrt{\phi(s)+\varepsilon_s}, \quad K = 0,$$
$$F = -\frac{3}{\rho^2}\sqrt{\phi+\varepsilon_s} + \frac{\varphi_{22}}{2\sqrt{\phi+\varepsilon_s}}, \qquad (52)$$
$$R = \frac{\varphi_{33}}{2\sqrt{\phi+\varepsilon_s}}, \quad N = Q = 0$$

式（52）的系数亦满足正交条件式（38），可见其变量亦是分离的。

鉴于 $\chi = 0$，故令 $u = p_2$，$v = p_3$，最后主轨迹方程可表示为

$$\varphi_2 = \frac{2(\phi+\varepsilon_s)}{\rho} \qquad (53)$$

曲近轴方程为

$$\frac{d}{ds}\left(\sqrt{\phi+\varepsilon_s}\frac{dp_2}{ds}\right) = \left(-\frac{3\sqrt{\phi+\varepsilon_s}}{\rho^2} + \frac{\varphi_{22}}{2\sqrt{\phi+\varepsilon_s}}\right)p_2,$$
$$\frac{d}{ds}\left(\sqrt{\phi+\varepsilon_s}\frac{dp_3}{ds}\right) = \frac{\varphi_{33}}{2\sqrt{\phi+\varepsilon_s}}p_3 \qquad (54)$$

当把以 s 表示的主轨迹坐标曲线，换成以坐标 (r,z) 表示的形式时，则由式（53）和式（54），便可得到我们曾经在文献［14］中建议的形式。

主轨迹方程

$$r'' = \frac{1+r'^2}{2[\varphi(z,r)+\varepsilon_s]}\left(\frac{\partial\varphi}{\partial r} - r'\frac{\partial\varphi}{\partial z}\right) \qquad (55)$$

曲近轴轨迹方程

$$p_2'' + F_1 p_2' + F_2 p_2 = 0,$$
$$p_3'' + G_1 p_3' + G_2 p_3 = 0 \qquad (56)$$

式中，

$$F_1 = G_1 = \frac{1+r'^2}{2[\varphi(z,r)+\varepsilon_s]}\frac{\partial\varphi}{\partial z} \qquad (57a)$$

$$F_2 = \frac{3r''^2}{(1+r'^2)^2} + \frac{1}{2[\varphi(z,r)+\varepsilon_s]}\left(-\frac{\partial^2\varphi}{\partial r^2} + 2r'\frac{\partial^2\varphi}{\partial r\partial z} - r'^2\frac{\partial^2\varphi}{\partial z^2}\right) \qquad (57b)$$

$$G_2 = -\frac{1+r'^2}{2[\varphi(z,r)+\varepsilon_s]} \frac{1}{r}\frac{\partial\varphi}{\partial r} \tag{57c}$$

再提醒一下，系数 F_1、G_1、F_2、G_2 中的各项应取主轨迹上的值。式（55）~（57）各式中的撇号乃是对于 z 的导数。

对于轴上点逸出的电子束，由于主轨迹为旋转对称轴，故 $r=r'=r''=0$，且

$$\left(\frac{1}{r}\frac{\partial\varphi}{\partial r}\right)_{r=0} = \left(\frac{\partial^2\varphi}{\partial r^2}\right)_{r=0} = -\frac{1}{2}\phi''(z) \tag{58}$$

令 $p = p_2 + \mathrm{i}p_3$，$\varepsilon_s = \varepsilon_z$。于是方程式（56）便化为如下的形式

$$p'' + \frac{1}{2}\frac{\phi'}{\phi(z)+\varepsilon_z}p' + \frac{1}{4}\frac{\phi''}{\phi(z)+\varepsilon_z}p = 0 \tag{59}$$

这便是众所周知的静电情况下的近轴方程，与轴外点逸出电子束所对应的曲近轴方程式（56）是协调的。式（59）自然亦可由方程式（51）略去磁场项而导得。

5 曲傍轴方程与曲近轴方程的异同点

现在扼要谈谈本文提出的宽电子束聚焦的曲近轴方程与细电子束聚焦的曲傍轴方程的异同点。

（1）当逸出电子束满足曲傍轴条件，且物面处于高电位下 $\phi(s=0)\neq 0$ 时，若令 $\varepsilon_s = \varepsilon_0$，则本文给出的曲线坐标系下的主轨迹方程式（26）和曲近轴轨迹方程式（27）便转化为 Г. А. Гринберг 方程[1][12]，使变量分离的曲近轴轨迹方程式（36）转化为 P. A. Sturrock 方程[4]，曲近轴系统的正交条件式（38）转化为 P. A. Sturrock 提出的细电子束聚焦的正交条件[4]。同样，上节实例中所有表达式均适用于细电子束聚焦的情况，只要令 $\varepsilon_s = \varepsilon_0$，$\phi(s=0) \gg \varepsilon_0$。

因此，从两类方程形式相类似来看，曲傍轴方程可以视为曲近轴方程的一个特例。

（2）在曲傍轴情况下，一般假定 $\phi(s) \gg \varepsilon_0$，故文献［1］、文献［12］中，通常令 $n = \sqrt{\phi}$。

在求解曲傍轴方程时，其线性无关的特解一般假定与式（39）不同。通常，若令 u_α、u_β 和 v_α、v_β 乃是满足正交条件下曲傍轴轨迹方程的特解，则它们满足如下初始条件：

$$\begin{cases} u_\alpha(0) = v_\alpha(0) = 0, \ u'_\alpha(0) = v'_\alpha(0) = 1, \\ u_\beta(0) = v_\beta(0) = 1, \ u'_\beta(0) = v'_\beta(0) = 0 \end{cases} \tag{60}$$

当给出的场使 $F - R = N - Q = 0$，即 $U = V$ 时，系统还同时满足正交条件，于是细束电子将在 $s = s_i$ 处形成点聚焦：

$$u_\alpha(s_i) = v_\alpha(s_i) = 0, \ u'_\alpha(s_i) = v'_\alpha(s_i) \tag{61}$$

故放大率之模可表示为

$$M_u = M_v = M = \left|\frac{1}{u'_\alpha(s_i)}\sqrt{\frac{\phi(0)}{\phi(s_i)}}\right| \tag{62}$$

当满足正交条件，且 $U \neq V$ 时，于是

$$u_\alpha(s_{1i}) = 0, \ v_\alpha(s_{2i}) = 0 \tag{63}$$

电子将在不同位置处 $(s_{1i} \neq s_{2i})$ 理想聚焦（形成线聚焦），其放大率之模为

$$M_u = \left| \frac{1}{u'_\alpha(s_{1i})} \sqrt{\frac{\phi(0)}{\phi(s_{1i})}} \right|,$$
$$M_v = \left| \frac{1}{v'_\alpha(s_{2i})} \sqrt{\frac{\phi(0)}{\phi(s_{2i})}} \right|$$
(64)

(3) 在傍轴细电子束聚焦下，可令 $\phi(s)$ = 常数，便出现纯磁场的聚焦。但在宽电子束聚焦下，假定 $\phi(s=0)=0$，电子需要受到加速而聚焦，故 $\phi(s=0)>0$。因此，不存在类似傍轴细束情况下的纯磁场的聚焦。

(4) 对于正交曲傍轴系统，当 $F-R=N-Q=0$ 时，系统在某一位置处将能形成无像散的几何相似的图像，此即所谓"高斯成像"。满足这一条件的系统称为高斯（曲轴）电子光学系统。当仅满足正交条件时，系统便有可能在两个不同位置处相互垂直的方向上形成线聚焦的图像。

对于正交曲近轴系统，便要考虑主轨迹携带的逸出电子初电位 ε_s，即使 $F-R=N-Q=0$，也只有 ε_s 相同的电子才能形成点聚焦。通常依据其阴极物面发射的初能分布选取与某一初电位 $\varepsilon_s=\varepsilon_{s_1}$ 相对应的聚焦位置作为像面位置。同样，当仅满足正交条件时，以 $\varepsilon_s=\varepsilon_{s_1}$ 逸出的电子束，将有可能在两个不同的位置处相互垂直的方向上形成线聚焦的图像。

由式（26）、式（27）与式（36）可见，无论主轨迹或是相邻轨迹（曲近轴轨迹），都取与主轨迹逸出电子初电位 ε_s 相同的值。从物理意义上来说若使相邻轨迹具有与主轨迹相同的主电子速度，即使相邻近轴轨迹的电子初电位沿主轨迹的分量亦等于 ε_s 时，正交曲近轴系统才具有点聚焦或线聚焦的性质。

(5) 正交条件下的曲傍轴轨迹方程乃是分离了变量 u、v 的二阶线性齐次微分方程组，其求解并不困难。而对于正交条件下的曲近轴方程组，其 $n=\sqrt{\phi(s)+\varepsilon_s}$，鉴于在阴极物面处，$\phi(s=0)=0$，若 $\varepsilon_s=0$，便出现奇点，其求解的困难度，与研究阴极透镜的线性方程的情况是一样的。

结束语

(1) 本文研究了电子束聚焦的最一般情况——曲线坐标系下宽电子束聚焦的普遍理论，首次推导了曲线坐标系下的主轨迹方程与曲近轴轨迹方程，给出了曲近轴系统的正交条件，并研究了正交曲近轴系统的两个实例。

(2) 本文的结果可以推广到研究曲线坐标系下细电子束聚焦的问题。目前在实际中广泛应用的电磁复合聚焦阴极透镜、静电阴极透镜及其轴外电子束的聚焦，静电与电磁聚焦同心球系统等电子光学问题均属于本文提出的正交曲近轴系统的例子。

因此，本文发展了 Гринберг[1]、Sturrock[4]、Касьянков[9] 等人的关于曲线坐标系下细电子束聚焦的理论，并进一步推进了 Recknagel[15]、Арцимович[16]、Семан[6]、西门纪业[10] 与作者[13] 等人关于轴对称阴极透镜的宽电子束聚焦的理论。

(3) 对于曲近轴电子光学问题，无论是轴对称的直近轴系统，或者是满足正交条件的曲近轴系统，自阴极物面逸出的宽电子束，其主轨迹与相邻轨迹必须取同一 ε_s 值（这里 ε_s 为沿主轨迹切线方向逸出的电子初电位），系统才具有理想聚焦（点聚焦或线聚焦）的性质。

(4) 宽电子束聚焦理论与细电子束聚焦理论有相似之处和根本差别。细电子束聚焦理论运用于解决大角度偏转系统等电子光学问题;而宽电子束聚焦理论则可用来解决阴极透镜等电子光学问题。

参 考 文 献

[1] ГРИНБЕРГ Г А. Иэбранные вопросы математическои теории электрических и магнитных явлений иэд[M]. Москва-Ленинград, Изд, А Н СССР, 1948.

[2] ГРИНБЕРГ Г А. Основы обшей теории фокусируюшего действия электростатических и магнитных полеи II. Пространственные электростатические поля [J]. Изд, А Н СССР, 1942,37(9):295 − 303.

[3] ГРИНБЕРГ Г А. Основы обшей теории фокусируюшего действия статических электрических и магнитных полеи [J]. [s. n.], 1943 (13):361 − 368.

[4] STURROCK P A. Static and dynamic electron optics[M]. Cambridge: Cambridge University Press, 1955.

[5] STURROCK P A. The imaging properties of electron beams in arbitrary static electromagnetic field[J]. Philosophical Transactions, 1952, A245:155 − 187.

[6] Ceмāн О И. 电子光学理论基础[M]. 北京:高等教育出版社,1958.

[7] 安德列安戈. 电工电信工程师数学[M]. 陆志钢,等,译. 北京:人民邮电出版社,1979.

[8] КИЛЬЧВСКИИ Н А. Элементы тензорного исчисления и его применение в механике[M]. Москва:ГИТГЛ,1954.

[9] КАСЬЯНКОВ П П. Теория элекромагнитных систем с криволинеиной осыо[M]. Москва:ИЛУ,1956.

[10] 西门纪业. 复合浸没物镜的电子光学性质和像差理论[J]. 物理学报,1957,13(4):339 − 356.

[11] 周立伟,艾克聪,潘顺臣. 关于电磁复合聚焦阴极透镜的像差理论[M]. 物理学报,1983,32(3):376 − 392.

[12] ЦУККЕРМАН И И. Электронная оптика в телевидении [M]. Москва-Ленинград, ГЭИ, 1956.

[13] CHOU L W. Electron optics of concentric electromagnetic focusing systems[J]. Advances in Electronics and Electron Physics,1979,52:119 − 132.

[15] RECKNAGEL A. Theorie des elektrisohen elecktronen miktroskops fur selbstrakler [J]. Z. Angew. Physik,1941,117:679(in Germen).

[16] АРЦИМОВИЧ Л А Электростатические свойства эмисионных систем[J]. Физ., А Н СССР Сер., 1944,8(6):313.

5.4 A Generalized Theory of Wide Electron Beam Focusing
宽电子束聚焦的普遍理论

Abstract: In the present paper, based on the general electron trajectory equation in a curvilinear coordinate system, considering the fact that the object (cathode) is placed directly in the electrostatic and magnetic fields and the cathode surface is to be imaged with wide electron beams, thus the electrons leave the cathode with small velocities and large inclinations, equations of curvilinear "paraxial" trajectories in the Frenet local curvilinear coordinate system have been derived. The orthogonality conditions for curvilinear "paraxial" systems with wide electron beam focusing are given and the electron optical properties of orthogonal system are studied. Two special cases of systems with wide electron beam focusing around the principal trajectory which is a linear axis of symmetry or an axis for a two-dimensional field with a plane symmetry have been discussed, and a practical computational example for three-electrode electrostatic cathode lens has also been given in this paper.

摘要：本文由局部曲线坐标下的普遍轨迹方程出发，考虑阴极物面电位 $\phi(0)=0$ 且自阴极面逸出电子速度很小而斜率可能很大的宽电子束的情况，推导了 Frenet 局部曲线坐标系下的主轨迹方程与曲近轴轨迹方程，研究了曲轴宽电子束聚焦系统的正交条件。文中还讨论了主轨迹为直轴时轴对称场的宽电子束聚焦和平面对称的二维静电场的宽电子束聚焦系统的特例，并给出了三电极阴极透镜的计算实例。

Introduction

Electron beam focusing has long been one of the most fundamental problems in electron optics, and great attention has been paid to it. The general theory of electron beam focusing for conventional electron lenses in a curvilinear coordinate system was established by G. A. Greenberg[1] and P. A. Sturrock[2] separately in the 1940s and 1950s. G. A. Greenberg deduced the paraxial trajectory equations with an arbitrary curved optical axis, based on the fundamental equation of electron motion. P. A. Sturrock investigated electron-optical systems with a curved optical axis, where the orthogonality condition was satisfied, applying the variation principle. Since then, a great deal has been done during the past decades[3-6].

Strictly speaking, the above-mentioned theory of narrow electron beam focusing with a curved optical axis, while satisfying the curved paraxial condition, is inadequate for the study of wide

---

Zhou Liwei. Beijing Institute of Technology. Advances in Electronics and Electron Physics (电子学与电子物理学的进展) V. 64B 1985, 574–589.

electron beam focusing (such as in the case of cathode lenses), in which the object (cathode) is placed directly in the electrostatic and magnetic fields and the cathode surface is to be imaged with a wide electron beam, so that electrons leave the cathode with small velocity and large inclination. Therefore, it is necessary to have a further look at the general theory of wide electron beam focusing with curved optical axis.

Similarly to cathode lenses with axial symmetry, we assume that in the system with wide electron beam focusing the curvilinear trajectories, being adjacent to the principal trajectory, satisfy the following "curved 'paraxial' condition" everywhere:

$$p^2(s) \approx 0, \quad p(s) \ll \rho(s), \quad p(s) \ll \tau(s) \tag{1}$$

$$\dot{p}^2(s) \ll 1 \tag{2}$$

and $\phi(0) = 0$. It is not necessary to assume that the condition $p'^2(s) \ll 1$ is satisfied everywhere. Here, s denotes the arc length of the curved optical axis from the object point, $p(s)$ is the altitude of the neighboring trajectories measured from the curved optical axis, $\rho(s)$, $\tau(s)$ are the radius of curvature and radius of torsion of the curved axis, respectively. Dot and prime notations are adopted to denote derivatives with respect to the time t and arc length s, respectively. In fact, the so called "curved 'paraxial' conditions" for systems with wide electron beam focusing regard $p(s)$, $\dot{p}(s)$ as the first order terms, and p^2, $\dot{p}^2$ and $\dot{p}p$ as the second order terms.

In the present article, based on the general trajectory equations solved by the tensor method in a Frenet local coordinate system, equations of the principal trajectory and the curvilinear "paraxial" trajectories are deduced. The orthogonality condition of a curvilinear "paraxial" system and the electron-optical properties of orthogonal curvilinear "paraxial" systems are studied and some practical examples of wide electron beam focusing satisfying the orthogonality condition will be given.

Problems concerned with narrow electron beam focusing in a curvilinear coordinate system or those concerned with wide electron beam focusing with axial and plane symmetry can be regarded as special cases thereof.

1 Electron trajectory equation in a curvilinear coordinate system

Let $r = r(s)$ represents the principal trajectory of an electron emitted from an object point at the cathode surface. The normal plane through an arbitrary point N on the curved axis intersects the neighboring curvilinear trajectory at point N^*. It follows that the position of N^* can be determined by a vector p from N to N^*. If the expression $p = p(s)$ is known, the neighboring curvilinear trajectory will be completely defined.

Introduce a Frenet local coordinate system (s, p_2, p_3) and denote the unit vectors at the point N by t, n, b for the tangent, principal normal, and bi-normal directions, respectively. The curvature radius and torsion radius of a principal trajectory r are expressed by $\rho = \rho(s)$ and $\tau = \tau(s)$. As mentioned above, the vector p is placed in the normal plane orthogonal to the tangent of the principal trajectory, therefore its components at the principal normal and bi-normal directions are p_2 and p_3, respectively.

The vector $\boldsymbol{r}^*$ at point N^* on the neighboring trajectory can be expressed by
$$\boldsymbol{r}^* = \boldsymbol{r} + \boldsymbol{p} = \boldsymbol{r} + p_2 \boldsymbol{n} + p_3 \boldsymbol{b} \tag{3}$$

Differentiating Eq. (3), using the Serret-Frenet equation of a curve, remembering that $(ds^*)^2 = (d\boldsymbol{r}^*)^2$ and expressing it in tensor form, we may have the components of the covariant and contravariant metric tensors. Then, from the contravariant components of Lorentz's equation in the nonrelativistic case, we can derive general trajectory equations in a curvilinear coordinate system (s, p_2, p_3), which is convenient for the study of wide electron beam focusing and its aberrations.

$$\frac{2\varphi_*}{\left(1 - \frac{1}{\rho}p_2\right)^2 + \left(\frac{1}{\tau}p_2\right)^2 + \left(\frac{1}{\tau}p_3\right)^2 - 2\frac{1}{\tau}p_3 p_2' + 2\frac{1}{\tau}p_2 p_3' + p_2'^2 + p_3'^2} p_2'' +$$

$$p_2' \frac{d}{ds}\left[\frac{\varphi_*}{\left(1 - \frac{1}{\rho}p_2\right)^2 + \left(\frac{1}{\tau}p_2\right)^2 + \left(\frac{1}{\tau}p_3\right)^2 - 2\frac{1}{\tau}p_3 p_2' + 2\frac{1}{\tau}p_2 p_3' + p_2'^2 + p_3'^2}\right] +$$

$$\frac{2\varphi_*}{\left(1 - \frac{1}{\rho}p_2\right)^2 + \left(\frac{1}{\tau}p_2\right)^2 + \left(\frac{1}{\tau}p_3\right)^2 - 2\frac{1}{\tau}p_3 p_2' + 2\frac{1}{\tau}p_2 p_3' + p_2'^2 + p_3'^2} \times$$

$$\left(\frac{1}{\rho} - \frac{1}{\rho^2}p_2 - \frac{1}{\tau^2}p_2 + \frac{1}{\tau^2}\tau' p_3 + \frac{\frac{1}{\rho\tau^2}p_3^2 + \frac{1}{\rho^2\tau}\rho' p_2 p_3}{1 - \frac{1}{\rho}p_2} - \frac{2\frac{1}{\rho\tau}p_3 p_2'}{1 - \frac{1}{\rho}p_2} - 2\frac{1}{\tau}p_3'\right)$$

$$= \frac{1}{\left(1 - \frac{1}{\rho}p_2\right)^2}\left\{\frac{1}{\tau}p_3 \frac{\partial\varphi}{\partial s} + \left[\left(1 - \frac{1}{\rho}p_2\right)^2 + \left(\frac{1}{\tau}p_3\right)^2\right]\frac{\partial\varphi}{\partial p_2} - \frac{1}{\tau^2}p_2 p_3 \frac{\partial\varphi}{\partial p_3}\right\} +$$

$$\frac{1}{1 - \frac{1}{\rho}p_2}\sqrt{\frac{2e\varphi_*/m_0}{\left(1 - \frac{1}{\rho}p_2\right)^2 + \left(\frac{1}{\tau}p_2\right)^2 + \left(\frac{1}{\tau}p_3\right)^2 - 2\frac{1}{\tau}p_3 p_2' + 2\frac{1}{\tau}p_2 p_3' + p_2'^2 + p_3'^2}} \times$$

$$\left(\left(p_3' + \frac{1}{\tau}p_2\right)\frac{\partial\Omega}{\partial s} - \left\{\left[\left(1 - \frac{1}{\rho}p_2\right)^2 + \left(\frac{1}{\tau}p_2\right)^2 + \left(\frac{1}{\tau}p_3\right)^2\right] - \frac{1}{\tau}p_3 p_2' + \frac{1}{\tau}p_2 p_3'\right\}\frac{\partial\Omega}{\partial p_3}\right),$$

$$\frac{2\varphi_*}{\left(1 - \frac{1}{\rho}p_2\right)^2 + \left(\frac{1}{\tau}p_2\right)^2 + \left(\frac{1}{\tau}p_3\right)^2 - 2\frac{1}{\tau}p_3 p_2' + 2\frac{1}{\tau}p_2 p_3' + p_2'^2 + p_3'^2} p_3'' +$$

$$p_3' \frac{d}{ds}\left[\frac{\varphi_*}{\left(1 - \frac{1}{\rho}p_2\right)^2 + \left(\frac{1}{\tau}p_2\right)^2 + \left(\frac{1}{\tau}p_3\right)^2 - 2\frac{1}{\tau}p_3 p_2' + 2\frac{1}{\tau}p_2 p_3' + p_2'^2 + p_3'^2}\right] +$$

$$\frac{2\varphi_*}{\left(1 - \frac{1}{\rho}p_2\right)^2 + \left(\frac{1}{\tau}p_2\right)^2 + \left(\frac{1}{\tau}p_3\right)^2 - 2\frac{1}{\tau}p_3 p_2' + 2\frac{1}{\tau}p_2 p_3' + p_2'^2 + p_3'^2} \times$$

$$\left(-\frac{1}{\tau^2}\tau' p_2 - \frac{1}{\tau^2}p_3 - \frac{\frac{1}{\tau\rho^2}\rho' p_2^2 + \frac{1}{\tau^2\rho}p_2 p_3}{1 - \frac{1}{\rho}p_2} + \frac{2\frac{1}{\tau}p_2'}{1 - \frac{1}{\rho}p_2}\right)$$

$$= \frac{1}{\left(1 - \frac{1}{\rho}p_2\right)^2}\left\{-\frac{1}{\tau}p_2\frac{\partial\varphi}{\partial s} - \frac{1}{\tau^2}p_2 p_3\frac{\partial\varphi}{\partial p_2} + \left[\left(1 - \frac{1}{\rho}p_2\right)^2 + \left(\frac{1}{\tau}p_2\right)^2\right]\frac{\partial\varphi}{\partial p_3}\right\} +$$

$$\frac{1}{1 - \frac{1}{\rho}p_2}\sqrt{\frac{2e\varphi_*/m_0}{\left(1 - \frac{1}{\rho}p_2\right)^2 + \left(\frac{1}{\tau}p_2\right)^2 + \left(\frac{1}{\tau}p_3\right)^2 - 2\frac{1}{\tau}p_3 p_2' + 2\frac{1}{\tau}p_2 p_3' + p_2'^2 + p_3'^2}} \times$$

$$\left(\left\{\left[\left(1 - \frac{1}{\rho}p_2\right)^2 + \left(\frac{1}{\tau}p_2\right)^2 + \left(\frac{1}{\tau}p_3\right)^2\right] - \frac{1}{\tau}p_3 p_2' + \frac{1}{\tau}p_2 p_3'\right\}\frac{\partial\Omega}{\partial p_2} - \left(p_2' - \frac{1}{\tau}p_3\right)\frac{\partial\Omega}{\partial s}\right) \quad (4)$$

where e is the magnitude of electron charge; m_0 is the electron mass; and $\varphi_* = \varphi + \varepsilon_0$, where ε_0 is the initial potential of electrons emitted from the cathode surface, and φ is the potential of the electrostatic field. The magnetic field H can be defined by a scalar magnetic potential Ω: $H = -\mathrm{grad}\,\Omega$.

2 Equations of the principal trajectory and its neighboring "paraxial" trajectories

In an electron-optical system with wide electron beam focusing, we consider the cathode with null electrostatic potential to be where the electron has small velocity. Now, we introduce the "curved 'paraxial' conditions", Eq. (1) and Eq. (2), for wide electron beam focusing into Eq. (4), regarding p_2, $\dot{p}_2 p_3$, and $\dot{p}_3$ as the first order terms and using the method given by Ximen J Ye and Zhou LiWei[7,8] investigate each of the coefficients in Eq. (4).

The electrostatic potential φ and magnetic scalar potential Ω are expanded in power series in p_2 and p_3 with coefficients that are functions of s, we have

$$\varphi(s,p_2,p_3) = \phi(s) + p_2\varphi_2(s) + p_3\varphi_3(s) + \frac{1}{2}p_2^2\varphi_{22}(s) + p_2 p_3\varphi_{23}(s) + \frac{1}{2}p_3^2\varphi_{33}(s) + \cdots \quad (5)$$

$$\Omega(s,p_2,p_3) = \Omega(s) + p_2\Omega_2(s) + p_3\Omega_3(s) + \frac{1}{2}p_2^2\Omega_{22}(s) + p_2 p_3\Omega_{23}(s) + \frac{1}{2}p_3^2\Omega_{33}(s) + \cdots$$

$$(6)$$

where the symbols introduced are

$$\varphi\bigg|_{\substack{p_2=0\\p_3=0}} = \varphi(s,0,0) = \phi(s),\quad \Omega\bigg|_{\substack{p_2=0\\p_3=0}} = \Omega(s,0,0) = \Omega(s) \quad (7)$$

$$\frac{\partial\varphi}{\partial p_i}\bigg|_{\substack{p_2=0\\p_3=0}} = \varphi_i,\ \frac{\partial\Omega}{\partial p_i}\bigg|_{\substack{p_2=0\\p_3=0}} = \Omega_i,\ \frac{\partial\varphi}{\partial p_i \partial p_j}\bigg|_{\substack{p_2=0\\p_3=0}} = \varphi_{ij},\ \frac{\partial^2\Omega}{\partial p_i \partial p_j}\bigg|_{\substack{p_2=0\\p_3=0}} = \Omega_{ij} \quad (8)$$

where

$$\varphi_* = \varphi(s,p_2,p_3) + \varepsilon_0,\ \varphi(0,0,0) = \phi(0) = 0,\ \varphi_1 = \frac{d\varphi}{ds}$$

Now, we examine the common coefficient of the first terms on the right-hand side of Eq.

(4) which is related to the tangent velocity $\dot{s}$. It may be seen that this coefficient at the initial point can be written as

$$\left[\frac{\varphi_*}{\left(1-\frac{1}{\rho}p_2\right)^2+\left(\frac{1}{\tau}p_2\right)_2+\left(\frac{1}{\tau}p_3\right)^2-2\frac{1}{\tau}p_3p_2'+2\frac{1}{\tau}p_2p_3'+p_2'^2+p_3'^2}\right]_{s=0,p_2=0,p_3=0} \quad (9)$$

$$=\left(\frac{\varepsilon_0}{1+p_2'^2+p_3'^2}\right)_{s=0,p_2=0,p_3=0}=\varepsilon_s$$

Therefore, its the first order approximation can be expressed as

$$\left[\frac{\varphi_*}{\left(1-\frac{1}{\rho}p_2\right)^2+\left(\frac{1}{\tau}p_2\right)_2+\left(\frac{1}{\tau}p_3\right)^2-2\frac{1}{\tau}p_3p_2'+2\frac{1}{\tau}p_2p_3'+p_2'^2+p_3'^2}\right]_{0+1} \quad (10)$$

$$=\varepsilon_s+\phi+p_2\varphi_2+p_3\varphi_3+(\varepsilon_s+\phi)\frac{2}{\rho}p_2$$

Similarly, after a series of rather complicated manipulations, we obtain the trajectory equations of the zeroth order approximation (in which $p_2=p_3=0$) as follows:

$$\frac{2[\phi+\varepsilon_s]}{\rho}=\varphi_2-\sqrt{2e/m_0}\sqrt{\phi+\varepsilon_s}\Omega_3$$

$$\varphi_3=-\sqrt{2e/m_0}\sqrt{\phi+\varepsilon_s}\Omega_2 \quad (11)$$

Eq. (11) are called the equations of the principal trajectroy (the axis of the electron beam or the axis of the system).

When Eq. (11) are satisfied, we may obtain the trajectory equations of the first order approximation, which represent equations for the neighboring trajectory:

$$\frac{d}{ds}\left(n\frac{dp_2}{ds}\right)=Fp_2+Np_3+2K\frac{dp_3}{ds}$$

$$\frac{d}{ds}\left(n\frac{dp_3}{ds}\right)=Rp_3-Qp_2-2K\frac{dp_2}{ds} \quad (12)$$

where n, K, F, R, N, and Q express the following functions at the curve of principal trajectory, similar to those given by I. I. Tsukermann[9]:

$$n=\sqrt{\phi+\varepsilon_s},\quad K=\frac{\sqrt{\phi+\varepsilon_s}}{\tau}+\sqrt{\frac{e}{8m_0}}\Omega_1,$$

$$F=\sqrt{\phi+\varepsilon_s}\left(\frac{1}{\tau^2}-\frac{1}{\rho^2}\right)+\frac{\varphi_{22}}{2\sqrt{\phi+\varepsilon_s}}-\frac{\varphi_2}{\rho\sqrt{\phi+\varepsilon_s}}+\sqrt{\frac{e}{2m_0}}\left[\frac{\Omega_1}{\tau}-\Omega_{23}-\frac{\Omega_3\varphi_2}{2(\phi+\varepsilon_s)}\right],$$

$$R=\frac{\sqrt{\phi+\varepsilon_s}}{\tau^2}+\frac{\varphi_{33}}{2\sqrt{\phi+\varepsilon_s}}+\sqrt{\frac{e}{2m_0}}\left[\frac{\Omega_1}{\tau}+\Omega_{23}+\frac{\Omega_2\varphi_3}{2(\phi+\varepsilon_s)}\right],$$

$$N=\frac{\varphi_1}{2\tau\sqrt{\phi+\varepsilon_s}}-\frac{\sqrt{\phi+\varepsilon_s}}{\tau^2}\frac{d\tau}{ds}-\frac{\varphi_3}{\rho\sqrt{\phi+\varepsilon_s}}+\frac{\varphi_{23}}{2\sqrt{\phi+\varepsilon_s}}-\sqrt{\frac{e}{2m_0}}\left[\Omega_{33}+\frac{\Omega_3\varphi_3}{2(\phi+\varepsilon_s)}\right],$$

$$Q=\frac{\varphi_1}{2\tau\sqrt{\phi+\varepsilon_s}}-\frac{\sqrt{\phi+\varepsilon_s}}{\tau^2}\frac{d\tau}{ds}-\frac{\varphi_{23}}{2\sqrt{\phi+\varepsilon_s}}-\sqrt{\frac{e}{2m_0}}\left[\Omega_{22}+\frac{\Omega_2\varphi_2}{2(\phi+\varepsilon_s)}\right] \quad (13)$$

It must be pointed out that the following relationship holds among the coefficients N, Q and K in Eq. (13):

$$(N+Q)/2 = \frac{dK}{ds} \tag{14}$$

Eq. (12) can be used to describe the neighboring trajectories with rather large initial slope, emitted from the cathode surface with potential $\phi(0) = 0$. We call them curvilinear "paraxial" trajectory equations.

It is to be noted that the value of the coefficients n, K, F, R, N, and K must be taken from the principal trajectory when solving Eq. (12). Thus, we usually solve Eq. (11) and Eq. (12) simultaneously. Besides, in the principal trajectory equations and in the curvilinear "paraxial" trajectory equations the same value for ε_s must be taken.

3 The orthogonality condition of systems with wide electron beam focusing

Now we derive the orthogonality condition for systems with wide electron beam focusing. Introducing the coordinate transformation

$$p_2 + ip_3 = (u + iv)e^{i\chi} \tag{15}$$

to Eq. (12) and supposing that

$$\chi' = \frac{d\chi}{ds} = -K/n \tag{16}$$

we obtain the curvilinear "paraxial" trajectory equations in a rotating coordinate system (u, v), but the variables u and v in these two equations are still not separated. We can prove that the variables will be completely separated if

$$\tan 2\chi = (N-Q)/(F-R) \tag{17}$$

Finally from Eq. (12) and using Eq. (15) - Eq. (17) we have curvilinear "paraxial" trajectory equations, in which the functions u, v are separated:

$$\frac{d}{ds}n\left(\frac{du}{ds}\right) = Uu,$$

$$\frac{d}{ds}n\left(\frac{dv}{ds}\right) = Vv \tag{18}$$

where

$$U = \frac{1}{2}(F+R) + \frac{1}{2}\sqrt{(F-R)^2 + (N-Q)^2} - (K^2/n)$$

$$V = \frac{1}{2}(F+R) - \frac{1}{2}\sqrt{(F-R)^2 + (N-Q)^2} - (K^2/n) \tag{19}$$

From Eq. (16) and Eq. (17), we can derive the conditional relationship, which the coefficients n, K, F, R, N and Q must satisfy, when the variables are separated:

$$n[(N-Q)(F'-R') - (F-R)(N'-Q')] = 2K[(F-R)^2 + (N-Q)^2] \tag{20}$$

This is the orthogonality condition for systems with wide electron beam focusing. Systems, which satisfy the orthogonality condition (20), will be termed orthogonal curvilinear "paraxial" systems. Equations (18) are the second order homogeneous linear differential equations in u and v, respectively. The general solutions can be expressed by two linear independent special solutions u_α, u_β and v_α, v_β, which satisfy the following conditions:

$$u_\alpha(0) = v_\alpha(0) = 0, \quad \sqrt{\varepsilon_s}u'_\alpha(0) = \sqrt{\varepsilon_s}v'_\alpha(0) = 1,$$
$$u_\beta(0) = v_\beta(0) = 1, \quad \sqrt{\varepsilon_s}u'_\beta(0) = \sqrt{\varepsilon_s}v'_\beta(0) = 0 \quad (21)$$

In order to obtain an image of a point, it is obviously necessary to make U and V in Eq. (18) equal to each other, which becomes possible only when

$$F - R = N - Q = 0 \quad (22)$$

Then, at the image plane, $s = s_i$, which corresponds to $\varepsilon_s = \varepsilon_{s_1}$, that is, the magnifications in the two directions will be the same.

Eq. (22) shows that the electron beam emitted from the cathode surface with $\varepsilon_s = \varepsilon_{s_1}$ will have ideal point focusing (stigmatic imaging), therefore the system will have properties similar to those of an electron-optical system with axial symmetry.

If a system with wide electron beam focusing only satisfies the orthogonality condition Eq. (20), but does not satisfy the condition Eq. (22) (thus $U \neq V$), the special solutions u_α, u_β and v_α, v_β are different. This means anisotropy in magnification and the positions s_{1i} and s_{2i} do not coincide with each other even if the initial energies ($\varepsilon_s = \varepsilon_{s_1}$) are the same. It turns out that a focusing segment is formed in the bi-normal direction b at the location $s = s_{1i}$ and another focusing segment is formed in the normal direction n at the location $s = s_{2i}$. Thus, the difference between the two locations can be defined as the astigmatism of the system.

4 Some Practical examples with Wide Electron Beam Focusing

4.1 A system in which the principal trajectory is a linear axis of symmetry

When the principal trajectory is a linear axis of symmetry, $1/\rho = 1/\tau = 0$. In axially symmetric fields, the potential on the axis takes the extreme value, and there is no difference between the normal and bi-normal directions.

Let $p = p_2 + ip_3$, $s = z$, $\varepsilon_s = \varepsilon_z$, $\phi(s) = \phi(z)$, $\Omega_1 = d\Omega/ds = -H(z)$, we then obtain

$$n = \sqrt{\phi(z) + \varepsilon_z}, \quad K = -\sqrt{\frac{e}{8m_0}}H(z)$$
$$F = R = -\frac{1}{4}\frac{\phi''}{\sqrt{\phi(z) + \varepsilon_z}}, \quad N - Q = 0 \quad (23)$$

Then,

$$U = V = -\frac{1}{4}\frac{\phi''}{\sqrt{\phi(z) + \varepsilon_z}} - \frac{e}{8m_0}\frac{H^2(z)}{\sqrt{\phi(z) + \varepsilon_z}} \quad (24)$$

It is obvious from Eq. (23) that a system with a linear axis of symmetry not only satisfies the condition Eq. (20), but also satisfies the condition Eq. (22). Thus, this system has ideal focusing properties (point focusing). Let

$$\boldsymbol{u} = u + \mathrm{i}v = \boldsymbol{p}\mathrm{e}^{-\mathrm{i}\chi} \tag{25}$$

where

$$\chi' = \sqrt{\frac{e}{8m_0}} \frac{H(z)}{\sqrt{\phi(z) + \varepsilon_z}} \tag{26}$$

Eq. (18) will then take the following form:

$$[\phi(z) + \varepsilon_z]\boldsymbol{u}'' + \frac{1}{2}\phi'(z)\boldsymbol{u}' + \frac{1}{4}\left[\phi'' + \frac{e}{2m_0}H^2(z)\right]\boldsymbol{u} = 0 \tag{27}$$

This is the linear differential equation of a system with wide electron beam focusing[7], where the principal trajectory is a straight axis having axial symmetry. Using Eq. (27), we may then solve a concentric spherical system of combined electrostatic and magnetic fields [10].

4.2 Electrostatic cathode lenses

For electrostatic cathode lenses with axial symmetry, the off-axis principal trajectory emitted from the cathode surface will be a plane curve which is normal to the cathode surface. Thus, $\Omega = 0$, $1/\tau = 0$, $\varphi_{23} = 0$. Then from Eq. (13) we have

$$n = \sqrt{\phi(s) + \varepsilon_s}, \quad K = 0,$$

$$F = -\frac{3}{\rho^2}\sqrt{\phi(s) + \varepsilon_s} + \frac{\varphi_{22}}{2\sqrt{\phi(s) + \varepsilon_s}},$$

$$R = \frac{\varphi_{33}}{2\sqrt{\phi(s) + \varepsilon_s}}, \quad N = Q = 0 \tag{28}$$

The coefficients in Eq. (28) also satisfy the orthogonality condition Eq. (20). It is obvious that the variables u and v are also separated.

Because $\chi = 0$, suppose that $u = p_2$, $v = p_3$. Then, the principal trajectory equation can be written as

$$\varphi_2 = \frac{2[\phi(s) + \varepsilon_s]}{\rho} \tag{29}$$

The curvilinear neighboring trajectory equations would become

$$\frac{\mathrm{d}}{\mathrm{d}s}\left(\sqrt{\phi(s) + \varepsilon_s}\frac{\mathrm{d}p_2}{\mathrm{d}s}\right) = \left\{-\frac{3\sqrt{\phi + \varepsilon_s}}{\rho^2} + \frac{\varphi_{22}}{2\sqrt{\phi(s) + \varepsilon_s}}\right\}p_2,$$

$$\frac{\mathrm{d}}{\mathrm{d}s}\left(\sqrt{\phi(s) + \varepsilon_s}\frac{\mathrm{d}p_3}{\mathrm{d}s}\right) = \frac{\varphi_{33}}{2\sqrt{\phi(s) + \varepsilon_s}}p_3 \tag{30}$$

The arc length s, which defines a coordinate curve of the principal trajectory, can be transformed into a form with cylindrical coordinates (r, z). From Eq. (29) and Eq. (30), we obtain the principal trajectory equation

$$r' = \frac{1 + r'^2}{2[\varphi(z,r) + \varepsilon_s]}\left(\frac{\partial \varphi}{\partial r} - r'\frac{\partial \varphi}{\partial z}\right) \tag{31}$$

and the curvilinear neighboring trajectory equations

$$p_2'' + F_1 p_2' + F_2 p_2 = 0 \tag{32a}$$

$$p_3'' + G_1 p_3' + F_2 p_3 = 0 \tag{32b}$$

where

$$F_1 = G_1 = \frac{1 + r'^2}{2[\varphi(z,r) + \varepsilon_s]} \frac{\partial \varphi}{\partial z},$$

$$F_2 = \frac{3r''^2}{(1 + r'^2)^2} + \frac{1}{2[\varphi(z,r) + \varepsilon_s]} \left(-\frac{\partial^2 \varphi}{\partial r^2} + 2r' \frac{\partial^2 \varphi}{\partial r \partial z} - r'^2 \frac{\partial^2 \varphi}{\partial z^2} \right),$$

$$G_2 = \frac{-(1 + r'^2)}{2[\varphi(z,r) + \varepsilon_s]} \frac{1}{r} \frac{\partial \varphi}{\partial r} \tag{33}$$

It is necessary to remember that the terms in the coefficients F_1, G_1, F_2, G_2 should be taken to be the values on the principal trajectory, and that the prime notations in Eq. (31), Eq. (32) and Eq. (33) denote derivatives with respect to the axial coordinate z.

For an electron beam emitted from the axial point of the cathode surface, because the principal trajectory is actually the axis of rotational symmetry, we will have $r = r' = r'' = 0$, and

$$\left(\frac{1}{r} \frac{\partial \varphi}{\partial r} \right)_{r=0} = \left(\frac{\partial^2 \varphi}{\partial r^2} \right)_{r=0} = -\frac{1}{2} \phi''(z) \tag{34}$$

Let $p = p_2 + \mathrm{i} p_3$, $\varepsilon_z = \varepsilon_s$, then Eq. (32) takes the following form:

$$p'' + \frac{1}{2} \frac{\phi'}{\phi(z) + \varepsilon_z} p' + \frac{1}{4} \frac{\phi''}{\phi(z) + \varepsilon_z} p = 0 \tag{35}$$

which is the well-known "paraxial" equation for electrostatic cathode lenses.

4.3 A two-dimensional field with the plane symmetry

Let the principal trajectory be an axis of symmetry for a two-dimensional field with the plane symmetry. Then, the curvilinear coordinate system (p_2, p_3, s) can be replaced by the Cartesian coordinate system (x, y, z), and the coordinate z, the axis of symmetry for the system, coincides with the principal trajectory. Therefore, $1/\rho = 1/\tau = 0$.

For a two-dimensional field, which is independent of the coordinate x, we have

$$\varphi_2 = \varphi_{22} = \varphi_{23} = 0, \quad \Omega_2 = \Omega_{22} = \Omega_{23} = 0 \tag{36}$$

In the plane of symmetry, the derivatives of electrostatic and magnetic scalar potentials with respect to y equal zero, thus $\varphi_3 = \Omega_3 = 0$,

$$\varphi_{33} = -\varphi_{11} = -\phi'', \quad \Omega_{33} = -\Omega_{11} = -\Omega'' \tag{37}$$

It will be seen from Eq. (20) that this system does not satisfy the orthogonality condition. Let $\Omega_1 = \Omega' = -H_z = -H$, then the curvilinear neighboring trajectory Eq. (12) can be written as

$$\left(\sqrt{\phi + \varepsilon_z} x' + \sqrt{\frac{e}{2m_0}} H y \right)' = 0, \tag{38a}$$

$$(\sqrt{\phi + \varepsilon_z} y')' + \frac{\phi''}{2 \sqrt{\phi + \varepsilon_z}} y - \sqrt{\frac{e}{2m_0}} H x' = 0 \tag{38b}$$

Integrating Eq. (38a) and substituting into Eq. (38b), we have

$$x' = -\sqrt{\frac{e}{2m_0}}\frac{1}{\sqrt{\phi+\varepsilon_z}}Hy + \frac{C}{\sqrt{\phi+\varepsilon_z}},$$

$$y'' + \frac{\phi'}{2(\phi+\varepsilon_z)}y' + \left[\frac{\phi''}{2(\phi+\varepsilon_z)} + \frac{eH^2}{2m_0(\phi+\varepsilon_z)}\right]y = \sqrt{\frac{e}{2m_0}}C\frac{H}{\phi+\varepsilon_z} \quad (39)$$

where

$$C = \sqrt{\varepsilon_z}x'_0 + \sqrt{e/2m_0}H_0 y_0 \quad (40)$$

From Eq. (39), we can derive the solution for a two-dimensional field with wide electron beam focusing. For the case of an electrostatic field of plane symmetry with wide electron beam focusing, we obtain

$$x' = \frac{x'_0\sqrt{\varepsilon_z}}{\sqrt{\phi(z)+\varepsilon_z}},$$

$$y'' + \frac{\phi'}{2[\phi(z)+\varepsilon_z]}y' + \frac{\phi''}{2[\phi(z)+\varepsilon_z]}y = 0 \quad (41)$$

It is obvious that the variables and y are separated. This is sufficient to prove that this kind of focusing system, which is actually the so-called cylindrical cathode lens, will focus only a plane electron beam located in the plane (y, z). In the direction perpendicular to the plane (y, z), there is no applied force due to the electric field. Thus, an electron beam emitted from the axial point of the cathode surface will be affected only in a direction perpendicular to the plane of symmetry. It will not be a point focusing as in a field with axial symmetry, but will produce a line focus in the direction perpendicular to the plane (y, z).

5 A practical computational example

We now give actual computations for three electrode cathode lens with electrostatic focusing[11] using the principal trajectory Eq. (31) and the curvilinear "paraxial" neighboring trajectory Eq. (32). The electrode structure and applied voltages are shown in Fig. 1.

For the image position and magnification of the system, we take $\varepsilon_s = 0.09$ V. According to the "paraxial" trajectory Eq. (35) we have $z_i = 73.127, 3$ mm, $M = -0.713, 16$.

Table 1 and Fig. 1 give the positions of tangential focus and sagittal focus and its corresponding astigmatism at different altitudes.

Using Eq. (31) and Eq. (32) we may compute some results for the principal trajectory and the tangential and sagittal "paraxial" neighboring trajectories at different altitudes (see Table 2).

The following initial conditions apply:

for the principal trajectory,

$$z = z_0, \quad r = r_0, \quad r' = r'_0$$

for the tangential trajectory,

$$(\bar{p}_2)_0 = (\bar{p}_3)_0 = 0$$

$$(\bar{p}'_2)_0 = \tan\alpha_0 \sqrt{1+r_0'^2}, \quad (\bar{p}'_3)_0 = 0 \quad (42)$$

Table 1 Positions of tangential focus and sagittal focus and the corresponding astigmatism at different altitudes   mm

Initial positions		Tangential focus		Sagittal focus		Astigmatism
r	z	z_T	r_T	z_s	r_s	η
0	0	73.127,3	0	73.127,3	0	0
2	0.07,026,2	73.148,2	−1.428,39	73.170,9	−1.429,24	0.022,77
4	0.282,098	72.678,1	−2.832,99	72.964,5	−2.854,23	0.287,20
6	0.638,736	72.317,1	−4.237,68	72.534,4	−4.262,06	0.218,67
8	1.145,841	71.811,3	−5.624,01	72.298,0	−5.697,72	0.492,29
10	1.811,988	71.087,9	−6.972,77	71.912,8	−7.131,58	0.840,10
12	2.649,468	69.856,0	−8.193,15	71.483,1	−8.578,02	1.672,07

for the sagittal trajectory,

$$(\bar{\bar{p}}_2)_0 = (\bar{\bar{p}}_3)_0 = 0,$$
$$(\bar{\bar{p}}'_3)_0 = \tan\alpha_0 \sqrt{1+r_0'^2}, \quad (\bar{\bar{p}}'_2)_0 = 0 \tag{43}$$

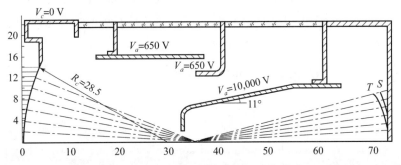

Fig. 1 The tangential and sagittal foci of a three-electrode cathode lens with electrostatic focusing

Table 2 The computed results for the principal trajectory and the tangential and sagittal "paraxial" neighboring trajectories   mm

Z	r	r'	$\bar{p}_2$	$\bar{p}'_2$	$\bar{\bar{p}}_3$	$\bar{\bar{p}}'_3$
1.811,99	10.000,00	−0.374,70	0	3.392,09	0	3.392,09
5.046,09	8.822,10	−0.353,26	0.537,44	0.077,26	0.536,18	0.076,38
11.046,09	6.790,55	−0.327,18	0.826,77	0.029,11	0.817,69	0.027,58
17.046,09	4.857,74	−0.319,72	0.924,17	0.003,66	0.907,50	0.003,04
23.046,09	2.958,39	−0.308,95	0.870,63	−0.021,06	0.854,79	−0.020,40
29.046,09	1.241,15	−0.253,61	0.690,17	−0.035,62	0.677,64	−0.035,40

Continued

Z	r	r'	$\bar{p}_2$	$\bar{p}_2'$	$\bar{\bar{p}}_3$	$\bar{\bar{p}}_3'$
35.171,09	-0.058,49	-0.192,67	0.539,44	-0.015,37	0.527,12	-0.014,70
41.171,09	-1.213,37	-0.192,49	0.448,89	-0.015,02	0.440,61	-0.014,34
47.296,09	-2.392,63	-0.192,50	0.356,97	-0.015,01	0.352,80	-0.014,33
53.296,09	-3.547,68	-0.192,51	0.266,94	-0.015,00	0.266,81	-0.014,33
59.296,09	-4.702,73	-0.192,51	0.176,92	-0.015,00	0.180,82	-0.014,33
71.087,90	-6.972,77	-0.192,51	0	-0.015,00	0.011,83	-0.014,33
71.912,80	-7.131,58	-0.192,51	-0.012,37	-0.015,00	0	-0.014,33

Conclusions

In this article, a generalized theory of wide electron beam focusing in a curvilinear coordinate system is investigated. Equations of the principal trajectory and the curvilinear "paraxial" trajectories are deduced. The orthogonality condition for a curvilinear "paraxial" system with wide electron beam focusing is given and the electron-optical properties of orthogonality systems are studied. Some practical examples of wide electron beam focusing satisfying the orthogonality condition including the computational results for a three-electrode electrostatic cathode lens are also given.

Problems involving narrow electron beam focusing in curvilinear coordinate systems or wide electron beam focusing with axial symmetry can be regarded as special cases of this generalized treatment.

Thus, this article tries to develop further the theory of wide electron beam focusing with axial or spherical symmetry, given by A. Recknagel[12], Ximen Ji Ye[7], and Zhou LiWei.

For problems involving wide electron beam focusing, whether it is a system with axial symmetry or a curvilinear "paraxial" system satisfying the orthogonality condition, the principal trajectory and the curvilinear "paraxial" neighboring trajectories must take the same value for ε_s, the initial energy of the principal trajectory in the tangential direction. Only then will the system give point focusing or line focusing.

There are similarities and differences between systems with narrow electron beam focusing and systems with wide electron beam focusing. The theory of narrow electron beam focusing is suitable for the problems of conventional lenses, while the theory of wide electron beam focusing is suitable to solve problems pertaining to cathode lenses.

References

[1] ГРИНБЕРГ Г А. Избранные Вопросы Математической Теории Элеских и

Магмитных Явлений[M]. Изд, А Н СССР,1948:507 - 535.

[2] STURROCK P A. Static and dynamic electron optics[M]. Cambridge: Cambridge University Press,1955.

[3] VANDAKUROV Y V. The theory of aberrations of electron-optical systems with a curvilinear axes [J]. Sov. Phys. -Tech. Phys. ,1957,2:2259.

[4] KAS' YANKOV P P. Theory of electromagnetic system with curvilinear axis [M]. Leningrad:Press of Leningrad University, 1956:50 - 57 (in Russian).

[5] HUTTER R G E. Deflection defocusing effects in cathode-ray tubes at large deflection angles[J]. IEEE Transactions on Electron Devices, 1970, 17(12):1022 - 1031.

[6] PLIES E, TYPKE D. Dreidimensional abbildende Elektronenmikroskope II. Theorie elektronenoptischer Systeme mit gekrümmter Achse[J]. Ztschrift für Naturforschung A, 1978, 33a:1361.

[7] XIMEN J Y. Electron-optical properties and aberration theory of combined immersion objectives[J]. Acta Physica Sinica, 1957,13:339 - 356.

[8] ZHOU L W, AI K C, PAN S C. On aberration theory of the combined electromagnetic focusing cathode lenses [J]. Acta Physica Sinica, 1983,32:376 - 392.

[9] TSUKERMANN I I. Electron optics in television[M]. Oxford:Pergamon Press,1961.

[10] CHOU L W. Electron optics of concentric spherical electromagnetic focusing systems [J]. Advances in Electronics and Electron Physics, 1979,52:119 - 132.

[11] ZHOU L W. Optics of wide electron beam focusing, electron optical systems [J]. Electron Optical Systems, SEM,Inc. ,1984:45 - 52.

[12] RECKNAGEL A. Theorie des elektrisohen elektronen miktroskops fur selbstrakler[J]. Z, Angew. Physik,1941,119:689 - 708(in Germen).

5.5 曲轴宽电子束聚焦的理论研究
A Theoretical Study of Wide Electron Beam Focusing Having Curvilinear Optical Axes

摘要：本文假定空间静电电位 φ 和磁感应强度矢量 B 在实验室坐标系中是已知的，利用沿主轨迹运动的非正交和正交两种曲线局部坐标系，研究了曲轴宽电子束聚焦的普遍理论。本文的数学处理对于计算机辅助设计宽电子束成像系统来说是适宜和方便的。

Abstract: A generalized theory of wide electron beam focusing with curvilinear optical axis is studied in the present paper using the non-orthogonal and orthogonal local coordinate systems moving along the principal trajectory, supposing that the spatial electrostatic potential φ and magnetic induction vector B in the laboratory coordinate system are known. The mathematical processing of this paper appears to be suitable and convenient for computer aided design of wide electron beam imaging systems.

引 言

带电粒子束的聚焦是电子光学的一个基本问题。曲线坐标系下细电子束聚焦的普遍理论已由 Г. А. Гринберг 和 P. A. Sturrock 于 20 世纪 40—50 年代分别用电动力学法和变分方法建立[1~5]。1970 年 R. G. E. Hutter[6] 用曲轴光学理论研究了彩色显像管中大角度偏转的问题。但是这些结果并不适用于研究阴极透镜中物面 $\varphi(0)=0$，且自阴极物面逸出的电子速度很小而斜率可能很大的宽电子束聚焦的问题。

近年来，在前人工作的基础上，作者利用张量方法，由曲线坐标系下的洛伦兹方程出发，导出了普遍形式的轨迹方程；并由此引入曲近轴条件，推导了适用于研究宽电子束聚焦的主轨迹方程和曲近轴轨迹方程，给出了曲近轴系统能够理想聚焦的正交条件，讨论了一些正交系统的实例，从而为宽电子束聚焦问题的研究开辟了一条新的途径[7~10]。

但是，作者以前的研究，如文献 [8] 中磁场以曲光轴上的磁标位表示，通常对于实际计算亦不方便；文献 [7] 对于由实验室坐标系的磁场分量求曲线坐标系的磁场分量并没有阐述得很清楚。因此，本文假定空间静电电位 φ 和磁感应强度矢量 B 在实验室坐标系中是已知函数，从而利用非正交和正交两种曲线局部坐标系，进一步研究宽电子束聚焦的普遍理论，以适合实际计算的要求。

1 曲线局部坐标系及场分量

如图 1 所示，以 r_N 表示主轨迹矢量，自 M_1 点逸出的主轨迹 M_1M_2 上的点 N 处作弗莱

周立伟，金伟其，倪国强．北京理工大学．光电子学技术（Optoelectronic Technology），No. 4, 1988, 8-22.

纳（Frenet）局部坐标系（t, n, b）。设在 N 点的法平面内，相邻轨迹的交点 N^* 与 N 点的矢量为 p，则相邻轨迹矢量 r^* 可以表示为

$$r^* = r_N + n = r_N + p_2 n + p_3 b \tag{1}$$

式中，p_2、p_3 为 p 在 Frenet 三面体的单位主法矢 n 和单位副法矢 b 上的投影。显然，只要求出 r_N 和 p，便能确定自 M_1 点逸出的电子束的运动规律。

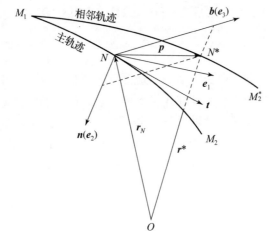

图 1 曲线坐标系的主轨迹与相邻轨迹，弗莱纳局部坐标系（t, n, b）与坐标基（e_1, e_2, e_3）

由微分式（1），利用 Frenet-Serret 公式，按文献[7]、文献[9]，我们得到

$$d r^* = ds e_1 + dp_2 e_2 + dp_3 e_3 \tag{2}$$

式中，坐标基的基矢 $e_i (i=1,2,3)$ 分别为：

$$e_1 = (1 - kp_2) t - \chi p_3 n + \chi p_2 b \tag{3a}$$

$$e_2 = n \tag{3b}$$

$$e_3 = b \tag{3c}$$

即建立了一个由弗莱纳局部坐标系的基矢（t, n, b）组合而成的非正交基，式中 k、χ 分别表示主轨迹的曲率和挠率。

由式（2）和式（3），可以得到度量张量 g_{ij} 如下[7]：

$$(g_{ij}) = \begin{pmatrix} (1-kp_2)^2 + \chi^2 p_2^2 + \chi^2 p_3^2 & -\chi p_3 & \chi p_2 \\ -\chi p_3 & 1 & 0 \\ \chi p_2 & 0 & 1 \end{pmatrix} \tag{4}$$

于是，对应的互逆基 $e^i (i=1,2,3)$ 为：

$$e^1 = \frac{1}{1-kp_2} t, \quad e^2 = \frac{\chi p_3}{1-kp_2} t + n, \quad e^3 = -\frac{\chi p_2}{1-kp_2} t + b \tag{5}$$

$$(g^{ij}) = -\frac{1}{(1-kp_2)^2} \begin{pmatrix} 1 & \chi p_3 & -\chi p_2 \\ \chi p_3 & (1-kp_2)^2 + \chi^2 p_3^2 & -\chi^2 p_2 p_3 \\ -\chi p_2 & -\chi^2 p_2 p_3 & (1-kp_2)^2 + \chi^2 p_2^2 \end{pmatrix} \tag{6}$$

按张量分析，利用上述的坐标基和互逆基，便可以表示有关的场矢量，如电场强度：

$$\boldsymbol{E}(s,p_2,p_3) = -\operatorname{grad}\varphi = -\frac{\partial \varphi}{\partial s}\boldsymbol{e}^1 - \frac{\partial \varphi}{\partial p_2}\boldsymbol{e}^2 - \frac{\partial \varphi}{\partial p_3}\boldsymbol{e}^3 \tag{7}$$

磁感应强度：

$$\boldsymbol{B}(s,p_2,p_3) = \boldsymbol{B}^i(s,p_2,p_3)\boldsymbol{e}_i = \boldsymbol{B}_i(s,p_2,p_3)\boldsymbol{e}^i \tag{8}$$

磁矢位：

$$\boldsymbol{A}(s,p_2,p_3) = \boldsymbol{A}^i(s,p_2,p_3)\boldsymbol{e}_i = \boldsymbol{A}_i(s,p_2,p_3)\boldsymbol{e}^i \tag{9}$$

上述这些在 $\boldsymbol{e}_i$ 和 $\boldsymbol{e}^i$ 前的量分别对应于该场的逆变和协变分量，它们之间的变换可以通过度量张量 $\boldsymbol{g}_{ij}$ 或逆变度量张量 $\boldsymbol{g}^{ij}$ 进行。顺便指出，上述最后两个表示式我们采用了张量分析中的爱因斯坦求和约定。

假设系统的静电电位 φ 和磁感应强度矢量 $\boldsymbol{B}$ 在实验室笛卡儿坐标系 (x,y,z) 中是已知函数，即

$$\varphi = \varphi(x,y,z),\quad \boldsymbol{B} = \boldsymbol{B}(x,y,z) = B_x(x,y,z)\boldsymbol{i} + B_y(x,y,z)\boldsymbol{j} + B_z(x,y,z)\boldsymbol{k} \tag{10}$$

式中，$\boldsymbol{i}$、$\boldsymbol{j}$、$\boldsymbol{k}$ 乃是笛卡儿直角坐标系的单位矢量。

在实验室坐标系下，可得弗莱纳局部坐标系的 $\boldsymbol{t}$、$\boldsymbol{n}$、$\boldsymbol{b}$ 在 $\boldsymbol{i}$、$\boldsymbol{j}$、$\boldsymbol{k}$ 上的投影：

$$\begin{aligned} &t_x = x'_N,\ t_y = y'_N,\ t_z = z'_N, \\ &n_x = \frac{1}{k}x''_N,\ n_y = \frac{1}{k}y''_N,\ n_z = \frac{1}{k}z''_N, \\ &b_x = \frac{1}{k}(z''_N y'_N - y''_N z'_N),\ b_y = \frac{1}{k}(x''_N z'_N - z''_N x'_N),\ b_z = \frac{1}{k}(y''_N x'_N - x''_N y'_N) \end{aligned} \tag{11}$$

主轨迹的曲率 k 和挠率 χ 亦可表示为

$$\begin{aligned} k &= (x''^2_N + y''^2_N + z''^2_N)^{1/2}, \\ \chi &= \frac{1}{k^2}\begin{vmatrix} x'_N & y'_N & z'_N \\ x''_N & y''_N & z''_N \\ x'''_N & y'''_N & z'''_N \end{vmatrix} \end{aligned} \tag{12}$$

上述各式中撇号"′"表示对主轨迹弧长 s 的微商。

将 N^* 点处的电位 φ 在曲光轴的 N 点处以 p_2、p_3 的幂次展开，其系数乃是弧长 s 的函数，则其表达式为

$$\varphi(s,p_2,p_3) = \sum_{i=0}^{\infty}\sum_{j=0}^{\infty}\frac{\varphi_{0ij}}{i!j!}p_2^i p_3^j \tag{13}$$

令 $\varphi(0,0,0) = \phi(s)$，即沿主轨迹曲光轴的电位，$\varphi_{100} = \phi'(s) = \mathrm{d}\phi/\mathrm{d}s$。式（13）中的量 φ_{0ij} 可以用实验室笛卡儿坐标系的已知偏微商来表达：

$$\begin{aligned} \varphi_{0ij} = \varphi_{oij}(s) &= (\boldsymbol{n}\cdot\nabla)^i(\boldsymbol{b}\cdot\nabla)^j\varphi|_{x_N,y_N,z_N} \\ &= \left(n_x\frac{\partial}{\partial x} + n_y\frac{\partial}{\partial y} + n_z\frac{\partial}{\partial z}\right)^i\left(b_x\frac{\partial}{\partial x} + b_y\frac{\partial}{\partial y} + b_z\frac{\partial}{\partial z}\right)^j\varphi|_{x_N,y_N,z_N} \end{aligned} \tag{14}$$

由于电位 φ 满足 Laplace 方程，故 $\varphi_{0ij}(i,j=0,1,2)$ 各系数之间并不是独立的。对于研究宽电子束聚焦，只须考虑到二次幂 $(i+j=2)$，则有

$$\varphi_{002} + \varphi_{020} + \phi''(s) - k\varphi_{010} = 0 \tag{15}$$

同样，对于 N^* 点处的磁感应强度 $\boldsymbol{B}(s,p_2,p_3)$，我们将它展开成 p_2、p_3 的幂级数：

$$\boldsymbol{B}(s,p_2,p_3) = \sum_{i=0}^{\infty}\sum_{j=0}^{\infty}\frac{(\boldsymbol{n}\cdot\nabla)^i(\boldsymbol{b}\cdot\nabla)^j}{i!j!}\boldsymbol{B}\bigg|_{p_2=p_3=0}p_2^i p_3^j \tag{16}$$

其逆变分量为 $B^1 = \boldsymbol{B}\cdot\boldsymbol{e}^1$，$B^2 = \boldsymbol{B}\cdot\boldsymbol{e}^2$，$B^3 = \boldsymbol{B}\cdot\boldsymbol{e}^3$，故逆变物理分量为 $B^{*i} = \sqrt{g_{ij}}B^i$ （$i=1, 2, 3$；不对 i 求和）。若以 B_t、B_n、B_b 来表示逆变物理分量 $B^{*i}(i=1,2,3)$，则有

$$B_t = \sqrt{(1-kp_2)^2 + \chi^2 p_2^2 + \chi^2 p_3^2}\, B^1 = \frac{\sqrt{(1-kp_2)^2+\chi^2 p_2^2+\chi^2 p_3^2}}{1-kp_2}\boldsymbol{B}\cdot\boldsymbol{t} \tag{17a}$$

$$= B_{t00} + B_{t10}p_2 + B_{t01}p_3 + B_{t20}p_2^2 + B_{t11}p_2 p_3 + B_{t02}p_3^2$$

$$B_n = \boldsymbol{B}^2 = \frac{\chi p_3}{1-kp_2}\boldsymbol{B}\cdot\boldsymbol{t} + \boldsymbol{B}\cdot\boldsymbol{n} = B_{n00} + B_{n10}p_2 + B_{n01}p_3 + B_{n20}p_2^2 + B_{n11}p_2 p_3 + B_{n02}p_3^2 \tag{17b}$$

$$B_b = \boldsymbol{B}^3 = -\frac{\chi p_3}{1-kp_2}\boldsymbol{B}\cdot\boldsymbol{t} + \boldsymbol{B}\cdot\boldsymbol{b} = B_{b00} + B_{b10}p_2 + B_{b01}p_3 + B_{b20}p_2^2 + B_{b11}p_2 p_3 + B_{b02}p_3^2 \tag{17c}$$

为了用实验室坐标分量来表示上述的系数，我们首先考虑 $\boldsymbol{B}$ 在 $\boldsymbol{t}$、$\boldsymbol{n}$、$\boldsymbol{b}$ 上的投影分量：

$$\boldsymbol{B}\cdot\boldsymbol{t} = \bar{B}_{t00} + \bar{B}_{t10}p_2 + \bar{B}_{t01}p_3 + \bar{B}_{t20}p_2^2 + \bar{B}_{t11}p_2 p_3 + \bar{B}_{t02}p_3^2 \tag{18a}$$

$$\boldsymbol{B}\cdot\boldsymbol{n} = \bar{B}_{n00} + \bar{B}_{n10}p_2 + \bar{B}_{n01}p_3 + \bar{B}_{n20}p_2^2 + \bar{B}_{n11}p_2 p_3 + \bar{B}_{n02}p_3^2 \tag{18b}$$

$$\boldsymbol{B}\cdot\boldsymbol{b} = \bar{B}_{b00} + \bar{B}_{b10}p_2 + \bar{B}_{b01}p_3 + \bar{B}_{b20}p_2^2 + \bar{B}_{b11}p_2 p_3 + \bar{B}_{b02}p_3^2 \tag{18c}$$

式中，

$$\bar{B}_{tk\lambda} = x'_N B_{xk\lambda} + y'_N B_{yk\lambda} + z'_N B_{zk\lambda} \tag{19a}$$

$$\bar{B}_{bk\lambda} = b_x B_{xk\lambda} + b_y B_{yk\lambda} + b_z B_{zk\lambda} \tag{19b}$$

$$\bar{B}_{nk\lambda} = n_x B_{xk\lambda} + n_y B_{yk\lambda} + n_z B_{zk\lambda} \tag{19c}$$

式中，

$$B_{xk\lambda} = \frac{1}{i!\,j!}(\boldsymbol{n}\cdot\nabla)^i(\boldsymbol{b}\cdot\nabla)^j B_x\big|_{x_N,y_N,z_N} \tag{20}$$

同样 $B_{yk\lambda}$、$B_{zk\lambda}$ 有类似的表达式，只须将 B_y、B_z 代以 B_x 即可。

将式（18）中各式代入式（17）中，便可得到 $B_{tk\lambda}$、$B_{nk\lambda}$、$B_{bk\lambda}$ 与 $\bar{B}_{tk\lambda}$、$\bar{B}_{nk\lambda}$、$\bar{B}_{bk\lambda}$ 等系数之间的关系式：

$$\begin{aligned}
&B_{t00} = \bar{B}_{t00},\ B_{t10} = \bar{B}_{t10},\ B_{t01} = \bar{B}_{t01},\ B_{t11} = \bar{B}_{t11},\\
&B_{t20} = \bar{B}_{t20} + \frac{1}{2}\chi^2 \bar{B}_{t00},\ B_{t02} = \bar{B}_{t02} + \frac{1}{2}\chi^2 \bar{B}_{t00},\\
&B_{n00} = \bar{B}_{n00},\ B_{n10} = \bar{B}_{n10},\ B_{n01} = \bar{B}_{n01} + \chi \bar{B}_{t00},\ B_{n20} = \bar{B}_{n20},\\
&B_{n11} = \bar{B}_{n11} + k\chi \bar{B}_{t00} + \chi \bar{B}_{t10},\ B_{n02} = \bar{B}_{n02} + \chi \bar{B}_{t00},\\
&B_{b00} = \bar{B}_{b00},\ B_{b10} = \bar{B}_{b10} - \chi \bar{B}_{t00},\ B_{b01} = \bar{B}_{b01},\\
&B_{b20} = \bar{B}_{b20} - k\chi \bar{B}_{t00} - \chi \bar{B}_{t10},\ B_{b11} = \bar{B}_{b11} - \chi \bar{B}_{t01},\ B_{b02} = \bar{B}_{b02}
\end{aligned} \tag{21}$$

由于静磁场满足无散和无旋条件，即 $\nabla\cdot\boldsymbol{B} = 0$，$\nabla\times\boldsymbol{B} = 0$，因而 $B_{tk\lambda}$、$B_{nk\lambda}$、$B_{bk\lambda}$（k,

$\lambda = 0,1,2$)等 18 个系数不都是独立的,它们之间存在着如下的关系式:

$$\begin{aligned}
&B'_{t00} + B_{n10} + B_{b01} - kB_{n00} = 0, \\
&B_{b10} - B_{n01} + 2\chi B_{t00} = 0, \\
&B_{t01} - B'_{b00} - \chi B_{n00} = 0, \\
&B'_{n00} - B_{t10} + kB_{t00} - \chi B_{b00} = 0, \\
&B'_{t10} + 2B_{n20} - 2kB_{n01} + B_{b11} - kB_{b02} = 0, \\
&B'_{t01} + B_{n11} - kB_{n01} + 2B_{b02} = 0, \\
&B_{b11} - 2B_{n02} + 3\chi B_{t01} = 0, \\
&2B_{b20} - B_{n11} + 3\chi(B_{t10} + kB_{t00}) = 0, \\
&B_{t11} - kB_{t01} - B'_{b10} + \chi(B_{b01} - B_{n10}) - \chi'B_{t00} - \chi B'_{t00} = 0, \\
&2B_{t02} - B'_{b01} - 2\chi B_{n01} + \chi^2 B_{t00} = 0, \\
&-B'_{n10} + 2B_{t20} - 2kB_{t10} + 2\chi B_{b10} + \chi^2 B_{t00} = 0, \\
&-B'_{n10} + B_{t11} + \chi'B_{t00} + \chi B'_{t00} - kB_{t01} - \chi B_{n10} + \chi B_{b01} = 0
\end{aligned} \tag{22}$$

由此可见,与磁感应强度有关的 $B_{tk\lambda}$、$B_{nk\lambda}$、$B_{bk\lambda}$ ($k,\lambda = 0,1,2$) 共有 18 个系数,只有 6 个系数是独立的。

实际上,在计算主轨迹和曲近轴轨迹时,磁感应的量只用到 7 个,即 B_{t00}、B_{n00}、B_{n10}、B_{n01}、B_{b00}、B_{b10}、B_{b01},其中只有 5 个量是独立的,其之间关系的表达式见式(22)的前两个式子。同样,电位的量用到 6 个,即 φ_{000}、φ_{001}、φ_{010}、φ_{020}、φ_{011}、φ_{002}。其中,由于表达式(15)成立,故也是 5 个量是独立的。

顺便指出,φ_{0ij} 是 $(i+j)$ 阶协变张量,$B_{tk\lambda}$、$B_{nk\lambda}$、$B_{bk\lambda}$ 乃是一阶逆变、$(k+\lambda)$ 阶协变张量。

2 曲线坐标系下的曲近轴方程与正交条件

在文献 [9] 中,我们由洛伦兹力和牛顿方程的协变形式出发,利用能量守恒定律,便可得到非相对论情况曲线坐标系下的电子轨迹方程:

$$2I[p''_2 + k - (k^2 + \chi^2)p_2 - \chi'p_3 - 2\chi p'_3] + (p'_2 - \chi p_3)\frac{dI}{ds} =$$
$$\frac{\partial \varphi}{\partial p_2} - \sqrt{\frac{2e}{m_0}} I^{1/2}(1 - kp_2) \left[\frac{p'_3}{\sqrt{(1-kp_2)^2 + \chi^2 p_2^2 + \chi^2 p_3^2}} B_t - B_b \right] \tag{23a}$$

$$2I[p''_3 - \chi^2 p_3 + \chi'p_2 + 2\chi p'_2] + (p'_3 - \chi p_2)\frac{dI}{ds} =$$
$$\frac{\partial \varphi}{\partial p_3} + \sqrt{\frac{2e}{m_0}} I^{1/2}(1 - kp_2) \left[\frac{p'_2}{\sqrt{(1-kp_2)^2 + \chi^2 p_2^2 + \chi^2 p_3^2}} B_t - B_b \right] \tag{23b}$$

式中,

$$I = \frac{\varphi_*}{(1-kp_2)^2 + (p'_2 - \chi p_3)^2 + (p'_3 + \chi p_2)^2} \tag{24}$$

m_0、e 分别为电子的静止质量和电荷;$\varphi_* = \varphi + \varepsilon_0$,称为规范化电位;$e\varepsilon_0$ 为电子逸出初能

量；B_t、B_n、B_b 为磁感逆变物理分量。

对于研究宽电子束聚焦问题，假定主轨迹的相邻轨迹处处满足如下曲近轴条件[8]：

$$p^2(s) \approx 0,$$
$$\frac{m_0}{2e} \frac{\dot{p}^2(s) - \dot{p}^2(0)}{\phi(s) + \varepsilon_s} \ll 1 \tag{25}$$

且 $\phi(0) = 0$，但是并不一定处处满足 $p'^2(s) \ll 1$ 的条件。

由轨迹方程式（23）出发，利用上述曲近轴条件，便可求得零级近似轨迹方程，即主轨迹方程：

$$\varphi_{010} - 2kn^2 + \sqrt{\frac{2e}{m_0}} nB_{b00} = 0 \tag{26a}$$

$$\varphi_{001} - \sqrt{\frac{2e}{m_0}} nB_{n00} = 0 \tag{26b}$$

和一级近似轨迹方程，即曲近轴轨迹方程：

$$\frac{d}{ds}\left(n\frac{dp_2}{ds}\right) = Fp_2 + Np_3 + 2K\frac{dp_3}{ds} \tag{27a}$$

$$\frac{d}{ds}\left(n\frac{dp_3}{ds}\right) = Rp_3 - Qp_2 - 2K\frac{dp_2}{ds} \tag{27b}$$

这里，将各系数定义为主轨迹曲线 s 的如下函数：

$$\begin{aligned}
& n = \sqrt{\phi + \varepsilon_s}, \quad K = n\chi - \sqrt{e/(8m_0)} B_{t00}, \\
& F = n(\chi^2 - k^2) + \frac{\varphi_{020}}{2n} - \frac{k\varphi_{010}}{n} + \sqrt{\frac{e}{2m_0}}\left(B_{b10} + \frac{\varphi_{010}}{2n^2}B_{b00}\right), \\
& R = n\chi^2 + \frac{\varphi_{002}}{2n} - \sqrt{\frac{e}{2m_0}}\left(B_{n01} + \frac{\varphi_{001}}{2n^2}B_{n00}\right), \\
& N = \frac{\chi\varphi_{100}}{2n} + n\chi' + \frac{\varphi_{011}}{2n} - \frac{k\varphi_{001}}{n} + \sqrt{\frac{e}{2m_0}}\left(B_{b01} + \frac{\varphi_{001}}{2n^2}B_{b00}\right), \\
& Q = \frac{\chi\varphi_{100}}{2n} + n\chi' - \frac{\varphi_{011}}{2n} + \sqrt{\frac{e}{2m_0}}\left(B_{n10} + \frac{\varphi_{010}}{2n^2}B_{n00}\right)
\end{aligned} \tag{28}$$

以上的推导过程同文献［7］、文献［8］。同样，不难证明：

$$\frac{dK}{ds} = \frac{N+Q}{2} \tag{29}$$

曲近轴轨迹方程式（27）乃是变量互相耦合的二阶线性微分方程组。在文献［7］中，我们曾证明，由方程式（27）所确定的曲近轴系统，一般情况下，并不具有理想聚焦性质，只有当系统满足如下的正交条件：

$$n[(N-Q)(F'-R') - (F-R)(N'-Q')] = 2K[(F-R)^2 + (N-Q)^2] \tag{30}$$

可将方程式（27）变为变量 u、v 分离的二阶线性齐次微分方程组：

$$\frac{d}{ds}\left(n\frac{du}{ds}\right) = Uu \tag{31a}$$

$$\frac{d}{ds}\left(n\frac{dv}{ds}\right) = Vv \tag{31b}$$

式中，
$$U = \frac{1}{2}(F+R) + \frac{1}{2}\sqrt{(F-R)^2 + (N-Q)^2} - \frac{K^2}{n} \quad (32a)$$

$$V = \frac{1}{2}(F+R) - \frac{1}{2}\sqrt{(F-R)^2 + (N-Q)^2} - \frac{K^2}{n} \quad (32b)$$

这里变量 u、v 是通过以下变换

$$(p_2 + ip_3) = (u + iv)e^{i\tau} \quad (33)$$

求得。式中，

$$\tau' = -(K/n) \quad (34)$$

或

$$\tan 2\tau = \frac{N-Q}{F-R} \quad (35)$$

由方程式（31）不难证明[8]，当 $U=V$ 时，系统可获得无像散点聚焦的像；当 $U \neq V$ 时，则将在两个位置形成线聚焦的理想像，它们与 $\varepsilon_s = \varepsilon_{s_1}$ 相对应。

3 曲近轴轨迹方程的求解

通常，以弧长 s 作为变量表示的方程，对于求解并不方便，我们可将它化为实验室笛卡儿直角坐标系下的形式，利用关系式

$$ds = \sqrt{1 + x'^2 + y'^2}\, dz \quad (36)$$

可将上面导得的方程对 s 求微商变为对 z 求微商。

于是，由主轨迹方程式（26），可以得到

$$\frac{d}{dz}\left(\frac{\sqrt{\phi + \varepsilon_s}}{\sqrt{1 + x'^2 + y'^2}} x'\right) = \frac{\sqrt{1 + x'^2 + y'^2}}{2\sqrt{\phi + \varepsilon_s}} \frac{\partial \phi}{\partial x} + \sqrt{\frac{e}{2m_0}}(B_{y00} - y' B_{z00}) \quad (37a)$$

$$\frac{d}{dz}\left(\frac{\sqrt{\phi + \varepsilon_s}}{\sqrt{1 + x'^2 + y'^2}} y'\right) = \frac{\sqrt{1 + x'^2 + y'^2}}{2\sqrt{\phi + \varepsilon_s}} \frac{\partial \phi}{\partial y} + \sqrt{\frac{e}{2m_0}}(B_{x00} - x' B_{z00}) \quad (37b)$$

如文献[7]所指出的，当进行正设计时，主轨迹方程式（26a）作上述变换时，另一主轨迹方程（26b）是自动满足的。

对于曲近轴轨迹方程，式（27）变为

$$\frac{d}{dz}(\sqrt{\phi + \varepsilon_s}\, p_2') = (1 + x'^2 + y'^2)(Fp_2 + Np_3) + 2K\sqrt{1 + x'^2 + y'^2}\, p_3' + \sqrt{\phi + \varepsilon_s}\, \frac{x'x'' + y'y''}{1 + x'^2 + y'^2} p_2'$$
$$(38a)$$

$$\frac{d}{dz}(\sqrt{\phi + \varepsilon_s}\, p_3') = (1 + x'^2 + y'^2)(Rp_3 + Qp_2) + 2K\sqrt{1 + x'^2 + y'^2}\, p_2' + \sqrt{\phi + \varepsilon_s}\, \frac{x'x'' + y'y''}{1 + x'^2 + y'^2} p_3'$$
$$(38b)$$

式中，ϕ、K、F、R、N、Q 等均应取主轨迹上的值。

在求解式（38）时，通常令

$$w_1 = p_2,\ w_2 = p_2',\ w_3 = p_3,\ w_4 = p_3' \quad (39)$$

将方程式（38）变为一阶联立方程组：

$$\begin{aligned}
w_1' &= w_2, \\
w_2' &= g_1 w_1 + g_2 w_3 + f_1 w_2 + f_2 w_4, \\
w_3' &= w_4, \\
w_4' &= h_1 w_1 + h_2 w_3 + f_1 w_4 - f_2 w_2
\end{aligned} \tag{40}$$

式中，

$$\begin{aligned}
g_1 &= \frac{1 + x'^2 + y'^2}{\sqrt{\phi + \varepsilon_s}} F, \\
g_2 &= \frac{1 + x'^2 + y'^2}{\sqrt{\phi + \varepsilon_s}} N, \\
f_1 &= \frac{x' x'' + y' y''}{1 + x'^2 + y'^2} - \frac{\phi'}{2(\phi + \varepsilon_s)}, \\
f_2 &= \frac{2 \sqrt{1 + x'^2 + y'^2}}{\sqrt{\phi + \varepsilon_s}} K, \\
h_1 &= -\frac{1 + x'^2 + y'^2}{\sqrt{\phi + \varepsilon_s}} Q, \\
h_2 &= \frac{1 + x'^2 + y'^2}{\sqrt{\phi + \varepsilon_s}} R
\end{aligned} \tag{41}$$

利用数值方法，可以求得方程组（40）的四组特解：

$$w_1^{(i)}, \quad w_2^{(i)}, \quad w_3^{(i)}, \quad w_4^{(i)} \quad (i = 1, 2, 3, 4)$$

设它们满足以下初始条件：

I	1	2	3	4
$w_1^{(i)} = p_2^{(i)}$	1	0	0	0
$w_2^{(i)} = (p_2^{(i)})'$	0	$1/\sqrt{\varepsilon_s}$	0	0
$w_3^{(i)} = p_3^{(i)}$	0	0	1	0
$w_4^{(i)} = (p_3^{(i)})'$	0	0	0	$1/\sqrt{\varepsilon_s}$

于是方程式（38）的通解如下：

$$\begin{aligned}
p_2(z) &= p_{20} w_1^{(1)} + \sqrt{\varepsilon_s} p_{20}' w_1^{(2)} + p_{30} w_1^{(3)} + \sqrt{\varepsilon_s} p_{30}' w_1^{(4)}, \\
p_3(z) &= p_{20} w_3^{(1)} + \sqrt{\varepsilon_s} p_{20}' w_3^{(2)} + p_{30} w_3^{(3)} + \sqrt{\varepsilon_s} p_{30}' w_3^{(4)}
\end{aligned} \tag{42}$$

4 转动的曲线坐标下的曲近轴轨迹方程

仿照文献 [6]、文献 [10] 在曲轴主轨迹的点 N 处引入转动的坐标系 $(\boldsymbol{t}, \boldsymbol{n}_f, \boldsymbol{b}_f)$，它与该点处的局部弗莱纳坐标系 $(\boldsymbol{t}, \boldsymbol{n}, \boldsymbol{b})$ 有以下的变换关系：

$$(p_2 + \mathrm{i} p_3) = (u + \mathrm{i} v) \mathrm{e}^{-\mathrm{i} \gamma} \tag{43}$$

式中，

$$\gamma = \gamma(s) = \int_0^s \chi(s)\,ds \tag{44}$$

按图 2，则有

$$\boldsymbol{n}_f = \boldsymbol{n}\cos\gamma - \boldsymbol{b}\sin\gamma,$$
$$\boldsymbol{b}_f = \boldsymbol{n}\sin\gamma - \boldsymbol{b}\cos\gamma \tag{45}$$

$(\boldsymbol{t},\ \boldsymbol{n}_f,\ \boldsymbol{b}_f)$ 在直角坐标系中的分量为：

$$t_{fxi} = t_{xi} = x_i',$$
$$n_{fxi} = \frac{1}{k}\left[x_i''\cos\gamma - (x_j'x_k'' - x_k'x_j'')\sin\gamma\right], \tag{46}$$
$$b_{fxi} = \frac{1}{k}\left[x_i''\sin\gamma + (x_j'x_k'' - x_k'x_j'')\cos\gamma\right]$$

式中，$i,\ j,\ k = 1,\ 2,\ 3$，$i,\ j,\ k$ 依次以 1，2，3 顺序循环排列；$x_1 = x,\ x_2 = y,\ x_3 = z$；撇号表示对弧长 s 的微商。

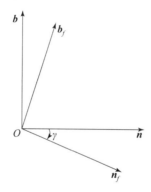

图 2　转动的正交曲线坐标系 $(\boldsymbol{t},\ \boldsymbol{n}_f,\ \boldsymbol{b}_f)$ 与弗莱纳局部坐标系 $(\boldsymbol{t},\ \boldsymbol{n},\ \boldsymbol{b})$

于是，相邻轨迹的微分 $d\boldsymbol{r}^*$ 可表示为[10]

$$d\boldsymbol{r}^* = (1 - ku\cos\gamma - kv\sin\gamma)ds\boldsymbol{t} + du\boldsymbol{n}_f + dv\boldsymbol{b}_f \tag{47}$$

显然，转动的 Frenet 局部坐标系乃是正交曲线坐标系，故以 $(u,\ v)$ 作为变量的电子轨迹方程可表示为

$$2I[u'' + k\cos\gamma(1 - ku\cos\gamma - kv\sin\gamma)] + u'\frac{dI}{ds} = \frac{\partial\varphi}{\partial u} + \sqrt{\frac{2e}{m_0}}I^{1/2}\left[(1 - ku\cos\gamma - kv\sin\gamma)B_v - v'B_s\right] \tag{48a}$$

$$2I[v'' + k\sin\gamma(1 - ku\cos\gamma - kv\sin\gamma)] + v'\frac{dI}{ds} = \frac{\partial\varphi}{\partial v} + \sqrt{\frac{2e}{m_0}}I^{1/2}\left[(1 - ku\cos\gamma - kv\sin\gamma)B_u - u'B_s\right] \tag{48b}$$

式中，

$$I = \frac{\varphi_*}{(1 - ku\cos\gamma - kv\sin\gamma)^2 + u'^2 + v'^2} \tag{49}$$

其中，B_s、B_u、B_v 分别表示 $\boldsymbol{B}$ 在 $\boldsymbol{t}$、$\boldsymbol{n}_f$、$\boldsymbol{b}_f$ 上的物理分量。

同样，我们可将静电电位 φ 和磁感应强度 $\boldsymbol{B}$ 在主轨迹 $(x_N,\ y_N,\ z_N)$ 处展成 u、v 的

幂级数：

$$\varphi(s,u,v) = \sum_{i=0}^{\infty}\sum_{j=0}^{\infty} \frac{\overline{\varphi}_{0ij}(s)}{i!j!} u^i v^j \tag{50}$$

$$B(s,u,v) = \sum_{i=0}^{\infty}\sum_{j=0}^{\infty} \frac{1}{i!j!} (\boldsymbol{n}_f \cdot \nabla)^i (\boldsymbol{b}_f \cdot \nabla)^j \boldsymbol{B}\big|_{x_N,y_N,z_N} u^i v^j \tag{51}$$

对于 $\overline{\varphi}_{0ij}$，它仍可用式 (14) 表示，只须将式 (14) 中 n_x、n_y、n_z 和 b_x、b_y、b_z 分别以 n_{fx}、n_{fy}、n_{fz} 和 b_{fx}、b_{fy}、b_{fz} 代替。同样，$\overline{\varphi}_{0ij}$ ($i,j = 0,1,2$) 各系数之间并不是独立的。当考虑到二次幂 ($i+j=2$) 时，则有关系式：

$$\overline{\varphi}_{002} + \overline{\varphi}_{020} + \phi''(s) - k\overline{\varphi}_{010}\cos\gamma - k\overline{\varphi}_{001}\sin\gamma = 0 \tag{52}$$

式中，$\overline{\varphi}_{000} = \phi(s)$，$\overline{\varphi}_{100} = \phi'(s)$。

若对式 (51) 乘 $\boldsymbol{n}_f$、$\boldsymbol{b}_f$ 和 $\boldsymbol{t}$，便不难得到以下分量表达式：

$$B_u = B_{u00} + B_{u10}u + B_{u01}v + B_{u20}u^2 + B_{u11}uv + B_{u02}v^2 \tag{53}$$

式中

$$B_{uk\lambda} = n_{fx}B_{xk\lambda} + n_{fy}B_{yk\lambda} + n_{fz}B_{zk\lambda} \tag{54}$$

式 (54) 中的 $B_{xk\lambda}$ 仍可用式 (20) 表示，只须将 $\boldsymbol{n}$、$\boldsymbol{b}$ 置换为 $\boldsymbol{n}_f$、$\boldsymbol{b}_f$，类似可以求得 $B_{yk\lambda}$ 和 $B_{zk\lambda}$。同样，B_v、B_s 以及相应的 $B_{vk\lambda}$、$B_{sk\lambda}$ 有类似式 (53) 和式 (54) 的表达式，这里不细述了。

同样，$B_{uk\lambda}$、$B_{vk\lambda}$、$B_{sk\lambda}$ ($k,\lambda = 0,1,2$) 的 18 个系数，它们之间也不是独立的，由 $\nabla \cdot \boldsymbol{B} = 0$ 和 $\nabla \times \boldsymbol{B} = 0$，可得

$$\begin{aligned}
&B_{u10} + B_{v01} + B'_{s00} - k\cos\gamma B_{u00} - k\sin\gamma B_{v00} = 0, \\
&B_{v10} - B_{u01} = 0, \\
&B_{s01} - B'_{v00} - k\sin\gamma B_{s00} = 0, \\
&B'_{u00} - B_{s10} + k\cos\gamma B_{s00} = 0, \\
&B'_{s10} + B_{v11} + 2B_{u20} - 2k\cos\gamma B_{u10} - k\cos\gamma B_{v01} - k\sin\gamma B_{v10} = 0, \\
&B'_{s01} + B_{u11} + 2B_{v02} - 2k\sin\gamma B_{v01} - k\sin\gamma B_{u10} - k\cos\gamma B_{u01} = 0, \\
&B'_{u10} - 2B_{s20} + 2k\cos\gamma B_{s10} = 0, \\
&B'_{v01} - 2B_{s02} + 2k\sin\gamma B_{s01} = 0, \\
&B_{s11} - B'_{v10} - k\cos\gamma B_{s01} - k\sin\gamma B_{s10} = 0, \\
&B_{s11} - B'_{u01} - k\sin\gamma B_{s10} - k\cos\gamma B_{s01} = 0, \\
&B_{u11} - 2B_{v20} = 0, \\
&B_{v11} - 2B_{u02} = 0
\end{aligned} \tag{55}$$

由于计算主轨迹和曲近轴轨迹，磁感应的量只用到 7 个，即 B_{s00}、B_{u00}、B_{u10}、B_{u01}、B_{v00}、B_{v10}、B_{v01}，故式 (55) 中只有前两个式子与这些系数有关。

由轨迹方程式 (48)，引入曲近轴条件式 (25)，便得到在 (s,u,v) 坐标下的轨迹方程：

$$\bar{\varphi}_{010} - 2kn^2\cos\gamma + \sqrt{\frac{2e}{m_0}}nB_{v00} = 0 \tag{56a}$$

$$\bar{\varphi}_{001} - 2kn^2\sin\gamma - \sqrt{\frac{2e}{m_0}}nB_{u00} = 0 \tag{56b}$$

和曲近轴轨迹方程：

$$\frac{\mathrm{d}}{\mathrm{d}s}\left(n\frac{\mathrm{d}u}{\mathrm{d}s}\right) = \bar{F}u + \bar{N}v + 2\bar{K}\frac{\mathrm{d}v}{\mathrm{d}s} \tag{57a}$$

$$\frac{\mathrm{d}}{\mathrm{d}s}\left(n\frac{\mathrm{d}v}{\mathrm{d}s}\right) = \bar{R}v - \bar{Q}u + 2\bar{K}\frac{\mathrm{d}u}{\mathrm{d}s} \tag{57b}$$

式中各系数当考虑主轨迹方程式（56）时，则有

$$n = \sqrt{\phi + \varepsilon_s}, \quad \bar{K} = -\sqrt{\frac{e}{8m_0}}B_{s00},$$

$$\bar{F} = \frac{\bar{\varphi}_{020}}{2n} - \frac{3}{4}\frac{(\bar{\varphi}_{010})^2}{n^3} - \frac{e}{2m_0}\frac{B_{v00}^2}{n} + \sqrt{\frac{e}{2m_0}}\left(B_{v10} - \frac{3}{2}\frac{\bar{\varphi}_{010}}{n^2}B_{v00}\right),$$

$$\bar{R} = \frac{\bar{\varphi}_{002}}{2n} - \frac{3}{4}\frac{(\bar{\varphi}_{001})^2}{n^3} - \frac{e}{2m_0}\frac{B_{u00}^2}{n} - \sqrt{\frac{e}{2m_0}}\left(B_{u01} - \frac{3}{2}\frac{\bar{\varphi}_{001}}{n^2}B_{u00}\right), \tag{58}$$

$$\bar{N} = \frac{\bar{\varphi}_{011}}{2n} - \frac{3}{4}\frac{\bar{\varphi}_{010}\bar{\varphi}_{001}}{n^3} + \frac{e}{2m_0}\frac{B_{v00}B_{u00}}{n} + \sqrt{\frac{e}{2m_0}}\left(B_{v10} + \frac{\bar{\varphi}_{010}}{2n^2}B_{u00} - \frac{\bar{\varphi}_{001}}{n^2}B_{v00}\right),$$

$$\bar{Q} = -\frac{\bar{\varphi}_{011}}{2n} + \frac{3}{4}\frac{\bar{\varphi}_{010}\bar{\varphi}_{001}}{n^3} - \frac{e}{2m_0}\frac{B_{v00}B_{u00}}{n} + \sqrt{\frac{e}{2m_0}}\left(B_{u10} + \frac{\bar{\varphi}_{001}}{2n^2}B_{v00} - \frac{\bar{\varphi}_{010}}{n^2}B_{u00}\right)$$

同样，不难证明

$$\frac{\mathrm{d}\bar{K}}{\mathrm{d}s} = \frac{\bar{N} + \bar{Q}}{2} \tag{59}$$

关于方程式（57）的求解以及在 (s, u, v) 坐标下内近轴系统的正交条件，可以仿照第 2 节求出，这里不再细述了。

5 讨论

（1）在研究宽电子束理想聚焦或一级近似曲近轴轨迹时，表征电位与磁场的系数各有 5 个是独立的（电位只须考虑到二次幂，磁感应强度到一次幂）。实际上，在进行逆设计时，考虑到还应该满足主轨迹方程，故只有 8 个系数是独立的，要按轨迹对于系统的要求来确定这些独立的任意系数。而在正设计时，由于系统的结构以及场都已确定，不难由实验室坐标系下的场系数确定这 10 个场系数。

（2）本文研究的两种曲线局部坐标系求解宽束成像系统，对于问题的描述是等价的，但是各有特点。试做以下比较：

对于前一种方法在引入 Frenet 局部坐标系后，又引入坐标基 e_i 和互逆基 e^i，而 e_i 或 e^i 的确定不是很方便，因它们不但与 t、n、b 有关，还与 p_2、p_3 有关。

但是，p_2、p_3 是在随主轨迹运动的 Frenet 局部坐标系 (t, n, b) 中定义的，因此，

就正设计而言，由主轨迹方程式（26）确定主轨迹，由曲近轴轨迹方程式（27）确定 p_2、p_3，计算是方便的，物理意义是明确的。同样，对于逆设计，即先给出主轨迹和相邻轨迹，由此确定场系数，继而确定系统的结构与电磁参数。在给出主轨迹时，相应的 Frenet 局部坐标系随之确定，便可相应地定义 p_2、p_3。由此可见，坐标基 e_i 和逆基 e^i 的引入仅是作为计算场系数的桥梁而已。

对于后一种方法，它引入转动的正交局部曲线坐标系 (t, n_f, b_f)，这一坐标基不随相邻轨迹变化，它的不便之处是在主轨迹的每一点处要对主轨迹挠率 χ 积分，并相对于 Frenet 局部坐标系 (t, n, b) 做相应转动，从而在计算场系数方面引入了工作量。尽管如此，应用这种转动的局部正交坐标系，在数学求解上并不困难。

总的说来，前一种方法更为方便和直观。

（3）本文所给出的数学处理对于计算机辅助求解宽电子束聚焦问题，特别是对于复合电磁聚焦成像系统是方便和适宜的，因为可以利用通常的方法计算系统在实验室坐标系中的场分布以及场系数。

（4）直轴复合电磁聚焦阴极透镜的近轴方程和转角公式，可以视为本文的特例。若注意到宽电子束与细电子束问题特解选取的差别，取 $\varepsilon_s = \varepsilon_0$，$\phi(0) \neq 0$，则本文的结果可以推广到研究细电子束聚焦问题。若令 $\phi(s) = \phi(s_0) = $ 常数，并令所有涉及电位的系数都等于零，则可研究 R. G. E. Hutter 提出的大角度偏转问题[6]。

（5）对前面所给出的普遍轨迹方程，若在展开时，进一步保留相邻轨迹的二、三级高幂次项，则可研究系统的二级和三级几何像差。

结束语

本文进一步发展了作者以前的工作，文中利用两种曲线局部坐标系，给出了磁场以磁感应强度 B 的逆变物理分量表示的普遍轨迹方程。文中给出静电电位 φ 和磁感应强度 B 沿曲轴展开的表示式，它们的场系数可以用实验室坐标系下场系数来表达。由此推导了宽电子束聚焦的主轨迹方程和曲近轴轨迹方程，讨论了系统的正交条件。

本文所给出的实验室坐标系下场系数表示式以及用变量 z 描写的主轨迹方程和曲近轴轨迹方程，对于计算机辅助设计求解是很有用的。

本文的结果可以推广到研究细电子束聚焦和大角度偏转的问题。

参 考 文 献

[1] ГРИНБЕРГ Г А Избранные вопросы математическои теории электрических явлений[M]. Москва-Ленинград：Изд, А Н СССР,1948.

[2] ГРИНБЕРГ Г А. Основы обшей теории фокусируюшего действия электростатических и магнитных полеи II. пространственные электростатические поля[J]. Изд, А Н СССР,1942, 37(9)：295 – 303.

[3] ГРИНБЕРГ Г А. Основы обшей теории фокусируюшего действия статических электрических и магнитнюх полей[J]. Ж. Т. Ф. , 1943, 13：361 – 388.

[4] STURROCK P A. Static and dynamic electron optics[M]. Cambridge：Cambridge

University Press, 1955.

[5] STURROCK P A. The imaging properties of electron beams in arbitrary static electromagnetic field[J]. Philosophical Transactions, 1952, A245:155-187.

[6] HUTTER R G E. Deflection defocusing effects in cathode ray tubes at large deflection angles[J]. IEEE-ED 1970,1:1022-1031.

[7] ZHOU L W. Optic of wide eletron beam focusing[J]. Eletron Optical Systens, SEM, Inc., 1984:45-62.

[8] ZHOU L W. A generalized theory of wide electron beam focusing[J]. Advances in Electronics and Electron Physics, 1985,64B:575-589.

[9] ZHOU L W, NI G Q, QIU B C. Tensor analysis of electron motion in curvilinear coordinate systems(Ⅰ)[J]. Optik,1988, 79(2):53-66.

[10] ZHOU L W, NI G Q, QIU B C. Tensor analysis of electron motion in curvilinear coordinate systems(Ⅱ)[J]. Optik,1988,78(3):101-107.

5.6 宽电子束聚焦的变分理论
Variation Theory of Wide Electron Beam Focusing

摘要：本文利用变分方法研究了正交和非正交两种曲线局部坐标基下宽电子束聚焦的普遍理论，若假定空间静电电位 φ 的磁感应强度矢量 **B** 在实验室坐标系中的分布是已知的。将变分原理下的细电子束聚焦推广到宽电子束聚焦，证明了其结果与用电动力学方法研究宽电子束聚焦的一致性。

Abstract: A generalized theory of wide electron beam focusing with curvilinear optical axis on both the orthogonal and the non-orthogonal curvilinear local coordinate bases has been studied by using the variation principle, in which the electrostatic potential φ and the magnetic induction vector **B** in the laboratory coordinate system are assumed to be known. The present paper represents a step further from the study of narrow electron beam focusing to that of wide electron beam focusing using the variation principle, and fully verifies the concordance between the variation method and the electrodynamic method in studying wide electron beam focusing.

引　言

用曲线局部坐标系来研究电子束聚焦问题已被证明是一种行之有效的方法，曲线坐标系下细电子束聚焦的普遍理论是由 P. A. Sturrock 和 Г. А. Гринберг 于 20 世纪 40—50 年代分别用轨迹法和变分法建立的[1~2]。此后人们又不断进行了研究，1970 年 R. G. E. Hutter 用曲轴光学理论研究了彩色显像管中大角度偏转问题[3]。1987 年 H. Rose 由变分原理出发，利用复数表示法，研究了曲轴细电子束聚焦的理论[4]。曲线坐标系下宽电子束聚焦的普遍理论是由本文作者近年来建立的[5~9]，它不仅适用于研究从 $\varphi(0)=0$ 的物面上逸出的电子初速度很小而初斜率可能很大的宽电子束聚焦问题，而且所得结果可将细电子束聚焦作为它的一个特例。文献 [5~9] 中利用张量表示法，由曲线坐标系下洛伦兹方程出发，导出了普遍轨迹方程，并引入曲近轴条件，推导了宽电子束聚焦的主轨迹方程和曲近轴轨迹方程，给出了曲近轴系统理想聚焦的正交条件。文献 [9] 中假定实验室直角坐标系下空间电位 φ 和磁感应强度 **B** 的分布已知，给出了适合计算机处理的宽电子束曲近轴轨迹方程。这些工作奠定了研究宽电子束聚焦问题的基础。

电子光学的研究方法可分为电动力学法（即轨迹法）和光程函数法（即变分法）。前者基于电子在电磁场中受到洛伦兹力作用，用电动力学方法来研究电子运动，其特点是比较直观，物理意义清晰；而后者基于与光学的相似性，用变分原理来研究电子运动，它的特点是应用简便，但较抽象。自然，这两种方法的结果是完全等效的[9]。

金伟其，周立伟，倪国强. 北京理工大学. 北京理工大学学报 (Journal of Beijing Institute of Technology)，V. 11, No. 2, 1991, 33 - 41.

在文献 [7~8] 中，推导了曲线坐标系下协变形式的轨迹方程。本文仍假定 φ 和 $\boldsymbol{B}$ 是已知的，利用变分原理研究宽电子束聚焦的普遍理论，证明也可获得与文献 [9] 相同的结果。

1　非正交曲线局部坐标基下的变分函数和普遍轨迹方程

如文献 [5~9] 所述，利用曲线坐标系研究电子在电磁场中的运动的基本思想，就是在由物点逸出的电子束中选取一条主轨迹 $\boldsymbol{r}_N$，在 $\boldsymbol{r}_N$ 上建立弗莱纳（Frenet）局部坐标系 $(\boldsymbol{t},\boldsymbol{n},\boldsymbol{b})$，然后在此运动的局部坐标系中，度量主轨迹法平面内电子束相邻轨迹 $\boldsymbol{r}^*$ 的偏离 $\boldsymbol{p}(=p_2\boldsymbol{n}+p_3\boldsymbol{b})$。显然，只要确定出 $\boldsymbol{r}_N$ 和 $\boldsymbol{p}$ 的变化规律，就确定了整个电子束的运动规律。

按照这一思想，利用张量表示法，由文献 [9]，可得非正的曲线局部坐标基 $(\boldsymbol{e}_1, \boldsymbol{e}_2, \boldsymbol{e}_3)$

$$\boldsymbol{e}_1 = (1-kp_2)\boldsymbol{t} - \chi p_3 \boldsymbol{n} + \chi p_2 \boldsymbol{b}, \quad \boldsymbol{e}_2 = \boldsymbol{n}, \quad \boldsymbol{e}_3 = \boldsymbol{b} \tag{1}$$

及其互逆基 $(\boldsymbol{e}^1, \boldsymbol{e}^2, \boldsymbol{e}^3)$

$$\boldsymbol{e}^1 = \frac{1}{1-kp_2}\boldsymbol{t}, \quad \boldsymbol{e}^2 = \frac{\chi p_3}{1-kp_2}\boldsymbol{t} + \boldsymbol{n}, \quad \boldsymbol{e}^3 = -\frac{\chi p_2}{1-kp_2}\boldsymbol{t} + \boldsymbol{b} \tag{2}$$

由此可以表示有关的场矢量。于是，相邻轨迹相对于主轨迹弧长的微商为

$$\frac{\mathrm{d}\boldsymbol{r}^*}{\mathrm{d}s} = \boldsymbol{e}_1 + p_2'\boldsymbol{e}_2 + p_3'\boldsymbol{e}_3 \tag{3}$$

电场强度为

$$\boldsymbol{E}(s,p_2,p_3) = -\mathrm{grad}\varphi = -\frac{\partial\varphi}{\partial s}\boldsymbol{e}^1 - \frac{\partial\varphi}{\partial p_2}\boldsymbol{e}^2 - \frac{\partial\varphi}{\partial p_3}\boldsymbol{e}^3 \tag{4}$$

磁感应强度为

$$\boldsymbol{B}(s,p_2,p_3) = B^i(s,p_2,p_3)\boldsymbol{e}_i = B_i(s,p_2,p_3)\boldsymbol{e}^i \tag{5}$$

磁矢位为

$$\boldsymbol{A}(s,p_2,p_3) = A^i(s,p_2,p_3)\boldsymbol{e}_i = A_i(s,p_2,p_3)\boldsymbol{e}^i \tag{6}$$

式中，s 为自阴极面算起的主轨迹弧长；p_2、p_3 分别为 $\boldsymbol{p}$ 在 $\boldsymbol{n}$、$\boldsymbol{b}$ 上的投影；k、χ 分别为主轨迹的曲率和挠率；$\varphi(s,p_2,p_3)$ 为系统的电位分布；在 $\boldsymbol{e}_i$ 和 $\boldsymbol{e}^i$ 前的量分别对应于该场量的逆变和协变分量，它们之间的转换可以通过度量张量 g_{ij} 或逆变度量张量 g^{ij} 进行[7~9]。

根据文献 [7~8] 可知，在此非正交曲线坐标基下，轨迹的变分函数为

$$\mu = \varphi_*^{1/2}[(1-kp_2)^2 + (p_2'-\chi p_3)^2 + (p_3'+\chi p_2)^2]^{1/2} - \eta(A_1 + p_2'A_2 + p_3'A_3) \tag{7}$$

式中，$\eta = \sqrt{e/(2m_0)}$，e/m_0 为电子的荷质比，$\varphi_*(s,p_2,p_3) = \varphi(s,p_2,p_3) + \varepsilon_0$，$\varepsilon_0$ 为电子逸出的初电位，利用欧拉（Euler）方程

$$\frac{\mathrm{d}}{\mathrm{d}s}\left(\frac{\partial\mu}{\partial p_i'}\right) - \frac{\partial\mu}{\partial p_i} = 0 \quad (i=2,3) \tag{8}$$

可以得到磁场以磁矢位 $\boldsymbol{A}$ 表示的电子运动普遍轨迹方程

$$2I[p_2'' + k(1-kp_2) - \chi^2 p_2 - \chi'p_3 - 2\chi p_3'] + (p_2' - \chi p_3)\frac{\mathrm{d}I}{\mathrm{d}s} =$$

$$\frac{\partial\varphi}{\partial p_2} - 2\eta I^{1/2}\left[\left(\frac{\partial A_1}{\partial p_2} - \frac{\partial A_2}{\partial s}\right) + \left(\frac{\partial A_3}{\partial p_2} - \frac{\partial A_2}{\partial p_3}\right)p_3'\right] \tag{9a}$$

$$2I[p_3'' - \chi^2 p_3 + \chi' p_2 - 2\chi p_2'] + (p_3' - \chi p_2)\frac{\mathrm{d}I}{\mathrm{d}s} = \frac{\partial\varphi}{\partial p_3} - 2\eta I^{1/2}\left[\left(\frac{\partial A_1}{\partial p_3} - \frac{\partial A_3}{\partial s}\right) + \left(\frac{\partial A_2}{\partial p_3} - \frac{\partial A_3}{\partial p_2}\right)p_2'\right]$$
(9b)

式中，A_i（$i = 1, 2, 3$）乃是磁矢量 A 的协变分量。参量 I 以下式表示

$$I = \frac{\varphi_*}{(1-kp_2)^2 + (p_2' - \chi p_3)^2 + (p_3' + \chi p_2)^2}$$
(10)

2 各级变分函数及轨迹方程

在电子光学中，用变分原理研究系统的理想聚焦性质和几何像差时，通常将变分函数 μ 按 p_2、p_3、p_2'、p_3' 的幂次展开，研究各级变分函数对电子轨迹的作用和影响。

为此，需要将 μ 中的有关项展成幂级数。对于电位 φ 可表示为[9]

$$\varphi(s, p_2, p_3) = \sum_{i=0}^{\infty}\sum_{j=0}^{\infty}\frac{\varphi_{0ij}(s)}{i!j!}p_2^i p_3^j$$
(11)

式中，φ_{0ij} 的意义以及相互关系见文献 [9]；对于磁矢量 A，虽然也可以将其分量直接展开为 p_2、p_3 的幂级数，然后根据其所满足的库仑规范 $\nabla \cdot A = 0$ 和磁矢量 A 的拉普拉辛 $\nabla^2 A = 0$ 来求出各系数之间的关系，但是这样对于实际计算是不方便的。因此，在本文中将利用文献 [9] 中磁感应强度 B 的展开式系数来表示 A 的幂级数展开式。

磁感应强度 B 和磁矢位 A 满足

$$B = \nabla \times A$$
(12)

对于现有坐标系，上式可写为分量形式

$$\begin{aligned}B_t &= \frac{\sqrt{(1-kp_2)^2 + \chi^2 p_2^2 + \chi^2 p_3^2}}{1-kp_2}\left(\frac{\partial A_3}{\partial p_2} - \frac{\partial A_2}{\partial p_3}\right),\\ B_n &= \frac{1}{1-kp_2}\left(\frac{\partial A_1}{\partial p_3} - \frac{\partial A_3}{\partial s}\right),\\ B_b &= \frac{1}{1-kp_2}\left(\frac{\partial A_2}{\partial s} - \frac{\partial A_1}{\partial p_2}\right)\end{aligned}$$
(13)

式中，B_t、B_n、B_b 是 B 的逆变物理分量。

我们知道，只有磁感应强度本身才有直接的物理意义，而磁矢位只是一个辅助参量，它可以有一定的任意性。因此，可以选择下述最简单的规范：

$$\frac{\partial A_2}{\partial p_3} = -\frac{\partial A_3}{\partial p_2}$$
(14)

$$A(s, 0, 0) = 0$$
(15)

于是，由式（13）以及 $\nabla \cdot B = 0$，可以得到

$$\begin{aligned}A_1 &= \frac{1}{2}\left\{\int_0^{p_3}\left[(1-kp_2)B_n + (B_n)_{p_2=0}\right]\mathrm{d}p_3 - \int_0^{p_2}\left[B_b + (B_b)_{p_3=0}\right](1-kp_2)\mathrm{d}p_2\right\},\\ A_2 &= -\frac{1}{2}\int_0^{p_3}\frac{1-kp_2}{(1-kp_2)^2 + \chi^2 p_2^2 + \chi^2 p_3^2}B_t\mathrm{d}p_3,\\ A_3 &= \frac{1}{2}\int_0^{p_2}\frac{1-kp_2}{(1-kp_2)^2 + \chi^2 p_2^2 + \chi^2 p_3^2}B_t\mathrm{d}p_2\end{aligned}$$
(16)

利用文献 [9] 中 B_t、B_n、B_b 的展开式

$$B_w = \sum_{i=0}^{\infty} \sum_{j=0}^{\infty} B_{wij} p_2^i p_3^j \quad (w = t, n, b) \tag{17}$$

便可得到

$$\begin{aligned}
A_1 &= -B_{b00}p_2 + B_{n00}p_3 - \frac{1}{2}(B_{b10} - kB_{b00})p_2^2 + \frac{1}{2}(B_{n10} - B_{b01} - kB_{n00})p_2 p_3 + \frac{1}{2}B_{n01}p_3^2 + \cdots \\
A_2 &= -\frac{1}{2}(B_{t00}p_3 + B_{t10}p_2 p_3 + \frac{1}{2}B_{t01}p_3^2 + \cdots) \\
A_3 &= \frac{1}{2}(B_{t00}p_2 + \frac{1}{2}B_{t10}p_2^2 + B_{t01}p_2 p_3 + \cdots)
\end{aligned} \tag{18}$$

如文献 [9] 中所指出的,系数 φ_{0ij} 和 B_{tij}、B_{nij}、B_{bij} 都是独立的,应满足一些关系式。若 φ 保留到二次幂,**B** 保留到一次幂,则只有电位和磁场各 5 个独立系数,当已知实验室直角坐标系中电位 φ 和磁感应强度 **B** 的分布时,利用文献 [9] 的方法,可以求出这些曲轴场系数。

将式 (11) 和式 (18) 代入式 (7),并按文献 [6] 的方法展开,得到

$$\mu = \mu_0 + \mu_1 + \mu_2 + \cdots \tag{19}$$

式中,

$$\mu_0 = n \tag{20a}$$

$$\mu_1 = \left(\frac{\varphi_{010}}{2n} - kn + \eta B_{b00}\right)p_2 + \left(\frac{\varphi_{001}}{2n} - \eta B_{n00}\right)p_3 \tag{20b}$$

$$\begin{aligned}
\mu_2 &= \frac{n}{2}(p_2'^2 + p_3'^2) + K(p_2 p_3' - p_3 p_2') + \frac{1}{2}Fp_2^2 + \frac{1}{2}Rp_3^2 + (N - K')p_2 + \\
&\quad \frac{\varepsilon_s}{2n}[k_0^2 p_{20}^2 + (p_{20}' - \chi_0 p_{30})^2 + (p_{30}' + \chi_0 p_{20})^2] \\
&= \frac{n}{2}(p_2'^2 + p_3'^2) + K(p_2 p_3' - p_3 p_2') + \frac{1}{2}Fp_2^2 + \frac{1}{2}Rp_3^2 - (Q - K')p_2 p_3 + \\
&\quad \frac{\varepsilon_s}{2n}[k_0^2 p_{20}^2 + (p_{20}' - \chi_0 p_{30})^2 + (p_{30}' + \chi_0 p_{20})^2]
\end{aligned} \tag{20c}$$

式中,

$$\begin{aligned}
n &= \sqrt{\phi(s) + \varepsilon_s}, \\
K &= n\chi - \frac{1}{2}\eta B_{t00}, \\
F &= n(\chi^2 - k^2) + \frac{\varphi_{020}}{2n} - \frac{k\varphi_{010}}{n} + \eta\left(B_{b10} + \frac{\varphi_{010}}{2n^2}B_{b00}\right), \\
R &= n\chi^2 + \frac{\varphi_{002}}{2n} - \eta\left(B_{n01} + \frac{\varphi_{001}}{2n^2}B_{n00}\right), \\
N &= \frac{\chi\varphi_{100}}{2n} + n\chi' + \frac{\varphi_{011}}{2n} - \frac{k\varphi_{001}}{n} + \eta\left(B_{b01} + \frac{\varphi_{001}}{2n^2}B_{b00}\right), \\
Q &= \frac{\chi\varphi_{100}}{2n} + n\chi' - \frac{\varphi_{011}}{2n} + \eta\left(B_{n10} + \frac{\varphi_{010}}{2n^2}B_{n00}\right)
\end{aligned} \tag{21}$$

且均取自主轨迹上的值；式中 ϕ 表示主轨迹上的电位；ε_s 为电子自阴极逸出的轴向初电位；$\varphi_{100} = \dfrac{d\phi}{ds}$，不难证明

$$\frac{dK}{ds} = \frac{N+Q}{2} \tag{22}$$

于是，由 $\mu = \mu_1$，将其代入式（8），得曲线坐标系下零级近似轨迹方程，即主轨迹方程

$$\begin{aligned}\varphi_{010} - 2kn^2 + 2\eta n B_{b00} &= 0, \\ \varphi_{001} - 2\eta n B_{n00} &= 0\end{aligned} \tag{23}$$

由 $\mu = \mu_1 + \mu_2$，当考虑到主轨迹方程式（23），便得到一级近似轨迹方程，即曲近轴轨迹方程

$$\begin{aligned}\frac{d}{ds}\left(n\frac{dp_2}{ds}\right) &= Fp_2 + Np_3 + 2K\frac{dp_3}{ds}, \\ \frac{d}{ds}\left(n\frac{dp_3}{ds}\right) &= Rp_3 - Qp_2 - 2K\frac{dp_2}{ds}\end{aligned} \tag{24}$$

可以看出，由变分原理得到的主轨迹方程和曲近轴轨迹方程与用轨迹法的结果完全一致[6~9]。关于方程求解方法和理想聚焦的正交条件等的讨论与文献 [5~9] 相同，这里不再细述。

当 μ 中进一步保留二、三级次幂，可以讨论系统的二、三级几何像差。另外，由 μ 也可以考虑由于电位（例如轴向初电位 ε_s）波动所产生的像差，我们将在以后的文章中讨论。

3 正交曲线局部坐标基下的变分函数及轨迹方程

若 p 不是向 n、b 方向投影，而是向转动坐标系 (t, n_f, b_f) 的 n_f、b_f 方向投影，则相当于做坐标变换

$$(p_2 + ip_3) = (u + iv)e^{-i\gamma} \tag{25}$$

式中，u、v 分别为 p 在 n_f、b_f 上的投影：

$$\gamma = \gamma(s) = \int_0^s \chi\,ds \tag{26}$$

$$n_f = n\cos\gamma - b\sin\gamma, \quad b_f = n\sin\gamma + b\cos\gamma \tag{27}$$

则此时曲线局部基是由 (t, n_f, b_f) 构成的正交基。其变分函数为[8]

$$\bar{\mu} = \varphi_*^{1/2}(h_1^2 + u'^2 + v'^2)^{1/2} - \eta(A_s + u'A_u + v'A_v) \tag{28}$$

式中，$h_1 = 1 - ku\cos\gamma - kv\sin\gamma$。将 $\bar{\mu}$ 代入欧拉方程，便得到转动的正交曲线坐标系 (s, u, v) 中的普遍轨迹方程

$$\begin{aligned}2\bar{I}(u'' + kh_1\cos\gamma) + u'\frac{d\bar{I}}{ds} &= \frac{\partial\varphi}{\partial u} - 2\eta\bar{I}^{1/2}\left[\left(\frac{\partial A_s}{\partial u} - \frac{\partial A_u}{\partial s}\right) + \left(\frac{\partial A_v}{\partial u} - \frac{\partial A_u}{\partial v}\right)v'\right], \\ 2\bar{I}(v'' + kh_1\sin\gamma) + v'\frac{d\bar{I}}{ds} &= \frac{\partial\varphi}{\partial v} - 2\eta\bar{I}^{1/2}\left[\left(\frac{\partial A_s}{\partial v} - \frac{\partial A_v}{\partial s}\right) + \left(\frac{\partial A_u}{\partial v} - \frac{\partial A_v}{\partial u}\right)u'\right]\end{aligned} \tag{29}$$

式中，

$$\bar{I} = \varphi_*/(h_1^2 + u'^2 + v'^2) \tag{30}$$

同样，将空间电位做如下展开

$$\varphi(s,u,v) = \sum_{i=0}^{\infty}\sum_{j=0}^{\infty}\frac{\overline{\varphi}_{0ij}(s)}{i!j!}u^i v^j \tag{31}$$

同时，由于式（12）满足分量式

$$\begin{aligned} B_s &= \frac{\partial A_v}{\partial u} - \frac{\partial A_u}{\partial v}, \\ B_u &= \left(\frac{\partial A_s}{\partial v} - \frac{\partial A_v}{\partial s}\right)\!\Big/h_1, \\ B_v &= \left(\frac{\partial A_u}{\partial s} - \frac{\partial A_s}{\partial u}\right)\!\Big/h_1 \end{aligned} \tag{32}$$

采用规范 $\dfrac{\partial A_u}{\partial v} = -\dfrac{\partial A_v}{\partial u}$，$\boldsymbol{A}(s,0,0)=0$，由式（32）可得

$$\begin{aligned} A_s &= \frac{1}{2}\left\{\int_0^v[h_1 B_u + (h_1 B_u)_{u=0}]\mathrm{d}v - \int_0^u[h_1 B_v + (h_1 B_v)_{v=0}]\mathrm{d}u\right\}, \\ A_u &= -\frac{1}{2}\int_0^v B_s \mathrm{d}v, \\ A_v &= \frac{1}{2}\int_0^u B_s \mathrm{d}u \end{aligned} \tag{33}$$

利用 $\boldsymbol{B}$ 按 u、v 展开的幂级数系数[9]，可以得到磁矢位的展开式

$$\begin{aligned} A_s =\ & -B_{v00}u + B_{u00}v - \frac{1}{2}(B_{v10} - kB_{v00}\cos\gamma)u^2 + \frac{1}{2}(B_{u10} - B_{v01} - kB_{u00}\cos\gamma + kB_{v00}\sin\gamma)uv + \\ & \frac{1}{2}(B_{u01} - kB_{u00}\sin\gamma)v^2 + \cdots, \\ A_u =\ & -\frac{1}{2}\left(B_{s00}v + B_{s10}uv + \frac{1}{2}B_{s01}v^2 + \cdots\right), \\ A_v =\ & \frac{1}{2}\left(B_{s00}u + \frac{1}{2}B_{s10}u^2 + B_{s01}uv + \cdots\right) \end{aligned} \tag{34}$$

系数 φ_{0ij} 和 B_{sij}、B_{uij}、B_{vij}（$i,j=0,1,2,\cdots$）同样满足一些关系式[9]，若已知实验室直角坐标系中 φ 和 $\boldsymbol{B}$ 的分布，则按文献[9]的方法可确定这些系数。于是，变分函数可表示为

$$\overline{\mu} = \overline{\mu}_0 + \overline{\mu}_1 + \overline{\mu}_2 + \cdots \tag{35}$$

式中，

$$\overline{\mu}_0 = n \tag{36a}$$

$$\overline{\mu}_1 = \left(\frac{\overline{\varphi}_{010}}{2n} - nk\cos\gamma + \eta B_{v00}\right)u + \left(\frac{\overline{\varphi}_{001}}{2n} - nk\sin\gamma - \eta B_{u00}\right)v \tag{36b}$$

$$\begin{aligned} \overline{\mu}_2 =\ & \frac{n}{2}(u'^2 + v'^2) + \overline{K}(uv' - vu') + \frac{1}{2}\overline{F}u^2 + \frac{1}{2}\overline{R}v^2 + \\ & (\overline{N} - \overline{K}')uv + \frac{\varepsilon_s}{2n}[k_0^2(u_0\cos\gamma_0 + v_0\sin\gamma_0)^2 + u_0'^2 + v_0^2] \end{aligned}$$

$$= \frac{n}{2}(u'^2+v'^2)+\bar{K}(uv'-vu')+\frac{1}{2}\bar{F}u^2+\frac{1}{2}\bar{R}v^2$$
$$-(\bar{Q}-\bar{K}')uv+\frac{\varepsilon_s}{2n}[k_0^2(u_0\cos\gamma_0+v_0\sin\gamma_0)^2+u_0'^2+v_0^2] \tag{36c}$$

式中，

$$n=\sqrt{\phi(s)+\varepsilon_s},$$

$$\bar{K}=-\frac{1}{2}\eta B_{s00},$$

$$\bar{F}=\frac{\bar{\varphi}_{020}}{2n}-\frac{3(\bar{\varphi}_{010})^2}{4n^3}-\eta^2\frac{B_{v00}^2}{n}+\eta\left(B_{v10}-\frac{3}{2}\frac{\bar{\varphi}_{010}}{n^2}B_{v00}\right),$$

$$\bar{R}=\frac{\bar{\varphi}_{002}}{2n}-\frac{3(\bar{\varphi}_{001})^2}{4n^3}-\eta^2\frac{B_{u00}^2}{n}-\eta\left(B_{u01}-\frac{3}{2}\frac{\bar{\varphi}_{001}}{n^2}B_{u00}\right), \tag{37}$$

$$\bar{N}=\frac{\bar{\varphi}_{011}}{2n}-\frac{3\bar{\varphi}_{010}\bar{\varphi}_{001}}{4n^3}+\eta^2\frac{B_{u00}B_{v00}}{n}+\eta\left(B_{v01}+\frac{\bar{\varphi}_{010}}{2n^2}B_{u00}-\frac{\bar{\varphi}_{001}}{n^2}B_{v00}\right),$$

$$\bar{Q}=-\frac{\bar{\varphi}_{011}}{2n}+\frac{3\bar{\varphi}_{010}\bar{\varphi}_{001}}{4n^3}-\eta^2\frac{B_{u00}B_{v00}}{n}+\eta\left(B_{u10}+\frac{\bar{\varphi}_{001}}{2n^2}B_{v00}-\frac{\bar{\varphi}_{010}}{n^2}B_{u00}\right)$$

同样不难得出 $\bar{K}'=(\bar{N}+\bar{Q})/2$。

由 $\bar{\mu}=\bar{\mu}_1$，即得主轨迹方程

$$\varphi_{010}-2n^2k\cos\gamma+2n\eta B_{v00}=0$$
$$\varphi_{001}-2n^2k\sin\gamma-2n\eta B_{u00}=0 \tag{38}$$

由 $\bar{\mu}=\bar{\mu}_1+\bar{\mu}_2$，并考虑方程式（38），便可得曲近轴轨迹方程

$$\frac{\mathrm{d}}{\mathrm{d}s}\left(n\frac{\mathrm{d}u}{\mathrm{d}s}\right)=\bar{F}u+\bar{N}v+2\bar{K}\frac{\mathrm{d}v}{\mathrm{d}s}$$
$$\frac{\mathrm{d}}{\mathrm{d}s}\left(n\frac{\mathrm{d}v}{\mathrm{d}s}\right)=\bar{R}v-\bar{Q}u-2\bar{K}\frac{\mathrm{d}u}{\mathrm{d}s} \tag{39}$$

上述公式与轨迹法所得到的结果也是一致的[9]，可由此讨论系统的聚焦性质，并可进一步研究几何像差或色像差。

结束语

本文用变分原理研究了宽电子束聚焦的普遍理论，利用两种曲线局部坐标系，给出了对应的变分函数和磁场以磁矢位表示的普遍轨迹方程；给出了用磁感应强度沿曲轴展开式系数表示的磁矢位沿轴展开式；并由一、二级变分函数给出了主轨迹方程和曲近轴轨迹方程，它们与轨迹法的结果完全一致，从而证明了在曲线坐标系下轨迹法和变分法结果的一致性。同时本文也为进一步研究宽电子束曲轴系统的像差提供了一条新的途径。

参 考 文 献

[1] ГРИНБЕРГ Г А. Избранные вопросы математическои теории электрических

явлений[M]. Москва-Ленинград: Изд, А Н СССР, 1948.

[2] STURROCK P A. Static and dynamic electron optics [M]. Cambridge: Cambridge University Press, 1955.

[3] HUTTER R G E. Deflection defocusing effects in cathode ray tubes at large deflection angles[J]. IEEE-ED, 1970, 17: 1022 – 1031.

[4] ROSE H. Hamiltonion magnetic optics[J]. Nucl. Instrum. & Methods in Phys. Research, 1987, A258: 374 – 401.

[5] ZHOU L W. Optics of wide electron beam focusing, electron optical systems[J]. Electron Optical Systems, SEM, Inc., 1984: 45 – 62.

[6] ZHOU L W. A generalized theory of wide electron beam focusing[J]. Advances in Electronics and Electron Physics, 1985: 575 – 589.

[7] ZHOU L W, NI G Q, QIU B C. Tensor analysis of electron motion in curvilinear coordinate systems(Ⅰ)[J]. Optik, 1988, 79(2): 53 – 66.

[8] ZHOU L W, NI G Q, QIU B C. Tensor analysis of electron motion in curvilinear coordinate systems(Ⅱ)[J]. Optik, 1988, 78(3): 101 – 107.

[9] 周立伟, 金伟其, 倪国强. 曲轴宽电子束聚焦理论的研究[J]. 光电子学技术, 1988, 8(4): 8 – 22.

5.7 相对论修正下的宽电子束聚焦的普遍理论
On the Relativistic Generalized Theory of
Wide Electron Beam Focusing

摘要：本文由 Frenet 局部曲线坐标系下用张量方法对相对论修正的普遍轨迹方程出发，考虑包含阴极物面的宽电子束聚焦的情况，推导了曲线坐标系下的主轨迹方程与曲近轴轨迹方程，研究了满足正交条件的近轴系统的特性。在曲线坐标系下的细电子束聚焦问题以及非相对论修正的宽电子束聚焦问题均可视为本文的特例。

Abstract: Based on the relativistic general trajectory equations solved by the tensor method in a Frenet local coordinate system, considering the characteristics of wide electron beam focusing possessing a cathode surface, equations of the principal trajectory and the curvilinear "paraxial" trajectory are deduced. The characteristics of the curvilinear "paraxial" system satisfying the orthogonal condition are studied. Problems concerning narrow electron beam focusing or non-relativistic wide electron beam focusing in a curvilinear coordinate system can be regarded as special cases thereof.

引 言

带电粒子束的聚焦是电子光学基本问题之一，长期以来受到人们的普遍重视。曲线坐标系下细电子束聚焦的普遍理论是由 Г. A. Гринберг[1~3] 和 P. A. Sturrock[4~5] 于 20 世纪 40—50 年代用电动力学方法和光程函数法分别建立的。而适用于研究阴极物面电位 $\phi(s=0)=0$，且自阴极逸出的电子速度很小而斜率很大的宽电子束问题的普遍理论，则是由周立伟等人[6~8]在近年来建立的。在文献 [6~8] 中，由曲线坐标系下的 Lorentz 方程出发，采用张量表示法，以逆变形式和协变形式导出了非相对论修正的普遍轨迹方程，给出了曲近轴轨迹方程和理想聚焦的正交条件，讨论了满足正交条件下曲近轴轨迹的理想聚焦性质，最后讨论了若干实例。

在实际电子光学系统中，当加速电压高达 5 万伏以上，甚至 10 万伏时，电子运动的相对论效应将不能忽略，必须考虑相对论修正。本文将宽电子束聚焦的普遍理论[7]推广到相对论情况，研究相对论修正下宽电子束聚焦的规律。

在研究宽束电子光学时，与探讨轴对称阴极透镜类似[10]，可以假定相邻轨迹处处满足如下曲近轴条件：

$$p^2(s) = 0, \dot{p}^2(s) \ll 1 \tag{1}$$

或严格写为

周立伟，金伟其，倪国强，史万宏. 北京理工大学. 电子科学学刊 Journal of Electronic (China), V. 10, No. 6, 1988, 520–527.

$$\frac{m_0(\dot{p}^2 - \dot{p}_0^2)/(2e)}{\varepsilon_s + \phi(1+\eta\phi)} \ll 1 \qquad (2)$$

且 $\phi(0)=0$，但并不一定处处要满足 $p'^2(s) \ll 1$。这里 s 为曲线轴自物点算起的弧长，$p(s)$ 为相邻轨迹相对于曲轴的偏离，点号和撇号分别表示对时间 t 和弧长 s 的微商。

1 曲线坐标系下相对论修正的电子普遍轨迹方程

设空间任意曲线 $\boldsymbol{r}=\boldsymbol{r}(x^1)$ 为自物面上 A 点逸出的电子束的曲线轴（主轨迹的矢径），x^1 为由 A 点算起的弧长，在点 N 作主轨迹曲线的法平面，其与相邻轨迹 AB^* 的交点 N^* 的位置由点 N 到点 N^* 的矢量 $\boldsymbol{p}$ 来表达。如果给定 $\boldsymbol{p}(s)$，则相邻轨迹的曲线便可完全确定。

在 N 点引入曲线坐标系 (x^1, x^2, x^3)，其单位矢量 $\boldsymbol{t}$、$\boldsymbol{n}$、$\boldsymbol{b}$ 为主轨迹在 N 处的切矢量、主法矢和副法矢。矢量 $\boldsymbol{p}$ 在 $\boldsymbol{n}$ 和 $\boldsymbol{b}$ 方向上的投影分量为 x^2 和 x^3，如图 1 所示。

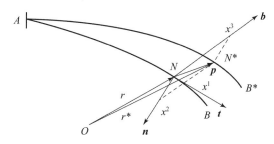

图 1 曲线坐标系、主轨迹和相邻轨迹

于是点 N^* 的相邻轨迹矢径 $\boldsymbol{r}^*$ 可表为
$$\boldsymbol{r}^* = \boldsymbol{r} + \boldsymbol{p} = \boldsymbol{r} + x^2\boldsymbol{n} + x^3\boldsymbol{b} \qquad (3)$$

设 $k=k(s)$，$\chi=\chi(s)$ 分别为主轨迹的曲率和挠率，利用微分几何中的 Frenet-Serret 公式，可得到弧长元的张量表达式为
$$(ds^*)^2 = g_{ij}dx^i dx^j, \quad (i,j=1,2,3) \qquad (4)$$
其中 g_{ij} 是协变度量张量的元素，且有
$$(g_{ij}) = \begin{bmatrix} (1-kx^2)^2 + (\chi x^2)^2 + (\chi x^3)^2 & -\chi x^3 & \chi x^2 \\ -\chi x^3 & 1 & 0 \\ \chi x^2 & 0 & 1 \end{bmatrix} \qquad (5)$$

度量张量的行列式值：
$$G = \det|(g_{ij})| = (1-kx^2)^2 \qquad (6)$$

由式（5），可得 (g_{ij}) 的逆矩阵 $(g_{ij})^{-1} = (g^{ij})$，将其称为逆变度量张量。

由电动力学中的 Lorentz 方程：
$$d(m\dot{\boldsymbol{r}}^*)/dt = -e\boldsymbol{E} - e(\dot{\boldsymbol{r}}^* \times \boldsymbol{B}) \qquad (7)$$

（这里 $\boldsymbol{E}$ 和 $\boldsymbol{B}$ 分别为电场强度和磁感应强度矢量，且 $\boldsymbol{E} = -\text{grad}\varphi$，$\boldsymbol{B} = -\mu_0 \text{grad}\Omega$，$\mu_0$ 为磁导率，φ 为空间电位，Ω 为空间磁标位）。可以得到相对论修正下逆变形式的 Lorentz 方程：
$$d(m\dot{x}^l)/dt + m\begin{Bmatrix} l \\ i \ j \end{Bmatrix}\dot{x}^i\dot{x}^j = e(A^l + B^l), \ (l=1,2,3) \qquad (8)$$

式中，$m = m_0 / \sqrt{1-(v/c)^2}$ 为相对论修正下的电子质量，m_0 和 e 分别为电子的静止质量和电荷；c 为光速；v 为电子速度；$\begin{Bmatrix} l \\ i \ j \end{Bmatrix}$ 为第二类 Christoffel 符号，且

$$\begin{Bmatrix} l \\ i \ j \end{Bmatrix} = \frac{1}{2} g^{kl} \left[\frac{\partial g_{kj}}{\partial x^i} + \frac{\partial g_{ik}}{\partial x^j} - \frac{\partial g_{ij}}{\partial x^k} \right] \tag{9}$$

而式（8）中 A^l 和 $B^l (l=1,2,3)$ 可表示为

$$A^l = g^{lj} \frac{\partial \varphi}{\partial x^j}, \quad B^l = \mu_0 \left(\dot{x}_j \frac{\partial \Omega}{\partial x^k} - \dot{x}_k \frac{\partial \Omega}{\partial x^j} \right) \Big/ \sqrt{G}, (l,j,k=1,2,3)$$

$$\dot{x}_l = g_{lj} \dot{x}^j \tag{10}$$

这里 $x_l (l=1,2,3)$ 为协变坐标。

若令 $\eta = e/2m_0 c^2$，$\varphi_* = \varphi + \varepsilon_0$ 为规范化电位，由电子运动的能量守恒定律，可得：

$$\begin{aligned} m &= m_0 (1 + 2\eta \varphi_*), \\ mv &= \sqrt{2em_0 \varphi_* (1 + \eta \varphi_*)} \end{aligned} \tag{11}$$

为方便起见，下面用另一种坐标符号 (s, p_2, p_3) 代替 (x^1, x^2, x^3)，则由式（8），考虑式（9）~式（11），可得电子在 (s, p_2, p_3) 坐标系中的运动方程：

$$(m_0/e)(1+2\eta\varphi_*) \{ \ddot{s} + [(k\chi p_3 - k'p_2)/(1-kp_2) + 2\eta (d\varphi/ds)/(1+2\eta\varphi_*)]\dot{s}^2 -$$
$$2k\dot{s}\dot{p}_2/(1-kp_2) \} = [(\partial\varphi/\partial s) + \chi p_3 (\partial\varphi/\partial p_2) - \chi p_2 (\partial\varphi/\partial p_3)]/(1-kp_2)^2 +$$
$$\mu_0 [(\dot{p}_2 - \chi p_3 \dot{s})(\partial\Omega/\partial p_3) - (\dot{p}_3 + \chi p_2 \dot{s})(\partial\Omega/\partial p_2)]/(1-kp_2) \tag{12a}$$

$$(m_0/e)(1+2\eta\varphi_*)\{\ddot{p}_2 + [k-(k^2+\chi^2)p_2 - \chi'p_3 + (k\chi p_3 - k'p_2)\chi p_3/(1-kp_2)]\dot{s}^2 -$$
$$2\chi\dot{s}\dot{p}_3 - [2k\chi p_3/(1-kp_2) - 2\eta(d\varphi/ds)/(1-2\eta\varphi_*)]\dot{s}\dot{p}_2\}$$
$$= \{\chi p_3 [(\partial\varphi/\partial s) + \chi p_3 (\partial\varphi/\partial p_2) - \chi p_2 (\partial\varphi/\partial p_3)] + (1-kp_2)^2 (\partial\varphi/\partial p_2)\}/(1-kp_2)^2 +$$
$$\mu_0 \{(\dot{p}_3 + \chi p_2 \dot{s})(\partial\Omega/\partial s) - \{[(1-kp_2)^2 + \chi^2 p_2^2 + \chi^2 p_3^2]\dot{s} -$$
$$\chi p_3 \dot{p}_2 + \chi p_2 \dot{p}_3\}(\partial\Omega/\partial p_3)\}/(1-kp_2) \tag{12b}$$

$$(m_0/e)(1+2\eta\varphi_*)\{\ddot{p}_3 - [\chi^2 p_3 - \chi'p_2 + (k\chi p_3 - k'p_2)\chi p_2/(1-kp_2)]\dot{s}^2 +$$
$$2\chi\dot{s}\dot{p}_2/(1-kp_2) - [2\eta(\partial\varphi/\partial s)/(1+2\eta\varphi_*)]\dot{s}\dot{p}_3\}$$
$$= \{-\chi p_2 [(\partial\varphi/\partial s) + \chi p_3 (\partial\varphi/\partial p_2) - \chi p_2 (\partial\Omega/\partial p_3)] +$$
$$(1-kp_2)^2 (\partial\varphi/\partial p_3)\}/(1-kp_2)^2 - \mu_0 \{(\dot{p}_2 - \chi p_3 \dot{s})(\partial\Omega/\partial s) -$$
$$\{[(1-kp_2)^2 + \chi^2 p_2^2 + \chi^2 p_3^2]\dot{s} - \chi p_3 \dot{p}_2 + \chi p_2 \dot{p}_3\}(\partial\Omega/\partial p_2)]\}/(1-kp_2) \tag{12c}$$

利用能量守恒关系，可得如下的变换：

$$m\dot{x}^l = mv(\text{d}s/\text{d}s^*)(\text{d}x^l/\text{d}s)$$

$$m\text{d}(m\dot{x}^l)/\text{d}t = (mv)^2 (\text{d}s/\text{d}s^*)^2 \text{d}^2 x^l/\text{d}s^2 + \frac{1}{2}(\text{d}x^l/\text{d}s)\{\text{d}[(mv)^2 (\text{d}s/\text{d}s^*)^2]/\text{d}s\} \tag{13}$$

这里，当 $l=1$ 时 $\text{d}x^l = \text{d}s$，且

$$(\text{d}s/\text{d}s^*)^2 = (1-kp_2)^2 + (p_2' - \chi p_3)^2 + (p_3' + \chi p_2)^2 \tag{14}$$

现在我们假定：

$$I = \varphi_* (1+\eta\varphi_*)/[(1-kp_2)^2 + (p_2' - \chi p_3)^2 + (p_3' + \chi p_2)^2] \tag{15}$$

由式（8）乘 m 以及式（11）、式（13）~式（15），经过繁复的推导，式（12a）可变换成如下的形式：

$$dI/ds = -2I(k\chi p_3 - k'p_2 - 2kp_2')/(1-kp_2) + (1+2\eta\varphi_*) \times \\ [(\partial\varphi/\partial s) + \chi p_3(\partial\varphi/\partial p_2) - \chi p_2(\partial\varphi/\partial p_3)]/(1-kp_2)^2 + \\ \sqrt{2eI/m_0}\mu_0[(p_2'-\chi p_3)(\partial\Omega/\partial p_2) - (p_3'+\chi p_2)(\partial\Omega/\partial p_2)]/(1-kp_2) \quad (16)$$

而式（12b）和式（12c）利用式（16）可得到两种完全等价的形式：

$$2I(p_2'' + k - (k^2+\chi^2)p_2 - \chi'p_3 - 2\chi p_3') + (p_2'-\chi p_3)dI/ds \\ = (1+2\eta\varphi_*)(\partial\varphi/\partial p_2) + \sqrt{2eI/m_0}\mu_0\{(p_3'+\chi p_2)[a\Omega/\partial s + \\ \chi p_3(\partial\Omega/\partial p_2) - \chi p_2(\partial\Omega/\partial p_3)] - (1-kp_2)^2(\partial\Omega/\partial p_3)\}/(1-kp_2) \quad (17a)$$

$$2I(p_3'' - \chi^2 p_3 + \chi'p_2 + 2\chi p_2') + (p_3'+\chi p_2)dI/ds \\ = (1+2\eta\varphi_*)(\partial\varphi/\partial p_3) - \sqrt{2eI/m_0}\mu_0\{(p_3'-\chi p_2)(\partial\Omega/\partial s + \\ \chi p_3(\partial\Omega/\partial p_2) - \chi p_2(\partial\Omega/\partial p_3)) - (1-kp_2)^2(\partial\Omega/\partial p_2)\}/(1-kp_2) \quad (17b)$$

或者：

$$2I[k + p_2'' - (k^2+\chi^2)p_2 - \chi'p_3 - 2\chi p_3' - (k\chi p_3 - k'p_2 - 2kp_2')(p_2'-\chi p_3)/(1-kp_2)] \\ = (1+2\eta\varphi_*)\{(1-kp_2)^2(\partial\varphi/\partial p_2) - (p_2'-\chi p_3)[\partial\varphi/\partial s + \chi p_3(\partial\varphi/\partial p_2) - \chi p_2(\partial\varphi/\partial p_3)]\}/ \\ (1-kp_2)^2 + \sqrt{2eI/m_0}\mu_0\{(p_3'+\chi p_2)[a\Omega/\partial s + p_2'(\partial\Omega/\partial p_2) - \chi p_2(\partial\Omega/\partial p_3)] - \\ [(1-kp_2)^2 + (p_2'-\chi p_3)^2](\partial\Omega/\partial p_3)\}/(1-kp_2) \quad (18a)$$

$$2I[p_3'' - \chi^2 p_3 + \chi'p_2 + 2\chi p_2' - (k\chi p_3 - k'p_2 - 2kp_2')(p_3'+\chi p_2)/(1-kp_2)] \\ = (1+2\eta\varphi_*)\{(1-kp_2)^2(\partial\varphi/\partial p_3) - (p_3'+\chi p_2)[\partial\varphi/\partial s + \chi p_3(\partial\varphi/\partial p_2) - \\ \chi p_2(\partial\varphi/\partial p_3)]\}/(1-kp_2)^2 - \sqrt{2eI/m_0}\mu_0\{(p_2'-\chi p_3)(\partial\Omega/\partial s) - p_3'(\partial\Omega/\partial p_3) + \\ \chi p_3(\partial\Omega/\partial p_2) - [(1-kp_2)^2 + (p_3'+\chi p_3)^2](\partial\Omega/\partial p_2)\}/(1-kp_2) \quad (18b)$$

以上即为曲线坐标系下相对论修正的普遍轨迹方程。

若在式（9）中以磁矢位 A 来表示磁感应强度 $B = \text{rot}A$，也可以类似文献[8]得到相应的相对论修正下的普遍轨迹方程。本文下面的推导将以磁标位 Ω 的形式来进行。

2 相对论修正下的主轨迹方程和曲近轴轨迹方程

与文献[10]、文献[11]类似，我们在普遍轨迹方程中引入曲近轴条件式（1）、式（2），将 p_2、p_3、$\dot p_2$、$\dot p_3$ 作为一级小量来考察式（18a）和式（18b）的各项系数。注意到当加速电压高达 10^5 V 时，$m/m_0 = 1 + 2\eta\varphi_* \approx 1.196$，故仍可将 $2\eta\varphi_*$ 作为小量来展开。

首先将 N^* 点处的电位 φ 和磁标位 Ω 在曲线轴的 N 点处展开，其表达式为：

$$\varphi(s, p_2, p_3) = \sum_{i=0}^{\infty}\sum_{j=0}^{\infty} \varphi_{0ij} p_2^i p_3^j/(i!j!), \\ \Omega(s, p_2, p_3) = \sum_{i=0}^{\infty}\sum_{j=0}^{\infty} \Omega_{0ij} p_2^i p_3^j/(i!j!) \quad (19)$$

式中，

$$\varphi_{0ij}(s) = \left(\frac{\partial^{i+j}\varphi}{\partial p_2^i \partial p_3^j}\right)\bigg|_{p_2=p_3=0},$$
$$\Omega_{0ij}(s) = \left(\frac{\partial^{i+j}\Omega}{\partial p_2^i \partial p_3^j}\right)\bigg|_{p_2=p_3=0} \tag{20}$$

令 $\phi(s) = \varphi_{000}(s)$，$\Omega(s) = \Omega_{000}(s)$ 分别为曲轴上的电位和磁标位分布。

现在考虑系数 I，由于在阴极面处 $\varphi(s=0) = 0$，$1 + \eta\varphi_* = 1 + \eta\varepsilon_0 \approx 1$。故在起点处，它可表示为

$$(I)|_{s=p_2=p_3=0} = \varepsilon_0/(1 + p_{20}'^2 + p_{30}'^2) = \varepsilon_s \tag{21}$$

因此，I 的 $0+1$ 级近似可表示为

$$(I)|_{0+1} = [\varepsilon_s + \phi(1 + \eta\phi)](1 + 2kp_2) + (p_2\varphi_{010} + p_3\varphi_{001})(1 + 2\eta\phi) \tag{22}$$

且式（22）成立的条件是式（2）的严格表达式成立。式中 ε_s 为电子沿曲轴的轴向初电位。

利用式（19）、式（20），将式（18a）和式（18b）（或者式（17a）和式（17b））展开为小量的各阶次项，考虑到曲线轴本身也是电子轨迹，则由零级项可得主轨迹方程：

$$2kn^2 - \varphi_{010}(1 + 2\eta\phi) + \sqrt{2e/m_0}\mu_0 n\Omega_{001} = 0 \tag{23a}$$

$$\varphi_{001}(1 + 2\eta\phi) + \sqrt{2e/m_0}\mu_0 n\Omega_{010} = 0 \tag{23b}$$

由一级项可得曲近轴轨迹方程：

$$\mathrm{d}(np_2')/\mathrm{d}s = Fp_2 + Np_3 + 2Kp_3' \tag{24a}$$

$$\mathrm{d}(np_3')/\mathrm{d}s = Rp_3 - Qp_2 - 2Kp_2' \tag{24b}$$

这里 n、F、K、R、N 和 Q 表示为主轨迹曲线 s 的如下函数：

$$n = \sqrt{\varepsilon_s + \phi(1+\eta\phi)}, \quad K = n\chi + \sqrt{e/(8m_0)}\mu_0\Omega_{100},$$
$$F = n(\chi^2 - k^2) + (1+2\eta\phi)\varphi_{020}/(2n) - (1+2\eta\varphi)k\varphi_{010}/n + \eta\varphi_{010}^2/n +$$
$$\quad \sqrt{e/(2m_0)}\mu_0[\chi\Omega_{100} - \Omega_{011} - (1+2\eta\phi)\varphi_{010}\Omega_{001}/(2n^2)],$$
$$N = (1+2\eta\phi)\chi\varphi_{100}/(2n) + n\chi' - (1+2\eta\varphi)k\varphi_{001}/n +$$
$$\quad (1+2\eta\varphi)\varphi_{011}/(2n) + \eta\varphi_{001}\varphi_{010}/n - \sqrt{e/(2m_0)}\mu_0[\Omega_{002} + (1+2\eta\phi)\varphi_{001}\Omega_{001}/(2n^2)],$$
$$R = n\chi^2 + (1+2\eta\phi)\varphi_{002}/(2n) + \eta\varphi_{001}^2/n +$$
$$\quad \sqrt{e/(2m_0)}\mu_0[\chi\varphi_{100} + \Omega_{011} + (1+2\eta\phi)\varphi_{001}\Omega_{010}/(2n^2)],$$
$$Q = (1+2\eta\phi)\chi\varphi_{100}/(2n) - (1+2\eta\phi)\varphi_{011}/(2n) + n\chi' - \eta\varphi_{010}\varphi_{001}/n -$$
$$\quad \sqrt{e/(2m_0)}\mu_0[\Omega_{020} + (1+2\eta\phi)\varphi_{010}\Omega_{010}/(2n^2)]$$

$$\tag{25}$$

φ_{ij}、Ω_{ijk} 实际对应的是协变微商的轴上值，只有当 $i=0$ 时式（20）才成立，即寻常微商等于协变微商。

利用式（23a）和式（23b），不难证明系数 N、Q 和 K 之间仍存在下列关系式：

$$\mathrm{d}K/\mathrm{d}s = (N + Q)/2 \tag{26}$$

应该指出：在求解时，相对论修正下的主轨迹方程式（23a）和式（23b）与曲近轴轨迹方程式（24a）和式（24b）的各系数必须取相同的 ε_s 值。

式（24a）和式（24b）式的求解可以采用降阶法，即化为四元一次方程组来求解[7]。可以证明，在普遍情况下，曲近轴系统并不是理想聚焦（点聚焦或线聚焦）系统，即不存在高斯光学性质。因此下节我们将限于讨论具有理想聚焦性质的曲近轴系统。

另外，由于电位和磁标位均满足拉氏方程以及主轨迹方程式（23），使得电位和磁标位的展开以及 k、χ 之间存在下列关系：

$$\varphi_{200} + \varphi_{020} + \varphi_{002} = 0, \quad \Omega_{200} + \Omega_{020} + \Omega_{002} = 0,$$

$$\varphi_{200} = \mathrm{d}^2\phi/\mathrm{d}s^2 - k\varphi_{010}, \quad \Omega_{200} = \mathrm{d}^2\Omega/\mathrm{d}s^2 - k\Omega_{010},$$

$$\mathrm{d}(2kn^2)/\mathrm{d}s = \varphi_{110} - k\varphi_{100} + \chi\varphi_{001} - \sqrt{2e/m_0}\mu_0 n [\Omega_{101} - \chi\Omega_{010} - (1 + 2\eta\phi)\varphi_{100}\Omega_{001}/(2n^2)],$$

$$\varphi_{101} - \chi\varphi_{010} = -\sqrt{2e/m_0}\mu_0 n [\Omega_{110} - k\Omega_{100} + \chi\Omega_{001} + (1 + 2\eta\phi)\varphi_{100}\Omega_{010}/(2n^2)]$$

(27)

式（27）表明，电位和磁标位及其微商之间的联系，无论是正设计或是逆设计，都必须满足上述关系。

上面的结果也可利用文献［8］所给的协变形式的轨迹方程得到。

3 相对论修正下的曲近轴系统的正交条件

与文献［5］、文献［7］类似，在 (p_2, p_3) 平面中引入旋转坐标系 (u, v)：

$$\begin{pmatrix} p_2 \\ p_3 \end{pmatrix} = \begin{pmatrix} \cos\gamma & -\sin\gamma \\ \sin\gamma & \cos\gamma \end{pmatrix} \begin{pmatrix} u \\ v \end{pmatrix} \tag{28}$$

式中，γ 为此二坐标系间的夹角。当系数 n、K、F、R、N 和 Q 满足如下曲近轴系统的正交条件

$$n[(N-Q)(F'-R') - (F-R)(N'-Q')] = 2K[(F-R)^2 + (N-Q)^2] \tag{29}$$

时，曲近轴轨迹方程式（24a）和式（24b）的变量便可分离，即

$$\mathrm{d}(nu')/\mathrm{d}s = Uu \tag{30a}$$

$$\mathrm{d}(nv')/\mathrm{d}s = Vv \tag{30b}$$

式中，

$$U = (F+R)/2 + \sqrt{(F-R)^2 + (N-Q)^2}/2 - K^2/n,$$

$$V = (F+R)/2 - \sqrt{(F-R)^2 + (N-Q)^2}/2 - K^2/n$$

(31)

式中，

$$\gamma' = -K/n, \quad \tan 2\gamma = (N-Q)/(F-R) \tag{32}$$

满足式（29）的正交条件所对应的系统称为正交曲近轴系统。

正如 P. A. Sturrock[5] 所指出的，满足正交条件的曲轴系统可以形成点聚焦或线聚焦的图像。因此正交条件式（29）的意义在于，它为判断系统是否能够形成理想聚焦给出了一个明确的判据。

式（30a）和式（30b）分别为 u 和 v 的二阶线性齐次微分方程，尽管方程在阴极面处存在奇异点，但求解方法与轴对称阴极透镜的线性方程并无根本差异。因此 u 和 v 的解可用两个线性独立的特解来表示。令 u_α、u_β 和 v_α、v_β 为式（30a）和式（30b）的两组特解，

它们满足如下初始条件：

$$u_\alpha(0) = v_\alpha(0) = 0, \sqrt{\varepsilon_s}u'_\alpha(0) = \sqrt{\varepsilon_s}v'_\alpha(0) = 1,$$
$$u_\beta(0) = v_\beta(0) = 1, \sqrt{\varepsilon_s}u'_\beta(0) = \sqrt{\varepsilon_s}v'_\beta(0) = 0 \tag{33}$$

式（30a）和式（30b）的通解可分别表示为 u_α、u_β 和 v_α、v_β 的线性组合，从而可讨论曲近轴轨迹的成像性质。

为了使系统获得无像散点聚焦的像，必须使 $U = V$，即有

$$F - R = N - Q = 0 \tag{34}$$

或者

$$(1+2\eta\phi)(\varphi_{002} - \varphi_{020} + 2k\varphi_{010})/n + nk^2 + \eta(\varphi_{001}^2 - \varphi_{010}^2)/n +$$
$$\sqrt{e/(2m_0)}\mu_0[\Omega_{011} + (1+2\eta\phi)(\varphi_{001}\Omega_{010} + \varphi_{010}\Omega_{001})/(4n^2)] = 0 \tag{35a}$$

$$(1+2\eta\phi)(\varphi_{011} - k\varphi_{001})/n + \sqrt{e/(2m_0)}\mu_0[\Omega_{020} - \Omega_{002} +$$
$$(1+2\eta\phi)(\Omega_{010}\varphi_{010} - \Omega_{001}\varphi_{001})/(2n^2)] = 0 \tag{35b}$$

由式（29）知，系统必然满足正交条件。

于是在 $\varepsilon_s = \varepsilon_{s_1}$ 对应的像面 $s = s_i$ 处，有

$$u_\alpha(\varepsilon_{s_1}, s_i) = v_\alpha(\varepsilon_{s_1}, s_i) = 0,$$
$$u'_\alpha(\varepsilon_{s_1}, s_i) = v'_\alpha(\varepsilon_{s_1}, s_i) \tag{36}$$

放大率之模可表达为

$$M_u = M_v = M = \left| 1/u'_\alpha(\varepsilon_{s_1}, s_i) \sqrt{\varepsilon_{s_1} + \phi(s_i)[1 + 2\eta\phi(s_i)]} \right| \tag{37}$$

这类系统以 $\varepsilon_s = \varepsilon_{s_1}$ 逸出的电子束将在其像面上理想聚焦（点聚焦），具有轴对称电子光学系统的性质。

若曲近轴系统仅满足正交条件式（29），并不满足条件式（34），则 $U \ne V$，因而 u_α、u_β 和 v_α、v_β 是不同的，系统聚焦性能各向异性。对于 $\varepsilon_s = \varepsilon_{s_1}$，两个方向的放大率之模为

$$M_u = \left| 1/u'_\alpha(\varepsilon_{s_1}, s_{i1}) \sqrt{\varepsilon_{s_1} + \phi(s_{i1})[1 + 2\eta\phi(s_{i1})]} \right|,$$
$$M_v = \left| 1/v'_\alpha(\varepsilon_{s_1}, s_{i2}) \sqrt{\varepsilon_{s_1} + \phi(s_{i2})[1 + 2\eta\phi(s_{i2})]} \right| \tag{38}$$

并且像的位置 s_{i1} 与 s_{i2} 不重合，即正交电子光学场可能有两个理想聚焦像，它们都以 $\varepsilon_s = \varepsilon_{s_1}$ 相对应。在 s_{i1} 处电子束在 $\xi_2(s_{i1})$ 方向形成（弧矢）焦线，而在 s_{i2} 处则在 $\xi_1(s_{i2})$ 方向形成（子午）焦线，此两者位置之差，表征了曲近轴系统的像散。其中

$$(\xi_1(s), \xi_2(s)) = (\mathbf{n}(s), \mathbf{b}(s)) \begin{pmatrix} \cos\gamma & -\sin\gamma \\ \sin\gamma & \cos\gamma \end{pmatrix} \tag{39}$$

即实际 $\xi_1(s_{i2})$ 和 $\xi_2(s_{i1})$ 不一定正交。

当加速电压不是很大时，可以忽略相对论效应，则在前面的所有公式中略去包含 η 的项，就得到文献 [7] 所得的非相对论形式的宽电子束聚焦的公式，即文献 [7] 所得结果是本文的一个特例。并且本文的结果也可以推广到研究细电子束聚焦的问题。

与文献 [7] 类似，可以证明：对于主轨迹为直轴且旋转对称的情况，由本文公式可以推得文献 [1] 所给的电磁复合聚焦阴极透镜近轴轨迹的线性方程和转角公式。对于静电阴极透镜的轴外轨迹可推得文献 [11] 中所给的轴外主轨迹方程和曲近轴轨迹方程，并

且这些系统均为正交曲近轴系统。

结束语

（1）本文研究了曲线坐标系下相对论修正的宽电子束聚焦的普遍理论，推导了电子束的主轨迹方程与曲近轴轨迹方程，给出了曲近轴系统的正交条件。

（2）本文的研究说明在相对论形式下，同样只有满足正交条件的曲近轴系统，且自阴极物面逸出的宽电子束，其主轨迹与相邻轨迹取同一 ε_s 值时，才具有理想聚焦（点或线聚焦）的性质。

（3）非相对论形式的宽电子束问题可以看作本文结果的一个特例，并且本文结果还可推广到曲轴细电子束聚焦问题。因而本文可基本包含目前实际中广泛使用的电子光学问题，发展了电子光学理论。

参 考 文 献

［1］ГРИНБЕРГ Г А. Избранные вопросы математическои теории электрических явлений［M］. Москва-Ленинград：Изд，А Н СССР，1948.

［2］ГРИНБЕРГ Г А. Основы обшей теории фокусируюшего действия электростатических и магнитных полеи II. пространственные электростатические поля［J］. Изд，А Н СССР，1942，37（9）：295 – 303.

［3］ГРИНБЕРГ Г А. Основы обшей теории фокусируюшего действия статических электрических и магнитнюх полей［J］. Ж. Т. Ф. , 1943, 13：361 – 388.

［4］STURROCK P A. Static and dynamic electron optics［M］. Cambridge：Cambridge University Press，1955.

［5］STURROCK P A. The imaging properties of electron beams in arbitrary static electromagnetic fields［J］. Philosophical Transactions, 1952, A245：155 – 187.

［6］ZHOU L W. Optics of wide electron beam focusing, electron optical systems［J］. Electron Optical Systems, SEM,Inc.,1984：45 – 62.

［7］ZHOU L W. A generalized theory of wide electron beam focusing［J］. Advances in Electronics and Electron Physics, 1985：575 – 589.

［8］ZHOU L W, NI G Q, QIU B C. Tensor analysis of electron motion in curvilinear coordinate systems［C］. Proceeding of the International Symposium on Electron Optics, Beijing, Sept. 9 – 13, 1986.

［9］西门纪业.复合浸没物镜的电子光学性质和像差理论［J］.物理学报,1957,13(4)：339 – 356.

［10］周立伟,艾克聪,潘顺臣.关于电磁复合聚焦阴极透镜的像差理论［J］.物理学报,1983,32：376 – 391.

［11］方二伦,周立伟.像管电子光学设计软件简介［C］.第四届全国电子光学讨论会,辽宁兴城,1989.

5.8 曲近轴方程组确定静电聚焦成像系统的场曲和像散的研究

A Study of Determining Astigmatism and Curvature of Field in Electrostatic Imaging Systems by Using a Set of Curvilinear "Paraxial" Equations

摘要：在大物面宽电子束聚焦成像系统的计算中，场曲和像散的确定是一个必须解决的实际问题。鉴于采用实际轨迹方程或运动方程解决这一问题在确定的真实的子午焦点时有一定困难，本文提出用曲近轴方程组确定成像系统的场曲和像散。本文研究和推导了在静电成像系统中沿主轨迹运动的 Frenet 局部坐标系和笛卡儿直角坐标系中两种形式的曲近轴轨迹方程组和曲近轴运动方程组，并给出了相应的计算初始条件以及计算实例。

Abstract: In the computation of imaging systems with a large object surface which emits angularly wide and transversely large electron beams, determination of astigmatism and curvature of field is a practical problem to be solved. Owing to the difficulties of practical electron trajectory equations or equations for the motion of electrons in solving problems concerning the determination of actual meridional focus, a set of curvilinear "paraxial" equations have been suggested to determine astigmatism and curvature of field for the imaging systems. In this paper, two forms for a set of curvilinear "paraxial" trajectory equations and equations of electron motion for electrostatic imaging systems have been studied and deduced, in which the Frenet local coordinate system and the Cartesian local coordinate system moving along the curved "principal" trajectory have been introduced. Initial conditions for computation and a practical computational example are also given.

引 言

在静电聚焦成像系统的场曲与像散的计算中，直接由实际轨迹方程（或运动方程）出发求解存在着两个问题。其一，实际轨迹方程并不是线性方程，它不能确定单元电子束以及系统的理想聚焦特性，诸如确定子午焦点与弧矢焦点、子午焦线与弧矢焦线等问题。其二，即使自同一物点发出的初能量和初角度相同但初方向相反的两条子午轨迹，由于各自所经历的场是不同的，它们与主轨迹并不会聚于一点，通常将有两个交点。因此，使用实际轨迹方程或运动方程计算时只能人为地或近似地确定子午焦点和弧矢焦点。我们在前两

周立伟[a]，方二伦[b]. a) 北京工业学院，b) 西安现代化学研究所. 第二届全国电子光学学术讨论会（The Second Domestic Symposium on Electron Optics），四川成都，1982.

篇文章[1,2]中建立的曲近轴方程组则基本上解决了上述的问题。本文将论证静电聚焦成像系统中曲近轴方程组的理想聚焦（线聚焦）的性质；从而确定子午焦点、弧矢焦点与相应的场曲和像散，以及子午焦线与弧矢焦线，并给出相应的计算初始条件。

此外，本文还推导了运动的直角坐标系表示的曲近轴运动方程与曲近轴轨迹方程，以及用该方程组确定场曲与像散等问题。

1 正交曲线坐标系下的曲近轴轨迹方程组

我们在文献[1]中曾证明了，适合大物面静电聚焦成像系统计算，以正交曲线坐标系表示的曲近轴轨迹方程组可表示如下。

主轨迹方程

$$r'' = \frac{1+r'^2}{2[\varphi(z,r)+\varepsilon_s]}\left(\frac{\partial \varphi}{\partial r} - r'\frac{\partial \varphi}{\partial z}\right) \tag{1}$$

曲近轴轨迹方程

$$\begin{aligned} p_2'' + F_1 p_2' + F_2 p_2 &= 0, \\ p_3'' + G_1 p_3' + G_2 p_3 &= 0 \end{aligned} \tag{2}$$

式中，

$$\begin{aligned} F_1 &= G_1 = \frac{1+r'^2}{2[\varphi(z,r)+\varepsilon_s]}\frac{\partial \varphi}{\partial z}, \\ F_2 &= \frac{3r''^2}{(1+r'^2)^2} + \frac{1}{2[\varphi(z,r)+\varepsilon_s]}\left(-\frac{\partial^2 \varphi}{\partial r^2} + 2r'\frac{\partial^2 \varphi}{\partial r\partial z} - r'^2\frac{\partial^2 \varphi}{\partial z^2}\right), \\ G_2 &= -\frac{1+r'^2}{2[\varphi(z,r)+\varepsilon_s]}\frac{1}{r}\frac{\partial \varphi}{\partial r} \end{aligned} \tag{3}$$

式中，z轴是系统的旋转对称轴，r为离轴距离，$\varphi = \varphi(z,r)$为空间静电电位，ε_s为垂直于阴极面逸出的主电子相应的初电位，$p_2 = p_2(z)$，$p_3 = p_3(z)$，撇号表示对z的导数。注意，系数F_1、G_1、F_2、G_2中的各量应取主轨迹上相应的值。

曲近轴轨迹方程组曾在文献[3]中被引用和叙述过，并编制了程序，进行了一系列像管的实际计算。

如图1所示，设α_0为逸出电子的初速v_0与过阴极面上的$P_0(z_0,r_0)$的法线之夹角，β_0为v_0在过P_0点并与阴极面相切的平面上的投影与$z-r$子午面和该切面的交线的夹角。α_0称为逸出角，它自阴极面的法线算起，由0°至90°；β_0称为方位角，它自子午面与阴极面在P_0点处的切面的交线算起，由0°至360°；ψ_0为P_0点的法线与系统轴的夹角，它自z轴算起，逆时针为正。

当给出初角度α_0、β_0以及逸出电子的初电位ε_0（它们唯一地确定初速v_0），便可决定ε_s值，$\varepsilon_s = \varepsilon_0\cos^2\alpha$。于是由方程式（1）和式（2）可确定主轨迹和相邻轨迹，从而求得子午焦点与弧矢焦点以及相应的场曲与像散值。

下面我们来确定系统的子午焦点与弧矢焦点以及相应的场曲与像散值。在给出初始条件时，应该使相邻轨迹的初电位沿主轨迹的分量等于主轨迹的初电位ε_s。因此，逸出电子应满足如下的初始条件：

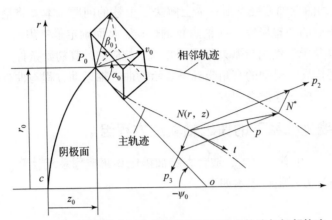

图1 静电阴极透镜中正交曲线坐标系下的主轨迹与相邻轨迹

对主轨迹方程：
$$z = z_0,\ r = r_0,\ r'_0 = \tan\psi_0 \tag{4}$$

对曲近轴轨迹方程：
$$(p_2) = (p_3)_0 = 0,$$
$$(p'_2)_0 = \frac{\tan\alpha_0 \cos\beta_0}{\cos\psi_0}, \tag{5}$$
$$(p'_3)_0 = \frac{\tan\alpha_0 \sin\beta_0}{\cos\psi_0}$$

这里
$$\cos\psi_0 = \frac{1}{\sqrt{1+r'^2_0}}$$

现论证曲近轴轨迹方程的线聚焦性质。

设子午（近轴）轨迹（$\beta_0 = 0°$）给出的初始条件为：
$$(\bar{p}_2)_0 = (\bar{p}_3)_0 = 0,$$
$$(\bar{p}'_2)_0 = \tan\alpha_0\sqrt{1+r'^2_0},\ (\bar{p}'_3)_0 = 0 \tag{6}$$

弧矢（近轴）轨迹（$\beta_0 = 90°$）给出的初始条件为：
$$(\bar{\bar{p}}_2)_0 = (\bar{\bar{p}}_3)_0 = 0,$$
$$(\bar{\bar{p}}'_2)_0 = \tan\alpha_0\sqrt{1+r'^2_0},\ (\bar{\bar{p}}'_3)_0 = 0 \tag{7}$$

圆括弧外的下标"0"表示 $z = z_0$ 处的值。

若我们所选择的解 $(\bar{p}_2,\ \bar{p}_3)$ 与 $(\bar{\bar{p}}_2,\ \bar{\bar{p}}_3)$ 是线性独立的。则由于曲近轴轨迹方程相对于 p_2、p_3 及其导数的线性性质，故满足条件
$$(p_2)_0 = (p_3)_0 = 0$$

的通解将可表示为：
$$p_2(z) = \cos\beta_0\,\bar{p}_2(z),$$
$$p_3(z) = \sin\beta_0\,\bar{\bar{p}}_3(z) \tag{8}$$

若我们假定，当 $z = z_T$ 时，$\bar{p}_2(z_T) = 0$，于是

$$p_2(z_T) = \cos\beta_0 \bar{p}_2(z_T) = 0,$$
$$p_3(z_T) = \sin\beta_0 \bar{\bar{p}}_3(z_T) \tag{9}$$

(z_T, r_T) 为子午焦点。由式（9）可见，在子午焦点处形成一段弧矢焦线，它位于副法线方向。

同样，当 $z = z_s$ 时，若 $\bar{\bar{p}}_3(z_s) = 0$，于是

$$p_2(z_s) = \cos\beta_0 \bar{p}_2(z_s),$$
$$p_3(z_s) = \sin\beta_0 \bar{\bar{p}}_3(z_s) = 0 \tag{10}$$

(z_s, r_s) 为弧矢焦点。由式（10）可见，在弧矢焦点处形成一段子午焦线，它位于主法线方向。

综上所述，实际计算时，只须给出的初始条件如下：
对主轨迹方程：

$$z = z_0, \ r = r_0, \ r_0' = \tan\psi_0$$

对曲近轴轨迹方程：

$$(\bar{p}_2)_0 = \bar{\bar{p}}_{30} = 0,$$
$$(\bar{p}_2')_0 = (\bar{\bar{p}}_3')_0 = \tan\alpha_0 \sqrt{1 + r_0'^2} \tag{11}$$

方程中 $\varepsilon_s = \varepsilon_0 \cos^2\alpha_0$。

由式（8），我们可确定相邻轨迹与主轨迹的偏离，而由式（9）、式（10）两式则可相应地确定子午焦点和弧矢焦线以及弧矢焦点与子午焦线。轴外不同高度逸出的电子（子午与弧矢方向）所形成焦点的总和就相应地构成子午和弧矢像场弯曲；而在主轨迹上子午束与弧矢束交点的差异的线段便可用来表示像散，如图 2 所示。

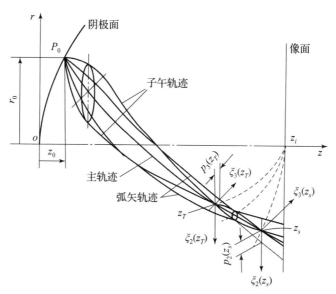

图 2　由曲近轴方程确定子午焦线与弧矢焦线

设像面位置 z_i 与某一 ε_{z_1} 相对应,则

$$F_T = z_i - z_T, \quad F_s = z_i - z_s \tag{12}$$

和

$$\eta = \sqrt{(z_T - z_s)^2 + (r_T - r_s)^2} \tag{13}$$

由它们可以分别确定子午场曲、弧矢场曲与像散。有时我们还将用 F_T 和 F_s 求中间值而取平均场曲 $\bar{F}$：

$$\bar{F} = \frac{1}{2}(F_T + F_s) \tag{14}$$

顺便指出,如果轴外物点 P_0 移到轴上,则作为单元电子束的参考轴即系统的光轴,于是主轨迹 r、r'、r'' 都成为零。此时曲近轴轨迹方程便化为众所周知的中心近轴轨迹方程。由此可见,轴外曲近轴方程与中心近轴方程是统一的,后者只是前者的一个特例。

2 正交曲线坐标系下的曲近轴运动方程组

文献［2］中我们曾导出了适用于静电阴极透镜计算的正交曲线坐标系下的曲近轴运动方程组的表达式,它们可表示为：

对主电子运动方程：

$$\ddot{r} = \frac{e}{m_0}\frac{\partial \varphi}{\partial r} \tag{15a}$$

$$\ddot{z} = \frac{e}{m_0}\frac{\partial \varphi}{\partial z} \tag{15b}$$

对曲近轴运动方程：

$$\ddot{\bar{p}}_2 = \frac{e}{m_0}\left\{-\frac{3}{2}\frac{1}{(1+\dot{r}^2/\dot{z}^2)}\frac{1}{[\varphi(z,r)+\varepsilon_s]}\left(\frac{\partial \varphi}{\partial r} - \frac{\dot{r}}{\dot{z}}\frac{\partial \varphi}{\partial z}\right)^2 + \frac{1}{1+\dot{r}^2/\dot{z}^2}\left[\frac{\partial^2 \varphi}{\partial r^2} - 2\frac{\dot{r}}{\dot{z}}\left(\frac{\partial \varphi}{\partial r\partial z}\right) + \frac{\dot{r}^2}{\dot{z}^2}\left(\frac{\partial^2 \varphi}{\partial z^2}\right)\right]\right\}\bar{p}_2 \tag{16a}$$

$$\ddot{\bar{p}}_3 = \frac{e}{m_0}\frac{1}{r}\frac{\partial \varphi}{\partial r}\bar{p}_3 \tag{16b}$$

此时,计算的初始条件可表示如下。

对主电子运动方程：

$$t = 0, \quad z = z_0, \quad r = r_0,$$
$$\dot{z}_0 = v_0\cos\alpha_0\cos\psi_0, \quad \dot{r}_0 = v_0\cos\alpha_0\sin\psi_0 \tag{17}$$

对曲近轴运动方程：

$$t = 0,$$
$$(\bar{p}_2)_0 = (\bar{\bar{p}}_3)_0 = 0,$$
$$(\dot{\bar{p}}_2)_0 = (\dot{\bar{\bar{p}}}_3)_0 = v_0\sin\alpha_0 \tag{18}$$

这里令 $\varepsilon_s = \frac{m_0}{2e}v_0^2\cos^2\alpha_0$。

方程式（16）的通解可表示为：

$$p_2(t) = \cos\beta_0 \, \bar{p}_2(t),$$
$$p_3(t) = \sin\beta_0 \, \bar{\bar{p}}_3(t)$$
(19)

设 t_T、t_s 分别所对应的 (z_T, r_T)、(z_s, r_s) 为子午焦点与弧矢焦点，则有

$$\bar{p}_2(t_T) = 0, \quad p_3(t_T) = \sin\beta_0 \, \bar{\bar{p}}_3(t_T),$$
$$\bar{\bar{p}}_3(t_s) = 0, \quad p_2(t_s) = \cos\beta_0 \, \bar{p}_2(t_s)$$
(20)

说明在 (z_T, r_T) 与 (z_s, r_s) 处分别形成弧矢焦线与子午焦线。

3 以运动的直角坐标系表示的曲近轴运动方程

设自阴极物面上的轴外某点 P_0 出发随着主轨迹以某一 "主" 电子速度运动着另一直角坐标系 (x, y, z) 的原点。在运动过程中，该运动的直角坐标系的 x 轴始终平行于 Or，z 轴平行于 Oz，而 y 轴始终垂直于子午面，如图 3 所示。

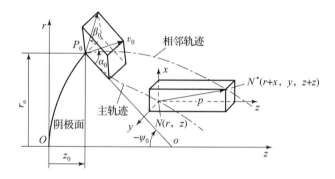

图 3 运动的直角坐标系下的主轨迹与曲近轴轨迹

如果自起始点 P_0 射出另一条电子轨迹，其初速度为 v_0，逸出角为 α_0，方位角为 β_0，且逸出电子初电位沿主轨迹的分量 $\varepsilon_0 \cos^2\alpha_0$ 与主轨迹射出的初电位 ε_s 相同，轨迹的坐标位置以 $(\bar{r}, \bar{y}, \bar{z})$ 表示。显然，此轨迹的空间位置可以用主轨迹和运动的直角坐标系离主轨迹的偏移来确定，即：

$$x = \bar{r} - r_\text{主} \tag{21a}$$
$$y = \bar{y} \tag{21b}$$
$$z = \bar{z} - z_\text{主} \tag{21c}$$

式中，$r_\text{主}$、$z_\text{主}$ 为主轨迹的坐标，它由主电子运动方程式（15）来决定。为方便起见，以下的表达式中各去掉了下标。

运动的直角坐标系下的曲近轴运动方程可由正交曲线坐标系下的曲近轴运动方程与主轨迹方程求得。由图 3 可见，正交曲线坐标 p_2 与运动和直角坐标 x、z 有以下关系：

$$x = \frac{p_2}{\sqrt{1+r'^2}}, \quad z = \frac{-p_2 r'}{\sqrt{1+r'^2}} \tag{22}$$

将式（21a）和式（21c）二次对 t 求导，并利用下述关系式

$$\begin{cases} \bar{r} = \dot{s}\dfrac{r'}{\sqrt{1+r'^2}} + \dot{p}_2 \dfrac{1}{\sqrt{1+r'^2}} + p_2 \dfrac{\mathrm{d}}{\mathrm{d}t}\left(\dfrac{1}{\sqrt{1+r'^2}}\right) \\ \bar{z} = \dot{s}\dfrac{1}{\sqrt{1+r'^2}} - \dot{p}_2 \dfrac{r'}{\sqrt{1+r'^2}} - p_2 \dfrac{\mathrm{d}}{\mathrm{d}t}\left(\dfrac{r'}{\sqrt{1+r'^2}}\right) \end{cases} \quad (23)$$

不难求得

$$\ddot{x} = \ddot{s}\dfrac{r'}{\sqrt{1+r'^2}} + \dot{s}\dfrac{\mathrm{d}}{\mathrm{d}t}\left(\dfrac{r'}{\sqrt{1+r'^2}}\right) + \ddot{p}_2\dfrac{1}{\sqrt{1+r'^2}} + 2\dot{p}_2\dfrac{\mathrm{d}}{\mathrm{d}t}\left(\dfrac{1}{\sqrt{1+r'^2}}\right) + p_2\dfrac{\mathrm{d}}{\mathrm{d}t}\left(\dfrac{-r'r''}{(1+r'^2)^{3/2}}\dfrac{\mathrm{d}z}{\mathrm{d}t}\right) - \dfrac{e}{m_0}\dfrac{\partial \varphi}{\partial r} \tag{24a}$$

$$\ddot{z} = \ddot{s}\dfrac{1}{\sqrt{1+r'^2}} + \dot{s}\dfrac{\mathrm{d}}{\mathrm{d}t}\left(\dfrac{1}{\sqrt{1+r'^2}}\right) - \ddot{p}_2\dfrac{r'}{\sqrt{1+r'^2}} - 2\dot{p}_2\dfrac{\mathrm{d}}{\mathrm{d}t}\left(\dfrac{r'}{\sqrt{1+r'^2}}\right) - p_2\dfrac{\mathrm{d}}{\mathrm{d}t}\left(\dfrac{-r'r''}{(1+r'^2)^{3/2}}\dfrac{\mathrm{d}z}{\mathrm{d}t}\right) - \dfrac{e}{m_0}\dfrac{\partial \varphi}{\partial z} \tag{24b}$$

在文献 [2] 中，我们曾导得

$$\dot{s} = \sqrt{\dfrac{2e}{m_0}}\sqrt{\varphi(z,r)+\varepsilon_s}\left(1+\dfrac{2}{\rho}p_2\right) \tag{25}$$

于是

$$\ddot{s} = \dfrac{e}{m_0}\left\{\dfrac{\partial \varphi}{\partial s}\left(1+\dfrac{4}{\rho}p_2\right) - \dfrac{4}{\rho^2}p_2\dfrac{\mathrm{d}\rho}{\mathrm{d}z}\dfrac{\mathrm{d}z}{\mathrm{d}s}[\varphi(z,r)+\varepsilon_s] + \dfrac{4}{\rho}\dot{p}_2\dfrac{1}{\dot{s}}[\varphi(z,r)+\varepsilon_s]\right\} \tag{26}$$

对于轴对称电场，以下表达式是成立的：

$$\begin{aligned}\dfrac{\mathrm{d}\varphi}{\mathrm{d}s} &= \dfrac{\partial \varphi}{\partial r}\dfrac{r'}{\sqrt{1+r'^2}} + \dfrac{\partial \varphi}{\partial z}\dfrac{1}{\sqrt{1+r'^2}}, \\ \dfrac{\mathrm{d}s}{\mathrm{d}z} &= \sqrt{1+r'^2}, \\ \dfrac{1}{\rho} &= \dfrac{r''}{(1+r'^2)^{3/2}}\end{aligned} \tag{27}$$

将以上表达式代入式（24）中，经过繁复的运算，并考虑到 $p_3 = y$，最后便可得到运动的直角坐标系下的曲近轴运动方程为：

$$\begin{aligned}\ddot{x} &= \dfrac{e}{m_0}\left(\dfrac{\partial^2 \varphi}{\partial r^2}x + \dfrac{\partial^2 \varphi}{\partial r \partial z}z\right), \\ \ddot{z} &= \dfrac{e}{m_0}\left(\dfrac{\partial^2 \varphi}{\partial r \partial z}x + \dfrac{\partial^2 \varphi}{\partial z^2}z\right), \\ \ddot{y} &= \dfrac{e}{m_0}\left(\dfrac{1}{r}\dfrac{\partial \varphi}{\partial r}y\right)\end{aligned} \tag{28}$$

于是，由式（23）和式（28），便可求得 t、$r_{主}(t)$、$z_{主}(t)$、$x(t)$、$y(t)$、$z(t)$。应该指出的是，在求解时，轨迹与电位及其偏导数都是取主轨迹上的值。

由方程式（28）可见，x、z 的方程中 x 与 z 是耦合的。这给求解与讨论系统的理想聚焦性质带来了一定的困难。但是，我们知道，$\ddot{x}$、$\ddot{z}$ 方程实际上乃是 $\ddot{p}_2$ 方程反映到运动的直角坐标系的表现形式，而 p_2 在运动的直角坐标系下的分量乃是 x、z，即

$$x^2 + z^2 = p_2^2 \tag{29}$$

于是，我们可给出以下的计算初始条件：

对主电子运动方程：
$$t=0,\ z=z_0,\ r=r_0,$$
$$\dot{z}_0 = v_0\cos\alpha_0\cos\psi_0,\ \dot{r}_0 = v_0\cos\alpha_0\sin\psi_0,\qquad(30)$$

对曲近轴运动方程：
$$t=0,\ \bar{x}_0 = \bar{\bar{y}}_0 = \bar{z}_0 = 0,$$
$$\dot{\bar{x}}_0 = v_0\sin\alpha_0\cos\psi_0,$$
$$\dot{\bar{z}}_0 = -v_0\sin\alpha_0\sin\psi_0,\qquad(31)$$
$$\dot{\bar{\bar{y}}}_0 = v_0\cos\alpha_0$$

方程式（28）的通解可表示为
$$x(t) = \bar{x}(t)\cos\beta_0,$$
$$z(t) = \bar{z}(t)\cos\beta_0,\qquad(32)$$
$$y(t) = \bar{\bar{y}}(t)\sin\beta_0$$

子午焦点应与 $\bar{x}(t_T)=0$，$\bar{z}(t_T)=0$ 相对应（或可由 $\bar{x}$、$\bar{z}$ 在相继点处的变号来确定），而与 t_T 相对应的主轨迹坐标 (z_{tT}, r_{tT}) 即为子午焦点，弧矢焦线为
$$y(t_T) = \sin\beta_0\,\bar{\bar{y}}(t_T)\qquad(33)$$

同样，弧矢焦点与 $\bar{\bar{y}}(t_s)=0$ 相对应，而与 t_s 相对应的主轨迹坐标 (z_{ts}, r_{ts}) 即弧矢焦点。子午焦线为
$$x(t_s) = \cos\beta_0\,\bar{x}(t_s),$$
$$z(t_s) = \cos\beta_0\,\bar{z}(t_s)\qquad(34)$$

子午焦线的长度 p_{2s} 为
$$p_{2s} = \sqrt{[\bar{x}(t_s)]^2 + [\bar{z}(t_s)]^2}\cos\beta_0\qquad(35)$$

4 以运动的直角坐标系表示的曲近轴轨迹方程

如果我们把上述以时间 t 为参量的运动方程转换成以坐标 z 为参量的轨迹方程的形式，则可得主轨迹方程
$$r'' = \frac{1+r'^2}{2[\varphi(z,r)+\varepsilon_s]}\left(\frac{\partial\varphi}{\partial r} - r'\frac{\partial\varphi}{\partial z}\right)\qquad(36)$$

和运动的直角坐标系下的曲近轴轨迹方程
$$x'' + \frac{1+r'^2}{2[\varphi(z,r)+\varepsilon_s]}\frac{\partial\varphi}{\partial z}x' = \frac{1+r'^2}{2[\varphi(z,r)+\varepsilon_s]}\left(\frac{\partial^2\varphi}{\partial r^2}x + \frac{\partial^2\varphi}{\partial r\partial z}z\right),$$
$$z'' + \frac{1+r'^2}{2[\varphi(z,r)+\varepsilon_s]}\frac{\partial\varphi}{\partial z}z' = \frac{1+r'^2}{2[\varphi(z,r)+\varepsilon_s]}\left(\frac{\partial^2\varphi}{\partial z^2}z + \frac{\partial^2\varphi}{\partial r\partial z}x\right),\qquad(37)$$
$$y'' + \frac{1+r'^2}{2[\varphi(z,r)+\varepsilon_s]}\frac{\partial\varphi}{\partial z}y' = \frac{1+r'^2}{2[\varphi(z,r)+\varepsilon_s]}\left(\frac{1}{r}\frac{\partial^2\varphi}{\partial r}y\right)$$

这里，$\varepsilon_s = \varepsilon_0\cos^2\alpha_0$，于是，可得 $z_主$、$r_主$、$x(z_主)$、$y(z_主)$、$z(z_主)$。以下的表达式中各去掉了下标"主"字。

同样，给出的计算初始条件如下：

对主轨迹方程：

$$z = z_0, \quad r = r_0, \quad r'_0 = \tan\psi_0 \tag{38}$$

对曲近轴轨迹方程：

$$\begin{aligned}
&\bar{x}_0 = \bar{\bar{y}}_0 = z_0 = 0, \\
&\bar{x}'_0 = \tan\alpha_0, \\
&\bar{z}'_0 = -\tan\alpha_0 \tan\psi_0, \\
&\bar{\bar{y}}'_0 = \tan\alpha_0/\cos\psi_0
\end{aligned} \tag{39}$$

方程式（37）的通解可表示为

$$\begin{aligned}
x(z) &= \bar{x}(z)\cos\beta_0, \\
z(z) &= \bar{z}(z)\cos\beta_0, \\
y(z) &= \bar{\bar{y}}(z)\sin\beta_0
\end{aligned} \tag{40}$$

由此不难确定子午焦点（z_T, r_T）、弧矢焦点（z_s, r_s）以及相应的弧矢焦线

$$y(z_T) = \sin\beta_0 \bar{\bar{y}}(z_T) \tag{41}$$

与子午焦线

$$\begin{aligned}
x(z_s) &= \cos\beta_0 \bar{x}(z_s), \\
z(z_s) &= \cos\beta_0 \bar{z}(z_s)
\end{aligned} \tag{42}$$

5 计算实例

现用曲线坐标系下的主轨迹方程与曲近轴轨迹方程计算某一三电极静电聚焦成像系统，其电极结构尺寸、相互位置与施加的电压，如图4所示。

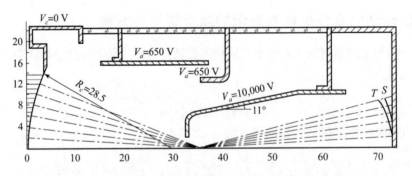

图4 某一三电极静电聚焦成像系统的子午焦点与弧矢焦点

计算的主要参数为：阴极曲率半径 $R_c = 28.5$ mm，$\varepsilon_0 = 1$ V，$\alpha_0 = 72.525°$（$\varepsilon_s = 0.09$ V），按近轴轨迹方程计算所得的像面位置与放大率为：

$$z_i = 73.127\,322\,4 \text{ mm}, \quad M = -0.713\,165\,92$$

表1给出了不同高度 r_0 下所对应的子午焦点和弧矢焦点的位置以及相应的像散值。

五、曲轴宽电子束聚焦理论

表1 不同高度 r_0 下的子午焦点与弧矢焦点的位置以及像散值 mm

逸出电子的初始位置		子午焦点		弧矢焦点		像散
r_0	z_0	z_T	r_T	z_s	r_s	η
0	0	73.127 3	0	73.127 3	0	0
2	0.070 262	73.148 2	−1.428 39	73.170 9	−1.429 24	0.022 77
4	0.282 097 8	72.678 1	−2.832 99	72.964 5	−2.854 23	0.287 20
6	0.638 736 5	72.317 1	−4.237 68	72.534 4	−4.262 06	0.218 67
8	1.145 841 3	71.811 3	−5.624 01	72.298 0	−5.697 72	0.492 29
10	1.811 987 7	71.087 9	−6.972 77	71.912 8	−7.131 58	0.840 10
12	2.649 468 1	69.856 0	−8.193 15	71.483 1	−8.578 02	1.672 07

表2列出了按式（1）与式（2）以及初始条件式（4）、（6）、（7）计算高度 $r_0 = 10$ mm, $\varepsilon_0 = 1$ V, $\alpha_0 = 72.525°$ ($\varepsilon_s = 0.09$ V) 下的主轨迹、子午曲近轴轨迹与弧矢曲近轴轨迹的结果。

表2 高度 $r_0 = 10$ mm 下的主轨迹、子午曲近轴轨迹与弧矢曲近轴轨迹的计算结果

z	R	r'	$\bar{p}_2$	$\bar{p}_2'$	$\bar{p}_3$	$\bar{p}_3'$
1.811 99	10.000 00	−0.374 70	0	3.392 09	0	3.392 09
5.046 09	8.822 10	−0.353 26	0.537 44	0.077 26	0.536 18	0.076 38
11.046 09	6.790 56	−0.327 18	0.826 77	0.029 11	0.817 69	0.027 58
17.046 09	4.857 74	−0.319 72	0.924 17	0.003 66	0.907 50	0.003 04
23.046 09	2.958 39	−0.308 95	0.870 63	−0.021 06	0.854 79	−0.020 40
29.046 09	1.241 15	−0.253 61	0.690 17	−0.035 62	0.677 64	−0.035 40
35.171 09	−0.058 49	−0.192 67	0.539 44	−0.015 37	0.527 12	−0.014 70
41.171 09	−1.213 37	−0.192 49	0.448 89	−0.015 02	0.440 61	−0.014 34
47.296 09	−2.392 63	−0.192 50	0.356 97	−0.015 01	0.352 80	−0.014 33
53.296 09	−3.547 68	−0.192 51	0.266 94	−0.015 00	0.266 81	−0.014 33
59.296 09	−4.702 73	−0.192 51	0.176 92	−0.015 00	0.180 82	−0.014 33
71.087 90	−6.972 77	−0.192 51	0	−0.015 00	0.011 83	−0.014 33
71.912 80	−7.131 58	−0.192 51	−0.012 37	−0.015 00	0	−0.014 33

结束语

（1）本文论述了以主轨迹为曲线轴而给出的曲近轴轨迹方程组或曲近轴运动方程组可

单一地给定与逸出主电子的初电位 ε_s 相对应的子午焦点与弧矢焦点的位置，证明了曲近轴方程是解决大物面成像系统的场曲和像散的一条有效途径。

（2）鉴于正交曲线坐标系下曲近轴方程的线性性质，故实际只需计算子午方向与弧矢方向的两条特殊轨迹，便可确定单元电子束的空间行进轨迹以及形成的子午焦线与弧矢焦线。

（3）本文给出了静电阴极透镜中计算子午焦点与弧矢焦点以及相应的焦线的初始条件。初始条件须保证相邻电子与主电子具有相同的"主电子初速度"，即使其相邻轨迹的初电位沿主轨迹的分量等于主轨迹的初电位 ε_s。

（4）本文由正交曲线坐标系下的 $\ddot{p}_2$ 方程过渡到运动的直角坐标系下的 x、z 方程。这两类方程各有特点，x、z 方程的形式虽然简单，但微分方程中 x、z 是耦合的。因此，通常在确定场曲与像散时以正交曲线坐标系下的曲近轴方程组更为合适。

参 考 文 献

[1] ZHOU L W. A generalized theory of wide electron beam focusing [J]. Advances in Electronics and Electron Physics, 1985,64(B):575 – 589.

[2] 周立伟. 曲线坐标系下近轴运动方程的研究[C]. 第二届全国电子光学学术讨论会,四川成都,1982.

[3] 方二伦,冯炽焘,周立伟. 文献名不详[J]. 光电技术,1980(2 – 3):71 – 81.

六、阴极透镜的几何横向像差与近轴横向像差理论

Theory of Geometrical Lateral Aberrations and Paraxial Lateral Aberrations for Cathode Lenses

6.1 关于电磁复合聚焦阴极透镜的几何横向像差理论
On Theory of Geometrical Aberrations for the Combined Electromagnetic Focusing Cathode Lenses

摘要：本文对电磁复合聚焦阴极透镜的几何像差的理论进行评述与研究。我们的研究表明，电磁复合聚焦阴极透镜的全部横向像差应由近轴横向像差与几何横向像差两部分组成。前者是由于逸出电子的轴向初能量的差异所形成，后者是由于实物几何形状的差异所形成，但本文将着重讨论三级几何横向像差。文中着重分析了西门纪业[1]和 Бонштедт[2] 的两篇代表性的论文，详细推导了两文的几何横向像差方程及其像差系数表示式。研究表明，尽管两文的三级几何横向像差系数在形式上迥然不同，但实质上是完全一致的，是可以互相转化的，从而证明 Бонштедт 对西门纪业的批评是没有根据的。

Abstract: In this article, theory of the geometrical aberration of the combined electromagnetic focusing cathode lenses has been commented and studied. Our investigation shows that the whole lateral aberration of the combined electromagnetic focusing cathode lenses should be composed of two parts: the paraxial lateral aberration and the geometrical lateral aberration, the former is formed by the differences of axial initial velocities of electrons emitted from the cathode surface; the latter is induced by the differences of geometrical figure of object. But in this paper, we emphatically discuss the theory of the third order geometrical lateral aberrations. Furthermore, we have emphatically analyzed and commented two previous representative papers concerning the geometrical lateral aberration theory of cathode lenses, which was first contributed by Ximen Jiye (1957)[1] and second by B. E. Bonschtedt (1964)[2]. It was shown that the formulae of aberration coefficients of the third order geometric lateral aberration for the electromagnetic cathode lenses given by Ximen Jiye and Bonschtedt, although different in form, it can be transformed from one to another. It is also enough to prove that the criticism of B. E. Bonschtedt addressed to Ximen Jiye is improper.

引 言

众所周知，阴极透镜中由阴极物面逸出速度很小而初斜率取任意值（可趋于无穷大）的电子参与成像。因此，细束电子光学像差理论并不适用于阴极透镜。在研究阴极透镜的电子光学性质与像差理论时，通常选择逸出点处电子的离轴高度 r_0 与逸出电子的初速度 $\dot{r}_0$ 作为一级小量，而将轨迹与像差用这两个量来描写。这样所获得的像差公式不仅包含了因几何形状的差异所形成的几何像差，也包含了因逸出电子初速度差异所形成的色像差。

周立伟，艾克聪ª，潘顺臣. 北京理工大学. 物理学报（Acta Physica Sinica），V. 32, No. 3, 1983, 376–391.

应该指出，静电聚焦阴极透镜与电磁复合聚焦阴极透镜的几何像差理论是由 Recknagel[3] 和西门纪业[1] 完成的。在 Recknagel 和西门纪业工作的基础上，文献 [4~7] 和文献 [2]、文献 [8] 继续讨论了静电阴极透镜与电磁复合聚焦阴极透镜的几何像差理论，推进了成像电子光学理论的研究与发展。但是，对于电磁复合聚焦阴极透镜像差理论的研究尚存在争论。如 Куликов 等人[8] 怀疑总的三级几何横向像差 Δu_3 内由于旋转坐标系的附加旋转而引起的像差项 $\Delta \chi (k \times u)$ 的存在。而 Бонштедт[2] 尽管承认采用西门纪业[1] 研究的方法，并再一次推导了三级几何横向像差系数的表示式①，但是，他却批评文献 [1] 中像差方程的自由项 F 的推导有误，从而导致所得到的像差系数也是不正确的。此文发表不久，本文作者之一在一篇论文报告稿中曾分析了文献 [1]、文献 [2] 中像差方程的自由项 F 的表达式，发现两文中的 F 实质上是一致的，初步指出了 Бонштедт 的批评是不正确的。1983 年，我们再一次对两文的像差方程和像差系数进行考察。其结论是，虽然两篇文章中的像差方程自由项 F 以及像差系数表达式形式上似乎差异很大，但实际是完全一致的，是可以互相转化的，从而证明 Бонштедт 对西门纪业的批评是没有根据的。此外，我们还证实了西门纪业[1] 文中像差系数的推导全部无误；相反，文献 [2] 给出的像差系数表达式却有一些错误与遗漏之处，需要加以补正。

早在 20 世纪六七十年代，作者关于电子光学像差理论的研究表明[10]，对阴极透镜等成像电子光学系统，像面应定义在某一轴向初能 ε_{z_1} 所对应的理想成像位置处，从而考察因逸出电子初能量的差异所形成的近轴横向像差，以及由于几何形状的差异所构成的几何横向像差①。

本文将对电磁复合聚焦阴极透镜的像差理论进行严密的推导与评述，提出了成像电子光学系统横向像差应是近轴横向像差与几何横向像差的合成。

这里，应该指出，在研究阴极透镜的像差理论时，我们曾对 Бонштедт 与西门纪业关于研究电磁复合阴极透镜的三级几何横向像差的理论进行了严密的考察。当时，一个有争论的问题是，三级几何横向应该包含 $v(z_i, \sqrt{\varepsilon_z})$ 的积分项否？此问题一直没有解决。在本文中，我们证明了在静电或电磁电子光学系统中，$v(z_i, \sqrt{\varepsilon_z})$ 为一级小量，在研究三级像差理论时，其相关的项都应略去，否则将出现四级小量。

1 实际轨迹方程

文献 [1]、文献 [2] 业已证明，旋转坐标系下轴对称电磁聚焦复合场中的实际轨迹方程可表示为

$$\frac{\varphi}{1+r'^2}u'' + \frac{1}{2}u'\frac{d}{dz}\left(\frac{\varphi}{1+r'^2}\right) - u\frac{\partial}{\partial(u^2)}(\varphi - 4A^{*2}) + 4\left(\frac{\varphi}{1+r'^2}\right)^{1/2}u(u\times u')k\frac{\partial}{\partial(u^2)}\left(\frac{A^*}{r}\right) = 0 \tag{1}$$

式中，$\varphi = \varphi(z, r)$ 为空间电位分布，$A^* = \sqrt{\frac{e}{8m_0}}A$，$A = A(z, r)$ 为旋转对称磁场中磁矢位方

---

① 这是周立伟 1965 年年初在列宁格勒电工学院电物理系的报告摘要：Чжоу ЛиВэй（周立伟），О выводах уравнения аберраций в комбинированной электро-магнитной катодной линзе, Тезисы доклада на ЭМА, ЛЭТИ, Ленинград, 1965.

位角分量的空间分布，e/m_0 为电子的荷质比，$\boldsymbol{k}$ 是 z 轴正方向的单位矢量，撇号"'"表示对 z 的导数。$\boldsymbol{r} = \boldsymbol{r}(z)$，$\boldsymbol{u} = \boldsymbol{u}(z)$ 分别表示固定坐标系 (X, Y) 和旋转坐标系 (x, y) 内垂直于 z 轴的平面矢量，两者之间的关系为

$$\boldsymbol{r} = \boldsymbol{u}\mathrm{e}^{\mathrm{i}\chi(z)} \tag{2}$$

式中，$\chi(z)$ 是由于磁场引起轨迹旋转而引入的旋转坐标系 (x, y) 相对于固定坐标系 (X, Y) 转过的角度。

$$\chi(z) = 2\int_0^z \frac{A^*}{r}\sqrt{\frac{1+r'^2}{\varphi}}\mathrm{d}z \tag{3}$$

这里，假定 $\chi(z=0) = 0$。

将式（1）叉乘 $\boldsymbol{u}$，不难证明旋转对称场中转动动量矩守恒：

$$\sqrt{\frac{\varphi}{1+r'^2}}(\boldsymbol{u}\times\boldsymbol{u}') = \left(\sqrt{\frac{\varphi}{1+r'^2}}\right)_0(\boldsymbol{u}_0\times\boldsymbol{u}_0') = 常数 \tag{4}$$

下标"0"表示在 $z=0$ 处的值。为了进一步简化方程式（1），西门纪业[1]引入了约化电位 $\tilde{\phi}$：

$$\tilde{\phi} = \varphi - 4A^{*2} - 4\left(\sqrt{\frac{\varphi}{1+r'^2}}\right)_0(\boldsymbol{u}_0\times\boldsymbol{u}_0')\boldsymbol{k}\frac{A^*}{r} \tag{5}$$

并证明了以下关系式成立：

$$\frac{\varphi}{1+r'^2} = \tilde{\phi} - \boldsymbol{u}'^2\frac{\varphi}{1+r'^2} \tag{6}$$

$$\frac{\mathrm{d}}{\mathrm{d}z}\left(\frac{\varphi}{1+r'^2}\right) = \frac{\partial\tilde{\phi}}{\partial z} \tag{7}$$

最后将式（1）简化为

$$\frac{2\varphi}{1+r'^2}\boldsymbol{u}'' + \boldsymbol{u}'\frac{\mathrm{d}}{\mathrm{d}z}\left(\frac{\varphi}{1+r'^2}\right) - 2\boldsymbol{u}\frac{\partial\tilde{\phi}}{\partial(\boldsymbol{u}^2)} = 0 \tag{8}$$

轨迹方程式（8）是由西门纪业[1]首先推导的，由式（7）可见，方程式（8）亦可表达成如文献［2］中给出的形式：

$$\frac{2\varphi}{1+r'^2}\boldsymbol{u}'' + \boldsymbol{u}'\frac{\partial\tilde{\phi}}{\partial z} - 2\boldsymbol{u}\frac{\partial\tilde{\phi}}{\partial(\boldsymbol{u}^2)} = 0 \tag{9}$$

方程式（8）或式（9）乃是我们研究电子光学成像系统的理想成像与像差的出发点。

2 近轴方程与理想成像

如上所述，对于大物面宽电子束成像系统，通常将电子轨迹和像差以 $\boldsymbol{r}_0$ 和 $\dot{\boldsymbol{r}}_0$ 作为一级小量展开来描写，令

$$\begin{aligned}&\boldsymbol{r}_0 = \boldsymbol{i}\chi_0 = \boldsymbol{i}r_0, \\ &\dot{\boldsymbol{r}}_0 = \boldsymbol{i}\sqrt{\frac{2e}{m_0}\varepsilon_\perp}\cos\beta_0 + \boldsymbol{j}\sqrt{\frac{2e}{m_0}\varepsilon_\perp}\sin\beta_0\end{aligned} \tag{10}$$

这里 $\varepsilon_\perp = \varepsilon_0\sin^2\alpha_0$，$\varepsilon_z = \varepsilon_0\cos^2\alpha_0$ 分别为逸出电子初速度的垂轴分量和轴向分量所对应的初

电位，ε_0 为逸出电子的初电位，电子逸出的初速度方向由逸出角 α_0 与方位角 β_0 所决定。

在非相对论、无空间电荷、磁场中无自由空间电流的情形下，φ 和 A^* 可以展开为

$$\varphi = \varepsilon_0 + \phi(z) - \frac{1}{4}\phi''(z)u^2 + \frac{1}{64}\phi^N(z)u^4 - \cdots \quad (11)$$

$$\frac{A^*}{r} = \frac{1}{2}h(z) - \frac{1}{16}h''(z)u^2 + \cdots \quad (12)$$

式中，$\phi(z)$ 为静电电位沿轴分布；$h(z) = \sqrt{\dfrac{e}{8m_0}}B(z)$，其中 $B(z)$ 为磁感应强度沿轴分布。

在实际轨迹方程式（8）或式（9）中，将各系数展开时，保留零级及一级的量，忽略二级以及更高级的量，从而得到电磁复合聚焦阴极透镜的线性方程，简称电子光学近轴方程：

$$Vu'' + \frac{1}{2}\phi'u' + Tu = 0 \quad (13)$$

这里

$$V = V(z) = \phi(z) + \varepsilon_z \quad (14a)$$

$$T = T(z) = \frac{1}{4}\phi''(z) + h^2(z) \quad (14b)$$

同样，展开式（3）的系数，取零级近似，则有

$$(\chi(z))_0 = \int_0^z \frac{h(z)}{\sqrt{V}}dz \quad (15a)$$

$$\chi'(z) = \frac{h(z)}{\sqrt{V}} \quad (15b)$$

近轴方程式（13）乃是关于 u 的二阶线性齐次常微分方程，可以用通常的方法求解，所得的解可以完整地描述系统的全部近轴光学性质。

设近轴方程式（13）的标量形式的两线性无关的特解分别为 $v(z)$ 和 $w(z)$，它们在 $z = 0$ 处满足以下的初始条件：

$$\begin{aligned} v(0) = 0, \; v'(0)\sqrt{\varepsilon_z} = 1, \\ w(0) = 1, \; w'(0)\sqrt{\varepsilon_z} = 0 \end{aligned} \quad (16)$$

此二特解满足的朗斯基行列式为

$$\sqrt{V}(v'w - vw') = 1 \quad (17)$$

假定在初始点 $z = 0$ 处旋转坐标与固定坐标重合，即 $\chi_0 = 0$，则 u 在阴极面的初始条件由式（2）、式（10）与式（15b）可表示为

$$u_0 = r_0, \; u_0' = r_0' - \chi_0'(\boldsymbol{k} \times \boldsymbol{r}_0) \quad (18)$$

于是方程式（13）的通解可以表示为

$$\boldsymbol{u} = \boldsymbol{r}_0 w + \left[\sqrt{\frac{m_0}{2e}}\dot{\boldsymbol{r}}_0 - (\boldsymbol{k} \times \boldsymbol{r}_0)h_0\right]v \quad (19)$$

在阴极透镜中，通常在以某一轴向初电位为 ε_{z_1} 的近轴电子所对应的像面上来讨论系统的理想成像的性质。特解 $v(z, \varepsilon_{z_1})$ 将决定像面位置 z_i，即

$$v(z_i, \varepsilon_{z_1}) = 0 \tag{20}$$

而另一特解 $w(z, \varepsilon_{z_1})$ 在 z_i 像面上将决定系统的放大率 M（当 $\chi(z_i) = m\pi$ 时）：

$$w(z_i, \varepsilon_{z_1}) = (-1)^m M \tag{21}$$

式中，m 为聚焦圈数。

3 横向像差的定义，近轴与几何横向像差的确定

在研究静电聚焦与电磁聚焦同心球系统时[10]，我们曾证实了系统的横向像差与所定义的轴向初电位 ε_{z_1} 对应的像面位置有关。如文献 [10] 所述，横向像差 $\Delta \boldsymbol{r}(z_i)$ 定义为在 z_i 像面上实际轨迹的落点与理想像点之间的偏离，即

$$\Delta \boldsymbol{r}(z_i) = \boldsymbol{r}_{\text{real}}(z_i, \varepsilon_z) - \boldsymbol{r}_*(z_i, \varepsilon_{z_1}) \tag{22}$$

式中，$\boldsymbol{r}_{\text{real}}(z_i, \varepsilon_z)$ 是高度为 r_0、初电位为 ε_0、初速度方向的角度为 α_0、β_0，其轴向初电位为 ε_z 的电子自阴极面逸出的实际轨迹在 $z = z_i$ 像面上的平面矢量；$\boldsymbol{r}_*(z_i, \varepsilon_{z_1})$ 则是同一高度为 r_0、轴向初电位为 ε_{z_1} 自阴极面发出的近轴电子轨迹在该像面上的平面矢量，且

$$\boldsymbol{r}_*(z_i, \varepsilon_{z_1}) = \boldsymbol{r}_0 M \tag{23}$$

按文献 [10]，可将式（22）表示为

$$\Delta \boldsymbol{r} = \Delta \boldsymbol{r}_* + \delta \boldsymbol{r} \tag{24}$$

式中，

$$\Delta \boldsymbol{r}_* = \boldsymbol{r}_*(z_i, \varepsilon_z) - \boldsymbol{r}_*(z_i, \varepsilon_{z_1}) \tag{25a}$$

$$\delta \boldsymbol{r} = \boldsymbol{r}_{\text{real}}(z_i, \varepsilon_z) - \boldsymbol{r}_*(z_i, \varepsilon_z) \tag{25b}$$

式（25a）实际上乃是具有不同的轴向初电位 ε_z 与 ε_{z_1} 的两条近轴轨迹的落点在 z_i 像面上的偏离（见图1）；$\Delta \boldsymbol{r}_*$ 称为近轴横向像差，它可表示为

$$\Delta \boldsymbol{r}_* = [w(z_i, \varepsilon_z) e^{i\chi(z_i, \varepsilon_z)} - M] \boldsymbol{r}_0 - (\boldsymbol{k} \times \boldsymbol{r}_0) h_0 v(z_i, \varepsilon_z) e^{i\chi(z_i, \varepsilon_z)} + \sqrt{\frac{m_0}{2e}} \dot{\boldsymbol{r}}_0 v(z_i, \varepsilon_z) e^{i\chi(z_i, \varepsilon_z)} \tag{26}$$

式（26）的前两项与初速度及方位角参量无关，而与 $\boldsymbol{r}_0$ 一次方成正比，称为径向与旋转方向的一级位移。最后一项与 $\dot{\boldsymbol{r}}_0$ 相关，乃是二级色差，即二级近轴横向色差。我们将专门讨论，这里就不细述了。

式（25b）乃是具有相同的轴向初电位 ε_z 的实际轨迹与近轴轨迹的落点在 z_i 像面上的偏离（见图1），$\delta \boldsymbol{r}$ 为三级横向像差，或称几何横向像差。由式（2），它可表示为

$$\delta \boldsymbol{r} = [(\boldsymbol{u} + \Delta \boldsymbol{u}) e^{i(\chi + \Delta \chi)} - \boldsymbol{u} e^{i\chi}]_{z_i} \tag{27}$$

将上式中的 $e^{i\Delta\chi}$ 展开成级数形式，略去高阶小量，最后可得

$$\delta \boldsymbol{r} = [\Delta \boldsymbol{u} + \Delta \chi (\boldsymbol{k} \times \boldsymbol{u})]_{z_i} e^{i\chi(z_i, \varepsilon_z)} = \Delta \boldsymbol{u}_3 e^{i\chi(z_i, \varepsilon_z)} \tag{28}$$

这里 $\Delta \boldsymbol{u}$ 乃是 $\boldsymbol{u}$ 的实际轨迹与近轴轨迹之差所引起的三级横向像差项，$\Delta \chi$ 乃是三级近似的轨迹与一级近似的轨迹的两个旋转坐标系的角度之差，$\Delta \chi (\boldsymbol{k} \times \boldsymbol{u})$ 乃是由于磁场产生旋转而引起的三级横向像差项。

总的三级横向像差 $\Delta \boldsymbol{u}_3$ 内除了 $\Delta \boldsymbol{u}$ 外还应包括 $\Delta \chi (\boldsymbol{k} \times \boldsymbol{u})$ 的问题早已在文献 [1] 的式（23）进行了论证，但文献 [8] 怀疑 $\Delta \chi (\boldsymbol{k} \times \boldsymbol{u})$ 的像差项的存在，式（28）再一次给出了该项存在的证明。由此可见，求几何横向像差 $\delta \boldsymbol{r}$ 便归结于求 $\Delta \boldsymbol{u}$ 和 $\Delta \chi (\boldsymbol{k} \times \boldsymbol{u})$。

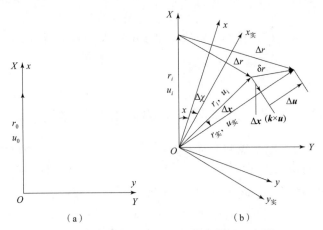

图1 近轴横向像差与几何横向像差示意图

将实际轨迹方程式（9）中的各系数展开，保留到二级量，略去四级以及更高级的量，便得到二级近似的各式

$$\left(\frac{\varphi}{1+r'^2}\right)_{0+2} = \varepsilon_z + \phi(z) - \frac{\phi''(z)}{4}u^2 - u^2 h^2 + u_0^2 h_0^2 - 2(h-h_0)\sqrt{\varepsilon_z}(\boldsymbol{u}_0 \times \boldsymbol{u}_0')\boldsymbol{k} -$$
$$u'^2[\varepsilon_z + \phi(z)] + u_0'^2 \varepsilon_z \tag{29a}$$

$$\left[\frac{\mathrm{d}}{\mathrm{d}z}\left(\frac{\varphi}{1+r'^2}\right)\right]_{0+2} = \phi'(z) - u^2 \frac{\mathrm{d}}{\mathrm{d}z}\left[\frac{\phi''(z)}{4}+h^2\right] - 2h'\sqrt{\varepsilon_z}(\boldsymbol{u}_0 \times \boldsymbol{u}_0')\boldsymbol{k} - \frac{\mathrm{d}}{\mathrm{d}z}(u'^2(\varepsilon_z+\phi(z)))$$
$$\tag{29b}$$

$$\left[\frac{\partial \phi}{\partial(\boldsymbol{u}^2)}\right]_{0+2} = -\frac{1}{4}\phi''(z) - h^2 + \frac{\phi'''(z)}{32}u^2 + \frac{hh''}{2}u^2 + \frac{1}{4}h''\sqrt{\varepsilon_z}(\boldsymbol{u}_0 \times \boldsymbol{u}_0')\boldsymbol{k} \tag{29c}$$

将以上三式代入式（9），整理后，便得到近轴方程式（13）与三级几何横向像差 $\Delta\boldsymbol{u}$ 的方程

$$V\Delta\boldsymbol{u}'' + \frac{1}{2}\phi'\Delta\boldsymbol{u}' + T\Delta\boldsymbol{u} = \boldsymbol{F}(\boldsymbol{u},\boldsymbol{u}',z) \tag{30}$$

式中，F 是像差方程的自由项，其中的 $\boldsymbol{u}$、$\boldsymbol{u}'$ 可用求解近轴轨迹方程式（13）所得到的平面轨迹矢量 $\boldsymbol{u}_1$、$\boldsymbol{u}_1'$ 来代替。在下面的讨论中，为简单起见，仍用 $\boldsymbol{u}$、$\boldsymbol{u}'$ 的符号，则可得到

$$\boldsymbol{F} = \boldsymbol{u}''S + \frac{1}{2}\boldsymbol{u}'\frac{\mathrm{d}S}{\mathrm{d}z} + \frac{1}{2}\boldsymbol{u}P \tag{31}$$

式中

$$S = T\boldsymbol{u}^2 - h_0^2\boldsymbol{u}_0^2 + 2\sqrt{\varepsilon_z}(h-h_0)(\boldsymbol{u}_0 \times \boldsymbol{u}_0')\boldsymbol{k} + \boldsymbol{u}'^2 V - \boldsymbol{u}_0'^2 \varepsilon_z \tag{32}$$

$$P = \left(\frac{1}{16}\phi'''(z) + hh''\right)\boldsymbol{u}^2 + \frac{1}{2}\sqrt{\varepsilon_z}h''(\boldsymbol{u}_0 \times \boldsymbol{u}_0')\boldsymbol{k} \tag{33}$$

由式（30）出发，利用变动任意常数法和朗斯基行列式（17），便可得到三级几何横向像差 $\Delta\boldsymbol{u}$ 值为

$$\Delta\boldsymbol{u} = v(z,\varepsilon_z)\int_0^z \boldsymbol{F}\frac{w}{\sqrt{V}}\mathrm{d}z - w(z,\varepsilon_z)\int_0^z \boldsymbol{F}\frac{v}{\sqrt{V}}\mathrm{d}z \tag{34}$$

而 $\Delta\chi$ 可由式（3）和式（15）得到

$$\Delta\chi(z) = \frac{1}{2}\int_0^z \left(\frac{hS}{V} - \frac{h''\boldsymbol{u}^2}{4}\right)\frac{\mathrm{d}z}{\sqrt{V}} \tag{35}$$

由式（34）和式（35），并考虑到三级近似和一级近似的轨迹条件之差，便可得到在 $z = z_i$ 像面上总的三级几何横向像差为

$$\Delta \boldsymbol{u}_3 = \Delta \boldsymbol{u}_3(z_i) = \frac{1}{8} h_0''(\boldsymbol{k} \times \boldsymbol{r}_0) \boldsymbol{r}_0^2 v(z_i, \varepsilon_z) + v(z_i, \varepsilon_z) \int_0^{z_i} \boldsymbol{F} \frac{w}{\sqrt{V}} \mathrm{d}z - w(z_i, \varepsilon_z) \int_0^{z_i} \boldsymbol{F} \frac{v}{\sqrt{V}} \mathrm{d}z +$$
$$\frac{1}{2} (\boldsymbol{k} \times \boldsymbol{u}(z_i)) \int_0^{z_i} \left(\frac{hS}{V} - \frac{h''\boldsymbol{u}^2}{4} \right) \frac{\mathrm{d}z}{\sqrt{V}} \tag{36}$$

将式（31）代入式（36），并利用分部积分和近轴方程式（13）进行化简，便有

$$\Delta \boldsymbol{u}_3(z_i) = \frac{1}{8} h_0''(\boldsymbol{k} \times \boldsymbol{r}_0) \boldsymbol{r}_0^2 v(z_i, \varepsilon_z) + \frac{1}{2} v(z_i, \varepsilon_z) \int_0^{z_i} \frac{Pw}{\sqrt{V}} \boldsymbol{u} \mathrm{d}z - \frac{1}{2} w(z_i, \varepsilon_z) \int_0^{z_i} \frac{Pv}{\sqrt{V}} \boldsymbol{u} \mathrm{d}z -$$
$$\frac{1}{2} v(z_i, \varepsilon_z) \int_0^{z_i} \frac{S}{V^{3/2}} (Tw\boldsymbol{u} + Vw'\boldsymbol{u}') \mathrm{d}z - \frac{1}{2} w(z_i, \varepsilon_z) \int_0^{z_i} \frac{S}{V^{3/2}} (Tv\boldsymbol{u} + Vv'\boldsymbol{u}') \mathrm{d}z +$$
$$\frac{1}{2} (\boldsymbol{k} \times \boldsymbol{u}(z_i)) \int_0^{z_i} \left(\frac{hS}{V} - \frac{h''\boldsymbol{u}^2}{4} \right) \frac{\mathrm{d}z}{\sqrt{V}} \tag{37}$$

注意在给出式（29a）时，已假定 $\phi''(z_0) = 0$，故仅讨论平面阴极情况。

应该指出，总的三级几何横向像差 $\Delta \boldsymbol{u}_3$ 的表达式是由西门纪业首先导出的[1]，虽然在形式上与式（37）有差异，但实质上是一致的。

顺便指出，20 世纪 80 年代以前，在研究三级几何横向像差时，有一个疑惑的问题一直困扰着当时的电子光学学术界，即要不要在几何横向像差的表示式中考虑与 $v(z_i, \varepsilon_z)$ 相关的积分项。在这里，我想将这个问题说得更明白一些，那就是，若近轴电子 $v(z, \varepsilon_z)$ 自阴极面物点逸出时，其轴向初能为 $\varepsilon_z = \varepsilon_{z_1}$，令它到达像面所在的轴上点 z_i 为理想像点，即 $v(z_i, \varepsilon_{z_1}) = 0$。那么，自同一物点逸出的另一相邻的近轴电子 $v(z, \varepsilon_z)$，其 $\varepsilon_z \neq \varepsilon_{z_1}$，它将到达理想像点的近旁，即 $v(z_i, \varepsilon_z) \neq 0$。现在的问题是，此 $v(z_i, \varepsilon_z)$ 所形成的离轴偏离将有多大，它具有什么样的数量级，主要受什么因素影响？当时，各执一词，都说不清楚。

对于这一问题，我们在电磁聚焦同心球系统的电子光学的研究中弄清楚了，并且严格证明了，无论是静电还是电磁复合的成像电子光学系统，下面的公式是普遍成立的：

$$v(z_i, \varepsilon_z) = \frac{2M}{E_c} \left[(\sqrt{\varepsilon_z} - \sqrt{\varepsilon_{z_1}}) - \frac{1}{2\sqrt{\varphi_i}} (\varepsilon_z - \varepsilon_{z_1}) \right] \tag{38}$$

这一研究清晰地回答了成像电子光学科学界长期困惑的问题：此 $v(z_i, \varepsilon_z)$ 的值乃是一个一级小量，仅与阴极面的电场强度 E_c、线放大率 M 相关，但与磁感应强度 B 无关。因此，当研究成像电子光学的三级几何横向像差时，积分号前出现的与 $v(z_i, \varepsilon_z)$ 相关的项都应该略去，否则它将构成四级小量。这就解决了长期以来研究成像电子光学三级几何横向像差表示式中与 $v(z_i, \varepsilon_z)$ 相关的积分要否取舍的问题。而且，由所导得的特解 $v(z_i, \varepsilon_z)$ 表示式，可直接推导出成像系统普遍成立的、决定系统极限鉴别率的 R – A 公式。这样，我们为 R – A 公式的溯源找到了理论依据。于是

$$\Delta \boldsymbol{u}_3(z_i) = -\frac{1}{2} w(z_i, \varepsilon_z) \int_0^{z_i} \frac{Pv}{\sqrt{V}} \boldsymbol{u} \mathrm{d}z - \frac{1}{2} w(z_i, \varepsilon_z) \int_0^{z_i} \frac{S}{V^{3/2}} (Tv\boldsymbol{u} + Vv'\boldsymbol{u}') \mathrm{d}z +$$
$$\frac{1}{2} (\boldsymbol{k} \times \boldsymbol{u}(z_i)) \int_0^{z_i} \left(\frac{hS}{V} - \frac{h''\boldsymbol{u}^2}{4} \right) \frac{\mathrm{d}z}{\sqrt{V}} \tag{39}$$

综上所述，我们的研究提出了在静态成像电子光学中，无论是静电聚焦，或是复合电磁聚焦成像系统，其横向像差应由近轴横向像差与几何横向像差所构成。近轴横向像差通

常由二级+三级近轴色差，以及三级近轴放大率色差与近轴各向异性彗差等所构成。几何横向像差即通常电子光学的三级几何横向像差，如球差、像散、场曲、彗差、畸变等。这样，就把成像电子光学系统的横向像差的构成说清楚了。

4 各个特殊类型三级几何横向像差系数的确定

由前所述，在像面 z_i 处，只有轴向初电位 $\varepsilon_z = \varepsilon_{z_1}$ 的电子才能理想成像，其特解 $v(z_i, \varepsilon_{z_1}) = 0$，$w(z_i, \varepsilon_{z_1}) = (-1)^m M$。将式（32）、式（33）表示的 S、P 代入式（39）中，经过繁复的运算，便可得到与 ε_{z_1} 相对应的 z_i 像面上总的三级横向像差 Δu_3 的 x、y 的分量表示式

$$\Delta u_{3x} = Ew(z_i)r_0^3 + \{[(C+D)\cos\beta_0 + c\sin\beta_0]w(z_i)\}r_0^2\sqrt{\frac{\varepsilon_\perp}{V_i}} +$$
$$[(G + F\cos 2\beta_0 + f\sin 2\beta_0)w(z_i)]r_0\left(\frac{\varepsilon_\perp}{V_i}\right) + [B\cos\beta_0 w(z_i)]\left(\frac{\varepsilon_\perp}{V_i}\right)^{3/2} \quad (40a)$$

$$\Delta u_{3y} = ew(z_i)r_0^3 + \{[(D-C)\sin\beta_0 + c\cos\beta_0]w(z_i)\}r_0^2\sqrt{\frac{\varepsilon_\perp}{V_i}} +$$
$$[(g - f\cos 2\beta_0 + F\sin 2\beta_0)w(z_i)]r_0\left(\frac{\varepsilon_\perp}{V_i}\right) + [B\sin\beta_0 w(z_i)]\left(\frac{\varepsilon_\perp}{V_i}\right)^{3/2} \quad (40b)$$

式中，各像差系数可表达如下：

$$B = -\frac{1}{2}V_i^{3/2}\int_0^{z_i}\left[Wv^4 - \frac{K_{vv}(K_{vv}-1)}{V}\right]\frac{dz}{\sqrt{V}},$$

$$f = -\frac{h_0}{\sqrt{V_i}}B + \frac{1}{2}V_i\int_0^{z_i}\left[\frac{h-h_0}{V}K_{vv} - \frac{h''}{4}v^2\right]\frac{dz}{\sqrt{V}},$$

$$g = 2f + \frac{1}{2}V_i\int_0^{z_i}\frac{h_0 K_{vv} - h}{V^{3/2}}dz,$$

$$F = -\frac{1}{2}V_i\int_0^{z_i}\left[Wv^3w - \frac{K_{vv}K_{vw}}{V}\right]\frac{dz}{\sqrt{V}},$$

$$G = 2F - \frac{1}{2}V_i\int_0^{z_i}\frac{K_{vw}}{V^{3/2}}dz,$$

$$c = -2\frac{h_0}{\sqrt{V_i}}F + \sqrt{V_i}\int_0^{z_i}\left[\frac{h}{V}K_{vw} - \frac{h''}{4}vw\right]\frac{dz}{\sqrt{V}},$$

$$D = -\sqrt{V_i}\int_0^{z_i}\left[W(h_0^2 v^2 + w^2)v^2 - \frac{1}{2}h_0 h''v^2 - \frac{(K_{ww} + h_0^2 K_{vv} - 2hh_0)}{V}K_{vv} + \frac{\phi''}{8V}\right]\frac{dz}{\sqrt{V}},$$

$$C = -\frac{1}{2}D - \sqrt{V_i}\int_0^{z_i}\left[Wv^2w^2 - \frac{K_{vw}^2}{V} - \frac{\phi''}{16V}\right]\frac{dz}{\sqrt{V}},$$

$$e = -\frac{h_0}{2\sqrt{V_i}}D - \frac{h_0}{16}\int_0^{z_i}\frac{\phi''}{V^{3/2}}dz + \frac{1}{2}\int_0^{z_i}\left[\frac{h}{V}(K_{ww} + h_0^2 K_{vv} - 2hh_0) - \frac{h''}{4}(h_0^2 v^2 + w^2)\right]\frac{dz}{\sqrt{V}},$$

$$E = -\frac{1}{2}\int_0^{z_i}\left[W(h_0^2 v^2 + w^2)vw - \frac{1}{2}h_0 h''vw - \frac{(K_{ww} + h_0^2 K_{vv} - 2hh_0)}{V}K_{vw}\right]\frac{dz}{\sqrt{V}}$$

$$(41)$$

式中，V_i 为像面 $z=z_i$ 处的电位；W、K_{vv} 等的表达式如下：

$$W = \frac{1}{16}\phi^{\mathrm{IV}} + hh'', \quad K_{vv} = Tv^2 + Vv'^2, \tag{42}$$
$$K_{vw} = Tvw + Vv'w', \quad K_{ww} = Tw^2 + Vw'^2,$$

式（40）和式（41）中，E 和 e 表示三级畸变项；C、D、c 表示三级场曲与像散项；G、F、g 和 f 表示三级彗差项；B 表示三级色球差项；$w(z_i)$ 即系统的放大率 M。

应该指出，本文导出的式（41）可转化为西门纪业[1]与 Бонштедт[2]给出的各像差系数表达式，证明了这两篇文章中的总的三级横向像差表达式实质上是一致的。如果将文献[2]中式（31）中各式做一些补正，则两文的像差系数表达式可以互相转化。这说明，由轨迹方程式（8）或式（9）出发研究阴极透镜的像差，其结果是完全一样的，因此，Бонштедт 对于西门纪业的文章的批评是不正确的。

5 静电阴极透镜各种特殊类型的三级几何横向像差系数的确定

静电阴极透镜在对应于 $v(z_i, \varepsilon_{z_1}) = 0$ 像面上的三级几何横向像差，早已由 Recknagel[3]导出。西门纪业[1]根据导出的电磁复合聚焦阴极透镜的像差公式，若不考虑磁场的情况，利用分部积分，消去像差系数中含有的高于二阶以上的电位导数，亦可化为 Recknagel 的情况。从我们导出的三级几何横向像差公式（40a）和式（40b）出发，若令 $h(z)=0$，亦可得到上述相同的结论。

在静电情况下，$h(z)=0$，式（14）与式（42）各量便化为如下形式：

$$T = \frac{1}{4}\phi'', \quad W = \frac{1}{16}\phi^{\mathrm{IV}}, \quad K_{vv} = \frac{1}{4}\phi''v^2 + Vv'^2, \tag{43}$$
$$K_{vw} = \frac{1}{4}\phi''vw + Vv'w', \quad K_{ww} = \frac{1}{4}\phi''w^2 + Vw'^2$$

于是式（40a）、式（40b）中所有含有 $h(z)$ 以小写英文字母表示的像差系数如 f、g 等都等于零。同样，类似于文献[1]、文献[3]、文献[11]，将像差系数的积分表达式化成不含有电位高阶导数的情况，于是有

$$\begin{aligned}
B = &-\frac{1}{32}V_i^{3/2}\left|\left(\frac{\phi'''}{\sqrt{V}} + \frac{1}{2}\frac{\phi''\phi'}{V^{3/2}}\right)v^4 - \frac{4\phi''v^3v'}{\sqrt{V}}\right|_0^{z_i} - \\
&\frac{1}{2}V_i^{3/2}\int_0^{z_i}\left[\left(-\frac{3}{32}\frac{\phi''^2}{V} + \frac{3}{64}\frac{\phi''\phi'^2}{V^2}\right)v^4 - \frac{3}{8}\frac{\phi''\phi'}{V}v^3v' + \frac{3}{4}\phi''v^2v'^2 - \frac{K_{vv}(K_{vv}-1)}{V}\right]\frac{\mathrm{d}z}{\sqrt{V}},
\end{aligned}$$

$$\begin{aligned}
F = &-\frac{1}{32}V_i\left|\left(\frac{\phi'''}{\sqrt{V}} + \frac{1}{2}\frac{\phi''\phi'}{V^{3/2}}\right)v^3w - \frac{3\phi''v^2v'w}{\sqrt{V}} - \frac{\phi''v^3w'}{\sqrt{V}}\right|_0^{z_i} - \\
&\frac{1}{2}V_i^{3/2}\int_0^{z_i}\left[\left(-\frac{3}{32}\frac{\phi''^2}{V} + \frac{3}{64}\frac{\phi''\phi'^2}{V^2}\right)v^3w - \frac{9}{32}\frac{\phi''\phi'}{V}\left(v^2v'w + \frac{1}{3}v^3w'\right) + \right. \\
&\left.\frac{3}{8}\phi''(vv'^2w + v^2v'w') - \frac{K_{vv}K_{vw}}{V}\right]\frac{\mathrm{d}z}{\sqrt{V}}
\end{aligned}$$

$$\tag{44}$$

$$D = -\frac{1}{16}\sqrt{V_i}\left|\left(\frac{\phi'''}{\sqrt{V}} + \frac{1}{2}\frac{\phi''\phi'}{V^{3/2}}\right)v^2w^2 - \frac{2\phi''vv'w^2}{\sqrt{V}} - \frac{2\phi''v^2ww'}{\sqrt{V}}\right|_0^{z_i} -$$

$$\sqrt{V_i}\int_0^{z_i}\left[\left(-\frac{3}{32}\frac{\phi''^2}{V} + \frac{3}{64}\frac{\phi''\phi'^2}{V^2}\right)v^2w^2 - \frac{3}{16}\frac{\phi''\phi'}{V}(vv'w^2 + v^2ww') +\right.$$

$$\left.\frac{3}{8}\phi''(v'^2w^2 + 4vv'ww' + v^2w'^2) - \frac{K_{ww}K_{vv}}{V} + \frac{\phi''}{8V}\right]\frac{\mathrm{d}z}{\sqrt{V}},$$

$$G = 2F - \frac{1}{2}V_i\int_0^{z_i}\frac{K_{vw}}{V^{3/2}}\mathrm{d}z, \qquad\qquad\qquad (44\,\text{续})$$

$$C = \frac{1}{2}D + \sqrt{V_i}\int_0^{z_i}\left(\frac{K_{vw}^2 - K_{w\omega}K_{vv}}{V} + \frac{3}{16}\frac{\phi''}{V}\right)\frac{\mathrm{d}z}{\sqrt{V}},$$

$$E = -\frac{1}{32}\left|\left(\frac{\phi'''}{\sqrt{V}} + \frac{1}{2}\frac{\phi''\phi'}{V^{3/2}}\right)vw^3 - \frac{3\phi''w^2w'v}{\sqrt{V}} - \frac{\phi''w^3v'}{\sqrt{V}}\right|_0^{z_i} -$$

$$\frac{1}{2}\int_0^{z_i}\left[\left(-\frac{3}{32}\frac{\phi''^2}{V} + \frac{3}{64}\frac{\phi''\phi'^2}{V^2}\right)vw^3 - \frac{9}{32}\frac{\phi''\phi'}{V}\left(w^2w'v + \frac{1}{3}w^3v'\right) +\right.$$

$$\left.\frac{3}{8}\phi''(ww'^2v + w^2w'v') - \frac{K_{ww}K_{vw}}{V}\right]\frac{\mathrm{d}z}{\sqrt{V}}$$

6 实例

我们在考察静电与电磁复合电子光学系统时，其结论是，系统横向像差应是近轴横向像差与几何横向像差的合成。现以均匀平行电磁复合聚焦系统为例，考察该系统的近轴横向像差与几何横向像差。

令

$$h(z) = h_0, \quad \phi'(z) = \frac{V_i}{l} = \text{常数}, \quad V(z) = \frac{V_i}{l}z + \varepsilon_z$$

因此可得

$$\phi'' = \phi''' = \phi^{IV} = 0, \quad h' = h'' = 0, \quad T = h_0^2, \quad W = 0$$

线性方程式（13）的标量形式的特解可表示为

$$v(z) = \frac{1}{h_0}\sin\chi(z), \quad w(z) = \cos\chi(z) \qquad\qquad (45)$$

式中，

$$\chi(z) = k\left(\sqrt{z + \frac{\varepsilon_z}{V_i}l} - \sqrt{\frac{\varepsilon_z}{V_i}l}\right) \qquad\qquad (46)$$

式中，

$$k = 2h_0\sqrt{\frac{l}{V_i}} \qquad\qquad (47)$$

设电子在 $\varepsilon_z = \varepsilon_{z_1}$ 对应的像面上理想成像，即 $\chi(z_i = l)|_{\varepsilon_{z_1}} = m\pi$ 便可导出成像位置与放大率为

$$z_i = l = \left(\frac{m\pi}{k}\right)^2 + \frac{2m\pi}{k}\sqrt{\frac{\varepsilon_{z_1}}{V_i}}l \qquad\qquad (48)$$

$$M = (-1)^m \cos(m\pi) = 1 \tag{49}$$

此系统内运动的电子轨迹,按式(2)、式(10)与式(19),并考虑到 $\theta_0 = 0$,则可表达成以下的指数形式[13]:

$$r(z) = r_0 + \sqrt{\varepsilon_\perp} \frac{\sin\chi}{h_0} e^{i(\beta_0 + \chi)} \tag{50}$$

现讨论系统的近轴横向像差与几何横向像差,由式(40)与式(43)可知

$$K_{vv} = 1, \quad K_{ww} = h_0^2, \quad K_{vw} = 0$$

将以上关系式均代入三级横向像差系统表达式(39)中,不难发现所有的各种特殊类型的像差系数均等于零。此乃由于在均匀平行复合场中以 ε_z 逸出的实际轨迹与近轴轨迹完全相同所致。因此,在 z_i 像面上系统只剩下二级横向像差。

系统的近轴横向像差即二级+三级横向色差 $\Delta r_*(z_i)$ 按式(24)与式(50)可表示为

$$\Delta r_*(z_i) = \sqrt{\varepsilon_\perp} \frac{\sin\chi(z_i)}{h_0} e^{i[\beta_0 + \chi(z_i)]} \tag{51}$$

式中,$\chi(z_i)$ 乃是轴向初电位 ε_z 的近轴电子使旋转坐标系在 $z_i = l$ 的像面上多或少旋转的角度,它可表示为

$$\chi(z_i = l)\big|_{\varepsilon_z} = m\pi - \frac{2h_0}{\phi_0'}(\sqrt{\varepsilon_z} - \sqrt{\varepsilon_{z_1}}) + \frac{h_0}{\phi_0'\sqrt{V_i}}(\varepsilon_z - \varepsilon_{z_1}) \tag{52}$$

将上式代入式(51),并以 $-E_c$ 代以 ϕ_0' 便可求得近轴横向像差的表达式为

$$\Delta r_* = \left[\frac{2}{E_c}\sqrt{\varepsilon_\perp}(\sqrt{\varepsilon_z} - \sqrt{\varepsilon_{z_1}}) - \frac{1}{E_c\sqrt{V_i}}\sqrt{\varepsilon_\perp}(\varepsilon_z - \varepsilon_{z_1})\right] \times$$
$$\exp\left\{i\left[\beta_0 + \frac{2h_0}{E_c}(\sqrt{\varepsilon_z} - \sqrt{\varepsilon_{z_1}}) - \frac{h_0}{E_c\sqrt{V_i}}(\varepsilon_z - \varepsilon_{z_1})\right]\right\} \tag{53}$$

由此可见,均匀平行电磁复合聚焦系统的三级几何横向像差等于零,于是仅剩下二级近轴横向像差(即二级+三级横向色差)。这与文献[10]的结果是一致的。

结束语

由上述的研究可以得到以下几点:

(1) 西门纪业和 Бонштедт 两文的像差方程的自由项 F 及其像差系数表达式虽然形式上迥然不同,实质上是可以互相转化的,是完全一致的。本文证明了 Бонштедт 对西门纪业的批评是不正确的。相反,Бонштедт 给出的像差系数表达式却有一些遗漏与错误,需要加以补正。

(2) 无论对于静电聚焦或是电磁复合聚焦阴极透镜,以高度 r_0,初始条件为 ε_0、α_0、β_0(其轴向初电位为 ε_z)逸出的实际电子轨迹在轴向初电位 ε_{z_1} 对应的像面上与理想成像位置的差异定义为横向像差,它表现为二级+三级近轴横向像差与三级几何横向像差的矢量和。

(3) 近轴横向像差表现为二级+三级横向色差,以及放大率近轴像差等;几何横向像差表现为中心色球差、场曲与像散、彗差与畸变等,都是三级项。由于逸出电子 $\varepsilon_z \neq \varepsilon_{z_1}$,尽管 $v(z_i, \varepsilon_{z_1}) = 0$,$v(z_i, \varepsilon_z)$ 虽不等于零,但它是一级小量,因此,三级几何横向像差不应

包含 $v(z_i,\varepsilon_z)$ 有关的积分项。

(4) 均匀平行电磁复合聚焦系统的实例表明，在这个系统中，尽管不存在三级几何横向像差，却存在二级近轴横向像差。

作者对于北京大学无线电电子学系西门纪业教授在本课题研究中给予的热忱支持与帮助表示衷心的感谢。

参 考 文 献

[1] XIMEN J Y. Electron-optical properties and aberration theory of combined immersion objectives[J] Acta Physica Sinica, 1957,13:339 – 356.

[2] БОНШТЕДТ Б Э. К расчету аберации катодных линз. [J]. Радиотехника и Электроника,1964,9 (5):844 – 850.

[3] RECKNAGEL A. Theorie des elektrisohen elektronen miktroskops fur selbstrakler[J]. Z, Angew. Physik,1941,119:689 – 708(in Germen).

[4] АРЦИМОВИЧ Л А. Электростатические свойства эмисионных систем[J]. А Н СССР Сер. Физ. , 1944, 8(6):313.

[5] ВОРОБЪЕВ Ю В. Фигуры рассеяния в электростатических иммерсионных линзах [J]. Ж. Т. Ф. , 1956, 26:2269 – 2280.

[6] 周立伟. 宽束电子光学 [M]. 北京：北京理工大学出版社,1993.

[7] КУЛИКОВ Ю В, МОНАСТЕРСКИЙ М А,УШАКОВ В А, ШЕЛЕВ М Я. Теория и расчет временных аберраций катодных линз[M]. Труды ФИАН 1985 – Т:153.

[8] КУЛИКОВ Ю В, МОНАСТЕРСКИЙ М А, ФЕЙГИН Х И. Теория аберрации третьего порядка катодных линз [J]. Ралиотехника и Электроника, 1978, 23 (1): 167 – 174.

[9] ВЛАСОВ А В. ШАПИРО Ю А. Методы расчета эмиссонных электрнно-оптических систем[M]. Машиностроение, 1974.

[10] CHOU L W. Electron optics of concentric spherical electromagnetic focusing systems [J]. Advances in Electronics and Electron Physics, 1979,52: 119 – 132.

[11] CHOU L W. Electron optics of concentric spherical electromagnetic focusing systems [J]. Advances in Electronics and Electron Physics, 1979,52:119 – 132.

[12] 周立伟. 变像管与像增强器的电子光学[M]. 北京：北京理工大学,1975.

6.2 Variation Theory of Geometrical Aberrations in Cathode Lenses
阴极透镜几何像差的变分理论

Abstract: In this paper, the aberration theory for cathode lenses with combined electrostatic and magnetic focusing fields is discussed in detail. The more general condition, when cathode is immersed in both magnetic and transversal electric fields, has also been considered. Basing on the variation principle, we have derived the general formulae of aberration coefficients of cathode lenses at an arbitrary image plane, and transformed them into the linear combination of the well-known aberration coefficients given by W. Glaser. It follows that the aberration for both the wide electron beam system (cathode lens) and the narrow electron beam system (conventional lens) can be treated by the universal variation theory. The paraxial electron trajectories of cathode lenses are described in the vector form. The aberrations are expressed in the matrix form, these are therefore suitable for the computer calculation.

摘要：本文运用变分原理研究了电磁复合聚焦阴极透镜的像差理论，导出了在任意像平面中像差系数的普遍公式，它们可以用 Glaser 导出的像差系数的线性组合来表示。这就表明了宽束与细束电子光学系统的像差都可以用统一的变分方法处理，并建立了两者之间的联系。本文采用矢量形式描写阴极透镜的近轴轨迹。并采用矩阵形式描写像差，本文所得的像差系数较之文献上已有结果形式更为简洁且适用于计算机计算。本文考虑了阴极面上磁场和横向电场不为零的情况，因而所得结果较为普遍。

Introduction

The cathode lens is usually considered as such a typical electron optical system, where the cathode is directly placed in the electrostatic and magnetic fields and the cathode surface is also the object surface to be imaged with wide electron beam. In 1957, one of the authors first established the strict aberration theory for cathode lenses with combined electric and magnetic focusing fields [1]. According to the results and method of Ref. [1], the aberration of cathode lenses was later calculated again by Ref. [2] in 1964. Then the aberration of cathode lenses at an arbitrary image plane has been derived by Ref. [3].

In present paper, by means of the variation principle we have studied the aberration for cathode lenses with combined electrostatic and magnetic fields and have derived the general

---

Ximen Jiye[a], Zhou Liwei[b], Ai Kecong[b]. a) Peking University, b) Beijing Institute of Technolgy. Optik, V. 66, No. 1, 1983, 19 – 34. Advances in Electronics and Electron Physics, V. 64B, 1985, 561 – 573.

formulae of aberration coefficients of cathode lenses at an arbitrary image plane. It has been shown that these aberration coefficients can be transformed into the linear combination of the well-known Glaser's geometrical and chromatic aberration coefficients [4]. It follows that the aberrations for both the wide electron beam system (cathode lens) and the narrow electron beam system (conventional lens) can be treated by the universal variation theory. Relationship between these two systems is thus established.

The paraxial trajectories of the wide electron beam of cathode lens are described in the vector form. The aberrations are expressed in the matrix form. This formalism is suitable for computer calculation. Therefore, we will generalize Glaser's aberration theory of the narrow electron beam system to the wide electron beam system. As an extension, the more general condition when cathode is immersed in both magnetic and transversal electrical fields, has also been considered.

1 Variation function and ray function for cathode lenses

In the rotationally symmetrical cathode lenses, let z-axis be its axis of rotation. We select the cathode plane $z = z_0$ (which is also the object plane) as the zero of the axial electrostatic potential $\phi = (z)$, i.e., $\phi(z_0) = 0$. Now, it should be noted that the normalization of the axial electrostatic potential for a cathode lens is quite different form that for a conventional lens. In fact, we use the special normalization for a cathode lens and the ordinary normalization for a conventional lens, which are defined respectively:

$$\varepsilon = \varepsilon_z + \varepsilon_\perp,$$
$$V(z) = \varepsilon_z + \phi(z) \text{ (for cathode lens)},$$
$$V(z) = \varepsilon + \phi(z) \text{ (for conventional lens)} \tag{1}$$

let ε_z, $\varepsilon_\perp$ and ε be the axial, transversal and total initial energy (in electron volt) respectively, which are referred to the electron emitted from the cathode.

The paraxial trajectory equation of the wide electron beam in cathode lens is given by[1]

$$V\boldsymbol{u}'' + \frac{1}{2}V'\boldsymbol{u}' + \frac{1}{4}(V'' + \mathscr{H})\boldsymbol{u} = 0 \tag{2}$$

Let the z-component of the axial magnetic induction be $H(z)$ which is related to the $\mathscr{H}(z)$ in script

$$\mathscr{H}(z) = \sqrt{\eta/2}H(z), \quad \eta = e/m_0 \tag{3}$$

where e is the absolute charge and m_0 is the mass of an electron. Let the prime denote the differentiation with respect to z, $v(z)$ and $w(z)$ are two indepedent solutions of Eq. (2), satisfying the initial conditions at the cathode plane $z = z_0$:

$$v(z_0) = 0, \quad v'(z_0)\sqrt{\varepsilon_z} = 1$$
$$w(z_0) = 1, \quad w'(z_0)\sqrt{\varepsilon_z} = 0 \tag{4}$$

According to Ref. [1, 3], the paraxial trajectory $\boldsymbol{u}$ and its slope $\boldsymbol{u}'$ may be expressed by means of $v(z)$ and $w(z)$ as follows

$$\boldsymbol{u}(z) = \boldsymbol{u}_0 w(z) + (2\eta)^{-1/2}\dot{\boldsymbol{u}}_0 v(z), \quad \boldsymbol{u}'(z) = \boldsymbol{u}_0 w'(z) + (2\eta)^{-1/2}\dot{\boldsymbol{u}}_0 v'(z) \tag{5}$$

It is to be noted that the trajectory vector **u** in the rotating Cartesian coordinate system rotates through an angle $\chi(z)$ with respect to the same trajectory vector **r** in the fixed Cartesian coordinate system. Hence, the relation between **u** and **r** is given by

$$\boldsymbol{r} = \boldsymbol{u} e^{i\chi(z)}, \quad \boldsymbol{u} = \boldsymbol{r} e^{-i\chi(z)} \tag{6}$$

where $\chi(z)$ is written by

$$\chi(z) = \frac{1}{2} \int_{z_0}^{z} \frac{\mathscr{H}}{V^{1/2}} dz \tag{7}$$

Using Eq. (6) and Eq. (7) we get the initial position $\boldsymbol{u}_0$ and initial velocity $\dot{\boldsymbol{u}}_0$:

$$\begin{aligned}(x_0, y_0) &= \boldsymbol{u}_0 = \boldsymbol{u}(z_0) = \boldsymbol{r}_0, \\ (\dot{x}_0, \dot{y}_0) &= \dot{\boldsymbol{u}}_0 = \dot{\boldsymbol{u}}(z_0) = \dot{\boldsymbol{r}}_0 - \dot{\chi}_0 \boldsymbol{r}_0^*\end{aligned} \tag{8}$$

In Eq. (8) $\dot{\chi}_0$ and $\boldsymbol{r}_0^*$ are defined by

$$\begin{aligned}\dot{\chi}_0 &= \sqrt{\eta/2}\, \mathscr{H}(z_0) = (\eta/2) H(z_0), \\ \boldsymbol{r}_0^* &= \boldsymbol{k} \times \boldsymbol{r}_0\end{aligned} \tag{9}$$

Let the dot denote the differentiation with respect to time, **k** be a unit vector in the direction of the rotationally symmetrical axis z, $\boldsymbol{r}_0^*$ be an orthogonal of $\boldsymbol{r}_0$. Hereafter an orthogonal vector is denoted by $\boldsymbol{r}^*$, which is rotated 90° with respect to the vector **r**.

Now it should be noted that the trajectory representations for a cathode lens are quite different from those for a conventional lens. In fact, for a cathode lens, the initial position $\boldsymbol{u}_0$ and initial velocity $\dot{\boldsymbol{u}}_0$ as the first order quantities are chosen to describe trajectory [1,3]. While for a conventional lens, the initial position $\boldsymbol{u}_0$ and initial slope $\boldsymbol{u}_0'$ as the first order quantities are used to describe trajectory [4-6].

We will generalize Glaser's aberration theory to the wide electron beam system. In order to calculate the thirdorder geometrical aberration, we consider the fourth order variational functions [4,6]

$$-F_4 = (2\eta)^{1/2} \left\{ \frac{1}{4} L (\boldsymbol{u} \cdot \boldsymbol{u})^2 + \frac{1}{2} M (\boldsymbol{u} \cdot \boldsymbol{u})(\boldsymbol{u}' \cdot \boldsymbol{u}') + \frac{1}{4} N (\boldsymbol{u}' \cdot \boldsymbol{u}')^2 + \right.$$
$$\left. PV^{1/2} (\boldsymbol{u} \cdot \boldsymbol{u})(\boldsymbol{u} \times \boldsymbol{u}')\boldsymbol{k} + QV^{1/2} (\boldsymbol{u}' \cdot \boldsymbol{u}')(\boldsymbol{u} \times \boldsymbol{u}')\boldsymbol{k} + KV [(\boldsymbol{u} \times \boldsymbol{u}')\boldsymbol{k}]^2 \right\} \tag{10}$$

where **u**, **u**' are given by Eq. (5), K, L, M, N, P, Q are the functions of $V(z)$ and $\mathscr{H}(z)$.

$$K = \frac{1}{8} \frac{\mathscr{H}^2}{V^{3/2}},$$

$$L = \frac{1}{32 V^{1/2}} \left[\frac{1}{V} (V'' + \mathscr{H}^2)^2 - V^{IV} - 4\mathscr{H}\mathscr{H}'' \right],$$

$$M = \frac{1}{8 V^{1/2}} (V'' + \mathscr{H}^2),$$

$$N = \frac{1}{2} V^{1/2},$$

$$P = \frac{1}{16 V^{1/2}} \left[\frac{\mathscr{H}}{V} (V'' + \mathscr{H}^2) - \mathscr{H}'' \right],$$

$$Q = \frac{1}{4V^{1/2}} \tag{11}$$

Then the fourth order ray function between the cathode plane z_0 (also the object plane) and an arbitrary image plane z_i is defined by

$$-\int_{z_0}^{z_i} F_4 dz = \frac{1}{4} A (\boldsymbol{u}_0 \cdot \boldsymbol{u}_0)^2 + \frac{1}{4} B (\dot{\boldsymbol{u}}_0 \cdot \dot{\boldsymbol{u}}_0)^2 + C (\boldsymbol{u}_0 \cdot \dot{\boldsymbol{u}}_0)^2 +$$
$$\frac{1}{2} D(\boldsymbol{u}_0 \cdot \boldsymbol{u}_0)(\dot{\boldsymbol{u}}_0 \cdot \dot{\boldsymbol{u}}_0) + E(\boldsymbol{u}_0 \cdot \boldsymbol{u}_0)(\boldsymbol{u}_0 \cdot \dot{\boldsymbol{u}}_0) +$$
$$F(\dot{\boldsymbol{u}}_0 \cdot \dot{\boldsymbol{u}}_0)(\boldsymbol{u}_0 \cdot \dot{\boldsymbol{u}}_0) + 2c(\boldsymbol{u}_0 \cdot \dot{\boldsymbol{u}}_0)(\boldsymbol{u}_0 \times \dot{\boldsymbol{u}}_0)\boldsymbol{k} +$$
$$e(\boldsymbol{u}_0 \cdot \boldsymbol{u}_0)(\boldsymbol{u}_0 \times \dot{\boldsymbol{u}}_0)\boldsymbol{k} + f(\dot{\boldsymbol{u}}_0 \cdot \dot{\boldsymbol{u}}_0)(\boldsymbol{u}_0 \times \dot{\boldsymbol{u}}_0)\boldsymbol{k} \tag{12}$$

Where the twin – and triple – scalar product are defined by
$$(\boldsymbol{u}_0 \cdot \dot{\boldsymbol{u}}_0) = (x_0 \dot{x}_0 + y_0 \dot{y}_0),$$
$$(\boldsymbol{u}_0 \times \dot{\boldsymbol{u}}_0)\boldsymbol{k} = (\boldsymbol{u}_0 \times \dot{\boldsymbol{u}}_0)\boldsymbol{k} = (x_0 \dot{y}_0 - y_0 \dot{x}_0) \tag{13}$$

In Eq. (12) coefficients A, B, C, D, E, F and c, e, f are the third order aberration coefficients suitable for cathode lens and are given as follows

$$A = (2\eta)^{1/2} \int_{z_0}^{z_i} [Lw^4 + 2Mw^2 w'^2 + Nw'^4] dz,$$

$$B = (2\eta)^{-3/2} \int_{z_0}^{z_i} [Lv^4 + 2Mv^2 v'^2 + Nv'^4] dz,$$

$$C = (2\eta)^{-1/2} \int_{z_0}^{z_i} [Lv^2 w^2 + 2Mvv' ww' + Nv'^2 w'^2 - K] dz,$$

$$D = (2\eta)^{-1/2} \int_{z_0}^{z_i} [Lv^2 w^2 + M(v^2 w'^2 + w^2 v'^2) + Nv'^2 w'^2 + 2K] dz,$$

$$E = \int_{z_0}^{z_i} [Lvw^3 + Mww'(vw)' + Nv'w'^3] dz,$$

$$F = (2\eta)^{-1} \int_{z_0}^{z_i} [Lv^3 w + Mvv'(vw)' + Nw'v'^3] dz,$$

$$c = (2\eta)^{-1/2} \int_{z_0}^{z_i} [Pvw + Qv'w'] dz,$$

$$e = \int_{z_0}^{z_i} [Pw^2 + Qw'^2] dz,$$

$$f = (2\eta)^{-1} \int_{z_0}^{z_i} [Pv^2 + Qv'^2] dz \tag{14}$$

Obviously, these aberration coefficients for cathode lens correspond to those given by Glaser[4] for the conventional lens.

Calculating the gradients of the fourth order ray function with respect to $\boldsymbol{u}_0$ and $\dot{\boldsymbol{u}}_0$, we obtain the third order geometrical aberration at an arbitrary image plane z_i

$$\Delta u_3(z_i) = (2\eta)^{-1/2} v(z_i) \frac{\partial}{\partial \boldsymbol{u}_0} \left[\int_{z_0}^{z_i} F_4 dz \right] - w(z_i) \frac{\partial}{\partial \dot{\boldsymbol{u}}_0} \left[\int_{z_0}^{z_i} F_4 dz \right] + (2\eta)^{-1/2} v(z_i) \left\{ \frac{\partial}{\partial \boldsymbol{u}'} F_4 \right\}_{z_0} \tag{15}$$

It should be pointed out that when Prof. Ximen Jiye discussed the Eq. 15 with Zhou Liwei in

1983, we could not confirmed the dimension of $v(z_i)$. Therefore, we temporary kept the related item with $v(z_i)$. Later, when we studied the electron optics related with electromagnetic focusing concentric spherical system, we found the following formula

$$v(z_i, \varepsilon_z) = \frac{2M}{E_c}\left[(\sqrt{\varepsilon_z} - \sqrt{\varepsilon_{z_1}}) - \frac{1}{2\sqrt{\phi_i}}(\varepsilon_z - \varepsilon_{z_1})\right] \tag{16}$$

is tenable. Thus, the term with $v(z_i)$ in Eq. (15) should be omitted. It shows that the term with $v(z_i)$ is only a small quantity of the first order. Therefore, we obtain the third order geometrical aberration at an arbitrary image plane z_i

$$\Delta \boldsymbol{u}_3(z_i) = -w(z_i)\frac{\partial}{\partial \dot{\boldsymbol{u}}_0}\left[\int_{z_0}^{z_i} F_4 \mathrm{d}z\right] \tag{17}$$

Inserting Eq. (10) and Eq. (12) into Eq. (15), we get the third order geometrical aberration $\Delta \boldsymbol{u}_3(z_i)$ as in Eq. (17)

$$\Delta \boldsymbol{u}_3(z_i) = w(z_i) \begin{bmatrix} \dot{\boldsymbol{u}}_0 \cdot \dot{\boldsymbol{u}}_0 \\ 2\dot{\boldsymbol{u}}_0 \cdot \dot{\boldsymbol{u}}_0 \\ 2(\boldsymbol{u}_0 \cdot \dot{\boldsymbol{u}}_0)k \\ \boldsymbol{u}_0 \cdot \boldsymbol{u}_0 \end{bmatrix}^{\mathrm{T}} \begin{bmatrix} B & F & f \\ F & C & c \\ f & c & o \\ D & E & e \end{bmatrix} \begin{bmatrix} \dot{\boldsymbol{u}}_0 \\ \boldsymbol{u}_0 \\ \boldsymbol{u}_0^* \end{bmatrix}$$

where T denotes the transposed matrix. The last term in the braces { } can be found in Ref. [1, 7].

In order to calculate the first order chromatic aberration we write the second order variation function[6]:

$$F_{2c} = (2\eta)^{-1/2}\left[\frac{1}{16V^{3/2}}(V''\mathscr{H})u^2 + \frac{1}{4V^{1/2}}u'^2 + \frac{\mathscr{H}}{4V}(\boldsymbol{u} \times \boldsymbol{u}')k\right] \tag{18}$$

Then, the second order ray function between the cathode plane z_0 and an arbitrary image plane z_i is given by[6]

$$\int_{z_0}^{z_i} F_{2c}\mathrm{d}z = \frac{1}{2}C_d(\dot{\boldsymbol{u}}_0 \cdot \dot{\boldsymbol{u}}_0) + C_m(\boldsymbol{u}_0 \cdot \dot{\boldsymbol{u}}_0) + \frac{1}{2}C_a(\boldsymbol{u}_0 \cdot \boldsymbol{u}_0) + C_r(\boldsymbol{u}_0 \times \dot{\boldsymbol{u}}_0)k \tag{19}$$

Calculating the gradients of the second order ray function with respect to $\boldsymbol{u}_0$ and $\dot{\boldsymbol{u}}_0$, we obtain the first order chromatic aberration at an arbitrary image plane z_i:

$$\Delta \boldsymbol{u}_c(z_i) = (2\eta \Delta \varepsilon)\left[-w(z_i)\frac{\partial}{\partial \dot{\boldsymbol{u}}_0}\int_{z_0}^{z_i} F_{2c}\mathrm{d}z\right] = (2\eta \Delta \varepsilon)\{-w(z_i)[C_d\dot{\boldsymbol{u}}_0 + C_m\boldsymbol{u}_0 - C_r\boldsymbol{u}_0^*]\} \tag{20}$$

In Eq. (19), $\Delta \varepsilon$ is the variation of the initial energy, C_m, C_d, C_a and C_r are the first order chromatic aberration coefficients[4,6]

$$C_m = \frac{1}{2}(2\eta)^{-1}\int_{z_0}^{z_i}\left[\frac{v'w'}{V^{1/2}} + \frac{V'' + \mathscr{H}^2}{4V^{3/2}}vw\right]\mathrm{d}z,$$

$$C_d = \frac{1}{2}(2\eta)^{-3/2}\int_{z_0}^{z_i}\left[\frac{v'^2}{V^{1/2}} + \frac{V'' + \mathscr{H}^2}{4V^{3/2}}v^2\right]\mathrm{d}z,$$

$$C_a = \frac{1}{2}(2\eta)^{-1/2}\int_{z_0}^{z_i}\left[\frac{w'^2}{V^{1/2}} + \frac{V'' + \mathscr{H}^2}{4V^{3/2}}w^2\right]\mathrm{d}z,$$

$$C_a = \frac{1}{2}(2\eta)^{-1/2} \int_{z_0}^{z_i} \frac{\mathscr{H}}{V^{3/2}} dz \tag{21}$$

Thus, nine third order geometrical aberration coefficients are given in Eq. (14) and four first order chromatic aberration coefficients are given in Eq. (21), which are suitable for cathode lenses with the wide electron beam. These results correspond to those given by Glaser for the conventional lenses with narrow electron beam.

According to Ref. [2, 3], we introduce the following notations:

$$h = \mathscr{H}/2, \quad T = \frac{1}{4}V'' + h^2, \quad W = \frac{1}{16}V^{IV} + hh'',$$
$$K_{vv} = Tv^2 + Vv'^2,$$
$$K_{vw} = Tvw + Vv'w',$$
$$K_{ww} = Tw^2 + Vw'^2 \tag{22}$$

Then, the aberration coefficients in Eq. (14) and Eq. (21) are rewritten as follows

$$A = -\frac{1}{2}(2\eta)^{1/2} \int_{z_0}^{z_i} \left[Ww^4 - \frac{K_{ww}^2}{V} \right] \frac{dz}{V^{1/2}},$$

$$B = -\frac{1}{2}(2\eta)^{-3/2} \int_{z_0}^{z_i} \left[Wv^4 - \frac{K_{vv}^2}{V} \right] \frac{dz}{V^{1/2}},$$

$$C = -\frac{1}{2}(2\eta)^{-1/2} \int_{z_0}^{z_i} \left[Wv^2w^2 - \frac{K_{vw}^2}{V} + \frac{h^2}{V} \right] \frac{dz}{V^{1/2}},$$

$$D = -\frac{1}{2}(2\eta)^{-1/2} \int_{z_0}^{z_i} \left[Wv^2w^2 - \frac{K_{vv}K_{ww}}{V} - \frac{2h^2}{V} \right] \frac{dz}{V^{1/2}},$$

$$E = -\frac{1}{2} \int_{z_0}^{z_i} \left[Wvw^3 - \frac{K_{vw}K_{ww}}{V} \right] \frac{dz}{V^{1/2}},$$

$$F = -\frac{1}{2}(2\eta)^{-1} \int_{z_0}^{z_i} \left[Wv^3w - \frac{K_{vv}K_{vw}}{V} \right] \frac{dz}{V^{1/2}},$$

$$c = -\frac{1}{2}(2\eta)^{-1/2} \int_{z_0}^{z_i} \left[\frac{h''vw}{4} - \frac{hK_{vw}}{V} \right] \frac{dz}{V^{1/2}},$$

$$e = -\frac{1}{2} \int_{z_0}^{z_i} \left[\frac{h''w^2}{4} - \frac{hK_{ww}}{V} \right] \frac{dz}{V^{1/2}},$$

$$f = -\frac{1}{2}(2\eta)^{-1} \int_{z_0}^{z_i} \left[\frac{h''v^2}{4} - \frac{hK_{vv}}{V} \right] \frac{dz}{V^{1/2}},$$

$$C_m = \frac{1}{2}(2\eta)^{-1} \int_{z_0}^{z_i} \frac{hK_{vw}}{V^{3/2}} dz,$$

$$C_d = \frac{1}{2}(2\eta)^{-3/2} \int_{z_0}^{z_i} \frac{K_{vv}}{V^{3/2}} dz,$$

$$C_a = \frac{1}{2}(2\eta)^{-1/2} \int_{z_0}^{z_i} \frac{K_{ww}}{V^{3/2}} dz,$$

$$C_r = \frac{1}{2}(2\eta)^{-1} \int_{z_0}^{z_i} \frac{h}{V^{3/2}} dz \tag{23}$$

2 Calculation of the third-order aberration for cathode lenses

Using the above-mentioned the third order geometrical and the first order chromatic aberration coefficient formulae, we can further derive the electron optical aberration for cathode lenses with the wide electron beam. The calculation procedure can be divided into three steps.

In the first step, the apparent initial position u_0 and initial velocity $\dot{u}_0$ in the rotating Cartesian coordinate system are transformed into the actual initial position r_0 and initial velocity $\dot{r}_0$ in the fixed Cartesian coordinate system. Taking the linear transformation of u_0, $\dot{u}_0$ into r_0, $\dot{r}_0$ [see Eq. (8)], we can derive the modified the third order aberration coefficients [5].

In the second step, taking into consideration of equality $\varepsilon = \varepsilon_z + \varepsilon_\perp$ in Eq. (1), we subtract the first order chromatic aberration (corresponding to $\varepsilon_\perp$) from the third order geometrical aberration (corresponding to ε) and then obtain the third order aberration (corresponding to ε_z) of a wide electron beam system in cathode lens.

In the third step, we assume that the cathode is immersed in both magnetic and transversal electrostatic fields, i. e., $\mathscr{H}(z_0) = \mathscr{H}_0 \neq 0$ and $V''(z_0) = V''_0 \neq 0$. In this case, the cathode plane $z = z_0$ is not an equipotential plane. Then, an electron emitted from the non-equipotential cathode plane at the initial position r_0, undergoes a variation of the initial energy equal to $V''_0 r_0^2/4$, which originates some additional chromatic aberration and should be subtracted from the third order aberration.

3 The aberrations associated with *w*-matrix

We rewrite the aberrations associated with *w*-matrix in Eq. (16) as below:

$$\Delta u_w(z_i) = w(z_i) \begin{bmatrix} \dot{u}_0 \cdot \dot{u}_0 \\ 2u_0 \cdot \dot{u}_0 \\ 2(u_0 \times \dot{u}_0)k \\ u_0 \cdot u_0 \end{bmatrix}^T \begin{bmatrix} B & F & f \\ F & C & c \\ f & c & o \\ D & E & e \end{bmatrix} \begin{bmatrix} \dot{u}_0 \\ u_0 \\ u_0^* \end{bmatrix} \quad (24a)$$

In the first step, transforming u_0, $\dot{u}_0$ into r_0, $\dot{r}_0$, we obtain the modified aberrations [5] $\Delta u_w(z_i)$ of Eq. (24b)

$$\Delta u_w(z_i) = w(z_i) \begin{bmatrix} \dot{r}_0 \cdot \dot{r}_0 \\ 2r_0 \cdot \dot{r}_0 \\ 2(r_0 \times \dot{r}_0)k \\ r_0 \cdot r_0 \end{bmatrix}^T \times$$

$$\begin{bmatrix} B & 0 & F & f-\dot{\chi}_0 B \\ E & 0 & C+2\dot{\chi}_0 f-\dot{\chi}_0^2 B & c-\dot{\chi}_0 F \\ f-\dot{\chi}_0 B & 0 & c-\dot{\chi}_0 F & 0 \\ D-6\dot{\chi}_0 f+3\dot{\chi}_0^2 B & 0 & E-2\dot{\chi}_0 c+\dot{\chi}_0^2 F & e-\dot{\chi}_0 D+3\dot{\chi}_0^2 f-\dot{\chi}_0^3 B \end{bmatrix} \begin{bmatrix} \dot{r}_0 \\ \dot{r}_0^* \\ r_0 \\ r_0^* \end{bmatrix} \quad (24b)$$

In the second step, we get the chromatic aberration corresponding to $\varepsilon_\perp$

$$\Delta u_w(z_i)|_{\varepsilon_\perp} = -w(z_i)(\dot{r}_0 \cdot \dot{r}_0)[C_d \dot{r}_0 + C_m r_0 + (C_r - \dot{\chi}_0 C_d) r_0^*] \quad (25)$$

In the third step, we get the chromatic aberration corresponding to $V_0'' r_0^2/4$

$$\Delta u_w(z_i)|_{V_0''} = -w(z_i)(V_0'' r_0 \cdot r_0/4)[C_d \dot{r}_0 + C_m r_0 + (C_r - \dot{\chi}_0 C_d) r_0^*] \quad (26)$$

The sum of Eq. (24) – Eq. (26) equals

$$\Delta u_{3w}(z_i) = \Delta u_w(z_i) + \Delta u_w(z_i)|_{\varepsilon_\perp} + \Delta u_w(z_i)|_{V_0''} \quad (27)$$

By summation and rearrangement of the w-matrix elements we obtain the final aberration formula associated with a generalized w-matrix $\Delta u_{3w}(z_i)$ in Eq. (28):

$$\Delta u_{3w}(z_i) = w(z_i) \begin{bmatrix} \dot{r}_0 \cdot \dot{r}_0 \\ 2r_0 \cdot \dot{r}_0 \\ 2(r_0 \times \dot{r}_0)k \\ r_0 \cdot r_0 \end{bmatrix}^T \times$$

$$\begin{bmatrix} B-C_d & 0 & F-C_m & f-\dot{\chi}_0 B - C_r + \dot{\chi}_0 C_d \\ F & 0 & C+2\dot{\chi}_0 f-\dot{\chi}_0^2 B & c-\dot{\chi}_0 F \\ f-\dot{\chi}_0 B & 0 & c-\dot{\chi}_0 F & 0 \\ \left\{\begin{array}{l}D-6\dot{\chi}_0 f+3\dot{\chi}_0^2 B \\ -(2\eta V_0''/4)C_d\end{array}\right\} & 0 & \left\{\begin{array}{l}E-2\dot{\chi}_0 c+\dot{\chi}_0^2 F \\ -(2\eta V_0''/4)C_m\end{array}\right\} & \left\{\begin{array}{l}e-\dot{\chi}_0 D+3\dot{\chi}_0^2 f-\dot{\chi}_0^3 B \\ -(2\eta V_0''/4)(C_r-\dot{\chi}_0 C_d)\end{array}\right\} \end{bmatrix} \begin{bmatrix} \dot{r}_0 \\ \dot{r}_0^* \\ r_0 \\ r_0^* \end{bmatrix} \quad (28)$$

If the electron beam emitted from the cathode has homogeneous axial initial energy ε_z, then the corresponding image plane z_i and the magnification m (referring to the n-th image) are given by

$$v(z_i, \varepsilon_z) = 0, \quad w(z_i, \varepsilon_z) = (-1)^n m \quad (29)$$

It is apparent that only w-matrix aberrations appear at z_i image plane corresponding to $v(z_i, \varepsilon_z) = 0$. When $V_0'' = 0$, the aberration expression Eq. (28) coincides with that already derived by Refs. [1-3].

Thus, for the summation of Eq. (28) gives the total third order aberration at an arbitrary image plane z_i for cathode lenses

$$\Delta u_3(z_i) = \Delta u_{3w}(z_i) \quad (30)$$

There are altogether nine third order geometrical and four first order chromatic aberration coefficients, whose linear combination gives the third order aberration coefficients associated with the generalized w-matrix. These aberration coefficients will be given in compact and canonical form suitable for computer calculation, and have been proved to be non-divergent whether ε_z, $\mathcal{H}_0$, V''_0 and $v(z_i)$ equal zero or not.

So far, the trajectories of cathode lenses are described by vectors $\boldsymbol{r}_0$ and $\dot{\boldsymbol{r}}_0$. It is clear that the cathode lenses can be conveniently represented by complex representation [3,7]. Therefore, we introduce the following transformation formulae.

$$\boldsymbol{r}_0 = r_0 e^{i\theta_0}, \quad \boldsymbol{r}_0^* = ir_0 e^{i\theta_0}, \quad \boldsymbol{r}_0 \cdot \boldsymbol{r}_0 = |\boldsymbol{r}_0|^2 = r_0^2,$$

$$\dot{\boldsymbol{r}}_0 = \sqrt{2\eta\varepsilon_\perp} e^{i\beta_0}, \quad \dot{\boldsymbol{r}}_0^* = i\sqrt{2\eta\varepsilon_\perp} e^{i\beta_0}, \quad \dot{\boldsymbol{r}}_0 \cdot \dot{\boldsymbol{r}}_0 = |\dot{\boldsymbol{r}}_0|^2 = 2\eta\varepsilon_\perp,$$

$$(\boldsymbol{r}_0 \cdot \dot{\boldsymbol{r}}_0) = r_0 \sqrt{2\eta\varepsilon_\perp} \cos(\beta_0 - \theta_0),$$

$$(\boldsymbol{r}_0 \times \dot{\boldsymbol{r}}_0)\boldsymbol{k} = r_0 \sqrt{2\eta\varepsilon_\perp} \sin(\beta_0 - \theta_0) \tag{31}$$

Thus, we obtain the complex aberration representation of cathode lens associated with the generalized w-and v-matrixes.

$$\Delta u_3(z_i) = \begin{bmatrix} \varepsilon_\perp/V_i \\ 2r_0\sqrt{\varepsilon_\perp/V_i}\cos(\beta_0-\theta_0) \\ 2r_0\sqrt{\varepsilon_\perp/V_i}\sin(\beta_0-\theta_0) \\ r_0^2 \end{bmatrix}^T \left\{ w(z_i) \left\{ \begin{bmatrix} B_w & 0 & G_w-F_w & g_w-f_w \\ E_w & 0 & C_w & c_w/2 \\ f_w & 0 & c_w/2 & 0 \\ D_w-C_w & 0 & E_w & e_w \end{bmatrix} \begin{bmatrix} \sqrt{\varepsilon_\perp/V_i}e^{i\beta_0} \\ i\sqrt{\varepsilon_\perp/V_i}e^{i\beta_0} \\ r_0 e^{i\theta_0} \\ ir_0 e^{i\theta_0} \end{bmatrix} \right\} \right\} \tag{32}$$

In Eq. (32) $V_i = V(z_i)$, while B_w, $\cdots$, e_w; are the third order aberration coefficients, which coincide with those given in Ref. [3] except for $V''_0 \neq 0$ in present paper. Thus, we derive the following aberration coefficients formulae

$$B_w = (2\eta V_i)^{3/2}(B - C_d),$$

$$G_w - F_w = (2\eta V_i)(F - C_m),$$

$$g_w - f_w = (2\eta V_i)(f - \dot{\chi}_0 B - C_r + \dot{\chi}_0 C_d),$$

$$F_w = (2\eta V_i)F,$$

$$C_w = (2\eta V_i)^{1/2}(C + 2\dot{\chi}_0 f - \dot{\chi}_0^2 B),$$

$$c_w/2 = (2\eta V_i)^{1/2}(c - \dot{\chi}_0 F),$$

$$f_w = (2\eta V_i)(f - \dot{\chi}_0 B),$$

$$D_w - C_w = (2\eta V_i)^{1/2}\{D - 6\dot{\chi}_0 f + 3\dot{\chi}_0^2 B - (2\eta V''_0/4)C_d\},$$

$$E_w = E - 2\dot{\chi}_0 c + \dot{\chi}_0^2 F - (2\eta V''_0/4)C_m,$$

$$e_w = e - \dot{\chi}_0 D + 3\dot{\chi}_0^2 f - \dot{\chi}_0^3 B - (2\eta V''_0/4)(C_r - \dot{\chi}_0 C_d) \tag{33}$$

By performing matrix multiplications in Eq. (32), we get the complex representation as follows:

$$\Delta \boldsymbol{u}_3(z_i) = r_0^3 e^{i\theta_0}[w(z_i)(E_w + ie_w)] \text{ (distortion) } +$$

$$r_0^2 \sqrt{\varepsilon_\perp/V_i} e^{i\beta_0}[w(z_i)D_w] \text{ (field curvature) } +$$

$$r_0^2 \sqrt{\varepsilon_\perp/V_i} e^{i(2\theta_0-\beta_0)}[w(z_i)(C_w + ic_w)] \text{ (astigmatism) } +$$

$$r_0(\varepsilon_\perp/V_i)e^{i\theta_0}[w(z_i)(G_w + ig_w)] \text{ (coma length) } +$$

$$r_0(\varepsilon_\perp/V_i)e^{i(2\beta_0-\theta_0)}[w(z_i)(F_w - if_w)] \text{ (coma radius) } +$$

$$(\varepsilon_\perp/V_i)^{3/2} e^{i\beta_0}[w(z_i)B_w] \text{ (chromatic-spherical- aberr)} \tag{34}$$

Obviously, Eq. (31) and Eq. (34) are general formulae of aberration coefficients for cathode lenses and describe the aberrations classification and their figures. The general Eq. (32) covers the basic results given in Ref. [1-3].

Finally, we can change the aberration $\Delta u_3(z_i)$ in the rotating coordinate system into the fixed coordinate system. Using Eq. (6) we get

$$\begin{aligned} r(z_i) &= e^{i\chi(z_i)} u(z_i) = e^{i\chi(z_i)}[u_g(z_i) + \Delta u_3(z_i)] \\ &= e^{i\chi(z_i)}\{\sqrt{\varepsilon_\perp} e^{i\beta_0} v(z_i) + r_0 e^{i\beta_0}[w(z_i) - i\dot{\chi}_0 v(z_i)(2\eta)^{-1/2}] + \Delta u_3(z_i)\} \end{aligned} \tag{35}$$

where $\chi(z_i)$ is given by Eq. (7), u_g by Eq. (5) and $\Delta u_3(z_i)$ by Eq. (34).

Conclusions

In the present paper, basing on the variation principle, we have discussed the aberration theory for cathode lenses with combined electrostatic and magnetic focusing fields. The paraxial trajectories are described in the vector form, and aberrations in the matrix form. The total third order aberrations at an arbitrary image plane are given in Eq. (28), which can also be transformed into the linear combination of Glaser's nine geometrical and four chromatic coefficients. Thus, we have derived the third order aberration coefficients of cathode lens, which are in compact and canonical form suitable for compute calculation.

As an extension, the cathode is assumed to be immersed in both magnetic and transversal electric fields. The aberration coefficients formulae referred to this more general condition are derived and proved to be non-divergent whether ε_z, $\mathcal{H}_0$, V_0'' and $v(z_i)$ equal zero or not.

Therefore, we have already applied the variation method to cathode lenses and derived the aberration coefficients, which formally coincide with those given by Glaser. It follows that the aberration for both the wide electron beam system and the narrow electron beam system can be treated by the universal variation theory. Relationship between these two systems is thus established.

Obviously, these conclusions may be of theoretical and practical importance.

References

[1] 西门纪业. 复合浸没物镜的电子光学性质和像差理论 [J]. 物理学报, 1957, 13 (4): 339-356.

[2] БОНШТЕДТ Б Э. К расчету аберации катодных линз [J]. Радиотехника и Электроника, 1964, 9 (5): 844-850.

[3] ZHOU L W, AI K C, PAN S C. On aberration theory of the combined electromagnetic focusing cathode lenses [J]. Acta Physica Sinica, 1983, 32: 376-392.

[4] GLASER W. Grundlagen der Elektronen optik [M]. Wien: Springer, 1952.

[5] XIMEN J Y. On the linear transformations of Gaussian trajectory parameters and their

influence upon electron optical aberrations [J]. Acta Physica Sinica, 1981, 30 (4): 472-477.

[6] XIMEN J Y. Principles of electron and ion optics and an introduction to the aberration theory [M]. Beijing: Science Press, 1983.

[7] LI Y, XIMEN J Y. On the relativistic aberration theory of a combined focusing-deflection system with multi-stage deflectors [J]. Acta Physica Sinica, 1982, 31 (5): 604-614.

6.3 静电成像电子光学近轴横向像差理论
Theory of Paraxial Lateral Aberrations of Electrostatic Imaging Electron Optics

摘要：在成像电子光学的研究中，备受关注的是二级近轴横向像差（即二级近轴色差）和三级几何横向像差，几乎忽略了三级近轴色差和三级近轴放大率色差的存在，理论上也没有给出近轴横向像差的一般形式。本文基于近轴方程的渐近解研究近轴横向像差的一般理论，探讨了近轴方程渐近解系数之间的关系式，证实了它们之间是相互关联的。本文的结果为研究简明形式的近轴横向像差提供了理论基础。

Abstract: In imaging electron optics, study of geometrical lateral aberrations of the third order and paraxial lateral aberrations of the second order (i. e., paraxial chromatic aberrations of second order) has traditionally been paid more attention to, but the existence of paraxial chromatic aberrations of the third order and paraxial chromatic aberrations of magnification of the third order was almost ignored, the general form of paraxial lateral aberration has not been studied theoretically. In the present paper, the paraxial lateral aberrations expressed in general form have been derived emphatically by using asymptotic solutions of paraxial equation. The relationship between the coefficients of asymptotic solutions of paraxial equation has been investigated, which proves that the coefficients of asymptotic solutions are related each other. Results of the present paper will have theoretical significance for studying the simple and clear form of paraxial lateral aberrations.

引 言

在成像电子光学的研究中，通常备受关注的是三级几何横向像差和二级近轴横向像差（即二级近轴色差），几乎忽略了三级近轴色差和三级近轴放大率色差的存在，也没有从理论上给出近轴横向像差的一般表示式。众所周知，决定空间分辨率极限的二级近轴色差在全部近轴横向像差中占有最主要的部分，它被称为众所周知的莱克纳格尔-阿尔齐莫维奇（Recknagel-Artimovich）公式[1-2]，其正确性只是从一些简单模型中得到了检验，并没有给出理论上的解释。迄今为止，近轴横向像差的一般表达式并没有给出。本文尝试在近轴方程的渐近解的基础上推导近轴横向像差（包括二级和三级近轴色差和三级近轴放大率色差）普遍形式的表达式，探讨渐近解的系数之间的关系式。

周立伟. 北京理工大学. 北京理工大学学报（Journal of Beijing Institute of technology）, V. 29, No. 11, 2009, 941–946.

1 静电电子光学成像系统的空间像差概述

按照近轴成像电子光学理论,所有自光阴极逸出具有给定的相同的轴向初能 $\sqrt{\varepsilon_{z_1}}$,初始高度为 r_0 的近轴电子轨迹,都将会聚于同一成像面 $z = z_i$ 的 $r^*(z_i, \varepsilon_{z_1}^{\frac{1}{2}}, r_0)$ 处,而与径向初能 $\sqrt{\varepsilon_r}$ 无关。成像电子光学系统的横向像差被定义为在轴向初电位为 $\sqrt{\varepsilon_{z_1}}$ 的近轴电子轨迹所决定的理想像面 $z = z_i$ 上实际成像点和理想成像点之间的差异。它可表示如下[3]:

$$\Delta r = \Delta r(z_i) = r(z_i, \varepsilon_z^{\frac{1}{2}}, \varepsilon_r^{\frac{1}{2}}, r_0) - r^*(z_i, \varepsilon_{z_1}^{\frac{1}{2}}, r_0) \tag{1}$$

这里,r、r^* 分别表示实际成像和理想成像在理想像面 $z = z_i$ 处的落点高度,它是由阴极面射出的实际电子和近轴电子经过系统后在理想像面上形成的。*号表示近轴轨迹。取圆柱坐标系 (z, r),轴向坐标 z 自阴极面 $z_0 = 0$ 算起,r_0 是电子的初始径向矢量,z_i 对应于理想成像面的位置,$\sqrt{\varepsilon_z}$、$\sqrt{\varepsilon_r}$ 是从光阴极逸出的电子的轴向初电位和径向初电位,$\sqrt{\varepsilon_{z_1}}$ 乃是所给定的电子的轴向初电位,它与理想成像位置 z_i 相对应。

公式 (1) 可表示为

$$\Delta r = \Delta r(z_i) = \Delta r^* + \delta r \tag{2}$$

式中,

$$\Delta r^* = r^*(z_i, \varepsilon_z^{\frac{1}{2}}, \varepsilon_r^{\frac{1}{2}}, r_0) - r^*(z_i, \varepsilon_{z_1}^{\frac{1}{2}}, r_0) \tag{3}$$

$$\delta r = r(z_i, \varepsilon_z^{\frac{1}{2}}, \varepsilon_r^{\frac{1}{2}}, r_0) - r^*(z_i, \varepsilon_z^{\frac{1}{2}}, \varepsilon_r^{\frac{1}{2}}, r_0) \tag{4}$$

这里,Δr^* 被称为近轴横向像差,它是由光阴极面同一高度 r_0,而逸出的轴向初电位 $(\sqrt{\varepsilon_z}, \sqrt{\varepsilon_r})$ 与 $(\sqrt{\varepsilon_{z_1}}, \sqrt{\varepsilon_r})$ 不同的两条近轴电子轨迹在 $\sqrt{\varepsilon_{z_1}}$ 所决定的理想像面 z_i 上落点之间的差异。δr 被称为几何横向像差,它是由光阴极面逸出的高度 r_0、轴向初电位 $(\sqrt{\varepsilon_z}, \sqrt{\varepsilon_r})$ 相同的实际电子轨迹和近轴电子轨迹在 $\sqrt{\varepsilon_{z_1}}$ 所决定的理想像面 z_i 上落点之间的差异。公式 (2) 表明,成像电子光学系统的横向像差 Δr 可以表示成近轴横向像差 Δr^* 与几何横向像差 δr 之组合;这在成像电子光学像差理论研究中是一个十分重要的概念。

众所周知,近轴轨迹 $r^*(z)$ 乃是下面的近轴方程

$$r^{*\prime\prime}(z) + \frac{1}{2} \frac{\phi'(z)}{\phi(z) + \varepsilon_z} r^{*\prime}(z) + \frac{1}{4} \frac{\phi''(z)}{\phi(z) + \varepsilon_z} r^*(z) = 0 \tag{5}$$

的解。此处,$\phi(z)$ 为轴上电位分布,撇号($'$) = d/dz,表示对 z 的导数。

实际轨迹 $r(z)$ 乃是下面的实际轨迹方程

$$r''(z) = \frac{1 + r'^2}{2\varphi_*} \left(\frac{\partial \varphi}{\partial r} - r' \frac{\partial \varphi}{\partial z} \right) \tag{6}$$

的解。式中,$\varphi_* = \varphi(z, r) + \varepsilon_0$,$\varepsilon_0$ 为自光阴极逸出的电子初电位,$\varepsilon_0 = \varepsilon_r + \varepsilon_z$;$\varphi = \varphi(z, r)$ 为空间电位分布,它可以由谢尔赤(Scherzer)级数展开成如下的形式:

$$\varphi(z, r) = \phi(z) - \frac{1}{4} \phi''(z) r^2 + \frac{1}{64} \phi^{\text{IV}}(z) r^4 - \cdots \tag{7}$$

近轴方程式 (5) 的通解可写成

$$r^*(z) = r_0 w(z) + \sqrt{\frac{m_0}{2e}} \dot{r}_0 v(z) \tag{8}$$

式中，m_0/e 为电子荷质比；r_0 和 $\dot{r}_0$ 分别为由阴极面逸出的初始径向矢量和径向初速。方程式（8）的两个特解 $v = v(z)$，$w = w(z)$ 满足如下初始条件：

$$v(z=0) = 0, \ v'(z=0) = \frac{1}{\sqrt{\varepsilon_z}},$$
$$w(z=0) = 1, \ w'(z=0) = 0 \tag{9}$$

现假设，所有自阴极面轴上点以相同的轴向初能 $\sqrt{\varepsilon_{z_1}}$ 逸出的近轴电子都聚焦于理想像点 $z = z_i$ 处，即

$$v(z_i, \sqrt{\varepsilon_{z_1}}) = 0,$$
$$w(z_i, \sqrt{\varepsilon_{z_1}}) = M \tag{10}$$

此处，M 为线放大率。此时，很显然

$$v(z_i, \sqrt{\varepsilon_z}) \neq 0, \ w(z_i, \sqrt{\varepsilon_z}) \neq M$$

由公式（3），我们便能得到位于 z_i 的理想像面上的近轴横向像差：

$$\Delta r^*(z_i) = [w(z_i, \sqrt{\varepsilon_z}) - w(z_i, \sqrt{\varepsilon_{z_1}})] r_0 + \sqrt{\frac{m_0}{2e}} \dot{r}_0 [v(z_i, \sqrt{\varepsilon_z}) - v(z_i, \sqrt{\varepsilon_{z_1}})]$$
$$= [w(z_i, \sqrt{\varepsilon_z}) - M] r_0 + \sqrt{\frac{m_0}{2e}} \dot{r}_0 v(z_i, \sqrt{\varepsilon_z}) \tag{11}$$

公式（11）是我们研究近轴横向像差的出发点。关于几何横向像差 δr，可以展开实际轨迹方程式（6）的各项，代入方程式（4），并利用公式（8），我们便能得到三级几何像差 δr_3 的如下方程式：

$$[\phi(z) + \varepsilon_z] \delta r_3'' + \frac{1}{2} \phi'(z) \delta r_3' + \frac{1}{4} \phi''(z) \delta r_3 = F(r^*, r^{*'}, z) \tag{12}$$

式（12）乃是二阶非齐次微分方程。式中，$F(r^*, r^{*'}, z)$ 是右端项，该式中的 r^* 是近轴轨迹的解。三级几何像差 δr_3 的求解在一系列的文献和著作[4-8]中已有叙述，这里就不细谈了。

2 渐近解求解近轴电子轨迹

文献[9-10]给出了一种利用渐近解求解近轴方程的特解 $v(z, \sqrt{\varepsilon_z})$、$w(z, \sqrt{\varepsilon_z})$ 的方法，它可以表述如下：

$$v(z, \sqrt{\varepsilon_z}) = \sqrt{\phi(z)} \xi_0(z) + \sqrt{\varepsilon_z} \eta_0(z) + \varepsilon_z \left[\frac{\xi_0(z)}{2\sqrt{\phi(z)}} + \xi_1(z) \sqrt{\phi(z)} \right] + \cdots \tag{13}$$

$$w(z, \sqrt{\varepsilon_z}) = \omega_0(z) + \sqrt{\varepsilon_z} \sqrt{\phi(z)} \zeta_0(z) + \varepsilon_z \omega_1(z) + \cdots \tag{14}$$

式（13）、式（14）中的系数 ξ_0、η_0、ω_0、ζ_0、ξ_1、ω_1 乃是下列微分方程

$$L_0 \begin{pmatrix} \xi_k \\ \zeta_k \end{pmatrix} = - \begin{pmatrix} \xi''_{k-1} \\ \zeta''_{k-i} \end{pmatrix} \tag{15a}$$

$$M_0 \begin{pmatrix} \eta_k \\ \omega_k \end{pmatrix} = - \begin{pmatrix} \eta''_{k-1} \\ \omega''_{k-i} \end{pmatrix} \tag{15b}$$

的解。式中，$\xi_{-1} = \zeta_{-1} = \eta_{-1} = \omega_{-1} = 0$。$L_0$、$M_0$ 为线性微分算子，它可表示为

$$L_0 = \phi(z)\frac{d^2}{dz^2} + \frac{3}{2}\phi'(z)\frac{d}{dz} + \frac{3}{4}\phi''(z) \tag{16}$$

$$M_0 = \phi(z)\frac{d^2}{dz^2} + \frac{1}{2}\phi'(z)\frac{d}{dz} + \frac{1}{4}\phi''(z) \tag{17}$$

方程式（15a）、式（15b）的初始条件可以按照下列次序决定：

$$\begin{aligned}\frac{\xi_0(0)\phi'(0)}{2} &= 1 \quad \rightarrow \quad \xi_0(0) + \eta_0(0) = 0 \\ &\downarrow \\ \frac{\xi_1(0)\phi'(0)}{2} &+ \xi_0'(0) + \eta_0'(0) = 0 \rightarrow \cdots\end{aligned} \tag{18}$$

$$\begin{aligned}\omega_0(0) &= 1 \quad \zeta_0(0) + \omega_1(0) = 0 \\ \downarrow \quad &\uparrow \quad \downarrow \\ \frac{\zeta_0(0) + \phi'(0)}{2} &+ \omega_0'(0) = 0 \rightarrow \cdots\end{aligned} \tag{19}$$

这里箭头表示计算的顺序。

由渐近解表达式（13）、式（14）和式（11）的定义，如果我们能求得位于理想像面 $z = z_i$（即屏面）处系数 ξ_0、η_0、ω_0、ζ_0、ξ_1、ω_1 的各值，我们便能求得近轴横向像差 $\Delta r^*(z_i)$：

$$\Delta r^*(z_i) = \left[\sqrt{\phi(z_i)}\zeta_0(z_i)(\sqrt{\varepsilon_z} - \sqrt{\varepsilon_{z_1}}) + \omega_1(z_i)(\varepsilon_z - \varepsilon_{z_1})\right] r_0 + \sqrt{\frac{m_0}{2e}}\dot{r}_0\left\{\eta_0(z_i)(\sqrt{\varepsilon_z} - \sqrt{\varepsilon_{z_1}}) + \left[\frac{\xi_0(z_i)}{2\sqrt{\phi(z_i)}} + \xi_1(z_i)\sqrt{\phi(z_i)}\right](\varepsilon_z - \varepsilon_{z_1})\right\} \tag{20}$$

上述 $\Delta r^*(z_i)$ 值的正负取决于近轴轨迹 $r^*(z_i, \sqrt{\varepsilon_{z_1}})$，由它算起，自下至上为正，反之为负。式（20）中的第一项表示近轴放大率色差，第二项表示近轴色差。

按照上述的计算顺序，我们可以由求解微分方程（15a）、式（15b）的一系列运算以获得系数 ξ_0、η_0、ω_0、ζ_0、ξ_1、ω_1 在理想像面 $z = z_i$ 的值，从理论上我们便能求得近轴横向像差 $\Delta r_*(z_i)$。但是，电子通过阳极后行进抵达像面这一区域通常乃是无场空间，此二特解 v、w 表示的电子轨迹乃是直线。故由渐近解直接求得位于 $z = z_i$ 的这些系数值谈何容易。因此，由式（20）求解近轴横向像差 $\Delta r^*(z_i)$ 的问题依然没有解决。

3 渐近解求解近轴横向像差

众所周知，对电子光学成像系统，轴上电位分布 $\phi(z)$ 在阳极段的某点 $z = z_a$ 后的区域通常是一常数，电子轨迹在这一区域以直线行进。这样，我们便可通过获取系数 ξ_0、η_0、ω_0、ζ_0、ξ_1、ω_1 在 $z = z_a$ 点的解求得位于 $z = z_i$ 处理想像面上的近轴横向像差 $\Delta r^*(z_i)$ 值。

按照式（11）的定义，近轴横向像差 $\Delta r^*(z_i)$ 可以视为近轴色差 $\Delta r_v^*(z_i)$ 以及近轴放大率色差 $\Delta r_w^*(z_i)$ 之和。即

$$\Delta r^*(z_i) = \Delta r_v^*(z_i) + \Delta r_w^*(z_i) = \sqrt{\frac{m_0}{2e}}\dot{r}_0\Delta v(z_i) + r_0\Delta w(z_i) \tag{21}$$

因 $z=z_i$ 处，$v(z_i, \sqrt{\varepsilon_{z_1}})=0$，故

$$\Delta v(z_i) = v(z_i, \sqrt{\varepsilon_z}),$$
$$\Delta w(z_i) = w(z_i, \sqrt{\varepsilon_z}) - M \tag{22}$$

假定，在 $z \leqslant z_a$ 处，$\phi(z) \neq \text{const}$；在 $z \geqslant z_a$ 处，$\phi(z) = \text{const} = \phi(z_a) = \phi(z_i)$。我们便可由表达式（13）的渐近解特解求得在 $z=z_a$ 处的值 $v(z_a, \sqrt{\varepsilon_z})$、$v'(z_a, \sqrt{\varepsilon_z})$，即

$$v(z_a, \sqrt{\varepsilon_z}) = a + b\sqrt{\varepsilon_z} + c\varepsilon_z \tag{23}$$
$$v'(z_a, \sqrt{\varepsilon_z}) = a' + b'\sqrt{\varepsilon_z} + c'\varepsilon_z \tag{24}$$

此处

$$a = \sqrt{\phi(z_a)}\xi_0(z_a),\ b = \eta_0(z_a),\ c = \frac{\xi_0(z_a)}{2\sqrt{\phi(z_a)}} + \xi_1(z_a)\sqrt{\phi(z_a)},$$
$$a' = \sqrt{\phi(z_a)}\xi'_0(z_a) + \frac{1}{2\sqrt{\phi(z_a)}}\phi'(z_a)\xi_0(z_a),\ b' = \eta'_0(z_a), \tag{25}$$
$$c' = \frac{\xi'_0(z_a)}{2\sqrt{\phi(z_a)}} - \frac{\phi'(z_a)\xi_0(z_a)}{4[\phi(z_a)]^{\frac{3}{2}}} + \xi'_1(z_a)\sqrt{\phi(z_a)} + \frac{\phi'(z_a)\xi_1(z_a)}{2\sqrt{\phi(z_a)}}$$

同样，我们可以由表达式（14）的渐近解特解求得在 $z=z_a$ 处的值 $w(z_a, \sqrt{\varepsilon_z})$、$w'(z_a, \sqrt{\varepsilon_z})$，即：

$$w(z_a, \sqrt{\varepsilon_z}) = d + e\sqrt{\varepsilon_z} + f\varepsilon_z \tag{26}$$
$$w'(z_a, \sqrt{\varepsilon_z}) = d' + e'\sqrt{\varepsilon_z} + f'\varepsilon_z \tag{27}$$

式中，

$$d = \omega_0(z_a),\ e = \sqrt{\phi(z_a)}\zeta_0(z_a),\ f = \omega_1(z_a),$$
$$d' = \omega'_0(z_a),\ e' = \sqrt{\phi(z_a)}\zeta'_0(z_a) + \frac{1}{2\sqrt{\phi(z_a)}}\phi'(z_a)\zeta_0(z_a),\ f' = \omega'_1(z_a) \tag{28}$$

首先，让我们先求 Δw 的解。由方程式（10），由于 $z=z_i$ 处，$v(z_i, \sqrt{\varepsilon_{z_1}})=0$，便有

$$w(z_i, \sqrt{\varepsilon_{z_1}}) = w(z_a, \sqrt{\varepsilon_{z_1}}) - \frac{v(z_a, \sqrt{\varepsilon_{z_1}})}{v'(z_a, \sqrt{\varepsilon_{z_1}})}w'(z_a, \sqrt{\varepsilon_{z_1}}) \tag{29}$$

相似地，我们也可得到

$$w(z_i, \sqrt{\varepsilon_z}) = w(z_a, \sqrt{\varepsilon_z}) - \frac{v(z_a, \sqrt{\varepsilon_{z_1}})}{v'(z_a, \sqrt{\varepsilon_{z_1}})}w'(z_a, \sqrt{\varepsilon_z}) \tag{30}$$

把表达式（23）、式（24）和式（26）、式（27）代入式（29）、式（30）中，经过一系列的变换，我们可以得到：

$$w(z_i, \sqrt{\varepsilon_z}) = \frac{1}{a'}(a'd - ad') + \frac{1}{a'}(a'e - ae')\sqrt{\varepsilon_z} - \frac{1}{a'}\left[\frac{d'}{a'}(a'b - ab')\right]\sqrt{\varepsilon_{z_1}} + \frac{1}{a'}(a'f - af')\varepsilon_z -$$
$$\frac{1}{a'}\left[\frac{e'}{a'}(a'b - ab')\right]\sqrt{\varepsilon_z}\sqrt{\varepsilon_{z_1}} - \frac{1}{a'}\frac{d'}{a'}\left[(a'c - ac') - \frac{b'}{a'}(a'b - ab')\right]\varepsilon_{z_1}$$

$$\tag{31}$$

而
$$w(z_i, \sqrt{\varepsilon_{z_1}}) = M = \frac{1}{a'}(a'd - ad') + \frac{1}{a'}\left[(a'e - ae') - \frac{d'}{a'}(a'b - ab')\right]\sqrt{\varepsilon_{z_1}} + \frac{1}{a'}\left\{a'f - af' - \frac{e'}{a'}(a'b - ab') - \frac{d'}{a'}\left[(a'c - ac') - \frac{b'}{a'}(a'b - ab')\right]\right\}\varepsilon_{z_1} \quad (32)$$

因此，由式（11）的定义，便有

$$\Delta w = w(z_i, \sqrt{\varepsilon_z}) - w(z_i, \sqrt{\varepsilon_{z_1}})$$
$$= \frac{1}{a'}\left[(a'e - ae')(\sqrt{\varepsilon_z} - \sqrt{\varepsilon_{z_1}}) + (a'f - af')(\varepsilon_z - \varepsilon_{z_1}) - \frac{e'}{a'}(a'b - ab')(\sqrt{\varepsilon_z} - \sqrt{\varepsilon_{z_1}})\sqrt{\varepsilon_{z_1}}\right] \quad (33)$$

现在我们讨论式（33）中的第一项，它可表示为

$$a'e - ae' = \phi(z_a)\left[\xi_0'(z_a)\zeta_0(z_a) - \xi_0(z_a)\zeta_0'(z_a)\right] \quad (34)$$

由方程式（15a）可见，求解 $\xi_0(z)$、$\zeta_0(z)$ 的微分方程乃是同一个方程式（16），而其初始条件也是相同的，故所求得的 $\xi_0(z)$ 和 $\zeta_0(z)$ 的解之间仅差一个常数值，于是有

$$\frac{\zeta_0'(z_a)}{\zeta_0(z_a)} = \frac{\xi_0'(z_a)}{\xi_0(z_a)} \quad (35)$$

因此

$$a'e - ae' = 0 \quad (36)$$

于是，式（32）的线放大率 M 和式（33）的 $\Delta w(z_i)$ 可以表示为

$$M = w(z_i, \sqrt{\varepsilon_{z_1}}) = \frac{1}{a'}(a'd - ad') - \frac{1}{a'}\left[\frac{d'}{a'}(a'b - ab')\right]\sqrt{\varepsilon_{z_1}} + \frac{1}{a'}\left[a'f - af' - \frac{1}{a'^2}(a'e' - b'd')(a'b - ab') - \frac{d'}{a'}(a'c - ac')\right]\varepsilon_{z_1} \quad (37)$$

$$\Delta w(z_i) = \frac{1}{a'}\left[(a'f - af')(\varepsilon_z - \varepsilon_{z_1}) - \frac{e'}{a'}(a'b - ab')(\sqrt{\varepsilon_z} - \sqrt{\varepsilon_{z_1}})\sqrt{\varepsilon_{z_1}}\right] \quad (38)$$

由于式（38）的像差项应与线放大率 M 有关，故将式（37）引入式（38），我们便得到

$$\Delta w(z_i) = -M\left[\frac{e'}{a'}\frac{(a'b - b'a)}{(a'd - ad')}(\sqrt{\varepsilon_z} - \sqrt{\varepsilon_{z_1}})\sqrt{\varepsilon_{z_1}} - \frac{a'f - af'}{a'd - ad'}(\varepsilon_z - \varepsilon_{z_1})\right] \quad (39)$$

实际上，在推导式（39）的过程中，线放大率 M 表达式（37）的展开式只需要精确到 $\sqrt{\varepsilon_{z_1}}$ 的零级近似。

式（39）的 $\Delta w(z_i)$ 还能进一步简化。在成像电子光学中，有两条途径可以求得线放大率 M 之值。第一条途径就是求特解 $w(z, \sqrt{\varepsilon_{z_1}})$ 在 $z = z_i$ 的值，即 $M = w(z_i, \sqrt{\varepsilon_{z_1}})$，表达式（37）就是这样求得的。第二条途径是利用拉格朗日 - 亥姆霍兹（Lagrange - Helmholtz）关系式，它可表示如下：

$$v'(z_i, \sqrt{\varepsilon_{z_1}}) = \frac{1}{\sqrt{\phi(z_i) + \varepsilon_{z_1}}M} = \frac{1}{\sqrt{\phi(z_a) + \varepsilon_{z_1}}M} = v'(z_a, \sqrt{\varepsilon_{z_1}}) \quad (40)$$

这样，由式（24），线放大率 M 便具有以下形式：

$$M = \frac{1}{\sqrt{\phi(z_a) + \varepsilon_{z_1}}v'(z_a, \varepsilon_{z_1})} = \frac{1}{\sqrt{\phi(z_a)}a'}\left[1 - \frac{b'}{a'}\sqrt{\varepsilon_{z_1}} - \left(\frac{b'}{a'} + \frac{1}{2\phi(z_a)} - \frac{b'^2}{a'^2}\right)\varepsilon_{z_1}\right] \quad (41)$$

当然，这两条途径应是等效的，它们应该得到相同的结果。将式（41）与式（37）相比较，其 $\sqrt{\varepsilon_{z_1}}$ 展开式中系数的同幂次项应该相等。因此，比较 $\sqrt{\varepsilon_{z_1}}$ 的零次幂项和一次幂项，渐近解系数之间便有如下对应的关系式：

$$a'd - ad' = \frac{1}{\sqrt{\phi(z_a)}} \tag{42}$$

$$\frac{d'(a'b - ab')}{b'(a'd - ad')} = 1 \tag{43}$$

此外，我们发现，在式（39）中，b、b' 和 f、f' 的表达式是和 $\eta_0(z_a)$、$\omega_1(z_a)$ 及其导数分别相联系的，而它们又是由同一微分方程式（17）求解的，故关系式 $b'/b = f'/f$ 成立。于是，有

$$\frac{a'b - ab'}{a'f - af'} = \frac{b\left(a' - a\dfrac{b'}{b}\right)}{f\left(a' - a\dfrac{f'}{f}\right)} = \frac{b}{f} = \frac{\eta_0(z_a)}{\omega_1(z_a)}$$

因此

$$\frac{f}{b}(a'b - ab') = a'f - af' \tag{44}$$

相似地，表示式 $\dfrac{e'}{a'}$ 可通过式（36）进一步简化如下：

$$\frac{e'}{a'} = \frac{e}{a} = \frac{\zeta_0(z_a)}{\xi_0(z_a)} \tag{45}$$

所有上述公式（42）、式（43）、式（44）和式（45）的正确性将在下一篇文章得到验证。

将式（42）和式（44）代入式（39）中，可以得到

$$\Delta w(z_i) = -M\sqrt{\phi(z_a)}(a'b - ab')\left[\frac{e'}{a'}(\sqrt{\varepsilon_z} - \sqrt{\varepsilon_{z_1}})\sqrt{\varepsilon_{z_1}} - \frac{f}{b}(\varepsilon_z - \varepsilon_{z_1})\right] \tag{46}$$

因此，近轴放大率色差 $\Delta r_w^*(z_i)$ 的一般形式可以表达如下：

$$\Delta r_w^*(z_i) = r_0 \Delta w(z_i) = -r_0 M\left\{\frac{1}{2}\phi'(z_a)\xi_0(z_a)\eta_0(z_a) + \phi(z_a)[\xi_0'(z_a)\eta_0(z_a) - \xi_0(z_a)\eta_0'(z_a)]\right\} \times \left\{\frac{\zeta_0(z_a)}{\xi_0(z_a)}(\sqrt{\varepsilon_z} - \sqrt{\varepsilon_{z_1}})\sqrt{\varepsilon_{z_1}} - \frac{\omega_1(z_a)}{\eta_0(z_a)}(\varepsilon_z - \varepsilon_{z_1})\right\} \tag{47}$$

现在，我们来求 Δv 的解。由定义式（11），有

$$\Delta v(z_i) = v(z_i, \sqrt{\varepsilon_z}) = v(z_a, \sqrt{\varepsilon_z}) - \frac{v'(z_a, \sqrt{\varepsilon_z})}{v'(z_a, \sqrt{\varepsilon_{z_1}})}v(z_a, \sqrt{\varepsilon_{z_1}}) \tag{48}$$

将式（23）、式（24）代入式（48），经过一系列的变换，可得到

$$\Delta v(z_i) = \frac{1}{a'}\left[(a'b - ab')(\sqrt{\varepsilon_z} - \sqrt{\varepsilon_{z_1}}) + (a'c - ac')(\varepsilon_z - \varepsilon_{z_1}) - \frac{b'}{a'}(a'b - ab')(\sqrt{\varepsilon_z} - \sqrt{\varepsilon_{z_1}})\sqrt{\varepsilon_{z_1}}\right] \tag{49}$$

在式（49）中引入线放大率 M 表达式（41），便有

$$\Delta v(z_i) = M\left[\sqrt{\phi(z_a)}(a'b - ab')(\sqrt{\varepsilon_z} - \sqrt{\varepsilon_{z_1}}) + \sqrt{\varphi(z_a)}(a'c - ac')(\varepsilon_z - \varepsilon_{z_1})\right] \quad (50)$$

实际上，在推导式（50）的过程中，线放大率 M 表达式（41）的展开式只需要精确到 $\sqrt{\varepsilon_{z_1}}$ 的一次幂项。

将式（25）的所有系数 a、b、c 的表达式代入式（50）中，经过一系列的变换，可以得到：

$$\Delta v(z_i) = M\left\{\left[\frac{\phi'(z_a)}{2}\xi_0(z_a)\eta_0(z_a) + \varphi(z_a)(\xi_0'(z_a)\eta_0(z_a) - \xi_0(z_a)\eta_0'(z_a))\right](\sqrt{\varepsilon_z} - \sqrt{\varepsilon_{z_1}}) + \right.$$
$$\left.\left[\frac{\phi'(z_a)}{2\sqrt{\phi(z_a)}}\xi_0^2(z_a) + (\phi(z_a)^{\frac{3}{2}}(\xi_0'(z_a)\xi_1(z_a) - \xi_0(z_a)\xi_1'(z_a)))\right](\varepsilon_z - \varepsilon_{z_1})\right\}$$
(51)

近轴横向色差的一般形式 $\Delta r_v^*(z_i)$ 便可写成如下形式：

$$\Delta r_v^*(z_i) = \sqrt{\frac{m_0}{2e}}\dot{r}_0 \Delta v(z_i) = \sqrt{\frac{m_0}{2e}}\dot{r}_0 M\left\{\left[\frac{\phi'(z_a)}{2}\xi_0(z_a)\eta_0(z_a) + \right.\right.$$
$$\phi(z_a)(\xi_0'(z_a)\eta_0(z_a) - \xi_0(z_a)\eta_0'(z_a))\right](\sqrt{\varepsilon_z} - \sqrt{\varepsilon_{z_1}}) + \quad (52)$$
$$\left.\left[\frac{\phi'(z_a)}{2\sqrt{\phi(z_a)}}\xi_0^2(z_a) + (\phi(z_a)^{\frac{3}{2}}(\xi_0'(z_a)\xi_1(z_a) - \xi_0(z_a)\xi_1'(z_a))\right](\varepsilon_z - \varepsilon_{z_1})\right\}$$

最后，我们得到近轴横向像差 $\Delta r^*(z_i)$ 一般形式的表达式，它由二级和三级近轴横向色差 $\Delta r_v^*(z_i)$ 和三级近轴放大率横向色差 $\Delta r_w^*(z_i)$ 所组成；它们是基于近轴方程的渐近解，其系数 ξ_0、η_0、ω_0、ζ_0、ξ_1、ω_1 和轴上电位分布 $\phi(z)$ 都取 $z = z_a$ 处之值。

我们在上面已经指出，近轴方程的渐近解给出的系数 ξ_0、η_0、ω_0、ζ_0、ω_1 并不是彼此独立的，它们是互相关联的。现在我们研究系数 ξ_1 和 ξ_0、η_0、ω_0、ζ_0、ω_1 之间的关系。比较式（41）与式（37），表达式中系数 $\sqrt{\varepsilon_{z_1}}$ 的二次幂项应该相等，于是有

$$\frac{1}{a'}\left[(a'f - af') - \frac{d'}{a'}(a'c - ac') - \frac{1}{a'^2}(a'e' - b'd')(a'b - ab')\right]$$
$$= \frac{1}{\sqrt{\phi(z_a)}a'}\left[-\frac{c'}{a'} - \frac{1}{2\phi(z_a)} + \left(\frac{b'}{a'}\right)^2\right] \quad (53)$$

上式可以变换为

$$\left(\frac{f}{b} - \frac{e'}{a'}\right)(a'b - ab') - \frac{d'}{a'}(a'c - ac') + \frac{1}{\sqrt{\phi(z_a)}}\frac{c'}{a'} + \frac{1}{2\phi(z_a)\sqrt{\phi(z_a)}} = 0 \quad (54)$$

最后，我们可以得到系数 $\xi_1(z_a)$ 应满足的方程：

$$\left[\omega_0'(z_a)\phi(z_a)\xi_0(z_a) + 1\right]\xi_1'(z_a) + \left[\frac{\phi'(z_a)}{2\phi(z_a)} - \omega_0'(z_a)\phi(z_a)\xi_0'(z_a)\right]\xi_1(z_a) +$$
$$\left[\frac{\xi_0'(z_a)}{\phi(z_a)} - \frac{\omega_0'(z_a)}{2\phi(z_a)}\phi'(z_a)\xi_0^2(z_a)\right] + \left[\frac{\omega_1(z_a)}{\eta_0(z_a)} - \frac{\zeta_0(z_a)}{\xi_0(z_a)}\right]\left[\sqrt{\phi(z_a)}\xi_0'(z_a) + \frac{\phi'(z_a)}{2\sqrt{\phi(z_a)}}\xi_0(z_a)\right] \times$$
$$\left\{\frac{\phi'(z_a)}{2\sqrt{\phi(z_a)}}\xi_0(z_a)\eta_0(z_a) + \sqrt{\varphi(z_a)}\left[\xi_0'(z_a)\eta_0(z_a) - \xi_0(z_a)\eta_0'(z_a)\right]\right\} = 0$$

式（55）表明，近轴方程渐近解给出的所有系数 ξ_0、η_0、ω_0、ζ_0、ω_1、ξ_1 并不是孤立的，它们应由方程式（16）或式（17）求解，而且应满足关系式（42）～式（45）和式（55）。

结束语

本文由近轴方程渐近解提出求解成像电子光学系统近轴横向像差的途径，推导了近轴横向像差普遍形式的表达式，研究了近轴方程渐近解系数之间的关系式，证明了它们之间是相互关联的。文中表达的一个重要信息是，成像电子光学系统近轴横向像差不仅仅应考虑二级近轴横向色差，虽则它占近轴横向像差的主要部分，而在研究全部横向像差时，也应考虑三级近轴色差和三级近轴放大率色差。本文证明了它们的存在，并首次导出了此两项像差的普遍形式的表达式。本文的结果为研究简明形式的近轴横向像差提供了理论基础。

致谢

本文系国家自然科学基金资助项目（60771070），高等学校博士点专项科研基金资助课题（B-117），特此致谢。

参 考 文 献

[1] RECKNAGEL A. Theorie des elektrisohen elektronen miktroskops fur selbstrakler[J]. Z. Angew. Physik, 1941, 119: 689-708.

[2] ARTIMOVICH L A. Electrostatic properties of emission systems[J]. Information of Academy od Sciences, USSR, Series of physics, 1944, 8(6): 313-328.

[3] 周立伟. 宽束电子光学[M]. 北京: 北京理工大学出版社, 1993.

[4] BANSHITET B A. On computation of aberrations in cathode lenses[J]. Radiotechnics and Electronics, 1964, 9(5): 844-850.

[5] KELMAN V M. Theory of aberrations of third order in cathode lenses, chromatic aberrations[J]. Radiotechnics and Electronics, 1978.

[6] MONASTYRSKI M A, KYLIKOV Y V. Theory of aberrations of third order in cathode lenses, chromatic aberrations[J]. Radiotechnics and Electronics, 1978, 23(3): 644-647.

[7] 周立伟, 艾克聪, 潘顺臣. 复合电磁聚焦阴极透镜的像差理论[J]. 物理学报, 1983, 32(3): 376-392.

[8] XIMEN J Y, ZHOU L W, AI K C. Variation theory of aberrations in cathode lenses[J]. Optik, 1983, 66(1): 19-34.

[9] MONASTYRSKI M A. On asymptotic solutions of paraxial equation of electron optics[J]. Journal of Technical Physics, 1978. 48(6): 1117-1122.

[10] ELIN B P, KATISHAOV B A, KYLIKOV Y V, et al. Numerical methods of optimization for mission electron-optical systems[M]. Novosibilsk: Science Press, 1987.

6.4 On Verification for Paraxial Lateral Aberrations of Imaging Electrostatic Electron Optics Based on Asymptotic Solutions

基于渐近解的静电成像电子光学近轴横向像差理论及其验证

Abstract: In imaging electron optics, study of geometrical lateral aberrations of third order and paraxial lateral aberrations of second order (i. e., paraxial chromatic aberrations of second order) has traditionally been paid more attention to, but the existence of paraxial chromatic aberrations of third order and paraxial chromatic aberrations of magnification of third order was almost ignored, and the general form of paraxial lateral aberration has not been studied theoretically. In the present paper, the paraxial lateral aberrations expressed in general form have been derived emphatically by using asymptotic solutions of paraxial equation. The relationship between the coefficients of asymptotic solutions has been investigated, which proves that the coefficients of asymptotic solutions are related each other. Through a two-electrode electrostatic spherical concentric system model, the two special solutions expressed by asymptotic solutions and accurate solutions in a two-electrode electrostatic spherical concentric system model have been deduced, and the paraxial lateral aberrations have been verified and tested, in which the aberration coefficients are solved by asymptotic solutions of paraxial equation. Result completely proves that the approach based on asymptotic solutions to solve the paraxial lateral aberrations are correct and practicable. The paraxial chromatic aberration of magnification of third order and the paraxial chromatic aberration of third order have been firstly given, and the R-A formula of paraxial chromatic aberration of second order has been deduced and confirmed which possess an greatest part in the whole paraxial lateral aberrations. A simple and clear form for expressing paraxial lateral aberrations of imaging electron optics is suggested for practical use. Results of the present paper will have theoretical and practical significance for the design of image tubes.

摘要：在成像电子光学的研究中，备受关注的是二级近轴横向色差和三级几何横向像差，人们几乎忽略了三级近轴色差和三级近轴放大率色差的存在，理论上也没有给出近轴横向像差的普遍形式。本文用近轴方程的渐近解导得近轴横向像差的普遍表示式，探讨近轴方程渐近解系数之间的关系式，并证实它们之间是相互关联的。通过两电极静电同心球系统，推导了近轴轨迹特解的渐近解和精确解的表示式，并对渐近解表示的近轴横向像差进行了验证，结果证明，基于渐近解求解成像电子光学近轴横向像差的方法和途径是可行的和足够精确的。文中导得了在横向像差中占有主要部分的二级近轴色差表示式，即莱克纳格尔–阿尔齐莫维奇（Recknagel-Artimovich）公式，并首次给出了三级近轴放大率色差和三级近轴色差的表达式。文中最后给出了以简明形式表示的近轴横向像差的表示式，它对于研究成像电子光学的像差理论具有理论价值和对于像管设计具有实际意义。

Introduction

In imaging electron optics, study of geometrical lateral aberrations of third order and paraxial lateral aberrations of second order has traditionally been paid more attention to, but the existence of paraxial lateral aberration of magnification and the paraxial chromatic aberration of third order was almost ignored which is the same order of magnitude compared to geometrical lateral aberrations of third order, and the general form of paraxial lateral aberration has not been studied theoretically.

As is known and recognized, the paraxial lateral aberrations of the second order, i. e., the paraxial chromatic aberration of second order which determines the limitation of spatial resolution, occupies a greatest part in the whole paraxial lateral aberrations. But till now the general expression of whole paraxial lateral aberration has not been given. The present paper is an attempt to derive the expression of the paraxial aberrations in general form on the basis of the asymptotic solutions, including the paraxial chromatic aberration of second order and third order, as well as the paraxial chromatic aberration of magnification of the third order. The relationship between the coefficients of asymptotic solutions will also be investigated.

The first aim of present paper is to deduce the paraxial lateral aberrations of imaging electron optics expressed in general form based on the asymptotic solution of paraxial equation given by Monastyrski[1,2]. The second aim of present paper is to test and verify the correctness of expression of paraxial lateral aberrations given here by a bi-electrode electrostatic spherical concentric system model. Besides, an expression of paraxial lateral aberration of imaging electron optics written in a simple and clear form will be suggested for the practical use.

1 Definition of lateral aberrations in electron optical imaging systems

According to theory of paraxial electron optics, all of paraxial electron rays emitted from photocathode with same axial initial potential $\varepsilon_{z_1}^{1/2} \sqrt{\varepsilon_{z_1}}$ and initial object height vector $\boldsymbol{r}_0$ will be focused at an ideal image plane at $z=z_i$, $\boldsymbol{r}^*(z_i, \varepsilon_{z_1}^{\frac{1}{2}}, \boldsymbol{r}_0)$, which does not depend on radial initial potential $\sqrt{\varepsilon_r}$. The lateral aberration of imaging electron optical systems is defined by spatial difference between practical imaging height vector and ideal imaging height vector at the ideal image plane $z=z_i$ determined by axial initial potential $\varepsilon_{z_1}^{1/2} \sqrt{\varepsilon_{z_1}}$ of paraxial electron rays, which can be expressed as [3]

$$\Delta \boldsymbol{r} = \Delta \boldsymbol{r}(z_i) = \boldsymbol{r}(z_i, \varepsilon_z^{\frac{1}{2}}, \varepsilon_r^{\frac{1}{2}}, \boldsymbol{r}_0) - \boldsymbol{r}^*(z_i, \varepsilon_{z_1}^{\frac{1}{2}}, \boldsymbol{r}_0) \tag{1}$$

where $r(z_i, \varepsilon_z^{\frac{1}{2}}, \varepsilon_r^{\frac{1}{2}}, r_0)$, $r^*(z_i, \varepsilon_{z_1}^{\frac{1}{2}}, r_0)$ express the practical imaging vector and the ideal imaging vector at the ideal image plane $z = z_i$, respectively, which are formed by the practical rays and the paraxial rays of electrons emitted from the photocathode go through the system. Here the star mark " * " represents the paraxial ray. Take the cylindrical coordinate system (z, r), the axial coordinate z is chosen from the photocathode $z_0 = 0$, r_0 is the initial object vector, z_i is the position of ideal image plane, which corresponds to a given axial initial potential $\sqrt{\varepsilon_{z_1}}$. $\sqrt{\varepsilon_z}$ and $\sqrt{\varepsilon_r}$ are the axial initial potential and radial initial potential of electrons emitted from the photocathode, respectively.

Thus, formula (1) can be expressed as

$$\Delta r = \Delta r(z_i) = \Delta r^* + \delta r \tag{2}$$

where

$$\Delta r^* = r^*(z_i, \varepsilon_z^{\frac{1}{2}}, \varepsilon_r^{\frac{1}{2}}, r_0) - r^*(z_i, \varepsilon_{z_1}^{\frac{1}{2}}, r_0) \tag{3}$$

$$\delta r = r(z_i, \varepsilon_z^{\frac{1}{2}}, \varepsilon_r^{\frac{1}{2}}, r_0) - r^*(z_i, \varepsilon_z^{\frac{1}{2}}, \varepsilon_r^{\frac{1}{2}}, r_0) \tag{4}$$

where Δr^* in Eq. (3) is called paraxial lateral aberration, which is the difference between the image height vector at the ideal image plane z_i determined by $\sqrt{\varepsilon_{z_1}}$, formed by two paraxial rays with different initial axial potential ($\sqrt{\varepsilon_z}$, $\sqrt{\varepsilon_r}$) and ($\sqrt{\varepsilon_{z_1}}$, $\sqrt{\varepsilon_r}$), emitted from same object vector r_0 at photocathode. δr in formula (4) is called geometrical lateral aberration, which is the difference at the ideal image plane z_i between the practical image height vector formed by practical rays and paraxial image height vector formed by paraxial rays, but both with same initial potential ($\sqrt{\varepsilon_z}$, $\sqrt{\varepsilon_r}$) emitted from same object height vector r_0 at photocathode. Eq. (2) shows that the lateral aberration Δr can be regarded as the composition of paraxial lateral aberration Δr^* and geometrical lateral aberration δr. It is a very important concept for studying the aberration theory in imaging electron optics.

As is known, the paraxial trajectory $r^*(z)$ is the solution of following paraxial ray equation:

$$r^{*\prime\prime}(z) + \frac{1}{2}\frac{\phi'(z)}{\phi(z) + \varepsilon_z} r^{*\prime}(z) + \frac{1}{4}\frac{\phi''(z)}{\phi(z) + \varepsilon_z} r^*(z) = 0 \tag{5}$$

where $\phi(z)$ is axial potential distribution, $' = d/dz$ expresses the differential with respect to z.

The general solution of paraxial ray Eq. (5) can be written as

$$r^*(z) = r_0 w(z) + \sqrt{\frac{m_0}{2e}} \dot{r}_0 v(z) \tag{6}$$

where m_0/e is the ratio of electron mass to charge; r_0 and $\dot{r}_0$ are initial object vector and initial radial velocity of electron emitted from the photocathode, respectively.

Two special solutions $v = v(z)$, $w = w(z)$ of Eq. (6) satisfy following initial conditions:

$$\begin{aligned} v(z=0) &= 0, \quad v'(z=0) = \frac{1}{\sqrt{\varepsilon_z}}; \\ w(z=0) &= 1, \quad w'(z=0) = 0 \end{aligned} \tag{7}$$

Suppose that all of electrons emitted from photocathode at initial axial point with same axial initial potential $\sqrt{\varepsilon_{z_1}}$ will be focused at the ideal image plane $z = z_i$, i. e. ,

$$v(z_i, \sqrt{\varepsilon_{z_1}}) = 0, \quad w(z_i, \sqrt{\varepsilon_{z_1}}) = M \tag{8}$$

and it is quite evident that $v(z_i, \sqrt{\varepsilon_z}) \neq 0$, $w(z_i, \sqrt{\varepsilon_z}) \neq M$, where M is the linear magnification.

From definition (3), we can define the paraxial lateral aberration at the ideal image plane ez_i, where $v(z_i, \sqrt{\varepsilon_{z_1}}) = 0$:

$$\Delta r^*(z_i) = \{w(z_i, \sqrt{\varepsilon_z}) - w(z_i, \sqrt{\varepsilon_{z_1}})\} r_0 + \sqrt{\frac{m_0}{2e}} \dot{r}_0 \{v(z_i, \sqrt{\varepsilon_z}) - v(z_i, \sqrt{\varepsilon_{z_1}})\}$$
$$= \{w(z_i, \sqrt{\varepsilon_z}) - M\} r_0 + \sqrt{\frac{m_0}{2e}} \dot{r}_0 v(z_i, \sqrt{\varepsilon_z}) \tag{9}$$

Eq. (9) can be written as:

$$\Delta r^*(z_i) = \Delta r_v^*(z_i) + \Delta r_w^*(z_i) = \sqrt{\frac{m_0}{2e}} \dot{r}_0 \Delta v(z_i) + r_0 \Delta w(z_i) \tag{10}$$

where $\Delta v(z_i) = v(z_i, \sqrt{\varepsilon_z})$, $\Delta w(z_i) = w(z_i, \sqrt{\varepsilon_z}) - M$.

Eq. (10) shows that the paraxial lateral aberration $\Delta r^*(z_i)$ can be regarded as the sum of the paraxial chromatic aberration $\Delta r_v^*(z_i)$ and the paraxial chromatic aberration of magnification $\Delta r_w^*(z_i)$. It is our starting point for studying paraxial aberration theory of imaging electron optics.

Concerning the geometrical aberrations δr, we can study it from the practical ray equation:

$$r''(z) = \frac{1 + r'^2}{2\varphi_*} \left(\frac{\partial \varphi}{\partial r} - r' \frac{\partial \varphi}{\partial z} \right) \tag{11}$$

where $\varphi_* = \varphi(z, r) + \varepsilon_0$. ε_0—initial potential of electron emitted from the photocathode $\varepsilon_0 = \varepsilon_r + \varepsilon_z$, $-\varphi = \varphi(z, r)$—distribution of spatial potential, which can be expanded by Scherzer series as following:

$$\varphi(z, r) = \phi(z) - \frac{1}{4} \phi''(z) r^2 + \frac{1}{64} \phi^{IV}(z) r^4 - \cdots \tag{12}$$

Expanding Eq. (11) and substituting Eq. (12) in it, from definition Eq. (4) we can obtain the equation of the geometrical lateral aberration of third – order δr_3 as following:

$$[\phi(z) + \varepsilon_z] \delta r_3'' + \frac{1}{2} \phi'(z) \delta r_3' + \frac{1}{4} \phi''(z) \delta r_3 = F(r^*, r^{*\prime}, z) \tag{13}$$

This is a non – homogenous differential equation of the second order, where $F(r^*, r^{*\prime}, z)$ is a term at right hand of the equation, and r^* is the solution of paraxial ray. Solution of the geometrical lateral aberrations of the third order δr_3 had been described in a lot of monographs and literature [4–8], and here we shall not discuss any more.

2 Paraxial lateral aberration solved by asymptotic solutions of paraxial equation

A method for solving the special solutions of $v(z, \sqrt{\varepsilon_z})$, $w(z, \sqrt{\varepsilon_z})$ of paraxial equation by using asymptotic solutions has been given in literature [2], which can be described as following:

$$v(z, \sqrt{\varepsilon_z}) = \sqrt{\phi(z)}\xi_0(z) + \sqrt{\varepsilon_z}\eta_0(z) + \varepsilon_z\left[\frac{\xi_0(z)}{2}\frac{\phi'(z)}{\sqrt{\phi(z)}} + \xi_1(z)\sqrt{\phi(z)}\right] + \cdots \quad (14)$$

$$w(z, \sqrt{\varepsilon_z}) = \omega_0(z) + \sqrt{\varepsilon_z}\sqrt{\phi(z)}\zeta_0(z) + \varepsilon_z\omega_1(z) + \cdots \quad (15)$$

where coefficients ξ_0, η_0, ω_0, ζ_0; ξ_1, ω_1 in Eq. (14), Eq. (15) are the solutions of following differential equations:

$$L_0\begin{pmatrix}\xi_K\\\zeta_K\end{pmatrix} = -\begin{pmatrix}\xi''_{k-1}\\\zeta''_{k-i}\end{pmatrix} \quad (16a)$$

$$M_0\begin{pmatrix}\eta_K\\\omega_K\end{pmatrix} = -\begin{pmatrix}\eta''_{k-1}\\\omega''_{k-i}\end{pmatrix} \quad (16b)$$

where $\xi_{-1} = \zeta_{-1} = \eta_{-1} = \omega_{-1} = 0$; L_0, M_0 is the linear differential operator, which can be expressed by

$$L_0 = \phi(z)\frac{\mathrm{d}^2}{\mathrm{d}z^2} + \frac{3}{2}\phi'(z)\frac{\mathrm{d}}{\mathrm{d}z} + \frac{3}{4}\phi''(z) \quad (17)$$

$$M_0 = \phi(z)\frac{\mathrm{d}^2}{\mathrm{d}z^2} + \frac{1}{2}\phi'(z)\frac{\mathrm{d}}{\mathrm{d}z} + \frac{1}{4}\phi''(z) \quad (18)$$

The initial condition of Eq. (16a) and Eq. (16b) can be determined according to the following procedure:

$$\begin{array}{c}\dfrac{\xi_0(0)\phi'(0)}{2} = 1 \rightarrow \xi_0(0) + \eta_0(0) = 0\\\downarrow\\\dfrac{\xi_1(0)\phi'(0)}{2} + \xi'_0(0) + \eta'_0(0) = 0 \rightarrow \text{and so on}\end{array} \quad (19)$$

$$\begin{array}{ccc}\omega_0(0) = 1 & \zeta_0(0) + \omega_1(0) = 0 & \\\downarrow & \uparrow & \downarrow \\\dfrac{\zeta_0(0) + \phi'(0)}{2} + \omega'_0(0) = 0 & \rightarrow \text{and so on} &\end{array} \quad (20)$$

where the arrow indicates the sequence of calculation.

From asymptotic solutions (14), solutions (15) and definition of (9), we may obtain ξ_0, η_0, ω_0, ζ_0; ξ_1, ω_1 at $z = z_i$ and the paraxial lateral aberrations. $\Delta r^*(z_i)$. According to the abovementioned procedure, we should carry on some operations for solving differential equations to obtain the value of coefficients ξ_0, η_0, ω_0, ζ_0, ξ_1, ω_1 at $z = z_i$. It is a complicated process and it is not easy to solve the value of coefficients, so the problem of paraxial lateral aberrations $\Delta r^*(z_i)$ given by Eq. (10) is still not solved.

As is known, for the electron optical imaging system, the axial potential distribution $\phi(z)$ at the anode-screen section behind a certain point $z = z_a$, usually is a constant, the trajectory moving in this section is a straight line. Thus, we could obtain the solutions of the coefficients ξ_0, η_0, ω_0, ζ_0; ξ_1, ω_1 at the position $z = z_a$, and to find the paraxial lateral aberrations $\Delta r^*(z_i)$ at image plane $z = z_i$.

Now we suggest the following process and procedures. Suppose that $\phi(z) \neq$ const $at \ z < z_a$, $\phi(z) =$ const $= \phi(z_a) = \phi(z_i)$ at $z \geq z_a$. This assumption is in accordance with the practical situation. Thus we can find the value of two special solutions v, w and their derivatives v', w' at $z = z_a$ from expression (14) and Expression (15) of asymptotic solutions. Thus, the expressions of $\Delta w(z_i)$, $\Delta v(z_i)$ in formula (10) can be written as:

$$\Delta w(z_i) = [w(z_a, \sqrt{\varepsilon_z}) - w(z_a, \sqrt{\varepsilon_{z_1}})] - \frac{v(z_a, \sqrt{\varepsilon_{z_1}})}{v'(z_a, \sqrt{\varepsilon_{z_1}})}[w'(z_a, \sqrt{\varepsilon_z}) - w'(z_a, \sqrt{\varepsilon_{z_1}})] \tag{21}$$

$$\Delta v(z_i) = v(z_i, \sqrt{\varepsilon_z}) = v(z_a, \sqrt{\varepsilon_z}) - \frac{v'(z_a, \sqrt{\varepsilon_z})}{v'(z_a, \sqrt{\varepsilon_{z_1}})} v(z_a, \sqrt{\varepsilon_{z_1}}) \tag{22}$$

Substituting all of values of $v(z_a, \sqrt{\varepsilon_z})$, $v'(z_a, \sqrt{\varepsilon_z})$, $w(z_a, \sqrt{\varepsilon_z})$, $w'(z_a, \sqrt{\varepsilon_z})$ and $v(z_a, \sqrt{\varepsilon_{z_1}})$, $v'(z_a, \sqrt{\varepsilon_{z_1}})$ given by asymptotic solution (14) and solution (15) into expressions (21) and expression (22), through a complicated transformation and introducing the linear magnification $M = w(z_i, \sqrt{\varepsilon_{z_1}})$ in it, we may obtain a general form of the paraxial chromatic aberration of magnification $\Delta r_w^*(z_i)$:

$$\Delta r_w^*(z_i) = r_0 \Delta w(z_i) = -r_0 M \left\{ \frac{1}{2} \phi'(z_a) \xi_0(z_a) \eta_0(z_a) + \phi(z_a) \right.$$
$$\left. [\xi_0'(z_a) \eta_0(z_a) - \xi_0(z_a) \eta_0'(z_a)] \right\} \times \left\{ \frac{\zeta_0(z_a)}{\xi_0(z_a)} (\sqrt{\varepsilon_z} - \sqrt{\varepsilon_{z_1}}) \sqrt{\varepsilon_{z_1}} - \frac{\omega_1(z_a)}{\eta_0(z_a)} (\varepsilon_z - \varepsilon_{z_1}) \right\} \tag{23}$$

and the paraxial chromatic aberration $\Delta r_v^*(z_i)$:

$$\Delta r_v^*(z_i) = \sqrt{\frac{m_0}{2e}} \dot{r}_0 \Delta v(z_i) = \sqrt{\frac{m_0}{2e}} \dot{r}_0 M \left\{ \left[\frac{\phi'(z_a)}{2} \xi_0(z_a) \eta_0(z_a) + \right. \right.$$
$$\phi(z_a)(\xi_0'(z_a) \eta_0(z_a) - \xi_0(z_a) \eta_0'(z_a)) \left] (\sqrt{\varepsilon_z} - \sqrt{\varepsilon_{z_1}}) + \right. \tag{24}$$
$$\left. \left[\frac{\phi'(z_a)}{2\sqrt{\phi(z_a)}} \xi_0^2(z_a) + (\phi(z_a))^{\frac{3}{2}} (\xi_0'(z_a) \xi_1(z_a) - \xi_0(z_a) \xi_1'(z_a)) \right] (\varepsilon_z - \varepsilon_{z_1}) \right\}$$

Finally, we obtain a general form of paraxial lateral aberrations $\Delta r^*(z_i)$, which is composed of the paraxial chromatic aberration $\Delta r_v^*(z_i)$ and paraxial chromatic aberration of magnification $\Delta r_w^*(z_i)$, based on asymptotic solutions of paraxial ray equation, where coefficients ξ_0, η_0, ω_0, ζ_0; ξ_1, ω_1 of asymptotic solutions and axial potential distribution $\phi(z)$ take the value at $z = z_a$. Regarding to the relationship between these coefficients, please read article[3] for reference.

3 Solving coefficients of asymptotic solutions in a bi-electrode electrostatic concentric spherical system model

According to the procedure for exploring asymptotic solution of paraxial equation, we can find out the solutions of coefficients ξ_0, η_0, ω_0, ζ_0; ξ_1, ω_1 at the position z_a which is based on the known axial potential distributions $\phi(z)$. It seems that it is not so complicated to find out the numerical solution of these coefficients, but the problem is that these coefficients cannot be showed by mathematical forms, thus it is difficult to master the physical characteristics and regularity from the point of view for designing imaging electron optical systems.

Now we try to seek for the solutions of coefficients ξ_0, η_0, ω_0, ζ_0; ξ_1, ω_1 given by asymptotic solution of paraxial ray equation through a bi-electrode electrostatic concentric spherical system model, the axial potential distribution $\phi(z)$ of which can be expressed by

$$\phi(z) = E_c \frac{-R_c z}{z + R_c} (0 \leq z \leq (R_a - R_c)) \tag{25}$$

where E_c is the strength of electric-field at the photocathode, R_c, R_a are radii of curvature of spherical photocathode and spherical anode, respectively, all of these quantities take negative value.

Let's first seek for the coefficients $\eta_0(z)$、$\xi_0(z)$ and $\xi_1(z)$. Suppose that the coefficient $\eta_0(z)$ can be expressed by

$$\eta_0(z) = \frac{-2z}{\phi(z)} \tag{26}$$

Substituting formula (26) into Eq. (16b):

$$\phi(z)\frac{d^2\eta_0}{dz^2} + \frac{1}{2}\phi'(z)\frac{d\eta_0}{dz} + \frac{1}{4}\phi''(z)\eta_0 = -\eta''_{-1} = 0 \tag{27}$$

we can get the following expression:

$$\frac{1}{\phi(z)}\phi'(z) + \frac{z}{2\phi(z)}\phi''(z) - \frac{z\phi'^2(z)}{\phi^2(z)} = 0 \tag{28}$$

Eq. (28) will be satisfied when the axial potential distribution $\phi(z)$ is expressed by Eq. (25). Thus, expression (26) is a solution of Eq. (16b).

Similarly, suppose that the coefficient $\xi_0(z)$ is expressed by

$$\xi_0(z) = \frac{2z}{\phi(z)} \tag{29}$$

Substituting it into Eq. (16a),

$$\phi(z)\frac{d^2\xi_0}{dz^2} + \frac{3}{2}\phi'(z)\frac{d\xi_0}{dz} + \frac{3}{4}\phi''(z)\xi_0 = -\xi''_{-1} = 0 \tag{30}$$

we can also get the expression similar to Eq. (28):

$$\frac{-1}{\phi(z)}\phi'(z) - \frac{z}{2\phi(z)}\phi''(z) + \frac{z\phi'^2(z)}{\phi^2(z)} = 0 \tag{31}$$

It is evident that Eq. (31) is also satisfied when the axial potential distribution $\phi(z)$ is expressed by Eq. (25). Thus, expression (29) is a solution of Eq. (16a).

Now, we seek for the expression of coefficient $\xi_1(z)$. When $z = 0$, no matter what the photocathode is spherical or planar electrode, formulae Eq. (26) and Eq. (29) at $z = 0$ can be expressed as follows:

$$\xi_0(z=0) = \frac{2}{\phi_0'}, \quad \eta_0(z=0) = -\frac{2}{\phi_0'},$$
$$\xi_0'(z=0) = 0, \quad \eta_0'(z=0) = 0 \tag{32}$$

which satisfy the initial conditions of Eq. (19):
$$\frac{\xi_0(0)\phi'(0)}{2} = 1, \quad \xi_0(0) + \eta_0(0) = 0 \tag{33}$$

and from the other initial condition of Eq. (19): $\frac{\xi_1(0)\phi'(0)}{2} + \xi_0'(0) + \eta_0'(0) = 0$, we found that $\xi_1(0) = 0$.

From (16a), $\xi_1(z)$ should satisfy following equation:
$$\phi(z)\frac{d^2}{dz^2}\xi_1 + \frac{3}{2}\phi'(z)\frac{d}{dz}\xi_1 + \frac{3}{4}\phi''(z)\xi_1 = -\xi_0'' \tag{34}$$

ξ_0'' can be found out by twice differential to Eq. (29):
$$\xi_0''(z) = \frac{2}{\phi^2(z)}\left[-2\phi'(z) - z\phi''(z) + \frac{2z\phi'^2(z)}{\phi(z)}\right] \tag{35}$$

Therefore, from Eq. (34) and Eq. (35), we may obtain non-homogeneous differential equation of second order for solving $\xi_1(z)$:
$$\frac{d^2\xi_1}{dz^2} + \frac{3}{2}\frac{\phi'(z)}{\phi(z)}\frac{d\xi_1}{dz} + \frac{3}{4}\frac{\phi''(z)}{\phi(z)}\xi_1 = \frac{2}{\phi^3(z)}\left[2\phi'(z) + z\phi''(z) - 2z\frac{\phi'^2(z)}{\phi(z)}\right] \tag{36}$$

Now we try to seek for the expression of $\xi_1(z)$, rewrite Eq. (36) to the following form:
$$\frac{d^2\xi_1}{dz^2} + \frac{3}{2}\frac{\phi'(z)}{\phi(z)}\frac{d\xi_1}{dz} + \frac{3}{4}\frac{\phi''(z)}{\phi(z)}\xi_1 = f(z) \tag{37}$$

where
$$f(z) = \frac{2}{\phi^3(z)}\left[2\phi'(z) + z\phi''(z) - 2z\frac{\phi'^2(z)}{\phi(z)}\right] \tag{38}$$

Eq. (37) is a non-homogeneous differential equation of second order. As is known, when $f(z) = 0$, it is a homogeneous differential equation of second order for solving $\xi_1(z)$:
$$\frac{d^2\xi_1}{dz^2} + \frac{3}{2}\frac{\phi'(z)}{\phi(z)}\frac{d\xi_1}{dz} + \frac{3}{4}\frac{\phi''(z)}{\phi(z)}\xi_1 = 0 \tag{39}$$

Eq. (39) has a special solution $y_1(z)$ like this
$$y_1(z) = \frac{2z}{\phi(z)} \tag{40}$$

The other special solution $y_2(z)$ can be found as
$$y_2(z) = y_1(z)\int \frac{e^{-\int \frac{3}{2}\frac{\phi'(z)}{\phi(z)}dz}}{y_1^2(z)}dz \tag{41}$$

If $\phi(z)$ satisfies expression (25), we may obtain
$$y_2(z) = \frac{z}{[\phi(z)]^{\frac{3}{2}}} \tag{42}$$

The condition which should satisfy Eq. (39) can be written as:

$$2\phi'(z)\left[\frac{z\phi'(z)}{\phi(z)} - 1\right] - z\phi''(z) = 0 \tag{43}$$

Eq. (43) is also satisfied when the axial potential distribution $\phi(z)$ is expressed by Eq. (25).

From expressions $y_1(z)$, $y_2(z)$ of Eq. (40), Eq. (42) and expression $f(z)$ of Eq. (38), we may obtain the solution of $\xi_1(z)$ of the non-homogeneous differential equation of second order Eq. (37):

$$\xi_1(z) = \int_{z=0}^{z} \frac{y_1(z)(-y_2(x)) + y_2(z)y_1(x)}{y_1(x)y_2'(x) - y_2(x)y_1'(x)} f(x) \, dx \tag{44}$$

i. e.,

$$\xi_1(z) = \int_{z=0}^{z} \left\{ \frac{2z}{\phi(z)} \frac{\phi^2(x)}{\phi'(x)x} - \frac{z}{\phi^{\frac{3}{2}}(z)} \frac{2\phi^{\frac{5}{2}}(x)}{\phi'(x)x} \right\} \left\{ \frac{2}{\phi^3(x)} \left[2\phi'(x) + x\phi''(x) - 2x \frac{\phi'^2(x)}{\varphi(x)} \right] \right\} dx \tag{45}$$

Now we try to seek for the expressions of coefficients $\omega_0(z)$, $\zeta_0(z)$, $\omega_1(z)$. From Eq. (16b), $\omega_0(z)$ should satisfy the following equation:

$$\phi(z) \frac{d^2}{dz^2} \omega_0 + \frac{1}{2} \phi'(z) \frac{d}{dz} \omega_0 + \frac{1}{4} \phi''(z) \omega_0 = -\omega_{-1}'' = 0 \tag{46}$$

From the special solution: $w(z=0) = 1$, we can get the initial condition: $\omega_0(z=0) = 1$. Let

$$\omega_0(z) = 1 + \frac{1}{R_c} z \tag{47}$$

Substituting (47) into Eq. (46), we may have

$$\frac{1}{2R_c} \left[\phi'(z) + \frac{R_c + z}{2} \phi''(z) \right] = 0 \tag{48}$$

It is not difficult to prove that Eq. (48) is also satisfied when the axial potential distribution $\phi(z)$ is expressed by Eq. (25).

Now we explore the expression of coefficient $\zeta_0(z)$. From Eq. (16a)

$$\phi(z) \frac{d^2}{dz^2} \zeta_0 + \frac{3}{2} \phi'(z) \frac{d}{dz} \zeta_0 + \frac{3}{4} \phi''(z) \zeta_0 = -\zeta_{-1} = 0 \tag{49}$$

Similar to abovementioned method, we may get a solution of Eq. (49):

$$\zeta_0(z) = \frac{2z}{(-R_c)\phi(z)} \tag{50}$$

Here we introduce the constant term $2/(-R_c)$ in order to satisfy the physical condition of system. When $z=0$, from Eq. (50), we have

$$\zeta_0(z=0) = \frac{-2}{(-R_c)E_c} \tag{51}$$

Now we explore the coefficient $\omega_1(z)$. From the initial condition (20), because $\omega_0'(0) = 1/R_c$, so the condition $\zeta_0(0)\phi_0'/2 + \omega_0'(0) = 0$ is satisfied. Then from $\zeta_0(0) + \omega_1(0) = 0$, we may obtain the following initial condition:

$$\omega_1(0) = \frac{2}{(-R_c)E_c} \tag{52}$$

Because $\omega_0''(z) = 0$, from Eq. (16b), we may have

$$\phi(z)\frac{d^2}{dz^2}\omega_1 + \frac{1}{2}\phi'(z)\frac{d}{dz}\omega_1(z) + \frac{1}{4}\phi''(z)\omega_1(z) = -\omega_0''(z) = 0 \quad (53)$$

similar to abovementioned method, we may have the solution of Eq. (53):

$$\omega_1(z) = \frac{-2z}{(-R_c)\phi(z)} \quad (54)$$

It is not difficult to prove that Eq. (49) and Eq. (53) are also satisfied when the axial potential distribution $\phi(z)$ is expressed by Eq. (25). Solution Eq. (54) just satisfies the condition of Eq. (52) at $z = 0$. Thus, we have obtained analytical forms of coefficients of asymptotic solutions ξ_0, η_0, ω_0, ζ_0; ξ_1, ω_1 in a bi-electrode electrostatic concentric spherical system model.

4 Asymptotic solutions of paraxial trajectories and paraxial aberrations in bi-electrode electrostatic concentric spherical system model

Now we try to find out the solution of paraxial trajectories and the paraxial lateral aberration in a bi-electrode electrostatic concentric spherical system model by using the expressions of coefficients given above. Substituting all of these solved expressions of coefficients $\xi_0(z)$, $\eta_0(z)$, $\omega_0(z)$, $\zeta_0(z)$ and $\omega_1(z)$ into two special solutions of paraxial Eq. (14) and Eq. (15), we have

$$v(z,\sqrt{\varepsilon_z}) = \frac{2z}{\sqrt{\phi(z)}} - \frac{2z}{\phi(z)}\sqrt{\varepsilon_z} + \frac{z\varepsilon_z}{\phi(z)\sqrt{\phi(z)}} = \frac{2z}{\sqrt{\phi(z)}}\left\{1 - \sqrt{\frac{\varepsilon_z}{\phi(z)}} + \frac{\varepsilon_z}{2\phi(z)}\right\} \quad (55)$$

$$w(z,\sqrt{\varepsilon_z}) = 1 + \frac{1}{R_c}z - \frac{2z}{R_c\sqrt{\phi(z)}}\sqrt{\varepsilon_z} + \frac{2z}{R_c\phi(z)}\varepsilon_z = 1 + \frac{1}{R_c}z - \frac{2z\sqrt{\varepsilon_z}}{R_c\sqrt{\phi(z)}}\left\{1 + \frac{\sqrt{\varepsilon_z}}{\sqrt{\phi(z)}}\right\}$$
$$(56)$$

According to Eq. (14), usually, it should have an expression of $\xi_1(z)$ in Eq. (55), because expression (45) of $\xi_1(z)$ contains a function $f(z)$ expressed by Eq. (38). But when the axial potential distribution $\phi(z)$ is expressed by Eq. (25) of a electrostatic concentric spherical system model, from Eq. (38), we get $f(z) = 0$. Therefore, the integral of expression Eq. (45) gives a zero solution $\xi_1(z) = 0$.

In the following paragraph, we shall prove that two special solutions of paraxial ray equation expressed by Eq. (55) and Eq. (56) on the basis of asymptotic solutions will have enough accuracy to solve the problem of paraxial lateral aberrations, thus it completely proves that the method and procedure given by Monastyrski to solve the paraxial equation of electron optics by asymptotic solutions are correct and practicable.

Now, we shall derive the expressions of paraxial lateral aberrations in a bi-electrode electrostatic concentric spherical system model on the basis of asymptotic solutions. By using the equation $\phi(z)$ of Eq. (25) and expressions of coefficients Eq. (26), Eq. (29), Eq. (47), Eq. (50) and Eq. (54), we can find out the value of $\eta_0(z)$, $\xi_0(z)$, $\omega_0(z)$, $\zeta_0(z)$, $\omega_1(z)$ at $z = z_a = R_a - R_c$ and its derivative of these coefficients:

$$\xi_0(z_a) = \frac{2(R_a - R_c)}{\phi_{ac}}, \quad \xi_0'(z_a) = -\frac{2n-2}{\phi_{ac}}, \quad \eta_0(z_a) = \frac{-2(R_a - R_c)}{\phi_{ac}}$$

$$\eta_0'(z_a) = \frac{2n-2}{\phi_{ac}}, \quad \zeta_0(z_a) = \frac{2(R_a - R_c)}{(-R_c)\phi_{ac}}, \quad \zeta_0'(z_a) = -\frac{2n-2}{(-R_c)\phi_{ac}}$$

$$\omega_0(z_a) = \frac{1}{n}, \quad \omega_0'(z_a) = \frac{1}{R_c}, \quad \omega_1(z_a) = \frac{-2(R_a - R_c)}{(-R_c)\phi_{ac}}, \quad \omega_1'(z_a) = \frac{2n-2}{(-R_c)\phi_{ac}} \quad (57)$$

and

$$\phi'(z_a) = -n^2 E_c, \quad E_c = \frac{\phi_{ac}}{R_c(n-1)} \quad (58)$$

where $n = R_c/R_a$, $\phi(z_a) = \phi_{ac}$; ϕ_{ac} indicates the voltage of anode with respect to photocathode.

Firstly, we shall derive the expression of linear magnification M, which can be obtained by two different approaches: from definition of special solution $M = w(z_i, \sqrt{\varepsilon_{z_1}})$, or from the relationship of Lagrange-Helmholtz:

$$M = \frac{1}{\sqrt{\varphi(z_a) + \varepsilon_{z_1}} v'(z_a, \sqrt{\varepsilon_{z_1}})} \quad (59)$$

where $v'(z_a, \sqrt{\varepsilon_{z_1}})$ can be obtained by expression (55) when $\sqrt{\varepsilon_z} = \sqrt{\varepsilon_{z_1}}$ and $z = z_a$. We can also prove that the two approaches for obtaining the linear magnification M have got same result as following:

$$M = w(z_i, \sqrt{\varepsilon_{z_1}}) = -\frac{1}{n-2}\left\{1 + \frac{2(n-1)}{(n-2)\sqrt{\phi_{ac}}}\sqrt{\varepsilon_{z_1}} + \frac{2n(n-1)}{(n-2)^2\phi_{ac}}\varepsilon_{z_1}\right\} \quad (60)$$

Thus the paraxial chromatic aberration of magnification $\Delta r_w^*(z_i)$ [Eq. (23)] and the paraxial chromatic aberration $\Delta r_v^*(z_i)$ [Eq. (24)] will be simplified as following:

$$\Delta r_w^*(z_i) = -r_0 M \frac{1}{2}\varphi'(z_a)\zeta_0(z_a)\eta_0(z_a)(\sqrt{\varepsilon_z} - \sqrt{\varepsilon_{z_1}})\sqrt{\varepsilon_z} \quad (61)$$

$$\Delta r_v^*(z_i) = -\sqrt{\frac{m_0}{2e}}\dot{r}_0 M \frac{\phi'(z_a)}{2}\xi_0^2(z_a)\left\{(\sqrt{\varepsilon_z} - \sqrt{\varepsilon_{z_1}}) - \frac{1}{\sqrt{\phi(z_a)}}(\varepsilon_z - \varepsilon_{z_1})\right\} \quad (62)$$

Substituting the coefficients of Eq. (57) into Eq. (61) and Eq. (62), introducing formula Eq. (58) of the electric field intensity at the photocathode E_c and Eq. (60) of linear magnification M in it, through a series of transformations, we may obtain the paraxial chromatic aberration of magnification $\Delta r_w^*(z_i)$ and the paraxial chromatic aberration $\Delta r_v^*(z_i)$ as following:

$$\Delta r_w^*(z_i) = r_0 \Delta w(z_i) = r_0\left[-\frac{2(M-1)}{\phi_{ac}}(\sqrt{\varepsilon_z} - \sqrt{\varepsilon_{z_1}})\sqrt{\varepsilon_z}\right] \quad (63)$$

$$\Delta r_v^*(z_i) = \sqrt{\frac{m_0}{2e}}\dot{r}_0 \Delta v(z_i) = \sqrt{\frac{m_0}{2e}}\dot{r}_0 \frac{2M}{E_c}\left\{(\sqrt{\varepsilon_z} - \sqrt{\varepsilon_{z_1}}) - \frac{1}{\sqrt{\phi_{ac}}}(\varepsilon_z - \varepsilon_{z_1})\right\} \quad (64)$$

Eq. (64) is called the paraxial chromatic aberrations of second order and third order; the first item of which was called as the well-known R-A formula in imaging electron optics [11,12], the second item is lesser one order of magnitude compared to first item in Eq. (64). Eq. (63) is called the paraxial chromatic aberration of magnification of third order.

5 Test and verification of solution of paraxial ray and paraxial aberrations by a bi-electrode electrostatic concentric spherical system model

Now we shall deduce the expressions of paraxial lateral aberrations on the basis of accurate solutions of paraxial ray equation in a bi-electrode electrostatic concentric spherical system model. The aim is to test and verify the correctness and accuracy of the expressions solved by asymptotic solutions.

Under the circumstance of a electrostatic concentric spherical system model, in which the axial potential distribution $\phi(z)$ is expressed by Eq. (25), fortunately, we have derived the accurate expressions of two special solutions of the paraxial ray equation (5) as following[9,10]:

$$v(z,\sqrt{\varepsilon_z}) = \frac{2z}{\sqrt{\phi(z)}}\left\{\sqrt{1+\frac{\varepsilon_z}{\varphi(z)}} - \sqrt{\frac{\varepsilon_z}{\phi(z)}}\right\} \tag{65}$$

$$w(z,\sqrt{\varepsilon_z}) = 1 + \frac{1}{R_c}z - \frac{\sqrt{\varepsilon_z}}{R_c}\frac{2z}{\phi(z)}[\sqrt{\phi(z)+\varepsilon_z} - \sqrt{\varepsilon_z}] \tag{66}$$

It is not difficult to prove that two special solutions (65) and (66) also satisfy the initial condition (7).

The difference between Eq. (65), Eq. (66) and Eq. (55), Eq. (56) is that the former is the accurate solutions of paraxial ray equation of Eq. (5) solved by direct method, the latter is the approximate solutions of paraxial ray equation solved by asymptotic method. But it should be pointed out that Eq. (55) and Eq. (56) are accurate to the second order approximation of $\sqrt{\varepsilon_z}$, that have sufficient accuracy for studying the paraxial lateral aberrations of the second and the third order and the geometrical lateral aberrations of third order. Therefore, expanding Eq. (65), Eq. (66) to the second order approximation of $\sqrt{\varepsilon_z}$, we find out that their expressions of $v(z, \sqrt{\varepsilon_z})$, $w(z, \sqrt{\varepsilon_z})$ have no difference with Eq. (55), Eq. (56). Their derivative with respect to z can be written as

$$v'(z,\sqrt{\varepsilon_z}) = \frac{(R_c+2z)}{(R_c+z)\sqrt{\phi(z)}}\left[1 - \frac{2z}{R_c+2z}\sqrt{\frac{\varepsilon_z}{\phi(z)}} + \frac{(2z-R_c)\varepsilon_z}{2(R_c+2z)\varphi(z)}\right] \tag{67}$$

$$w'(z,\sqrt{\varepsilon_z}) = \frac{1}{R_c} - \frac{\sqrt{\varepsilon_z}}{R_c\sqrt{\phi(z)}}\left[2 - \frac{z}{\phi(z)}\phi'(z)\right] + \frac{2\varepsilon_z}{R_c\phi(z)}\left[1 - \frac{z}{\phi(z)}\phi'(z)\right] \tag{68}$$

Now we derive the paraxial chromatic aberration $\Delta r_v^*(z_i)$ and paraxial chromatic aberration of magnification $\Delta r_w^*(z_i)$ at the image plane $z = z_i$. $\Delta v(z_i)$ and $\Delta w(z_i)$ have been defined by Eq. (21) and Eq. (22). From Eq. (55), Eq. (56), Eq. (67) and Eq. (68), we can get the values of special solutions v, w, $v'w'$ at $z = z_a = (R_a - R_c)$. Substituting all of these expressions into Eq. (23) and Eq. (24), we may obtain

$$\Delta r_w^*(z_i) = r_0 \Delta w(z_i) = r_0\left\{-\frac{2(n-1)}{(n-2)\phi_{ac}}[(\sqrt{\varepsilon_z} - \sqrt{\varepsilon_{z_1}})\sqrt{\varepsilon_{z_1}} - (\varepsilon_z - \varepsilon_{z_1})]\right\} \tag{69}$$

$$\Delta r_v^*(z_i) = \sqrt{\frac{m_0}{2e}} \dot{r}_0 \Delta v(z_i) = \sqrt{\frac{m_0}{2e}} \dot{r}_0 \left\{ \frac{2n(R_a - R_c)}{(n-2)\phi_{ac}} \right\} (\sqrt{\varepsilon_z} - \sqrt{\varepsilon_{z_1}}) - \frac{1}{\sqrt{\phi_{ac}}} (\varepsilon_z - \varepsilon_{z_1}) + \frac{2(n-1)}{(n-2)\sqrt{\phi_{ac}}} (\sqrt{\varepsilon_z} - \sqrt{\varepsilon_{z_1}}) \sqrt{\varepsilon_{z_1}} \right\} \quad (70)$$

Introducing E_c and M into Eq. (69) and Eq. (70), we finally obtain

$$\Delta r_w^*(z_i) = r_0 \Delta w(z_i) = r_0 \left[-\frac{2(M-1)}{\phi_{ac}} (\sqrt{\varepsilon_z} - \sqrt{\varepsilon_{z_1}}) \sqrt{\varepsilon_z} \right] \quad (71)$$

$$\Delta r_v^*(z_i) = \sqrt{\frac{m_0}{2e}} \dot{r}_0 \Delta v(z_i) = \sqrt{\frac{m_0}{2e}} \dot{r}_0 \frac{2M}{E_c} \left\{ (\sqrt{\varepsilon_z} - \sqrt{\varepsilon_{z_1}}) - \frac{1}{\sqrt{\phi_{ac}}} (\varepsilon_z - \varepsilon_{z_1}) \right\} \quad (72)$$

which are fully identical with Eq. (63) and Eq. (64). It completely proves that we use the method of asymptotic solutions of paraxial ray equation of imaging electron optics given by Monastyrski to solve the problem of paraxial lateral aberration is correct.

6 Discussions

Let's discuss the first item in Eq. (72). This is the paraxial chromatic aberration of second order, which is the only one of aberration of second order and occupies a greatest part in whole paraxial lateral aberration, and it exits everywhere at the whole image plane including the axial point. Usually, expression of paraxial chromatic aberration of second order is called as R-A formula, which has wide application in designing image tubes for evaluating image quality such as image resolution. The great advantage of this formula is that it does not depend on system structure and axial potential distribution, but only depends on initial radial potential and initial axial potential of electrons emitted from the photocathode, the strength of electric-field at the photocathode and the linear magnification of system.

It should be pointed out, although the derivation of Eq. (72) is based on a bi-electrode electrostatic concentric spherical system model, it still has universal significance for practical use. Let's explain it in detail. For an electrostatic electron optical system, whatever its object surface is planar or spherical, we can imagine that an equipotential surface at $z = z_m$ which is similar to mesh anode located in front of the object surface, a virtual bi-electrode electrostatic concentric spherical system or a virtual system of uniform field will be constructed When electrons emit from the object surface with initial potential ($\sqrt{\varepsilon_z}$, $\sqrt{\varepsilon_r}$) and pass through the virtual system, it will form a paraxial chromatic aberration at a virtual image plane $z = -z_{i*}$, which is behind the object surface, and which is determined by $\sqrt{\varepsilon_{z_1}}$.

Suppose that a short lens is connected in front of the above-mentioned virtual system, the potential distribution of which is composed of the field

$$\phi(z(-z_i, z_m)) = \phi_m, \quad \phi''(z(z_m, z_a)) \neq 0, \quad \phi(z(z_a, z_i)) = \phi_i$$

We assume that an virtual object having height $\Delta r_v^*(z = -z_{i*})$

$$\Delta r_v^*(z = -z_{i*}) = \sqrt{\frac{m_0}{2e}} \dot{r}_0 \frac{2M_1}{E_c} (\sqrt{\varepsilon_z} - \sqrt{\varepsilon_{z_1}}) \quad (73)$$

where M_1 is the linear magnification of this virtual system, E_c is the field intensity at photocathode surface.

We assume that an virtual object having height Δr_v^* ($z = -z_{i*}$) is placed at the position $z = -z_{i*}$. Seeing that the short lens can be regarded as an ideal electron lens, that does not produce additional aberration. Therefore, when the virtual object having height Δr_v^* ($z = -z_{i*}$) goes through the short lens, a magnified or reduced image with linear magnification M_2 will be formed at the image position $z = z$; Then we have

$$\Delta r_v^* (z = z_i) = \sqrt{\frac{m_0}{2e}} \dot{r}_0 \frac{2M_1 M_2}{E_c} (\sqrt{\varepsilon_z} - \sqrt{\varepsilon_{z_1}}) \tag{74}$$

where M_2 is a linear magnification formed by the short lens, $M_1 M_2 = M$. Thus the first item in Eq. (72) still holds, but M in Eq. (72) should be understood as an overall linear magnification of whole system.

Now we sum up the abovementioned description. Though the paraxial lateral aberrations can be solved by obtaining the coefficients of asymptotic solutions of paraxial ray equation, but it seems too complicated for solving these coefficients. It shows in this research that the formula of paraxial chromatic aberration of second order derived by a bi-electrode electrostatic concentric spherical system, which has a very simple and clear form and which possess a greatest part in whole paraxial lateral aberrations, is universal applicable for practical use. Thus we fully prove that the R-A formula of paraxial chromatic aberration of second order in a concrete system holds. Besides, we firstly have obtained concrete expression of paraxial chromatic aberrations of third order, as well as the paraxial chromatic aberration of magnification of third order which are lesser one order of magnitude compared to the paraxial chromatic aberrations of second order.

In general case, in practical calculation and design for image tubes, the paraxial chromatic aberration of second order should only be considered, since it is good enough to evaluate the image quality of the system, it is unnecessary to consider the paraxial chromatic aberrations of third order and paraxial chromatic aberrations of magnification of third order. But when the geometrical lateral aberration of third order is studied, it should consider not only the paraxial chromatic aberration of second order, but also the paraxial chromatic aberrations of third order and paraxial chromatic aberrations of magnification of third order, because they are same order of magnitude with the geometrical lateral aberration of third order. In this case, we would like to recommend a formula of the paraxial lateral aberration expressed in a simple and clear form as following:

$$\Delta r^* (z_i) = \sqrt{\frac{m_0}{2e}} \dot{r}_0 \left\{ \frac{2M}{E_c} (\sqrt{\varepsilon_z} - \sqrt{\varepsilon_{z_1}}) + \frac{2M}{E_c \sqrt{\phi(z_i)}} (\varepsilon_z - \varepsilon_{z_1}) \right\} + r_0 \frac{-2(M-1)}{\phi(z_i)} (\sqrt{\varepsilon_z} - \sqrt{\varepsilon_{z_1}}) \sqrt{\varepsilon_z} \tag{75}$$

It should be pointed that the first item in expression (75) is an accurate formula which is good enough for practical use, the second and third items are approximate formulae, but they possess only a very little part in whole paraxial lateral aberrations. Expression (75) is very convenient for practical use, since it does not depend on the system structure and the axial potential distribution. We believe that the more accurate expressions for paraxial lateral aberrations of the third order for practical use will be found out.

Summary

In the present paper, we have studied an approach based on the asymptotic solutions of paraxial equation for solving paraxial lateral aberrations of imaging electron optics. The paraxial lateral aberrations expressed in general form have been derived emphatically. The relationship between the coefficients of asymptotic solutions has been investigated, which proves that the coefficients of asymptotic solutions are related each other. Through a bi-electrode electrostatic spherical concentric system model, the coefficients of asymptotic solutions of paraxial equation have been solved in this paper, the two special solutions expressed by asymptotic solutions and accurate solutions in this system model have been deduced, and the paraxial lateral aberrations have been verified and tested. Result completely proves that the approach based on asymptotic solutions to solve the paraxial lateral aberrations is correct and practicable. At the same time, we fully prove that the well-known R-A formula of paraxial chromatic aberration of second order in a concrete system holds. Besides, we firstly have obtained concrete expressions of paraxial chromatic aberration of third order, as well as the paraxial chromatic aberration of magnification of third order. Although these aberrations are lesser one order of magnitude compared to the paraxial chromatic aberrations of second order, but it should not be ignored when the geometrical lateral aberration of third order is studied. Finally, a simple and clear form for expressing paraxial lateral aberrations of imaging electron optics which does not depend on the system structure and axial potential distribution is suggested for practical use. Results of the present paper will have theoretical significance for the aberration theory of electron optics and practical significance for the design of image tubes.

Acknowledgements

This project was supported by National Natural Science Foundation of China (No: 60771070) and the Special Science Fund of Doctoral Discipline for Institutions of Higher Learning (No: B117).

References

[1] ELIN B P, KATISHAOV B A, MONASTYRSKI M A, et al. Numerical methods of optimization for emission electron-optical systems [M]. Novosibilsk: Science Press, 1987.

[2] MONASTYRSKI M A. On asymptotic solutions of paraxial equation of electron optics [J]. Journal of Technical Physics, 1978, 48 (6): 1117 – 1122.

[3] ZHOU L W. On theory of paraxial lateral aberrations in imaging electrostatic electron optics based on asymptotic solutions [C]// International Symposium on Photoelectronic Detection

and Imaging 2007: Photoelectronic Imaging and Detection SPIE, 2008, 6621 662101 -1 -12.

[4] BANSHITET B A. On computation of aberrations in cathode lenses [J]. Radiotechnics and Electronics 1964, 9 (5) 844 -850.

[5] KELMAN V M. Theory of cathode lenses, II: Electrostatic lenses with axial symmetrical fields [J]. Journal of Technical Physics 1973, 43 (1): 52 -56.

[6] MONASTYRSKI M A, KYLIKOV Y V. Theory of aberrations of third order in cathode lenses, chromatic aberrations. Radiotechnics and Electronics 1978, 23 (3): 644 -647.

[7] ZHOU L W, AI K C, PAN S. C. On theory of geometrical lateral aberrations for the combined electromagnetic focusing cathode lenses [J]. Acta Physica Sinica 1983, 32 (3): 376 -391 (in Chinese).

[8] XIMEN J Y. ZHOU L W, AI K C. Variation theory of aberrations in cathode lenses [J]. Optik. 1983, 66 (1): 19 -34.

[9] ZHOU L W. Electron optics with wide electron beam focusing [M]. Beijing: Beijing Institute of Technology Press, 1983.

[10] ZHOU L W. Electron optics of concentric spherical electromagnetic focusing systems [J]. Advances in Electronics and Electron Physics, 1979, 52: 119 -132.

[11] RECKNAGEL A. Theorie des elektrisohen elektronen miktroskops fur selbstrakler [J]. Z. Angew. Physik, 1941, 119: 689 -708.

[12] ARTIMOVICH L A. Electrostatic properties of emission systems [J]. Information of Academy of Sciences, USSR, Series of Physics, 1944, 8 (6): 313 -328.

6.5 Theory of Paraxial Lateral Aberrations in Imaging Electrostatic Electron Optics Based on Asymptotic Solutions
渐近解求解静电成像电子光学近轴横向像差的理论

Abstract: In the present paper, the paraxial lateral aberrations of the second order and the third order expressed in a general form have been derived emphatically by using asymptotic solutions of paraxial equation of electron optics. The relationship between the coefficients of asymptotic solutions of paraxial equation has been investigated, which proves that the coefficients of asymptotic solutions are related to each other. It has proven that the paraxial lateral chromatic aberration of the second order possesses the greatest part in the whole paraxial lateral aberrations, and the paraxial chromatic aberration of magnification is only related to infinitesimal quantities of the third order. Results of the present paper will have theoretical significance for studying theory of imaging electron optics.

摘要：本文重点研究利用成像电子光学的渐近解导出的二级近轴横向像差（即二级近轴色差）和三级几何横向像差，探讨了近轴方程渐近解系数之间的关系，证明了在近轴横向像差中二级近轴横色差居主要地位，而近轴放大率色差居次要地位。本文基于近轴方程的渐近解研究近轴横向像差的一般理论，其结果为研究简明形式的近轴横向像差提供了理论基础。

Introduction

In imaging electron optics, study of geometrical lateral aberrations of the third order and paraxial lateral aberrations of the second order has traditionally been paid more attention to, but the general form of paraxial lateral aberration has not been studied theoretically. The paraxial chromatic aberration of the second order was given in a simple and clear form which was summarized as a well-known R-A formula. but the existence of paraxial lateral aberration of magnification and the paraxial chromatic aberration of the third order was often ignored which is the same order of magnitude compared to geometrical lateral aberrations of the third order.

As is known and recognized, the paraxial chromatic aberration of the second order which determines the limitation of spatial resolution, occupies the greatest part in the whole paraxial lateral aberrations. But in studying paraxial lateral aberrations, though the solution of electron trajectory with accuracy to the infinitesimal of the second order and the asymptotic solutions of paraxial electron optics have been given[1-2], but the expressions of the whole paraxial lateral

---

Zhou Liwei (周立伟). Beijing Institute of Technology. Optik, V. 122, 2011, 300 – 306.

aberrations have not been expressed in a general form. The first aim of the present paper is to find the paraxial lateral aberrations of the second and the third orders expressed in general form based on the asymptotic solutions of paraxial equation of electron optics given by M. A. Monastyrski[2]. The second aim is to find the relationship between the coefficients of asymptotic solutions of paraxial equation, which proves that the coefficients of asymptotic solutions are related to each other.

1 On spatial aberration theory of electrostatic electron optical imaging systems

According to our study of aberration theory for electron optical imaging systems and study of electron optics for a bi-electrode concentric spherical system of electrostatic focusing, the lateral aberrations of electron optical imaging systems are expressed by spatial difference between practical imaging height and ideal imaging height at the ideal image plane determined by paraxial electron trajectory emitted from photocathode with axial initial energy $\sqrt{\varepsilon_{z_1}}$, which can be defined as[3]

$$\Delta \boldsymbol{r} = \Delta \boldsymbol{r}(z_i) = \boldsymbol{r}_{\text{real}}\left(z_i, \varepsilon_z^{\frac{1}{2}}, \varepsilon_r^{\frac{1}{2}}, r_0\right) - \boldsymbol{r}_{\text{para}}\left(z_i, \varepsilon_{z_1}^{\frac{1}{2}}, \varepsilon_r^{\frac{1}{2}}, r_0\right) \tag{1}$$

where $\boldsymbol{r}_{\text{real}}$, $\boldsymbol{r}_{\text{para}}$ express the practical imaging height and the ideal imaging height at an ideal image plane $z = z_i$, formed by the practical rays and the paraxial rays of electrons emitted from the photocathode go through the system, respectively. For the cylindrical coordinate system (z, r), the axial coordinate z is chosen from the photocathode $z_0 = 0$, r_0 is the initial radial vector of electron, z_i is the distance from the photocathode to the ideal imaging plane. $\sqrt{\varepsilon_z}$, $\sqrt{\varepsilon_r}$ are the initial axial energy and initial radial energy of electrons emitted from the photocathode; $\sqrt{\varepsilon_{z_1}}$ is a given initial energy of electron emission along the axial direction, which corresponds the ideal image position z_i.

According to theory of paraxial electron optics, all of paraxial electron trajectories emitted from photocathode with the same $\sqrt{\varepsilon_{z_1}}$ and initial height r_0 will be focused at the same image position $z = z_i(\varepsilon_{z_1})$, $r = r_*(z_i)$, which does not depend on $\sqrt{\varepsilon_r}$. The position $z = z_i$, $r = r_*(z_i)$ is called the ideal image position.

Substituting r, r_* for r_{real}, r_{para}, respectively, then Eq. (1) can be expressed as

$$\Delta \boldsymbol{r} = \Delta \boldsymbol{r}(z_i) = \boldsymbol{r}_{\text{real}}\left(z_i, \varepsilon_z^{\frac{1}{2}}, \varepsilon_r^{\frac{1}{2}}, r_0\right) - \boldsymbol{r}_*\left(z_i, \varepsilon_{z_1}^{\frac{1}{2}}, \varepsilon_r^{\frac{1}{2}}, r_0\right) = \Delta \boldsymbol{r}_* + \delta \boldsymbol{r} \tag{2}$$

We may express Eq. (2) as following:

$$\Delta \boldsymbol{r}_* = \boldsymbol{r}_*\left(z_i, \varepsilon_z^{\frac{1}{2}}, \varepsilon_r^{\frac{1}{2}}, r_0\right) - \boldsymbol{r}_*\left(z_i, \varepsilon_{z_1}^{\frac{1}{2}}, r_0\right) \tag{3}$$

$$\delta \boldsymbol{r} = \boldsymbol{r}\left(z_i, \varepsilon_z^{\frac{1}{2}}, \varepsilon_r^{\frac{1}{2}}, r_0\right) - \boldsymbol{r}_*\left(z_i, \varepsilon_z^{\frac{1}{2}}, \varepsilon_r^{\frac{1}{2}}, r_0\right) \tag{4}$$

where $\Delta \boldsymbol{r}_*$ in Eq. (3) is called paraxial lateral aberration, which is the difference between the image height and the ideal image height formed by two paraxial rays emitted from photocathode with the same height r_0, but at image plane z_i determined by $\sqrt{\varepsilon_{z_1}}$, δr in formula (4) is called geometrical lateral aberration, which is the difference between the practical image height and paraxial image height with the same initial energy ($\sqrt{\varepsilon_z}$, $\sqrt{\varepsilon_r}$) and the same height r_0 and $\sqrt{\varepsilon_z} = \sqrt{\varepsilon_{z_1}}$, as shown in Fig. 1.

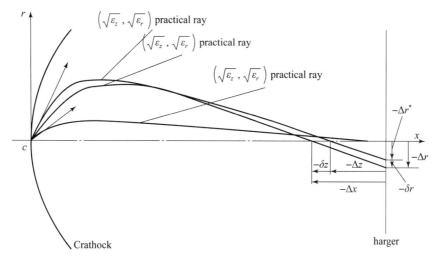

Fig. 1 The lateral aberrations Δr is the sum of paraxial lateral aberration Δr_* and geometrical lateral aberration δr

The paraxial trajectory $r_*(z)$ is the solution of following paraxial ray equation:

$$r''(z) + \frac{1}{2}\frac{\phi'(z)}{\phi(z)+\varepsilon_z}r'(z) + \frac{1}{4}\frac{\phi''(z)}{\phi(z)+\varepsilon_z}r(z) = 0 \tag{5}$$

where $\phi(z)$ is axial potential distribution, $' = \frac{d}{dz}$ expresses the differential with respect to z

The practical trajectory $r(z)$ is the solution of following practical ray equation:

$$r''(z) = \frac{1+r'^2}{2\varphi_*}\left(\frac{\partial\varphi}{\partial r} - r'\frac{\partial\varphi}{\partial z}\right) \tag{6}$$

where $\varphi_* = \varphi(z,r) + \varepsilon_0$, ε_0—initial energy of electron emitted from photocathode, $\varepsilon_0 = \varepsilon_r + \varepsilon_z$, $\varphi = \varphi(z,r)$—spatial potential distribution which can be expanded by Scherzer series expansions as following:

$$\varphi(z,r) = \phi(z) - \frac{1}{4}\phi''(z)r^2 + \frac{1}{64}\phi^{IV}(z)r^4 - \cdots \tag{7}$$

The general solution of paraxial trajectory equation (5) can be written as

$$r_*(z) = r_0 w(z) + \sqrt{\frac{m_0}{2e}}\dot{r}_0 v(z) \tag{8}$$

where m_0/e is the ratio of electron mass to charge.

Two special solutions $v = v(z)$, $w = w(z)$ will satisfy the following initial conditions:

$$v(z=0) = 0, \ v'(z=0) = \frac{1}{\sqrt{\varepsilon_z}},$$
$$w(z=0) = 1, \ w'(z=0) = 0 \tag{9}$$

Suppose that all of electrons emitted from photocathode at initial axial point with same axial initial energy will be focused at the ideal image plane $z = z_i$, i. e. ,

$$v(z_i, \sqrt{\varepsilon_{z_1}}) = 0, \quad w(z_i, \sqrt{\varepsilon_{z_1}}) = M \tag{10}$$

where M is the magnification, and it is quite evident that $v(z_i, \sqrt{\varepsilon_z}) \neq 0$, $w(z_i, \sqrt{\varepsilon_z}) \neq M$.

From formula (1-3), we can get the paraxial lateral aberration at the ideal image plane z_i:

$$\Delta \boldsymbol{r}_*(z_i) = \{w(z_i, \sqrt{\varepsilon_z}) - w(z_i, \sqrt{\varepsilon_{z_1}})\} \boldsymbol{r}_0 + \sqrt{\frac{m_0}{2e}} \dot{\boldsymbol{r}}_0 \{v(z_i,) - v(z_i, \sqrt{\varepsilon_{z_1}})\}$$

$$= \{w(z_i, \sqrt{\varepsilon_z}) - M\} \boldsymbol{r}_0 + \sqrt{\frac{m_0}{2e}} \dot{\boldsymbol{r}}_0 v(z_i, \sqrt{\varepsilon_z}) \tag{11}$$

Formula (11) is our starting point for studying the paraxial lateral aberrations. Regarding to geometrical lateral aberration of third order, expanding each term in practical trajectory equation (6) and substituting them into formula (4), then by using formula (8), we can get the equation of geometrical lateral aberration of the third order δr:

$$[\phi(z) + \varepsilon_z]\delta r'' + \frac{1}{2}\phi'(z)\delta r' + \frac{1}{4}\phi''(z)\delta r = F(\boldsymbol{r}_*, \boldsymbol{r}'_*, z) \tag{12}$$

Eq. (12) is a non-homogeneous equation of the second order, where $F(\boldsymbol{r}_*, \boldsymbol{r}'_*, z)$ is an item at the right hand. Solution of the equation of geometrical lateral aberration of the third order δr was studied in a lot of literatures and monographs[4-8]. We shall not describe it in the following.

2 On paraxial lateral aberrations solved by asymptotic solutions of paraxial equation

A method of solving the special solutions of $v(z, \sqrt{\varepsilon_z})$, $w(z, \sqrt{\varepsilon_z})$ of paraxial equation by using asymptotic solutions has been given in Ref. [2], which can be described as following:

$$v(z, \sqrt{\varepsilon_z}) = \sqrt{\phi(z)}\xi_0(z) + \sqrt{\varepsilon_z}\eta_0(z) + \varepsilon_z\left[\frac{\xi_0(z)}{2\sqrt{\phi(z)}} + \xi_1(z)\sqrt{\phi(z)}\right] + \cdots \tag{13}$$

$$w(z, \sqrt{\varepsilon_z}) = \omega_0(z) + \sqrt{\varepsilon_z}\sqrt{\phi(z)}\zeta_0(z) + \varepsilon_z\omega_1(z) + \cdots \tag{14}$$

where coefficients ξ_0, η_0, ω_0, ζ_0, ξ_1, ω_1 in formulae (13), (14) are the solutions of the following differential equations:

$$L_0\begin{pmatrix}\xi_K \\ \zeta_K\end{pmatrix} = -\begin{pmatrix}\xi''_{K-1} \\ \zeta''_{K-i}\end{pmatrix}, \quad M_0\begin{pmatrix}\eta_K \\ \omega_K\end{pmatrix} = -\begin{pmatrix}\eta''_{K-1} \\ \omega''_{K-i}\end{pmatrix} \tag{15a, 15b}$$

where

$$\xi_{K-1} = \zeta_{K-1} = \eta_{K-1} = \omega_{k-1} = 0$$

L_0, M_0 are linear differential operators, which can be expressed by

$$L_0 = \phi(z)\frac{\mathrm{d}^2}{\mathrm{d}z^2} + \frac{3}{2}\phi'(z)\frac{\mathrm{d}}{\mathrm{d}z} + \frac{3}{4}\phi''(z) \tag{16}$$

$$M_0 = \phi(z)\frac{\mathrm{d}^2}{\mathrm{d}z^2} + \frac{1}{2}\phi'(z)\frac{\mathrm{d}}{\mathrm{d}z} + \frac{1}{4}\phi''(z) \tag{17}$$

The initial condition of Eq. (15a), Eq. (15b) can be determined according to the following procedure:

$$\frac{\xi_0(0)\phi'(0)}{2} = 1 \rightarrow \xi_0(0) + \eta_0(0) = 0 \tag{18}$$

$$\downarrow$$

$$\frac{\xi_1(0)\phi'(0)}{2} + \xi'_0(0) + \eta'_0(0) = 0 \rightarrow \text{and so on}$$

$$\omega_0 = 1;\ \zeta_0(0) + \omega_1(0) = 0 \tag{19}$$

$$\downarrow \quad \uparrow \quad \downarrow$$

$$\frac{\zeta_0(0) + \phi'(0)}{2} + \omega'_0(0) = 0 \rightarrow \text{and so on}$$

where the arrows indicate the sequence of calculation.

From asymptotic solution (13), solution (14) and the definition of (11), we may obtain the paraxial lateral aberrations $\Delta r_*(z_i)$ at the ideal image plane $z = z_i$, if we can get the value of coefficients ξ_0, η_0, ω_0, ζ_0, ξ_1, ω_1 at the ideal image plane $z = z_i$ (i.e., at the screen):

$$\Delta r_*(z_i) = \{ \sqrt{\phi(z_i)}\zeta_0(z_i)(\sqrt{\varepsilon_z} - \sqrt{\varepsilon_{z_1}}) + \omega_1(z_i)(\varepsilon_z - \varepsilon_{z_1})\}r_0 +$$

$$\sqrt{\frac{m_0}{2e}}\dot{r}_0 \left\{ \eta_0(z_i)(\sqrt{\varepsilon_z} - \sqrt{\varepsilon_{z_1}}) + \left[\frac{\xi_0(z_i)}{2\sqrt{\phi(z_i)}} + \xi_1(z_i)\sqrt{\phi(z_i)}\right](\varepsilon_z - \varepsilon_{z_1})\right\} \tag{20}$$

The positive or negative value of above-mentioned $\Delta r_*(z_i)$ depend on the $r_*(z_i, \sqrt{\varepsilon_{z_1}})$ of paraxial ray, which take as a criterion, from bottom to top, it will take positive, otherwise it takes negative. In formula (20), the first item expresses the paraxial chromatic aberration of magnification, and the second item expresses the paraxial chromatic aberration.

3 On aberration coefficients of paraxial lateral aberrations solved by asymptotic solutions of paraxial equation

According to the above-mentioned procedure, we should carry on some operations for solving differential equations to obtain the value of coefficients ξ_0, η_0, ω_0, ζ_0, ξ_1, ω_1 at the ideal image plane $z = z_i$, it is a complicated process and it is not easy to understand the image quality characteristics of the system, so the problem of paraxial lateral aberrations $\Delta r_*(z_i)$ given by formula (20) is still not solved.

As is known, for the electron optical imaging systems, the axial potential distribution $\phi(z)$ at the anode-screen section behind a certain point $z = z_a$ usually is a constant, the trajectory moving in this section is a straight line. Thus, we could obtain the solutions of the coefficients ξ_0, η_0, ω_0, ζ_0, ξ_1, ω_1 at the position $z = z_a$, and to find the paraxial lateral aberrations $\Delta r_*(z_i)$ at image plane $z = z_i$.

According to the definition (11), the paraxial lateral aberrations $\Delta r_*(z_i)$ can be regard as the sum of the paraxial chromatic aberration $\Delta r_* v(z_i)$ and the paraxial chromatic aberration of magnification $\Delta r_* w(z_i)$.

$$\Delta r_*(z_i) = \Delta r_* v(z_i) + \Delta r_* w(z_i) = \sqrt{\frac{m_0}{2e}} \dot{r}_0 \Delta v(z_i) + r_0 \Delta w(z_i) \tag{21}$$

where
$$\Delta v(z_i) = v(z_i, \sqrt{\varepsilon_z}) , \quad \Delta w(z_i) = w(z_i, \sqrt{\varepsilon_z}) - M \tag{22}$$

Suppose at $z \leqslant z_0$, $\phi(z) \neq \text{const}$; at $z > z_a$, $\phi(z) = \text{const} = \phi(z_a) = \phi(z_i)$, we can get special solutions at $z = z_a$ from expressions (13) and (14): $v(z_a, \sqrt{\varepsilon_z})$, $v'(z_a, \sqrt{\varepsilon_z})$ and $w(z_a, \sqrt{\varepsilon_z})$, $w'(z_a, \sqrt{\varepsilon_z})$ which can be expressed as following:

$$v(z_a, \sqrt{\varepsilon_z}) = a + b\sqrt{\varepsilon_z} + c\varepsilon_z \tag{23}$$

$$v'(z_a, \sqrt{\varepsilon_z}) = a' + b'\sqrt{\varepsilon_z} + c'\varepsilon_z \tag{24}$$

where
$$a = \sqrt{\phi(z_a)} \xi_0(z_a) , \quad b = \eta_0(z_a) , \quad c = \frac{\xi_0(z_a)}{2\sqrt{\phi(z_a)}} + \xi_1(z_a) \sqrt{\phi(z_a)} ,$$

$$a' = \sqrt{\phi(z_a)} \xi_0'(z_a) + \frac{1}{2\sqrt{\phi(z_a)}} \phi'(z_a) \xi_0(z_a) , \quad b' = \eta_0'(z_a) ,$$

$$c' = \frac{\xi_0'(z_a)}{2\sqrt{\phi(z_a)}} - \frac{\phi'(z_a) \xi_0(z_a)}{4[\phi(z_a)]^{\frac{3}{2}}} + \xi_1'(z_a)\sqrt{\phi(z_a)} + \frac{\phi'(z_a)\xi_1(z_a)}{2\sqrt{\phi(z_a)}} \tag{25}$$

and
$$w(z_a, \sqrt{\varepsilon_z}) = d + e\sqrt{\varepsilon_z} + f\varepsilon_z \tag{26}$$

$$w'(z_a, \sqrt{\varepsilon_z}) = d' + e'\sqrt{\varepsilon_z} + f'\varepsilon_z \tag{27}$$

where
$$d = \omega_0(z_a) , \quad e = \sqrt{\phi(z_a)} \zeta_0(z_a) , \quad f = \omega_1(z_a)$$

$$d' = \omega_0'(z_a) , \quad e' = [\sqrt{\phi(z_a)} \zeta_0'(z_a) + \frac{1}{2\sqrt{\phi(z_a)}} \phi'(z_a) \zeta_0(z_a)] , \quad f' = \omega_1'(z_a) \tag{28}$$

Firstly, let's find the solution of Δw. From Eq. (10), at the ideal image position $z = z_i$, $v(z_i, \sqrt{\varepsilon_{z_1}}) = 0$

$$w(z_i, \sqrt{\varepsilon_{z_1}}) = w(z_a, \sqrt{\varepsilon_{z_1}}) - \frac{v(z_a, \sqrt{\varepsilon_{z_1}})}{v'(z_a, \sqrt{\varepsilon_{z_1}})} w'(z_a, \sqrt{\varepsilon_{z_1}}) \tag{29}$$

Similarly, we also can get

$$w(z_i, \sqrt{\varepsilon_z}) = w(z_a, \sqrt{\varepsilon_z}) - \frac{v(z_a, \sqrt{\varepsilon_z})}{v'(z_a, \sqrt{\varepsilon_z})} w'(z_a, \sqrt{\varepsilon_z}) \tag{30}$$

Substituting all of the expressions (23), (24) and (26), (27) into (29), (30), through a series of transformation, we may obtain

$$w(z_i, \sqrt{\varepsilon_z}) = \frac{1}{a'}[a'd - ad'] + \frac{1}{a'}[a'e - ae']\sqrt{\varepsilon_z} - \frac{1}{a'}\left[\frac{d'}{a'}(a'b - ab')\right]\sqrt{\varepsilon_{z_1}} + \frac{1}{a'}[a'f - af']\varepsilon_z -$$
$$\frac{1}{a'}\left[\frac{e'}{a'}(a'b - ab')\right]\sqrt{\varepsilon_z}\sqrt{\varepsilon_{z_1}} - \frac{1}{a'}\frac{d'}{a'}\left[(a'c - ac') - \frac{b'}{a'}(a'b - ab')\right]\varepsilon_{z_1} \tag{31}$$

and

$$w(z_i, \sqrt{\varepsilon_z}) = M = \frac{1}{a'}[a'd - ad'] + \frac{1}{a'}\left[(a'e - ae') - \frac{d'}{a'}(a'b - ab')\right]\sqrt{\varepsilon_{z_1}} +$$
$$\frac{1}{a'}\left\{a'f - af' - \frac{e'}{a'}(a'b - ab') - \frac{d'}{a'}\left[(a'c - ac') - \frac{b'}{a'}(a'b - ab')\right]\right\}\varepsilon_{z_1} \quad (32)$$

Therefore, from the definition (11), we obtain

$$\Delta w = w(z_i, \sqrt{\varepsilon_z}) - w(z_i, \sqrt{\varepsilon_{z_1}})$$
$$= \frac{1}{a'}\left\{(a'e - ae')(\sqrt{\varepsilon_z} - \sqrt{\varepsilon_{z_1}}) + (a'f - af')(\varepsilon_z - \varepsilon_{z_1}) - \frac{e'}{a'}(a'b - ab')(\sqrt{\varepsilon_z} - \sqrt{\varepsilon_{z_1}})\sqrt{\varepsilon_{z_1}}\right\}$$
$$(33)$$

Now we discuss the first item in expression (33), which can be expressed as following:

$$[a'e - ae'] = \phi(a_z)\{\xi_0'(z_a)\zeta_0(z_a) - \xi_0(z_a)\zeta_0'(z_a)\} \quad (34)$$

From Eq. (15a), we know that the differential equation for solving $\xi_0(z)$, $\zeta_0(z)$ is the same equation (16), and their initial conditions are identical, so the solutions of $\xi_0(z)$ and $\zeta_0(z)$ to be obtained will have only a difference of constant. We may have

$$\frac{\zeta_0'(z_a)}{\zeta_0(z_a)} = \frac{\xi_0'(z_a)}{\xi_0(z_a)\zeta_0'} \quad (35)$$

Therefore,

$$a'e - ae' = 0 \quad (36)$$

The magnification M of expression (32) and $\Delta w(z_i)$ of expression (33) may be written as

$$M = w(z_i, \sqrt{\varepsilon_z}) = \frac{1}{a'}[a'd - ad'] - \frac{1}{a'}\left[\frac{d'}{a'}(a'b - ab')\right]\sqrt{\varepsilon_{z_1}} +$$
$$\frac{1}{a'}\left[a'f - af' - \frac{1}{a'^2}(a'e' - b'd')(a'b - ab') - \frac{d'}{a'}(a'c - ac')\right]\varepsilon_{z_1} \quad (37)$$

$$\Delta w(z_i) = \frac{1}{a'}\left\{(a'f - af')(\varepsilon_z - \varepsilon_{z_1}) - \frac{e'}{a'}(a'b - ab')(\sqrt{\varepsilon_z} - \sqrt{\varepsilon_{z_1}})\sqrt{\varepsilon_{z_1}}\right\} \quad (38)$$

In virtue of that the paraxial lateral aberration is related to the magnification M, introducing (37) into (38), we obtain

$$\Delta w(z_i) = M\left\{\frac{e'}{a'}\frac{(a'b - ab')}{(a'd - ad')}(\sqrt{\varepsilon_z} - \sqrt{\varepsilon_{z_1}})\sqrt{\varepsilon_{z_1}} - \frac{a'f - af'}{a'd - ad'}(\varepsilon_z - \varepsilon_{z_1})\right\} \quad (39)$$

Actually, in the process of derivation for formula (39), in the expression of magnification M, it only needs to have an accuracy to the approximation of the zero the order of $\sqrt{\varepsilon_{z_1}}$ in the power series of Eq. (37).

Expression (39) of $\Delta w(z_i)$ can be further simplified. In imaging electron optics, there are two different approaches to obtain the value of magnification M. The first approach is to find the value of special solution $w(z, \sqrt{\varepsilon_{z_1}})$ at $z = z_i$: $M = w(z_i, \sqrt{\varepsilon_z})$, as we have done in formula (37). The second approach is to make use of the Lagrange-Helmholtz relationship as following:

$$v'(z_i, \sqrt{\varepsilon_z}) = \frac{1}{\sqrt{\phi(z_i) + \varepsilon_{z_1}}M} = \frac{1}{\sqrt{\phi(z_a) + \varepsilon_{z_1}}M} = v'(z_a, \sqrt{\varepsilon_z}) \quad (40)$$

Therefore, from (24), the magnification M has the following form:

$$M = \frac{1}{\sqrt{\phi(z_a) + \varepsilon_{z_1}} Mv'(z_a, \sqrt{\varepsilon_z})} = \frac{1}{\sqrt{\phi_{ac}} a'} \left[1 - \frac{b'}{a'} \sqrt{\varepsilon_{z_1}} - \left(\frac{c'}{a'} + \frac{1}{2\phi(z_a)} - \frac{b'^2}{a'^2} \right) \varepsilon_{z_1} \right] \quad (41)$$

Of course, the two approaches of M are equivalent, they should give identical results. Comparing Eq. (41) to Eq. (37), the coefficients of same order of $\sqrt{\varepsilon_{z_1}}$ in the power series of expressions should be identical, for the items of the zeroth and the first order of $\sqrt{\varepsilon_{z_1}}$, we may have

$$[a'd - ad'] = \frac{1}{\sqrt{\phi(z_a)}} \quad (42)$$

$$\frac{d'(a'b - ab')}{b'(a'd - ad')} = 1 \quad (43)$$

Furthermore, we found in formula (39) that expressions of b, b' and f, f' are related to $\eta_0(z_a)$, $\omega_1(z_a)$ and their derivative respectively, which are solved by the same differential equation (17), so the relationship of $\frac{b'}{b} = \frac{f'}{f}$ holds. Then, we may have

$$\frac{(a'b - ab')}{(a'f - af')} = \frac{b\left(a' - a\frac{b'}{b}\right)}{f\left(a' - \frac{f'}{f}\right)} = \frac{b}{f} = \frac{\eta_0(z_a)}{\omega_1(z_a)}$$

Therefore,

$$\frac{f}{b}(a'b - ab') = a'f - af' \quad (44)$$

Similarly, expression $\frac{e'}{a'}$ can be further simplified by expression (36) as following:

$$\frac{e'}{a'} = \frac{e}{a} = \frac{\zeta_0(z_a)}{\xi_0(z_a)} \quad (45)$$

The correctness of all of these formula, Eq. (42), Eq. (43), Eq. (44) and Eq. (45) will be verified and tested in the following paper.

Substituting Eq. (42) and Eq. (44) into expression (39), we may obtain

$$\Delta w(z_i) = -M \sqrt{\phi(z_a)}(a'b - ab') \left\{ \frac{e'}{a'}(\sqrt{\varepsilon_z} - \sqrt{\varepsilon_{z_1}}) \sqrt{\varepsilon_{z_1}} - \frac{f}{b}(\varepsilon_z - \varepsilon_{z_1}) \right\} \quad (46)$$

Therefore, the general form of paraxial chromatic aberration of magnification $\Delta \boldsymbol{r}_* w(z_i)$ can be written as

$$\Delta \boldsymbol{r}_* w(z_i) = \boldsymbol{r}_0 \Delta w(z_i) = -\boldsymbol{r}_0 M \left\{ \frac{1}{2} \phi'(z_a) \xi_0(z_a) \eta_0(z_a) + \phi(z_a) [\xi_0'(z_a) \eta_0(z_a) - \xi_0(z_a) \eta_0'(z_a)] \right\} \times \left\{ \frac{\zeta_0(z_a)}{\xi_0(z_a)} (\sqrt{\varepsilon_z} - \sqrt{\varepsilon_{z_1}}) \sqrt{\varepsilon_{z_1}} - \frac{w_1(z_a)}{\eta_0(z_a)} (\varepsilon_z - \varepsilon_{z_1}) \right\} \quad (47)$$

Now, let's find the solution of Δv. From definition (11), we have

$$\Delta v(z_i) = v(z_i, \sqrt{\varepsilon_z}) = v(z_a, \sqrt{\varepsilon_z}) - \frac{v'(z_a, \sqrt{\varepsilon_z})}{v'(z_a, \sqrt{\varepsilon_{z_1}})} v(z_a, \sqrt{\varepsilon_{z_1}}) \quad (48)$$

Substituting all of the expression (23), expression (24) into expression (48), through a series of transformation, we obtain

$$\Delta v(z_i) = \frac{1}{a'} \left\{ (a'b - ab')(\sqrt{\varepsilon_z} - \sqrt{\varepsilon_{z_1}}) + (a'c - ac') \right.$$
$$\left. (\varepsilon_z - \varepsilon_{z_1}) - \frac{b'}{a'}(a'b - ab')(\sqrt{\varepsilon_z} - \sqrt{\varepsilon_{z_1}})\sqrt{\varepsilon_{z_1}} \right\} \quad (49)$$

Introducing the magnification M of Eq. (41) into Eq. (49), we may obtain

$$\Delta v(z_i) = M\{\sqrt{\phi(z_a)}(a'b - ab')(\sqrt{\varepsilon_z} - \sqrt{\varepsilon_{z_1}}) + \sqrt{\phi(z_a)}(a'c - ac')(\varepsilon_z - \varepsilon_{z_1})\} \quad (50)$$

Actually, for derivation of formula (50), in magnification M, we only use the zero the and the first order of $\sqrt{\varepsilon_{z_1}}$ in the power series of (41).

Substituting all of expressions (25) of coefficients a, b, c into Eq. (50), through a series of transformation, we may obtain

$$\Delta v(z_i) = M \left\{ \begin{array}{l} \left[\dfrac{\phi'(z_a)}{2} \xi_0(z_a) \eta_0(z_a) + \phi(z_a)(\xi_0'(z_a)\eta_0(z_a) - \xi_0(z_a)\eta_0'(z_a)) \right] (\sqrt{\varepsilon_z} - \sqrt{\varepsilon_{z_1}}) + \\ \left[\dfrac{\phi'(z_a)}{2\sqrt{\phi(z_a)}} \xi_0^2(z_a) + (\phi(z_a))^{\frac{3}{2}} (\xi_0'(z_a)\xi_1(z_a) - \xi_0(z_a)\xi_1'(z_a)) \right] (\varepsilon_z - \varepsilon_{z_1}) \end{array} \right\}$$
(51)

The general form of paraxial chromatic aberration $\Delta r_* v(z_i)$ can be written as

$$\Delta r_* v(z_i) = \sqrt{\frac{m_0}{2e}} \dot{r}_0 \Delta v(z_i) = \sqrt{\frac{m_0}{2e}} \dot{r}_0 M$$

$$\left\{ \begin{array}{l} \left[\dfrac{\phi'(z_a)}{2} \xi_0(z_a) \eta_0(z_a) + \varphi(z_a)(\xi_0'(z_a)\eta_0(z_a) - \xi_0(z_a)\eta_0'(z_a)) \right] (\sqrt{\varepsilon_z} - \sqrt{\varepsilon_{z_1}}) + \\ \left[\dfrac{\phi'(z_a)}{2\sqrt{\phi(z_a)}} \xi_0^2(z_a) + (\phi(z_a))^{\frac{3}{2}} (\xi_0'(z_a)\xi_1(z_a) - \xi_0(z_a)\xi_1'(z_a)) \right] (\varepsilon_z - \varepsilon_{z_1}) \end{array} \right\} (52)$$

Finally, we obtain the general form of paraxial lateral aberrations Δr_* which is composed of the paraxial chromatic aberration $\Delta r_* v$ of the second and the third order and the paraxial chromatic aberration of magnification $\Delta r_* w$ of the third order, based on asymptotic solutions of paraxial equation, where coefficients ξ_0, η_0, ω_0, ζ_0, ω_1 and axial potential distribution $\phi(z)$ take the value at $z = z_a$.

We have mentioned before that the coefficients of ξ_0, η_0, w_0, ζ_0, ω_1 given by asymptotic solutions of paraxial equation, are not independent, they are connected with each other. Now, we study the relationship between ξ_1 and ξ_0, η_0, ω_0, ζ_0, ω_1. Comparing Eq. (41) to Eq. (37), the coefficients of the second order of $\sqrt{\varepsilon_{z_1}}$ in the power series of expressions should be identical, we may have

$$\frac{1}{a'}\left\{(a'f - af') - \frac{d'}{a'}(a'c - ac') - \frac{1}{a'^2}(a'e' - b'd')\right\} = \frac{1}{\sqrt{\phi_{ac}} a'}\left\{-\frac{c'}{a'} - \frac{1}{2\phi_{ac}}\left(\frac{b'}{a'}\right)^2\right\} \quad (53)$$

Expression (53) can be transformed to

$$\left(\frac{f}{b} - \frac{e'}{a'}\right)(a'b - ab') - \frac{d'}{a'}(a'c - ac') + \frac{1}{\sqrt{\phi_{ac}}}\frac{c'}{a'} + \frac{1}{2\phi_{ac}\sqrt{\phi_{ac}}} = 0 \quad (54)$$

Finally, we can get an equation in which $\xi_1(z_a)$ should satisfy

$$\{\omega_0'(z_a)\phi(z_a)\xi_0'(z_a)+1\}\xi_1'(z_a)+\left\{\frac{\phi'(z_a)}{2\phi(z_a)}-\omega_0'(z_a)\phi(z_a)\xi_0'(z_a)\right\}\xi_1(z_a)+$$

$$\left[\frac{\xi_0'(z_a)}{\phi(z_a)}-\frac{\omega_0'(z_a)}{2\phi(z_a)}\phi'(z_a)\xi_0^2(z_a)\right]+\left(\frac{\omega_1(z_a)}{\eta_0(z_a)}-\frac{\zeta_0(z_a)}{\xi_0(z_a)}\right)\left[\sqrt{\phi(z_a)}\xi_0'(z_a)+\frac{\phi'(z_a)}{2\sqrt{\phi(z_a)}}\xi_0(z_a)\right]\times$$

$$\left\{\frac{\phi'(z_a)}{2\sqrt{\phi(z_a)}}\xi_0(z_a)\eta_0(z_a)+\sqrt{\phi(z_a)}\left[\xi_0'(z_a)\eta_0(z_a)-\xi_0(z_a)\eta_0'(z_a)\right]\right\}=0$$

Formula (55) shows that all of coefficients of ξ_0, η_0, ω_0, ζ_0, ω_1, ξ_1 given by asymptotic solutions of paraxial equation are not isolated, they should be solved by Eq. (16) or Eq. (17), and they should also satisfy relationship Eq. (42), Eq. (43), Eq. (44), Eq. (45) and Eq. (55).

Conclusions

In the present paper, a clear and definite definition for the spatial aberrations of imaging electron optical systems has been given. The spatial aberrations are composed of paraxial lateral aberrations and geometrical lateral aberrations. The paraxial lateral aberrations are combined by paraxial sphero-chromatic aberration the of the second and the third orders and paraxial chromatic aberration of magnification of the third order. The paraxial chromatic aberration of magnification has been firstly deduced and which proves that it only relates to infinitesimal quantities of the third order. The expressions of paraxial lateral aberrations in a general form based on asymptotic solutions of paraxial equation of electron optics have been derived, the relationship between the coefficients of asymptotic solutions of paraxial equation has been deduced which was proven that the coefficients of asymptotic solutions are related to each other.

Acknowledgements

This project was supported by National Science Foundation of China (NSFC), Russian Foundation for Basic Research (RFBR) and the Special Science Fund of Doctoral Discipline for Institutions of Higher Learning.

References

[1] ELIN B P, KATISHAOV B A, KYLIKOV Y V, et al. Numerical methods of optimization for emission electron-optical systems [M]. Novosibilsk: Science Press, 1987.

[2] MONASTYRSKI M A. On asymptotic solutions of paraxial equation of electron optics [J]. Journal of Technical Physics, 1978, 48 (6): 1117-1122.

[3] ZHOU L W. Electron optics with wide beam focusing [M]. Beijing: Beijing Institute of Technology Press, 1993.

[4] BANSHITET B A. On computation of aberrations in cathode lenses [J]. Radiotechnics and Electronics 1964, 9 (5): 844-850 (in Russian).

[5] KELMAN V M, et al. Theory of cathode lenses II, Electrostatic lenses with axial symmetrical fields [J]. Journal of Technical Physics, 1973, 43 (1): 52-56.

[6] MONASTYRSKI M A KYLIKOV Y V. Theory of aberrations of third order in cathode lenses, chromatic aberrations [J]. Radiotechnics and Electronics, 1978, 23 (3): 644-647.

[7] ZHOU L W, AI K C, P S C. On aberration theory of cathode lenses with combined electromagnetic focusing [J]. Acta Physica Sinica, 1983, 32 (3): 376-392.

[8] XIMEN J Y, ZHOU L W, AI K C. Variation theory of aberrations in cathode lenses [J]. Optik, 1983, 66 (1): 19-34.

6.6 Test and Verification of Theory for Paraxial Lateral Aberrations by a Bi-electrode Electrostatic Concentric Spherical System Model
两电极静电同心球系统模型验证近轴横向像差理论的研究

Abstract: In the present paper, the paraxial lateral aberrations of the second and the third order, in which the aberration coefficients are solved by asymptotic solutions of the paraxial equation, have been verified and tested. The analytical expressions of asymptotic solutions in a bi-electrode electrostatic spherical concentric system model have been obtained. Result completely proves that the method and procedure given by M. A. Monastyrski to solve the paraxial equation of electron optics by asymptotic solutions are correct and practicable. The paraxial chromatic aberration of magnification has been firstly deduced and proved that it is only related to infinitesimal quantities of the third order. The R-A formula of paraxial chromatic aberration of the second order has been deduced and confirmed which possess the greatest part in the whole paraxial lateral aberrations. Results of the present paper will have theoretical and practical significance for the design of image tubes.

摘要：本文对像差系数通过近轴方程的渐近解求得的二级和三级近轴横向像差进行验证。渐近解的解析表示式是通过一个两电极同心球系统的模型求得的。结果完全证实 M. A. Monastyrski 提出用渐近解来求解电子的光学近轴方程的方法和步骤是正确和符合实际的。本文首次导出和证实了近轴放大率色差仅与非常小的三级小量相联系。文中推导了占全部近轴像差的绝大部分二级近轴色差的 R – A 公式。本文结果对于像管的设计具有理论和实际意义。

Introduction

In previous paper[1], the paraxial lateral aberrations of the second order and the third order expressed in a general form has been deduced, which is based on the asymptotic solutions of paraxial equation of electron optics suggested by M. A. Monastyrski[2,3]. But it still has two problems that should be solved. The first is how to test and verify the theory of paraxial lateral aberrations, to prove the method to be correct; the second is how to find a simple and clear form of paraxial lateral aberrations for the practical use. In this paper, a bi-electrode electrostatic

---

Zhou Liwei, Gong Hui. Beijing Institute of Technology. SPIE Proceedings 2007, 6621: 66212A – 1 – 13.

spherical concentric system model in which the potential distribution and the paraxial ray can be expressed by analytical solutions is applied to test and verify the correctness of given expressions of paraxial lateral aberrations. An expression of paraxial lateral aberrations written in a simple and clear form is suggested for the practical use.

1 Theory of paraxial lateral aberrations solved by asymptotic solutions of paraxial equation

As is well known, the paraxial ray equation in imaging electron optics can be expressed as following:

$$r''(z) + \frac{1}{2}\frac{\phi'(z)}{\phi(z)+\varepsilon_z}r'(z) + \frac{1}{4}\frac{\phi''(z)}{\phi(z)+\varepsilon_z}r(z) = 0 \tag{1}$$

where $r(z)$ is the paraxial ray, $\phi(z)$ is axial potential distribution, $\sqrt{\varepsilon_z}$ is the initial axial energy of electrons emitted from the photocathode; $' = \frac{d}{dz}$ expresses the differential with respect to z.

The general solution of paraxial ray equation (1) can be written as

$$r(z) = r_0 w(z) + \sqrt{\frac{m_0}{2e}}\dot{r}_0 v(z) \tag{2}$$

where $\frac{m_0}{e}$ is the ratio of electron mass to charge; r_0 and $\dot{r}_0$ are the initial radial vector and radial velocity of electron emitted from the cathode surface, respectively; $v = v(z)$, $w = w(z)$ are two special solutions which satisfy the following initial conditions:

$$\begin{aligned} v(z=0) &= 0, & v'(z=0) &= \frac{1}{\sqrt{\varepsilon_z}}, \\ w(z=0) &= 1, & w'(z=0) &= 0 \end{aligned} \tag{3}$$

Suppose that all of electrons emitted from photocathode with the same axial initial energy $\sqrt{\varepsilon_{z_1}}$ will be focused at the ideal image plane $z = z_i$, i. e,

$$v(z_i, \sqrt{\varepsilon_{z_1}}) = 0 \qquad w(z_i, \sqrt{\varepsilon_{z_1}}) = M \tag{4}$$

and it is quite evident that $v(z_i, \sqrt{\varepsilon_z}) \neq 0$, $w(z_i, \sqrt{\varepsilon_z}) \neq M$, where M is the linear magnification of system.

From formula (2), we can define the paraxial lateral aberrations at the ideal image plane z_i:

$$\Delta r_*(z_i) = \{w(z_i,\sqrt{\varepsilon_z}) - w(z_i,\sqrt{\varepsilon_{z_1}})\}r_0 + \sqrt{\frac{m_0}{2e}}\dot{r}_0\{v(z_i,\sqrt{\varepsilon_z}) - v(z_i,\sqrt{\varepsilon_{z_1}})\} \tag{5}$$

Formula (5) shows that the paraxial lateral aberrations $\Delta r_*(z_i)$ can be regard as the sum of the paraxial chromatic aberration $\Delta r_{*v}(z_i)$ and the paraxial chromatic aberration of magnification $\Delta r_{*w}(z_i)$, which can be written as following:

$$\Delta r_*(z_i) = \Delta r_* v(z_i) + \Delta r_* w(z_i) = \sqrt{\frac{m_0}{2e}}\dot{r}_0 \Delta v(z_i) + r_0 \Delta w(z_i) \tag{6}$$

where

$$\Delta v(z_i) = v(z_i, \sqrt{\varepsilon_z}), \quad \Delta w(z_i) = w(z_i, \sqrt{\varepsilon_z}) - M \tag{7}$$

Formula (6) is our starting point for studying the paraxial lateral aberrations.

A method of solving the special solutions of $v(z, \sqrt{\varepsilon_z})$, $w(z, \sqrt{\varepsilon_z})$ of paraxial equation by using asymptotic solutions has been given in Ref. [2], which can be described as following:

$$v(z, \sqrt{\varepsilon_z}) = \sqrt{\phi(z)}\xi_0(z) + \sqrt{\varepsilon_z}\eta_0(z) + \varepsilon_z\left[\frac{\xi_0(z)}{2}\frac{\phi'(z)}{\sqrt{\phi(z)}} + \xi_1(z)\sqrt{\phi(z)}\right] + \cdots \tag{8}$$

$$w(z, \sqrt{\varepsilon_z}) = \omega_0(z) + \sqrt{\varepsilon_z}\sqrt{\phi(z)}\zeta_0(z) + \varepsilon_z\omega_1(z) + \cdots \tag{9}$$

where coefficients ξ_0, η_0, ω_0, ζ_0; ξ_1, ω_1 in Eq. (8), Eq. (9) are the solutions of the following differential equations:

$$L_0\begin{pmatrix}\xi_K \\ \zeta_K\end{pmatrix} = -\begin{pmatrix}\xi''_{k-1} \\ \zeta''_{k-i}\end{pmatrix} \tag{10a}$$

$$M_0\begin{pmatrix}\eta_K \\ \omega_K\end{pmatrix} = -\begin{pmatrix}\eta''_{k-1} \\ \omega''_{k-i}\end{pmatrix} \tag{10b}$$

where $\xi_{-1} = \zeta_{-1} = \eta_{-1} = \omega_{-1} = 0$, L_0, M_0 is the linear differential operator, which can be expressed by

$$L_0 = \phi(z)\frac{d^2}{dz^2} + \frac{3}{2}\phi'(z)\frac{d}{dz} + \frac{3}{4}\phi''(z) \tag{11}$$

$$M_0 = \phi(z)\frac{d^2}{dz^2} + \frac{1}{2}\phi'(z)\frac{d}{dz} + \frac{1}{4}\phi''(z) \tag{12}$$

The initial condition of Eq. (10a) and Eq. (10b) can be determined according to the following procedures:

$$\frac{\xi_0(0)\phi'(0)}{2} = 1 \quad \rightarrow \quad \xi_0(0) + \eta_0(0) = 0$$
$$\downarrow \tag{13}$$
$$\frac{\xi_1(0)\phi'(0)}{2} + \xi'_0(0) + \eta'_0(0) = 0 \rightarrow \text{and so on}$$

$$\omega_0(0) = 1 \quad \zeta_0(0) + \omega_1(0) = 0$$
$$\downarrow \quad \uparrow \quad \downarrow \tag{14}$$
$$\frac{\zeta_0(0) + \phi'(0)}{2} + \omega'_0(0) = 0 \rightarrow \text{and so on}$$

where arrows indicate the sequence of calculation.

From asymptotic solution (8), solution (9) and the definition of (6), we may obtain the paraxial lateral aberrations $\Delta r_*(z_i)$ at the ideal image plane $z = z_i$, if we can get the value of coefficients ξ_0, η_0, ω_0, ζ_0; ξ_1, ω_1 at the ideal image plane $z = z_i$, but it is a complicated process for solving the differential equation (8), equation (9). Furthermore, it is not easy to understand the aberration characteristics of the system, so the problem of paraxial lateral aberrations $\Delta r_*(z_i)$ given by (6) is still not solved.

Now, we suggest following processing and procedure. Suppose $\phi(z) \neq$ const at $z \leqslant z_a$; $\phi(z) = $ const $= \phi(z_a) = \phi(z_i)$ at $z_i \geqslant z \geqslant z_a$, we can get the value of two special solutions at $z = z_a$ from expression (8) and expression (9). Therefore, the paraxial lateral aberrations can be obtained by the following expressions:

$$\Delta w(z_i) = [w(z_a, \sqrt{\varepsilon_z}) - w(z_a, \sqrt{\varepsilon_{z_1}})] - \frac{v(z_a, \sqrt{\varepsilon_{z_1}})}{v'(z_a, \sqrt{\varepsilon_{z_1}})}[w'(z_a, \sqrt{\varepsilon_z}) - w'(z_a, \sqrt{\varepsilon_{z_1}})]$$
(15)

$$\Delta v(z_i) = v(z_i, \sqrt{\varepsilon_z}) = v(z_a, \sqrt{\varepsilon_z}) - \frac{v'(z_a, \sqrt{\varepsilon_z})}{v'(z_a, \sqrt{\varepsilon_{z_1}})} v(z_a, \sqrt{\varepsilon_{z_1}})$$
(16)

Substituting all of the values of $v(z_a, \sqrt{\varepsilon_z})$, $v'(z_a, \sqrt{\varepsilon_z})$, $w(z_a, \sqrt{\varepsilon_z})$, $w'(z_a, \sqrt{\varepsilon_z})$ and $v(z_a, \sqrt{\varepsilon_{z_1}})$, $v'(z_a, \sqrt{\varepsilon_{z_1}})$ given by expression (8), expression (9) into expression (15) and expression (16), through a series of transformation, and introducing the linear magnification $M = w(z_i, \sqrt{\varepsilon_{z_1}})$ into expression (15) and expression (16), we may obtain the general form of paraxial chromatic aberration of magnification $\Delta r_{*w}(z_i)$:

$$\Delta r_{*w}(z_i) = r_0 \Delta w(z_i) = -r_0 M \left\{ \frac{1}{2}\phi'(z_a)\xi_0(z_a)\eta_0(z_a) + \phi(z_a)[\xi_0'(z_a)\eta_0(z_a) - \xi_0(z_a)\eta_0'(z_a)] \right\} \times$$
$$\left\{ \frac{\zeta_0(z_a)}{\xi_0(z_a)}(\sqrt{\varepsilon_z} - \sqrt{\varepsilon_{z_1}})\sqrt{\varepsilon_{z_1}} - \frac{\omega_1(z_a)}{\eta_0(z_a)}(\varepsilon_z - \varepsilon_{z_1}) \right\}$$
(17)

and the general form of paraxial sphero-chromatic aberration $\Delta r_{*v}(z_i)$:

$$\Delta r_{*v}(z_i) = \sqrt{\frac{m_0}{2e}} \dot{r}_0 \Delta v(z_i) = \sqrt{\frac{m_0}{2e}} \dot{r}_0 M \left\{ \left[\frac{\phi'(z_a)}{2} \xi_0(z_a)\eta_0(z_a) + \right. \right.$$
$$\phi(z_a)(\xi_0'(z_a)\eta_0(z_a) - \xi_0(z_a)\eta_0'(z_a)) \left] (\sqrt{\varepsilon_z} - \sqrt{\varepsilon_{z_1}}) + \right.$$
$$\left. \left[\frac{\phi'(z_a)}{2\sqrt{\phi(z_a)}} \xi_0^2(z_a) + (\phi(z_a))^{\frac{3}{2}}(\xi_0'(z_a)\xi_1(z_a) - \xi_0(z_a)\xi_1'(z_a)) \right](\varepsilon_z - \varepsilon_{z_1}) \right\}$$
(18)

Finally, we obtain the general form of paraxial lateral aberrations Δr_* which is composed of the paraxial chromatic aberration Δr_{*v} of the second and third orders and the paraxial chromatic aberration of magnification Δr_{*w} of the third order, based on asymptotic solutions of paraxial equation, where coefficients ξ_0, η_0, ω_0, ζ_0; ξ_1, ω_1 and axial potential distribution $\phi(z)$ take the value at $z = z_a$.

2 Solving coefficients of asymptotic solutions in a bi-electrode electrostatic concentric spherical system model

According to the procedure for exploring asymptotic solution of paraxial equation, we can find out the solutions of coefficients ξ_0, η_0, ω_0, ζ_0; ξ_1, ω_1 at the position z, which is based on the known axial potential distributions $\phi(z)$. It seems that to find out the numerical solution of these coefficients is not so complicated, but they can not be showed by mathematical formula, thus it is difficult to master the physical characteristics and regularity for designing imaging electron optical systems.

Now, we try to seek for the solutions of coefficients ξ_0, η_0, ω_0, ζ_0; ξ_1, ω_1 given by asymptotic solution of paraxial ray equation through a bi-electrode electrostatic concentric spherical system model, the axial potential distribution $\phi(z)$ of which can be expressed by

$$\phi(z) = E_c \frac{-R_c z}{z + R_c}, \quad (0 \leqslant z \leqslant (R_a - R_c)) \tag{19}$$

where E_c is the electric field strength at the photocathode; R_c, R_a are radius of curvature of spherical photocathode and spherical anode, all of these quantities take negative value.

Let's first seek for the coefficients $\eta_0(z)$, $\xi_0(z)$ and $\xi_1(z)$. Suppose that the coefficient $\eta_0(z)$ can be expressed by

$$\eta_0(z) = \frac{-2z}{\phi(z)} \tag{20}$$

Substituting formula (20) into Eq. (10b) and Eq. (12):

$$\phi(z)\frac{d^2\eta_0}{dz^2} + \frac{1}{2}\phi'(z)\frac{d\eta_0}{dz} + \frac{1}{4}\phi''(z)\eta_0 = -\eta''_{-1} = 0 \tag{21}$$

we can get the following expression:

$$\frac{1}{\phi(z)}\phi'(z) + \frac{z}{2\phi(z)}\phi''(z) - \frac{z\phi'^2(z)}{\phi^2(z)} = 0 \tag{22}$$

Eq. (22) will be satisfied when the axial potential distribution $\phi(z)$ is expressed by formula (19), where $0 \leqslant z \leqslant (R_a - R_c)$. Thus, expression (20) is a solution of coefficient $\eta_0(z)$. Similarly, suppose that the coefficient $\xi_0(z)$ is expressed by

$$\xi_0(z) = \frac{2z}{\phi(z)} \tag{23}$$

Substituting it into Eq. (10a) and Eq. (11):

$$\phi(z)\frac{d^2\xi_0}{dz^2} + \frac{3}{2}\phi'(z)\frac{d\xi_0}{dz} + \frac{3}{4}\phi''(z)\xi_0 = -\xi''_{-1} = 0 \tag{24}$$

we can also get the expression similar to Eq. (22):

$$\frac{-1}{\phi(z)}\phi'(z) - \frac{z}{2\phi(z)}\phi''(z) + \frac{z\phi'^2(z)}{\phi^2(z)} = 0 \tag{25}$$

It is evident that Eq. (25) is also satisfied when the axial potential distribution $\phi(z)$ is expressed by formula (19), where $0 \leqslant z \leqslant (R_a - R_c)$.

Now, we try to seek for the expression of coefficient $\xi_1(z)$. When $z = 0$, no matter what the photocathode is a spherical or planar electrode, Eq. (23) and Eq. (20) at $z = 0$ can be expressed as following:

$$\xi_0(z=0) = \frac{2}{\phi'_0}, \quad \eta_0(z=0) = -\frac{2}{\phi'_0}, \tag{26}$$
$$\xi'_0(z=0) = 0, \quad \eta'_0(z=0) = 0$$

which satisfies the initial conditions of equation (13):

$$\frac{\xi_0(0)\phi'(0)}{2}=1, \quad \xi_0(0)+\eta_0(0)=0 \tag{27}$$

and from the other initial condition (13): $\frac{\xi_1(0)\phi'(0)}{2}+\xi_0'(0)+\eta_0'(0)=0$, we found that $\xi_1(0)=0$. From Eq. (10a) and (11), $\xi_1(z)$ should satisfy the following equation:

$$\phi(z)\frac{d^2}{dz^2}\xi_1+\frac{3}{2}\phi'(z)\frac{d}{dz}\xi_1+\frac{3}{4}\phi''(z)\xi_1=-\xi_0'' \tag{28}$$

ξ_0'' can be found out by twice differential to formula (23):

$$\xi_0''(z)=\frac{2}{\phi^2(z)}\left[-2\phi'(z)-z\phi''(z)+\frac{2z\phi'^2(z)}{\phi(z)}\right] \tag{29}$$

Therefore, from Eq. (28) and Eq. (29), we may have a non-homogeneous differential equation of the second order for solving $\xi_1(z)$:

$$\frac{d^2\xi_1}{dz^2}+\frac{3}{2}\frac{\phi'(z)}{\phi(z)}\frac{d\xi_1}{dz}+\frac{3}{4}\frac{\phi''(z)}{\phi(z)}\xi_1=\frac{2}{\phi^3(z)}\left[2\phi'(z)+z\phi''(z)-2z\frac{\phi'^2(z)}{\phi(z)}\right] \tag{30}$$

Rewrite Eq. (30) to the following form:

$$\frac{d^2\xi_1}{dz^2}+\frac{3}{2}\frac{\phi'(z)}{\phi(z)}\frac{d\xi_1}{dz}+\frac{3}{4}\frac{\phi''(z)}{\phi(z)}\xi_1=f(z) \tag{31}$$

where

$$f(z)=\frac{2}{\phi^3(z)}\left[2\phi'(z)+z\phi''(z)-2z\frac{\phi'^2(z)}{\phi(z)}\right] \tag{32}$$

From Eq. (31) we know that when $f(z)=0$, it is a homogeneous differential equation of the second order

$$\frac{d^2\xi_1}{dz^2}+\frac{3}{2}\frac{\phi'(z)}{\phi(z)}\frac{d\xi_1}{dz}+\frac{3}{4}\frac{\phi''(z)}{\phi(z)}\xi_1=0 \tag{33}$$

which has a special solution like this

$$y_1(z)=\frac{2z}{\phi(z)} \tag{34}$$

The other special solution $y_2(z)$ can be found as

$$y_2(z)=y_1(z)\int\frac{e^{-\int\frac{3}{2}\frac{\phi'(z)}{\phi(z)}dz}}{y_1^2(z)}dz \tag{35}$$

when $\phi(z)$ is expressed by Eq. (19), we may obtain

$$y_2(z)=\frac{z}{[\phi(z)]^{\frac{3}{2}}} \tag{36}$$

The condition which should satisfy Eq. (33) can be written as

$$2\phi'(z)\left[\frac{z\phi'(z)}{\phi(z)}-1\right]-z\phi''(z)=0 \tag{37}$$

Eq. (37) is really satisfied bi-electrode electrostatic concentric spherical system model. From expressions $y_1(z)$, $y_2(z)$ of Eq. (34), Eq. (36) and expression $f(z)$ of Eq. (32), we may obtain the solution of $\xi_1(z)$ of the non-homogeneous differential equation of the second order Eq. (31)

$$\xi_1(z) = \int_{z=0}^{z} \frac{y_1(z)(-y_2(x)) + y_2(z)y_1(x)}{y_1(x)y_2'(x) - y_2(x)y_1'(x)} f(x) dx \tag{38}$$

i. e.,

$$\xi_1(z) = \int_{z=0}^{z} \left\{ \frac{2z}{\phi(z)} \frac{\phi^2(x)}{\phi'(x)x} - \frac{z}{\phi^{\frac{3}{2}}(z)} \frac{2\phi^{\frac{5}{2}}(x)}{\phi'(x)x} \right\} \left\{ \frac{2}{\phi^3(x)} \left[2\phi'(x) + x\phi''(x) - 2x\frac{\phi'^2(x)}{\phi(x)} \right] \right\} dx \tag{39}$$

Now, we try to seek for the expressions of coefficients $\omega_0(z)$, $\zeta_0(z)$, $\omega_1(z)$. From Eq. (10b) and Eq. (12), $\omega_0(z)$ should satisfy the following equation:

$$\phi(z)\frac{d^2}{dz^2}\omega_0 + \frac{1}{2}\phi'(z)\frac{d}{dz}\omega_0 + \frac{1}{4}\phi''(z)\omega_0 = -\omega''_{-1} = 0 \tag{40}$$

From the special solution: $w(z=0)=1$, we can get the initial condition: $\omega_0(z=0)=1$. Let

$$\omega_0(z) = 1 + \frac{1}{R_c}z \tag{41}$$

substituting Eq. (41) into Eq. (40), we may have

$$\frac{1}{2R_c}\left[\phi'(z) + \frac{R_c + z}{2}\phi''(z)\right] = 0 \tag{42}$$

It is not difficult to prove that formula (42) is also satisfied by expression (19) two-electrode electrostatic concentric spherical system model at $0 \leq z \leq (R_a - R_c)$.

Now, we explore coefficient $\zeta_0(z)$. From Eq. (10a) and Eq. (11)

$$\phi(z)\frac{d^2}{dz^2}\zeta_0 + \frac{3}{2}\phi'(z)\frac{d}{dz}\zeta_0 + \frac{3}{4}\phi''(z)\zeta_0 = -\zeta_{-1} = 0 \tag{43}$$

Similar to above-mentioned method, we may get a solution of Eq. (43):

$$\zeta_0(z) = \frac{2z}{(-R_c)\phi(z)} \tag{44}$$

Here we introduce the constant term $2/(-R_c)$ in order to satisfy the physical condition of system. When $z=0$, from Eq. (44), we get

$$\zeta_0(z=0) = \frac{-2}{(-R_c)E_c} \tag{45}$$

Therefore, from the initial condition (14), because $\omega'_0(0) = 1/R_c$, so the condition $\zeta_0(0)\phi'_0/2 + \omega'_0(0) = 0$ is satisfied. Then, from $\zeta_0(0) + \omega_1(0) = 0$, we may obtain the following initial condition:

$$\omega_1(0) = \frac{2}{(-R_c)E_c} \tag{46}$$

Now, we explore coefficient $\omega_1(z)$. Because $\omega''_0(z) = 0$, from Eq. (10b), we have

$$\phi(z)\frac{d^2}{dz^2}\omega_1 + \frac{1}{2}\phi'(z)\frac{d}{dz}\omega_1(z) + \frac{1}{4}\phi''\omega_1(z) = -\omega''_0(z) = 0 \tag{47}$$

Similar to above-mentioned method, we may have the solution of Eq. (47):

$$\omega_1(z) = \frac{-2z}{(-R_c)\phi(z)} \tag{48}$$

Solution (48) just satisfies condition (46) at $z = 0$.

Thus we have obtained the asymptotic solutions of coefficients ξ_0, η_0, ω_0, ζ_0; ξ_1, ω_1 in a bi-electrode electrostatic concentric spherical system model.

3 Paraxial lateral aberrations in a bi-electrode electrostatic concentric spherical system model solved by asymptotic solution

Now, we try to find out the paraxial trajectories and the paraxial lateral aberration in a two-electrode electrostatic concentric spherical system model by using the expressions of coefficients given above. Substituting all of theses solved expressions of coefficients $\xi_0(z)$, $\eta_0(z)$, $\omega_0(z)$, $\zeta_0(z)$ and $\omega_1(z)$ into the asymptotic solutions of paraxial Eq. (8) and Eq. (9), we have

$$v(z, \sqrt{\varepsilon_z}) = \frac{2z}{\sqrt{\varphi(z)}} - \frac{2z}{\phi(z)}\sqrt{\varepsilon_z} + \frac{z\varepsilon_z}{\phi(z)\sqrt{\phi(z)}} = \frac{2z}{\sqrt{\phi(z)}}\left\{1 - \sqrt{\frac{\varepsilon_z}{\phi(z)}} + \frac{\varepsilon_z}{2\phi(z)}\right\} \tag{49}$$

$$w(z, \sqrt{\varepsilon_z}) = 1 + \frac{1}{R_c}z - \frac{2z}{R_c\sqrt{\phi(z)}}\sqrt{\varepsilon_z} + \frac{2z}{R_c\phi(z)}\varepsilon_z = 1 + \frac{1}{R_c}z - \frac{2z\sqrt{\varepsilon_z}}{R_c\sqrt{\phi(z)}}\left\{1 + \frac{\sqrt{\varepsilon_z}}{\sqrt{\phi(z)}}\right\} \tag{50}$$

According to expression (8), it should have an expression of $\xi_1(z)$ in formula (49), because expression (39) of $\xi_1(z)$ contains a function $f(z)$ expressed by Eq. (32). But when the axial potential distribution $\phi(z)$ is expressed by formula (19) of a bi-electrode electrostatic concentric spherical system model, from Eq. (32), we get $f(z) = 0$. Therefore, the integral of expression (38) gives a zero solution: $\xi_1(z) = 0$.

In the following paragraph, we shall prove that formula (49) and formula (50) are the approximate expressions of two special solutions of paraxial ray equation (1), in which the potential distribution $\phi(z)$ is expressed by formula (19). Formula (49), formula (50) also satisfy the initial conditions (3) of two special solutions simultaneously. We will prove that formula (49) and formula (50), expressing two special solutions of paraxial rays solved by asymptotic solutions, have enough accuracy to investigate the problem of paraxial lateral aberration, thus it shows that the method and the procedure given by M. A. Monastyrski to solve the paraxial equation of electron optics by asymptotic solutions are correct and practicable.

Now, we shall test and verify the expressions of paraxial lateral aberrations in a two-electrode electrostatic concentric spherical system model on the basis of asymptotic solutions given by Monastyrski. By using the formula (19) of $\phi(z)$ and expressions of coefficients (20), (23), (41), (44) and (48), we can find out the value of these coefficients $\eta_0(z)$, $\xi_0(z)$, $\omega_0(z)$, $\zeta_0(z)$, $\omega_1(z)$ at $z = z_a = R_a - R_c$:

$$\xi_0(z_a) = \frac{2(R_a - R_c)}{\phi_{ac}}, \quad \xi_0'(z_a) = -\frac{2n-2}{\phi_{ac}}, \quad \eta_0(z_a) = \frac{-2(R_a - R_c)}{\phi_{ac}},$$

$$\eta'_0(z_a) = \frac{2n-2}{\phi_{ac}}, \quad \zeta_0(z_a) = \frac{2(R_a - R_c)}{(-R_c)\phi_{ac}}, \quad \zeta'_0(z_a) = -\frac{2n-2}{(-R_c)\phi_{ac}}$$

$$\omega_0(z_a) = \frac{1}{n}, \quad \omega'_0(z_a) = \frac{1}{R_c}, \quad \omega_1(z_a) = \frac{-2(R_a - R_c)}{(-R_c)\phi_{ac}}, \quad \omega'_1(z_a) = \frac{2n-2}{(-R_c)\phi_{ac}} \quad (51)$$

and

$$\phi'(z_a) = -n^2 E_c, \quad E_c = \frac{\phi_{ac}}{R_c(n-1)} \quad (52)$$

where $n = \frac{R_c}{R_a}$; $\phi(z_a) = \phi_{ac}$, ϕ_{ac} indicates the voltage of anode with respect to photocathode.

Firstly, we shall test and verify the linear magnification M, which was given by two different approaches: $M = w(z_i, \sqrt{\varepsilon_{z_1}})'$ or the Lagrange-Helmholtz relationship of imaging electron optics

$$M = \frac{1}{\sqrt{\phi(z_a) + \varepsilon_{z_1}} v'(z_a \cdot \varepsilon_{z_1})} \quad (53)$$

Substituting all of these expressions (51) of $\eta_0(z)$, $\xi_0(z)$, $\omega_0(z)$, $\zeta_0(z)$, $\omega_1(z)$ and expression (52) of E_c, into expression (53), we can prove that the two approaches for obtaining the magnification M have the same result as following:

$$M = w(z_i, \sqrt{\varepsilon_{z_1}}) = -\frac{1}{n-2}\left[1 + \frac{2(n-1)}{(n-2)\sqrt{\phi_{ac}}}\sqrt{\varepsilon_{z_1}} + \frac{2n(n-1)}{(n-2)^2\phi_{ac}}\varepsilon_{z_1}\right] \quad (54)$$

Secondly, we shall test and verify the relationships between these asymptotic coefficients of $\eta_0(z)$, $\xi_0(z)$, $\omega_0(z)$, $\zeta_0(z)$, $\omega_1(z)$ which is deduced from the two approaches for obtaining M. Similar to above treatment, it is not difficult to prove that expressions the relationships are tenable in the two-electrode electrostatic concentric spherical system model.

Now, we shall discuss the expression of paraxial lateral aberration in the circumstance of two-electrode electrostatic concentric system model. From Eq. (51), we found that

$$\xi_0(z_a) = -\eta_0(z_a), \quad \xi'_0(z_a) = -\eta'_0(z_a), \quad \frac{\omega_1(z_a)}{\eta_0(z_a)} = \frac{\zeta_0(z_a)}{\xi_0(z_a)}$$

and

$$\xi_1(z_a) = 0, \quad \xi'_1(z_a) = 0$$

Thus, the paraxial chromatic aberration of magnification Δr_{*w} [expression (17)] and the paraxial chromatic aberration $\Delta r_{*v}(z_i)$ (expression (18)) will be simplified as following:

$$\Delta r_{*w}(z_i) = -r_0 M \frac{1}{2}\phi'(z_a)\zeta_0(z_a)\eta_0(z_a)(\sqrt{\varepsilon_z} - \sqrt{\varepsilon_{z_1}})\sqrt{\varepsilon_z} \quad (55)$$

$$\Delta r_{*v}(z_i) = -\sqrt{\frac{m_0}{2e}}\dot{r}_0 M \frac{\phi'(z_a)}{2}\xi_0^2(z_a)\left[(\sqrt{\varepsilon_z} - \sqrt{\varepsilon_{z_1}}) - \frac{1}{\sqrt{\phi(z_a)}}(\varepsilon_z - \varepsilon_{z_1})\right] \quad (56)$$

Substituting the coefficients of (51) into expression (55) and expression (56), and introducing the formula (52) of E_c and formula (54) of M into expression (55) and expression (56), through a series of transformations, we may obtain the paraxial chromatic aberration of magnification Δr_{*w} of the third order and the paraxial chromatic aberration Δr_{*v} of the second and the third order as following:

$$\Delta r_{*w}(z_i) = r_0 \Delta w(z_i) = r_0 \left[-\frac{2(M-1)}{\phi_{ac}} (\sqrt{\varepsilon_z} - \sqrt{\varepsilon_{z_1}}) \sqrt{\varepsilon_z} \right] \quad (57)$$

$$\Delta r_{*v}(z_i) = \sqrt{\frac{m_0}{2e}} \dot{r}_0 \Delta v(z_i) = \sqrt{\frac{m_0}{2e}} \dot{r}_0 \frac{2M}{E_c} \left\{ (\sqrt{\varepsilon_z} - \sqrt{\varepsilon_{z_1}}) - \frac{1}{\sqrt{\phi_{ac}}} (\varepsilon_z - \varepsilon_{z_1}) \right\} \quad (58)$$

Formula (58) is called the paraxial chromatic aberration of the second order and the third order; the first item of which was called as the well known R-A formula in imaging electron optics[4,5]. Formula (57) is called the paraxial chromatic aberration of magnification of the third order, which is a lesser magnitude of one order compared to the first item in formula (58).

4 Testing and verification of paraxial lateral aberrations by a bi-electrode electrostatic concentric spherical system model

Now we shall find the expressions of paraxial lateral aberrations in a actual bi-electrode electrostatic concentric spherical system model on the basis of exact solution of two special solutions. The aim is to verify and test the correctness and accuracy of the expressions solved by asymptotic solutions.

Under the circumstance of a two electrode electrostatic concentric spherical system model, in which the axial potential distribution $\phi(z)$ is expressed by formula (19), fortunately, we have derived the exact expressions of two special solutions of the paraxial trajectory equation (1) as following:

$$v(z, \sqrt{\varepsilon_z}) = \frac{2z}{\sqrt{\phi(z)}} \left\{ \sqrt{1 + \frac{\varepsilon_z}{\phi(z)}} - \sqrt{\frac{\varepsilon_z}{\phi(z)}} \right\} \quad (59)$$

$$w(z, \sqrt{\varepsilon_z}) = 1 + \frac{1}{R_c} z - \frac{\sqrt{\varepsilon_z}}{R_c} \frac{2z}{\phi(z)} [\sqrt{\phi(z) + \varepsilon_z} - \sqrt{\varepsilon_z}] \quad (60)$$

where $\phi(z)$ is axial potential distribution, expressed by formula (19). It is not difficult to prove that two special solution (59) and solution (60) also satisfy the initial conditions (3).

The difference between expression (59), expression (60), expression (49), and expression (50) is that the former is the accurate solutions of paraxial ray equation of (1) solved by direct method, the latter is the approximate solutions of paraxial ray equation solved by asymptotic method. But it should be pointed out that expressions (49) and expression (50) are accurate to the second order approximation of $\sqrt{\varepsilon_z}$, that have sufficient accuracy for studying paraxial lateral aberrations of the second and the third order and geometrical lateral aberrations of the third order. Therefore, in the following, we shall use expression (49) and expression (50) to study paraxial lateral aberrations, their derivative with respect to z, which is accurate to the second order approximation of $\sqrt{\varepsilon_z}$, can be written as

$$v'(z, \sqrt{\varepsilon_z}) = \frac{R_c + 2z}{(R_c + z)\sqrt{\phi(z)}} \left[1 - \frac{2z}{R_c + 2z} \sqrt{\frac{\varepsilon_z}{\phi(z)}} + \frac{(2z - R_c)\varepsilon_z}{2(R_c + 2z)\phi(z)} \right] \quad (61)$$

$$w'(z, \sqrt{\varepsilon_z}) = \frac{1}{R_c} - \frac{\sqrt{\varepsilon_z}}{R_c \sqrt{\varphi(z)}} \left[2 - \frac{z}{\varphi(z)} \phi'(z) \right] + \frac{2\varepsilon_z}{R_c \phi(z)} \left[1 - \frac{z}{\phi(z)} \phi'(z) \right] \quad (62)$$

Now we shall discuss the paraxial chromatic aberration $\Delta r_{*v}(z_i)$ and paraxial chromatic aberration of magnification $\Delta r_{*w}(z_i)$ at the image plane $z = z_i$. As we discussed above, $\Delta w(z_i)$ and $\Delta v(z_i)$ are defined by Eq. (15) and Eq. (16). From expression (49), expression (50), expression (61) and expression (62), we can obtain the value of v, w, $v'w'$ at $z = z_a = (R_a - R_c)$:

$$v(z_a, \sqrt{\varepsilon_z}) = \frac{2(R_a - R_c)}{\sqrt{\phi_{ac}}}\left(1 - \sqrt{\frac{\varepsilon_z}{\phi_{ac}}} + \frac{\varepsilon_z}{2\phi_{ac}}\right) \tag{63}$$

$$w(z_a, \sqrt{\varepsilon_z}) = \frac{1}{n}\left[1 + \frac{2(n-1)\sqrt{\varepsilon_z}}{\sqrt{\phi_{ac}}} - \frac{2(n-1)\varepsilon_z}{\phi_{ac}}\right] \tag{64}$$

$$v'(z_a, \sqrt{\varepsilon_z}) = -\frac{n-2}{\sqrt{\phi_{ac}}}\left[1 - \frac{2(n-1)}{n-2}\sqrt{\frac{\varepsilon_z}{\phi_{ac}}} + \frac{(3n-2)\varepsilon_z}{2(n-2)\phi_{ac}}\right] \tag{65}$$

$$w'(z_a, \sqrt{\varepsilon_z}) = \frac{1}{R_c} + \frac{(n-2)\sqrt{\varepsilon_z}}{R_c\sqrt{\phi_{ac}}} - \frac{2(n-1)\varepsilon_z}{R_c\phi_{ac}} \tag{66}$$

Substituting all of these quantities of v, w, v', w' at $z = (R_a - R_c)$ into Eq. (15) and Eq. (16), finally we have

$$\Delta r_{*w}(z_i) = r_0 \Delta w(z_i) = r_0 \left[-\frac{2(n-1)}{(n-2)\phi_{ac}}(\sqrt{\varepsilon_z} - \sqrt{\varepsilon_{z_1}})\sqrt{\varepsilon_{z_1}} - (\varepsilon_z - \varepsilon_{z_1})\right] \tag{67}$$

$$\Delta r_{*v}(z_i) = \sqrt{\frac{m_0}{2e}}\dot{r}_0 \Delta v(z_i) = \sqrt{\frac{m_0}{2e}}\dot{r}_0 \left\{\frac{2n(R_a - R_c)}{(n-2)\phi_{ac}}(\sqrt{\varepsilon_z} - \sqrt{\varepsilon_{z_1}}) - \frac{1}{\sqrt{\phi_{ac}}}(\varepsilon_z - \varepsilon_{z_1}) + \frac{2(n-1)}{(n-2)\sqrt{\phi_{ac}}}(\sqrt{\varepsilon_z} - \sqrt{\varepsilon_{z_1}})\sqrt{\varepsilon_{z_1}}\right\} \tag{68}$$

Similar to above-mentioned treatment, we found that the formula (54) of M holds and introduce M and E_c into Eq. (67) and Eq. (68), finally we obtain the formulae that are fully identical with expressions (57) and expression (58). It completely proves that the asymptotic solutions of paraxial equation of electron optics given by M. A. Monastyrski for solving the problem of paraxial lateral aberration are correct.

5 Discussions

Let's discuss the first item in formula (58), this is the paraxial sphero-chromatic aberration of the second order, which occupies the greatest part in the whole paraxial lateral aberration. Formula (58) has a wide application in designing image tubes for evaluating image quality, such as image resolution. The great advantage of this formula is that it does not depend on the system structure and the axial potential distribution, but only depends on initial radial energy and initial axial energy of electrons emitted from the photocathode, the electric-field strength at the photocathode and the linear magnification of the system.

It should be pointed out, although the derivation of expression (58) is based on a two-electrode electrostatic concentric spherical system model, it still has universal significance for practical use. Let's give an example. For a cathode lens, whatever its object surface is planar or spherical, we can imagine that the object surface and an equipotential surface which likes a transparent mesh at $z = z_m$, and which is located near the object surface, will construct a virtual two-electrode electrostatic concentric spherical system or a virtual system of uniform field. When electrons emit from the object surface with initial condition ($\sqrt{\varepsilon_z}$, $\sqrt{\varepsilon_r}$) and pass through the virtual system, it will form a paraxial sphero-chromatic aberration at a virtual image plane $z = -z_{i*}$, which is behind the object surface, and which is determined by $\sqrt{\varepsilon_{z_1}}$. According to above-mentioned theory, we may find that a paraxial chromatic aberration of the second order

$$\Delta r_{*v}(z = -z_{i*}) = \sqrt{\frac{m_0}{2e}} \dot{r}_0 \frac{2M_1}{E_c}(\sqrt{\varepsilon_z} - \sqrt{\varepsilon_{z_1}}) \qquad (69)$$

is constructed at the virtual image plane $z = -z_{i*}$, where M_1 is the linear magnification of this virtual system. constructed by the cathode and the mesh; E_c is the electric field strength at the photocathode.

Suppose that a short lens is located in front of the above-mentioned virtual system, which is composed of the field $\phi(z(-z_{i*},z_m)) = \phi_m$, $\phi''(z(z_m,z_a)) \neq 0$, $\phi(z(z_a,z_i)) = \phi_i$. We assume that an virtual object having height $\Delta v(z = -z_{i*})$ is placed at the virtual object plane $z = -z_{i*}$, when electrons emitted from the virtual object goes through the short lens, since the short lens is an ideal electron lens, which does not produce aberrations. Thus a magnified or reduced image with magnification M_2 will be formed at the image position $z = z_i$. Then, we have

$$\Delta r_{*v}(z = z_i) = \sqrt{\frac{m_0}{2e}} \dot{r}_0 \frac{2M_1 M_2}{E_c}(\sqrt{\varepsilon_z} - \sqrt{\varepsilon_{z_1}}) \qquad (70)$$

where M_2 is a linear magnification formed by the short lens. Let $M_1 M_2 = M$. Thus, the first item in formula (58) still holds, but M should be understood as an overall linear magnification of the whole system.

Now, we sum up the above-mentioned description, though the paraxial lateral aberrations can be solved by obtaining the coefficients of asymptotic solutions of paraxial equation, but it seems too complicated for solving these coefficients. It shows that the R-A formula of paraxial sphero-chromaric aberration of the second order which possess the greatest part in the whole paraxial lateral aberrations is universal applicable for electrostatic cathode lenses.

Although the conclusion given above is not new, but we fully prove that the R-A formula of paraxial sphero-chromaric aberration of the second order in a concrete electron optical system holds, and we firstly have obtained concrete forms of paraxial lateral aberrations of the third order, which are a lesser magnitude of one order compared to the R-A formula.

Therefore, in practical calculation, in general case, the R-A formula of paraxial sphero-chromaric aberration of the second order is good enough to evaluate the image quality of the system. But when the geometrical lateral aberration of the third order is studied, it should consider the paraxial lateral aberrations not only its second order, but also its third order. In this case, we would like to recommend a formula of the paraxial lateral aberration expressed in a simple and clear form as following:

$$\Delta r_*(z_i) = \sqrt{\frac{m_0}{2e}} \dot{r}_0 \left\{ \frac{2M}{E_c}(\sqrt{\varepsilon_z} - \sqrt{\varepsilon_{z_1}}) + \frac{2M}{E_c \sqrt{\phi(z_i)}}(\varepsilon_z - \varepsilon_{z_1}) \right\} + r_0 \frac{-2(M-1)}{\phi_{ac}}(\sqrt{\varepsilon_z} - \sqrt{\varepsilon_{z_1}})\sqrt{\varepsilon_z} \quad (71)$$

It should be pointed that the first item in Eq. (71) is accurate formula which is good enough for practical use, the second and third items in Eq. (71) are approximate formulae, but they are possess only a little part in the whole paraxial lateral aberrations. We believe that more accurate expressions of paraxial lateral aberrations of the third order expressed by simple and clear form will be found out.

Conclusions

In the present paper, we have verified and tested the paraxial lateral aberrations given by asymptotic solutions, by using analytical solutions of a two-electrode electrostatic spherical concentric system model. Result shows that the asymptotic solutions of paraxial equation of electron optics for solving the problem of paraxial lateral aberrations are correct. Through a two-electrode electrostatic spherical concentric system model, we have proved that the well-known R-A formula is tenable, in which the paraxial chromatic aberration of the second order expressed in a simple and clear form, is only related to the electric field strength at the photocathode, the linear magnification of system, and the initial radial energy and initial axial energy of electrons emitted from the photocathode, but does not depend on the concrete structure of system and the practical axial potential distribution. We have also suggested an expression of paraxial lateral aberrations written in a simple and clear form when the geometrical lateral aberrations of the third order will be studied.

It seems that results of the present paper have theoretical significance for imaging electron optics, and it may be useful for the design of image tubes and in studying aberrations in imaging electron optics.

Acknowledgements

This project was supported by National Natural Science Foundation of China, Russian Foundation for Basic Research (RFBR) and the Special Science Fund of Doctoral Discipline for Institutions of Higher Learning.

References

[1] ZHOU L W. On theory of paraxial lateral aberrations of imaging electrostatic electron optical systems based on asymptotic solutions [C]. SPIE Proceedings 6621, 2007, 6621: 1 – 12.

[2] ELIN B P, KATISHAOV B A, KYLIKOV Y V, MONASTYRSKI M A. Numerical

methods of optimization for emission electron-opctical systems [J]. Novosibilsk: Science Press, (in Russian) 1987.

[3] MONASTYRSKI M A. On asymptotic solutions of paraxial equation of electron optics. Journal of Technical Physics 1978, 48 (6): 1117 - 1122 (in Russian).

[4] RECKNAGEL A. Theorie des elektrisohen elektronen miktroskops fur selbstrakler [J]. Z. Angew Physik, 1941, 119: 689 - 708 (in Germen).

[5] ARTIMOVICH L A. Electrostatic properties of emission systems [J]. Information of Academy od Sciences, USSR, Series of Physics, 1944, 8 (6): 313 - 328 (in Russian).

[6] ZHOU L W. Electron optics with wide beam focusing [M]. Beijing: Beijing Institute of Technology Press, 1993 (in Chinese).

[7] CHOU L W. Electron optics of concentric spherical electromagnetic focusing systems. Advances in Electronics and Electron Physics, 1979, 52: 119 - 132.

七、成像系统的电子光学传递函数
Electron Optical Transfer Function of Imaging Systems

7.1 成像系统的电子光学调制传递函数与均方根半径的研究
Study of Electron optical Modulation Transfer Function and Root Mean Square Radius in Imaging Systems

摘要：本文研究了逸出光电子的初能量为余弦分布、麦克斯韦分布和贝塔分布下近贴聚焦与锐聚焦系统的电子光学调制传递函数（MTF）与均方根半径，证明了以指数函数形式表示的解析式 $T(f) = \exp[-(\pi\rho f)^2]$ 足以精确地表达成像系统的 MTF，这里 ρ 是成像系统的均方根（RMS）半径 $\bar{r}$，具有明确的物理意义。文中给出了上述三种光电子初能分布下成像系统的均方根半径表达式，最后给出了像管 MTF 计算的若干实例。

Abstract: The electron-optical modulation transfer function (MTF) and root mean square (RMS) radius in the proximity focusing and sharp focusing systems for the photoelectronic image devices are studied in this paper, where the initial energy of emitted photoelectron is supposed as having cosine, Maxwell and beta distributions. It has been proved that the formula expressed by the exponential function as $T(f) = \exp[-(\pi\rho f)^2]$ is accurate enough to express the MTF of the image focusing systems, where the focusing error coefficient ρ, will be confirmed as the RMS radius $\bar{r}$ of image systems, so it has a definite physical meaning. Under consideration of these three initial energy distributions of photoelectrons, the MTF and its RMS radius of the image systems are given. Finally, some practical examples in MTF calculations of image tubes are given for illustrative purposes.

引 言

成像系统的 MTF 由于能对光电成像器件的像质作出全面准确的评定，近十年来已引起人们的广泛注意，国内外发表了不少理论分析文章，这些文章大多是由静电和电磁聚焦系统中普遍成立了二级近轴横向色差公式出发讨论的。代表性的工作有：Csorba[1,2] 求得了麦克斯韦分布和余弦分布下近贴聚焦系统以及余弦分布下锐聚焦系统的全色 MTF 解析表达式；Ezard[3] 和 Mulder[4] 分别求出了近贴聚焦和锐聚焦系统的单色 MTF 解析表达式；Hartly[5] 仅求出若干贝塔分布下极限像面的点扩散函数（PSF）和线扩散函数（LSF）的解析表达式，而其他像面上的 MTF 则是用数值方法求出的；周立伟和方二伦[6] 研究了成像系统各种初能分布下全色点扩散函数的确定；朱克正[7] 在 Mulder[4] 的基础上，通过单色 MTF 对初能分布求统计平均值得出了贝塔分布和麦克斯韦分布下全色 MTF 的解析表达式。这里"全色"系指电子初能量的不同分布，"单色"系指电子初能量相同。

周立伟[a]，艾克聪[a]，方二伦[b]. a) 北京工业学院，b) 西安现代化学研究所. 北京工业学院学报（Journal of Beijing Institute of Technology），No. 3, 1982, 36–51.

应该指出,在这些工作中,Csorba 提出以指数函数形式表示系统的 MTF 具有简便实用的特点。按照文献 [1],求 MTF 将归结于求成像系统的聚焦误差系数 ρ。但如何获得该系数及其物理意义都是不够明确的;加之他所给出的余弦分布下近贴聚焦系统的 ρ 值有误。而锐聚焦系统的 ρ 值与实际计算值也有较大出入,从而降低了公式的实用性。

还应指出,目前的文献对于均方根半径及其与 MTF 之间的联系和应用均阐述得不够。Beurle[8]最早讨论了麦克斯韦分布下锐聚焦系统的均方根半径,但并没有将其与 MTF 联系起来。Csorba[1]尽管给出了指数形式的 MTF 解析表达式,也没有将它与均方根半径联系起来。朱克正在文献 [7] 中研究了贝塔分布下锐聚焦系统中均方根半径与 MTF 的联系,但并没有从实际系统的计算中进一步证明,而且文献 [1]、文献 [7] 仅只讨论了最佳像面的情况,对其他像面的情况并未加以讨论。

本文研究和推导了各种初能分布下近贴聚焦和锐聚焦系统的全色 MTF 和均方根半径解析表达式,证明了成像系统 MTF 的指数表达式 $T(f) = \exp[-(\pi\rho f)^2]$ 中的 ρ 正是系统的均方根半径 $\bar{r}$,具有明确的物理意义。同时,我们由各种初能分布下 MTF 和均方根半径的解析表达式,以及由静电聚焦系统和倾斜型电磁聚焦系统[9]的实际计算都证明了:均方根半径 $\bar{r}$ 的大小与 MTF 的优劣在各个像面上都是一一对应的;均方根半径 $\bar{r}$ 最小的像面正是 MTF 的最佳像面。因此均方根半径不仅可以用来表征系统的鉴别率特性,而且可能作为计算 MTF 的一个辅助评价指标,具有计算量小、直观方便等优点。从物理意义上讲,均方根半径的大小正反映了逸出电子束在像面上落点密集的程度,而密集的电子落点必将使 MTF 得以改善,两者自然是密切相关的。本文最后举例说明了整管 MTF 的计算,说明以指数函数形式表示的 MTF 具有计算速度快、形式简洁、使用方便,且有令人满意的精度等优点。

1 光电子初角度和初能量分布、点扩散函数、调制传递函数与均方根半径

1.1 初角度和初能量分布

在成像系统中,一般假定逸出光电子的初角度服从朗伯定律[10,11],即电子以初角度 α 逸出的概率为

$$G(\alpha) = \cos\alpha \tag{1}$$

而常用的逸出光电子的初能分布函数有以下三种形式:

(1) 归一化的余弦分布:

$$N(\xi) = \frac{\pi^2}{2(\pi-2)}\xi\cos\left(\frac{\pi}{2}\xi\right) \tag{2}$$

(2) 归一化的贝塔分布:

$$N(\xi) = \frac{(k+l+1)!}{k!\,l!}\xi^k(1-\xi)^l \tag{3}$$

在以上两种分布中: $\xi = \frac{\varepsilon}{\varepsilon_m}$。$\varepsilon$ 和 ε_m 分别是逸出电子的发射初电位和最大发射初电

位，均以伏特表示。

（3）归一化的麦克斯韦分布：

$$N(\eta) = \eta\exp(-\eta) \tag{4}$$

式中，$\xi = \dfrac{\varepsilon}{\varepsilon_p}$，$\varepsilon_p$ 是逸出电子的最可几发射初电位，以伏特表示。

1.2 点扩散函数（PSF）

在我们讨论的成像系统内，设在阴极面中心，ΔA 为光阴极面上无限小的微面元，I 为发射电流密度。若逸出光电子是单能量分布且角度服从朗伯定律，则在逸出角 $\alpha \sim \alpha + \mathrm{d}\alpha$、方位角 $\beta \sim \beta + \mathrm{d}\beta$ 之间由 ΔA 逸出的光电子将落在 $r\mathrm{d}r\mathrm{d}\beta$ 的像面微元内，此微元内的单色点像电流密度可表示为

$$j(r) = \frac{I\Delta A \sin\alpha\cos\alpha\, \mathrm{d}\alpha\,\mathrm{d}\beta}{\pi r \mathrm{d}r\mathrm{d}\beta} \tag{5}$$

式（5）乃是我们讨论成像系统在各种初能分布下全色 PSF 和全色 MTF 的出发点。

由单色点像电流密度 $j(r)$，在考虑前述逸出电子初能量分布的情况下，可求得全色点像电流密度 $J(r)$：

$$J(r) = \int_D j(r)N(\xi)\mathrm{d}\xi \text{（余弦分布或贝塔分布）} \tag{6}$$

或

$$J(r) = \int_D j(r)N(\eta)\mathrm{d}\eta \text{（麦克斯韦分布）} \tag{7}$$

积分域 D 与对 $r \sim (r+\mathrm{d}r)$ 有贡献的 α 和 ξ（或 η）有关。对此全色点像电流密度做归一化处理：

$$2\pi \int_0^\infty CrJ(r)\mathrm{d}r = 1 \tag{8}$$

便可求得归一化的全色点像电流密度分布。令

$$P(r) = CJ(r)$$

$P(r)$ 通常称为全色点扩散函数，C 是归一化系数，随不同系统和不同初能分布而异。

1.3 调制传递函数（MTF）

目前，求成像系统的 MTF 通常有两种途径。对于近贴聚焦系统，则可对全色点扩散函数 $P(r)$ 做傅里叶–贝塞尔变换（也称零阶汉克尔变换）而求得系统的全色 MTF：

$$T(f) = 2\pi\int_0^\infty rP(r)\mathrm{J}_0(2\pi fr)\mathrm{d}r \tag{9}$$

这里 $\mathrm{J}_0(2\pi fr)$ 是零阶贝塞尔函数，且

$$\mathrm{J}_0(2\pi fr) = \sum_{n=0}^{\infty} \frac{(-1)^n (2\pi fr)^{2n}}{2^{2n}(n!)^2} \tag{10}$$

式中，f 为空间频率。

对于锐聚焦系统，则由式（5）先求系统的单色点像电流密度 $j(r)$，然后直接对 $j(r)$ 做傅里叶–贝塞尔变换，并做归一化处理，便可求得单色 MTF[4]：

$$T_{\text{mono}}(f) = \frac{\int_0^\infty r j(r) J_0(2\pi f r) \, dr}{\int_0^\infty r j(r) \, dr} \tag{11}$$

对式（11）求某一初能分布下初电位的统计平均值，便可求得全色调制传递函数 $T(f)$：

$$T(f) = \int_0^1 T_{\text{mono}}(f) N(\xi) \, d\xi \quad (\text{余弦或贝塔分布}) \tag{12}$$

或

$$T(f) = \int_0^\infty T_{\text{mono}}(f) N(\eta) \, d\eta \quad (\text{麦克斯韦分布}) \tag{13}$$

将积分化为和数形式，取前两项，略去 $(2\pi f r)^4$ 以上的高次项，则可将上述 $T(f)$ 表示成指数形式：

$$T(f) = \exp[-(\pi \rho f)^2] \tag{14}$$

将式（14）对 $\sqrt{\dfrac{\varepsilon_{z_1}}{\varepsilon_{\text{m}}}}$ 或 $\sqrt{\dfrac{\varepsilon_{z_1}}{\varepsilon_{\text{p}}}}$ 求导取极值，便可求得锐聚焦位置与相应的最佳 MTF。

1.4 均方根（RMS）半径

当假定逸出光电子的初角度服从朗伯定律、初能量服从某一分布的情况下，成像系统的均方根半径 $\bar{r}$ 可表示为

$$\bar{r}^2 = \int_0^1 \int_0^{\frac{\pi}{2}} r^2 \sin 2\alpha N(\xi) \, d\xi d\alpha \quad (\text{余弦或贝塔分布}) \tag{15}$$

或

$$\bar{r}^2 = \int_0^\infty \int_0^{\frac{\pi}{2}} r^2 \sin 2\alpha N(\eta) \, d\eta d\alpha \quad (\text{麦克斯韦分布}) \tag{16}$$

以上二式中的 r 值即是成像平面上的二级近轴横向像差（锐聚焦系统）或散射圆半径（近贴聚焦系统）。

由式（15）或式（16）分别对 $\sqrt{\dfrac{\varepsilon_{z_1}}{\varepsilon_{\text{m}}}}$ 或 $\sqrt{\dfrac{\varepsilon_{z_1}}{\varepsilon_{\text{p}}}}$ 求导取极值，则可求得最小均方根半径值 $\bar{r}_{\min}$ 及其相应的像面位置。

2 近贴聚焦系统的调制传递函数与均方根半径

对于近贴聚焦系统，在文献［12］中我们已经导得在该系统中以初电位 ε、初角度 α 逸出的电子在阳极平面上形成的散射圆半径 r 可表示为

$$r = \frac{2L\sqrt{\varepsilon}}{\phi_{ac}} \sin \alpha \left\{ \sqrt{\phi_{ac} + \varepsilon \cos^2 \alpha} - \sqrt{\varepsilon} \cos \alpha \right\} \tag{17}$$

式中，L 为阴极至阳极之间的距离，ϕ_{ac} 为极间加速电位（V）。

考虑到 $\sqrt{\dfrac{\varepsilon}{\phi_{ac}}} \ll 1$，故上式可简化为

$$r = \frac{2L\sqrt{\varepsilon}}{\sqrt{\phi_{ac}}} \sin \alpha \tag{18}$$

式（18）乃是我们计算近贴聚焦系统 MTF 和均方根半径的出发点。

将式（18）及其对 α 的导数代入式（5），就可求得近贴聚焦系统单色点像电流密度的解析表达式为

$$j(r) = \frac{I\Delta A\phi_{ac}}{4\pi L^2 \varepsilon} \tag{19}$$

下面将从式（19）出发，讨论前述三种初能分布下近贴聚焦系统的归一化全色 PSF、MTF 与均方根半径。

2.1 余弦分布

将式（19）与式（2）代入式（6），积分限取 $\int_{\xi_p}^{1}$，$\xi_p = \frac{1}{4}\frac{\phi_{ac}}{\varepsilon_m}\frac{r^2}{L^2}$，则可求得全色点像电流密度为

$$J(r) = \frac{\Delta A I'}{4(\pi-2)L^2}\frac{\phi_{ac}}{\varepsilon_m}\left[1 - \sin\left(\frac{\pi}{8}\frac{\phi_{ac}}{\varepsilon_m}\frac{r^2}{L^2}\right)\right] \tag{20}$$

式中，I' 为发射点多能量电流密度，此式经过式（8）的归一化处理，就可求得归一化的全色点扩散函数为

$$P(r) = \frac{\phi_{ac}}{4(\pi-2)L^2 \varepsilon_m}\left[1 - \sin\left(\frac{\pi}{8}\frac{\phi_{ac}}{\varepsilon_m}\frac{r^2}{L^2}\right)\right] \tag{21}$$

将式（21）代入式（9）做傅里叶-贝塞尔变换，便可求得归一化的全色 MTF 的解析表达式为

$$T(f) = \frac{2\pi}{\pi-2}\sum_{n=0}^{\infty}(-1)^n \frac{\left(2\pi fL\sqrt{\frac{\varepsilon_m}{\phi_{ac}}}\right)^{2n}}{(n!)}\left[\frac{1}{2(n+1)} - \sum_{i=0}^{\infty}(-1)^i \frac{\left(\frac{\pi}{2}\right)^{2i+1}}{(2i+1)!}\frac{1}{2(2i+n+2)}\right] \tag{22}$$

在上式中，取 $n=0$、1 两项，略去 $\left(2\pi fL\sqrt{\frac{\varepsilon_m}{\phi_{ac}}}\right)^4$ 及以上高次项，则式（22）便可表示为如下形式：

$$T(f) = \exp\left[-10.290\,072\,27\frac{\varepsilon_m}{\phi_{ac}}L^2 f^2\right] \tag{23}$$

或可表示成式（14）的形式，式中

$$\rho = 1.021\,078\,989 L\left(\frac{\varepsilon_m}{\phi_{ac}}\right)^{1/2} \tag{24}$$

应该指出：我们求得的 ρ 值与 Csorba[1] 给出的式（13a）

$$\rho = 1.48 L\left(\frac{\varepsilon_p}{\phi_{ac}}\right)^{1/2}, (\varepsilon_p = 0.548\varepsilon_m)$$

有差别，究其原因在于：按 Csorba 在文献 [1] 中称，调制传递函数是点扩散函数归一化的傅里叶变换。但是由文献 [13] 我们知道，对圆对称的点扩散函数应该做傅里叶-贝塞尔变换（即零阶汉克尔变换）才能求得系统的 MTF。经过检验，Csorba 上述的 ρ 值似乎是

将点扩散函数作为线扩散函数进行傅里叶变换的，由此导致了不正确的结果。还应指出，Csorba 对导得的指数表达式中 ρ 的物理意义的阐述也是不明确的，他称 ρ 为聚焦误差系数，实际 ρ 是逸出电子在该像面上的均方根半径 $\bar{r}$。我们将在下面予以证明。

将式（18）代入式（15），便可求得余弦分布下近贴聚焦系统的均方根半径为

$$\bar{r} = \left[\frac{2(\pi^2 - 8)}{\pi(\pi - 2)} \frac{\varepsilon_m L^2}{\phi_{ac}}\right]^{1/2} = 1.021\,078\,989 \left(\frac{\varepsilon_m}{\phi_{ac}}\right)^{1/2} L \tag{25}$$

将式（25）与式（24）进行比较可以看出，MTF 指数表达式中的 ρ 正是该像面上的均方根半径 $\bar{r}$。

2.2 麦克斯韦分布

将式（19）与式（4）代入式（7），积分限取 $\int_{\eta_p}^{\infty}$，而 $\eta_p = \frac{1}{4}\frac{\phi_{ac}}{\varepsilon_p}\frac{r^2}{L^2}$，由式（8）进行归一化处理，并由式（9）对 $P(r)$ 做傅里叶-贝塞尔变换，便可求得麦克斯韦分布下 $P(r)$ 与 $T(f)$ 分别为

$$P(r) = \frac{\phi_{ac}}{4\pi L^2 \varepsilon_p} \exp\left[-\frac{1}{4}\frac{\phi_{ac}}{\varepsilon_p}\left(\frac{r}{L}\right)^2\right] \tag{26}$$

$$T(f) = \exp\left[-4\frac{\varepsilon_p}{\phi_{ac}}(\pi L f)^2\right] = \exp\left[-(\pi\rho f)^2\right] \tag{27}$$

式中 ρ 同样可证明乃是均方根半径 $\bar{r}$，即

$$\rho = \bar{r} = 2L\left(\frac{\varepsilon_p}{\phi_{ac}}\right)^{1/2} \tag{28}$$

上式曾在文献 [1] 中给出过。

2.3 贝塔分布

在贝塔分布式（3）下，用与前述相同的方法，不难求得全色点扩散函数 $P(r)$ 为

$$P(r) = \frac{\phi_{ac}}{4\pi L^2 \varepsilon_m} \frac{(k+l+1)!}{k!l!} \sum_{i=0}^{l}(-1)^i \frac{l!}{(l-i)!i!}\frac{1}{k+i}\left[1 - \left(\frac{1}{4}\frac{\phi_{ac}}{\varepsilon_m}\frac{r^2}{L^2}\right)^{k+i}\right] \tag{29}$$

全色调制传递函数 $T(f)$ 则可表示为

$$T(f) = \sum_{n=0}^{\infty} \frac{(-1)^n \left(2\pi f L\sqrt{\frac{\varepsilon_m}{\phi_{ac}}}\right)^{2n}}{n!(n+1)!} \frac{(k+l+1)!(n+k)!}{k!(n+k+l+1)!} \tag{30}$$

其指数形式仍可用式（14）近似表示，式中 ρ 即均方根半径 $\bar{r}$，它可表示为

$$T(f) = \exp\left[-(\pi\rho f)^2\right] \tag{31a}$$

$$\bar{r} = \rho = \left(\frac{2\varepsilon_m L^2}{\phi_{ac}}\frac{k+1}{k+l+2}\right)^{1/2} \tag{31b}$$

现将各种初能分布下 MTF 的指数表达式和均方根半径 $\bar{r}$ 的解析式列在表 1 中。

表 1　三种初能分布下近贴聚焦系统的 MTF 与均方根半径

初能分布 $N(\xi)$ 或 $N(\eta)$		MTF 与均方根半径	$T(f)=\exp[-(\pi\rho f)^2]$	均方根半径 $\bar{r}=\rho$
余弦分布		$\dfrac{\pi^2}{2(\pi-2)}\xi\cos\left(\dfrac{\pi}{2}\xi\right)$	$\exp\left[-10.290\,072\,27\,\dfrac{\varepsilon_m}{\phi_{ac}}L^2f^2\right]$	$1.021\,078\,989\left(\dfrac{\varepsilon_m}{\phi_{ac}}\right)^{1/2}L$
麦克斯韦分布		$\eta\exp(-\eta)$	$\exp\left[-4\,\dfrac{\varepsilon_m}{\phi_{ac}}(\pi Lf)^2\right]$	$2\left(\dfrac{\varepsilon_p}{\phi_{ac}}\right)^{1/2}L$
贝塔分布	$\beta_{2,1}$	$12\xi^2(1-\xi)$	$\exp\left[-1.2\,\dfrac{\varepsilon_m}{\phi_{ac}}(\pi Lf)^2\right]$	$1.095\,445\,115\left(\dfrac{\varepsilon_m}{\phi_{ac}}\right)^{1/2}L$
	$\beta_{1,1}$	$6\xi(1-\xi)$	$\exp\left[-\dfrac{\varepsilon_m}{\phi_{ac}}(\pi Lf)^2\right]$	$\left(\dfrac{\varepsilon_m}{\phi_{ac}}\right)^{1/2}L$
	$\beta_{1,2}$	$12\xi(1-\xi)^2$	$\exp\left[-0.8\,\dfrac{\varepsilon_m}{\phi_{ac}}(\pi Lf)^2\right]$	$0.894\,427\,191\left(\dfrac{\varepsilon_m}{\phi_{ac}}\right)^{1/2}L$
	$\beta_{1,3}$	$20\xi(1-\xi)^2$	$\exp\left[-0.666\,\dfrac{\varepsilon_m}{\phi_{ac}}(\pi Lf)^2\right]$	$0.816\,496\,58\left(\dfrac{\varepsilon_m}{\phi_{ac}}\right)^{1/2}L$
	$\beta_{1,4}$	$30\xi(1-\xi)^4$	$\exp\left[-0.571\,428\,5\,\dfrac{\varepsilon_m}{\phi_{ac}}(\pi Lf)^2\right]$	$0.755\,928\,946\left(\dfrac{\varepsilon_m}{\phi_{ac}}\right)^{1/2}L$
	$\beta_{1,8}$	$90\xi(1-\xi)^8$	$\exp\left[-0.363\,636\,\dfrac{\varepsilon_m}{\phi_{ac}}(\pi Lf)^2\right]$	$0.603\,022\,688\left(\dfrac{\varepsilon_m}{\phi_{ac}}\right)^{1/2}L$

3　锐聚焦系统的调制传递函数与均方根半径

文献 [12] 和文献 [14] 中已经证明，对于锐聚焦系统，二级近轴横向色球差公式

$$\Delta r = \frac{2M\sqrt{\varepsilon_r}}{E_c}(\sqrt{\varepsilon_z} - \sqrt{\varepsilon_{z_1}}) \tag{32}$$

是普遍成立的。式中，M 是系统的放大率，E_c 是系统的阴极面场强（伏特/毫米），M 和 E_c 均取正值。ε_{z_1} 是决定某一理想聚焦像面的轴向初电位，ε_z、ε_r 分别是逸出电子轴向初能和径向初电位，均以伏特表示。

在单色情况下，二级近轴横向色差公式可表示为

$$\Delta r = \frac{M\varepsilon}{E_c}\left(\sin 2\alpha - 2\sqrt{\frac{\varepsilon_{z_1}}{\varepsilon}}\sin\alpha\right) \tag{33}$$

对于单色 MTF，则可由式 (33)、式 (5) 与式 (11) 直接求得[4]

$$T_{\text{单}}(f) = \int_0^{\frac{\pi}{2}} \sin 2\alpha J_0\left[\frac{2\pi fM\varepsilon}{E_c}\left(\sin 2\alpha - 2\sin\alpha\sqrt{\frac{\varepsilon_{z_1}}{\varepsilon}}\right)\right]d\alpha \tag{34}$$

引用二项式定理

$$(a-b)^n = \sum_{m=0}^{n}(-1)^m C_n^m a^{n-m} b_m$$

对上式进行积分，便可求得锐聚焦系统单色 MTF 的表达式：

$$T_{\text{单}}(f) = \sum_{n=0}^{\infty}\frac{(-1)^n}{(n!)^2}\left(\frac{2\pi fM\varepsilon}{E_c}\right)^{2n}\sum_{m=0}^{2n}\frac{(-1)^m}{(n!)^2}C_{2n}^m\left(\frac{\varepsilon_{z_1}}{\varepsilon}\right)^{\frac{m}{2}}\beta\left(n+1, n-\frac{m}{2}+1\right) \tag{35}$$

式中，$\beta\left(n+1, n-\dfrac{m}{2}+1\right)$ 是贝塔函数，可由伽马函数求出

$$\beta(m,n) = \frac{\Gamma(m)\Gamma(n)}{\Gamma(m+n)} \tag{36}$$

下面我们将从式（36）出发，讨论各种初能分布下锐聚焦系统的全色 MTF。

3.1 余弦分布

由式（12）对单色 MTF 式（35）求初能分布函数的统计平均值，就可求得余弦分布下锐聚焦系统归一化的全色 MTF 表达式：

$$T(f) = \frac{\pi^2}{2(\pi-2)} \sum_{n=0}^{\infty} \frac{(-1)^n F^{2n}}{(n!)^2} \sum_{i=0}^{2n} \left[(-1)^i C_{2n}^i \left(\frac{\varepsilon_{z_1}}{\varepsilon_m}\right)^{\frac{1}{2}} \cdot \right.$$

$$\left. \beta\left(n+1, n-\frac{i}{2}+1\right) \sum_{m=0}^{\infty} (-1)^m \frac{\left(\frac{\pi}{2}\right)^{2m}}{(2m)!} \frac{1}{2n+2m+2-\frac{i}{2}} \right] \tag{37}$$

式中，

$$F = \frac{2\pi f M \varepsilon_m}{E_c} \tag{38}$$

在式（37）中，略去 F^4 以上的高次项，也可将 $T(f)$ 表达成式（14）的指数形式：式中 ρ 亦可证明即为均方根半径 $\bar{r}$，它可表示为

$$\rho = \bar{r} = \frac{M\varepsilon_m}{E_c}\left(0.213\,486\,657 - 0.858\,658\,572\sqrt{\frac{\varepsilon_{z_1}}{\varepsilon_m}} + 1.042\,602\,298\,\frac{\varepsilon_{z_1}}{\varepsilon_m}\right)^{1/2} \tag{39}$$

由式（39），不难求得最佳 MTF 对应的像面位置为

$$\left.\sqrt{\frac{\varepsilon_{z_1}}{\varepsilon_m}}\right|_{opt} = 0.411\,786\,247 \tag{40}$$

于是与最佳 MTF 对应的 ρ_{min} 便可表示为

$$\rho_{min} = 0.191\,558\,767\,\frac{M\varepsilon_m}{E_c} \tag{41}$$

下面，我们以本文所导出的余弦分布下全色 MTF 的均方根半径表达式（39）和式（41）与 Csorba[1] 给出的如下公式

$$T(f) = \exp\left[-0.06\left(\frac{\pi f M \varepsilon_p}{E_c}\right)^2\right] = \exp[-(\pi\rho f)^2]$$

$$\rho = 0.245\,\frac{M\varepsilon_p}{E_c} = 0.134\,\frac{M\varepsilon_m}{E_c}$$

以及由二级近轴横向色差公式，通过将初角度分为 6 等份、初能量分为 10 等份计算 PSF 和 LSF，最后求得的 MTF，三者进行对比，如表 2 所示。由表 2 的比较可以看出：我们导出的公式比 Csorba 导出的公式更为接近实际计算值。

表 2　余弦分布下三种 MTF 计算值的比较（$E_c = 80$ V/mm，$\varepsilon_m = 0.9124$ V，$M = 1$）

$f/(\text{lp} \cdot \text{mm}^{-1})$	$T(f)$（本文（14）式）	$T(f)$（实际计算）	$T(f)$（Csorba（15）式）
10	0.995 300 278	0.994 987 52	0.997 697 554
20	0.981 333 615	0.980 171 05	0.990 821 977
30	0.958 489 855	0.956 197 14	0.979 467 815
40	0.927 399 171	0.924 090 79	0.963 790 241
50	0.888 902 648	0.885 181 38	0.944 001 522
60	0.844 014 841	0.841 008 83	0.920 366 241
70	0.793 878 938	0.793 217 39	0.893 195 408
80	0.739 719 070	0.743 450 71	0.862 839 659
90	0.682 790 842	0.693 244 79	0.829 681 738
100	0.624 333 873	0.643 947 55	0.794 128 493
110	0.565 528 436	0.596 651 88	0.756 602 623

3.2　麦克斯韦分布

在麦克斯韦分布下，与上相似，不难求得归一化的全色 MTF 解析表达式为

$$T(f) = \sum_{n=0}^{\infty} \frac{(-1)^n}{(n!)^2} \left(\frac{2\pi f M \varepsilon_p}{E_c} \right)^{2n} \sum_{m=0}^{2n} (-1)^m C_{2n}^m \left(\sqrt{\frac{\varepsilon_{z_1}}{\varepsilon_p}} \right)^m \beta\left(n+1, n - \frac{m}{2} + 1\right) \Gamma\left(2n - \frac{m}{2} + 2\right) \tag{42}$$

同样取 $n = 0$、1 两项，则全色 MTF 的指数表达式仍可用式（14）表示。式中的 ρ 仿前可证即是系统的均方根半径 $\bar{r}$，于是可得：

$$\rho = \bar{r} = \frac{2M\varepsilon_p}{E_c} \left(1 - \sqrt{\frac{\varepsilon_{z_1}}{\varepsilon_p}} \sqrt{\pi} + \frac{\varepsilon_{z_1}}{\varepsilon_p} \right)^{1/2} \tag{43}$$

最佳 MTF 对应的像面位置与最小均方根半径便可表示如下：

$$\left. \sqrt{\frac{\varepsilon_{z_1}}{\varepsilon_p}} \right|_{\text{opt}} = \frac{\sqrt{\pi}}{2} \tag{44}$$

$$\rho_{\min} = \bar{r}_{\min} = 0.926\,502\,75 \frac{M\varepsilon_p}{E_c} \tag{45}$$

3.3　贝塔分布

在贝塔分布下，系统的全色 MTF 与均方根半径 $\bar{r} = \rho$ 可分别表示如下：

$$T(f) = \frac{(k+l+1)!}{k!\,l!} \sum_{n=0}^{\infty} \frac{(-1)^n F^{2n}}{(n!)^2} \sum_{m=0}^{2n} \left[(-1)^m C_{2n}^m \left(\sqrt{\frac{\varepsilon_{z_1}}{\varepsilon_m}} \right)^m \cdot \right.$$

$$\left. \beta\left(n+1, n - \frac{m}{2} + 1\right) \beta\left(2n + k - \frac{m}{2} + 1, l + 1\right) \right] \tag{46}$$

$$\bar{r} = \rho = \frac{M\varepsilon_m}{E_c} \left[\frac{2}{3} \frac{(k+2)(k+1)}{(k+l+2)(k+l+3)} - \frac{32}{15} \frac{(k+l+1)!}{k!} \frac{\Gamma(2.5+k)}{\Gamma(3.5+k+l)} \sqrt{\frac{\varepsilon_{z_1}}{\varepsilon_m}} + \frac{2(k+1)}{(k+l+2)} \frac{\varepsilon_{z_1}}{\varepsilon_m} \right]^{1/2}$$
(47)

最佳 MTF 所对应的像面位置及最小均方根半径便可表示为

$$\sqrt{\frac{\varepsilon_{z_1}}{\varepsilon_m}}\bigg|_{opt} = \frac{8}{15} \frac{(k+l+2)!}{(k+1)!} \frac{\Gamma(k+2.5)}{\Gamma(k+l+3.5)} \quad (48)$$

$$\rho_{min} = \bar{r}_{min} = \frac{M\varepsilon_m}{E_c} \left[\frac{2}{3} \frac{(k+2)(k+1)}{(k+l+2)(k+l+3)} - \frac{128}{225} \frac{(k+l+1)!(k+l+2)!}{k!(k+1)!} \left(\frac{\Gamma(k+2.5)}{\Gamma(k+l+3.5)} \right)^2 \right]^{1/2}$$
(49)

将不同的 k、l 值代入上两式,便可求得各种贝塔分布下最佳 MTF 的位置与最小均方根半径值。表 3 列出了各种初能分布下的最佳 MTF、最小均方根半径及其相对应的像面位置。

表 3 锐聚焦系统的 MTF 与均方根半径

初能分布 $N(\xi)$ 或 $N(\eta)$	MTF 与均方根半径	MTF $T_{opt}(f) = \exp[-(\pi\rho f)^2]$	均方根半径 $\bar{r}_{min} = \rho_{min}$	T_{opt} 与 $\bar{r}_{min}$ 的位置 $\sqrt{\frac{\varepsilon_{z_1}}{\varepsilon_m}}$ 或 $\sqrt{\frac{\varepsilon_{z_1}}{\varepsilon_p}}$
余弦分布	$\frac{\pi^2}{2(\pi-2)}\xi\cos\left(\frac{\pi}{2}\xi\right)$	$\exp\left[-0.036\,702\,665\left(\frac{\pi f M \varepsilon_m}{E_c}\right)^2\right]$	$0.191\,579\,392\left(\frac{M\varepsilon_m}{E_c}\right)$	$0.411\,779\,733$
麦克斯韦分布	$\eta\exp(-\eta)$	$\exp\left[-0.858\,407\,346\left(\frac{\pi f M \varepsilon_m}{E_c}\right)^2\right]$	$0.926\,502\,75\left(\frac{M\varepsilon_p}{E_c}\right)$	$0.886\,226\,925$
贝塔分布 $\beta_{2.1}$	$12\xi^2(1-\xi)$	$\exp\left[-0.043\,777\,845\left(\frac{\pi f M \varepsilon_m}{E_c}\right)^2\right]$	$0.209\,231\,568\left(\frac{M\varepsilon_m}{E_c}\right)$	$0.430\,976\,431$
贝塔分布 $\beta_{1.1}$	$6\xi(1-\xi)$	$\exp\left[-0.034\,880\,322\left(\frac{\pi f M \varepsilon_m}{E_c}\right)^2\right]$	$0.186\,762\,743\left(\frac{M\varepsilon_m}{E_c}\right)$	$0.406\,349\,206$
贝塔分布 $\beta_{1.2}$	$12\xi(1-\xi)^2$	$\exp\left[-0.024\,163\,296\left(\frac{\pi f M \varepsilon_m}{E_c}\right)^2\right]$	$0.155\,445\,482\left(\frac{M\varepsilon_m}{E_c}\right)$	$0.369\,408\,369$
贝塔分布 $\beta_{1.3}$	$20\xi(1-\xi)^2$	$\exp\left[-0.017\,720\,91\left(\frac{\pi f M \varepsilon_m}{E_c}\right)^2\right]$	$0.133\,119\,908\left(\frac{M\varepsilon_m}{E_c}\right)$	$0.340\,992\,341$
贝塔分布 $\beta_{1.4}$	$30\xi(1-\xi)^4$	$\exp\left[-0.013\,549\,073\left(\frac{\pi f M \varepsilon_m}{E_c}\right)^2\right]$	$0.116\,400\,488\left(\frac{M\varepsilon_m}{E_c}\right)$	$0.318\,259\,518$
贝塔分布 $\beta_{1.8}$	$90\xi(1-\xi)^8$	$\exp\left[-0.006\,002\,198\left(\frac{\pi f M \varepsilon_m}{E_c}\right)^2\right]$	$0.077\,477\,386\left(\frac{M\varepsilon_m}{E_c}\right)$	$0.258\,509\,74$

从表 3 和表 2 可以看出:$\beta_{1.8}$ 分布的低频特性最好,其次是 $\beta_{1.4} \sim \beta_{1.1}$ 分布,再其次是余弦分布和 $\beta_{2.1}$ 分布,麦克斯韦分布的低频特性亦比较差。应该指出:我们推导出的近贴聚焦和锐聚焦系统的 MTF 指数表达式是在略去 F^4 以上的高次项后求出的。因此,在低频情况下与实际符合得较好。由表 3 可见,初能分布不同,最佳 MTF 所对应的像面位置也不同,且随着低频特性的变好而逐渐向极限像面移动。

表 4 给出了初能分布为 $\beta_{1.8}$ 分布,由实际计算求得的 MTF 值与本文 MTF 理论公式所求值的对比结果。由此可见,两者是很接近的。

由以上分析可见，全色 MTF 指数形式的表达式与实际计算 MTF 值是很接近的，具有令人满意的精度。同时我们通过理论分析以及由表1和表3的比较，证明了 MTF 指数表达式中的 ρ 正是系统的均方根半径 $\bar{r}$，最佳 MTF 的像面位置也正与最小均方根半径的对应，而且只要给定 $\sqrt{\dfrac{\varepsilon_{z_1}}{\varepsilon_m}}$，则 MTF 的优劣与均方根半径的大小在各个像面上都将是一一对应的。

表4　锐聚焦系统最佳像面处理论 MTF 与实际 MTF 的比较
($\beta_{1.8}$ 分布，$\varepsilon_m = 2$ V，$M = 1$，$E_c = 80$ V/mm)

$f(\text{lp}\cdot\text{mm}^{-1})$	$T_{\text{opt}}(f)$（本文）	$T_{\text{opt}}(f)$（实际计算）
10	0.996 304 337	0.995 045 604 6
20	0.985 299 128	0.980 543 442 4
30	0.967 226 578	0.957 516 888 6
40	0.942 480 546	0.927 488 053 0
50	0.911 592 234	0.892 251 311 2
60	0.875 211 244	0.853 636 804 0
70	0.834 082 884	0.813 311 862 4
80	0.789 022 848	0.772 650 416 8
90	0.740 890 454	0.732 677 487 0
100	0.690 561 668	0.694 076 294 6

4　整管 MTF 的计算

在求出系统的电子光学 MTF 后，如果知道像管光学纤维面板和荧光屏的 MTF，就可以很方便地遵照线性元件 MTF 相乘的原理：

$$T_{整管}(f) = T_{电光}(f) \times T_{屏}(f) \times T^2_{面板}(f) \tag{50}$$

来求出整管的 MTF。式（50）中假定像管具有两块纤维学光学面板。

Johnson 曾经指出[15]，大多数光电成像器件的 MTF 可以用如下公式来描述：

$$T(f) = \exp[-(f/f_c)^n] \tag{51}$$

式中，f 是空间频率（lp/mm），f_c 是空间频率常数，n 是器件特性指数。只要给出 f_c 和 n 两个参数，记为 (f_c, n)，就可以唯一地确定其调制传递函数。

对于 $P-20$ 荧光屏，文献 [15] 中给出了两种测量数据：a. $(f_c, n) = (46, 1.1)$ 和 b. $(f_c, n) = (24, 1.1)$。在 5% 的调制度下，分别给出极限鉴别率为 125 (lp/mm) 和 65 (lp/mm)。对于光纤面板，文献 [15] 给出：a. 丝径为 5.8 μm 时，$(f_c, n) = (96, 1.7)$；丝径为 6 μm 时，$(f_c, n) = (80, 1.8)$。下面我们对荧光屏取 $(f_c, n) = (46, 1.1)$，对于光学纤维面板取 $(f_c, n) = (80, 1.8)$ 来计算第一代微光像增强器的 MTF。于是由式（50）可求得整管 MTF 的解析表达式为

$$T_{整管}(f) = \exp[-(\pi\rho f)^2 - (f/46)^{1.1} - 2(f/80)^{1.8}] \tag{52}$$

同文献 [1]，若取 $\varepsilon_m = 0.912\ 4$ V，$E_c = 80$ V/mm，$M = 1$，$f = 64$ lp/mm 时，由本文

式 (14)、式 (41) 可求得余弦分布下电子光学系统的 MTF 为：$T_{opt}(f) \approx 0.824$。而面板和屏的 MTF 则分别为 0.512 与 0.237。于是可求得第一代微光倒像管整管的 MTF 为

$$T(f) = 0.824 \times 0.237 \times (0.512)^2 = 0.051\ 2$$

若我们以 5% 的调制度定义极限鉴别率时，则上述整管的极限鉴别率高于 64 lp/mm。这与实际测试结果是很接近的。

对于第一代三级级联像增强器，可以得级联管的 MTF 的解析表达式为

$$T_{级联}(f) = \exp[-3(\pi\rho f)^2 - 3(f/46)^{1.1} - 6(f/80)^{1.8}] \qquad (53)$$

式中的 ρ、f_c、n 的意义仍如前述。

当 $f = 30$ lp/mm 时，三级级联管的 MTF 为

$$T_{级联}(f) = 0.048\ 388\ 662$$

由以上计算可以看出，当知道像管光学纤维面板和屏的 (f_c, n) 后，使用 MTF 的指数表达式估算整管或三级级联管的 MTF 是非常方便的。

结束语

本文导出了近贴聚焦和锐聚焦系统在各种初能分布下的全色调制传递函数与均方根半径的解析表达式，并通过理论推导和实际计算说明成像系统的均方根半径与 MTF 之间有着密切的联系。证明 MTF 的指数表达式中的 ρ 正是系统的均方根半径 $\bar{r}$ 的大小，不仅可以表征系统鉴别率特性的好坏，而且和 MTF 的优劣在各个像面上都是一一对应的。本文还对不同初能分布下系统的锐聚焦特性和最佳像面位置的变化做了探讨，证明了用指数函数形式表示系统的 MTF 具有计算速度快、形式简洁、使用方便且有令人满意的精度等优点。

参 考 文 献

[1] CSORBA I P. Modulation transfer function of image tube lenses [J]. Applied Optics, 1977, 16(10): 2647 – 2650.

[2] CSORBA I P. Resolution limitations of electromagnetically focused image-intensifier tubes [J]. RCA Review, 1970, 31: 534.

[3] EZARD L A. Measurement of the effective photo-electron emission energy [J]. RCA Review, 1975, 36(4): 711 – 721.

[4] MULDER H. Electron-optical Transfer functions of image intensifies [J]. Advances in Electronics and Electron Physics, 1972, 33A: 563 – 570.

[5] HARTLY K F. On the electron of magnetically focused image tubes. [J]. J. Phys. D. Appl. Phys., 1974, 7(12): 1612 – 1620.

[6] 周立伟, 方二伦. 关于阴极透镜的点扩散函数[J]. 工程光学, 1979(2): 24.

[7] 朱克正. 阴极透镜电子光学传递函数[J]. 电子学报, 1980 (3): 25.

[8] BEURLE R L, WREATHALL W M. Aberration in magnetic focus systems [J]. Advances in Electronics and Electron Physics, 1962, 16: 333 – 339.

[9] 周立伟,方二伦. 倾斜型电磁聚焦系统的电子光学[J]. 工程光学,1980 (2).

[10] 吴全德. 在发射式电子光学系统中光电子速度分布对像质量的影响(Ⅰ) [J]. 物理学报,1956,12(5):419.

[11] ZACHAROV B. A demagnifying image tube for nuclear physics applications [J]. Advances in Electronics and Electron Physics,1962,16:71.

[12] 周立伟. 变像管与像增强器的电子光学[M]. 北京:北京理工大学出版社,1975.

[13] GOODMAN J W. Introduction to fourier optics [M]. New York:McGraw-Hill,1968:2.

[14] CHOU L W. Electron optics of concentric electromagnetic focusing systems [J]. Advances in Electronics and Electron Physics,1979,52:119 – 132.

[15] JOHNSON C B. Classification of electron-optical device modulation transfer functions [J]. Advances in Electronics and Electron Physics,1972,33B:579 – 584.

7.2 On Determination of Poly-energetic Point Spread Function and Modulation Transfer Function in Electron-optical Imaging Systems Using Three Dimensional Coordinates
用三维坐标对电子光学成像系统点扩散函数和调制传递函数的研究

Abstract: A study for determining the poly-energetic point spread function (PSF), root mean square (RMS) radius of aberration and electron optical modulation transfer function (MTF) through the second order transverse chromatic aberrations of imaging electron optical system by means of a geometric figure of three dimensional coordinates has been fully described in the paper.

摘要：本文通过建立成像电子光学系统二级横向色球差的三维曲面图，用形象的手段全面探讨了阴极透镜的点扩散函数（PSF）、横向色差的均方根（RMS）半径和电子光学调制传递函数（MTF）的确定。

Introduction

In the last twenty years, the point spread function and the modulation transfer function in imaging electron optical systems have been investigated by many authors[1-12]. All these works have brought about developments in the study of MTF and evaluation of image quality for the cathode lenses. But up to now there are still a series of essential problems which should be further investigated and solved.

For example,

(1) with the exception of few cases under special conditions, the general analytic expressions of poly-energetic PSF and MTF in terms of the the second order transverse aberration formula have not yet been obtained;

(2) the identity of poly-energetic MTFs obtained from two methods[6-9] has not been proved strictly;

(3) the physical meaning of the "focusing error coefficient"[1] in the MTF's approximate expression of image tube MTF = exp $[-(\pi\rho f)^2]$ [1,2,4,5,10,11] is not clear;

(4) to determine which method of calculating PSF and MTF of imaging systems is more simple and convenient;

Ni Guoqiang[a], Zhou Liwei[a], Fang Erlun[b]. a) Beijing Institute of Technology, b) Xi'an Research Institute of Modern Chemistry. Acta Armamentarii English Edition（兵工学报英文版）No. 1, 1989, 77 – 86.

(5) the poly-energetic PSF expression and the choice of bounds for its integral have not been fully solved.

In view of the above-mentioned facts, in the present paper, the cubic curved surface graph of the second order paraxial lateral aberration formula using a three-dimensional coordinate system in built up. From this we can find out the answers to these problems thoroughly.

1 Fundamental formulae

For the cathode lenses of electron optical imaging systems, it is generally accepted that the angle α between the direction of photoelectrons emitted from axial point of cathode and the axis of the system obeys the Lambert law, and the initial energy distribution of photoelectrons usually complies with the normalized cosine's, beta's, Maxwell, or Reyleigh distribution.

For a rotationally symmetric imaging system, a cylindrical coordinate system is applied. Its origin is located at the central point of the cathode and the z-axis coincides with the axis of the imaging system. Then, the normalized mono-energetic current from the cathode central point, for photoelectrons located between two circular cones of half vertex angles α and $\alpha + d\alpha$ and with normalized energies between ξ and $\xi + d\xi$, can be expressed by

$$dI(\xi,\alpha) = N(\xi)\sin 2\alpha \, d\alpha \, d\xi \tag{1}$$

where $N(\xi)$ is the density distribution function of initial energy of photoelectrons; $\xi = \varepsilon/\varepsilon_m$ ($0 \leq \xi \leq 1$), ε and ε_m are the equivalent initial potential of photoelectrons and its maximum value ($\varepsilon_m = \varepsilon_{max}$), respectively; ε is thus the initial energy of electrons. After moving through the image tube, all these electrons strike at a circular ring with inner radius r and outer radius $r + dr$ on the image plane, in which r is the second order transverse aberration radius (for sharply focusing systems) or the radius of diffusing circle (for proximity systems).

The second order transverse aberration formula of the axial point is universally set up for electrostatic or electromagnetic cathode lenses. For sharply focusing imaging systems this formula can be expressed as follows:

$$\Delta r = r = R_m | \xi \sin 2\alpha - 2\xi_{z_1}^{1/2} \xi^{1/2} \sin \alpha | \tag{2}$$

where $R_m = M\varepsilon_m/E_c$; M is the radial magnification of system; E_c is the electric field intensity at the cathode plane, both of them taking the absolute values; $\xi_{z_1} = \varepsilon_{z_1}/\varepsilon_m$, ε_{z_1} is the axial component of initial potential corresponding to the electrons which can be ideally focused on the image plane under consideration.

For proximity systems, the radius of diffusing circle can be approximately expressed by

$$\Delta r = r = R_m \xi^{1/2} \sin \alpha \tag{3}$$

where $R_m = 2L(\varepsilon_m/\phi_{ac})^{1/2}$; L is the distance between two planar electrodes; ϕ_{ac} is the potential applied to the electrodes.

Form Eq. (1) the normalized mono-energetic PSF in the circular ring with inner radius r and outer radius $r + dr$ can be obtained:

$$j(r) = \sum_i \frac{\sin 2\alpha_i \, d\alpha_i}{2\pi r |dr|} \tag{4}$$

where α_i is the initial angle with which the electron emitted from photocathode will have contributions to the current in the circular ring.

According to Eq. (4), a probability statistics on the initial energy distribution of electrons which will have contributions to the current in the circular ring is carried out, then the normalized poly-energetic PSF can be expressed as follows:

$$P(r) = \int_{\xi_j}^1 j(r) N(\xi) \, d\xi \tag{5}$$

where ξ_j is the minimum value of ξ, which will have contributions to the current density at r.

Applying the Fourier-Bessel conversion on $P(r)$, we can obtain the poly-energetic MTF, which is denoted as $\mathrm{MTF}_1(f)$:

$$\mathrm{MTF}_1(f) = 2\pi \int_0^\infty \left[\int_{\xi_j}^1 j(r) N(\xi) \, d\xi \right] r J_0(2\pi fr) \, dr \tag{6}$$

where f is the spatial frequency, and $J_0(x)$ is the zeroth order Bessel function. Eq. (6) is a sequential integral for r and ξ, and it can be considered as double integral of a certain integral region $\sum$ on the (r, ξ) plane. If $\sum$ satisfies the Dirichlet condition which allows the integral order to exchange, the above integral can be changed into the following form:

$$\mathrm{MTF}_1(f) = 2\pi \int_{\xi_0}^1 \left[\int_{r_0}^\infty r j(r) J_0(2\pi fr) \, dr \right] N(\xi) \, d\xi \tag{7}$$

where r_0, ξ_0 are new integral lower bounds which need to be determined.

If we apply Fourier-Bessel conversion on $j(r)$, to obtain the mono-energetic MTF_ξ, then apply statistics on it about the initial energy distribution of photoelectrons, the MTF of system can be obtained and denoted as $\mathrm{MTF}_2(f)$:

$$\mathrm{MTF}_2(f) = 2\pi \int_0^1 \left[\int_0^\infty r j(r) J_0(2\pi fr) \, dr \right] N(\xi) \, d\xi \tag{8}$$

By comparing Eq. (7) with Eq. (8), it is obvious that if $\sum$ satisfies the Dirichlet condition and $\xi_0 = 0$, $r_0 = 0$, (the Dirichlet condition follows that on the (r, ξ) two-dimensional coordinate plane the intersection points of each straight line of $r = $ const or $\xi = $ const and the boundaries of $\sum$ aren't more than two), the order of integration can be changed, then the following relation can be proved immediately:

$$\mathrm{MTF}_1(f) = \mathrm{MTF}_2(f)$$

All above-mentioned formulae are applicable when the initial energy distribution of electrons complies with the cosine, or beta, or Rayleigh distribution. For the Maxwell distribution, provided ε_m is substituted by ε_p, the most probable initial potential of electrons emitted from photocathode, and the higher bound of integral 1 by ∞, the formulae still hold true, and the integration will be convergent.

2 Study of the sharply focusing system using the three-dimensional coordinates

From the aberration Eq. (2), the three-dimensional coordinate aberration curved surfaces $r = r(\xi, \alpha)$ at the imaging planes corresponding to $\xi_{z_1}^{1/2}$ can be drawn up.

Fig. 1 shows five curved surfaces at the corresponding typical image planes
$$(\xi_{z_1}^{1/2} = 0; \quad 0.25; \quad 0.300,28; \quad 0.5; \quad 1)$$

Because the optimal image plane of sharply focusing system is always located between the image planes corresponding to $\xi_{z_1}^{1/2} = 0$ and $\xi_{z_1}^{1/2} = 1$ respectively, it need only to discuss the cases for $0 \leqslant \xi_{z_1}^{1/2} \leqslant 1$.

The following facts are obvious from Fig. 1.

(1) When $\alpha = 0$ or $\xi = 0$, $r \equiv 0$.

(2) Generally, there are two peak regions: one is an arch peak region (marked as region-1) and the other is sharp peak region (region-2) whose blade corresponds to $\alpha = \pi/2$. The demarcation line between the two regions at $r = 0$ on the (ξ, α) plane satisfies the following equation:
$$\xi^{1/2} \cos \alpha = \xi_{z_1}^{1/2} \quad (\xi \geqslant \xi_{z_1}) \tag{9}$$

(3) At the image plane corresponding to $\xi_{z_1}^{1/2} = 0$, only region-1 exists, which is symmetrical about $\alpha = \pi/4$, and at the image plane of $\xi_{z_1}^{1/2} = 1$, only region-2 exists. Along with the increase of $\xi_{z_1}^{1/2}$ from 0 to 1, the area and height of region-1 are gradually reduced, while those of region-2 become larger.

(4) There is an extreme value r_{1m} in region-1 on each cross section of ξ = const. The equation for the α-coordinate of these extreme value points is $\dfrac{dr}{d\alpha} = 0$, that is
$$\cos \alpha = [(\xi_{z_1} + 8\xi)^{1/2} + \xi_{z_1}^{1/2}]/4\xi^{1/2} \quad (\xi \geqslant \xi_{z_1}) \tag{10}$$

thus,
$$r_{1m}/R_m = 0.5\xi^{1/2}[(\xi_{z_1} + 8\xi)^{1/2} - 3\xi_{z_1}^{1/2}]\{1 - [(\xi_{z_1} + 8\xi)^{1/2} + \xi_{z_1}^{1/2}]^2/16\xi\}^{1/2} \quad (\xi \geqslant \xi_{z_1}) \tag{11}$$

When $\xi = 1$, r_{1m} takes the maximum value r_{1mm}:
$$r_{1mm}/R_m = 0.5[(\xi_{z_1} + 8)^{1/2} - 3\xi_{z_1}^{1/2}]\{1 - [(\xi_{z_1} + 8)^{1/2} + \xi_{z_1}^{1/2}]^2/16\}^{1/2} \tag{12}$$

(5) There is an extreme value r_{2m} for each value of ξ in region-2:
$$r_{2m}/R_m = 2\xi_{z_1}^{1/2}\xi^{1/2} \tag{13}$$

which is located at $\alpha = \pi/2$. When $\xi = 1$, r_{2m} achieves the maximum value r_{2mm}:
$$r_{2mm}/R_m = 2\xi_{z_1}^{1/2} \tag{14}$$

(6) It can be easily shown that: $\dfrac{dr_{2m}}{d\xi} > 0$, $\dfrac{d^2 r_{2m}}{d\xi^2} < 0$; $\dfrac{dr_{1m}}{d\xi} > 0$, $\dfrac{d^2 r_{1m}}{d\xi^2} > 0$; which indicate that the increase of r_{1m} is more rapid than r_{2m} along with the increase of ξ. If there is a value of ξ_α,

at which relation $r_\alpha = r_{1m} = r_{2m}$ exists, there must be the relation $r_{1m} > r_{2m}$ when $\xi > \xi_\alpha$.

Let $r_m = \max(r_{1m}, r_{2m})$, which represents the maximum aberration radius of mono-energetic electrons at the image plane of $\xi_{z_1}^{1/2}$. Let $r_{mm} = \max(r_{1mm}, r_{2mm})$, which is the maximum aberration radius of poly-energetic electrons on this image plane. Then $j(r) = 0$ when $r > r_m$, and $P(r) = 0$ when $r > r_{mm}$.

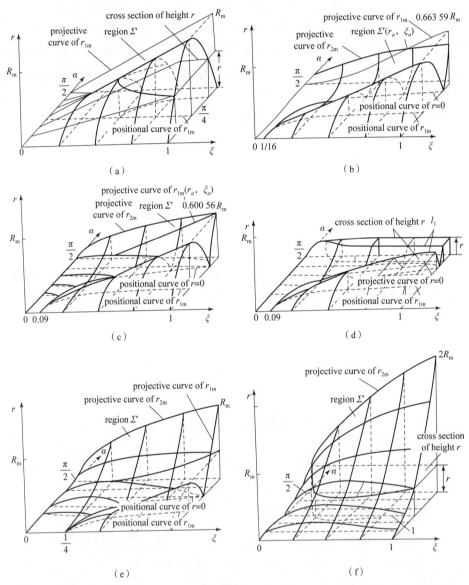

Fig. 1 Three dimensional aberration curved surfaces (r, ξ, α) at the image plane corresponding to $\xi_{z_1}^{1/2}$

(a) $\sqrt{\xi_{z_1}} = 0$; (b) $\sqrt{\xi_{z_1}} = 0.25$; (c) $\sqrt{\xi_{z_1}} = 0.30028$; (d) $\sqrt{\xi_{z_1}} = 0.5$; (e) $\sqrt{\xi_{z_1}} = 1$

(7) From Eq. (11) and Eq. (13) we can obtain: $\xi_\alpha^{1/2} = \xi_{z_1}^{1/2}/0.300,28$. It exists only at those image planes with $0 \leq \xi_{z_1}^{1/2} \leq 0.300,28$.

(8) When $0.300,28 < \xi_{z_1}^{1/2} \leq 1$, the relation $r_{1m} < r_{2m}$ is always set up and the projective

region $\sum'$ of the curved surface on the (r, ξ) plane is shown in Fig. 2 (a).

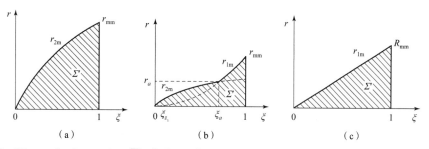

Fig. 2 The projection region $\sum'$ of aberration curved surface on the coordinate plane (r, ξ)

(a) $0.300, 28 \leq \sqrt{\xi_{z_1}} \leq 1$; (b) $0 < \sqrt{\xi_{z_1}} < 0.300, 28$; (c) $\sqrt{\xi_{z_1}} = 0$

(9) When $0 < \xi_{z_1}^{1/2} < 0.300, 28$, the $\sum'$ of the curved surface is drawn up as in Fig. 2 (b) and

$$r_\alpha = 2R_m \xi_{z_1}/0.300,28 \quad (0 < \xi_{z_1}^{1/2} < 0.300,28) \tag{15}$$

From Eq. (2) it is known that ξ_α corresponds to the coordinate value of $\alpha = \alpha_0$, where $\alpha_0 = 38°10'$ ($\sin\alpha_0 = (\sqrt{5}-1)/2$) and it does not depend on the position of image plane.

(10) When $\xi_{z_1}^{1/2} = 0$, then $\xi_\alpha = 0$, $r_\alpha = 0$; and the aberration equation can be expressed as

$$r = R_m \xi \sin 2\alpha \tag{16}$$

r_{1m} is located at $\alpha = \pi/4$, $r_{1m} = R_m \xi$. Its projective region $\sum'$ is given in Fig. 3 (c).

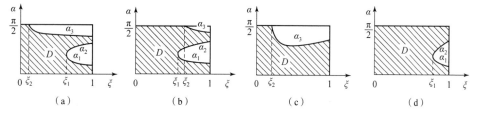

Fig. 3 Projection regions D

From Eq. (2) we know that at the image plane of $\xi_{z_1}^{1/2}$, those electrons which are emitted from cathode central point and possess certain initial conditions of ξ and α fall in the disc with radius r. If the total current $I(r)$ in the disc is considered as a probability distribution function of the random variable r, then the current of the image in the ring with inner radius r and outer radius $r + dr$ can be taken as a probability density function of the random variable r, namely

$$P(r) = \frac{1}{2\pi r} \frac{dI(r)}{dr} \tag{17}$$

Cutting an aberration curved surface with a plane of height $r(r < r_{mm})$ which is parallel to the (ξ, α) plane, we obtain the intersection curve l_1. Projecting l_1 in the (ξ, α) plane, we obtain the curve l, which cuts apart a region-D that is the projection of the part of the aberration curved surface under the plane with height r on the (ξ, α) plane. All electrons whose initial conditions take the values of ξ and α in region-D will fall in the disc with radius r and we have

$$I(r) = \iint_D dI(\xi, \alpha) \tag{18}$$

Usually l consists of three curves (see Fig. 3). $\alpha_i = \alpha_i (r, \xi, \xi_{z_1}^{1/2})$ ($i = 1, 2, 3$), which are the solutions of Eq. (2), with $\alpha_1 \leq \alpha_2 \leq \alpha_3$ (equal-sign- will be taken when $\xi = \xi_1$). In Fig. 3, ξ_1 and ξ_2 are the ξ-coordinates of the intersection points in which a plane with height r cuts off the arch peak r_{1m} of region-1 and the sharp peak r_{2m} of region-2, respectively. They can be expressed as follows:

$$\xi_1^{1/2} = \frac{\xi_{z_1}^{1/2} \sin\alpha + [\xi_{z_1} \sin^2\alpha + (r/R_m) \sin 2\alpha]^{1/2}}{\sin 2\alpha}, \tag{19}$$

$$\cos\alpha = [\xi_{z_1}^{1/2} + (\xi_{z_1} + 8\xi_1)^{1/2}]/4\xi_1^{1/2},$$

$$\xi_2 = r^2/(4\xi_{z_1} R_m^2) \tag{20}$$

For all the four cases in Fig. 3, we evaluate the integration of Eq. (18) on the region D which is divided into some parts. Noting that $\alpha_1(\xi_1) = \alpha_2(\xi_2)$, $\alpha_3(\xi_2) = \pi/2$, and using Eq. (17), we have the following general expression for all the four cases.

$$P(r) = \frac{1}{2\pi r} \int_{\xi_i}^1 \sum_i \sin 2\alpha_i \left|\frac{d\alpha_i}{dr}\right| N(\xi) d\xi \tag{21}$$

where $d\alpha_1/dr > 0$, $d\alpha_2/dr < 0$, $d\alpha_3/dr > 0$; $\xi_i = \min(\xi_1, \xi_2)$. Thus, we have achieved Eq. (4) and Eq. (5) from strict derivation.

It is clear that the projective region $\sum'$ in Fig. 2 are just the integral region $\sum$ that need to be determined. They all satisfy the above-mentioned two conditions about $\sum$. As a result, in sharply focusing systems, the MTFs derived by the above-mentioned two methods are identical.

In fact, $P(r)$ is the sum of the following two terms or one of them:

$$P_1(r) = \frac{1}{2\pi r} \int_{\xi_1}^1 \left(\sin 2\alpha_1 \frac{d\alpha_1}{dr} - \sin 2\alpha_2 \frac{d\alpha_2}{dr}\right) N(\xi) d\xi,$$

$$P_2(r) = \frac{1}{2\pi r} \int_{\xi_2}^1 \sin 2\alpha_3 \frac{d\alpha_3}{dr} N(\xi) d\xi \tag{22}$$

which represent parts of the poly-energetic current densities at distance r from the axis, contributed by the electrons whose initial conditions are in region-1 or region-2, respectively. $P(r)$ can be computed using Table 1.

Table 1 Table for computing PSF $0 < r < r_{mm}$

$\xi_{z_1}^{1/2}$	0	$0 < \xi_{z_1}^{1/2} < 0.300,28$		$0.300,28$	$0.300,28 < \xi_{z_1}^{1/2} < 1$		1
R	—	$r < r_{2mm}$	$r \geq r_{2mm}$	—	$r < r_{1mm}$	$r \geq r_{1mm}$	—
α_1, α_2 and $P_1(r)$	exist	exist	exist	exist	exist	—	—
α_3 and $P_2(r)$	—	exist	—	exist	exist	exist	exist
$P(r)$	$P_1(r)$	$P_1(r) + P_2(r)$	$P_1(r)$	$P_1(r) + P_2(r)$	$P_1(r) + P_2(r)$	$P_2(r)$	$P_2(r)$

Of all the cross sections of the curved surface with ξ = const there are only three kinds of curves (see Fig. 4). Ref. [1] only gave a curve similar to Fig. 4 (c). But it was a two-dimensional graph (mono-energetic) rather than a three-dimensional (poly-energetic) one, so the poly-energetic PSF with initial energy distribution could not be further investigated.

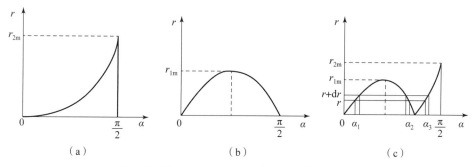

Fig. 4 The planar curves of $r - \alpha$ (ξ = const)

In accordance with the above-mentioned facts, the mono-energetic can be expressed as:

$$\mathrm{MTF}_\xi(f) = \int_0^{r_m} \sum_i \sin 2\alpha_i \left| \frac{d\alpha_i}{dr} \right| J_0(2\pi f r) dr$$

$$= \int_0^{\pi/2} \sin 2\alpha J_0 [2\pi f R_m(\xi \sin 2\alpha - 2\xi_{z_1}^{1/2} \xi^{1/2} \sin\alpha)] d\alpha \tag{23}$$

$$= \sum_{n=0}^{\infty} \frac{(-1)^n}{(n!)^2} (2\pi f R_m)^{2n} \sum_{m=0}^{2n} (-1)^m C_{2n}^m \beta\left(n+1, n-\frac{m}{2}+1\right) \xi_{z_1}^{m/2} \xi^{2n-m/2} \tag{24}$$

where C_M^N is the combination symbol and $\beta(m,n)$ is beta function. Then, we have

$$\mathrm{MTF}_2(f) = \int_0^1 \mathrm{MTF}_\xi(f) N(\xi) d\xi = \int_0^1 \left[\int_0^{\pi/2} \sin\alpha J_0(2\pi f r) dr \right] N(\xi) d\xi \tag{25}$$

Eq. (23) was given by Ref. [7]. Eq. (25) was shown in Ref. [14], but its physical meaning was not clear.

Substituting Eq. (24) into Eq. (25) we may have the analytic expressions of MTF under various initial energy distributions. The MTF_2 under cosine, Beta and Maxwell distributions have been given by Ref. 8. Here, the MTF under Rayleigh distribution will be given in addition. Its initial energy distribution function of electrons will be

$$N(\xi) = K\xi \exp(-4\xi) \tag{26}$$

where $K = 16/[1 - 5\exp(-4)] = 17.612,963,39$, $\varepsilon_m = 4\varepsilon_p$, then

$$\mathrm{MTF}(f) = K \sum_{n=0}^{\infty} \frac{(-1)^n}{n!} (2\pi f R_m) 2n \left[\exp(-4) \sum_{m=0,2,4,\cdots}^{2n} C_{2n}^m \left(n - \frac{m}{2}\right)! \xi_{z_1}^{1/2} \times \right.$$

$$\left. \sum_{i=1}^{\infty} \frac{4^{i-1}}{\left(2n - \frac{m}{2} + i + 1\right)!} - \sum_{m=1,3,5,\cdots}^{2n-1} C_{2n}^m \frac{\left(n - \frac{m}{2}\right)!}{\left(2n - \frac{m}{2} + 1\right)!} \xi_{z_1}^{m/2} \sum_{i=0}^{\infty} \frac{(-1)^i 4^i}{i!\left(2n - \frac{m}{2} + i + 2\right)} \right]$$

$$\tag{27}$$

For lower spatial frequencies, the progression of Eq. (27) can be truncated, since only terms for $n=0$ and $n=1$ are significant:

$$\text{MTF}(f) = 1 - (\pi f \rho)^2 \qquad (28)$$

where $\rho = (0.155,910,444 - 0.651,092,922\xi_{z_1}^{1/2} + 0.838,703,601\xi_{z_1})^{1/2} R_m$.

The optimal image plane, the minimum value of $\rho(\rho_{\min})$ and optimal MTF($\text{MTF}|_{\text{opt}}$) on this plane thus are given as follows respectively:

$$\xi_{z_1}^{1/2}|_{\text{opt}} = 0.388,154,361 \quad (\varepsilon_{z_1}/\varepsilon_p)^{1/2}|_{\text{opt}} = 0.776,308,723),$$

$$\rho_{\min} = 0.171,895,797 M\varepsilon_m/E_c = 0.687,583,189 M\varepsilon_p/E_c,$$

$$\text{MTF}(f)|_{\text{opt}} = 1 - 0.029,548,165 (\pi f R_m)^2 \qquad (29)$$

For Maxwell distribution, $(\varepsilon_{z_1}/\varepsilon_p)^{1/2}|_{\text{opt}} = 0.886,226,917$, $\rho_{\min} = 0.926,502,750 M\varepsilon_p/E_c$ [8]. By comparison it is obvious that by cutting off the tail of high energy of Maxwell distribution the optimal image plane moves toward to the plane $\xi_{z_1}^{1/2} = 0$, $\rho_{\min}$ reduces and the MTF of the imaging system becomes better.

The resolution of imaging systems under Rayleigh distribution was discussed in Ref. [15 – 16], where the planes for $\xi_{z_1}^{1/2} = 0$ and $(\varepsilon_{z_1}/\varepsilon_p)^{1/2} = 1$ were considered as the optimal planes respectively, but in fact the author applied the Maxwell distribution during calculation. It is considered to be inappropriate and incorrect[17].

On the plane $\xi_{z_1}^{1/2} = 0$, the results become very simple (see Fig. 2 (c) and Fig. 5):

$$\begin{aligned} \alpha_1 &= \frac{1}{2}\sin^{-1}(r/R_m\xi), \\ \alpha_2 &= \frac{\pi}{2} - \frac{1}{2}\sin^{-1}(r/R_m\xi) \end{aligned} \qquad (30)$$

$$j(r) = 1/[2\pi R_m^2 \xi (\xi^2 - \xi_1^2)^{1/2}] \quad (\xi_1 = r/R_m) \qquad (31)$$

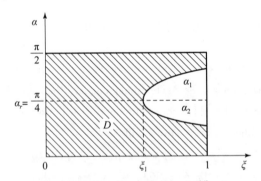

Fig. 5　Projection region D of $\xi_{z_1}^{1/2} = 0$

Substituting these two equations and the expressions of $N(\xi)$ into Eq. (21), we can obtain expressions of $P(r)$ under various initial energy distributions.

For example, when beta distributions are taken, $P(q)$ (where $q = r/R_m$ is the dimensionless aberration radius) will be written as:

$$P(q)|_{\beta_{1,1}} = \frac{3}{\pi R_m^2}\left[\ln\frac{1+(1-q^2)^{1/2}}{q} - (1-q^2)^{1/2}\right] \qquad (32)$$

$$P(q)|_{\beta_{1,2}} = \frac{6}{\pi R_m^2}\left[\left(1+\frac{1}{2}q^2\right)\ln\frac{1+(1-q^2)^{1/2}}{q} - \frac{3}{2}(1-q^2)^{1/2}\right] \quad (33)$$

$$P(q)|_{\beta_{2,1}} = \frac{3}{\pi R_m^2}\left[(1-q^2)^{1/2} - q^2\ln\frac{1+(1-q^2)^{1/2}}{q}\right] \quad (34)$$

$$P(q)|_{\beta_{1,3}} = \frac{10}{\pi R_m^2}\left[\left(1+\frac{3}{2}q^2\right)\ln\frac{1+(1-q^2)^{1/2}}{q} - \frac{11}{6}(1-q^2)^{1/2} - \frac{2}{3}q^2(1-q^2)^{1/2}\right] \quad (35)$$

$$P(q)|_{\beta_{1,4}} = \frac{15}{\pi R_m^2}\left[\left(1+3q^2+\frac{3}{8}q^4\right)\ln\frac{1+(1-q^2)^{1/2}}{q} - \frac{25}{12}(1-q^2)^{1/2} - \frac{55}{24}q^2(1-q^2)^{1/2}\right] \quad (36)$$

$$P(q)|_{\beta_{1,8}} = \frac{45}{\pi R_m^2}\left[\left(1+14q^2+\frac{105}{4}q^2+\frac{34}{4}q^6+\frac{35}{128}q^8\right)\ln\frac{1+(1-q^2)^{1/2}}{q} + 6(1-q^2)^{1/2} - \frac{77}{6}(1-q^2)^{3/2} + \frac{14}{5}(1-q^2)^{5/2} + \frac{221}{168}(1-q^2)^{7/2} - \frac{189}{4}q^2(1-q^2)^{1/2} + \frac{91}{6}q^2(1-q^2)^{3/2} + \frac{591}{80}q^2(1-q^2)^{5/2} - \frac{35}{4}q^4(1-q^2)^{1/2} + \frac{1763}{192}q^4(1-q^2)^{3/2} - \frac{35}{128}q^6(1-q^2)^{1/2}\right] \quad (37)$$

The above expressions for $P(r)$ under $\beta_{1,1}$ distributions are all divergent at $r=0$, but their MTFs obtained from Fourier-Bessel conversion are convergent. For instance:

$$\text{MTF}_1(f)|_{\beta_{1,1},\xi_{z_1}^{1/2}=0} = 6\sum_{n=0}^{\infty}\frac{(-1)^n}{(n!)^2}(\pi fR_m)^{2n} \times \left[\frac{1}{2(n+1)}\sum_{m=0}^{n}C_n^m B(m+1,n+2) + \frac{1}{4(n+1)^2} - \frac{1}{2}B\left(\frac{3}{2},n+1\right)\right] \quad (38)$$

The following expression was given by Ref. [8]:

$$\text{MTF}_2(f)|_{\beta_{1,1},\xi_{z_1}^{1/2}=0} = 6\sum_{n=0}^{\infty}(-1)^n\frac{(2\pi fR_m)^{2n}}{(2n+3)!} \quad (39)$$

Although the two forms are different, their expansions are identical:

$$\text{MTF}(f) = 1 - \frac{1}{20}(2\pi fR_m)^2 + \frac{6}{7!}(2\pi fR_m)^4 - \frac{6}{9!}(2\pi fR_m)^6 + \cdots \quad (40)$$

Similarly, for $\beta_{2,1}$ distribution, we can write MTF_1 and MTF_2 respectively:

$$\text{MTF}_1(f)|_{\beta_{2,1},\xi_{z_1}^{1/2}=0} = 6\sum_{m=0}^{\infty}\frac{(-1)^n}{(n!)^2}(\pi fR_m)^{2n} \times \left[\frac{1}{2}B\left(\frac{3}{2},n+1\right) - \frac{1}{4(n+1)^2} - \frac{1}{2(n+2)}\sum_{m=0}^{n=1}C_{n+1}^m B(m+1,n+3)\right] \quad (41)$$

$$\text{MTF}_1(f)|_{\beta_{2,1},\xi_{z_1}^{1/2}=0} = 6\sum_{m=0}^{\infty}\frac{(-1)^n}{(n!)^2}(\pi fR_m)^{2n} \quad (42)$$

Again the two forms are not alike, but their expansions are the same:

$$\text{MTF}(f) = 1 - \frac{1}{15}(2\pi fR_m)^2 + \frac{7}{9!}(2\pi fR_m)^4 - \frac{96}{10!}(2\pi fR_m)^6 + \cdots \quad (43)$$

It is evident that the forms of $\text{MTF}_2(f)$ are succinct and the expansions [see Eq. (40) and Eq. (43)] attenuate very quickly for lower spatial frequencies.

At lower frequencies, the zeroth order Bessel function should only take the first and second terms. Substituting them into Eq. (25), and using the definition of the root mean square (RMS) radius of imaging system's aberration, we obtain immediately:

$$\text{MTF}(f) = 1 - (\pi f)^2 \int_0^1 \int_0^{\pi/2} r^2 \sin 2\alpha N(\xi) \, d\alpha d\xi = 1 - (\pi \rho f)^2 \qquad (44)$$

$$\rho = (\overline{r^2})^{1/2} = (\int_0^1 \int_0^{\pi/2} r^2 \sin 2\alpha N(\xi) \, d\alpha d\xi)^{1/2} \qquad (45)$$

It shows that the quantity ρ, which is called the "focusing error coefficient" in Ref. [1], is just the RMS radius of aberration and it does not depend on initial energy distributions of photoelectrons. For other initial angle distribution form of electrons this conclusion also holds true.

3 Study of the proximity focusing system using three-dimensional coordinates

Form Eq. (3) and Eq. (4) the mono-energetic PSF of the system can be easily written as:

$$j(r) = 1/(\pi R_m^2 \xi) \qquad (46)$$

In Fig. 6 the cubic curved surface of aberration Eq. (3) for proximity systems is drawn up and from it we can see that $r_m = R_m \xi^{1/2}$ (when $\alpha = \pi/2$), $r_{mm} = R_m$ (when $\xi = 1$). The following formula is obtained easily:

$$P(r) = \frac{1}{\pi R_m^2} \int_{\xi_1}^1 \frac{N(\xi)}{\xi} d\xi \quad (\xi_1 = r^2/R_m^2) \qquad (47)$$

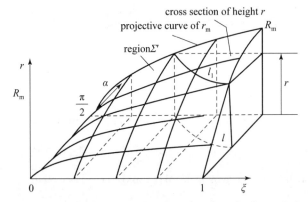

Fig. 6 Three dimensional aberration curved surface (r, ξ, α) of proximity focusing system

It can be seen from this that the projective region of aberration curved surface on the (r, ξ) plane is just the integral region $\sum$ which satisfies the above-mentioned two conditions. So for proximity systems the following relation still holds true: $\text{MTF}_1(f) \equiv \text{MTF}_2(f)$, $\rho = (\overline{r^2})^{1/2}$. Here we shall give the PSF and MTF under Rayleigh distribution to complement Ref. [8]:

$$P(r) = \frac{K}{4\pi R_m^2} [\exp(-4r^2/R_m^2) - \exp(-4)] \qquad (48)$$

$$\text{MTF}(f) = \frac{K}{\exp(4)} \sum_{n=0}^\infty \frac{(-1)^n}{n!} (\pi f R_m)^{2n} \sum_{m=1}^\infty \frac{4^{m-1}}{(n+m+1)!} \qquad (49)$$

At lower frequencies, we may take only the terms of $n = 0$ and $n = 1$ in Eq. (49), then we have

$$\text{MTF}(f) = 1 - 0.209{,}675{,}915\,(\pi f R_m)^2 \tag{50}$$

$$\rho = 0.457{,}903{,}827 R_m = 1.831{,}615{,}311 L\,(\varepsilon_p/\phi_{ac})^{1/2} \tag{51}$$

By comparing it with $\rho = 2L\,(\varepsilon_p/\phi_{ac})^{1/2}$ under Maxwell distribution[8], it is obvious that the value of ρ under Rayleigh distribution is smaller and the system's MTF becomes better.

Conclusions

(1) In the present paper, by using three-dimensional coordinate system (r, ξ, α) the cubic geometrical graphs of the first order aberration radius have been built up, the poly-energetic PSF and MTF of sharply focusing systems and proximity systems have been discussed.

(2) From strict mathematic derivation the mono-energetic and poly-energetic PSF expressions have been obtained. Their physical meaning is very clear. All parameters and integral bounds in the PSF formulae have definite meanings and their expressions to be solved in the graphs of aberration curved surfaces are also given.

(3) In this paper the identity of MTFs calculated using two methods has been proved strictly. It is thus possible to avoid the complicated and dwell on the simple, enabling as to obtain the theoretical and analytic expressions of MTF at different focusing image planes of cathode lenses under various initial energy distributions of photoelectrons.

(4) In the paper, we have proved that at all focusing image planes of cathode lenses the so called "focusing error coefficient" ρ is just the RMS radius of the first order transverse aberration. Thus, it is theoretically justified that the RMS radius of aberration can be used to express the performance of resolution of the system as an auxiliary evaluating index for calculating MTF. Its advantages lie in that the computation is much easier and the physical meaning is very clear.

(5) The results show that the suggested method of the three-dimensional coordinates may be of practical importance for deeply understanding the physical meaning of aberration formulae of photoelectronic imaging systems and for making further improvements on PSF and MTF of the image tubes.

References

[1] CSORBA I P. Resolution limitations of electro-magnetically focused image-intensifier tubes [J]. RCA Review, 1970, 31:534

[2] CSORBA I P. Modulation transfer function of image tube lenses[J]. Applied Optics, 1977, 16(10): 2647 – 2650.

[3] HARTLEY K F. On the electron optics of magnetically focused image tubes. [J]. J. Phys. D: Appl. Phys., 1974, 7:1612.

[4] СЕМЕНОВ Е П. Зависимость разрешаюшеи способности иммерсионных электрических линз от величины напряженности электрстического поля у катода. [J]. РиЭ., 1968, 13:334 – 339.

[5] СЕМЕНОВ Е П. Разрешаюая способность катоднои системы с однородным

полем[J]. ОМП., 1971,11:4 −6.

[6] EZARD L A. Measurement of the effective photo-electron emission energy [J]. RCA Review, 1975,36:711 −721.

[7] КУЛИКОВ Ю В. Об оценке качества изображения катодных электронно-оптических систем с учетом аберации третьего порядка[J]. Радиотехника и Электроника, 1975,20 (6):1249 −1254.

[8] ZHOU L W, AI K C, FANG E L. Study of electron optical modulation transfer function and root mean square radius in the photo electronics image focusing systems [J]. Journal of Beijing Institute of Technology,1982,3: 36 −51 (in Chinese).

[9] ZHU K Z. Cathode lens electron optics transfer function[J]. Acta Electronica,1980,3: 25 (in Chinese).

[10] JOHNSON C B. Classification of electron-optical device modulation transfer functions [J]. Advances in Electronics and Electron Physics,1972,33B:579 −584.

[11] BEURLE R L, WREATHALL W M. Aberration in magnetic focus systems [J]. Advances in Electronics and Electron Physics,1962,16:333.

[12] SONG K C. The effects of performance feedback, self-esteem, performance standard on feedback recipient's responses: an attributional analysis [D]. Columbus: Ohio State University,1985.

[13] ЦЫГАНЕНКО В В, ЛАЧАЩВИЛИ Р А, БОБРОВСКИЙ М А. Разрешаюшая способность бипнарных электронно-оптических систем[J]. ОМП, 1972,12:12 −15.

[14] WENDT G. Sur le pouvoir separateur du convertissenr d'images a champs homogenes electrostatique et magnetique. [J]. Annales de Radioelectricite, 1955, 39:74 −82 (in French).

[15] БОНЩТЕДТ Б, ДМИТРИЕВА Т, ЦУККЕРМАН И И. К расчету разрешающей способности электронно-оптического преобразователя с однородными полями[J]. Ж. Т. Ф., 1956, 26(9):1966 −1968.

[16] WU Q D. The influence of velocity distribution of photoelectron upon image qualities in emission electron optical systems(II)[J]. Acta Physica Sinica,1957,13: 90 (in Chinese).

[17] WU Q D. The influence of velocity distribution of photoelectron upon image qualities in emission electron optical systems (I)[J]. Acta Physica Sinica,1956,12: 419(in Chinese).

7.3 电子光学成像系统全色点扩散函数的研究
A Study of Multi-energy Point Spread Function in Electron Optical Imaging Systems

摘要：本文建立新的二级近轴横向色差半径的三维坐标曲面图形，用形象的手段讨论阴极透镜的全色点扩散函数，给出了在一般像面上归一化全色点扩散函数的表达式及若干实例。

Abstract: A new geometric figure of the three dimensional coordinate for studying the second order paraxial lateral chromatic aberration radius in cathode lenses has been suggested. The multi-energy point spread function (PSF) considering angular distribution and initial energy distribution for the cathode lenses has vividly been depicted and the expressions of normalized multi-energy PSF at arbitrary image plane have also been given, in which both the various parameters and the limits of different integrals have a clear and definite physical and mathematical meaning. Some practical examples for determining multi-energy PSF in the sharp focusing systems and proximity focusing systems have also been described.

引 言

长期以来，由阴极透镜普遍成立的二级近轴横向色差公式出发计算成像系统全色点扩散函数（PSF），是各国电子光学工作者颇为关心的问题[1~3]。早在1956年，吴全德[4~5]在研究光电子发射初能量呈贝塔（beta）分布下成像系统的分辨距离时，已从原理上提出了求成像系统全色PSF的方法。尽管文献[4]没有具体说明所求的是PSF以及积分限的由来，结果还是正确的，却一直没有得到应有的重视。以后文献[2~3]曾探讨了求PSF的方案，但问题并没有圆满地解决。

在锐聚焦系统中，二级近轴横向色差半径 Δr 可表示为

$$\Delta r = R_m \left| \xi \sin 2\alpha - 2\sqrt{\xi_{z_1}}\sqrt{\xi}\sin\alpha \right| \tag{1}$$

式中，$R_m = M\varepsilon_m/E_c$；M 为系统横向放大率；E_c 为阴极面处场强，均取绝对值；α 为电子发射的初角度；$\xi = \varepsilon/\varepsilon_m (0 \leqslant \xi \leqslant 1)$，$\varepsilon$、$\varepsilon_m$ 分别为电子发射的初电位和最大初电位；$\xi_{z_1} = \varepsilon_{z_1}/\varepsilon_m$，$\varepsilon_{z_1}$ 是在所讨论的像面上理想聚焦的电子所具有的轴向初电位，$\sqrt{\xi_{z_1}}$ 值对应于确定的像面位置。

近贴系统的散射圆半径 Δr 的近似表达式为

倪国强[a]，周立伟[a]，方二伦[b]. a）北京理工大学，b）西安现代化学研究所. 北京理工大学学报（Journal of Beijing Institute of Technology），V.8, No.1, 1988, 31–42.

$$\Delta r = R_m \sqrt{\xi} \sin\alpha \tag{2}$$

式中，$R_m = 2L(\varepsilon_m/\phi_{ac})^{1/2}$，$L$ 为两平面间距离，ϕ_{ac} 为极间加速电位。

我们在文献 [1] 中，由式 (1) 与式 (2) 出发，分别建立了锐聚焦系统的 $\sqrt{\xi_{z_1}}$ 像面上和近贴聚焦系统的像面上像差半径 $r = \Delta r = \Delta r(\xi, \alpha)$ 三维坐标曲面。由像差方程得知，在像面上，满足一定 ξ、α 条件的电子打在离轴半径为 r 的圆盘内。把该圆盘内的归一化总电流值 $I(r)$ 看作是对 r 的概率分布函数，则落在半径 $r \to (r+dr)$ 圆环内的点像电流值就相应于对 r 的概率密度函数，于是归一化的全色 PSF 可由下式求得

$$P(r) = \frac{1}{2\pi r}\frac{dI(r)}{dr} \tag{3}$$

用高度为 r（$r < r_{mm}$，r_{mm} 为像面上最大像差半径）的平面截像差曲面，将截线 l_1 向 (ξ, α) 坐标平面投影，得 l 曲线，它分划出在 r 平面以上的像差曲面部分在 (ξ, α) 平面上的投影区域 D。以 D 内的 (ξ, α) 值为初始条件的电子均落在半径为 r 的圆盘内。在电子初角度分布假定满足朗伯 (Lambert) 定律，初能量分布密度函数为 $N(\xi)$ 时，有

$$I(r) = \iint_D N(\xi) \sin 2\alpha \, d\alpha \, d\xi \tag{4}$$

由此，文献 [1] 给出了 $P(r)$ 的积分表达式，各参数及积分限在坐标系中有明确的物理意义和几何意义，并进而讨论了电子光学调制传递函数 (MTF) 的确定，从理论上比较全面系统地解决了电子初角度和初能量分布时的 PSF 和 MTF 的计算问题。

但是，对锐聚焦系统的一般像面，上述方法还有两个困难：一是积分下限之一的 ξ_1 求解繁复，很难用显式表达；二是积分函数与像差方程有关，该方程实际上是一个四次方程；积分 $P(r)$ 时，随着积分变量 ξ 值的变化，要反复求解该方程，计算量大。

为克服上述困难，对 (r, ξ, α) 坐标系做形如

$$\xi = \xi, \quad v_r = \sqrt{\xi} \sin\alpha \tag{5}$$

的坐标变换，通过雅可比 (Jacobi) 行列式，式 (4) 就成为

$$I(q) = \iint_{D_n} 2v_r \frac{N\xi}{\xi} dv_r d\xi \tag{6}$$

式中，$q = r/R_m$ 为无量纲像差半径，v_r 实际上是电子发射的无量纲径向初速度。于是 $I(q)$ 就变换为在 (q, ξ, v_r) 坐标系中，被积函数在区域 D_n 上的积分，D_n 是像差方程 $q = q(\xi, v_r)$ 的曲面被高度为 q 的平面截在平面以下的曲面部分在 (ξ, v_r) 坐标平面上的投影区域。

以上公式对电子初能量分布为余弦、贝塔、瑞利 (Rayleigh) 分布均适用；对电子初能量服从麦克斯韦 (Maxwell) 分布，只要做一定的代换，公式依然成立[1]。

下面我们用新的三维坐标系分别研究锐聚焦系统与近贴系统的全色 PSF，为计算全色 PSF 提供切实方便的途径。

1 锐聚焦系统像差半径曲面图及参数确定

在新的坐标系中，像差方程式 (1) 成为

$$q = 2v_r \left| \sqrt{\xi - v_r^2} - \sqrt{\xi_{z_1}} \right| \tag{7}$$

由此可画出在以 $\sqrt{\xi_{z_1}}$ 为参变量的各个像面上像差半径的三维曲面图。这里选择 $\sqrt{\xi_{z_1}} = 0$、0.300 28、0.5 及 1 这四个典型像面，如图 1 所示。由图可见：

a. (ξ, v_r) 定义域：$0 \leq \xi \leq 1$，$0 \leq v_r \leq 1$，$v_r \leq \sqrt{\xi}$。

b. $v_r = 0$ 时，$q \equiv 0$。

c. 一般存在两个峰区：拱形峰区（峰区 1）与以 $v_r = \sqrt{\xi}$（定义域边界线）为劈刃的弯尖劈峰区）（峰区 2），其分界线（$q = 0$）方程为

$$\xi = v_r^2 + \xi_{z_1} \tag{8}$$

它的两个端点坐标为 $(\xi_{z_1}, 0)$ 与 $(1, \sqrt{1-\xi_{z_1}})$。

d. 在 $\sqrt{\xi_{z_1}} = 0$ 像面上，峰区 2 消失。随着 $\sqrt{\xi_{z_1}}$ 增大，两峰区分界线逐渐向右平移，峰区 1 所占区域减小，高度趋低，峰区 2 变化趋势相反，直到 $\sqrt{\xi_{z_1}} = 1$ 像面，峰区 1 消失。

e. 峰区 1 在每个 ξ 为常数的截面上有极值 q_{1m}，极值点在 (ξ, v_r) 坐标平面上的位置曲线方程由 $\dfrac{\mathrm{d}q}{\mathrm{d}v_r} = 0$ 导得

$$\xi = 2v_r^2 + \sqrt{\xi_{z_1}} \sqrt{\xi - v_r^2} \tag{9}$$

可解得

$$v_r\big|_{q_{1m}} = \left[(4\xi - \xi_{z_1} - \sqrt{\xi_{z_1}} \sqrt{\xi_{z_1} + 8\xi})/8\right]^{1/2} \quad (\xi \geq \xi_{z_1}) \tag{10}$$

$$q_{1m} = \left[\xi - \sqrt{\xi_{z_1}}(\sqrt{\xi_{z_1}} + \sqrt{\xi_{z_1} + 8\xi})/4\right]^{1/2} \cdot$$
$$\left\{\left[\xi + \sqrt{\xi_{z_1}}(\sqrt{\xi_{z_1}} + \sqrt{\xi_{z_1} + 8\xi})/4\right]^{1/2} - \sqrt{2\xi_{z_1}}\right\} \quad (\xi \geq \xi_{z_1}) \tag{11}$$

当 $\xi = 1$ 时，q_{1m} 达到最大值 q_{1mm}：

$$q_{1mm} = \left[1 - \sqrt{\xi_{z_1}}(\sqrt{\xi_{z_1}} + \sqrt{\xi_{z_1} + 8})/4\right]^{1/2} \cdot$$
$$\left\{\left[1 + \sqrt{\xi_{z_1}}(\sqrt{\xi_{z_1}} + \sqrt{\xi_{z_1} + 8})/4\right]^{1/2} - \sqrt{2\xi_{z_1}}\right\} \quad (\xi \geq \xi_{z_1}) \tag{12}$$

f. 峰区 2 的最大值 q_{2m}（对每个 ξ 值）位于 $\xi = v_r^2$ 边界线上，峰值是

$$q_{2m} = 2\sqrt{\xi_{z_1}} v_r \tag{13}$$

q_{2m} 与 v_r 成正比，其空间曲线在 (q, v_r) 坐标平面上的投影是一条通过 $(0, 0)$ 点、斜率为 $2\sqrt{\xi_{z_1}}$ 的直线，记为 "q_p 线"。

在 $v_r = 1$ 处，q_{2m} 达到最大值 q_{2mm}：

$$q_{2mm} = 2\sqrt{\xi_{z_1}} \tag{14}$$

令 $q_{mm} = \max(q_{1mm}, q_{2mm})$，表示该像平面上的最大像差半径。

g. 在 $\sqrt{\xi_{z_1}}$ 像面上，在 $\xi = 1$ 截面上像差方程为

$$v_r^4 - (1 - \xi_{z_1})v_r^2 \pm \sqrt{\xi_{z_1}} q v_r + q^2/4 = 0 \tag{15}$$

由式（8）知，$v_r\big|_{q=0, \xi=1} = \sqrt{1-\xi_{z_1}}$，故上式中 "＋" 号对应于 $v_r \in (0, \sqrt{1-\xi_{z_1}})$，"－" 号对应于 $v_r \in (\sqrt{1-\xi_{z_1}}, 1)$。由式（15）确定的 $\xi = 1$ 截面上的像差方程曲线，记为 "q_s 线"。

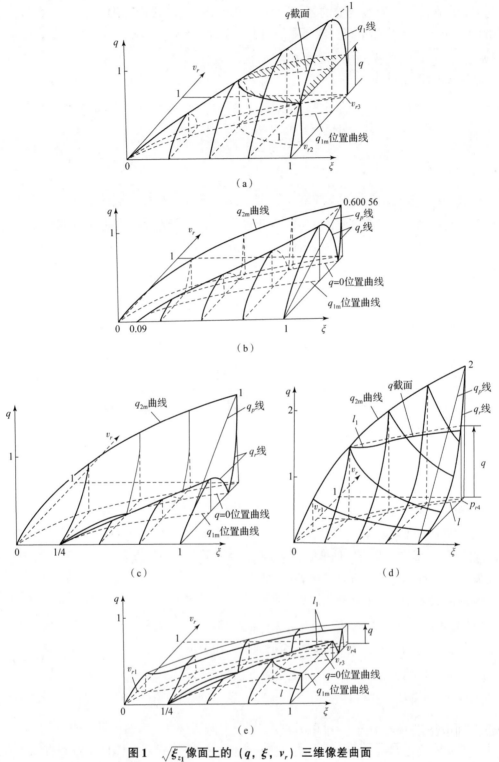

图1 $\sqrt{\xi_{z_1}}$ 像面上的 (q, ξ, v_r) 三维像差曲面

(a) $\sqrt{\xi_{z_1}}=0$；(b) $\sqrt{\xi_{z_1}}=0.30028$；(c) $\sqrt{\xi_{z_1}}=0.5$；(d) $\sqrt{\xi_{z_1}}=1$；(e) $\sqrt{\xi_{z_1}}=0.5$，q 截面

2 锐聚焦系统全色 PSF 表达式

对 $\sqrt{\xi_{z_1}}$ 像面,用高度为 $q(q<q_{mm})$ 的平面截 (q,ξ,v_r) 曲面(见图1),得 D_n 区域(见图2)。一般情况下,l 有两条曲线:$\xi_1(q,v_r)$ 与 $\xi_2(q,v_r)$,分别位于峰区1和峰区2,它们的方程是:

$$\xi_{1,2}=v_r^2+[\xi_{z_1}\pm q/(2v_r)]^2 \tag{16}$$

ξ_1 取"+"号,ξ_2 取"-"号。$\xi_2(q,v_r)$ 与区域边界线 $\xi=v_r^2$ 的交点的 v_r 坐标记为 v_{r1},则

$$\xi_2(q,v_{r1})=v_{r1}^2 \tag{17}$$

而 $\xi_1(q,v_r)$ 和 $\xi_2(q,v_r)$ 与 $\xi=1$ 的交点分别记为 v_{r2}、v_{r3}、v_{r4}(见图2),则

$$\xi_1(q,v_{r2})=1,\quad \xi_1(q,v_{r3})=1,\quad \xi_2(q,v_{r4})=1 \tag{18}$$

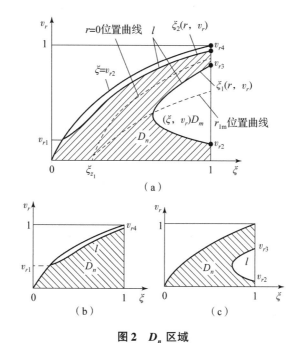

图 2 D_n 区域

设 $\xi_1(q,v_r)$ 曲线与 q_{1m} 极值位置曲线交点坐标为 $(\bar{\xi},\bar{v}_r)$,容易证明,当 $v_r=\bar{v}_r$ 时,$\dfrac{d\xi_1}{dv_r}=0$;当 $v_r>\bar{v}_r$ 时,$\dfrac{d\xi_1}{dv_r}>0$;而当 $v_r<\bar{v}_r$ 时,$\dfrac{d\xi_1}{dv_r}<0$。同样易证,对 $\xi_2(q,v_r)$ 曲线,总有 $\dfrac{d\xi_2}{dv_r}>0$。因而 $\xi_1(q,v_r)$ 与 $\xi_2(q,v_r)$ 以及与 v_r 为常数的直线的交点至多都只有一个。

于是,$I(q)$ 可按下述过程积分:

$$\begin{aligned}I(q)&=\iint_{D_n}dI(\xi,v_r)=\left(\iint_{D_n+D_m}-\iint_{D_m}\right)dI(\xi,v_r)=\left(\int_0^{v_{r1}}\int_{v_{r2}}^l+\int_{v_{r1}}^{v_{r4}}\int_{\xi_2}^1-\int_{v_{r2}}^{v_{r3}}\int_{\xi_1}^1\right)2v_r\frac{N(\xi)}{\xi}d\xi dv_r\\&=f(l)(v_{r4}^2-v_{r3}^2+v_{r2}^2)-\int_0^{v_{r1}}2f(v_r^2)v_rdv_r-\int_{v_{r1}}^{v_{r4}}2f(\xi_2)v_rdv_r+\int_{v_{r2}}^{v_{r3}}2f(\xi_1)v_rdv_r\end{aligned}$$

$$\tag{19}$$

式中 D_m 区域见图2（a），$f(\xi)$ 是 $N(\xi)/\xi$ 的一个原函数。利用式（17）、式（18），由式（3），可得

$$P(q) = \frac{1}{\pi R_m^2 q}\int_{v_{r2}}^{v_{r3}}\frac{N(\xi_1)}{\xi_1}\left(\sqrt{\xi_{z_1}}+\frac{q}{2v_r}\right)dv_r + \frac{1}{\pi R_m^2 q}\int_{v_{r1}}^{v_{r4}}\frac{N(\xi_2)}{\xi_2}\left(\sqrt{\xi_{z_1}}-\frac{q}{2v_r}\right)dv_r \quad (20)$$

式中，$\xi_{1,2}$ 以式（16）代入，这便是（q,ξ,v_r）坐标系中，归一化的全色 PSF 表达式，与文献[4]的结果完全一致。这就从另一条途径证明吴全德的结果是完全正确的，并赋予其表达式中的积分限以十分明确的几何意义和物理意义。

将式（20）的前、后项分别记为 $Q(+q)|_{v_{r2}}^{v_{r3}}$ 与 $Q(-q)|_{v_{r1}}^{v_{r4}}$，分别代表具有峰区 1 与峰区 2 内初始条件的电子对 q 处电流密度的贡献。它们对应的积分微元分别记为 $dQ(+q)$、$dQ(-q)$。作为特殊情况，所获得的 D_n 区域还有图 2（b）、图 2（c）所示的两种，这时，式（20）就只有后一项没有前一项。

3 锐聚焦系统全色 PSF 积分限的确定

由上可知，v_{r1} 是 q_{2m} 峰值内线上高为 q 的点 v_r 的坐标值，$v_{r2,r3,r4}$ 是 $\xi=1$ 截面上像差为 q 时的 v_r 坐标值，即它们分别为 q_p 线与 q_s 线上高为 q 的点的 v_r 坐标值。故 v_{r1} 满足式（13），即

$$v_{r1} = q/(2\sqrt{\xi_{z_1}}) \quad (21)$$

v_{r2}、v_{r3}、v_{r4} 满足式（15），对 v_{r2}、v_{r3}，方程取"+"号，对 v_{r4}，取"-"号。对于不包含 v_r^3 项的四次方程式（15），可求得其解析解，它的根与二次方程

$$v_r^2 + Av_r/2 + (y\mp\sqrt{\xi_{z_1}}q/A) = 0 \quad (22)$$

的根重合，此处

$$A = \pm[8y+4(1-\xi_{z_1})]^{1/2} \quad (23)$$

而 y 是三次方程

$$8y^3 + 4(1-\xi_{z_1})y^2 - 2q^2 y - q^2 = 0 \quad (24)$$

的实根。$v_{rj}(j=1,2,3,4)$ 的存在性，见表 1。

表 1 v_{r1}、v_{r2}、v_{r3}、v_{r4} 的存在性（$0<q\leqslant q_{mm}$）

参数	$\sqrt{\xi_{z_1}}=0$	$0<\sqrt{\xi_{z_1}}<0.30028$		$\sqrt{\xi_{z_1}}=0.30028$	$0.30028<\sqrt{\xi_{z_1}}<1$		$\sqrt{\xi_{z_1}}=1$
		$q\geqslant q_{2mm}$	$q<1\leqslant q_{2mm}$		$q<q_{1mm}$	$q\geqslant q_{1mm}$	
v_{r2},v_{r3}	存在	存在	存在	存在	存在	存在	—
v_{r1},v_{r4}	—	—	存在	存在	存在	存在	存在

对于 $\sqrt{\xi_{z_1}}=0$ 的像面，v_{r1}、v_{r4} 不存在，而

$$v_{r2,r3} = [(1\mp\sqrt{1-q^2})/2]^{1/2} \quad (25)$$

对于 $\sqrt{\xi_{z_1}}=1$ 的像面，v_{r2}、v_{r3} 不存在，而

$$v_{r1} = q/2, v_{r4} = [\sqrt{2y}+(2q/\sqrt{2y}-2y)^{1/2}]/2 \quad (26)$$

式中

$$y = \left(\frac{q}{4}\right)^{2/3} \left[\left(1+\sqrt{1-\frac{4}{27}q^2}\right)^{1/3} + \left(1-\sqrt{1-\frac{4}{27}q^2}\right)^{1/3}\right] \quad (27)$$

将其代入式（26），用级数展开，保留有限项，得

$$v_{r4} = q^{\frac{1}{3}} - \frac{1}{12}q - \frac{1}{72}q^{\frac{5}{3}} - \frac{5}{1\,296}q^{\frac{7}{3}} - \frac{1}{768}q^3 - \frac{91}{186\,624}q^{\frac{11}{3}} - \frac{1\,021}{6\,718\,464}q^{\frac{13}{3}} \quad (28)$$

对于一般像面，也可遵循上述过程进行计算。

4 一般像面上全色 PSF 的解析表达式实例

贝塔分布下的全色 PSF，由式（20）可得

$$dQ(\pm q) = \frac{(k+l+1)!}{k!\,l!}\frac{1}{\pi R_m^2 q}\left[v_r^2 + \left(\sqrt{\xi_{z_1}}\pm\frac{q}{2v_r}\right)^2\right]^{k-1}\left[1 - v_r^2 - \left(\sqrt{\xi_{z_1}}\pm\frac{q}{2v_r}\right)^2\right]^l \left(\sqrt{\xi_{z_1}}\pm\frac{q}{2v_r}\right)dv_r \quad (29)$$

式中，k、l 为取正整数的常数。对之积分，可得到贝塔分布下全色 PSF 的具体表达式。例如：

$$Q(\pm q)|_{\beta_{1,1}} = \frac{6}{\pi R_m^2 q}\left[-\frac{1}{3}\sqrt{\xi_{z_1}}v_r^3 \mp \frac{1}{4}qv_r^2 + (1-\xi_{z_1})\sqrt{\xi_{z_1}}v_r \pm \frac{q}{2}(1-3\xi_{z_1})\ln v_r + \frac{3}{4}\sqrt{\xi_{z_1}}q^2\frac{1}{v_r} \pm \frac{1}{16}q^3\frac{1}{v_r^2}\right] \quad (30)$$

$$Q(\pm q)|_{\beta_{1,2}} = \frac{12}{\pi R_m^2 q}\left\{\frac{1}{5}\sqrt{\xi_{z_1}}v_r^5 \pm \frac{1}{8}qv_r^4 - \frac{2}{3}(1-\xi_{z_1})\sqrt{\xi_{z_1}}v_r^3 \mp \frac{1}{2}q(1-3\xi_{z_1})v_r^2 + \left[\frac{3}{2}q^2+(1-\xi_{z_1})^2\right]\sqrt{\xi_{z_1}}v_r \pm \left[\frac{1}{4}q^3+\frac{1}{2}q(1-\xi_{z_1})(1-5\xi_{z_1})\right]\ln v_r + \frac{1}{2}q^2\sqrt{\xi_{z_1}}(3-5\xi_{z_1})\frac{1}{v_r} \pm \frac{1}{8}q^3(1-5\xi_{z_1})\frac{1}{v_r^2} - \frac{5}{48}\sqrt{\xi_{z_1}}q^4\frac{1}{v_r^3} \mp \frac{1}{128}q^5\frac{1}{v_r^4}\right\} \quad (31)$$

$$Q(\pm q)|_{\beta_{2,1}} = \frac{12}{\pi R_m^2 q}\left[-\frac{1}{5}\sqrt{\xi_{z_1}}v_r^5 \mp \frac{1}{8}qv_r^4 + \frac{1}{3}\sqrt{\xi_{z_1}}(1-2\xi_{z_1})v_r^3 \pm \frac{1}{4}q(1-6\xi_{z_1})v_r^2 + \sqrt{\xi_{z_1}}\left(\xi_{z_1}-\xi_{z_1}^2-\frac{3}{2}q^2\right)v_r \pm q\left(\frac{3}{2}\xi_{z_1}-\frac{5}{2}\xi_{z_1}^2-\frac{1}{4}q^2\right)\ln v_r - q^2\sqrt{\xi_{z_1}}\left(\frac{4}{3}-\frac{5}{2}\xi_{z_1}\right)\frac{1}{v_r} \mp \frac{1}{16}q^3(1-10\xi_{z_1})\frac{1}{v_r^2} + \frac{5}{48}q^4\sqrt{\xi_{z_1}}\frac{1}{v_r^3} \mp \frac{1}{128}q^5\frac{1}{v_r^4}\right] \quad (32)$$

类似地，还可求得 $\beta_{1,n}$ 分布当 n 值由 2 至 8 的全色 PSF 解析表达式。

同样可求得余弦分布的 PSF：

$$dQ(\pm q) = \frac{1}{\pi R_m^2 q}\frac{\pi^2}{2(\pi-2)}\left(\sqrt{\xi_{z_1}}\pm\frac{q}{2v_r}\right)\left\{\sum_{n=0}^{\infty}(-1)^n\frac{1}{(2n)!}\left(\frac{\pi}{2}\right)^{2n}\left[v_r^2+\left(\sqrt{\xi_{z_1}}\pm\frac{q}{2v_r}\right)^2\right]^{2n}\right\}dv_r \quad (33)$$

和瑞利分布下的 PSF：

$$dQ(\pm q) = \frac{K}{\pi R_m^2 q}\exp(-4\xi_{z_1})\left(\sqrt{\xi_{z_1}}\pm\frac{q}{2v_r}\right)\cdot\left[\sum_{n=0}^{\infty}\frac{(-4)^n}{n!}\left(v_r^2\pm\frac{q}{v_r}\sqrt{\xi_{z_1}}+\frac{q^2}{4v_r^2}\right)^n\right]dv_r \quad (34)$$

其中，$K=16/[1-5\exp(-4)]=17.61296339$。式（33）中 n 取至 2 已有足够的精度，式（34）级数在 $\left(v_r^2 \pm \dfrac{q}{v_r}\sqrt{\xi_{z_1}}+\dfrac{q^2}{4v_r^2}\right)$ 较大时，收敛较慢，n 需取到 10，才有足够的精度。

将上节中所确定的积分限代入 $Q(\pm q)$ 表达式，便可得到 PSF 的具体表达式或数值。

例如，在 $\sqrt{\xi_{z_1}}=0$ 的像面，对 $\beta_{1,1}$、$\beta_{1,2}$、$\beta_{2,1}$，将式（25）代入式（30）、式（31）、式（32），就得到和文献 [1, 4] 完全一致的结果；对 $\beta_{1,3}$、$\beta_{1,4}$、$\beta_{1,8}$，有

$$P(q)\Big|_{\beta_{1,3},\sqrt{\xi_{z_1}}=0}=\dfrac{10}{\pi R_m^2}\left[\left(1+\dfrac{3}{2}q^2\right)\ln\dfrac{1+\sqrt{1-q^2}}{q}-\dfrac{3}{2}\sqrt{1-q^2}-\dfrac{1}{3}(1-q^2)^{3/2}-q^2\sqrt{1-q^2}\right] \tag{35}$$

$$P(q)\Big|_{\beta_{1,4},\sqrt{\xi_{z_1}}=0}=\dfrac{15}{\pi R_m^2}\left[\left(1+3q^2+\dfrac{3}{8}q^4\right)\ln\dfrac{1+\sqrt{1-q^2}}{q}-\sqrt{1-q^2}-\dfrac{27}{8}q^2\sqrt{1-q^2}-\dfrac{13}{12}(1-q^2)^{3/2}\right] \tag{36}$$

$$P(q)\Big|_{\beta_{1,8},\sqrt{\xi_{z_1}}=0}=\dfrac{45}{\pi R_m^2}\left[\left(1+14q^2+\dfrac{105}{4}q^4+\dfrac{35}{4}q^6+\dfrac{35}{128}q^8\right)\ln\dfrac{1+\sqrt{1-q^2}}{q}-\dfrac{255}{128}\sqrt{1-q^2}-\dfrac{73}{128}(1-q^2)^{3/2}-\dfrac{93}{640}(1-q^2)^{5/2}-\dfrac{9}{896}(1-q^2)^{7/2}-\dfrac{321}{16}q^2\sqrt{1-q^2}-\dfrac{17}{8}q^2(1-q^2)^{3/2}-\dfrac{9}{80}q^2(1-q^2)^{5/2}-\dfrac{777}{32}q^4\sqrt{1-q^2}-\dfrac{21}{32}q^4(1-q^2)^{3/2}-\dfrac{63}{16}q^6\sqrt{1-q^2}\right] \tag{37}$$

以上三式，形式上与文献 [1] 的结果略有差异，实际上是完全等同的。

将 $\sqrt{\xi_{z_1}}=1$ 像面上的 v_{r1}、v_{r4}，即式（26）代入 $Q(-q)$ 项，同样可得 PSF，如对 $\beta_{1,1}$ 有

$$P(q)\Big|_{\beta_{1,1},\sqrt{\xi_{z_1}}=1}=\dfrac{6}{\pi R_m^2}\left[\ln\left(q^{1/3}-\dfrac{1}{12}q-\dfrac{1}{72}q^{5/3}-\dfrac{5}{1296}q^{7/3}-\dfrac{1}{768}q^3-\dfrac{91}{186\,624}q^{11/3}-\dfrac{1\,021}{6\,718\,464}q^{13/2}\right)-\ln\dfrac{q}{2}-\dfrac{19}{12}+\dfrac{13}{12}q^{2/3}-\dfrac{5}{144}q^{4/3}-\dfrac{29}{1\,728}q^2+\dfrac{39}{20\,736}q^{8/3}+\dfrac{289}{559\,872}q^{10/3}+\dfrac{823}{2\,239\,488}q^4\right] \tag{38}$$

对麦克斯韦分布，区域延伸到无穷，$v_{r2}\to 0$，$v_{r3}=v_{r4}\to\infty$，v_{r1} 仍由式（21）确定，结果仍收敛。

5 计算锐聚焦系统全色 PSF 方法概述

计算锐聚焦系统全色 PSF 的步骤如下：

(1) 确定像面位置，即确定 $\sqrt{\xi_{z_1}}$ 值；

(2) 确定 q 值；

(3) 按式（12）、式（14）计算 q_{1mm}、q_{2mm}；

(4) 按表 1，对不同的 $\sqrt{\xi_{z_1}}$ 值及 q 值，做必要的判断，确定 v_{rj} 的存在性；

(5) 由式（21）、式（15）求解 v_{rj}；

(6) 代入式（20）或已积得的 $Q(\pm q)$ 表达式，计算 $P(q)$。

与 (r, ξ, α) 坐标系中 PSF 参数确定和求解过程[1]相比，这里更容易些，积分限 v_{rj} 计算简便，其方程不包含参变量 ξ，特别有利于数值计算。

6 在 (q, ξ, v_r) 坐标系中对近贴聚焦系统的研究

对近贴聚焦系统，将式（2）做如式（5）的坐标变换，则有

$$q = v_r \tag{39}$$

该像差方程在 (q, ξ, v_r) 坐标系中的曲面图如图 3 所示，其定义域与锐聚焦系统相同。其主要特点是 $v_r = 0$，$q \equiv 0$；曲面顶线，线面是平面；$q_m = \sqrt{\xi}$，$q_{mm} = 1$；q_p 线与 q_s 线重合，为 $\xi = 1$ 截面内斜率为 1 的直线。

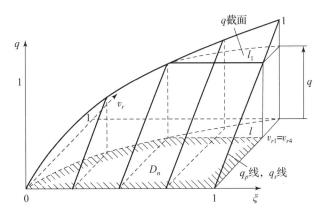

图 3 近贴聚焦系统 (q, ξ, v_r) 三维像差曲面

采用上述同样方法，可得

$$P(q) = \frac{1}{\pi R_m^2} \int_{\xi_1}^1 \frac{N(\xi)}{\xi} d\xi \quad (\xi_1 = v_r^2 = q^2) \tag{40}$$

这和文献［1］的结果完全一致。

结束语

本文由阴极透镜普遍成立的二级近轴横向色差公式出发，建立像差半径的三维坐标系，用形象的方法讨论锐聚焦系统与近贴聚焦系统的全色 PSF，由严格的数学推导，给出了归一化全色 PSF 的表达式，其物理意义明确，各参数及积分限在像差曲面图形中有确定的含义和求解表达式，从而较为系统完整地解决了在一般像面上计算电子具有初角度和初能量分布时的全色 PSF 的问题，文中还给出了若干初能分布下全色 PSF 的具体表达式。

结果表明，所建议的三维坐标法，对于加深理解像差公式的物理意义、深入研究和计算像管透镜的 PSF 具有实际意义。

参 考 文 献

［1］NI G Q, ZHOU L W, FANG E L. An inverse design of electrostatic focusing field for electrostatic and magnetic imaging［J］. Acta Armamentari Sinica, 1989, 1:77.

[2] CSORBA I P. Resolution limitations of electromagnetically focused image-intensifier tubes [J]. RCA Review, 1970, 31:534.

[3] HARTLEY K F. On the electron optics of magnetically focused image tubes [J]. J. Phys. D: Appl. Phys., 1974, 7:1612 – 1620.

[4] 吴全德. 在发射式电子光学系统中光电子速度分布对像品质的影响(Ⅰ)[J]. 物理学报, 1956, 12(5):419.

[5] 吴全德. 在发射式电子光学系统中光电子速度分布对像品质的影响(Ⅱ)[J]. 物理学报, 1957(01):78 – 89.

7.4 On Modulation Transfer Function of Cathode Lenses in Image Tubes
关于像管阴极透镜的调制传递函数

Abstract: Due to comprehensive evaluating image quality of systems or components, the modulation transfer function (MTF) has found wide applications in photoelectronic image devices, such as image converter tubes and image intensifiers. The approximate formula of MTF for cathode lenses of proximity focusing, electrostatic sharp focusing and electromagnetic focusing in image tubes has usually expressed by $\text{MTF}(f) = \exp[-(\pi \rho f)^2]$ where f is the spatial frequency (lp/mm); ρ is the focusing error coefficient, as called by I. P. Csorba, $\rho = k_1 L (\varepsilon_m/V_a)^{1/2}$ for the proximity focusing; $\rho = k_2 (M\varepsilon_m/E_c)$, for the electrostatic sharp focusing and electromagnetic focusing, where L is the distance between photocathode and phosphor screen (mm); V_a is the voltage at screen (V), M is the magnification; E_c is the field intensity at photocathode (V/mm); ε_m is the maximum initial potential (V), which is equivalent to maximum probably initial energy of electron emission; k_1, k_2 are the constants, depending on the initial angular distribution and initial energy distribution of electrons emitted from the photocathode surface.

In our investigation, we have proved that ρ has a definite physical meaning, it is better to be defined as root mean square (RMS) radius of the second order paraxial lateral aberration at the optimal image plane of cathode lenses. The more exact values of k_1, k_2 are also given in the present paper, while the initial angular distribution follows the Lambert law or cubic-cosine angular distribution and the initial energy distribution is expressed by cosine, Maxwell or beta distribution. The image quality of the image intensifiers of the first, second and third generations has been evaluated by using the given MTF formula.

摘要：调制传递函数（MTF）由于能全面综合地评价元件和系统的像质，在光电成像器件如变像管和像增强器中获得了广泛的应用。像管中近贴聚焦、静电锐聚焦和电磁聚焦的阴极透镜的 MTF 的近似式通常可表为 $\text{MTF}(f) = \exp[-(\pi \rho f)^2]$，式中，$f$ 是空间频率（lp/mm）；ρ 是近贴聚焦误差系数（I. P. Csorba），$\rho = k_1 L (\varepsilon_m/V_a)^{1/2}$，对静电锐聚焦或电磁聚焦，$\rho = k_2 (M\varepsilon_m/E_c)$；$L$ 是光阴极至荧光屏之距（mm）；V_a 是屏上电压（V），E_c 是阴极面上的场强（V/mm），ε_m 是最大初电位（V），它等同于电子发射的最大或然初能量，k_1，k_2 是常数，它们取决于由光阴极逸出电子的初角度分布和初能量

Zhou Liwei[a], Zhang Zhiquan[a], Ni Guoqiang[a], Fang Erlun[b], Jin Weiqi[a], Zhang Ling[a]. a) Beijing Institute of Technology; b) Xi'an Research Institute of Modern Chemistry. 1991 Symposium on Photoelectronic Image Devices, The Institute of Physics Conference Series, No. 121, 1991, 405–419.

分布。

在我们的研究中,我们证明了 ρ 具有明确的物理意义,它最好是被定义为阴极透镜的最佳像面上的二级近轴横向像差的均方根半径。本文给出了当初角度遵循朗伯定律或立方-余弦分布;初电位以余弦分布,麦克斯韦分布以及贝塔分布表示时 k_1、k_2 的更精确的值,并利用所给出的 MTF 公式对一代、二代以及三代像增强器的像质进行了评价。

Introduction

Three types of focusing cathode lenses most commonly used in image tubes are classified by their modes of focusing.

(1) Proximity focusing, i. e, a bi-planar cathode lens, consisting of a homogeneous axial field.

It has found wide applications in wafer tubes of the second and third generations. Proximity focusing is formed between the photocathode and input surface of the microchannel plate (MCP), as well as between the output surface of MCP and the phosphor screen.

(2) Electrostatic focusing, i. e., electrostatic cathode lens, consisting of axial symmetrical electrostatic field. It has found wide applications in infrared image converter tubes, cascade image intensifiers of the first generation and image-inverting tubes of the second generation.

(3) Electromagnetic focusing, consisting of axial symmetrical electrostatic and magnetic fields, used for transferring the electron image from the photocathode to the phosphor screen. It has found wide applications in electron photographic image tubes and magnetic image intensifiers with high resolution and small geometrical distortion.

The last two types of focusing are sometimes referred to as sharp focusing.

Modulation transfer function, as a result of its comprehensive assessment of image quality, has a special significance in evaluating cathode lenses and image tubes. In the last two decades, MTF of cathode lenses and image tubes has been investigated by many authors[1-11]. All these works have brought about developments in the study of MTF and assessment of image quality for cathode lenses and image tubes. Studies on the analytical solutions of the poly-energetic point spread function (PSF) and electron-optical modulation transfer function in cathode lenses have been fully described in papers[12,13,14], in which we have given a geometric figure in three dimensional coordinates by using the second order paraxial transverse aberration formula.

This paper will further investigate the approximate formulae of poly-energetic MTF expressed as exponential function for the proximity focusing and sharp focusing in image tubes under conditions of various initial emission angular distributions and initial emission energy distributions of the photoelectrons. An assessment of image quality for image intensifiers of the first, second and third generations will also be given.

1 Basic formulae

1.1 Initial angular distributions of electron emission

For cathode lenses of image tubes, when the S-type photocathode is adopted, it is generally accepted that the initial angular distribution of electron emission obeys Lambert law, that is, the probability of initial emission angle a of emitted photoelectrons can be given as:

$$G(\alpha) = \cos\alpha \tag{1}$$

For the negative electron affinity (NEA) photocathode used in the third generation wafer tubes, assuming that the initial emission angle of photoelectrons obeys cubic-cosine angular distribution[11] owing to concentration of emission of emitted electrons, we have

$$G(\alpha) = \cos^3\alpha \tag{2}$$

1.2 Initial energy distributions of electron emission

For the initial emission energy distribution of photoelectrons emitted from the photocathode, the following three kinds of distributions are usually accepted:

(1) Normalized cosine distribution[3].

$$N(\xi) = \frac{\pi^2}{2(\pi-2)}\xi\cos\left(\frac{\pi}{2}\xi\right) \tag{3}$$

(2) Normalized beta distribution[5].

$$N(\xi) = \frac{(k+l+1)!}{k!\,l!}\xi^k(1-\xi)^l \tag{4}$$

(3) Normalized Maxwell distribution[3].

$$N(\eta) = \eta\exp(-\eta) \tag{5}$$

where $\xi = \varepsilon/\varepsilon_m$, $\eta = \varepsilon/\varepsilon_p$, ε is the initial potential equivalent of initial emission energy of emitted electrons; ε_m is the maximum potential equivalent of maximum emission energy; ε_p is the potential equivalent of the most probable of emission energy.

1.3 The second order paraxial lateral aberration

For the assessment of MTF of image tubes, it is generally accepted that the cathode lens is free of third order geometrical aberrations and the chromatic aberration present is caused by the variation of axial emission energy of the photoelectrons[6]. For the sharp focusing systems, the second order paraxial transverse aberration formula can be expressed as follows:

$$\Delta r = \frac{M\varepsilon_m}{E_c}|\xi\sin 2\alpha - 2\xi_{z_1}^{1/2}\xi^{1/2}\sin\alpha| \tag{6}$$

where M is the radial magnification of system, E_c is the electrical field strength at the cathode, both of which taking the absolute values. $\xi_{z_1} = \varepsilon_{z_1}/\varepsilon_m$, ε_{z_1} is the axial component of initial potential corresponding to the electrons which can be ideally focused on the image plane under consideration.

For proximity focusing systems, the second order paraxial transverse aberration, i.e., the radius of diffusing circle at the plane of the phosphor screen (or target), can be expressed by

$$\Delta r = 2L \left(\frac{\varepsilon_m}{V}\right)^{1/2} \xi^{1/2} \sin \alpha \tag{7}$$

where L is the spacing between two planar electrodes, V is the voltage applied to the planar electrodes.

1.4 Point spread function

For axial symmetrical imaging systems, the normalized monoenergetic PSF in the circular ring with inner radius r and outer radius $r + dr$ can be expressed by

$$j(r) = \sum_i \frac{CG(\alpha_i) \sin \alpha_i d\alpha_i}{2\pi r |dr|} \tag{8}$$

where C is a constant introduced for normalization, $C = 2$ and $C = 4$ for Lambert and cubic-cosine angular distributions respectively, α_i are these initial angles with which the electrons emitted from the photocathode will have contributions to the current in the circular ring.

According to Eq. (8), a probability statistics on the energy distribution of electron emission which will have contributions to the current in the circular ring is carried out, we can have the normalized poly-energetic PSF.

1.5 Modulation transfer function

As have been discussed in Ref. [12, 13], the mathematical description for solving MTF can be derived as:

$$\text{MTF}_\xi(f) = 2\pi \int_0^\infty r j(r) J_0(2\pi f r) dr \tag{9}$$

$$\text{MTF}(f) = \int_0^1 \text{MTF}_\xi(f) N(\xi) d\xi \tag{10}$$

where f is the spatial frequency, $J_0(2\pi f r)$ is the zeroth order Bessel function, $\text{MTF}_\xi(f)$ is the mono-energetic MTF, $\text{MTF}(f)$ expresses the poly-energetic MTF.

If we express the integration of Eq. (10) in term of a form of summation, the progression of summation formula can be truncated at lower spatial frequencies. Neglecting $(2\pi f r)^4$ and the other higher order terms, the MTF of formula (10) can be expressed in an exponential form:

$$\text{MTF}(f) = \exp[-(\pi \rho f)^2] \tag{11}$$

where ρ has been defined as the focusing error coefficient in Ref. [3, 4], but we have proved that ρ should better be defined as a root mean square (RMS) radius of system, i.e., $\rho = \bar{r}$, as will be proved later.

2 Polyenergetic MTF with proximity focusing in image tubes

Using Eq. (8) and Eq. (9), the mono-energetic MTF with proximity focusing can be expressed as

$$\text{MTF}_\xi(f) = \sum_{n=0}^\infty (-1)^n \frac{\frac{C}{2}}{\left(n + \frac{C}{2}\right)! n!} \left(2\pi f L \sqrt{\frac{\varepsilon}{V}}\right)^{2n} \tag{12}$$

Statistical treatment of Eq. (12) regarding the initial energy distributions of photoelectrons will give from Eq. (10) analytical expressions of the normalized poly-energetic MTF of proximity focusing under different initial energy distributions, when the initial emission angle follows Lambert distribution or cubic-cosine distribution, as shown in Table 1.

Table 1 Normalized poly-energetic MTF of proximity focusing under different emission energy distributions and emission angular distributions

Initial emission energy distribution	Normalized poly-energetic MTF	$C = 2$ for Lambert distribution $C = 4$ for cubic-cosine angular distribution
cosine distribution $N(\xi) = \frac{\pi^2}{2(\pi-2)}\xi\cos\left(\frac{\pi}{2}\xi\right)$		$\mathrm{MTF}(f) = \frac{\pi^2}{2(\pi-2)}\sum_{n=0}^{\infty}(-1)^n \frac{C}{2n!\left(n+\frac{C}{2}\right)!}\left(2\pi fL\sqrt{\frac{\varepsilon_m}{V}}\right)^{2n} \times \left\{\frac{2}{\pi} + \left[(-1)^{\frac{n+2}{2}}\frac{(n+1)!}{\left(\frac{\pi}{2}\right)^{n+2}}\right]_{0,2,4,\cdots} + \sum_{k=1}^{k=\left(\frac{n+1}{2}\right)}(-1)^k \frac{(n+1)!}{(n-2k+1)!\left(\frac{\pi}{2}\right)^{2k+1}}\right\}$
beta distribution $N(\xi) = \frac{(k+l+1)!}{k!\, l!}\xi^k(1-\xi)^l$		$\mathrm{MTF}(f) = \sum_{n=0}^{\infty}(-1)^n \frac{\frac{C}{2}}{n!\left(n+\frac{C}{2}\right)!}\left(2\pi fL\sqrt{\frac{\varepsilon_m}{V}}\right)^{2n}\frac{(k+l+1)!(k+n)!}{k!(k+n+l+1)!}$
Maxwell distribution $N(\eta) = \eta\exp(-\eta)$		$\mathrm{MTF}(f) = \sum_{n=0}^{\infty}(-1)^n \frac{\frac{C}{2}(n+1)}{\left(n+\frac{C}{2}\right)!}\left(2\pi fL\sqrt{\frac{\varepsilon_p}{V}}\right)^{2n}$ when $C = 2$, $\mathrm{MTF}(f) = \exp\left[-\left(2\pi fL\sqrt{\frac{\varepsilon_p}{V}}\right)^2\right]$

For lower spatial frequencies, the progression in Table 1 can be truncated, since only terms for $n = 0$, and $n = 1$ are significant, then we have the approximate formula of MTF expressed by an exponential function as Eq. (11). The value of ρ is shown in Table 2.

Table 2 The approximate formula of normalized poly-energetic MTF and its ρ value of proximity focusing under different emission energy distributions and emission angular distributions

Initial emission energy distribution	Approximate formula of normalized polyenergetic MTF $\mathrm{MTF}(f) = \exp[-(\pi\rho f)^2]$	
	$C = 2$ for Lambert distribution	$C = 4$ for cubic-cosine angular distribution
	ρ	ρ
cosine distribution $N(\xi) = \frac{\pi^2}{2(\pi-2)}\xi\cos\left(\frac{\pi}{2}\xi\right)$	$\rho = 1.021,078,988L\sqrt{\frac{\varepsilon_m}{V}}$	$\rho = 0.833,707,502L\sqrt{\frac{\varepsilon_m}{V}}$

Initial emission energy distribution	Approximate formula of normalized polyenergetic MTF $\mathrm{MTF}(f) = \exp[-(\pi\rho f)^2]$	
	$C = 2$ for Lambert distribution	$C = 4$ for cubic-cosine angular distribution
	ρ	ρ
beta distribution $N(\xi) = \dfrac{(k+l+1)!}{k!\,l!}\xi^k(1-\xi)^l$	$\rho\|_{\beta_{l,k}} = \left(\dfrac{2}{k+l+2}(k+1)\right)^{1/2} L\sqrt{\dfrac{\varepsilon_m}{V}}$ $\rho\|_{\beta_{1,1}} = L\sqrt{\dfrac{\varepsilon_m}{V}}$ $\rho\|_{\beta_{1,4}} = 0.755,928,946 L\sqrt{\dfrac{\varepsilon_m}{V}}$ $\rho\|_{\beta_{1,8}} = 0.603,022,688 L\sqrt{\dfrac{\varepsilon_m}{V}}$	$\rho\|_{\beta_{l,k}} = \left(\dfrac{4}{3}\dfrac{k+1}{k+l+2}\right)^{1/2} L\sqrt{\dfrac{\varepsilon_m}{V}}$ $\rho\|_{\beta_{1,1}} = 0.816,496,58 L\sqrt{\dfrac{\varepsilon_m}{V}}$ $\rho\|_{\beta_{1,4}} = 0.617,213,399 L\sqrt{\dfrac{\varepsilon_m}{V}}$ $\rho\|_{\beta_{1,8}} = 0.492,365,936 L\sqrt{\dfrac{\varepsilon_m}{V}}$
Maxwell distribution $N(\eta) = \eta\exp(-\eta)$	$\rho = 2L\sqrt{\dfrac{\varepsilon_p}{V}}$	$\rho = 1.632,993,162 L\sqrt{\dfrac{\varepsilon_p}{V}}$

3 Poly-energetic MTF with sharp focusing in image tubes

For solving the poly-energetic MTF with sharp focusing systems, we shall confine our discussion only to cases with Lambert angular distribution. Similarly, starting from Eq. (1), Eq. (6) and Eq. (9), we can derive the expression of mono-energetic MTF with sharp focusing under the condition of Lambert distribution:

$$\mathrm{MTF}_\xi(f) = \sum_{n=0}^{\infty} \frac{(-1)^n}{(n!)^2} \left(\frac{2\pi f M\varepsilon}{E_c}\right)^{2n} \sum_{m=0}^{2n} (-1)^m C_{2n}^m \left(\frac{\varepsilon_{z_1}}{\varepsilon}\right)^{\frac{m}{2}} \beta\left(n+1, n-\frac{m}{2}+1\right) \quad (13)$$

where $\beta\left(n+1, n-\dfrac{m}{2}+1\right)$ is the beta function, which can be expressed by the Gamma function:

$$\beta(m,n) = \frac{\Gamma(m)\Gamma(n)}{\Gamma(m+n)} \quad (14)$$

Statistical treatment of Eq. (13) regarding the initial emission energy distributions of photoelectrons will give from Eq. (10) analytical expressions of normalized poly-energetic MTF of sharp focusing under different initial energy distributions and Lambert angular distribution, as shown in Ref. [12, 13]. Similarly, the approximate formula of poly-energetic MTF can be expressed as a form of exponential function, in which the ρ value depends on the image plane corresponding to $\sqrt{\xi_{z_1}}\|_{\mathrm{opt}}$. If we make search for minimal value of ρ, we can find the optimal image plane corresponding to $\sqrt{\xi_{z_1}}\|_{\mathrm{opt}}$ or $\sqrt{\eta_{z_1}}\|_{\mathrm{opt}}$, the $\rho_{\min}$ and optimal $\mathrm{MTF}\|_{\mathrm{opt}}$ on this plane, as given in Table 3.

Table 3 Approximate formula of poly-energetic MTF with sharp focusing for ρ_{min} and $\sqrt{\xi_{z_1}}|_{opt}$ or $\sqrt{\eta_{z_1}}|_{opt}$ corresponding to optimal image plane

Initial emission energy distribution	Approximate formula of MTF $MTF(f) = \exp[-(\pi\rho f)^2]$ ρ	ρ_{min} and $\sqrt{\xi_{z_1}}\|_{opt}$ or $\sqrt{\eta_{z_1}}\|_{opt}$ corresponding to optimal image plane
cosine distribution $N(\xi) = \dfrac{\pi^2}{2(\pi-2)}\xi\cos\left(\dfrac{\pi}{2}\xi\right)$	$\rho = \dfrac{M\varepsilon_m}{E_c}\{0.213,486,657 - 0.858,658,572\xi_{z_1}^{1/2} + 1.042,602,298\xi_{z_1}\}^{1/2}$	$\rho_{min} = 0.191,558,767\dfrac{M\varepsilon_m}{E_c}$ $\sqrt{\xi_{z_1}}\|_{opt} = 0.411,786,247$
Beta distribution $N(\xi) = \dfrac{(k+l+1)!}{k!\,l!}\xi^k(1-\xi)^l$	$\rho = \dfrac{M\varepsilon_m}{E_c}\left\{\dfrac{2}{3}\dfrac{(k+2)(k+1)}{(k+l+2)(k+l+3)} - \dfrac{32}{15}\dfrac{(k+l+1)!}{k!}\dfrac{\Gamma(2.5+k)}{\Gamma(3.5+k+l)}\xi_{z_1}^{1/2} + \dfrac{2(k+1)}{k+l+2}\xi_{z_1}\right\}^{1/2}$	$\rho_{min} = \dfrac{M\varepsilon_m}{E_c}\left\{\dfrac{2}{3}\dfrac{(k+2)(k+1)}{(k+l+2)(k+l+3)} - \dfrac{128}{225}\dfrac{(k+l+1)!(k+l+2)!}{k!(k+l)!}\times\left[\dfrac{\Gamma(2.5+k)}{\Gamma(k+l+3.5)}\right]^2\right\}^{1/2}$ $\sqrt{\xi_{z_1}}\|_{opt} = \dfrac{8}{15}\dfrac{(k+l+2)!}{(k+1)!}\dfrac{\Gamma(2.5+k)}{\Gamma(k+l+3.5)}$
$\beta_{1,1}$: $N(\xi) = 6\xi(1-\xi)$	$\rho\|_{\beta_{1,1}} = \dfrac{M\varepsilon_m}{E_c}\left\{\dfrac{1}{5} - \dfrac{256}{315}\xi_{z_1}^{1/2} + \xi_{z_1}\right\}^{1/2}$	$\beta_{1,1}$: $\rho_{min} = 0.186,762,743\dfrac{M\varepsilon_m}{E_c}$ $\sqrt{\xi_{z_1}}\|_{opt} = 0.406,349,220,6$
$\beta_{1,4}$: $N(\xi) = 30\xi(1-\xi)^4$	$\rho\|_{\beta_{1,4}} = \dfrac{M\varepsilon_m}{E_c}\left\{\dfrac{1}{14} - \dfrac{491.52}{135,135}\xi_{z_1}^{1/2} + \dfrac{4}{7}\xi_{z_1}\right\}^{1/2}$	$\beta_{1,4}$: $\rho_{min} = 0.116,400,488\dfrac{M\varepsilon_m}{E_c}$ $\sqrt{\xi_{z_1}}\|_{opt} = 0.318,259,518$
$\beta_{1,8}$: $N(\xi) = 90\xi(1-\xi)^8$	$\rho\|_{\beta_{1,8}} = \dfrac{M\varepsilon_m}{E_c}\left\{\dfrac{1}{33} - \dfrac{125,829.12}{669,278.61}\xi_{z_1}^{1/2} + \dfrac{4}{11}\xi_{z_1}\right\}^{1/2}$	$\beta_{1,8}$: $\rho_{min} = 0.077,477,386\dfrac{M\varepsilon_m}{E_c}$ $\sqrt{\xi_{z_1}}\|_{opt} = 0.258,509,74$
Maxwell distribution $N(\eta) = \eta\exp(-\eta)$	$\rho = 2\dfrac{M\varepsilon_p}{E_c}\{1 - \eta_{z_1}^{1/2}\sqrt{\pi} + \eta_{z_1}\}^{1/2}$	$\rho_{min} = 0.926,502,75\dfrac{M\varepsilon_p}{E_c}$ $\sqrt{\eta_{z_1}}\|_{opt} = 0.886,226,925$

I. P. Csorba [3,4] has given the approximate formula of poly-energetic MTF of sharp focusing, expressed in the form of exponential function (11), under cosine energy distribution and Lambert angular distribution. The value of ρ has been derived as [4]

$$\rho = 0.134\frac{M\varepsilon_m}{E_c} \tag{15}$$

In our paper, we have derived

$$\rho_{opt} = 0.191,558,767 \frac{M\varepsilon_m}{E_c} \qquad (16)$$

and

$$\sqrt{\xi_{z_1}}\big|_{opt} = 0.411,786,247 \qquad (17)$$

Starting from the second order paraxial transverse aberration formula (6), dividing the initial emission angle and initial emission energy into 6 and 10 equal parts respectively, we have computed the point spread functions and the line spread function, and finally the modulation transfer function for a typical sharp focusing system ($E_c = 80\text{V/mm}$, $\varepsilon_m = 0.912$, 4V, $M = 1$). Comparing the computed value of MTF with that from the approximate formula (11) for MTF, in which ρ is expressed by the above two values, we obtain the computational results, as shown in Table 4.

Table 4 A comparison of three computations for the MTF in sharp focusing system in case of cosine emission energy distribution

f (lp/mm)	Computed value MTF(f)	MTF(f) = exp[$-(\pi\rho f)^2$] ρ—Eq. (16) in present paper	MTF(f) = exp[$-(\pi\rho f)^2$] ρ—Eq. (6.25) in Ref. [4]
10	0.994,987,52	0.995,300,278	0.997,697,554
20	0.980,171,05	0.981,333,61	0.990,821,977
30	0.956,197,14	0.958,489,855	0.979,467,815
40	0.924,090,79	0.927,399,171	0.963,790,241
50	0.885,181,38	0.888,902,648	0.944,001,522
60	0.841,008,83	0.844,014,841	0.920,366,241
70	0.793,217,39	0.793,878,938	0.893,195,408
80	0.743,450,71	0.739,719,070	0.862,839,659
90	0.693,244,79	0.682,790,842	0.829,681,738
100	0.643,947,55	0.624,333,873	0.794,128,493

It will be seen from Table 4 that the MTF given in the present paper (Eq. (16)) is in closer agreement with computed values of MTF than the other one, given by I. P. Csorba [Eq. (6-25)], and that the approximate MTF formula expressed as an exponential function, in which ρ is defined to be RMS radius of the second order paraxial transverse aberration at the optimal image plane of cathode lenses, has sufficient accuracy at lower frequencies.

If we compare the figures given in Table 2 and Table 3, it is obvious that the MTF characteristics having the $\beta_{1,8}$ distribution is the best one, the one of $\beta_{1,4}$ distribution comes next, then the $\beta_{1,1}$ and cosine distributions, while those of the Maxwell distribution behaves worst.

4 Root mean square (RMS) radius of cathode lenses in image tubes

At lower frequencies, the zeroth order Bessel function can only take the first and second terms ($n = 0;\ 1$), substituting them into Eq. (12), considering Eq. (16) and Eq. (18), and using the definition of the root mean square (RMS) radius of imaging system's aberrations, we obtain immediately

$$\text{MTF}(f) = 1 - (\pi f)^2 \int_0^1 \int_0^{\frac{\pi}{2}} r^2 CG(\alpha) \sin\alpha N(\xi) \mathrm{d}\xi \mathrm{d}\alpha = 1 - (\pi \rho f)^2$$

where

$$\rho = \bar{r} = \left\{ \int_0^1 \int_0^{\frac{\pi}{2}} r^2 CG(\alpha) \sin\alpha N(\xi) \mathrm{d}\xi \mathrm{d}\alpha \right\}^{1/2}$$

(for the cosine, beta energy distributions) (18a)

or

$$\text{MTF}(f) = 1 - (\pi f)^2 \int_0^{\infty} \int_0^{\frac{\pi}{2}} r^2 CG(\alpha) \sin\alpha N(\eta) \mathrm{d}\eta \mathrm{d}\alpha = 1 - (\pi \rho f)^2$$

where

$$\rho = \bar{r} = \left\{ \int_0^{\infty} \int_0^{\frac{\pi}{2}} r^2 CG(\alpha) \sin\alpha N(\eta) \mathrm{d}\eta \mathrm{d}\alpha \right\}^{1/2}$$

(for the Maxwell energy distribution) (18b)

where r is the second order paraxial transverse aberration at the image plane (for sharp focusing) or the radius of diffusing circle on the screen (for proximity focusing).

For sharp focusing systems, differentiating Eq. (18a) or Eq. (18b) with respect to $\sqrt{\xi_{z_1}}$ or $\sqrt{\eta_{z_1}}$ respectively, taking the extreme value we can obtain the value of minimal RMS radius $\rho_{\min} = \bar{r}_{\min}$ and the optimal image plane, corresponding to $\sqrt{\xi_{z_1}}\big|_{\text{opt}}$ or $\sqrt{\eta_{z_1}}\big|_{\text{opt}}$ respectively.

It is not difficult to prove that ρ and $\bar{r}$ are identical. Substituting Eq. (1), Eq. (3), Eq. (7) into Eq. (18a) and supposing $C = 2$, and integrating Eq. (18a), we may have the RMS radius of proximity focusing systems in case of cosine energy distribution, which is equal to ρ:

$$\bar{r} = \left\{ \frac{2(\pi^2 - 8)}{\pi(\pi - 2)} \frac{\varepsilon_m L^2}{V} \right\}^{1/2} = 1.021,078,989 L \left(\frac{\varepsilon_m}{V} \right)^{1/2} = \rho \quad (19)$$

The above description shows that the exponential approximate formula for MTF of cathode lenses in image tubes approaches to practical MTF with sufficient accuracy at lower frequencies, where ρ in the formula is just the RMS radius of system, the optimal image plane corresponding to the optimal MTF which is correspondent to the minimal RMS radius $\bar{r}_{\min} = \rho_{\min}$.

Furthermore, it has been discovered that the size of RMS radius corresponds to the characteristics of MTF at each image plane. Such a correspondence ensures that the parameter of RMS radius can be used as an index for the evaluation of MTF of the system in designing electron optical systems of image tubes[15].

5 MTF computation of image tubes

As is well known, the MTF of the image tube is a product of the sine-wave response (MTF) of the optical components of the image tube. The major components contributing to the MTF are the input and output fiber-optic faceplates, fiber-optic twister, phosphor screen, microchannel plate (MCP) and the electron optics of sharp focusing, electron optics of proximity focusing for the input and output MCP spacing. The MTF of electron optics has been described in the previous paragraphs.

In has been pointed out that for most of the photoelectronic image devices and their components, the MTF can be expressed by the following formula[10]:

$$\text{MTF}(f) = \exp\left[-\left(\frac{f}{f_c}\right)^n\right] \quad (20)$$

where f is the spatial frequency (lp/mm); f_c is a constant of the spatial frequency; n is the characteristic index of the device or component. If two parameters f_c, n (written as (f_c, n)) are given, its MTF can be uniquely defined. In Table 5, the (f_c, n) values of the major optical components in the image tube are given. This is a result of numerical simulation for the data given in chapter 6 of the Ref [4].

Table 5 Values of (f_c, n) for the optical components in image tubes

Optical components of image tube	$\text{MTF}(f) = \exp\left[-\left(\frac{f}{f_c}\right)^n\right]$				
	Input and output fiber-optic $\text{MTF}_{\text{FO}}(f)$	Fiber-optic twister $\text{MTF}_{\text{FOT}}(f)$	Phosphor screen 1 $\text{MTF}_{\text{PS1}}(f)$	Phosphor screen 2 $\text{MTF}_{\text{PS2}}(f)$	Microchannel plate $\text{MTF}_{\text{MCP}}(f)$
(f_c, n)	(66, 2.21)	(38, 2.9)	(51, 1.34)	(42, 1.49)	(29, 1.8)
Remarks	For 6 μm fiber diameter $\text{MTF}_{\text{FO}} = 0.033$ at 115 lp/mm	for 10 μm fiber diameter $\text{MTF}_{\text{FOT}} = 0.033$ at 58 lp/mm	for a better screen $\text{MTF}_{\text{PS1}} = 0.03$ at 130 lp/mm	for a usual screen $\text{MTF}_{\text{PS2}} = 0.03$ at 98 lp/mm	$\text{MTF}_{\text{MCP}} = 0.03$ at 58 lp/mm

5.1 The first generation cascade image intensifier

The first generation cascade image intensifier adopts electrostatic sharp focusing system. Suppose that the initial emission angle obeys Lambert law, photocathode is S-20, its initial emission energy distribution takes $\beta_{1,4}$, $E_c = 80$ V/mm, $\varepsilon_m = 1$ V, $M = 1$, according to Table 3, we obtain the RMS radius of the sharp focusing:

$$\rho_{\min}\big|_{\text{SF}} = 0.116,400,488 \frac{E\varepsilon_m}{E_c} = 0.001,455,006 \text{ mm} \quad (21)$$

Therefore, the MTF of the first generation image intensifier tube (1st gen. tube) and the first generation cascade image intensifier (1st gen. intensifier) will be, respectively:

$$\text{MTF}(f)|_{\text{1st gen. tube}} = \text{MTF}_{\text{SF}}(f) \times \text{MTF}_{\text{PS}}(f) \times [\text{MTF}_{\text{FO}}(f)]^2 \quad (22)$$

and

$$\text{MTF}(f)|_{\text{1st gen. intensifier}} = \{\text{MTF}(f)|_{\text{1st gen. tube}}\}^3 \quad (23)$$

where $\text{MTF}_{\text{SF}}(f)$ is the MTF of sharp focusing system. According to the assessment of the limiting resolution at 3% response, we have attained the limiting resolutions of the first generation image intensifier tube and the first generation cascade image intensifier respectively, as shown in Table 6.

$$\text{MTF}_{\text{SF}}(f) = \exp[-(0.001,455,006\pi f)^2], \quad \text{MTF}_{\text{FO}}(f) = \exp\left[-\left(\frac{f}{66}\right)^{2.21}\right]$$

Table 6 A predication of the limiting resolutions of 1st gen. image intensifier tube and 1st gen. cascade image intensifier

$\text{MTF}_{\text{PS1}}(f) = \exp\left[-\left(\frac{f}{51}\right)^{1.34}\right]$		$\text{MTF}_{\text{PS2}}(f) = \exp\left[-\left(\frac{f}{42}\right)^{1.49}\right]$	
1st gen. image intensifier tube	1st gen. cascade image intensifier	1st gen. image intensifier tube	1st gen. cascade image intensifier
$f = 65$ lp/mm	$f = 36$ lp/mm	$f = 60$ lp/mm	$f = 32$ lp/mm
$\text{MTF}(f)\|_{\text{1st gen. tube}}$ $= 0.915,499,972$ $\times 0.250,555,266$ $\times (0.380,289,768)^2$ $= 0.033,173$	$\text{MTF}(f)\|_{\text{1st gen. intensifier}}$ $= (0.973,284,246)^3$ $\times (0.534,165,927)^3$ $\times (0.769,541,132)^6$ $= 0.029,183,851,73$	$\text{MTF}(f)\|_{\text{1st gen. tube}}$ $= 0.927,539,648$ $\times 0.182,429,738$ $\times (0.444,827,245)^2$ $= 0.033,481,96$	$\text{MTF}(f)\|_{\text{1st gen. intensifier}}$ $= (0.978,831,44)^3$ $\times (0.513,319,95)^3$ $\times (0.817,156,492)^6$ $= 0.037,767,615$

5.2 The second generation image-inverting tube

The MTF of the second generation image-inverting tube (2nd gen. IIT) can be written as

$$\begin{aligned}\text{MTF}(f)|_{\text{2nd gen. IIT}} = &\text{MTF}_{\text{SF}}(f) \times \text{MTF}_{\text{MCP}}(f) \times \text{MTF}_{\text{MSPF}}(f) \\ &\times \text{MTF}_{\text{PS}}(f) \times [\text{MTF}_{\text{FO}}(f)]^2\end{aligned} \quad (24)$$

where $\text{MTF}_{\text{MSPF}}(f)$ is the MTF of proximity focusing of the MCP—phosphor screen spacing. Now we evaluate the limiting resolution of the 2nd generation image-inverting tube. The ρ value of sharp focusing in Eq. (17) remains the same. For the proximity focusing of MCP—phosphor screen spacing, according to Ref. [3], suppose that the initial emission angle obeys Lamber law, the initial emission energy distribution takes Maxwell distribution, the most probable initial potential $\varepsilon_p = 0.2$V, the MCP is the phosphor screen spacing $L = (0.75 - 1)$ mm, the applied voltage $V = 6,000$ V, from Table 2, we have

$$\rho|_{\text{MSPF}} = 2L\sqrt{\frac{\varepsilon_p}{V}} = 0.008,660,254 - 0.011,547,005 \text{ mm} \quad (25)$$

The evaluated value of limiting resolution for the 2nd generation image-inverting tube is given in Table 7.

Table 7 A predication of the limiting resolution of 2nd gen. image-inverting tube

$\mathrm{MTF}_{SF}(f) = \exp[-(0.001,455,006\pi f)^2]$	
$\mathrm{MTF}_{FO}(f) = \exp\left[-\left(\frac{f}{66}\right)^{2.21}\right]$ $\mathrm{MTF}_{MCP}(f) = \exp\left[-\left(\frac{f}{29}\right)^{1.8}\right]$	
$\mathrm{MTF}_{PS1}(f) = \exp\left[-\left(\frac{f}{51}\right)^{1.34}\right]$	$\mathrm{MTF}_{PS2}(f) = \exp\left[-\left(\frac{f}{42}\right)^{1.49}\right]$
$\mathrm{MTF}_{MSPF1}(f) = \exp[-(0.008,660,254\pi f)^2]$	$\mathrm{MTF}_{MSPF2}(f) = \exp[-(0.011,547,005\pi f)^2]$
$f = 35$ lp/mm	$f = 32$ lp/mm
$\mathrm{MTF}(f=35)\vert_{\text{2nd gen. IIT}} = 0.974,729,188$ $\times 0.245,899,256 \times 0.403,826,521$ $\times 0.546,719,441 \times (0.781,805,636)^2$ $= 0.032,347,531$	$\mathrm{MTF}(f=32)\vert_{\text{2nd gen. IIT}} = 0.978,831,43$ $\times 0.303,049,097 \times 0.259,881,402$ $\times 0.513,319,951 \times (0.817,156,492)^2$ $= 0.032,442,94$

5.3 The second generation wafer tube

$$\mathrm{MTF}(f)\vert_{\text{wafer tube}} = \mathrm{MTF}_{FO}(f) \times \mathrm{MTF}_{CMPF}(f) \times \mathrm{MTF}_{MCP}(f) \\ \times \mathrm{MTF}_{MSPF}(f) \times \mathrm{MTF}_{FS}(f) \times \mathrm{MTF}_{FOT}(f) \tag{26}$$

Now, we evaluate the limiting resolution of 2nd generation wafer tube. The values of $\mathrm{MTF}_{FO}(f)$, $\mathrm{MTF}_{MCP}(f)$, $\mathrm{MTF}_{FS}(f)$ and $\mathrm{MTF}_{FOT}(f)$ still takes the data in Table 5. The ρ value of $\mathrm{MTF}_{MSPF}(f)$ in Eq. (21) remains the same for 2nd generation wafer tube. For the proximity focusing of photocathode—MCP spacing, we suppose that the initial emission angle obeys Lambert law, the initial emission energy distribution takes $\beta_{1,4}$, the photocathode—MCP spacing $L = (0.1-0.2)$ mm, the applied voltage $V = 200$ V, $\varepsilon_m = 1$ V, from Table 2, we have

$$\rho\vert_{CMPF} = 0.755,928,946L\sqrt{\frac{\varepsilon_m}{V}} = 0.005,352,248 - 0.010,690,449 \text{ mm} \tag{27}$$

The evaluated value of the limiting resolution for the 2nd generation wafer tube is shown in Table 8.

Table 8 A predication of the limiting resolution of 2nd generation wafer tube

$\mathrm{MTF}_{FO}(f) = \exp\left[-\left(\frac{f}{66}\right)^{2.21}\right]$	
$\mathrm{MTF}_{MCP}(f) = \exp\left[-\left(\frac{f}{29}\right)^{1.8}\right]$ $\mathrm{MTF}_{FOT}(f) = \exp\left[-\left(\frac{f}{38}\right)^{2.9}\right]$	
$\mathrm{MTF}_{PS1}(f) = \exp\left[-\left(\frac{f}{51}\right)^{1.34}\right]$	$\mathrm{MTF}_{PS2}(f) = \exp\left[-\left(\frac{f}{42}\right)^{1.49}\right]$
$\mathrm{MTF}_{CMPF1}(f) = \exp[-(0.005,345,224,8\pi f)^2]$	$\mathrm{MTF}_{CMPF2}(f) = \exp[-(0.010,690,449\pi f)^2]$
$\mathrm{MTF}_{MSPF1}(f) = \exp[-(0.008,660,254\pi f)^2]$	$\mathrm{MTF}_{MSPF2}(f) = \exp[-(0.011,547,005\pi f)^2]$

Continued

$\mathrm{MTF}_{\mathrm{FO}}(f) = \exp\left[-\left(\dfrac{f}{66}\right)^{2.21}\right]$	
$\mathrm{MTF}_{\mathrm{MCP}}(f) = \exp\left[-\left(\dfrac{f}{29}\right)^{1.8}\right]$ $\mathrm{MTF}_{\mathrm{FOT}}(f) = \exp\left[-\left(\dfrac{f}{38}\right)^{2.9}\right]$	
$f = 31$ lp/mm	$f = 26$ lp/mm
$\mathrm{MTF}(f)\big|_{\text{2nd gen. wafer tube}} = 0.828,413,939$ $\times 0.762,623,319 \times 0.323,826,63$ $\times 0.490,980,075 \times 0.598,513,18$ $\times 0.574,597,712 = 0.034,578,45$	$\mathrm{MTF}(f)\big|_{\text{2nd gen. wafer tube}} = 0.880,192,289,2$ $\times 0.466,499,961 \times 0.439,748,158$ $\times 0.410,828,147 \times 0.612,990,708$ $\times 0.716,984,462 = 0.032,603,08$

5.4 The third generation wafer tube

The MTF of the third generation wafer tube can be described with the following formula:

$$\mathrm{MTF}(f)\big|_{\text{3rdgen. wafer tube}} = \mathrm{MTF}_{\mathrm{CMPF}}(f) \times \mathrm{MTF}_{\mathrm{MCP}}(f) \times \mathrm{MTF}_{\mathrm{MSPF}}(f) \\ \times \mathrm{MTF}_{\mathrm{PS}}(f) \times \mathrm{MTF}_{\mathrm{FO}}(f) \tag{28}$$

Now, we discuss the limiting resolution of 3rd generation wafer tube. For the proximity focusing of photocathode—MCP spacing, suppose that the initial emission angle takes cubic-cosine distribution and the initial emission energy takes $\beta_{1,8}$ distribution, the photocathode—MCP spacing $L = 0.1$ mm, the applied voltage $V = 200$ V, $\varepsilon_\mathrm{m} = 0.5$ V, from Table 2, we have

$$\rho_{\mathrm{CMPF}}(f) = 0.492,365,936 L \sqrt{\dfrac{\varepsilon_\mathrm{m}}{V}} = 0.002,461,829,68 \text{ mm} \tag{29}$$

For the proximity focusing of MCP—phosphor screen spacing, suppose that the MCP—phosphor screen spacing $L = 0.75$ mm, the other data remain the same as in the second generation wafer tube, then we have

$$\rho_{\mathrm{MSPF}}(f) = 2L \sqrt{\dfrac{\varepsilon_\mathrm{p}}{V}} = 0.008,660,254 \text{ mm} \tag{30}$$

The evaluated value of the limiting resolution for 3rd generation wafer tube is given in Table 9.

Table 9 A predication of the limiting resolution of 3rd generation wafer tube

$\mathrm{MTF}_{\mathrm{CMPF2}}(f) = \exp[-(0.002,461,829,7\pi f)^2]$ $\mathrm{MTF}_{\mathrm{MCP}}(f) = \exp\left[-\left(\dfrac{f}{29}\right)^{1.3}\right]$	
$\mathrm{MTF}_{\mathrm{MSPF2}}(f) = \exp[-(0.008,660,254\pi f)^2]$ $\mathrm{MTF}_{\mathrm{FS}}(f) = \exp\left[-\left(\dfrac{f}{51}\right)^{1.4}\right]$	
$\mathrm{MTF}_{\mathrm{FO}}(f) = \exp\left[-\left(\dfrac{f}{66}\right)^{2.21}\right]$	$\mathrm{MTF}_{\mathrm{FOT}}(f) = \exp\left[-\left(\dfrac{f}{38}\right)^{2.9}\right]$
$f = 36$ lp/mm	$f = 33$ lp/mm
$\mathrm{MTF}(f)\big|_{\text{2nd gen. wafer tube}} = 0.925,407,401$ $\times 0.228,595,311 \times 0.383,151,219$ $\times 0.534,165,927 \times 0.769,541,132$ $= 0.033,317,968$	$\mathrm{MTF}(f)\big|_{\text{2nd gen. wafer tube}} = 0.936,936,855$ $\times 0.284,237,265 \times 0.446,596,473$ $\times 0.572,330,296 \times 0.514,670,074$ $= 0.035,033,399,4$

Conclusions

In the present paper, we have verified that the value of ρ in the approximate exponential formula of the MTF for the cathode lenses of image tubes is just the RMS radius $\bar{r}$, which has a clear and definite physical meaning. The analytical solution and the approximate expression of the MTF and its RMS radius have also been given in cases of different initial angular distributions and initial emission energy distributions. If the values of RMS radius of sharp focusing and proximity focusing are given, and the values of (f_c, n) for the optical components of image tube are known, one can then predict the characteristic of the image tube.

Acknowledgements

The work is supported by Chinese Natural Science Foundation and the Science and Technology Foundation of the Academy of Ordnance Science, China.

References

[1] CSORBA I P. Resolution limitations of electromagnetically focused image-intensifier tubes[J]. RCA Review, 1970, 31: 534.

[2] CSORBA I P. Modulation transfer function calculation of electrostatic electron lenses [J]. RCA Review, 1972, 33: 393.

[3] CSORBA I P. Modulation transfer function of image tube lenses[J]. Applied Optics, 1977, 16(10): 2647 – 2650.

[4] CSORBA I P. Image tubes[M]. Howard W. Sams & Co. Inc., 1985.

[5] Hartley K F. On the electron optics of magnetically focused image tubes[J]. J. Phys. D, 1974, 7: 1612.

[6] ZHOU L W, ZHANG Z Q, NI G Q, et al. On modulation transfer function of cathode lenses in image tubes[J]. Institute of Conference Series 121: 405 – 419, 1991.

[7] SIMENOV Y P. Unknown title [J]. Optical Mechanical Industry, 1971, 11: 4 (in Russian).

[8] EZARD L A. Measurement of the effective photo-electron emission energ[J]. RCA Review, 1975, 36(4): 711 – 721.

[9] ZHU K Z. Cathode lens electron optics transfer function[J]. Acta Electronica, 1980, 3: 25 (in Chinese).

[10] JOHNSON C B. A Magnetically focused image intensifier employing evaporated field electrodes [J]. Advances in Electronics and Electron Physics, 1972, 33B: 579.

[11] KULIKOV Y V, MONASTYRSKI M A. On chromatic aberrations in cathode lenses [J]. Radiotechnique and Electronics, 1976 (1): 2251 – 2254.

[12] 周立伟,艾克聪,方二伦. 成像系统的电子光学调制传递函数与均方根半径的研

究[J]. 北京理工大学出版社, 1982(3):36.

[13] NI G Q, ZHOU L W, FANG E L. An inverse design of electrostatic focusing field for electrostatic and magnetic imaging[J]. Acta Armamentari Sinica, 1989,1:77 (in English).

[14] NI G Q, ZHOU L W, FANG E L. A study of multi-energy point spread function in electron optical imaging systems [J]. Journal of Beijing Institute of Technology, 1988 (01): 35 – 46.

[15] ZHOU L W, FANG E L. Electron optics of oblique electromagnet- ic focusing systems [J]. Journal of Beijing Institute of Technology, 1990, 10(S1):14 (in English).

[16] ZHOU L W. A generalized theory of wide electron beam focusing[J]. Advances in Electronics and Electron Physics, 1985,64B:575 – 589.

7.5 一种用曲轴轨迹确定轴外点扩散分布的方法
A Method for Determining the Off-axis Point Spread Distribution by Using Trajectories Around Curved Axes

摘要：本文根据曲轴宽电子束聚焦理论，研究了一种用曲轴轨迹确定轴外点扩散分布的方法。这种新方法可在保持与传统方法相同的精度条件下，减小电子光学传递函数的计算量，提高计算速度，对大物面成像系统的计算具有实际意义。

Abstract: Based on the theory of wide electron beam focusing with curved axis, a method for determining the off-axis point spread distribution is studied using the trajectories around curved axes. The result shows that the new method of calculating the off-axis electron optical transfer function can reduce the amount of computation and increase the speed of calculation, yet retaining the same degree of accuracy when compared with the traditional method. It will have practical meaning in computing imaging systems with a large photocathode surface.

引 言

在大物面电子光学成像系统的设计计算中，一项主要的任务是确定系统的轴外电子轨迹的成像质量，尤其是关于传递函数的计算，需要首先确定轴外物点的点扩散函数。以往的计算方法和手段大都是通过直接追迹实际轨迹来实现的。由于要追迹上万条的电子轨迹，且每条轨迹的追迹都要逐点数值求解场分布及其导数，因此计算量很大。在文献[1]中根据静电阴极透镜直轴三级几何像差和二级近轴横向色差的形式，通过追迹150条电子轨迹拟合像差系数，进而衍算出点扩散函数（10 800个电子轨迹落点），从而减少了计算量。但由于以轴上点电位及特解来推算轴外点的电子光学像差，其可靠性与准确性仍是一个问题。此外，它仍然需花费不少的计算时间。

采用曲轴宽电子束聚焦理论是解决这一问题的一条有效途径。其特点是只需确定曲轴主轨迹及主轨迹上的场系数，通过求解曲近轴轨迹和曲轴像差系数来得到整个曲近轴电子束及像差，因而大大下降计算工作量。但由于一般曲轴像平面 s_i 与系统像平面 z_i 并不重合，故存在确定 z_i 像面上轨迹落点的问题。作者曾利用几何投影关系研究了静电阴极透镜像平面上点扩散分布的确定方法[2]，而对于电磁聚焦成像系统，由于存在着图像的旋转，不能简单地采用几何投影关系。本文将研究一种更具有普遍意义的轴外点扩散分布确定的方法。

金伟其，倪国强，周立伟. 北京理工大学. 北京理工大学学报（Journel of Beijing Institute of Technology），V.11, No.4, 1991, 90-93.

1 曲轴轨迹

如图 1 所示，设 $r_主(s)$ 为自 M_1 点逸出的主轨迹 $\overline{M_1M_2}$ 位置矢量，s 为主轨迹弧长，在 N 点以主轨迹的切向矢量 t、主法矢 n 和副法矢 b 构成 Frenet 曲线坐标基 (t, n, b)，并由 Frenet 坐标基绕 t 旋转 $\gamma(s)$ 角得到旋转曲线坐标基 (t, n_f, b_f)，即

$$\begin{bmatrix} t \\ n_f \\ b_f \end{bmatrix} = \begin{bmatrix} 1 & 0 & 0 \\ 0 & \cos\gamma & -\sin\gamma \\ 0 & \sin\gamma & \cos\gamma \end{bmatrix} \begin{bmatrix} t \\ n \\ b \end{bmatrix} \quad (1)$$

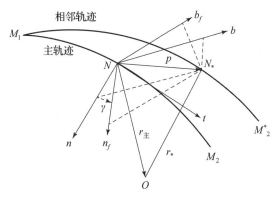

图 1 曲线坐标系

利用 Frenet-Serret 公式[3]

$$t' = kn, \quad n' = -kt + \chi b, \quad b' = -\chi n \quad (2)$$

可得到旋转坐标基满足

$$\begin{aligned} t' &= k(n_f\cos\gamma + b_f\sin\gamma), \\ n_f' &= -kt\cos\gamma - (\chi - \gamma')b_f, \\ b_f' &= -kt\sin\gamma - (\chi - \gamma')n_f \end{aligned} \quad (3)$$

式中，k 和 χ 分别为曲轴主轨迹的曲率和挠率。

设在 N 点的法平面内，斜出轨迹的交点 N_* 与 N 点的距离为 p，则斜出轨迹矢量 r_* 可表示为[4]

$$r_* = r_主 + p = r_主 + un_f + vb_f \quad (4)$$

式中，u 和 v 分别是 p 向 n_f 和 b_f 上的投影。因此，只要求出 $r_主(s)$ 和 $p(s)$ 便能确定整个电子束的运动规律。

根据曲轴宽电子束聚焦理论，曲轴轨迹 u、v 可表示如下[2]：

$$\begin{aligned} u &= u_g + \Delta u_c^{(0)} + \Delta u_c^{(1)} + \delta u^{(2)} + \cdots \\ v &= v_g + \Delta v_c^{(0)} + \Delta v_c^{(1)} + \delta v^{(2)} + \cdots \end{aligned} \quad (5)$$

式中，u_g、v_g 为曲近轴轨迹，$\delta u^{(2)}$、$\delta v^{(2)}$ 为曲轴二级几何像差，$\Delta u_c^{(0)}$、$\Delta v_c^{(0)}$ 和 $\Delta u_c^{(1)}$、$\Delta v_c^{(1)}$ 分别为曲轴主轨迹色差和曲近轴色差。

当已知实验室坐标系下的系统场分布后，主轨迹方程可用直角坐标系下的普遍轨迹方程来计算

$$\begin{cases} \dfrac{d}{dz}\left[\dfrac{(\varphi+\varepsilon_{s_1})^{1/2}}{(1+x'^2+y'^2)^{1/2}}x'\right]=\dfrac{(1+x'^2+y'^2)^{1/2}}{2(\varphi+\varepsilon_{s_1})^{1/2}}\dfrac{\partial\varphi}{\partial x}+\left(\dfrac{e}{2m_0}\right)^{1/2}(B_y-y'B_z)\\ \dfrac{d}{dz}\left[\dfrac{(\varphi+\varepsilon_{s_1})^{1/2}}{(1+x'^2+y'^2)^{1/2}}y'\right]=\dfrac{(1+x'^2+y'^2)^{1/2}}{2(\varphi+\varepsilon_{s_1})^{1/2}}\dfrac{\partial\varphi}{\partial y}-\left(\dfrac{e}{2m_0}\right)^{1/2}(B_x-x'B_z) \end{cases} \quad (6)$$

式中，ε_{s_1} 为曲轴主轨迹的初电位，φ 为主轨迹上的电位分布，B_x、B_y、B_z 分别为主轨迹上磁感应强度在直角坐标轴上的投影分量，e/m_0 为电子荷质比。

由主轨迹可确定曲率 k、挠率 χ、曲线坐标基（$\boldsymbol{t}$, $\boldsymbol{n}$, $\boldsymbol{b}$）和（$\boldsymbol{t}$, $\boldsymbol{n}_f$, $\boldsymbol{b}_f$）在直角坐标轴上的分量，并可进一步确定曲轴场系数、曲近轴轨迹和各种像差系数[2]。于是，在曲轴像平面 s_i 上的曲轴轨迹落点坐标（5）是曲轴轨迹初始条件（u_0, v_0, $(\varepsilon_s)^{1/2}u'_0$, $(\varepsilon_s)^{1/2}v'_0$）和轴向初电位差 $\Delta\varepsilon_s=\varepsilon_s-\varepsilon_{s_1}$ 的函数。换句话说，通过改变轨迹初始条件，可确定整个曲轴电子束在 s_i 像面上的落点分布。

2　系统像面上点扩散分布的确定

在电磁聚焦成像系统中，曲轴像平面 s_i 不仅在子午面内对于系统轴上像平面 z_i 有一个转角[2]，而且在弧矢面内也有一个转角，故几何投影关系很复杂。

实际上，利用式（3）和式（4），可以把曲轴法平面内斜出轨迹 $\boldsymbol{r}_*$ 及其对 $z_\text{主}$ 的导数 $d\boldsymbol{r}_*/dz_\text{主}$ 表示为直角坐标系中的分量形式

$$w=w_\text{主}+un_{fw}+vb_{fw}$$
$$w'=w'_\text{主}+[u'-(1+x'^2_\text{主}+y'^2_\text{主})^{1/2}\chi_1 v]n_{fw}+[v'+(1+x'^2_\text{主}+y'^2_\text{主})^{1/2}\chi_1 v]b_{fw}- \quad (7)$$
$$(1+x'^2_\text{主}+y'^2_\text{主})^{1/2}k(u\cos\gamma+v\sin\gamma)t_w \quad (z'_\text{主}=1,w=x,y,z)$$

式中，$\chi_1=\chi-\gamma'$，t_w、n_{fw}、b_{fw} 分别为基矢 $\boldsymbol{t}$、$\boldsymbol{n}_f$、$\boldsymbol{b}_f$ 在 x、y、z 方向的分量，撇"'"表示对 $z_\text{主}$ 的导数。

设对应 $z_\text{主}=z_i$ 的曲轴像平面为 s_i，则在 s_i 上（u_i, v_i）点离 z_i 平面的距离为

$$\Delta z=z(s_i)-z_i=u_i n_{fz}+v_i b_{fz} \quad (8)$$

一般情况下，Δz 是很小的，且像平面电子运动速度较快（轴向速度），故可假设该斜出轨迹继续运动的轨迹近似为直线，则其在 z_i 平面上的落点坐标（x_i, y_i）为

$$w_i=w(s_i)+w'(s_i)\Delta z=w_\text{主}(s_i)+u_i n_{fw}+v_i b_{fw}+$$
$$\{w'_\text{主}(s_i)+[u'_i-(1+x'^2_\text{主}+y'^2_\text{主})^{1/2}\chi_1 v_i]n_{fw}+$$
$$[v'+(1+x'^2_\text{主}+y'^2_\text{主})^{1/2}\chi_1 u_i]b_{fw}- \quad (9)$$
$$(1+x'^2_\text{主}+y'^2_\text{主})^{1/2}k(u_i\cos\gamma+v_i\sin\gamma)\Delta z$$
$$(w_i=x_i,y_i)$$

式中，t_w、n_{fw}、b_{fw}、$x_\text{主}$、$y_\text{主}$、χ_1、k 均取 s_i 处的值。

设系统轴上的中心放大率为 M_0，则得到轴外轨迹在像平面 z_i 上相对于理想像点的像差

$$\Delta w_i=[w_\text{主}(s_i)-M_0 w_0]+(u_i n_{fw}+v_i b_{fw})+w'(s_i)\Delta z \quad (10)$$

对于系统轴外曲轴物点来说（$u_0=v_0=0$），式（10）的第一项相当于直轴像差理论中的畸变，第二项相当于场曲引起的弥散。第三项是其他像差的影响。因此，当计算到曲轴

二级几何像差时，已基本包含了直轴三级几何像差的影响。

将式（5）代入式（10），经初始条件同幂次归纳就可得到 z_i 像平面上的像差系数，然后利用不同的轨迹初始条件，就可快速计算轴外物点的点扩散分布，进而计算电子光学传递函数。

实际计算时，还可根据计算类型，决定曲轴轨迹式（5）中的像差取舍。例如，计算轨迹时，可选 $u=u_g$ 和 $v=v_g$，由式（7）确定 ε_{s_1} 的曲轴电子束中的任意一条电子轨迹所对应的直角坐标及其斜率，这对于电子以直角坐标形式的输出或者绘制是很有效的；若选取曲近轴轨迹附加某一项曲轴像差，则可分析其对成像质量的影响。

结束语

根据曲轴宽电子束聚焦理论，研究了一种新的用曲轴轨迹确定大物面成像系统轴外点扩散分布的方法。其特点是速度快、精度高。这一方法已在本文的电磁聚焦成像系统设计计算软件包中得到应用。由于可以提高电子光学传递函数的计算速度，因此对大物面成像系统的设计计算，特别是考虑轴外像质的优化设计中具有实际意义。

参 考 文 献

[1] 方二伦,冯炽焘,周立伟. 变像管及像增强器电子学系统的计算机分析与设计[J]. 光电技术,1988(2~3):71-81.

[2] 金伟其,周立伟,倪国强. 面对称静电场中曲轴宽电子束聚焦的像差理论[J]. 北京理工大学学报,1990,10(4):34-44.

[3] SYNGE J L, SCHILD A. Tensor calculus [M]. New York：Dover Publications. Inc.,1969.

[4] 周立伟,金伟其,倪国强. 曲轴宽电子束聚焦理论的研究[J]. 光电子学技术,1988,8(4):8-22.

八、成像电子光学系统的计算与设计

Computation and Design of Imaging Electron Optical Systems

8.1 变像管及像增强器电子光学系统的计算机辅助设计

CAD of Electron Optical Systems in Image Converter Tubes and Image Intensifiers

摘要：本文叙述了运用电子计算机对变像管和像增强器的电子光学系统进行分析和设计的相关内容，包括静电场和电子轨迹计算的基本公式、电子光学特性和像差的确定，以及算法与程序设计。为了计算和分析不同类型像管的电子光学系统，程序采用填写"边界数据表"的方法以输入电极结构参数和加在电极上的电压参数以及由光阴极逸出的电子束的初始条件。由此"边界数据表"，程序便自动形成边界点参数。像管电子光学系统计算与设计的主要算法有：网格点上的静电电位计算和迭代采用有限差分法（FDM）；逐次超松弛法（SOR），场中流动点的电位及其一、二偏导数的计算采用 Lagrange 插值法。各种不同的轨迹微分方程如曲轴轨迹方程、实际轨迹方程和近轴轨迹方程被应用于计算电子轨迹。最后，文中讨论了电子光学特性和像差的确定以及系统的像质评定的问题。

Abstract: This paper describes a work on the design and analysis for electron optical systems in image converter tubes and image intensifiers using a computer. It includes basic formulae of computations for electrostatic field and electron trajectories, determination of electron optical properties and aberrations, algorithm and programming. In order to compute and analyze electron optical systems of different types, the programming adopts a method of filling in a "table of boundary data" for the input of electrodes' configuration parameters, electrical parameters, and initial data of electron beam emitted from the photocathode. From the "table of boundary data", the program forms parameters of boundary points automatically. Algorithms adopted comprise, among them: the finite difference method (FDM) and the successive over-relaxation (SOR) method are adopted for the electrostatic potential computation and iteration at the mesh points. Lagrange interpolation is used for computations of potential at flowing points in the field and for their partial derivatives of the first and second orders. The different differential equations of trajectories, such as trajectory equation based on curvilinear optical axis, practical trajectory equation and "paraxial" trajectory equation are used for computing the electron trajectories. Finally, determination of electron optical imaging properties and aberrations, as well as problem of evaluation of image quality for image systems have been discussed in the present paper.

方二伦[a]，冯炽焘[b]，周立伟[c]. a) 西安现代化学研究所，b) 昆明技术物理研究所，c) 北京工业学院. 光电子技术（Photoelectronic Technology），No. 2-3, 1980, 71-81.

引 言

变像管与像增强器在近红外及微光夜视、微光电视、红外显微观察、天文观察、高速摄影、公安侦破以及工业和医学等许多领域中，都有广泛和重要的应用。静电聚焦型的像管具有结构简单、使用方便、体积小、质量轻等优点，特别适用于研制各种军用夜视仪和其他小型仪器之用。随着社会主义四个现代化建设事业的飞跃发展，迫切要求我们能够自行设计和研制各种新型优质的像管。

众所周知，静电聚焦型像管的电子光学系统，属于浸没物镜式的宽电子束成像系统，它的阴极面本身就是物面，并且直接处于电子透镜的场中。对于这种系统，各种模拟装置，甚至模拟计算机，由于精度的限制或其他原因，一般只能得到极为近似的结果。然而，这种电子光学系统的分析设计所要求解决的问题具有较完整的数学描述和精密的结果。20 世纪 60 年代以来，由于计算技术的迅猛发展，像管电子光学系统的计算机分析，已逐渐为人们所熟悉。现在，各单位研制某一管型，几乎无例外地要进行数值分析的工作。这就客观地表明，运用电子计算机进行计算，是分析设计静电阴极透镜成像系统的精确而有效的方法[1]。

概括地说，用数值方法分析计算静电阴极透镜，需首先求出电极系统内部的电位分布；作为数学问题，是求在给定边界条件下拉普拉斯偏微分方程的解。这通常以与系统轴线相平行和相垂直的两族直线，将系统半子午面上由电极边界所包围的区域划分成网格；然后在全部网格内节点上，以有限差商近似代替拉氏偏微分方程中的偏导数，建立起有限差分方程，并对这些方程所构成的方程组，进行迭代求解。其次则依据所求出的电位分布，在给定的初始条件下，求电子轨迹微分方程或运动微分方程的解。最后由各种电子轨迹的计算结果，便可确定系统的电子光学成像参量和像差，以及进一步计算电子光学鉴别率和电子光学传递函数等。下面分别做具体的论述。

1 静电场的计算

通常，用网格法计算像管电子光学系统的电位分布 $\varphi(z,r)$，可以归结为在给定的边界所包围的封闭区域内（实际系统的边界在不同电极之间往往是非封闭的，对此，可以在适当的位置作所谓"补充边界线"，并采取适当的插补计算，求出其上网格点的电位值并予固定之），求解拉普拉斯（Laplace）偏微分方程：

$$\frac{\partial^2 \varphi}{\partial z^2} + \frac{1}{r}\frac{\partial \varphi}{\partial r} + \frac{\partial^2 \varphi}{\partial r^2} = 0 \tag{1}$$

式中，z、r 分别为子午面上轴向和径向坐标。

一般地说，用数值方法求解方程式（1）有各种不同的方法。例如，D. R. Cruise[2] 等提出用积分形式的泊松（Poisson）方程，通过计算电极面上的电荷分布，来计算系统中任意点的电位和感应分析（Induction Analysis），或称电荷分布方法（Charge Distribution）[2,3]；F. Schaff[4,5] 探索将电位按本征函数展开，从而用解析函数求和的形式，来确定静电场的电位分布；以及使用三角形网格的"有限元素法"等。这些方法都各有自己的特点，但对于我们所要处理的全部计算内容，都还存在一定的问题（例如用任意的三角形网格，如何精确计算任意流动点的电位及其一、二阶偏导数;以及当电极形状尺寸有改变时,如何很快地形成边界点参数等)。

我们采用最简单和通用的"五点差分公式",对方程式(1)进行数值求解。此时,任一内点的电位值 φ_0,可以用它上、下、左、右四个点的电位 φ_1、φ_2、φ_3、φ_4 表示,其一般形式的公式为

$$\varphi_0 = \frac{1}{A}(A_1\varphi_1 + A_2\varphi_2 + A_3\varphi_3 + A_4\varphi_4) \tag{2}$$

对于四个方向的点与中心点不等距的一般情况,可导出

$$\begin{cases} A_1 = (2r_0 + t_2h)/t_1(t_1+t_2)r_0 \\ A_2 = (2r_0 - t_1h)/t_2(t_1+t_2)r_0 \\ A_3 = 2/t_3(t_3+t_4) \\ A_4 = 2/t_4(t_3+t_4) \end{cases} \tag{3}$$

式中,h 为基本步长,四个方向的点与中心点的距离分别为 $h_1=t_1h$,$h_2=t_2h$,$h_3=t_3h$,$h_4=t_4h$;r_0 为中心点与 z 轴的距离。

当 $r_0=0$ 时,即对应于轴上点,计算 φ_0 时只引用 φ_1、φ_3、φ_4,此时可导出:

$$\begin{cases} A_1 = 4/t_1^2 \\ A_2 = 0 \\ A_3 = 2/t_3(t_3+t_4) \\ A_4 = 2/t_4(t_3+t_4) \end{cases} \tag{4}$$

而当 $t_1=t_2=t_3=t_4$ 时,即对应于等距网格的普通内点,则得

$$\begin{cases} A_1 = 1 + h/2r_0 \\ A_2 = 1 - h/2r_0 \\ A_3 = A_4 = 1 \end{cases} \tag{5}$$

对于等距网格的轴上点,则化为

$$A_1 = 4, \quad A_2 = 0, \quad A_3 = A_4 = 1 \tag{6}$$

并且对于所用的情况都有

$$A = A_1 + A_2 + A_3 + A_4 \tag{7}$$

因此,公式(2)可以改写为

$$\varphi_0 = \sum_{i=1}^{4} A_i\varphi_i / \sum_{i=1}^{4} A_i \tag{8}$$

按上述方法处理,对于任一系统,我们都可以得到与内节点的数目相等的差分方程所构成的方程组。在通用程序中,对它们采用 Young – Frankel 超松弛法(Successive Over – Relaxation method)进行迭代求解;对每一点的计算,用"Seidel – Liebmann"迭代,并按照 B. A. Carre 的方法[6],由程序过程自动选求超松弛加速因子。

实际计算中,对于一个系统的任一节点,如何确定 t_1、t_2、t_3、t_4 和 A_1、A_2、A_3、A_4,需由具体的电极结构来确定;这对于各种不同的系统,情况往往是很复杂的。为了使计算程序具有通用性和灵活性,我们成功地实现了"自动形成边界点参数"的程序设计,用这

一方法通过一次扫描，便定出内点中所有邻近边界的不规则点的四个电位系数 A_1、A_2、A_3、A_4，并把它们依序存放起来，直接提供电位迭代计算之用。

关于电位初值的给定，理论上可以是任意的。实际计算时应对它做适当的选取，但无须追求十分精确。现在，我们的通用程序中采用的方法是：在电极之间，按电极电位值在 z 方向做 4/3 次方插值；即若阴极与屏的电位分别为 V_c 和 V_s，屏的 z 坐标为 z_m，则轴向坐标为 z 的任一内点的电位初值可取为：

$$\varphi_0 = V_c + (V_s - V_c)(z/z_m)^{4/3} \tag{9}$$

我们的通用程序允许采用大、中、小三种步长，并在必要时划取子区间做精密计算。根据我们自己的经验，初步计算一个管型，只要取 1 500 个左右的网格点，便可得到大体符合的结果；而若取到 4 000～6 000 个点，则可以给出比较理想的结果（现在我们在 DJS-6 和 TQ-16 计算机上的通用程序，留给电位矩阵的内存单元分别为 8 192 和 12 288）。一个大约 1 000 个点的系统，迭代计算约 150 遍，便可收敛到相对残差为 10^{-8}～10^{-9}；在 DJS-6 和 TQ-16 计算机上，需时分别为 2 分 30 秒和 1 分 15 秒左右；在此基础上将网格折半，得网格点数约 4 000，迭代约 250 遍又可收敛到同样精度，在两种机器上，需时分别约 15 分钟和 7 分 30 秒。

为了对所计算电子光学系统的静电场做形象、直观的描绘，可以基于已算出的电位分布数据，计算并绘出等位线和电力线。

在 $z-r$ 子午面上，等位线就是 "$\varphi(z,r) = \text{Const}$" 的曲线族，而电力线则是等位线的正交轨线族。从物理上说，某一条电力线在其上任一点的正向切线（定义从阴极指向阳极的方向为正），就可确定电子在该点时，所受电场力作用的方向；而电子轨迹在其上任一点的正向切线，所确定的乃是运动电子在该点的瞬时速度的方向。

我们的通用程序中使用"扫描搜索法"计算等电位线，即通过有序的搜索过程，按所计算的等电位线电位值 φ_k，通过插值计算，求出等位线与网格线相交之各点的坐标值，便可绘出等电位曲线。

电力线的计算使用"微分方程追迹法"[7]。这时，导出确定它的方程组为：

$$\begin{aligned} \frac{dz}{ds} &= \frac{\partial \varphi}{\partial z} \bigg/ \left[\left(\frac{\partial \varphi}{\partial z}\right)^2 + \left(\frac{\partial \varphi}{\partial r}\right)^2 \right], \\ \frac{dr}{ds} &= \frac{\partial \varphi}{\partial r} \bigg/ \left[\left(\frac{\partial \varphi}{\partial z}\right)^2 + \left(\frac{\partial \varphi}{\partial r}\right)^2 \right] \end{aligned} \tag{10}$$

方程中 s 是作为自变量的弧长；任意流动点的 $\partial \varphi/\partial z$、$\partial \varphi/\partial r$ 可根据网格点的电位分布，由拉格朗日（Lagrange）公式确定。方程式 (10) 的解的形式为

$$z = z(s), \quad r = r(s) \tag{11}$$

求数值解时，应首先给出初始条件 $z|_{s=s_0} = z_0$ 和 $r|_{s=s_0} = r_0$，通常选择 (z_0, r_0) 为阴极面上的点，与此相对应的 $s_0 = 0$。

此外，由所计算出来的网格点电位，还可以计算阴极面上的场强分布、系统轴线上的加密电位及其 1~4 阶导数、轴上电位分布曲线扭转点，以及在计算电力线和电子轨迹时，用 4×4 拉格朗日二向插值和求导公式，来计算子午面上任意点的电位及其一、二阶偏导数。

2 电子轨迹和电子光学参量的计算[8,9]

在求得系统的电位分布 $\varphi(z,r)$ 之后，就可以通过各种电子轨迹的计算，来确定电子光学成像参量和电子光学像差。为此，根据不同的要求，需选用不同形式的方程，计算各种初始条件的轨迹，这在许多文献资料中略而不提，而本文则对它们做较详细的说明。

总的说来，计算电子运动的轨迹，可以引用运动微分方程或轨迹微分方程，二者在原理上是等效的，我们曾对此做过有关的论述和比较；通用程序中则选用轨迹微分方程做实际计算。我们基于最一般的矢量形式的方程：

$$\frac{2V_*}{1+r'^2}r'' + \frac{\partial V}{\partial z}r' - \frac{\partial V}{\partial r} = 0 \tag{12}$$

式中，$V_* = V(z,r) + \varepsilon$，其中 $V(z,r)$ 为空间电位分布，ε 为与电子的初能量相对应的加速电位（简称初电位）。用它便可导出用于实际计算的各种形式的轨迹方程。

2.1 基本电子光学参量的计算

分析和设计某一像管的电子光学系统，都必须首先确定基本的参量：像面位置 z_s、近轴放大率 M_0 和近轴色球差 δ_c（通常折算到阴极面上定义）。这里，z_s 不但用来确定实际像管荧光屏的位置，而且是计算各种电子光学像差的基准面。

由方程式（12）做近轴的假定，便可导出大家熟知的近轴轨迹方程：

$$[\varepsilon_z + \phi(z)]r'' + \frac{1}{2}\phi'(z)r' + \frac{1}{4}\phi''(z)r = 0 \tag{13}$$

式中，$\phi(z)$ 为轴上电位分布，$\varepsilon_z = \varepsilon\cos^2\alpha$ 为电子的轴向初电位。根据此方程的线性齐次特性可知，由阴极面中心以不同初角度 α 发出的、所有具有相同的轴向初电位 ε_{z_1} 的近轴轨迹都将重新聚焦于同一点，由此焦点的轴向坐标 z_s 所确定的像面，即为对应于轴向初电位 ε_{z_1} 的理想成像面；同时，离轴高度为 r_c、轴向初电位仍为 ε_{z_1} 近轴轨迹，在 z_s 像面上的落点与轴线的距离 r_s 对 r_c 之比值，即为对应于 z_s 像面的近轴放大率。按此，如果电子由阴极面发出时的最大初电位为 $\varepsilon_{\max}$，则 ε_{z_1} 取 $0 \sim \varepsilon_{\max}$ 所对应的近轴轨迹的像点，便确定了实际像面所在的区间。当 $\varepsilon_{z_1} = 0$ 时，所对应的像面称为极限像面（简记为 t 面）；当 $\varepsilon_{z_1} = \varepsilon_{\max}$ 时，所对应的像面称为高斯像面（简记为 g 面）。同时，根据二级横向色球差的"R - A"公式

$$\Delta r = -2M_0 \frac{\sqrt{\varepsilon_r}}{E_c}(\sqrt{\varepsilon_{z_1}} - \sqrt{\varepsilon_z}) \tag{14}$$

（式中，E_c 为阴极面中心点的场强，ε_z 为决定像面位置的轴向初电位，ε_z、ε_r 为决定弥散圆半径的轴向和径向初电位）可以证明：在 t 面和 g 面之间，存在一弥散圆半径最小的最佳像面（简记为 m 面），它对应于 $\varepsilon_{z_1} = 0.09\varepsilon_{\max}$；通常就用它确定实际成像面的位置。

实际上，我们计算初电位 $\varepsilon_{z_1} = \varepsilon_{\max}$，初角度 $\alpha = 72.525°$、$90°$、$45°$ 和 $\varepsilon = 2\varepsilon_{\max}$、$\alpha = 45°$ 的四条轨迹，就可以确定 m、t、g 三个像面和近轴色球差；并可以由确定像面位置的轨迹，用由朗斯基（Wronski）行列式导出的公式

$$M_0 = \sqrt{\varepsilon_r}/r'(z_s)\sqrt{\phi(z)+\varepsilon_z} \tag{15}$$

来计算相应的近轴放大率 M_0（式中，ε_r、ε_z 分别为所计算轨迹的径向和轴向初电位，$\phi(z_s)$、$r'(z_s)$ 为此轴上点轨迹到达像面时的电位和斜率）。

在程序中，为了进行对比和作为计算轴外实际轨迹的基面，我们还按照实际轨迹方程（详后），引用与上面确定 m、t、g 三个面相同的初始条件，计算相应的三个面 m'、t'、g'。此时 m'、t'、g' 面与对应的 m、t、g 面的位置在 z 轴方向之差，反映了纵向几何像差。

2.2 轴外放大率与畸变的确定

以上所述 M_0 是系统的近轴放大率。实际由系统的中心逐渐移向轴外，不同高度 r_c 所对应的实际放大率并不一样，于是就导致像的几何形状的失真，即产生了畸变。

由于主轨迹（即阴极面上沿法线方向反射出来的电子轨迹）可以视为由一点发出的单元电子束的参考轴，因此通常定义阴极面上高度为 r_c 的主轨迹在给定的像面（平面的或球面的）上的落点的径向高度 r_s 与 r_c 之比，叫作相应点的轴外放大率 M_r。即

$$M_r = r_s/r_c \tag{16}$$

而畸变则定义为

$$\beta_r = \frac{M_r - M_0}{M_0} \times 100\% \tag{17}$$

下面要讨论到，在计算场曲和像散时，要用主轨迹方程与 p-q 曲轴方程联立求解，所以在实际计算时，可以计算场曲、像散的同时，就由主轨迹确定轴外放大率 M_r，并计算出畸变 β_r。

如果一个系统仅有畸变存在，则由阴极面上一点发出的电子，仍将重新聚焦于一点，这样在理想像面上所得到的仍是清晰的像，只是就整体而论，图像的尺寸比例有了失真。对于不同的系统，随着被映照的物点远离系统的轴线，像点对物点的比例，即像的放大率随之减小或增大，就相应产生"桶形"或"枕形"畸变。在静电阴极透镜中，经常出现的是枕形畸变，但我们可以设计带有校正电极的系统，以减小或消除这种像差。

2.3 场曲、像散和轴外像差的确定

场曲与像散产生的主要原因是：物点远离系统曲线和轴外的单元电子束相对于系统的轴线与倾斜。对于实际的系统，具体可归结为两点：①阴极面上轴外点的场强，以及轴外沿单元电子束参考轴的电位分布，与整个系统的轴上不尽相同，特别是轴外场强低于中心场强的情况；②围绕轴外单元电子束的场，对于主轨迹是非旋转对称的，即作用于单元电子束的场是各向异性的。

当与某一轴向初电位 ε_{z_1} 相对应的像面位置确定后，轴外物点以法向初速 $\varepsilon_n = \varepsilon_{z_1}$ 射出的斜轨迹，由于前一原因以及轴外主轨迹到达像面的光程较长，相邻轨迹通常在尚未到达像面位置时，就与主轨迹相交了；这样，在不同的入射高度 r_c，像点与确定的理想像面有不同的偏离，就造成像面的弯曲。再由于第二个原因，同一高度 r_c 且逸出角 α 相同的轴外子午轨迹与弧矢轨迹，也并不会聚于同一点。因此，与"子午焦点"和"弧矢焦点"相对应，就分别有"子午场曲"和"弧矢场曲"；而用同一单元电子束的子午与弧矢焦点的空间离散，就可以表征"像散"。

实际上，确定与轴外物点相对应的像点，即子午和弧矢焦点，从而确定场曲和像散；以及根据初始条件不同的曲近轴轨迹相对于理想像点的离散来定义色球差，在概念上应与通过中心近轴轨迹的计算来确定不同的像面和色球差是类似的。为此，可以将导出方程式（13）时的考虑推广到轴外：把由阴极面上轴外点发出的主轨迹，作为该点参与成像的单元电子束的参考轴，来讨论成像电子束中对于主轨迹是近轴射线的相邻轨迹。这时，由于轴外的主轨迹一般为曲线，故通常将相对于它的近轴方程，称为"曲近轴方程"。

基于以上的考虑，设与主轨迹相邻的子午与弧矢方向的曲近轴轨迹分别为 $p = p(z)$ 和 $q = q(z)$，则由基本方程式（12）可导出主轨迹方程为

$$r'' = \frac{1 + r'^2}{2V_n(z,r)}\left(\frac{\partial V}{\partial r} - r'\frac{\partial V}{\partial z}\right) \tag{18}$$

式中，$V_n(z,r) = V(z,r) + \varepsilon_n$，其中 $\varepsilon_n = \varepsilon_{\max}\cos^2\alpha$，确定于相邻轨迹沿主轨迹方向的初电位。而以相对于主轨迹的偏离量 p、q 为变量的子午和弧矢曲轴方程分别为：

$$\begin{aligned}p'' + F_1 p' + F_2 p &= 0,\\ q'' + G_1 q' + G_2 q &= 0\end{aligned} \tag{19}$$

式中，

$$F_1 = G_1 = \frac{1 + r'^2}{2V_n}\frac{\partial V}{\partial z}$$

$$F_2 = \frac{3r''^2}{(1 + r'^2)^2} - \frac{1}{2V_n}\left(\frac{\partial^2 V}{\partial r^2} - 2r'\frac{\partial^2 V}{\partial z \partial r} + r'^2\frac{\partial^2 V}{\partial z^2}\right)$$

$$G_2 = -\frac{1 + r'^2}{2V_n}\frac{1}{r}\frac{\partial V}{\partial r}$$

将方程式（18）与式（19）相联立，计算 $\varepsilon = \varepsilon_{\max}$，$\alpha = 72.525°$、$90°$、$45°$ 和 $\varepsilon = 2\varepsilon_{\max}$，$\alpha = 45°$ 的四组轨迹，便可以确定相对于 m、t、g 三个像面的子午焦点 $(z_p, r_p)m$、t、g 和弧矢焦点 $(z_q, r_q)m$、t、g，以及色球差和畸变。最后，用公式

$$F_p = z_s - z_p, \quad F_q = z_s - z_q \tag{20}$$

$$\eta = [(z_p - z_q)^2 + (r_p - r_q)^2]^{1/2} \tag{21}$$

分别确定子午、弧矢场曲及像散。

2.4 非对称像差和彗差的确定

前面讨论了由于轴外主轨迹的场在子午与弧矢方向的各向异性，而导致子午与弧矢焦点的离散，即像散。实际上，对于大物面的阴极透镜，即使是在子午面内，主轨迹两侧的场也是不对称的。所以，若用子午实际轨迹方程

$$r'' = \frac{1 + r'^2}{2V_*}\left(\frac{\partial V}{\partial r} - r'\frac{\partial V}{\partial z}\right) \tag{22}$$

（式中，$V_* = V(z,r) + \varepsilon$）计算主轨迹和相对于它具有逸出角为 $\pm\alpha$（对应于最大像差的计算，取 $\alpha = \pm 90°$）的两条子午轨迹，则可得到斜轨迹与主轨迹的两个交点 (z_+, r_+) 和 (z_-, r_-)，其空间离散

$$W = [(z_+ - z_-)^2 + (r_+ - r_-)^2]^{1/2} \tag{23}$$

便可定义为以纵像差形式表示的"子午非对称像差"；它与前面所讨论的像散 η 具有相同

的量纲，故可以相互比较。

这里应注意，如是二电极同心球的系统，纵然像散上述子午非对称像差都不存在。但由于轴外单元束相对于与系统轴线相垂直的平面屏是倾斜的，因而逸出时对于主轨迹为对称的两条子午轨迹在平面屏上的落点，对于主轨迹与屏的交点仍然是不对称的。

在实际系统中，我们把上述两种原因所造成的、初始条件对于主轨迹为对称的相邻轨迹在像面上的失对称性统称为彗形像差。按此，若 $\alpha = \pm 90°$ 的两条子午实际轨迹，在 z_s 像面上的落点的 r 坐标分别为 $r_{\max}$、$r_{\min}$，而主轨迹在同一像面上交点的 r 坐标为 r_s，则

$$\bar{W} = \frac{1}{2}(r_{\max} + r_{\min}) - r_s \tag{24}$$

即可作为以横向像差形式表示的彗差。

2.5 完全用实际轨迹确定各种像差的方法

以上我们着眼于将各种电子光学像差尽可能地分离开来，采用了各种形式的轨迹方程进行计算；但我们也可以采用与几何光学相类似的方法，完全用子午和弧矢实际轨迹的计算数据来确定各种像差。

为此，我们首先由基本方程式（12），导出适用于计算任意空间实际轨迹的方程组为

$$r'' = \frac{1 + r'^2 + y'^2}{2V_*}\left(\frac{1}{\rho}\frac{\partial V}{\partial \rho}r - \frac{\partial V}{\partial z}r'\right),$$
$$y'' = \frac{1 + r'^2 + y'^2}{2V_*}\left(\frac{1}{\rho}\frac{\partial V}{\partial \rho}y - \frac{\partial V}{\partial z}y'\right) \tag{25}$$

这里，使用 $z-r-y$ 空间直角坐标，$\rho = \sqrt{r^2 + y^2}$，$V_n = V(z,r,y) + \varepsilon$。一般情况下，电子自阴极面上某点 (z_0, r_0) 发射出来时，其初速 v_0 具有任意的逸出角 α 和方位角 β，所对应的便是一般空间轨迹。特殊情况时，若 $\beta = 0°$ 或 $180°$，则初速度没有与子午面相垂直的 y 方向的分量，同时也没有 y 方向的作用力，所对应的便是子午轨迹；此时由方程式（25）简化便得方程式（22）。而若 $\beta = 90°$ 或 $270°$，则所对应的空间轨迹便是通常所说的弧矢轨迹。

对于所讨论的问题，我们用中心实际轨迹所确定的像平面作为基面。例如，相对于 t' 面，可以用 $\alpha = 0°$ 的轴外主轨迹的落点，确定轴外放大率和畸变；用 $\alpha = \pm 90°$ 的两条子午轨迹的交点 (z_z, r_z) 作为子午焦点；用 $\alpha = 90°$ 的弧矢轨迹与子午面的交点 (z_y, r_y) 作为弧矢焦点；用 z_x、z_y 相对于 t 面的偏离分别确定子午和弧矢场曲，用子午与弧矢焦点的空间离散确定像散；用 r_x、r_y 相对于主轨迹的横向偏离，确定子午和弧矢彗差；并可用一系列的空间实际轨迹来确定像面上的最大弥散像斑。

3 像管的像质评定

前面我们讨论了静电阴极透镜电子光学像差的定义和计算的问题。一般来说，任一系统的各种像差越小，像的质量就越好；因此，在设计一个像管时，总是力图通过修改电极系统的结构尺寸，电极电位的适当取值，使之不但满足预定的结构形式外廓尺寸和基本电子光学参量的要求，还要尽可能地减小各种电子光学像差。但是一般来说，残留的像差总

是不可能完全消除，甚至有时会出现这样的情况：某一种像差减小了，又会出现别的像差反而增大。因此在实际使用中，就有必要选用某种参量，对系统的成像质量做综合的评定。

以往，实际用来作为像管像质评定的综合参量，是电子光学系统和整管的"鉴别率"。这种方法用于实际检验比较简单，至今还在使用。但是，后来人们逐渐发现，"鉴别率"并不是一个很确切的物理量，只用它还不能作为评定光学器件成像质量的充分判据。近年来，国外许多定型的像管和新的工作，大都给出调制传递函数的曲线或数据；国内许多单位，也在积极研究测量或计算光学和电子光学传递函数的问题，已经取得了可喜的成果。关于像管电子光学传递函数的理论研究以及计算，将另文叙述。

4 像管的方案计算与公差分析

在我们的通用程序中，先后采用了"自动形成逻辑尺"和"自动形成边界点参数"的结构处理方法，这不但能对已定系统进行较详尽和全面的分析，而且能够在按预定的要求设计新的管型时，有效地进行方案试算，从而挑选最佳结构，给出详细的指标参量和像差数据等；同时，还可以对几何尺寸和电参量的微扰，进行公差的和焦深的分析。

仅就上机计算而言，方案计算和公差分析计算是类似的，只是目的不同。为此，我们分析计算某一像管，在开始填"边界数据表"时，就先确定需做变动计算的尺寸参量和电参量，把这些量的数据定为"基本数据"，并通过一简单的"形成边界数据处理程序"，对基本数据进行加工而形成全部边界数据。这样，对于一个试算的系统，只要结构形式基本不变，就允许对边界参数做单个的修改；此时，只需临时更改基本数据，再执行处理程序，可迅速形成的边界参数，便可满足方案计算和公差分析的要求。

这里所谓电子光学公差，即系统的结构尺寸和电极电位所允许的误差值，它是依据这些量的变化影响于系统的聚焦成像性质和程度而确定的。通常，在电子光学系统中，当电极形状、尺寸、相对位置以及各电极的工作电位值有改变或起伏时，系统的电位分布、电子轨迹以及由它们所确定的电子光学参量和像差，就将随之而变。可是，在工艺上实现所设计的系统时，总不可避免地会有偏差。于是，就有必要确定这种偏差的允许范围——电子光学公差，使系统各指标参量的变动能够控制在许可的范围内。

通常，我们希望按照设计方案制造和装配的管子，其放大率的变化与一定视场内鉴别率或传递函数的降低，能够控制在某一容许的范围之内。实际上，当系统的边界条件只是发生不大的改变时，系统总的性能、各种像差的相对关系，并不会发生较大的质变。其主要影响表现为像面位置的移动及放大率的改变，而后者又与前者是相关的。所以，一般只需要用"成像面位置的偏移"这一指标，就足以表征几何尺寸和电参量微扰的影响。

我们借鉴于几何光学，引入"焦深"的概念，来对问题做综合的分析。这可以首先计算有关的每一几何尺寸和电参量改变时，所引起成像面位置改变的数据；然后根据零件加工和装配实际可能达到的精度，试给出各单项可变量的公差值，用统计的方法，推算成像面位置可能的移动量 $\Delta \tilde{z}$，用它与允许的焦深 $\Delta \tilde{z}_s$ 相比，来检验所给公差的合理性。很显然，若某一量的改变，引起像面的摆动较大，就应给以较小的公差；反之，则应放宽要求。这里可得到一个重要的结论是：在加工研制过程中，一个像管的像质是否合格，取决于各可变参量的起伏，视其所引起的像面位置的漂移，是否超出了允许的焦深，而不是只看荧光屏的位置是否符合设计的尺寸位置。某些情况下，根据公差计算的数据，适当变动个别的安装尺寸（如极间距离 l、像面位置 z_s），就能够补偿因零件制造等原因所引起的像面偏离。

参 考 文 献

[1] 冯炽焘,方二伦,周立伟. 变像管及像增强器电子光学系统的数值计算与设计[J]. 光电技术,1980,2-3:71-81.

[2] CRUISE D R. A Numerical method for the determination of an electric field about a complicated boundary[J]. Journal of Applied Physics, 1963, 34(12):3477-3479.

[3] Munro E. CAD of electron lenses by the finite element. Image processing and CAD in electron optics[M]. Academic Press,1973:284-323.

[4] SCHAFF F. Subminiature deflection circuit operates integrated sweep circuits in TV camera[J]. Z. Angew, Phys., 1967, Bd. 23, Heft. 2, S. 64.

[5] HARTH W, SCHAFF F. Cardinal elements and spherical aberration constant of a strong two-cylinder mesh lens[J]. IEEE Transactions on Electron Devices, 1968, 14(12):860-861.

[6] CARRE B A. The determination of the optimum accelerating factor for successive over-relaxation[J]. Computer Journal, 1961, 4(1):73-78.

[7] KULSRUD H E. A programming system for electron optical simulation [J]. RCA Review. 1967,28(2):351-365.

[8] 周立伟. 变像管与像增强器的电子光学[M]. 北京:北京理工大学出版社,1975.

[9] 周立伟. 夜视器件中的电子光学[M]. 北京:北京理工大学出版社,1977.

8.2 Optimization Design of Image Tubes with Electrostatic Focusing
静电聚焦像管的优化设计

Abstract: The present paper applies a constrained variable metric method (CVMM) for the optimization design of image tubes with electrostatic focusing, and a multigrid method (MGM) for the computation of the electrostatic field in designing image tubes, thus to improve the efficiency of field computation, and to reduce the time of computation. For the design of diode tubes, triode tubes, zoom tubes and gated tubes, we have investigated the objective function having a least-square-fit form with weight factors to carry out the optimization computation. The result of optimization has shown that the suggested method for designing image tubes given in the present paper is a practical and effective one.

摘要：本文将约束变尺度法（CVMM）用于静电聚焦像管的优化设计，将多重网格法（MGM）用于像管设计中静电场的计算，从而提高了电场计算的效率，减少了计算时间。对于二电极像管、三电极像管、变倍管与选通管，本文研究了具有权因子最小拟合形式的目标函数以进行优化计算，计算结果表明本文所给出的设计像管的方法是一种既实际又有效的方法。

Introduction

The optimization design of image tubes has begun since the 1970s, but as it is restricted by the numerical algorithm and optimization techniques, the problem has not been fully solved, especially for systems with wide electron beam focusing. For the optimization design of image tubes, it not only needs a fast convergence algorithm calling for lesser number of the objective function, but also needs a new numerical method which solves the electrostatic field very fast.

The present paper combines the CVMM with the MGM, and yields a new method of optimization, having high efficiency and fast convergence.

1 Principle

1.1 Optimization design of image tubes

The problem of optimization design of image tubes is a numerical one for solving the nonlinear

---

Zhang Zhiquan[a], Zhou Liwei[a], Jin Weiqi[a], Fang Erlun[b]. a) Beijing Institute of Technology, b) Xi'an Research Institute of Modern Chemistry. Proceedings of '93 Beijing International Symposium on Photoelectronic Detection and Imaging. SPIE V. 1982 (1993), 238–244.

programming that is given below:

$$\begin{aligned} \min \quad & f(\boldsymbol{x}), \quad & \boldsymbol{x} \in E^n \\ \text{s.t.} \quad & c_j(\boldsymbol{x}) = 0, \quad & j \in E \\ & c_j(\boldsymbol{x}) \geq 0, \quad & j \in I \end{aligned} \quad (1)$$

where $f(\boldsymbol{x})$ is the objective function, which represents the image characteristic parameters and aberrations, $c_j(\boldsymbol{x})$ are also image parameters, which should satisfy the restrained conditions during the optimizing process, $x_i \ (i = 1, 2, \cdots, n)$ are the geometric structural and electrical parameters of the image tube. So the optimization design of image tubes can be described as: to search a set of $x_i \ (i = 1, 2, \cdots, n)$, which makes the objective function $f(\boldsymbol{x})$ minimum under the constrained conditions as given in problem (1).

1.2 Multigrid method (MGM)

The most troublesome problem in studying the optimization design of image tubes is that it should spend plenty of time for the field computation. Besides the various numerical solution methods so far used, such as finite difference method (FDM), finite element method (FEM) and integral equation method (IEM), the multigrid method is a new one. Because the MGM has very fast rate of convergence and its convergence is almost not changed with the step of length h, it is very suitable for the computation in solving the boundary value problem which needs a large number of iterations.

The multigrid method combines the traditional iteration (such as Gauss-Seidel iteration with $\omega = 1$, Jacobi iteration, etc.) with the Coarse Grid Correction technique, and transfers the error of field value of the finer grid to the coarser grid. Because the error is not smooth to the coarser grid, so it will have fast convergence in the iteration at the coarser grid points. At last, put the error transferred to the finer grid and add it to the corresponding grid points, an approximate solution of the boundary value problem will be obtained. It shows that the MGM is a highly efficient and very fast iteration algorithm, thus it lays a good foundation for the optimization design of image tubes [1,2].

A flow chart of the algorithm of MGM is shown in Fig. 1 [3].

1.3 Constrained variable metric method (CVMM)

The constrained variable metric method given in this paper is an algorithm which is a reformation of the Powell's method. Because it adopts linear-searching technique and a newest watchdog technique, and so on, the CVMM is the best one among numerical optimization methods for a nonlinear programming algorithm, and has many advantages such as fast convergence, high efficiency, high reliability, good capability of adaptation and lesser calls for objective function. So it is very suitable for the optimization in image tube design, which is difficult in the evaluation of values of the objective function.

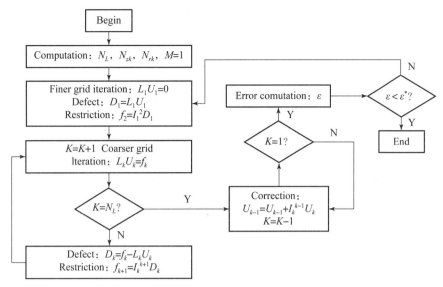

Fig. 1 Flow chart of the multigrid method

The procedures of CVMM are as follows[4].

Put the initial values: $\boldsymbol{x}_{(0)}$、$\boldsymbol{\lambda}_{(0)}$、$\boldsymbol{\alpha}_{(0)}$、$\boldsymbol{B}^{(0)}$ and $k=0$.

Compute the objective function and gradual value of the original problem (1) and construct a problem of quadratic programming QP.

$$\begin{aligned}
\min \quad & QP(\boldsymbol{d}) = \nabla f[\boldsymbol{x}]^{\mathrm{T}}\boldsymbol{d} + \frac{1}{2}\boldsymbol{d}^{\mathrm{T}}\boldsymbol{B}\boldsymbol{d} \quad && \boldsymbol{x},\boldsymbol{d} \in E^n \\
\text{s. t.} \quad & c_i[\boldsymbol{x}] + \nabla c_i[\boldsymbol{x}]^{\mathrm{T}}\boldsymbol{d} = 0 \quad && I \in E \\
& c_i[\boldsymbol{x}] + \nabla c_i[\boldsymbol{x}]^{\mathrm{T}}\boldsymbol{d} \geqslant 0 \quad && i \in I
\end{aligned} \quad (2)$$

Solve the quadratic programming QP, and determine a new Lagrange multiple vector $\boldsymbol{\lambda}^{(k+1)}$ and a searching direction $\boldsymbol{d}^{(k)}$.

Use the watchdog technique to determine the step length factor $\alpha_{(k)}$, and put a new approximate minimal point as

$$\boldsymbol{x}^{(k+1)} = \boldsymbol{x}^{(k)} + \alpha_{(k)}\boldsymbol{d}^{(k)} \quad (3)$$

where $\alpha_{(k)}$ is a searching step length, determined by the chosen linear searching tactics.

Give infinitesimal positive constants ε_1, ε_2, ε_3, if the following formula

or

$$\|\nabla_x L[\boldsymbol{x}^{(k+1)},\boldsymbol{\lambda}_{(k+1)}]\|_2 \leqslant \varepsilon_1 \quad (4)$$

$$c_j[\boldsymbol{x}^{(k+1)}] \geqslant \varepsilon_e, \quad i \in E \cup I \quad (5)$$

$$\frac{|f[\boldsymbol{x}^{(k+1)}] - f[\boldsymbol{x}^{(k)}]|}{|f[\boldsymbol{x}^{(k)}]|} \leqslant \varepsilon_3 \quad (6)$$

are satisfied, then stop computation and output the optimization results, otherwise go to problem (2).

Use BFGS formula

$$B^{(k+1)} = B^{(k)} - \frac{B^{(k)}\delta\delta^T B^{(k)}}{\delta^T B^{(k)}\delta} + \frac{\gamma\gamma^T}{\delta^T\gamma} \quad (7)$$

to the correct approximate Hesse matrix $B^{(k)}$, $B^{(k+1)}$ will be obtained. Put $k = k + 1$, return to problem (2). In equation (7), $\delta = x^{(k+1)} - x^{(k)}$ is the variable difference, $\gamma = \nabla_x L[x^{(k+1)}, \lambda^{(k)}] - \nabla_x L[x^{(k)}, \lambda^{(k)}]$ is the gradual difference of Lagrange function; $\eta = \theta\gamma + (1-\theta)B^{(k)}\delta (0 \leqslant \theta \leqslant 1)$ is the correct value of vector γ, and θ is a positive constant.

2 Optimization program

As discussed above, considering the characteristic of electrostatic image tubes, we use the concept of "fictitious grid points" and extend the standard MGM to the rotational symmetrical problem with arbitrary boundary, in order to replace the traditional methods of iteration for the field distribution computation[3].

We combine the MGM given above with CVMM, and adopt a least-square-fit method with weight factors to construct the objective function of the multi-objective optimization, choose some structural and electrical parameters of the system as independent variables, and some image characteristic parameters as constrained conditions. The optimization is to make the objective function $f(x)$ minimized under these restrictions.

A flow chart of the program of optimization design is shown in Fig. 2.

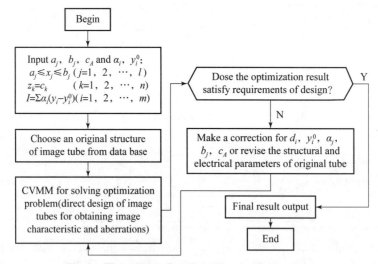

Fig. 2 Flow chart of optimization for image tubes

3 Examples and analysis

We have developed a package of software of CAD using the optimization method for image tube design. From the data library for image tube design in which data of different types of image tubes, such as diode tubes, triode tubes, zoom tubes and gated tubes, have been stored, we may choose one of them as an initial type of the tube to be designed. Its geometrical structural parameters and electrical parameters will be chosen as independent variables to be optimized, and the image position, magnification, as well as technical restrictions of variable parameters, will be selected as the restrained conditions. Some characteristic values, such as the chromatic aberration and off-axis aberrations including field curvature, astigmatism, distortion, etc., can be chosen to construct the objective function of the multi-objective optimization. The program will automatically make a search for a series of geometrical and electrical parameters, so that the objective function tends toward a minimum, and finally an optimal image tube satisfying the given requirements of design can be obtained.

Since the optimization design of image tubes is a problem of nonlinear multi-objective optimization, each of the aberrations of image tube is related to the geometrical and electrical parameters of the tube. The objective function of the multi-objective optimization can be constructed as a least-square-fit form with weight factors i. e. , $f = \sum \alpha_i (f_i - f_i^*)^2$, where α_i is the weight factor of the corresponding aberration, f_i, f_i^* are the computational value and the objective value, respectively.

Two examples of image tube optimization designing are given in the following.

Example 1:

We choose a triode image tube as the first example of optimization design, the construction of which is shown in Fig. 3. The following geometrical parameters are selected as the independent variables to be optimized.

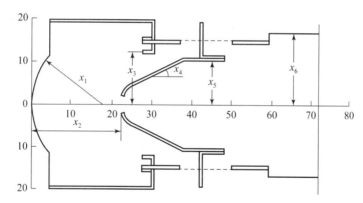

Fig. 3 Initial construction for the optimization design of triode image tube

x_1—radius of curvature of photocathode; x_2—distance between photocathode and anode;

x_3—radius of neck of cathode cylinder; x_4—slope of conical anode;

x_5—radius of sleeve of anode; x_6—radius of screen electrode.

The main characteristic values and aberrations which are to be optimized are:

Length of the image tube L;   Magnification at optimal image plane M;

Chromatic aberration f_3^*;   Tangential field curvature f_{14}^*;

Sagittal field curvature f_{15}^*;   Distortion f_{17}^*;

Off-axis chromatic aberration f_{18}^*;

where the off-axis aberrations are taken at 75% of the field of view.

For convenience, in the beginning, we suppose that objective values f_3^*, f_{14}^*, f_{15}^*, f_{17}^*, f_{18}^* are equal to zero, then the objective function will take the form $f = 100 \times (f_3)^2 + (f_{14})^2 + (f_{15})^2 + (f_{17})^2 + (f_{18})^2$. The iteration process and the result of optimization are shown in Table 1, where Nit represents the number of iteration times.

Table 1 The iteration process of optimization design under the objective function
$f = 100 \times (f_3)^2 + (f_{14})^2 + (f_{15})^2 + (f_{17})^2 + (f_{18})^2$

Nit	L/mm	M	f_3	f_{14}	f_{15}	f_{17}	f_{18}	$f = \Sigma \alpha_i (f_i)^2$
0	75.446,114	-1.666,936	0.020,086	4.154,992	-1.457,580	-3.548,831	0.119,781	32.037,392
1	71.276,787	-1.495,108	0.019,393	2.385,140	-1.229,897	-1.053,729	0.066,136	8.353,867
2	71.512,878	-1.496,113	0.019,868	2.573,591	-1.000,043	-0.828,408	0.068,236	8.353,887
3	72.023,018	-1.501,374	0.020,383	3.260,101	-0.044,014	-0.429,088	0.083,495	10.862,828
4	71.990,883	-1.499,182	0.020,394	3.175,735	0.029,223	0.013,531	0.078,631	10.134,104
5	71.980,293	-1.499,636	0.019,906	2.783,050	-0.473,936	0.012,621	0.068,745	8.014,494
6	71.997,757	-1.500,073	0.019,555	2.041,598	-1.309,270	-0.123,290	0.054,097	5.938,675
7	71.989,265	-1.499,556	0.018,069	1.844,299	-1.425,719	0.013,582	0.052,998	5.469,758
8	71.996,994	-1.499,959	0.017,955	1.746,294	-1.527,114	0.021,589	0.054,768	5.417,323
9	71.998,856	-1.499,971	0.017,969	1.708,356	-1.568,981	0.023,301	0.052,462	5.415,766
10	71.999,794	-1.499,995	0.017,964	1.722,174	-1.553,899	0.022,095	0.054,640	5.416,231

Example 2:

We make optimization design of a zoom tube, the construction of which is shown in Fig. 4, and the following electrical parameters are selected as the independent variables to be optimized:

V_g——applied voltage of zoom electrode;

V_a——applied voltage of anode.

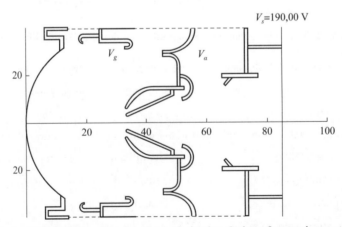

Fig. 4 Initial construction for the optimization design of zoom image tube

The main characteristic values and aberrations which are to be optimized are:

Length of the image tube $L = 85.00$ mm;

Magnification at optimal image plane $M = 0.3, 0.4, 0.5, 0.6, 0.7$;

Chromatic aberration f_3^*;

Tangential field curvature f_{14}^*;

Sagittal field curvature f_{15}^*;

Distortion f_{17}^*;

Off-axis chromatic aberration f_{18}^*;

where the off-axis aberrations are taken at 75% of the field of view.

During the optimization design of a zoom tube, letting L, M be the constrained parameters, screen voltage $V_s = 19,000$ V be fixed, and supposing that f_3^*, f_{14}^*, f_{15}^*, f_{17}^*, f_{18}^* are equal to zero. We have the objective function below:

$$f = (f_3)^2 + (f_{14})^2 + (f_{15})^2 + (f_{17})^2 + (f_{18})^2$$

The result of optimization of a zoom image tube is shown in Table 2.

Table 2 Optimization design of a zoom image tube

No.	1	2	3	4	5
M	-0.300,000	-0.400,004	-0.500,000	-0.599,993	-0.700,001
L (mm)	84.999,992	85.000,465	85.000,008	85.000,008	85.000,305
V_g (V)	169.539,948	-29.408,150	-205.674,759	-406.257,111	-652.927,856
V_a (V)	2,224.251,953	3,173.957,520	4,420.067,871	6,104.984,863	8,490.474,609
f_3	0.008,554	0.009,907	0.009,213	0.008,289	0.009,980
f_{14}	8.268,538	9.473,101	11.195,140	12.927,758	13.407,029
f_{15}	2.849,189	4.976,107	7.328,090	10.162,670	11.674,387
f_{17}	2.243,783	1.027,729	3.056,011	5.965,831	9.862,988
f_{18}	0.255,703	0.230,617	0.221,263	0.227,977	0.178,248

Conclusions

Practical computations of the image tube design have shown that:

(1) MGM is a fast convergence numerical method in field computation. The efficiency of MGM is over three times higher than that of FDM (finite difference method) in reaching the same accuracy of iteration. The more the number of mesh points the discrete system has, the higher the computational efficiency of MGM.

(2) The constrained variable metric method (CVMM), which uses new optimization techniques, such as the watchdog-technique and analysis of monotonicity, ensures the design to

have fast convergence, high efficiency and lesser number of times for calling the objective function. It appears that the CVMM is fitted for the optimization design of electrostatic image tubes.

(3) Results for designing practical image tubes using the software have proved that the method of optimization given in the present paper is an advanced, practical and effective one. It not only will promote the practicability of optimization design for image tubes, but also can be extended to the optimization design of electron optical systems with narrow electron beam focusing.

References

[1] HACKBUSCH W. Multigrid methods and applications[M]. Berlin:Springer, 1985.

[2] BRANDT A. Multi-level adaptive solutions to boundary-value problems [J]. Mathematics of Computation, V. 1977,31, 138:333 – 390.

[3] ZHANG Z Q, ZHOU L W, JIN W Q, et al, Multigrid method for computation of the rotational symmetrical electrostatic field [J]. Chinese Journal of Electronics, 1993, 2 (2): 43 – 49.

[4] YU J, ZHOU J. The OPB – 1 program library for optimization methods[M]. Beijing: Mechanical Industry Press,1989 (in Chinese).

8.3 Some Problems of Mathematical Simulation in Optimization Design of Electrostatic Image Tubes
静电像管优化设计中若干数值模拟问题的研究

Abstract: Three problems of mathematical simulation have been discussed in the present paper when optimizing the electron optical system design of electrostatic image tubes. The multigrid method (MGM) is suggested for solving the rotational symmetrical electrostatic field, which proves that the computational efficiency of MGM is 2 – 4 times better than the traditional finite difference method (FDM) when the same accuracy of iteration is reached. The constrained variable metric method (CVMM), which ensures the design to have fast convergence, and high efficiency and lesser number of times of calling for objective is recommended for optimization design. The objective function having least-square fit form with weight factors is investigated for the problem of nonlinear multi-objective optimization. The result of optimization appears that the suggested mathematical methods given in this paper for optimization design of electrostatic image tubes are practical and effective.

摘要：本文讨论了在优化静电像管的电子光学系统时数学模拟中的三个问题。本文建议用多重网格法（MGM）求解轴对称静电场，证明了在同样的迭代精度时，MGM 的计算效率 2~4 倍优于传统的有限差分法（FDM）。本文建议在优化设计时采用受限可变尺度法（CVMM），它能保证设计具有快速的收敛和高效率，以及在优化设计时能较少地呼唤目标的次数。本文对非线性、多目标的优化问题推荐了具有最小二乘方的权重因子的目标函数。实际优化的结果表明本文所建议的数值方法对于静电像管的优化设计是实用和有效的。

Introduction

The optimization design of electron optical systems is an inevitable outcome of direct design. During the process of design when using optimization method, the designer needs not change the geometrical and electrical parameters manually and the result of design will not very be restricted by experience and ability of the designer. Actually, it not only improves the efficiency and quality of the design, but also fully brings out latent potentialities of the system. So the optimization

---

Zhou Liwei[a], Zhang Zhiquan[b], Jin Weiqi[a], Fang Erlun[c], Ni Guoqiang[a], Zhang Liangzhong[a]. a) Beijing Institute of Technology, b) Armored Force Engineering Institute, c) Xi'an Research Institute of Modern Chemistry. Proc. Electron-Beam Sources and Charged-Particle Optics, SPIE 2522, 1995, 102.

design is one of the most interesting problems in electron optical system design, and much attention has been paid to it.

The optimization design of electron optical systems has begun since the 1970s. The axial potential distribution and magnetic induction of the system, in which the coefficient of spherical aberration reaches a minimal value, has been investigated by H. Rose[1] by using variation method and by M. Szilagyi[2] by using dynamic programming, but the problem is too difficult to find a practical optimum structure from the given field distribution. In 1975, E. Munro[3] engaged in optimization design of the magnetic focusing-deflection system by using Powell method.

In 1982, H. C. Chu and E. Munro[4] investigated computerized optimization of the electron-beam lithography system by using method of the damping least squares. In 1984, Gu Changxin et al. tried to apply the "simplex method[5]" as well as the "complex method[6]" to the optimization design of an extended field lens with a criterion of minimum objective parameter coefficient of spherical aberration. M. Scheinfein and A. Galantai[7] have chosen more suitable objective function and applied the multi-objective optimization technique, and introduced the constraint optimization technique and the simplex scaling method to the design. In 1988, J. Podbrdsky and O. L. Krivanek[8] studied the optimization problem for the diaphragm lens of the electron probe system from the modulation transfer function, and F. Lenz[9] has investigated the optimization problem for the electron optical system composed of four magnetic lenses having magnification $M = 0.01$. But all of above-mentioned investigations were aimed at electron optical systems with narrow beam focusing.

For the electron optical imaging systems, in 1990, V. P. Flin, et al.[10] firstly published their investigation on optimization design of emission-imaging electron optical systems, in which the numerical aspects of the main stage for computing and optimization of emission-imaging systems have been given; the solution of the first types of integral equations for the accurate evaluation of the potential and its perturbation have been considered in detail; the formulation and solution algorithms of cathode-lens parametric optimization problems have been specified; and the solution of two optimization problems for an imaging system has been given. In 1993, we investigated the optimization design of image tubes with electrostatic focusing[11], in which the multigrid method (MGM) has been described for the computation of rotational symmetrical electrostatic field[12] and the constrained variable metric method (CVMM) has been used for solving the nonlinear constraint optimization problem.

Three problems should be solved or answered when using optimization method for designing electrostatic image tubes. The first is to find a high efficient and very fast numerical algorithm for solving Laplace boundary problems. The second is to apply an advanced optimization algorithm having a series of advantages not only in fast convergency, high efficiency, high reliability and good capability of adaptation, but also in a few calls for evaluation of functions; the third is to investigate a suitable mathematical model, form of objective function and constrained conditions for the optimization design of image tubes.

In the present paper, the multigrid method for solving the Laplace boundary problem, the constrained variable metric method for solving the nonlinear constraint optimization problem and the objective function for the problem of nonlinear multi-objective optimization will be discussed in optimization design of electrostatic image tubes.

1 Multigrid method for solving boundary problem

1.1 Fundamental algorithm of the MGM

The multigrid method (MGM) is a new numerical method in solving the boundary-value problems[13,14], that is still in its development in the recent years. One of the great advantages of the MGM, when compared with the classical method of iteration, is that the rate of convergence of MGM is independent of the step length h adopted in the discrete domain of the system. It shows the superiority of high efficiency and high speed, especially when the step length h of system decreases. So this method has been broadly used in hydrodynamics, structural mechanics and eigenvalue problems.

For the engineering problem, as well as the electron-optical system design, it is often necessary to solve the boundary-value problem which has the following form:

$$\begin{aligned} LU(x,y) &= F(x,y), \quad (x,y) \in \Omega; \\ U &= F_0(x,y), \quad (x,y) \in \Gamma \end{aligned} \quad (1)$$

where L is a differential operator; F is a source function; U is the field distribution to be solved and Γ is the boundary of the domain Ω.

In general, the method of numerical computation for Eq. (1) is to simplify the above-mentioned problem in a discrete form as what follows:

$$\begin{aligned} L_k U_k &= F_k, \quad (x,y) \in \Omega_k; \\ U_k &= F_0, \quad (x,y) \in \Gamma_k \end{aligned} \quad (2)$$

and to use a certain numerical solution method for solving the problem, where Ω_k is the discrete domain of Ω; Γ_k is the discrete boundary of Γ. The form of operator L_k is related to the adapted difference form instead of the differential form and the numerical method for solving Eq. (2). Although the traditional method of iteration such as the Guass-Seidel iteration with "five points difference pattern" can successfully be applied to boundary problems with closed boundary and region of arbitrary shape, its velocity of convergence slows down when the number of iteration increases. The Fourier analysis of error function $U^{(i)} - U$ shows that the high-frequency part of the eigen-function of $U^{(i)} - U$ has properties of fast convergence, but the low-frequency part of eigen-function of $U^{(i)} - U$ has properties of slow convergence. Within the first few numbers of iteration, the convergence is fast, but when the number of iteration increases, the error function $U^{(i)} - U$ becomes a smooth function of the grid, and the convergence slows down very quickly (this phenomenon is so called "smoothness effect"), so its computational efficiency is very slow.

In contrast to the traditional method of iteration, the multigrid grid method[12] combines the finite difference method (FDM) with a "coarser grid correction" technique, and yields a highly efficient and very fast iteration algorithm.

The multigrid method can be described as following:

Supposing that the domain to be solved is a square region in the rectangular plane coordinate system:

$$0 \leq x \leq x_{max} = 1,$$
$$0 \leq y \leq y_{max} = 1 \tag{3}$$

for simplicity, we divide x_{max} and y_{max} into equal parts, respectively; that is

$$h_x = h_y = h_1 = x_{max}/2^N = y_{max}/2^N = 1/2^N$$

At first level, i. e., at the finest grids, the boundary problem is written as

$$L_1 U_1(x,y) = F_1(x,y), \quad (x,y) \in \Omega_1; \tag{4a}$$
$$U_1(x,y) = F_1(x,y), \quad (x,y) \in \Gamma_1 \tag{4b}$$

where Ω_1 is the discrete domain of Ω, Γ_1 is the discrete boundary of Γ.

Firstly, by using the traditional method, such as Jacobi, Gauss-Seidel method for iteration, the approximate solution of field distribution at finest grids (step length $h = h_1 = h_x = h_y$) will be obtained through J times of iteration. If suppose the accurate solution is U_1^*, the error or defect of the approximate solution V_1 will be

$$\Delta V_1 = U_1^* - V_1, \tag{5}$$

or
$$U_1^* = \Delta V_1 + V_1 \tag{6}$$

Apparently, ΔV_1 is the accurate value of correction.

Owing to the fact that the U_1^* satisfies Eq. (4) certainly, substituting the expression (6) in Eq. (4a), we have

$$L_1 U_1^* = L_1(V_1 + \Delta V_1) = F_1,$$
$$L_1 \Delta V_1 = F_1 - L_1 V_1 \tag{7}$$

Eq. (7) expresses the equation for which ΔV_1 should be satisfied. As we have been described above, ΔV_1 is a smoothness function of the grids. In order to evaluate ΔV_1 as fast as possible, we define a transformation I_1^2, and transform the defect ΔV_1 and the right-end term $F_1 - L_1 V_1$ of the equation at the finest grids ($k = 1$) into the coarser grids ($k = 2$, $h = 2h$). In substance, this transformation is a weighted restriction, that is, at a certain coarser grid point, the values of 8 surrounding grid points are concentrated into the coarser grid point:

$$I_1^2 L_1 \Delta V_1 = I_1^2 (F_1 - L_1 V_1) \tag{8}$$

At the coarser grids ($k = 2$), U_2 satisfies the following equation:

$$L_2 U_2(x,y) = F_2(x,y), \quad (x,y) \in \Omega_2;$$
$$U_2(x,y) = 0, \quad (x,y) \in \Gamma_2 \tag{9}$$

where $l_2 = l_1$ (or $l_2 = I_1^2 \times L_1 \times I_2^1$, I_2^1 is the adjoint matrix of I_1^2), is the differential operator at coarser grids, $U_2 = I_1^2 \Delta V_1$ is the defect at the coarser grids, $F_2 = I_1^2(F_1 - L_1 V_1)$ is the source function of the defect at coarser grids. Eq. (9) shows that in the case of coarser grids the defect at the domain's boundary is always equal to zero.

It is clear that Eq. (9) is identical with Eq. (4) in form, but the step length of the coarser grids is increased two times, the amount of coarser grids is decreased four times, thus it is easier to solve Eq. (9) compared to Eq. (4). But as described above, we can get only an approximate numerical solution of Eq. (9).

Suppose that the approximate value V_2 of defect U_2 at coarser grids will be obtained from Eq. (9) through j_2 times of iteration, the error will be

$$\Delta V_2 = U_2^* - V_2 \tag{10}$$

where U_2^* is the accurate value of the defect U_2. In the same way, the defect equation at more coarser grids ($k = 3$, $h = 4h_1$) can be written as

$$L_3 U_3(x,y) = F_3(x,y), \quad (x,y) \in \Omega_3;$$
$$U_3(x,y) = 0, \quad (x,y) \in \Gamma_3 \tag{11}$$

where $L_3 = L_2$ (or $L_3 = I_3^2 \times L_2 \times I_2^3$, I_3^2 is the adjoint matrix of I_2^3), is the differential operator at the more coarser grids, $U_3 = I_2^3 \Delta V_2$ is the defect at the more coarser grids, $F_3 = I_2^3 (F_2 - L_2 V_2)$ is the source function of U_3.

In this recurrent way, till the coarsest grid ($k = N$, $h = 2^{(N-1)} \times h_1 = 1/2$), we have

$$L_N U_N(x,y) = F_N(x,y), \quad (x,y) \in \Omega_N;$$
$$U_N(x,y) = 0, \quad (x,y) \in \Gamma_N \tag{12}$$

where $L_n = L_{N-1}$ (or $L_n = I_{N-1}^N \times L_N \times I_N^{N-1}$, I_N^{N-1} is the adjoint matrix of I_{N-1}^N), is the differential operator at the coarsest grids. In fact, only one internal grid point ($x = 1/2$, $y = 1/2$) exists at this step. It is easy to evaluate the solution $U_N = L^{-1}(F_N)$. After having obtained U_N, by using the reverse transformation I_N^{N-1} of the transformation I_{N-1}^N, the process of which is so called the "prolongation," we can interpolate the value at grid points of $k = N - 1$ level through the value of $k = N$ level:

$$\Delta V_{N-1} = I_N^{N-1} U_N \tag{13}$$

where I_N^{N-1} is the transformation operator from coarse grid level to fine grid level.

Adding the value of (13) to the iteration value at $K = N - 1$ level, we obtain the correction value of the defect at this level:

$$\overline{U}_{N-1} = V_{N-1} + \Delta V_{N-1} \tag{14}$$

Similarly, interpolating the value at grids of $k = N - 2$ level through the "prolongation" by using $\overline{U}_{N-1}$ at grids of $k = N - 1$ level, we may have the correction value of the defect at $k = N - 2$. The same process goes again till the finest level ($k = 1$, $h = h_1$), we may obtain the approximate correction value of the desired field:

$$\overline{U}_1 = V_1 + \Delta V_1 \tag{15}$$

During the above-mentioned process, the transformation operator from coarse grids to fine grids, which is called "prolongation[10]", is the bilinear interpolation, i. e.,

$$I_k^{k-1} = \frac{1}{4}\begin{bmatrix} 1 & 2 & 1 \\ 2 & 4 & 2 \\ 1 & 2 & 1 \end{bmatrix} \quad (16)$$

the transformation operator I_{k-1}^k from fine grids to coarse grids, which is called "restriction," is the self-adjoint operator, i. e. ,

$$I_k^{k-1} = \frac{1}{16}\begin{bmatrix} 1 & 2 & 1 \\ 2 & 4 & 2 \\ 1 & 2 & 1 \end{bmatrix} \quad (17)$$

1.2 MGM for an irregular domain

The standard MGM is carried out only for a rectangular domain. In order to apply the MGM to compute the rotational symmetrical electrostatic field with arbitrary boundary, we have adopted the following techniques:

(1) Imbedding the domain Ω in which the value U is to be solved and the boundaries of which have arbitrary shapes to a rectangular domain, we treat the maximum values of z and r directions of Ω as z_{max} and r_{max} of the rectangular domain, and give the points which are located between outside of the domain Ω to be solved and inside of the rectangular domain always zero (these points are called the "outside-domain points").

(2) Using a concept of "fictitious grid points," that is, increasing a row and/or a column of grid points on the top and/or the right-side of the rectangular domain, we assume that the requirement of grid coarser is always satisfied (the step length h_{i+1} of the coarser grids is two times that of h_i of the finer grids). All of the "fictitious grid points" can be regarded as "outside-domain points", and are also given as zero.

(3) During the computation of grid correction, for the transformation of the operator I_{k+1}^k, that is, for the prolongation from a coarser grid level to a finer grid level, we use the bilinear interpolation in the z and r direction of the rotational symmetrical system, thus the approximate values of the finer grid points are obtained by linear interpolation from the values of the coarser grid points, while for I_k^{k+1}, which is a restriction from finer grid level to coarser grid level, is a conjugate transposed matrix of I_{k+1}^k [13].

Besides the above-mentioned treatment, a "code of logic rule" technique is used to distinguish the grid points at different levels. Therefore, we can easily distinguish the "electrode points" (which include the inner electrode points and the boundary electrode points) and the "inner points" (which include the regular and irregular inner points), as well as the "outside-domain points," thus simplifying the logic computation of iteration and improving the efficiency of computation.

1.3 Program design mid computation

We have compiled a program of MGM for the computation of rotational symmetrical electrostatic field using C language. A flow chart for the program is shown in Fig. 1.

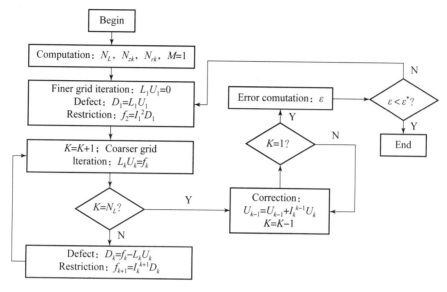

Fig. 1 A flow chart for the program of the multigrid method

Table 1 shows a result of some points of the axial potential for an immersion lens with two equal-diameter cylinders as computed by FDM (number of iterations: 208, total average relative error: 4.569×10^{-10}, time of computation: 9.77 s) and by MGM ($J_1 = 2$, $J_2 = 3$, $Nit = 11$, total average relative error: 6.990×10^{-10}, time of computation: 2.03 s, where $J_1 = 2$, $J_2 = 3$ are the numbers of iteration at the level of $k = 1$ and $k = 2, 3, 4$, Nit is the cycle number of iteration of the multigrid method), all of the computation being carried out on the same personal computer (type 486, 33 MHz).

Table 1 Axial potential computation of an immersion lens with two equal-diameter cylinders by FDM and MGM

z/mm	Values of the axial potential/V					T/s
	1	2	3	4	5	
FDM	29.132,885	58.673,203,9	89.033,285,1	120.635,297	153.916,147	9.77
MGM	29.132,889,0	58.673,206,2	89.033,288,5	120.635,301	153.916,152	2.03

Table 2 shows a result of axial potential computation for a certain typical triode image tube, as shown in Fig. 3 (The total number of finest grid points is 5,945) and gives a comparison between FDM (iteration times: 345, total average relative error: 4.780×10^{-10}, time of computation: 52.67 s) with MGM ($J_1 = 2$, $J_2 = 4$, $Nit = 14$, total average relative error: 3.985×10^{-10}, time of computation: 9.66 s). The time of computation of MGM is again 1/5.45 of FDM.

Table 2 Axial potential computation of a typical triode image tube by FDM and MGM

Values of the axial potential/V					
z/mm	FDM	MGM	z/mm	FDM	MGM
0	0.000,000	0.000,000	40	4 332.786,46	4 332.786,46
5	190.879,749	190.879,763	45	3 858.010,57	3 858.010,57
10	533.090,256	533.090,276	50	2 977.727,83	2 977.727,83
15	1 209.088,96	1 209.088,98	55	2 025.699,68	2 025.699,68
20	2 831.518,82	2 381.518,83	60	1 385.785,78	1 385.785,78
25	4 545.448,84	4 545.448,84	65	1 045.289,28	1 045.289,28
30	4 577.979,58	4 577.979,58	70	858.664,315	858.664,315
35	4 518.114,24	4 518.114,24	72	800.000,000	800.000,000

2 Optimization method

2.1 Optimization problem of image tube design

The problem of image tube design is a numerical one for solving the nonlinear programming such as that given below[10]:

$$\begin{aligned} \min \quad & f(\boldsymbol{x}), \quad \boldsymbol{x} \in E^l \\ \text{s.t.} \quad & c_j(\boldsymbol{x}), \quad j \in E; \\ & c_j(\boldsymbol{x}) > 0, \quad j \in I \end{aligned} \qquad (18)$$

where $f(\boldsymbol{x})$ is the objective function, which represents the image characteristic parameters and aberrations, $c_j(\boldsymbol{x})$ are also image parameters, which should satisfy the constrained conditions during the optimizing process, $x_j(j = l, 2, \cdots, j)$ are the geometrical structural and electrical parameters of the image tube. So the optimization problem of image tube design can be described as: to search a set of $x_j(j = l, 2, \cdots, j)$, which make the objective function $f(\boldsymbol{x})$ minimum under the constrained conditions as given in problem (18).

2.2 Constrained variable metric method (CVMM) for solving the nonlinear constraint optimization problem

The constrained variable metric method (CVMM)[15] given in this paper is an algorithm which is a reformation of the Powell's method. Because it adopts a linear-searching technique and a newest watchdog technique, and so on, the CVMM is the best one among numerical optimization methods for a programming algorithm, and has many advantages such as fast convergence, high efficiency, high reliability, good capability of adaptation and lesser calls for objective function. So it is very suitable for the optimization in image tube design, which is difficult in the evaluation of values of the objective function.

The basic thinking of CVMM is as following: for the optimization problem [Eq. (18)] we produce a Lagrange function, and use this function to construct a sub-problem of quadratic programming with a non-equality constrained conditions at each iteration point, and solve the extremum solution with the aid of numerical method. The extremum solution of each iteration for the sub-problem of quadratic programming is taken for the searching direction of this iteration, and the watchdog technique is used to determine the step-length factor, thus a new iteration point is produced. A series of this kind of iteration points will finally approximate the solution of the problem.

The procedures of CVMM are as follows[15]:

(1) Put the initial values: $X_{(0)}$, $\lambda_{(0)}$, $\alpha_{(0)}$, $\boldsymbol{B}^{(0)}$ and $k = 0$;

(2) Compute the objective function and gradual value of the original problem (18) and construct a problem of quadratic programming QP.

$$\min QP(\boldsymbol{d}) = \nabla f[\boldsymbol{x}]^T \boldsymbol{d} + \frac{1}{2}\boldsymbol{d}^T \boldsymbol{B} \boldsymbol{d} \quad x,d \in E^1$$
$$\text{s. t.} \quad c_j[\boldsymbol{x}] + \nabla c_j[\boldsymbol{x}]^T \boldsymbol{d} = 0, \quad j \in E \quad (19)$$
$$c_j[\boldsymbol{x}] + \nabla c_j[\boldsymbol{x}]^T \boldsymbol{d} \geq 0, \quad j \in I$$

(3) Solve the quadratic programming QP, and determine a new Lagrange multiple vector $\lambda_{(k-1)}$ and a searching direction $\boldsymbol{d}^{(k)}$.

(4) Use the watchdog technique to determine the step length factor $\boldsymbol{\alpha}_{(k)}$, and put a new approximate minimal point as

$$\boldsymbol{x}^{(k+1)} = \boldsymbol{x}^{(k)} + \alpha_{(k)} \boldsymbol{d}^{(k)} \quad (20)$$

Where $\boldsymbol{\alpha}_{(k)}$ is a searching step length, determined by the chosen linear searching tactics.

(5) Give infinitesimal positive ε_1, ε_2, ε_3, if the following formula

$$|\nabla_x \sim L[\boldsymbol{x}^{(k+1)}, \lambda_{(k+1)}]|_2 \leq \varepsilon_1 \quad (21)$$

or
$$c_j[\boldsymbol{x}^{(k+1)}] \geq \varepsilon_2, \quad j \in E \cup I \quad (22)$$

$$\frac{|f[\boldsymbol{x}^{(k+1)}] - f[\boldsymbol{x}^{(k)}]|}{|f[\boldsymbol{x}^{(k)}]|} \leq \varepsilon_3 \quad (23)$$

are satisfied, then stop the computation and output the optimization results, otherwise go to Eq. (6).

(6) Use the BFGS formula [16]

$$\boldsymbol{B}^{(k+1)} = \boldsymbol{B}^{(k)} - \frac{B^{(k)}\boldsymbol{\delta}\boldsymbol{\delta}^T B^{(k)}}{\boldsymbol{\delta}^T B^{(k)}\boldsymbol{\delta}} + \frac{\boldsymbol{\gamma}\boldsymbol{\gamma}^T}{\boldsymbol{\delta}^T \boldsymbol{\gamma}} \quad (24)$$

to correct the approximate Hesse matrix $\boldsymbol{B}^{(k)}$, $\boldsymbol{B}^{(k+1)}$ will be obtained. Put $k = k + 1$, return to problem (2). In the equation (24), $\boldsymbol{\delta} = \boldsymbol{x}^{(k+1)} - \boldsymbol{x}^{(k)}$ is the variable difference, $\boldsymbol{\gamma} = \nabla_x L[\boldsymbol{x}^{(k+1)}, \lambda_{(k+1)}] - \nabla_x L[\boldsymbol{x}^{(k)}, \lambda_{(k)}]$ is the gradual difference of Lagrange function; $\eta = \theta\boldsymbol{\gamma} + (1 - \theta)\boldsymbol{B}^{(k)}\boldsymbol{\delta}(0 \leq \theta \leq <1)$ is the correct value of vector $\boldsymbol{\gamma}$, and θ is a positive constant.

3 Optimization design

3.1 Mathematical model and objective function for optimizing image tubes

Since the optimization design of image tubes is a problem of nonlinear multi-objective optimization, each of the aberration is related to the geometrical and electrical parameters of the tube. The objective function of the multi-objective optimization can be constructed as a least-square-fit form with weight factors.

Suppose $x = (x_1, x_2, \cdots x_l)$ represent the geometrical structural parameters and electrical parameters of the electron optical system, which are chosen as independent variables to be optimized; $y_i(i=1,2,\cdots,m)$ be the objective values to be achieved which represent the image characteristics and aberrations; $z_k(k=1,2,\cdots,n)$ be also the objective values, but they are conditions to be satisfied during the optimization.

The optimization problem of electron-optical system for the image tube can be described as following:

Under the conditions

$$a_j \leqslant x_j \leqslant b_j, \quad (j=1,2,\cdots,l)$$
$$c_k \leqslant z_k \leqslant d_k, \quad (k=1,2,\cdots,n) \tag{25}$$

the problem is to make a search for a series of geometrical and electrical parameters so that the objective function

$$f(x) = \sum_{i=1}^{m} \alpha_j (y_i - y_i^0)^2 \tag{26}$$

tends toward a minimum; where α_j is the weight factor of the corresponding aberration; $y_i^0 (i=1, 2, \cdots m)$ is the objective value to be desired; a_j, b_j are usually the technical restrictions for the independent variables x_j, c_k, d_k are usually the requirements for the image characteristics of the system. A flow chart of the program of optimization design is shown in Fig. 2.

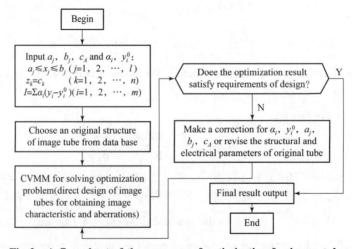

Fig. 2 A flow chart of the program of optimization for image tubes

3.2 Practical examples of optimization design

We have developed a package of software of CAD using the optimization for image tube design. From the data library for image tube design in which data of different types of image tubes, such as diode tubes, triode tubes, zoom tubes and gated tubes, have been stored, we may choose one of them as an initial type of the tube to be optimized.

1) A triode image tube

For the optimization design of a triode image tube, the following geometrical parameters are selected as the independent variables to be optimized.

X_1—radius of curvature of photocathode;

X_2—distance between photocathode and anode;

X_3—radius of neck of cathode cylinder;

X_4—slope of conical anode;

X_5—radius of sleeve of anode;

X_6—radius of screen electrode.

The main characteristic values and aberration which are to be optimized are:

Length of image tube L;   Magnification at optimal image plane M;

Chromatic aberration f_3^*;   Sagittal field curvature f_{14}^*;

Tangential field curvature f_{15}^*;   Distortion f_{17}^*;

Off-axis chromatic aberration f_{18}^*;

where the off-axis chromatic aberration are taken at 75% of the field of view.

For convenience, in the beginning, we select an initial construction of a triode image tube for the optimization design, as shown in Fig. 3, and suppose that the objective values f_3^*, f_{14}^*, f_{15}^*, f_{17}^*, f_{18}^* are equal to zero, then the objective function will take the form $f = 100(f_3^*)^2 + (f_{14}^*)^2 + (f_{15}^*)^2 + (f_{17}^*)^2 + (f_{18}^*)^2$.

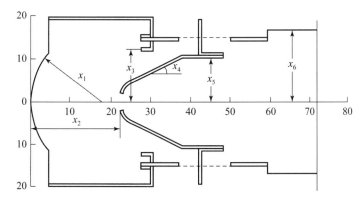

Fig. 3 The initial construction of a triode image tube for optimization design

The iteration process and the result of optimization of the triode image tube are shown in Table 3, where NIT represents the number of iteration times.

It is noteworthy that from Table 3 along with the decreasing of the radius of curvature of the photocathode, the symbols of tangential field curvature and sagittal field curvature become opposite, it means that it is possible to flatten the image plane.

Table 3 Optimization design of a triode image tube

Nit	L/mm	M	f_3^*	f_{14}^*	f_{15}^*	f_{17}^*	f_{18}^*	$f = \sum \alpha_j (f_j^*)^2$
0	75.446,114	-1.666,94	0.020,086	4.154,992	-1.457,580	-3.548,831	0.119,781	32.037,392
1	71.276,787	-1.495,11	0.019,393	2.385,140	-1.229,897	-1.053,729	0.066,136	8.353,867
2	71.512,878	-1.496,11	0.019,868	2.573,591	-1.000,043	-0.828,408	0.068,236	8.353,887
3	72.023,018	-1.501,37	0.020,383	3.260,101	-0.044,014	-0.429,088	0.083,495	10.862,828
4	71.990,883	-1.499,18	0.020,394	3.175,735	-0.029,223	0.013,531	0.078,631	10.134,104
5	71.980,293	-1.499,64	0.019,906	2.783,050	-0.473,936	0.012,621	0.068,745	8.014,494
6	71.997,757	-1.500,07	0.019,555	2.041,598	-1.309,270	0.123,290	0.054,097	5.938,675
7	71.989,265	-1.499,56	0.018,069	1.844,299	-1.425,719	0.013,582	0.052,998	5.469,758
8	71.996,994	-1.499,96	0.017,955	1.746,294	-1.527,114	0.021,589	0.054,768	5.417,323
9	71.998,856	-1.499,97	0.017,969	1.708,356	-1.568,981	0.023,301	0.052,462	5.415,766
10	71.999,794	-1.500,00	0.017,964	1.722,174	-1.553,899	0.022,095	0.054,640	5.416,231

2) A zoom image tube

The zoom image tube design seems not easy in the electron-optical system design, but it is quite simple by using optimization method. We assume that we have got an optimized design for an zoom tube with $M = 0.3$ and suppose that the geometrical construction of the zoom tube, as shown in Fig. 4, is not to be changed, and the two following electrical parameters will be selected as independent variables to be optimized:

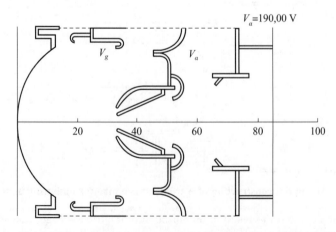

Fig. 4 The initial construction of a zoom image tube for optimization design

V_g—applied voltage of the zoom electrode;

V_α — applied voltage of the anode.

The main characteristic values which are to be optimized are:

Length of the zoom tube: $L = 85.00$ mm;

Magnification at optimal image plane: $M = -0.3, -0.4, -0.5, -0.6, -0.7$.

During the optimization design of the zoom tube, letting L, M be the constrained parameters, the screen voltage $V_\alpha = 19,000$ V be fixed, the aberrations f_3^*, f_{14}^*, f_{15}^*, f_{17}^*, f_{18}^*, as defined above, be equal to zero, the objective function has the form (26) with weight factors $\alpha_i = 1$. The result of optimization is shown in Table 4.

Table 4 Optimization design of a zoom image tube

No.	1	2	3	4	5
M	-0.300,000	-0.400,004	-0.500,000	-0.599,993	-0.700,001
L/mm	84.999,992	85.000,465	85.000,008	85.000,008	85.000,305
V_g/V	169.540,0	-29.408,2	-205.674,8	-406.257,1	-652.927,9
V_α/V	2 224.252,0	3 173.957,5	4 420.067,9	6 104.984,9	8 490.474,6
f_3^*	0.008,554	0.009,907	0.009,213	0.008,289	0.009,980
f_{14}^*	8.268,538	9.473,101	11.195,140	12.927,758	13.407,029
f_{15}^*	2.849,189	4.976,107	7.328,090	10.162,670	11.674,387
f_{17}^*	2.243,783	1.027,729	3.056,011	5.965,831	9.862,988
f_{18}^*	0.255,703	0.230,617	0.221,263	0.227,977	0.178,248

Summary

The present paper has proven that:

(1) The multigrid method (MGM) is really an excellent and fast numerical computation method, its computation efficiency is 2 - 4 times higher than that of the traditional finite difference method (FDM) in reaching the same accuracy of iteration. Furthermore, the more the number of grid points in the discrete system has. the higher the computational efficiency of MGM reaches.

(2) The constrained variable metric method (CVMM), which uses new optimization techniques such as the watchdog-technique and analysis of monotonicity, ensures the design to have fast convergence, high efficiency and lesser number of times of calling for the objective function. It appears that the CVMM is fitted for the optimization design of image tubes.

(3) Since the optimization design of electrostatic image tubes is a problem of nonlinear multi-objective optimization, each of aberrations of image tube is related to the geometrical and electrical parameters of the tube, the objective function having a least-square-fit form with weight factors has

been investigated to carry out the optimization computation.

Results for designing practical image tubes by using the software have proved that the suggested mathematical simulation given in the present paper is an advanced, practical and effective one.

Acknowledgements

The work is supported by National Natural Science Foundation of China and the Science and Technology Foundation of the Academy of Ordnance Science, China. We would like to thank Prof. M. A. Monastyrsky from General Physics Institute of Russian Academy of Science, Moscow, and Dr. Y. V. Kulikov from Institute of Electron Devices, Moscow, for helpful discussions.

References

[1] ROSE H. Electronenoptische Aplanate[J]. Optik, 1971,343:285 - 311.

[2] SZILAGYI M. Synthesis of electron lenses [C]//Electron Optical Systems for Microscopy, Microanalysis and Microlithography. SEM Inc., AMF O'Hare, 1984:75 - 84.

[3] MUNRO E. Design and optimization of magnetic lenses and deflection systems for electron beams[J]. Journal Of Vacuum Science & Technology, 1975,12(6):1146 - 1150.

[4] CHU H C, MUNRO E. Numerical analysis of electron beam lithography systems. Part 4: Computerized optimization of the electron optical performance of electron beam lithography systems using the damped least squares method [J]. Optik, 1982,61(3):213 - 236.

[5] GU C X, CHEN N Q, WU X M. The optimization design of an electron optical system [J]. Acta Electronica Sinica, 1984,12, (2):41 - 47 (in Chinese).

[6] GU C X, SHAN L Y. Constrained optimization design of an electron optical system [C]//Electron Optical Systems for Microscopy, Microanalysis and Microlithography. SEM Inc., AMF O'Hare 1984:91 - 96.

[7] SCHEINFEIN M, GALANTAI A. Multi-objective optimization techniques for design of electrostatic charged particle lenses[J]. Optik, 1986,74:152 - 164.

[8] PODBRDSKY J, KRIVANEK O L. A universal mesh for calculating the properties of highly saturated magnetic electron lenses by the finite element mothed[J]. Optik,1988,79(4): 177 - 182.

[9] LENZ F. Optimization of the imaging properties of an electron optical de-magnifying system[J]. Optik, 1988,78(4):135 - 140.

[10] KATESHOV V A, KULIKOV Y V, МОНАСТЕРСКИЙ M A, et al. Emission-imaging electron-optical system design [J]. Advances in Electronics and Electron Physics, 1990, 78, Supplement:155 - 278.

[11] ZHANG Z Q, ZHOU L W, JIN W Q, et al. Optimization design of image tubes with electrostatic focusing [J]. Proceedings of SPIE—The International Society for Optical Engineering,1982:238 - 244.

[12] ZHANG Z Q, ZHOU L W, JIN W Q, et al. A multigrid method for the computation of rotational symmetrical electrostatic fields [J]. Chinese Journal of Electronics, 1993, 2 (1): 15 – 19.

[13] HACKBUSCH W. Multigrid methods and applications [M]. Berlin: Springer, 1985.

[14] BRANDT A. Multilevel adaptive solutions to boundary-value problems [J]. Mathematics of Computation, 1977, 31(l38): 333 – 390.

[15] YU J, ZHOU J. Program Library of Optimization Methods OPB – 1 [M]. Beijing: Mechanical Industry Press, 1989 (in Chinese).

[16] FIELDING K. Algorithm 387 function minimization and linear search [J]. Communications of the ACM, 1970, 13: 509.

8.4 A Study on Electron Optical Systems for Conical Immersion Lenses
圆锥浸没透镜电子光学系统的研究

Abstract: In the present paper, computation of electrostatic field distribution for conical immersion lenses is studied by using the boundary element method (BEM), and the electron trajectories incident from the object and image spaces have been traced using a set of curvilinear paraxial trajectory equations and practical trajectory equations, then the electron optical parameters, such as cardinal points of lens, have been determined. The computational results coincide with the analytical asymptotical method are given by S. Y. Yavor, et al.

摘要：本文研究利用边界元素法计算圆锥透镜场分布，用曲傍轴轨迹方程组、实际轨迹方程追迹计算由物方和像方发出的电子轨迹，确定电子光学参量。其计算结果与 S. Y. Yavor 等人采用渐近解析方法的结果相符合。

Introduction

Electrostatic focusing and deflecting systems with energy dispersion are widely used in modern charged particle beam analytical instruments with high precision. Great attention has been paid to the conical electrostatic focal and deflection lenses (including conical lenses and conical mirrors) which are developed in recent years, since they have a series of advantages over other devices, such as high capability of focusing and deflection for the charged particles. Particularly, the angular distribution and energy distribution can be simultaneously obtained without moving the sample or the analyzer. Such an instrument now is one of the main technical measures for investigating surface structure of matter.

Differing from the traditional narrow beam systems, the principal trajectories in conical lenses are curves with quite a difference in the initial energy and initial angular distributions of particle beams. This is a kind of electron optical systems with curvilinear axes which is different from electron optical systems with linear axes and has special mass dispersion and energy dispersion and aberration properties. Brewer, et al (1979) firstly applied the mono-chromatometer and analyzer formed by a pair of coaxial conical electrodes to the electron spectrometer and gave a preliminary analysis of the field distribution and associated dispersing and focusing properties, as well as the

---

Zhang Ling[a], Zhou Liwei[a], Jin Weiqi[a], Fang Erlun[b]. a) Beijing Institute of Technology; b) Xi'an Research Institute of Modern Chemistry. Chinese Journal of Electronics (电子学报（英文）), V 2, 1993, 43–49.

angular aberration coefficients[1,2]. H. A. Engelhardt, W. Back (1981) and D. Menzel designed a novel electrostatic energy-dispersion particle analyzer which consists of a toroidal prism and a truncated conical lens. By using the spatial dispersion characteristics, such an analyzer can be used to measure the azimuth angle distribution of charged particles originating in a sample spot[3]. M. I. Yavor (1990) presented an asymptotic analysis of electrostatic field calculation for coaxial conical lens[4]. Since 1980s, S. Y. Yavor and L. A. Baranova have investigated the electron optical properties of both energy analyzer and lenses formed by conical electrodes with the approximate potential distribution[5,6].

In this paper, computation of electrostatic field distribution for the conical immersion lenses is studied by means of boundary element method and the electron trajectories incident from the object and image spaces have been traced using a set of curvilinear paraxial trajectory equations and practical trajectory equations. The electron-optical properties of conical immersion lenses are studied from the practical application aspects.

1 Basic equations of conical immersion lenses

A coaxial conical lens (shown as Fig. 1) is formed by the inner and outer electrodes whose surfaces are two cones with parallel generating lines. θ_0 is the half-angle of the cone. The distance between the inner cone and outer cone is $2b$. V_1 and V_2 are potentials applied to two ends of the immersion lens respectively. The charged particles move between the cones. The hollow particle beam through the conical lens will take shape of a circular form or disk annulus. By changing the potential applied to the electrodes, it is possible to create an additional radial field which makes the axial trajectory along coordinate x curve, and this changes the radius of image disk and contributes to the optical power.

1.1 Potential distribution

Let us introduce a new Cartesian coordinates (x, y) to the immersion lens, as shown in Fig. 1. The electrode potential is V_1 for $x < 0$, and V_2 for $x > 0$. r_c is the distance from symmetrical axis z to the origin O of the Cartesian coordinate system.

For the rotationally symmetrical electrostatic system, the potential distribution in the cylindrical coordinate system satisfies Laplace's equation:

$$\frac{\partial^2 \varphi}{\partial r^2} + \frac{\partial^2 \varphi}{\partial z^2} + \frac{1}{r}\frac{\partial \varphi}{\partial r} = 0 \tag{1}$$

If we assume the potential distribution on the plane z-r takes the form:

$$\varphi(x,y) = V_1 + (V_2 - V_1)\varphi_0(x,y,\varepsilon) \tag{2}$$

where $\varepsilon = b/r_c$; $\varphi_0(x,y,\varepsilon)$ may be called as factor of potential distribution. Consider the conversion relationship between z-r and z-y coordinates

$$\begin{aligned} z &= x\cos\theta_0 - y\sin\theta_0, \\ r &= x\sin\theta_0 + y\cos\theta_0 + r_c \end{aligned} \tag{3}$$

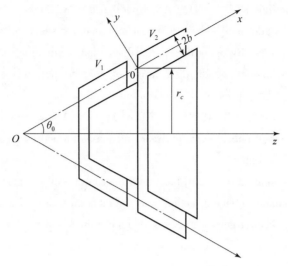

Fig. 1 Conical immersion lens

The Laplace Eq. (1) in the $x-y$ coordinate can be written as:

$$\frac{\partial^2 \varphi_0}{\partial x^2} + \frac{\partial^2 \varphi_0}{\partial y^2} + \frac{1}{r_c + x\sin\theta_0 + y\cos\theta_0}\left(\frac{\partial \varphi_0}{\partial x}\sin\theta_0 + \frac{\partial \varphi_0}{\partial y}\cos\theta_0\right) = 0 \quad (4)$$

To introduce dimensionless coordinates $\zeta = x/b$, $\eta = y/b$, we can rewrite Eq. (4) as follows:

$$\frac{\partial^2 \varphi_0}{\partial \zeta^2} + \frac{\partial^2 \varphi_0}{\partial \eta^2} + \frac{\varepsilon}{1 + \varepsilon(\zeta\sin\theta_0 + \eta\cos\theta_0)}\left(\frac{\partial \varphi_0}{\partial \zeta}\sin\theta_0 + \frac{\partial \varphi_0}{\partial \eta}\cos\theta_0\right) = 0 \quad (5)$$

where the boundary conditions are:

$$\begin{aligned}\varphi_0(\zeta, \pm 1, \varepsilon) &= 0, \quad \zeta < 0;\\ \varphi_0(\zeta, \pm 1, \varepsilon) &= 1, \quad \zeta > 0\end{aligned} \quad (6)$$

If we assume ε to be small: $\varepsilon \ll 1$, $\varphi_0(\zeta,\eta,\varepsilon)$ may be expanded as following:

$$\varphi_0(\zeta,\eta,\varepsilon) = \sum_{j=0}^{\infty} u_j(\zeta,\eta)\varepsilon^j \quad (7)$$

Substituting Eq. (7) into Eq. (5) and using the boundary condition (6), we obtain the potential distribution along the x-axis and the partial derivative of potential with respect to y on the x-axis[6]:

$$\varphi(\zeta,0) = V_1 + (V_2 - V_1)\left\{u_0(\zeta,0) + \frac{\varepsilon}{2}\sin\theta_0[g_1(\zeta) - \zeta u_0(\zeta,0)] + \right.$$
$$\left. \frac{\varepsilon^2}{8}[(3\zeta^2\sin^2\theta_0 - \cos^2\theta_0)u_0(\zeta,0) - 2\zeta\sin^2\theta_0 g_1(\zeta) - 2y_1(\zeta) - \sin^2\theta_0 v_1(\zeta)]\right\} \quad (8)$$

$$\left.\frac{\partial \varphi(\zeta,\eta)}{\partial \eta}\right|_{\eta=0} = (V_2 - V_1)\left\{\frac{\varepsilon}{2}\cos\theta_0\left[\frac{1}{2} + \frac{1}{2}\text{th}\frac{\pi\zeta}{2} - u_0(\zeta,0)\right] + \right.$$
$$\left.\frac{\varepsilon^2}{8}\sin\theta_0\cos\theta_0\left[6\zeta u_0(\zeta,0) - 2g_1(\zeta) - \zeta\left(2 + \text{th}\frac{\pi\zeta}{2}\right) - \frac{2}{\pi}\ln\left(2\text{ch}\frac{\pi\zeta}{2}\right)\right]\right\} \quad (9)$$

where th and ch are hyperbolic functions respectively, u_0, g_1, y_1 and v_1 can be expressed in following forms:

$$\left.\begin{array}{l} u_0(\zeta,0) = \dfrac{1}{2}y_0(-\zeta), \\[4pt] g_1(\zeta) = g_2(-\zeta), \\[4pt] y_1(\zeta) = -g_3(-\zeta) + \dfrac{\zeta}{2}g_2(-\zeta), \\[4pt] v_1(\zeta) = 2g_3(-\zeta) \end{array}\right\} \zeta < 0; \quad (10a)$$

$$\left.\begin{array}{l} u_0(\zeta,0) = 1 - \dfrac{1}{2}y_0(\zeta), \\[4pt] g_1(\zeta) = \zeta + g_2(\zeta), \\[4pt] y_1(\zeta) = -\dfrac{1}{2} + g_3(\zeta) + \dfrac{\zeta}{2}g_2(\zeta), \\[4pt] v_1(\zeta) = \zeta^2 + 1 - 2g_3(\zeta) \end{array}\right\} \zeta > 0; \quad (10b)$$

where

$$y_0(\zeta) = 1 - \frac{2}{\pi}\arctan\left(\operatorname{sh}\frac{\pi\zeta}{2}\right),$$

$$g_2(\zeta) = \frac{4}{\pi^2}\sum_{n=1}^{\infty}\frac{(-1)^{n-1}}{(2n-1)^2}\exp\left[-\pi\left(n-\frac{1}{2}\right)\zeta\right],$$

$$g_3(\zeta) = \frac{8}{\pi^3}\sum_{n=1}^{\infty}\frac{(-1)^{n-1}}{(2n-1)^3}\exp\left[-\pi\left(n-\frac{1}{2}\right)\zeta\right] \quad (11)$$

1.2 Paraxial trajectory equations

From the expressions of field distribution derived above, the paraxial trajectory equation in meridional plane where x-axis is reference axis can be written as:

$$y'' + \frac{\phi'(x)}{2\phi(x)}y' + \frac{1}{2\phi(x)}\left[\phi''(x) + \frac{1}{x}\phi'(x)\right]y = \frac{\phi_1(x)}{2\phi(x)} \quad (12)$$

where $\phi_1(x) = \left.\dfrac{\partial \varphi(x,y)}{\partial y}\right|_{y=0}$ is the potential derivative with respect to y on the x-axis, while primes "'" denote differentiation with respect to x.

Consequently, the paraxial trajectories are described by means of an inhomogeneous linear second order differential equation. When $\phi_1(x) = 0$, it coincides with the paraxial trajectory equation of a trans-axial lens in vertical plane. When $\theta_0 = 0$, which means that conical lenses transform into coaxial cylindrical ones, the paraxial equation takes another form (since the second term in the brackets $\phi'(x)/x$ is vanished).

$$y'' + \frac{\phi'(x)}{2\phi(x)}y' + \frac{\phi''(x)}{2\phi(x)}y = \frac{\phi_1(x)}{2\phi(x)} \quad (13)$$

Obviously, the left part of Eq. (13) coincides with the trajectory equation in two-dimensional lenses.

As is well known, the general solution of the inhomogeneous linear differential equation is a sum of a general solution of the homogeneous equation and a particular solution of the inhomogeneous one. The general solution of the homogeneous equation determines the focusing properties of lens. The particular solution of the inhomogeneous equation describes the deflection properties of the charged particle beam. Hence the conical lenses converge charged particles, thus forming circular images of point or circular sources. The term on the right-hand side of the trajectory Eq. (12) causes the deflection of a charged particle beam and also changes the circular image diameter.

2 Numerical calculation

We have discussed the field distribution along x-axis of the conical immersion lens and the paraxial trajectory equation by means of asymptotic method presented by Yavor, et al. If ε is much more smaller than 1 ($\varepsilon \ll 1$), Eq. (12) can be solved by numerical integral method and approximate focal parameters of conical immersion lens can be obtained. But there are some limitations for practical application. On the one hand, the analytical method is only suitable to simple and ideal structure model, which usually has many differences from practical system, thus leads to errors between the calculated results and the real case. On the other hand, the analytical method will not be used for designing and analyzing various practical systems because of its complex derivation process and mathematical difficulties. Therefore, most of actual conical lenses will be hardly solved by analytical method. The processing method given in the present paper will have no restrictions for the calculation of the various practical systems.

2.1 Computation of field distribution

We make use of boundary element method to calculate potential distributions and their derivatives of different orders, because conical lens is also a kind of rotationally symmetrical system with open boundary. Compared with the FEM and FDM, BEM has a series of advantages as below:

(1) It is suitable to the calculation of field distribution with open boundary;

(2) It requires to calculate the potential and its derivatives only at some specific points, which can save the memory of computer.

(3) By means of the surface charge density on the boundary elements, it can reduce the three dimensional problem to a two dimensional one, so that we can design and anlayze the new type electron-optical systems with three dimensional structure by using PC computers.

The basic computation formulae of electrostatic field of rotationally symmetric systems are[7]

$$\varphi(z_j, r_j) = \sum_{i=1}^{n} A_{ij} \sigma_i \qquad (14)$$

$$\frac{\partial \varphi(z_j, r_j)}{\partial r_j} = \sum_{i=1}^{n} B_{ij} \sigma_i \qquad (15)$$

$$\frac{\partial \varphi(z_j, r_j)}{\partial z_j} = \sum_{i=1}^{n} C_{ij} \sigma_i \qquad (16)$$

where

$$A_{ij} = \frac{1}{\pi\varepsilon_0} \int_{(\Delta l_i)} \frac{rK(k^2)}{S} \mathrm{d}l \tag{17}$$

$$B_{ij} = \frac{1}{\pi\varepsilon_0} \int_{(\Delta_i)} \frac{r(z_j - z)}{S^3} \frac{E(k^2)}{1-k^2} \mathrm{d}l \tag{18}$$

$$C_{ij} = \frac{1}{\pi\varepsilon_0} \int_{(\Delta l_i)} \frac{r}{2r_j S} \left\{ \left[1 - \frac{2r_j(r_j + r)}{S^2}\right] \frac{E(k^2)}{1-k^2} - K(k^2) \right\} \mathrm{d}l \tag{19}$$

where

$$S = \sqrt{(z_j - z)^2 + (r_j + r)^2} \tag{20}$$

$$K(k^2) = \int_0^{\pi/2} \frac{\mathrm{d}\theta}{\sqrt{(1 - k^2 \cos^2\theta)}} \tag{21}$$

$$E(k^2) = \int_0^{\pi/2} \sqrt{(1 - k^2 \cos^2\theta)} \mathrm{d}\theta \tag{22}$$

$$k^2 = \frac{4rr_j}{S^2}, \quad m = 1 - k^2 \tag{23}$$

where, n is the total number of boundary elements, σ_i is the charge density on the boundary elements, (z, r) is the coordinate of field point, (z_j, r_j) is the coordinate of source point, Δl_i is the width of each boundary element. The integral region is over the surface of all electrodes. $K(k^2)$ and $E(k^2)$ are elliptic integrals of the first kind and second kind, which can be calculated by the following formulae:

$$K(k^2) = \sum_{i=0}^{4} a_i m^i - \ln(m) \sum_{i=0}^{4} b_i m^i \tag{24}$$

$$E(k^2) = 1 + \sum_{i=1}^{4} c_i m^i - \ln(m) \sum_{i=1}^{4} d_i m^i \tag{25}$$

where the coefficients of a_i, b_i, c_i and d_i are as Table 1.

Table 1 Coefficients of a_i, b_i, c_i and d_i

i	a_i	b_i	c_i	d_i
0	1.386,294,361,12	0.500,000,000,00	—	—
1	0.096,663,442,59	0.124,985,935,97	0.443,251,414,63	0.249,983,683,10
2	0.035,900,923,83	0.068,802,485,76	0.062,606,012,20	0.092,001,800,37
3	0.037,425,637,13	0.033,283,553,46	0.047,573,835,46	0.040,696,975,26
4	0.014,511,962,12	0.004,417,870,12	0.017,365,064,51	0.005,264,496,39

It is easy to see that as $(z,r) \to (z_j, r_j)$, thus $k \to 1$ and logarithmic singularity appears. We process these singularity integral by using "modified Gauss integral formula", which can be written as follows:

$$I = \int_{-1}^{1} F(x) \mathrm{d}x = \sum_{i=1}^{3} G_u [F(-p_u) + F(p_u)] \tag{26}$$

where $F(x)$ is the integrated function; p_u is the nodal points; G_u is the weight factors

(Table 2).[8]

Table 2　Nodal points P_u and weight factors G_u

u	p_u	G_u
1	0. 006, 722, 304, 11	0. 211, 613, 092, 57
2	0. 441, 855, 063, 00	0. 470, 757, 464, 49
3	0. 870, 098, 761, 21	0. 317, 629, 442, 94

2.2　Determination of positions for cardinal points

From the above-mentioned BEM, it is not difficult to obtain the potential distribution along axis x, and the paraxial trajectory can be solved based on Eq. (12), in which the axis x is the reference axis. But it only gives meridional paraxial trajectory, the sagittal paraxial trajectory could not be solved using Eq. (12). Furthermore, since the electrostatic forces exist in the y direction, the principal trajectory moving along the x-axis will be forced out of the x-axis. We determine the parameters of Gaussian cardinal points for conical lens by means of paraxial trajectory equations having curvilinear optical axis (Eq. (12) with the x-axis as reference axis is not suitable to the computation of the parameters of the cardinal points for conical lens). The principal trajectory equation is written as:

$$r' = \frac{1+r'^2}{2\varphi_n(z,r)}\left(\frac{\partial \varphi}{\partial r} - r'\frac{\partial \varphi}{\partial z}\right) \tag{27}$$

The paraxial trajectory equations having curvilinear optical axis will have the form:

$$\begin{aligned} p'' + F_1 p' + F_2 p &= 0 \\ q'' + G_1 q' + G_2 q &= 0 \end{aligned} \tag{28}$$

where

$$F_1 = G_1 = \frac{1+r'^2}{2\varphi_n(z,r)}\frac{\partial \varphi}{\partial z}$$

$$F_2 = \frac{3r''^2}{(1+r'^2)^2} - \frac{1}{2\varphi_n}\left[\frac{\partial^2 \varphi}{\partial r^2} - 2r'\frac{\partial^2 \varphi}{\partial z \partial r} + r'^2\frac{\partial^2 \varphi}{\partial z^2}\right] \tag{29}$$

$$G_2 = -\frac{1+r'^2}{2\varphi_n(z,r)}\left(\frac{1}{r}\frac{\partial \varphi}{\partial r}\right)$$

where $\varphi_n(z,r) = \varphi(z,r) + \varepsilon_n$, ε_n is the initial potential of electron along the principal trajectory, p and q are the distances of meridional trajectory and sagittal trajectory to the principal trajectory, while the primes denote differentiation with respect to z. It should be pointed out that all the field parameters are taken along the principal trajectory. Tracing group of principal trajectories and neighboring trajectories, we can obtain the parameters of cardinal points in conical lens, as shown in Fig. 2.

Parameters of cardinal points for immersion lens in the image space:

We trace a principal trajectory along x-axis coming from object space and a neighboring trajectory that is parallel to the x-axis which is a distance p_0 apart from principal trajectory, then we can determine the parameters of cardinal points in the image space:

1) Focal point F_i

The focal point is a point of intersection of the principal trajectory and the neighboring trajectory (where $p = 0$). The position of focal point on x-axis is $x(F_i)$. Fig. 2 shows the focal point in image space.

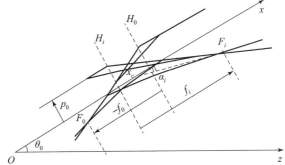

Fig. 2　The definition of cardinal point parameters

2) Principal point H_i

When the neighboring trajectory is extended backward in the focal point F_i, it will intersect the extension of original parallel ray $y = p_0$. The plane passing through the intersection and perpendicular to the x-axis is called "principal plane", the intersection of the principal plane with the x-axis is called "principal point", and the position of principal point on x-axis is $x(H_i)$. The slope of backward extension of the neighboring trajectory is:

$$K = \left(\frac{dp}{ds} + \frac{dR}{dz}\right) \bigg/ \left(1 - \frac{dp}{ds}\frac{dR}{dz}\right) \tag{30}$$

3) Focal length f_i

The distance which is apart from the focal point to the principal point is called the "focal length".

$$f_i = x(F_i) - x(H_i) \tag{31}$$

4) Deflection angle of principal trajectory α_i

The deflection angle α_i is the angle between the x-axis and the backward extension of principal trajectory at focal point, then we have

$$\tan \alpha_i = \left(\frac{dR}{dz} - \tan \theta_0\right) \bigg/ \left(1 + \frac{dR}{dz}\tan \theta_0\right) \tag{32}$$

5) Deflection center of the principal trajectory x_{ci}

The deflection center of the principal trajectory x_{ci} is the intersection of backward extension of the principal trajectory at focal point with the x-axis.

Similarly, we can also have the definitions of the parameters $x(F_0), x(H_0), f_0, \alpha_0$ and x_0 in the object space.

It should be pointed out that we can also determine the parameters of cardinal points by tracing practical trajectory with the practical trajectory equations. The computation for the actual system shows that the result is the same as by tracing trajectory with curvilinear optical axis. Here we shall not describe this procedure in detail.

3 Computation program ELCP

The computation program of electrostatic conical lens is compiled by C language and can be run on the IBM PC/XT/286/386/486 or their compatible computers. The pop-down menu is used to operate computing tasks, which is convenient for the user to operate. After inputting the structure parameters and potential parameters on the electrodes of electron-optical system, the program can be used to automatically calculate the cardinal points (positions of focal point and principal point), focal length, deflection center and deflection angle of principal trajectory. For the stronger or weaker lenses, the program can yield corresponding results in asymptotic cardinal point or real cardinal point. Fig. 3 is a flow chart of the computation program for the electrostatic conical lens.

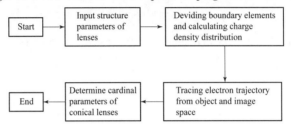

Fig. 3 Flow chart of the computation program of electrostatic conical lens

4 Practical example

As an example, we have computed the electron-optical properties of conical immersion lens, shown as Fig. 1. The structure parameters of this lens are $\theta_0 = 30°$, $r_c = 1$ and $b = 0.25 r_c$. The computation results of cardinal point parameters are listed in Table 3, in which f is focal length, $x(F)$ is the position of focal point, x_c and α are respectively the deflection center and deflection angle of principal trajectory, while the subscripts i and o indicate respectively the parameters in image space and in object space. These results are given in Fig. 4, in which the solid lines with denotation "○" are results calculated by the numerical method of our presentation, while the dashed lines with denotation "∗" are the analyzed results obtained by S. Y. Yavor and L. A. Baranova.

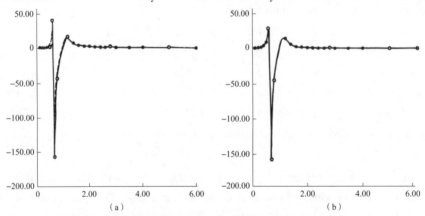

Fig. 4 Computation results of cardinal point parameters

(a) $f_i - V_2/V_1$; (b) $x(F_i) - V_2/V_1$;

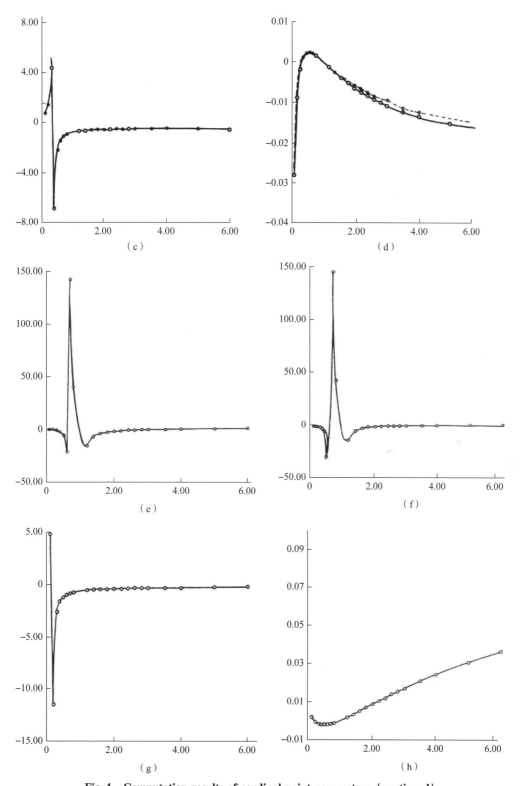

Fig. 4 Computation results of cardinal point parameters (continued)

(c) $x_{ci} - V_2/V_1$; (d) $\tan\alpha_i - V_2/V_1$; (e) $f_0 - V_2/V_1$; (f) $x(F_0) - V_2/V_1$; (g) $x_{c0} - V_2/V_1$; (h) $\tan\alpha_0 - V_2/V_1$

Table 3 Electron-optical parameters of conical immersion lens

V_2/V_1	f_i	f_0	$x(F_i)$	$x(F_0)$	x_{ci}	x_{c0}	$\tan \alpha_i$	$\tan \alpha_0$
0.20	0.487,1	−1.688,4	1.361,3	−0.578,5	1.440,3	−11.508,3	−0.008,8	−0.000,5
0.40	1.808,0	−6.134,2	4.197,1	−2.636,6	−6.804,9	1.634,5	0.000,9	−0.002,3
0.60	38.576,2	−15.556,9	28.070,0	−21.558,1	−1.396,0	−1.031,4	0.002,1	−0.002,2
0.8	−43.689,3	42.510,9	−45.809,8	40.527,2	−0.895,4	−0.819,9	0.001,4	−0.001,3
1.20	15.569,2	−14.870,8	14.544,3	−15.916,1	−0.618,1	−0.646,7	−0.001,5	0.001,6
1.40	4.886,0	−6.265,7	6.013,7	−7.162,5	−0.563,0	−0.602,8	−0.003,0	0.003,2
1.60	4.320,8	−3.737,2	3.541,1	−4.540,3	−0.527,0	−0.571,5	−0.004,3	0.005,0
1.80	3.152,4	−2.600,3	2.433,6	−3.338,3	−0.502,8	−0.546,9	−0.005,6	0.006,7
2.20	2.092,0	−1.578,3	1.446,7	−2.233,9	−0.473,8	−0.514,7	−0.007,9	0.010,1
2.60	1.613,6	−1.125,9	1.009,4	−1.731,0	−0.458,5	−0.493,7	−0.009,7	0.013,5
2.80	1.463,5	−0.986,6	0.873,2	−15,732	−0.454,0	−0.484,1	−0.010,5	0.015,1
3.00	1.347,1	−0.878,4	0.767,8	−1.448,4	−0.450,8	−0.476,9	−0.011,3	0.016,6

References

[1] BREWER D F C, NEWELL W R, SMITH A C H. A coaxial cone electrostatic velocity analyzer I: Analysis of electron optical properties[J]. Journal of Physics E, 1980, 13(1) 114 − 122.

[2] BREWER D F C, NEWELL W R, SMITH A C H. A coaxial cone electrostatic velocity analyzer II: Construction and use in high-resolution electron-scattering experiments[J]. Journal of Physics E, 1980, 13(1): 123 − 127.

[3] ENGELHARDT H A, BACK W, MENZEL D. Novel charge particle analyzer for momentum determination in the multi-channeling mode I: Design aspects and electron/ion optical properties[J]. Review of Scientific Instruments, 1981, 52: 835 − 839.

[4] YAVOR M I. Asymptotic methods applied to electrostatic field calculations for ion-optical systems[J]. Nuclear Instruments & Methods in Physics Research Section A, 1990, A 298: 415 − 420.

[5] YAVOR S Y. Optics of conical electrostatic analyzing and focusing systems[J]. Nuclear Instruments & Methods in Physics Research Section A, 1990, A 298: 421 − 425.

[6] BARANOVA L A, YAVOR M I, YAVOR S Y. Coaxial lenses with longitudinal field for hollow beam focusing[J]. Journal of Technical Physics, 1990, 60, (8): 16 − 22 (in Russian).

［7］ZHANG L. CAD and software for the rotationally and non-rotationally electron-optical system by using boundary element method［J］. Thesis, Kunming Research Institute of Physics, China, 1989 (in Chinese).

［8］KASPER E. On the solution of integral equations arising in electron optical field computations［J］. Optik, 1983, 64, (2):157-169.

8.5 A Multigrid Method for the Computation of Rotational Symmetrical Electrostatic Fields
多重网格法计算轴对称电场的研究

Abstract: The present paper describes a multigrid method which is used to calculate the electrostatic potential distributions for the rotational symmetrical electron-optical systems having internal electrodes and different boundaries. It appears that the computational efficiency and accuracy of the multigrid method are better than those obtainable with the traditional finite difference method (FDM). This method has been applied to design and analyze the electrostatic fields for electron lenses and image tubes with electrostatic focusing, as well as for the electron gun of camera tubes.

摘要：本文描述了多重网格法可用于区域形状任意和有"内电极"的实际电子光学系统轴对称电场的电位计算，其计算精度和效率均优于传统的有限差分法，故可广泛应用于分析设计静电聚焦像管的电子光学系统及摄像管电子枪的发射系统等。

Introduction

The multigrid method (MGM) is a new numerical method in solving the boundary-value problems, that is still in its development in the recent years. One of the greatest advantages of the MGM, when compared with the classical method of iteration, is that the rate of convergence of MGM is independent of the step length h adopted in the discrete domain of the system. It shows the superiority of high efficiency and high speed especially when the step length h of system decreases. So this method has been broadly used in hydrodynamics, structural mechanics and eigenvalue problems.

In view of the optimization design of electron optical imaging systems which normally needs plenty of time for the field computation, we have applied the MGM to solve the rotational symmetrical electrostatic field, and have tried to adapt it to the general domain having any shape boundaries.

Zhang Zhiquan[a], Zhou Liwei[a], Jin Weiqi[a], Fang Erlun[b]. a) Beijing Institute of Technology, b) Xi'an Research Institute of Modern Chemistry. Chinese Journal of Electronics (电子学报（英文）), V.2, No.1, 1993, 15–19.

1 Principle

1.1 Fundamental algorithm of the MGM

For the engineering design, it is often necessary to solve the boundary-value problem which has the following form:

$$\begin{aligned} LU(x,y) &= F(x,y), \quad (x,y) \in \Omega; \\ U &= F_0(x,y), \quad (x,y) \in \Gamma \end{aligned} \quad (1)$$

where L is a differential operator; F is a source function; U is the field distribution to be solved and Γ is the boundary of the domain Ω.

In general, the method of numerical computation for Eq. (1) is to simplify the above-mentioned problem in a discrete form as follows:

$$\begin{aligned} L_k U_k &= F_k, \quad (x,y) \in \Omega_k; \\ U_k &= F_0, \quad (x,y) \in \Gamma_k \end{aligned} \quad (2)$$

and to use a certain approximate1 method, such as the Gauss-Seidel iteration, for solving the problem, where Ω_k is the discrete domain of Ω, Γ_k is the discrete boundary of Γ.

Although the traditional method of iteration can be applied to boundary problems with closed boundary and region of arbitrary shape, its velocity of convergence slows down when the number of iteration increases. The Fourier analysis of the error function $U^{(i)} - U$ shows that the high-frequency part of the Eigen-function of $U^{(i)} - U$ has properties of fast convergence, but the low-frequency part of the Eigen-function of $U^{(i)} - U$ has properties of slow convergence. Within the first few numbers of iteration, the convergence is fast, but when the numbers of iteration increases, the error function $U^{(i)} - U$ becomes a smooth function of grid, and the convergence slows down very quickly (this phenomenon is the so called "smoothness effect"), so its computational efficiency is very low.

In contrast to the traditional method of iteration, the multigrid method uses a "coarser grid correction" technique, which transfers the accurate computation of the field value U on the fine grids to the computation of the deviation ΔV (which is called the defect or error function) of the field value on the coarse grids. Since the defect function ΔV is not smooth on the coarser grids, so the convergent rate of iteration is fast in the first few numbers of iteration, thus it accelerates the process in solving the field distribution. Therefore, the MGM combines the traditional finite difference method (FDM) with the "coarser grid correction" technique, and yields a highly efficient and very fast iteration algorithm.

A more detailed discussion about the MGM is given in Ref. [1].

1.2 MGM for an irregular domain

The standard MGM is carried out only for rectangular domain. In order to apply the MGM to compute the rotational symmetrical electrostatic field with arbitrary boundary, we have adopted the following techniques:

(1) Imbedding the domain Ω in which the value U is to be solved and the boundaries of which have arbitrary shapes to a rectangular domain, we treat the maximum values of the z and r directions of Ω as z_{MAX} and r_{MAX} of the rectangular domain, and give the points which are located outside of the domain Ω to be solved and inside of the rectangular domain always zero (these points are called the "outside-domain points").

(2) Using a concept of "fictitious grid points", that is, increasing a row and/or a column of grid points on the top and/or the right-side of the rectangular domain, we assume that the requirement of grid coarse is always satisfied (the step length h_{i+1} of the coarser grid is two times that of h_i of the finer grid). All of the "fictitious grid points" can be regarded as "outside-domain points", and are also given as zero.

(3) During the computation of grid correction, for the transformation of the operator I_{k+1}^k, that is, prolongation from the coarser grid level to a finer grid level, we use the bilinear-interpolation, i. e. in the z and r direction of the rotational symmetrical system, the approximate values of the finer grid points are obtained by linear interpolation from the values of the coarser grid points, while for I_{k+1}^k, which is a restriction from finer grid level to coarser grid level, is a conjugate transposed matrix of I_{k+1}^k[1].

Besides the above-mentioned treatment, a "code of logic rule" technique is used to distinguish the grid points at different levels. Therefore, we can easily distinguish the "electrode points" (which include the inner electrode points and the boundary electrode points) and the "inner points" (which include the regular and irregular inner points), as well as the "outside-domain points", thus simplifying the logic computation of iteration and improving the efficiency of computation.

2 Program design and computation

As discussed above, considering the characteristic of electrostatic image tubes, we use the concept of the "fictitious grid points" to extend the standard MGM algorithm to the rotational symmetrical problems having arbitrary boundaries, and to substitute MGM for FDM in solving the field distribution of electron optical systems.

We have compiled a program of MGM for the computation of rotational symmetrical electrostatic field using the C language. A flow chart for the program is shown in Fig. 1.

In order to test the efficiency and accuracy of MGM, we have taken the two examples low and compared them with FDM.

Example 1: an immersion lens with two equal-diameter cylinders, with radius of the cylinders 20 mm, the length of each cylinder 32 mm, the distance between two cylinders 6 mm, the applied voltages 0 V and 10,000 V, respectively. The system is separated by four grid levels as follows: N_{zk} and N_{rk} are the numbers of the grid points in the level k in the directions z and r: at the first level, $h_{z1} = h_{r1} = 1$, $N_{z1} = 71$, $N_{r1} = 21$, at the second level, $h_{z2} = h_{r2} = 2$, $N_{z2} = 36$, $N_{r2} = 11$,

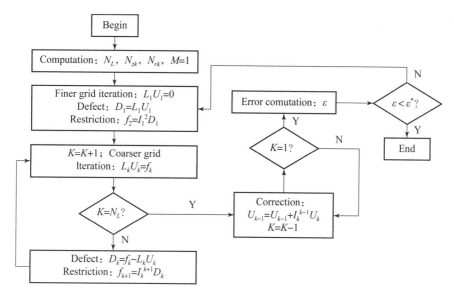

Fig. 1 A flow chart of the multigrid method

at the third level, $h_{z3} = h_{r3} = 4$, $N_{z3} = 19$, $N_{r3} = 6$, at the fourth level, $h_{z4} = h_{r4} = 8$, $N_{z4} = 10$, $N_{r4} = 4$ (The total number of grid points at $k = 1$ level is 1,491).

It should be pointed out that the first column of grid points on the right at the third level, and the first row of grid points on the top at the fourth level, as well as the first column of grid points on the right at the fourth level are taken as "fictitious grid points". All of these points are located outside of the domain Ω to be solved.

Table 1 shows the result of some points of the axial potential as computed by FDM (number of iterations: 208, total average relative error: 4.569×10^{-10}, time of computation: 9.77 s) and by MGM ($J_1 = 2$, $J_2 = 3$, $Nit = 11$, total average relative error: 6.990×10^{-10}, time of computation: 2.03 s, where J_1, J_2 are the numbers of iteration at the level of $k = 1$ and $k = 2, 3, 4$, Nit is the cycle number of iteration of the multigrid method), all of the computations being carried out on the same personal computer (type 486, 33 MHz).

It can be seen from Table 1 that the computational result by MGM are very close to FDM, and the time of computation of MGM is only 1/4.8 of FDM, so MGM is 3.8 times faster than FDM.

Table 1 Axial potential computation of the immersion lens with two equal-diameter cylinders by FDM and MGM

z/mm	Values of the axial potential/V					T/s
	1	2	3	4	5	
FDM	29.132,888,5	58.673,203,9	89.033,285,1	120.635,297	153.916,147	9.77
MGM	29.132,889,0	58.673,206,2	89.033,288,5	120.635,301	153.916,152	2.03

Example 2: a typical triode image tube, with tube length 72 mm, raidus of tube 20 mm,

radius of curvature of the photocathode 17 mm, radius of hole on anode 2 mm and applied voltage on the anode 4,600 V, applied voltage on the screen 800 V, as shown in Fig. 2. The system is separated by five grid levels as follows, at the first level, $h_{z1} = h_{r1} = 0.5$, $N_{z1} = 145$, $N_{r1} = 41$, at the second level, $h_{z2} = h_{r2} = 1$, $N_{z2} = 73$, $N_{r2} = 21$, at the third level, $h_{z3} = h_{r3} = 2$, $N_{z3} = 37$, $N_{r3} = 11$, at the fourth level, $h_{z4} = h_{r4} = 4$, $N_{z4} = 19$, $N_{r4} = 6$, at the fifth level, $h_{z5} = h_{r5} = 8$, $N_{z5} = 10$, $N_{r5} = 4$ (The total number of grid points at $k = 1$ level is 5,945).

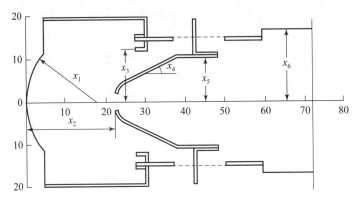

Fig. 2 Configuration of a triode image tube

The first two of grid points on the top at the fifth level are the "fictitious grid points", which are located outside of the domain Ω to be solved.

Table 2 shows a comparison of field computation by MGM with different J_1 and J_2, in which T is the time of computation reaching the accuracy of total relative error of 10^{-10}.

Table 2 A comparison of field computation by MGM with different J_1 and J_2

J_1	J_2	Nit	$z = 1$ mm	$z = 2$ mm	$z = 3$ mm	T/s
2	2	19	29.902,938,2	63.498,492,8	101.203,002	11.47
2	3	15	29.902,938,2	63.498,492,8	101.203,002	9.67
2	4	14	29.902,938,2	63.498,492,8	101.203,002	9.66
2	5	14	29.902,938,2	63.498,492,8	101.203,002	10.21
2	6	14	29.902,938,2	63.498,492,8	101.203,002	10.77
2	2	17	29.902,938,2	63.498,492,8	101.203,002	11.71
2	3	14	29.902,938,2	63.498,492,8	101.203,002	10.27
2	4	12	29.902,938,1	63.498,492,8	101.203,002	9.28
2	5	12	29.902,938,2	63.498,492,8	101.203,002	9.77
2	6	12	29.902,938,2	63.498,492,8	101.203,002	10.21

Table 3 shows the result of axial potential computation, and gives a comparison between FDM (iteration times: 345 s, total average relative error: 4.780×10^{-10}, time of computation:

52.67 s) with MGM ($J_1=2$, $J_2=4$, $Nit=14$, total average relative error: 3.985×10^{-10}, time of computation: 9.66 s). The time of computation of MGM is again only 1/5.45 of FDM.

If we compare the results obtained with ones having a higher accuracy, for example, 29.902,935,0 V, the potential at $z=1$ mm by FDM, the relative error of potential at $z=1$ mm will be (29.902,935,0 − 29.902,938,2)/29.902,935,0 = 1.07×10^{-7}. It shows that the accuracy is enough for a large number of engineering problems.

Table 3 Axial potential computation of a typical triode image tube by FDM and MGM

Values of the axial potential/V					
z/mm	FDM	MGM	z/mm	FDM	MGM
0	0.000,000	0.000,000	40	4 332.786,46	4 332.786,46
5	190.879,749	190.879,763	45	3 858.010,57	3 858.010,57
10	533.090,256	533.090,276	50	2 977.727,83	2 977.727,83
15	1 209.088,96	1 209.088,98	55	2 025.699,68	2 025.699,68
20	2 831.518,82	2 381.518,83	60	1 385.785,78	1 385.785,78
25	4 545.448,84	4 545.448,84	65	1 045.289,28	1 045.289,28
30	4 577.979,58	4 577.979,58	70	858.664,315	858.664,315
35	4 518.114,24	4 518.114,24	72	800.000,000	800.000,000

Conclusions

The present paper describes a multigrid method used to calculate the electrostatic potential distribution for electron-optical systems in electron lenses and image tubes. The "fictitious coarse gird points" and "imbedding" techniques for irregular regions, as well as the "code of logic rule" technique, have been introduced so that the method suggested not only enjoys the characteristics of high efficiency and high speed of the standard MGM, but also can be applied in the calculation of tasks like rotational symmetrical electron-optical systems in electron lenses and image tubes involving internal electrodes and irregular boundaries.

In the package of program for solving electrostatic fields in image tubes, the algorithm has been written in the C language. Results for a large number of practical computations of image tubes show that on the same condition of required accuracy of iterations, for a discrete system which has 1,400 grid-points, the speed of MGM is 3 times faster than FDM, and for a discrete system which has 5,000 grid-points, the MGM is even more efficient (about 5 times faster than FDM). This means that the computational efficiency of MGM is superior to FDM, so MGM is a new and fast computational method for electrostatic fields in the design of image tubes, especially for the optimization design of image tube which needs a large number of field computations, MGM can also be widely used not only for solving the Laplace field (such as in cathode lenses and electrostatic electron lenses), but also for solving the Poisson field (such as in camera tubes and emission systems).

References

[1] HACKBUSCH W. Multigrid methods and applications[M]. Berlin: Springer, 1985.

[2] ACHI B. Multi-level adaptive solutions to boundary-value problems, mathematics of computation [J]. Mathematics of computation. 1977, 31 (138): 333 –390.

[3] DENDY J E. Jr.: Black box multigrid[J]. Journal of Computational Physics, 1982, 48: 366 –386.

[4] MANFRED R, ULRICH T, GERD W. A note on MGR methods[J]. Linear Algebra & Its Applications, 1983, 49: 1 –26.

[5] ZHOU H, NI G Z. Multigrid method applied to the steady electromagnetic field numerical analysis[J]. Journal of Chinese Electrical Engineering, 1990, 10, (5): 20 –26(in Chinese).

8.6 一种计算轴对称磁场的边界元 – 有限元混合法的研究

A Combined Boundary Element-finite Element Method for Computing the Rotational Symmetrical Magnetic Field

摘要：本文由静磁场的基本方程和有限元的原理出发，研究了一种新的计算带非饱和磁介质轴对称磁场的边界元 – 有限元混合法，编制了计算电磁聚焦成像系统聚焦磁场的程序，这种方法解决了单纯使用有限元法的边界封闭问题，为成像系统及其他磁透镜的工程设计提供了一种有效方法。

Abstract: Based on the fundamental equations of static-magnetic field and the principle of finite element method, a new combined boundary element-finite element method for the computation of rotational symmetrical magnetic field with unsaturated iron medium is studied and a program for the design of electromagnetic focusing imaging systems is drawn up. The method shows that the question of the closed boundary, when using finite element method only can now be settled. It has also shown that the new method is very efficient for the engineering design of imaging systems, as well as for ordinary magnetic lenses.

引 言

电磁复合聚焦成像系统广泛应用于近代光电成像器件中，特别是近年来对 $M\neq 1$ 的图像无旋转的电磁聚焦成像系统理论和设计方法的研究工作[1-3]，突破了以往放大率只能为1的限制，这对系统性能的扩展以及进一步的应用无疑是很大的促进。

电磁聚焦成像系统的聚焦磁场是长磁透镜，它不同于一般短磁透镜，其场可分布在一个较大的区域。虽然可以用几个裸线圈的结合来逼近所需的场分布，但为了减小励磁电流，降低功耗、屏蔽外界电磁场的干扰以及提高系统边缘线圈所产生的磁感的峰值[3]，往往要加上铁芯外壳；同时，成像系统的输入输出视场又要求铁芯两端应有足够的开口尺寸。显然，这样的磁场是带铁介质的半开放磁场，若使用单纯的有限元程序[4]来计算这种半开放磁场，则存在边界封闭问题。由于受到计算机内存量的限制，若处理不当，将会造成很大的误差。特别在端面开口较大时，这个问题就更为突出。为了从根本上解决这一问

金伟其[a]，周立伟[a]，倪国强[a]，方二伦[b]. a) 北京理工大学, b) 西安现代化学研究所. 北京理工大学学报（Journal of Beijing Institute of Technology）, V. 11, No. 4, 1991, 37 – 44.

题，本文由静磁场的基本方程和有限元的原理出发，研究了一种适用于求解带铁介质半开放轴对称磁场的边界元-有限元混合法，为电磁聚焦成像系统和其他磁透镜的工程设计提供了一种可行的方法。

1 边界元-有限元混合法的基本原理

对于均匀且各向同性介质中的静磁场，由静磁场的基本方程：

$$\nabla \cdot \boldsymbol{B} = 0, \quad \nabla \times \boldsymbol{H} = 0 \tag{1}$$

在库仑规范$\nabla \cdot \boldsymbol{A} = 0$的条件下，其磁矢位$\boldsymbol{A}$满足

$$\nabla^2 \boldsymbol{A} = -\mu \boldsymbol{j} \tag{2}$$

式中，μ为介质的磁导率，$\boldsymbol{B}$为磁感应强度，$\boldsymbol{H}$为磁场强度，$\boldsymbol{j}$为电流密度矢量。

从式（1）和式（2）出发，利用矢量形式的格林恒等式，可以证明：在以正则面S为边界的一个闭合域V中，任意一点$\rho_1(x_1,y_1,z_1) \in V$的磁矢位$\boldsymbol{A}(\rho_1)$可表示为[6]：

$$\begin{aligned}\boldsymbol{A}(\rho_1) = & \frac{\mu}{4\pi}\iiint_V \frac{\boldsymbol{j}(\rho)}{|\rho-\rho_1|}\mathrm{d}V - \frac{1}{4\pi}\oiint_S \frac{\boldsymbol{n}\times\boldsymbol{B}}{|\rho-\rho_1|}\mathrm{d}s - \\ & \frac{1}{4\pi}\oiint_S (\boldsymbol{n}\times\boldsymbol{A})\times\nabla\frac{1}{|\rho-\rho_1|}\mathrm{d}s - \frac{1}{4\pi}\oiint_S (\boldsymbol{n}\cdot\boldsymbol{A})\nabla\frac{1}{|\rho-\rho_1|}\mathrm{d}s\end{aligned} \tag{3}$$

式中，$\boldsymbol{j}(\rho)$是V中流动点$\rho(x,y,z)$的电流密度；$\boldsymbol{n}$、$\boldsymbol{B}$和$\boldsymbol{A}$分别为边界S上ρ点处的外法向单位矢量、磁感应强度和磁矢位；$|\rho-\rho_1|$为ρ到ρ_1的距离。上式表明V中任意一点的磁矢位可分为两部分的贡献，即内部电流的贡献（表现为体积分）和外部的贡献（表现为边界积分）。

对于轴对称磁场，磁矢位只有旋转分量

$$\boldsymbol{A} = A\boldsymbol{\theta}°, \tag{4}$$
$$\boldsymbol{B} = -\frac{1}{r}\frac{\partial(rA)}{\partial z}\boldsymbol{r}° + \frac{1}{r}\frac{\partial(rA)}{\partial r}\boldsymbol{z}°$$

这里$(\boldsymbol{z}°, \boldsymbol{r}°, \boldsymbol{\theta}°)$为轴对称坐标系的坐标基。因此，利用磁矢位$\boldsymbol{A}$来求解轴对称磁场，可降低问题的维数，减少计算量。

如图1所示，在轴对称磁场中，设闭合曲面S是连续曲线l绕对称轴z的旋转曲面，它将全空间分割为V_1和V_2两个部分，且全部电流和磁介质都包含在V_1中，V_2是均匀的自由空间，即

$$\boldsymbol{j}(\rho) = 0 \quad (\rho \in V_2) \tag{5}$$

考虑到磁矢位的距离倒数衰减规律以及以下关系

$$|\rho-\rho_1| = [(z-z_1)^2 + r^2 + r_1^2 - 2rr_1\cos\theta]^{1/2},$$
$$(\boldsymbol{n}\cdot\boldsymbol{A}) = 0 \tag{6}$$

于是，由式（3）可求V_2中任意一点的磁矢位

$$A(r_1,z_1) = \frac{1}{4\pi}\int_l \left[G\frac{\partial A(r,z)}{\partial n} - A(r,z)\frac{\partial G}{\partial n}\right]r\mathrm{d}l \tag{7}$$

其中$\partial/\partial n$为方向导数，G为格林函数，且有

$$G = 2\int_0^\pi \frac{\cos\theta\mathrm{d}\theta}{\sqrt{(z-z_1)^2+r^2+r_1^2-2rr_1\cos\theta}} = \frac{4}{k}\left(\frac{1}{rr_1}\right)^{1/2}\left[\left(1-\frac{1}{2}k^2\right)K - E\right] \tag{8}$$

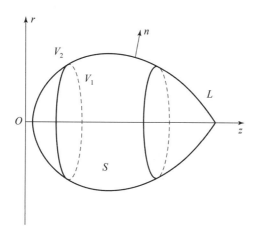

图 1　区域分割示意图

$$k^2 = \frac{4rr_1}{(z-z_1)^2 + (r+r_1)^2} \leq 1 \tag{9}$$

式中，K、E 分别为第一类和第二类完全椭圆积分。

式（7）表明：只要知道边界 l 上的磁矢位，就可求出 V_2 中的磁矢位。

1.2　有限元法

对于轴对称磁场，可得在 S 面积（子午面上）内的能量泛函为[5]

$$F = \iint_S \left\{ \frac{1}{2\mu}\left[\left(\frac{\partial A}{\partial z}\right)^2 + \left(\frac{\partial A}{\partial r} + \frac{A}{r}\right)^2\right] - JA \right\} 2\pi r dr dz \tag{10}$$

经过单元划分和插值近似，在能量最小的条件下，可得到有限元方程

$$[K][A] = [b] \tag{11}$$

式中，$[K]$ 是刚度矩阵（带状对称）；$[A]$ 为所求的磁矢位向量；$[b]$ 是激励向量。式（11）在封闭的边界条件下可以唯一地求解。

1.3　轴对称磁场边界元 – 有限元混合法

一般来说，边界元法较适合于求解开放边界问题；有限元法较适合于求解变介质场问题（如带铁介质磁场），但需要在封闭边界条件内求解[5]。因此，若将两者的优点结合，即在区域内用有限元划分，用边界元封闭边界（或给出一组关系式），就可构成一种适合于求解变介质开放场问题的方法，这就是边界元 – 有限元混合法的基本思想[7]。

图 2 是轴对称磁场子午面（r, z）内的分布示意图，电流和磁介质均集中于边界 l_C（即 S_1）内。为求解 l_C 内的磁矢位分布，再假设边界 l_B 比 l_C 稍大，在 l_B（即 S_1+S_2）内用有限元分割并建立方程组，且使 l_B 与 l_C 之间只有一层有限元（设 l_C 内有 N_I 个内部节点，l_C 和 l_B 上分别有 N_C 和 N_B 个节点）。此时由于 l_B 上节点磁矢位未知，边界未封闭，有限元方程还不能求解。因而我们进一步在 l_C 以外到无穷远 l_∞ 的区域内，利用边界元法建立 l_B 和 l_C 上磁矢位之间的关系，然后代入有限元方程就可求解磁矢位分布。

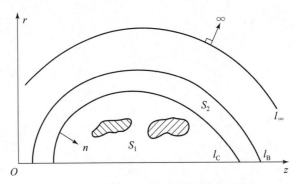

图 2 子午面内电流和介质的分布示意图

在 l_B 内经适当排序，有限元方程可写为

$$\begin{bmatrix} [K_{II}] & [K_{IC}] & 0 \\ [K_{IC}]^T & [K_{CC}] & [K_{CB}] \\ 0 & [K_{CB}]^T & [K_{BB}] \end{bmatrix} \begin{bmatrix} [A_I] \\ [A_C] \\ [A_B] \end{bmatrix} = \begin{bmatrix} [b_I] \\ 0 \\ 0 \end{bmatrix} \tag{12}$$

式中，$[K_{II}]$ 是与内部节点有关的 $N_I \times N_I$ 阶对称带状的刚度矩阵；$[K_{CC}]$ 和 $[K_{BB}]$ 分别是与 l_C 和 l_B 上节点有关的刚度矩阵；$[K_{IC}]$ 和 $[K_{CB}]$ 则分别表示内部与 l_C 以及 l_C 与 l_B 上节点之间的耦合刚度矩阵；$[A_I]$、$[A_C]$、$[A_B]$ 分别为内部、l_C 和 l_B 节点的磁矢位向量；$[b_I]$ 为激励向量。

由式（7）可知，在 l_C 与 l_∞ 之间的磁矢位可由 l_C 上的磁矢位及其偏微商的积分得到，因而我们可建立 $[A_C]$ 和 $[A_B]$ 之间的关系。由有限元单元的插值函数可知，在 l_C 与 l_B 之间（一层有限元）

$$[A] = [\alpha_C]^T [A_C] + [\alpha_B]^T [A_B] \tag{13}$$

式中，$[\alpha_C]$ 和 $[\alpha_B]$ 分别为插值函数向量。将场点设在 l_B 上的节点，则由式（7）可得

$$[A_B] = [P_C][A_C] + [P_B][A_B] \tag{14}$$

$$[A_B] = [P][A_C] \tag{15}$$

$$[P] = (I - [P_B])^{-1} [P_C] \tag{16}$$

式中，$[P]$ 是 $N_B \times N_C$ 阶变换矩阵，当设定了 l_C 和 l_B 上节点的位置，$[P]$ 可唯一确定。

将式（15）代入式（12），就得到边界元-有限元方程

$$\begin{bmatrix} [K_{II}] & [K_{IC}] \\ [K_{IC}]^T & [K_{CC}] + [K_{CB}][P] \end{bmatrix} \begin{bmatrix} [A_I] \\ [A_C] \end{bmatrix} = \begin{bmatrix} [b_I] \\ 0 \end{bmatrix} \tag{17}$$

在轴上边界条件 $A(0,z) = 0$ 下，式（17）可唯一求解磁矢位分布 $[A_I]$ 和 $[A_C]$，并可由式（15）或式（7）求出 $[A_B]$ 或者 l_C 以外任意点的磁矢位。这样就实现了半开放磁场计算的边界自动封闭。

2 程序的实现

根据上述基本原理，我们用 FORTRAN 语言编制了微机上使用的边界元-有限元混合法计算软件（程序框图如图 3 所示），它彻底解决了单纯用有限元法计算时的封闭边界问题，适用于计算电磁聚焦成像系统中的带铁介质半开放磁场，也可用于计算其他磁透镜场

（如单极靴磁透镜等）。

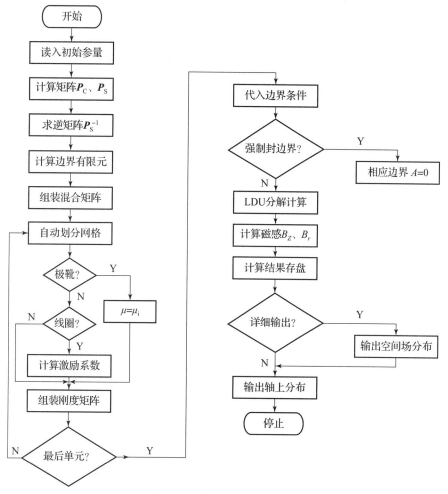

图 3　混合法计算程序框图

在计算软件中采用了如下具体方法：

（1）兼顾到计算精度、计算量和内存量，并考虑到电磁聚焦成像系统的特点，我们采用了四点四边形等参有限元，采用 7×7 阶高斯积分或解析积分方法求矩阵系数。

（2）根据实际要求和计算量限制，在磁介质以外选择一条矩形曲线作为 l_C 边界，在内部给出大网格点坐标及分割数，由程序自动细分，在 l_C 外部平行外延一层单元得到 l_B 边界。

（3）注意式（17）的刚度矩阵 $[K]$ 已不是对称矩阵，但仍保持稀疏和部分对称，其非零元素分布形状如图 4 阴影部分所示，我们采用 LDU 分解法进行矩阵计算，因而只需储存图中阴影部分。

（4）考虑到实际系统在磁介质以外和远离电流的部分 l_B 边界上，磁矢位往往已接近于零，程序可根据系统特点，在精度允许的条件下，对这些 l_B 边界进行强制边界条件 $A=0$ 的处理，从而大大减少计算量，显著提高计算速度。

（5）本程序虽是针对非饱和磁场编制，但混合法的思想也适用于饱和磁场计算，此时

只需将能量泛函式（10）变为[4]

$$F = \iint_S \left[\int_0^{|B|} H \cdot dB - JA\right] 2\pi r dr dz \quad (18)$$

式中，磁场强度 H 与 B 的关系由磁介质的磁化曲线 $B-H$ 确定。经过类似的推导，仍可得到类似式（17）的边界元–有限元方程。通过迭代求解就可求出饱和磁场分布。

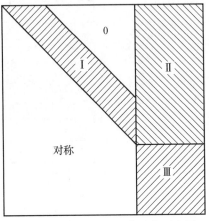

图 4　刚度矩阵非零元素分布示意图

3　计算实例

利用半开放轴对称磁场的边界元–有限元混合法计算软件，我们已实现了对电磁聚焦成像系统聚焦磁场的实际计算。为了比较混合法的精度，下面给出两个计算实例。

3.1　裸露线圈的磁场分布

首先对内、外径分别为 20 mm 和 30 mm，厚度为 3 mm 的裸露线圈用混合法和圆环电流磁场积分的方法[8]（简称解析法）分别进行了计算。计算中线圈电流密度取 0.8 A/mm²，混合法共用 1 057 个节点，在 IBM–C286（主频 12 MHz）机上耗时约 39 min。表 1 给出了这两种方法计算的部分轴上磁感应强度值 B 和 $r_0=4$ 处的磁矢位值 A 及两者之间的相对误差。由表可见，两者的相对误差小于 1.3‰，可以满足工程设计的精度要求。

表 1　解析法和混合法计算裸露线圈磁场的精度比较

z 坐标 /mm	$r=0$, $B/(\times 10^{-4}$ T$)$			$r=4$, $A/(\times 10^{-4}$ T·mm$^{-1})$		
	解析法	混合法	误差/‰	解析法	混合法	误差/‰
0.0	52.028	52.076	0.925 7	104.600	104.71	1.051 6
2.0	51.756	51.803	0.914 6	104.050	104.16	1.057 2
4.0	50.946	50.992	0.901 9	102.410	102.51	0.976 5
6.0	49.619	49.664	0.897 6	99.723	99.82	0.972 7
8.0	47.813	47.856	0.888 8	96.057	96.15	0.968 2

续表

z 坐标 /mm	$r=0, B/(\times 10^{-4}\, T)$			$r=4, A/(\times 10^{-4}\, T \cdot mm^{-1})$		
	解析法	混合法	误差/‰	解析法	混合法	误差/‰
10.0	45.581	45.622	0.903 9	91.520	91.61	0.983 4
12.0	42.993	43.035	0.975 7	86.258	86.35	1.066 6
14.0	40.137	40.180	1.058 9	80.447	80.54	1.156 0
16.0	37.109	37.151	1.138 3	74.290	74.38	1.211 4
18.0	34.007	34.047	1.188 0	67.991	68.07	1.161 9
20.0	30.924	30.960	1.154 4	61.745	61.82	1.214 7

3.2 单极靴磁透镜的计算

单极靴磁透镜是近年来发展起来的一种磁透镜，它可产生强的磁场和小的半宽度，从而可得到低像差图像，且由于只有一个极靴，可提供更大的样品空间，适宜在诸如电子束光刻等电子光学仪器中应用，故受到人们的极大重视[9]。

单极靴透镜是一种典型的带铁介质半开放磁透镜，若直接用有限元计算，则精度依赖于计算机内存和边界的选择，使用混合法程序则可完全解决这一问题。

图 5 给出了用混合法（1 512 个节点）和有限元法（1 000 个节点，分别在 $z=-65$ mm，150 mm，$R=200$ mm 处封边）对一个典型单极靴透镜的计算结果，由图 5 可见混合法得到的轴上磁感应强度分布比较光滑，而有限元法由于网格较粗及强制边界影响，产生了抖动和曲线下压，从而产生误差。

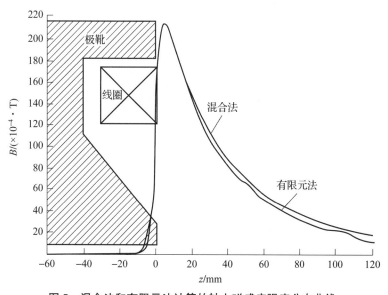

图 5 混合法和有限元法计算的轴上磁感应强度分布曲线

结 束 语

本文研究了适合计算带铁介质半开放轴对称磁场的边界元－有限元混合法，编制了相应的计算软件。对实际系统的计算表明：对于半开放问题，混合法具有比单纯有限元法更高的计算精度，可满足工程设计的要求，从而较为彻底地解决了计算半开放系统的边界封闭问题，可作为计算电磁聚焦成像系统和其他轴对称磁透镜的有效方法和工具。

参 考 文 献

［1］ZHOU L W, NI G Q, FANG E L. Electrostatic and magnetic imaging without image rotation［J］. Electron Optical Systems, SEM Inc, 1984, 37 – 43.

［2］周立伟,仇伯仓,倪国强. 电磁聚焦移像系统中聚焦磁场的逆设计［J］. 电子学报, 1989, 17(2):21 – 27.

［3］NI G Q, ZHOU L W, JIN W Q, et al. A study of new methods for designing electrostatic and magnetic imaging systems with variable magnification. International Conference on Optoelectronic Science and Engineering'90, SPIE, 1990, 1230:55 – 57.

［4］MUNRO E. Computer – aided – design methods in electron optics［D］. Cambridge: University of Cambridge, 1971.

［5］盛剑霓. 电磁场数值分析［M］. 北京:科学出版社,1984.

［6］STRATTON J A. Electromagnetic theory［M］. New – York: McGraw – Hill book company,1941.

［7］李忠元. 电磁场边界元素法［M］. 北京:北京理工大学出版社,1987.

［8］斯迈思 W R. 静电学和电动力学［J］. 戴世强,译. 北京:科学出版社,1982.

［9］MULVEY T. Magnetic electron lenses.［J］. Scanning Electron Microscopy, 1982, 18.

九、成像电子光学系统的逆设计

Inverse Design of Imaging Electron Optical Systems

9.1 静电聚焦成像系统电子光学逆设计的研究
A Study of Electron-optical Inverse Design for Electrostatic Focusing Imaging Systems

摘要：与通常的电子光学系统设计相反，电子光学逆设计是在给定的系统成像特性下，反求电子轨迹与电位分布，从而确定能实现所要求的成像特性的系统的结构参量与电参量。本文由简正形式的近轴轨迹方程出发，探讨了典型静电成像系统的电子轨迹与电位分布所应满足的要求，导出了近轴轨迹与轴上电位分布之间的关系式。文中提出了用多项式为数学模型模拟电子轨迹，并通过电子计算机进行求解的方法，最后给出了计算实例。

Abstract: Contrary to the normal way of electrostatic electron-optical system design, the process of inverse design, given image characteristics of a system, consists of three steps: first, to determine the electron trajectories; secondly, to solve the potential distribution, and thirdly, to calculate the structural and electrical parameters in order to fulfil the required image characteristics. In this paper, based on the Picht's "paraxial" equation, the requirements that the electron trajectories and potential distribution must realize for typical electrostatic focusing image system are discussed, and the relationship between "paraxial" trajectories and potential distribution along axis is also derived. It has been suggested in the paper that the polynomial as a mathematical model is used to simulate the electron trajectory, and some examples are shown. Finally, computer results of an example for such designs process are given.

引 言

众所周知，电子光学系统设计通常是在给定的结构参量与电参量下，利用所给的边界条件，通过解拉普拉斯方程求解电位分布，解运动方程或轨迹方程求电子轨迹，从而确定系统成像位置、像差与调制传递函数等。这种方法称为电子光学系统正设计。

电子光学逆设计一反上述的计算过程，它是在给定系统的成像特性——像面位置与放大率下，反求电子轨迹与电位分布，从而确定系统的结构参量与电参量。此外，还可通过寻求像差极小值，进行优化逆设计。

进行成像系统的逆设计，需要从理论与实际探讨以下几个问题：

（1）由所给定的成像参量求适应静电聚焦成像系统的轴上电位分布。

（2）由轴上电位分布求空间电位分布，并证明该问题的唯一性。

（3）由空间电位分布获得现实上可能的电极结构。

周立伟，潘顺臣，艾克聪. 北京工业学院. 北京工业学院学报（Journal of Beijing Institute of Technology），No. 1, 1983, 17-34.

电子光学逆设计的理论基础是由 Гринберг 奠定的[1]，以后 Касьянков 等人又进行了发展，对静电透镜的逆设计问题进行了探讨[2]。但是迄今为止的文献，并没有考虑静电成像系统的特点，所得到的电位分布与实际典型成像系统不相适应，故很难找到合适的电极结构，能保证实现系统的成像特性——物像位置与放大率的要求。本文作者之一曾于 1965 年在一篇报告[3]中考虑静电成像系统的特点，提出了静电成像系统逆设计的基本途径与方法。

本文是在该基础上的进一步发展。按照本文提出的方法，系统的成像参量能满足设计要求，且由于轴上电位分布考虑了静电成像系统的实际情况，能较容易获得所需要的电极结构。

1 静电聚焦成像系统逆设计求解问题的提出

在旋转对称静电成像系统中，近轴轨迹简正形式的线性方程可表示为

$$R''(z) + \frac{3}{16}\left[\frac{\phi'(z)}{\phi_*(z)}\right]^2 R(z) = 0 \tag{1}$$

式中，$\phi_*(z) = \phi(z) + \varepsilon_z$，其中 $\phi(z)$ 为轴上电位分布，ε_z 为轴向初速所对应的电子初电位。

若以 $R_\alpha(z)$ 作为式（1）的一个特解，则可得到轴上电位分布与简正轨迹特解之间的关系式：

$$\phi_*(z) = [\phi_*(z)] \exp\left(\int_{z_c}^{z} \sqrt{-\frac{16 R''_\alpha}{3 R_\alpha}} dz\right) \tag{2}$$

式中，z_c 为自阴极 $z_0 = 0$ 算起某一位置的纵坐标。考虑静电成像系统轴上电位分布的特征，上式积分号前取正号。

由于简正近轴轨迹方程的奇异性，式（2）的积分下限不能选在阴极面上。为此，在阴极面附近选取一点 z_c，使过此点的等位面与阴极面之间的场尽可能接近理想分布。于是理想系统的解析解便可用来表示该区域的轨迹和电位分布。众所周知，阴极面场强 E 与阴极面曲率半径 R_c 是静电成像系统的两个重要参量，前者将决定系统的极限鉴别率，后者将影响系统的轴外像差，特别是场曲与像散。这样，E 和 R_c 便可由设计者事先按要求给定。

利用静电聚焦成像系统普遍成立的拉亥关系式可导得[4]

$$M = \frac{1}{\sqrt{\phi_*(z_i)} r'_\alpha(z_i)} \tag{3}$$

式中，$r_\alpha(z)$ 为近轴方程的特解，它满足初始条件：

$$r_\alpha(0) = 0, \quad r'_\alpha(0) = \frac{1}{\varepsilon_z^{1/2}} \tag{4}$$

于是按照 Picht[5]，令

$$r_\alpha(z) = R_\alpha(z)[\phi_*(z)]^{-1/4} \tag{5}$$

将上式对 z 求导，并将它与式（2）代入式（3），于是放大率 M 便可表示为

$$M = \frac{[\phi_*(z_c)]^{-1/4}}{R'_\alpha(z_i)} \exp\left[-\int_{z_c}^{z_i} \sqrt{-\frac{1}{3}\frac{R''_\alpha(z)}{R_\alpha(z)}} dz\right] \tag{6}$$

式中，z_i 为成像位置，$R_\alpha(z_i) = 0$。

从原理上讲，逆设计的问题归结于在给定的 M 和 $l = z_i$ 下求解满足式（6）的简正轨迹的特解 R_α，并由式（2）确定系统的沿轴电位分布。因此，我们将由式（6）、式（2）出发进一步探讨静电成像系统的逆设计。

2 逆设计求解轨迹的限制条件

从表面上看，似乎只要找到 $R_\alpha(z)$，它满足式（6）的要求，便可实现系统的电子光学逆设计。实际上，这样获得的系统是很难实现的，且不能满足静电成像系统的要求。因此，我们需分析轴对称静电成像系统典型轴上电位分布及运动电子轨迹的特征，进而探讨逆设计求解对于轨迹 $R_\alpha(z)$ 的限制。

众所周知，典型静电成像系统轴上电位分布通常可以分为以下三个区域：

(1) 近阴极区 $[0, z_c]$，在该区域内，无论阴极面是球面或是平面，电位分布与电子轨迹都具有解析解。

(2) 聚焦 – 发散区 $[z_c, z_h]$，其间出现拐点 z_g，$\phi''(z_g) = 0$，在 z_g 点的左端，$\phi''(z) > 0$，在 z_g 的右端，$\phi''(z) < 0$。

(3) 等位区 $[z_h, z_i]$，$\phi'(z) = 0$，电子以不变的斜率做直线运动。因此，对于静电成像系统的逆设计，要尽可能使由电子轨迹 $R_\alpha(z)$ 导得的电位分布符合上述典型分布的特征。

在进行逆设计时，为了研究方便起见，取 $\varepsilon_z = 0$，即所研究的仅是极限轨迹的情况，这样的假定并不影响逆设计问题的普遍性。

在阴极区 $[0, z_c]$ 内，轴上分布 $\phi(z)$ 及其轨迹 $r_\alpha(z)$ 按文献 [4] 可近似表示如下：

$$\phi(z) = Ez\left(1 + \frac{z}{R_c}\right) \tag{7}$$

$$r_\alpha(z) = 2z^{1/2}E^{-1/2}\left(1 - \frac{z}{2R_c}\right) \tag{8}$$

式中，E 取正值；曲率半径为 R_c 的球面阴极凹向阳极，取正值。

于是，在 $[z_c, z_h]$ 区域内的轴上电位分布 $\phi(z)$ 按式（2）可表示为

$$\phi(z) = E\eta\left(1 + \frac{\eta}{R_c}\right)\exp\left(\int_{z_c}^{z_h}\sqrt{-\frac{16}{3}\frac{R''_\alpha}{R_\alpha}}\,dz\right) \tag{9}$$

由式（6）决定的系统放大率 M 亦可表示为

$$M = \frac{1}{R'_\alpha(z_h)}\left[E\eta\left(1 + \frac{\eta}{R_c}\right)\right]^{-1/4}\exp\left(\int_{z_c}^{z_h}\sqrt{-\frac{1}{3}\frac{R''_\alpha}{R_\alpha}}\,dz\right) \tag{10}$$

式中，$z_c = \eta$，积分限从 z_c 积到 z_h，是因为考虑到在 $[z_h, z_i]$ 区间内 $R'_\alpha(z_i) = R'_\alpha(z_h)$，$R''_\alpha(z) = 0$ 的原因。式（9）、式（10）乃是求解静电成像系统电子光学逆设计的出发点。

根据上述的电位分布与电子轨迹特征，我们进一步考虑进行逆设计时对 $R_\alpha(z)$ 在区间 $z_c \leq z \leq z_h$ 内所应满足的条件。

(1) 在 $z = z_c$ 处，$R_\alpha(z)$ 和 $R'_\alpha(z)$ 应是连续的。由式（7）和式（8）可得

$$R_\alpha(z) = 2E^{-1/4}\eta^{3/4}L_1,$$
$$R'_\alpha(z) = \frac{3}{2}E^{-1/4}\eta^{-1/4}L_2 \tag{11}$$

式中，
$$L_1 = 1 - \frac{\eta}{4R_c}, \quad L_2 = 1 - \frac{7\eta}{12R_c}$$

(2) 在区间 $[z_c, z_h]$ 内，$R_\alpha(z) > 0$。

(3) $R_\alpha(z)$ 应能保证物像距离的要求，即在 $z = z_h$ 处，轨迹的高度与斜率应使 $R_\alpha(z_i) = 0$。

(4) $R_\alpha(z)$ 的假定应满足在 $[z_c, z_h]$ 内，使 $R''_\alpha(z) \leq 0$。

(5) $R_\alpha(z)$ 的假定应使 $\phi'(z_h) = \phi''(z_h) = 0$，$R''_\alpha(z_h) = 0$。

(6) $R_\alpha(z)$ 的假定应使所求得的电位值在 $z = z_c$ 处连续，即在 $z = z_c$ 处，电位分布将由式 (7) 所确定。

(7) $R_\alpha(z)$ 的给定应使 $\phi(z)$ 在 $[z_c, z_h]$ 区间内有一个拐点 z_g，即 $\phi''(z_g) = 0$。

(8) $R_\alpha(z)$ 的假定应满足放大率 M 的要求。

如果所假定的 $R_\alpha(z)$ 能满足上述八点要求，则由式 (9) 便能确定满足要求的轴上电位分布 $\phi(z)$。

3 多项式模拟电子轨迹

选择能满足上述条件函数 $R_\alpha(z)$ 可以有许多种表达式，但需要遵循两条原则：

一是函数要尽可能地简单且能足以表达现实上存在的轨迹；二是能比较容易地满足上述所给出的限制条件。

下面我们用多项式模拟特解 $R_\alpha(z)$。假定 $R''_\alpha(z)$ 的形式可表示为

$$R''_\alpha(z) = 2NE^{-1/4}l^{-5/4}\left(\frac{z}{l} - \frac{h}{l}\right)^n Q(z) \quad (\eta \leq z \leq h) \tag{12}$$

式中，l 为所给定的系统长度，$l = z_i$；h 为从阴极面到进入等位区前的距离，$h = z_h$；N 为待定系数，它将保证 $R''_\alpha \leq 0$；幂次 n 取整数。

$Q(z)$ 假定是一个多项式，并且限制 $Q(z) > 0$，$\eta \leq z \leq h$，它可表示为

$$Q(z) = \left(\frac{z}{l} - a_1\right)\left(\frac{z}{l} - a_2\right)\cdots\left(\frac{z}{l} - a_m\right)\left[\left(\frac{z}{l}\right)^2 + p\left(\frac{z}{l}\right) + q\right] \tag{13}$$

现进一步分析 R_α 遵循以上限制条件下各待定系数应满足的关系式。

由式 (10) 可见，当 $z = h$ 时，$R''_\alpha(z) = 0$，条件 (5) 的要求部分满足。

对式 (10) 积分两次，使 $R_\alpha(z)$ 和 $R'_\alpha(z)$ 在 $z = z_c$ 处满足条件 (1)；同时我们限制 N，使其在 $\left(\frac{\eta}{l}, \frac{h}{l}\right)$ 区间内，$R''_\alpha(z) < 0$，这说明特解 $R_\alpha(z)$ 是向上凸的曲线。因为在区间的起点，$R_\alpha(z_c) > 0$，$R'_\alpha(z_c) > 0$，故若在 $z = h$ 处 $R_\alpha(z_h) > 0$，则条件 (2) 和 (4) 分别都被满足。

由上述条件 (2) 的限制，$R_\alpha(z) > 0$，故 a_1、a_2、$\cdots$、a_m 的值应位于区间 $\left[\frac{\eta}{l}, \frac{h}{l}\right]$ 之外，且参数 p、q 应使 $\left(\frac{z}{l}\right)^2 + p\left(\frac{z}{l}\right) + q$ 之根为复数，则其判别式在区间 $\left[\frac{\eta}{l}, \frac{h}{l}\right]$ 内应该小

于零，即

$$p^2 - 4q < 0 \tag{14}$$

现确定式（12）中幂次 n 的下限，由式（1）可得

$$\begin{cases} \dfrac{\phi'(z)}{\phi(z)} = \left(-\dfrac{16}{3}\dfrac{R''_\alpha}{R_\alpha}\right)^{1/2} \\ \dfrac{\phi''(z)}{\phi(z)} = -\dfrac{2}{\sqrt{3}}\left[\dfrac{R'''_\alpha}{R_\alpha\left(-\dfrac{R''_\alpha}{R_\alpha}\right)^{1/2}} + \dfrac{R'_\alpha}{R_\alpha}\left(-\dfrac{R''_\alpha}{R_\alpha}\right)^{1/2}\right] + \dfrac{16}{3}\left(-\dfrac{R''_\alpha}{R_\alpha}\right) \end{cases} \tag{15}$$

显然，若要满足条件（5），只须将 R_α、R'_α、R''_α 及 R'''_α 代入上式，不难看出，n 幂次必须满足：

$$n \geqslant 3$$

在等位区 $[z_h, z_i]$ 内，电子轨迹做直线运动，条件（3）要求

$$-\dfrac{R_\alpha(h)}{R'_\alpha(h)} = l - h \tag{16}$$

由电位连续条件（6），可求得在 $z = z_c$ 处，下述关系式成立：

$$-\dfrac{16 R''_\alpha(\eta)}{3 R_\alpha(\eta)} = \left[\dfrac{\phi'(\eta)}{\phi(\eta)}\right]^2 = \left(\dfrac{L_4}{L_3}\right)^2 \dfrac{1}{\eta^2} \tag{17}$$

式中：

$$L_3 = 1 + \dfrac{\eta}{R_c}, \qquad L_4 = 1 + \dfrac{2\eta}{R_c}$$

为了实现条件（7），所假设的特解 R_α 在某点 $z = z_g (\eta \leqslant z_g < h)$ 应满足下列关系式：

$$\left[-\dfrac{16}{3}\dfrac{R''_\alpha(z_g)}{R_\alpha(z_g)}\right]^{3/2} = \dfrac{8[R'''_\alpha(z_g) R_\alpha(z_g) - R'_\alpha(z_g) R''_\alpha(z_g)]}{3 R_\alpha^2(z_g)} \tag{18}$$

它由拐点条件 $\phi''(z_g) = 0$ 求得。若令：

$$x = \left[-\dfrac{16}{3}\dfrac{R''_\alpha(z)}{R_\alpha(z)}\right]^{3/2}, \qquad y = \dfrac{8}{3}\left[\dfrac{R'''_\alpha(z)}{R_\alpha(z)} - \dfrac{R''_\alpha(z) R'_\alpha(z)}{R_\alpha^2(z)}\right]$$

则式（18）可写为

$$x\big|_{z_g} = y\big|_{z_g}$$

若将 R_α、R'_α、R''_α 和 R'''_α 的表达式代入上式，便可发现：当 $z \to z_h$ 时，x 以 $(z - z_h)^{9/2}$ 趋于零，而 y 以 $(z - z_h)^3$ 趋于零，显然有

$$x\big|_{z \to z_h} < y\big|_{z \to z_h}$$

而当 $z = z_c$ 时，可有

$$x\big|_{z \to z_c} < y\big|_{z \to z_c} \tag{19}$$

由上面两个不等式可见，x 与 y 随着坐标 z 变化朝相反的趋势变化，则在 $[z_c, z_h]$ 区间内，由 x、y 所确定的两条曲线必然相交，该交点便是所求的拐点 z_g，于是式（19）即为电位分布具有拐点的条件。

到此，我们已假设了简正轨迹特解的二阶导数的表达式并由限制条件给出了该表达式中待定系数的一些限制与变化范围，由此，可通过确定上述系数进行系统的逆设计。

4 逆设计计算电聚焦成像系统的实例

下面我们将通过几个具体实例来进一步说明逆设计的方法。

4.1 实例1

令 $n=3$，有

$$Q(z) = \left(\frac{z}{l}\right)^2 + p\left(\frac{z}{l}\right) + q \tag{20}$$

则 $R''_\alpha(z)$ 有如下形式：

$$R''_\alpha(z) = 2N^{-1/4} l^{-5/4} \left(\frac{z}{l} - \frac{h}{l}\right)^3 \left[\left(\frac{z}{l}\right)^2 + p\left(\frac{z}{l}\right) + q\right] \tag{21}$$

对上式进行积分，并使 R_α、R'_α 在 $z=z_c$ 处满足式 (11)，便得到 $[z_c, z_h]$ 区间内简正轨迹的特解及其一阶导数表达式：

$$R'_\alpha(z) = 2NE^{-1/4} l^{-1/4} T(z) \tag{22}$$

$$R_\alpha(z) = 2NE^{-1/4} l^{3/4} P(z) \tag{23}$$

式中，

$$\begin{aligned} T(z) &= S_7 + pS_6 + qS_5 + \frac{3l^{1/4}L_2}{4N\eta^{1/4}}, \\ P(z) &= W_7 + pW_6 + qW_5 + \frac{3l^{1/4}L_2}{4N\eta^{1/4}}\left(\frac{z}{l} - \frac{\eta}{l}\right) + \frac{\eta^{3/4}L_1}{Nl^{3/4}} \end{aligned} \tag{24}$$

式中，

$$W_j = \frac{\left[\left(\frac{z}{l}\right)^j - \left(\frac{\eta}{l}\right)^j\right]}{j(j-1)} - \frac{3\left(\frac{h}{l}\right)\left[\left(\frac{z}{l}\right)^{j-1} - \left(\frac{\eta}{l}\right)^{j-1}\right]}{(j-1)(j-2)} +$$

$$\frac{3\left(\frac{h}{l}\right)^2\left[\left(\frac{z}{l}\right)^{j-2} - \left(\frac{\eta}{l}\right)^{j-2}\right]}{(j-2)(j-3)} - \frac{\left(\frac{h}{l}\right)^3\left[\left(\frac{z}{l}\right)^{j-3} - \left(\frac{\eta}{l}\right)^{j-3}\right]}{(j-3)(j-4)} -$$

$$\left(\frac{z}{l} - \frac{\eta}{l}\right)\left[\frac{\left(\frac{\eta}{l}\right)^{j-1}}{j-1} - \frac{3\left(\frac{h}{l}\right)\left(\frac{\eta}{l}\right)^{j-2}}{j-2} + \frac{3\left(\frac{h}{l}\right)^2\left(\frac{\eta}{l}\right)^{j-3}}{j-3} - \frac{\left(\frac{h}{l}\right)^3\left(\frac{\eta}{l}\right)^{j-4}}{j-4}\right] \tag{25a}$$

$$S_j = W'_j = \frac{\left[\left(\frac{z}{l}\right)^{j-1} - \left(\frac{\eta}{l}\right)^{j-1}\right]}{j-3} - \frac{3\left(\frac{h}{l}\right)\left[\left(\frac{z}{l}\right)^{j-2} - \left(\frac{\eta}{l}\right)^{j-2}\right]}{j-2} +$$

$$\frac{3\left(\frac{h}{l}\right)^2\left[\left(\frac{z}{l}\right)^{j-3} - \left(\frac{\eta}{l}\right)^{j-3}\right]}{j-3} - \frac{\left(\frac{h}{l}\right)^3\left[\left(\frac{z}{l}\right)^{j-4} - \left(\frac{\eta}{l}\right)^{j-4}\right]}{j-4}$$

$$(j=5,6,7) \tag{25b}$$

由式 (17)、式 (12) 并利用式 (10)，可以得到

$$\frac{1}{N} = \frac{-16\left(\frac{\eta}{l}\right)^{5/4} L_3^2}{3L_1 L_4^2}\left(\frac{\eta}{l} - \frac{h}{l}\right)^3\left[\left(\frac{\eta}{l}\right)^2 + p\left(\frac{\eta}{l}\right) + q\right] \tag{26}$$

由条件式（16）可得：

$$\left[\left(\frac{h}{l}\right)S_7(h)-W_7(h)-S_7(h)\right]+p\left[\left(\frac{h}{l}\right)S_6(h)-W_6(h)-S_6(h)\right]+$$

$$q\left[\left(\frac{h}{l}\right)S_5(h)-W_5(h)-S_5(h)\right]+\frac{3l^{1/4}L_2}{4N\eta^{1/4}}\left(\frac{\eta}{l}-1\right)-\frac{\eta^{3/4}L_1}{Nl^{3/4}}=0 \quad (27)$$

不难证明，当 $z=h$ 时，

$$\left(\frac{h}{l}\right)S_j(h)-W_j(h)=S_{j+1}(h) \quad (28)$$

于是式（27）可写为

$$[S_8(h)-S_7(h)]+p[S_7(h)-S_6(h)]+q[S_6(h)-S_5(S)]-\frac{\eta^{3/4}}{Nl^{3/4}}\left(L_1-\frac{3}{4}L_2\right)-\frac{3l^{1/4}L_2}{4N\eta^{1/4}}=0 \quad (29)$$

将式（26）代入式（29），便可解出 q 与 p 之间的关系式：

$$q=-p\frac{k_7}{k_6}-\frac{k_8}{k_6} \quad (30)$$

式中，

$$k_j=S_j(h)-S_{j-1}(h)+\frac{16L_3^2}{3L_1L_4^2}\left(\frac{\eta}{l}\right)^{j-5}\left(\frac{\eta}{l}-\frac{h}{l}\right)^3\left[\frac{\eta}{l}\left(L_1-\frac{3}{4}L_2\right)+\frac{3}{4}L_2\right] \quad (31)$$

式（30）的 p、q 值还应满足条件式（14），于是有

$$\frac{1}{4}p^2+\frac{k_7}{k_6}p+\frac{k_8}{k_6}<0 \quad (32)$$

由此可解出 p 的取值范围为：

$$p_2>p>p_1 \quad (33)$$

此外，

$$p_2=-2\frac{k_7}{k_6}+2\sqrt{\left(\frac{k_7}{k_6}\right)^2-\frac{k_8}{k_6}},$$

$$p_1=-2\frac{k_7}{k_6}-2\sqrt{\left(\frac{k_7}{k_6}\right)^2-\frac{k_8}{k_6}} \quad (34)$$

而且，在 p、q 取实数值的情况下，应使所选择的与电位分布有关的参数 η、h 满足不等式：

$$k_7^2>k_8k_6 \quad (35)$$

由拐点条件式（19），还应满足以下不等式：

$$\frac{L_4}{\eta L_3}>-\frac{1}{2l}\left[\frac{3}{\left(\frac{\eta}{l}-\frac{h}{l}\right)}+\frac{2\left(\frac{\eta}{l}\right)+p}{\left(\frac{\eta}{l}\right)^2+p\left(\frac{\eta}{l}\right)+q}\right]+\frac{3L_2}{8\eta L_1} \quad (36)$$

按照假定 $Q(z)=\left(\frac{z}{l}\right)^2+p\left(\frac{z}{l}\right)+q>0$，故由式（36）还可解出 p 值必须满足的一个下限不等式

$$p > p_3 \tag{37}$$

式中，

$$p_3 = \left\{ -\frac{2\eta}{l} - \left[\left(\frac{\eta}{l}\right)^2 - \frac{k_8}{k_6} \right] \left(\frac{2lL_4}{\eta L_3} + \frac{3l}{\eta - h} - \frac{3lL_2}{4\eta L_1} \right) \right\} \times \left[1 + \left(\frac{\eta}{l} - \frac{k_7}{k_4} \right) \left(\frac{2lL_4}{\eta L_3} + \frac{3l}{\eta - h} - \frac{3lL_2}{4\eta L_1} \right) \right]^{-1} \tag{38}$$

由上面所给出的 p、q、N 的表达式可见，当 E、η、h、R_c 的值给定后，系数 N、q 都取决于 p 值，而 p 的取值范围应由式（33）和式（37）两个不等式决定。

实际计算时，首先由给出的 M、l 选取 E、R_c、η、h 等值，用关系式（35）判断所选定的参数是否合理，然后由式（33）和式（37）确定 p 值的取值范围。在此范围内优选 p，使它能由式（10）求得的 M 恰好等于给定的放大率，最后由式（9）便可求得该静电成像系统的轴上电位分布。

按照上面所推导的公式与所叙述的步骤，我们在 DJS–8 机上编制了程序对该逆设计问题进行求解。程序只需输入系统的目标参量——物像距离 l 与放大率（取负值）以及系统的结构参数 R_c、E、η、h，便可求得简正轨迹 R_α 与轴上电位分布 ϕ 及其一、二阶导数。静电成像系统逆设计框图如图 1 所示。

图 2 乃是在给定的目标值 l 与 M 下用该程序计算求得的简正轨迹，即轴上电位分布及其导数的曲线。由图可见，它们与静电成像系统的典型分布是很接近的。

实际应用时，由于轨迹微分方程的齐次性，当 ϕ 变为 $K\phi$ 时，像面位置与放大率不变。因此可按实际需要对总加速电位 ϕ_{ac} 乘以一常数。自然，此时阴极面的场强亦将变化，需要重新复算，直到符合要求为止。

4.2 实例 2

令 $n = 3$，$m = 1$，

$$Q(z) = \left(\frac{z}{l} - a_1\right)\left[\left(\frac{z}{l}\right)^2 + p\left(\frac{z}{l}\right) + q\right] \tag{39}$$

则有

$$R''_\alpha(z) = 2NE^{-1/4}l^{-5/4}\left(\frac{z}{l} - \frac{h}{l}\right)^3 \left(\frac{z}{l} - a_1\right)\left[\left(\frac{z}{l}\right)^2 + p\left(\frac{z}{l}\right) + q\right] \tag{40}$$

与上例相同，$R'_\alpha(z)$、$R_\alpha(z)$ 仍可用式（22）、式（23）表示，但式中 $T(z)$、$P(z)$ 为：

$$T(z) = (S_8 - a_1 S_7) + p(S_7 - a_1 S_6) + q(S_6 - a_1 S_5) + \frac{3l^{1/4}L_2}{4N\eta^{1/4}} \tag{41a}$$

$$P(z) = (W_8 - a_1 W_7) + p(W_7 - a_1 W_6) + q(W_6 - a_1 W_5) + \frac{3l^{1/4}L_2}{4N\eta^{1/4}}\left(\frac{z}{l} - \frac{\eta}{l}\right) + \frac{\eta^{3/4}L_1}{Nl^{3/4}} \tag{41b}$$

式中，W_j、S_j 仍以式（25）表示。由条件式（17）可导得

$$\frac{1}{N} = -\frac{16\eta^{5/4}L_3^2}{3l^{5/4}L_1 L_4^2}\left(\frac{\eta}{l} - \frac{h}{l}\right)^3 \left(\frac{\eta}{l} - a_1\right)\left[\left(\frac{\eta}{l}\right)^2 + p\left(\frac{\eta}{l}\right) + q\right] \tag{42}$$

由上式可见，当 $a_1 > \frac{h}{l}$ 时，$\frac{1}{N} < 0$；当 $a_1 < \frac{h}{l}$ 时，$\frac{1}{N} > 0$。

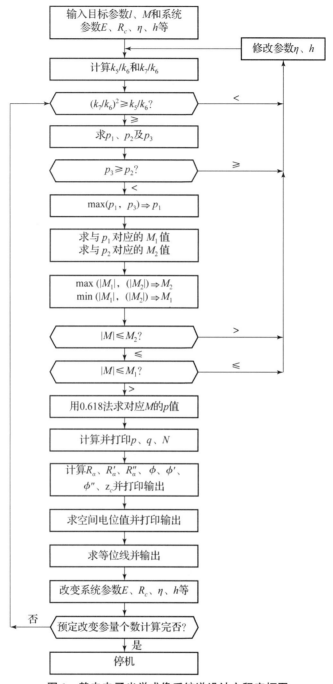

图 1 静电电子光学成像系统逆设计主程序框图

由条件式（16）与式（42），可以导得 a_1、q、p 之间的关系式：

$$q = -\frac{k_8 - a_1 k_7}{k_7 - a_1 k_6} p - \frac{k_9 - a_1 k_8}{k_7 - a_1 k_6} \tag{43}$$

式中的 k_j 表达式仍以式（31）表示。

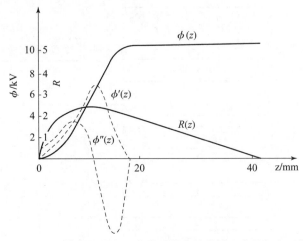

图 2 逆设计求得的静电电子光学成像系统的轴上电位与电子轨迹

($l = 42$ mm, $M = -1.2$, $E_c = 100$ V/mm, $R_c = \infty$, $\eta = 1.5$ mm, $h = 18$ mm, $\phi_{ac} = 10\,643$ V)

参数 p 在本例中应满足的不等式可表示为

$$p_2 > p > p_1 \tag{44}$$

式中，

$$p_2 = -2\frac{k_8 - a_1 k_7}{k_7 - a_1 k_6} + 2\sqrt{\left(\frac{k_8 - a_1 k_7}{k_7 - a_1 k_6}\right)^2 - \frac{k_9 - a_1 k_8}{k_7 - a_1 k_6}}$$

$$p_1 = -2\frac{k_8 - a_1 k_7}{k_7 - a_1 k_6} - 2\sqrt{\left(\frac{k_8 - a_1 k_7}{k_7 - a_1 k_6}\right)^2 - \frac{k_9 - a_1 k_8}{k_7 - a_1 k_6}} \tag{45}$$

为了使 p、q 取实数值，下列不等式应该成立：

$$a_1^2(k_7^2 - k_6 k_8) - a_1(k_7 k_8 - k_6 k_9) + (k_8^2 - k_7 k_9) \geqslant 0 \tag{46}$$

按要求，$R_\alpha(z)$ 在区间 $[\eta, h]$ 应大于零，因此，a_1 的取值由下面不等式决定。若

$$(k_9 k_6 - k_8 k_7)^2 - 4(k_7^2 - k_8 k_6)(k_8^2 - k_9 k_7) < 0 \tag{47}$$

则

$$a_1 < \frac{\eta}{l} \quad \text{或} \quad a_1 > \frac{\eta}{l} \tag{48}$$

若式 (47) 不成立，则 a_1 的取值除满足式 (48) 外，还应满足下列不等式：

$$a_1 < \frac{(k_7 k_8 - k_6 k_9) - \sqrt{(k_7 k_8 - k_6 k_9)^2 - 4(k_7^2 - k_6 k_8)(k_8^2 - k_7 k_9)}}{2(k_7^2 - k_6 k_8)} \tag{49a}$$

或

$$a_1 > \frac{(k_7 k_8 - k_6 k_9) - \sqrt{(k_7 k_8 - k_6 k_9)^2 - 4(k_7^2 - k_6 k_8)(k_8^2 - k_7 k_9)}}{2(k_7^2 - k_6 k_8)} \tag{49b}$$

顺便指出，在取值时，还必须使 $a_1 \neq k_7 k_6$。由拐点条件式 (19) 还可解得 p 的另一个下限值：

$$p > p_3 \tag{50}$$

式中，

$$p_3 = \left\{ -2\frac{\eta}{l} - \left[\left(\frac{\eta}{l}\right)^2 - \frac{(k_9 - a_1 k_8)}{(k_7 - a_1 k_6)}\right]\left(\frac{2lL_4}{\eta L_3} + \frac{3}{\frac{\eta}{l} - \frac{h}{l}} + \frac{1}{\frac{\eta}{l} - a_1} - \frac{3lL_2}{4\eta L_1}\right)\right\} \times$$

$$\left[1 + \left(\frac{\eta}{l} - \frac{k_8 - a_1 k_7}{k_7 - a_1 k_6}\right)\left(\frac{2lL_4}{\eta L_3} + \frac{3}{\frac{\eta}{l} - \frac{h}{l}} + \frac{1}{\frac{\eta}{l} - a_1} - \frac{3lL_2}{4\eta L_1}\right)\right]^{-1} \tag{51}$$

与上一例相比较，由于增加 $\left(\frac{z}{l} - a_1\right)$ 项，所得到的关系式亦较为复杂，除了可选择参量 p 外，还多了一个可选择的参量 a_1。根据线性代数的理论，这样求得的解将不是唯一的。这可应用两个变量的优化方法求得满足所要求的解。可在求解的过程中加入一些限制条件，如控制系统的某项像差（如场曲、畸变等）来选择最优解。

4.3 实例 3

将 m、n 取整数，$m \geq 1$，$n \geq 3$，

$$Q(z) = \left[\left(\frac{z}{l}\right)^2 + p\left(\frac{z}{l}\right) + q\right]\prod_{i=1}^{m}\left(\frac{z}{l} - a_i\right) \tag{52}$$

于是 $R''_\alpha(z)$ 可表示为

$$R''_\alpha(z) = 2NE^{-1/4}l^{-5/4}\left(\frac{z}{l} - \frac{h}{l}\right)^n\left[\left(\frac{z}{l}\right)^2 + p\left(\frac{z}{l}\right) + q\right]\prod_{i=1}^{m}\left(\frac{z}{l} - a_i\right) \tag{53}$$

显然，这是普遍情况的表达式。此时 $R'_\alpha(z)$、$R_\alpha(z)$ 同样可用式（22）、式（23）表示，但式中 $T(z)$、$P(z)$ 可表示成以下形式：

$$T(z) = A_{n+m+4} + pA_{n+m+3} + qA_{n+m+2} + \frac{3l^{1/4}L_2}{4N\eta^{1/4}} \tag{54}$$

$$P(z) = B_{n+m+4} + pB_{n+m+3} + qB_{n+m+2} + \frac{3l^{1/4}L_2}{4N\eta^{1/4}}\left(\frac{z}{l} - \frac{\eta}{l}\right) + \frac{\eta^{3/4}L_1}{Nl^{3/4}} \tag{55}$$

式中，

$$A_\lambda = S_\lambda + (-1)\sum_{i=1}^{m} a_i S_{\lambda-1} + (-1)^2 \sum_{j>i,i=1,2\cdots}^{m} a_i a_j S_{\lambda-2} + \cdots +$$

$$(-1)^{m-1}\sum_{f>\cdots>j>i,i=1,2\cdots}^{m}\overbrace{a_i,a_j\cdots a_f}^{(m-1)\uparrow}S_{\lambda-m+1} + (-1)^m a_1 a_2 \cdots a_m S_{\lambda-m} \tag{56a}$$

$$S_\lambda = \sum_{\mu=0}^{n}(-1)^\mu C_n^\mu \left(\frac{h}{l}\right)^\mu \frac{\left[\left(\frac{z}{l}\right)^{\lambda-\mu-1} - \left(\frac{h}{l}\right)^{\lambda-\mu-1}\right]}{\lambda - \mu - 1} \tag{56b}$$

$$B_\lambda = W_\lambda + (-1)\sum_{i=1}^{m} a_i W_{\lambda-1} + (-1)^2 \sum_{j>i,i=1,2\cdots}^{m} a_i a_j W_{\lambda-2} + \cdots +$$

$$(-1)^{m-1}\sum_{f>\cdots>j>i,i=1,2\cdots}^{m}\overbrace{a_i,a_j\cdots a_f}^{(m-1)\uparrow}W_{\lambda-m+1} + (-1)^m a_1 a_2 \cdots a_m W_{\lambda-m} \tag{57a}$$

$$W_\lambda = \sum_{\mu=0}^{n}(-1)^\mu C_n^\mu \left(\frac{h}{l}\right)^\mu \frac{\left[\left(\frac{z}{l}\right)^{\lambda-\mu} - \left(\frac{h}{l}\right)^{\lambda-\mu}\right]}{(\lambda-\mu)(\lambda-\mu-1)} - \left(\frac{z}{l} - \frac{\eta}{l}\right)\sum_{\mu=0}^{n}(-1)^\mu C_n^\mu \left(\frac{h}{l}\right)^\mu \frac{\left(\frac{\eta}{l}\right)^{\lambda-\mu-1}}{\lambda-\mu-1}$$

$$\tag{57b}$$

其中，$\lambda = n+m+2, n+m+3, n+m+4$。

C_n^μ 为 n 个元素中取 μ 个元素的组合。同样，我们还可以求得

$$\frac{1}{N} = -\frac{16L_3^2}{3L_1L_4^2}\left(\frac{\eta}{l}\right)^{5/4}\left(\frac{\eta}{l}-\frac{h}{l}\right)^n\left[\left(\frac{\eta}{l}\right)^2+p\left(\frac{\eta}{l}\right)+q\right]\prod_{i=1}^{m}\left(\frac{\eta}{l}-a_i\right) \tag{58}$$

$$q = -\frac{C_{n+m+4}}{C_{n+m+3}}p - \frac{C_{n+m+5}}{C_{n+m+3}} \tag{59}$$

式中，

$$C_{\lambda+1} = k_{\lambda+1}(h) + (-1)\sum_{i=1}^{m}a_i k_\lambda(h) + (-1)^2\sum_{j>i,i=1,2\cdots}^{m}a_i a_j k_{\lambda-1}(h) + \cdots +$$

$$(-1)^{m-1}\sum_{f>\cdots>j>i,i=1,2\cdots}^{m}\overbrace{a_i,a_j\cdots a_f}^{m-1\uparrow}k_{\lambda-m+2}(h) + (-1)^m a_1 a_2 \cdots a_m k_{\lambda-m+1}(h) \tag{60}$$

$$k_{\lambda+1}(h) = S_{\lambda+1}(h) - S_\lambda(h) + t_{\lambda-n-1} \tag{61}$$

$$t_{\lambda-n-1} = \frac{16L_3^2}{3L_1L_4^2}\left(\frac{\eta}{l}\right)^{\lambda-n-1}\left(\frac{\eta}{l}-\frac{h}{l}\right)^n\left[\frac{\eta}{l}\left(L_1-\frac{3}{4}L_2\right)+\frac{3}{4}L_2\right] \tag{62}$$

p 的取值范围同前，其中 p_1、p_2、p_3 为

$$p_1 = -2\frac{C_{n+m+4}}{C_{n+m+3}} - 2\sqrt{\left(\frac{C_{n+m+4}}{C_{n+m+3}}\right)^2 - \frac{C_{n+m+5}}{C_{n+m+3}}} \tag{63}$$

$$p_2 = -2\frac{C_{n+m+4}}{C_{n+m+3}} + 2\sqrt{\left(\frac{C_{n+m+4}}{C_{n+m+3}}\right)^2 - \frac{C_{n+m+5}}{C_{n+m+3}}} \tag{64}$$

$$p_3 = \left\{-2\frac{\eta}{l} - \left[\left(\frac{\eta}{l}\right)^2 - \frac{C_{n+m+5}}{C_{n+m+3}}\right]\left[2\frac{lL_4}{\eta L_3} + \frac{n}{\frac{\eta}{l}-\frac{h}{l}} + \sum_{i=1}^{m}\frac{1}{\frac{\eta}{l}-a_i} - \frac{3lL_2}{4\eta L_1}\right]\right\} \times$$

$$\left[1 + \left(\frac{\eta}{l} - \frac{C_{n+m+4}}{C_{n+m+3}}\right)\left(2\frac{lL_4}{\eta L_3} + \frac{n}{\frac{\eta}{l}-\frac{h}{l}} + \sum_{i=1}^{m}\frac{1}{\frac{\eta}{l}-a_i} - \frac{3lL_2}{4\eta L_1}\right)\right]^{-1} \tag{65}$$

同样 a_1 的取值除了满足

$$a_1 < \frac{\eta}{l} \quad \text{或} \quad a_1 > \frac{\eta}{l} \tag{66}$$

外，还应满足下列不等式

$$[C_{n+m+4}]^2 - [C_{n+m+3}][C_{n+m+5}] \geqslant 0 \tag{67}$$

由上可见，对于普遍情况，关系式更为复杂。但无论 n、m 取何值，变化是有规律的，只须使某些系数项做一些改变就可以了。自然，a_1、a_2、$\cdots$、a_m 等待定系数的引入将使轴上电位分布出现多种解，它们都能满足给定的物像距离和放大率的要求。解决这个问题的最好办法是引入一些像差控制项，并用最优化方法进行求解。

5 关于空间电位与电极系统的确定

在电动力学中已证明，在给定边界条件下，拉普拉斯（或泊松方程）的解是唯一的。同样文献中已经证明，对于轴对称系统，由已知轴上电位分布确定的空间电位分布也是唯

一的。

由此可见，当由聚焦特性条件确定了轴上电位分布 $\phi(z)$ 后，空间电位分布便完全为唯一性定理所限制。按此定理，只存在唯一的空间电位分布 $\phi(z,r)$，它取已给定的轴上电位值。因此，所求得的成像系统亦是唯一的，而且电位的求解与方法无关。

尽管从理论和实践上，很难找到一种电位分布，它与轴上预先给定的电位分布完全吻合。但是，这种限制并不妨碍实际的应用。首先，对于工程物理技术问题，需要求得的只是某一有限区域范围内的电位分布，而不一定是整个空间；其次，电位分布只要求近似值。物理量的确定，不可能绝对精确，只要能满足一定的精度即可，严格的重合对于工程技术问题并不重要。

按照 Weierstrass 定理：如果函数在实轴有限区间内是连续的，则在该区域内，它可以近似地用多项式表达到任意的精度。

由此定理，有文献中曾导出以下推论。

推论1：由该对称轴确定的区间内所求得的任何连续电位分布，可以用无数个任意阶次、任意精度的电位分布来近似表示。

推论2：这样的电位分布可以在无奇异点的平面上构成。

推论3：这样的电位分布可以由位于对称轴外任意数量点处预先计算求得的值所构成。

上述推论对于实现所需要的成像系统具有实际意义。由此所得到的主要结论是：在电子光学逆设计问题中，若要精确地实现预定的轴上电位分布，将有可能导致很不合理甚至不可能的电极结构形状。而轴上电位分布值的微小变化几乎将不显著影响电子轨迹，都有可能获得对于实际应用更合适的电极系统。这个结论正是按预先给定的轴上电位分界来确定电极结构的理论基础。

由轴上电位分布求空间电位分布的问题已在文献中详细叙述，可以有多种方法，如谢尔赤级数法与积分法，但比较简单适用的是网格法，它以阴极面中心为原点，在子午面上复盖步长为 h 的正方形网格。因此，网格点上某一点的电位值 φ_{ij} 与其相邻点的电位值 $\varphi_{i-1,j-1}$、$\varphi_{i-1,j}$、$\varphi_{i-1,j+1}$、$\varphi_{i-2,j}$ 的关系，可以通过将有限差分方程取代拉普拉斯方程得到。这样，我们可利用已知的轴上电位分布，先求第一行网格点上的电位值，以后各行的电位值，则可依次逐步求得。最后利用这些网格节点上的电位值 φ_{ij}，用扫描搜索法便能形成系统的等电位曲线，由此等位线便可考虑电极的配置。

自然，在求得的等位线中，我们只能选择有限的几个作为电极，并加上相应的工作电压。同时，还须考虑系统的边界电极的安置，构成一个封闭的成像系统。实际情况并非如此简单。首先，由 $\varphi(z,r)$ 所确定的等位面通常是复杂的曲面，需要以简单形状如平面、柱面或球面替代之。其次，取代所选择等位面的电极应当有一圆孔，以便使电子能通过孔阑而到达屏上。显然孔的存在将使电极系统所形成的场有别于计算得到的分布，从而有可能使实际电子轨迹与原来假设的有较大的差别。因此，在通常情况下，需要对给定的电极系统进行修正。

对于上述问题，可以有两种方法对电极系统进行修正。一种是以已知轴上电位分布作为目标函数，另一种是以系统的成像参量（放大率与像面位置）作为目标函数。鉴于这已属于优化设计的问题，就不一一叙述了。

结束语

(1) 本文提出了逆设计求解静电成像系统的新方法，即按已给定的系统的成像参量的要求，考虑实际成像系统轴上电位分布与电子轨迹的特点，给出静电成像系统对轨迹的限制条件，并且用多项式模拟电子轨迹。通过电子轨迹来计算轴上电位分布。系统的成像参量可以通过选择待定参数来确定。

(2) 对某一个实例编制了设计程序，并进行实例计算。计算结果证明，用本方法模拟求得的成像系统的轴上电位分布与典型系统是比较一致的。

逆设计用于成像系统的设计尚属于一种尝试。诚然，在反求电极结构参量上还存在不少困难，难度较大。从目前的情况来看，尚须正设计加以辅助。但对成像系统逆设计所进行的探讨，无疑将对电子光学系统设计具有意义。我们这里探讨的虽是静电成像系统的逆设计，但其思想和方法也适用于其他形式的电子透镜系统。

参 考 文 献

[1] Гринберг Г А. Избранные вопросы математическои теории электрических явлений[M]. Москва-Ленинград：Изд, А Н СССР,1948.

[2] КАСЬЯНКОВ П П. О расчете электронных линз по задым фокусирующим свойствам в параксиальной области. [J]. Ж. Т. Ф. , 1953,23(3):531-540.

[3] ЧЖОУ Л В. Прямой метод расчета электронно-оптической системы преобраэователей иэображеиия с электростатической Фокусировкой[J]. Доклад на ЭМАе ЛЭТИ,1965.

[4] 周立伟. 夜视器件的电子光学[M]. 北京:北京理工大学出版社,1977.

[5] PICHT J. The wave of a moving classical electron[J]. Progress in Optics, 1966, 5:351-370.

9.2 An Inverse Design of Magnetic Focusing Coil for Electrostatic and Magnetic Imaging
电磁聚焦成像系统中磁聚焦线圈的逆设计

Abstract: An inverse design of magnetic focusing coil for electrostatic and magnetic imaging is investigated in the present paper. For the inverse problem of designing magnetic focusing coil, a mathematical model has been developed, an optimization method called BFGS has been developed, an universal program using Fortran language has been drawn up in this article. On the basis of computational results, a magnetic focusing coil with four variable magnifications and a prototype tube of electrostatic and magnetic imaging without image rotation have been made. The results show that the computational and measuring values of axial magnetic induction coincide with the theoretical ones. It fully proves that the theory of electrostatic and magnetic imaging without image rotation and its method of inverse design described here are correct.

摘要：本文研究了电磁聚焦移像系统中磁聚焦线圈的逆设计。针对磁聚焦线圈系统设计的逆问题，作者建立了数学模型，应用 BFGS 优化方法，采用 Fortrain 语言编制了应用程序。在计算结果的基础上，制作了一个具有四个可变放大率的磁聚焦线圈和一个无图像旋转都得静电磁像管。计算结果和所要求的磁感应强度的轴上分布相当一致。文中还给出了实验测量和验证。

Introduction

Electrostatic and magnetic imaging systems have long been widely used in the photo-electronic image devices, such as image intensifiers, camera tubes and electronic photographic image tubes, in recent 30 years. As compared with pure electrostatic imaging systems, they have a series of advantages over them, for instance, large field of view, low distortion, uniformity of resolution in the whole field, high modulation transfer function, flat input and output windows and arbitrary choice of tube's length and diameter, etc. It is really a pity that such a sort of systems applied in practice have been confined to the parallel and uniform electrostatic and magnetic fields with magnification of unity, so that their outstanding performances have not been fully developed as yet in practical applications.

The theory and design of electrostatic and magnetic imaging without image rotation but with magnification $M > 1$ or $M < 1$ has been investigated in detail in previous papers[1-3], in which we have established two conditions for designing electrostatic and magnetic imaging systems without image rotation:

$$\frac{3(V+\varepsilon_{z_1}^*)}{G^2}\left(\frac{dG}{dz}\right)^2 - \frac{2(V+\varepsilon_{z_1}^*)}{G}\frac{d^2G}{dz^2} - \frac{1}{G}\frac{dV}{dz}\frac{dG}{dz} + \frac{d^2V}{dz^2} = 0 \qquad (1)$$

$$\lambda_n = \frac{e}{8m_0}\frac{B_0^2 l^2}{\phi_i} = \frac{n^2\pi^2}{\left[\int_0^1 \frac{G(z)}{\sqrt{V(z)+\varepsilon_{z_1}^*}}dz\right]^2} \qquad (2)$$

where e/m_0 is the ratio of electron charge to mass; e is the absolute magnitude of electron charge; m_0 is the electron mass; z is the axial coordinate measured in unit of distance l between cathode and screen; $\phi(z)$, axial electrostatic potential, at the cathode $\phi(0) = (0)$, at the image plane $\phi_i = \phi(z)$; $B(z)$ is the axial magnetic induction, at the cathode

$$B(0) = B_0, \ V = V(z) = \phi(z)/\phi_i, \ G = G(z) = B(z)/B_0, \ G_i = G(1), \ G_i = 1/M^2$$

where M is the radial magnification of the system; $e\varepsilon_{z_1}^*$ is a certain initial axial energy of the electron emitted from the cathode surface, all electrons having the same initial axial energy are ideally focused at the image plane $z = 1$; λ_n is the eigen value of n loops.

From above-mentioned two conditions (1) and (2), the design procedure for electrostatic and magnetic imaging systems without image rotation but with variable magnifications can be summed up as follows:

For given axial electrostatic potential $V(z)$, given magnetic induction $G(z)$, and some expected magnifications, one could use condition (1) to determine $G(z)$ from given $V(z)$ or $V(z)$ from given $G(z)$, then determine the value of n, evaluate the eigen value λ_n from condition (2), and choose suitable values of ϕ_i, B_0 and l, which must be in agreement with condition (2), and finally design electrodes and magnetic circuits in a way that they generate the fields $V(z)$ and $G(z)$ with the constants ϕ_i and B_0.

In fact, as we know, the above-mentioned method is an inverse design. It means that we conversely design electrode system and magnetic focusing coil from the axial field distributions $V(z)$ and $G(z)$, which satisfy two conditions (1) and (2) for obtaining an image without image rotation.

For the sake of convenience, we suppose that the axial electrostatic potential $V(z)$ is a linear distribution, then series of data of axial magnetic induction $G(z)$, which correspond to different magnifications, can be obtained. In view of designing and realizing an electrode system with linear distribution of axial electrostatic potential without any difficulties, then the inverse design of magnetic focusing coil from given data of axial magnetic induction becomes a problem of interest.

In the present paper, an inverse design of magnetic focusing coil with four variable magnifications will be discussed. For the problem of an inverse design of magnetic focusing coil, we construct a mathematical model suited for computation in a personal computer. An optimization method called BFGS for the inverse problem of designing magnetic focusing coil will be introduced. The results show that the computational and measuring values of axial magnetic induction coincide with the theoretical ones.

1 Mathematical model

Suppose we study a rotational symmetrical magnetic coil without iron medium, let its ampere-turn distribution along the axis z be $n(z)$, but along the radius r be uniform; the axial coordinate positions of coil for the left side and the right side be a and b, respectively; the inner and outer radii of coil be R_1 and R_2, respectively; as shown in Fig. 1, then the magnetic induction $\overline{B}(z)$ along the symmetrical axis z generated by the coil can be formulated as follows:

$$\overline{B}(z) = \frac{\mu_0}{2(R_2 - R_1)} \int_a^b \int_{R_1}^{R_2} \frac{r^2 n(x)}{[r^2 + (z-x)^2]^{3/2}} dr dx \quad (3)$$

where r, x are radial and axial coordinates, respectively, $r \in [R_1, R_2], x \in [a, b]$; μ_0 is the magnetic permeability.

Integrating Eq. (3) with respect to r, we obtain

$$\overline{B}(z) = \frac{\mu_0}{2(R_2 - R_1)} \int_a^b n(x) \left\{ \frac{R_1}{[R_1^2 + (z-x)^2]^{1/2}} - \frac{R_2}{[R_2^2 + (z-x)^2]^{1/2}} + \ln \frac{R_2 + [R_2^2 + (z-x)^2]^{1/2}}{R_1 + [R_1^2 + (z-x)^2]^{1/2}} \right\} dx \quad (4)$$

Suppose the axial magnetic induction distribution to be expected be $B(z)$, $z \in [a_0, b_0]$, where a_0, b_0 correspond to the axial coordinate positions of tube for the left side and the right side, we would like to get ampere-turn distribution $n(z)$ of the coil so well that the axial magnetic induction $\overline{B}(z)$ at the interval $[a_0, b_0]$ generated by the coil approaches $B(z)$ as far as possible. Using the definition of mean square deviation, we construct the following functional for solving a problem of the rotational symmetrical magnetic field:

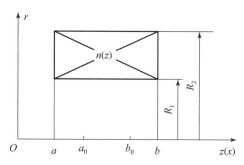

Fig. 1　A schematic diagram of magnetic focusing coil without iron medium

$$L = \frac{1}{b_0 - a_0} \int_{a_0}^{b_0} [B(z) - \overline{B}(z)]^2 dz$$

$$= \frac{1}{b_0 - a_0} \int_{a_0}^{b_0} \left\{ B(z) - \frac{\mu_0}{2(R_2 - R_1)} \int_a^b \left[\frac{R_1}{[R_1^2 + (z-x)^2]^{1/2}} - \frac{R_2}{[R_2^2 + (z-x)^2]^{1/2}} + \ln \frac{R_2 + [R_2^2 + (z-x)^2]^{1/2}}{R_1 + [R_1^2 + (z-x)^2]^{1/2}} \right] n(x) dx \right\}^2 dz \quad (5)$$

Now the inverse problem is changed into solving $n(z)$, which makes functional an extreme value.

Obviously, it is difficult for us to solve $n(z)$ analytically from Eq. (5). So we apply the Euler's finite difference method of the calculus of variations to solve $n(z)$. Let's take $m+1$ points in the interval $[a,b]$, which correspond to $z_0 = a$, z_1, z_2, $\cdots$, z_{m-1}, $z_m = b$, it means that the magnetic focusing coil is divided into m segments, suppose the current in each segment of coil be uniform, i.e., at i-th segment of coil, $[z_{i-1}, z_i]$, $n(z) = n_i = \text{const}(i=1,2,\cdots,m)$, then the current density J_i of i-th segment of coil can be expressed by

$$J_i = \frac{n_i}{R_2 - R_1} \quad (i=1,2,\cdots,m)$$

Evidently, the function in Eq. (5) to be integrated is so complicated that we have to expend a large amount of time in calculating the numerical integral. So we change the integral in Eq. (5) into a form of summation, then Eq. (5) will be discreted:

$$L = \frac{1}{K}\sum_{j=1}^{K}\left\{B(z_j) - \frac{\mu_0}{2}\sum_{i=1}^{m}J_i\left[(z_i - z_j)\ln\frac{R_2 + [R_2^2 + (z_i - z_j)^2]^{1/2}}{R_1 + [R_1^2 + (z_i - z_j)^2]^{1/2}} + \right.\right.$$
$$\left.\left.(z_j - z_{i-1})\ln\frac{R_2 + [R_2^2 + (z_j - z_{i-1})^2]^{1/2}}{R_1 + [R_1^2 + (z_j - z_{i-1})^2]^{1/2}}\right]\right\}^2 \quad (6)$$

where $z_j(j=1,2,\cdots,K)$ are axial positions of sampling points; $z_j \in [a_0, b_0]$; $B(z_j)$, the expected value of axial magnetic induction at these sampling points.

Considering that the requirements for the sampling points in Eq. (6) will be different (for example, in the electrostatic and magnetic imaging systems, owing to the low velocity of electron motion in the neighborhood of cathode surface, we hope that the error in this region for magnetic coil design is as small as possible, but in the other region far from the cathode, as at the middle, the error control may be relatively relaxed), we should weight each sampling point according to the error control in magnetic focusing coil design. Besides, we should take account of the thickness of the frame wall for the practical magnetic focusing coil and consider the fact that the inner and outer radii of every segment of coil may be different from each other. Then, Eq. (6) should take the following form:

$$L = \frac{1}{K}\sum_{j=1}^{K}A_j\left\{B(z_j) - \frac{\mu_0}{2}\sum_{i=1}^{m}J_i\left[(z_j - z_{i1})\ln\frac{R_{i2} + [R_{i2}^2 + (z_j - z_{i1})^2]^{1/2}}{R_{i1} + [R_{i1}^2 + (z_j - z_{i1})^2]^{1/2}} + \right.\right.$$
$$\left.\left.(z_{i2} - z_j)\ln\frac{R_{i2} + [R_{i2}^2 + (z_{i2} - z_j)^2]^{1/2}}{R_{i1} + [R_{i1}^2 + (z_{i2} - z_j)^2]^{1/2}}\right]\right\}^2 \quad (7)$$

where A_j is the weight factor, $\sum_{j=1}^{K}A_j = K(j=1,2,\cdots,K)$, $A_j > 0$, $z_{i1}, z_{i2}(i=1,2,\cdots,m)$ are the axial coordinates of the left side and right side for the i-th segment's coil, respectively; R_{i1}, R_{i2} are the inner and outer radii of the i-th segment's coil, respectively.

If z_{i1}, z_{i2}, R_{i1} and R_{i2} in Eq. (7) remain constant, i.e., the structure and position of each segment's coil are fixed, then L will be the function of current densities J_i ($i=1,2,\cdots,m$) of the magnetic focusing coil. From above-mentioned treatment, the problem for solving functional becomes an extreme value problem for solving multivariate function. In other words, we should explore a set of values for J_i so that the value of objective function in Eq. (7) tends toward a minimum.

In view of the fact that Eq. (7) is a nonlinear function, the inverse problem for designing magnetic focusing coil of electrostatic and magnetic imaging system may be included in the nonlinear programming:

$$\min_{J \in R^m} L(J) \tag{8}$$

and

$$|J_i| < J_{\max} \tag{9}$$

where $J = [J_1, J_2, \cdots, J_m]^T$ describes the independent variable vector in the m-dimensional linear space R^m; the superscript "T" indicates the transposed matrix. In order to beaccord with the demands, the constraint condition (9) of current density prescribes a limit to the current which flows through the coil.

The gradient vector of objective function L is

$$\nabla L = \left[\frac{\partial L}{\partial J_1}, \frac{\partial L}{\partial J_2}, \cdots, \frac{\partial L}{\partial J_m} \right]^T \tag{10}$$

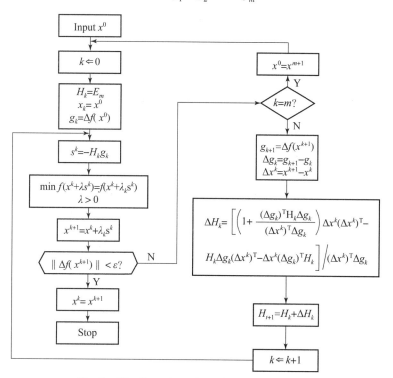

Fig. 2　The flow chart for using BFGS method

where $\dfrac{\partial L}{\partial J_1}$ can be obtained from Eq. (7):

$$\frac{\partial L}{\partial J_1} = -\frac{2}{K} \sum_{j=1}^{K} A_j \left\{ B(z_j) - \frac{\mu_0}{2} \sum_{i=1}^{m} J_i \left[(z_j - z_{i1}) \ln \frac{R_{i2} + [R_{i2}^2 + (z_j - z_{i1})^2]^{1/2}}{R_{i1} + [R_{i1}^2 + (z_j - z_{i1})^2]^{1/2}} + \right. \right.$$

$$(z_{i2} - z_j)\ln\frac{R_{i2} + [R_{i2}^2 + (z_{i2} - z_j)^2]^{1/2}}{R_{i1} + [R_{i1}^2 + (z_{i2} - z_j)^2]^{1/2}}\Big\} \times$$

$$\frac{\mu_0}{2}\Big\{(z_j - z_{l1})\ln\frac{R_{l2} + [R_{l2}^2 + (z_j - z_{l1})^2]^{1/2}}{R_{l1} + [R_{l1}^2 + (z_j - z_{l1})^2]^{1/2}} +$$

$$(z_{l2} - z_j)\ln\frac{R_{l2} + [R_{l2}^2 + (z_{l2} - z_j)^2]^{1/2}}{R_{l1} + [R_{l1}^2 + (z_{l2} - z_j)^2]^{1/2}}\Big\}, \quad (l = 1,2,\cdots,m) \tag{11}$$

2 An optimization method for inverse design of magnetic focusing coil

From the optimization methods, we choose one of the methods of scale change, which is called BFGS (Broyden, Fletcher, Goldfarb, Shanno) method and apply it to optimize the objective function (7). Now let's give a brief introduction for BFGS method [4].

Actually, the BFGS method is an optimization method, which directly uses the derivatives of function to search optimum.

For the problem of nonlinear programming

$$\min_{x \in R^m} f(\boldsymbol{x})$$

its iterative formula may be written as

$$x^{k+1} = x^k - \lambda_k H_k \nabla f(\boldsymbol{x}^k) \tag{12}$$

where $\boldsymbol{x} = [x_1, x_2, \cdots, x_m]^T$ is a vector in the linear space R^m:

$$\nabla f(\boldsymbol{x}^k) = \left[\frac{\partial f}{\partial x_1}, \frac{\partial f}{\partial x_2}, \cdots, \frac{\partial f}{\partial x_m}\right]^T$$

the gradient vector of function $f(\boldsymbol{x})$ at $\boldsymbol{x}^k$; λ_k the optimum step length; H_k is called the iterative matrix, which is a symmetrical positively definite matrix, its iterative formula can be written as

$$H_{k+1} = H_k + \Delta H_k \tag{13}$$

where ΔH_k is called the corrective matrix, it only depends on the present iterative values x^{k+1}, x^k and their corresponding gradient vectors $\nabla f(x^{k+1})$ and $\nabla f(x^k)$. ΔH_k can be expressed by

$$\Delta H_k = \left[\left(1 + \frac{(\Delta g_k)^T H_k \Delta g_k}{(\Delta x^k)^T \Delta g_k}\right)\Delta x^k (\Delta x^k)^T - H_k \Delta g_k (\Delta x^k)^T - \Delta x^k (\Delta g_k)^T H_k\right]/(\Delta x^k)^T \Delta g_k \tag{14}$$

where $g_k = \nabla f(\boldsymbol{x}^k)$, $\Delta g_k = g_{k+1} - g_k$, $\Delta x^k = x^{k+1} - x^k$. Expression (12) – Expression (14) are called BFGS formulae.

The iteration process of BFGS method can be summed up as follows:

(1) Give an initial point $x^0 \in R^m$ and the permissible error $\varepsilon > 0$, let $H_0 = E_m$ (unit matrix of m order), $g_0 = \nabla f(\boldsymbol{x}^0)$, $k = 0$;

(2) Let $s^k = -H_k g_k$;

(3) Take one dimensional searching along s^k direction from x^k, find λ_k:

$$\min_{\lambda > 0} f(x^k + \lambda s^k) = f(x^k + \lambda_k s^k)$$

(4) Let $x^{k+1} = x^k + \lambda_k s^k$;

(5) Check the convergence criteria:

$$\| \nabla f(x^{k+1}) \| < \varepsilon$$

if it is satisfied, then the iteration comes to an end, otherwise, goes to (6).

(6) Check $k = m$?, let $x^0 = x^{m+1}$, return to (1); otherwise, when $k < m$, calculate the iterative matrix H_{k+1} using formulae (13) and (14), then let $k \Leftarrow k+1$, return to (2).

During the iteration process, whenever $k = m$, we get x^{m+1} and substitute x^{m+1} for x^0. It means that we take the fastest declining direction in order to decrease harmful effects given by computational errors. The flow chart of above-mentioned iteration process is shown in Fig. 2.

When we design the magnetic focusing coil using BFGS method, Eq. (7) is taken as the objective function, its gradient vector is given by formulae (10) and (11). Thus, a universal program using Fortran language has been drawn up for the inverse design of magnetic focusing coil. The principal advantage of the program is not only convenient but also practical.

3 A practical example

We have designed a magnetic focusing coil for the electrostatic and magnetic imaging system without image rotation but with variable magnifications by using our universal programme. The schematic diagram of the magnetic focusing coil is shown in Fig. 3.

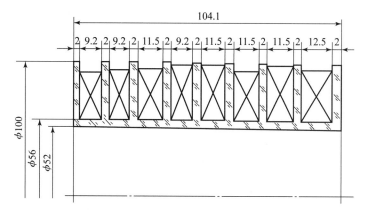

Fig. 3 A practical magnetic focusing coil without iron medium

Table 1 shows the computational data of design for the magnetic focusing coil, which can realize an electron optical system of zoom type corresponding to four magnifications such as 0.8, 0.9, 1.0 and 1.2.

The results show that the computational values of axial magnetic induction generated by magnetic focusing coil closely approach the expected theoretical ones. For the above-mentioned design of four variable magnifications, the mean relative error is between 0.23% – 0.96%, if fully satisfies the requirements of engineering design.

After having finished winding of the coil, we measured its magnetic induction along the axis. Result shows that the experimental data coincide with the computational ones within experimental error. Fig. 4 shows the theoretical, computational and experimental measured values of magnetic induction along axis z, in which the magnifications correspond to 0.8, 0.9, 1.0 and 1.2, respectively.

Table 1 Computational data of coil design

Coil No.	Turns	Ampere-turns			
		$M=0.8$	$M=0.9$	$M=1.0$	$M=1.2$
1	144	15.984	14.636	152.950	228.000
2	144	−31.185	−7.226	141.351	228.223
3	180	29.691	161.204	141.958	279.709
4	144	299.855	251.270	20.750	−119.804
5	180	203.543	55.834	123.309	80.196
6	180	−8.380	144.983	136.269	154.651
7	180	319.002	127.891	16.584	23.408
8	198	475.751	409.234	378.364	168.170

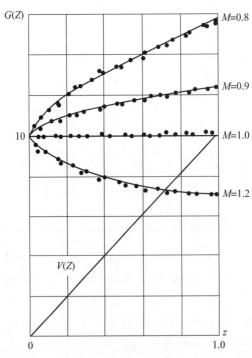

Fig. 4 Axial magnetic induction distributions for a practical system with four variable magnifications
(——theoretical curves, ······ computational data, ∗∗∗∗∗∗ measuring values)

Fig. 5 shows an experimental prototype of electron optical system, which is designed according to the above-mentioned method.

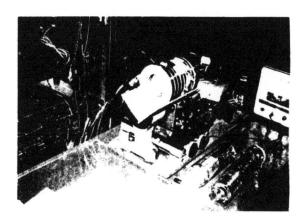

Fig. 5 An experimental prototype tube of electrostatic and magnetic imaging with four variable magnifications

It should be pointed out that when using our universal program the requirements for the structure parameters of the coil can be relaxed. As the errors appear in the structure of coil frame or in making the coil winding, one could change the input parameters to the computer, thus obtaining the corresponding current values of each segment's coil to correct the error effect.

Conclusions

In present paper, we have discussed an inverse design of magnetic focusing coil for the electrostatic and magnetic imaging systems. A mathematic model suited for computation has been constructed, a magnetic focusing coil for electron optical system with variable magnifications has been designed using BFGS method and an experimental prototype tube has been made. The result fully proves the theory of electrostatic and magnetic imaging without image rotation and its method of inverse design described here are correct.

The great advantage of the inverse design given by this paper lies in the fact that for a given magnetic focusing coil package, in which the structure of coil frame has been determined, one could change the current values of each segment coil to get different magnetic induction as required. It not only gives a practical and useful method for designing magnetic focusing coil of electrostatic and magnetic imaging systems without image rotation but with variable magnifications, but also provides a new approach for designing rotational symmetrical magnetic coil without iron medium. The results of present paper may be of practical importance for electron optical system design.

References

[1] ZHOU L W. On the theory of combined electrostatic and magnetic focusing imaging systems[J]. Journal of Beijing Institute of Technology, 1983(3): 12 - 24.

[2] ZHOU L W, NI G Q, FANG E L. A Study of electrostatic and magnetic imaging

transference systems without image rotation[J]. Acta Electronica, 1984, 12, (3), 33–40 (in Chinese).

[3] ZHOU L W, NI G Q, FANG E L. Electrostatic and magnetic imaging without image Rotation[J]. Electron Optical Systems, SEM. Inc., 1984: 37–43.

[4] FIELDING K. Algorithm 387 function minimization and linear search[J]. Communication of the ACM, 1970, 13(8), 509–510.

9.3 An Inverse Design of Electrostatic Focusing Field for Electrostatic and Magnetic Imaging
电磁聚焦成像中静电聚焦场的逆设计

Abstract: An inverse design of electrostatic focusing field for electrostatic and magnetic imaging is investigated. Using the potential superimposition theorem of electrostatic field in multi-electrode system, a mathematical model has been developed and an optimization method has been introduced into computation for designing the electrostatic focusing field of the imaging system.

摘要：本文研究了电磁成像中的静电聚焦场的逆设计。作者应用多电极系统中静电场的电位叠加原理，发展了一种数学模型，并将最优化方法引入成像系统静电场的设计与计算。

Introduction

Imaging systems with electrostatic and magnetic combined focusing have long been widely used in photo-electronic imaging devices. The theory and design of electrostatic and magnetic imaging without image rotation but with magnification $M > 1$ or $M < 1$ have been investigated in detail in previous papers[1-3], in which we have established two conditions for designing electrostatic and magnetic imaging systems without image rotation:

$$\frac{3(V + \varepsilon_{z_1}^*)}{G^2}\left(\frac{dG}{dz}\right)^2 - \frac{2(V + \varepsilon_{z_1}^*)}{G}\frac{d^2 G}{dz^2} - \frac{1}{G}\frac{dV}{dz}\frac{dG}{dz} + \frac{d^2 V}{dz^2} = 0 \tag{1}$$

$$\lambda_n = \frac{e}{8m_0}\frac{B_0^2 l^2}{\varphi_i} = \frac{n^2 \pi^2}{\left[\int_0^1 \frac{G(z)}{\sqrt{V(z) + \varepsilon_{z_1}^*}}\right]^2} \tag{2}$$

where e/m_0 is the ratio of electron charge to mass; e should be taken to be an absolute magnitude; z is the axial coordinate measured in unit of distance l between the photocathode and screen; $V(z)$ is the axial electrostatic potential, at the cathode $\varphi(0) = 0$, at the screen $\varphi_i = \varphi(z = 1)$; $B(z)$ is axial magnetic induction, at the cathode $B(0) = B_0$, $V = V(z) = \varphi(z)/\varphi_i$, $G = G(z) = B(z)/B_0$; $\varepsilon_{z_1}^* = \varepsilon_{z_1}/\varphi_i$, $e\varepsilon_{z_1}$ is a certain initial axial energy of the electron emitted from the cathode surface, all electrons having the same initial axial energy are ideally focused at the image plane $z = 1$; λ_n is the eigenvalue of n loops.

Ni Guoqiang[a], Zhou Liwei[a], Jin Weiqi[a], Fang Erlun[b]. a) Beijing Institute of Technology, b) Xi'an Research Institute of Modern Chemistry. Chinese Journal of Electronics English Edition (电子科学英文版), V. 7, No. 1, 1990, 6-14.

Actually, the system design from the above-mentioned condition (1) and condition (2) is an inverse design. It means that we conversely design the electrode system and magnetic focusing coil and determine the electric and structure parameters from the axial field distributions $V(z)$ and $G(z)$, which satisfy condition (1) and condition (2) for obtaining an image without image rotation.

In Ref. [4], based on the preceding formulae, we assumed that the axial electrostatic potential $V(z)$ is a linear distribution, then a series of data of axial magnetic induction $G(z)$ corresponding to different magnifications can be obtained and an inverse design of magnetic focusing coil from given data of axial magnetic induction can be carried out. Thus, a magnetic focusing coil without iron medium and the prototype tube have been made, which gives an image with magnifications 0.8, 0.9, 1.0, 1.2, but without image rotation.

According to the references [2, 3], for the axial electrostatic potential with linear distribution, the axial magnetic inductions corresponding to a variety of magnifications can be attained. In principle, it is possible to produce an identical magnetic coil for realizing all these magnetic induction distributions by means of the method given in the Ref. [4]. But the computation shows that the magnetic field produced by the coil is difficult to approach the required distributions, or the current through the coil given by computation will be too large to realize, when the magnification is deviated far from unity. As we know, however, for reducing the spherical and chromatic aberration, the second order derivative of axial magnetic induction distribution should keep positive in the whole lens, and just this fact leads to a substantive improving of the spectrometer, but for an axial potential with linear distribution, the magnetic field distribution does not satisfy this requirement when $M<1$. It may be seen that for the image system with electrostatic and magnetic combined focusing, an inverse design for the electrostatic and magnetic fields should be made at the same time, so that a better or an optimum practical system can be obtained. Therefore, how we do realize the inverse design of the electrostatic field of system becomes an imperative problem.

The theoretical foundation of inverse design for electron optics has been laid by Grinberg, et al[7]. In Ref [8], a method of inverse design for the electrostatic focusing imaging system was approached in detail. Since 1970, as the optimization design of the electron optical system has been developing in depth, some idea and methods of inverse design have emerged as the time went on[8]. In 1974, P. W. Hawkes[9] described an idea to construct a system from the axial field distributions, that is, he made a search after an electrode and an unsaturated pole system by using the method of surface charge so as to produce the given axial potential and magnetic induction distributions. The dynamic programming proposed by M. Szilagyi[10,11], and the variation method given by H. Rose[12], searched for the axial electrostatic and magnetic field distributions when the spherical aberration coefficient is minimum, but it could not obtain a practical structure. In Ref. [10], a segmental spline function of third order is used to approximate locally the given axial potential distribution. On the basis of the spatial potential distribution obtained by Scherzer's series expansion of axial symmetrical system, by selecting some of equipotential surfaces as the electrodes structure of system, and after repeated corrections, a realizable structure of electrodes can be obtained. The result shows that the reconstruction of electrodes or magnetic pole piece from the given axial field distributions is more complicated and more difficult than searching for field distributions from a given electron optical characteristic.

The present paper is concerned in the inverse design of electrostatic focusing field in the electrostatic and magnetic imaging system with variable magnifications but without image rotation, which put forward a new problem. It requires that various field distributions can be obtained in the same system, that is, it makes demands on obtaining different field distributions by changing electric parameters when the structure of electrodes has been fixed.

Actually, the design of electrostatic image tube of the zoom type deals with the same problem. But up to now, the design of this kind of image tube still depends mainly on the direct design of step by step process, as well as on the designer's experience.

In view of this fact, the present paper will explore the inverse design of electrostatic focusing field in the electrostatic and magnetic combined focusing system, and design the practical imaging system realizing a given electrostatic field distribution by using the optimization method.

1 Mathematical model

Starting with the axial potential distribution, the inverse design for the electrostatic and magnetic combined focusing system can be considered. We hope for finding a certain electrode system, in which $V(z)$, the dimensionless axial potential distribution in the interval $(0, 1)$ generated by the electrode system, measured in screen voltage φ_i at the image plane, approaches the $\overline{V}(z)$—dimensionless axial potential distribution to be expected—as far as possible. Using the definition of mean square deviation, we construct the following functional for solving a problem of rotational symmetric electrostatic field:

$$L = \int_0^1 [\overline{V}(z) - V(z)]^2 dz \tag{3}$$

which makes the functional an extreme value.

From the theorem of potential superimposition in the electrostatic field we know that the potential distribution generated by a multi-electrode system, can inevitably be obtained by the linear superimposition of fundamental potential distribution φ_1^*, φ_2^*, $\cdots$, $\varphi_{n_1}^*$, in which n_1 is equal to the number of electrodes, that is

$$\varphi(r,z) = \sum_{k=1}^{n_1} V_k \cdot \varphi_k^*(r,z) \tag{4}$$

where $\varphi(r,z)$ is the spatial potential distribution at the meridional plane (r, z) in the rotational symmetrical system; V_k are the values of potential applied to the k-th electrode $(k = 1, 2, \cdots, n)$, $\varphi_k^*(r,z)$ is the fundamental spatial potential distribution of k-th electrode, can be defined as follows: in a system formed by n_1-electrodes, let the potential of k-th electrode be 1, and the potential of other electrodes be zero, so the attained spatial potential distribution is called fundamental spatial distribution. Obviously, φ_k^* depends only on the electrodes structure and location with each other.

For the axial potential distribution, similarly, we have

$$V(z) = \sum_{k=1}^{n_1} V_k^* \cdot \varphi_k^*(z) \tag{5}$$

where $\varphi_k^*(z)$ is the fundamental axial potential distribution of k-th electrode; $\varphi_k^*(z) = \varphi_k^*(0,z)$, and $V_k^* = V_k/\varphi_i$. For short, herein after we shall neglect all star marks $*$ in the symbols of V_k^* and $\varphi_k^*(z)$, which are simplified as V_k and $\varphi_k(z)$.

After the structure and mutual location of electrodes have been determined, the computational potential distribution obtained by the computer are always discrete. At the same time, for the sake of saving time, we change the integral of Eq. (3) into a form of summation, then Eq. (3) will be discrete. Furthermore, for the cathode lens, the cathode potential is always identically equal to zero, so the cathode should not be counted in the number of electrodes. As we know, for system with n_1 electrodes (not including the cathode), only $n_1 - 1$ fundamental potential distribution are independent with each other, and the potential at the image plane is identically equal to φ_i, that is, $V_i \equiv 1$.

Considering that the requirements for the discrete sampling points will not be the same, (For example, in the electrostatic and magnetic imaging systems, owing to the low velocity of electron motion in the neighborhood of cathode surface, as well as in the neighborhood of screen, at which electrons strikes) we hope that the error in these regions for the electrostatic field design is as small as possible. But in the other region far from the cathode and the screen, as at the middle, the error control may be relatively relaxed. Therefore, we should weight each sampling point according to the error control in the electrostatic focusing field design. Then, Eq. (3) should take the following form:

$$L = \frac{1}{m} \sum_{j=1}^{m} A_j \left[\sum_{k=1}^{n} V_k \cdot \varphi_k(z_j) + \varphi_i(z_j) - V(z_j) \right]^2 \tag{6}$$

where $z_j (j=1,2,\cdots,m)$ is the coordinate of axial sampling point, $z_j \in (0,1)$; $\varphi_k(z_j)(k=1,2,\cdots,n)$, values of fundamental potential formed by the k-th intermediate electrode or screen electrode at point z_j respectively; $V(z_j)$ is the value of potential to be expected; n is the number of intermediate electrodes, not including cathode and screen; A_j is the weight factor. $\sum_{j=1}^{m} A_j = m$, $A_j > 0$.

When the structure parameters of electrodes are fixed, L will be the function of each intermediate electrode $V_k(k=1,2,\cdots,n)$. From the above-mentioned treatment, the problem for solving functional becomes an extreme value problem for solving the multivariate function. In other words, we should explore a set of values for V_k so that the objective function, that is, the L value in Eq. (6), tends toward a minimum:

$$\min_{V \in R^n} L(V) \tag{7}$$

where $V = [V_1, V_2, \cdots, V_n]^T$ describes the independent variable vector in the n-dimensional linear space R^n; the superscript "T" indicates the transposed matrix. The gradient vector of objective function is

$$\Delta L = \left[\frac{\partial L}{\partial V_1}, \frac{\partial L}{\partial V_2}, \cdots, \frac{\partial L}{\partial V_n}\right]^T \tag{8}$$

where $\frac{\partial L}{\partial V_l}$ can be obtained from Eq. (6):

$$\frac{\partial L}{\partial V_1} = \frac{2}{m}\sum_{j=1}^{m} A_j \left[\sum_{k=1}^{n} V_k \varphi_k(z_j) + \varphi_i(z_j) - V(z_j)\right] \cdot \varphi_l(z_j), \quad (l = 1, 2, \cdots, n) \tag{9}$$

where the values of $\varphi_k(z_j)$, $\varphi_i(z_j)$ in Eq. (6) and Eq. (9) will be provided by the direct design for computational program of electrostatic field in the image tube.

From the optimization methods—methods of scale change, we choose one of them, which is called BFGS (Broyden, Fletcher, Goldfarb, Shanno) method and apply it to optimizing the objective function Eq. (6). A brief introduction of the computational idea and the iteration procedures for the BFGS method has been given in our previous paper[4], here it would not be described any more.

We have designed the electrostatic focusing field for the electrostatic and magnetic imaging system. A universal program using Fortran language has been drawn up for the inverse design of electrostatic focusing field. The principal advantage of the program is universal, convenient and practically functional.

2 Practical examples

The electrode structure of an experimental prototype tube for the electrostatic and magnetic imaging system developed in our previous paper[4] is shown in Fig. 1. Besides the plane photocathode and the plane phosphor screen, the prototype tube has 5 intermediate electrodes, and its structure is very simple.

Using the above-mentioned method, we compute this experimental prototype tube which will realize the axial potential distribution as the following:

$$V(z) = \gamma z / [1 - (1 - \gamma)z] \tag{10}$$

Actually, formula (10) is an expression of the axial potential distribution for a typical concentric spherical system. This electrostatic field combined with the magnetic field expressed by the following axial magnetic induction distribution[13]

$$G(z) = 1/[1 - (1 - \gamma)z]^2 \tag{11}$$

has formed an electrostatic and magnetic imaging system, which can give an ideal focusing image without image rotation[2,3]. It is not difficult to prove that the parameter γ in formula (11) is just the lateral magnification M of the system. When $\gamma = 1.0$, formula (10), formula (11) describe a homogeneous and parallel combined electrostatic and magnetic field.

Fig. 2 shows the curves of axial fundamental potential distribution for each of the inter-mediate electrodes and the image plane in this prototype tube, while some curves of axial potential distribution expressed by formula (10) are shown in Fig. 3 when the parameter γ corresponds to different values.

Computational result shows that the main source of the error of approach accuracy is determined by the plane input and output windows used in the image tube. The second order derivative of actual axial potential distribution at two ends of the tube is equal to zero, i. e., $V''(0) = 0$, and $V''(1) = 0$. All these fundamental axial potential distributions have the same characteristics. It makes the practical field distribution generated by the prototype tube to be no identical spherical system in the neighborhood of the cathode and screen. The more the value of γ diverges from unity, the bigger the error.

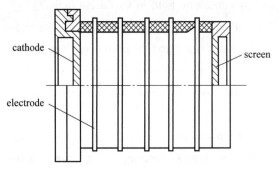

Fig. 1 Schematic diagram of electrode structure of the experimental prototype tube in the electrostatic and magnetic imaging system

If we make an interpolation treatment of the third order natural spline function (SPLM-Spline method) for the required axial potential distribution in the neighborhood of the cathode and screen, satisfying the natural boundary conditions ($V''(0) = 0$, $V''(1) = 0$), and do a computation, the convergence of iteration will become faster, and the approach accuracy will be improved.

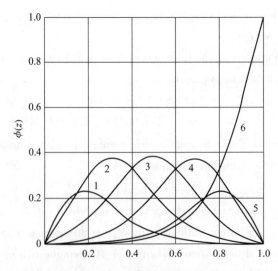

Fig. 2 Curves of the axial fundamental potential distribution for each of electrodes in an experimental prototype tube

Table 1 and Table 2 give part of computational results for the axial potential distribution expressed by formula (10) when γ is from 0.5 to 3.0. Computation shows that it can not be distinguished by using the curve plotting for the computational value and the required value, as shown in Fig. 3.

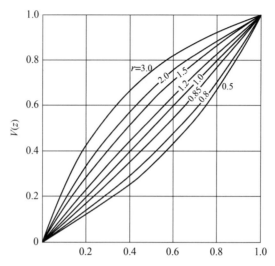

Fig. 3 Curves of axial potential distribution expressed by formula (10)

Table 1 Part of computational results for formula (10): Effects of number of iteration and accuracy by using spline method

Parameter γ	No correction by the spline method at the beginning and the end			Correction by the spline method at the beginning and the end		
	Number of iteration	Value of objective function	Mean relative error	Number of iteration	Value of objective function	Mean relative error
0.5	12	$6.847, 3 \times 10^{-7}$	$1.916, 2 \times 10^{-3}$	5	$4.159, 8 \times 10^{-7}$	$1.189, 6 \times 10^{-3}$
0.6	31	$2.144, 4 \times 10^{-7}$	$1.334, 7 \times 10^{-3}$	5	$1.263, 4 \times 10^{-7}$	$7.380, 6 \times 10^{-4}$
0.85	19	$1.032, 1 \times 10^{-8}$	$4.610, 7 \times 10^{-4}$	5	$5.002, 7 \times 10^{-9}$	$2.708, 5 \times 10^{-4}$
1.0	16	$1.196, 7 \times 10^{-10}$	$3.180, 8 \times 10^{-5}$	5	$2.985, 6 \times 10^{-11}$	$2.332, 5 \times 10^{-5}$
1.2	22	$1.737, 1 \times 10^{-8}$	$5.815, 0 \times 10^{-4}$	5	$6.894, 6 \times 10^{-9}$	$3.331, 0 \times 10^{-4}$
1.6	34	$2.620, 4 \times 10^{-7}$	$1.847, 4 \times 10^{-3}$	5	$1.052, 3 \times 10^{-7}$	$1.106, 9 \times 10^{-3}$
20	28	$1.140, 4 \times 10^{-6}$	$3.218, 1 \times 10^{-3}$	5	$4.680, 0 \times 10^{-7}$	$1.974, 8 \times 10^{-3}$
3.0	17	$9.753, 1 \times 10^{-6}$	$6.967, 7 \times 10^{-3}$	5	$4.042, 4 \times 10^{-6}$	$4.408, 0 \times 10^{-3}$

Both theory and computation show that the approach accuracy will be improved obviously if we increase the number of intermediate electrodes. For instance, for the preceding prototype tube, if the number of electrodes increases from 5 to 7, 9, or 11, but the other structure parameters remain unchanged, it makes an order or several orders of magnitude improvement in approach accuracy for the field distribution expressed by formula (10). In practice, the number of electrodes need not take too many, then a quite good accuracy will be obtained, which satisfies the requirements of engineering design.

Table 2 Part of computational results for formula (10) at different electrode potential values

Parameter γ	Potential at intermediate electrodes					Potential V_6
	V_1	V_2	V_3	V_4	V_5	
0.5	0.028,07	0.174,42	0.285,47	0.518,74	0.654,53	1.000,0
0.6	0.043,24	0.201,26	0.340,07	0.562,34	0.739,69	1.000,0
0.85	0.083,90	0.269,32	0.447,54	0.663,64	0.851,47	1.000,0
1.0	0.111,50	0.307,24	0.499,19	0.694,31	0.887,13	1.000,0
1.2	0.152,52	0.352,72	0.557,34	0.736,47	0.917,68	1.000,0
1.6	0.245,92	0.426,04	0.647,99	0.792,00	0.952,50	1.000,0
20	0.349,77	0.478,08	0.716,61	0.824,18	0.972,91	1.000,0
3.0	0.632,81	0.635,72	0.841,82	0.855,00	1.006,28	1.000,0

In addition, as is known, the hyperboloid field is a fundamental and simple one which is capable of focusing a wide electron beam, and which is having several ideal properties. In the early literature, for example, in Ref. [14], the hyperboloid field was studied as a focusing system of narrow electron beam. Recently, it has been investigated as a focusing system of wide electron beam, given in Ref. [15].

For the electrostatic and magnetic imaging, the axial potential distribution of the hyperboloid field can be expressed by

$$V(z) = (1-\beta)z^2 + \beta z \quad (12)$$

If parameter β takes different values, it can simulate a variety of hyperboloid fields. When $\beta = 1.0$, formula (10) expresses a uniform and parallel field. Computations for the field distribution given by Eq. (12) in which $0.3 \leq \beta \leq 2.0$, show that the approach accuracy is quite good (part of computational results is given in Table 3). It provides a practical and effective approach for further studying hyperboloid field.

Table 3 Part of computational results for formula (12) (number of iteration: 5)

Parameter β	Mean relative error	Potential at intermediate electrodes					Potential V_6
		V_1	V_2	V_3	V_4	V_5	
0.3	$3.031,2 \times 10^{-3}$	-0.022,62	0.136,48	0.276,96	0.523,43	0.753,27	1.000,0
0.6	$1.064,4 \times 10^{-3}$	0.035,19	0.209,30	0.372,54	0.596,37	0.810,88	1.000,0

续表

Parameter β	Mean relative error	Potential at intermediate electrodes					Potential V_6
		V_1	V_2	V_3	V_4	V_5	
0.8	$4.388,3 \times 10^{-4}$	0.073,74	0.257,85	0.436,27	0.644,99	0.849,29	1.000,0
1.0	$2.446,4 \times 10^{-5}$	0.112,28	0.306,40	0.499,99	0.693,61	0.887,70	1.000,0
1.2	$2.863,4 \times 10^{-4}$	0.150,82	0.354,95	0.563,72	0.742,24	0.926,11	1.000,0
1.5	$6.187,8 \times 10^{-4}$	0.208,63	0.427,78	0.659,31	0.815,17	0.983,73	1.000,0
1.8	$8.615,7 \times 10^{-4}$	0.266,45	0.500,61	0.754,89	0.888,11	1.041,35	1.000,0
2.0	$9.913,7 \times 10^{-4}$	0.304,99	0.549,16	0.818,62	0.936,74	1.079,76	1.000,0

3 Discussions

For the axial symmetrical system, once the axial potential distribution required is determined, usually one reconstructs the spatial potential distribution from the axial potential distribution on the basis of the finite difference method with live point mesh, the Scherzer series expansion method and the Scherzer integral method etc, and scan equipotential lines, select some of them as electrodes to be disposed.

In fact, the practical case is much more complicated. Firstly, we can only select a limited number of equipotential lines as the electrodes, and apply corresponding operating voltages. Secondly, only the ideal transparent mesh can really and truly approach the form of the original potential line. Obviously, it can not be accomplished. If we use a few of diaphragms with apertures instead of meshes, the equipotential line will be sure to penetrate aperture of the diaphragm, spreads out from the dense region to the sparse region and bends up, so that the field distribution is deviated from the former distribution, as required. Furthermore, as we consider the electrodes from the figure of equipotential line, usually it should have appropriate correction shape, so that a reasonable technologically structure will be ensured. But this correction brings about new error, and, even in that case, the electrodes will not be sure to have a simplified geometrical form. Moreover, by using this method, the computational result can only satisfy requirements of one field distribution. When the field distribution required is changed, it is difficult to obtain this field distribution in such a way that the potentials applied at the electrodes are changed. Obviously, it is not appropriate for the electrostatic and magnetic imaging systems with variable magnifications.

In fact, there is free surface charge distribution on surface of the metal electrode, and it is the singular point in electrostatic field. Therefore, if we dispose the electrode system according to the figure of equipotential line, the spatial potential distribution reconstructed by computation from axial potential distribution will certainly be destroyed, because it is actually contrary to the premise, on which the computation depends than there are no free charge distributions in the space. The spatial potential distribution is only determined from the axial potential distribution, for a certain point at the axis in an axial symmetrical system, its effective radius does not go beyond the distance between this axial point and closest point of the electrode. Therefore, this paper presents a new method for designing the electrostatic field, which is totally different from the method of traditional design, in which one reconstructs distribution and selects several of these equipotential lines to dispose the electrodes. Apparently, it is fully profitable for designing the electrostatic field.

Conclusions

(1) The present paper has explored and investigated the inverse design of electrostatic focusing field in electrostatic and magnetic imaging systems. According to the theorem of potential superimposition in the electrostatic field composed by a multi-electrode system, a minimum duplicative functional mathematical model has been set up. Based on the computational data given by the method of direct design, a practical and effective inverse design has been made for the multi-electrode's system by use of an optimization method of non-linear programming.

(2) Computational results in a wide spread range show that for the axial potential distribution satisfying with the general demands of electrostatic and magnetic imaging system the computation is stable and convergent, and its accuracy is good enough to satisfy the requirements of engineering design.

(3) The method given in the present paper shows that the electrode structure which is as simple as possible can be used to realize various different field distributions in electrostatic and magnetic imaging systems. The outstanding advantage lies in that the requirements for the structure parameters of the electrode system could be relaxed, and for the electrode system, the structure of which has been determined, and the potential value of each intermediate electrode can be changed to obtain different field distributions. The present paper provides an effective and practical method for designing electrostatic and magnetic imaging systems with variable magnifications but without image rotation, as well as gives a new approach for designing and computing other kind of electrostatic focusing systems.

References

[1] 周立伟. 电磁聚焦移像系统理论的研究[J]. 北京理工大学学报,1983,3(9):12-24.

[2] ZHOU L W, NI G Q, FANG E L. Electrostatic and magnetic imaging without image rotation[J]. Electron Optical Systems, SEM, Inc., 1984, 37-43.

[3] 周立伟,倪国强,方二伦. 图像无旋转的电磁聚焦移像系统的研究[J]. 电子学报,1984(06):33-34.

[4] 周立伟,倪国强,仇伯仓. 曲线坐标系下电子运动的张量分析[J]. 电子学报,1988(05):57-69.

[5] GLASER W. Über den öffnungsfehler der elektronenlinsen[J] Zeitschrift für Physik,1940,116: 19-33.

[6] ZHANG Z Q, ZHOU L W, JIN W Q, et al. Multigrid method for computation of the rotational symmetrical electrostatic field[J]. Chinese Journel of Electronics,1993,2(2):43-49.

[7] GRINBERG G A. Selected problems of mathematical theory of electrical and magnetic phenomena[M]. New York: Press of Academic Sciences, 1948:507-537.

[8] 周立伟,潘顺臣,艾克聪.静电聚焦成像系统电子光学逆设计的研究[J].北京理工大学学报,1983, 1:1-18.

[9] HAWKES P W. Image processing and computer aided design in electron optics[J]. Optikt,1973,38: 589-591.

[10] SZILAGYI M. A dynamic programming search for electrostatic immersion lenses with minimum spherical aberration[J]. Optik, 1978,50:35-51.

[11] SZILAGYI M. Synthesis of electron lenses, electron optical systems[J]. Electron Optical Systems, SEM,Inc.,1984:75-84.

[12] ROSE H, MOSES R W. Jr. Minimaler offnungsfehler magnetischer rund-und zylinderlinsen bei feldfreiem objektraum[J]. Optik,1973, 37(3): 316-336.

[13] 周立伟.电磁聚焦同心球系统的电子光学[J].兵工学报,1979,1:66-79.

[14] 谢曼 О И.电子光学理论基础[M].朱宜,译.北京:高等教育出版社,1958:125-127.

[15] 周立伟,史万宏.两电极双曲场静电聚焦成像系统的电子光学[C].第三次全国电子光学学术讨论会,咸阳,1984.

十、动态成像电子光学
Dynamic Imaging Electron Optics

10.1 On Paraxial Chromatic Aberrations Limited Temporal Resolution and Spatial Resolution in Image Tubes
论像管中近轴色差对时间分辨率与空间分辨率的限制

Abstract: In the present paper, the influence of paraxial chromatic aberrations of electrostatic cathode lenses on temporal resolution and spatial resolution in image tubes with state-mode operation and dynamic-mode operation is described. Based on an ideal model—a bi-electrode concentric spherical electrostatic focusing system, it is proven that the secondorder paraxial spatial (lateral) aberration, and the firstorder paraxial temporal aberration, have the similar form. The root mean square (RMS) radius of spatial resolution and the value of time-transit-spread are discussed, and the optimal image positions for dynamic-mode and static-mode operation are also calculated and compared.

摘要：本文对静态与动态工作方式下的像管中静电阴极透镜的近轴色差对时间分辨率和空间分辨率的影响进行了讨论。基于两电极同心球静电聚焦系统的模型，证明了其二级空间横向近轴色差与一级时间近轴像差具有类似的形式。本文讨论了空间分辨率与时间离散的均方根半径（RMS），计算和比较了静态和动态模式下的最佳成像面的位置。

Introduction

In the coming of new era, development of the ultrafast technology is worth paying attention. Research on the ultrafast technology is mainly divided into two fields: the one is study of the ultrafast pulses, the other is study of the streak tubes. Ultrafast pulses are those in the picosecond (10^{-12} s) and femtosecond (10^{-15} s) realms-so quick that the light takes several hundred femtoseconds just to traverse the diameter of a human hair (human hair varies from 10 to 100 μm in diameter). A femtosecond laser delivers a relatively high-energy pulse in a very short time period so that any effects of thermal diffusion and related damage are minimized compared to other techniques. In fact, an ultrafast laser can cut safely through pressed pellets of common explosives, and can drill very tiny holes in steel or otherwise micro-machine material without causing collateral damage. High speed image techniques such as streak tubes can provide the researcher with an understanding of very fast physical processes.

The ultrafast laser technology was evident when Professor Ahmed Zewail of the California Institute of Technology was awarded winner of the 1999 Nobel prize for chemistry for his work with

---

Zhou Liwei. Beijing Institute of Technology. Proceedings of the Sino-Russia Academic Conference. Beijing Institute of Technology Press, 2000, 30 – 34.

femtosecond pluses, and for his studies of the transition states of chemical reactions using femtosecond spectroscopy. Zewail's work involves the use of ultrafast laser spectroscopy to better understanding how chemical bonds form and break and he has been credited with bringing about a new chemical field of study femtosecond chemistry. Recently, a device was demonstrated that produces a femtosecond-resolution series of images-akin to a movie-showing areas of instantaneous birefringence induced by a focused ultrafast laser pulse in air[1]. So it has a practical meaning to study the limitation of temporal resolution, as well as the limitation of spatial resolution, for the pico-femtosecond high speed photographic camera.

1 Problems are Advanced

For an electron optical imaging system, spatial-trajectory-spread and transit-time-spread produced by the emission energy and emission angle of photoelectrons are two of the fundamental problems of high-speed image-tube camera. In the static mode operation, the spatial-trajectory-spread of electron trajectories at the image plane determines limited spatial resolution. In the streak mode operation, because of time-transit-spread, a point image of a photo event of very short duration is imaged by the camera as a line. The length of the line is proportional to the time-transit-spread. Obviously, the shorter the length of the line, the better the time resolution of the camera.

It was proven that the spatial-trajectory-spread of an imaging system is mainly determined by the second order lateral aberration, i.e., chromatic lateral aberration Δr, which can be expressed by Recknagel-Artimovich formula as following[2-3]:

$$\Delta r = \frac{2M}{E_c} \sqrt{\varepsilon_r} (\sqrt{\varepsilon_z} - \sqrt{\varepsilon_{z_1}}) \qquad (1)$$

Where M is the magnification of the system; E_c is the field strength of electric field; ε_r, ε_z represent the radial and axial initial energy respectively; ε_{z_1} is a certain axial initial energy corresponding to the image position.

Formula (1) shows that the spatial lateral aberration in the second order approximation does not depend on the concrete electrode structure and potential distributions, but depends on the emission angle and emission energy, the field strength of electric field at the photocathode, the magnification of the system and the position of image plane.

It should be mentioned that the time transit spread ΔT, which can be called the first order temporal chromatic aberration at the image plane, was given by Savoisy-Fanchenko[4] formula in 1950s, and then was proven by I. P. Csorba[5] in 1970s as following:

$$\Delta T = \sqrt{\frac{2m_0}{e}} \frac{1}{E_c} \sqrt{\varepsilon_z} \qquad (2)$$

where m_0/e is the ratio of mass to charge of electron.

Formula (2) shows that the temporal aberration in the first order approximation depends only on the emission angle and emission energy, the field strength of electric field at the photocathode, but does not depend on the concrete electrode structure and potential distributions, as well as the magnification of the system and the position of image plane.

The difference between two formulae is obviously. We may find the non-harmonious and non-symmetry between the two formulae if we think it over. Actually, the spatial-trajectory-spread and the time-transit-spread at an image plane is produced by the emission energy and emission angle of the same photoelectrons. It is strange that the former depends on the position of image plane, but the later does not. This paper will relate the static problem of imaging electron optics to the dynamic problem, and clears up the theoretical difference between them.

2 Bi-electrode Concentric Spherical System to Study Temporal Resolution and Spatial Resolution

The best model for investigating the theoretical problem is a bi-electrode concentric spherical electrostatic focusing system. This is because the potential distribution and electron trajectories can be written as analytical forms, and the focusing characteristics and aberrations (spatial and temporal) can be investigated quantitatively. With a concentric spherical electrostatic focusing system, whether realized accurately or approximately, a good image quality can be obtained.

For the static optics of the concentric spherical electrostatic system, the previous works were given by E. Ruska (1933)[6], P. Schagen (1952)[7]. but all of these works remained at the zerotborder approximation of imaging, only P. Schagen discussed the chromatic aberration at a specified image position. Formula (1) has long been strictly proven by us in Ref. [8], and the formula of the secondorder lateral chromatic aberration has also been deduced based on the concentric spherical system. For the dynamic optics of the concentric spherical electrostatic system, the work of I. P. Csorba[5] only confirmed the conclusion of Savoisky-Fanchenko formula.

In the joint project "Fundamental principles of photoelectron image formation with time resolution of $10^{-13} - 10^{-14}$ s", supported by Russian Foundation for Basic Research (RFBR) and National Natural Science Foundation of China (NSFC) collaborated by General Physics Institute (GPI), Russian Academy of Sciences and Beijing Institute of Technology (BIT), headed by M. Y Schelev and Zhou Liwei. For the investigation temporal resolution and spatial resolution in image tubes working in static mode operation and dynamic mode operation, we have carried out following studies.

(1) Static and dynamic electron optics for a concentric spherical system of electrostatic focusing.

(2) Static and dynamic paraxial electron optics for a concentric spherical system of electrostatic focusing.

(3) On time-transit-spread of a concentric spherical system in streak mode operation.

(4) On root mean square (RMS) radius of lateral chromatic aberration and root mean square (RMS) value of time-transit-spread.

In the first two papers, we have discussed chromatic aberration and temporal aberration by two approaches—practical ray equation and motion equation, and paraxial ray equation and motion equation. We have proved that the secondorder lateral chromatic aberration Δr, as shown in formula (1) and the first order temporal chromatic aberration ΔT is as following:

$$\Delta T = \sqrt{\frac{2m_0}{e}} \frac{1}{E_c} (\sqrt{\varepsilon_z} - \sqrt{\varepsilon_{z_1}}) \tag{3}$$

We made sure that the two approaches which gave same expressions are rigorous, since the expressions are expressed by analytical forms.

It is natural that we relate the static problem of imaging electron optics to the dynamic problem. Actually, it is the same thing which reflects temporal and spatial characteristics, that is, the spatial aberration and the time aberration are produced by the divergence of initial emission energy at the image plane. So it is not strange that the expressions of aberration, either spatial or temporal, have the similar form.

In the third paper, we have proven that the firstorder time transit-spread exits everywhere, either within the field produced by two concentric sphere or out of the field in the region beyond the anode. The expression of the first order time-transit-spread which occupies the main part in the whole time-transit-spread, has a very compact form. The first order time-transit-spread is only related to the field strength at the photocathode, the axial initial energy of electron emission and a certain axial initial energy to be chosen, but it is not related to the concrete structure of the system, and the lateral magnification of the system.

In the fourth paper, the root mean square (RMS) radius of lateral chromatic aberration $\Delta \bar{r}$ and the root mean square (RMS) value of transit-time-spread ΔT are discussed, suppose that the initial angular distribution of emitted electron follows the Lambertian distribution, and the initial velocity distribution follows the beta distribution or cosine distribution, the optimal image position for dynamic-mode operation and for static mode operation are also calculated and compared. From the calculations we can make conclusion that either for the spatial frequency characteristics or the time-transit-spread characteristics, $\beta_{1.8}$ distribution seems the best one among the distributions listed, then in order, $\beta_{1.4}$, $\beta_{1.2}$, $\beta_{1.1}$, and Cosine, the $\beta_{2.1}$ is the worst one. The given results show a direction or principle to choose a photocathode, in which the spatial resolution and temporal resolution will satisfy the requirements for the streak tubes.

Summary

In the present paper, the first order time-transit-spread which influences the temporal resolution exits everywhere is strictly deduced in a concentric spherical system of electro static focusing. It is proven that the second order lateral chromatic aberration which limits the spatial resolution has the similar form as the first order time-transit-spread. The root mean square (RMS) radius of lateral chromatic aberration $\Delta \bar{r}$ and the root mean square (RMS) value of transit-time-spread $\Delta \bar{T}$ have been discussed. The position of an optimal image for dynamic mode operation is very close to the optimal position for static mode operation.

References

[1] WALLACE J. Imager sees femtosecond-sale birefringence[J]. Laser Focus World,1999, 35(12):15-16.

[2] RECKNAGEL A. Theorie des elektrisohen elektronen miktroskops fur selbstrakler[J]. Z. Angew Physik,1941,117:689-708.

[3] ARTIMOVICH L A. Electrostatic properties of mission systems[J]. Bulletin of Academy of Sciences,Soviet Union,Series of Physics,1944,8(6):131-338. (in Russian).

[4] SAVOISY Y K,FANCHENKO S D. Physical basis of electron-optical characteristics[J]. Reports of Academy of Sciences,Soviet Union,1956,108(2):218-221. (in Russian).

[5] CSORBA I P. Transit-time-spread limited time resolution of image tubes in streak operation[J]. RCA Review,1971,32:650-659.

[6] RUSKA E. Zur fokussierbarkeit von kathoden-Strahlbundeln grosser ausgangsquerchnitte [J]. Z. Angew. Physik,1933,83(9):684-687.

[7] SCHAGEN P,BRUINING H,FRANKEN L C. A simple electrostatic electron-optical system with only one voltage[J]. Philips Research Reports,1952,7(2):119-130.

[8] ZHOU L W. Electron optics with wide be am focusing[M]. Beijing Institute of Technology Press,1993(in Chinese).

10.2 论两电极静电同心球系统的近轴电子光学及其空间-时间像差

On Paraxial Electron Optics and Its Spatial-temporal Aberrations for a Bi-electrode Electrostatic Concentric Spherical System

摘要：本文由近轴轨迹方程和近轴运动方程出发，研究两电极静电同心球系统的空-时轨迹及其像差。首先求解两电极静电同心球系统中自光阴极逸出的运动电子的近轴空间-时间轨迹，然后讨论此系统的静动态电子光学及其空间-时间像差，揭示近轴电子光学成像的一般规律。文中定义和推导了近轴空间像差和近轴时间像差，表明完全可以直接由近轴轨迹方程和近轴运动方程出发来研究电子光学成像及其空间-时间像差。

Abstract: Start from the paraxial trajectory equation and paraxial motion equation, the spatial-temporal trajectory and its aberrations in a bi-electrode electrostatic concentric system have been investigated. First, the paraxial spatial-temporal trajectory emitted from photocathode in the bi-electrode electrostatic concentric system has been solved, then the static and dynamic electron optics and its spatial-temporal aberrations have been discussed, and the common rule of paraxial imaging electron optics has been revealed. In the present paper, the paraxial spatial-temporal aberrations have been defined and deduced. It shows that we can investigate the electron optical imaging and its spatial-temporal aberrations directly from the paraxial trajectory equation or paraxial motion equation.

引言

在成像电子光学的研究中，众多研究者包括我们在内[1~6]，对三级几何横向像差给予了很大的注意，而对近轴横向像差却重视不够。实际上，由于近轴横向像差与自阴极面逸出电子的轴向初速的分散相联系，它不但是成像的中心点最主要的像差，其二级近轴横向色差决定了系统的极限分辨率，而且它在整个像面上处处存在。文献[1]严格证明了在像面中心点处，其三级几何横向像差与三级近轴横向像差属于同一数量级，却远小于二级近轴横向色差。对于近轴时间像差，同样也有类似的结论，即二级几何时间像差与二级近轴时间像差属于同一数量级，却远小于一级近轴时间色差；而一级近轴时间色差决定了系统的时间分辨率，因此，近轴横向像差和近轴时间像差的研究是一个值得关切的课题。在

周立伟[a]，公慧[a]，张智诠[b]，张轶飞[b]. a) 北京理工大学，b) 装甲兵工程学院. 物理学报（Acta Physica Sinica），V. 59，No 8，2010，5450-5459.

文献［1］中，我们由实际轨迹方程出发，讨论了两电极静电同心球系统的静态电子光学和空间像差；由电子运动方程出发，讨论了此系统的动态电子光学和时间像差。结果表明，不但空间像差可以视为近轴空间像差和几何空间像差的合成，时间像差也可以视为近轴时间像差和几何时间像差的合成。这是成像电子光学一个十分重要的概念，将给电子光学的研究带来一些新的内容。

正如文献［1］的研究所认识到的，二级近轴横向色差与一级近轴时间色差在全部横向像差和时间像差中占有绝大部分，但它们仅与近轴解相联系。故由近轴轨迹方程和近轴运动方程的角度出发研究近轴横向像差和近轴时间像差自然引起了人们的兴趣。本文的工作将再一次证实，近轴像差确实是以不同轴向初速 $v_z(\sim\sqrt{\varepsilon_z})$ 和 $v_{z_1}(\sim\sqrt{\varepsilon_{z_1}})$ 逸出电子的两个近轴解（空间解或时间解）之间的差异，其中携有 $\sqrt{\varepsilon_{z_1}}$ 的电子决定理想成像的位置。本文从近轴解出发，证明了近轴（空间或时间）像差可以由近轴（轨迹或运动）方程的解求得，而求解近轴（轨迹或运动）方程远比求解实际（轨迹或运动）方程来得简单和容易得多。

1 两电极静电同心球系统的近轴轨迹的求解及其空间像差

1.1 近轴轨迹的求解及近轴成像

图 1 所示的是两电极静电同心球系统。设球面阴极 C 和栅状球面阳极 A 的曲率半径分别为 R_c 和 R_a，系统的共同曲率中心为 O。设阴极 C 的电位 $\phi_c=0$，栅状阳极 A 对于阴极 C 的电位为 ϕ_{ac}，电子以初速度 v_0、初角度 α_0 自阴极原点射出。

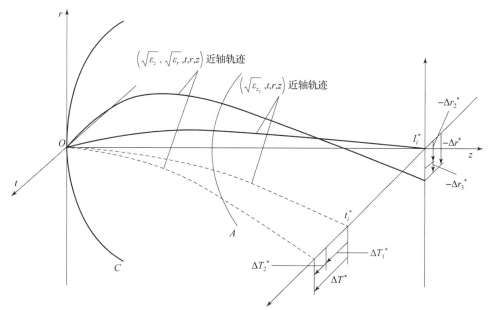

图 1 两电极静电同心球系统（$n=R_c/R_a>2$）空间与时间的近轴轨迹、近轴空间像差与时间像差

众所周知，在成像系统的电子光学中，标量形式的近轴轨迹方程可表示如下：

$$r''(z) + \frac{1}{2}\frac{\phi'(z)}{\phi(z)+\varepsilon_z}r'(z) + \frac{1}{4}\frac{\phi''(z)}{\phi(z)+\varepsilon_z}r(z) = 0 \tag{1}$$

式中，$r(z)$ 为电子射线的径向高度，$\phi(z)$ 为轴上电位分布，"'"、"''" 分别表示对 z 的一阶和二阶导数，ε_z 对应于由阴极射出的电子的轴向初电位。

在两电极静电同心球系统中，轴上电位分布 $\phi(z)$ 可以表示为

$$\phi(z) = \frac{-z}{(n-1)(z+R_c)}\phi_{ac} \tag{2}$$

式中，$n = R_c/R_a$，R_c、R_a 取负值。当电子射线自阴极原点射出时，轨迹的初始条件为

$$r_0 = r(z=0) = 0, \quad r_0' = \tan\alpha_0 = \frac{\sqrt{\varepsilon_r}}{\sqrt{\varepsilon_z}} \tag{3}$$

将式（2）代入式（1），不难导得式（1）在初始条件式（3）下的解析解为

$$r^*(z) = 2z\sqrt{\frac{\varepsilon_r}{\phi(z)}}\left(\sqrt{1+\frac{\varepsilon_z}{\phi(z)}} - \sqrt{\frac{\varepsilon_z}{\phi(z)}}\right) \tag{4}$$

式中，星号 * 表示近轴状况。由能量守恒定律，可以求得近轴电子的轴向速度为

$$\dot{z}^* = \sqrt{\frac{2e}{m_0}[\phi(z)+\varepsilon_z]} \tag{5}$$

由式（4），可得到近轴电子轨迹的径向速度及其斜率的表示式[8]：

$$\dot{r}^*(z) = \frac{1}{R_c+z}\sqrt{\frac{2e}{m_0}\varepsilon_r}\left\{R_c + 2z\sqrt{1+\frac{\varepsilon_z}{\phi(z)}}\left[\sqrt{1+\frac{\varepsilon_z}{\phi(z)}} - \sqrt{\frac{\varepsilon_z}{\phi(z)}}\right]\right\} \tag{6}$$

$$r'^*(z) = \frac{1}{R_c+z}\sqrt{\frac{\varepsilon_r}{\phi(z)}}\left\{2z\left[\sqrt{1+\frac{\varepsilon_z}{\phi(z)}} - \sqrt{\frac{\varepsilon_z}{\phi(z)}}\right] + \frac{R_c}{\sqrt{1+\frac{\varepsilon_z}{\phi(z)}}}\right\} \tag{7}$$

现确定近轴轨迹在轴上的理想成像位置 R_i^* 的表示式。令 $\sqrt{\varepsilon_z} = \sqrt{\varepsilon_{z_1}}$，在阳极处，$z = R_a - R_c$，$r(z) = R_a - R_c = r_a$，$\phi(z = R_a - R_c) = \phi_{ac}$，由式（4）和式（7），可以得到电子轨迹在阳极处的径向高度及其斜率，其近似式可表示为：

$$r_a^* = 2(R_a - R_c)\sqrt{\frac{\varepsilon_r}{\phi_{ac}}}\left(1 - \sqrt{\frac{\varepsilon_{z_1}}{\phi_{ac}}} + \frac{1}{2}\frac{\varepsilon_{z_1}}{\phi_{ac}}\right) \tag{8}$$

$$r_a'^* = r_i'^* = \tan\alpha_i^* = -(n-2)\sqrt{\frac{\varepsilon_r}{\phi_{ac}}}\left[1 - \frac{2(n-1)}{n-2}\sqrt{\frac{\varepsilon_{z_1}}{\phi_{ac}}} + \frac{3n-2}{2(n-2)}\frac{\varepsilon_{z_1}}{\phi_{ac}}\right] \tag{9}$$

设电子射线的理想成像位置离曲率中心之距为 R，它可表示为

$$R_i^* = \frac{r_a^*}{-r_a'^*} + R_a \tag{10}$$

将式（8）和式（9）代入式（10）中，略去阶次高于 $\varepsilon_{z_1}/\phi_{ac}$ 的远小于 1 的所有项，可得到理想成像位置 R_i^* 的表示式为

$$R_i^* = \frac{-R_c}{n-2}\left[1 + \frac{2(n-1)}{n-2}\sqrt{\frac{\varepsilon_{z_1}}{\phi_{ac}}} + \frac{2n(n-1)}{(n-2)^2}\frac{\varepsilon_{z_1}}{\phi_{ac}}\right] \tag{11}$$

这表明，由近轴解的途径所导得的理想成像位置及其斜率的表示式（11）和式（9）与文献［1］由实际解的途径所导得的式（14）和式（15）是完全一样的。

由式（11），可以讨论电子光学系统的理想成像性质。如系统的线放大率 M 可以表示为

$$M = \frac{R_i^*}{R_c} = -\frac{1}{n-2}\left[1 + \frac{2(n-1)}{n-2}\sqrt{\frac{\varepsilon_{z_1}}{\phi_{ac}}} + \frac{2n(n-1)}{(n-2)^2}\frac{\varepsilon_{z_1}}{\phi_{ac}}\right] \quad (12)$$

显然，当 $n > 2$ 时形成实像，而当 $n < 2$ 时形成虚像。

在高斯光线光学中，物像之间关系式遵循 Lagrange-Helmholtz 不变式。在近轴电子光学中，这个关系式依然存在。它表示了一个实际电子光学成像系统在近轴条件下的一种普遍性质。对近轴成像，如果定义角放大率为 $\Gamma = r_i'^*/r_0'^*$，则由式（11）和式（9），可以获得角放大率 Γ 和线放大率 M 的乘积，如下式所示：

$$M\Gamma = \sqrt{\frac{\varepsilon_{z_1}}{\phi_{ac}}}\left[1 - \frac{\varepsilon_{z_1}}{2\phi_{ac}} + \frac{3}{8}\left(\frac{\varepsilon_{z_1}}{\phi_{ac}}\right)^2\right] = \sqrt{\frac{\varepsilon_{z_1}}{\phi_{ac} + \varepsilon_{z_1}}} \quad (13)$$

式（13）被称为成像电子光学中的 Lagrange-Helmholtz 不变式。于是，轴向放大率 M 便可表示为

$$M_L = \frac{\Delta z_i}{\Delta z_0} = \frac{\Delta r_i}{\tan\alpha_i} \bigg/ \frac{\Delta r_0}{\tan\alpha_0} = \frac{\Delta r_i}{\Delta r_0} \bigg/ \frac{\tan\alpha_i}{\tan\alpha_0} = \frac{M}{\Gamma} = M^2\frac{\sqrt{\phi_{ac} + \varepsilon_{z_1}}}{\sqrt{\varepsilon_{z_1}}} \quad (14)$$

这样就严格地推导了成像电子光学的两个重要关系式（13）和式（14）。

在两电极静电同心球系统中，有两个参数能定量地表示电子束的会聚性质[8]：

（1）由整个阴极面射出的电子束所形成的交叉点直径，称为交叉颈，以 $2r_{cr}$ 表示，如图 1 所示，它可表示为

$$2r_{cr} = -2R_i^*\frac{r_i'^*}{\cos\theta - \sin\theta \cdot r_i'^*} \approx -\frac{4r_0}{\sin 2\theta}\sqrt{\frac{\varepsilon_r}{\phi_{ac}}} \quad (15)$$

式中，r_0 为物高；θ 为半个阴极面的有效尺寸与中心轴之间的夹角，取负值。

（2）由物点以初电位和初角度（$\sqrt{\varepsilon_0}$，α_0）发射的电子束在阳极上形成的弥散圆半径 r_a^*，如图 1 所示。它可表示为

$$r_a^* \approx 2(R_a - R_c)\sqrt{\frac{\varepsilon_r}{\phi_{ac}}} \quad (16)$$

这表明，在近轴条件下，阳极上的弥散圆半径 r_a^* 的大小主要取决于与阴极和阳极之间的距离 $l = R_a - R_c$ 和电子的径向初电位和总电位的比值 $\sqrt{\varepsilon_r/\phi_{ac}}$，而与电极的形状无关。这一性质有助于像管电子光学系统设计时对阳极孔径大小的估计。

1.2 近轴空间像差

现推导近轴空间像差。因为得到的是近轴轨迹的解，故仅讨论近轴空间像差。按照文献［1］的像差定义，由式（11），可以得到近轴纵向色差 Δz^* 按阶次划分的如下表示式：

$$\Delta z^* = R_i^*(\sqrt{\varepsilon_z}) - R_i^*(\sqrt{\varepsilon_{z_1}}) = \Delta z_1^* + \Delta z_2^* \quad (17)$$

式中，

$$\Delta z_1^* = \frac{2M^2}{-E_c}\sqrt{\phi_{ac}}(\sqrt{\varepsilon_z} - \sqrt{\varepsilon_{z_1}}) \tag{18}$$

$$\Delta z_2^* = \frac{2M^2}{-E_c}\left[\frac{n}{n-2}(\varepsilon_z - \varepsilon_{z_1}) - \frac{4(n-1)}{n-2}\sqrt{\varepsilon_{z_1}}(\sqrt{\varepsilon_z} - \sqrt{\varepsilon_{z_1}})\right] \tag{19}$$

式中，E_c 为阴极面上的电场强度，取负值。

近轴横向色差 Δr^* 便可表示为

$$\Delta r^* = -\Delta z^* \tan\alpha_i = r^*(z_i^*, \sqrt{\varepsilon_z}, \sqrt{\varepsilon_r}) - r^*(z_i^*, \sqrt{\varepsilon_{z_1}}) = \Delta r_2^* + \Delta r_3^* \tag{20}$$

式中，

$$\Delta r_2^* = \frac{2M}{E_c}\sqrt{\varepsilon_r}(\sqrt{\varepsilon_z} - \sqrt{\varepsilon_{z_1}}) \tag{21}$$

$$\Delta r_3^* = \frac{2M}{-E_c}\sqrt{\phi_{ac}}\sqrt{\varepsilon_r}(\sqrt{\varepsilon_z} - \sqrt{\varepsilon_{z_1}}) \tag{22}$$

Δz_1^*、Δz_2^* 和 Δr_2^*、Δr_3^* 的定义已在文献 [1] 中给出，其表示式（18）、式（19）、式（21）和式（22）已通过实际轨迹解的途径求得[1]。式（21）所示的二级近轴横向色差，通常称为 Recknagel-Aritimovich 公式[9,10]，它的确是由于近轴轨迹的轴向初速之间的差异形成的。这样，通过近轴解的途径导出了近轴空间像差。

2 两电极静电同心球系统的近轴飞行时间的求解及近轴时间像差

2.1 近轴飞行时间的求解

在两电极静电同心球系统的理想模型中，电子离阴极的飞行时间的精确表示式 $t(z)$ 已被导出过[1,13]。现推导飞行时间的近轴表示式 $t^*(z)$。若对近轴电子的轴向速度表示式（5）进行积分，可得

$$t^*(z) = \int_0^z \frac{dz}{\left\{\frac{2e}{m_0}[\phi(z) + \varepsilon_z]\right\}^{1/2}} \tag{23}$$

将 $\phi(z)$ 的表示式（2）代入式（23）中，便能得到电子自阴极逸出在初始条件 $z=0$，$t=0$，$\dot{z}_0 = \sqrt{2e\varepsilon_z/m_0}$ 下飞行时间的近轴表示式 $t^*(z)$ 为

$$t^*(z) = \frac{-R_c}{\left\{\frac{2e}{m_0}\left[\frac{\phi_{ac} - (n-1)\varepsilon_z}{(n-1)}\right]\right\}^{1/2}} \times$$

$$\left\{\sqrt{1 + \frac{z}{R_c}}\sqrt{\frac{(n-1)\varepsilon_z}{\phi_{ac} - (n-1)\varepsilon_z}} - \frac{z}{R_c} - \sqrt{\frac{(n-1)\varepsilon_z}{\phi_{ac} - (n-1)\varepsilon_z}} + \left[1 + \frac{(n-1)\varepsilon_z}{\phi_{ac} - (n-1)\varepsilon_z}\right] \times\right.$$

$$\left.\left[\arcsin\sqrt{\frac{(n-1)\varepsilon_z}{\phi_{ac}} - \frac{[\phi_{ac} - (n-1)\varepsilon_z]z}{R_c\phi_{ac}}} - \arcsin\sqrt{\frac{(n-1)\varepsilon_z}{\phi_{ac}}}\right]\right\}$$

$$\tag{24}$$

$t^*(z)$ 可称为近轴飞行时间。式（23）和式（24）表明，$t^*(z)$ 与电子的径向初速 $\sqrt{\varepsilon_r}$

无关。

为简化式（24），首先计算电子自光阴极到达阳极的近轴飞行时间，即 t_{C-A}^*。此时，阳极离阴极之距为 $z = -R_c + R_a$。由式（24），因为 $(R_a - R_c)/R_c = (n-1)/n$，便得到

$$t_{C-A}^* = \frac{-R_c}{\left\{\frac{2e}{m_0}\left[\frac{\phi_{ac} - (n-1)\varepsilon_z}{(n-1)}\right]\right\}^{1/2}} \times$$

$$\left\{\sqrt{\frac{R_a}{R_c}}\sqrt{\frac{(n-1)[\phi_{ac} + \varepsilon_z]}{[\phi_{ac} - (n-1)\varepsilon_z]n}} - \sqrt{\frac{(n-1)\varepsilon_z}{\phi_{ac} - (n-1)\varepsilon_z}} + \left[1 + \frac{(n-1)\varepsilon_z}{\phi_{ac} - (n-1)\varepsilon_z}\right] \times\right.$$

$$\left.\left[\arcsin\sqrt{\frac{(n-1)(\phi_{ac} + \varepsilon_z)}{n\phi_{ac}}} - \arcsin\sqrt{\frac{(n-1)\varepsilon_z}{\phi_{ac}}}\right]\right\} \quad (25)$$

展开式（25），略去其中高于 ε_z/ϕ_{ac} 阶次等高次小项，通过一系列的变换，便得到由阴极到阳极的近轴飞行时间 t_{C-A}^* 的近似表示式：

$$t_{C-A}^* = \left(\frac{2m_0}{e}\right)^{1/2}\left\{\frac{R_a - R_c}{2\sqrt{\phi_{ac}}} - \frac{1}{2}R_c\sqrt{\frac{n-1}{\phi_{ac}}}\arcsin\sqrt{\frac{(n-1)}{n}} - n\frac{R_a - R_c}{\sqrt{\phi_{ac}}}\sqrt{\frac{\varepsilon_z}{\phi_{ac}}} + \right.$$

$$\left. n\frac{R_a - R_c}{\sqrt{\phi_{ac}}}\frac{\varepsilon_z}{\phi_{ac}}\left[\frac{3n-1}{4n} + \frac{3}{4}\sqrt{n-1}\arcsin\sqrt{\frac{(n-1)}{n}}\right]\right\} \quad (26)$$

由此可见，由近轴解的途径导得的式（26）与文献［1］中由实际解的途径导得的式（38）是完全一样的，只要令 $\sqrt{\varepsilon_z} = \sqrt{\varepsilon_r}$。

其次，要计算携有 $\sqrt{\varepsilon_z}$ 的近轴电子从阳极飞出经历无场空间到达中心轴的近轴飞行时间 t_{C-A}^*。即令 v_a^* 为电子离阳极时的近轴轴向初速，此时近轴电子轨迹乃是一条直线，它到达中心轴的距离为 l，近轴飞行时间 t_{C-A}^* 可表示为

$$t_{C-A}^* = \frac{l}{v_a^*} \quad (27)$$

式中，$l = [r_a^2 + (-R_a + R_i)^2]^{1/2}$，$v_a^* = \sqrt{v_z^2 + 2e\phi_{ac}/m_0}$，代入式（27），并考虑近轴条件，便得到由阳极到像点的近轴飞行时间 t_{C-A}^* 的近似表示式：

$$t_{C-A}^* = -\sqrt{\left(\frac{2m_0}{e}\right)}\frac{R_a(n-1)}{(n-2)\sqrt{\phi_{ac}}} \times \left[1 + \frac{n}{n-2}\sqrt{\frac{\varepsilon_z}{\phi_{ac}}} + \frac{n^2 + 4n - 4}{2(n-2)^2}\frac{\varepsilon_z}{\phi_{ac}}\right] \quad (28)$$

式（28）与文献［1］中的式（43）是一样的。

2.2 近轴时间像差

近轴时间像差 ΔT^* 是在由轴向初速 $\sqrt{\varepsilon_{z_1}}$ 确定的像面上度量的，它被定义为自阴极面逸出的两条携有 $\sqrt{\varepsilon_z}$ 和 $\sqrt{\varepsilon_{z_1}}$ 的近轴轨线落在像面上在飞行时间上的差异。如图 1 所示，图中除空间坐标轴外，还假设有一个时间坐标轴。ΔT^* 由两部分组成：①携有 $\sqrt{\varepsilon_z}$ 和 $\sqrt{\varepsilon_{z_1}}$ 的两条近轴电子射线从阴极到达阳极的时间渡越弥散；②此两条近轴电子射线由阳极到达确定的理想像面上的时间渡越弥散。于是，可得

$$\Delta T^* = t_{C-A}^*(\sqrt{\varepsilon_z}) + (t^* + \Delta t)_{A-i^*}^*(\sqrt{\varepsilon_z}) - t_{C-A}^*(\sqrt{\varepsilon_{z_1}}) - t_{A-i^*}^*(\sqrt{\varepsilon_{z_1}}) \quad (29)$$

这里 $t_{C-A}^*(\sqrt{\varepsilon_z})$ 同样以式(26)表示,但 $t_{A-i^*}^*(\sqrt{\varepsilon_z})$ 以式(28)表示,两式中的 $\sqrt{\varepsilon_z}$ 均以 $\sqrt{\varepsilon_{z_1}}$ 代之,而

$$(t^* + \Delta t)_{A-i^*}^*(\sqrt{\varepsilon_z}) = t_{A-i^*}^*(\sqrt{\varepsilon_z}) + \Delta t_{i-i^*}^*(\sqrt{\varepsilon_z}) \tag{30}$$

现求解式(30)携有 $\sqrt{\varepsilon_z}$ 电子的近轴射线自阳极出发经无场空间首先到达中心轴,然后抵达由 $\sqrt{\varepsilon_{z_1}}$ 确定的理想像面,令电子自中心轴到理想像面的时间为 $\Delta t_{i-i^*}^*(\sqrt{\varepsilon_z})$。与文献[1]的推导和处理相同,可得

$$\Delta t_{i-i^*}^*(\sqrt{\varepsilon_z}) = R_a \frac{n}{n-2} \sqrt{\frac{m_0}{2e}} \frac{1}{\sqrt{\phi_{ac}}} \times \left[\frac{2n(n-1)}{(n-2)\sqrt{\phi_{ac}}} (\sqrt{\varepsilon_z} - \sqrt{\varepsilon_{z_1}}) + \frac{2n(n-1)}{(n-2)^2 \phi_{ac}} (\varepsilon_z - \varepsilon_{z_1}) \right] \tag{31}$$

将式(28)和式(31)代入式(30)中,便有

$$(t^* + \Delta t)_{A-i^*}^*(\sqrt{\varepsilon_z}) = -\sqrt{\frac{2m_0}{e}} \frac{R_a(n-1)}{(n-2)\sqrt{\phi_{ac}}} \times \left[1 + \frac{n}{(n-2)} \sqrt{\frac{\varepsilon_z}{\phi_{ac}}} + \frac{n^2 + 4n - 4}{2(n-2)^2} \frac{\varepsilon_z}{\phi_{ac}} \right] +$$

$$R_a \frac{n}{(n-2)} \sqrt{\frac{m_0}{2e}} \frac{1}{\sqrt{\phi_{ac}}} \left[\frac{2(n-1)}{(n-2)\sqrt{\phi_{ac}}} (\sqrt{\varepsilon_z} - \sqrt{\varepsilon_{z_1}}) + \frac{2n(n-1)}{(n-2)^2 \phi_{ac}} (\varepsilon_z - \varepsilon_{z_1}) \right] \tag{32}$$

最后,由式(29),可得以级次划分的近轴时间像差 ΔT^* 如下:

$$\Delta T^* = \Delta T_1^* + \Delta T_2^* \tag{33}$$

式中,

$$\Delta T_1^* = -\left(\frac{2m_0}{e}\right)^{1/2} \frac{\sqrt{\varepsilon_z} - \sqrt{\varepsilon_{z_1}}}{E_c} \tag{34}$$

$$\Delta T_2^* = \left(\frac{2m_0}{e}\right)^{1/2} \frac{\varepsilon_z - \varepsilon_{z_1}}{E_c \sqrt{\phi_{ac}}} \times \left[-\frac{3n^2 - 9n + 2}{4n(n-2)} - \frac{3}{8}\sqrt{n-1} \frac{1}{n} \left(\pi - \arcsin \frac{2}{n}\sqrt{n-1} \right) \right] \tag{35}$$

ΔT_1^* 和 ΔT_2^* 分别称为一级和二级近轴时间像差,其定义和式(34)、式(35)已在文献[1]中给出过,但这里是从近轴解的途径导出的。应该指出,Savoisy、Fanchenko[11] 和 Csorba[12] 所导出的 ΔT_1^* 表示式乃是 $\sqrt{\varepsilon_{z_1}} = 0$ 时(即在极限像面位置上)的式(34)。

结论

由以上研究可以得到以下结论:

(1) 本文给出了求解两电极静电同心球系统的近轴空间像差和近轴时间像差的另一条途径——近轴解的途径。这一工作使我们确信,无论是实际解或是近轴解的途径,只要遵循我们所给出的像差定义,所得结果将是完全一样的。

(2) 通过近轴轨迹方程和近轴运动方程推导了两电极静电同心球系统中运动电子的近轴轨迹解和近轴飞行时间解,以及近轴空间像差和近轴时间像差。结果表明,不论是二级近轴横向像差 Δr_2^* 或者是一级近轴时间像差 ΔT_1^*,前者决定像管空间分辨的基本限制,而后者决定高速摄影变像管时间分辨的基本限制,二者都取决于近轴解。这意味着,在研究成像电子光学系统时,应十分注意近轴解。

（3）通过近轴解途径，再一次推导了两电极静电同心球系统中三级近轴横向像差和二级近轴时间像差，它们分别与三级几何横向像差和二级几何时间像差属于同一数量级。这表明，在成像电子光学系统中，当研究几何像差（横向像差或时间像差）δr_3 或 δT_2 时，必须同时考虑近轴像差（横向像差或时间像差）Δr_3^* 或 ΔT_2^*。本文和文献［1］清晰地指出，三级近轴横向像差 Δr_3^* 和二级近轴时间像差 ΔT_2^* 是存在的，而且再一次给出了这两个像差的具体表达式。

参 考 文 献

［1］周立伟,公慧,张智诠.两电极静电同心球系统的近轴电子光学及其空间－时间像差[J].物理学报,2010,59(8):5450－5458.

［2］西门纪业.复合浸没物镜的电子光学性质和像差理论[J].物理学报,1957,13(4):339－356.

［3］周立伟,艾克聪,潘顺臣.关于电磁复合聚焦阴极透镜的像差理论[J].物理学报,1983(3):376－392.

［4］西门纪业,周立伟,艾克聪.阴极透镜像差的变分理论[J].物理学报,1983,32(12):1536－1546.

［5］艾克聪,周立伟,西门纪业.宽束和细束电磁复合聚焦球面阴极透镜的像差理论[J].物理学报,1986,35(009):1199－1209.

［6］艾克聪,西门纪业,周立伟.电磁复合聚焦－偏转球面阴极透镜的相对论像差理论[J].物理学报,1986,35(9):1210－1222.

［7］周立伟.宽电子束光学[M].北京:北京理工大学出版社,1993.

［8］周立伟,宽电子束光学[J]//周立伟电子光学学术论文选.北京:北京理工大学出版社,1994:11－29.

［9］ARTIMOVICH L A. Electrostatic properties of emission systems [J]. Bulletin of Academy of Sciences, Physics Series, 1944,8(6): 313－328(in Russian).

［10］RECKNAGEL A. Theorie des elektrisohen elecktronen miktroskops fur selbstrakler [J]. Z. Angew. Physik, 1941,117:689(in Germen).

［11］SAVOISKY Y K, FANCHENKO S D. Physical basis of electron optical chronograph [J]. Report of Academy of Sciences, 1956,108(2):218－221 (in Russian).

［12］CSORBA I P. Transit-time-spread-limited time resolution of image tubes in streak operation[J]. RCA Review, 1971,32:650－659.

［13］周立伟,李元,MONASTYRSKI M A,等.静电聚焦同心球系统验证电子光学成像系统的时间像差理论[J].物理学报,2005,54(8):3596.

10.3 Paraxial Imaging Electron Optics and Its Spatial-temporal Aberrations for a Bi-electrode Concentric Spherical System with Electrostatic Focusing
静电聚焦同心球系统的近轴成像电子光学及其空间－时间像差

Abstract: As is known to us, the paraxial solutions play an important role in studying electron optical imaging system and its spatial-temporal aberrations, but investigation of a bi-electrode concentric spherical system with electrostatic focusing directly from paraxial ray equation and paraxial motion equation has not been done before. In this paper, we shall use the paraxial equations to study the spatial-temporal trajectories and their aberrations for a bi-electrode concentric spherical system with electrostatic focusing.

摘要：本文由近轴轨迹方程和近轴运动方程出发，研究静电聚焦同心球系统的空间－时间轨迹及其像差。作者首先求解两电极静电同心球系统中自光阴极逸出的运动电子的近轴空间－时间轨迹，然后讨论此系统的静动态电子光学及其空间－时间像差，揭示近轴电子光学成像的一般规律。文中定义和推导了近轴空间像差和近轴时间像差，由近轴轨迹方程和近轴运动方程出发来研究电子光学成像及其空间－时间像差。

Introduction

In the previous paper[1], for a bi-electrode concentric spherical electrostatic system, we had discussed the static electron optics and spatial aberrations by practical electron ray equation, the dynamic electron optics and temporal aberrations by practical electron motion equation. The result shows that not only the spatial aberration can be regard as composition of the paraxial spatial aberration and the geometrical spatial aberration, but also the temporal aberration can be regard as composition of the paraxial temporal aberration and the geometrical temporal aberration. This is a very important concept which will bring something new for the electron optical study.

As we recognized from previous study, the paraxial lateral aberration of the second order and the paraxial temporal aberration of the first order occupy the greatest part in the whole spatial and temporal aberrations, but they are only related to the paraxial solutions. So it is interesting to study the paraxial spatial and temporal aberrations from the point of view of paraxial ray equation

---

Zhou Liwei[a], GongHui[a], Zhang Zhiquan[b], Zhang Yifei[b]. a) Beijing Institute of Technology, b) Institute of Armored Force Engineering. Optik, No. 122, 2011, 295－299.

and paraxial motion equation. This work will once again confirm that the paraxial aberrations actually are the difference between two paraxial solutions (spatial or temporal) of emitted electrons with different initial axial potential $\sqrt{\varepsilon_z}$ and $\sqrt{\varepsilon_{z_1}}$, in which the latter determines the ideal image position. The present paper shows another approach from paraxial solutions, which proves that the paraxial aberrations (spatial or temporal) can be found by solutions of paraxial (ray or motion) equation, that is greatly simpler than solutions of practical (ray or motion) equations.

In the present paper, start from the paraxial ray equation and paraxial motion equation, the paraxial spatial-temporal trajectory of moving electron emitted from the photocathode has been solved for a bi-electrode concentric spherical system with electrostatic focusing. The paraxial static and dynamic electron optics, as well as the paraxial spatial-temporal aberrations in this system are then discussed, the general regularity of imaging in paraxial optical system has been explored. The paraxial spatial aberrations, as well as the paraxial temporal aberrations with different orders, have been defined and deduced, that are classified by the order of $(\varepsilon_z/\phi_{ac})^{1/2}$ and $(\varepsilon_r/\phi_{ac})^{1/2}$. Thus, we get same conclusions about paraxial spatial and temporal aberrations as we have given in the previous paper and it completely shows that the paraxial spatial-temporal aberrations can be investigated directly from the paraxial ray equation and paraxial motion equation.

1 Paraxial ray solution and its spatial aberrations for a bi-electrode concentric spherical system with electrostatic focusing

1.1 Paraxial ray solution and paraxial imaging

As is known, in electron optics of electrostatic imaging systems, the paraxial ray equation can be written as

$$r''(z) + \frac{1}{2}\frac{\phi'(z)}{\phi(z)+\varepsilon_z}r'(z) + \frac{1}{4}\frac{\phi''(z)}{\phi(z)+\varepsilon_z}r(z) = 0 \tag{1}$$

where $r(z)$ is the radial height of emitted electron ray; $\phi(z)$ is the axial potential distribution; "′" represents the differentiation to z; $\sqrt{\varepsilon_z}$ is the axial initial potential of electron emitted from the cathode.

For simplicity, Eq. (1) is expressed in a scaler form.

In the bi-electrode concentric spherical system, the axial potential distribution $\phi(z)$ can be expressed by[1]

$$\phi(z) = \frac{-z}{(n-1)(z+R_c)}\phi_{ac} \tag{2}$$

where $n = R_c/R_a$; R_c, R_a are the curvature radii of spherical cathode C and of the mesh-spherical anode A, respectively; ϕ_{ac} is the potential of the anode A with respect to cathode C; $\phi_c = 0$.

When the electron ray emits from the axial point at photocathode with initial emission energy and initial emission angle ($\sqrt{\varepsilon_0}$, α_0), the initial condition of trajectory will take the form:

$$r_0 = r(z=0) = 0, \quad r'_0 = \tan\alpha_0 = \frac{\sqrt{\varepsilon_r}}{\sqrt{\varepsilon_z}} \tag{3}$$

Substituting Eq. (2) into Eq. (1), it is not difficult to derive an analytical solution of Eq. (1) under the condition (3) as following[2,3]:

$$r^*(z) = 2z\sqrt{\frac{\varepsilon_r}{\phi(z)}}\left\{\sqrt{1+\frac{\varepsilon_z}{\phi(z)}} - \sqrt{\frac{\varepsilon_z}{\phi(z)}}\right\} \tag{4}$$

where the star mark * indicates the paraxial situation. The axial velocity $\dot{z}^*$ of paraxial electron can be found from the law of conversation of energy[3]:

$$\dot{z}^* = \sqrt{\frac{2e}{m_0}[\phi(z)+\varepsilon_z]} \tag{5}$$

From Eq. (4), we may get expressions for radial velocity and slope of the paraxial electron trajectory:

$$\dot{r}^*(z) = \frac{1}{R_c+z}\sqrt{\frac{2e}{m_0}\varepsilon_r}\left\{R_c + 2z\sqrt{1+\frac{\varepsilon_z}{\phi(z)}}\left(\sqrt{1+\frac{\varepsilon_z}{\phi(z)}} - \sqrt{\frac{\varepsilon_z}{\phi(z)}}\right)\right\} \tag{6}$$

$$r'^*(z) = \frac{1}{R_c+z}\sqrt{\frac{\varepsilon_r}{\phi(z)}}\left\{2z\left(\sqrt{1+\frac{\varepsilon_z}{\phi(z)}} - \sqrt{\frac{\varepsilon_z}{\phi(z)}}\right) + \frac{R_c}{\sqrt{1+\frac{\varepsilon_z}{\phi(z)}}}\right\} \tag{7}$$

Now, we determine the ideal image position R_i^* of paraxial trajectory at axis. Let $\sqrt{\varepsilon_z} = \sqrt{\varepsilon_{z_1}}$, $z = R_a - R_c$, $r(z = R_a - R_c) = r_a$, $\phi(z = R_a - R_c) = \phi_{ac}$, from Eq. (4) and Eq. 7 we may obtain expressions for the radial height of electron ray and its slope at the anode, their approximations can be written as

$$r_a^* = 2(R_a - R_c)\sqrt{\frac{\varepsilon_r}{\phi_{ac}}}\left\{1 - \sqrt{\frac{\varepsilon_{z_1}}{\phi_{ac}}} + \frac{1}{2}\frac{\varepsilon_{z_1}}{\phi_{ac}}\right\} \tag{8}$$

$$r_a'^* = r_i'^* = -(n-2)\sqrt{\frac{\varepsilon_r}{\phi_{ac}}}\left\{1 - \frac{2(n-1)}{n-2}\sqrt{\frac{\varepsilon_{z_1}}{\phi(z)}} + \frac{3n-2}{2(n-2)}\frac{\varepsilon_{z_1}}{\phi_{ac}}\right\} \tag{9}$$

Suppose the distance of ideal image position of paraxial electron ray of $\sqrt{\varepsilon_{z_1}}$ from the curvature center O is R_i^*, which can be expressed by

$$R_i^* = \frac{r_a^*}{-r_a'^*} + R_a \tag{10}$$

Substituting Eq. (8) and Eq. (9) in Eq. (10), neglecting the higher order than terms of $\varepsilon_{z_1}/\phi_{ac}$, we obtain

$$R_i^* = \frac{-R_c}{n-2}\left\{1 + \frac{2(n-1)}{n-2}\sqrt{\frac{\varepsilon_{z_1}}{\phi_{ac}}} + \frac{2n(n-1)}{(n-2)^2}\frac{\varepsilon_{z_1}}{\phi_{ac}}\right\} \tag{11}$$

It shows that formulae (11) and formula (9), which express the ideal image position and its slope, derived from paraxial approach and from practical approach in previous paper[1].

Relationship for object-image space in Gaussian geometrical optics follows the Lagrange-Helmholtz invariant, as well as in paraxial electron optics. It represents an universal characteristics of a practical electron optical imaging system at paraxial condition. From formula (11), we can discuss the electron optical ideal imaging characteristics, such as linear magnification and Lagrange-Helmholtz invariant. The linear magnification M of the system can be expressed by

$$M = \frac{R_i^*}{R_c} = -\frac{1}{n-2}\left\{1 + \frac{2(n-1)}{n-2}\sqrt{\frac{\varepsilon_{z_1}}{\phi_{ac}}} + \frac{2n(n-1)}{(n-2)^2}\frac{\varepsilon_{z_1}}{\phi_{ac}}\right\} \quad (12)$$

Obviously, it forms a real image when $n > 2$ or a virtual image when $n < 2$.

For paraxial imaging, if we define angular magnification $\Gamma = r_i'^*/r_0'^*$, from formula (11) and formual (9) we may obtain the product of angular magnification Γ and linear magnification M as following:

$$M\Gamma = \sqrt{\frac{\varepsilon_{z_1}}{\phi_{ac}}}\left\{1 - \frac{\varepsilon_{z_1}}{2\phi_{ac}} + \frac{3}{8}\left(\frac{\varepsilon_{z_1}}{\phi_{ac}}\right)^2\right\} = \sqrt{\frac{\varepsilon_{z_1}}{\phi_{ac} + \varepsilon_{z_1}}} \quad (13)$$

Formula (13) is the Lagrange-Helmholtz invariant in imaging electron optics. The longitudinal magnification M_L can be expressed as following:

$$M_L = \frac{\Delta z_i}{\Delta z_0} = \frac{\frac{\Delta r_i}{\tan \alpha_i}}{\frac{\Delta r_0}{\tan \alpha_0}} = \frac{\frac{\Delta r_i}{\Delta r_0}}{\frac{\tan \alpha_i}{\tan \alpha_0}} = \frac{M}{\Gamma} = M^2 \frac{\sqrt{\phi_{ac} + \varepsilon_{z_1}}}{\sqrt{\varepsilon_{z_1}}} \quad (14)$$

Thus, we firstly strictly deduced these two important formula (13) and formula (14) in imaging electron optics.

There are two parameters quantitatively to express the convergence of electron beam in a bi-electrode spherical concentric system[3].

1) Radius of crossover r_{cr} formed by electron beams emitted from the whole cathode surface

Let θ be the angle between the effective size of half cathode surface and the central axis, we may obtain the expression for radius of crossover r_{cr}:

$$r_{cr} = -R_i^* \frac{r_i'^*}{\cos\theta - \sin\theta \, r_i'^*} \quad (15)$$

its first order approximation will be

$$r_{cr} \approx -\frac{2r_0}{\sin 2\theta}\sqrt{\frac{\varepsilon_r}{\phi_{ac}}} \quad (16)$$

where r_0 is the radial height of the object.

2) Radius of diffusing circle r_a^* at mesh-anode formed by electron rays from the object point with initial potential and initial angle ($\sqrt{\varepsilon_0}$, α_0).

From Eq. (6), at the anode, we may have

$$r_a^* = 2(R_a - R_c)\sqrt{\frac{\varepsilon_r}{\phi_{ac}}}\left\{\left(1 + \frac{\varepsilon_z}{\phi_{ac}}\right)^{1/2} - \left(\frac{\varepsilon_z}{\phi_{ac}}\right)^{1/2}\right\} \tag{17}$$

its first order approximation will be

$$r_a^* = 2(R_a - R_c)\sqrt{\frac{\varepsilon_r}{\phi_{ac}}} \tag{18}$$

It means that in paraxial condition the diffusing circle at mesh-anode is only related to the distance between cathode and anode $l = R_a - R_c$ and the radial initial potential of electron $\sqrt{\varepsilon_r/\phi_{ac}}$, but have nothing to do with the form of electrodes.

1.2 Paraxial spatial aberration

Now, we derive the paraxial spatial aberrations. Since we have obtained the paraxial ray solutions, only the paraxial spatial aberrations will be discussed, see Fig. 1. According to the definition of aberrations in the previous paper[1], from Eq. (11), we may obtain the expression for paraxial longitudinal aberration Δz^* as following:

$$\Delta z^* = R_i^*(\sqrt{\varepsilon_z}) - R_i^*(\sqrt{\varepsilon_{z_1}}) = \Delta z_1^* + \Delta z_2^* \tag{19}$$

where

$$\Delta z_1^* = \frac{2M^2\sqrt{\phi_{ac}}}{-E_c}(\sqrt{\varepsilon_z} - \sqrt{\varepsilon_{z_1}}) \tag{20}$$

$$\Delta z_2^* = \frac{2M^2}{-E_c}\left\{\frac{n}{n-2}(\varepsilon_z - \varepsilon_{z_1}) - \frac{4(n-1)}{n-2}\sqrt{\varepsilon_{z_1}}(\sqrt{\varepsilon_z} - \sqrt{\varepsilon_{z_1}})\right\} \tag{21}$$

where E_c is the strength of electric-field at the photocathode.

The paraxial lateral aberration Δr^* can be expressed by

$$\begin{aligned}\Delta r^* &= -\Delta z^* \tan \alpha_i^* \\ &= r^*(z_i^*, \sqrt{\varepsilon_z}, \sqrt{\varepsilon_r}) - r^*(z_i^*, \sqrt{\varepsilon_{z_1}}) = \Delta r_2^* + \Delta r_3^*\end{aligned} \tag{22}$$

where

$$\Delta r_2^* = \frac{2M}{E_c}\sqrt{\varepsilon_r}(\sqrt{\varepsilon_z} - \sqrt{\varepsilon_{z_1}}) \tag{23}$$

$$\Delta r_3^* = \frac{2M}{-E_c\sqrt{\phi_{ac}}}\sqrt{\varepsilon_r}(\varepsilon_z - \varepsilon_{z_1}) \tag{24}$$

The definitions of Δz_1^*, Δz_2^* and Δr_2^*, Δr_3^* and its Eq. (20), Eq. (21) and Eq. (23), (24) have been derived through the approach of practical ray solution in the previous paper[1]. Expression (23) is just the Recknagel-Artimovich formula, which is formed by the difference of paraxial ray solution[4,5]. Thus, we have derived the paraxial spatial aberrations from the approach of paraxial ray solution.

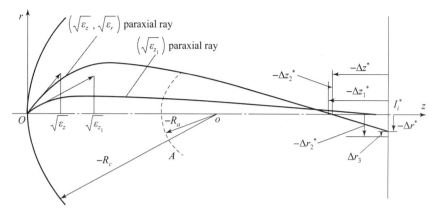

Fig. 1 The paraxial lateral and longitudinal aberrations in a bi-electrode concentric spherical system with electrostatic focusing

2 Paraxial temporal solution and its temporal aberrations for a bi-electrode concentric spherical system with electrostatic focusing

2.1 Paraxial temporal solution

In an ideal model of bi-electrode concentric spherical system the accurate expression of flight time $t(z)$ for electrons leaving from the cathode has been derived [1,6], now we deduce the paraxial expression of time $t^*(z)$. Integrating the expression of axial velocity of paraxial electron (5), we have

$$t^*(z) = \int_0^z \frac{\mathrm{d}z}{\left\{\frac{2e}{m_0}[\phi(z) + \varepsilon_z]\right\}^{1/2}} \tag{25}$$

Substituting the expression (2) of $\phi(z)$ in Eq. (25), we can obtain the paraxial expression of flight time $t^*(z)$, when an electron emits from photocathode under the following initial condition: $z=0$, $t=0$, and $\dot{z}_0 = \sqrt{2e\varepsilon_z/m_0}$.

$$t^*(z) = \frac{-R_c}{\left(\frac{2e}{m_0}\left[\frac{\phi_{ac} - (n-1)\varepsilon_z}{(n-1)}\right]\right)^{1/2}} \left\{ \sqrt{1 + \frac{z}{R_c}}\sqrt{\frac{(n-1)\varepsilon_z}{\phi_{ac} - (n-1)\varepsilon_z} - \frac{z}{R_c}} - \sqrt{\frac{(n-1)\varepsilon_z}{\phi_{ac} - (n-1)\varepsilon_z}} + \right.$$

$$\left. \left(1 + \frac{(n-1)\varepsilon_z}{\phi_{ac} - (n-1)\varepsilon_z}\right)\left[\arcsin\sqrt{\frac{(n-1)\varepsilon_z}{\phi_{ac}} - \frac{[\phi_{ac} - (n-1)\varepsilon_z]z}{R_c\phi_{ac}}} - \arcsin\sqrt{\frac{(n-1)\varepsilon_z}{\phi_{ac}}}\right] \right\}$$

(26)

$t^*(z)$ is called paraxial flight time. Formula (25) or formula (26) means that $t^*(z)$ does not depend on the initial radial potential of electron $\sqrt{\varepsilon_r}$.

Firstly, we determine the paraxial flight time from photocathode to anode, i.e., t^*_{C-A}, the distance from photocathode to anode is $z = -R_c + R_a$. From (2-2), because $(R_c - R_a)/R_c = (n-1)/n$, we have

$$t_{C-A}^* = \frac{-R_c}{\left\{\frac{2e}{m_0}\left[\frac{\phi_{ac}-(n-1)\varepsilon_z}{(n-1)}\right]\right\}^{1/2}} \left\{\sqrt{\frac{R_a}{R_c}}\sqrt{\frac{(n-1)(\phi_{ac}+\varepsilon_z)}{[\phi_{ac}-(n-1)\varepsilon_z]n}} - \sqrt{\frac{(n-1)\varepsilon_z}{\phi_{ac}-(n-1)\varepsilon_z}} + \right.$$

$$\left. \left(1+\frac{(n-1)\varepsilon_z}{\phi_{ac}-(n-1)\varepsilon_z}\right)\left[\arcsin\sqrt{\frac{(n-1)(\phi_{ac}+\varepsilon_z)}{n\phi_{ac}}} - \arcsin\sqrt{\frac{(n-1)\varepsilon_z}{\phi_{ac}}}\right]\right\} = t_{C-A}^*|_1 + t_{C-A}^*|_2 \quad (27)$$

Now, we simplify Eq. (27). Firstly, expanding the expression of $t_{C-A}^*|_1$, neglecting the higher order than ε_z/ϕ_{ac} terms which are greatly smaller than 1, we have

$$t_{C-A}^*|_1 = \frac{-R_c}{\left\{\frac{2e}{m_0}\left[\frac{\phi_{ac}-(n-1)\varepsilon_z}{(n-1)}\right]\right\}^{1/2}} \left\{\sqrt{\frac{R_a}{R_c}}\sqrt{\frac{(n-1)(\phi_{ac}+\varepsilon_z)}{[\phi_{ac}-(n-1)\varepsilon_z]n}} - \sqrt{\frac{(n-1)\varepsilon_z}{\phi_{ac}-(n-1)\varepsilon_z}}\right\} \quad (28)$$

$$= \left(\frac{m_0}{2e}\right)^{1/2}\frac{R_a-R_c}{\sqrt{\phi_{ac}}}\left\{1 - n\sqrt{\frac{\varepsilon_z}{\phi_{ac}}} + \frac{1}{2}(2n-1)\frac{\varepsilon_z}{\phi_{ac}}\right\}$$

Secondly, the expression $t_{C-A}^*|_2$ can be written as:

$$t_{C-A}^*|_2 = \frac{-R_c}{\left\{\frac{2e}{m_0}\left[\frac{\phi_{ac}-(n-1)\varepsilon_z}{(n-1)}\right]\right\}^{1/2}} \left\{\left(1+\frac{(n-1)\varepsilon_z}{\phi_{ac}-(n-1)\varepsilon_z}\right)\right.$$

$$\left.\left[\arcsin\sqrt{\frac{(n-1)(\phi_{ac}+\varepsilon_z)}{n\phi_{ac}}} - \arcsin\sqrt{\frac{(n-1)\varepsilon_z}{\phi_{ac}}}\right]\right\} \quad (29)$$

Now, we simplify expression of Eq. (29). Let $x = \sqrt{(n-1)(\phi_{ac}+\varepsilon_z)/n\phi_{ac}}$, $y = \sqrt{(n-1)\varepsilon_z/\phi_{ac}}$, because of $x^2+y^2 \leq 1$, by using following formula:

$$\arcsin x - \arcsin y = \arcsin(x\sqrt{1-y^2} - y\sqrt{1-x^2}),$$

thus terms in the brackets of Eq. (29) containing anti-trigonometric function can be simplified as

$$\arcsin\sqrt{\frac{(n-1)(\phi_{ac}+\varepsilon_z)}{n\phi_{ac}}} - \arcsin\sqrt{\frac{(n-1)\varepsilon_z}{\phi_{ac}}} = \arcsin\sqrt{\frac{n-1}{n}} - \sqrt{n-1}\sqrt{\frac{\varepsilon_z}{\phi_{ac}}} + \frac{1}{2}\sqrt{n-1}\frac{\varepsilon_z}{\phi_{ac}} \quad (30)$$

Therefore, expression 27 can be written as

$$t_{C-A}^*|_2 = -R_c\sqrt{\frac{m_0}{2e}}\sqrt{\frac{n-1}{\phi_{ac}}}\left(1+\frac{3}{2}\frac{(n-1)\varepsilon_z}{\phi_{ac}}\right)\left\{\arcsin\sqrt{\frac{n-1}{n}} - \sqrt{n-1}\sqrt{\frac{\varepsilon_z}{\phi_{ac}}} + \frac{1}{2}\sqrt{n-1}\frac{\varepsilon_z}{\phi_{ac}}\right\} \quad (31)$$

Thus, we obtain the approximate expression of paraxial flight time t_{C-A}^* as following:

$$t_{C-A}^* = \left(\frac{2m_0}{e}\right)^{1/2}\left\{\frac{R_a-R_c}{2\sqrt{\phi_{ac}}} - \frac{1}{2}R_c\sqrt{\frac{n-1}{\phi_{ac}}}\arcsin\sqrt{\frac{n-1}{n}} - n\frac{R_a-R_c}{\sqrt{\phi_{ac}}}\sqrt{\frac{\varepsilon_z}{\phi_{ac}}} + \right.$$

$$\left. n\frac{R_a-R_c}{\sqrt{\phi_{ac}}}\frac{\varepsilon_z}{\phi_{ac}} + n\frac{R_a-R_c}{\sqrt{\phi_{ac}}}\left[\frac{1}{4}\frac{3n-1}{n} + \frac{3}{4}\sqrt{n-1}\arcsin\sqrt{\frac{n-1}{n}}\right]\right\} \quad (32)$$

It shows that from the approach of paraxial solution, the expression (32) derived coincides with the expression derived from practical solution in the previous paper[1].

Secondly, we should calculate the paraxial flight time from anode to the axis, i.e., $t_{A-i^*}^*$ for the paraxial electron with initial axial potential $\sqrt{\varepsilon_z}$ leaving from the anode in the region free from field. Let v_a^* be the paraxial axial velocity of electron leaving from the anode, the paraxial electron ray is a straight line, which reaches the central axis from anode, its distance is l, the time interval is $t_{A-i^*}^*$, so we have

$$t_{A-i^*}^* = \frac{l}{v_a^*} \tag{33}$$

where $l = [r_a^2 + (-R_a + R_i)^2]^{1/2}$; $v_a^* = \sqrt{v_z^2 + \dfrac{2e}{m_0}\phi_{ac}}$.

Considering the paraxial condition, their approximations can be written as

$$v_a^* = \sqrt{\frac{2e}{m_0}\phi_{ac}}\left(1 + \frac{1}{2}\frac{\varepsilon_z}{\phi_{ac}}\right),$$

$$l \approx -R_a + R_i = \frac{-2R_c(n-1)}{n(n-2)}\left\{1 + \frac{n}{n-2}\sqrt{\frac{\varepsilon_z}{\phi_{ac}}} + \frac{n^2}{(n-2)^2}\frac{\varepsilon_z}{\phi_{ac}}\right\}$$

Substituting above formula (33), we obtain the approximate value of $t_{A-i^*}^*$:

$$t_{A-i^*}^* = -\sqrt{\frac{2m_0}{e}}\frac{R_a(n-1)}{(n-2)\sqrt{\phi_{ac}}}\left\{1 + \frac{n}{n-2}\sqrt{\frac{\varepsilon_z}{\phi_{ac}}} + \frac{n^2+4n-4}{2(n-2)^2}\frac{\varepsilon_z}{\phi_{ac}}\right\} \tag{34}$$

The expression (34) coincides with the expression (19), where $\sqrt{\varepsilon_z} = \sqrt{\varepsilon_{z_1}}$ in the previous paper.

2.2 Paraxial temporal aberration

The paraxial temporal aberration ΔT^* is measured in the ideal image plane determined by paraxial ray with $\sqrt{\varepsilon_{z_1}}$, and defined by the flight time difference between the two paraxial rays travelling with ($\sqrt{\varepsilon_z}$) and ($\sqrt{\varepsilon_{z_1}}$). As shown in Fig. 2, ΔT^* consists of two parts, the one is the time-transit-spread of the two paraxial electron rays from cathode to anode, the other is the time-transit-spread of this two paraxial electron rays arriving at the ideal image plane determined by paraxial ray with $\sqrt{\varepsilon_{z_1}}$, that is

$$\Delta T^* = t^*|_{C-A}(\sqrt{\varepsilon_z}) + (t^* + \Delta t)^*|_{A-i^*}(\sqrt{\varepsilon_z}) - t^*|_{C-A}(\sqrt{\varepsilon_{z_1}}) - t^*|_{A-i^*}(\sqrt{\varepsilon_{z_1}}) \tag{35}$$

where $t^*|_{C-A}(\sqrt{\varepsilon_z})$, $t^*|_{C-A}(\sqrt{\varepsilon_{z_1}})$ are expressed by Eq. (8), $t^*|_{A-i^*}(\sqrt{\varepsilon_{z_1}})$ by Eq. (9), and

$$(t^* + \Delta t)^*|_{A-i^*}(\sqrt{\varepsilon_z}) = t^*|_{A-i}(\sqrt{\varepsilon_z}) + \Delta t^*|_{i-i^*}(\sqrt{\varepsilon_z}) \tag{36}$$

Now we solve Eq. (36). The paraxial ray travelling with ($\sqrt{\varepsilon_z}$) from the anode through the region free from field firstly arrives the central axis, then reaches the ideal image plane determined by paraxial ray with $\sqrt{\varepsilon_{z_1}}$, the time difference is $\Delta t^*|_{i-i^*}(\sqrt{\varepsilon_z})$. Similar to the derivation and treatment in the previous paper, we may obtain

$$\Delta t^* \big|_{i-i^*}(\sqrt{\varepsilon_z}) = R_a \frac{n}{n-2}\sqrt{\frac{m_0}{2e}} \frac{1}{\sqrt{\phi_{ac}}} \left\{ \frac{2(n-1)}{(n-2)\sqrt{\phi_{ac}}}(\sqrt{\varepsilon_z}-\sqrt{\varepsilon_{z_1}}) + \frac{2n(n-1)}{(n-2)^2 \phi_{ac}}(\varepsilon_z - \varepsilon_{z_1}) \right\}$$
(37)

Substituting Eq. (34) and Eq. (37) in Eq. (36), we have

$$(t^* + \Delta t)^* \big|_{A-i^*}(\sqrt{\varepsilon_z}) = -\sqrt{\frac{2m_0}{e}} \frac{R_a(n-1)}{(n-2)\sqrt{\phi_{ac}}} \left\{ 1 + \frac{n}{n-2}\sqrt{\frac{\varepsilon_z}{\phi_{ac}}} + \frac{n^2+4n-4}{2(n-2)^2}\frac{\varepsilon_z}{\phi_{ac}} \right\} +$$

$$R_a \frac{n}{n-2}\sqrt{\frac{m_0}{2e}} \frac{1}{\sqrt{\phi_{ac}}} \left\{ \frac{2(n-1)}{(n-2)\sqrt{\phi_{ac}}}(\sqrt{\varepsilon_z}-\sqrt{\varepsilon_{z_1}}) + \frac{2n(n-1)}{(n-2)^2 \phi_{ac}}(\varepsilon_z - \varepsilon_{z_1}) \right\}$$
(38)

Finally, from Eq. (35), we may obtain the paraxial temporal aberration ΔT^* which is classified in order-sequence as following:

$$\Delta T^* = \Delta V T_1^* + \Delta V T_2^* \tag{39}$$

where
$$\Delta T_1^* = \left(\frac{2m_0}{e}\right)^{1/2} \frac{\sqrt{\varepsilon_z}-\sqrt{\varepsilon_{z_1}}}{E_c} \tag{40}$$

$$\Delta T_2^* = \left(\frac{2m_0}{e}\right)^{1/2} \frac{\varepsilon_z - \varepsilon_{z_1}}{E_c \sqrt{\phi_{ac}}} \times \left\{ -\frac{3n^2-9n+2}{4n(n-2)} - \frac{3}{8}\sqrt{n-1}\frac{1}{n}\left(\pi - \arcsin\frac{2}{n}\sqrt{n-1}\right) \right\} \tag{41}$$

where ΔT_1^* and ΔT_2^* are called the paraxial temporal aberrations of the first order and second order, respectively. Formula (40) and formulae (41) we have deduced in the previous paper, but here the paraxial temporal aberrations have been derived from the approach of paraxial solutions.

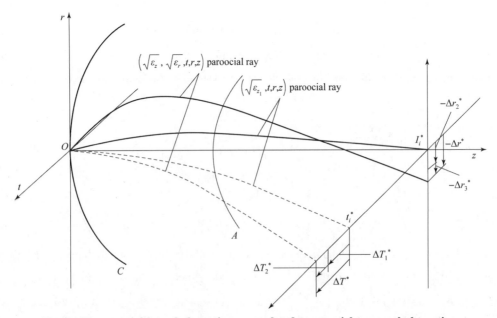

Fig. 2 The paraxial lateral aberrations are related to paraxial temporal aberrations

Conclusions

From the above investigation, we can reach the following conclusions:

(1) In present paper, we have given another approach—the approach of paraxial solution, to solve the paraxial spatial aberrations and paraxial temporal aberration in a bi-electrode concentric spherical system with electrostatic focusing. The work convinced us that either the approach of paraxial solution or the approach of practical solution, if follows the given definition of aberrations, the results will be fully identical.

(2) We have deduced the paraxial ray solution and paraxial temporal solution, as well as the paraxial spatial aberration and paraxial temporal aberration in a bi-electrode concentric spherical system with electrostatic focusing through the paraxial ray equation and paraxial motion equation. The result shows that either the paraxial lateral aberration of the second order Δr_2^* or the paraxial temporal aberration of the first order ΔT_1^*, in which the former determines the fundamental limitation of spatial resolution for image tubes,, the latter determines the fundamental limitation of temporal resolution of high-speed photographic image converter tube, depend only on the paraxial solutions. It means that we should lay emphasis on the investigation of paraxial solutions in further study of imaging electron optics.

(3) We have once again derived the expressions of the paraxial lateral aberration of the third order Δr_3^*, as well as the paraxial temporal aberration of the second order ΔT_2^*, in a bi-electrode concentric spherical system with electrostatic focusing, which are same magnitude with the geometrical lateral aberration of the third order δr_3 and the geometrical temporal aberration of the second order δT_2, respectively. It means that for imaging electron optics, while study the geometrical (lateral or temporal) aberrations: δr_3, δT_2, the paraxial (spatial or temporal) aberrations: Δr_3^*, ΔT_2^* should be considered at the same time. The present paper and previous paper clearly indicate that the paraxial lateral aberration of the third order Δr_3^* and paraxial temporal aberration of the second order ΔT_2^* exit, and the concrete expressions of these two aberrations have firstly been given.

Acknowledgements

This project was supported by National Natural Science Foundation of China and the Special Science Fund of Doctoral Discipline for Institutions of Higher Learning.

References

[1] ZHOU L W, GONG H, ZHANG Y F, et al. Static and dynamic imaging electron optics and spatial-temporal aberrations in a bi-electrode spherical concentric system with electrostatic focusing[J]. International Journal for Light and Electron Optics, 2011, 122(4):287-294.

[2] ZHOU L W. Electron optics with wide beam focusing[M]. Beijing: Beijing Institute of

Technology Press, 1993 (in Chinese).

[3] ZHOU L W. Electron optics of concentric spherical electrostatic focusing systems with two electrodes [M]//Focusing and Imaging of Wide Electron Beams—Selected Papers on Electron Optics by Zhou Liwei. Beijing: Beijing Institute of Technology Press, 1994: 11 – 29 (in Chinese).

[4] ARTIMOVICH L A. Electrostatic properties of emission systems [J]. Bulletin of Academy of Sciences, Physics Series, 1944, 8(6): 313 – 328 (in Russian).

[5] RECKNAGEL A. Theorie des elektrisohen elecktronen miktroskops fur selbstrakler [J]. Z. Angew. Physik, 1941, 117: 689 – 708 (in German).

[6] ZHOU L W, LI Y, ZHANG Z Q, et al. Test and verification of temporal aberration theory for electron optical imaging systems by an electrostatic concentric spherical system [J]. Acta Physica Sinica, 2005, 54(8): 3597 – 3603 (in Chinese).

10.4 Static and Dynamic Imaging Electron Optics and Spatial-temporal Aberrations in a Bi-electrode Spherical Concentric System with Electrostatic Focusing
两电极静电聚焦同心球系统的静动态成像电子光学及其空间－时间像差

Abstract: In the present paper, based on the practical electron ray equation and electron motion equation, for a bi-electrode concentric spherical system with electrostatic focusing, the spatial-temporal trajectories of moving electrons emitted from the photocathode have been solved, the exact and approximate formulae for image position and flight time of electrons, have been deduced. Starting from analytic solutions of spatial-temporal trajectories, the electron optical spatial-temporal properties of this system are then discussed. According to the definitions of spatial-temporal aberrations, the paraxial and geometrical lateral aberrations, as well as the paraxial and geometrical temporal aberrations, have been deduced.

摘要：基于实际电子轨迹方程和电子运动方程，本文求解了两电极同心球静电聚焦系统中自光阴极逸出的运动电子的空间－时间轨迹，导出了成像位置以及电子飞行时间的精确和近似表示式。作者由空间－时间轨迹的解析解，讨论了系统电子光学的空间－时间特性。按照空间－时间像差的定义，导出了近轴和几何横向（空间）像差，以及近轴和几何时间像差。

Introduction

For an electron optical imaging system, spatial-trajectory-spread and transit-time-spread produced by initial emission energy and initial emission angle of photoelectrons are the fundamental problems of image tubes and high-speed image converter tubes. In the static mode operation for these two tubes, the spatial-trajectory-spread of electron trajectories at the image plane determines limited spatial resolution. In the streak mode operation for high-speed image converter tube, the point image of a photo event of very short duration is imaged by the camera as a line because of transit-time-spread. The length of the line is proportional to the transit-time-

---

Zhou Liwei[a], Gong Hui[a], Zhang Zhiquan[b], Zhang Yifei[b]. a) Beijing Institute of Technology, b) Institute of Armored Force Engineering.
Optik, V. 122, No. 4, 2011, 287–294.

spread. Obviously, the shorter the length of the line, the better the time resolution of the camera.

It was proven that the spatial-trajectory-spread of an imaging system is mainly determined by the paraxial lateral chromatic aberration of the second order, which is usually summarized as Recknagel-Artimovich formula[1,2]. This formula shows that the paraxial lateral chromatic aberration of the second order does not depend on the concrete electrode structure and potential distributions, but depends on the initial electron emission energy, the strength of electric field at the photocathode, the linear magnification of the system and the position of image plane.

Similarly, the transit-time-spread of an imaging system is mainly determined by the paraxial temporal aberration of first order, which is expressed by Savoisky-Fanchenko formula[3]. This formula shows that the temporal aberration of the first order depends only on the initial emission energy, the strength of electric field at the photocathode, but does not depend on the concrete electrode structure and potential distribution. Besides, it does not depend on the linear magnification of the system and the position of image plane.

Thus, we may find many similarities between Recknagel-Artimovich formula and Savoisky-Fanchenko formula, but the difference exists. We can also find the non-harmonious and non-symmetry between them if we think it over. Actually, either the spatial-trajectory-spread or the transit-time-spread at an image plane, which are produced by spatial-temporal trajectories emitted from photocathode with initial emission energy and initial emission angle of same photoelectrons. It is strange that the former depends on the position of image plane, but the later does not.

The best model for investigating the difference is a bi-electrode concentric spherical electrostatic system with electrostatic focusing. This is because in this system the potential distribution and electron trajectory can be written as analytical forms, the focusing and imaging characteristics and the spatial-temporal aberrations can be investigated quantitatively. With a concentric spherical electrostatic system with electrostatic focusing, whether realized accurately or approximately, a good image quality can be obtained.

For the static electron optics of a bi-electrode concentric spherical electrostatic system with electrostatic focusing, the previous work was given by E. Ruska (1933)[5] and P. Schagen (1952)[6], but all of these works remained at the zero the order approximation of imaging. The more accurate formulae for the solution of electron trajectory and lateral aberrations of this system were given by Zhou Liwei in his monographs[7,8]. For the dynamic electron optics of a bi-electrode concentric spherical electrostatic system with electrostatic focusing, the work of I. P. Csorba[4] has only confirmed the conclusion of Savoisy-Fanchenko formula[3], and the more accurate formula of temporal aberration has been developed by Zhou Liwei, et al[9].

In the present paper, for a bi-electrode concentric spherical electrostatic system, the static imaging electron optics and spatial aberrations will be discussed in part I, the dynamic imaging electron optics and temporal aberrations will be discussed in part II.

1. Static electron optics and spatial aberrations for a bi-electrode concentric spherical system with electrostatic focusing

1.1 Ray equation and its solution

Fig. 1 shows a bi-electrode electrostatic concentric spherical system. Suppose that the curvature radii of spherical cathode C and of the mesh-spherical anode A are R_c and R_a, respectively, the common curvature center is O. Because of spherical symmetry of the system, we can use the polar coordinate (ρ, φ) to describe the electron trajectory. Let the original point of the polar coordinate is located at the center O of curvature, the potential of the photocathode is $\phi_c = 0$, the potential of mesh-anode A with respect to the photocathode C is ϕ_{ac}, the photoelectron with initial emission velocity v_0 and initial emission angle α_0 is emitted from the photocathode. Suppose that the moving direction of electron ray is from left to right, we stipulate that the line segment is positive from left to right or from bottom to top, otherwise is negative. R_c, R_a, ρ, R_i, R_i^* are calculated from the curvature center O to the corresponding spherical surface; Δz is calculated from the ideal image position I_i^*; Δr is calculated from the central axis $\varphi = 0$.

Suppose that all of the angles be measured by the acute angle, we stipulate that rotation turns round counter clockwise is positive, otherwise is negative. Similar to the line segments, we should stipulate the beginning axis for the angle calculation, α_0, φ, γ, α_i are calculated from central axis $\varphi = 0$, ζ, τ are calculated from the joint line connecting the center O and the point concerned as a beginning axis.

The field between two concentric spheres is equivalent to the electrostatic field produced by a point charge which is concentrated in spherical center. $\phi_{\rho c}$ is the potential at coordinate ρ with respect to photocathode C, can be expressed as

$$\phi_{\rho c} = E_c R_c \left(\frac{R_c}{\rho} - 1 \right) \tag{1}$$

where E_c is the strength of electric-field at the photocathode. When $\rho = R_a$, we have

$$E_c = \frac{\phi_{ac}}{R_c(n-1)} \tag{2}$$

where $n = R_c/R_a$. For the concave cathode-anode system, R_c, R_a take negative value, $n > 1$; for the convex cathode-anode system, R_c, R_a take positive value, $n < 1$. E_c always takes negative value n whether $n > 1$ or $n < 1$.

In a bi-electrode concentric spherical system, consider an electron leaving photocathode with initial velocity v_0 and initial angle α_0 as indicated in Fig. 1, the motion of electron path follows the law of conversation of energy and the law of conversation of angular momentum:

$$\frac{1}{2}m_0 v^2 = \frac{1}{2}m_0 v_0^2 + e\phi_{\rho c} \tag{3}$$

$$\rho^2 \dot{\varphi} = R_c^2 \dot{\varphi}_0 = \text{const} \tag{4}$$

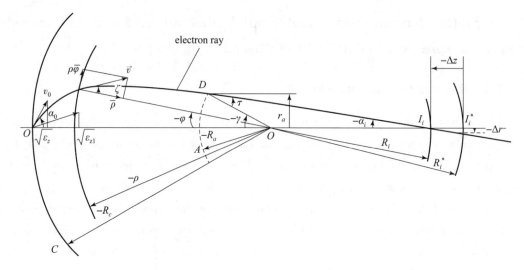

Fig. 1 A bi-electrode concentric spherical system with electrostatic focusing ($n = R_c/R_a > 2$)

where m_0, e are electron mass and charge, e takes absolute value; v is the velocity at polar coordinate (ρ, φ), $\dot{\varphi} = d\varphi/dt$.

From Eq. (3) and Eq. (4), three kinds of expressions for ray equation have been obtained. One is the E. Ruska[5] equation, in which φ is the variable, i.e., $\rho = f(\varphi)$ which can be written as following:

$$\mu = \left(1 - \frac{p}{2\sin^2\alpha_0}\right)\cos\varphi - \frac{1}{\tan\alpha_0}\sin\varphi + \frac{p}{2\sin^2\alpha_0} \tag{5}$$

The other one is P. Schagen[6] equation, in which ρ is the variable, i.e., $\varphi = f(\rho)$, which can be written as following:

$$\tan\frac{\varphi}{2} = \frac{-\cos\alpha_0 + [1 + p(\mu - 1) - \mu^2\sin^2\alpha_0]^{1/2}}{-\dfrac{p}{\sin\alpha_0} + (\mu + 1)\sin\alpha_0} \tag{6}$$

where $\mu = R_c/\rho$, $p(n-1) = \phi_{ac}/\varepsilon_0$, $\varepsilon_0 = m_0 v_0^2/2e$. ε_0 is the applied potential required for a rest electron to obtain the emission energy, which is called initial potential, p is only a parameter, but $p(n-1)$ has clear physical meaning, which represents the ratio of applied potential to the initial potential. Usually, for an ordinary image tube, $p(n-1) = 10^3 - 10^4$.

It should be noted that the E. Ruska and P. Schagen equations only give the flight position of electron emitted from photocathode, but the flight direction of electron at any position has not been clearly given. Furthermore, these two equations were not convenient for calculating the moving electron trajectory in this system, for example, the half-angle expression in P. Schagen equation. For solving this problem, the third equations $\varphi = f(\rho)$, $\zeta = \psi(\rho)$ were given by Zhou Liwei, expressed in the form of $\sin\varphi$, $\cos\varphi$ and $\tan\zeta$[7,8]:

$$\sin\varphi = -\frac{(\mu - 1)(c + d_0)}{bd_0 + b_0 c}\sin\alpha_0 \tag{7a}$$

$$\cos\varphi = \frac{bc + b_0 d_0}{bd_0 + b_0 c} \quad (7b)$$

$$\tan\zeta = \frac{\mu\sin\alpha_0}{b} \quad (8)$$

where
$$b_0 = \cos\alpha_0, \quad b = \left[1 + \frac{\phi_{ac}}{\varepsilon_0}\frac{(\mu-1)}{(n-1)} - \mu^2\sin^2\alpha_0\right]^{\frac{1}{2}},$$

$$c = 1 - 2\mu(n-1)\frac{\varepsilon_0}{\phi_{ac}}\sin^2\alpha_0, \quad d_0 = 1 - 2(n-1)\frac{\varepsilon_0}{\phi_{ac}}\sin^2\alpha_0$$

Formula (7a), formula (7b), and formula (8) are generally appropriate whether $n > 1$ or $n < 1$, which is very simple and convenient for calculation. The great advantage of formula of (7a), formula (7b) and formula (8) is that it has analogical property when solving the problem of ray tracing for multi-electrode concentric spherical system with electrostatic focusing.

1.2 Determination of ideal image position R_i^* and its paraxial imaging

From Eq. (7a), Eq. (7b) and Eq. (8), it is not difficult to trace the electron ray emitted from photocathode with initial parameters ($\sqrt{\varepsilon_0}$, α_0) and to define the position D at the anode A, thus we can define angle τ and γ. We assume that the mesh-anode is transparent for electrons passing through, and after the mesh-anode there is a region free from the field. When the electron goes through the mesh-anode, the magnitude and direction of electron velocity will not be changed, the solution and its slope of electron ray are continuous. Therefore, the trajectory after the mesh-anode is a straight line tangentially to the trajectory at point D, and intersects the axis $\varphi = 0$ at point I_i.

According to the triangle $\triangle ODI_i$ given in Fig. 1, it is not difficult to obtain the expression of R_i at the intersection I_i:

$$R_i = R_a\frac{\sin\tau}{\sin\alpha_i} = R_a\frac{\tan\tau}{\sin\gamma + \tan\tau\cos\gamma} \quad (9)$$

where α_i is the angle of electron ray at intersection I_i with axis $\varphi = 0$; R_i is the distance from curvature center O to intersection I_i.

Obviously, if $\mu = n$, $\varphi = \gamma$, $\zeta = \tau$, $b = b_1$, $c = c_0$, from formula (7a), formula (7b) and formula (8) we can get the expressions of $\sin\gamma$, $\cos\gamma$ and $\tan\tau$. Using formula (9), we may obtain exact expressions of trajectory intersection and its slope at I_i for electron ray emitted from photocathode with initial parameters ($\sqrt{\varepsilon_0}$, α_0) in the bi-electrode electrostatic concentric spherical system:

$$R_i = -R_a\frac{n(b_1 d_0 + b_0 c_0)}{b_1(n-1)(c_0 + d_0) - n(b_1 c_0 + b_0 d_0)} \quad (10)$$

$$\tan\alpha_i = -\frac{b_1(n-1)(c_0 + d_0) - n(b_1 c_0 + b_0 d_0)}{b_1(b_1 c_0 + b_0 d_0) + n(n-1)(c_0 + d_0)\sin^2\alpha_0}\sin\alpha_0 \quad (11)$$

where

$$b_1 = \left[1 + \frac{\phi_{ac}}{\varepsilon_0} - n^2 \sin^2\alpha_0\right]^{\frac{1}{2}}, \quad c_0 = 1 - 2n(n-1)\frac{\varepsilon_0}{\phi_{ac}}\sin^2\alpha_0$$

Now, we simplify the above-mentioned expressions for R_i and $\tan\alpha_i$. Let $\sqrt{\varepsilon_z} = \sqrt{\varepsilon_0}\cos\alpha_0$, $\sqrt{\varepsilon_r} = \sqrt{\varepsilon_0}\sin\alpha_0$ and using binomial theorem to make power series expansion for the terms Eq. (10) and Eq. (11), neglecting $(\varepsilon_0/\phi_{ac})^{3/2}$ and its higher order terms which is greatly smaller than 1, we may obtain the approximate expressions of R_i and its slope $\tan\alpha_i$ at trajectory intersection I_i:

$$R_i = -R_a \frac{n}{n-2}\left\{1 + \frac{2(n-1)}{n-2}\sqrt{\frac{\varepsilon_z}{\phi_{ac}}} + \frac{2n(n-1)}{(n-2)^2}\frac{\varepsilon_z}{\phi_{ac}} - \frac{2(n-1)^2}{n-2}\frac{\varepsilon_r}{\phi_{ac}}\right\} \quad (12)$$

$$\tan\alpha_i = -(n-2)\sqrt{\frac{\varepsilon_r}{\phi_{ac}}}\left\{1 - \frac{2(n-1)}{n-2}\sqrt{\frac{\varepsilon_z}{\phi_{ac}}} + \frac{3n-2}{2(n-2)}\frac{\varepsilon_z}{\phi_{ac}} + \frac{(n-1)(n^2-n+2)}{2(n-2)}\frac{\varepsilon_r}{\phi_{ac}}\right\} \quad (13)$$

From this, we can study the electron optical imaging properties and spatial aberrations for the bi-electrode concentric spherical system.

Considering the paraxial imaging condition for imaging systems in expression (12) and expression (13), let $\sqrt{\varepsilon_z} = \sqrt{\varepsilon_{z_1}}$, $\sqrt{\varepsilon_{z_1}}$ is corresponding to initial axial potential that paraxial electron carries it to paraxial image position at I_i^*. Neglecting term of ε_r/ϕ_{ac} in the brackets of expressions, we may obtain the approximate expressions of R_i^* and $\tan\alpha_i^*$ at I_i^* for the paraxial ray leaving the photocathode with the initial axial potential $\sqrt{\varepsilon_{z_1}}$:

$$R_i^* = -R_a \frac{n}{n-2}\left\{1 + \frac{2(n-1)}{n-2}\sqrt{\frac{\varepsilon_{z_1}}{\phi_{ac}}} + \frac{2n(n-1)}{(n-2)^2}\frac{\varepsilon_{z_1}}{\phi_{ac}}\right\} \quad (14)$$

$$\tan\alpha_i^* = -(n-2)\sqrt{\frac{\varepsilon_r}{\phi_{ac}}}\left\{1 - \frac{2(n-1)}{n-2}\sqrt{\frac{\varepsilon_{z_1}}{\phi_{ac}}} + \frac{3n-2}{2(n-2)}\frac{\varepsilon_{z_1}}{\phi_{ac}}\right\} \quad (15)$$

where the star mark (*) expresses the paraxial condition. Formula (14) shows that at the paraxial condition the ideal imaging will be formed so long as the axial initial velocity of electron $\sqrt{\varepsilon_{z_1}}$, remains the same, and if $\sqrt{\varepsilon_{z_1}}$ is changed, the position of ideal imaging will also be changed.

The linear magnification M of the system can be expressed by

$$M = \frac{R_i^*}{R_c} = -\frac{1}{n-2}\left\{1 + \frac{2(n-1)}{n-2}\sqrt{\frac{\varepsilon_{z_1}}{\phi_{ac}}} + \frac{2n(n-1)}{(n-2)^2}\frac{\varepsilon_{z_1}}{\phi_{ac}}\right\} \quad (16)$$

Obviously, it forms a real image upside down when $n > 2$ or a virtual image erect stand when $n < 2$.

1.3 Spatial (lateral) aberration—which is composed of paraxial spatial (lateral) aberration and geometrical spatial (lateral) aberration

In the following, we divide the spatial aberration into two parts, the paraxial aberration and the geometrical aberration. As shown in Fig. 2, the focusing position I_i^* of electrons emitted from the cathode with $\sqrt{\varepsilon_{z_1}}$ in paraxial condition is defined as an ideal image position, and the practical rays from the same object point with initial condition ($\sqrt{\varepsilon_z}$, $\sqrt{\varepsilon_r}$), will intersect the axial at I_i, it forms a displacement of image position along the axial direction, which is called longitudinal aberration as following:

$$\Delta z = R_i(\sqrt{\varepsilon_r}, \sqrt{\varepsilon_z}) - R_i^*(\sqrt{\varepsilon_{z_1}}) \\ = R_i^*(\sqrt{\varepsilon_r}, \sqrt{\varepsilon_z}) - R_i^*(\sqrt{\varepsilon_{z_1}}) + R_i(\sqrt{\varepsilon_r}, \sqrt{\varepsilon_z}) - R_i^*(\sqrt{\varepsilon_r}, \sqrt{\varepsilon_z}) \quad (17)$$

Substituting Eq. (14) and Eq. (12) into Eq. (17), we get the expression for longitudinal aberration Δz:

$$\Delta z = \Delta z^* + \delta z \\ = \Delta z_1^* + \Delta z_2^* + \delta z_1 + \delta z_2 \quad (18)$$

where

$$\Delta z_1^* = \frac{2M^2 \sqrt{\phi_{ac}}}{-E_c}(\sqrt{\varepsilon_z} - \sqrt{\varepsilon_{z_1}}) \quad (19)$$

$$\Delta z_2^* = \frac{2M^2}{-E_c}\left\{\frac{n}{n-2}(\varepsilon_z - \varepsilon_{z_1}) - \frac{4(n-1)}{n-2}\sqrt{\varepsilon_{z_1}}(\sqrt{\varepsilon_z} - \sqrt{\varepsilon_{z_1}})\right\} \quad (20)$$

$$\delta z_1 = 0, \quad \delta z_2 = \frac{2M^2}{E_c}(n-1)\varepsilon_r \quad (21)$$

Similarly, at the ideal image plane determined in paraxial condition by $\sqrt{\varepsilon_{z_1}}$, the lateral aberration Δr can be expressed as following:

$$\Delta r = -\Delta z \tan \alpha_i \quad (22)$$

where $\tan \alpha_i$ is the slope of practical electron ray of ($\sqrt{\varepsilon_z}$, $\sqrt{\varepsilon_r}$) at point I_i composed with the axis $\varphi = 0$, it is determined by Eq. (13).

Substituting Eq. (18) and Eq. (13) in Eq. (22), we may obtain the lateral aberration Δr as following:

$$\Delta r = r(z_i^*, \sqrt{\varepsilon_z}, \sqrt{\varepsilon_r}) - r^*(z_i^*, \sqrt{\varepsilon_{z_1}}) \\ = r^*(z_i^*, \sqrt{\varepsilon_z}, \sqrt{\varepsilon_r}) - r^*(z_i^*, \sqrt{\varepsilon_{z_1}}) + r(z_i^*, \sqrt{\varepsilon_z}, \sqrt{\varepsilon_r}) - r^*(z_i^*, \sqrt{\varepsilon_z}, \sqrt{\varepsilon_r}) \quad (23) \\ = \Delta r^* + \delta r = \Delta r_2^* + \Delta r_3^* + \delta r_2 + \delta r_3$$

where

$$\Delta r_2^* = \frac{2M}{E_c}\sqrt{\varepsilon_r}(\sqrt{\varepsilon_z} - \sqrt{\varepsilon_{z_1}}) \quad (24)$$

$$\Delta r_3^* = \frac{2M}{-E_c\sqrt{\phi_{ac}}}\sqrt{\varepsilon_r}(\varepsilon_z - \varepsilon_{z_1}) \quad (25)$$

$$\delta r_2 = 0, \quad \delta r_3 = \frac{2M}{-E_c\sqrt{\phi_{ac}}}(n-1)\varepsilon_r^{\frac{3}{2}} \quad (26)$$

The definition of spatial aberrations is as following (see Fig. 2).

Δr^*—paraxial lateral aberration, that is the radial displacement between paraxial ray of

($\sqrt{\varepsilon_z}$, $\sqrt{\varepsilon_r}$) and paraxial ray of $\sqrt{\varepsilon_{z_1}}$ at the ideal image plane which is determined by paraxial ray of $\sqrt{\varepsilon_{z_1}}$.

Δz^* — paraxial longitudinal aberration, that is the axial displacement between paraxial ray of ($\sqrt{\varepsilon_z}$, $\sqrt{\varepsilon_r}$) and paraxial ray of $\sqrt{\varepsilon_{z_1}}$.

δr — geometrical lateral aberration, that is the radial displacement between paraxial ray of ($\sqrt{\varepsilon_z}$, $\sqrt{\varepsilon_r}$) and practical ray of ($\sqrt{\varepsilon_z}$, $\sqrt{\varepsilon_r}$) at the ideal image plane.

δz — geometrical longitudinal aberration, that is the axial displacement between paraxial ray of ($\sqrt{\varepsilon_z}$, $\sqrt{\varepsilon_r}$) and practical ray of ($\sqrt{\varepsilon_z}$, $\sqrt{\varepsilon_r}$) at the ideal image plane.

The lower index numeral of aberration is given according to the power of $\sqrt{\varepsilon_z}/\sqrt{\phi_{ac}}$ and $\sqrt{\varepsilon_r}/\sqrt{\phi_{ac}}$, which represents the order of this aberration.

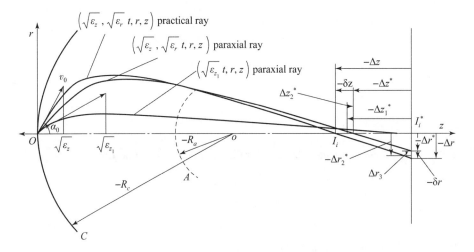

Fig. 2 Spatial (lateral) aberration can be regarded as composition of paraxial spatial (lateral) aberration and geometrical spatial (lateral) aberration

2 Dynamic electron optics and temporal aberrations for a bi-electrode concentric spherical system with electrostatic focusing

2.1 Solution of flight time of electron motion

Now, we discuss the dynamic electron optics and temporal aberrations for a bi-electrode concentric spherical system with electrostatic focusing. Firstly, we should solve the flight time of electron motion in this system.

Starting from the initial condition of electron emitted from photocathode: $R_c \dot{\varphi}_0 = v_0 \sin \alpha_0$, and from Eq. (4), we may have

$$\rho^2 \dot{\varphi} = R_c v_0 \sin \alpha_0 \tag{27}$$

The electron velocity v in the polar coordinate (ρ, φ) can be written as

$$v^2 = \dot{\rho}^2 + (\rho\dot{\varphi})^2 \tag{28}$$

From Eq. (27) and Eq. (28), we may obtain

$$\dot{\rho} = \frac{d\rho}{dt} = \sqrt{v^2 - (R_c^2 v_0^2 \sin^2\alpha_0/\rho^2)} \tag{29}$$

Substituting

$$v^2 = v_0^2 + \frac{2e}{m_0}\frac{\phi_{ac}}{n-1}\left(\frac{R_c}{\rho} - 1\right) \tag{30a}$$

in Eq. (29), we have

$$\dot{\rho} = \sqrt{\frac{2e\varepsilon_0}{m_0}}[1 + p(\mu - 1) - \mu^2\sin^2\alpha_0]^{\frac{1}{2}} \tag{30b}$$

By using transformation:

$$\dot{\rho} = \frac{d\rho}{dt} = -\frac{R_c}{\mu^2}\frac{d\mu}{dt} \tag{31}$$

Eq. (30b) will take this form:

$$dt = \frac{-R_c d\mu}{\sqrt{\frac{2e\varepsilon_0}{m_0}}\mu^2\sqrt{1 + p(\mu - 1) - \mu^2\sin^2\alpha_0}} \tag{32}$$

Integrating Eq. (32) and considering $\rho = R_c$, $\mu = 1$, when $t = 0$, we may obtain the formula of flight time t in the space between cathode and anode spheres:

$$t = \frac{-R_c}{\sqrt{\frac{2e\varepsilon_0}{m_0}}}\left\{-\frac{\sqrt{1 + p(\mu - 1) - \mu^2\sin^2\alpha_0}}{(1-p)\mu} + \frac{\cos\alpha_0}{1-p} - \right.$$

$$\left. \frac{p}{2(1-p)\sqrt{p-1}}\left[\arcsin\frac{p(\mu-2)+2}{\mu\sqrt{p^2 + 4(1-p)\sin^2\alpha_0}} + \arcsin\frac{p-2}{\sqrt{p^2 + 4(1-p)\sin^2\alpha_0}}\right]\right\} \tag{33}$$

2.2 Determination of flight time for the practical ray and paraxial ray arriving at anode

From Eq. (33), let $\varepsilon_0 = \varepsilon_r + \varepsilon_z$, $\rho = R_a$, i.e., $\mu = n$, we can get an exact expression of t_{C-A} which is the flight time arriving at the anode for the practical ray of ($\sqrt{\varepsilon_r}$, $\sqrt{\varepsilon_z}$) emitted from photocathode:

$$t_{C-A} = \sqrt{\frac{m_0}{2e}}\frac{-R_c}{\left[\frac{\phi_{ac}}{n-1} - \varepsilon_z - \varepsilon_r\right]}\left\{\frac{1}{n}\sqrt{\phi_{ac} + \varepsilon_z + (1-n^2)\varepsilon_r} - \sqrt{\varepsilon_z}\right\} - \sqrt{\frac{m_0}{2e}}\frac{\frac{\phi_{ac}}{n-1}R_c}{2\left(\frac{\phi_{ac}}{n-1} - \varepsilon_z - \varepsilon_r\right)^{3/2}} \times$$

$$\left\{\arcsin\frac{\left(1 - \frac{2}{n}\right) + \frac{2}{n}\frac{(n-1)}{\phi_{ac}}(\varepsilon_z + \varepsilon_r)}{\sqrt{1 - 4\left(\frac{\phi_{ac}}{n-1} - \varepsilon_z - \varepsilon_r\right)\varepsilon_r\left(\frac{n-1}{\phi_{ac}}\right)^2}} + \arcsin\frac{1 - 2\frac{n-1}{\phi_{ac}}(\varepsilon_z + \varepsilon_r)}{\sqrt{1 - 4\left(\frac{\phi_{ac}}{n-1} - \varepsilon_z - \varepsilon_r\right)\varepsilon_r\left(\frac{n-1}{\phi_{ac}}\right)^2}}\right\}$$

$$\tag{34}$$

Expanding Eq. (34) and neglecting all of these terms higher than orders of ε_z/ϕ_{ac} and ε_r/ϕ_{ac}, we may obtain an approximate expression of t_{C-A}:

$$t_{C-A} = -\sqrt{\frac{m_0}{2e}} \frac{R_c(n-1)}{n} \frac{1}{\sqrt{\phi_{ac}}} \left\{ 1 + \frac{2n-1}{2\phi_{ac}} \varepsilon_z + \frac{(n-1)^2}{2\phi_{ac}} \varepsilon_r \right\} +$$

$$\sqrt{\frac{m_0}{2e}} \frac{R_c(n-1)}{\sqrt{\phi_{ac}}} \sqrt{\varepsilon_z} - \sqrt{\frac{m_0}{2e}} \frac{R_c}{2} \frac{\sqrt{n-1}}{\sqrt{\phi_{ac}}} \left(1 + \frac{3}{2} \frac{n-1}{\phi_{ac}} \varepsilon_z + \frac{3}{2} \frac{n-1}{\phi_{ac}} \varepsilon_r \right) \times \quad (35)$$

$$\left\{ \arcsin \left[\left(1 - \frac{2}{n}\right) + \frac{2(n-1)}{n\phi_{ac}} \varepsilon_z + \frac{2(n-1)^2}{n\phi_{ac}} \varepsilon_r \right] + \arcsin \left[1 - 2\frac{n-1}{\phi_{ac}} \varepsilon_z \right] \right\}$$

The last term of expression (35) should be further simplified. Through a series of transformation, it can be written as

$$\left\{ \arcsin \left[\left(1 - \frac{2}{n}\right) + \frac{2(n-1)\varepsilon_z}{n\phi_{ac}} + \frac{2(n-1)^2}{n\phi_{ac}} \varepsilon_r \right] + \arcsin \left[1 - 2\frac{n-1}{\phi_{ac}} \varepsilon_z \right] \right\}$$

$$= \left\{ 2\arcsin \sqrt{\frac{n-1}{n}} - 2\sqrt{n-1} \sqrt{\frac{\varepsilon_z}{\phi_{ac}}} + \sqrt{n-1} \frac{\varepsilon_z}{\phi_{ac}} + 2(n-1)\sqrt{n-1} \frac{\varepsilon_r}{\phi_{ac}} \right\} \quad (36)$$

Therefore, the approximate expression of t_{C-A} can be expressed by

$$t_{C-A} = \left(\frac{2m_0}{e}\right)^{\frac{1}{2}} \left\{ \frac{-R_c(n-1)}{2\sqrt{\phi_{ac}}} \frac{1}{n} + \frac{-R_c}{2} \sqrt{\frac{n-1}{\phi_{ac}}} \arcsin \sqrt{\frac{n-1}{n}} + \right.$$

$$\frac{R_c(n-1)}{\sqrt{\phi_{ac}}} \sqrt{\frac{\varepsilon_z}{\phi_{ac}}} + \frac{-R_c(n-1)}{\sqrt{\phi_{ac}}} \left(\frac{3n-1}{4n} + \frac{3}{4} \sqrt{n-1} \arcsin \sqrt{\frac{n-1}{n}} \right) \frac{\varepsilon_z}{\phi_{ac}} + \quad (37)$$

$$\left. \frac{-R_c(n-1)}{\sqrt{\phi_{ac}}} \left[\frac{(n-1)(3n-1)}{4n} + \frac{3}{4} \sqrt{n-1} \arcsin \sqrt{\frac{n-1}{n}} \right] \frac{\varepsilon_r}{\phi_{ac}} \right\}$$

For the paraxial ray, when neglect the terms of ε_r/ϕ_{ac}, and let $\sqrt{\varepsilon_z} = \sqrt{\varepsilon_{z_1}}$, we have

$$t_{C-A}^* = \left(\frac{2m_0}{e}\right)^{\frac{1}{2}} \left\{ \frac{-R_c(n-1)}{2\sqrt{\phi_{ac}}} \frac{1}{n} + \frac{-R_c}{2} \sqrt{\frac{n-1}{\phi_{ac}}} \arcsin \sqrt{\frac{n-1}{n}} + \frac{R_c(n-1)}{\sqrt{\phi_{ac}}} \sqrt{\frac{\varepsilon_{z_1}}{\phi_{ac}}} + \right.$$

$$\left. \frac{-R_c(n-1)}{\sqrt{\phi_{ac}}} \left(\frac{3n-1}{4n} + \frac{3}{4} \sqrt{n-1} \arcsin \sqrt{\frac{n-1}{n}} \right) \frac{\varepsilon_{z_1}}{\phi_{ac}} \right\} \quad (38)$$

2.3 Calculation of flight time of electron in the region free from field behind the mesh-anode

When the electron arrives at the anode from the photocathode, its velocity v_a can be expressed by

$$v_a = \sqrt{v_0^2 + \frac{2e}{m_0} \phi_{ac}} \quad (39)$$

In the region free from field behind the anode, the electron ray is a straight line. Let the distance of electron ray from the mesh-anode to the axis is l, the flight time is t_{A-i}, we have

$$t_{A-i} = \frac{l}{v_a} \quad (40)$$

From the triangle ΔODI_i in Fig. 1, Eq. (40) can be expressed by

$$t_{A-i} = R_i \frac{\sin(-\gamma)}{v_a \sin\tau} \tag{41}$$

where R_i can be expressed by (12); $\sin(-\gamma)$ can be obtained from Eq. (7) when $\varphi = \gamma$; From Eq. (8), when $\zeta = \tau$, $\sin\tau$ will take the following form:

$$\sin\tau = \frac{n\sin\alpha_0}{\sqrt{b_1^2 + n^2\sin^2\alpha_0}} \tag{42}$$

Expanding R_i, $\sin(-\gamma)$ and $\sin\tau$, and neglecting all of these terms higher than orders of ε_z/ϕ_{ac} and ε_r/ϕ_{ac}, and from Eq. (41), for the practical ray from the mesh-anode to the axis $\varphi = 0$ in the region free from field, we may obtain the approximate expression of flight time t_{A-i} as following:

$$t_{A-i} = -\sqrt{\frac{2m_0}{e}} \frac{R_a(n-1)}{(n-2)\sqrt{\phi_{ac}}} \left\{ 1 + \frac{n}{n-2}\sqrt{\frac{\varepsilon_z}{\phi_{ac}}} + \frac{n^2+4n-4}{2(n-2)^2}\frac{\varepsilon_z}{\phi_{ac}} - \frac{(n-1)(n^2-n+2)}{2(n-2)}\frac{\varepsilon_r}{\phi_{ac}} \right\} \tag{43}$$

If neglect the terms of ε_r/ϕ_{ac} in Eq. (43), and let $\sqrt{\varepsilon_z} = \sqrt{\varepsilon_{z_1}}$, for the paraxial ray of $\sqrt{\varepsilon_{z_1}}$ from the mesh-anode to the axis $\varphi = 0$ in the region free from field, we may obtain the approximate expression of flight time $t_{A-i^*}^*$:

$$t_{A-i^*}^* = -\sqrt{\frac{2m_0}{e}} \frac{R_a(n-1)}{(n-2)\sqrt{\phi_{ac}}} \left[1 + \frac{n}{n-2}\sqrt{\frac{\varepsilon_{z_1}}{\phi_{ac}}} + \frac{n^2+4n-4}{2(n-2)^2}\frac{\varepsilon_{z_1}}{\phi_{ac}} \right] \tag{44}$$

2.4 Temporal aberrations

The temporal aberration ΔT is measured in the ideal image plane determined by paraxial ray of $\sqrt{\varepsilon_{z_1}}$, which is defined by the difference of flight time between the practical electron of ($\sqrt{\varepsilon_z}$, $\sqrt{\varepsilon_r}$) and paraxial electron of $\sqrt{\varepsilon_{z_1}}$ arriving at the image plane. It consists of two parts, the one is the time-transit-spread of these two electrons from photocathode to anode, the other one is the time-transit-spread of these two electrons from anode to the ideal image plane, as shown in Fig. 2, that is

$$\begin{aligned}\Delta T &= t_{C-i^*}(\sqrt{\varepsilon_z},\sqrt{\varepsilon_r}) - t_{C-i^*}^*(\sqrt{\varepsilon_{z_1}}) \\ &= t_{C-A}(\sqrt{\varepsilon_z},\sqrt{\varepsilon_r}) + t_{A-i^*}(\sqrt{\varepsilon_z},\sqrt{\varepsilon_r}) - t_{C-A}^*(\sqrt{\varepsilon_{z_1}}) - t_{A-i^*}^*(\sqrt{\varepsilon_{z_1}})\end{aligned} \tag{45}$$

where $t_{C-A}(\sqrt{\varepsilon_z},\sqrt{\varepsilon_r})$, $t_{C-A}^*(\sqrt{\varepsilon_{z_1}})$, $t_{A-i^*}^*(\sqrt{\varepsilon_{z_1}})$ are expressed by Eq. (37), Eq. (38), Eq. (44), respectively. Solution of $t_{A-i^*}(\sqrt{\varepsilon_z},\sqrt{\varepsilon_r})$ can be regarded as composition of two parts:

$$t_{A-i^*}(\sqrt{\varepsilon_z},\sqrt{\varepsilon_r}) = t_{A-i}(\sqrt{\varepsilon_z},\sqrt{\varepsilon_r}) + \Delta t_{i-i^*}(\sqrt{\varepsilon_z},\sqrt{\varepsilon_r}) \tag{46}$$

where $t_{A-i}(\sqrt{\varepsilon_z},\sqrt{\varepsilon_r})$ is the flight time of the practical electron of ($\sqrt{\varepsilon_z}$, $\sqrt{\varepsilon_r}$) from anode to central axis $\varphi = 0$, which is expressed by Eq. (43). Suppose that the practical electron flies with velocity v_a from central axis $\varphi = 0$ to the ideal image plane, the spatial distance between them is

Δl, the time difference is $\Delta t_{i-i^*}(\sqrt{\varepsilon_z}, \sqrt{\varepsilon_r})$, then we have

$$\Delta t_{i-i^*}(\sqrt{\varepsilon_z}, \sqrt{\varepsilon_r}) = \frac{\Delta l}{v_a} = \frac{-\Delta z}{\cos\alpha_i \cdot v_a} = -\Delta z \sqrt{1+\tan^2\alpha_i}\frac{1}{v_a} \tag{47}$$

where $\tan\alpha_i$, v_a are expressed by Eq. (13) and Eq. (39) respectively. $\Delta z = R_i - R_i^*$, R_i, R_i^* are expressed by Eq. (12) and Eq. (14) respectively.

Expanding Eq. (47), and neglecting all of these terms higher than orders of ε_z/ϕ_{ac} and ε_r/ϕ_{ac}, we may obtain

$$\Delta t_{i-i^*}(\sqrt{\varepsilon_z}, \sqrt{\varepsilon_r}) = R_a \frac{n}{n-2}\sqrt{\frac{m_0}{2e}}\frac{1}{\sqrt{\phi_{ac}}}\left\{\frac{2(n-1)}{(n-2)\sqrt{\phi_{ac}}}(\sqrt{\varepsilon_z}-\sqrt{\varepsilon_{z_1}}) + \frac{2n(n-1)}{(n-2)^2\phi_{ac}}(\varepsilon_z-\varepsilon_{z_1}) - \frac{2(n-1)^2}{(n-2)}\frac{\varepsilon_r}{\phi_{ac}}\right\} \tag{48}$$

Thus, $t_{A-i^*}(\sqrt{\varepsilon_z}, \sqrt{\varepsilon_r})$ expressed by (46) can be calculated by the following formula:

$$t_{A-i^*}(\sqrt{\varepsilon_z}, \sqrt{\varepsilon_r}) = -\sqrt{\frac{2m_0}{e}}\frac{R_a(n-1)}{(n-2)\sqrt{\phi_{ac}}}\left[1 + \frac{n}{n-2}\sqrt{\frac{\varepsilon_z}{\phi_{ac}}} + \frac{n^2+4n-4}{2(n-2)^2}\frac{\varepsilon_z}{\phi_{ac}} - \frac{(n-1)(n^2-n+2)}{2(n-2)}\frac{\varepsilon_r}{\phi_{ac}}\right] + R_a\frac{n}{n-2}\sqrt{\frac{m_0}{2e}}\frac{1}{\sqrt{\phi_{ac}}}\left[\frac{2(n-1)}{(n-2)\sqrt{\phi_{ac}}}(\sqrt{\varepsilon_z}-\sqrt{\varepsilon_{z_1}}) + \frac{2n(n-1)}{(n-2)^2\phi_{ac}}(\varepsilon_z-\varepsilon_{z_1}) - \frac{2(n-1)^2}{(n-2)}\frac{\varepsilon_r}{\phi_{ac}}\right] \tag{49}$$

Finally, we can also classify the temporal aberration by order-sequence, as we have done in the treatment with spatial aberration. As shown in Fig. 2, from Eq. (45), the temporal aberration ΔT can be regarded as composition of paraxial temporal aberration ΔT^* and geometrical temporal aberration δT, it can be expressed as following:

$$\begin{aligned}\Delta T &= \Delta T^* + \delta T \\ &= \Delta T_1^* + \Delta T_2^* + \delta T_1 + \delta T_2\end{aligned} \tag{50}$$

where

$$\Delta T_1^* = \left(\frac{2m_0}{e}\right)^{\frac{1}{2}}\frac{\sqrt{\varepsilon_z}-\sqrt{\varepsilon_{z_1}}}{E_c} \tag{51}$$

$$\Delta T_2^* = \left(\frac{2m_0}{e}\right)^{\frac{1}{2}}\frac{\varepsilon_z-\varepsilon_{z_1}}{E_c\sqrt{\phi_{ac}}}\times\left[-\frac{3n^2-9n+2}{4n(n-2)} - \frac{3}{8}\sqrt{n-1}\frac{1}{n} - \frac{3}{8}\sqrt{n-1}\left(\pi-\arcsin\frac{2}{n}\sqrt{n-1}\right)\right] \tag{52}$$

$$\delta T_1 = 0,$$

$$\delta T_2 = \left(\frac{2m_0}{e}\right)^{\frac{1}{2}}\frac{\varepsilon_r}{E_c\sqrt{\phi_{ac}}}\times\left\{\frac{(n-1)[-(n-2)^2(3n-1)+2(n^2-n+2)]}{4n(n-2)^2} - \frac{3}{8}\sqrt{n-1}\left(\pi-\arcsin\frac{2}{n}\sqrt{n-1}\right)\right\} \tag{53}$$

The definition of temporal aberrations is as following (see Fig. 3):

ΔT^* — paraxial temporal aberration, that is the difference of flight time between paraxial electron of ($\sqrt{\varepsilon_z}$, $\sqrt{\varepsilon_r}$) and paraxial electron of $\sqrt{\varepsilon_{z_1}}$, which is measured at the ideal image plane;

δT — geometrical temporal aberration, that is the difference of flight time between paraxial electron and practical electron having same ($\sqrt{\varepsilon_z}$, $\sqrt{\varepsilon_r}$), which is measured at the ideal image plane.

The lower index numeral of aberration is given according to the power of $\sqrt{\varepsilon_z}/\sqrt{\phi_{ac}}$, which represents the order of this aberration.

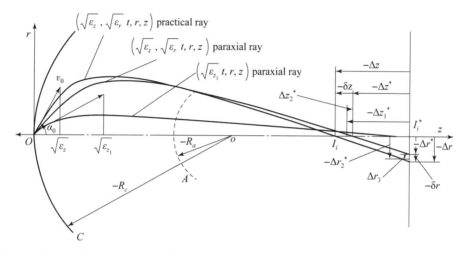

Fig. 3 Temporal aberrations can be regarded as composition of paraxial temporal aberration and geometrical temporal aberration

Summary

From the above investigation, we can reach the following conclusions.

(1) Our study naturally relates the static problem of imaging electron optics to the dynamic problem. Actually, electrons emitted from photocathode flying in space carry not only the spatial information, but also the temporal information, both of which display on the image plane as spatial-temporal characteristics, such as spatial aberration and temporal aberration.

(2) From the study of electron optics and aberrations for a bi-electrode concentric spherical system with electrostatic focusing, we get a very important concept. Whether the spatial aberration or the temporal aberration, both can be expressed in the combination of paraxial aberration and geometrical aberration, that is, the lateral aberration can be regarded as composition of paraxial lateral aberration and geometrical lateral aberration, the temporal aberration can be regarded as composition of paraxial temporal aberration and geometrical temporal aberration.

(3) In a bi-electrode concentric spherical system working in static mode operation, we have

confirmed that the spatial resolution is mainly determined by the paraxial lateral aberration of the second order Δr_2^*, i.e., Recknagel-Artimovich formula, and we have firstly derived the concrete expressions of the paraxial lateral aberration of the third order Δr_3^* and geometrical lateral aberration of the third order δr_3. Similarly, in a bi-electrode concentric spherical system working in streak mode operation, we have confirmed that the temporal aberration is mainly determined by the paraxial temporal aberration of the first order ΔT_1^* i.e., Savoisy-Fanchenko formula, we have also firstly derived the concrete expressions of the second order paraxial temporal aberration of the second order ΔT_2^* and geometrical temporal aberration of the secondorder δT_2. It should be noted that either the paraxial lateral aberration of the second order Δr_2^* or the paraxial temporal aberration of the first order ΔT_1^*, both of them occupy greatest part of spatial-temporal aberrations and these two aberrations are only related to the paraxial rays, so the emphasis of further study for imaging electron optics should be directed to paraxial optics and its paraxial aberrations.

(4) All of the formulae in this paper derived from a bi-electrode concentric spherical system with electrostatic focusing for the electron ray tracing and flight time, the paraxial lateral aberration and geometrical lateral aberration, as well as the paraxial temporal aberration and geometrical temporal aberration, can be used to test and verify the theory, method and algorithm in the research of imaging electron optics.

Acknowledgements

This project was supported by National Natural Science Foundation of China and the Special Science Fund of Doctoral Discipline for Institutions of Higher Learning.

References

[1] ARTIMOVICH L A. Electrostatic properties of emission systems[J]. Bulletin of Academy of Sciences USSR. Physics series, 1944, 8(6): 313 - 328.

[2] RECKNAGEL A. Theorie des elektrisohen elecktronen miktroskops fur selbstrakler. Z. Angew Physik, 1941, 117: 689 - 708 (in German).

[3] SAVOISKY Y K, FANCHENKO S D. Physical basis of electron optical chronograph[R]. Report of Academy of Sciences, USSR 108, 2: 218 - 221, 1956.

[4] CSORBA I P. Transit-time-spread-limited time resolution of image tubes in streak operation[J]. RCA Review 32 650 - 659, 1971.

[5] RUSKA E. Zur fokussierbarkeit von kathoden - strahlbundeln grosser ausgangsquerchnitte[J]. Z. Angew. Physik. 1933, 83(9): 684 - 687 (in German).

[6] SCHAGEN P, BRUINING H, FRANCKEN J C. A simple electrostatic electron - optical system with only one voltage[J]. Philips Research Report, 1952, 7(2): 119 - 130.

[7] ZHOU L W. Electron optics with wide beam focusing[M]. Beijing: Beijing Institute of Technology Press, 1993 (in Chinese).

[8] ZHOU L W. Electron optics of concentrate spherical electrostatic focusing systems with

two electrodes(in Chinese)[J]//Focusing and imaging of wide electron beams—Selected papers on electron optics by Zhou Liwei. Beijing:Beijing Institute of Technology Press,1994:11-29.

[9] ZHOU L W,LI Y,ZHANG Z Q, et al. Test and verification of temporal abbreviation theory for electron optical imaging systems by an electrostatic concentric spherical [J]. Acta Physical Sinica,2005,54(8):3597-3603(in Chinese).

10.5 两电极静电同心球系统的成像电子光学及其空间-时间像差
Imaging Electron Optics and Spatial-temporal Aberrations in a Bi-electrode Spherical Concentric System with Electrostatic Focusing

摘要：本文基于实际轨迹方程和电子运动方程，研究了两电极静电同心球系统中由光阴极逸出的运动电子的空间-时间轨迹，推导了成像位置和电子运动时间的精确和近似表达式，并由空间和时间轨迹的解析解出发讨论系统的空间-时间成像特性。文中对空间-时间像差给出统一的定义，即横向像差可视为近轴横向像差与几何横向像差的合成，时间像差可视为近轴时间像差与几何时间像差的合成，并导出了所有相应的像差表示式。

Abstract: In the present paper, based on the practical electron ray equation and electron motion equation, for a bi-electrode concentric spherical system with electrostatic focusing, the spatial-temporal trajectories of moving electrons emitted from the photocathode are obtained, the exact and the approximate formulae for image position and moving time of electrons are deduced. Starting from the solutions of spatial-temporal trajectories, the electron optical spatial-temporal properties of this system are then discussed. A unified definition of spatial-temporal aberrations is given, in which the lateral aberration can be regarded as a combination of paraxial lateral aberration and geometrical lateral aberration, and the temporal aberration can be regarded as a combination of paraxial temporal aberration and geometrical temporal aberration. All expressions of these corresponding aberrations are deduced.

引言

对成像电子光学系统，由光电子发射的初能量和初角度所形成的空间成像离散和渡越时间弥散是像管和高速摄影变像管的基本问题，在静态模式下，电子轨迹在像面上的空间成像离散决定了系统的极限分辨率；而在高速摄影变像管的条纹模式下，极短暂留光子事件的点像被相机成像，由于渡越时间弥散成一条线状。此线的长度比例于渡越时间弥散。显然，线的长度越短，相机的时间分辨率越好。

业已证明，成像系统的空间成像离散主要决定于二级近轴横向像差，它通常被表示为 Recknagel-Artimovich 公式[1,2]。公式表明，二级近轴横向像差（亦称二级近轴色差）与电极的具体结构和电位分布无关，而与电子发射的初能量、阴极面上的电场强度、系统的线

周立伟[a]，公慧[a]，张智诠[b]，张轶飞[b]. a) 北京理工大学，b) 装甲兵工程学院. 物理学报, 2010, V. 59, No. 8, 5450-5459.

放大率以及像面位置有关。同样，成像系统的时间渡越弥散主要决定于一级近轴时间像差，它通常被表示为 Savoisky-Fanchenko 公式[3]。公式表明，一级近轴时间像差（亦称一级时间色差）同样与电极的具体结构和电位分布无关，而与电子发射的初能量、阴极面上的电场强度有关；此外，它还与系统的线放大率以及像面位置无关。关于阴极透镜的三级像差理论，西门纪业、周立伟及艾克聪等人有较为详细的报道[4~8]。

Recknagel-Artimovich 公式和 Savoisky-Fanchenko[3] 公式之间有很多相似点，但其差异也很明显，仔细考察这两个公式，就可发现此二公式之间的不和谐性和不对称性。实际上，无论是空间成像离散或是渡越时间弥散，都是由阴极面逸出电子发射的初能量和初角度的空－时轨迹所形成的，它们之间必然有联系。但奇怪的是，在一些文献中，Recknagel-Artimovich 公式与像面位置有关，而 Savoisky-Fanchenko 公式与像面位置无关。两电极静电同心球系统是研究这一电子光学特性较好的模型，这是因为，系统的电位分布和电子轨迹都能表示成解析形式，能定量地研究系统的聚焦成像性质以及空间－时间像差。对两电极静电同心球系统的研究，前人的工作见文献 [9]、文献 [10]，但这些工作停留在成像的零级近似上。求解此系统的电子轨迹和横向像差的较为精确的公式可见文献 [11]、文献 [12]。对两电极静电同心球系统的动态光学，文献 [13] 仅是证实 Savoisky-Fanchenko 的结论；更为精确的时间像差的公式见文献 [14]。

1　两电极静电同心球系统的静态电子光学及其空间像差

1.1　轨迹方程及其求解

图 1 所示是两电极静电同心球系统。设球面阴极 C 和栅状球面阳极 A 的曲率半径分别为 R_c 和 R_a，系统的共同曲率中心为 O。由于系统的球对称性，故可用极坐标 (ρ, φ) 来描述电子轨迹。令极坐标的原点位于系统的曲率中心 O，并设阴极 C 的电位 $\phi_c = 0$，栅状阳极 A 对于阴极 C 的电位为 ϕ_{ac}，电子以初速度 v_0、初角度 α_0 自阴极原点射出。设射线行进方向自左至右，故规定线段由左向右为正，由下向上为正，反之为负。线段 R_c、R_a、ρ、R_i、R_i^* 均以曲率中心 O 算起到相应球面的顶点；Δz 自理想像面位置 I_i^* 点算起；Δr 由 $\varphi = 0$ 轴线算起。角度一律以锐角来度量，规定逆时针转为正，顺时针转为负。和线段参量要规定计算起点一样，角度也要规定起始轴。图中 α_0、φ、γ、α_i 均以 $\varphi = 0$ 轴线为起始轴，ζ、τ 均以过该点与中心 O 的连线为起始轴，逆时针转为正，顺时针转为负。

同心球系统两电极间的场等效于集中球心的点电荷所产生的场。故不难求得矢径坐标 ρ 处相对于阴极 C 的电位 $\phi_{\rho c}$ 的表达式为

$$\phi_{\rho c} = E_c R_c \left(\frac{R_c}{\rho} - 1 \right) \tag{1}$$

式中，E_c 为阴极面上的电场强度。当 $\rho = R_a$ 时，则有

$$E_c = \frac{\phi_{ac}}{R_c(n-1)} \tag{2}$$

式中，$n = R_c/R_a$，对于凹面阴极－阳极系统，R_c、R_a 为负值，$n > 1$；对于凸面阴极－阳极系统，R_c、R_a 为正值，$n < 1$。由此可见，无论 $n > 1$ 或 $n < 1$，E_c 永为负值。

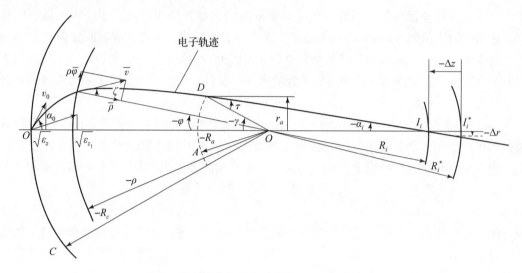

图1 两电极静电同心球系统 ($n = R_c/R_a > 2$)

在两电极同心球系统中,电子在初速度为 v_0、初角度为 α_0 的逸出方向与中心轴线所构成的平面内的运动遵循能量守恒定律和有心力场角动量守恒定律,即

$$\frac{1}{2}m_0 v^2 = \frac{1}{2}m_0 v_0^2 + e\phi_{\rho c} \tag{3}$$

$$\rho^2 \dot{\varphi} = R_c^2 \dot{\varphi}_0 = \text{const} \tag{4}$$

式中,m_0、e 分别为电子的质量和电荷,e 取绝对值,v 为电子在极坐标 (ρ, φ) 处的速度;$\dot{\varphi} = d\varphi/dt$。由式(3)和式(4),可导得两种形式的轨迹方程。

(1)Ruska 方程,它以 φ 为变量,即 $\rho = f(\varphi)$,表示如下:

$$\mu = \left(1 - \frac{p}{2\sin^2\alpha_0}\right)\cos\varphi - \frac{1}{\tan\alpha_0}\sin\varphi + \frac{p}{2\sin^2\alpha_0} \tag{5}$$

(2)Schagen[10] 方程,它以 ρ 为变量,即 $\varphi = f(\rho)$,表示如下:

$$\tan\frac{\varphi}{2} = \frac{-\cos\alpha_0 + [1 + p(\mu-1) - \mu^2\sin^2\alpha_0]^{1/2}}{-\dfrac{p}{2\sin\alpha_0} + (\mu+1)\sin\alpha_0} \tag{6}$$

式中,$\mu = R_c/\rho$,$p(n-1) = \phi_{ac}/\varepsilon_0$,$\varepsilon_0 = m_0 v_0^2/2e$。$\varepsilon_0$ 为静止的光电子为获得其发射能量所要求的加速电位,称为初电位;p 仅是一个参量,但 $p(n-1)$ 具有明显的物理意义,表示系统的加速电位与初电位的比值。在像管中,一般 $p(n-1) = 10^4 \sim 10^5$。

应该指出,Ruska 和 Schagen 给出的方程仅给出自光阴极逸出电子的飞行位置,而没有清晰地给出电子在任一位置的飞行方向。在实际计算时,应用 Schagen[10] 方程和 Ruska[9] 方程求解轨迹行进的方向和位置等参量,都很不方便。为解决这一问题,周立伟给出了此系统中求解电子的飞行轨迹位置及其行进方向的方程[11,12],它以 $\sin\varphi$、$\cos\varphi$、$\tan\zeta$ 的形式表示为:

$$\sin\varphi = -\frac{(\mu-1)(c + d_0)}{bd_0 + b_0 c}\sin\alpha_0 \tag{7a}$$

$$\cos\varphi = \frac{bc + b_0 d_0}{bd_0 + b_0 c} \tag{7b}$$

$$\tan\zeta = \frac{\mu \sin\alpha_0}{b} \tag{8}$$

式中，

$$b_0 = \cos\alpha_0,$$

$$d_0 = 1 - 2(n-1)\frac{\varepsilon_0}{\phi_{ac}}\sin^2\alpha_0,$$

$$b = \left[1 + \frac{\phi_{ac}}{\varepsilon_0}\frac{(\mu-1)}{(n-1)} - \mu^2 \sin^2\alpha_0\right]^{1/2},$$

$$c = 1 - 2\mu(n-1)\frac{\varepsilon_0}{\phi_{ac}}\sin^2\alpha_0$$

式（7a）、式（7b）、式（8）在计算上更为简捷方便，无论对于 $n > 1$ 或 $n < 1$ 的系统都是普遍适用的。特别是在求解多电极静电同心球系统的轨迹时，式（7a）、式（7b）、式（8）具有类推性，这是其最大优点。

1.2 理想成像位置 R_i 的确定及其近轴成像

由式（7a）、式（7b）和式（8），不难求得自阴极上原点以初始条件（$\sqrt{\varepsilon_0}$，α_0）射出的电子到达阳极 A 上 D 点的轨迹，由此可以定出角度 τ 和 γ，如图 1 所示。假定阳极是对于电子透明等电位的栅状电极，在栅状阳极后面的空间为无场空间。当电子通过栅状阳极时，假定电子速度的大小和方向都不变，轨迹的解及其斜率都是连续的。因此，通过栅状阳极后的轨迹将是一条切于过 D 点轨迹的直线，交于 $\varphi = 0$ 的轴线为 I_i 点。

按照图 1 所给出的 $\triangle ODI_i$，不难求得交点 I_i 位置处 R_i 的表达式

$$R_i = R_a \frac{\sin\tau}{\sin\alpha_i} = R_a \frac{\tan\tau}{\sin\gamma + \tan\tau\cos\gamma} \tag{9}$$

式中，α_i 为电子轨迹在交点 I_i 处与轴线 $\varphi = 0$ 的交角；R_i 为曲率中心 O 至交点 I_i 的距离。

显然，若 $\mu = n$，$\varphi = \gamma$，$\zeta = \tau$，$b = b_1$，$c = c_0$，由式（7a）、式（7b）和式（8）可以获得 $\sin\gamma$、$\cos\gamma$、$\tan\tau$ 的表达式，利用式（9），便得到两电极同心球系统中以初始条件（$\sqrt{\varepsilon_0}$，α_0）自阴极原点射出轨迹的交轴位置 I_i 及其斜率的精确表达式：

$$R_i = -R_a \frac{n(b_1 d_0 + b_0 c_0)}{b_1(n-1)(c_0 + d_0) - n(b_1 c_0 + b_0 d_0)} \tag{10}$$

$$\tan\alpha_i = -\frac{b_1(n-1)(c_0 + d_0) - n(b_1 c_0 + b_0 d_0)}{b_1(b_1 c_0 + b_0 d_0) + n(n-1)(c_0 + d_0)\sin^2\alpha_0}\sin\alpha_0 \tag{11}$$

式中，

$$b_1 = \left(1 + \frac{\phi_{ac}}{\varepsilon_0} - n^2\sin^2\alpha_0\right)^{1/2},$$

$$c_0 = 1 - 2n(n-1)\frac{\varepsilon_0}{\phi_{ac}}\sin^2\alpha_0$$

现简化上述的 R_i 和 $\tan\alpha_i$ 的表达式。令 $\sqrt{\varepsilon_z} = \sqrt{\varepsilon_0}\cos\alpha_0$，$\sqrt{\varepsilon_r} = \sqrt{\varepsilon_0}\sin\alpha_0$，用二项式

定理按乘幂展开式（10）和式（11）各项，略去较 1 小得多的 $(\varepsilon_0/\phi_{ac})^{3/2}$ 项及其以上高阶乘幂项，便得到轨迹交轴位置 I_i 处的 R_i 及其斜率 $\tan\alpha_i$ 的近似式为

$$R_i = -R_a \frac{n}{n-2}\left[1 + \frac{2(n-1)}{n-2}\sqrt{\frac{\varepsilon_z}{\phi_{ac}}} + \frac{2n(n-1)}{(n-2)^2}\frac{\varepsilon_z}{\phi_{ac}} - \frac{2(n-1)^2}{n-2}\frac{\varepsilon_r}{\phi_{ac}}\right] \quad (12)$$

$$\tan\alpha_i = -(n-2)\sqrt{\frac{\varepsilon_r}{\phi_{ac}}}\left[1 - \frac{2(n-1)}{n-2}\sqrt{\frac{\varepsilon_z}{\phi_{ac}}} + \frac{3n-2}{2(n-2)}\frac{\varepsilon_z}{\phi_{ac}} + \frac{(n-1)(n^2-n+2)}{2(n-2)}\frac{\varepsilon_r}{\phi_{ac}}\right] \quad (13)$$

由此出发，可以研究两电极静电同心球系统的电子光学成像性质及空间像差。在式（12）和式（13）中，令 $\sqrt{\varepsilon_z} = \sqrt{\varepsilon_{z_1}}$，$\varepsilon_{z_1}$ 为近轴电子到达近轴成像位置 I_i^* 所对应的轴向初速。考虑成像系统的近轴条件，略去上两式括弧内含有 ε_r/ϕ_{ac} 的各项，则可得到自阴极面原点以轴向初速 $\sqrt{\varepsilon_{z_1}}$ 逸出的近轴轨迹到达近轴成像位置处的 R_i^* 和 $\tan\alpha_i^*$ 的表达式：

$$R_i^* = -R_a\frac{n}{n-2}\left[1 + \frac{2(n-1)}{n-2}\sqrt{\frac{\varepsilon_{z_1}}{\phi_{ac}}} + \frac{2n(n-1)}{(n-2)^2}\frac{\varepsilon_{z_1}}{\phi_{ac}}\right] \quad (14)$$

$$\tan\alpha_i^* = -(n-2)\sqrt{\frac{\varepsilon_r}{\phi_{ac}}}\left[1 - \frac{2(n-1)}{n-2}\sqrt{\frac{\varepsilon_{z_1}}{\phi_{ac}}} + \frac{3n-2}{2(n-2)}\frac{\varepsilon_{z_1}}{\phi_{ac}}\right] \quad (15)$$

式（14）表明，在近轴条件下，只有 $\sqrt{\varepsilon_{z_1}}$ 相同电子束才能理想聚焦，当 $\sqrt{\varepsilon_{z_1}}$ 改变时，理想成像的位置也随着变动。

系统的线放大率 M 可以表示为

$$M = \frac{R_i^*}{R_c} = -\frac{1}{(n-2)}\left[1 + \frac{2(n-1)}{n-2}\sqrt{\frac{\varepsilon_{z_1}}{\phi_{ac}}} + \frac{2n(n-1)}{(n-2)^2}\frac{\varepsilon_{z_1}}{\phi_{ac}}\right] \quad (16)$$

很明显，在 $n>2$ 时形成倒立的实像，在 $n<2$ 时形成正立的虚像。

1.3 空间像差近轴横向像差和几何横向像差

下面把空间像差分为两部分：近轴像差和几何像差，如图 2 所示。定义从阴极面原点以 $\sqrt{\varepsilon_{z_1}}$ 发出的电子按近轴条件聚焦的位置 I_i^* 为理想像面位置，而从同一点发出的 $(\sqrt{\varepsilon_z}, \sqrt{\varepsilon_r})$ 的实际电子轨迹交于 I_i 处，则产生像的轴向位移，即纵向像差

$$\begin{aligned}\Delta z &= R_i(\sqrt{\varepsilon_r}, \sqrt{\varepsilon_z}) - R_i^*(\sqrt{\varepsilon_{z_1}}) \\ &= R_i^*(\sqrt{\varepsilon_r}, \sqrt{\varepsilon_z}) - R_i^*(\sqrt{\varepsilon_{z_1}}) + R_i(\sqrt{\varepsilon_r}, \sqrt{\varepsilon_z}) - R_i^*(\sqrt{\varepsilon_r}, \sqrt{\varepsilon_z}) \\ &= \Delta z^* + \delta z\end{aligned} \quad (17)$$

将式（12）和式（14）代入式（17）中，便得到纵向像差 Δz 的表达式

$$\Delta z = \Delta z^* + \delta z = \Delta z_1^* + \Delta z_2^* + \delta z_1 + \delta z_2 \quad (18)$$

$$\Delta z_1^* = \frac{2M^2\sqrt{\phi_{ac}}}{-E_c}(\sqrt{\varepsilon_z} - \sqrt{\varepsilon_{z_1}}) \quad (19)$$

$$\Delta z_2^* = \frac{2M^2}{-E_c}\left[\frac{n}{n-2}(\sqrt{\varepsilon_z} - \sqrt{\varepsilon_{z_1}}) - \frac{4(n-1)}{n-2}\sqrt{\varepsilon_{z_1}}(\sqrt{\varepsilon_z} - \sqrt{\varepsilon_{z_1}})\right] \quad (20)$$

$$\delta z_1 = 0, \quad \delta z_2 = \frac{2M^2}{E_c}(n-1)\varepsilon_r \tag{21}$$

通常，人们关心的是横向像差。在近轴条件下，在 $\sqrt{\varepsilon_{z_1}}$ 所确定的理想像面上，横向像差 Δr 可表示为

$$\Delta r = -\Delta z \tan\alpha_i \tag{22}$$

式中，$\tan\alpha_i$ 是 ($\sqrt{\varepsilon_z},\sqrt{\varepsilon_r}$) 的实际电子轨线在 I_i 点与 $\varphi=0$ 轴线所成的斜率，由式（13）确定。将式（13）和式（18）代入式（22）中，便得到横向像差 Δr 的表达式：

$$\begin{aligned}\Delta r &= r(z_i^*,\sqrt{\varepsilon_z},\sqrt{\varepsilon_r}) - r^*(z_i^*,\sqrt{\varepsilon_{z_1}}) \\ &= r^*(z_i^*,\sqrt{\varepsilon_z},\sqrt{\varepsilon_r}) - r^*(z_i^*,\sqrt{\varepsilon_{z_1}}) + r(z_i^*,\sqrt{\varepsilon_z},\sqrt{\varepsilon_r}) - r^*(z_i^*,\sqrt{\varepsilon_z},\sqrt{\varepsilon_r}) \\ &= \Delta r^* + \delta r = \Delta r_2^* + \Delta r_3^* + \delta r_2 + \delta r_3 \end{aligned} \tag{23}$$

式中，

$$\Delta r_2^* = \frac{2M}{E_c}\sqrt{\varepsilon_r}(\sqrt{\varepsilon_z}-\sqrt{\varepsilon_{z_1}}) \tag{24}$$

$$\Delta r_3^* = \frac{2M}{-E_c\sqrt{\phi_{ac}}}\sqrt{\varepsilon_r}(\sqrt{\varepsilon_z}-\sqrt{\varepsilon_{z_1}}) \tag{25}$$

$$\delta r_2 = 0, \quad \delta r_3 = \frac{2M}{-E_c\sqrt{\phi_{ac}}}(n-1)\varepsilon_r^{3/2} \tag{26}$$

横向像差的符号和定义如下：

Δr^* 为近轴横向像差（即近轴横向色差），是在 $\sqrt{\varepsilon_{z_1}}$ 确定的理想像面上 ($\sqrt{\varepsilon_z},\sqrt{\varepsilon_r}$) 的近轴轨迹与 $\sqrt{\varepsilon_{z_1}}$ 的近轴轨迹间的径向偏离；Δr_2^*，Δr_3^* 分别称为二级、三级近轴横向色差。δr 为几何横向球差，是在 $\sqrt{\varepsilon_{z_1}}$ 确定的理想像面上 ($\sqrt{\varepsilon_z},\sqrt{\varepsilon_r}$) 的近轴轨迹与 ($\sqrt{\varepsilon_z}$, $\sqrt{\varepsilon_r}$) 实际轨迹之间的径向偏离；δr_3 称为三级几何横向球差。像差的下标数字按 $\sqrt{\varepsilon_z}/\sqrt{\phi_{ac}}$ 和 $\sqrt{\varepsilon_r}/\sqrt{\phi_{ac}}$ 确定的幂次给出，表示该像差的阶次。

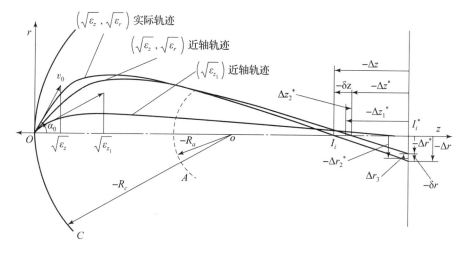

图2 空间像差可以视为近轴横向像差和几何像横向差的合成

2 两电极静电同心球系统的动态电子光学及时间像差

2.1 电子运动的飞行时间解

现在讨论两电极静电同心球系统的动态电子光学及时间像差。首先求解此系统中运动电子的飞行时间。由光阴极逸出的初始条件 $R_c\dot{\varphi} = R_c v_0 \sin\alpha_0$ 出发,并由式 (4) 有

$$\rho^2 \dot{\varphi} = R_c v_0 \sin\alpha_0 \tag{27}$$

在极坐标 (ρ, φ) 下,电子速度 v 可以写成

$$v^2 = \dot{\rho}^2 + (\rho\dot{\varphi})^2 \tag{28}$$

由式 (27) 和式 (28),可得

$$\dot{\rho} = \frac{d\rho}{dt} = \sqrt{v^2 - R_c^2 v_0^2 \sin^2\alpha_0 / \rho^2} \tag{29}$$

将

$$v^2 = v_0^2 + \frac{2e}{m_0}\frac{\phi_{ac}}{(n-1)}\left(\frac{R_c}{\rho} - 1\right) \tag{30}$$

代入式 (29) 中,便有

$$\dot{\rho} = \sqrt{\frac{2e\varepsilon_0}{m_0}}[1 + \rho(\mu-1) - \mu^2 \sin^2\alpha_0]^{1/2} \tag{31}$$

利用变换

$$\dot{\rho} = \frac{d\rho}{dt} = -\frac{R_c}{\mu^2}\frac{d\mu}{dt} \tag{32}$$

式 (31) 便能表达成以下的形式:

$$dt = \frac{-R_c d\mu}{\sqrt{\frac{2e\varepsilon_0}{m_0}}\mu^2 \sqrt{1 + \rho(\mu-1) - \mu^2 \sin^2\alpha_0}} \tag{33}$$

积分式 (33),$t=0$ 时,$\rho = R_c$,$\mu = 1$,便得到电子在阴极和阳极球面空间中的飞行时间 t 的表示式:

$$t = -\frac{R_c}{\sqrt{\frac{2e\varepsilon_0}{m_0}}}\left\{-\frac{\sqrt{1+\rho(\mu-1)-\mu^2\sin^2\alpha_0}}{(1-\rho)\mu} + \frac{\cos\alpha_0}{1-\rho} - \frac{\rho}{2(1-\rho)\sqrt{\rho-1}} \times \right.$$
$$\left.\left[\arcsin\frac{\rho(\mu-2)+2}{\mu\sqrt{\rho^2+4(1-\rho)\sin^2\alpha_0}} + \arcsin\frac{\rho-2}{\sqrt{\rho^2+4(1-\rho)\sin^2\alpha_0}}\right]\right\} \tag{34}$$

2.2 实际轨迹和近轴轨迹抵达阳极的飞行时间的确定

由式 (34),令 $\varepsilon_0 = \varepsilon_r + \varepsilon_z$,$\rho = R_a$,即 $\mu = n$,我们便能得到自光阴极逸出 ($\sqrt{\varepsilon_r}$,$\sqrt{\varepsilon_z}$) 的电子的实际轨迹抵达阳极的飞行时间 t_{C-A} 的精确式为

$$t_{C-A} = \frac{-R_c(n-1)}{\sqrt{\frac{2e}{m_0}}[\phi_{ac} - (n-1)\varepsilon_z - (n-1)\varepsilon_r]} \left[\frac{1}{n}\sqrt{\phi_{ac} + \varepsilon_z + (1-n^2)\varepsilon_r} - \sqrt{\varepsilon_z}\right] +$$

$$\frac{-R_c}{\sqrt{\frac{2e}{m_0}}} \frac{\frac{\phi_{ac}}{n-1}}{2\left(\frac{\phi_{ac}}{n-1} - \varepsilon_z - \varepsilon_r\right)\sqrt{\frac{\phi_{ac}}{n-1} - \varepsilon_z - \varepsilon_r}} \times$$

$$\left\{\arcsin \frac{\left(1 - \frac{2}{n}\right) + \frac{2(n-1)(\varepsilon_z + \varepsilon_r)}{n\phi_{ac}}}{\sqrt{1 - 4\left(\frac{\phi_{ac}}{n-1} - \varepsilon_z - \varepsilon_r\right)\varepsilon_r\left(\frac{n-1}{\phi_{ac}}\right)^2}} + \arcsin \frac{1 - 2\frac{n-1}{\phi_{ac}}(\varepsilon_z + \varepsilon_r)}{\sqrt{1 - 4\left(\frac{\phi_{ac}}{n-1} - \varepsilon_z - \varepsilon_r\right)\varepsilon_r\left(\frac{n-1}{\phi_{ac}}\right)^2}}\right\} \quad (35)$$

展开式（35），略去所有高于 ε_z/ϕ_{ac} 和 ε_r/ϕ_{ac} 阶次的项，我们便能得到 t_{C-A} 的近似式为

$$t_{C-A} = \frac{-R_c(n-1)}{\sqrt{\frac{2e}{m_0}\phi_{ac}}} \frac{1}{n}\left[1 + \frac{(2n-1)}{2\phi_{ac}}\varepsilon_z + \frac{(n-1)^2}{2\phi_{ac}}\varepsilon_r\right] + \frac{R_c(n-1)}{\sqrt{\frac{2e}{m_0}\phi_{ac}}}\sqrt{\varepsilon_z} +$$

$$\frac{-R_c}{\sqrt{\frac{2e}{m_0}}} \frac{\sqrt{n-1}}{2\sqrt{\phi_{ac}}}\left(1 + \frac{3}{2}\frac{n-1}{\phi_{ac}}\varepsilon_z + \frac{3}{2}\frac{n-1}{\phi_{ac}}\varepsilon_r\right) \times$$

$$\left\{\arcsin\left[\left(1 - \frac{2}{n}\right) + \frac{2(n-1)\varepsilon_z}{n\phi_{ac}} + \frac{2(n-1)^2}{n\phi_{ac}}\varepsilon_r\right] + \arcsin\left(1 - 2\frac{n-1}{\phi_{ac}}\varepsilon_z\right)\right\} \quad (36)$$

式（36）的最后一项还应该进一步简化。经过一系列的变换，此式可变换为

$$\arcsin\left[\left(1 - \frac{2}{n}\right) + \frac{2(n-1)\varepsilon_z}{n\phi_{ac}} + \frac{2(n-1)^2}{n\phi_{ac}}\varepsilon_r\right] + \arcsin\left(1 - 2\frac{n-1}{\phi_{ac}}\varepsilon_z\right)$$

$$= 2\arcsin\sqrt{\frac{n-1}{n}} - 2\sqrt{n-1}\sqrt{\frac{\varepsilon_z}{\phi_{ac}}} + \sqrt{n-1}\frac{\varepsilon_z}{\phi_{ac}} + 2(n-1)\sqrt{n-1}\frac{\varepsilon_r}{\phi_{ac}} \quad (37)$$

对近轴轨迹，在式（36）中，略去含有 ε_r/ϕ_{ac} 的项，并令 $\sqrt{\varepsilon_z} = \sqrt{\varepsilon_{z_1}}$，便得到自光阴极逸出的 $\sqrt{\varepsilon_{z_1}}$ 近轴电子到达阳极的近轴飞行时间为

$$t_{C-A}^* = \left(\frac{2m_0}{e}\right)^{1/2}\left\{\frac{-R_c(n-1)}{2\sqrt{\phi_{ac}}}\frac{1}{n} + \frac{-R_c}{2}\sqrt{\frac{n-1}{\phi_{ac}}}\arcsin\sqrt{\frac{n-1}{n}} + \frac{R_c(n-1)}{\sqrt{\phi_{ac}}}\sqrt{\frac{\varepsilon_{z_1}}{\phi_{ac}}} + \right.$$

$$\left. \frac{-R_c(n-1)}{\sqrt{\phi_{ac}}}\left(\frac{3n-1}{4n} + \frac{3}{4}\sqrt{n-1}\arcsin\sqrt{\frac{n-1}{n}}\right)\frac{\varepsilon_{z_1}}{\phi_{ac}}\right\} \quad (38)$$

2.3 栅状阳极后无场空间区域的电子飞行时间的计算

当电子由阴极抵达栅状阳极时，其速度 v_a 可表示为

$$v_a = \sqrt{v_0^2 + \frac{2e}{m_0}\phi_{ac}} \quad (39)$$

在阳极后无场空间的区域，电子轨迹乃是一条直线。令此轨迹由阳极到轴线的距离为 l，

飞行时间为 t_{A-i}，于是由图中 ΔODI_i，它可以表示为

$$t_{A-i} = \frac{l}{v_a} = R_i \frac{\sin(-\gamma)}{v_a \sin\tau} \tag{40}$$

这里 R_i 以式（12）表示，$\sin(-\gamma)$ 可由式（7a）当 $\varphi = -\gamma$ 时获得。由式（8），当 $\zeta = \tau$ 时，$\sin\tau$ 可表示为

$$\sin\tau = \frac{n\sin\alpha_0}{\sqrt{b_1^2 + n^2\sin^2\alpha_0}} \tag{41}$$

展开 R_i、$\sin(-\gamma)$ 和 $\sin\tau$ 等式，略去所有高于 ε_z/ϕ_{ac} 和 ε_r/ϕ_{ac} 阶次的项，由式（40），便得到实际电子在无场空间由阳极抵达轴线 $\varphi = 0$ 的飞行时间 t_{A-i} 的近似式：

$$t_{A-i} = -\sqrt{\frac{2m_0}{e}} \frac{R_a(n-1)}{(n-2)\sqrt{\phi_{ac}}} \left[1 + \frac{n}{n-2}\sqrt{\frac{\varepsilon_z}{\phi_{ac}}} + \frac{n^2+4n-4}{2(n-2)^2}\frac{\varepsilon_z}{\phi_{ac}} - \frac{(n-1)(n^2-n+2)}{2(n-2)}\frac{\varepsilon_r}{\phi_{ac}}\right] \tag{42}$$

若略去式（42）中的 ε_r/ϕ_{ac} 项，并令 $\sqrt{\varepsilon_z} = \sqrt{\varepsilon_{z_1}}$，便可得到 $\sqrt{\varepsilon_{z_1}}$ 的近轴电子在无场空间由阳极抵达轴线 $\varphi = 0$ 的飞行时间 $t_{A-i^*}^*$ 近似式：

$$t_{A-i^*}^* = -\sqrt{\frac{2m_0}{e}} \frac{R_a(n-1)}{(n-2)\sqrt{\phi_{ac}}} \left[1 + \frac{n}{n-2}\sqrt{\frac{\varepsilon_{z_1}}{\phi_{ac}}} + \frac{n^2+4n-4}{2(n-2)^2}\frac{\varepsilon_{z_1}}{\phi_{ac}}\right] \tag{43}$$

2.4 时间像差（近轴时间像差和几何时间像差）

时间像差 ΔT 是在 $\sqrt{\varepsilon_{z_1}}$ 的近轴轨迹所对应的理想像面上衡量的，被定义为（$\sqrt{\varepsilon_z}$，$\sqrt{\varepsilon_r}$）的实际电子与 $\sqrt{\varepsilon_{z_1}}$ 的近轴电子到达此理想像面上的飞行时间的差异。它由两部分组成：一是此二电子由阴极出发到达阳极时的时间弥散，二是此二电子由阳极出发到达理想像面时的时间弥散，即

$$\begin{aligned}\Delta T &= t_{C-i^*}(\sqrt{\varepsilon_z},\sqrt{\varepsilon_r}) - t_{C-i^*}^*(\sqrt{\varepsilon_{z_1}}) \\ &= t_{C-A}(\sqrt{\varepsilon_z},\sqrt{\varepsilon_r}) + t_{A-i^*}(\sqrt{\varepsilon_z},\sqrt{\varepsilon_r}) - t_{C-A}^*(\sqrt{\varepsilon_{z_1}}) - t_{A-i^*}^*(\sqrt{\varepsilon_{z_1}})\end{aligned} \tag{44}$$

式中，$t_{C-A}(\sqrt{\varepsilon_z},\sqrt{\varepsilon_r})$、$t_{C-A}^*(\sqrt{\varepsilon_{z_1}})$、$t_{A-i^*}^*(\sqrt{\varepsilon_{z_1}})$ 分别以式（36）、式（38）和式（43）表示。而 $t_{A-i^*}(\sqrt{\varepsilon_z},\sqrt{\varepsilon_r})$ 的求解可视为由两部分组成：

$$t_{A-i^*}(\sqrt{\varepsilon_z},\sqrt{\varepsilon_r}) = t_{A-i}(\sqrt{\varepsilon_z},\sqrt{\varepsilon_r}) + \Delta t_{i-i^*}(\sqrt{\varepsilon_z},\sqrt{\varepsilon_r}) \tag{45}$$

式中，$t_{A-i}(\sqrt{\varepsilon_z},\sqrt{\varepsilon_r})$ 乃是（$\sqrt{\varepsilon_z}$，$\sqrt{\varepsilon_r}$）的实际电子由阳极抵达中心轴时的飞行时间，以式（42）表示。设此实际电子以速度 v_a 飞行由中心轴到达理想像面，其间的空间距离为 Δl，时间差异为 $\Delta t_{i-i^*}(\sqrt{\varepsilon_z},\sqrt{\varepsilon_r})$，则有

$$\Delta t_{i-i^*}(\sqrt{\varepsilon_z},\sqrt{\varepsilon_r}) = \frac{\Delta l}{v_a} = \frac{-\Delta z}{\cos\alpha_i \cdot v_a} = -\Delta z \sqrt{1+\tan^2\alpha_i}\frac{1}{v_a} \tag{46}$$

式中，$\tan\alpha_i$、v_a 分别以式（13）和式（39）表示。$\Delta z = R_i - R_i^*$，R_i、R_i^* 分别以式（12）和式（14）表示。

展开式（46）中的各表示式，略去所有高于 ε_z/ϕ_{ac} 和 ε_r/ϕ_{ac} 阶次的项，可以得到

$$\Delta t_{i-i^*}(\sqrt{\varepsilon_z},\sqrt{\varepsilon_r}) = R_a \frac{n}{n-2}\sqrt{\frac{m_0}{2e}}\frac{1}{\sqrt{\phi_{ac}}}\left[\frac{2(n-1)}{(n-2)\sqrt{\phi_{ac}}}(\sqrt{\varepsilon_z}-\sqrt{\varepsilon_{z_1}})+\right.$$
$$\left.\frac{2n(n-1)}{(n-2)^2\phi_{ac}}(\varepsilon_z-\varepsilon_{z_1}) - \frac{2(n-1)^2}{(n-2)}\frac{\varepsilon_r}{\phi_{ac}}\right] \tag{47}$$

于是式（45）表示的飞行时间 $t_{A-i^*}(\sqrt{\varepsilon_z},\sqrt{\varepsilon_r})$ 便可由以下式算得

$$t_{A-i^*}(\sqrt{\varepsilon_z},\sqrt{\varepsilon_r}) = -\sqrt{\frac{2m_0}{e}}\frac{R_a(n-1)}{(n-2)\sqrt{\phi_{ac}}}\left[1+\frac{n}{n-2}\sqrt{\frac{\varepsilon_z}{\phi_{ac}}}+\frac{n^2+4n-4}{2(n-2)^2}\frac{\varepsilon_z}{\phi_{ac}}-\frac{(n-1)(n^2-n+2)}{2(n-2)}\frac{\varepsilon_r}{\phi_{ac}}\right]+$$
$$R_a\frac{n}{n-2}\sqrt{\frac{m_0}{2e}}\frac{1}{\sqrt{\phi_{ac}}}\left[\frac{2(n-1)}{(n-2)\sqrt{\phi_{ac}}}(\sqrt{\varepsilon_z}-\sqrt{\varepsilon_{z_1}})+\frac{2n(n-1)}{(n-2)^2\phi_{ac}}(\varepsilon_z-\varepsilon_{z_1})-\frac{2(n-1)^2}{(n-2)}\frac{\varepsilon_r}{\phi_{ac}}\right] \tag{48}$$

最后，也可以将时间像差按阶次分类，就像我们在处理空间像差时一样；在电子行进的途径加一时间坐标，在记录电子轨迹的同时，记录下电子的飞行时间，如图3所示。

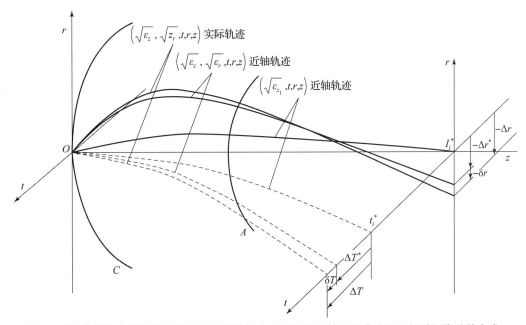

图3　时间像差和空间像差示意图（时间像差也可以视为近轴时间像差和几何时间像差的合成）

由式（44）可知，时间像差 ΔT 可以视为近轴时间像差 ΔT^* 和几何时间像差 δT 的合成，它可以表示如下：

$$\Delta T = \Delta T^* + \delta T = \Delta T_1^* + \Delta T_2^* + \delta T_1 + \delta T_2 \tag{49}$$

式中，

$$\Delta T_1^* = \left(\frac{2m_0}{e}\right)^{1/2}\frac{\sqrt{\varepsilon_z}-\sqrt{\varepsilon_{z_1}}}{E_c} \tag{50}$$

$$\Delta T_2^* = \left(\frac{2m_0}{e}\right)^{1/2}\frac{\varepsilon_z-\varepsilon_{z_1}}{E_c\sqrt{\phi_{ac}}}\times\left[-\frac{3n^2-9n+2}{4n(n-2)}-\frac{3}{8}\sqrt{n-1}\times\frac{1}{n}\left(\pi-\arcsin\frac{2}{n}\sqrt{n-1}\right)\right] \tag{51}$$

$$\delta T_1 = 0,$$

$$\delta T_2 = \left(\frac{2m_0}{e}\right)^{1/2} \frac{\varepsilon_r}{E_c \sqrt{\phi_{ac}}} \times \left\{ \frac{(n-1)[-(n-2)^2(3n-1)+2(n^2-n+2)]}{4n(n-2)^2} - \frac{3}{8}\sqrt{n-1}\left(\pi - \arcsin\frac{2}{n}\sqrt{n-1}\right) \right\} \tag{52}$$

以上的时间像差的定义如下：

ΔT^* 为近轴时间像差，它是在理想像面上 ($\sqrt{\varepsilon_z}$, $\sqrt{\varepsilon_r}$) 的近轴电子与 $\sqrt{\varepsilon_{z_1}}$ 的近轴电子的飞行时间的差异；ΔT_1^* 和 ΔT_2^* 分别称为一级、二级近轴时间像差。δT 为几何时间像差，它是在理想像面上具有相同 ($\sqrt{\varepsilon_z}$, $\sqrt{\varepsilon_r}$) 的近轴电子与实际电子的飞行时间的差异；δT_2 为二级几何时间像差。

结论

由以上研究，我们可以得到以下结论：

（1）本文的研究很自然地将成像电子光学的静态问题与动态问题联系起来。实际，自光阴极逸出的电子在空间飞行时携带的不仅是空间信息，还有时间信息。此二信息在像面上显示出其空间-时间特性，如横向像差和时间像差。

（2）由两电极静电同心球系统电子光学和像差的研究，我们得到一个重要的概念：不管是横向像差或者是时间像差，都可以表示为近轴像差和几何像差的组合；即横向像差可以视为近轴横向像差和几何横向像差的合成，时间像差可以视为近轴时间像差和几何时间像差的合成。

（3）当两电极静电同心球系统以静态模式工作时，我们证实了空间分辨率主要取决于二级近轴横向色差 Δr_2^*，即 Recknagel-Artimovich 公式，并且首次导得三级近轴横向像差 Δr_3^* 和三级几何横向像差 δr_3 的具体表示式。同样，当两电极静电同心球系统以动态模式工作时，我们证实了时间分辨率主要取决于一级近轴时间像差 ΔT_1^*，即 Savoisky-Fanchenko 公式；而且也首次导得二级近轴时间像差 ΔT_2^* 和二级几何时间像差 δT_2 的具体表示式。应该指出，占有空间像差和时间像差绝大部分的二级近轴横向色差 Δr_2^* 和一级近轴时间像差 ΔT_1^* 仅与近轴轨迹相联系，因此，今后的研究宜将重点放在近轴电子光学及其近轴像差上。

（4）本文所导得的两电极静电同心球系统的所有公式，如电子追迹（轨迹和时间）、近轴横向像差与几何横向像差、近轴时间像差和几何时间像差，均可以用来检验和校正成像电子光学的理论、方法和算法。

参 考 文 献

[1] ARTIMOVICH L A. Electrostatic properties of emission systems[J]. Information of Academy of Sciences, Series of Physics, 1944, 8(6):313-328.

[2] RECKNAGEL A. Theorie des elektrisohen elecktronen miktroskops fur selbstrakler[J]. Z. Angew. Physik, 1941, 117:689-708 (in German).

[3] SAVOISKY Y K, FANCHENKO S D. Physical basis of electron optical chronograph[J]. Report of Academy of Sciences, 1956, 108(2):218-221 (in Russian).

[4] 西门纪业. 复合浸没物镜的电子光学性质和像差理论[J]. 物理学报, 1957, 13(4):339-356.

[5] 周立伟, 艾克聪, 潘顺臣. 关于电磁复合聚焦阴极透镜的像差理论[J]. 物理学报, 1983(3):376-392.

[6] 西门纪业, 周立伟, 艾克聪. 阴极透镜像差的变分理论[J]. 物理学报, 1983, 32(12):1536-1546.

[7] 艾克聪, 周立伟, 西门纪业. 宽束和细束电磁复合聚焦球面阴极透镜的像差理论[J]. 物理学报, 1986, 35(9):1199-1209.

[8] 艾克聪, 西门纪业, 周立伟. 电磁复合聚焦-偏转球面阴极透镜的相对论像差理论[J]. 物理学报, 1986, 35(9):1210-1222.

[9] RUSKA E. Zur fukussierkeit von Kathoden strahlbundeln grosser Ausgangsquerchnitte[J]. Z. Angew. Physik, 1973, 83(9):684-687.

[10] SCHAGEN P, BRUINING H, FRANCKEN J C. A simple electrostatic electron-optical with only one voltage[J]. Philips Research Report, 1952, 7(2):119-130.

[11] 周立伟. 宽束电子光学[M]. 北京:北京理工大学出版社, 1993.

[12] 周立伟. 宽电子束聚焦与成像——周立伟电子光学学术论文选[M]. 北京:北京理工大学出版社, 1994.

[13] CSORBA I P. Transit-time-spread-limited time resolution of image tubes in streak operation[J]. RCA Review, 1971, 32:650-659.

[14] 周立伟, 李元, 张智诠, 等. 静电聚焦同心球系统验证电子光学成像系统的时间像差理论[J]. 物理学报, 2005, 54(8):3597-3603.

10.6 静电聚焦同心球系统的静态和动态近轴光学
Static and Dynamic Paraxial Optics of Electrostatic Focusing Concentric Spherical Systems

摘要：本文由近轴电子轨迹方程和近轴电子运动方程出发，研究了静电聚焦两电极同心球系统中电子运动轨迹和电子运动时间的解析解，探讨了电子光学的理想成像性质，推导了横向色差与时间色差。本文的结果进一步证实上文的结论是正确的，即时间色差和横向色差正是初始条件参量 $(\varepsilon_r, \varepsilon_z)$ 与 $(\varepsilon_r, \varepsilon_{z_1})$ 的两条近轴电子轨迹在成像位置处在时间和空间上的差异。

Abstract: Based on the paraxial electron trajectory equation and paraxial electron motion equation, the analytical solutions for electron trajectory and the moving time of electrons in a concentric spherical system composed of two electrodes with electrostatic focusing have been studied in the present paper. The electron optical imaging characteristics has been explored, the lateral chromatic aberration and temporal chromatic aberration have been derived and the time-transit-spread of electrons through the space have also been investigated. The results show that the conclusion we have done in the previous paper was correct, i.e., the lateral chromatic aberration and temporal chromatic aberration are the spatial difference and temporal difference of two paraxial electron trajectories at the ideal image plane formed by initial emitted parameters $(\varepsilon_r, \varepsilon_z)$ and $(\varepsilon_r, \varepsilon_{z_1})$.

引言

在上文中我们由普遍轨迹方程和电子运动方程研究了静电聚焦两电极同心球系统的成像性质以及各级电子光学几何像差与时间像差。所得到的结论是，横向色差和时间色差乃是初始条件参量 $(\varepsilon_r, \varepsilon_z)$ 与 $(\varepsilon_r, \varepsilon_{z_1})$ 的两条近轴电子轨迹在成像位置处在空间和时间上的差异，而且它们占据横向像差和时间像差的主要部分。因此，如果就研究色差而言，只需要研究近轴电子光学即可。本文将从近轴轨迹方程和近轴运动方程出发，研究静电聚焦两电极同心球系统中电子运动轨迹的电子运动时间的解析解，从而探讨电子光学的成像性质及其横向色差与时间色差。所得到的结论与上文是一致的。

1 同心球系统的静态近轴光学

1.1 由近轴轨迹方程求解电子轨迹

众所共知，在静电阴极透镜的电子光学中，近轴轨迹方程可表示为

$$r''(z) + \frac{1}{2}\frac{\phi'(z)}{\phi(z)+\varepsilon_z}r' + \frac{1}{4}\frac{\phi''(z)}{\phi(z)+\varepsilon_z}r = 0 \tag{1}$$

在同心球系统中，轴上电位分布 $\phi(z)$ 可表示成

$$\phi(z) = \frac{-z}{(n-1)(z+R_c)}\phi_{ac} \tag{2}$$

式中，R_c 为球面阴极 C 的曲率半径；R_a 为栅状球面阳极 A 的曲率半径，阳极 A 对于阴极 C 的电位为 ϕ_{ac}，$n = \frac{R_c}{R_a}$，设阴极 C 的电位 $\phi_c = 0$。按照前文的约定，R_c、R_a 均取负值，z 自阴极面算起，取正值。

不难证明，在阴极面轴上点逸出的 $(\varepsilon_0, \alpha_0)$ 的电子轨迹，初始条件为

$$r_0 = r(z=0) = 0, \quad r_0' = \tan\alpha_0 = \frac{\sqrt{\varepsilon_r}}{\sqrt{\varepsilon_z}}$$

情况下，方程式（1）在电位分布式（2）下的解析解为

$$r^*(z) = 2z\left(\frac{\varepsilon_r}{\phi(z)}\right)^{\frac{1}{2}}\left[\sqrt{1+\frac{\varepsilon_z}{\phi(z)}} - \sqrt{\frac{\varepsilon_z}{\phi(z)}}\right] \tag{3}$$

式中，星号 $*$ 表示近轴情况。由式（3）和近轴电子的轴向速度表达式

$$\dot{z}^* = \sqrt{\frac{2e}{m_0}[\phi(z)+\varepsilon_z]}$$

不难求得近轴电子轨迹的斜率和径向速度表示式：

$$r'^* = \frac{1}{R_c+z}\sqrt{\frac{\varepsilon_r}{\phi(z)}}\left\{2z\left[\sqrt{1+\frac{\varepsilon_z}{\phi(z)}} - \sqrt{\frac{\varepsilon_z}{\phi(z)}}\right] + R_c\frac{1}{\sqrt{1+\frac{\varepsilon_z}{\phi(z)}}}\right\} \tag{4}$$

$$\dot{r}^* = \frac{1}{R_c+z}\sqrt{\frac{2e}{m_0}\varepsilon_r}\left\{R_c + 2z\sqrt{1+\frac{\varepsilon_z}{\phi(z)}}\left[\sqrt{1+\frac{\varepsilon_z}{\phi(z)}} - \sqrt{\frac{\varepsilon_z}{\phi(z)}}\right]\right\} \tag{5}$$

1.2 近轴轨迹的交轴位置表达式

当 $z = R_a - R_c$，$r(z = R_a - R_c) = r_a$，$\phi(z = R_a - R_c) = \phi_{ac}$，可得到电子轨迹在阳极处的径向高度的表达式：

$$r_a = 2(R_a - R_c)\left(\frac{\varepsilon_r}{\phi_{ac}}\right)^{\frac{1}{2}}\left(\sqrt{1+\frac{\varepsilon_z}{\phi_{ac}}} - \sqrt{\frac{\varepsilon_z}{\phi_{ac}}}\right) \tag{6}$$

与斜率表达式：

$$r_a' = \frac{1}{R_a}\sqrt{\frac{\varepsilon_r}{\phi_{ac}}}\left[2(R_a - R_c)\left(\sqrt{1+\frac{\varepsilon_z}{\phi_{ac}}} - \sqrt{\frac{\varepsilon_z}{\phi_{ac}}}\right) + R_c\frac{1}{\sqrt{1+\frac{\varepsilon_z}{\phi_{ac}}}}\right] \tag{7}$$

其近似式为

$$r_a = 2(R_a - R_c)\left(\frac{\varepsilon_r}{\phi_{ac}}\right)^{\frac{1}{2}}\left(1 - \sqrt{\frac{\varepsilon_z}{\phi_{ac}}} + \frac{1}{2}\frac{\varepsilon_z}{\phi_{ac}}\right) \tag{8}$$

$$r'_a = -\sqrt{\frac{\varepsilon_r}{\phi_{ac}}}(n-2)\left[1 - \frac{2(n-1)}{n-2}\sqrt{\frac{\varepsilon_z}{\phi(z)}} + \frac{3n-2}{2(n-2)}\frac{\varepsilon_z}{\phi_{ac}}\right] \tag{9}$$

于是可求得电子到达轴上点的位置离曲率中心的距离 R_i^* 为

$$R_i^* = \frac{r_a}{-r'_a} + R_a$$

对以上各式进行展开，略去高于 $\frac{\varepsilon_z}{\phi_{ac}}$ 的幂次项，便得到

$$R_i^* = \frac{-R_c}{n-2}\left[1 + \frac{2(n-1)}{n-2}\sqrt{\frac{\varepsilon_z}{\phi_{ac}}} + \frac{2n(n-1)}{(n-2)^2}\frac{\varepsilon_z}{\phi_{ac}}\right] \tag{10}$$

这与前面一篇文章所导出的公式是一致的。

1.3 电子光学性质与近轴几何像差

与前文类似，我们可讨论系统的近轴电子光学性质及几何像差：

$$M = \frac{R_i^*}{R_c} = -\frac{1}{n-2}\left[1 + \frac{2(n-1)}{n-2}\left(\frac{\varepsilon_{z_1}}{\phi_{ac}}\right)^{\frac{1}{2}} + \frac{2n(n-1)}{(n-2)^2}\frac{\varepsilon_{z_1}}{\phi_{ac}}\right] \tag{11}$$

由于我们研究的是近轴情况，故仅有近轴像差。按照前文的定义和处理，不难求得近轴纵向色差 Δz^* 的表达式为

$$\Delta z^* = \sum_{j=1}^{2} \Delta z_j^* \tag{12}$$

式中，

$$\Delta z_1^* = \frac{2M^2 \sqrt{\phi_{ac}}}{-E_c}(\sqrt{\varepsilon_z} - \sqrt{\varepsilon_{z_1}}) \tag{13}$$

$$\Delta z_2^* = \frac{2M^2}{-E_c}\left[\frac{n}{n-2}(\varepsilon_z - \varepsilon_{z_1}) - \frac{4(n-1)}{n-2}\sqrt{\varepsilon_{z_1}}(\sqrt{\varepsilon_z} - \sqrt{\varepsilon_{z_1}})\right] \tag{14}$$

近轴横向色差 Δr^* 的表达式为

$$\Delta r^* = \sum_{j=1}^{2} \Delta r_j^* \tag{15}$$

式中，

$$\Delta r_1^* = \frac{2M}{E_c}\sqrt{\varepsilon_r}(\sqrt{\varepsilon_z} - \sqrt{\varepsilon_{z_1}}) \tag{16}$$

$$\Delta r_2^* = \frac{2M}{-E_c \sqrt{\phi_{ac}}}\sqrt{\varepsilon_r}(\varepsilon_z - \varepsilon_{z_1}) \tag{17}$$

2 条纹工作方式下同心球系统的动态近轴光学

2.1 由近轴运动方程求解时间 t

由近轴电子轴向速度的表达式

$$\dot{z} = \frac{dz}{dt} = \left\{\frac{2e}{m_0}[\phi(z) + \varepsilon_z]\right\}^{\frac{1}{2}} \tag{18}$$

出发，积分上式，由

$$t(z) = \int_0^z \frac{\mathrm{d}z}{\left\{\frac{2e}{m_0}[\phi(z) + \varepsilon_z]\right\}^{\frac{1}{2}}}$$

可以得到电子自阴极面（$t=0$，$z_0=0$，$r_0=0$）上发出的初始条件为（ε_0，α_0）、在任一位置 z 的时间 t 的近轴情况下的精确表达式：

$$t(z) = t(z,0,\varepsilon_z^{\frac{1}{2}},0) = \frac{-R_c}{\left\{\frac{2e}{m_0}\left[\frac{\phi_{ac}-(n-1)\varepsilon_z}{(n-1)}\right]\right\}^{\frac{1}{2}}}\left\{\sqrt{1+\frac{z}{R_c}}\sqrt{\frac{(n-1)\varepsilon_z}{\phi_{ac}-(n-1)\varepsilon_z}}-\frac{z}{R_c}-\sqrt{\frac{(n-1)\varepsilon_z}{\phi_{ac}-(n-1)\varepsilon_z}}+\right.$$
$$\left.\left(1+\frac{(n-1)\varepsilon_z}{\phi_{ac}-(n-1)\varepsilon_z}\right)\left[\arcsin\sqrt{\frac{(n-1)\varepsilon_z}{\phi_{ac}}-\frac{[\phi_{ac}-(n-1)\varepsilon_z]z}{R_c\phi_{ac}}}-\arcsin\sqrt{\frac{(n-1)\varepsilon_z}{\phi_{ac}}}\right]\right\}$$
(19)

2.2 电子到达阳极的时间近似式

先求当 $z = -R_c + R_a$ 时电子到达阳极的时间 t_{C-A} 之值。因为 $\frac{R_c - R_a}{R_c} = \frac{n-1}{n}$，故有

$$t_{C-A} = \frac{-R_c}{\left\{\frac{2e}{m_0}\left[\frac{\phi_{ac}-(n-1)\varepsilon_z}{(n-1)}\right]\right\}^{\frac{1}{2}}}\left\{\sqrt{\frac{R_a}{R_c}}\sqrt{\frac{(n-1)(\phi_{ac}+\varepsilon_z)}{[\phi_{ac}-(n-1)\varepsilon_z]n}}-\sqrt{\frac{(n-1)\varepsilon_z}{\phi_{ac}-(n-1)\varepsilon_z}}+\right.$$
$$\left.\left[1+\frac{(n-1)\varepsilon_z}{\phi_{ac}-(n-1)\varepsilon_z}\right]\left[\arcsin\sqrt{\frac{(n-1)(\phi_{ac}+\varepsilon_z)}{n\phi_{ac}}}-\arcsin\sqrt{\frac{(n-1)\varepsilon_z}{\phi_{ac}}}\right]\right\}$$
$$= t_{C-A}|_1 + t_{C-A}|_2$$
(20)

现简化上面的公式，先求第一项，对上式展开，略去高于 $\frac{\varepsilon_z}{\phi_{ac}}$ 的幂次项，得

$$t_{C-A}|_1 = \left(\frac{m_0}{2e}\right)^{\frac{1}{2}}\frac{R_a - R_c}{\sqrt{\phi_{ac}}}\left[1 - n\sqrt{\frac{\varepsilon_z}{\phi_{ac}}} + \frac{1}{2}(2n-1)\frac{\varepsilon_z}{\phi_{ac}}\right]$$
(21)

再求 $t_{C-A}|_2$ 的后一部分，因为 $x^2 + y^2 \leq 1$，$\arcsin x - \arcsin y = \arcsin(x\sqrt{1-y^2} - y\sqrt{1-x^2})$，故带有反三角函数项可简化为

$$\arcsin\sqrt{\frac{(n-1)[\phi_{ac}+\varepsilon_z]}{n\phi_{ac}}} - \arcsin\sqrt{\frac{(n-1)\varepsilon_z}{\phi_{ac}}} = \arcsin\sqrt{\frac{n-1}{n}} - \sqrt{n-1}\sqrt{\frac{\varepsilon_z}{\phi_{ac}}} + \frac{1}{2}\sqrt{n-1}\frac{\varepsilon_z}{\phi_{ac}}$$

于是

$$t_{C-A}|_2 = -R_c\sqrt{\frac{m_0}{2e}}\sqrt{\frac{n-1}{\phi_{ac}}}\left[1 + \frac{3}{2}\frac{(n-1)\varepsilon_z}{\phi_{ac}}\right]\left(\arcsin\sqrt{\frac{n-1}{n}} - \sqrt{n-1}\sqrt{\frac{\varepsilon_z}{\phi_{ac}}} + \frac{1}{2}\sqrt{n-1}\frac{\varepsilon_z}{\phi_{ac}}\right)$$
(22)

最后，可得电子到达阳极时的时间 t_{C-A} 的近似表达式：

$$t_{C-A} = \left(\frac{2m_0}{e}\right)^{\frac{1}{2}} \left[\frac{R_a - R_c}{2\sqrt{\phi_{ac}}} - \frac{1}{2}R_c\sqrt{\frac{n-1}{\phi_{ac}}}\arcsin\sqrt{\frac{n-1}{n}} - n\frac{R_a - R_c}{\sqrt{\phi_{ac}}}\sqrt{\frac{\varepsilon_z}{\phi_{ac}}} + \right.$$
$$\left. n\frac{R_a - R_c}{\sqrt{\phi_{ac}}}\frac{\varepsilon_z}{\phi_{ac}}\left(\frac{1}{4}\frac{3n-1}{n} + \frac{3}{4}\sqrt{n-1}\arcsin\sqrt{\frac{n-1}{n}}\right)\right] \quad (23)$$

这与前文所导出的是一致的。

2.3 近轴轨迹在无场区经历的时间的表达式

在无场空间中，电子自阳极射出的速度为 v_a，它到达轴上点的长度为 l，时间为 t_{A-i}，则有

$$t_{A-i} = \frac{l}{v_a} \quad (24)$$

式中，

$$l = \left[r_a^2 + (-R_a + R_i^*)^2\right]^{\frac{1}{2}}, \quad v_a = \sqrt{v_0^2 + \frac{2e}{m_0}\phi_{ac}}$$

其中，r_a、R_i^* 以式（8）、式（10）表示，v_a、l 可近似表示为如下：

$$v_a \approx \sqrt{\frac{2e}{m_0}\phi_{ac}}\left(1 + \frac{1}{2}\frac{\varepsilon_z}{\phi_{ac}}\right),$$

$$l \approx -R_a + R_i = \frac{-2R_c(n-1)}{n(n-2)}\left[1 + \frac{n}{n-2}\sqrt{\frac{\varepsilon_z}{\phi_{ac}}} + \frac{n^2}{(n-2)^2}\frac{\varepsilon_z}{\phi_{ac}}\right]$$

在式（24）中代入以上各式，即可求得电子自阳极出发到达轴上点的时间 t_{A-i} 的近似值：

$$t_{A-i} = -\sqrt{\frac{2m_0}{e}}\frac{R_a(n-1)}{(n-2)\sqrt{\phi_{ac}}}\left[1 + \frac{n}{n-2}\sqrt{\frac{\varepsilon_z}{\phi_{ac}}} + \frac{n^2 + 4n - 4}{2(n-2)^2}\frac{\varepsilon_z}{\phi_{ac}}\right] \quad (25)$$

这与前文所导出的公式是一致的。

2.4 近轴时间像差

近轴横向时间像差 ΔT^* 是在成像面上衡量的。它定义为在由 ε_{z_1} 确定的成像面上以初始条件为 $(\varepsilon_r, \varepsilon_z)$ 逸出的近轴电子与以初始条件为 $(\varepsilon_r, \varepsilon_{z_1})$ 的近轴电子到达位于 ε_{z_1} 所确定的像面上的时间差异。它由两部分组成：一是此二近轴电子到达阳极时的时间弥散；二是此二电子由阳极出发到达由 ε_{z_1} 确定的近轴成像位置的像面上的时间弥散，即

$$\Delta T^* = t|_{C-A}(\varepsilon_z) + (t + \Delta t)|_{A-i^*}(\varepsilon_z) - t|_{C-A}(\varepsilon_{z_1}) - t|_{A-i^*}(\varepsilon_{z_1}) \quad (26)$$

式中，$t|_{C-A}(\varepsilon_z)$、$t|_{C-A}(\varepsilon_{z_1})$ 以式（23）表示。$t|_{A-i^*}(\varepsilon_z)$ 以式（25）表示。而

$$(t + \Delta t)|_{A-i^*}(\varepsilon_z) = t|_{A-i}(\varepsilon_z) + \Delta t|_{i-i^*}(\varepsilon_z) \quad (27)$$

现求式（27），$(\varepsilon_0, \alpha_0)$ 的轨迹到达轴上再到达像面上，其时间差异为 $\Delta t|_{i-i^*}(\varepsilon_z, \varepsilon_r)$，按照前文的推导，不难求得

$$\Delta t|_{i-i^*}(\varepsilon_z, \varepsilon_r) = R_a\frac{n}{n-2}\sqrt{\frac{m_0}{2e}}\frac{1}{\sqrt{\phi_{ac}}}\left[\frac{2(n-1)}{(n-2)\sqrt{\phi_{ac}}}(\sqrt{\varepsilon_z} - \sqrt{\varepsilon_{z_1}}) + \frac{2n(n-1)}{(n-2)^2\phi_{ac}}(\varepsilon_z - \varepsilon_{z_1})\right]$$

$$(28)$$

故在等位区内，$(\varepsilon_z, \varepsilon_r)$ 的电子到达位于 ε_{z_1} 的像面的时间表达式可表示为

$$(t+\Delta t)^*|_{A-i^*}(\varepsilon_z) = -\sqrt{\frac{2m_0}{e}}\frac{R_a(n-1)}{(n-2)\sqrt{\phi_{ac}}}\left[1+\frac{n}{n-2}\sqrt{\frac{\varepsilon_z}{\phi_{ac}}}+\frac{n^2+4n-4}{2(n-2)^2}\frac{\varepsilon_z}{\phi_{ac}}\right]+$$

$$R_a\frac{n}{n-2}\sqrt{\frac{m_0}{2e}}\frac{1}{\sqrt{\phi_{ac}}}\left[\frac{2(n-1)}{(n-2)\sqrt{\phi_{ac}}}(\sqrt{\varepsilon_z}-\sqrt{\varepsilon_{z_1}})+\frac{2n(n-1)}{(n-2)^2\phi_{ac}}(\varepsilon_z-\varepsilon_{z_1})\right]$$

(29)

故自阴极面逸出的 $(\varepsilon_z, \varepsilon_r)$ 的近轴电子到达位于 ε_{z_1} 确定的像面的时间与 ε_{z_1} 近轴电子到达的时间之差，即近轴时间像差 ΔT^* 以阶次分类最后可表示为

$$\Delta T^* = \sum_{j=1}^{2}\Delta T_j^* \tag{30}$$

式中，

$$\Delta T_1^* = \left(\frac{2m_0}{e}\right)^{\frac{1}{2}}\frac{\sqrt{\varepsilon_z}-\sqrt{\varepsilon_{z_1}}}{E_c} \tag{31}$$

$$\Delta T_2^* = \left(\frac{2m_0}{e}\right)^{\frac{1}{2}}\frac{\varepsilon_z-\varepsilon_{z_1}}{E_c\sqrt{\phi_{ac}}}\times\left[-\frac{3n^2-9n+2}{4n(n-2)}-\frac{3}{8}\sqrt{n-1}\frac{1}{n}\left(\pi-\arcsin\frac{2}{n}\sqrt{n-1}\right)\right] \tag{32}$$

在上面的文章中普遍运动方程导得的时间像差，若考虑近轴条件，其结果与式（31）和式（32）是一致的。这就证明了近轴时间像差只需要通过近轴运动方程就可以求得。

结束语

本文从近轴轨迹方程和近轴运动方程出发，研究静电聚焦两电极同心球系统中电子运动轨迹的电子运动时间的解析解，从而探讨电子光学的成像性质及其横向色差与时间色差。所得到的结论与上文是一致的。

参 考 文 献

[1] ZHOU L W, GONG H, ZHANG Z Q, et al. Paraxial imaging electron optics and its spatial – temporal aberrations for a bi – electrode concentric spherical system with electrostatic focusing[J]. Optik, 2011,122（4）：295－299.

[2] 周立伟. 宽束电子光学[M]. 北京：北京理工大学出版社，1993.

[3] 周立伟. 宽电子束聚焦与成像[J]//周立伟电子光学学术论文选. 北京：北京理工大学出版社，1994.

[4] ARTIMOVICH L A. Electrostatic properties of emission systems[J]. Information of Academy of Sciences, Series of Physics, 1944, 8（6）：689－708.

[5] RECKNAGEL A. Theorie des elektrisohen elecktronen miktroskops fur selbstrakler[J]. Z. Angew. Physik, 1941,117：689－708（in German）.

[6] 周立伟，李元，张智铨，等. 静电聚焦同心球系统验证电子光学成像系统的时间像差理论[J]. 物理学报，2005，54(8)：3597－3603.

10.7 On Electron-optical Spatial-temporal Aberrations in a Bi-electrode Spherical Concentric System with Electrostatic Focusing

两电极静电聚焦同心球电子光学系统的空间-时间像差的研究

Abstract: In the present paper, based on the practical electron ray equation and electron motion equation for a bi-electrode concentric spherical system with electrostatic focusing, the spatial-temporal trajectory of moving electron emitted from the photocathode is solved, the exact and approximate formulae for image position and arriving time, have been deduced. From the solution of spatial-temporal trajectory, the electron optical spatial and temporal properties of this system are then discussed, the paraxial and geometrical spatial and temporal aberrations with different orders are defined and deduced.

摘要：本文基于实际轨迹方程和电子运动方程，对两电极静电同心球系统研究了由光阴极逸出的运动电子的空间和时间轨迹，推导了成像位置和电子飞行时间的精确和近似表达式，并由空间和时间轨迹的解析解出发讨论了系统的空间-时间成像特性，导出了各级近轴与几何空间-时间像差的表示式。

Introduction

For an electron optical imaging system, spatial-trajectory-spread and transit-time-spread produced by the emission energy and emission angle of photoelectrons are the fundamental problems of image tubes and high-speed image converter tubes. In the static mode operation for these two tubes, the spatial-trajectory-spread of electron trajectories at the image plane determines limited spatial resolution. In the streak mode operation for high-speed image converter tube, a point image of a photo event of very short duration is imaged by the camera as a line because of transit-time-spread. The length of the line is proportional to the transit-time-spread. Obviously, the shorter the length of the line, the better the time resolution of the camera.

It was proven that the spatial-trajectory-spread of an imaging system is mainly determined by the second order paraxial lateral aberration, which can be expressed by Recknagel-Artimovich formula[1,2]. Formula shows that the second order lateral aberration does not depend on the

---

Zhou Liwei[a], Gong Hui[a], Zhang Zhiquan[b], Zhang Yifei[b]. a) Beijing Institute of Technology, b) Institute of Armored Force Engineering. Proc. of SPIE 2009, V. 7384, No. 35, 1–12.

concrete electrode structure and potential distributions, but depends on the electron emission energy, the strength of electric field at the photocathode, the magnification of the system and the position of image plane.

It should be mentioned that the transit-time-spread, which can be called the first order paraxial temporal aberration, was given by Savoisky-Fanchenko formula in 1956[3], and then was proven by I. P. Csorba in 1970[4]. The formula shows that the first order temporal aberration depends only on the axial emission energy, the strength of electric field at the photocathode, but does not depend on the concrete electrode structure and potential distributions, as well as the magnification of the system and the position of image plane.

The difference between Recknagel-Artimovich formula and Savoisky-Fanchenko formula is obviously. We may find the non-harmonious and non-symmetry between the two formulae if we think it over. Actually, the spatial-trajectory-spread and the transit-time spread at an image plane is produced by the emission energy and emission angle of the same photoelectrons. It is strange that the former depends on the position of image plane, but the later does not.

The best model for investigating the difference is a bi-electrode concentric spherical electrostatic focusing system. This is because in this system the potential distribution and electron trajectory can be written as analytical forms, and the focusing and imaging characteristics and aberrations (spatial and temporal) can be investigated quantitatively. With a concentric spherical electrostatic focusing system, whether realized accurately or approximately, a good image quality can be obtained.

For the static electron optics of the concentric spherical electrostatic system, the previous works were given by E. Ruska (1933)[5], P. Schagen (1952)[6], but all of these works remained at zero-order approximation of imaging. The more accurate formulae for the solution of electron trajectory and lateral aberrations were given by Zhou Liwei in his monographs[7,8]. For the dynamic electron optics of the concentric spherical electrostatic system, the work of I. P. Csorba[4] has only confirmed the conclusion of Savoisky-Fanchenko formula[3], and the more accurate formula of temporal aberration has been developed by Zhou Liwei, et al[9].

In the present paper, for a bi-electrode concentric spherical electrostatic system, the static imaging electron optics and spatial aberrations will be discussed in part I, the dynamic imaging electron optics and temporal aberrations will be discussed in part II.

1 Static imaging electron optics and spatial aberrations for a bi-electrode concentric spherical system with electrostatic focusing

1.1 Ray equation and its solution

Fig. 1 shows a bi-electrode concentric spherical electrostatic system. Suppose that the curvature radii of spherical cathode C and of the mesh-spherical anode A are R_c and R_a, the common curvature center is O. Because of spherical symmetry of the system, we can use the polar coordinate (ρ, φ) to describe the electron trajectory. The original point of the polar coordinate is located at the center O of curvature, the potential of the photocathode is $\phi_c = 0$, the potential of mesh-anode A with respect to the photocathode C is ϕ_{ac}, the photoelectron with initial emission velocity v_0 and initial emission angle α_0 is emitted from the photocathode. R_c, R_a, ρ, R_i, R_i^* are calculated from the curvature center O to the corresponding spherical surface; Δz is calculated from the ideal image position I_i^*; Δr is calculated from the central axis $\varphi = 0$. Suppose that the moving direction of electron ray is from left to right, we stipulate that the line segment is positive from left to right or from bottom to top, otherwise is negative. All of the angles should be measured by the acute angle, we stipulate that rotation turns round counterclockwise is positive, otherwise is negative. Similar to the line segments, we should stipulate the beginning axis for the angle calculation. α_0, φ, γ, α_i are calculated from central axis $\varphi = 0$. ζ, τ are calculated from the joint line connecting the center O and the point concerned as a beginning axis.

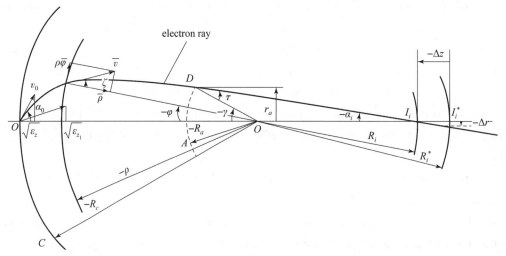

Fig. 1 A bi-electrode concentric spherical system with electrostatic focusing ($n = R_c/R_a > 2$)

The field between two concentric spheres is equivalent to the electrostatic field produced by a point charge which is concentrated in spherical center. The formula of $\phi_{\rho c}$, the potential at coordinate ρ with respect to photocathode C, can be expressed as

$$\phi_{\rho c} = E_c R_c \left(\frac{R_c}{\rho} - 1 \right) \tag{1}$$

where E_c electric-field strength at the photocathode. When $\rho = R_a$, we have

$$E_c = \frac{\phi_{ac}}{R_c(n-1)} \tag{2}$$

where $n = R_c/R_a$. For the concave cathode-anode system, R_c, R_a take negative value, $n > 1$; for the convex cathode-anode system, R_c, R_a take positive values, $n < 1$. E_c always takes negative value whether $n > 1$ or $n < 1$.

In a bi-electrode concentric spherical system, consider an electron leaving cathode sphere with an emission velocity v_0 and emission angle α_0 as indicated in Fig. 1, the motion of electron path follows the law of conversation of energy and the law of conversation of angular momentum:

$$\frac{1}{2}m_0 v^2 = \frac{1}{2}m_0 v_0^2 + e\phi_{pc} \tag{3}$$

$$\rho^2 \dot{\varphi} = R_c^2 \dot{\varphi}_0 = \text{const.} \tag{4}$$

where m_0, e are electron mass and charge, e takes absolute value. v is the velocity at polar coordinate (ρ, φ), $\dot{\varphi} = d\varphi/dt$.

From Eq. (3) and Eq. (4), we can get the ray expressions: the E. Ruska[5] equation $\rho = f(\varphi)$ or the P. Schagen[6] equation $\varphi = f(\rho)$ which can be written as following

$$\mu = \left(1 - \frac{p}{2\sin^2\alpha_0}\right)\cos\varphi - \frac{1}{\tan\alpha_0}\sin\varphi + \frac{p}{2\sin^2\alpha_0} \tag{5}$$

$$\tan\frac{\varphi}{2} = \frac{-\cos\alpha_0 + [1 + p(\mu - 1) - \mu^2\sin^2\alpha_0]^{1/2}}{-\dfrac{p}{\sin\alpha_0} + (\mu + 1)\sin\alpha_0} \tag{6}$$

where $\mu = R_c/\rho$, $p(n-1) = \varphi_{ac}/\varepsilon_0$, $\varepsilon_0 = m_0 v_0^2/2e$. ε_0 is the applied potential required for a rest electron to obtain the emission energy, which is called initial potential, p is only a parameter, but $p(n-1)$ has physical meaning, and for an ordinary image tube, $p(n-1) = 10^3 - 10^4$.

It should be noted that the Ruska[5] and Schagen[6] equations only give the flying position of electron emitted from photocathode, but the flying direction of electron has not been clearly given. Furthermore, their equations were not convenient for calculating the moving electron trajectory in this system. For solving this problem, the new equations $\varphi = f(\rho)$, $\zeta = \psi(\rho)$ have been given by . Zhou Liwei, expressed in the form of $\sin\varphi$, $\cos\varphi$ and $\tan\zeta$ [7,8]:

$$\sin\varphi = -\frac{(\mu - 1)(c + d_0)}{bd_0 + b_0 c}\sin\alpha_0 \tag{7a}$$

$$\cos\varphi = \frac{bc + b_0 d_0}{bd_0 + b_0 c} \tag{7b}$$

$$\tan\zeta = \frac{\mu\sin\alpha_0}{b} \tag{8}$$

where
$$b_0 = \cos\alpha_0, \quad b = \left[1 + \frac{\phi_{ac}(\mu - 1)}{\varepsilon_0(n-1)} - \mu^2\sin^2\alpha_0\right]^{\frac{1}{2}}$$

$$c = 1 - 2\mu(n-1)\frac{\varepsilon_0}{\phi_{ac}}\sin^2\alpha_0, \quad d_0 = 1 - 2(n-1)\frac{\varepsilon_0}{\phi_{ac}}\sin^2\alpha_0$$

Eq. (7a), Eq. (7b), and Eq. (8) are generally appropriate whether $n > 1$ or $n < 1$.

1.2 Determination of ideal image position R_i^* and its paraxial imaging

From Eq. (7a), Eq. (7b) and Eq. (8), it is not difficult to trace the electron ray emitted from photocathode with ($\sqrt{\varepsilon_0}$, α_0) and to define the position D at the anode A, thus we can determine angle τ and γ. We assume that the anode is a mesh-electrode which will be transparent for electron passing through, and after the mesh-anode there is a free from field region. When the electron goes through the mesh-anode, the magnitude and direction of electron velocity will not be changed. Therefore, the trajectory after the anode is a straight line tangentially to the trajectory at point D, and intersects the axis $\varphi = 0$ at point I_i.

According to the triangle $\triangle ODI_i$ given in Fig. 1, we can get the expression of R_i at the intersection I_i:

$$R_i = R_a \frac{\sin \tau}{\sin \alpha_i} = R_a \frac{\tan \tau}{\sin \gamma + \tan \tau \cos \gamma} \tag{9}$$

where α_i is the angle of electron ray at intersection I_i with axis $\varphi = 0$; R_i is the distance from curvature center O to trajectory intersection I_i.

Obviously, if $\mu = n$, $\varphi = \gamma$, $\zeta = \tau$, $b = b_1$, $c = c_0$, from expressions (7a), expression (7b) and expression (8) we can get the formulae of $\sin \gamma$, $\cos \gamma$ and $\tan \tau$. Using formula (9), we may obtain the exact expressions of trajectory intersection and its slope at I_i for electron ray emitted from photocathode with initial condition ($\sqrt{\varepsilon_0}$, α_0) in the bi-electrode concentric spherical system:

$$R_i = -R_a \frac{n(b_1 d_0 + b_0 c_0)}{b_1(n-1)(c_0 + d_0) - n(b_1 c_0 + b_0 d_0)} \tag{10}$$

$$\tan \alpha_i = -\frac{b_1(n-1)(c_0 + d_0) - n(b_1 c_0 + b_0 d_0)}{b_1(b_1 c_0 + b_0 d_0) + n(n-1)(c_0 + d_0)\sin^2 \alpha_0} \sin \alpha_0 \tag{11}$$

where $b_1 = \left(1 + \frac{\phi_{ac}}{\varepsilon_0} - n^2 \sin^2 \alpha_0\right)^{\frac{1}{2}}$, $c_0 = 1 - 2n(n-1)\frac{\varepsilon_0}{\phi_{ac}}\sin^2 \alpha_0$

Now, we simplify the above-mentioned expressions for R_i and $\tan \alpha_i$. Let $\sqrt{\varepsilon_z} = \sqrt{\varepsilon_0}\cos \alpha_0$, $\sqrt{\varepsilon_r} = \sqrt{\varepsilon_0}\sin \alpha_0$ and using binomial theorem to make power series expansion for the terms in Eq. (10) and Eq. (11), neglecting $(\varepsilon_0/\phi_{ac})^{3/2}$ and its higher order terms which is greatly smaller than 1, we may obtain the approximate expressions for the trajectory intersection R_i and its slope $\tan \alpha_i$ at I_i:

$$R_i = -R_a \frac{n}{n-2}\left[1 + \frac{2(n-1)}{n-2}\sqrt{\frac{\varepsilon_z}{\phi_{ac}}} + \frac{2n(n-1)}{(n-2)^2}\frac{\varepsilon_z}{\phi_{ac}} - \frac{2(n-1)^2}{n-2}\frac{\varepsilon_r}{\phi_{ac}}\right] \tag{12}$$

$$\tan \alpha_i = -(n-2)\sqrt{\frac{\varepsilon_r}{\phi_{ac}}}\left[1 - \frac{2(n-1)}{n-2}\sqrt{\frac{\varepsilon_z}{\phi_{ac}}} + \frac{3n-2}{2(n-2)}\frac{\varepsilon_z}{\phi_{ac}} + \frac{(n-1)(n^2-n+2)}{2(n-2)}\frac{\varepsilon_r}{\phi_{ac}}\right] \tag{13}$$

From this, we can study the electron optical properties and spatial aberrations for the bi-electrode concentric spherical system.

Considering the paraxial imaging condition for cathode lenses, let $\sqrt{\varepsilon_z} = \sqrt{\varepsilon_{z_1}}$, neglecting ε_r/ϕ_{ac} term which is greatly smaller than 1 in the brackets of expression (12) and expression (13), we may obtain the approximate expressions of R_i^* and $\tan \alpha_i^*$ at I_i^* for the paraxial ray leaving the photocathode with the initial condition $\sqrt{\varepsilon_{z_1}}$:

$$R_i^* = -R_a \frac{n}{n-2}\left[1 + \frac{2(n-1)}{n-2}\sqrt{\frac{\varepsilon_{z_1}}{\phi_{ac}}} + \frac{2n(n-1)}{(n-2)^2}\frac{\varepsilon_{z_1}}{\phi_{ac}}\right] \tag{14}$$

$$\tan\alpha_i^* = -(n-2)\sqrt{\frac{\varepsilon_r}{\phi_{ac}}}\left[1 - \frac{2(n-1)}{n-2}\sqrt{\frac{\varepsilon_{z_1}}{\phi_{ac}}} + \frac{3n-2}{2(n-2)}\frac{\varepsilon_{z_1}}{\phi_{ac}}\right] \quad (15)$$

where the mark (*) expresses the paraxial condition. $\sqrt{\varepsilon_{z_1}}$ is corresponding to the initial axial velocity that electron carries to paraxial image position at I_i^*. Formula (14) means that the ideal imaging will be formed so long as the axial initial velocity of electron $\sqrt{\varepsilon_{z_1}}$ remains the same.

The lateral magnification M of the system can be expressed by

$$M = \frac{R_i^*}{R_c} = -\frac{1}{n-2}\left[1 + \frac{2(n-1)}{n-2}\left(\frac{\varepsilon_{z_1}}{\phi_{ac}}\right)^{\frac{1}{2}} + \frac{2n(n-1)}{(n-2)^2}\frac{\varepsilon_{z_1}}{\phi_{ac}}\right] \quad (16)$$

Obviously, it forms a real image when $n > 2$ or a virtual image when $n < 2$.

1.3 Spatial aberrations—paraxial lateral aberration and geometrical lateral aberration

In the following, we divide the spatial aberration into two parts, the paraxial aberration and the geometrical aberration. As shown in Fig. 2, the focusing position I_i^* of electrons emitted from cathode with $\sqrt{\varepsilon_{z_1}}$ in paraxial condition is defined as an ideal image position, and the practical rays from the same object point with ($\sqrt{\varepsilon_0}$, α_0), corresponding to ($\sqrt{\varepsilon_z}$, $\sqrt{\varepsilon_r}$), will intersect the axial at I_i, it forms a displacement of image position along the axial direction, which is called longitudinal aberration as following:

$$\begin{aligned}\Delta z &= R_i(\sqrt{\varepsilon_r},\sqrt{\varepsilon_z}) - R_i^*(\sqrt{\varepsilon_{z_1}}) \\ &= R_i(\sqrt{\varepsilon_r},\sqrt{\varepsilon_z}) - R_i^*(\sqrt{\varepsilon_r},\sqrt{\varepsilon_z}) + R_i^*(\sqrt{\varepsilon_r},\sqrt{\varepsilon_z}) - R_i^*(\sqrt{\varepsilon_{z_1}}) \\ &= \Delta z^* + \delta z\end{aligned} \quad (17)$$

Substituting Eq. (14) and Eq. (12) into Eq. (17), we get the expression for longitudinal aberration Δz:

$$\Delta z = \Delta z^* + \delta z = \Delta z_1^* + \Delta z_2^* + \delta z_1 + \delta z_2 \quad (18)$$

where

$$\Delta z_1^* = \frac{2M^2\sqrt{\varphi_{ac}}}{-E_c}(\sqrt{\varepsilon_z} - \sqrt{\varepsilon_{z_1}}) \quad (19)$$

$$\Delta z_2^* = \frac{2M^2}{-E_c}\left\{\frac{n}{n-2}(\varepsilon_z - \varepsilon_{z_1}) - \frac{4(n-1)}{n-2}\sqrt{\varepsilon_{z_1}}(\sqrt{\varepsilon_z} - \sqrt{\varepsilon_{z_1}})\right\} \quad (20)$$

$$\delta z_1 = 0, \quad \delta z_2 = \frac{2M^2}{E_c}(n-1)\varepsilon_r \quad (21)$$

Similarly, at the ideal image plane determined in paraxial condition by $\sqrt{\varepsilon_{z_1}}$, we may obtain the lateral aberration Δr as following:

$$\Delta r = -\Delta z\tan\alpha_i \quad (22)$$

where $\tan\alpha_i$ is expressed by Eq. (13).

Substituting Eq. (18) and Eq. (13) in Eq. (22), the lateral aberration Δr will have the following form:

$$\Delta r = \Delta r^* + \delta r = r(z_i^*, \sqrt{\varepsilon_z}, \sqrt{\varepsilon_r}) - r^*(z_i^*, \sqrt{\varepsilon_{z_1}})$$
$$= r(z_i^*, \sqrt{\varepsilon_z}, \sqrt{\varepsilon_r}) - r^*(z_i^*, \sqrt{\varepsilon_z}, \sqrt{\varepsilon_r}) + r^*(z_i^*, \sqrt{\varepsilon_z}, \sqrt{\varepsilon_r}) - r^*(z_i^*, \sqrt{\varepsilon_{z_1}})$$
$$= \Delta r_2^* + \Delta r_3^* + \delta r_2 + \delta r_3$$

(23)

where

$$\Delta r_2^* = \frac{2M}{E_c} \sqrt{\varepsilon_r}(\sqrt{\varepsilon_z} - \sqrt{\varepsilon_{z_1}}) \tag{24}$$

$$\Delta r_3^* = \frac{2M}{-E_c \sqrt{\phi_{ac}}} \sqrt{\varepsilon_r}(\varepsilon_z - \varepsilon_{z_1}) \tag{25}$$

$$\delta r_2 = 0, \quad \delta r_3 = \frac{2M}{-E_c \sqrt{\phi_{ac}}}(n-1)\varepsilon_r^{\frac{3}{2}} \tag{26}$$

The definition of aberrations is as following (see Fig. 2).

Δr^*—paraxial lateral aberration, that is the displacement of paraxial ray with ($\sqrt{\varepsilon_z}$, $\sqrt{\varepsilon_r}$) at the ideal image plane which is corresponding to paraxial ray with $\sqrt{\varepsilon_{z_1}}$.

Δz^*—paraxial longitudinal aberration in axial direction.

δr— geometrical lateral aberration, that is the displacement of paraxial ray and practical ray with same ($\sqrt{\varepsilon_z}$, $\sqrt{\varepsilon_r}$) at ideal image plane, δz is the geometrical longitudinal aberration in axial direction;

Δz_j^* ($j = 1, 2$), Δr_j^* ($j = 2, 3$)—the j-th order paraxial longitudinal aberration and paraxial lateral aberration, respectively;

δz_j ($j = 1, 2$), δr_j ($j = 2, 3$)—the j-th order geometrical longitudinal aberration and geometrical lateral aberration, respectively.

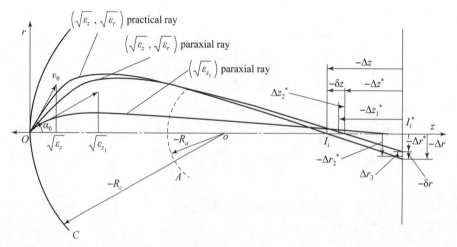

Fig. 2 Spatial aberrations are composed of paraxial aberration and geometrical aberration

2 Dynamic imaging electron optics and temporal aberrations for a bi-electrode concentric spherical system with electrostatic focusing

2.1 Motion equation and its time solution

Now, we discuss the dynamic imaging electron optics and temporal aberrations for a bi-electrode concentric spherical system with electrostatic focusing.

Starting from the initial condition of electron emitted from photocathode: $R_c \dot{\varphi}_0 = v_0 \sin \alpha_0$, and from Eq. (2) – Eq. (4), we may have

$$\rho^2 \dot{\varphi} = R_c v_0 \sin \alpha_0 \tag{27}$$

The electron velocity v in the polar coordinate (ρ, φ) can be written as

$$v^2 = \dot{\rho}^2 + (\rho \dot{\varphi})^2 \tag{28}$$

From Eq. (27) and Eq. (28), we may obtain

$$\dot{\rho} = \frac{d\rho}{dt} = \sqrt{v^2 - (R_c^2 v_0^2 \sin^2 \alpha_0 / \rho^2)} \tag{29}$$

Substituting

$$v^2 = v_0^2 + \frac{2e}{m_0} \frac{\phi_{ac}}{(n-1)} \left(\frac{R_c}{\rho} - 1 \right) \tag{30}$$

In Eq. (29), we have

$$\dot{\rho} = \sqrt{\frac{2e\varepsilon_0}{m_0}} [1 + p(\mu - 1) - \mu^2 \sin^2 \alpha_0]^{\frac{1}{2}} \tag{31}$$

From

$$\dot{\rho} = \frac{d\rho}{dt} = -\frac{R_c}{\mu^2} \frac{d\mu}{dt} \tag{32}$$

The expression (31) will take this form:

$$dt = \frac{-R_c d\mu}{\sqrt{\frac{2e\varepsilon_0}{m_0}} \mu^2 \sqrt{1 + p(\mu - 1) - \mu^2 \sin^2 \alpha_0}} \tag{33}$$

Integrating Eq. (33) and considering $\rho = R_c$, $\mu = 1$, when $t = 0$, we may obtain the flying time t at the space between cathode and anode spheres

$$t = \frac{-R_c}{\sqrt{\frac{2e\varepsilon_0}{m_0}}} \left\{ -\frac{\sqrt{1 + p(\mu - 1) - \mu^2 \sin^2 \alpha_0}}{(1-p)\mu} + \frac{\cos \alpha_0}{1-p} - \frac{p}{2(1-p)\sqrt{p-1}} \left[\arcsin \frac{p(\mu - 2) + 2}{\mu \sqrt{p^2 + 4(1-p)\sin^2 \alpha_0}} + \arcsin \frac{p - 2}{\sqrt{p^2 + 4(1-p)\sin^2 \alpha_0}} \right] \right\} \tag{34}$$

2.2 Determination of flying time for the practical ray and paraxial ray emitted from photocathode to anode

From Eq. (34), let $\rho = R_a$, i.e., $\mu = n$, $\varepsilon_0 = \varepsilon_r + \varepsilon_z$, we can get an exact expression of

t_{C-A} which is the time arriving at the anode for the practical ray emitted from photocathode with ($\sqrt{\varepsilon_r}$, $\sqrt{\varepsilon_z}$):

$$t_{C-A} = \sqrt{\frac{m_0}{2e}} \frac{-R_c}{\left[\frac{\varphi_{ac}}{(n-1)} - \varepsilon_z - \varepsilon_r\right]} \left[\frac{1}{n}\sqrt{\phi_{ac} + \varepsilon_z + (1-n^2)\varepsilon_r} - \sqrt{\varepsilon_z}\right] - \sqrt{\frac{m_0}{2e}} \frac{\frac{\phi_{ac}}{(n-1)}R_c}{2\left[\frac{\phi_{ac}}{(n-1)} - \varepsilon_z - \varepsilon_r\right]^{3/2}} \times$$

$$\left\{\arcsin \frac{\left(1 - \frac{2}{n}\right) + \frac{2}{n}\frac{(n-1)}{\phi_{ac}}(\varepsilon_z + \varepsilon_r)}{\sqrt{1 - 4\left[\frac{\phi_{ac}}{(n-1)} - \varepsilon_z - \varepsilon_r\right]\varepsilon_r\left(\frac{n-1}{\phi_{ac}}\right)^2}} + \arcsin \frac{1 - 2\frac{n-1}{\phi_{ac}}(\varepsilon_z + \varepsilon_r)}{\sqrt{1 - 4\left(\frac{\phi_{ac}}{(n-1)} - \varepsilon_z - \varepsilon_r\right)\varepsilon_r\left(\frac{n-1}{\phi_{ac}}\right)^2}}\right\}$$

(35)

Expanding (35) and neglecting all of these terms higher than ε_z/ϕ_{ac} and ε_r/ϕ_{ac} orders which are greatly less than 1, we may obtain an approximate expression of t_{C-A}:

$$t_{C-A} = -\sqrt{\frac{m_0}{2e}}\frac{R_c(n-1)}{n\sqrt{\phi_{ac}}}\left[1 + \frac{(2n-1)}{2\phi_{ac}}\varepsilon_z + \frac{(n-1)^2}{2\phi_{ac}}\varepsilon_r\right] +$$

$$\sqrt{\frac{m_0}{2e}}\frac{R_c(n-1)}{\sqrt{\phi_{ac}}}\sqrt{\varepsilon_z} - \sqrt{\frac{m_0}{2e}}\frac{R_c}{2}\frac{\sqrt{n-1}}{\sqrt{\phi_{ac}}}\left(1 + \frac{3}{2}\frac{n-1}{\phi_{ac}}\varepsilon_z + \frac{3}{2}\frac{n-1}{\phi_{ac}}\varepsilon_r\right) \times \quad (36)$$

$$\left\{\arcsin\left[\left(1 - \frac{2}{n}\right) + \frac{2(n-1)}{n\phi_{ac}}\varepsilon_z + \frac{2(n-1)^2}{n\phi_{ac}}\varepsilon_r\right] + \arcsin\left(1 - 2\frac{n-1}{\phi_{ac}}\varepsilon_z\right)\right\}$$

The last term of above-mentioned formula can be written as

$$\left\{\arcsin\left[\left(1 - \frac{2}{n}\right) + \frac{2(n-1)\varepsilon_z}{n\phi_{ac}} + \frac{2(n-1)^2}{n\phi_{ac}}\varepsilon_r\right] + \arcsin\left(1 - 2\frac{n-1}{\phi_{ac}}\varepsilon_z\right)\right\} =$$

$$\left\{2\arcsin\sqrt{\frac{n-1}{n}} - 2\sqrt{n-1}\sqrt{\frac{\varepsilon_z}{\phi_{ac}}} + \sqrt{n-1}\frac{\varepsilon_z}{\phi_{ac}} + 2(n-1)\sqrt{n-1}\frac{\varepsilon_r}{\phi_{ac}}\right\}$$

(37)

Therefore, the approximate expression of t_{C-A} can be expressed by

$$t_{C-A} = \sqrt{\frac{2m_0}{e}}\left\{\frac{-R_c(n-1)}{2n\sqrt{\phi_{ac}}} - \frac{R_c}{2}\sqrt{\frac{n-1}{n}}\arcsin\sqrt{\frac{n-1}{n}} + \frac{R_c(n-1)}{\sqrt{\phi_{ac}}}\sqrt{\frac{\varepsilon_z}{\phi_{ac}}} - \frac{R_c(n-1)}{\sqrt{\phi_{ac}}} \times\right.$$

$$\left.\left[\left(\frac{3n-1}{4n} + \frac{3}{4}\sqrt{n-1}\arcsin\sqrt{\frac{n-1}{n}}\right)\frac{\varepsilon_z}{\phi_{ac}} + \left[\frac{(n-1)(3n-1)}{4n} + \frac{3}{4}\sqrt{n-1}\arcsin\sqrt{\frac{n-1}{n}}\right]\frac{\varepsilon_r}{\phi_{ac}}\right]\right\}$$

(38)

For the paraxial ray, when neglect the terms of ε_r/ϕ_{ac}, and let $\sqrt{\varepsilon_z} = \sqrt{\varepsilon_{z_1}}$, we have

$$t^*_{C-A} = \left(\frac{2m_0}{e}\right)^{\frac{1}{2}}\left[\frac{-R_c(n-1)}{2\sqrt{\phi_{ac}}}\frac{1}{n} + \frac{-R_c}{2}\sqrt{\frac{n-1}{n}}\arcsin\sqrt{\frac{n-1}{n}} + \frac{R_c(n-1)}{\sqrt{\phi_{ac}}}\sqrt{\frac{\varepsilon_{z_1}}{\phi_{ac}}} + \right.$$

$$\left.\frac{-R_c(n-1)}{\sqrt{\phi_{ac}}}\left(\frac{3n-1}{4n} + \frac{3}{4}\sqrt{n-1}\arcsin\sqrt{\frac{n-1}{n}}\right)\frac{\varepsilon_{z_1}}{\phi_{ac}}\right]$$

(39)

2.3 Calculation of flying time in the region free from field behind the mesh-anode

When the electron arrives at the anode from the cathode, its velocity v_a can be expressed by

$$v_a = \sqrt{v_0^2 + \frac{2e}{m_0}\phi_{ac}} \tag{40}$$

In the region free from field, the length of electron ray leaving from the mesh-anode and arriving at the axis is l, thus the time will be t_{A-i}, we have

$$t_{A-i} = \frac{l}{v_a} \tag{41}$$

From the triangle $\triangle ODI_i$ in Fig. 1, Eq. (41) takes the form:

$$t_{A-i} = R_i \frac{\sin(-\gamma)}{v_a \sin\tau} \tag{42}$$

where R_i can be expressed by Eq. (12); $\sin(-\gamma)$ from Eq. (7) when $\varphi = \gamma$, and from Eq. (8), $\sin\tau$ will have the form:

$$\sin\tau = \frac{n\sin\alpha_0}{\sqrt{b_1^2 + n^2\sin^2\alpha_0}} \tag{43a}$$

Expanding R_i, $\sin(-\gamma)$ and $\sin\tau$, and neglecting all of these terms higher than ε_z/ϕ_{ac} and ε_r/ϕ_{ac} orders which are greatly less than 1, we may obtain approximate expressions for $\sin\tau$, $\sin(-\gamma)$ and v_a in Eq. (42), then we may obtain the approximate expression of time t_{A-i} for the electron ray leaving from the mesh-anode to the axis in the region free from field:

$$t_{A-i} = -\sqrt{\frac{2m_0}{e}} \frac{R_a(n-1)}{(n-2)\sqrt{\phi_{ac}}} \left\{1 + \frac{n}{n-2}\sqrt{\frac{\varepsilon_z}{\phi_{ac}}} + \frac{n^2+4n-4}{2(n-2)^2}\frac{\varepsilon_z}{\phi_{ac}} - \frac{(n-1)(n^2-n+2)}{2(n-2)}\frac{\varepsilon_r}{\phi_{ac}}\right\} \tag{43b}$$

For the paraxial imaging with $\sqrt{\varepsilon_{z_1}}$, neglecting the terms of ε_r/ϕ_{ac} in Eq. (43), the approximate expression of time $t_{A-i^*}^*$ will be

$$t_{A-i^*}^* = -\sqrt{\frac{2m_0}{e}} \frac{R_a(n-1)}{(n-2)\sqrt{\phi_{ac}}} \left\{1 + \frac{n}{n-2}\sqrt{\frac{\varepsilon_{z_1}}{\phi_{ac}}} + \frac{n^2+4n-4}{2(n-2)^2}\frac{\varepsilon_{z_1}}{\phi_{ac}}\right\} \tag{44}$$

2.4 Temporal aberrations

The temporal aberration ΔT is measured in the ideal image plane determined by paraxial ray with $\sqrt{\varepsilon_{z_1}}$, which is defined by the time difference between the practical ray travelling with $(\sqrt{\varepsilon_z}, \sqrt{\varepsilon_r})$ and paraxial ray travelling with $\sqrt{\varepsilon_{z_1}}$. It consists two parts, the one is the time-transit-spread of these two electron rays from cathode to anode, the other one is the time-transit-spread of these two electron rays arriving at the ideal image plane determined by paraxial ray with $\sqrt{\varepsilon_{z_1}}$, as shown in Fig. 3, that is

$$\begin{aligned}\Delta T &= t_{c-i^*}(\sqrt{\varepsilon_z}, \sqrt{\varepsilon_r}) - t_{C-i^*}^*(\sqrt{\varepsilon_{z_1}}) \\ &= t_{C-A}(\sqrt{\varepsilon_z}, \sqrt{\varepsilon_r}) + t_{A-i^*}(\sqrt{\varepsilon_z}, \sqrt{\varepsilon_r}) - t_{C-A}^*(\sqrt{\varepsilon_{z_1}}) - t_{A-i^*}^*(\sqrt{\varepsilon_{z_1}})\end{aligned} \tag{45}$$

where $t_{C-A}(\sqrt{\varepsilon_z}, \sqrt{\varepsilon_r})$, $t_{C-A}^*(\sqrt{\varepsilon_{z_1}})$ are expressed by Eq. (38) and Eq. (39) respectively, and $t_{A-i^*}^*(\sqrt{\varepsilon_{z_1}})$ is expressed by Eq. (44).

The term of $t_{A-i^*}(\sqrt{\varepsilon_z}, \sqrt{\varepsilon_r})$ can be solved by

$$t_{A-i^*}(\sqrt{\varepsilon_z}, \sqrt{\varepsilon_r}) = t_{A-i}(\sqrt{\varepsilon_z}, \sqrt{\varepsilon_r}) + \Delta t_{i-i^*}(\sqrt{\varepsilon_z}, \sqrt{\varepsilon_r}) \tag{46}$$

where $t_{A-i}(\sqrt{\varepsilon_z}, \sqrt{\varepsilon_r})$ is expressed by Eq. (43). Suppose that the practical ray with ($\sqrt{\varepsilon_z}, \sqrt{\varepsilon_r}$) arrives at the central axis and then to the ideal image plane determined by paraxial ray with $\sqrt{\varepsilon_{z_1}}$, the spatial difference between them is Δl, temporal difference is $\Delta t_{i-i^*}(\sqrt{\varepsilon_z}, \sqrt{\varepsilon_r})$, we have

$$\Delta t_{i-i^*}(\sqrt{\varepsilon_z}, \sqrt{\varepsilon_r}) = \frac{\Delta l}{v_a} = \frac{-\Delta z}{\cos\alpha_i \cdot v_a} = -\Delta z \sqrt{1 + \tan^2\alpha_i} \frac{1}{v_a} \tag{47}$$

where $\tan\alpha_i$, v_a are expressed by Eq. (13) and Eq. (40) respectively. $\Delta z = R_i - R_i^*$, R_i, R_i^* are expressed by Eq. (12) and Eq. (14) respectively. When $\sqrt{\varepsilon_z} > \sqrt{\varepsilon_{z_1}}$, $\Delta z = R_i - R_i^*$ is positive, and when $\sqrt{\varepsilon_z} < \sqrt{\varepsilon_{z_1}}$, $\Delta z = R_i - R_i^*$ is negative. The beginning point is from the ideal image position determined by paraxial ray with ε_{z_1}, the value from left to right is positive.

Expanding all of expressions in Eq. (47), and neglecting all of these terms higher than ε_z/ϕ_{ac} and ε_r/ϕ_{ac} orders which are greatly less than 1, we may obtain

$$\Delta t_{i-i^*}(\sqrt{\varepsilon_z}, \sqrt{\varepsilon_r}) = \sqrt{\frac{2m_0}{e}} \frac{n(n-1)}{n-2} \frac{R_a}{\sqrt{\phi_{ac}}} \left\{ \frac{(\sqrt{\varepsilon_z} - \sqrt{\varepsilon_{z_1}})}{\sqrt{\phi_{ac}}} + \frac{n}{(n-2)} \frac{(\varepsilon_z - \varepsilon_{z_1})}{\phi_{ac}} - (n-1)\frac{\varepsilon_r}{\phi_{ac}} \right\} \tag{48}$$

So the expression (46) can be calculated by the following formula:

$$t_{A-i^*}(\sqrt{\varepsilon_z}, \sqrt{\varepsilon_r}) = -\sqrt{\frac{2m_0}{e}} \frac{(n-1)R_a}{(n-2)\sqrt{\phi_{ac}}} \left\{ 1 + \frac{n}{n-2}\sqrt{\frac{\varepsilon_z}{\phi_{ac}}} + \frac{n^2+4n-4}{2(n-2)^2}\frac{\varepsilon_z}{\phi_{ac}} - \frac{(n-1)(n^2-n+2)}{2(n-2)}\frac{\varepsilon_r}{\phi_{ac}} \right\} + \sqrt{\frac{2m_0}{e}} \frac{n(n-1)}{(n-2)^2} \frac{R_a}{\sqrt{\phi_{ac}}} \left\{ \frac{(\sqrt{\varepsilon_z} - \sqrt{\varepsilon_{z_1}})}{\sqrt{\phi_{ac}}} + \frac{n}{(n-2)} \frac{(\varepsilon_z - \varepsilon_{z_1})}{\phi_{ac}} - (n-1)\frac{\varepsilon_r}{\phi_{ac}} \right\} \tag{49}$$

Finally, from Eq. (45), we can also classify the temporal aberration by order-sequence, as we have done in the treatment with spatial aberration. As shown in Fig. 3, the temporal aberration ΔT can be composed of paraxial temporal aberration ΔT^* and geometrical temporal aberration δT, it can be expressed as following:

$$\Delta T = \Delta T^* + \delta T = \Delta T_1^* + \Delta T_2^* + \delta T_1 + \delta T_2 \tag{50}$$

where

$$\Delta T_1^* = \left(\frac{2m_0}{e}\right)^{\frac{1}{2}} \frac{\sqrt{\varepsilon_z} - \sqrt{\varepsilon_{z_1}}}{E_c} \tag{51}$$

$$\Delta T_2^* = \left(\frac{2m_0}{e}\right)^{\frac{1}{2}} \frac{\varepsilon_z - \varepsilon_{z_1}}{E_c \sqrt{\varphi_{ac}}} \times \left\{ -\frac{3n^2 - 9n + 2}{4n(n-2)} - \frac{3}{8}\sqrt{n-1}\frac{1}{n}\left(\pi - \arcsin\frac{2}{n}\sqrt{n-1}\right) \right\} \tag{52}$$

$$\delta T_1 = 0, \quad \delta T_2 = \left(\frac{2m_0}{e}\right)^{\frac{1}{2}} \frac{\varepsilon_r}{E_c \sqrt{\varphi_{ac}}} \times \left\{ \frac{(n-1)[-(n-2)^2(3n-1) + 2(n^2-n+2)]}{4n(n-2)^2} - \frac{3}{8}\sqrt{n-1}\left(\pi - \arcsin\frac{2}{n}\sqrt{n-1}\right) \right\}$$

(53)

where ΔT^* is the paraxial temporal aberration, that is the time-transit-spread of paraxial ray with ($\sqrt{\varepsilon_z}$, $\sqrt{\varepsilon_r}$), diverging from paraxial ray with $\sqrt{\varepsilon_{z_1}}$ at the ideal image plane; δT is the geometrical temporal aberration, that is the time difference of paraxial ray and practical ray with same ($\sqrt{\varepsilon_z}$, $\sqrt{\varepsilon_r}$) at the ideal image plane; ΔT_j^* ($j = 1, 2$), δT_j ($j = 1, 2$) are the j-th order paraxial temporal aberration and geometrical temporal aberration, respectively.

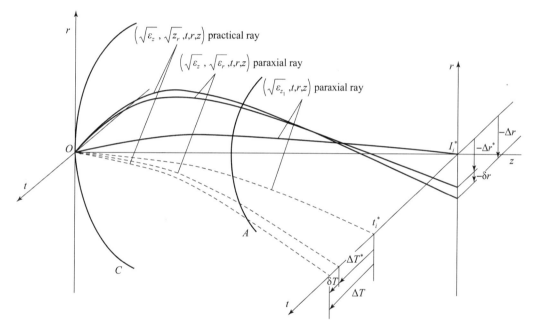

Fig. 3 Temporal aberrations can be composed of paraxial temporal aberration and geometrical temporal aberration

Conclusions

From the above investigation, we can reach the following conclusions.

(1) In electron optics of a bi-electrode concentric spherical system with electrostatic focusing, whether the expression of spatial aberration or the expression of temporal aberration, both can be expressed in the combination of paraxial aberration and geometrical aberration, that is, the spatial aberration can be expressed in combination of paraxial spatial aberration and geometrical spatial aberration, the temporal aberration can be expressed in combination of paraxial temporal aberration and geometrical temporal aberration.

(2) In electron optical imaging systems working in static mode operation, the lateral

aberration of axial point caused by spatial-trajectory-spread produced by the emission energy and emission angle of photoelectrons is the fundamental limitation of spatial resolution, which is mainly determined by the second order paraxial lateral aberration Δr_2^*, as is shown in Eq. (24), that is related to the electric-field strength at the photocathode, the lateral magnification of the system and the initial energy of electron, as well as the ideal image position determined by $\sqrt{\varepsilon_{z_1}}$. Similarly, in electron optical imaging systems working in streak mode operation, the temporal aberration caused by time-transit-spread produced by the emission energy and emission angle of photoelectrons is also the fundamental limitation of temporal resolution of high-speed photographic image converter tube, which is mainly determined by the first order paraxial temporal aberration ΔT_1^*, as shown in Eq. (51), that is also related to the electric-field strength at the photocathode, and the axial initial energy of electron, as well as the ideal image position which is determined by $\sqrt{\varepsilon_{z_1}}$, but it is really not related to the lateral magnification of the system.

Start from Eq. (51) and Eq. (24) we can study the time modulation transfer function of the image tube in streak mode operation and the modulation transfer function of the image tube in static mode operation.

(3) It is natural that our study relates the static problem of imaging electron optics to the dynamic problem. Actually, the flying electrons emitted from cathode carries not only the space information, but also the time information, both of which reflects spatial and temporal characteristics, displayed on the image plane as spatial aberration and temporal aberration. So it is not strange that their formulae of aberrations have the similar form. It should be noted that either the second order paraxial lateral aberration or the first order paraxial temporal aberration are only related to the paraxial rays, so the emphasis of further study should be directed to paraxial electron optics and their paraxial aberrations.

References

[1] ARTIMOVICH L A. Electrostatic properties of emission systems [J]. Bulletin of Academy of Sciences, USSR, Physics Series, 1944, 8(6):313-328(in Russian).

[2] RECKNAGEL A. Theorie des elektrisohen elecktronen miktroskops fur selbstrakler[J]. Z. Angew. Physik, 1941, 117:689-708 (in German).

[3] SAVOISKY Y K, FANCHENKO S D. Physical basis of electron optical chronograph[J]. Report of Academy of Sciences,1956,108, 2: 218-221 (in Russian).

[4] CSORBA I P. Transit-time-spread-limited time resolution of image tubes in streak operation[J]. RCA Review,1971,32:650-659.

[5] RUSKA E. Zur fokussierbarkeit von kathoden-strahlbundeln grosser ausgangsquerchnitte [J]. Z. Angew. Physik,1933, 83(9):684-687 (in German).

[6] SCHAGEN P, BRUINING H, FRANCKEN J C. A simple electrostatic electron-optical

system with only one voltage[J]. Philips Research Report, 1952, 7(2):119-130.

[7] ZHOU L W. Electron optics with wide beam focusing[M]. Beijing: Beijing Institute of Technology Press, 1993 (in Chinese).

[8] ZHOU L W. Focusing and imaging of wide electron beams[J]//Selected papers on electron optics by Zhou Liwei[M]. Beijing: Beijing Institute of Technology Press, 1994:11-29 (in Chinese).

[9] ZHOU L W, LI Y, MONASTYRSKI M A, et al. Test and verification of temporal aberration theory for electron optical imaging systems by an electrostatic concentric spherical system[J]. Acta Physica Sinica 2005, 54(8): 3597-3603 (in Chinese).

10.8 Paraxial Imaging Electron Optics and Its Spatial-temporal Aberrations for a Bi-electrode Concentric Spherical System with Electrostatic Focusing

两电极静电聚焦同心球系统的近轴成像电子光学及其空间－时间像差

Abstract: As is known, the paraxial solutions play an important role in studying electron optical imaging system and its aberrations, but the investigation of a bi-electrode concentric spherical system with electrostatic focusing directly from paraxial electron ray equation and paraxial electron motion equation has not been done before. In this paper, we shall use the paraxial electron ray equation and paraxial electron motion equations to study the spatial-temporal trajectories and their aberrations for a bi-electrode concentric spherical system with electrostatic focusing. The paraxial static and dynamic imaging electron optics, as well as the paraxial spatial-temporal aberrations in this system are then discussed, the regularity of paraxial imaging optical properties has been given. The paraxial spatial aberrations, as well as the paraxial temporal aberrations with different orders, have been defined and deduced, that are classified by the order of $(\varepsilon_z/\phi_{ac})^{\frac{1}{2}}$ and $(\varepsilon_r/\phi_{ac})^{\frac{1}{2}}$.

摘要：众所周知，在研究电子光学成像系统及其像差时近轴解起着重要的作用，但直接由近轴轨迹方程或是近轴运动方程出发研究两电极静电聚焦同心球系统的工作并没有进行。在本文中，我们将从近轴运动方程和近轴轨迹方程出发研究两电极静电聚焦同心球系统的电子运动的空－时轨迹及其像差，并谈论近轴静态与动态成像电子光学特性及其空－时间像差，推导并定义各级近轴空间像差和近轴时间像差，它们是以$(\varepsilon_z/\phi_{ac})^{\frac{1}{2}}$和$(\varepsilon_r/\phi_{ac})^{\frac{1}{2}}$的序列划分的。

Introduction

In the previous paper [1], for a bi-electrode concentric spherical electrostatic system, we had discussed the static electron optics and spatial aberrations by practical electron ray equation, the dynamic electron optics and temporal aberrations by practical electron motion equation. The result shows that not only the spatial aberration can be regard as composition of the paraxial spatial aberration and the geometrical spatial aberration, but also the temporal aberration can be regard as composition of the paraxial temporal aberration and the geometrical temporal aberration.

As is known, the second-order paraxial lateral aberration and the first-order paraxial temporal aberration, which are only related to the paraxial solutions, occupy an important place in the whole spatial and temporal aberrations. So it is interesting to study the paraxial spatial and temporal aberrations from paraxial electron ray equation and paraxial electron motion equation. This work will once again confirm that the paraxial aberrations actually are the difference between two paraxial solutions (spatial or temporal) with different initial velocities $\sqrt{\varepsilon_z}$ and $\sqrt{\varepsilon_{z_1}}$, in which the latter determines the ideal image position. The present paper shows another approach to prove that the paraxial aberrations can be found by solutions of paraxial (ray or motion) equation, which is much simpler than solutions of practical (ray or motion) equations.

1 Paraxial ray solution and its spatial aberrations for a bi-electrode concentric spherical system with electrostatic focusing

1.1 Paraxial ray solution and paraxial imaging

As is known, in electron optics of electrostatic cathode lenses, the paraxial ray equation can be written as

$$r''(z) + \frac{1}{2}\frac{\phi'(z)}{\phi(z)+\varepsilon_z}r'(z) + \frac{1}{4}\frac{\phi''(z)}{\phi(z)+\varepsilon_z}r(z) = 0 \quad (1)$$

where $r(z)$ is the electron ray; $\phi(z)$ is the axial potential distribution; " ' " represents the differentiation to z. $\sqrt{\varepsilon_z}$ corresponds to the axial initial velocity of electron emitted from the cathode.

In the bi-electrode concentric spherical system, $\phi(z)$ can be expressed by[1]

$$\phi(z) = \frac{-z}{(n-1)(z+R_c)}\phi_{ac} \quad (2)$$

where $n = R_c/R_a$; R_c, R_a are curvature radii of spherical cathode C and of the mesh-spherical anode A respectively; ϕ_{ac} is the potential of the anode A with respect to cathode C; $\phi_c = 0$.

When the electron ray emits from the axial point at photocathode with emission energy and emission angle ($\sqrt{\varepsilon_0}$, α_0), the initial condition of trajectory will take the form:

$$r_0 = r(z=0) = 0, \quad r'_0 = \tan\alpha_0 = \frac{\sqrt{\varepsilon_r}}{\sqrt{\varepsilon_z}} \quad (3)$$

Substituting Eq. (2) in Eq. (1), it is not difficult to derive an analytical solution of Eq. (1) under the condition (3) as following [2,3]:

$$r^*(z) = 2z\sqrt{\frac{\varepsilon_r}{\phi(z)}}\left[\sqrt{1+\frac{\varepsilon_z}{\phi(z)}} - \sqrt{\frac{\varepsilon_z}{\phi(z)}}\right] \quad (4)$$

where the star mark $*$ indicates the paraxial situation. The axial velocity $\dot{z}^*$ of paraxial electron can be found from the law of conversation of energy [3]:

$$\dot{z}^* = \sqrt{\frac{2e}{m_0}[\phi(z) + \varepsilon_z]} \tag{5}$$

From Eq. (4), we may get expressions for radial velocity and slope of the paraxial electron trajectory:

$$\dot{r}^*(z) = \frac{1}{R_c + z}\sqrt{\frac{2e}{m_0}}\varepsilon_r\left[R_c + 2z\sqrt{1 + \frac{\varepsilon_z}{\phi(z)}}\left(\sqrt{1 + \frac{\varepsilon_z}{\phi(z)}} - \frac{\varepsilon_z}{\phi(z)}\right)\right] \tag{6}$$

$$r'^*(z) = \frac{1}{R_c + z}\sqrt{\frac{\varepsilon_r}{m_0}}\left[2z\left(\sqrt{1 + \frac{\varepsilon_z}{\phi(z)}} - \sqrt{\frac{\varepsilon_z}{\phi(z)}}\right) + \frac{R_c}{\sqrt{1 + \frac{\varepsilon_z}{\phi(z)}}}\right] \tag{7}$$

Now we determine the ideal image position R_i^* of paraxial trajectory at axis. Let $\sqrt{\varepsilon_z} = \sqrt{\varepsilon_{z_1}}$, $z = R_a - R_c$, $r(z = R_a - R_c) = r_a$, $\phi(z = R_a - R_c) = \phi_{ac}$, from Eq. (4) and Eq. (7) we may obtain expressions for the radial height of electron ray and its slope at the anode, their approximations are

$$r_a^*(z) = 2(R_a - R_c)\sqrt{\frac{\varepsilon_r}{\phi_{ac}}}\left(1 - \sqrt{\frac{\varepsilon_{z_1}}{\phi_{ac}}} + \frac{1}{2}\frac{\varepsilon_{z_1}}{\phi_{ac}}\right) \tag{8}$$

$$r'^*_a(z) = r'^*_i(z) = -\sqrt{\frac{\varepsilon_r}{\phi_{ac}}}(n-2)\left[1 - \frac{2(n-1)}{n-2}\sqrt{\frac{\varepsilon_{z_1}}{\phi(z)}} + \frac{3n-2}{2(n-2)}\frac{\varepsilon_{z_1}}{\phi_{ac}}\right] \tag{9}$$

The ideal image position of electron ray from the curvature center R_i^* is

$$R_i^* = R_a - \frac{r_a^*}{r'^*_a} \tag{10}$$

Substituting Eq. (8) and Eq. (9) in Eq. (10), neglecting the higher order than $\varepsilon_{z_1}/\phi_{ac}$ terms which is much smaller than 1, we obtain

$$R_i^* = \frac{-R_c}{n-2}\left[1 + \frac{2(n-1)}{n-2}\sqrt{\frac{\varepsilon_{z_1}}{\phi_{ac}}} + \frac{2n(n-1)}{(n-2)^2}\frac{\varepsilon_{z_1}}{\phi_{ac}}\right] \tag{11}$$

It shows that from this approach, the expression (11) derived coincides with the expression in Ref [1].

From Eq. (11), we can discuss the electron optical imaging characteristics, such as paraxial imaging, lateral magnification and Lagrange-Helmholtz invariant. The lateral magnification M of the system can be expressed by

$$M = \frac{R_i^*}{R_c} = -\frac{1}{n-2}\left[1 + \frac{2(n-1)}{n-2}\sqrt{\frac{\varepsilon_{z_1}}{\phi_{ac}}} + \frac{2n(n-1)}{(n-2)^2}\frac{\varepsilon_{z_1}}{\phi_{ac}}\right] \tag{12}$$

Obviously, it forms a real image when $n > 2$ or a virtual image when $n < 2$.

It shows that the invariant for object-image space in Gaussian geometrical optics follows the Lagrange-Helmholtz invariant, as well as in paraxial imaging electron optics. It represents an universal characteristics of a practical electron optical imaging system at paraxial condition. For a paraxial image, if we define angular magnification $\Gamma = r'^*_i/r'_0$, from formula (11) and formula (9) we may obtain the product of angular magnification Γ and lateral magnification M as following:

$$M\Gamma = \sqrt{\frac{\varepsilon_{z_1}}{\phi_{ac}}}\left\{1 + \frac{\varepsilon_{z_1}}{2\phi_{ac}} + \frac{3}{8}\left(\frac{\varepsilon_{z_1}}{\phi_{ac}}\right)^2\right\} = \sqrt{\frac{\varepsilon_{z_1}}{\phi_{ac} + \varepsilon_{z_1}}} \tag{13}$$

Formula (13) is the Lagrange-Helmholtz invariant in imaging electron optics. Thus, the longitudinal magnification M_L can be expressed as following:

$$M_L = \frac{\Delta z_i}{\Delta z_0} = \frac{\frac{\Delta r_i}{\tan \alpha_i}}{\frac{\Delta r_0}{\tan \alpha_0}} = \frac{\frac{\Delta r_i}{\Delta r_0}}{\frac{\tan \alpha_i}{\tan \alpha_0}} = \frac{M}{\Gamma} = M^2 \sqrt{\frac{\phi_{ac} + \varepsilon_{z_1}}{\varepsilon_{z_1}}} \tag{14}$$

It shows that we firstly strictly deduced these two important formulae (13) and Eq. (14) in imaging electron optics.

There are two parameters quantitatively to express the convergence of electron beam in a bi-electrode spherical concentric system [3]:

1) Radius of crossover r_{cr} formed by electron beams emitted from whole cathode surface

Let θ be the angle between the effective size of half cathode surface and the central axis, we may obtain the expression for radius of crossover r_{cr}:

$$r_{cr} = -R_i^* \frac{r_i'^*}{\cos\theta - \sin\theta r_i'^*} \tag{15}$$

its first order approximation will be

$$r_{cr} \approx -\frac{2r_0}{\sin 2\theta}\sqrt{\frac{\varepsilon_r}{\phi_{ac}}} \tag{16}$$

where r_0 is the height of object.

2) Radius of diffusing circle r_a^* at mesh-anode formed by electron rays from the object point with emission parameters ($\sqrt{\varepsilon_0}$, α_0)

From Eq. (8), we may have

$$r_a^* = 2(R_a - R_c)\sqrt{\frac{\varepsilon_r}{\phi_{ac}}}\left[\left(1 + \frac{\varepsilon_z}{\phi_{ac}}\right)^{\frac{1}{2}} - \left(\frac{\varepsilon_z}{\phi_{ac}}\right)^{\frac{1}{2}}\right] \tag{17}$$

It means that in paraxial condition the diffusing circle at mesh-anode is only related to the distance between cathode and anode $l = R_a - R_c$ and the initial energy in radial direction $\sqrt{\varepsilon_r/\phi_{ac}}$, but have nothing to do with the form of electrodes.

1.2 The paraxial spatial aberration

Now, we derive the paraxial spatial aberrations. Since we investigate the paraxial ray solutions, only the paraxial spatial aberrations exist, see Fig. 1. From Eq. (11), according to the definition and treatment in the previous paper [1], we may obtain the expression for paraxial longitudinal aberration Δz^* as following:

$$\Delta z^* = R_i^*(\sqrt{\varepsilon_z}) - R_i^*(\sqrt{\varepsilon_{z_1}}) = \Delta z_1^* - \Delta z_2^* \tag{18}$$

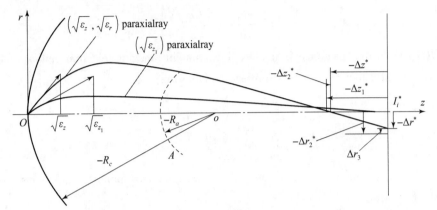

Fig. 1 The paraxial lateral and longitudinal aberrations in a bi-electrode concentric spherical system with electrostatic focusing

where

$$\Delta z_1^* = \frac{2M^2}{-E_c} \sqrt{\phi_{ac}} (\sqrt{\varepsilon_z} - \sqrt{\varepsilon_{z_1}}) \tag{19}$$

$$\Delta z_2^* = \frac{2M^2}{-E_c} \left\{ \frac{n}{n-2}(\varepsilon_z - \varepsilon_{z_1}) - \frac{4(n-1)}{n-2}\sqrt{\varepsilon_{z_1}}(\sqrt{\varepsilon_z} - \sqrt{\varepsilon_{z_1}}) \right\} \tag{20}$$

where E_c is the electric-field strength at the photocathode.

The paraxial lateral aberration Δr^* can be expressed by

$$\Delta r^* = -\Delta z^* \tan \alpha_i^* = r^*(\Delta z_i^*, \sqrt{\varepsilon_z}, \sqrt{\varepsilon_r}) - r^*(\Delta z_i^*, \sqrt{\varepsilon_{z_1}}) = \Delta r_2^* - \Delta r_3^* \tag{21}$$

where

$$\Delta r_2^* = \frac{2M}{E_c}\sqrt{\varepsilon_r}(\sqrt{\varepsilon_z} - \sqrt{\varepsilon_{z_1}}) \tag{22}$$

$$\Delta r_3^* = \frac{2M}{-E_c \sqrt{\phi_{ac}}}\sqrt{\varepsilon_r}(\varepsilon_z - \varepsilon_{z_1}) \tag{23}$$

The definitions of Δz_1^*, Δz_2^* and Δr_2^*, Δr_3^* and its Eq. (19), Eq. (20) and Eq. (22), Eq. (23) have been derived in the previous paper [1]. Expression (22) is just the Recknagel-Artimovich formula, which is formed by the difference of paraxial solutions [4,5]. Thus, we have derived the paraxial spatial aberrations from the approach of paraxial solutions.

2 Paraxial temporal solution and its temporal aberrations for a bi-electrode concentric spherical system with electrostatic focusing

2.1 Paraxial temporal solution

In an ideal model of bi-electrode concentric spherical system the accurate expression of time $t(z)$ for electrons leaving from the cathode has been derived [1,6], now we deduce the paraxial expression of time $t^*(z)$. From the axial velocity of paraxial electron expression (5), integrating it, we have

$$t^*(z) = \int_0^z \frac{dz}{\left\{\frac{2e}{m}[\phi(z) + \varepsilon_z]\right\}^{\frac{1}{2}}} \tag{24}$$

and substituting Eq. (2) in Eq. (24), we can obtain the paraxial expression of time $t^*(z)$ for the electron ray leaving from the cathode, when an electron emits from photocathode under the following initial conditions: $z=0$, $t=0$, and $\dot{z}_0 = \sqrt{2e\varepsilon_z m_0}$

$$t^*(z) = \frac{-R_c}{\left[\frac{2e}{m_0}\left[\frac{\phi_{ac} - (n-1)\varepsilon_z}{(n-1)}\right]\right]^{1/2}}\left\{\sqrt{1 + \frac{z}{R_c}}\sqrt{\frac{(n-1)\varepsilon_z}{\phi_{ac} - (n-1)\varepsilon_z} - \frac{z}{R_c}} - \sqrt{\frac{(n-1)\varepsilon_z}{\phi_{ac} - (n-1)\varepsilon_z}} + \left(1 + \frac{(n-1)\varepsilon_z}{\phi_{ac} - (n-1)\varepsilon_z}\right)\left[\arcsin\sqrt{\frac{(n-1)\varepsilon_z}{\phi_{ac}} - \frac{[\phi_{ac} - (n-1)\varepsilon_z]z}{R_c \phi_{ac}}} - \arcsin\sqrt{\frac{(n-1)\varepsilon_z}{\phi_{ac}}}\right]\right\} \tag{25}$$

Formula (24) or formula (25) means that $t^*(z)$ does not depend on $\sqrt{\varepsilon_r}$.

Now we determine the time of the paraxial ray arriving in anode from the photocathode. Calculating t^*_{C-A}, which is the time of paraxial ray arriving at the anode from the cathode where $z = -R_c + R_a$. From formula (25), because of $(R_c - R_a)R_c = (n-1)n$, we have

$$t^*_{C-A} = \frac{-R_c}{\left\{\frac{2e}{m_0}\left[\frac{\phi_{ac} - (n-1)\varepsilon_z}{n-1}\right]\right\}^{1/2}}\left\{\sqrt{\frac{R_a}{R_c}}\sqrt{\frac{(n-1)(\phi_{ac} + \varepsilon_z)}{[\phi_{ac} - (n-1)\varepsilon_z]n}} - \sqrt{\frac{(n-1)\varepsilon_z}{\phi_{ac} - (n-1)\varepsilon_z}} + \left(1 + \frac{(n-1)\varepsilon_z}{\phi_{ac} - (n-1)\varepsilon_z}\right)\left[\arcsin\sqrt{\frac{(n-1)(\phi_{ac} + \varepsilon_z)}{n\phi_{ac}}} - \arcsin\sqrt{\frac{(n-1)\varepsilon_z}{\phi_{ac}}}\right]\right\} = t^*_{C-A}\big|_1 + t^*_{C-A}\big|_2 \tag{26}$$

Now we simplify expression (26). Firstly, expanding the expression of $t^*_{C-A}\big|_1$, neglecting the higher order than $\frac{\varepsilon_z}{\phi_{ac}}$ terms which is much smaller than 1, we have

$$t^*_{C-A}\big|_1 = \frac{-R_c}{\left[\frac{2e}{m_0}\left(\frac{\phi_{ac} - (n-1)\varepsilon_z}{n-1}\right)\right]^{1/2}}\left\{\sqrt{\frac{R_a}{R_c}}\sqrt{\frac{(n-1)(\phi_{ac} + \varepsilon_z)}{[\phi_{ac} - (n-1)\varepsilon_z]n}} - \sqrt{\frac{(n-1)\varepsilon_z}{\phi_{ac} - (n-1)\varepsilon_z}}\right\}$$
$$= \left(\frac{m_0}{2e}\right)^{1/2}\frac{R_a - R_c}{\sqrt{\phi_{ac}}}\left\{1 - n\sqrt{\frac{\varepsilon_z}{\phi_{ac}}} + \frac{1}{2}(2n-1)\frac{\varepsilon_z}{\phi_{ac}}\right\} \tag{27}$$

Secondly, $t^*_{C-A}\big|_2$ has the form:

$$t^*_{C-A}\big|_2 = \frac{-R_c}{\left\{\frac{2e}{m_0}\left[\frac{\phi_{ac} - (n-1)\varepsilon_z}{n-1}\right]\right\}^{1/2}}\left\{\left(1 + \frac{(n-1)\varepsilon_z}{\phi_{ac} - (n-1)\varepsilon_z}\right)\left[\arcsin\sqrt{\frac{(n-1)(\phi_{ac} + \varepsilon_z)}{n\phi_{ac}}} - \arcsin\sqrt{\frac{(n-1)\varepsilon_z}{\phi_{ac}}}\right]\right\} \tag{28}$$

Now, we simplify expression (28). Let $x = \sqrt{\frac{(n-1)(\phi_{ac} + \varepsilon_z)}{n\phi_{ac}}}$, $y = \sqrt{\frac{(n-1)\varepsilon_z}{\phi_{ac}}}$, because of

$x^2 + y^2 \leqslant 1$, by using following formula: $\arcsin x - \arcsin y = \arcsin(x\sqrt{1-y^2} - y\sqrt{1-x^2})$, thus, terms in the brackets of Eq. (28) containing anti-trigonometric function can be simplified as

$$\arcsin\sqrt{\frac{(n-1)(\phi_{ac}+\varepsilon_z)}{n\phi_{ac}}} - \arcsin\sqrt{\frac{(n-1)\varepsilon_z}{\phi_{ac}}} = \arcsin\sqrt{\frac{n-1}{n}} - \sqrt{n-1}\sqrt{\frac{\varepsilon_z}{\phi_{ac}}} + \frac{1}{2}\sqrt{n-1}\frac{\varepsilon_z}{\phi_{ac}} \tag{29}$$

Therefore,

$$t^*_{C-A}\Big|_2 = -R_c\sqrt{\frac{m_0}{2e}}\sqrt{\frac{n-1}{\phi_{ac}}}\left[1 + \frac{3}{2}\frac{(n-1)\varepsilon_z}{\phi_{ac}}\right]\left(\arcsin\sqrt{\frac{n-1}{n}} - \sqrt{n-1}\sqrt{\frac{\varepsilon_z}{\phi_{ac}}} + \frac{1}{2}\sqrt{n-1}\frac{\varepsilon_z}{\phi_{ac}}\right) \tag{30}$$

Finally, we obtain the approximate expression of time t^*_{C-A} as following:

$$t^*_{C-A} = \left(\frac{2m_0}{e}\right)^{\frac{1}{2}}\left[\frac{R_a - R_c}{2\sqrt{\phi_{ac}}} - \frac{1}{2}R_c\sqrt{\frac{n-1}{\phi_{ac}}}\arcsin\sqrt{\frac{n-1}{n}} - n\frac{R_a - R_c}{\sqrt{\phi_{ac}}}\sqrt{\frac{\varepsilon_z}{\phi_{ac}}} + n\frac{R_a - R_c}{\sqrt{\phi_{ac}}}\frac{\varepsilon_z}{\phi_{ac}}\left(\frac{1}{4}\frac{3n-1}{n} + \frac{3}{4}\sqrt{n-1}\arcsin\sqrt{\frac{n-1}{n}}\right)\right] \tag{31}$$

It shows that from this approach, the expression (31) derived coincides with the expression in the previous paper[1].

Now, we calculate the flying time for the paraxial ray in the region free from field after the mesh-anode. Let v^*_a be the electron velocity leave from the anode, the paraxial electron ray is a straight line, which reaches the central axis from anode, and its distance is l, time interval is $t^*_{A-i^*}$, so we have

$$t^*_{A-i} = \frac{l}{v^*_a}$$

$$l = [r_a^2 + (-R_a + R_i)^2]^{\frac{1}{2}}, \quad v_a = \sqrt{v_0^2 + \frac{2e}{m_0}\phi_{ac}} \tag{32}$$

Considering the paraxial condition, their approximations can be written as

$$v^*_a = \sqrt{\frac{2e}{m_0}\phi_{ac}}\left(1 + \frac{1}{2}\frac{\varepsilon_z}{\phi_{ac}}\right),$$

$$l \approx -R_a + R_i = \frac{-2R_c(n-1)}{n(n-2)}\left[1 + \frac{n}{n-2}\sqrt{\frac{\varepsilon_z}{\phi_{ac}}} + \frac{n^2}{(n-2)^2}\frac{\varepsilon_z}{\phi_{ac}}\right]$$

Substituting above formulae in (32), we obtain the approximate value of t^*_{A-i}:

$$t^*_{A-i} = -\sqrt{\frac{2m_0}{e}}\frac{R_a(n-1)}{(n-2)\sqrt{\phi_{ac}}}\left[1 + \frac{n}{n-2}\sqrt{\frac{\varepsilon_z}{\phi_{ac}}} + \frac{n^2 + 4n - 4}{2(n-2)^2}\frac{\varepsilon_z}{\phi_{ac}}\right] \tag{33}$$

2.2 Paraxial temporal aberration

The paraxial temporal aberration ΔT^* is measured in the ideal image plane determined by paraxial ray with $\sqrt{\varepsilon_{z_1}}$, and defined by the time difference between the two paraxial rays travelling with ($\sqrt{\varepsilon_z}$) and ($\sqrt{\varepsilon_{z_1}}$). As shown in Fig. 2, it consists two parts, the one is the time-transit-spread of the two paraxial electron rays from cathode to anode, the other one is the time-transit-spread of this two paraxial electron rays arriving at the ideal image plane determined by paraxial ray with $\sqrt{\varepsilon_{z_1}}$, that is

$$\Delta T^* = t^*|_{C-A}(\sqrt{\varepsilon_z}) + (t^* + \Delta t)^*|_{A-i^*}(\sqrt{\varepsilon_z}) - t^*|_{C-A}(\sqrt{\varepsilon_{z_1}}) - t^*|_{A-i^*}(\sqrt{\varepsilon_{z_1}}) \quad (34)$$

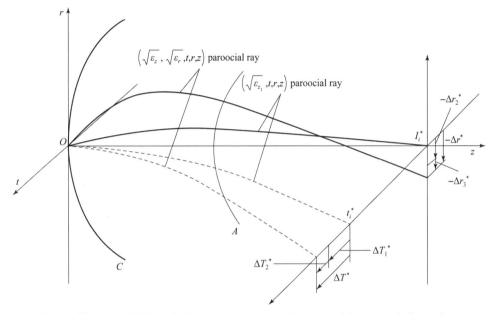

Fig. 2 The paraxial lateral aberrations are related to paraxial temporal aberrations

where $t^*|_{C-A}(\sqrt{\varepsilon_z})$, $t^*|_{C-A}(\sqrt{\varepsilon_{z_1}})$ are expressed by formula (31), $t^*|_{A-i^*}(\sqrt{\varepsilon_{z_1}})$ by formula (33), and

$$(t^* + \Delta t)^*|_{A-i^*}(\sqrt{\varepsilon_z}) = t^*|_{A-i}(\sqrt{\varepsilon_z}) + \Delta t^*|_{i-i^*}(\sqrt{\varepsilon_z}) \quad (35)$$

Now we solve Eq. (34). The paraxial ray travelling with ($\sqrt{\varepsilon_z}$) in the region free from field firstly arrives the central axis, then reaches the ideal image plane determined by paraxial ray with ($\sqrt{\varepsilon_{z_1}}$), the time difference is $\Delta t^*|_{i-i^*}(\sqrt{\varepsilon_z})$.

Similar to the derivation and treatment in the previous paper, we may obtain

$$\Delta t^*|_{i-i^*}(\sqrt{\varepsilon_z}) = R_a \frac{n}{n-2}\sqrt{\frac{m_0}{2e}} \frac{1}{\sqrt{\phi_{ac}}}\left[\frac{2(n-1)}{(n-2)\sqrt{\phi_{ac}}}(\sqrt{\varepsilon_z} - \sqrt{\varepsilon_{z_1}}) + \frac{2n(n-1)}{(n-2)^2\phi_{ac}}(\varepsilon_z - \varepsilon_{z_1})\right]$$

(36)

Substituting Eq. (33) and Eq. (36) in Eq. (35),

$$(t^* + \Delta t)^*|_{A-i^*}(\sqrt{\varepsilon_z}) = -\sqrt{\frac{2m_0}{e}}\frac{R_a(n-1)}{(n-2)\sqrt{\phi_{ac}}}\left[1 + \frac{n}{n-2}\sqrt{\frac{\varepsilon_z}{\phi_{ac}}} + \frac{n^2+4n-4}{2(n-2)^2}\frac{\varepsilon_z}{\phi_{ac}}\right] + \frac{n(n-1)}{(n-2)^2}\sqrt{\frac{2m_0}{e}}\frac{R_a}{\sqrt{\phi_{ac}}}\left[\frac{(\sqrt{\varepsilon_z} - \sqrt{\varepsilon_{z_1}})}{\sqrt{\phi_{ac}}} + \frac{n(\varepsilon_z - \varepsilon_{z_1})}{(n-2)\phi_{ac}}\right]$$

(37)

Finally, from Eq. (34), we may obtain the paraxial temporal aberration ΔT^* which is classified in order sequence as following:

$$\Delta T^* = \Delta T_1^* + \Delta T_2^* \tag{38}$$

$$\Delta T_1^* = \left(\frac{2m_0}{e}\right)^{\frac{1}{2}} \frac{\sqrt{\varepsilon_z} - \sqrt{\varepsilon_{z_1}}}{E_c} \tag{39}$$

$$\Delta T_2^* = \left(\frac{2m_0}{e}\right)^{\frac{1}{2}} \frac{\varepsilon_z - \varepsilon_{z_1}}{E_c \sqrt{\phi_{ac}}} \times \left[-\frac{3n^2 - 9n + 2}{4n(n-2)} - \frac{3}{8}\sqrt{n-1}\frac{1}{n}\left(\pi - \arcsin\frac{2}{n}\sqrt{n-1}\right) \right] \tag{40}$$

The definitions of ΔT_1^* and ΔT_2^* were given in the previous papers. we have deduced formula (39) and formula (40) in the previous papers. Thus, we have derived the paraxial temporal aberrations from the approach of paraxial solutions.

Conclusions

From the above investigation, we can reach the following conclusions.

(1) In the present paper, we have given another approach to solve the spatial aberrations and temporal aberration in a bi-electrode concentric spherical system with electrostatic focusing, and we made sure that the two approaches which gave the same expressions are rigorous, since all of the expressions are expressed by analytical forms.

(2) We have deduced the paraxial ray solution and paraxial temporal solution in a bi-electrode concentric spherical system with electrostatic focusing through the paraxial ray equation and paraxial motion equation, and by using the paraxial ray solution and paraxial temporal solution, the paraxial spatial aberration and paraxial temporal aberration have been derived. The result shows that the second order paraxial lateral aberration Δr_2^*, which is the fundamental limitation of spatial resolution for image tubes, as well as the first order paraxial temporal aberration ΔT_1^*, which is the fundamental limitation of temporal resolution of high-speed photographic image converter tube, depend only on the paraxial solutions. It means that we should lay emphasis on the investigation of paraxial solutions in further study of imaging electron optics.

(3) We have once again derived the expressions of the third order paraxial lateral aberration Δr_3^* as well as the second order paraxial temporal aberration ΔT_2^*, in a bi-electrode concentric spherical system with electrostatic focusing, which his the same magnitude with the third order geometrical lateral aberration δr_3 and the second order geometrical temporal aberration δT_2. It means that for imaging electron optics while study the geometrical (lateral or temporal) aberrations: δr_3, δT_2, the paraxial (spatial or temporal) aberrations: Δr_3^*, ΔT_2^* should be considered at the same time. The present paper and previous paper clearly indicate that the third order paraxial lateral aberration Δr_3^* and the second order paraxial temporal aberration ΔT_2^* exist, and the concrete form of these two aberrations have been given.

Acknowledgements

This project was supported by National Natural Science Foundation of China and the Special Science Fund of Doctoral Discipline for Institutions of Higher Learning.

References

[1] ZHOU L W, GONG H, ZHANG Z Q, et al. On electron – optical spatial and temporal aberrations in a bi – electrode spherical concentric system with electrostatic focusing [C]. Proceedings of International Symposium on Photoelectronic Detection and Imaging 2009, 7384, 35: 1 – 12.

[2] ZHOU L W. Electron optics with wide beam focusing [M]. Beijing: Beijing Institute of Technology Press, 1993 (in Chinese).

[3] ZHOU L W. Electron optics of concentric spherical electrostatic focusing systems with two electrodes [J]//Focusing and Imaging of Wide Electron Beams—Selected Papers on Electron Optics by Zhou Liwei. Beijing: Beijing Institute of Technology Press, 1994: 11 – 29 (in Chinese).

[4] ARTIMOVICH L A. Electrostatic properties of emission systems [J]. Bulletin of Academy of Sciences, Physics Series, 1944, 8(6): 313 – 328 (in Russian).

[5] RECKNAGEL A. Theorie des elektrisohen elecktronen miktroskops fur selbstrakler. Z. Angew. Physik, 1941, 117: 689 – 708 (in German).

[6] ZHOU L W, LI Y, ZHANG Z Q, et al. Test and verification of temporal aberration theory for electron optical imaging systems by an electrostatic concentric spherical system [J]. Acta Physica Sinica, 2005, 54(8): 3597 – 3603 (in Chinese).

十一、动态成像电子光学的时间像差理论

Temporal Aberration Theory of Dynamic Imaging Electron Optics

11.1 关于动态电子光学时间像差理论的研究
On the Theory of Temporal Aberrations for Dynamic Electron Optics

摘要：本文提出了"直接积分法"研究动态电子光学成像系统时间像差的新理论。文中以阴极面逸出的某一轴向电子初能为 ε_{z_1} 的近轴电子轨迹为比较基准，给出了时间像差的新定义，导出了动态电子光学成像系统时间像差系数的积分表达式。利用静电聚焦同心球系统理想模型的解析解检验了直接积分法（DIM），研究了"τ-变分法"求解成像系统时间像差系数的表达式，证明了两者的正确性与等价性。与"τ-变分法"比较，"直接积分法"给出的二级几何时间像差系数以积分形式表示，计算更为简便，更适于成像系统的实际计算与设计。文中同时还从方法论的角度叙述了研究动态电子光学时间像差理论的体会。

Abstract: A new theory for studying temporal aberrations of dynamic electron optical imaging systems by the direct integral method (DIM) is put forward in this paper. A new definition of temporal aberration is given taking a certain initial energy of electron emission emitted from the cathode surface along the axial direction ε_{z_1} as a criterion. New expressions of the temporary aberration coefficients expressed in integral forms for the dynamic electron optical imaging system are deduced. An electrostatic concentric spherical system model is used to test and verify the expressions of temporal aberrations given by DIM and the "τ-Variation Method", the analytical solutions prove that both of the two methods are correct and equivalent. Compared to the "τ-Variation Method", DIM needs only to carry out the integral calculation for the three of temporal geometrical aberration coefficients of the second order, and is therefore more convenient and suitable for computation in practical design operations. Finally, what we have learned from the study of the theory of temporal aberrations for dynamic electron optics from the point of view of methodology was elaborated.

引言

1980 年，Monastyrski 和 Schelev 首先提出"τ-变分法"探讨电子光学成像系统的时间像差理论[1]。其核心是从变分的角度考察动态电子光学成像系统的时间像差，该文以轴向初能 $\varepsilon_{z_1}=0$ 的电子作为基准给出了计算电子光学成像系统一级和二级时间像差系数的途径，导出的几何时间像差系数的表示式涉及微分方程的求解。这一工作使电子光学工作者有可能在高速摄影变像管和条纹管的设计与计算中考虑二级时间像差项对时间分辨力的影

周立伟. 北京理工大学. 北京理工大学学报（Beijing Institute of Technology Press），V. 26, No. 5, 2006, 377 – 382.

响,具有科学意义和实际价值。

关于电子光学成像系统的空间像差或时间像差,早在 20 世纪 40—50 年代的研究中就确定了二级横向像差或一级时间像差的特性,其结论如下。

电子光学成像系统的空间弥散特性主要由二级近轴横向色差 Δr_1^* 决定,它以 Recknagel-Artimovich 公式表示[2,3]:

$$\Delta r_1^* = \frac{2M}{E_c}\sqrt{\varepsilon_r}(\sqrt{\varepsilon_z} - \sqrt{\varepsilon_{z_1}}) \tag{1}$$

电子光学成像系统的时间弥散特性主要由一级时间色差 ΔT_1 以 Savoisky-Fanchenko 公式表示[4]:

$$\Delta T_1 = \sqrt{\frac{2m_0}{e}}\frac{1}{E_c}\sqrt{\varepsilon_z} \tag{2}$$

式中, ε_z、ε_r 分别为电子从光阴极发射时的轴向初能量和径向初能量;M 为系统放大率;E_c 为阴极面上的场强(取负值);e/m_0 为电子的荷质比。式(1)中, $\varepsilon_{z_1}(0 \leq \varepsilon_{z_1} \leq \varepsilon_{0\max})$ 为与理想成像位置相对应的某电子轴向初能量, Δr_1^* 是在该理想成像面上考察的。

不难看出,表示一级时间像差 ΔT_1 的式(2)与表示一级横向像差 Δr_1^* 的式(1)之间存在某种不协调性与不对称性。这种不协调性唯一可以统一起来解释的是,目前普遍采用的评价时间渡越弥散或时间像差是以某一轴向初能量 $\varepsilon_{z_1} = 0$ 的电子作为比较基准探讨轴向初能量 $\varepsilon_{z_1} \neq 0$ 的电子所构成的时间离散或形成的时间像差。

考察电子光学成像系统的空间像差与时间像差可以发现,这本来是同一事物,即由阴极面逸出的光电子的发射初能量分散在某一成像面上所显现的空间弥散特性,或是在空间某一位置(包括成像面位置)处所显现的时间弥散特性。如果以阴极面逸出的某一轴向电子初能 $\varepsilon_{z_1}(0 \leq \varepsilon_{z_1} \leq \varepsilon_{0\max})$ 作为比较基准考察一级时间像差 ΔT_1,则式(2)应表示为

$$\Delta T_1 = \sqrt{\frac{2m_0}{e}}\frac{1}{E_c}(\sqrt{\varepsilon_z} - \sqrt{\varepsilon_{z_1}}) \tag{3}$$

作者曾用静电聚焦同心球系统的理想模型严格证明了式(3)的成立[5~7]。由式(3)可见,研究时间像差及其理论必然应考虑其与 ε_{z_1} 的联系。

1 时间像差定义

按照静电聚焦和电磁聚焦以及同心球系统成像电子光学系统的像差理论,空间像差(横向像差)Δr 被定义为[8]

$$\Delta r = \Delta r(z_i) = r_{实际}(z, \varepsilon_r^{1/2}, \varepsilon_z^{1/2}, r_0) - r_{近轴}(z, \varepsilon_{r_1}^{1/2}, \varepsilon_{z_1}^{1/2}, r_0) \tag{4}$$

式中, $r_{实际}$、$r_{近轴}$ 分别为由光阴极发射的实际电子轨迹和近轴电子轨迹到达某一成像位置 z_i 处的空间位置;$\varepsilon_{z_1}(0 \leq \varepsilon_{z_1} \leq \varepsilon_{0\max})$ 为与理想成像位置相对应的某一电子轴向初能量,它乃是横向像差的比较基准。取圆柱坐标系 (z, r),轴向坐标 z 自阴极面 $z_0 = 0$ 算起,r_0 为电子出射的径向初始矢量。

在文献[9]中,曾证明式(4)可以表示为

$$\Delta r = \Delta r^* + \delta r = \Delta r_1^* + \Delta r_2^* + \delta r \tag{5}$$

即横向像差 Δr 被定义为由近轴横向像差 Δr^* 和几何横向像差 δr 两部分所组成,而近轴横

向像差 Δr^* 可细分为一级近轴横向像差 Δr_1^* 和二级近轴横向像差 Δr_2^*，Δr_2^* 与 δr 为同一数量级。

鉴于电子光学成像系统的横向像差或时间像差是同一事物（即由光阴极逸出的光电子的发射初能量分散）在某一成像面上所显现的空间弥散特性，或在空间某一位置处所显现的时间弥散特性，故时间像差也应表达为与式（4）相似的形式，其定义式为

$$\Delta t = \Delta t(z) = t_{实际}(z, \varepsilon_r^{1/2}, \varepsilon_z^{1/2}, r_0) - t_{近轴}(z, \varepsilon_{z_1}^{1/2}, \varepsilon_{z_1}^{1/2}, r_0) \tag{6}$$

式中，$t_{实际}$、$t_{近轴}$ 分别为由光阴极发射的实际电子轨迹和近轴电子轨迹到达某一位置 z 处所经历的时间；ε_{z_1} 为时间像差作为比较基准的某一近轴电子轨迹对应的轴向初能量。

式（4）与式（6）之差异在于，电子光学成像系统的空间像差是在 ε_{z_1} 对应的成像面 z_i（实像面或虚像面）上衡量的，而时间像差是在系统的任一位置 z（包括实像面的位置 z_i）上度量的。

将 $t_{实际}$、$t_{近轴}$ 以 t、t^* 代之，可以证明，近轴电子轨迹所经历的时间 t^* 与电子的径向初能 $\varepsilon_r^{1/2}$ 和电子逸出高度 r_0 无关。于是式（6）便可表示为

$$\Delta t = t(z, \varepsilon_r^{1/2}, \varepsilon_z^{1/2}, r_0) - t^*(z, \varepsilon_{z_1}^{1/2}) \tag{7}$$

将式（7）表达成类似式（5）的形式

$$\Delta t = \Delta T(z, \varepsilon_z^{1/2}, \varepsilon_{z_1}^{1/2}) + \Delta \tau(z, \varepsilon_r^{1/2}, \varepsilon_z^{1/2}, r_0) = \Delta T + \Delta \tau \tag{8}$$

式中，

$$\Delta T = t^*(z, \varepsilon_z^{1/2}) - t^*(z, \varepsilon_{z_1}^{1/2}) = \Delta T_1 + \Delta T_2 \tag{9}$$

$$\Delta \tau = t(z, \varepsilon_r^{1/2}, \varepsilon_z^{1/2}, r_0) - t^*(z, \varepsilon_z^{1/2}) \tag{10}$$

这里，ΔT 称为近轴时间像差或时间色差，它表示轴向初能不同的两条近轴电子轨迹的时间差异。ΔT 又可细分为一级时间色差 ΔT_1 和二级时间色差 ΔT_2；$\Delta \tau$ 称为几何时间像差，它表示轴向初能相同的实际电子轨迹与近轴电子轨迹的时间差异。ΔT_2 与 $\delta \tau$ 为同一数量级。式（9）、式（10）的时间像差定义是研究的出发点。

2 直接积分法求时间像差表达式[10~12]

求时间像差最简便、最直接的途径是对轴对称静电场下电子运动方程

$$\ddot{z} = \frac{e}{m_0} \frac{\partial \varphi}{\partial z} \tag{11}$$

进行积分，求得轴向速度 $\dot{z}$ 后，再对 $\frac{1}{\dot{z}} \frac{dt}{dz}$ 积分，便可求得时间 t 的表达式。

当把空间电位分布 $\varphi = \varphi(z, r)$ 的谢尔赤级数展开式

$$\varphi(z, r) = \phi(z) - \frac{r^2}{4} \phi''(z) + \frac{r^4}{64} \phi^{\text{IV}}(z) - \cdots \tag{12}$$

代入式（11）进行积分后，出现了求解 $\int \frac{r^2}{4} \phi'''(z) dz$，而出现双重积分的问题。这是研究过程中的主要困难。研究中，利用近轴轨迹方程及其两个特解，解决了这个难题。

于是，通过一系列变换，可得到自光阴极面发射的电子在位置矢量 r、方向角 θ_0、逸出角 α_0、方位角 β_0、初速度 v_0、电子初能 ε_0（$v_0 = \sqrt{2e\varepsilon_0/m_0}$，$\varepsilon_z^{1/2} = \varepsilon_0^{1/2} \cos \alpha_0$，$\varepsilon_r^{1/2} = \varepsilon_0^{1/2} \sin \alpha_0$）

的初始条件下，在系统中所行进的时间：

$$t(z,\varepsilon_r^{1/2},\varepsilon_z^{1/2},r_0) = \int_0^z \frac{\mathrm{d}z}{\left(\frac{2e}{m_0}\right)^{1/2}[\phi(z)+\varepsilon_z]^{1/2}}\left\{1 + \frac{1}{2}\frac{1}{[\phi(z)+\varepsilon_z]}\cdot\right.$$

$$\left[\varepsilon_r\left(v'^2\phi_* + \frac{1}{4}\phi''v^2 - 1\right) + r_0^2\left(w'^2\phi_* + \frac{1}{4}\phi''w^2 - \frac{1}{4}\phi_0''\right) + \right.$$

$$\left.\left. 2r_0\sqrt{\varepsilon_r}\cos(\theta_0-\beta_0)\left(v'w'\phi_* + \frac{1}{4}\phi''vw\right)\right]\right\} \tag{13}$$

式中，$\phi(z)$为轴上电位分布；$\phi'(z)$、$\phi''(z)$为$\phi(z)$对z的一阶、二阶导数；$v=v(z)$，$w=w(z)$为近轴轨迹方程的两个线性无关的特解，它满足初始条件：

$$v(z_0=0)=0,\ v'(z_0=0)=\frac{1}{\sqrt{\varepsilon_z}},$$
$$w(z_0=0)=1,\ w'(z_0=0)=0 \tag{14}$$

考虑近轴条件，展开式（13）的第一项，由近轴时间像差的定义式（9），便可得到近轴时间像差ΔT及像差系数表达式：

$$\Delta T = a_2(\varepsilon_z^{1/2} - \varepsilon_{z_1}^{1/2}) + A_{22}(\varepsilon_z - \varepsilon_{z_1}) \tag{15}$$

式中，a_2为一级近轴时间像差系数或称一级时间色差系数，且有

$$a_2 = -\sqrt{\frac{2m_0}{e}}\frac{1}{\phi'(0)} = \sqrt{\frac{2m_0}{e}}\frac{1}{E_c} \tag{16}$$

A_{22}为二级近轴时间像差系数或称二级时间色差系数，且有

$$A_{22} = \frac{1}{2}\sqrt{\frac{2m_0}{e}}\left\{\frac{1}{\sqrt{\phi(z)}}\frac{1}{\phi'(z)} + \int_0^z \frac{\phi''(z)}{\sqrt{\phi(z)}[\phi'(z)]^2}\mathrm{d}z\right\} \tag{17}$$

同样展开式（13）的二阶项，由几何时间像差的定义式（10），便可得到几何时间像差$\Delta\tau$及像差系数表达式：

$$\Delta\tau = A_{11}\varepsilon_r + 2A_{13}\varepsilon_r^{1/2}r_0 + A_{33}r_0^2 \tag{18}$$

式中，A_{11}为二级时间球差系数，且有

$$A_{11} = \int_0^z \frac{1}{2\left(\frac{2e}{m_0}\right)^{1/2}[\phi(z)+\varepsilon_z]^{3/2}}\left(v'^2\phi_* + \frac{1}{4}\phi''v^2 - 1\right)\mathrm{d}z \tag{19}$$

A_{13}为二级时间场曲系数，且有

$$A_{13} = \int_0^z \frac{1}{2\left(\frac{2e}{m_0}\right)^{1/2}[\phi(z)+\varepsilon_z]^{3/2}}\cos(\theta_0-\beta_0)\left(v'w'\phi_* + \frac{1}{4}\phi''vw\right)\mathrm{d}z \tag{20}$$

A_{33}为二级时间畸变系数，且有

$$A_{33} = \int_0^z \frac{1}{2\left(\frac{2e}{m_0}\right)^{1/2}[\phi(z)+\varepsilon_z]^{3/2}}\left(w'^2\phi_* + \frac{1}{4}\phi''w^2 - \frac{1}{4}\phi_0''\right)\mathrm{d}z \tag{21}$$

3 直接积分法时间像差理论的验证[13]

静电聚焦两电极同心球系统是一个理想模型，由此系统的电位场分布和行进的电子轨

迹可以获得明晰的解析表达式，若能求得电子在此系统行进时间的解析表达式，则可用它作为标准考察直接积分法的正确性。

设同心球系统的球面阴极 C 和栅状球面阳极 A 曲率半径分别为 R_c 和 R_a，栅状阳极 A 对于阴极 C 的电位为 ϕ_{ac}，取极坐标 (ρ, φ)，则由能量守恒定律和角动量守恒定律

$$\frac{1}{2}m_0 v^2 = \frac{1}{2}m_0 v_0^2 + e\phi_{\rho c} \tag{22}$$

$$\rho^2 \dot{\varphi} = R_c^2 \dot{\varphi} = R_c v_0 \sin \alpha_0 \tag{23}$$

出发，可以求得自阴极面发射的电子经历时间的解析表达式：

$$t = \sqrt{\frac{m_0}{2e}} \frac{1}{\frac{\phi_{ac}}{n-1} - \varepsilon_0} \left[\sqrt{-\rho^2 \left(\frac{\phi_{ac}}{n-1} - \varepsilon_0\right) + \frac{\phi_{ac}}{n-1} R_c \rho - \varepsilon_0 R_c^2 \sin^2 \alpha_0} - \right.$$

$$\frac{1}{2} \frac{\frac{\phi_{ac}}{n-1}}{\sqrt{\frac{\phi_{ac}}{n-1} - \varepsilon_0}} R_c \arctan \frac{-\frac{\phi_{ac}}{n-1} R_c + 2\rho \left(\frac{\phi_{ac}}{n-1} - \varepsilon_0\right)}{2\sqrt{\frac{\phi_{ac}}{n-1} - \varepsilon_0} \sqrt{-\rho^2 \left(\frac{\phi_{ac}}{n-1} - \varepsilon_0\right) + \frac{\phi_{ac}}{n-1} R_c \rho - \varepsilon_0 R_c^2 \sin^2 \alpha_0}} +$$

$$\left. R_c \sqrt{\varepsilon_0 \cos^2 \alpha_0} + \frac{1}{2} \frac{\frac{\phi_{ac}}{n-1}}{\sqrt{\frac{\phi_{ac}}{n-1} - \varepsilon_0}} R_c \arctan \frac{2\varepsilon_0 - \frac{\phi_{ac}}{n-1}}{2\sqrt{\frac{\phi_{ac}}{n-1} - \varepsilon_0} \sqrt{\varepsilon_0 \cos^2 \alpha_0}} \right] \tag{24}$$

式中，$n = R_c / R_a$。

展开式 (24)，并按式 (9)、式 (10) 的定义把它表达成式 (15) 和式 (18) 的形式，则可得近轴时间像差系数和几何时间像差系数如下：

$$a_2 = \sqrt{\frac{2m_0}{e}} \frac{1}{E_c} \tag{25}$$

$$A_{22} = -\sqrt{\frac{2m_0}{e}} \frac{1}{E_c} \left(\frac{n-1}{\phi_{ac}}\right)^{1/2} \left[\frac{-(R_c + z)(2R_c - z)}{4R_c \sqrt{-z(R_c + z)}} + \frac{3}{4} \arctan \sqrt{\frac{-z}{R_c}}\right] \tag{26}$$

$$A_{11} = \frac{3}{4} \sqrt{\frac{2m_0}{e}} \left(\frac{n-1}{\phi_{ac}}\right)^{1/2} \frac{1}{E_c R_c} \left(-\sqrt{z}\sqrt{-z - R_c} - R_c \arctan \frac{\sqrt{z}}{\sqrt{-z - R_c}}\right) \tag{27}$$

$$A_{13} = -\cos(\theta_0 - \beta_0) \frac{1}{2} \sqrt{\frac{2m_0}{e}} \frac{1}{R_c^2 E_c} z \tag{28}$$

$$A_{33} = \sqrt{\frac{2m_0}{e}} \sqrt{\frac{n-1}{\phi_{ac}}} \frac{3}{8R_c^2} \left(\sqrt{z}\sqrt{-z - R_c} - R_c \arctan \frac{\sqrt{z}}{\sqrt{-z - R_c}}\right) \tag{29}$$

现在对直接积分法的正确性进行检验。将同心球系统轴上电位分布 $\phi(z)$ 表达式

$$\phi(z) = \frac{\phi_{ac}}{n-1} \frac{-z}{z + R_c} \tag{30}$$

及其近轴轨迹方程的两个特解

$$v(z) = \frac{2z}{\phi(z)} \left[\sqrt{\phi(z) + \varepsilon_z} - \sqrt{\varepsilon_z}\right] \tag{31}$$

$$w(z) = 1 + \frac{1}{R_c}z - \frac{\sqrt{\varepsilon_z}}{R_c}\frac{2z}{\phi(z)}\left[\sqrt{\phi(z)+\varepsilon_z} - \sqrt{\varepsilon_z}\right] \tag{32}$$

代入式（16）、式（17）、式（19）、式（20）、式（21），亦可得到与式（25）~式（29）同样的结果。由此证明了直接积分法时间像差理论的正确性。

4 直接积分法与τ变分法研究时间像差理论的比较[14]

研究表明，用τ变分法所导得的一级时间色差系数 a_2 的表达式与式（16）完全一致；其二级时间色差系数 A_{22} 和二级几何时间像差系数 A_{11}、A_{13}、A_{33} 的表达式如下：

$$A_{22} = -\frac{\ddot{z}}{2\dot{z}^3}z_{\alpha_2}^2 + \frac{\dot{z}_{\alpha_2}z_{\alpha_2}}{\dot{z}^2} - \frac{z_{\alpha_2\alpha_2}}{2\dot{z}}, \quad T^{(z)}(z_{\alpha_2\alpha_2}) = \phi\phi'''\frac{2}{\phi_0'^2} \tag{33}$$

$$A_{11} = -\frac{z_{\alpha_1\alpha_1}}{2\dot{z}}, \quad T^{(z)}(z_{\alpha_1\alpha_1}) = -\frac{1}{4}\phi'''v^2 \tag{34}$$

$$A_{13} = -\frac{z_{\alpha_1\alpha_3}}{2\dot{z}}, \quad T^{(z)}(z_{\alpha_1\alpha_3}) = -\frac{1}{4}\phi'''wv\cos(\beta_0-\theta_0) \tag{35}$$

$$A_{33} = -\frac{z_{\alpha_3\alpha_3}}{2\dot{z}}, \quad T^{(z)}(z_{\alpha_3\alpha_3}) = -\frac{1}{4}\phi'''w^2 \tag{36}$$

式（33）~式（36）中，$z_{\alpha_i\alpha_j}$ 是微分方程

$$T^{(z)}(z_{\alpha_i\alpha_j}) = \frac{\mathrm{d}}{\mathrm{d}z}\left(\phi\frac{\mathrm{d}}{\mathrm{d}z}z_{\alpha_i\alpha_j} - \frac{1}{2}\phi'z_{\alpha_i\alpha_j}\right), \quad (i=1,2,3; j=1,2,3) \tag{37}$$

的解，且式（33）中各值可表示为

$$z_{\alpha_2} = \frac{2}{\phi_0'}\sqrt{\phi}, \quad \dot{z} = \sqrt{\frac{2e\phi}{m_0}}, \quad \dot{z}_{\alpha_2} = \sqrt{\frac{2e}{m_0}}\frac{\phi'}{\phi_0'} = \sqrt{\frac{2e}{m_0}}\frac{R_c^2}{(-R_c-z)^2}, \quad \ddot{z} = \frac{e}{m_0}\phi' \tag{38}$$

将同心球系统的场表达式（30）和轨迹的特解表达式（31）、式（32）代入式（33）~式（36），得到的二级时间色差系数 A_{22} 和二级几何时间像差系数 A_{11}、A_{13}、A_{33} 的表达式与式（26）~式（29）是完全一样的，由此也证明了τ变分时间像差理论的正确性以及它与直接积分法的等价性。

5 本研究科学方法的讨论

英国著名科学家卡尔·雷蒙·波普（K. R. Popper）提出演绎检验法（又称试错法）作为科学方法，它可归纳为如下4个一组的图解表示：

P1（Problem1，问题1）→TT（Tentative Theory，尝试性理论）→EE（Elimination of Error，排除错误）→P2（Problem2，问题2）……

波普关于科学研究过程的这个图解表明，可以从某个问题 P1 开始，无论是理论问题还是历史问题；接着做出尝试性解答 TT（推测的或假设性的解答，一种尝试性理论）；然后将其提交，按照证据进行批评性讨论 EE（即排除错误）；如果可得到确认，结果出现新的问题 P2。

下面结合本研究讲一点学习波普科学方法的体会。关于动态电子光学的时间像差理论，俄罗斯科学家提出了一种名为τ变分时间像差理论，20余年来没有人提出疑问。当作者着手研究时间像差理论时，脑海中有两个问题：一是τ变分时间像差理论的正确性，有

无其他更为简捷的途径研究时间像差理论;二是变分理论求得的时间像差系数的准确性(精确性),要找一个理想模型来检验和比较。

这两个问题(即理论的正确性和解决问题的准确性)一直没有被怀疑过,只被简单系统证实过。提出的第1个问题,实际是试图考验变分理论,或否定、或证实、或寻找一个更好的理论;提出的第2个问题,是考验这个理论的可信程度及其适用性。这种思考就是波普演绎检验法中的问题1(Problem1)。研究结果是提出了一种名为直接积分法的时间像差理论,提出了一种新的时间像差定义,即时间像差应由近轴时间像差和几何时间像差两部分组成,其比较基准是轴向电子初能为 $\varepsilon_{z_1}(0 \leqslant \varepsilon_{z_1} \leqslant \varepsilon_{0\max})$ 的近轴电子轨迹,这一过程便是演绎检验法中尝试性理论(Tentative Theory)。这一理论的研究结果是:时间像差系数的求解可以直接用积分形式表示,并不需要用变分理论求解微分方程,而且变分理论仅适用于轴向电子初能为 $\varepsilon_{z_1}=0$ 的情况。研究到这一步,这两种理论孰是孰非并没有解决,故必须寻找一种途径进行严格检验。

下一步就是演绎检验法中的排除错误(elimination of error)。作者找到了一种静电聚焦两电极同心球系统的理想模型,并且也找到了电子在此系统中行进时间的解析解。检验结果表明,这两种理论不但是正确的,而且是精确的。所谓正确是指从两种不同途径出发获得了完全一致的结果,而且,τ 变分理论的结果经过变换也可表达成直接积分法的形式。所谓精确是指两条途径的计算结果与理想模型的解析解是精确一致的。

研究表明,新的理论——直接积分法时间像差理论包容了旧的理论——τ 变分时间像差理论。由此可见,直接积分法时间像差理论是一种更好的理论,于是便进入了问题2(Problem2),推动了科学的进步。

本文中所提出的动态电子光学时间像差理论满足了科学上一个良好理论模式的条件,它具有以下几个特点:

①新理论(直接积分法时间像差理论)。从最基本的物理学定律——电子光学运动方程出发,前提清楚,概念清晰,数学推导正确,得到的数学形式较原理论(τ 变分时间像差理论)更为简洁。

②原理论所能解释的现象,新理论不但都能解释,而且扩展了应用范围:它将原理论以轴向初能量 $\varepsilon_{z_1}=0$ 的比较基准推广到 $0 \leqslant \varepsilon_{z_1} \leqslant \varepsilon_{0\max}$ 的情况。

③新理论经过静电同心球系统的理想模型解析解的检验,证明是正确的和精确的。同时,通过这一理想模型也证明了原理论(τ 变分时间像差理论)的正确性和精确性,说明这两条途径殊途同归。

④新理论中提出了时间像差的新定义。这一理论的出发点是时间像差与空间像差乃是同一现象的时空表现。空间像差是由近轴空间像差和几何空间像差所组成;时间像差由近轴时间像差和几何时间像差所组成。新理论引入的这一假设是电子光学科学理论的一个进步,它把动态电子光学与静态电子光学有机地联系起来了。

⑤新理论具有预测性,它可对目前研究的电磁聚焦成像系统的时间像差理论进行推断和预测:可以证明,无论静电聚焦电子光学成像系统还是电磁聚焦电子光学成像系统,其时间色差仅与静电场有关。

从这一科学问题的研究,可以看出方法论的指导作用。①从逻辑上发现电子光学成像系统现有的空间像差理论与时间像差理论之间的矛盾,即其不协调性,提出了切入点。

②把宽束电子光学的静态问题和动态问题联系起来，可以看到，这是同一事物（即由光阴极面逸出的光电子的发射初能量分散）在某一基准面或成像面上表现的时空特性，并把空间像差和时间像差在定义上统一起来。③采用类比的方法，把静态宽束电子光学的那一套移植到动态电子光学上，给出了新的时间像差定义。④采用理想模型检验的方法，验证原理论和新理论的正确性与精确性，通过一系列比较后做出了科学的结论。

研究表明，直接积分法研究动态电子光学时间像差理论所得结果使计算更简便，更适于电子光学成像系统的实际计算和设计。

参 考 文 献

[1] MONASTYRSKI M A, SCHELEV M Y. Theory of temporal aberrations of cathode lenses [M]. Moscow: Lebedeev Institute of Physics, 1980: 1 – 38.

[2] RECKNAGEL A. Theorie des elektrisohen elecktronen miktroskops fur selbstrakler [J]. J. Z Angew Physik, 1941: 117: 689 – 708.

[3] ARTIMOVICH L A. Electrostatic properties of emission systems [J]. Bulletin of Academy of Sciences USSR Physics Series, 1944, 8(6): 313 – 328.

[4] SAVOISKY Y K, FANCHENKO S D. Physical foundation of electron-optical chronograph [J]. Report of Academy of Sciences, 1956, 108(2): 218 – 221.

[5] 周立伟. 宽电子束聚焦与成像[J]//周立伟电子光学学术论文选. 北京: 北京理工大学出版社, 1994.

[6] ZHOU L W. Electron optics of concentric spherical electromagnetic focusing systems [J]. Advances in Electronics and Electron Physics, 1979, 52: 119 – 132.

[7] 周立伟. 关于像管时间渡越弥散表达式的研究[G]//中国工程院第五次院士大会学术报告汇编, 北京, 信息与电子工程部, 2000: 16 – 20.

[8] 周立伟, 艾克聪, 潘顺臣. 关于电磁复合聚焦阴极透镜的像差理论[J]. 物理学报, 1983, 32(3): 376 – 392.

[9] 周立伟. 宽电子光学[M]. 北京: 北京理工大学出版社, 1993.

[10] 周立伟, 李元, 张智诠, 等. 直接积分法研究电子光学成像系统的时间像差理论[J]. 物理学报, 2005, 54(8): 3591 – 3596.

[11] ZHOU L W, LI Y, ZHANG Z Q, et al. On the theory of temporal aberrations for electron optical imaging systems by using "Direct Integral Method" [J]. SPIE, 2005, 5580: 710 – 724.

[12] ZHOU L W, LI Y, ZHANG Z Q, et al. On the theory of temporal aberrations for cathode lenses [J]. Optik, 2005, 116(4): 175 – 184.

[13] 周立伟, 李元, 张智诠, 等. 静电聚焦同心球系统验证电子光学成像系统的时间像差理论[J]. 物理学报, 2005, 54(8): 3597 – 3603.

[14] 周立伟, MONASTYRSKI M A, SCHELEV M Y, 等. 关于 τ 变分法研究电子光学成像系统的时间像差理论[J]. 电子学报, 2006, 34(2): 193 – 197.

[15] 赖辉亮, 金太军. 波普传[M]. 石家庄: 河北人民出版社, 1998.

11.2 关于 τ-变分法研究动态电子光学成像系统的时间像差理论

On the Temporal Aberration Theory of Dynamic Electron Optical Imaging Systems by τ-Variation Method

摘要：本文采用新的时间像差定义，考察了"τ-变分法"研究动态电子光学成像系统的时间像差理论。结果表明，"τ-变分法"所给出的以微分方程形式表示的几何时间像差系数，亦可获得以积分形式的表示式，证明了"直接积分法"与"τ-变分法"这两条途径殊途同归。

Abstract: By the use of a novel definition of temporal aberration, the temporal aberration theory of dynamic electron optical imaging system was again investigated by the "τ-Variation Method". Results show that the expressions in integral form can also be obtained by the "τ-Variation Method" which usually needs to solve the differential equations for the geometrical temporal aberration coefficients of the second order. It also proves that the Direct Integral Method and "τ-Variation Method" reach the same goal by different ways.

引言

1980 年，Monastyrski 和 Schelev 首先提出"τ-变分法"[3]探讨电子光学成像系统的时间像差理论，其核心是以变分方法考察电子光学成像系统的时间像差，该文以轴向初能量 $\varepsilon_{z_1}=0$ 的电子作为基准给出了计算电子光学成像系统一级和二级时间像差系数的途径，但导得的几何时间像差系数的表示式涉及微分方程的求解。文献[3]的工作使得有可能在高速摄影变像管和条纹管的设计与计算中考虑二级时间像差项对时间分辨率的影响。

本文的目的是由所给出的时间像差定义[1,2]进一步考察"τ-变分法"[3]研究电子光学成像系统的时间像差理论，并证明由"τ-变分法"出发亦可获得以积分形式表示的二级几何时间像差系数表达式。

1 时间像差的定义

文献[1]给出的时间像差的定义如下：

周立伟[a]，M. A. Monastyrski[b]，M. Y. Schelev[b]，张智诠[c]，李元[a]. a) 北京理工大学，b) 俄罗斯科学院普罗霍洛夫普通物理研究所，c) 装甲兵工程学院. 电子学报（Acta Electronica Sinica），V. 34 No. 2, 2006, 193-197.

$$\Delta t = t(z, \varepsilon_r^{1/2}, \varepsilon_z^{1/2}, r_0) - t^*(z, \varepsilon_{z_1}^{1/2})$$
$$= \Delta T(z, \varepsilon_z^{1/2}, \varepsilon_{z_1}^{1/2}) + \Delta \tau(z, \varepsilon_r^{1/2}, \varepsilon_z^{1/2}, r_0)$$
$$= \Delta T + \Delta \tau \tag{1}$$

式中，t、t^* 分别表示由光阴极发射的实际电子轨迹和近轴电子轨迹所经历的时间；ε_z、ε_r 分别为电子从光阴极发射时的轴向初能量和径向初能量；ε_{z_1} 为作为比较基准的近轴电子轨迹的轴向初能，取圆柱坐标系 (z, r)，轴向坐标 z 自阴极面 $z_0 = 0$ 算起，r_0 为电子出射的径向初始矢量。

按式（1）的定义，总的时间像差可分为近轴时间像差（或称近轴时间色差）和几何时间像差两部分。近轴时间像差 ΔT 可表示为：

$$\Delta T = t^*(z, \varepsilon_z^{1/2}) - t^*(z, \varepsilon_{z_1}^{1/2}) = a_2(\varepsilon_z^{1/2} - \varepsilon_{z_1}^{1/2}) + A_{22}(\varepsilon_z - \varepsilon_{z_1}) \tag{2}$$

几何时间像差 $\Delta \tau$ 可表示为：

$$\Delta \tau = t(z, \varepsilon_r^{1/2}, \varepsilon_z^{1/2}, r_0) - t^*(z, \varepsilon_z^{1/2}) = A_{11}\varepsilon_r + 2A_{13}\varepsilon_r^{1/2} r_0 + A_{33} r_0^2 \tag{3}$$

式（2）表明，近轴时间像差 ΔT 乃是两条不同的 ε_z 下近轴轨迹在同一轴向位置 z 处的时间差异。式（3）表明，几何时间像差 $\Delta \tau$ 乃是在相同的 ε_z 下实际轨迹与近轴轨迹在同一轴向位置 z 处的时间差异。

于是，由式（1），总的时间像差可表示为：

$$\Delta t = a_2(\varepsilon_z^{1/2} - \varepsilon_{z_1}^{1/2}) + A_{22}(\varepsilon_z - \varepsilon_{z_1}) + A_{11}\varepsilon_r + 2A_{13}\varepsilon_r^{1/2} r_0 + A_{33} r_0^2 \tag{4}$$

式中，a_2 为一级近轴时间像差系数，A_{22} 为二级近轴时间像差系数，A_{11} 为二级几何时间球差系数，A_{33} 为二级几何时间畸变系数，A_{13} 为二级几何时间场曲系数。

2 变分法求时间像差系数的理论

如图 1 所示，电子自阴极平面 M_0 点发射的初始位置矢量为 r_0，方向角为 θ_0，初始电子的逸出角为 α_0，方位角为 β_0，电子逸出的初速度为 v_0，$v_0 = \sqrt{2e\varepsilon_0/m_0}$，$\varepsilon_z^{1/2} = \varepsilon_0^{1/2} \cos\alpha_0$，$\varepsilon_r^{1/2} = \varepsilon_0^{1/2} \sin\alpha_0$。

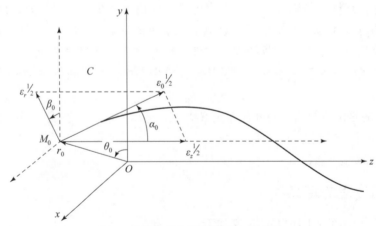

图 1 电子自阴极面发射的初始状态

众所周知，电子在轴对称静电场中的运动由 Lorentz 方程描述：

$$\ddot{r} = \frac{e}{m_0}\frac{\partial \varphi}{\partial r},$$
$$\ddot{z} = \frac{e}{m_0}\frac{\partial \varphi}{\partial z} \tag{5}$$

式中，$\varphi = \varphi(z,r)$ 为空间电位分布，将位置矢量 r 以复数形式表示，$r = x + iy$，空间电位分布便可表示为

$$\varphi(z,r) = \phi(z) - \frac{r \cdot r^*}{4}\phi''(z) + \frac{(r \cdot r^*)^2}{64}\phi^{IV}(z) - \cdots \tag{6}$$

式中，$\phi = \phi(z)$ 为轴上电位分布，$\phi'' = \phi''(z)$，$\phi^{IV} = \phi^{IV}(z)$ 分别为轴上电位分布的二阶和四阶导数。

求解式（5）的初始条件为

$$r(0) = r_0 e^{i\theta_0},\ \dot{r}(0) = \sqrt{\frac{2e}{m_0}}\varepsilon_r^{1/2}e^{i\beta_0},\ \theta_0 = \arg r(0),\ z(0) = 0,\ \dot{z}(0) = \sqrt{\frac{2e}{m_0}}\varepsilon_z^{1/2} \tag{7}$$

于是，式（5）的解可表示为

$$r = r(\varepsilon_r^{1/2},\varepsilon_z^{1/2},r_0),$$
$$z = z(\varepsilon_r^{1/2},\varepsilon_z^{1/2},r_0) \tag{8}$$

如果已知电位分布 φ 的同时，给出了参量 $\varepsilon_r^{1/2}$、$\varepsilon_z^{1/2}$、r_0 的值，则在时间 t 的任一时刻唯一地确定了带电粒子的轨迹。参量 $\varepsilon_r^{1/2}$、$\varepsilon_z^{1/2}$、r_0 在我们考察的情况下被认为是一级小量。

由函数 $t = t(z_1,\varepsilon_r^{1/2},\varepsilon_z^{1/2},r_0)$ 的定义，可得对任一固定的 $z_1 > 0$：

$$z[t(z_1,\varepsilon_r^{1/2},\varepsilon_z^{1/2},r_0),\varepsilon_r^{1/2},\varepsilon_z^{1/2},r_0] = z_1 \tag{9}$$

此式对于所有参量 $\varepsilon_r^{1/2}$、$\varepsilon_z^{1/2}$、r_0 值都是成立的。由于 z_1 坐标的任意性，故下标 1 我们有时便略去。

由于变分法并不为从事电子光学工作者所熟悉，故这里较为详细地介绍其在求解时间像差表达式中的各级像差系数中的应用。

现引入矢量符号 $\alpha(\alpha_1 = \varepsilon_r^{1/2},\alpha_2 = \varepsilon_z^{1/2},\alpha_3 = r_0)$，将式（8）简写为如下形式：

$$z[t(z_1,\alpha),\alpha] = z_1 \tag{10}$$

将式（10）对 α_i、α_j 进行微分，可得：

$$\frac{dz}{d\alpha_i} = \frac{\partial z}{\partial t}\frac{\partial t}{\partial \alpha_i} + \frac{\partial z}{\partial \alpha_i} = \dot{z}t_{\alpha_i} + z_{\alpha_i} = 0 \tag{11}$$

$$\frac{d}{d\alpha_j}(\dot{z}t_{\alpha_i} + z_{\alpha_i}) = (\ddot{z}t_{\alpha_j} + \dot{z}_{\alpha_j})t_{\alpha_i} + \dot{z}t_{\alpha_i\alpha_j} + \dot{z}_{\alpha_i}t_{\alpha_j} + z_{\alpha_i\alpha_j} = 0 \tag{12}$$

在式（12）的推导中我们应用了关系式 $\dot{z}\frac{\partial}{\partial t}\left[\frac{\partial t}{\partial \alpha_i}\right]t_{\alpha_j} = 0$。

由式（11）、式（12）可解得

$$t_{\alpha_i} = -\frac{z_{\alpha_i}}{\dot{z}} \tag{13}$$

$$t_{\alpha_i\alpha_j} = -\frac{\ddot{z}}{\dot{z}^3}z_{\alpha_i}z_{\alpha_j} + \frac{z_{\alpha_i}\dot{z}_{\alpha_j} + \dot{z}_{\alpha_i}z_{\alpha_j}}{\dot{z}^2} - \frac{z_{\alpha_i}z_{\alpha_j}}{\dot{z}} \tag{14}$$

由式（4）可得

$$\alpha_2 = \frac{\partial t}{\partial \alpha_2} = t_{\alpha_2} \tag{15}$$

$$A_{ij} = \frac{1}{2}\frac{\partial^2 t}{\partial \alpha_i \partial \alpha_j} = \frac{1}{2} t_{\alpha_i \alpha_j} \tag{16}$$

式中，$(i,j) = (1,1)$，$(1,3)$，$(3,3)$，$(2,2)$。

这样，求时间像差系数 α_2、A_{ij} 实际归结于计算式（13）、式（14）的右端项。我们现在就对它进行分析。

由电子运动方程式（5）的第一式，利用关系式（6），取其一阶项，可以得到

$$\ddot{z} - \frac{e}{m_0}\frac{\partial \varphi}{\partial z} = \ddot{z} - \frac{e}{m_0}\phi' = 0 \tag{17}$$

于是

$$\frac{\partial \varphi}{\partial \alpha_i}\left(\ddot{z} - \frac{e}{m_0}\phi'\right) = \frac{\partial}{\partial \alpha_i}(\ddot{z}) - \frac{e}{m_0}\frac{\partial \phi'}{\partial z}\frac{\partial z}{\partial \alpha_i} = 0 \tag{18}$$

此即一阶偏导数（变分）的方程，它具有如下形式

$$T^{(i)}(z_{\alpha_i}) = \ddot{z}_{\alpha_i} - \frac{e}{m_0}\phi'' z_{\alpha_i} = 0 \quad (i = 1, 2, 3) \tag{19}$$

式中，$T^{(i)}$ 为二阶微分算符，它与时间 t 相关。$T^{(i)}$ 可表示为

$$T^{(i)} = \frac{d^2}{dt^2} - \frac{e}{m_0}\phi'' \tag{20}$$

同样求二阶偏导数（变分）方程时，需要考虑式（6）的二阶项，则有

$$\ddot{z} - \frac{e}{m_0}\frac{\partial \phi}{\partial z} = \ddot{z} - \frac{e}{m_0}\phi' + \frac{e}{m_0}\frac{r \cdot r^*}{4}\phi'''(z) = 0 \tag{21}$$

于是

$$\frac{\partial}{\partial \alpha_i}\left[\ddot{z} - \frac{e}{m_0}\phi' + \frac{e}{m_0}\frac{r \cdot r^*}{4}\phi'''(z)\right] = \frac{\partial}{\partial \alpha_i}(\ddot{z}) - \frac{e}{m_0}\frac{d\phi'}{dz}\frac{\partial z}{\partial \alpha_i} + \frac{1}{4}\frac{e}{m_0}\phi'''(r^* \cdot r_{\alpha_i} + r \cdot r^*_{\alpha_i}) = 0$$

即

$$\ddot{z}_{\alpha_i} - \frac{e}{m_0}\phi'' z_{\alpha_i} + \frac{1}{4}\frac{e}{m_0}\phi'''(r^* \cdot r_{\alpha_i} + r \cdot r^*_{\alpha_i}) = 0 \tag{22}$$

再对上述方程求偏导数：

$$\ddot{z}_{\alpha_i \alpha_j} - \frac{e}{m_0}\phi'' z_{\alpha_i \alpha_j} + \frac{1}{4}\frac{e}{m_0}\phi'''(r_{\alpha_i} r^*_{\alpha_j} + r^*_{\alpha_i} r_{\alpha_j}) = 0$$

因此

$$T^{(i)}(z_{\alpha_i}, \alpha_j) = \ddot{z}_{\alpha_i \alpha_j} - \frac{e}{m_0}\phi'' z_{\alpha_i \alpha_j} = \frac{e}{m_0}\phi'''\left[z_{\alpha_i} z_{\alpha_j} - \frac{1}{4}(r_{\alpha_i} r^*_{\alpha_j} + r^*_{\alpha_i} r_{\alpha_j})\right] \tag{23}$$

这里，电子轨迹 $r = r(z)$ 应满足轴对称电子光学成像系统的近轴轨迹方程：

$$\phi_*(z) r'' + \frac{1}{2}\phi'(z) r' + \frac{1}{4}\phi''(z) r = 0 \tag{24}$$

式中，$\phi_*(z) = \phi(z) + \varepsilon_z$。式（24）的通解可表示为

$$r(z) = r_0 w(z) + \sqrt{\frac{m_0}{2e}}\gamma_0 v(z) \tag{25}$$

式中，特解 $v = v(z)$，$w = w(z)$ 满足如下初始条件：

$$v(z_0 = 0) = 0, \quad v'(z_0 = 0) = \frac{1}{\sqrt{\varepsilon_z}};$$
$$w(z_0 = 0) = 1, \quad w'(z_0 = 0) = 0 \tag{26}$$

现利用能量关系式：

$$\frac{m_0}{2}\dot{z}^2 = e\phi_* \tag{27}$$

将式（19）的 $T^{(i)}(z_{\alpha_i})$ 中的 $\ddot{z}_{\alpha_i}$ 展开为

$$\ddot{z}_{\alpha_i} = \frac{d^2}{dt^2}(z_{\alpha_i}) = \frac{d}{dz}\left(\frac{dz_{\alpha_i}}{dz}\frac{dz}{dt}\right)\frac{dz}{dt} = \frac{2e}{m_0}\phi_* \frac{d^2}{dz^2}(z_{\alpha_i}) + \frac{d}{dz}(z_{\alpha_i})\frac{e}{m_0}\phi' \tag{28}$$

将它代入式（19），可得

$$\phi_* \frac{d^2}{dz^2}(z_{\alpha_i}) + \frac{1}{2}\phi' \frac{d}{dz}(z_{\alpha_i}) - \frac{1}{2}\phi'' z_{\alpha_i} = 0 \tag{29}$$

令 $T^{(z)}$ 表示二阶线性微分算符，且

$$T^{(z)} = \phi_* \frac{d^2}{dz^2} + \frac{1}{2}\phi'\frac{d}{dz} - \frac{1}{2}\phi'' = 0 \tag{30}$$

于是有

$$T^{(z)}(z_{\alpha_i}) = 0 \tag{31}$$

同样，可将式（23）中的 $\ddot{z}_{\alpha_i\alpha_j}$ 展开为

$$\ddot{z}_{\alpha_i\alpha_j} = \frac{d}{dt^2}(z_{\alpha_i\alpha_j}) = \frac{2e}{m_0}\left[\phi_* \frac{d^2}{dz^2}(z_{\alpha_i\alpha_j}) + \frac{1}{2}\phi'\frac{d}{dz}(z_{\alpha_i\alpha_j})\right] \tag{32}$$

故有

$$\ddot{z}_{\alpha_i\alpha_j} - \frac{e}{m_0}\phi'' z_{\alpha_i\alpha_j} = \frac{2e}{m_0}\left[\phi_* \frac{d^2}{dz^2}(z_{\alpha_i\alpha_j}) + \frac{1}{2}\phi'\frac{d}{dz}(z_{\alpha_i\alpha_j}) - \frac{1}{2}\phi'\frac{d}{dz}(z_{\alpha_i\alpha_j})\right] = \frac{2e}{m_0}T^{(z)}(z_{\alpha_i\alpha_j}) \tag{33}$$

故由式（23）可得

$$T^{(i)}(z_{\alpha_i\alpha_j}) = \frac{1}{2}\phi''\left[z_{\alpha_i}z_{\alpha_j} - \frac{1}{4}(r_{\alpha_i}r_{\alpha_j}^* + r_{\alpha_i}^* r_{\alpha_j})\right] \tag{34}$$

3 一级近轴时间像差系数的求解

现在我们求齐次方程式（31）即 $T^{(z)}(z_{\alpha_i}) = 0$ 在 $i = 1, 2, 3$ 时的解。如文献 [1] 所述，$\varepsilon_r^{1/2}$ 和 r_0 并不引入近轴运动方程中，故

$$z_{\alpha_i} = 0, \quad i = 1, 3 \tag{35}$$

因此，由式（13），$t_{\alpha_i} = 0$，$i = 1, 3$，于是 $a_1 = 0$，$a_3 = 0$，故像差公式中不出现 a_1、a_3 系数项。

对于 $i = 2$，则 $\alpha_2 = \varepsilon_z^{1/2}$。不难证明，满足条件式 $\dot{z}(0) = \sqrt{2e\varepsilon_z/m_0}$ 下，方程式（31）的解为

$$z_{\alpha_2} = \frac{2}{\phi_0'}\sqrt{\phi_*} \tag{36}$$

其初始条件可表示为

$$z_{\alpha_2}|_{z=0} = \frac{2}{\phi_0'}\sqrt{\varepsilon_z}, \quad \dot{z}_{\alpha_2}|_{z=0} = \sqrt{\frac{2e}{m_0}} \tag{37}$$

这里，$\phi_0' = \phi'(0)$ 是阴极面中心的电位梯度。令 $E_c = -\phi_0'$ 为阴极面中心的场强，取负值。由式（13）表示的一级近轴像差系数或称一级时间色差系数可表示为

$$\alpha_2 = t_{\alpha_2} = \frac{\partial t}{\partial \alpha_2} = \frac{\partial z}{\partial \alpha_2}\frac{1}{\frac{\partial z}{\partial t}} = \frac{-z_{\alpha_2}}{\dot{z}} = \frac{-\frac{2}{\phi_0'}\sqrt{\phi_*}}{\sqrt{\frac{2e}{m_0}\phi_*}} = -\frac{1}{\phi_0'}\sqrt{\frac{2m_0}{e}} = \frac{1}{E_c}\sqrt{\frac{2m_0}{e}} \tag{38}$$

式（38）表明，系数 α_2 将取负值，其物理意义是：具有某一轴向初速 $\varepsilon_z^{1/2} \neq 0$ 的带电粒子到达 z_1 位置的时间较另一轴向初速 $\varepsilon_z^{1/2} = 0$ 的带电粒子要短一些。式（38）与文献 [5]、文献 [6] 中获得的阴极透镜时间渡越弥散公式相一致。

4 二级几何时间像差系数的求解

现在我们转到分析具有二级变分值的方程式（34）的右端项。$r_{\alpha_i}^{(z)}$ 值按初始条件式（7）及特解式（25），便有

$$r_{\alpha_i}^{(z)} = \begin{cases} ve^{i\beta_0}, & i=1 \\ 0, & i=2 \\ we^{i\theta_0}, & i=3 \end{cases} \tag{39}$$

利用式（39）及式（35）、式（36）所有可能的组合，便可将式（34）写成三阶矩阵的形式：

$$F = \begin{vmatrix} -\frac{1}{4}\phi'''v^2 & 0 & -\frac{1}{4}\phi'''vw\cos(\theta_0-\beta_0) \\ 0 & \frac{2}{\phi_0^2}\phi'''\phi_* & 0 \\ -\frac{1}{4}\phi'''vw\cos(\theta_0-\beta_0) & 0 & -\frac{1}{4}\phi'''w^2 \end{vmatrix} \tag{40}$$

由于式（7）的初始条件对于 $\varepsilon_r^{1/2}$、$\varepsilon_z^{1/2}$、r_0 的线性关系，故变分形式的方程的初始条件都等于零，对于 $(i,j)=(1,3)$ 的情况：

$$z_{\alpha_i\alpha_j}|_{z=0} = 0, \quad \dot{z}_{\alpha_i\alpha_j}|_{z=0} = 0 \tag{41}$$

由式（34）与式（40）便可得函数 $z_{\alpha_1\alpha_3}$ 满足方程

$$T^{(i)}(z_{\alpha_i\alpha_j}) = -\frac{1}{4}\phi'''vw\cos(\theta_0-\beta_0) \tag{42}$$

应用洛毕达法则，得初始条件

$$z_{\alpha_1\alpha_3}|_{z=0} = 0, \quad z'_{\alpha_1\alpha_3}|_{z=0} = 0 \tag{43}$$

将下列的关系式：

$$\frac{d}{dz}\left(v'w'\phi_* + \frac{1}{4}\phi''vw\right) = \frac{1}{4}\phi'''vw \tag{44}$$

$$\frac{\mathrm{d}}{\mathrm{d}z}\left(\phi_* z'_{\alpha_1\alpha_3} - \frac{1}{2}\phi'_{\alpha_1\alpha_3}\right) = \phi_* z''_{\alpha_1\alpha_3} + \frac{1}{2}\phi' z'_{\alpha_1\alpha_3} - \frac{1}{2}\phi'' z_{\alpha_1\alpha_3} = T^{(z)}(z_{\alpha_1\alpha_3}) \tag{45}$$

置换式（42）的两端并进行积分得

$$\phi_* z'_{\alpha_1\alpha_3} - \frac{1}{2}\phi'_{\alpha_1\alpha_3} = -\cos(\theta_0 - \beta_0)\left(v'w\phi_* + \frac{1}{4}\phi'''vw\right) + C_1$$

由式（43）的初始条件及特解初始条件式（26），可得 $C_1 = 0$，于是

$$\phi_* z'_{\alpha_1\alpha_3} - \frac{1}{2}\phi' z_{\alpha_1\alpha_3} = -\cos(\theta_0 - \beta_0)\left(v'w\phi_* + \frac{1}{4}\phi'''vw\right) \tag{46}$$

引入线性微分算符

$$R^{(z)} = \phi_* \frac{\mathrm{d}}{\mathrm{d}z} - \frac{1}{2}\phi' \tag{47}$$

可将方程式（46）表达成以下形式：

$$R^{(z)}(z_{\alpha_1\alpha_3}) = -\left(v'w'\phi_* + \frac{1}{4}\phi''vw\right)\cos(\theta_0 - \beta_0) \tag{48}$$

对于 $(i,j) = (1,1)$ 情况，由式（34）与式（40），便可得函数 $z_{\alpha_1\alpha_3}$ 满足方程

$$T^{(z)}(z_{\alpha_1\alpha_3}) = -\frac{1}{4}\phi''' v^2 \tag{49}$$

仿照式（44）、式（45）的处理，积分得

$$\phi_* z'_{\alpha_1\alpha_3} - \frac{1}{2}\phi' z_{\alpha_1\alpha_3} = -\left(v'^2\phi_* + \frac{1}{4}\phi''v^2\right) + C_2$$

常数 C_2 可由特解 v 的初始条件求得 $C_2 = 1$。

最后可得

$$R^{(z)}(z_{\alpha_1\alpha_1}) = -v'^2\phi_* - \frac{1}{4}\phi''v^2 + 1 \tag{50}$$

对于 $(i,j) = (3,3)$ 情况仿照上述，将 $v \to w$、下标 $1 \to 3$ 做相应的置换，可得

$$\phi_* z'_{\alpha_3\alpha_3} - \frac{1}{2}\phi' z_{\alpha_3\alpha_3} = -\left(w^2\phi_* + \frac{1}{4}\phi''w^2\right) + C_3$$

常数 C_3 可由特解 w 的初始条件求得 $C_3 = \frac{1}{4}\phi''_0$，则

$$R^{(z)}(z_{\alpha_3\alpha_3}) = -w'^2\phi_* - \frac{1}{4}\phi''w'^2 + \frac{1}{4}\phi''_0 \tag{51}$$

将式（48）、式（50）、式（51）求得的 $z_{\alpha_i\alpha_j}$ 代入式（14），并考虑式（35）：$z_{\alpha_1} = 0$，$z_{\alpha_3} = 0$，$\dot{z}_{\alpha_1} = 0$，$\dot{z}_{\alpha_3} = 0$，则由式（14）可得时间像差系数 A_{ij} 表达式如下：

$$A_{ij} = \frac{1}{2}t_{\alpha_i\alpha_j} = \frac{1}{2}\left(-\frac{z_{\alpha_i\alpha_j}}{\dot{z}}\right) = -\frac{1}{4}\sqrt{\frac{2m_0}{e}}\frac{z_{\alpha_i\alpha_j}}{\sqrt{\phi_*}} \quad (i=1, j=3), (i=1, j=1), (i=3, j=3) \tag{52}$$

由此可见，式（52）的 A_{ij} 值是通过求解式（48）、式（50）、式（51）的微分方程的 $z_{\alpha_i\alpha_j}$ 求得。这是比较繁复的，我们现在把它化为积分形式，若令

$$z_{\alpha_i\alpha_j} = \sqrt{\phi_*}\, y_{ij}(z) \tag{53}$$

则式（48）变为

$$\phi_*^{3/2} y'_{13}(z) = -\cos(\theta_0 - \beta_0)\left(v'w'\phi_* + \frac{1}{4}\phi''vw\right) \tag{54}$$

由此可解得

$$y_{13}(z) = -\int_0^z \frac{1}{[\phi(z)+\varepsilon_z]^{3/2}} \cos(\theta_0-\beta_0)\left(v'w'\phi_* + \frac{1}{4}\phi''vw\right)dz \tag{55}$$

于是，由式（52），便得到 A_{13} 的积分表示式：

$$A_{13} = \frac{1}{4}\sqrt{\frac{2m_0}{e}} \int_0^z \frac{1}{[\phi(z)+\varepsilon_z]^{3/2}} \cos(\theta_0-\beta_0)\left(v'w'\phi_* + \frac{1}{4}\phi''vw\right)dz \tag{56}$$

与此完全类似，利用式（53）的转换，由方程式（50）、式（51）分别可解得

$$y_{11}(z) = -\int_0^z \frac{1}{[\phi(z)+\varepsilon_z]^{3/2}}\left[-\left(v'^2\phi_* + \frac{1}{4}\phi''v^2\right)+1\right]dz \tag{57}$$

$$y_{33}(z) = -\int_0^z \frac{1}{[\phi(z)+\varepsilon_z]^{3/2}}\left[-\left(w'^2\phi_* + \frac{1}{4}\phi''w^2\right)+\frac{1}{4}\phi_0''\right]dz \tag{58}$$

于是，由式（52）便得到 A_{11} 和 A_{33} 的积分表示式：

$$A_{11} = \frac{1}{4}\sqrt{\frac{2m_0}{e}}\int_0^z \frac{1}{[\phi(z)+\varepsilon_z]^{3/2}}\left[-\left(v'^2\phi_* + \frac{1}{4}\phi''v^2\right)-1\right]dz \tag{59}$$

$$A_{33} = \frac{1}{4}\sqrt{\frac{2m_0}{e}}\int_0^z \frac{1}{[\phi(z)+\varepsilon_z]^{3/2}}\left[w'^2\phi_* + \frac{1}{4}\phi''w^2 - \frac{1}{4}\phi_0''\right]dz \tag{60}$$

式（56）、式（59）和式（60）所表示的二级几何时间像差系数表达式与"直接积分法"所导出的公式完全一致。

5 二级近轴时间像差系数的求解

对于 $(i,j)=(2,2)$ 的情况，由式（34）与式（40），便可得函数 $z_{\alpha_2\alpha_2}$ 满足方程

$$T^{(z)}(z_{\alpha_2\alpha_2}) = \frac{2}{\phi_0'^2}\phi'''\phi_* \tag{61}$$

它在初始条件

$$z_{\alpha_2\alpha_2}|_{z=0} = 0, \quad z'_{\alpha_2\alpha_2}|_{z=0} = \frac{2}{\phi_0'^2}\phi_0''$$

下积分可得

$$\phi_* z'_{\alpha_2\alpha_2} - \frac{1}{2}\phi' z_{\alpha_2\alpha_2} = \frac{2}{\phi_0'^2}\left(\phi_*\phi'' - \frac{1}{2}\phi'^2\right) + C_4 \tag{62}$$

由初始条件得 $C_4=1$，于是得

$$\phi_* z'_{\alpha_2\alpha_2} - \frac{1}{2}\phi' z_{\alpha_2\alpha_2} = \frac{2}{\phi_0'^2}\left(\phi_*\phi'' - \frac{1}{2}\phi'^2\right) + 1 \tag{63}$$

即

$$R^{(z)}(z_{\alpha_2\alpha_2}) = \frac{2}{\phi_0'^2}\left(\phi_*\phi'' - \frac{1}{2}\phi'^2\right) + 1 \tag{64}$$

其解便是 $z_{\alpha_2\alpha_2}$。其初始条件应满足式（62）。

由式（14）的 $t_{\alpha_i\alpha_j}$ 表达式，令 $i=2$, $j=2$，代入 $z_{\alpha_2\alpha_2}$ 及

$$\ddot{z} = \frac{e}{m_0}\phi', \quad \dot{z} = \sqrt{\frac{2e}{m_0}\phi_*}, \quad z_{\alpha_2} = \frac{2}{\phi_0'}\sqrt{\phi_*}$$

$$\dot{z}_{\alpha_2} = \frac{\mathrm{d}}{\mathrm{d}t}(z_{\alpha_2}) = \frac{\mathrm{d}z}{\mathrm{d}t}\frac{\mathrm{d}}{\mathrm{d}z}\left(\frac{2}{\phi_0'}\sqrt{\phi_*}\right) = \sqrt{\frac{2e}{m_0}\frac{\phi'}{\phi_0'}}$$

各式,并利用式(16)的 A_{ij} 表达式可以导得

$$A_{22} = \frac{1}{2} t_{\alpha_2\alpha_2} = \frac{1}{2}\sqrt{\frac{2m_0}{e}}\left(\frac{\phi'}{\sqrt{\phi_*}\phi_0'^2} - \frac{z_{\alpha_2\alpha_2}}{2\sqrt{\phi_*}}\right) \tag{65}$$

若令

$$z_{\alpha_2\alpha_2} = \frac{2\phi'}{\phi_0'^2} - y_{22} \tag{66}$$

代入式(63),得到

$$\phi_* \frac{2\phi''}{\phi_0'^2} - \phi_* y_{22}' - \frac{1}{2}\frac{2\phi'^2}{\phi_0'^2} + \frac{1}{2}\phi' y_{22} = \frac{2}{\phi_0'^2}\left(\phi_*\phi'' - \frac{1}{2}\phi'^2\right) + 1$$

由此可得

$$\phi_* y_{22}' - \frac{1}{2}\phi' y_{22} + 1 = 0 \tag{67}$$

其解为

$$y_{22}'(z) = \frac{2}{\phi'} + 2\sqrt{\phi_*}\int_0^z \frac{\phi''}{\sqrt{\phi_*}\phi'^2}\mathrm{d}z \tag{68}$$

将式(68)代入式(66),便可证明 $z'_{\alpha_2\alpha_2}$ 确实满足式(62)。我们便可把式(65)写成如下形式:

$$A_{22} = \frac{1}{4}\sqrt{\frac{2m_0}{e}}\frac{1}{\sqrt{\phi_*}}\left(\frac{2}{\phi'} + 2\sqrt{\phi_*}\int_0^z \frac{\phi''}{\sqrt{\phi_*}\phi_0'^2}\mathrm{d}z\right) \tag{69}$$

为了避免 A_{22} 的计算,当 $\varepsilon_z = 0$ 时,式(69)在表面上似乎趋于无穷大的困难。文献[1]提出了足以精确描述式(69)的表示式:

$$A_{22} = -\frac{1}{4}\sqrt{\frac{2m_0}{e}}\int_0^z \frac{1}{\phi(z)^{3/2}}\mathrm{d}z \tag{70}$$

结束语

本文由时间像差的定义出发,以轴向电子初能为 $\varepsilon_{z_1}(0 \leqslant \varepsilon_{z_1} \leqslant \varepsilon_{0\max})$ 作为参考基准,用变分法导出了一级近轴时间像差表示式、二级近轴时间像差系数的积分表达式和以微分方程形式表示的二级几何时间像差系数表达式。本文推广了文献[3]的应用范围,并将变分法以微分方程形式表示的二级几何时间像差系数表达式转换为积分形式。本文证明,尽管"直接积分法"与"τ-变分法"在数学处理上有很大的差异,但殊途同归,其时间像差系数的表达式实质是完全一致的。

参 考 文 献

[1]周立伟,李元,张智诠,等.直接积分法研究电子光学成像系统的时间像差理论[J].物理学报,2005,54(8):3591-3596.

[2]周立伟,李元,张智诠,等.静电聚焦同心球系统验证电子光学成像系统的时间像差

理论[J]. 物理学报,2005,54(8):3597-3603.

[3] MONASTYRSKI M A, SCHELEV M Y. Theory of temporal aberrations of cathode lenses [M]. Moscow: Lebedeev Institute of Physics, 1980:1-38.

[4] 周立伟. 宽束电子光学[M]. 北京:北京理工大学出版社,1993.

[5] SAVOISKY Y K, FANCHENKO S D. Physical basis of electron optical chronograph [J]. Report of Academy of Sciences, 1956,108(2):218-221.

[6] ZHOU L W. Electron optics of concentric spherical electromagnetic focusing systems [J]. Advances in Electronics and Electron Physics, 1979, 52:119-132.

11.3 Theory of Temporal Aberrations for Cathode Lenses
阴极透镜的时间像差理论

Abstract: A new approach to the theory of temporal aberration for the cathode lenses is given in the present paper. A definition of temporal aberration is given, in which a certain initial energy of electron emission along the axial direction ε_{z_1} ($0 \leqslant \varepsilon_{z_1} \leqslant \varepsilon_{0\max}$) is considered. A new method to calculate the temporal aberration coefficients of cathode lenses named "Direct Integral Method" is also presented. The "Direct Integral Method" gives new expressions of the temporal aberration coefficients which are expressed in integral forms. The difference between "Direct Integral Method" and "τ-Variation Method" is that the "τ-Variation Method" needs to solve the differential equations for the three of temporal geometrical aberration coefficients of the second order, while the "Direct Integral Method" needs only to carry out the integral calculation for all of these temporal aberration coefficients of the second order. All of the formulae of the temporal aberration coefficients deduced from "Direct Integral Method" and "τ-Variation Method" have been verified by an electrostatic concentric spherical system model, and contrasted with the analytical solutions. Results show that these two methods have got identical solutions and the solutions of temporal aberration coefficients of the first and second orders are the same with the analytical solutions. Although some forms of the results seem different, but they can be transformed into the same form. Thus, it can be concluded that these two methods given by us are equivalent and correct, but the "Direct Integral Method" is related to solving integral formula, which is more convenient for computation and could be suggested to use in the practical design.

摘要：本文给出了研究阴极透镜时间像差理论的新途径。本文将时间像差定义在某一电子发射的轴向初能量为 $\varepsilon_{z_1}(0 \leqslant \varepsilon_{z_1} \leqslant \varepsilon_{0\max})$ 所决定的像面上。本文提出一种名为"直接积分法"计算时间像差系数的新方法。"直接积分法"的时间像差系数的新表示式以积分形式表示。"直接积分法"与"τ-变分法"的差异在于，"τ-变分法"需要求解3个二级时间几何像差系数的微分方程，而"直接积分法"只需要对3个二级时间几何像差系数进行直接的积分运算就可以了。本文对"直接积分法"与"τ-变分法"导得的公式用静电同心球系统的模型进行检验，并用解析解进行了对照。结果表明，这两种方法可获得同样的解，且其一级与二级时间像差系数都能获得解析解。这表明，俄方与中方所提出的方法

Zhou Liwei[a], Li Yuan[a], Zhang Zhiquan[c], M. A. Monastyrski[b], M. Y. Schelev[b]. a) Beijing Institute of Technology, b) Prokhovov General Institute of Physics, Russian Academy of Sciences, c) Institute of Armored Force Engineering, Beijing. Optik V. 116, 2005, 175-184.

The present paper have been published in 26th International Congress on High-Speed Photography and Photonics, 2004, Proceedings of SPIE 5580 (17 March 2005). Title of this paper is On the Theory of Temporal Aberrations for Electron Optical Imaging Systems by using "Direct Integral Method".

是等同的,而且是正确的。但"直接积分法"只需要进行求解积分公式,其计算更为方便,故建议在实际设计中采用。

Introduction

Theory of temporal aberrations which occupies an important place in dynamic electron optics with wide beam focusing and its system design was first studied by M. A. Monastyrski and M. Y. Schelev in 1980, who brought forward the "τ-Variation Method" [1] to calculate the temporal aberrations of the cathode lenses, gave the expressions of temporal aberration coefficients of the first and second orders and verified the correctness of the theory in some ideal models which may get analytical solutions.

It should be mentioned that the transit-time-spread for the cathode lenses, which can be called the temporal aberration of the first order, or the temporal chromatic aberration of the first order was given by Savoisy-Fanchenko in 1956[2], and then proven by I. P. Csorba in 1971[3] and by M. A. Monastyrski and Schelev M. Y in 1980 [1]. It can be expressed as following:

$$\Delta T_1 = \sqrt{\frac{2m_0}{e}} \frac{1}{E_c} \sqrt{\varepsilon_z} \tag{1}$$

It was proven[4] that the spatial-trajectory-spread of cathode lenses is mainly determined by the lateral chromatic aberration of the second order, which can be expressed by Recknagel-Artimovich formula:

$$\Delta r_1^* = \frac{2M}{E_c} \sqrt{\varepsilon_r}(\sqrt{\varepsilon_z} - \sqrt{\varepsilon_{z_1}}) \tag{2}$$

where ε_z, ε_r are the initial axial energy and initial radial energy of electron emitted from the photocathode, respectively; ε_{z_1} is the initial axial energy of electron corresponding to paraxial image position; M is the magnification of system; E_c is the strength of electric-field at the photocathode, E_c takes negative value; e/m_0 is the the ratio of electron charge to mass.

The difference between these two formulae is obviously. We may find the non-harmonious and non-symmetry between them, if we think it over. It may be seen that both of the first order temporal aberration and the second order spatial aberration do not depend on the concrete electrode structure and axial potential distribution. It is strange that the later depends on ε_{z_1} that corresponds to the position of image plane, but the former does not. In fact, regarding to temporal aberration and the spatial aberration, it is actually the same thing about the initial energy spread of electrons emitted from the photocathode expressed as spatial-trajectory-spread characteristics at the image plane and the transit-time spread characteristics at a certain z plane. Formula (2) shows that the spatial aberration should be measured at an image plane corresponding to ε_{z_1}, but the temporal aberration can be measured at any z position (including image plane). As we know, to define the temporal aberration, it should have a criterion to evaluate the time difference between two moving electrons. The non-harmonious between formula (1) and formula (2) can only be explained that the theory of temporal aberrations at present is taken $\varepsilon_{z_1} = 0$ as a criterion to explore the temporal aberrations formed by initial axial energies $\varepsilon_z \neq 0$ of electrons, emitted from the photocathode.

The aim of this paper is to investigate a new approach to the theory of temporal aberrations for the cathode lenses. A new definition of temporal aberrations is given at a certain position z along the axial direction which is determined by $0 \leq \varepsilon_{z_1} \leq \varepsilon_{0\max}$, where ε_{z_1} is a given initial energy of electron emission along the axial direction. A new method to calculate the temporal aberration coefficients of cathode lenses, which is named "Direct Integral Method" is also presented. All of the formulae of the temporal aberration coefficients deduced from "Direct Integral Method" and "τ Variation Method" have been verified by an electrostatic concentric spherical system model, and contrasted them with the analytical solutions.

1 Definition of temporal aberrations

According to our study of electron optics for a concentric spherical system of electrostatic focusing, the temporal aberrations of cathode lenses can be defined as[5]

$$\Delta t = \Delta t(z) = t_{\text{practical}}(z, \varepsilon_r^{1/2}, \varepsilon_z^{1/2}, r_0) - t_{\text{paraxial}}(z, \varepsilon_{r1}^{1/2}, \varepsilon_{z_1}^{1/2}, r_0) \tag{3}$$

where $t_{\text{practical}}$, t_{paraxial} express the time of which the practical ray and paraxial ray of electrons emitted from the photocathode go through the system, respectively; ε_z, ε_r are the initial axial energy and initial radial energy of electrons emitted from the photocathode, respectively; ε_{z_1} is a given initial energy of electron emission along the axial direction; r_0 is the initial height of electron emission. Take the cylindrical coordinate system (z, r), the axial coordinate z is chosen from the photocathode $z_0 = 0$.

In the following, we shall prove that t_{paraxial} does not depend on $\varepsilon_{r1}^{1/2}$ and r_0. Substituting t, t^* for $t_{\text{practical}}$, t_{paraxial} respectively, then formula (3) can be expressed as

$$\Delta t = t(z, \varepsilon_r^{1/2}, \varepsilon_z^{1/2}, r_0) - t^*(z, \varepsilon_{z_1}^{1/2}) \tag{4}$$

We may express formula 4) as following:

$$\Delta t = \Delta T(z, \varepsilon_z^{1/2}, \varepsilon_{z_1}^{1/2}) + \Delta \tau(z, \varepsilon_r^{1/2}, \varepsilon_z^{1/2}, r_0) = \Delta T + \Delta \tau \tag{5}$$

where

$$\Delta T = t^*(z, \varepsilon_z^{1/2}) - t^*(z, \varepsilon_{z_1}^{1/2}) \tag{6}$$

$$\Delta \tau = t(z, \varepsilon_r^{1/2}, \varepsilon_z^{1/2}, r_0) - t^*(z, \varepsilon_z^{1/2}) \tag{7}$$

where ΔT, $\Delta \tau$ can be called temporal chromatic aberrations and temporal geometrical aberrations, respectively.

If the temporal aberration which is introduced by the initial velocity distribution of photoelectrons emitted from the photocathode can be reduced to the study of a function Δt at a certain position z with the relationship of small parameters $\varepsilon_r^{1/2}$, $\varepsilon_z^{1/2}$, $\varepsilon_{z_1}^{1/2}$, r_0, then the temporal aberrations expression (6) and expression (7) can be written as

$$\Delta T = a_2(\varepsilon_z^{1/2} - \varepsilon_{z_1}^{1/2}) + A_{22}(\varepsilon_z - \varepsilon_{z_1}) \tag{8}$$

$$\Delta \tau = a_1 \varepsilon_r^{1/2} + a_3 r_0 + A_{11}\varepsilon_r + 2A_{12}\varepsilon_r^{1/2}\varepsilon_z^{1/2} + 2A_{13}\varepsilon_r^{1/2}r_0 + 2A_{23}\varepsilon_z^{1/2}r_0 + A_{33}r_0^2 \tag{9}$$

where $a_i = a_i(z)$, $i = 1, 2, 3$; $A_{ij} = A_{ij}(z)$; $i, j = 1, 2, 3$ are called the temporal aberration coefficients of the first and second orders for the cathode lenses, respectively.

2 Temporal aberration coefficients of cathode lenses solved by "Direct Integral Method"

As shown in Fig. 1, r_0 is the the initial position vector from the photocathode, θ_0 is the the initial direction angle, α_0 is the the initial emission angle, β_0 is the initial position angle, v_0 is the the initial emission velocity, where

$$v_0 = \sqrt{\frac{2e}{m_0}\varepsilon_0}, \quad \varepsilon_z^{1/2} = \varepsilon_0^{1/2}\cos\alpha_0, \quad \varepsilon_r^{1/2} = \varepsilon_0^{1/2}\sin\alpha_0$$

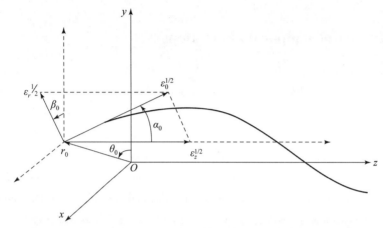

Fig. 1 The initial condition of electron emitted from the photocathode

The electron motion equation in the electrostatic field with axial symmetry can be written as

$$\ddot{z} = \frac{e}{m_0}\frac{\partial\varphi}{\partial z} \tag{10}$$

Using the Scherzer series expansions of spatial potential distribution $\varphi = \varphi(z,r)$:

$$\varphi(z,r) = \phi(z) - \frac{r^2}{4}\phi(z) + \frac{r^4}{64}\phi^{\text{IV}}(z) - \cdots \tag{11}$$

and substituting formula (11) to formula (10), we get

$$\ddot{z} = \frac{e}{m_0}\left[\phi'(z) - \frac{r^2}{4}\phi'''(z)\right] \tag{12}$$

where $\phi(z)$ is the axial potential distribution, $\phi'(z)$, $\phi''(z)$, $\phi'''(z)$, $\phi^{\text{IV}}(z)$ are the derivatives of $\phi(z)$ with respect to z.

Formula (12) can be transformed into the following form:

$$d(\dot{z}^2) = \frac{2e}{m_0}\left[\phi'(z) - \frac{r^2}{4}\phi'''(z)\right]dz$$

Integrate it, we have

$$\dot{z}^2 = \frac{2e}{m_0}\int\phi'(z)\,dz - \frac{2e}{m_0}\int\frac{r^2}{4}\phi'''(z)\,dz + \frac{2e}{m_0}\varepsilon_z \tag{13}$$

In order to solve the second integration in formula (13), we shall make the following transformation.

The paraxial electron motion equation in the cathode lenses with axial symmetry has the following form:

$$[\phi(z) + \varepsilon_z]r'' + \frac{1}{2}\phi'(z)r' + \frac{1}{4}\phi''(z)r = 0 \qquad (14)$$

Its general solution can be written as

$$r(z) = r_0 w(z) + \sqrt{\frac{m_0}{2e}}\dot{r}_0 v(z) \qquad (15)$$

where the two special solutions $v = v(z)$, $w = w(z)$ satisfy the following initial conditions:

$$v(z_0 = 0) = 0, \quad v'(z_0 = 0) = \frac{1}{\sqrt{\varepsilon_z}},$$
$$w(z_0 = 0) = 1, \quad w'(z_0 = 0) = 0 \qquad (16)$$

Therefore, we have

$$r^2(z) = r_0^2 w^2(z) + 2w(z)v(z)r_0\sqrt{\varepsilon_r}\cos(\theta_0 - \beta_0) + \varepsilon_r v^2(z) \qquad (17)$$

Substituting expression (17) to the second term at the right hand of formula (13), we get

$$\int \frac{r^2}{4}\phi'''(z)\mathrm{d}z = \int \frac{1}{4}r_0^2 w^2(z)\phi'''(z)\mathrm{d}z +$$
$$\int \frac{1}{2}r_0\sqrt{\varepsilon_r}\cos(\theta_0 - \beta_0)w(z)v(z)\phi'''(z)\mathrm{d}z + \int \frac{1}{4}\varepsilon_r v^2(z)\phi'''(z)\mathrm{d}z \qquad (18)$$

Now, we calculate the second term at the right hand of formula (18). From

$$\frac{\mathrm{d}}{\mathrm{d}z}\left(v'w'\phi_* + \frac{1}{4}\phi''vw\right) = \phi'v'w' + \phi_*v''w' + \phi_*v'w'' + \frac{1}{4}\phi'''vw + \frac{1}{4}\phi''v'w + \frac{1}{4}\phi''vw'$$

and using the following paraxial ray equations

$$\phi_* v'' = -\frac{1}{2}\phi'v' - \frac{1}{4}\phi''v, \text{ and } \phi_* w'' = -\frac{1}{2}\phi'w' - \frac{1}{4}\phi''w$$

we obtain

$$\frac{\mathrm{d}}{\mathrm{d}z}\left(v'w'\phi_* + \frac{1}{4}\phi''vw\right) = \frac{1}{4}\phi'''vw \qquad (19)$$

where $\phi_* = \phi(z) + \varepsilon_z$.

Therefore, integrating formula (19), we have

$$\int \frac{1}{4}w(z)v(z)\phi'''(z)\mathrm{d}z = \left(v'w'\phi_* + \frac{1}{4}\phi''vw\right) + C_1$$

where the constant $C_1 = 0$ from the initial conditions (16).

The second term at the right hand of formula (18) can be written as

$$\int \frac{1}{2}r_0\sqrt{\varepsilon_r}\cos(\theta_0 - \beta_0)w(z)v(z)\phi'''(z)\mathrm{d}z = 2r_0\sqrt{\varepsilon_r}\cos(\theta_0 - \beta_0)\left(v'w'\phi_* + \frac{1}{4}\phi''vw\right) \qquad (20)$$

where θ_0 is the initial direction angle, β_0 is the initial position angle, as shown in Fig. 1.

Now, we calculate the first term at the right hand of formula (18). Similar to the deduction of formula (19), we may have

$$\frac{\mathrm{d}}{\mathrm{d}z}\left(w'^2\phi_* + \frac{1}{4}\phi''w^2\right) = \frac{1}{4}\phi'''w^2 \tag{21}$$

Integrate formula (21), we have

$$\int \frac{1}{4}w^2(z)\phi'''(z)\,\mathrm{d}z = \left(w'^2\phi_* + \frac{1}{4}\phi''w^2\right) + C_2$$

where the constant $C_2 = -\frac{1}{4}\phi_0''$ from the initial conditions (16).

The first term at the right hand of formula (18) can be written as

$$\int \frac{1}{4}r_0^2 w^2(z)\phi'''(z)\,\mathrm{d}z = r_0^2\left(w'^2\phi + \frac{1}{4}\phi''w^2 - \frac{1}{4}\phi_0''\right) \tag{22}$$

Similarly, for the third term at the right hand of formula (18), we obtain

$$\frac{\mathrm{d}}{\mathrm{d}z}\left(v'^2\phi_* + \frac{1}{4}\phi''v^2\right) = \frac{1}{4}\phi'''v^2 \tag{23}$$

Integrate formula (23), we have

$$\int \frac{1}{4}v^2(z)\phi'''(z)\,\mathrm{d}z = \left(v'^2\phi_* + \frac{1}{4}\phi''v^2\right) + C_3$$

where the constant $C_3 = -1$ from the initial conditions (16).

The third term at the right hand of formula (18) can be written as

$$\int \frac{1}{4}\varepsilon_r v^2(z)\phi'''(z)\,\mathrm{d}z = \varepsilon_r\left(v'^2\phi_* + \frac{1}{4}\phi''v^2 - 1\right) \tag{24}$$

Substituting expression (20), expression (22) and expression (24) to the formula (13), we get

$$\dot{z}^2 = \frac{2e}{m_0}[\phi(z) + \varepsilon_z] - \frac{2e}{m_0}\Big[\varepsilon_r\left(v'^2\phi_* + \frac{1}{4}\phi''v^2 - 1\right) + r_0^2\left(w'^2\phi + \frac{1}{4}\phi''w^2 - \frac{1}{4}\phi_0''\right) + 2r_0\sqrt{\varepsilon_r}\cos(\theta_0 - \beta_0)\left(v'w'\phi_* + \frac{1}{4}\phi''vw\right)\Big] \tag{25}$$

At last, from $\mathrm{d}t = \frac{\mathrm{d}z}{\dot{z}}$ we may obtain the time expression of an electron which reaches a certain position z in the system:

$$t(z,\varepsilon_r^{1/2},\varepsilon_z^{1/2},r_0) = \int_0^z \frac{\mathrm{d}z}{\left(\frac{2e}{m_0}\right)^{1/2}[\phi(z)+\varepsilon_z]^{1/2}}\Big\{1 + \frac{1}{2}\frac{1}{[\phi(z)+\varepsilon_z]}\Big[\varepsilon_r\left(v'^2\phi_* + \frac{1}{4}\phi''v^2 - 1\right) + r_0^2\left(w'^2\phi_* + \frac{1}{4}\phi''w^2 - \frac{1}{4}\phi_0''\right) + 2r_0\sqrt{\varepsilon_r}\cos(\theta_0 - \beta_0)\left(v'w'\phi_* + \frac{1}{4}\phi''vw\right)\Big]\Big\} \tag{26}$$

For the paraxial electrons, if neglect the term r^2 in formula (13), we have

$$\dot{z}^2 = \frac{2e}{m_0}\int \phi'(z)\,\mathrm{d}z + \frac{2e}{m_0}\varepsilon_z \tag{27}$$

Therefore,

$$t^*(z,\varepsilon_z^{1/2}) = \int_0^z \frac{\mathrm{d}z}{\left(\frac{2e}{m_0}\right)^{1/2}[\phi(z)+\varepsilon_z]^{1/2}} \tag{28}$$

Formula (28) can also be obtained from formula (26) when neglect the quadric term of $\varepsilon_r^{1/2}$ and r_0.

From the definition of formula (6) and from formula (28), we have

$$\Delta T = a_2(\varepsilon_z^{1/2} - \varepsilon_{z_1}^{1/2}) + A_{22}(\varepsilon_z - \varepsilon_{z_1}) + 0(\varepsilon_z^{3/2} - \varepsilon_{z_1}^{3/2})$$
$$= \int_0^z \frac{dz}{\left(\frac{2e}{m_0}\right)^{1/2}[\phi(z) + \varepsilon_z]^{1/2}} - \int_0^z \frac{dz}{\left(\frac{2e}{m_0}\right)^{1/2}[\phi(z) + \varepsilon_{z_1}]^{1/2}} \quad (29)$$

where a_2 is the temporal aberration coefficient of the first order, which may be called as temporal chromatic aberration coefficient of the first order, A_{22} is the temporal chromatic aberration coefficient of the second order, they are the coefficients of $\sqrt{\varepsilon_z}$ and ε_z of power series expansion of formula (28), respectively.

From the definition of formula (7), comparing formula (9) to formula (26), we have

$$a_1 = a_3 = 0, \quad A_{12} = A_{23} = 0$$

Therefore,

$$\Delta \tau = A_{11}\varepsilon_r + 2A_{13}\varepsilon_r^{1/2}r_0 + A_{33}r_0^2 \quad (30)$$

where

$$A_{11} = \int_0^z \frac{1}{2\left(\frac{2e}{m_0}\right)^{1/2}[\phi(z) + \varepsilon_z]^{3/2}}\left(v'^2\phi_* + \frac{1}{4}\phi''v^2 - 1\right)dz \quad (31)$$

$$A_{13} = \int_0^z \frac{1}{2\left(\frac{2e}{m_0}\right)^{1/2}[\phi(z) + \varepsilon_z]^{3/2}}\left(v'w'\phi_* + \frac{1}{4}\phi''vw\right)dz \quad (32)$$

$$A_{33} = \int_0^z \frac{1}{2\left(\frac{2e}{m_0}\right)^{1/2}[\phi(z) + \varepsilon_z]^{3/2}}\left(w'^2\phi_* + \frac{1}{4}\phi''w^2 - \frac{1}{4}\phi_0''\right)dz \quad (33)$$

where A_{11} is the temporal spherical aberration coefficient of the second order; A_{13} is the temporal aberration coefficient of field of curvature of the second order; A_{33} is the temporal distortion aberration coefficient of the second order.

3 Determination of temporal chromatic aberration coefficients a_2 and A_{22}

Formula (29) only gives an approach to determine temporal chromatic aberration coefficients a_2 and A_{22}, but their concrete expressions have not been given. Now, we suggest two ways to solve the temporal chromatic aberration coefficients a_2 and A_{22}.

The first way for solving the temporal chromatic aberration coefficients is by using Taylor series expansion. From expression (28) we know that the temporal chromatic aberration coefficients a_2 and A_{22} only depend on axial potential distribution $\phi(z)$, which can be expressed by Taylor series expansion as following:

$$\phi(z) = \sum_1^m \frac{E_m}{m!} z^m \quad (34)$$

Substituting the Taylor series expansion (34) when $m = 1$ to expression (28), and integrating it, we have

$$t^*(z, \varepsilon_z^{1/2}) = \sqrt{\frac{2m_0}{e}} \frac{1}{E_1}(\sqrt{E_1 z + \varepsilon_z} - \sqrt{\varepsilon_z}) \qquad (35)$$

Expanding the expression (35) according to a power of series of $\varepsilon_z^{1/2}$, we can get

$$t^*(z, \varepsilon_z^{1/2}) = \sqrt{\frac{2m_0}{e}} \frac{1}{E_1}\left[\sqrt{E_1 z} - \sqrt{\varepsilon_z} + \frac{1}{2}\frac{1}{\sqrt{E_1 z}}\varepsilon_z + 0(\varepsilon_z)^{3/2}\right] \qquad (36)$$

From the definition of coefficients a_2, A_{22}, we may obtain

$$a_2 = -\sqrt{\frac{2m_0}{e}}\frac{1}{E_1} = \sqrt{\frac{2m_0}{e}}\frac{1}{E_c} \qquad (37)$$

$$A_{22}|_{m=1} = \sqrt{\frac{2m_0}{e}}\frac{1}{2E_1\sqrt{E_1 z}} \qquad (38)$$

We shall prove that the expression (37) of coefficient a_2 is valid in any circumstances.

Similarly, if we take $m = 2$ in the Taylor series expansion (34), we may also obtain the accurate analytical solution of a_2, A_{22}, where a_2 is also expressed by formula (37), and A_{22} can be expressed by

$$A_{22}|_{m=2} = \sqrt{\frac{2m_0}{e}}\frac{E_1 + E_2 z}{E_1^2 \sqrt{2z(2E_1 + E_2 z)}} \qquad (39)$$

Similarly, we may get the approximate analytical solutions of A_{22} when $m = 3, 4$:

$$A_{22}|_{m=3} \approx \sqrt{\frac{2m_0}{e}}\left\{\frac{1}{E_1^2\sqrt{2z(2E_1 + E_2 z)}}\left[E_1 + zE_2 + \frac{z^2 E_1 E_3}{6(2E_1 + zE_2)}\right]\right\}$$

$$A_{22}|_{m=4} \approx \sqrt{\frac{2m_0}{e}}\left\{\frac{1}{E_1^2\sqrt{2z(2E_1+E_2 z)}}\left[E_1 + zE_2 + \frac{z^2 E_1 E_3}{6(2E_1 + zE_2)}\right] + \frac{E_4}{24\sqrt{2}E_2^{5/2}}\left[3\ln\frac{E_1 + E_2 z + \sqrt{E_2}\sqrt{2E_1 z + E_2 z^2}}{E_1} - \frac{4z^2\sqrt{E_2}(3E_1 + 2E_2 z)}{(2E_1 z + E_2 z^2)^{3/2}}\right]\right\}$$

The second way for solving the temporal chromatic aberration coefficients is by using integral forms. Expanding the expression (28) according to a power of series of $\varepsilon_z^{1/2}$, we can get

$$\frac{1}{2}\sqrt{\frac{2m_0}{e}}\int_0^z \frac{dz}{[\phi(z) + \varepsilon_z]^{1/2}} = \text{Constant term} + a_2 \varepsilon_z^{1/2} + A_{22}\varepsilon_z + 0(\varepsilon_z^{3/2}) \qquad (40)$$

Because

$$[\sqrt{\phi(z) + \varepsilon_z}]' = \frac{\phi'(z)}{2\sqrt{\phi(z) + \varepsilon_z}}$$

we may express Eq. (40) as following:

$$\int_0^z \frac{dz}{[\phi(z) + \varepsilon_z]^{\frac{1}{2}}} = \int_0^z \frac{2[\sqrt{\phi(z)+\varepsilon_z}]'}{\phi'(z)} = 2\frac{\sqrt{\phi(z)+\varepsilon_z}}{\phi'(z)}\bigg|_0^z + 2\int_0^z \frac{\phi''(z)\sqrt{\phi(z)+\varepsilon_z}}{[\phi'(z)]^2}dz$$

$$= 2\frac{\sqrt{\phi(z)}}{\phi'(z)} + 2\int_0^z \frac{\phi''(z)\sqrt{\phi(z)}}{[\phi'(z)]^2}dz - 2\frac{\sqrt{\varepsilon_z}}{\phi'(0)} + \left\{\frac{1}{\phi'(z)\sqrt{\phi(z)}} + \int_0^z \frac{\phi''(z)}{[\phi'(z)]^2\sqrt{\phi(z)}}dz\right\}\varepsilon_z$$

Comparing it to Eq. (34), we obtain

$$\text{Constant term} = \sqrt{\frac{2m_0}{e}} \left\{ \frac{\sqrt{\phi(z)}}{\phi'(z)} + \int_0^z \frac{\phi''(z)}{[\phi'(z)]^2} \sqrt{\phi(z)} dz \right\}$$

$$a_2 = -\sqrt{\frac{2m_0}{e}} \frac{1}{\phi'(0)} = \sqrt{\frac{2m_0}{e}} \frac{1}{E_c} \quad (41)$$

$$A_{22} = \frac{1}{2} \sqrt{\frac{2m_0}{e}} \left\{ \frac{1}{\sqrt{\phi(z)}} \frac{1}{\phi'(z)} + \int_0^z \frac{\phi''(z)}{\sqrt{\phi(z)} [\phi'(z)]^2} dz \right\} \quad (42)$$

Expression (41), expression (42) of the temporal chromatic aberration coefficients a_2 and A_{22}, have been derived by using "τ Variation Method[1]."

To find simpler expression for solving A_{22}, we make the following transformation for the expression (42):

$$\begin{aligned} A_{22} &= \frac{1}{2} \sqrt{\frac{2m_0}{e}} \left\{ \frac{1}{\sqrt{\phi(z)}} \frac{1}{\phi'(z)} - \int_0^z \left\{ \frac{1}{\sqrt{\phi(z)}} \frac{\mathrm{d}}{\mathrm{d}z} \left[\frac{1}{\phi'(z)} \right] \right\} \mathrm{d}z \right\} \\ &= \frac{1}{2} \sqrt{\frac{2m_0}{e}} \left\{ \frac{1}{\sqrt{\phi(z)}} \frac{1}{\phi'(z)} - \frac{1}{\sqrt{\phi(z)} \phi'(z)} + \frac{1}{\sqrt{\phi(0)} \phi'(0)} + \int_0^z \frac{1}{\phi'(z)} \mathrm{d} \left[\frac{1}{\sqrt{\phi(z)}} \right] \right\} \\ &= \frac{1}{2} \sqrt{\frac{2m_0}{e}} \left\{ \frac{1}{\sqrt{\phi(0)} \phi'(0)} + \int_0^z \frac{1}{\phi'(z)} \mathrm{d} \left[\frac{1}{\sqrt{\phi(z)}} \right] \right\} \end{aligned}$$

(43)

In the above-mentioned formula, the first term at the right hand tends to infinitive, but it could be canceled out each other with the integral value of the second term at the right hand when $z = 0$. So the integral of formula (43) will have a definite value, therefore we may write (43) as following:

$$A_{22} = -\frac{1}{4} \sqrt{\frac{2m_0}{e}} \int \frac{1}{\phi(z)^{3/2}} \mathrm{d}z \quad (44)$$

Substituting the Taylor series expansion (34) when $m = 1, 2$ to the expression (42), expression (44) and integrating it, the results obtained are identical with expression (38), expression (39). It shows that the two methods "Direct Integral Method" and "τ Variation Method" are equivalent. We shall further test and verify expression (42) and expression (44) by an electrostatic concentric spherical system model.

4 Verification of temporal chromatic aberration coefficients by an electrostatic concentric spherical system model

4.1 Analytical solutions of the temporal chromatic aberration coefficients a_2 and A_{22} in an electrostatic concentric spherical system model[5]

Fig. 2 shows an electrostatic concentric spherical electrostatic system model. Suppose that the curvature radii of spherical cathode C and of the mesh-spherical anode A are R_c and R_a, the common curvature center is O. Because of spherical symmetry of the system, we can use the polar coordinate (ρ, φ) to describe the electron trajectory. Let the original point of the polar coordinate is located at the center O of curvature, the potential of the photocathode is $\phi_c = 0$, the potential of mesh-anode A with respect to the photocathode C is ϕ_{ac}, the photoelectron with initial emission velocity v_0 and initial emission angle α_0 is emitted from the photocathode. The expression of ϕ_{pc}, the potential at polar coordinate ρ with respect to photocathode, is known as:

$$\phi_{pc} = E_c R_c \left(\frac{R_c}{\rho} - 1 \right) \qquad (45)$$

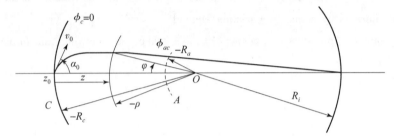

Fig. 2 Electron motion in an electrostatic concentric spherical system

where E_c is the electric-field strength at the photocathode. When $\rho = R_a$, we have

$$E_c = \frac{\phi_{ac}}{R_c(n-1)} \qquad (46)$$

where $n = R_c/R_a$. For the concave cathode-anode system, R_c, R_a take negative values.

The motion of electron path follows the law of conversation of energy and the law of conversation of angular momentum:

$$\frac{1}{2} m_0 v^2 = \frac{1}{2} m_0 v_0^2 + e\phi_{pc},$$

$$\rho^2 \dot\varphi = R_c^2 \dot\varphi_0 = R_c v_0 \sin\alpha_0$$

Obviously, the electron velocity can be described as:

$$v^2 = \dot\rho^2 + (\rho\dot\varphi)^2$$

Therefore,

$$\dot\rho = \frac{d\rho}{dt} = \sqrt{v^2 - R_c^2 v_0^2 \sin^2\alpha_0 / \rho^2} \qquad (47)$$

Substituting

$$v^2 = v_0^2 + \frac{2e}{m_0} \frac{\phi_{ac}}{(n-1)} \left(\frac{R_c}{\rho} - 1 \right)$$

to the expression (47), we have

$$\dot\rho = \frac{d\rho}{dt} = \sqrt{v_0^2 + \frac{2e}{m_0} \frac{\phi_{ac}}{n-1} \left(\frac{R_c}{\rho} - 1 \right) - \frac{R_c^2 v_0^2 \sin^2\alpha_0}{\rho^2}} \qquad (48)$$

From $\varepsilon_0 = \frac{1}{2} \frac{m_0}{e} v_0^2$ and from expression (48), we can get the differential expression for time t:

$$dt = \sqrt{\frac{m_0}{2e}} \frac{-\rho d\rho}{\sqrt{\left(\frac{\phi_{ac}}{n-1} - \varepsilon_0\right)}\sqrt{-\rho^2 + \frac{\phi_{ac}}{(n-1)\left(\frac{\phi_{ac}}{n-1} - \varepsilon_0\right)}R_c\rho - \frac{\varepsilon_0 R_c^2 \sin^2\alpha_0}{\left(\frac{\phi_{ac}}{n-1} - \varepsilon_0\right)}}} \quad (49)$$

Integrating Eq. (49) and considering boundary condition when $\rho = R_c$, $t = 0$, we obtain the analytical solution of time t, in which the electron emitted from the photocathode goes through:

$$t = \sqrt{\frac{m_0}{2e}} \frac{1}{\left(\frac{\phi_{ac}}{n-1} - \varepsilon_0\right)} \left\{ \sqrt{-\rho^2\left(\frac{\phi_{ac}}{n-1} - \varepsilon_0\right) + \frac{\phi_{ac}}{n-1}R_c\rho - \varepsilon_0 R_c^2 \sin^2\alpha_0} \right.$$

$$-\frac{1}{2} \frac{\frac{\phi_{ac}}{n-1}}{\sqrt{\left(\frac{\phi_{ac}}{n-1} - \varepsilon_0\right)}} R_c \arctan \frac{-\frac{\phi_{ac}}{n-1}R_c + 2\rho\left(\frac{\phi_{ac}}{n-1} - \varepsilon_0\right)}{2\sqrt{\frac{\phi_{ac}}{n-1} - \varepsilon_0}\sqrt{-\rho^2\left(\frac{\phi_{ac}}{n-1} - \varepsilon_0\right) + \frac{\phi_{ac}}{n-1}R_c\rho - \varepsilon_0 R_c^2 \sin^2\alpha_0}} +$$

$$\left. R_c\sqrt{\varepsilon_0\cos^2\alpha_0} + \frac{1}{2}\frac{\frac{\phi_{ac}}{(n-1)}}{\sqrt{\left(\frac{\phi_{ac}}{n-1} - \varepsilon_0\right)}}R_c \arctan \frac{2\varepsilon_0 - \frac{\phi_{ac}}{n-1}}{2\sqrt{\frac{\phi_{ac}}{n-1} - \varepsilon_0}\sqrt{\varepsilon_0\cos^2\alpha_0}} \right\} \quad (50)$$

If we consider the paraxial condition of electron emission, and neglect the quadric terms of $\varepsilon_r^{1/2}$ and r_0, then $\varepsilon_0 = \varepsilon_z$, $\rho = R_c + z$. Expression (50) can be transformed to

$$t^*(z, \varepsilon_z^{\frac{1}{2}}) = \sqrt{\frac{m_0}{2e}} \frac{1}{\left(\frac{\phi_{ac}}{n-1} - \varepsilon_z\right)} \left\{ \sqrt{-(R_c+z)^2\left(\frac{\phi_{ac}}{n-1} - \varepsilon_z\right) + \frac{\phi_{ac}}{n-1}R_c(R_c+z)} - \right.$$

$$\frac{1}{2}\frac{\frac{\phi_{ac}}{n-1}}{\sqrt{\left(\frac{\phi_{ac}}{n-1} - \varepsilon_z\right)}}R_c\arctan\frac{-\frac{\phi_{ac}}{n-1}R_c + 2(R_c+z)\left(\frac{\phi_{ac}}{n-1} - \varepsilon_z\right)}{2\sqrt{\frac{\phi_{ac}}{n-1} - \varepsilon_z}\sqrt{-(R_c+z)^2\left(\frac{\phi_{ac}}{n-1} - \varepsilon_z\right) + \frac{\phi_{ac}}{n-1}R_c(R_c+z)}} +$$

$$\left. R_c\sqrt{\varepsilon_z} + \frac{1}{2}\frac{\frac{\phi_{ac}}{n-1}}{\sqrt{\left(\frac{\phi_{ac}}{n-1} - \varepsilon_z\right)}}R_c\arctan\frac{2\varepsilon_z - \frac{\phi_{ac}}{n-1}}{2\sqrt{\frac{\phi_{ac}}{n-1} - \varepsilon_z}\sqrt{\varepsilon_z}} \right\} \quad (51)$$

Expanding the expression (51) of $t^*(z, \varepsilon_z^{1/2})$ according to a power of series of $\varepsilon_z^{1/2}$, we can get the expressions for temporal chromatic aberration coefficients of the first order a_2 and of the second order A_{22}:

$$a_2 = R_c\sqrt{\frac{m_0}{2e}\frac{2(n-1)}{\phi_{ac}}} = \sqrt{\frac{2m_0}{e}}\frac{1}{E_c} \quad (52)$$

$$A_{22} = -\sqrt{\frac{2m_0}{e}}\frac{1}{E_c}\left(\frac{n-1}{\phi_{ac}}\right)^{1/2}\left[\frac{(-2R_c^2 - R_c z + z^2)}{4R_c\sqrt{-z(R_c+z)}} - \frac{3}{8}\arctan\frac{-(R_c+2z)}{2\sqrt{-z(R_c-z)}} + \frac{3}{16}\pi\right] \quad (53)$$

4.2 Verification of the temporal chromatic aberration coefficients a_2 and A_{22} by "Direct Integral Method"

From formula (45), the expression of axial potential distribution $\phi(z)$ in the electrostatic concentric spherical electrostatic system model can be written as

$$\phi(z) = \frac{\phi_{ac}}{n-1} \frac{-z}{z + R_c} \tag{54}$$

Substituting formula (54) to the formula (28) and integrating it, we may obtain the analytical solution of time $t^*(z)$ at any position of z along the axial direction, in which the electron emitted from the photocathode goes through: at the initial condition ($t = 0$, $z_0 = 0$, $r_0 = 0$) and at the paraxial ray condition:

$$t^*(z, \varepsilon_z^{1/2}) = \frac{-R_c}{\left[\frac{2e}{m_0}\left[\frac{\phi_{ac} - (n-1)\varepsilon_z}{n-1}\right]\right]^{1/2}} \left\{ \sqrt{1 + \frac{z}{R_c}} \sqrt{\frac{(n-1)\varepsilon_z}{\phi_{ac} - (n-1)\varepsilon_z} - \frac{z}{R_c}} - \sqrt{\frac{(n-1)\varepsilon_z}{\phi_{ac} - (n-1)\varepsilon_z}} + \left(1 + \frac{(n-1)\varepsilon_z}{\phi_{ac} - (n-1)\varepsilon_z}\right)\left[\arcsin\sqrt{\frac{(n-1)\varepsilon_z}{\phi_{ac}} - \frac{[\phi_{ac} - (n-1)\varepsilon_z]z}{R_c \phi_{ac}}} - \arcsin\sqrt{\frac{(n-1)\varepsilon_z}{\phi_{ac}}}\right] \right\} \tag{55}$$

Expanding formula (55), and expressing it as following form:

$$t^*(z, \varepsilon_z^{1/2}) = t^*(z, 0) + a_2 \varepsilon_z^{1/2} + A_{22} \varepsilon_z + 0(\varepsilon_z)^{3/2}$$

then, we have

$$t^*(z, 0) = \left(\frac{2m_0}{e}\right)^{1/2} \frac{-R_c}{\sqrt{\phi_{ac}}} \frac{\sqrt{n-1}}{2} \sqrt{\frac{z}{-R_c}\left(1 + \frac{z}{R_c}\right)} - R_c \frac{1}{2} \sqrt{\frac{2m_0}{e}} \sqrt{\frac{n-1}{\phi_{ac}}} \arcsin\sqrt{\frac{-z}{R_c}},$$

$$a_2 = \sqrt{\frac{2m_0}{e}} \frac{1}{E_c},$$

$$A_{22} = -\sqrt{\frac{2m_0}{e}} \frac{1}{E_c} \left(\frac{n-1}{\phi_{ac}}\right)^{1/2} \left\{ \frac{-(R_c + z)(2R_c - z)}{4R_c \sqrt{-z(R_c + z)}} + \frac{3}{4}\left(\arcsin\sqrt{\frac{-z}{R_c}}\right) \right\} \tag{56}$$

Formula (56) can be transformed into the form of formula (53). Thus, we have proven that the approach of direct integral method is correct.

5 Comparison between "direct integral method" and "τ-Variation Method" for solving temporal aberration coefficients

5.1 Direct integral method

In the following, we shall use the electrostatic concentric spherical electrostatic system model to find the temporal aberration coefficients of the second order by using the "Direct Integral Method" as before and make a comparison with "τ-Variation Vethod."

For an electrostatic concentric spherical electrostatic system model, the two special solutions of paraxial ray equation are as following[5]:

$$v(z) = \frac{2z}{\phi(z)}[\sqrt{\phi(z) + \varepsilon_z} - \sqrt{\varepsilon_z}] \tag{57}$$

$$w(z) = 1 + \frac{1}{R_c}z - \frac{\sqrt{\varepsilon_z}}{R_c}\frac{2z}{\phi(z)}[\sqrt{\phi(z) + \varepsilon_z} - \sqrt{\varepsilon_z}] \tag{58}$$

Substituting formula (54), formula (57), formula (58) of $\phi(z)$, v, w to the formula (31), formula (32), formula (33) and formula (44) of temporal aberration coefficients of the second order A_{11}, A_{13}, A_{33} and A_{22}, integrating it, we may obtain

$$A_{11} = \frac{3}{4}\sqrt{\frac{2m_0}{e}}\left(\frac{n-1}{\phi_{ac}}\right)^{1/2}\frac{1}{E_c R_c}\left[-\sqrt{z}\sqrt{-z-R_c} - R_c \arctan \frac{\sqrt{z}}{\sqrt{-z-R_c}}\right] \tag{59}$$

$$A_{13} = -\cos(\theta_0 - \beta_0)\frac{1}{2}\sqrt{\frac{2m_0}{e}}\frac{1}{R_c^2 E_c}z \tag{60}$$

$$A_{33} = \sqrt{\frac{2m_0}{e}}\sqrt{\frac{n-1}{\phi_{ac}}}\frac{3}{8R_c^2}\left(\sqrt{z}\sqrt{-z-R_c} - R_c \arctan \frac{\sqrt{z}}{\sqrt{-z-R_c}}\right) \tag{61}$$

$$A_{22} = \frac{1}{4}\sqrt{\frac{2m_0}{e}}\left(\frac{n-1}{\phi_{ac}}\right)^{3/2}\left[\frac{-(R_c+z)(-2R_c+z)}{\sqrt{z}\sqrt{-(R_c+z)}} - 3R_c \arctan\left(\frac{\sqrt{z}}{\sqrt{-(R_c+z)}}\right)\right] \tag{62}$$

Formula (62) can be transformed into the form of formula (56) and formula (53).

5.2 τ-Variation Method

According to "τ-Variation Method[1]," expressions for solving the temporal aberration coefficients can be described as below.

For the temporal aberration coefficients of the first order,

$$a_1 = a_3 = 0, \quad a_2 = \sqrt{2m_0/e}(1/E_c),$$

for the temporal aberration coefficients of the second order,

$$A_{12} = A_{23} = 0,$$

and A_{11}, A_{13}, A_{33} and A_{22} can be solved by the following operations:

$$A_{11} = -\frac{z_{\alpha_1 \alpha_1}}{2\dot{z}} \tag{63}$$

where $z_{\alpha_1 \alpha_1}$ is a solution of the differential equation of

$$T^{(z)}(z_{\alpha_1 \alpha_1}) = -\frac{1}{4}\phi''' v^2 \tag{64}$$

$$A_{13} = -\frac{z_{\alpha_1 \alpha_3}}{2\dot{z}} \tag{65}$$

where $z_{\alpha_1 \alpha_3}$ is a solution of the differential equation

$$T^{(z)}(z_{\alpha_1 \alpha_3}) = -\frac{1}{4}\phi''' vw \cos(\beta_0 - \theta_0) \tag{66}$$

$$A_{33} = -\frac{z_{\alpha_3 \alpha_3}}{2\dot{z}} \tag{67}$$

where $z_{\alpha_3 \alpha_3}$ is a solution of the differential equation

$$T^{(z)}(z_{\alpha_3\alpha_3}) = -\frac{1}{4}\phi''' w^2 \tag{68}$$

$$A_{22} = -\frac{\ddot{z}}{2\dot{z}^3}z_{\alpha_2}^2 + \frac{\dot{z}_{\alpha_2}z_{\alpha_2}}{\dot{z}^2} - \frac{z_{\alpha_2\alpha_2}}{2\dot{z}} \tag{69}$$

where $z_{\alpha_2\alpha_2}$ is a solution of the differential equation

$$T^{(z)}(z_{\alpha_2\alpha_2}) = \phi\phi''' \frac{2}{\phi_0'^2} \tag{70}$$

The differential operator $T^{(z)}$ can be defined as

$$T^{(z)}(z_{\alpha_i\alpha_j}) = \frac{1}{2}\phi''' \left[z_{\alpha_i}z_{\alpha_j} - \frac{1}{4}(r_{\alpha_i}r_{\alpha_j}^* + r_{\alpha_i}^* r_{\alpha_j}) \right] \tag{71}$$

In the above formulae, a vector $\boldsymbol{\alpha}$ is introduced, its three components will represent

$$\alpha_1 = \varepsilon_r^{1/2}, \quad \alpha_2 = \varepsilon_z^{1/2}, \quad \alpha_3 = r_0 \tag{72}$$

and $z_{\alpha_i} = \frac{\partial z}{\partial \alpha_i}$, $i = 1, 2, 3$.

Firstly, we calculate the coefficient A_{11}. Integrating the differential Eq. (64):

$$T^{(z)}(z_{\alpha_1\alpha_1}) = \frac{\mathrm{d}}{\mathrm{d}z}\left(\phi \frac{\mathrm{d}}{\mathrm{d}z}z_{\alpha_1\alpha_1} - \frac{1}{2}\phi' z_{\alpha_1\alpha_1} \right) = -\frac{1}{4}\phi''' v^2 \tag{73}$$

we have

$$\phi z'_{\alpha_1\alpha_1} - \frac{1}{2}\phi' z_{\alpha_1\alpha_1} = -\left(v'^2 \phi + \frac{1}{4}\phi'' v^2 \right) + 1 \tag{74}$$

Substituting the axial potential distribution $\phi(z)$ of formula (54) and two special solutions v and w of formula (57), formula (58) of an electrostatic concentric spherical electrostatic system to the formula (74), we obtain

$$z'_{\alpha_1\alpha_1} - \frac{1}{2}\frac{R_c}{z(z+R_c)}z_{\alpha_1\alpha_1} = \frac{n-1}{\phi_{ac}}\frac{3z}{(z+R_c)} \tag{75}$$

Thus, $z_{\alpha_1\alpha_1}$ can be solved. From Eq. (63), A_{11} can be expressed by

$$A_{11} = -\frac{3}{4}\sqrt{\frac{2m_0}{e}}\sqrt{\frac{n-1}{\phi_{ac}}}\frac{1}{E_c R_c}\left(R_c \arctan\frac{\sqrt{z}}{\sqrt{-z-R_c}} + \sqrt{z}\sqrt{-z-R_c} \right) \tag{76}$$

Regarding to coefficient A_{13}, integrating the differential Eq. (66):

$$T^{(z)}(z_{\alpha_1\alpha_3}) = \frac{\mathrm{d}}{\mathrm{d}z}\left(\phi \frac{\mathrm{d}}{\mathrm{d}z}z_{\alpha_1\alpha_3} - \frac{1}{2}\phi' z_{\alpha_1\alpha_3} \right) = -\frac{1}{4}\phi''' wv\cos(\beta_0 - \theta_0) \tag{77}$$

we have

$$\phi z'_{\alpha_1\alpha_3} - \frac{1}{2}\phi' z_{\alpha_1\alpha_3} = -\cos(\beta_0 - \theta_0)\left(v'w'\phi + \frac{1}{4}\phi'' vw \right) \tag{78}$$

Similarly, substituting expressions $\phi(z)$, v, w of formula (54), formula (57), formula (58) to the formula (78), we obtain

$$z'_{\alpha_1\alpha_3} + \frac{1}{2}\frac{R_c}{z(-z-R_c)}z_{\alpha_1\alpha_3} = \cos(\beta_0 - \theta_0)\sqrt{\frac{n-1}{\phi_{ac}}}\frac{2}{R_c}\frac{z^{\frac{1}{2}}}{(-z-R_c)^{\frac{1}{2}}} \tag{79}$$

Thus, $z_{\alpha_1\alpha_3}$ can be solved. From Eq. (65), A_{13} can be expressed by:

$$A_{13} = -\cos(\beta_0 - \theta_0)\frac{1}{2}\sqrt{\frac{2m_0}{e}}\frac{1}{R_c^2 E_c}z \tag{80}$$

Regarding to coefficient A_{33}, integrating the differential Eq. (68):

$$T^{(z)}(z_{\alpha_3\alpha_3}) = \frac{d}{dz}\left(\phi\frac{d}{dz}z_{\alpha_3\alpha_3} - \frac{1}{2}\phi' z_{\alpha_3\alpha_3}\right) = -\frac{1}{4}\phi''' w^2 \tag{81}$$

we have

$$\phi z'_{\alpha_3\alpha_3} - \frac{1}{2}\phi' z_{\alpha_3\alpha_3} = -\left(w'^2\phi + \frac{1}{4}\phi'' w^2\right) + \frac{1}{4}\phi_0'' \tag{82}$$

Similarly, substituting expressions $\phi(z)$, v, w of formula (54), formula (57), formula (58) to the formula (82), we obtain

$$z'_{\alpha_3\alpha_3} - \frac{1}{2}\frac{R_c}{z(z+R_c)}z_{\alpha_3\alpha_3} = -\frac{3}{2R_c^2} \tag{83}$$

Thus, $z_{\alpha_3\alpha_3}$ can be solved. From formula (67) and A_{33} can be expressed by

$$A_{33} = \sqrt{\frac{2m_0}{e}}\sqrt{\frac{n-1}{\phi_{ac}}}\frac{3}{8R_c^2}\left(\sqrt{z}\sqrt{-z-R_c} - R_c \arctan\frac{\sqrt{z}}{\sqrt{-z-R_c}}\right) \tag{84}$$

Regarding to coefficient A_{22}, integrating the differential Eq. (70):

$$T^{(z)}(z_{\alpha_2\alpha_2}) = \frac{d}{dz}\left(\phi\frac{d}{dz}z_{\alpha_2\alpha_2} - \frac{1}{2}\phi' z_{\alpha_2\alpha_2}\right) = \phi'''\frac{2}{\phi_0'^2}\phi \tag{85}$$

at the initial condition $z_{\alpha_2\alpha_2}|_{z=0} = 0$, $z'_{\alpha_2\alpha_2}|_{z=0} = 0$, we have

$$\phi z'_{\alpha_2\alpha_2} - \frac{1}{2}\phi' z_{\alpha_2\alpha_2} = \frac{2}{\phi_0'^2}\left(\phi\phi'' - \frac{1}{2}\phi'^2\right) + 1 \tag{86}$$

Similarly, substituting expressions $\phi(z)$, v, w of formula (54), formula (57), formula (58) to the formula (86), we obtain

$$z'_{\alpha_2\alpha_2} - \frac{1}{2}\frac{R_c}{z(z+R_c)}z_{\alpha_2\alpha_2} = -\frac{(n-1)z(z^2 + 4zR_c + 6R_c^2)}{(z+R_c)^3 \phi_{ac}} \tag{87}$$

Thus $z_{\alpha_2\alpha_2}$ can be solved.

$$z_{\alpha_2\alpha_2} = -\frac{\sqrt{z}}{\sqrt{-z-R_c}}\frac{n-1}{\varphi_{ac}}\left[-3R_c \arctan\frac{\sqrt{z}}{\sqrt{-z-R_c}} + \frac{\sqrt{z}}{\sqrt{-z-R_c}}\frac{z^2 - 3R_c^2}{-z-R_c}\right] \tag{88}$$

Other parameters in formula (69) can be expressed by

$$z_{\alpha_2} = \frac{2}{\varphi_0'}\sqrt{\varphi}, \quad \ddot{z} = \frac{e}{m_0}\varphi', \quad \dot{z} = \sqrt{\frac{2e\varphi}{m_0}}, \quad \dot{z}_{\alpha_2} = \sqrt{\frac{2e}{m_0}}\frac{\varphi'}{\varphi_0'} = \sqrt{\frac{2e}{m_0}}\frac{R_c^2}{(-R_c - z)^2} \tag{89}$$

Substituting all of these expressions in formula (69), we can get

$$A_{22} = -\sqrt{\frac{2m_0}{e}}\frac{1}{E_c}\left(\frac{n-1}{\phi_{ac}}\right)^{1/2}\left[\frac{-2R_c^2 - R_c z + z^2}{4R_c\sqrt{z}\sqrt{-(z+R_c)}} + \frac{3}{4}\arctan\frac{\sqrt{z}}{\sqrt{-(z+R_c)}}\right] \tag{90}$$

It can be seen from comparison between Eq. (59), Eq. (60), Eq. (61), Eq. (62)) and Eq. (76), Eq. (80), Eq. (84), Eq. (90) for A_{11}, A_{13}, A_{33}, A_{22} that in an electrostatic concentric spherical electrostatic system model the expressions of temporal aberration coefficients solved by "Direct Integral Method" and "τ-Variation Method" have complete identical form. It verifies that the two methods are equivalent and correct.

It should be pointed out that for solving coefficients A_{11}, A_{13}, A_{33} and A_{22} the "Direct Integral Method" is only related to solve integral expressions, but the "τ-Variation Method" is related to solving differential equations for coefficients A_{11}, A_{13}, A_{33} and to solve integral expression for coefficient A_{22}, it seems that the "Direct Integral Method" is simpler and more convenient than the "τ-Variation Method" for the practical computation of temporal aberration coefficients.

Conclusions

A new approach to the theory of temporal aberration for the cathode lenses which is named "Direct Integral Method" is put forward in the present paper. A new definition of temporal aberration is given in which a certain initial energy of electron emission along the axial direction $\varepsilon_{z_1}(0 \leq \varepsilon_{z_1} \leq \varepsilon_{0max})$ is considered. New expressions of the temporal geometrical aberration coefficients of the second order expressed in integral forms have been obtained by the "Direct Integral Method". All of the formulae of the temporal aberration coefficients deduced from "Direct Integral Method" and "τ-Variation Method" have been verified by an electrostatic concentric spherical system model and contrasted their results with analytical solutions. Results show that these two methods have got complete identical solutions. It can be concluded that both of two methods given by us are equivalent and correct.

It should be pointed out that the expressions for solving temporal aberration coefficients of the second order A_{11}, A_{13}, A_{33} and A_{22} given by the "Direct Integral Method" are related to solve integral expressions, which are more convenient for computation and could be suggested to use in the practical design.

Acknowledgements

This project was supported by National Natural Science Foundation of China and Russian Foundation for Basic Research Russia (RFBR).

References

[1] MONASTYRSKI M A, SCHELEV M Y. Theory of temporal aberrations of cathode lenses [M]. Moscow: Lebedeev Institute of Physics, 1980: 1 - 38.

[2] SAVOISKY Y K, FANCHENKO S D. Physical basis of electron optical chronograph[J]. Report of Academy of Sciences, 1956, 108, 2: 218 - 221 (in Russian).

[3] CSORBA I P. Transit-time-spread-limited time resolution of image tubes in streak operation[J]. RCA Review 1971, 32: 650 - 659.

[4] CHOU L W. Electron optics of concentric spherical electromagnetic focusing systems[J]. Advances in Electronics and Electron Physics, 1979, 52: 119 - 132.

[5] ZHOU L W. Electron optics with wide beam focusing[M]. Beijing: Beijing Institute of Technology Press, 1993.

11.4 静电聚焦同心球系统验证电子光学成像系统的时间像差理论

Verification of Temporal Aberration Theory of Imaging Electron Optical Systems by Electrostatic Focusing Concentric Spherical System

摘要：关于动态电子光学成像系统的时间像差理论，我们提出了计算时间像差系数的两种方法——"τ-变分法"与"直接积分法"。它们的差别在于："τ-变分法"计算二级几何时间像差系数时必须求解微分方程，而直接积分法仅需进行积分运算。本文采用静电同心球系统的理想模型对这两种方法的正确性进行检验。计算表明：这两种方法求解电子光学成像系统的时间像差系数的结果完全一致，所求得的时间色差系数与理想模型的解析解完全相同，从而证明两种方法是等价并且正确的。本文的工作表明，直接积分法的计算更为简便，更适于实际系统的计算与设计。

Abstract: Regarding the aberration theory of dynamic electron optical systems, we have suggested two methods to calculate the temporal aberration coefficients—one is the "τ-Variation Method," the other is direct integral method. The difference between them is that the "τ-Variation Method" for solving the aberration coefficients requires to solve the differential equation, but the direct integral method needs only to solve integral computation. Computation shows that the results given by two methods for solving the temporal aberration coefficients are identical, and their analytical solutions of the temporal chromatic coefficients are the same with the ideal model. It shows that the two methods are equivalent and correct. Our work shows that computation by the direct integral method is simpler than "τ-Variation Method," and it is more applicable for the practical system design and computation.

引言

关于动态电子光学成像系统的时间像差理论，我们提出了计算时间像差系数的两种方法——"τ-变分法"[1]和"直接积分法"[2]。这两种方法的主要差别在于："τ-变分法"计算二级几何时间像差系数时必须求解微分方程，而直接积分法仅需进行积分运算。本文利用静电聚焦同心球系统的理想模型，其电位场和电子轨迹均可获得解析解，系统的空间和时间特性亦可求得明晰的解析表达式，并对"τ-变分法"与"直接积分法计"算成像

周立伟[a]、李元[a]、张智诠[b], Monastyrski M A[c], Schelev M Y[c]. a) 北京理工大学，b) 装甲兵工程学院，c) 俄罗斯科学院普罗霍洛夫普通物理研究所. 物理学报, V. 54, No. 8, 2005, 3597–3603.

系统时间像差系数的正确性进行验证。

1 "τ-变分法"与"直接积分法"求解近轴时间像差系数的正确性的验证

关于静电聚焦同心球系统模型的电子光学成像特性和横向像差,文献[3]、文献[4]中已做了详细的叙述。由静电聚焦同心球系统的理想模型,我们可以求解自阴极面逸出的电子的运动轨迹,便能求得任一时刻下的电子所在的位置的解析表达式或任一位置处电子所经历的时间的解析表达式。这就为研究系统的电子光学成像性质、空间像差和时间像差提供了方便。本文首先利用静电聚焦同心球系统的时间表达式求相应的近轴时间像差系数即时间色差系数 a_2、A_{22} 的解析解,然后利用静电聚焦同心球系统场的表达式求直接积分法及 τ-变分法的时间色差系数 a_2、A_{22},它们亦以解析形式表示。通过结果的比较,说明 τ 变分法与直接积分法两种方法计算成像系统近轴时间像差系数的正确性。

1.1 求相应的时间色差系数 a_2、A_{22} 的解析解

对于静电聚焦同心球系统,如果设球面阴极 C 和栅状球面阳极 A 的曲率半径为 R_c 和 R_a,系统的共同曲率中心为 O。对于凹面阴极和凸面阳极情况,R_c 和 R_a 取负值。由于系统的球对称性,故可用极坐标 (ρ, φ) 来描述电子轨迹。令极坐标的原点位于系统的曲率中心 O,并设阴极 C 的电位 $\phi_c = 0$,栅状阳极 A 对于阴极 C 的电位为 ϕ_{ac},电子以初速度 v_0、初角度 α_0 自阴极上某点 z_0 射出。同时规定线段由左向右、由下向上为正,反之为负;角度以轴线或法线为轴,逆时针为正,顺时针为负。对于凹面阴极-凸面阳极系统,各参数定义如图 1 所示。

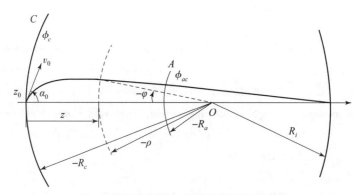

图 1 静电聚焦同心球系统中的电子运动

已知两电极同心球系统中任一点处的电位表达式 $\phi_{\rho c}$ 为

$$\phi_{\rho c} = \frac{\phi_{ac}}{n-1}\left(\frac{R_c}{\rho} - 1\right) \tag{1}$$

式中,$n = R_c/R_a$。根据能量守恒定律和角动量守恒定律,有

$$\frac{1}{2}m_0 v^2 = \frac{1}{2}m_0 v_0^2 + e\phi_{\rho c} \tag{2}$$

$$\rho^2 \dot{\varphi} = R_c^2 \dot{\varphi}_0 = R_c v_0 \sin\alpha_0 \tag{3}$$

在极坐标系下，电子速度可表示为

$$v^2 = \dot{\rho}^2 + (\rho\dot{\varphi})^2$$

于是有

$$\dot{\rho} = \frac{d\rho}{dt} = \sqrt{v^2 - R_c^2 v_0^2 \sin^2\alpha_0/\rho^2} \tag{4}$$

将

$$v^2 = v_0^2 + \frac{2e}{m_0}\frac{\phi_{ac}}{n-1}\left(\frac{R_c}{\rho} - 1\right)$$

代入式（4），得

$$\dot{\rho} = \frac{d\rho}{dt} = \sqrt{v_0^2 + \frac{2e}{m_0}\frac{\phi_{ac}}{n-1}\left(\frac{R_c}{\rho} - 1\right) - \frac{R_c^2 v_0^2 \sin^2\alpha_0}{\rho^2}} \tag{5}$$

定义 $\varepsilon_0 = \frac{1}{2}\frac{m_0}{e}v_0^2$，$m_0/e$ 为电子的荷质比。由式（5），可得时间 t 的微分表达式

$$dt = \sqrt{\frac{m_0}{2e}}\frac{-\rho d\rho}{\sqrt{\left(\frac{\phi_{ac}}{n-1} - \varepsilon_0\right)}\sqrt{-\rho^2 + \frac{\phi_{ac}}{(n-1)\left(\frac{\phi_{ac}}{n-1} - \varepsilon_0\right)}R_c\rho - \frac{\varepsilon_0 R_c^2 \sin^2\alpha_0}{\left(\frac{\phi_{ac}}{n-1} - \varepsilon_0\right)}}} \tag{6}$$

积分式（6），并考虑边界条件 $\rho = R_c$ 时 $t = 0$，便得到自阴极面发射的电子经历时间的解析表达式

$$t = \sqrt{\frac{m_0}{2e}}\frac{1}{\frac{\phi_{ac}}{n-1} - \varepsilon_0}\left[\sqrt{-\rho^2\left(\frac{\phi_{ac}}{n-1} - \varepsilon_0\right) + \frac{\phi_{ac}}{n-1}R_c\rho - \varepsilon_0 R_c^2 \sin^2\alpha_0} - \right.$$

$$\frac{1}{2}\frac{\frac{\phi_{ac}}{n-1}}{\sqrt{\frac{\phi_{ac}}{n-1} - \varepsilon_0}}R_c \arctan\frac{-\frac{\phi_{ac}}{n-1}R_c + 2\rho\left(\frac{\phi_{ac}}{n-1} - \varepsilon_0\right)}{2\sqrt{\frac{\phi_{ac}}{n-1} - \varepsilon_0}\sqrt{-\rho^2\left(\frac{\phi_{ac}}{n-1} - \varepsilon_0\right) + \frac{\phi_{ac}}{n-1}R_c\rho - \varepsilon_0 R_c^2 \sin^2\alpha_0}} +$$

$$\left. R_c\sqrt{\varepsilon_0 \cos^2\alpha_0} + \frac{1}{2}\frac{\frac{\phi_{ac}}{n-1}}{\sqrt{\frac{\phi_{ac}}{n-1} - \varepsilon_0}}R_c \arctan\frac{2\varepsilon_0 - \frac{\phi_{ac}}{n-1}}{2\sqrt{\frac{\phi_{ac}}{n-1} - \varepsilon_0}\sqrt{\varepsilon_0 \cos^2\alpha_0}}\right] \tag{7}$$

若考虑电子发射的近轴情况，略去 $\varepsilon_r^{\frac{1}{2}}$、$r_0$ 的二阶小量项，于是 $\varepsilon_0 = \varepsilon_z$，$\rho = R_c + z$，式（7）变为

$$t^*(z, \varepsilon_z^{\frac{1}{2}}) = \sqrt{\frac{m_0}{2e}}\frac{1}{\frac{\phi_{ac}}{n-1} - \varepsilon_z}\left[\sqrt{-(R_c + z)^2\left(\frac{\phi_{ac}}{n-1} - \varepsilon_z\right) + \frac{\phi_{ac}}{n-1}R_c(R_c + z)} - \right.$$

$$\frac{1}{2}\frac{\frac{\phi_{ac}}{n-1}}{\sqrt{\frac{\phi_{ac}}{n-1} - \varepsilon_z}}R_c \arctan\frac{-\frac{\phi_{ac}}{n-1}R_c + 2(R_c + z)\left(\frac{\phi_{ac}}{n-1} - \varepsilon_z\right)}{2\sqrt{\frac{\phi_{ac}}{n-1} - \varepsilon_z}\sqrt{-(R_c + z)^2\left(\frac{\phi_{ac}}{n-1} - \varepsilon_z\right) + \frac{\phi_{ac}}{n-1}R_c(R_c + z)}} +$$

$$R_c\sqrt{\varepsilon_z} + \frac{1}{2}\frac{\frac{\phi_{ac}}{n-1}}{\sqrt{\frac{\phi_{ac}}{n-1}-\varepsilon_z}}R_c\arctan\frac{2\varepsilon_z - \frac{\phi_{ac}}{n-1}}{2\sqrt{\frac{\phi_{ac}}{n-1}-\varepsilon_z}\sqrt{\varepsilon_z}}\right] \quad (8)$$

文献 [2] 给出时间像差的定义如下：

$$\Delta t = \Delta T(z,\varepsilon_z^{\frac{1}{2}},\varepsilon_{z_1}^{\frac{1}{2}}) + \Delta\tau(z,\varepsilon_r^{\frac{1}{2}},\varepsilon_z^{\frac{1}{2}},r_0) = \Delta T + \Delta\tau \quad (9)$$

式中，ΔT、$\Delta\tau$ 分别称为近轴时间像差（或时间色差）和几何时间像差。时间色差 ΔT 以下式表示：

$$\Delta T = t^*(z,\varepsilon_z^{\frac{1}{2}}) - t^*(z,\varepsilon_{z_1}^{\frac{1}{2}}) = a_2(\varepsilon_z^{\frac{1}{2}} - \varepsilon_{z_1}^{\frac{1}{2}}) + A_{22}(\varepsilon_z - \varepsilon_{z_1}) \quad (10)$$

把 $t^*(z,\varepsilon_z^{\frac{1}{2}})$ 的表达式（8）按 $\varepsilon_z^{\frac{1}{2}}$ 的幂次展开，便得到常数项以及与 $\varepsilon_z^{\frac{1}{2}}$、$\varepsilon_z$ 相关的一、二级小量。同样，展开 $t^*(z,\varepsilon_{z_1}^{\frac{1}{2}})$，亦可得到相同的常数项以及与 $\varepsilon_{z_1}^{\frac{1}{2}}$、$\varepsilon_{z_1}$ 相关的一、二级小量。由式（10），其相应的一级和二级时间色差系数 a_2、A_{22} 表达式为：

$$a_2 = R_c\sqrt{\frac{m_0}{2e}\frac{2(n-1)}{\phi_{ac}}} = \sqrt{\frac{2m_0}{e}}\frac{1}{E_c} \quad (11)$$

$$A_{22} = -\sqrt{\frac{2m_0}{e}}\frac{1}{E_c}\left(\frac{n-1}{\phi_{ac}}\right)^{\frac{1}{2}}\left[\frac{(-2R_c^2-R_cz+z^2)}{4R_c\sqrt{-z(R_c+z)}} - \frac{3}{8}\arctan\frac{-(R_c+2z)}{2\sqrt{-z(R_c+z)}} + \frac{3}{16}\pi\right] \quad (12)$$

式（11）表示的电子渡越弥散是 Savoisky-Fanchenko 在 1956 年首先提出的[5]，并为文献 [1]、文献 [6] 所证实。

1.2 利用"直接积分法"求时间色差系数 a_2、A_{22}

在文献 [2] 中我们已求得在轴对称静电场中运动的近轴电子的时间表达式

$$t^*(z,\varepsilon_z^{\frac{1}{2}}) = \int_0^z \frac{\mathrm{d}z}{\left(\frac{2e}{m_0}\right)^{\frac{1}{2}}[\phi(z)+\varepsilon_z]^{\frac{1}{2}}} \quad (13)$$

将式（13）代入式（10）中，便可得到时间色差表达式

$$\Delta T = a_2(\varepsilon_z^{\frac{1}{2}} - \varepsilon_{z_1}^{\frac{1}{2}}) + A_{22}(\varepsilon_z - \varepsilon_{z_1})$$

$$= \int_0^z \frac{\mathrm{d}z}{\left(\frac{2e}{m_0}\right)^{\frac{1}{2}}[\phi(z)+\varepsilon_z]^{\frac{1}{2}}} - \int_0^z \frac{\mathrm{d}z}{\left(\frac{2e}{m_0}\right)^{\frac{1}{2}}[\phi(z)+\varepsilon_{z_1}]^{\frac{1}{2}}} \quad (14)$$

按静电聚焦同心球系统，轴上电位分布 $\phi(z)$ 的表达式为

$$\phi(z) = \frac{\phi_{ac}}{n-1}\frac{-z}{z+R_c} \quad (15)$$

将式（15）代入式（13）中，便可求得电子自阴极面上发出（$t=0$，$z_0=0$，$r_0=0$）初始条件为 (ε_0,α_0) 的轨迹在任一位置的时间 $t^*(z,\varepsilon_z^{\frac{1}{2}})$ 的近轴表达式

$$t^*(z,\varepsilon_z^{\frac{1}{2}}) = \frac{-R_c}{\left\{\frac{2e}{m_0}\left[\frac{\phi_{ac}-(n-1)\varepsilon_z}{n-1}\right]\right\}^{\frac{1}{2}}}\left\{\sqrt{1+\frac{z}{R_c}}\sqrt{\frac{(n-1)\varepsilon_z}{\phi_{ac}-(n-1)\varepsilon_z}} - \frac{z}{R_c} - \sqrt{\frac{(n-1)\varepsilon_z}{\phi_{ac}-(n-1)\varepsilon_z}} + \right.$$

$$\left[1+\frac{(n-1)\varepsilon_z}{\phi_{ac}-(n-1)\varepsilon_z}\right]\left[\arcsin\sqrt{\frac{(n-1)\varepsilon_z}{\phi_{ac}}-\frac{[\phi_{ac}-(n-1)\varepsilon_z]z}{R_c\phi_{ac}}}-\arcsin\sqrt{\frac{(n-1)\varepsilon_z}{\phi_{ac}}}\right]\right\}$$
(16)

展开式（16），并把它表达成如下形式：

$$t^*(z,\varepsilon_z^{\frac{1}{2}}) = t^*(z,0) + a_2\varepsilon_z^{\frac{1}{2}} + A_{22}\varepsilon_z + 0(\varepsilon_z)^{\frac{3}{2}} \tag{17}$$

可得

$$t^*(z,0) = \left(\frac{2m_0}{e}\right)^{\frac{1}{2}}\frac{-R_c}{\sqrt{\phi_{ac}}}\frac{\sqrt{n-1}}{2}\sqrt{\frac{z}{-R_c}\left(1+\frac{z}{R_c}\right)} - R_c\frac{1}{2}\sqrt{\frac{2m_0}{e}}\sqrt{\frac{n-1}{\phi_{ac}}}\arcsin\sqrt{\frac{-z}{R_c}}$$

$$a_2 = \sqrt{\frac{2m_0}{e}}\frac{1}{E_c} \tag{18}$$

$$A_{22} = -\sqrt{\frac{2m_0}{e}}\frac{1}{E_c}\left(\frac{n-1}{\phi_{ac}}\right)^{\frac{1}{2}}\left\{\frac{-(R_c+z)(2R_c-z)}{4R_c\sqrt{-z(R_c+z)}} + \frac{3}{4}\left(\arcsin\sqrt{\frac{-z}{R_c}}\right)\right\} \tag{19}$$

式（18）与式（11）完全相同，式（19）通过变换可以表达成式（12）的形式，说明直接积分法求解时间像差系数的方法是正确的。文献[2]利用分部积分法对式（13）进行变换求得 a_2 和 A_{22} 的表达式。a_2 的表达式与式（18）一致，A_{22} 可以表达如下：

$$A_{22} = -\frac{1}{4}\sqrt{\frac{2m_0}{e}}\int\frac{1}{\phi(z)^{\frac{3}{2}}}dz \tag{20}$$

将式（15）代入式（20）中，同样亦可求得 A_{22} 的表达式（12）。

1.3 利用"τ-变分法"求时间色差系数 a_2、A_{22}

关于"τ-变分法"研究电子光学成像系统的时间像差理论，请参阅文献[1]，这里就不细述了。τ-变分法所导得的一级时间色差系数 a_2 的表达式与式（18）完全一致，其二级时间色差系数求解的表达式如下：

$$A_{22} = -\frac{\ddot{z}}{2\dot{z}^3}z_{\alpha_2}^2 + \frac{\dot{z}_{\alpha_2}z_{\alpha_2}}{\dot{z}^2} - \frac{z_{\alpha_2\alpha_2}}{2\dot{z}} \tag{21}$$

式中，$z_{\alpha_2\alpha_2}$ 是微分方程 $T^{(z)}(z_{\alpha_2\alpha_2}) = \phi\phi'''\frac{2}{\phi_0'^2}$ 的解，且式中各值可表示为

$$z_{\alpha_2} = \frac{2}{\phi_0'}\sqrt{\phi}, \quad \dot{z} = \sqrt{\frac{2e\phi}{m_0}}, \quad \dot{z}_{\alpha_2} = \sqrt{\frac{2e}{m_0}}\frac{\phi'}{\phi_0'} = \sqrt{\frac{2e}{m_0}}\frac{R_c^2}{(-R_c-z)^2}, \quad \ddot{z} = \frac{e}{m_0}\phi' \tag{22}$$

在上述公式中引入矢量 $\boldsymbol{\alpha}$，其三个分量表示为

$$\alpha_1 = \varepsilon_r^{\frac{1}{2}}, \quad \alpha_2 = \varepsilon_z^{\frac{1}{2}}, \quad \alpha_3 = r_0 \tag{23}$$

式中，$z_{\alpha_i} = \frac{\partial z}{\partial \alpha_i}$，$i = 1, 2, 3$。

为了求解 $z_{\alpha_2\alpha_2}$，首先求解下列微分方程

$$T^{(z)}(z_{\alpha_2\alpha_2}) = \frac{d}{dz}\left(\phi\frac{d}{dz}z_{\alpha_2\alpha_2} - \frac{1}{2}\phi'z_{\alpha_2\alpha_2}\right) = \phi'''\frac{2}{\phi_0'^2}\phi \tag{24}$$

在初始条件 $z_{\alpha_2\alpha_2}|_{z=0} = 0$，$z'_{\alpha_2\alpha_2}|_{z=0} = 0$ 下积分，可得

$$\phi z'_{\alpha_2\alpha_2} - \frac{1}{2}\phi' z_{\alpha_2\alpha_2} = \frac{2}{\phi_0'^2}\left(\phi\phi'' - \frac{1}{2}\phi'^2\right) + 1 \tag{25}$$

将式（15）的 $\phi(z)$ 及其导数代入式（25），经过整理，得

$$z'_{\alpha_2\alpha_2} - \frac{1}{2}\frac{R_c}{z(z+R_c)}z_{\alpha_2\alpha_2} = -\frac{(n-1)z(z^2+4zR_c+6R_c^2)}{(z+R_c)^3\phi_{ac}} \tag{26}$$

由此解得

$$z_{\alpha_2\alpha_2} = -\frac{\sqrt{z}}{\sqrt{-z-R_c}}\frac{n-1}{\phi_{ac}}\left(-3R_c\arctan\frac{\sqrt{z}}{\sqrt{-z-R_c}} + \frac{\sqrt{z}}{\sqrt{-z-R_c}}\frac{z^2-3R_c^2}{-z-R_c}\right) \tag{27}$$

将上式以及式（22）的量代入式（21）中，可得

$$A_{22} = -\sqrt{\frac{2m_0}{e}}\frac{1}{E_c}\left(\frac{n-1}{\phi_{ac}}\right)^{\frac{1}{2}}\left[\frac{-2R_c^2-R_cz+z^2}{4R_c\sqrt{-z(z+R_c)}} + \frac{3}{4}\arctan\frac{\sqrt{z}}{\sqrt{-(z+R_c)}}\right] \tag{28}$$

经过验证，式（28）通过变换可以表达成式（19）或式（12）的形式，这说明 τ – 变分法求解时间色差系数的方法也是正确的。这样，我们通过静电聚焦同心球系统证明了 τ – 变分法与直接积分法的正确性。

2 同心球系统验证 "τ – 变分法" 与 "直接积分法" 求解几何时间像差系数的等价性

上面我们用静电聚焦同心球系统模型证明了 "τ – 变分法" 与直接积分法计算时间色差系数 a_2、A_{22} 的结果是一致的。下面我们由几何时间像差的表达式[2]

$$\Delta\tau = A_{11}\varepsilon_r + 2A_{13}\varepsilon_r^{\frac{1}{2}}r_0 + A_{33}r_0^2 \tag{29}$$

出发，依然利用静电聚焦同心球系统的场和轨迹特解表达式证明 "直接积分法" 和 "τ – 变分法" 求解二级几何时间像差系数的结果也是一致的，从而证明它们之间的等价性。

2.1 利用直接积分法求几何时间像差系数

在文献 [2] 中，我们用直接积分法导得的二级几何时间像差系数的公式为

$$A_{11} = \int_0^z \frac{1}{2\left(\frac{2e}{m_0}\right)^{\frac{1}{2}}[\phi(z)+\varepsilon_z]^{\frac{3}{2}}}\left(v'^2\phi_* + \frac{1}{4}\phi''v^2 - 1\right)dz \tag{30}$$

$$A_{13} = \int_0^z \frac{1}{2\left(\frac{2e}{m_0}\right)^{\frac{1}{2}}[\phi(z)+\varepsilon_z]^{\frac{3}{2}}}\cos(\theta_0-\beta_0)\left(v'w'\phi_* + \frac{1}{4}\phi''vw\right)dz \tag{31}$$

$$A_{33} = \int_0^z \frac{1}{2\left(\frac{2e}{m_0}\right)^{\frac{1}{2}}[\phi(z)+\varepsilon_z]^{\frac{3}{2}}}\left(w'^2\phi_* + \frac{1}{4}\phi''w^2 - \frac{1}{4}\phi_0''\right)dz \tag{32}$$

将同心球系统轴上电位分布 $\phi(z)$ 表达式（15）和近轴轨迹方程的两个特解[3,4]如下：

$$v(z) = \frac{2z}{\phi(z)}[\sqrt{\phi(z)+\varepsilon_z} - \sqrt{\varepsilon_z}] \tag{33}$$

$$w(z) = 1 + \frac{1}{R_c}z - \frac{\sqrt{\varepsilon_z}}{R_c}\frac{2z}{\phi(z)}[\sqrt{\phi(z)+\varepsilon_z} - \sqrt{\varepsilon_z}] \tag{34}$$

代入时间像差系数 A_{11}、A_{13}、A_{33} 的表达式（30）、式（31）、式（32）中，积分之，即可求得：

$$A_{11} = \frac{3}{4}\sqrt{\frac{2m_0}{e}}\left(\frac{n-1}{\phi_{ac}}\right)^{\frac{1}{2}}\frac{1}{E_c R_c}\left(-\sqrt{z}\sqrt{-z-R_c} - R_c \arctan\frac{\sqrt{z}}{\sqrt{-z-R_c}}\right) \tag{35}$$

$$A_{13} = -\cos(\theta_0 - \beta_0)\frac{1}{2}\sqrt{\frac{2m_0}{e}}\frac{1}{R_c^2 E_c}z \tag{36}$$

$$A_{33} = \sqrt{\frac{2m_0}{e}}\sqrt{\frac{n-1}{\phi_{ac}}}\frac{3}{8R_c^2}\left(\sqrt{z}\sqrt{-z-R_c} - R_c \arctan\frac{\sqrt{z}}{\sqrt{-z-R_c}}\right) \tag{37}$$

2.2 利用"τ-变分法"求几何时间像差系数

依照"τ-变分法"[1]二级几何时间像差系数 A_{11}、A_{13}、A_{33} 的计算步骤如下：

$$A_{11} = -\frac{z_{\alpha_1\alpha_1}}{2\dot{z}} \tag{38a}$$

$$T^{(z)}(z_{\alpha_1\alpha_1}) = -\frac{1}{4}\phi'''v^2 \tag{38b}$$

$$A_{13} = -\frac{z_{\alpha_1\alpha_3}}{2\dot{z}} \tag{39a}$$

$$T^{(z)}(z_{\alpha_1\alpha_3}) = -\frac{1}{4}\phi'''vw\cos(\beta_0-\theta_0) \tag{39b}$$

$$A_{33} = -\frac{z_{\alpha_3\alpha_3}}{2\dot{z}} \tag{40a}$$

$$T^{(z)}(z_{\alpha_3\alpha_3}) = -\frac{1}{4}\phi'''w^2 \tag{40b}$$

式中，$z_{\alpha_i\alpha_j}$ 乃是微分方程 $T^{(z)}(z_{\alpha_i\alpha_j})$（$i=1,3;j=1,3$）的解。

首先计算 A_{11}，由方程（38b）

$$T^{(z)}(z_{\alpha_1\alpha_1}) = \frac{\mathrm{d}}{\mathrm{d}z}\left(\phi\frac{\mathrm{d}}{\mathrm{d}z}z_{\alpha_1\alpha_1} - \frac{1}{2}\phi'z_{\alpha_1\alpha_1}\right) = -\frac{1}{4}\phi'''v^2 \tag{41}$$

积分，得到

$$\phi z'_{\alpha_1\alpha_1} - \frac{1}{2}\phi' z_{\alpha_1\alpha_1} = -\left(v'^2\phi + \frac{1}{4}\phi''v^2\right) + 1 \tag{42}$$

在式（42）中代入同心球系统的轴上电位分布 $\phi(z)$ 表达式（15）及近轴轨迹的特解 v 的表达式（33），经过整理，得

$$z'_{\alpha_1\alpha_1} - \frac{1}{2}\frac{R_c}{z(z+R_c)}z_{\alpha_1\alpha_1} = \frac{n-1}{\phi_{ac}}\frac{3z}{(z+R_c)} \tag{43}$$

由此解得 $z_{\alpha_1\alpha_1}$：

$$z_{\alpha_1\alpha_1} = \frac{3(n-1)\sqrt{z}}{\phi_{ac}\sqrt{-z-R_c}}\left(R_c \arctan\frac{\sqrt{z}}{\sqrt{-z-R_c}} + \sqrt{z}\sqrt{-z-R_c}\right) \tag{44}$$

因 $\dot{z} = \sqrt{\frac{2e\phi}{m_0}}$，于是，由式（38a）可得

$$A_{11} = -\frac{z_{\alpha_1\alpha_1}}{2\dot{z}} = -\frac{3}{4}\sqrt{\frac{2m_0}{e}}\sqrt{\frac{n-1}{\phi_{ac}}}\frac{1}{E_c R_c}\left(R_c \arctan\frac{\sqrt{z}}{\sqrt{-z-R_c}} + \sqrt{z}\sqrt{-z-R_c}\right) \quad (45)$$

关于 A_{13} 的计算，同样，由方程 (39b)

$$T^{(z)}(z_{\alpha_1\alpha_3}) = \frac{\mathrm{d}}{\mathrm{d}z}\left(\phi\frac{\mathrm{d}}{\mathrm{d}z}z_{\alpha_1\alpha_3} - \frac{1}{2}\phi' z_{\alpha_1\alpha_3}\right) = -\frac{1}{4}\phi''' wv\cos(\beta_0 - \theta_0) \quad (46)$$

积分得

$$\phi z'_{\alpha_1\alpha_3} - \frac{1}{2}\phi' z_{\alpha_1\alpha_3} = -\cos(\beta_0 - \theta_0)\left(v'w'\phi + \frac{1}{4}\phi'' vw\right) \quad (47)$$

同样，代入 $\phi(z)$ 表达式（15）及特解 v、w 的表达式（33）、式（34），经过整理，得

$$z'_{\alpha_1\alpha_3} + \frac{1}{2}\frac{R_c}{z(-z-R_c)}z_{\alpha_1\alpha_3} = \cos(\beta_0 - \theta_0)\sqrt{\frac{n-1}{\phi_{ac}}}\frac{2}{R_c}\frac{z^{\frac{1}{2}}}{(-z-R_c)^{\frac{1}{2}}} \quad (48)$$

由此解得 $z_{\alpha_1\alpha_3}$：

$$z_{\alpha_1\alpha_3} = \frac{2\cos(\beta_0 - \theta_0)(n-1)z\sqrt{\phi}}{R_c \phi_{ac}} \quad (49)$$

于是，由式（39a）可得

$$A_{13} = -\frac{1}{2}\frac{z_{\alpha_1\alpha_3}}{\dot{z}} = -\cos(\theta_0 - \beta_0)\frac{1}{2}\sqrt{\frac{2m_0}{e}}\frac{1}{R_c^2 E_c}z \quad (50)$$

关于 A_{33} 的计算，同样，由方程 (40b)

$$T^{(z)}(z_{\alpha_3\alpha_3}) = \frac{\mathrm{d}}{\mathrm{d}z}\left(\phi\frac{\mathrm{d}}{\mathrm{d}z}z_{\alpha_3\alpha_3} - \frac{1}{2}\phi' z_{\alpha_3\alpha_3}\right) = -\frac{1}{4}\phi''' w^2 \quad (51)$$

积分，并考虑初始条件，得

$$\phi z'_{\alpha_3\alpha_3} - \frac{1}{2}\phi' z_{\alpha_3\alpha_3} = -\left(w'^2\phi + \frac{1}{4}\phi'' w^2\right) + \frac{1}{4}\phi''_0 \quad (52)$$

代入式（15）、式（34），经过整理后，得

$$z'_{\alpha_3\alpha_3} - \frac{1}{2}\frac{R_c}{z(z+R_c)}z_{\alpha_3\alpha_3} = -\frac{3}{2R_c^2} \quad (53)$$

由此解得 $z_{\alpha_3\alpha_3}$：

$$z_{\alpha_3\alpha_3} = -\frac{3z}{2R_c^2} + \frac{3\sqrt{z}}{2R_c\sqrt{-z-R_c}}\arctan\frac{\sqrt{z}}{\sqrt{-z-R_c}} \quad (54)$$

于是，由式（40a）可得

$$A_{33} = -\frac{1}{2}\frac{z_{\alpha_3\alpha_3}}{\dot{z}} = \frac{3}{4R_c^2}\sqrt{\frac{m_0}{2e}}\sqrt{\frac{n-1}{\phi_{ac}}}\left(\sqrt{z}\sqrt{-z-R_c} - R_c \arctan\frac{\sqrt{z}}{\sqrt{-z-R_c}}\right) \quad (55)$$

比较式（35）、式（36）、式（37）与式（45）、式（50）、式（55）可以看出，在静电聚焦同心球系统中，用直接积分法和 τ - 变分法所求解的二级几何时间像差系数表达式具有完全相同的形式。经过三角变换，可以发现二级时间色差系数表达式（19）与式（28）也是等价的，从而说明这两种方法在计算结果上都是非常准确的。不过在计算中，τ - 变分法由于涉及对微分方程的求解，计算难度较大，而且各中间量的求解也稍显艰涩，不如直接积分法求解那么便捷。

结束语

本文通过静电聚焦同心球系统的理想模型检验了直接积分法与"τ-变分法"求解成像系统时间像差系数的正确性与等价性，所得到的结果说明两种途径殊途同归，是完全正确的。应该指出，直接积分法给出的二级时间像差系数表达式 A_{11}、A_{13}、A_{33} 和 A_{22} 都是以积分形式表示的，便于计算，可以在实际系统的计算与设计中应用。

参 考 文 献

[1] MONASTYRSKI M A, SCHELEV M Y. Theory of temporal aberrations of cathode lenses[M]. Moscow: Lebedeev Institute of Physics, 1980.

[2] 周立伟, 李元, 张智诠, 等. 直接积分法研究电子光学成像系统的时间像差理论[J]. 物理学报, 2005, 54(8): 3591-3596.

[3] 周立伟. 宽束电子光学[M]. 北京: 北京理工大学出版社, 1993.

[4] 周立伟. 宽电子束聚焦与成像[J]//周立伟电子光学学术论文选. 北京: 北京理工大学出版社, 1994.

[5] SAVOISKY Y K, FANCHENKO S D. Physical basis of electron optical chro nograph[J]. Reporp of Acaclemy of sciences. 1956: 108, 2: 218-221.

[6] CSORBA I P. Transit-time-spread-limited time resolution of image tubes in streak operation[J]. RCA Review, 1971, 32: 650-659.

11.5 静电同心球电子光学系统时间像差系数的计算
Computation of Temporal Aberration Coefficients of Electrostatic Concentric Spherical Electron Optical System

摘要：使用"直接积分法"计算电子光学系统的时间像差系数是我们近期研究取得的进展。本文以直接积分法为基础，利用所开发的 ODESI-SD 软件包对电子光学静电成像同心球系统理想模型的时间像差系数进行了计算，并与基于"τ-变分法"理论的 ELIM 软件包的计算结果进行比较，发现二者存在一定差异。利用此二软件包对同心球系统的时间像差系数的解析解进行验算，证实 ODESI-SD 软件包的计算更为精确。计算结果表明，直接积分法可以应用于电子光学成像系统的实际计算与设计中。

Abstract: Using the "Direct Integral Method" to calculate the temporal aberration coefficients of dynamic electron optical imaging systems is a recent development studied by us. Based on the direct integral method, the temporal aberration coefficients of an electron optical static imaging system—a concentric spherical ideal model were calculated by ODESI-SD software package developed by us. Compared with the temporal aberration coefficients calculated by "τ - Variation Method" based on ELIM software package, the results between them have some differences. By the use of concentric spherical theory of electron optics, the analytical solutions of the ideal model have been deduced. The results verify that the temporal aberration coefficients calculated by ODESI-SD software package is more precise, at the same time, the results show that the direct integral method can be applied to the computation in the practical calculation and the design of the electron optical imaging system.

引言

电子光学时间像差理论在高速摄影条纹管的计算与设计中占有重要的地位，但研究的机构仅有俄罗斯科学院普通物理研究所（GPI）光电子部与北京理工大学（BIT）光学工程的电子光学研究课题组。GPI 的 Monastyrski 等人于 1980 年提出了"τ-变分法"[1]求解电子光学系统的时间像差系数，为从理论的角度分析条纹管动态成像特性提供了可能。近期，我们与他们合作提出了"直接积分法"[2]求解成像电子光学系统的时间像差理论并对时间像差进行了重新定义，使时间像差理论更加完善与严谨。尽管在文献 [3] 中利用静电聚焦同心球

李元[1]，周立伟[1]，方二伦[2]. 1) 北京理工大学，2) 西安现代化学研究所. 光电技术与系统文选——中国光学学会光电技术专业委员会成立二十周年暨第 11 届全国光电技术与系统学术会议论文集，2005.

系统的理想模型从理论上对"τ-变分法"与"直接积分法"计算成像系统时间像差系数的正确性进行了验证,但是截至目前,直接积分法尚未应用于实际计算与设计中,其实际可行性与精确性尚需得到检验。本文以"直接积分法"的时间像差系数积分表达式作为计算公式,使用 ODESI-SD 软件包对一同心球电子光学系统的时间像差系数进行了计算。

1 时间像差系数的计算公式

按照文献[2]对时间像差的定义,由光电子初速度分布引起的时间像差 Δt 也可以表示为近轴时间像差 ΔT 与几何时间像差 $\Delta \tau$ 之和,即

$$\Delta t = \Delta T + \Delta \tau \tag{1}$$

它们可用以下公式表示:

$$\Delta T = a_2(\sqrt{\varepsilon_z} - \sqrt{\varepsilon_{z_1}}) + A_{22}(\varepsilon_z - \varepsilon_{z_1}) \tag{2}$$

$$\Delta \tau = A_{11}\varepsilon_r + 2A_{13}\sqrt{\varepsilon_r}r_0 + A_{33}r_0^2 \tag{3}$$

式中,a_2、A_{22} 分别称为一、二级近轴时间像差系数;A_{11}、A_{13} 和 A_{33} 分别称为二级时间球差、场曲和畸变系数。以上述定义为基础,文献[2]提出一种计算电子光学成像系统时间像差系数的新方法——直接积分法,所给出的计算公式如下。(括号内为单位)

$$a_2 = -\sqrt{\frac{2m_0}{e}} \frac{1}{\phi'(0)} \quad (\mathrm{s}/\sqrt{\mathrm{V}}) \tag{4}$$

$$A_{11} = \int_0^z \frac{1}{2\sqrt{2e/m_0}[\phi(z) + \varepsilon_z]^{3/2}} (v'^2\phi_* + \frac{1}{4}\phi''v^2 - 1)\mathrm{d}z \quad (\mathrm{s}/\mathrm{V}) \tag{5}$$

$$A_{13} = \int_0^z \frac{1}{2\sqrt{2e/m_0}[\phi(z) + \varepsilon_z]^{3/2}} (v'w'\phi_* + \frac{1}{4}\phi''vw)\mathrm{d}z \quad \left(\frac{\mathrm{s}}{\sqrt{\mathrm{V}} \cdot \mathrm{mm}}\right) \tag{6}$$

$$A_{33} = \int_0^z \frac{1}{2\sqrt{2e/m_0}[\phi(z) + \varepsilon_z]^{3/2}} (w'^2\phi_* + \frac{1}{4}\phi''w^2 - \frac{1}{4}\phi_0'')\mathrm{d}z \quad \left(\frac{\mathrm{s}}{\mathrm{mm}^2}\right) \tag{7}$$

$$A_{22} = \frac{1}{2}\sqrt{\frac{2m_0}{e}}\left\{\frac{1}{\sqrt{\phi(z)}\sqrt{\phi'(z)}} + \int_0^z \frac{\phi''(z)}{\sqrt{\phi(z)}[\phi'(z)]^2}\right\}\mathrm{d}z \quad (\mathrm{s}/\mathrm{V}) \tag{8}$$

式中,$\phi(z)$ 表示轴上电位分布,e/m_0 为电子的荷质比,v、w 为轴对称成像系统的近轴轨迹方程的特解。从式(4)~式(8)中可以发现,直接积分法的主要特点是计算二级时间像差系数时,不必求解微分方程,仅需进行积分运算,这就大大地减轻了数值计算难度。

2 ODESI 软件包与 ELIM 软件包对时间像差系数的计算比较

在两电极同心球理想模型中,取阳极电位 $\phi_{ac} = 10\,000$ V,阴极半径 $R_c = -30$ mm,阳极半径 $R_a = -10$ mm。分别使用基于直接积分法理论的 ODESI 软件包与基于"τ-变分法"的 ELIM 软件包[6]求解上述系统的时间像差系数,所得结果如表 1 所示。

表 1 同心球系统的近轴与几何时间像差系数的数值解 (在 $z = 20$ mm 处)

像差系数软件包	$a_2/(\mathrm{s}\cdot\mathrm{V}^{-\frac{1}{2}})$	$A_{22}/(\mathrm{s}\cdot\mathrm{V}^{-1})$	$A_{11}/(\mathrm{s}\cdot\mathrm{V}^{-1})$	$A_{13}/(\mathrm{s}\cdot\mathrm{V}^{-\frac{1}{2}}\cdot\mathrm{mm}^{-1})$	$A_{33}/(\mathrm{s}\cdot\mathrm{mm}^{-2})$
ODESI-SD	$-2.023\,3\mathrm{E}-11$	$3.399\,0\mathrm{E}-13$	$1.038\,5\mathrm{E}-13$	$2.248\,1\mathrm{E}-13$	$8.502\,84\mathrm{E}-13$

续表

像差系数软件包	$a_2/(\text{s}\cdot\text{V}^{-\frac{1}{2}})$	$A_{22}/(\text{s}\cdot\text{V}^{-1})$	$A_{11}/(\text{s}\cdot\text{V}^{-1})$	$A_{13}/(\text{s}\cdot\text{V}^{-\frac{1}{2}}\cdot\text{mm}^{-1})$	$A_{33}/(\text{s}\cdot\text{mm}^{-2})$
ELIM	−2.023 29E−11	3.397 97E−13	1.038 53E−13	−2.248E−13	−9.367 5E−14
解析解	−2.023 29E−11	3.398 99E−13	1.038 48c−13	2.248 1E−13	8.504 92E−13

表 1 中 ODESI-SD 与 ELIM 软件包的计算结果略有不同，但非常接近。其差异是因为研究的是二级像差，在计算系统空间电位时谢尔赤级数展开式 ODESI-SD 取到二阶小项，即

$$\varphi(z,r) = \phi(z) - \frac{1}{4}\phi''(z)r^2 + \cdots \tag{9}$$

其中，$\phi(z)$ 是系统轴上电位分布。在同心球系统中，

$$\phi(z) = \phi_{ac}\frac{-z}{(n-1)(z+R_c)} \tag{10}$$

将式（9）代入式（10）并整理得到

$$\varphi(r,z) = \phi(z) + \frac{R_c r^2}{2(z+R_c)^2 z}\phi(z) \tag{11}$$

式（9）就是 ODESI-SD 在计算同心球系统时所用的空间电位表达式。而 ELIM 软件在计算同心球系统的空间电位时直接使用解析表达式：

$$\varphi(r,z) = \phi(z) + \frac{1}{2}\left[R_c + \frac{R_c(R_c+z)}{\sqrt{r^2+(z+R_c)^2}}\right]\phi(z) \tag{12}$$

式（12）实际上包含了高次项，即将高阶时间像差引入公式中，从而造成 ELIM 软件计算得到的轴外几何时间像差系数与 ODESI-SD 的结果不同。

为了对 ODESI-SD 软件包的结果做进一步的验证，下面使用式（9）推导同心球理想模型时间像差系数的解析解。

3 同心球系统时间像差系数的解析解

在文献［3］中，已求得同心球系统近轴时间像差系数的解析解

$$a_2 = R_c\sqrt{\frac{m_0}{2e}\frac{2(n-1)}{\phi_{ac}}} \tag{13}$$

$$A_{22} = -R_c\sqrt{\frac{2m_0}{e}}\left(\frac{n-1}{\phi_{ac}}\right)^{3/2}\left[\frac{(-2R_c^2-R_c z+z^2)}{4R_c\sqrt{-z(R_c+z)}} - \frac{3}{8}\arctan\frac{-(R_c+2z)}{z\sqrt{-z(R_c+z)}} + \frac{3}{16}\pi\right] \tag{14}$$

下面求几何时间像差系数的解析解。

利用轴对称静电场下电子运动方程及谢尔赤级数展开式（9），可得

$$\dot{z}^2 = \frac{2e}{m_0}\int_0^z \phi'(z)\mathrm{d}z - \frac{2e}{m_0}\int_0^z \frac{r^2}{4}\phi''(z)\mathrm{d}z + \frac{2e}{m_0}\varepsilon_z \tag{15}$$

对于同心球系统，有

$$r(z) = r_0\left\{1 + \frac{z}{R_c} - \frac{\sqrt{\varepsilon_z}}{R_c}\frac{2z}{\phi(z)}\left[\sqrt{\phi(z)+\varepsilon_z} - \sqrt{\varepsilon_z}\right]\right\} + \frac{2z}{\phi(z)}\sqrt{\varepsilon_r}\left[\sqrt{\phi(z)+\varepsilon_z} - \sqrt{\varepsilon_z}\right] \tag{16}$$

将式（16）代入式（15）中，考虑到单位为毫米时电位导数数量级的变化，整理并且积分，可求得同心球系统的几何时间像差系数的解析解如下：

$$A_{11} = \frac{3}{2}\sqrt{\frac{m_0}{2e}}\left(\frac{n-1}{\phi_{ac}}\right)^{3/2}\left[-z\sqrt{-(z+R_c)} - R_c \arctan\frac{\sqrt{z}}{\sqrt{-(z+R_c)}}\right] \quad (17)$$

$$A_{13} = -\frac{\sqrt{2}}{2}\sqrt{\frac{m_o}{e}}\frac{(n-1)z}{R_c\phi_{ac}}\times 10^{-3} \quad (18)$$

$$A_{33} = \frac{3\sqrt{2}}{8R_c^2}\sqrt{\frac{m_0}{e}}\sqrt{\frac{n-1}{\phi_{ac}}}\left[\sqrt{-z(z+R_c)} - R_c \arctan\frac{\sqrt{z}}{\sqrt{-(z+R_c)}}\right]\times 10^{-6} \quad (19)$$

将同心球系统的结构参数代入式（13）、式（14）、式（17）、式（18）、式（19）中，得到解析解（见表1）。从表中可以看出，ODESI-SD软件包的计算结果是十分精确的。

结束语

本文基于所开发的ODESI-SD软件包对"直接积分法"推导的理论公式用同心球系统进行了验算。结果表明，"直接积分法"所导得的宽束电子光学静电成像系统时间像差系数表达式具有很高的精确度，可以应用在实际电子光学系统的计算与设计中。

参 考 文 献

[1] MONASTYRSKI M A, SCHELEV M Y. Theory of temporal aberrations of cathode lenses [M]. Moscow: Lebedeev Institute of Physics, 1980.

[2] 周立伟,李元,张智诠,等. 直接积分法研究电子光学成像系统的时间像差理论[J]. 物理学报, 2005, 54(8): 3591-3596.

[3] 周立伟,李元,张智诠,等. 静电聚焦同心球系统验证电子光学成像系统的时间像差理论[J]. 物理学报, 2005, 54(8): 3597-3603.

[4] ZHOU L W, LI Y, ZHANG Z Q, et al. On the theory of temporal aberrations for electron optical imaging systems by using Direct Integral Method[C]. At the 26th International congress on High-Speed Photography and Photonics, Sept. 21, 2004, Alexandria Virginia USA, Proc. of SPIE, V. 5580, 2004: 710-724.

[5] ZHOU L W, LI Y, ZHANG Z Q, et al. On the theory of temporal aberrations for cathode lenses[J]. Optik, 2005, 116(4): 175-184.

[6] MONASTYRSKI M A, ANDREEV S V, KOMEEVA M V, et al. Numerical techniques and software for static and dynamic image tubes design. Electron-Beam Sources and Charged-Particle Optics Meeting[J]. SPIE, 1995, 2522: 167-177.

11.6 On the Theory of Temporal Aberrations for Dynamic Electron Optics
关于动态电子光学的时间像差理论的研究

Abstract: A new theory for temporal aberrations of dynamic optics by applying the "Direct Integral Method" (DIM) is put forward in the present paper. A new definition of temporal aberration is given, in which a certain initial energy of electron emission emitted from a photocathode along the axial direction was taken as a criterion. New expressions of the temporary aberration coefficients in integral forms for the electron optical imaging systems have been deduced. An electrostatic concentric spherical system model is used to test and verify expressions of the coefficients given by the "Direct Integral Method" and "τ-Variation Method" (τVM). The analytical solutions prove that both methods are correct and equivalent. Compared with the "τ-Variation Method", the "Direct Integral Method" only needs to carry out the integral calculation for computation in the practical design. Finally, results of the study for the theory of temporal aberrations of electron optical imaging systems, from the point of view of methodology, have been elaborated.

摘要: 本文提出了应用"直接积分法"研究动态电子光学的时间像差理论,给出了时间像差的新定义,即它是以从光阴极逸出的某一电子沿轴向发射初能作为评判标准。本文导出了以积分形式表示电子光学系统新的时间像差系数表示式,并用一个静电同心球系统的模型检查与验证"τ-变分法"与"直接积分法"所给出的像差系数表示式。解析解证明了这两种方法是正确的和等同的。与"τ-变分法"比较,"直接积分法"在实际设计中只需要进行积分运算。最后,作者从方法论的观点出发研究了成像电子光学系统的时间像差理论。

Introduction

The theory of temporal aberrations, which holds an important place in the study of electron optics with its wide beam focusing and system design, was first investigated by M. A. Monastyrski and M. Y. Schelev in 1980 [1]. This initial study brought forward the "τ-Variation Method" to calculate the temporal aberration coefficients of the first and second orders, in which a certain initial energy of electron emission emitted from the cathode surface along the axial direction $\sqrt{\varepsilon_{z_1}} = 0$ as a criterion is considered. However, the derived expressions of geometrical temporal aberration coefficients have been related to solve differential equations. Work on the "τ-Variation Method" raised possibilities for scientists working on electron optics, particularly to consider the influence of temporal aberrations of the first order in designing image intensifiers and high-speed photographic image tubes that had scientific significance and practical applications.

The study of lateral aberrations and temporal aberrations of electron optical imaging systems in the 1950s and 1970s started to investigate the paraxial lateral chromatic aberration of the second order and the paraxial temporal chromatic aberration of the first order. Conclusions were as following.

The spatial-trajectory-spread of electron optical imaging systems is mainly determined by the second order paraxial lateral aberration, which can be expressed by the Recknagel-Artimovich formula [2,3]

$$\Delta r_2^* = \frac{2M}{E_c}\sqrt{\varepsilon_r}(\sqrt{\varepsilon_z} - \sqrt{\varepsilon_{z_1}}) \tag{1}$$

The transit-time-spread of electron optical imaging systems is mainly determined by paraxial temporal chromatic aberration of the first order as

$$\Delta T_1 = \sqrt{\frac{2m_0}{e}}\frac{1}{E_c}\sqrt{\varepsilon_z} \tag{2}$$

which has been given by Savoisky-Fanchenko[4] and proven by Csorba[5].

In Eq. (1) and Eq. (2), $\sqrt{\varepsilon_z}$ and $\sqrt{\varepsilon_r}$ are the initial axial energy and initial radial energy of electrons emitted from the photocathode, respectively, M is the magnification of the system, E_c is the strength of the electric-field at the photocathode, E_c takes a negative value, and e/m_0 is the ratio of electron charge to mass. In Eq. (1), $\sqrt{\varepsilon_{z_1}}$ is a given initial axial energy of electron corresponding to paraxial image position, and lateral aberrations are inspected only at this ideal image plane.

The difference between Eq. (1) expressing paraxial lateral aberration of the second order and Eq. (2) expressing the first order paraxial temporal aberration is obvious. However, we are able to find out the non-harmonious and non-symmetry between the two equations. It may be seen that neither the lateral aberration (in the second order approximation) nor the temporal aberration (in the first order approximation) depends on the concrete electrode structure and axial potential distribution. It is strange to find that the former depends on $\sqrt{\varepsilon_{z_1}}$, which corresponds to the position of the ideal image plane, but the latter does not.

In fact, in terms of temporal and spatial aberrations, the initial energy spread of electrons emitted from the photocathode expressed as the spatial-trajectory-spread characteristics at the image plane are actually the same as the transit-time spread characteristics at a certain z plane. To define the temporal aberration, there should be a criterion to evaluate the time difference between two moving electrons. The non-harmonious relationship between Eq. (1) and Eq. (2) can only be explained by Eq. (1) taking $\sqrt{\varepsilon_{z_1}} = 0$ as a criterion to explore the temporal aberrations formed by initial axial energies $\sqrt{\varepsilon_{z_1}} \neq 0$ of electrons emitted from the photocathode.

If we take $\sqrt{\varepsilon_{z_1}}$ ($0 \leq \sqrt{\varepsilon_{\pi 1}} \leq \sqrt{\varepsilon_{\pi_{\max}}}$) a certain initial energy of electron emission emitted from the photocathode along the axial direction, as a criterion to explore ΔT_1, i.e., the paraxial temporal chromatic aberration of the first order, then Eq. (2) should be written as follows:

$$\Delta T_1 = \sqrt{\frac{2m_0}{e}} \frac{1}{E_c} (\sqrt{\varepsilon_z} - \sqrt{\varepsilon_{z_1}}) \tag{3}$$

Eq. (3) has been strictly verified by an electrostatic concentric spherical system model [6,7]. It is evident from the equation that temporal aberrations and its theory should include the relation with $\sqrt{\varepsilon_{z_1}}$.

1 Definition of temporal aberrations

According to our study of aberration theory of electron optics with wide beam focusing, as well as the aberration theory for a concentric spherical system of electrostatic focus, the lateral aberrations Δr of electron optical imaging systems can be defined as[8,9]:

$$\Delta r = \Delta r(z_i) = r_{\text{practical}}(z_i, \varepsilon_r^{1/2}, \varepsilon_z^{1/2}, r_0) - r_{\text{paraxial}}(z_i, \varepsilon_r^{1/2}, \varepsilon_z^{1/2}, r_0) \tag{4}$$

where $r_{\text{practical}}$, r_{paraxial} express the lateral height at an image plane z_i, formed by the practical ray and paraxial ray of electrons emitted from the photocathode going through the system, respectively; and r_0 is the initial height of electron emission. Taking the cylindrical coordinate system (z, r), the axial coordinate z is chosen from the photocathode $z_0 = 0$.

In Ref. [8, 10] we have proven that Eq. (4) can be expressed by

$$\Delta r = \Delta r^* + \delta r = \Delta r_2^* + \Delta r_3^* + \delta r_3 \tag{5}$$

For example, lateral aberrations Δr are defined as the sum of paraxial lateral aberration and geometrical lateral aberration δr, and the paraxial lateral aberration Δr^* is composed of paraxial lateral aberrations of the second order Δr_2^* and of third order Δr_3^*. The paraxial lateral aberration of the third order Δr_3^* and the geometrical lateral aberration of the third order δr_3 are in the same order of magnitude. δr_3 has been investigated in a lot in Ref. [11 – 13].

Since the lateral aberrations and the temporal aberrations of electron optical imaging systems actually are the same, with the initial energy spread and the initial angular spread of electrons emitted from the photocathode expressed as spatial-trajectory-spread characteristics at the ideal image plane or as transit-time spread characteristics at a certain z plane, including the ideal image plane.

The temporal aberrations can be similarly defined as Eq. (4):

$$\Delta t = \Delta t(z) = t_{\text{practical}}(z, \varepsilon_r^{1/2}, \varepsilon_z^{1/2}, r_0) - t_{\text{paraxial}}(z, \varepsilon_r^{1/2}, \varepsilon_z^{1/2}, r_0) \tag{6}$$

where $t_{\text{practical}}$, t_{paraxial} express the time when the practical ray and paraxial ray of electrons emitted from the photocathode going through the system, respectively.

The difference between Eq. (4) and Eq. (6) is that the lateral aberration of electron optical imaging systems is measured at an ideal image plane z_i (real or virtual) and corresponds to $\sqrt{\varepsilon_{z_1}}$, while the temporal aberration of such systems is measured at a certain z plane, including the real image plane. The lateral aberrations are inspected only at this ideal image plane.

We have proven that t_{paraxial} does not depend on $\varepsilon_r^{1/2}$ and r_0[14]. Substituting t, t^* for $t_{\text{practical}}$, t_{paraxial}, respectively; then Eq. (6) can be expressed as

$$\Delta t = t(z, \varepsilon_r^{1/2}, \varepsilon_z^{1/2}, r_0) - t^*(z, \varepsilon_{z_1}^{1/2}) \tag{7}$$

Similarly to Eq. (5), we may express Eq. (7) as

$$\Delta t = \Delta T(z, \varepsilon_z^{1/2}, \varepsilon_{z_1}^{1/2}) + \Delta\tau(z, \varepsilon_r^{1/2}, \varepsilon_z^{1/2}, r_0) = \Delta T + \Delta\tau \tag{8}$$

where

$$\Delta T = t^*(z, \varepsilon_z^{1/2}) - t^*(z, \varepsilon_{z_1}^{1/2}) = \Delta T_1 + \Delta T_2 \tag{9}$$

$$\Delta\tau = t(z, \varepsilon_r^{1/2}, \varepsilon_z^{1/2}, r_0) - t^*(z, \varepsilon_z^{1/2}) = \Delta\tau_2 \tag{10}$$

in which ΔT is called the paraxial temporal aberration, which expresses time difference between two paraxial rays with different initial axial energies. ΔT can be divided into two parts, the paraxial temporal aberration of the first order ΔT_1 and the paraxial temporal aberration of the second order ΔT_2. $\Delta\tau(\Delta\tau_2)$ is called the geometrical temporal aberration of the second order, which expresses the time difference between practical ray and paraxial ray with the same initial axial energies. Here, ΔT_2 and $\Delta\tau_2$ are in the same order of magnitude. The definition of temporal aberrations expressed by Eq. (9) and Eq. (10) will be the starting point of our investigation.

2 Expressions for temporal aberration coefficients solved by "Direct Integral Method" [14-16]

The most direct and simplest approach for solving temporal aberration coefficients is undoubtedly to integrate the electron motion equation in the electrostatic field with axial symmetry

$$\ddot{z} = \frac{e}{m_0}\frac{\partial \varphi}{\partial z} \tag{11}$$

After the axial velocity $\dot{z}$ is obtained, we may integrate $\frac{1}{\dot{z}} = \frac{dt}{dz}$, then we can derive the expression of time t.

Using the Scherzer series expansions of spatial potential distribution $\varphi = \varphi(z,r)$

$$\varphi(z,r) = \phi(z) - \frac{r^2}{4}\phi''(z) + \frac{r^4}{64}\phi^{IV}(z) - \cdots \tag{12}$$

and substituting Eq. (12) to Eq. (11), we get

$$\dot{z}^2 = \frac{2e}{m_0}\int \phi'(z)dz - \frac{2e}{m_0}\int \frac{r^2}{4}\phi'''(z)dz + \frac{2e}{m_0}\varepsilon_z$$

To solve the integral $\frac{r^2}{4}\phi'''(z)dz$ in the expression above, we will meet a problem for solving double integrals — the most difficult part of the investigation. The problem was settled by using the paraxial electron motion equation and its two special solutions. Through a series of complicated transformations, we may obtain the time expression of an electron which reaches a certain position z in the system under the following initial conditions: the initial position vector r_0, the initial direction angle θ_0, the initial emission angle α_0, the initial position angle β_0, and the initial emission velocity v_0, i.e., initial energy of electron

$$\varepsilon_z^{1/2} = \varepsilon_0^{1/2}\cos\alpha_0, \quad \varepsilon_r^{1/2} = \varepsilon_0^{1/2}\sin\alpha_0,$$

$$t(z,\varepsilon_r^{1/2},\varepsilon_z^{1/2},r_0) = \int_0^z \frac{dz}{\left(\frac{2e}{m_0}\right)^{1/2}[\phi(z)+\varepsilon_z]^{1/2}}\left\{1 + \frac{1}{2}\frac{1}{[\phi(z)+\varepsilon_z]}\left[\varepsilon_r\left(v'^2\phi_* + \frac{1}{4}\phi''v^2 - 1\right) + r_0^2\left(w'^2\phi_* + \frac{1}{4}\phi''w^2\frac{1}{4}\phi''_0 - \frac{1}{4}\phi''_0\right) + 2r_0\sqrt{\varepsilon_r}\cos(\theta_0-\beta_0)\left(v'w'\phi_* + \frac{1}{4}\phi''vw\right)\right]\right\} \quad (13)$$

where $\phi(z)$ is the axial potential distribution, $\phi'(z)$; $\phi''(z)$ are the first and second orders derivatives of $\phi(z)$ with respect to z. $v = v(z)$; $w = w(z)$ are two linear independent special solutions of the paraxial electron motion equation satisfying the following initial conditions:

$$v(z_0=0) = 0, \quad v'(z_0=0) = \frac{1}{\sqrt{\varepsilon_z}}, \quad (14)$$

$$w(z_0=0) = 1, \quad w'(z_0=0) = 0$$

Considering the paraxial condition of electron trajectories and expanding the first order term in expression (13), from the definition of paraxial temporal aberration (9), we can get the expressions of paraxial temporal aberrations ΔT and its aberration coefficients:

$$\Delta T = a_2(\varepsilon_z^{1/2} - \varepsilon_{z_1}^{1/2}) + A_{22}(\varepsilon_z - \varepsilon_{z_1}) \quad (15)$$

where a_2 is the paraxial temporal chromatic aberration coefficient of the first order

$$a_2 = -\sqrt{\frac{2m_0}{e}}\frac{1}{\phi'(0)} = \sqrt{\frac{2m_0}{e}}\frac{1}{E_c} \quad (16)$$

A_{22} is the paraxial temporal chromatic aberration coefficient of the second order

$$A_{22} = \frac{1}{2}\sqrt{\frac{2m_0}{e}}\left\{\frac{1}{\sqrt{\phi(z)}}\frac{1}{\phi'(z)} + \int_0^z \frac{\phi'(z)}{\sqrt{\phi(z)}[\phi'(z)]^2}dz\right\} \quad (17)$$

Similarly, expanding the second order term in expression (13), from the definition of geometrical temporal aberrations of expression (10), we can get the expressions of geometrical temporal aberrations $\Delta\tau$ and its aberration coefficients as follows:

$$\Delta\tau = A_{11}\varepsilon_r + 2A_{13}\varepsilon_r^{1/2}r_0 + A_{33}r_0^2 \quad (18)$$

where

$$A_{11} = \int_0^z \frac{1}{2(2e/m_0)^{1/2}[\phi(z)+\varepsilon_z]^{3/2}}\left(v'^2\phi_* + \frac{1}{4}\phi''v^2 - 1\right)dz \quad (19)$$

$$A_{13} = \int_0^z \frac{1}{2(2e/m_0)^{1/2}[\phi(z)+\varepsilon_z]^{3/2}}\cos(\theta_0-\beta_0) \times \left(v'w'\phi_* + \frac{1}{4}\phi''vw\right)dz \quad (20)$$

$$A_{33} = \int_0^z \frac{1}{2(2e/m_0)^{1/2}[\phi(z)+\varepsilon_z]^{3/2}} \times \left(w'^2\phi_* + \frac{1}{4}\phi''w^2 - \frac{1}{4}\phi''_0\right)dz \quad (21)$$

where A_{11} is the temporal spherical aberration coefficient, A_{13} is the temporal aberration coefficient of field of curvature, and A_{33} is the temporal distortion aberration coefficient of the second order.

3 Verification of temporal aberration coefficients deduced from "direct integral method" by an electrostatic concentric spherical system model [7]

The correctness of expressions for temporal aberration coefficients deduced from the "Direct Integral Method" should be tested and verified. We take a bi-electrode electrostatic concentric spherical system as an ideal model, in which the electrostatic potential distribution and the moving electron rays can be expressed by distinct analytical expressions. Although the system has been investigated by many authors[17-19], the time of moving electrons expressed by analytical solution has not been given. If we can get the analytical expression of time for moving electron rays in this system, it may be used as a criterion to test and verify the correctness of the "Direct Integral Method".

Suppose that in the bi-electrode electrostatic concentric spherical system the curvature radii of spherical cathode C and of the mesh-spherical anode A are R_c and R_a, respectively, the common curvature center is O. Because of spherical symmetry, we can use the polar coordinate (ρ, φ) to describe the electron ray. Locating the origin of the polar coordinate at the curvature center O, the potential of the photocathode is $\phi_c = 0$, the potential of mesh-anode A with respect to the photocathode is ϕ_{ac}, and the potential at polar coordinate ρ with respect to photocathode is $\phi_{\rho c}$.

The motion of electron path follows the law of energy conservation and the law of angular momentum conservation:

$$\frac{1}{2}m_0 v^2 = \frac{1}{2}m_0 v_0^2 + e\phi_{\rho c} \qquad (22)$$

$$\rho^2 \varphi = R_c^2 \varphi_0 = R_c v_0 \sin\alpha_0 \qquad (23)$$

Starting from Eq. (22) and Eq. (23), we may obtain the analytical solution of time t, when the electron ray emitted from the photocathode goes through:

$$t = \sqrt{\frac{m_0}{2e\phi_{ac}/(n-1) - \varepsilon_0}} \frac{1}{} \times \Bigg\{ \sqrt{-\rho^2[\phi_{ac}/(n-1) - \varepsilon_0] + \phi_{ac}/(n-1)R_c\rho - R_c^2\varepsilon_0 \sin^2\alpha_0} - $$

$$\frac{1}{2}\frac{\phi_{ac}/(n-1)}{\sqrt{\phi_{ac}/(n-1) - \varepsilon_0}} \times $$

$$R_c \arctan \frac{-\phi_{ac}/(n-1)R_c + 2\rho[\phi_{ac}/(n-1) - \varepsilon_0]}{2\sqrt{\phi_{ac}/(n-1) - \varepsilon_0}\sqrt{-\rho^2[\phi_{ac}/(n-1) - \varepsilon_0] + \phi_{ac}/(n-1)R_c\rho - R_c^2\varepsilon_0 \sin^2\alpha_0}} + $$

$$\frac{1}{2}\frac{\phi_{ac}/(n-1)}{\sqrt{\phi_{ac}/(n-1) - \varepsilon_0}} R_c \arctan \frac{R_c\sqrt{\varepsilon_0 \cos^2\alpha_0} + 2\varepsilon_0 - \phi_{ac}/(n-1)}{2\sqrt{\phi_{ac}/(n-1) - \varepsilon_0}\sqrt{\varepsilon_0 \cos^2\alpha_0}} \Bigg\} \qquad (24)$$

where $n = R_c/R_a$.

Expanding Eq. (24) to the forms of expressions Eq. (15) and Eq. (18) according to definitions Eq. (9) and Eq. (10), we can get the expressions for paraxial temporal chromatic aberration coefficients of the first order a_2 and of the second order A_{22}:

$$a_2 = \sqrt{\frac{2m_0}{e}} \frac{1}{E_c} \tag{25}$$

$$A_{22} = -\sqrt{\frac{2m_0}{e}} \frac{1}{E_c} \left(\frac{n-1}{\phi_{ac}}\right)^{1/2} \left[\frac{-(R_c+z)(2R_c-z)}{4R_c\sqrt{-z(R_c+z)}} + \frac{3}{4}\left(\arcsin\sqrt{\frac{-z}{R_c}}\right) \right] \tag{26}$$

Now, we shall test and verify the correctness of expressions for temporal aberration coefficients deduced from "Direct Integral Method". Substituting the expression of axial potential distribution $\phi(z)$ in the electrostatic concentric spherical electrostatic system model

$$\phi(z) = \frac{\phi_{ac}}{n-1} \frac{-z}{(z+R_c)} \tag{27}$$

and the two special solutions of paraxial ray equation[5,9]

$$v(z) = \frac{2z}{\phi(z)} [\sqrt{\phi(z)+\varepsilon_z} - \sqrt{\varepsilon_z}] \tag{28}$$

$$w(z) = 1 + \frac{1}{R_c}z - \frac{\sqrt{\varepsilon_z}}{R_c} \frac{2z}{\phi(z)} [\sqrt{\phi(z)+\varepsilon_z} - \sqrt{\varepsilon_z}] \tag{29}$$

into Eq. (16) and Eq. (17), we can get the same results as Eq. (25) and Eq. (26). Thus, we have proven that the "Direct Integral Method" is correct.

Similarly, from Eq. (19) and Eq. (21), by using the analytical solutions of electrostatic concentric spherical electrostatic system models Eq. (27) and Eq. (29), we can also obtain the solution of coefficients of the geometrical temporal aberrations

$$A_{11} = \frac{3}{4}\sqrt{\frac{2m_0}{e}} \left(\frac{n-1}{\phi_{ac}}\right)^{1/2} \frac{1}{E_c R_c} \left(-\sqrt{z}\sqrt{-z-R_c} - R_c \arctan\frac{\sqrt{z}}{\sqrt{-z-R_c}} \right) \tag{30}$$

$$A_{13} = -\cos(\theta_0 - \beta_0) \frac{1}{2} \sqrt{\frac{2m_0}{e}} \frac{1}{R_c^2 E_c} z \tag{31}$$

$$A_{33} = \sqrt{\frac{2m_0}{e}} \sqrt{\frac{n-1}{\phi_{ac}}} \frac{3}{8R_c^2} \left(\sqrt{z}\sqrt{-z-R_c} - R_c \arctan\frac{\sqrt{z}}{\sqrt{-z-R_c}} \right) \tag{32}$$

4 Comparison for studying temporal aberration theory between "Direct Integral Method" and "τ-Variation Method"[1,19]

Our study has shown that the expression of coefficients for paraxial temporal chromatic aberration of the first order a_2 deduced from "τ-Variation Method" is completely identical with Eq. (16). The expressions of coefficients for paraxial temporal chromatic aberration of the second order A_{22} and coefficients for geometrical temporal aberrations of the second order A_{11}, A_{13}, A_{33} can be written as follows:

$$\begin{aligned} A_{22} &= -\frac{\ddot{z}}{2\dot{z}^3} z^2_{\alpha_2} + \frac{\dot{z}_{\alpha_2} z_{\alpha_2}}{\dot{z}^2} - \frac{z_{\alpha_2\alpha_2}}{2\dot{z}}, \\ T^{(z)}(z_{\alpha_2\alpha_2}) &= \phi\phi''' \frac{2}{\phi_0'^2} \end{aligned} \tag{33}$$

$$A_{11} = -\frac{z_{\alpha_1\alpha_1}}{2\dot{z}},$$
$$T^{(z)}(z_{\alpha_1\alpha_1}) = -\frac{1}{4}\phi'''v^2 \tag{34}$$

$$A_{13} = -\frac{z_{\alpha_1\alpha_3}}{2\dot{z}},$$
$$T^{(z)}(z_{\alpha_1\alpha_3}) = -\frac{1}{4}\phi'''vw\cos(\theta_0 - \beta_0) \tag{35}$$

$$A_{33} = -\frac{z_{\alpha_3\alpha_3}}{2\dot{z}},$$
$$T^{(z)}(z_{\alpha_3\alpha_3}) = -\frac{1}{4}\phi'''w^2 \tag{36}$$

where in expressions Eq. (33) and Eq. (36), $z_{\alpha_i\alpha_j}$ is the solution of the following differential equations:

$$T^{(z)}(z_{\alpha_i\alpha_j}) = \frac{d}{dz}\left(\phi\frac{d}{dz}z_{\alpha_i\alpha_j} - \frac{1}{2}\phi' z_{\alpha_i\alpha_j}\right) \quad (i=1,2,3; j=1,2,3) \tag{37}$$

and each expression in Eq. (33) can be written as

$$z_{\alpha_2} = \frac{2}{\phi'_0}\sqrt{\phi},\quad \dot{z} = \sqrt{\frac{2e\phi}{m_0}},$$
$$\dot{z}_{\alpha_2} = \sqrt{\frac{2e}{m_0}}\frac{\phi'}{\phi'_0} = \sqrt{\frac{2e}{m_0}}\frac{R_c^2}{(-R_c-z)^2},\quad \ddot{z} = \frac{e}{m_0}\phi' \tag{38}$$

Substituting the expression Eq. (27) of axial potential distribution $\phi(z)$ of electrostatic concentric spherical electrostatic system model and the two special solutions Eq. (28) and Eq. (29) of its paraxial ray equation into Eq. (33) and Eq. (36), we can prove that the expressions of coefficients for paraxial temporal chromatic aberration of the second order A_{22} and coefficients for geometrical temporal aberrations of the second order A_{11}, A_{13}, A_{33} are completely identical with Eq. (26) and Eq. (30), Eq. (32). Thus, we fully proved the correctness and equivalence of the "τ Variation Method" and "Direct Integral Method" for studying temporal aberration theory.

5 Discussion of scientific method in this research[20]

A British philosopher of science K. R. Popper has put forward a deductive and inspection method as a scientific method (also called as trial and error method) for research, which can be summarized by the following route.

P_1(Problem1)→TT(Tentative Theory)→EE
(Elimination of Error)→P_2 (Problem 2) …

The route shows that we can start from a theoretical or historical problem P_1; then we make a tentative answer (a guess or a hypothetical answer), and give a tentative theory TT; after that we submit it to a critical discussion according to evidence collected as the step to eliminate error EE. If it can be confirmed, a new problem P_2 appears.

Now, I would like to talk about what we have learned from Popper's scientific method in this research.

Twenty eight years ago, two Russian scientists put forward a temporal aberration theory for studying electron optical imaging systems, named "τ-Variation Method", which no one has been skeptical of since then. When I began to work on the theory, two problems arose in my mind. The first was about the correctness of the "τ-Variation Method" and if I will be able to find a simpler and direct approach to study the temporal aberration theory. The second was about the accuracy of coefficients of temporal aberrations given by the "τ-ariation method", and if I will be able to find an ideal model to test and verify the coefficients of temporal aberrations. These two problems, the correctness of aberration theory and the accuracy of expressions of coefficients given by "τ Variation Method", have not been thoroughly studied. The first problem that I raised is actually to make an attempt to test the "τ-variation method", confirming or negating it, or find a better theory. The second problem that I raised is actually to verify the degree to trust the theory and its feasibility. Such kind of thought is actually Problem 1 in the deductive and inspection method. This thought gives out a so-called "Direct Integral Method" for studying temporal aberration theory, and a new definition of temporal aberrations has been produced, i.e., that temporal aberrations should be composed of paraxial temporal aberration and geometrical temporal aberration, in which the criterion of comparison is the paraxial electron ray with axial electron energy $\sqrt{\varepsilon_{z_1}}(0 \leq \sqrt{\varepsilon_{z_1}} \leq \sqrt{\varepsilon_{0\max}}$. This process is just the tentative theory in the deductive and inspection method. This research shows that the solution of coefficients of temporal aberrations can be directly expressed by integral forms, and that there is no need to solve the aberration coefficients via differential equations deduced from "τ-Variation Method".

Up to this step, the problem "which one is right and which one is wrong of these two methods has not been solved". We must figure out an approach to make a critical testing for these two methods. This step is the elimination of error in the deductive and inspection method. In this step, we have discovered an ideal model, i.e., a bi-electrode spherical concentric system with electrostatic focusing, and found the analytical solution of moving time of electrons going through the system. Result of this test and verification has shown that these two methods are not only correct, but also accurate. The correctness comes from the completely identical results obtained from two different approaches, and we have also proven that the result of τVM can be transformed to the form given by the DIM. The accuracy comes from the result that two approaches are in complete agreement with the analytical solutions given by the ideal model. The study shows that the new DIM for temporal aberration theory contains the former τVM. In view of this, we can say that the DIM for temporal aberration theory is a better one. Thus, we reach Problem 2, and progress of science has been promoted.

It seems that the DIM for temporal aberration theory given by the present paper satisfies a condition of good theoretical pattern in science, which has the following characteristics:

(1) The new method DIM starts from the motion equation of electron optics under the

fundamental laws of physics. The premises are clear, the concept is distinct, the mathematical deduction is correct, and the mathematical forms obtained by the new method are simpler and clearer than the former method τVM.

(2) The new method DIM not only can explain all phenomena which can be explained by the former one, but also extend the former method in which the criterion of comparison is based on axial energy $\sqrt{\varepsilon_{z_1}}=0$ to the new one in which the criterion of comparison is based on axial energy $0 \leqslant \sqrt{\varepsilon_{z_1}} \leqslant \sqrt{\varepsilon_{0\max}}$.

(3) The two methods have been verified and tested by using analytical solutions of a bi-electrode spherical concentric model, and it has been proven that both methods are correct and accurate. This means that the two approaches for solving temporal aberration theory reached the same goal by taking different routes.

(4) A new definition of temporal aberrations has been put forward in dynamic electron optics study with the starting point that temporal and lateral aberrations are actually the space-time expression of the same phenomenon. Since lateral aberrations are composed of paraxial lateral aberration and geometrical lateral aberration, the temporal aberrations should also be composed of paraxial temporal aberration and geometrical temporal aberration. The hypothesis of introducing the new definition advances electron optics theory, which connects an internal relationship between static electron optics and dynamic electron optics.

(5) The new method DIM has predictive property. It can make a prediction or deduction for dynamic electron optics with electro-magnetic focusing, in which the temporal aberration has had limited investigation. We shall prove that for either the electron optical imaging systems with electro-magnetic focusing, or the the electron optical imaging systems with electrostatic focusing, the paraxial temporal aberration of the first order is related only to electrostatic field strength at the photocathode, which will be still expressed by Eq. (16).

From the study of this scientific problem, we can find the directive function of methodology.

(1) Starting from a logical point of view, we found the contradiction between the spatial aberration theory and the temporal aberration theory, i.e., their non-harmonious relationship and their non-symmetry, and we put forward a point of breakthrough for studying temporal aberration theory.

(2) We have connected the spatial aberration problem with the temporal aberration problem, and found that the initial energy spread of electrons emitted from the photocathode expressed as spatial-trajectory-spread characteristics at the image plane can also be expressed as transit-time spread characteristics at a certain z plane. The two kinds of aberrations can be unified from the point of view of definition.

(3) We have applied a method of analogy in this research, transplanted the method for studying lateral aberration theory to the study of temporal aberration theory, and put forward a new definition for temporal aberrations.

(4) We have applied a testing method by using an ideal model, and verified that the new

method and the former method are correct and equivalent. Thus, we reached a scientific conclusion through a series of comparisons.

Our study shows that the "direct integral method" to study electron optical imaging systems is only related to solving integral expressions, which are more convenient for computation and could be recommended for the practical design.

Conclusions

In this paper. we described what we have obtained and what we have learned in studying the temporal aberration theory in terms of methodology. A new approach to the theory of temporal aberration for electron optical imaging systems called as "Direct Integral Method" is put forward. A new definition of temporal aberration is given, where a certain initial energy of electron mission along the axial direction $\sqrt{\varepsilon_{z_1}}(0 \leqslant \sqrt{\varepsilon_{z_1}} \leqslant \sqrt{\varepsilon_{0max}})$ as criterion is considered. New expressions of the temporal geometrical aberration coefficients of the second order in integral forms have been obtained by "DIM". All of the formulae for temporal aberration coefficients deduced from the "Direct Integral Method" and "τ-Variation Method" have been verified by an electrostatic concentric spherical system model. Contrasting their results with analytical solutions shows that these two methods have complete identical solutions. It can be concluded that both methods are equivalent and correct. It should be pointed out that the expressions for solving temporal aberration coefficients of the second order A_{11}, A_{13}, A_{33} and A_{22} given by the "Direct Integral Method" are related to solve integral expressions, which are more convenient for computation and could be suggested for use in the practical design of electron optical imaging systems. What we have learned from the study for the theory of temporal aberrations from the point of view of methodology has been summarized.

Acknowledgements

This project was supported by National Natural Science Foundation of China and International Bureau of Cooperation, as well as the Special Science Fund of Doctoral Discipline for Institutions of Higher Learning.

References

[1] MONASTYRSKI M A, SCHELEV M Y. Theory of temporal aberrations of cathode lenses [M]. Moscow: Lebedeev Institute of Physics, 1980.

[2] RECKNAGEL A. Theorie des elektrisohen elecktronen miktroskops fur selbstrakler[J]. Z. Angew. Physik, 1941, 117:689 - 708.

[3] ARTIMOVICH L A. Electrostatic properties of emission systems[J]. Bulletin of Academy of Sciences, USSR Physics Series, 1994, 8(6):313 - 328.

[4] SAVOISYK Y K, FANCHENKO S D. Physical foundation of electron-optical chronograph [J]. Report of Academy of Sciences, USSR, 1956, 108(2):218 - 221(in Russian).

[5] CSORBA I P. Chromatic aberration limited image transfer characteristics of image tube lenses of simple geometry[J]. RCA Review,1970,31(3):534-550

[6] ZHOU L W,LI Y,ZHANG Z Q,et al. Test and verification of temporal aberration theory for electron optical imaging systems by an electrostatic concentric spherical system[J]. Acta Physica Sinica,2005,54(8):3597-3603(in Chinese)

[7] ZHOU L W. Electron optics with wide beam focusing[M]. Beijing:Beijing Institute of Technology Press,1993(in Chinese)

[8] ZHOU L W,AI K C,Pan S C. On aberration theory of cathode lenses with combined electromagnetic focusing[J]. Acta Physica Sinica,1983,32(3),376-392(in Chinese)

[9] ZHOU L W. Electron optics of concentric spherical system composed of two electrodes [M]//Focusing and Imaging of Wide Electron Beams—Selected Papers on Electron Optics by Zhou Liwei. Beijing:Beijing Institute of Technology Press,1994,11-29(in Chinese)

[10] XIMEN J Y. Electron-optical properties and aberration theory of combined immersion objectives[J]. Acta Physica Sinica,1957,13(4):339-356(in Chinese)

[11] KULIKOV Y V,MONASTYRSKI M A,FEIDING H E. Aberration theory of third order of cathode lenses:aberrations of cathode lenses with combined electric and magnetic fields[J]. Radiotechnics and Electronics,1978,23(1):167-174

[12] XIMEN J Y,ZHOU L W,AI K C. Variation theory of aberrations in cathode lenses[J]. Optik,1983,66:19-34.

[13] ZHOU L W,LI Y,ZHANG Z Q,et al. On the theory of temporal aberrations for cathode lenses[J]. Optik,2005,116(4):175-184

[14] ZHOU L W,LI Y,ZHANG Z Q,et al. Theory of temporal aberrations for electron optical imaging systems by "Direct Integral Method"[J]. Acta Physica Sinica,2005,54(8):3591-3596 (in Chinese).

[15] ZHOU L W,LI Y,ZHANG Z Q,et al. On the theory of temporal aberrations for electron optical imaging systems by using "Direct Integral Method"[C]. Proceedings of SPIE,2005,5580: 710-724.

[16] RUSKA E. Zur fokussierbarkeit von kathoden-strahlbundeln grosser ausgangsquerchnitte [J]. Z. Angew. Physik,1993,83(9):684-687(in German).

[17] SCHAGEN P,BRUINING H,FRANCKEN J C. A simple electrostatic electron-optical system with only one voltage[J]. Philips Research Reports,1952,7(2):119-130.

[18] ZHOU L W. Electron optics of concentric spherical electro-magnetic focusing systems [J]. Advances in Electronics and Electron Physics,1979,52:119-132.

[19] ZHOU L W,MONASTYRSKI M A,SCHELEV M Y,et al. On the temporal aberration theory of electron optical imaging systems by "τ-Variation Method"[J]. Acta Electronica Sinica, 2006,34(2):193-197(in Chinese).

[20] LAI H L,JIN T J. Biography of K. Popper[M]. Shijiazhuang:Hebei People's Publicing House,1998.

11.7 直接积分法研究电子光学成像系统的时间像差理论
On Temporal Aberration Theory of Imaging Electron Optical System by Using Direct Integral Method

摘要：本文提出了计算动态电子光学成像系统时间像差系数的新方法——"直接积分法"。以从阴极面逸出的轴向电子初能为 ε_{z_1} 的近轴电子轨迹为比较基准，给出了时间像差的定义，详细叙述了直接积分法并给出了求解动态电子光学成像系统时间像差系数的积分表达式。研究表明，若采用"τ-变分法"求得的二级几何时间像差系数必须求解微分方程，而直接积分法求得的二级几何时间像差系数全部以积分形式表示，仅需进行积分运算，更适用于成像系统的实际计算与设计。

Abstract: A new approach to calculate the temporal aberration coefficients of dynamic electron optical imaging systems is put forward in the present paper. A new definition of temporal aberration is given in which a certain initial energy of electron emission along the axial direction ε_{z_1} as a criterion is considered. A new method to calculate the temporal aberration coefficients of dynamic electron optical imaging systems named "direct integral method" is presented which gives new expressions of the temporal aberration coefficients expressed in integral forms. The difference between "Direct Integral Method" and "τ-Variation Method" is that the "τ-Variation Method" needs to solve the differential equations for the three of temporal geometrical aberration coefficients of the second order, while the "direct integral method" needs only to carry out the integral calculation of them, which is more convenient and suitable for computation in the practical design.

引言

在宽束动态电子光学成像系统的计算与设计中，时间像差理论占有一个重要的位置。1957 年，Savoisky 和 Fanchenko[1]首先提出了时间渡越弥散即一级时间像差的表达式。1971 年，Csorba[2]对其进行了证明。1980 年，Monastyrski 和 Schelev[3]用"τ-变分法"提出了一种较为完整的时间像差理论，给出了计算动态电子光学成像系统一级和二级时间像差系数的表达式。

文献 [4] 中已经证明，电子光学成像系统的空间弥散特性主要由二级横向色差决定，它可由 Recknagel-Artimovich 公式表示

$$\Delta r_1^* = \frac{2M}{E_c}\sqrt{\varepsilon_r}(\sqrt{\varepsilon_z} - \sqrt{\varepsilon_{z_1}}) \tag{1}$$

周立伟[a]，李元[a]，张智诠[b]，M. A. Monastyrski[c]，M. Y. Schelev[c]. a) 北京理工大学，b) 装甲兵工程学院，c) 俄罗斯科学院普罗霍洛夫普通物理研究所. 物理学报, V. 54, No. 8, 2005, 3591–3596.

而成像系统的时间弥散特性主要由一级时间像差（即时间色差）ΔT_1 以 Savoisky-Fanchenko 公式[1]表示

$$\Delta T_1 = \sqrt{\frac{2m_0}{e}} \frac{1}{E_c} \sqrt{\varepsilon_z} \qquad (2)$$

式中，ε_z、ε_r 分别为电子从光阴极发射时的轴向初能量和径向初能量，ε_{z_1} 为与理想成像位置相对应的电子轴向初能量，M 为系统放大率，E_c 为阴极面上的场强（取负值），e/m_0 为电子的荷质比。

不难看出，式（1）、式（2）之间存在某种不协调性和不对称性。式（1）、式（2）表明，系统的横向色差在二级近似下，时间像差在一级近似下，与电极结构和电位分布无关，而与电子的轴向逸出初能 ε_z、阴极面上的场强 E_c 有关。这两个公式的差异在于，式（1）中含有放大率参量 M 以及与理想成像位置相对应的电子轴向初能量 ε_{z_1} 的参量；但式（2）中既不包含放大率的参量，也不包含相应的 ε_{z_1} 参量。同样，在"τ-变分法"的时间像差理论中，也并没有与 ε_z 相比较作为基准的 ε_{z_1} 参量。

关于电子光学成像系统的空间像差或时间像差，这里指的像差即所谓色差，它实际上是同一事物（即由阴极面逸出的光电子的发射初能量分散）在某一成像面上所显现的空间弥散特性或是在空间某一位置（包括成像面位置）处所显现的时间弥散特性。我们可以理解式（2）没有引入放大率的参量，因为时间像差乃是两条不同的电子轨迹（实际轨迹与近轴轨迹）在任一位置 z（包括成像面位置）处的时间离散，而空间像差必须在 ε_{z_1} 所对应的成像位置处衡量。但是在式（2）中没有引入 ε_{z_1} 的参量显然是不协调的。这种不协调性唯一可以统一起来的解释是：目前普遍采用的评价时间渡越弥散或时间像差理论是以某一轴向初能量 $\varepsilon_{z_1} = 0$ 的电子作为比较基准来探讨轴向初能量 ε_z 不等于零的电子所构成的时间离散或形成的时间像差。

本文的目的是探讨一种更为简便的，能直接求解电子光学成像系统时间像差系数的新途径，它不同于现有的"τ-变分法"。文中以阴极面逸出的轴向电子初能 ε_{z_1} 在 $0 \leq \varepsilon_{z_1} \leq \varepsilon_{0\max}$ 的近轴电子轨迹为基准，研究时间像差的定义，给出了计算动态成像电子光学系统的时间像差系数的新方法——直接积分法。

1 时间像差的定义

按照我们对静电聚焦和电磁聚焦成像电子光学系统的像差理论研究[5,6]以及对静电聚焦同心球系统的电子光学成像理论的研究[7]，轴对称成像电子光学系统的时间像差可定义为

$$\Delta t = \Delta t(z) = t_{\text{real}}(z, \varepsilon_r^{1/2}, \varepsilon_z^{1/2}, \boldsymbol{r}_0) - t_{\text{prax}}(z, \varepsilon_{r_1}^{1/2}, \varepsilon_{z_1}^{1/2}, \boldsymbol{r}_0) \qquad (3)$$

式中，t_{real}、t_{prax} 分别表示由光阴极发射的实际电子轨迹和近轴电子轨迹到达同一位置所经历的时间；ε_z、ε_r 分别为电子从光阴极发射时的轴向初能量和径向初能量；ε_{z_1} 为作为比较基准的近轴电子轨迹的轴向初能量。取圆柱坐标系 (z, r)，轴向坐标 z 自阴极面 $z_0 = 0$ 算起，$\boldsymbol{r}_0$ 为电子出射的径向初始矢量。

将 t_{real}、t_{prax} 分别以 t、t^* 代之，下面将证明，近轴电子轨迹所经历的时间 t^* 与电子的

径向初能 $\varepsilon_{r_1}^{1/2}$、电子逸出高度 r_0 值无关。于是式（3）便可表示为

$$\Delta t = t(z, \varepsilon_r^{1/2}, \varepsilon_z^{1/2}, r_0) - t^*(z, \varepsilon_{z_1}^{1/2}) \tag{4}$$

我们将式（4）表达成以下的形式：

$$\Delta t = \Delta T(z, \varepsilon_z^{1/2}, \varepsilon_{z_1}^{1/2}) + \Delta\tau(z, \varepsilon_r^{1/2}, \varepsilon_z^{1/2}, r_0) = \Delta T + \Delta\tau \tag{5}$$

式中，ΔT 称为近轴时间像差或时间色差，它表示轴向初能不同的两条近轴电子轨迹的时间差异，由下式表示：

$$\Delta T = t^*(z, \varepsilon_z^{1/2}) - t^*(z, \varepsilon_{z_1}^{1/2}) \tag{6}$$

$\Delta\tau$ 称为几何时间像差，它表示轴向初能相同的实际电子轨迹与近轴电子轨迹的时间差异，由下式表示：

$$\Delta\tau = t(z, \varepsilon_r^{1/2}, \varepsilon_z^{1/2}, r_0) - t^*(z, \varepsilon_z^{1/2}) \tag{7}$$

由光电子初速度分布引起的时间像差归结于研究函数 Δt 在某一轴向位置 z 处与小参量 $\varepsilon_r^{1/2}$、$\varepsilon_z^{1/2}$、$\varepsilon_{z_1}^{1/2}$、r_0 之间的关系式，时间像差可以表示为

$$\Delta T = a_2(\varepsilon_z^{1/2} - \varepsilon_{z_1}^{1/2}) + A_{22}(\varepsilon_z - \varepsilon_{z_1}) \tag{8}$$

$$\Delta\tau = a_1\varepsilon_r^{1/2} + a_3 r_0 + A_{11}\varepsilon_r + 2A_{12}\varepsilon_r^{1/2}\varepsilon_z^{1/2} + 2A_{13}\varepsilon_r^{1/2}r_0 + 2A_{23}\varepsilon_z^{1/2}r_0 + A_{33}r_0^2 \tag{9}$$

式中，$a_i = a_i(z)(i=1,2,3)$，$A_{ij} = A_{ij}(z)(i,j=1,2,3)$ 分别称为成像电子光学系统的一级、二级时间像差系数。

2 "直接积分法"求解成像电子光学系统的时间像差系数

2.1 几何时间像差系数 A_{11}、A_{13} 和 A_{33} 的确定

如图 1 所示，电子自阴极面发射的初始位置矢量为 r_0，方向角为 θ_0，初始电子的逸出角为 α_0，方位角为 β_0，电子逸出的初速度为 v_0，且有

$$v_0 = \sqrt{\frac{2e}{m_0}\varepsilon_0}, \quad \varepsilon_z^{1/2} = \varepsilon_0^{1/2}\cos\alpha_0, \quad \varepsilon_r^{1/2} = \varepsilon_0^{1/2}\sin\alpha_0$$

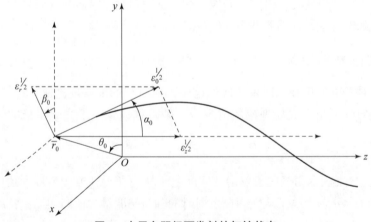

图 1 电子自阴极面发射的初始状态

轴对称静电场下电子运动方程为

$$\ddot{z} = \frac{e}{m_0}\frac{\partial \phi}{\partial z} \tag{10}$$

利用空间电位分布 $\varphi = \varphi(z, \boldsymbol{r})$ 的谢尔赤级数展开式：

$$\varphi(z, r) = \phi(z) - \frac{\boldsymbol{r}^2}{4}\phi''(z) + \frac{\boldsymbol{r}^4}{64}\phi^{(4)}(z) - \cdots \tag{11}$$

式中，$\phi(z)$ 表示轴上电位分布，$\phi'(z)$、$\phi''(z)$、$\phi'''(z)$、$\phi^{(4)}(z)$ 乃是 $\phi(z)$ 对 z 的导数。

将式（11）代入式（10），可得

$$\mathrm{d}(\dot{z}^2) = \frac{2e}{m_0}\left[\phi'(z) - \frac{\boldsymbol{r}^2}{4}\phi'''(z)\right]\mathrm{d}z \tag{12}$$

对式（12）进行积分，于是有

$$\dot{z}^2 = \frac{2e}{m_0}\int \phi'(z)\,\mathrm{d}z - \frac{2e}{m_0}\int \frac{\boldsymbol{r}^2}{4}\phi'''(z)\,\mathrm{d}z + \frac{2e}{m_0}\varepsilon_z \tag{13}$$

为了求得式（13）等号右端第二项的积分，对 $\int \frac{\boldsymbol{r}^2}{4}\phi'''(z)\,\mathrm{d}z$ 做如下变换。

轴对称成像系统的近轴轨迹方程[5,7]

$$[\varphi(z) + \varepsilon_z]\boldsymbol{r}'' + \frac{1}{2}\varphi'(z)\boldsymbol{r}' + \frac{1}{4}\varphi''(z)\boldsymbol{r} = 0 \tag{14}$$

其解可表示为

$$\boldsymbol{r}(z) = \boldsymbol{r}_0 w(z) + \sqrt{\frac{m_0}{2e}}\dot{\boldsymbol{r}}_0 v(z) \tag{15}$$

这里，特解 $v = v(z)$，$w = w(z)$ 满足如下初始条件：

$$\begin{aligned}
v(z_0 = 0) &= 0, \\
v'(z_0 = 0) &= \frac{1}{\sqrt{\varepsilon_z}}, \\
w(z_0 = 0) &= 1, \\
w'(z_0 = 0) &= 0
\end{aligned} \tag{16}$$

于是有

$$\boldsymbol{r}^2(z) = r_0^2 w^2(z) + 2w(z)v(z)r_0\sqrt{\varepsilon_r}\cos(\theta_0 - \beta_0) + \varepsilon_r v^2(z) \tag{17}$$

将式（17）代入式（13）等号右端的第二项，有

$$\int \frac{\boldsymbol{r}^2}{4}\phi'''(z)\,\mathrm{d}z = \int \frac{1}{4}r_0^2 w^2(z)\phi'''(z)\,\mathrm{d}z + \int \frac{1}{2}r_0\sqrt{\varepsilon_r}\cos(\theta_0 - \beta_0)w(z)v(z)\phi'''(z)\,\mathrm{d}z + \int \frac{1}{4}\varepsilon_r v^2(z)\phi'''(z)\,\mathrm{d}z \tag{18}$$

先求式（18）等号右端的第二项：$\int \frac{1}{2}r_0\sqrt{\varepsilon_r}\cos(\theta_0 - \beta_0)w(z)v(z)\phi'''(z)\,\mathrm{d}z$。

$$\frac{\mathrm{d}}{\mathrm{d}z}\left(v'w'\phi_* + \frac{1}{4}\phi''vw\right) = \phi'v'w' + \phi_*v''w' + \phi_*v'w'' + \frac{1}{4}\phi'''vw + \frac{1}{4}\phi''v'w + \frac{1}{4}\phi''vw'$$

式中，$\phi_* = \phi(z) + \varepsilon_z$，并利用

$$\phi_*v'' = -\frac{1}{2}\phi'v' - \frac{1}{4}\phi''v$$

$$\phi_* w'' = -\frac{1}{2}\phi' w' - \frac{1}{4}\phi'' w$$

可得

$$\frac{\mathrm{d}}{\mathrm{d}z}\left(v'w'\phi + \frac{1}{4}\phi''vw\right) = \frac{1}{4}\phi'''vw \tag{19}$$

于是，最后可得

$$\int \frac{1}{4}w(z)v(z)\phi'''(z)\mathrm{d}z = \left(v'w'\phi_* + \frac{1}{4}\phi''vw\right) + C_1$$

由初始条件式（16），可得常数 $C_1 = 0$。故有

$$\int \frac{1}{2}r_0\sqrt{\varepsilon_r}\cos(\theta_0 - \beta_0)w(z)v(z)\phi'''(z)\mathrm{d}z = 2r_0\sqrt{\varepsilon_r}\cos(\theta_0 - \beta_0)\left(v'w'\phi + \frac{1}{4}\phi''vw\right) \tag{20}$$

现求式（18）等号右端的第一项。类似式（19）的推导，可得

$$\frac{\mathrm{d}}{\mathrm{d}z}\left(w'^2\phi_* + \frac{1}{4}\phi''w^2\right) = \frac{1}{4}\phi'''w^2 \tag{21}$$

两边积分，得

$$\int \frac{1}{4}w^2(z)\phi'''(z)\mathrm{d}z = \left(w'^2\phi_* + \frac{1}{4}\phi''w^2\right) + C_2$$

由初始条件式（16），可得常数 $C_2 = -\frac{1}{4}\phi_0''$，故有

$$\int \frac{1}{4}r_0^2 w^2(z)\phi'''(z)\mathrm{d}z = r_0^2\left(w'^2\phi + \frac{1}{4}\phi''w^2 - \frac{1}{4}\phi_0''\right) \tag{22}$$

同样，对于式（18）等号右端的第三项，两边积分，得

$$\int \frac{1}{4}v^2(z)\phi'''(z)\mathrm{d}z = \left(v'^2\phi_* + \frac{1}{4}\phi''v^2\right) + C_3 \tag{23}$$

由初始条件式（16），可得常数 $C_3 = -1$，故有

$$\int \frac{1}{4}\varepsilon_r v^2(z)\phi'''(z)\mathrm{d}z = \varepsilon_r\left(v'^2\phi_* + \frac{1}{4}\phi''v^2 - 1\right) \tag{24}$$

将式（20）、式（22）、式（24）代入式（13），便有

$$\dot{z}^2 = \frac{2e}{m_0}[\phi(z) + \varepsilon_z] - \frac{2e}{m_0}\Big[\varepsilon_r\left(v'^2\phi_* + \frac{1}{4}\phi''v^2 - 1\right) + r_0^2\left(w'^2\phi + \frac{1}{4}\phi''w^2 - \frac{1}{4}\phi_0''\right) +$$

$$2r_0\sqrt{\varepsilon_r}\cos(\theta_0 - \beta_0)\left(v'w'\phi_* + \frac{1}{4}\phi''vw\right)\Big] \tag{25}$$

这样，由 $\mathrm{d}t = \dfrac{\mathrm{d}z}{\dot{z}}$，便得到电子到达系统内某点的时间表达式

$$t(z, \varepsilon_r^{1/2}, \varepsilon_z^{1/2}, r_0) = \int_0^z \frac{\mathrm{d}z}{\left(\frac{2e}{m}\right)^{1/2}[\phi(z) + \varepsilon_z]^{1/2}}\bigg\{1 + \frac{1}{2}\frac{1}{[\phi(z) + \varepsilon_z]}\Big[\varepsilon_r\left(v'^2\phi_* + \frac{1}{4}\phi''v^2 - 1\right) +$$

$$r_0^2\left(w'^2\phi + \frac{1}{4}\phi''w^2 - \frac{1}{4}\phi_0''\right) + 2r_0\sqrt{\varepsilon_r}\cos(\theta_0 - \beta_0)\left(v'w'\phi_* + \frac{1}{4}\phi''vw\right)\Big]\bigg\} \tag{26}$$

对于近轴电子，在式（13）中略去 r^2 项，则有

$$\dot{z}^2 = \frac{2e}{m_0}\int \phi'(z)\mathrm{d}z + \frac{2e}{m_0}\varepsilon_z \tag{27}$$

于是有

$$t^*(z,\varepsilon_z^{1/2}) = \int_0^z \frac{\mathrm{d}z}{\left(\frac{2e}{m_0}\right)^{1/2}[\phi(z)+\varepsilon_z]^{1/2}} \tag{28}$$

式（28）亦可由式（26）略去 $\varepsilon_r^{1/2}$ 和 r_0 的二次项求得。这表明，对于近轴电子轨迹，无论是从轴上点逸出的，或是从高度为 r_0 的轴外点逸出的，只要其轴向初能量 ε_z 相同，它所经历的时间是相同的。式（28）表明：t^* 与 $\varepsilon_r^{1/2}$、r_0 值无关。

由式（7）的定义，并由式（26）、式（28），对照式（9），便有 $a_1 = a_3 = 0$，$A_{12} = A_{23} = 0$。因此

$$\Delta\tau = A_{11}\varepsilon_r + 2A_{13}\varepsilon_r^{1/2}r_0 + A_{33}r_0^2 \tag{29}$$

式中，

$$A_{11} = \int_0^z \frac{1}{2\left(\frac{2e}{m_0}\right)^{1/2}[\phi(z)+\varepsilon_z]^{3/2}}\left(v'^2\phi_* + \frac{1}{4}\phi''v^2 - 1\right)\mathrm{d}z \tag{30}$$

$$A_{13} = \int_0^z \frac{1}{2\left(\frac{2e}{m_0}\right)^{1/2}[\phi(z)+\varepsilon_z]^{3/2}}\cos(\theta_0-\beta_0)\left(v'w'\phi_* + \frac{1}{4}\phi''vw\right)\mathrm{d}z \tag{31}$$

$$A_{33} = \int_0^z \frac{1}{2\left(\frac{2e}{m_0}\right)^{1/2}[\phi(z)+\varepsilon_z]^{3/2}}\left(w'^2\phi_* + \frac{1}{4}\phi''w^2 - \frac{1}{4}\phi_0''\right)\mathrm{d}z \tag{32}$$

这里，A_{11} 为二级时间球差系数，A_{13} 为二级时间场曲系数，A_{33} 为二级时间畸变系数。

2.2 时间色差系数 a_2 和 A_{22} 的确定

由式（6）的定义，并由式（28），可得

$$\begin{aligned}\Delta T &= a_2(\varepsilon_z^{1/2} - \varepsilon_{z_1}^{1/2}) + A_{22}(\varepsilon_z - \varepsilon_{z_1}) \\ &= \int_0^z \frac{\mathrm{d}z}{\left(\frac{2e}{m_0}\right)^{1/2}[\phi(z)+\varepsilon_z]^{1/2}} - \int_0^z \frac{\mathrm{d}z}{\left(\frac{2e}{m_0}\right)^{1/2}[\phi(z)+\varepsilon_{z_1}]^{1/2}}\end{aligned} \tag{33}$$

式中，a_2 为一级近轴时间像差系数或简称一级时间色差系数；A_{22} 为二级近轴时间像差系数或简称为二级时间色差系数，它们分别是式（33）按 $\sqrt{\varepsilon_z}$ 展开后与 $\sqrt{\varepsilon_z}$ 和 ε_z 项幂次相对应的系数。

式（33）仅给出了确定时间色差系数 a_2 和 A_{22} 的途径，但并未给出具体的表达式。下面我们用两种方法求时间色差系数 a_2 和 A_{22}。

2.3 用泰勒级数法求时间色差系数 a_2 和 A_{22}

由式（33）可见，时间色差系数 a_2 和 A_{22} 仅与轴上电位分布 $\phi(z)$ 有关。现我们将 $\phi(z)$ 以泰勒级数展开式表示为

$$\phi(z) = \sum_{1}^{m} \frac{E_m}{m!} z^m \tag{34}$$

式中，E_m 是与场有关的系数。当 $m=1$ 时，$E_1 = -E_c$。把 $\phi(z)$ 代入式 (32)，积分之，便得到

$$t^*(z, \varepsilon_z^{1/2}) = \sqrt{\frac{2m_0}{e}} \frac{1}{E_1} (\sqrt{E_1 z + \varepsilon_z} - \sqrt{\varepsilon_z}) \tag{35}$$

按常数项和 $\varepsilon_z^{\frac{1}{2}}$ 的幂次对上式展开，得

$$t^*(z, \varepsilon_z^{1/2}) = \sqrt{\frac{2m_0}{e}} \frac{1}{E_1} \left[\sqrt{E_1 z} - \sqrt{\varepsilon_z} + \frac{1}{2}\frac{1}{\sqrt{E_1 z}}\varepsilon_z + (\varepsilon_z)^{3/2} \right] \tag{36}$$

由式 (33)，系数 a_2、A_{22} 可以表示为

$$a_2 = -\sqrt{\frac{2m_0}{e}} \frac{1}{E_1} = \sqrt{\frac{2m_0}{e}} \frac{1}{E_c} \tag{37}$$

$$A_{22}|_{m=1} = \sqrt{\frac{2m_0}{e}} \frac{1}{2E_1 \sqrt{E_1 z}} \tag{38}$$

下面我们将证明，式 (37) 是处处成立的。

在式 (34) 的泰勒级数中，若取 $m=2$，用同样的方法可求得 a_2、A_{22} 值，a_2 值仍以式 (37) 表示，A_{22} 的精确解析式为

$$A_{22}|_{m=2} = \sqrt{\frac{2m_0}{e}} \frac{E_1 + E_2 z}{E_1^2 \sqrt{2z(2E_1 + E_2 z)}} \tag{39}$$

同样可以求得 $m=3,4$ 时的 A_{22} 的近似表达式：

$$A_{22}|_{m=3} \approx \sqrt{\frac{2m_0}{e}} \left\{ \frac{1}{E_1^2 \sqrt{2z(2E_1 + E_2 z)}} \left[E_1 + zE_2 + \frac{z^2 E_1 E_3}{6(2E_1 + zE_2)} \right] \right\} \tag{40}$$

$$A_{22}|_{m=4} \approx \sqrt{\frac{2m_0}{e}} \left\{ \frac{1}{E_1^2 \sqrt{2z(2E_1 + E_2 z)}} \left[E_1 + zE_2 + \frac{z^2 E_1 E_3}{6(2E_1 + zE_2)} \right] + \right.$$

$$\left. \frac{E_4}{24\sqrt{2}E_2^{\frac{5}{2}}} \left[3\ln\frac{E_1 + E_2 z + \sqrt{E_2}\sqrt{2E_1 z + E_2 z^2}}{E_1} - \frac{4z^2 \sqrt{E_2}(3E_1 + 2E_2 z)}{(2E_1 z + E_2 z^2)^{\frac{3}{2}}} \right] \right\} \tag{41}$$

文献 [3] 应用 "τ 变分法" 求得时间色差系数 a_2 和 A_{22} 的表达式：

$$a_2 = \sqrt{\frac{2m_0}{e}} \frac{1}{E_c} \tag{42}$$

$$A_{22} = \sqrt{\frac{2m_0}{e}} \left(\frac{1}{2} \frac{1}{\sqrt{\phi}} \frac{1}{\phi'} + \frac{1}{2} \int_0^z \frac{\phi''}{\sqrt{\phi}\phi'^2} dz \right) \tag{43}$$

当将 $m=1,2$ 的 $\phi(z)$ 的表达式 (34) 代入式 (43)，得到的结果与式 (38)、式 (39) 完全一致，从而证明泰勒级数法在轴上的电位分布 $\phi(z)$ 取级数展开的近似情况下与 "$\tau-$变分法" 是等效的。

2.4 用积分表达式求时间色差系数 a_2 和 A_{22}

对式 (32)，若按 $\varepsilon_z^{\frac{1}{2}}$ 的级数展开，我们可以得到

$$\frac{1}{2}\sqrt{\frac{2m_0}{e}}\int_0^z \frac{\mathrm{d}z}{[\phi(z)+\varepsilon_z]^{1/2}} = a_2 \varepsilon_z^{1/2} + A_{22}\varepsilon_z + 0(\varepsilon_z^{3/2}) \tag{44}$$

由于

$$[\sqrt{\phi(z)+\varepsilon_z}]' = \frac{\phi'(z)}{2\sqrt{\phi(z)+\varepsilon_z}}$$

所以

$$\int_0^z \frac{\mathrm{d}z}{[\phi(z)+\varepsilon_z]^{1/2}} = \int_0^z \frac{2[\sqrt{\phi(z)+\varepsilon_z}]'}{\phi'(z)} = 2\frac{\sqrt{\phi(z)+\varepsilon_z}}{\phi'(z)}\bigg|_0^z + 2\int_0^z \frac{\phi''(z)\sqrt{\phi(z)+\varepsilon_z}}{[\phi'(z)]^2}\mathrm{d}z$$

$$= 2\frac{\sqrt{\phi(z)}}{\phi'(z)} + 2\int_0^z \frac{\phi''(z)\sqrt{\phi(z)}}{[\phi'(z)]^2}\mathrm{d}z - 2\frac{\sqrt{\varepsilon_z}}{\phi'(0)} + \left\{\frac{1}{\phi'(z)\sqrt{\phi(z)}} + \int_0^z \frac{\phi''(z)}{[\phi'(z)]^2\sqrt{\phi(z)}}\mathrm{d}z\right\}\varepsilon_z$$

将上式与式（44）进行比较，我们可以得到

$$\mathrm{Const} = \sqrt{\frac{2m_0}{e}}\left\{\frac{\sqrt{\phi(z)}}{\phi'(z)} + \int_0^z \frac{\phi''(z)\sqrt{\phi(z)}}{[\phi'(z)]^2}\mathrm{d}z\right\}$$

$$a_2 = -\sqrt{\frac{2m_0}{e}}\frac{1}{\phi'(0)} = \sqrt{\frac{2m_0}{e}}\frac{1}{E_c} \tag{45}$$

$$A_{22} = \frac{1}{2}\sqrt{\frac{2m_0}{e}}\left\{\frac{1}{\sqrt{\phi(z)}}\frac{1}{\phi'(z)} + \int_0^z \frac{\phi''(z)}{\sqrt{\phi(z)}[\phi'(z)]^2}\mathrm{d}z\right\} \tag{46}$$

时间色差系数 a_2 与 A_{22} 的表达式（45）、式（46）与"τ - 变分法"中的结果表达式（42）、式（43）是完全一致的[3,6]。

对式（46）做进一步简化，

$$A_{22} = \frac{1}{2}\sqrt{\frac{2m_0}{e}}\left\{\frac{1}{\sqrt{\phi(z)}}\frac{1}{\phi'(z)} + \int_0^z \frac{\phi''(z)}{\sqrt{\phi(z)}[\phi'(z)]^2}\mathrm{d}z\right\}$$

$$= \frac{1}{2}\sqrt{\frac{2m_0}{e}}\left\{\frac{1}{\phi'(z)\sqrt{\phi(z)}} - \frac{1}{\sqrt{\phi(z)}}\frac{1}{\phi'(z)}\bigg|_0^z + \int_0^z \frac{1}{\phi'(z)}\mathrm{d}\left[\frac{1}{\sqrt{\phi(z)}}\right]\right\}$$

$$= -\frac{1}{2}\sqrt{\frac{2m_0}{e}}\left[\frac{1}{\phi'(0)\sqrt{\phi(0)}} + \frac{1}{2}\int_0^z \frac{1}{\phi(z)^{\frac{3}{2}}}\mathrm{d}z\right] \tag{47}$$

当 $z=0$ 时式（47）中的 $\frac{1}{\phi'(0)\sqrt{\phi(0)}}$ 与 $\frac{1}{2}\int_0^z \frac{1}{[\phi(z)]^{3/2}}\mathrm{d}z$ 会出现无穷大量，两相抵消后可以写成如下的形式：

$$A_{22} = -\frac{1}{4}\sqrt{\frac{2m_0}{e}}\int \frac{1}{[\phi(z)]^{3/2}}\mathrm{d}z \tag{48}$$

当将 $m=1$，2 的 $\phi(z)$ 表达式（34）代入式（48），得到的结果依然与式（38）、式（39）完全一致，从而证明此积分表达式（48）与"τ - 变分法"中的结果表达式（46）是等效的。我们将在下面的文章用静电聚焦同心球系统的模型证明式（48）与式（46）是等效的。

结束语

本文提出了一种计算电子光学成像系统时间像差系数的新途径——直接积分法，给出了自阴极面逸出的轴向电子初能 ε_{z_1} 在 $0 \leqslant \varepsilon_{z_1} \leqslant \varepsilon_{0\max}$ 的近轴电子轨迹作为比较基准的时间像差的定义，导出了时间像差系数表达式，其一级时间色差系数 a_2、二级时间色差系数 A_{22} 分别以式（37）、式（48）表示；二级时间球差系数 A_{11}、二级时间场曲系数 A_{13}、二级时间畸变系数 A_{33} 分别以式（30）、式（31）、式（32）表示。本文给出的二级时间像差系数表达式以积分形式表示，便于计算，可以在实际设计中应用。

参 考 文 献

[1] SAVOISKY Y K, FANCHENKO S D. Physical basis of electron optical chronograph[J]. Report of Academy of Sciences, 1956,108(2):218 – 221 (in Russian).

[2] CSORBA I P. Transit-time-spread-limited time resolution of image tubes in streak operation[J]. RCA Review 1971,32:650 – 659.

[3] MONASTYRSKI M A, SCHELEV M Y. Theory of temporal aberrations for cathode lenses[M]. Moscow: Lebedeev Institute of Physics,1980.

[4] CHOU L W. Electron optics of concentric spherical electromagnetic focusing systems [J]. Advances in Electronics and Electron Physics, 1979,52:119.

[5] 周立伟,艾克聪,潘顺臣. 关于电磁复合聚焦阴极透镜的像差理论[J]. 物理学报, 1983,32:376 – 391.

[6] 西门纪业, 周立伟, 艾克聪. 阴极透镜像差的变分理论[J]. 物理学报, 1983, 32(12):1536 – 1546.

[7] 周立伟. 宽电子束光学[M]. 北京:北京理工大学出版社, 1993.

11.8 静电同心球系统验证直接积分法的时间像差系数的研究

Testing and Verification for Temporal Aberration Coefficients by Direct Integral Method by Using an Electrostatic Concentric Spherical System

摘要：本文应用作者提出的"直接积分法"计算成像电子光学系统的时间像差系数，它的优点是计算时不需要求解微分方程，只需要对积分表示式求解即可，计算方便、实用，适于计算机求解。"直接积分法"为动态电子光学成像系统时间像差的计算提供了更新而简便的途径。但这一方法导得的积分表达式是否正确，必须经过检验。本文应用静电聚焦同心球系统的理想模型的解析解对其进行检验。结果表明，"直接积分法"求得的一级时间色差系数 a_2、二级时间色差系数 A_{22} 和二级时间球差系数 A_{11} 的积分值与解析值完全相同，从而证明了"直接积分法"计算时间像差系数的正确性。

Abstract: In the present paper, the "Direct Integral Method" (DIM) given by the author has been applied to calculate the temporal aberration coefficients for the imaging electron optical systems. The advantage of DIM is that it needs only to solve the integral expressions, but not needs to solve the differential equations, the computation is practical and applicable by using the computer. The DIM provides a newer and simpler approach for the computation of temporal aberration coefficients in a dynamic imaging electron optical system. But the integral expressions deduced by DIM is correct or not, that is a question? So it requires a critical testing and verification. In the present paper, an analytical solution given by an ideal model of an electrostatic concentric spherical system has been tested and verified. Result shows that the integral value of the first order temporal aberration coefficient a_2, the second order temporal aberration coefficient A_{22} and the second order temporal chromatic aberration coefficient A_{11} completely coincide with the analytical value. It proves the correctness of DIM for calculating the temporal aberration coefficients.

引言

时间像差定义为由坐标原点（$r_0 = 0$）、径向初速为零（$\varepsilon_\rho^{\frac{1}{2}} = 0$,）、轴向初速为 $\varepsilon_{z_1}^{\frac{1}{2}}$

周立伟[1]，李元[1]，张智诠[2]，方二伦[3]，M. A. Monastyrski[4]，M. Y. Schelev[4]. 1) 北京理工大学光电工程系，2) 装甲兵工程学院，3) 西安现代化学研究所，4) 俄罗斯科学院普通物理研究所光电子部. 物理学报，V. 54, No. 8, 2005, 3597–3603.

($\varepsilon_{z_1}^{\frac{1}{2}} \neq 0$ 或 $\varepsilon_{z_1}^{\frac{1}{2}} = 0$）逸出的近轴电子轨迹的飞行时间 $\bar{\tau}$，与初始条件参量（$\varepsilon_\rho^{\frac{1}{2}}$，$\varepsilon_z^{\frac{1}{2}}$，$r_0$）的带电粒子运动到达某一位置 z 的飞行时间 τ 之间的时间差异。成像电子光学系统由光电子初速分布所造成的时间像差可用下式表示[1,2,3]：

$$\Delta \tau = a_2(\varepsilon_z^{\frac{1}{2}} - \varepsilon_{z_1}^{\frac{1}{2}}) + A_{22}(\varepsilon_z - \varepsilon_{z_1}) + A_{11}\varepsilon_\rho + 2A_{13}\varepsilon_z^{\frac{1}{2}} r_0 + A_{33} r_0^2 \quad (1)$$

式中，a_2 为一级时间色差系数，A_{22} 为二级时间色差系数，A_{11} 为二级时间球差系数，A_{13} 为二级时间场曲系数，A_{33} 为二级时间畸变系数。

直接积分法所导得的时间像差系数的表达式可表示为

$$A_{11} = \int_0^z \frac{1}{2\left(\frac{2e}{m}\right)^{\frac{1}{2}} [\phi(z) + \varepsilon_z]^{\frac{3}{2}}} \left(v'^2 \phi_* + \frac{1}{4}\phi'' v^2 - 1\right) dz \quad (2)$$

$$A_{13} = \int_0^z \frac{1}{2\left(\frac{2e}{m}\right)^{\frac{1}{2}} [\phi(z) + \varepsilon_z]^{\frac{3}{2}}} \cos(\theta_0 - \beta_0) \left(v'w' \phi_* + \frac{1}{4}\phi'' vw\right) dz \quad (3)$$

$$A_{33} = \int_0^z \frac{1}{2\left(\frac{2e}{m}\right)^{\frac{1}{2}} [\phi(z) + \varepsilon_z]^{\frac{3}{2}}} \left(w'^2 \phi_* + \frac{1}{4}\phi'' w^2 - \frac{1}{4}\phi_0''\right) dz \quad (4)$$

实际上，直接积分法导得一级时间色差系数 a_2 和二级时间色差系数 A_{22} 的表达式，它已由文献［1］求得，可表示为

$$a_2 = -\frac{1}{\phi_0'}\sqrt{\frac{2m}{e}} \quad (5)$$

$$A_{22} = \frac{1}{4}\sqrt{\frac{2m}{e}} \frac{1}{\sqrt{\phi}} \left(\frac{2}{\phi'} + 2\sqrt{\phi} \int_0^z \frac{\phi''}{\sqrt{\phi}\phi'^2} dz\right) \quad (6)$$

问题是，直接积分法所导得的时间像差系数的积分表达式是否正确，必须经过检验。本文应用静电聚焦同心球系统的理想模型的解析解对一级时间色差系数 a_2、二级时间色差系数 A_{22} 和二级时间球差系数 A_{11} 进行检验。

1 同心球系统模型求时间色差及其系数

作为一种求取时间像差系数的方法，其正确与否直接影响时间像差的计算结果。但是由于电子光学系统的结构复杂，一般只能得到各参数的数值解，在这样的情况下，无法说明结果的正确性。鉴于这样的原因，我们选用可得到解析解的同心球系统，用解析的方法得到它的时间像差系数，再用直接法计算该系统的时间像差系数，比较所得的两个结果，就可以对直接法计算的正确性做出判断。

对于同心球系统，如果设球面阴极 C 和栅状球面阳极 A 的曲率半径为 R_c 和 R_a，系统的共同曲率中心为 O。由于系统的球对称性，故可用极坐标（ρ, φ）来描述电子轨迹。令极坐标的原点位于系统的曲率中心 O，并设阴极 C 的电位 $\phi_c = 0$，栅状阳极 A 对于阴极 C 的电位为 ϕ_{ac}，电子以初速度 v_0、初角度 α_0 自阴极上某点 C 射出。同时规定线段由左向右、由下向上为正，反之为负；角度逆时针转为正，顺时针转为负。在这样的规定下，对于凹面阴极－阳极系统，各参数定义如图 1 所示：

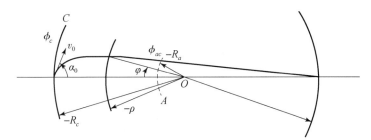

图 1　静电聚焦同心球系统

已知两电极同心球系统中任一点处的电位表达式为

$$\phi_{\rho c} = \frac{\phi_{ac}}{n-1}\left(\frac{R_c}{\rho} - 1\right) \tag{7}$$

其中，$n = R_c/R_a$。根据能量守恒定律和角动量守恒定律，有

$$\frac{1}{2}m_0 v^2 = \frac{1}{2}m_0 v_0^2 + e\phi_{\rho c} \tag{8}$$

$$\rho^2 \dot{\varphi} = R_c^2 \dot{\varphi}_0 = R_c v_0 \sin\alpha_0 \tag{9}$$

式中各参数如图 1 所示。显然，电子速度可表示为

$$v^2 = \dot{\rho}^2 + (\rho\dot{\varphi})^2$$

那么可得到下面的式子：

$$\dot{\rho} = \frac{d\rho}{dt} = \sqrt{v^2 - R_c^2 v_0^2 \sin^2\alpha_0/\rho^2} \tag{10}$$

将

$$v^2 = v_0^2 + \frac{2e}{m_0}\frac{\phi_{ac}}{(n-1)}\left(\frac{R_c}{\rho} - 1\right)$$

代入式（10），得

$$\dot{\rho} = \sqrt{\frac{2e\varepsilon_0}{m_0}}\left[1 + p(\mu-1) - \mu^2\sin^2\alpha_0\right]^{\frac{1}{2}} \tag{11}$$

这里，$p(n-1) = \dfrac{\phi_{ac}}{\varepsilon_0}$，$\mu = \dfrac{R_c}{\rho}$。

由

$$\dot{\rho} = \frac{d\rho}{dt} = -\frac{R_c}{\mu^2}\frac{d\mu}{dt}$$

式（11）可写成如下形式：

$$dt = \frac{-R_c d\mu}{\sqrt{\dfrac{2e\varepsilon_0}{m_0}}\mu^2\sqrt{1 + p(\mu-1) - \mu^2\sin^2\alpha_0}} \tag{12}$$

积分式（12），当 $t = 0$ 时，$\rho = R_c$，$\mu = 1$，可以得到

$$t = \frac{-R_c}{\sqrt{\dfrac{2e\varepsilon_0}{m_0}}}\left\{-\frac{\sqrt{1+p(\mu-1)-\mu^2\sin^2\alpha_0}}{(1-p)\mu} + \frac{\cos\alpha_0}{1-p} - \right.$$

$$\frac{p}{2(1-p)\sqrt{p-1}}\left[\arcsin\frac{p(\mu-2)+2}{\mu\sqrt{p^2+4(1-p)\sin^2\alpha_0}}+\arcsin\frac{p-2}{\sqrt{p^2+4(1-p)\sin^2\alpha_0}}\right]\right\} \quad (13)$$

现求由阴极到达阳极上的时间的精确表达式 t_{C-A}。

由式 (13)，令 $\rho = R_a$，$\mu = n$，我们便可获得自阴极逸出以初参量 $(\varepsilon_0, \alpha_0)$ 的电子到达阳极上的时间的精确表达式 t_{C-A}：

$$t_{C-A} = \frac{-R_c(n-1)}{\sqrt{\frac{2e}{m_0}}[\phi_{ac}-(n-1)\varepsilon_0]}\left[\frac{1}{n}\sqrt{\phi_{ac}+\varepsilon_0(1-n^2\sin^2\alpha_0)}-\sqrt{\varepsilon_0}\cos\alpha_0\right]+$$

$$\frac{-R_c}{\sqrt{\frac{2e}{m_0}}}\frac{\frac{\phi_{ac}}{n-1}}{2\left(\frac{\phi_{ac}}{n-1}-\varepsilon_0\right)\sqrt{\frac{\phi_{ac}}{n-1}-\varepsilon_0}}\left[\arcsin\frac{\frac{\phi_{ac}}{(n-1)}\left(1-\frac{2}{n}\right)+\frac{2\varepsilon_0}{n}}{\sqrt{\left(\frac{\phi_{ac}}{n-1}\right)^2-4\left(\frac{\phi_{ac}}{n-1}-\varepsilon_0\right)\varepsilon_0\sin^2\alpha_0}}+\right.$$

$$\left.\arcsin\frac{\frac{\phi_{ac}}{n-1}-2\varepsilon_0}{\sqrt{\left(\frac{\phi_{ac}}{n-1}\right)^2-4\left(\frac{\phi_{ac}}{n-1}-\varepsilon_0\right)\varepsilon_0\sin^2\alpha_0}}\right] \quad (14)$$

令 $\varepsilon_0 = \varepsilon_r + \varepsilon_z$，式 (14) 可表达成如下形式：

$$t_{C-A} = \frac{-R_c(n-1)}{\sqrt{\frac{2e}{m_0}}[\phi_{ac}-(n-1)\varepsilon_z-(n-1)\varepsilon_r]}\left[\frac{1}{n}\sqrt{\phi_{ac}+\varepsilon_z+(1-n^2)\varepsilon_r}-\sqrt{\varepsilon_z}\right]+$$

$$\frac{-R_c}{\sqrt{\frac{2e}{m_0}}}\frac{\frac{\phi_{ac}}{n-1}}{2\left(\frac{\phi_{ac}}{n-1}-\varepsilon_z-\varepsilon_r\right)\sqrt{\frac{\phi_{ac}}{n-1}-\varepsilon_z-\varepsilon_r}}\times$$

$$\left[\arcsin\frac{\left(1-\frac{2}{n}\right)+\frac{2(n-1)(\varepsilon_z+\varepsilon_r)}{n\phi_{ac}}}{\sqrt{1-4\left(\frac{\phi_{ac}}{n-1}-\varepsilon_z-\varepsilon_r\right)\varepsilon_r\left(\frac{n-1}{\phi_{ac}}\right)^2}}+\arcsin\frac{1-2\frac{n-1}{\phi_{ac}}(\varepsilon_z+\varepsilon_r)}{\sqrt{1-4\left(\frac{\phi_{ac}}{n-1}-\varepsilon_z-\varepsilon_r\right)\varepsilon_r\left(\frac{n-1}{\phi_{ac}}\right)^2}}\right] \quad (15)$$

现展开式 (15)，略去高于 $\frac{\varepsilon_z}{\phi_{ac}}$、$\frac{\varepsilon_r}{\phi_{ac}}$ 等远小于 1 的高阶项，便可获得自阴极逸出以初参量 $(\varepsilon_0, \alpha_0)$ 的电子到达阳极上的时间的近似表达式 t_{C-A}：

$$t_{C-A} = \frac{-R_c(n-1)}{\sqrt{\frac{2e}{m_0}\phi_{ac}}}\frac{1}{n}\left[1+\frac{(2n-1)}{2\phi_{ac}}\varepsilon_z+\frac{(n-1)^2}{2\phi_{ac}}\varepsilon_r\right]+\frac{R_c(n-1)}{\sqrt{\frac{2e}{m_0}\phi_{ac}}}\sqrt{\varepsilon_z}+$$

$$\frac{-R_c}{\sqrt{\frac{2e}{m_0}}}\frac{\phi_{ac}\sqrt{n-1}}{2(\phi_a)^{\frac{3}{2}}}\left(1+\frac{3}{2}\frac{n-1}{\phi_{ac}}\varepsilon_z+\frac{3}{2}\frac{n-1}{\phi_{ac}}\varepsilon_r\right)\times$$

$$\left\{\arcsin\left[\left(1-\frac{2}{n}\right)+\frac{2(n-1)\varepsilon_z}{n\phi_{ac}}+\frac{2(n-1)^2}{n\phi_{ac}}\varepsilon_r\right]+\arcsin\left(1-2\frac{n-1}{\phi_{ac}}\varepsilon_z\right)\right\} \quad (16)$$

式（16）的最后一项可以写成如下的形式：

$$\left\{\arcsin\left[\left(1-\frac{2}{n}\right)+\frac{2(n-1)\varepsilon_z}{n\phi_{ac}}+\frac{2(n-1)^2}{n\phi_{ac}}\varepsilon_r\right]+\arcsin\left(1-2\frac{n-1}{\phi_{ac}}\varepsilon_z\right)\right\}$$

$$=\pi-\arcsin\left(\frac{2}{n}\sqrt{n-1}\right)-2\sqrt{n-1}\sqrt{\frac{\varepsilon_z}{\phi_{ac}}}+\sqrt{n-1}\frac{\varepsilon_z}{\phi_{ac}}+2(n-1)\sqrt{n-1}\frac{\varepsilon_r}{\phi_{ac}} \quad (17)$$

因此，t_{C-A} 的近似式可以表示成以下的形式：

$$t_{C-A}=\left(\frac{2m_0}{e}\right)^{\frac{1}{2}}\left\{\frac{-R_c(n-1)}{2\sqrt{\phi_{ac}}}\frac{1}{n}+\frac{-R_c}{4}\sqrt{\frac{n-1}{\phi_{ac}}}\times\left[\pi-\arcsin\left(\frac{2}{n}\sqrt{n-1}\right)\right]+\right.$$

$$\frac{R_c(n-1)}{\sqrt{\phi_{ac}}}\sqrt{\frac{\varepsilon_z}{\phi_{ac}}}+\frac{-R_c(n-1)}{\sqrt{\phi_{ac}}}\left\{\frac{3n-1}{4n}+\frac{3}{8}\sqrt{n-1}\left[\pi-\arcsin\left(\frac{2}{n}\sqrt{n-1}\right)\right]\right\}\frac{\varepsilon_z}{\phi_{ac}}+$$

$$\left.\frac{-R_c(n-1)}{\sqrt{\phi_{ac}}}\left[\frac{(n-1)(3n-1)}{4n}+\frac{3}{8}\sqrt{n-1}\left(\pi-\arcsin\left(\frac{2}{n}\sqrt{n-1}\right)\right)\right]\frac{\varepsilon_r}{\phi_{ac}}\right\} \quad (18)$$

因为

$$\arcsin\sqrt{\frac{n-1}{n}}=\frac{1}{2}\left(\pi-\arcsin\frac{2}{n}\sqrt{n-1}\right)$$

故式（18）可以表示为

$$t_{C-A}=\left(\frac{2m_0}{e}\right)^{\frac{1}{2}}\left\{\frac{-R_c(n-1)}{2\sqrt{\phi_{ac}}}\frac{1}{n}+\frac{-R_c}{2}\sqrt{\frac{n-1}{\phi_{ac}}}\arcsin\sqrt{\frac{n-1}{n}}+\right.$$

$$\frac{R_c(n-1)}{\sqrt{\phi_{ac}}}\sqrt{\frac{\varepsilon_z}{\phi_{ac}}}+\frac{-R_c(n-1)}{\sqrt{\phi_{ac}}}\left(\frac{3n-1}{4n}+\frac{3}{4}\sqrt{n-1}\arcsin\sqrt{\frac{n-1}{n}}\right)\frac{\varepsilon_z}{\phi_{ac}}+$$

$$\left.\frac{-R_c(n-1)}{\sqrt{\phi_{ac}}}\left[\frac{(n-1)(3n-1)}{4n}+\frac{3}{4}\sqrt{n-1}\arcsin\sqrt{\frac{n-1}{n}}\right]\frac{\varepsilon_r}{\phi_{ac}}\right\} \quad (19)$$

对于近轴射线，当略去 $\dfrac{\varepsilon_r}{\phi_{ac}}$ 项时，便有

$$t^*_{C-A}=\left(\frac{2m_0}{e}\right)^{\frac{1}{2}}\left\{\frac{-R_c(n-1)}{2\sqrt{\phi_{ac}}}\frac{1}{n}+\frac{-R_c}{2}\sqrt{\frac{n-1}{\phi_{ac}}}\arcsin\sqrt{\frac{n-1}{n}}+\frac{R_c(n-1)}{\sqrt{\phi_{ac}}}\sqrt{\frac{\varepsilon_{z_1}}{\phi_{ac}}}+\right.$$

$$\left.\frac{-R_c(n-1)}{\sqrt{\phi_{ac}}}\left[\frac{3n-1}{4n}+\frac{3}{4}\sqrt{n-1}\arcsin\sqrt{\frac{n-1}{n}}\right]\frac{\varepsilon_{z_1}}{\phi_{ac}}\right\} \quad (20)$$

把 t 的表达式按 ε_z 展开，即可得到一级时间色差系数和二级时间色差系数为

$$a_2=R_c\sqrt{\frac{m_0}{2e}\frac{2(n-1)}{\phi_{ac}}}=-\sqrt{\frac{2m}{e}}\frac{1}{\phi'_0} \quad (21)$$

$$A_{22}=-\frac{R_c}{8}\sqrt{\frac{m_0}{2e}}\left(\frac{n-1}{\phi_{ac}}\right)^{\frac{3}{2}}\left[3\pi+\frac{4\rho^2}{R_c\sqrt{(R_c-\rho)\rho}}+\frac{12\sqrt{-\rho}}{\sqrt{\rho-R_c}}-6\arctan\frac{R_c-2\rho}{2\sqrt{(R_c-\rho)\rho}}\right] \quad (22)$$

令 $\rho=R_a$，就可以得到同心球系统中阳极上的像差系数为

$$A_{22}=\frac{R_c}{8}\sqrt{\frac{m_0}{2e}}\left(\frac{n-1}{\phi_{ac}}\right)^{\frac{3}{2}}\left(-3\pi+\frac{4}{n\sqrt{n-1}}-\frac{12}{\sqrt{n-1}}-6\arctan\frac{n-2}{2\sqrt{n-1}}\right) \quad (23)$$

2 "直接积分法"求解时间色差系数

根据"直接积分法",一级时间色差系数 a_2 和二级时间色差系数 A_{22} 分别是

$$a_2(\varepsilon_z^{\frac{1}{2}} - \varepsilon_{z_1}^{\frac{1}{2}}) + A_{22}(\varepsilon_z - \varepsilon_{z_1}) = \int_0^z \frac{\mathrm{d}z}{\left(\frac{2e}{m}\right)^{\frac{1}{2}}[\phi(z) + \varepsilon_z]^{\frac{1}{2}}} - \int_0^z \frac{\mathrm{d}z}{\left(\frac{2e}{m}\right)^{\frac{1}{2}}[\phi(z) + \varepsilon_{z_1}]^{\frac{1}{2}}} \quad (24)$$

的展开式中有关 $\sqrt{\varepsilon_z}$ 和 ε_z 的系数。

在同心球系统中:

$$\phi(z) = \frac{\phi_{ac}}{n-1} \frac{-z}{z+R_c} \quad (25)$$

将式 (25) 代入式 (24) 中,可得

$$\int_0^z \frac{\mathrm{d}z}{\left(\frac{2e}{m}\right)^{\frac{1}{2}}[\phi(z) + \varepsilon_z]^{\frac{1}{2}}} = \int_0^z \frac{\mathrm{d}z}{\left(\frac{2e}{m}\right)^{\frac{1}{2}}\left[\frac{\phi_{ac}}{n-1}\frac{-z}{z+R_c} + \varepsilon_z\right]^{\frac{1}{2}}}$$

$$= \sqrt{\frac{m}{2e}} \int_0^z \sqrt{\frac{(n-1)(z+R_c)}{-z\phi_{ac} + \varepsilon_z(n-1)(z+R_c)}} \mathrm{d}z = \sqrt{\frac{m}{2e}} \int_0^z \sqrt{\frac{-(z+R_c)}{\frac{z\phi_{ac}}{n-1} - \varepsilon_z(z+R_c)}} \mathrm{d}z$$

$$= \sqrt{\frac{m}{2e\left(\frac{\phi_{ac}}{n-1} - \varepsilon_z\right)}} \int_0^z \frac{\sqrt{-R_c - z}}{\sqrt{-\frac{\varepsilon_z R_c}{\frac{\phi_{ac}}{n-1} - \varepsilon_z} + z}} \mathrm{d}z$$

$$= \sqrt{\frac{m}{2e\left(\frac{\phi_{ac}}{n-1} - \varepsilon_z\right)}} \left[-\sqrt{-bR_c} + \sqrt{-R_c - z}\sqrt{b+z} + \frac{1}{2}(b - R_c)\arctan\frac{-b - R_c}{2\sqrt{-R_c b}} - \frac{1}{2}(b - R_c)\arctan\frac{-b - R_c - 2z}{2\sqrt{-R_c - z}\sqrt{b+z}} \right]$$

式中,$b = -\dfrac{\varepsilon_z R_c}{\dfrac{\phi_{ac}}{n-1} - \varepsilon_z}$。

对上式按 ε_z 展开,得一级时间色差系数 a_2 和二级时间色差系数 A_{22} 分别为:

$$a_2 = \sqrt{\frac{2m}{e}}\frac{(n-1)}{\phi_{ac}}R_c = -\sqrt{\frac{2m}{e}}\frac{1}{\phi_0'} \quad (26)$$

$$A_{22} = -\frac{1}{8}\sqrt{\frac{m}{2e}}\left(\frac{n-1}{\phi_{ac}}\right)^{\frac{3}{2}}\left[\frac{4(-2R_c^2 - R_c z + z^2)}{\sqrt{-R_c - z}\sqrt{z}} - 6R_c \arctan\frac{-R_c - 2z}{2\sqrt{-R_c - z}\sqrt{z}} + 3\pi R_c\right] \quad (27)$$

在阳极上,$z = R_a - R_c = \dfrac{R_c}{n} - R_c$,则有

$$A_{22} = \frac{R_c}{8}\sqrt{\frac{m}{2e}}\left(\frac{n-1}{\phi_{ac}}\right)^{\frac{3}{2}}\left[\frac{4\left(\frac{1}{n}-3\right)}{\sqrt{n-1}} - 6\arctan\frac{n-2}{2\sqrt{n-1}} - 3\pi\right] \quad (28)$$

结束语

比较式（21）、式（22）与式（26）、式（27）可以看出，直接积分法求得的一阶时间色差系数与解析值完全相同，比较式（23）与式（28），说明直接积分法求得的二阶时间色差系数在阳极上也是完全吻合的。这样，我们通过求解代表轴上色差的 a_2 与 A_{22}，说明直接积分法所得的结果在阳极上与解析值完全一致，从而通过理论推导证明了直接积分法计算像差系数的正确性与准确性。

参 考 文 献

[1]周立伟,李元,张智诠,等.直接积分法研究成像电子光学系统的时间像差理论[J].物理学报,2005,54(8):3591-3596.

[2]ZHOU L W,LI Y,ZHANG Z Q,et al. On the theory of temporal aberrations for electron optical imaging systems by using direct integral method[C]. Proceedings of SPIE 2005.

[3]周立伟.宽束电子光学[M].北京:北京理工大学出版社,1993.